Handbook of Cyclization Reactions

Edited by
Shengming Ma

Related Titles

Carreira, Erick M. / Kvaerno, Lisbet

Classics in Stereoselective Synthesis

2008

ISBN: 978-3-527-32452-1

Crabtree, R. H. (ed.)

Handbook of Green Chemistry Set 1: Green Catalysis

Series editor: Anastas, P. T. (ed.)

2009

ISBN: 978-3-527-31577-2

Dubois, P. / Coulembier, O. / Raquez, J.-M. (eds.)

Handbook of Ring-Opening Polymerization

2009

ISBN: 978-3-527-31953-4

Dupont, J. / Pfeffer, M. (eds.)

Palladacycles

Synthesis, Characterization and Applications

2008

ISBN: 978-3-527-31781-3

Handbook of Cyclization Reactions

Volume 1

Edited by
Shengming Ma

WILEY-VCH Verlag GmbH & Co. KGaA

The Editor

Prof. Dr. Shengming Ma
State Key Labor. of Organomet.
Shanghai Institute of Organo.
354 Fenglin Lu
Shanghai 200032
Peoples Republic of China

All books published by Wiley-VCH are carefully produced. Nevertheless, authors, editors, and publisher do not warrant the information contained in these books, including this book, to be free of errors. Readers are advised to keep in mind that statements, data, illustrations, procedural details or other items may inadvertently be inaccurate.

Library of Congress Card No.: applied for

British Library Cataloguing-in-Publication Data
A catalogue record for this book is available from the British Library.

Bibliographic information published by the Deutsche Nationalbibliothek
The Deutsche Nationalbibliothek lists this publication in the Deutsche Nationalbibliografie; detailed bibliographic data are available on the Internet at http://dnb.d-nb.de.

© 2010 WILEY-VCH Verlag GmbH & Co. KGaA, Weinheim

All rights reserved (including those of translation into other languages). No part of this book may be reproduced in any form – by photoprinting, microfilm, or any other means – nor transmitted or translated into a machine language without written permission from the publishers. Registered names, trademarks, etc. used in this book, even when not specifically marked as such, are not to be considered unprotected by law.

Printed in the Federal Republic of Germany
Printed on acid-free paper

Typesetting   Thomson Digital, Noida, India
Printing   betz-Druck GmbH, Darmstadt
Binding   Litges & Dopf Buchbinderei GmbH, Heppenheim
Cover Design   Formgeber, Eppelheim

ISBN: 978-3-527-32088-2

Contents

Contents to Volume 2 *XV*
Preface *XVII*
List of Contributors to Volume 1 *XIX*

1	**Asymmetric Catalysis of Diels–Alder Reaction** *1*	
	Haifeng Du and Kuiling Ding	
1.1	Introduction *1*	
1.2	Asymmetric Diels–Alder Reaction *2*	
1.2.1	Lewis Acid Catalyzed Asymmetric Diels–Alder Reaction *2*	
1.2.1.1	Chiral Boron, Aluminum, and Indium Complexes *2*	
1.2.1.2	Chiral Copper, Magnesium, and Zinc Complexes *7*	
1.2.1.3	Chiral Transition Metal Complexes *10*	
1.2.1.4	Chiral Rare Earth Metal Complexes *12*	
1.2.2	Organocatalysis of Asymmetric Diels–Alder Reaction *14*	
1.2.2.1	Chiral Secondary Amine Catalyzed Asymmetric Diels–Alder Reaction *14*	
1.2.2.2	Chiral Primary Amine Catalyzed Asymmetric Diels–Alder Reaction *16*	
1.2.2.3	Brønsted Acid Catalyzed Asymmetric Diels–Alder Reaction *17*	
1.2.2.4	Bifunctional Organocatalysis of Asymmetric Diels–Alder Reaction via Hydrogen Bonding *19*	
1.3	Asymmetric Oxa-Diels–Alder Reaction *21*	
1.3.1	Lewis Acid Catalyzed Asymmetric Oxa-Diels-Alder Reaction *21*	
1.3.1.1	Chiral Aluminum, Boron, and Indium Complexes *21*	
1.3.1.2	Chiral Titanium and Zirconium Complexes *23*	
1.3.1.3	Chiral Chromium Complexes *26*	
1.3.1.4	Chiral Copper and Zinc Complexes *28*	
1.3.1.5	Chiral Rhodium, Palladium, and Platinum Complexes *33*	
1.3.1.6	Chiral Rare Earth Metal Complexes *36*	
1.3.2	Organocatalysis of Asymmetric Oxa-Diels–Alder Reaction *37*	
1.3.2.1	Hydrogen Bonding Promoted Asymmetric Oxa-Diels–Alder Reaction *37*	

Handbook of Cyclization Reactions. Volume 1.
Edited by Shengming Ma
Copyright © 2010 WILEY-VCH Verlag GmbH & Co. KGaA, Weinheim
ISBN: 978-3-527-32088-2

1.3.2.2	Chiral Secondary Amine Catalyzed Asymmetric Oxa-Diels–Alder Reaction *39*	
1.4	Representative Applications in Total Synthesis *41*	
1.5	Conclusion *46*	
1.6	Experimental: Selected Procedures *46*	
1.6.1	Procedure for the Preparation of Chiral Boron Complex 8b (R = o-tol, Ar = Ph) and Its Application in Asymmetric Diels–Alder Reactions *46*	
1.6.2	Procedure for Chiral Secondary Amine 42 Catalyzed Asymmetric Diels–Alder Reaction of Cyclopentadiene with (E)-Cinnamaldehyde *47*	
1.6.3	Procedure for 65/Ti/70 Catalyzed Asymmetric Hetero Diels–Alder Reaction of Benzaldehyde with Danishefsky's Diene *47*	
1.6.4	Procedure for the Preparation of the Chiral Chromium Complex (1S, 2R)-76a and Its Application in the Asymmetric Hetero Diels–Alder Reaction of Ethyl Vinyl Ether with Crotonaldehyde *47*	
1.6.5	Procedure for Chiral Copper Complex 82a Catalyzed Reaction of Asymmetric Hetero Ethyl Pyruvate with Danishefsky's Diene *48*	
1.6.6	Procedure for the Preparation of Chiral Dirhodium Carboxamidate Complex 95 and Its Application in the Asymmetric Hetero Diels–Alder Reaction of Phenylpropargyl Aldehyde with Danishefsky's Diene *48*	
1.6.7	Procedure for TADDOL 52 Promoted Asymmetric Hetero Diels–Alder reaction of Benzaldehyde with 1-(N,N-Dimethylamino)-3-tert-butyldimethylsiloxy-1,3-Diene via Hydrogen Bonding Activation *49*	
1.6.8	Procedure for Chiral Amine 109 Promoted Inverse Electron Demand Hetero Diels–Alder Reaction of Butyraldehyde and Enone *49*	
	References *51*	
2	**Catalytic Asymmetric Aza Diels–Alder Reactions** *59*	
	Yasuhiro Yamashita and Shū Kobayashi	
2.1	Introduction *59*	
2.2	Chiral Lewis Acid Catalysis *59*	
2.3	Chiral Organocatalysis *76*	
2.4	Conclusion *82*	
2.5	Experimental: Selected Procedures *83*	
2.5.1	Typical Experimental Procedure for Asymmetric aza Diels–Alder Reactions Using A Chiral Zr Complex *83*	
2.5.2	Typical Experimental Procedure for Asymmetric Aza Diels–Alder Reactions Using a Chiral Ag Complex *83*	
	References *84*	
3	**1,3-Dipolar Cycloaddition** *87*	
	Takuya Hashimoto and Keiji Maruoka	
3.1	Introduction *87*	
3.2	Nitrones *88*	

3.2.1	Beginning of Lewis Acid Catalyzed 1,3-DC of Nitrones *89*	
3.2.2	Chiral Lewis Acid Catalyzed 1,3-DC of Nitrones and α,β-Unsaturated Carbonyl Compounds Bearing Ancillary Coordinating Group *91*	
3.2.3	Asymmetric 1,3-DC of Nitrones and α,β-Unsaturated Aldehydes *95*	
3.2.3.1	Organocatalysis *95*	
3.2.3.2	Lewis Acid Catalysis *96*	
3.2.4	Asymmetric 1,3-DC of Nitrones and Other Electron-Deficient Olefins *101*	
3.2.5	Inverse Electron Demand 1,3-DC of Nitrones *103*	
3.2.6	Kinugasa Reaction *105*	
3.3	Azomethine Ylides *107*	
3.3.1	Early Examples of Asymmetric 1,3-DC of *N*-Metalated Azomethine Ylides Derived from Aldimines of α-Amino Acid Esters *108*	
3.3.2	Development of Asymmetric 1,3-DC of *N*-Metalated Azomethine Ylides with α,β-Unsaturated Carbonyl Compounds *109*	
3.3.3	Asymmetric 1,3-DC of *N*-Metalated Azomethine Ylides with Vinyl Sulfones and Nitroalkenes *117*	
3.3.4	Organocatalytic Asymmetric 1,3-DC of Azomethine Ylides *119*	
3.3.5	Transition Metal Catalyzed 1,3-DC of Münchnones *121*	
3.3.6	Lewis Acid Catalyzed 1,3-DC of Azomethine Ylide Derived from the Imidate of 2-Aminomalonate *123*	
3.3.7	Lewis Acid Promoted Generation of Electrophilic Azomethine Ylides via C–C Bond Cleavage of Aziridines *123*	
3.3.8	Generation and 1,3-DC of Metal-Containing Azomethine Ylide via Electrophilic Activation of Alkynes *124*	
3.4	Azomethine Imines *126*	
3.4.1	Asymmetric 1,3-DC of Azomethine Imines: Use of 3-Oxopyrazolidin-1-Ium-2-Ides as a 1,3-Dipole *127*	
3.4.2	Asymmetric [3 + 2] Cycloaddition of *N*-Acylhydrazones with Olefins *130*	
3.5	Carbonyl Ylides *134*	
3.5.1	Dirhodium(II)-Catalyzed Asymmetric Tandem Carbonyl Ylide Formation-1,3-DC *134*	
3.5.2	Three-Component Tandem Carbonyl Ylide Formation-1,3-DC *139*	
3.5.3	Chiral Lewis Acid Catalyzed 1,3-DC of Carbonyl Ylides Generated by Dirhodium(II) Co-Catalyst *140*	
3.5.4	1,3-DC of Metal-Cotaining Carbonyl Ylides Generated via Electrophilic Activation of Alkynes *143*	
3.6	Nitrile Oxides and Nitrile Imines *147*	
3.6.1	Magnesium(II)-Associated Hydroxy-Directed 1,3-DC of Nitrile Oxides *148*	
3.6.2	Catalytic Asymmetric 1,3-DC of Nitrile Oxides *152*	
3.6.3	1,3-DC of Nitrile Imines *153*	
3.7	Diazo Compounds *154*	

3.7.1	Lewis Acid Catalyzed Asymmetric 1,3-DC of Diazo Compounds with α,β-Unsaturated Carbonyl Compounds *155*
3.7.2	Lewis acid Catalyzed 1,3-DC of Diazo Compounds with Alkynes *157*
3.8	Conclusions *158*
3.9	Experimental: Selected Experimental *159*
3.9.1	Preparation of *N*-Benzylidenebenzylamine *N*-oxide *159*
3.9.2	General Procedure for the Preparation of the Aldimine of *N*-Arylidene Amino Acid Ester *159*
3.9.3	Preparation of 1-Benzylidene-3-oxopyrazolidin-1-ium-2-ide *160*
3.9.4	Preparation of *tert*-Butyl Diazoacetate *160*
	References *160*

4 Intramolecular 1,2-Addition and 1,4-Addition Reactions   *169*
Xiyan Lu and Xiuling Han

4.1	Introduction *169*
4.2	Cyclization via Intramolecular 1,2-Addition Reactions *169*
4.2.1	1,2-Addition of Carbonyl Compounds *169*
4.2.1.1	1,2-Addition of Dicarbonyl Compounds (Aldol Reaction) *169*
4.2.1.2	Vinylogous Aldol Reaction *173*
4.2.1.3	1,2-Addition of Alkene-, Alkyne- or Arene-Substituted Carbonyl Compounds *177*
4.2.2	1,2-Addition of Imines *181*
4.2.2.1	1,2-Addition of Alkene or Alkyne Substituted Imino Compounds (Imino Ene Reactions) *182*
4.2.2.2	Nucleophilic Addition to Imino Groups *186*
4.2.3	1,2-Addition of Nitriles *191*
4.2.3.1	Acid-Catalyzed Reactions *191*
4.2.3.2	Base-Catalyzed Reactions *192*
4.2.3.3	Nucleophilic Addition to Nitriles *194*
4.2.3.4	Cycloaddition Mediated by a Stoichiometric Amount of Transition Metals *197*
4.3	Cyclization via Intramolecular 1,4-Addition Reactions *199*
4.3.1	General *199*
4.3.2	1,4-Addition of Carbanions (Michael Addition) *200*
4.3.2.1	1,4-Additions Mediated by Bases *200*
4.3.2.2	1,4-Additions Mediated by Lewis Acids *203*
4.3.2.3	Catalyzed 1,4-Addition Reactions *205*
4.3.3	1,4-Addition of Heteronucleophiles *210*
4.3.4	Other Acceptors for 1,4-Addition Reactions *214*
4.3.5	Catalytic Asymmetric Intramolecular Cyclization via 1,4-Addition *216*
4.4	Conclusion and Perspectives *217*
4.5	Experimental: Selected Procedures *217*
4.5.1	1,2-Addition of Alkyne Substituted Carbonyl Compounds *217*
4.5.1.1	General Procedure for Cyclization of 5-Yn-1-ones *217*

4.5.2	1,2-Addition of Arene Substituted Carbonyl Compounds	218
4.5.2.1	Representative Procedure for Asymmetric Cyclization of o-(Acylmethyloxy)-arylboronic Acid to Furnish the Optically Active Cycloalkanols	218
4.5.3	Nucleophilic Addition to Imino Groups	218
4.5.3.1	Typical Procedure for the Palladium-Catalyzed Indole Synthesis	218
4.5.4	1,2-Addition of Nitriles	218
4.5.4.1	Procedure for the 1.2-Addition of Nitrile	218
4.5.5	1,2-Addition of Nitriles	219
4.5.5.1	Synthesis of Enantiomerically Enriched Atropoisomers of 2-Arylpyridines	219
4.5.6	Catalyzed 1,4-Addition Reactions	219
4.5.6.1	Typical Procedure for the Phosphine-Catalyzed Tandem Nucleophilic Additions	219
4.5.7	Catalytic Asymmetric Intramolecular Cyclization via 1,4- Addition	219
4.5.7.1	General Procedure for Asymmetric 1,4-Addition	219
	References	220
5	**Cyclic Carbometallation of Alkenes, Arenes, Alkynes and Allenes**	**227**
	Ron Grigg and Martyn Inman	
5.1	Introduction	227
5.2	Carbometallation of Alkenes	228
5.2.1	Stoichiometric Intramolecular Carbometallation of Alkenes	228
5.2.2	Intramolecular Heck Reactions	230
5.2.3	Intramolecular Catalytic Alkene Carbometallation: Other Termination Steps	233
5.3	Alkyne Carbometallation	238
5.3.1	Stoichiometric Methods	238
5.3.2	Catalytic Alkyne Carbometallations	239
5.4	Carbometallations of Arenes	243
5.5	Carbometallations of Allenes	243
5.6	Multiple Cyclization Reactions	245
5.6.1	Stoichiometric Polycyclization Methods	247
5.6.2	Fused- and Bridged-Ring Forming Catalytic Cyclizations	247
5.6.3	Spiro-Mode Polycyclizations	249
5.6.4	Transannular or "Zipper" Mode Polycyclizations	250
5.6.5	Circular or "Dumbbell" Polycyclizations	251
5.6.5.1	2 + 2 + 2 Cyclization Reactions	251
5.6.5.2	Cyclopropanations	255
5.7	Acylpalladations	257
5.8	Conclusion	261
5.9	Experimental: Selected Procedures	262
5.9.1	Synthesis of 11	262
5.9.2	Synthesis of 41	262
5.9.3	Synthesis of 185	262

5.9.4	Synthesis of 207 263
	References 264

6	**Transition Metal-Catalyzed Intramolecular Allylation Reactions** 271
	Zhan Lu and Shengming Ma
6.1	Introduction 271
6.2	Palladium-Catalyzed Intramolecular Allylation Reactions 271
6.2.1	Cyclic Allylations of Carbonucleophiles 271
6.2.2	Cyclic Allylic Aminations 282
6.2.3	Cyclic O- or S-Allylation 294
6.3	Iridium-Catalyzed Intramolecular Allylation Reactions 301
6.4	Ni-Catalyzed Intramolecular Allylation Reactions 304
6.5	Rh-Catalyzed Intramolecular Allylation Reactions 304
6.6	Conclusion and Perspectives 305
6.7	Experimental: Selected Procedures 305
6.7.1	Synthesis of Optically Active Carbocycle 9 with Chiral Ligand S-7b 305
6.7.2	Synthesis of Lactone 15(n = 4) in an Aqueous–Organic Biphasic System 306
6.7.3	Synthesis of Lactone 25a in an Aqueous–Organic Biphasic system 306
6.7.4	Synthesis of (4S,5S)-3-oxo-5-vinyl-1-aza-bicyclo[2.2.2]octane-4-carboxylic Acid Ethyl Ester 307
6.7.5	Synthesis of trans-2-vinylcyclopentanecarbaldehyde 307
6.7.6	Synthesis of 3-vinylindan-1,1-dicarboxylic Acid Diethyl Ester 307
6.7.7	Synthesis of (3S,3Sa,6Ra)-5-methylene-3-(2-naphthoxy) tetrahydrocyclopenta[c]furan-1-One 307
6.7.8	Synthesis of 1-Ethenyl-2-trifluoroacetyl-6,7-dimethoxy-1,2,3,4-tetrahydro-isoquinoline 308
6.7.9	Synthesis of (R)-2-Isopropenyl-2,3-dihydrobenzofuran 308
6.7.10	Synthesis of Optically Active Cyclic Carbonate S-226 308
6.7.11	Synthesis of 2,3-dihydro-1,4-benzodioxine 309
6.7.12	Synthesis of Optically Active Carbocycles and Azacycles via Ir-Catalyzed Intramolecular Allylation 309
6.7.13	Synthesis of Compound 270 (n = 2) 309
6.7.14	Synthesis of N-(p-methoxyphenyl)-4-vinyl-2-oxazolidinone 309
	References 310

7	**Cyclic Coupling Reactions** 315
	Xuefeng Jiang and Shengming Ma
7.1	Introduction 315
7.2	Cyclic Coupling Involving Organoboron 315
7.3	Cyclic Coupling Involving Organotin 321
7.4	Cyclic Coupling Involving Organosilicon 328
7.5	Cyclic Coupling Involving Amines or Alcohols 330
7.6	Cyclic Coupling via the Heck Reaction 344
7.7	Cyclic Coupling Involving Terminal Alkynes 354

7.8	Conclusion and Perspectives	*357*
7.9	Experimental: Selected Procedures	*357*
7.9.1	Synthesis of Compound 4	*357*
7.9.2	Synthesis of Compound 17	*357*
7.9.3	Synthesis of Compound 44 (R = Ph)	*358*
7.9.4	Synthesis of Compound 86	*358*
7.9.5	Synthesis of Compound 91	*359*
7.9.6	Synthesis of Compound 100 (Z = Boc, n = 2)	*359*
7.9.7	Synthesis of Compound 135	*359*
7.9.8	Synthesis of Compound 261	*360*
7.9.9	Synthesis of Compound 278	*360*
7.9.10	Synthesis of Compound 291	*360*
	References	*362*
8	**Transition Metal-Mediated [2 + 2 + 2] Cycloadditions**	*367*
	David Lebœuf, Vincent Gandon, and Max Malacria	
8.1	Introduction	*367*
8.2	Benzene Derivatives	*368*
8.2.1	Rapid Construction of Functionalized Benzenes Using Heteroatom-Substituted Alkynes	*368*
8.2.1.1	Alkynylsilanes	*368*
8.2.1.2	Alkynylboronates	*369*
8.2.1.3	Ynamides	*370*
8.2.1.4	Alkynylphosphines	*370*
8.2.1.5	Alkynylhalides	*370*
8.2.2	Rapid Construction of the Naphthalene Core from Transient Benzynes	*372*
8.2.3	Rapid Construction of Polyphenylenes	*373*
8.2.3.1	Helicenes	*373*
8.2.3.2	Silafluorenes	*374*
8.2.3.3	Angular [N]-Phenylenes	*375*
8.2.4	Development of New Catalytic Systems for the [2 + 2 + 2] Cycloaddition	*377*
8.2.4.1	Rhodium Cationic Complexes	*377*
8.2.4.2	Inexpensive Catalytic Systems	*377*
8.2.4.3	Microwave and Solid-Supported Cycloadditions	*378*
8.2.5	Enantioselective Reactions	*379*
8.3	1,3-Cyclohexadienes and their Heterocyclic Counterparts	*381*
8.3.1	Emergence (or Re-emergence) of Peculiar Unsaturated Partners	*381*
8.3.1.1	C≡C Bonds	*381*
8.3.1.2	C=C Bonds	*382*
8.3.1.3	C=X Bonds	*385*
8.3.2	Enantioselective [2 + 2 + 2] Cycloaddition	*391*
8.4	Cyclohexenes and their Heterocyclic Counterparts	*391*
8.5	Conclusion and Perspectives	*396*

8.6	Experimental: Selected Procedures 397
8.6.1	Synthesis of Compound 9 397
8.6.2	Synthesis of Compound 31 397
8.6.3	Synthesis of Compound 67 397
8.6.4	Synthesis of Compounds 100 and 101 398
8.6.5	Synthesis of Compound 116 398
8.6.6	Synthesis of Compound 145 398
8.6.7	Synthesis of Compound 174 399
	References 400
9	**Cyclizations Based on Cyclometallation** 407
	Hao Guo and Shengming Ma
9.1	Introduction 407
9.2	Titanium-Catalyzed Cyclizations 407
9.3	Zirconium-Catalyzed Cyclizations 414
9.4	Ruthenium-Catalyzed Cyclizations 419
9.5	Cobalt-Catalyzed Cyclizations 425
9.6	Rhodium-Catalyzed Cyclizations 427
9.7	Iridium-Catalyzed Cyclizations 439
9.8	Nickel-Catalyzed Cyclizations 440
9.9	Palladium-Catalyzed Cyclizations 448
9.10	Conclusion and Perspectives 449
9.11	Experimental: Selected Procedures 450
9.11.1	Typical Procedure for the Titanium-Catalyzed Cyclizations Shown in Scheme 9.4 450
9.11.2	Typical Procedure for the Zirconium-Catalyzed Cyclizations Shown in Scheme 9.16 450
9.11.3	Typical Procedure for the Ruthenium-Catalyzed Cyclizations Shown in Scheme 9.22 450
9.11.4	Typical Procedure for the Cobalt-Catalyzed Cyclizations Shown in Scheme 9.34 451
9.11.5	Typical Procedure for the Rhodium-Catalyzed Cyclizations Shown in Scheme 9.47 451
9.11.6	Typical Procedure for the Iridium-Catalyzed Cyclizations Shown in Scheme 9.63 451
9.11.7	Typical Procedure for the Nickel-Catalyzed Cyclizations Shown in eq 9.3 451
9.11.8	Typical Procedure for the Palladium-Catalyzed Cyclizations Shown in Scheme 9.77 452
	References 453
10	**Transition Metal Catalyzed Cyclization Reactions of Functionalized Alkenes, Alkynes, and Allenes** 457
	Nitin T. Patil and Yoshinori Yamamoto
10.1	Introduction 457

10.2	Cyclization Reactions of Alkenes 457
10.2.1	Cyclization with Tethered Nucleophiles 457
10.2.2	Hydroarylation Reactions 460
10.2.3	Other Cascade Processes 462
10.3	Cyclization Reactions of Alkynes 466
10.3.1	Cyclization with Tethered Nucleophiles 466
10.3.2	Domino sp–sp^2 Coupling and Cyclization 475
10.3.3	Cyclization/Functional Group Migration Reactions 479
10.3.4	Cyclization with Carbonyls/Imines/Epoxides/Acetals/Thioacetals, and so on 483
10.3.4.1	Aliphatic Tether 483
10.3.4.2	Aromatic Tether 486
10.3.5	Hydroarylation Reactions 501
10.3.6	Miscellanious Cyclization Reactions 502
10.4	Cyclization Reactions of Allenes 504
10.4.1	Cyclization with Tethered Nucleophiles 504
10.4.2	Cyclization with Carbonyls 512
10.4.3	Hydroarylation Reactions 513
10.5	Conclusion and Perspective 515
10.6	Experimental: Selected Procedures 516
10.6.1	Synthesis of Compound 13 516
10.6.2	Synthesis of Compound 20 516
10.6.3	Synthesis of Compound 40 517
10.6.4	Synthesis of Compound 82 517
10.6.5	Synthesis of Compound 100 517
10.6.6	Synthesis of Compound 134 517
10.6.7	Synthesis of Compound 145 517
10.6.8	Synthesis of Compound 184 518
10.6.9	Synthesis of Compound 259 518
10.6.10	Synthesis of Compound 275 518
10.6.11	Synthesis of Compound 287 518
10.6.12	Synthesis of Compound 325 518
	References 519
11	**Ring-Closing Metathesis of Dienes and Enynes** 527
	Miwako Mori
11.1	Introduction 527
11.2	Ring-Closing Metathesis of Dienes 531
11.2.1	Synthesis of Carbocycles and Heterocycles Using Ring-Closing Metathesis of Dienes 531
11.2.2	Synthesis of Carbo- and Heterocycles Using Ring-Opening Metathesis–Ring-Closing Metathesis of Dienes 542
11.2.3	Catalytic Asymmetric Ring Closing Metathesis of Dienes 547
11.2.3.1	Asymmetric Synthesis Using a Chiral Molybdenum Catalyst 547

11.2.3.2	Asymmetric Olefin Metathesis Using Ruthenium Catalyst *552*
11.2.4	Polymer-Supported Catalysts *554*
11.2.5	Synthesis of Natural Products Using RCM *557*
11.3	Ring-Closing Metathesis of Enynes *562*
11.3.1	Synthesis of Carbo- and Hetero-cycles Using Ring-Closing Metathesis of Enynes *562*
11.3.2	Ring-Closing Metathesis of Dienynes *573*
11.3.3	Ring-Opening Metathesis–Ring-Closing Metathesis of Cycloalkene-ynes *578*
11.3.4	Synthesis of Natural Products Using Ring-Closing Metathesis of Enyne *586*
11.4	Perspective *592*
11.5	Experimental: Selected Procedure *592*
11.5.1	Typical Procedure for the Synthesis of a Cyclized Compound from Enyne Using RCM *592*
	References *593*

12	**Ring-Closing Metathesis of Alkynes** *599*
	Paul W. Davies
12.1	Introduction *599*
12.2	Alkyne Metathesis *599*
12.3	Ring-Closing Alkyne Metathesis as a Synthetic Strategy *600*
12.4	Catalyst Systems for Ring-Closing Alkyne Metathesis *602*
12.4.1	The Mortreux "Instant" System *602*
12.4.2	Tungsten Alkylidyne Catalysts *602*
12.4.3	Molybdenum-Based Catalysts *603*
12.4.4	Comparison of the Reaction Systems *604*
12.5	Complex Molecule Synthesis using Ring-Closing Alkyne Metathesis *605*
12.5.1	Alkyne–Alkyne Metathesis *605*
12.5.2	Enyne-Yne Metathesis *612*
12.6	Ring-Closing Alkyne Metathesis within Transition Metal Coordination Spheres *615*
12.7	Cyclo Dimerizations, Trimerizations, and Oligomerizations *616*
12.8	Conclusion and Perspectives *618*
12.9	Experimental: Selected Procedures *618*
12.9.1	RCAM Synthesis of Compound 26 with Complex 1 *618*
12.9.2	RCAM Synthesis of Compound 26 with Complex 1 under Microwave Conditions *619*
12.9.3	RCAM Synthesis of Compound 28 with Complex 3 *619*
12.9.4	RCAM Synthesis of Compound 35a with Complex 2 *619*
12.9.5	Enyne-yne RCAM Synthesis of Compound 54 with Complex 3 *619*
12.9.6	Polycyclization of 67 with Complex 2 *619*
12.9.7	Polycyclization of 67 with Complex 9 *620*
	References *620*

Contents to Volume 2

13   Transition-Metal-Catalyzed Cycloisomerizations and Nucleophilic Cyclization of Enynes   *625*
Elena Herrero-Gómez and Antonio M. Echavarren

14   Cyclopropanation, Epoxidation and Aziridination Reactions   *687*
Song Ye and Yong Tang

15   Cyclization of Cyclopropane- or Cyclopropene-Containing Compounds   *733*
Junliang Zhang and Yuanjing Xiao

16   Transition Metal-Catalyzed Ring Expansion Cyclization Reactions   *813*
Masahiro Yoshida and Yoshimitsu Nagao

17   Hydrometallation-Initiated Cyclization Reactions   *843*
Isamu Matsuda

18   Catalytic Dipolar Cycloadditions of Alkynes with Azides and Nitrile Oxides   *917*
Valery V. Fokin

19   Electrophilic Cyclizations   *951*
Félix Rodríguez and Francisco J. Fañanás

20   Cyclizations Based on C–H Activation   *991*
David A. Capretto, Zigang Li, and Chuan He

21   Friedel–Crafts Type Cyclizations   *1025*
Ryuji Hayashi and Gregory R. Cook

22   Macrolactones and Macrolactams   *1055*
Chang-Liang Sun, Bi-Jie Li, and Zhang-Jie Shi

Handbook of Cyclization Reactions. Volume 1.
Edited by Shengming Ma
Copyright © 2010 WILEY-VCH Verlag GmbH & Co. KGaA, Weinheim
ISBN: 978-3-527-32088-2

23	**Free Radical Cyclization Reactions** *1099*	
	Jake Zimmerman, Amanda Halloway, and Mukund P. Sibi	
24	**Photocyclization Reactions** *1149*	
	Axel G. Griesbeck	
25	**Asymmetric Organocatalyzed Cyclization Reactions** *1199*	
	Liu-Zhu Gong, Jun Jiang, Meng-Xia Xue, and Shi-Wei Luo	

Index *1243*

Preface

Cyclic compounds including carbo- and heterocycles are one of the most important types of organic compounds showing interesting potentials in organic synthesis, pharmaceuticals, fragrances, agricultural industry, and material science, etc. In addition, cyclic skeletons also exist widely in natural products showing interesting biological activities. Thus, much attention has been paid to the development of efficient methods for their synthesis. The cyclic scaffolds are constructed via the formation of carbon-carbon and carbon-heteroatom bonds. Generally speaking, the syntheses of three- or four-membered rings are not easy due to the intrinsic strain of the rings; the formations of 5~7-memebered rings are very easy; the construction of 8~12-membered rings, usually hampered by entropic/enthalpic factors, is challenging; the assembly of macrocyclic compounds is not very difficult providing the reactions are conducted under diluted conditions.

There are some monographs in this area, most of them are in the area of cycloaddition reactions. Recently, in addition to the traditional Friedel-Crafts-type cyclizations, electrophilic cyclizations, substitution reactions, and different types of cycloadditions reactions, due to the rapid development of organometallic chemistry, organocatalysis, radical chemistry, etc., many new methods have been developed for the efficient construction of cyclic compounds from readily available starting materials, thus, there is a need to have a handbook summarizing the most typical advances in this area. With this purpose in mind, Wiley-VCH launched a task to ask me to act as an Editor to invite some of the well-known chemists in this area to write a book entitled "Handbook of Cyclization Reactions".

It is not so easy to include every sound contribution in this area, thus, I had come up with an outline of 24 chapters to try to include some of the most important advances in this area. It is always a difficult job to invite the chemists who have made important contributions to this area to come to this book-writing task due to their busy schedules. However, I am happy to see that many authors had kindly accepted my invitation to write a chapter or more for this handbook. Even in cases the originally proposed authors are too busy for such a chapter by themselves, they kindly suggested the best alternative authors for me. I thank all these people. Many

Handbook of Cyclization Reactions. Volume 1.
Edited by Shengming Ma
Copyright © 2010 WILEY-VCH Verlag GmbH & Co. KGaA, Weinheim
ISBN: 978-3-527-32088-2

thanks go to Dr. Elke Maase, Dr. Martin Graf, Dr. Ina Wiedemann and Yvonne Eckstein at Wiley-VCH. Finally I would also like to thank Mrs. Gong Hua and Chunling Fu in my group for their efforts devoted to this project.

Shanghai
July, 2009

Shengming Ma

List of Contributors to Volume 1

Paul W. Davies
University of Birmingham
School of Chemistry
Edgbaston
Birmingham, B15 2TT
UK

Kuiling Ding
Chinese Academy of Sciences
Shanghai Institute of
Organic Chemistry
State Key Laboratory of
Organometallic Chemistry
345 Lingling Road
Shanghai 200032
China

Haifeng Du
Beijing National Laboratory
for Molecular Sciences
Institute of Chemistry
Research Center for Chemical Biology
CAS Key Laboratory of Molecular
Recognition and Function
No. 21st North Street Zhongguancun
Beijing 100190
China

Vincent Gandon
Université Pierre et Marie Curie-Paris 6
Institut de Chimie Moléculaire
(FR 2769)
Laboratoire de Chimie Organique
(UMR CNRS 7611)
Case 229, 4 Place Jussieu
75252 Paris Cedex 05
France

Ron Grigg
University of Leeds
School of Chemistry
Molecular Innovation, Diversity and
Automated Synthesis (MIDAS) Centre
Woodhouse Lane
Leeds, LS2 9JT
UK

Hao Guo
Chinese Academy of Sciences
Shanghai Institute of
Organic Chemistry
State Key Laboratory of
Organometallic Chemistry
354 Fenglin Lu
Shanghai 200032
China

Xiuling Han
Chinese Academy of Sciences
Shanghai Institute of
Organometallic Chemistry
State Key Laboratory of
Organometallic Chemistry
354 Feng Lin Lu
Shanghai 200032
China

Takuya Hashimoto
Kyoto University
Graduate School of Science
Department of Chemistry
Rigakubu-ichi-goukan
Sakyo Kyoto 606-8502
Japan

Martyn Inman
University of Leeds
School of Chemistry
Molecular Innovation, Diversity and
Automated Synthesis (MIDAS) Centre
Woodhouse Lane
Leeds LS2 9JT
UK

Xuefeng Jiang
Chinese Academy of Sciences
Shanghai Institute of
Organic Chemistry
State Key Laboratory of
Organometallic Chemistry
354 Fenglin Road
Shanghai 200032
China

and

East China Normal University
Department of Chemistry
Shanghai Key Laboratory of Green
Chemistry and Technology
3663 North Zhongshan Road
Shanghai 200062
China

Shū Kobayashi
The University of Tokyo
School of Science and Graduate School
of Pharmaceutical Sciences
Department of Chemistry
Hongo, Bunkyo-ku
Tokyo 113-0033
Japan

David Lebœuf
Université Pierre et Marie Curie-Paris 6
Institut de Chimie Moléculaire
(FR 2769)
Laboratoire de Chimie Organique
(UMR CNRS 7611)
Case 229, 4 Place Jussieu
75252 Paris Cedex 05
France

Bi-Jie Li
Peking University
College of Chemistry and Molecular
Engineering
Beijing National Laboratory of
Molecular Sciences (BNLMS)
PKU Green Chemistry Centre and Key
Laboratory of Bioorganic Chemistry and
Molecular Engineering of Ministry of
Education
100871 Beijing
China

Xiyan Lu
Chinese Academy of Sciences
Shanghai Institute of
Organometallic Chemistry
State Key Laboratory of
Organometallic Chemistry
354 Feng Lin Lu
Shanghai 200032
China

Zhan Lu
Zhejiang University
Department of Chemistry
Laboratory of Molecular
Recognition and Synthesis
Hangzhou 310027
Zhejiang
China

Shengming Ma
Zhejiang University
Department of Chemistry
Laboratory of Molecular
Recognition and Synthesis
Hangzhou 310027
Zhejiang
China

Max Malacria
Université Pierre et Marie Curie-Paris 6
Institut de Chimie Moléculaire
(FR 2769)
Laboratoire de Chimie Organique
(UMR CNRS 7611)
Case 229, 4 Place Jussieu
75252 Paris Cedex 05
France

Keiji Maruoka
Kyoto University
Graduate School of Science
Department of Chemistry
Rigakubu-ichi-goukan
Sakyo Kyoto 606-8502
Japan

Miwako Mori
Health Sciences University of Hokkaido
1757 Kanazawa
Tobetsu-cho
Ishikari-gun
Hokkaido 061-0293
Japan

Nitin T. Patil
Indian Institute of Chemical Technology
Organic Chemistry Division II
Uppal Road
Hyderabad 500007
India

Yoshinori Yamamoto
Tohoku University
Graduate School of Science
Department of Chemistry
Aramaki, Aoba-ku
Sendai 980-8578
Japan

Yasuhiro Yamashita
The University of Tokyo
School of Science and Graduate School
of Pharmaceutical Sciences
Department of Chemistry
Hongo, Bunkyo-ku
Tokyo 113-0033
Japan

1
Asymmetric Catalysis of Diels–Alder Reaction

Haifeng Du and Kuiling Ding

1.1
Introduction

The Diels–Alder (DA) reaction (or "diene synthesis") between a diene and a dienophile generates two σ bonds stereoselectively and up to four chiral centers in a single step to afford six-membered cyclic compounds. This cycloaddition reaction named after Professor Otto Paul Hermann Diels (1876–1954) and his student Kurt Alder (1902–1958) was discovered in 1928 during studies on the reaction of benzoquinone with cyclopentadiene (Scheme 1.1) [1, 2], and has become an extremely useful and classic methodology in organic synthesis. For their discovery, Diels and Alder received the Nobel Prize in chemistry in 1950. Since this landmark discovery, great progress in the area has been achieved and the historical development for the past 80 years can be divided into three periods: (i) from the discovery to the 1950s, the studies mainly focused on the substrate scope and mechanism research, and one of the representative works was the Alder endo rule; (ii) from the 1950s to the 1970s, the DA reaction was successfully applied to the total synthesis of many complex molecules, and two important theories [molecular orbital theory (Woodward-Hoffmann rule) and frontier orbital theory (Kenichi Fukui)] were applied to explain the mechanism; (iii) from the 1980s to the present day, the exploration and application of an enantioselective version of DA reactions have attracted tremendous interest from chemists [3].

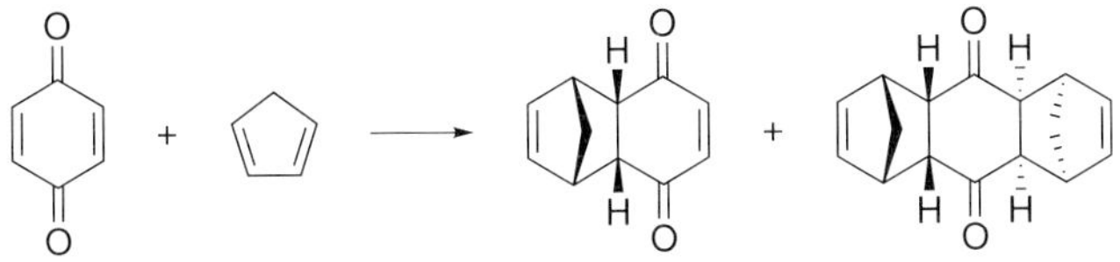

Scheme 1.1 The discovery of DA reaction.

Different types of DA reactions have been extensively reviewed, such as intramolecular DA reactions [4], dehydro-DA reactions [5], enzymatic catalysis of DA

Handbook of Cyclization Reactions. Volume 1.
Edited by Shengming Ma

ISBN: 978-3-527-32088-2

reactions [6], Lewis acid catalyzed enantioselective DA and the hetero Diels–Alder (HDA) reaction [7], organocatalysis of the DA and HDA reactions [8], DA reaction in aqueous media and so on [9]. For enantioselective DA reactions, chiral catalysts play a key role in the reactivity and enantioselectivity. In this chapter we will mainly discuss the development of chiral catalysts, including metal-based chiral Lewis acids and organocatalysts for DA and oxa HDA reactions, as well as their applications in the synthesis of natural products and biologically important compounds. The catalytic asymmetric aza Diels–Alder reaction is comprehensively reviewed in the following chapter by Kobayashi and will not be included here.

1.2
Asymmetric Diels–Alder Reaction

1.2.1
Lewis Acid Catalyzed Asymmetric Diels–Alder Reaction

Chiral Lewis acid catalyzed asymmetric reactions represent the most powerful methods to afford optically active compounds. For DA reactions, many excellent results have been achieved by applying various chiral Lewis acids as catalysts. In the following section, the development of catalytic asymmetric DA reactions will be highlighted, based on the category of Lewis acid catalyst.

1.2.1.1 Chiral Boron, Aluminum, and Indium Complexes

Chiral Boron Complexes In the field of asymmetric DA reactions, chiral boron complexes are among the most effective catalysts and have been thoroughly investigated [10]. In 1988, Yamamoto's group reported an efficient chiral (acyloxy) boron catalyst **1** (CAB) for the DA reaction between acrylic acid and cyclopentadiene to give the corresponding adduct in 93% yield and 78% ee (Scheme 1.2), which demonstrated for the first time the possibility of achieving useful enantioselectivities for DA reactions of simple dienophiles with a simple chiral controller ligand [11a]. Further expanding the substrate scope to α,β-unsaturated aldehydes and cyclic or acyclic dienes [11b] and studies on the mechanism [11c] make this catalyst more practical. In 1991, Corey and coworkers envisaged the possibility that the (S)-tryptophan-derived chiral oxazaborolindione **2** would facilitate the Diels–Alder pathway represented by the transition-state assembly (Scheme 1.2) through attractive interaction as well as the usual steric repulsion, and this surmise was confirmed by the reaction of cyclopentadiene and 2-bromoacrolein, affording the corresponding cycloadduct in 95% yield and 99% ee with the desired absolute configuration (Scheme 1.2) [12]. Allo-Threonine derived catalysts **3** were recently reported by Harada's group to catalyze the DA reactions of a variety of enone dienophiles with dienes, including cyclic, acyclic dienes or furan, in high yields, good diastereoselectivities and excellent enantioselectivities (Scheme 1.2) [13].

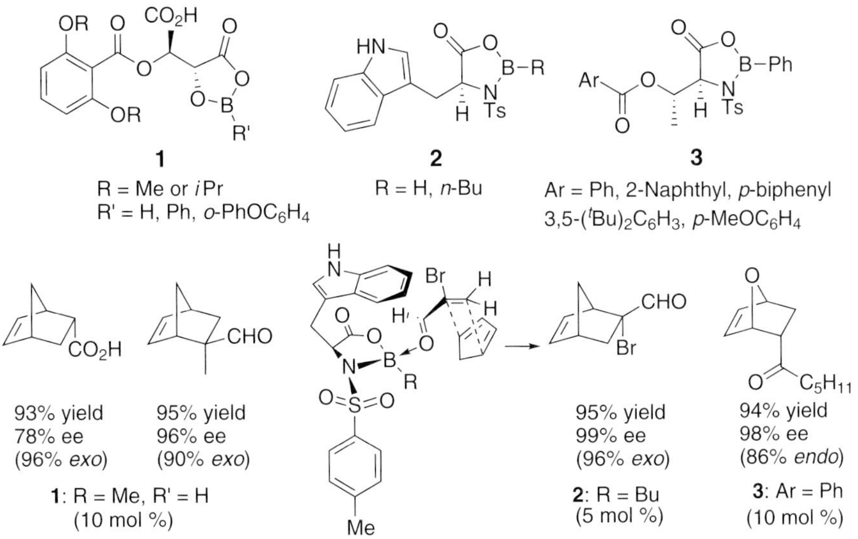

1.2 Asymmetric Diels–Alder Reaction

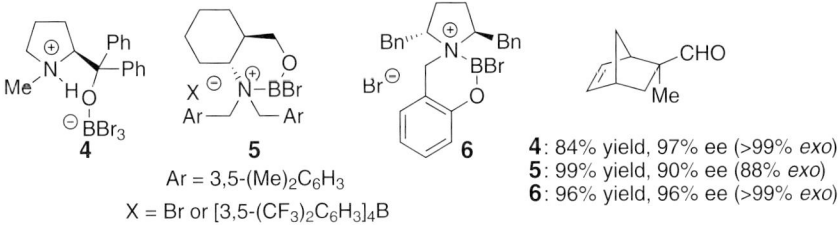

Scheme 1.2 Selected examples for enantioselective DA reactions catalyzed by **1–3**.

Cationic boron catalyst was found to be very effective for the asymmetric DA reaction due to its strong acidic property. Kobayashi's group and Aggarwal's group independently employed the chiral boron complex **4** (20 mol%) containing Lewis acid and Brønsted acid units as catalyst for the cycloaddition reaction between methacrolein and cyclopentadiene, up to 97% ee of the adduct was obtained (Scheme 1.3) [14]. In 1996, Corey's group developed a chiral super-Lewis acidic catalyst, excellent results (up to 98% ee) were obtained for the reaction of both active and inactive 1,3-dienes under the catalysis of **5** (10 mol%) [15a], and a similar catalyst **6** was developed in 2003 [15b].

Scheme 1.3 Selected examples for enantioselective DA reactions catalyzed by **4–6**.

Great progress for catalytic asymmetric DA reactions has been achieved using boron complexes by introducing a Brønsted acid or Lewis acid to assist or activate chiral boron complexes (Scheme 1.4). Yamamoto's group made pioneering contributions to this field [16]. In 1994, Yamamoto designed a chiral boron catalyst **7** via intramolecular Brønsted acid activation, and excellent reactivities and enantioselectivities were afforded for the reactions between α,β-unsaturated aldehydes and dienes (Scheme 1.4) [16a]. Later, a series of Brønsted-assisted Lewis acid (BLA)

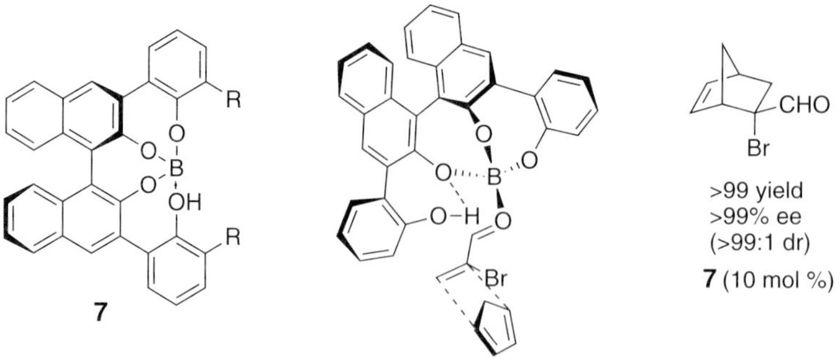

Scheme 1.4 Enantioselective DA reaction catalyzed by BLA catalyst.

catalysts were further synthesized and subjected to the asymmetric DA reactions. Intramolecular Brønsted acid was found to play an important role in accelerating the rate of DA reactions and in achieving a high level of enantioselectivity, due to intramolecular hydrogen bonding interaction and attractive π–π donor–acceptor interaction in the transition-state assembled by hydroxy aromatic groups in a chiral BLA catalyst (Scheme 1.4) [16b, c].

In 2002, Corey and coworkers developed a powerful chiral Lewis superacid **8a** which was easily generated from commercially available amino alcohol by condensation with an aryboroxine to afford an oxazaborolidine, followed by subsequent activation with triflic acid. Complex **8a** was found to be highly effective for the DA reactions of a variety of α,β-unsaturated aldehydes, ketones, and esters (Scheme 1.5). In order to avoid the decomposition of catalyst **8a**, triflimide was employed to generate a more stable catalyst **8b**. Under the catalysis of **8b**, DA reaction of a wide range of dienophiles, including diethyl fumarates, acrylate, enones and quinones, with cyclic or acyclic dienes gave the corresponding products in very high yields and ees (Scheme 1.5) [18].

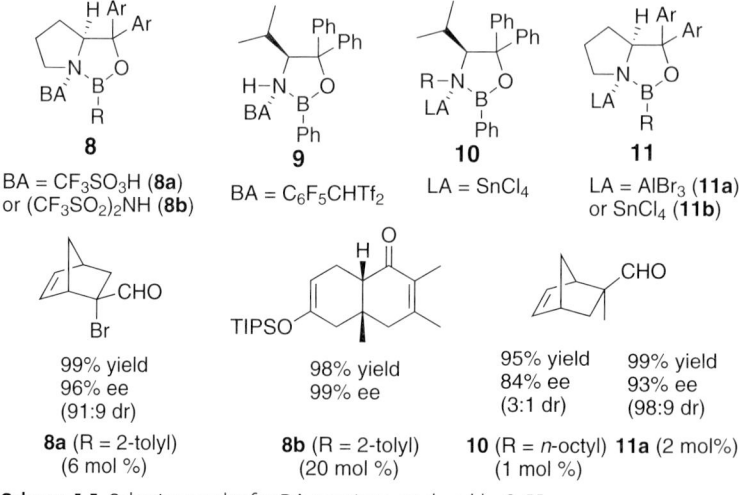

Scheme 1.5 Selective results for DA reactions catalyzed by **8–11**.

Yamamoto and coworkers reported the first regio- and enantioselective DA reactions of 1- and 2-substituted cyclopentadienes using Brønsted acid activated chiral oxazaborolindine **9** as catalysts. In the reactions of 1- and 2-substituted cyclopentadiene mixtures with ethyl acrylate, only 2-sustituted cyclopentadienes were found to be reactive, giving the corresponding products in high yields and ees (Scheme 1.6). A one-pot procedure for the preparation of the cycloadduct containing adjacent all-carbon quaternary stereocenters was also successfully achieved, whereby excess ethyl acrylate was employed to consume all the 2-substituted cyclopentadiene, and a more active dienophile (quinone) was added subsequently to react with the remaining 1-substituted cyclopentadiene [19].

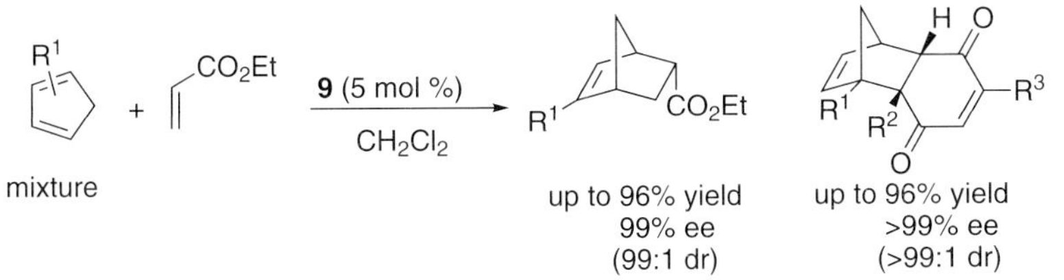

Scheme 1.6 Enantioselelective DA reactions catalyzed by **9**.

In 2005, Yamamoto and coworkers utilized a Lewis acid to activate chiral oxazaborolidine for asymmetric DA reactions and a type of moisture-tolerant catalysts **10** was developed. This catalyst system was effective for the reactions of a variety of dienophiles, including α,β-unsaturated aldehydes, esters, enones and quinones, affording the corresponding cycloadducts with a high level of reactivities and enantioselectivities (Scheme 1.5) [20a]. Corey's group also found that $AlBr_3$ could activate oxazaborolidine to generate useful catalysts **11a** for DA reactions of a very wide range of dienophiles and dienes (Scheme 1.5) [20b]. Very recently, Paddon-Row and coworkers reported the first computational investigations for oxazaborolidine catalyzed asymmetric DA reactions, which is very helpful for understanding the mechanistic aspects for the boron catalysts and the related transition state [20c].

Chiral Aluminum and Indium Complexes Corey's group reported a highly enantioselective DA reaction using a chiral aluminum complex **12** (10–20 mol%) derived from diamide ligand as catalyst, up to 95% ee of the adducts was obtained. Transition-state assembly was proposed and the absolute stereopreference in this DA reaction was believed to be the result of catalyst **12** binding to the acrylyl carbonyl of the dienophile at the lone pair anti to nitrogen, fixing the acrylyl group in the s-*trans* conformation (Scheme 1.7) [21]. Chiral aluminum complexes of biaryl ligands have also been exploited in asymmetric DA reactions. Wulff and coworkers developed catalyst **13** (0.5–10 mol%) derived from VAPOL, which was highly efficient for the reactions of cyclopentadiene and 2-methacrolein to give the *exo* cycloadduct in up to 100% yield with 98% ee (Scheme 1.8) [22]. In 2001, Yamamoto's group employed multinuclear chiral aluminum Lewis acids **14** (5–10 mol%), easily generated from the reaction of organoaluminum reagents with optically pure binaphthol derivatives, for DA reactions of cyclopentadiene and methyl acrylate, moderate to good enantioselectivies (Scheme 1.8) were attained [23]. Recently, Renaud and coworkers utilized hydroxamic acids as templates for chiral aluminum promoted enantioselective DA reactions, a stoichiometric amount of complex **15** was required in order to get high yields and enantiomeric excess of the adduct (Scheme 1.8). Facile coversion of the products to the corresponding alcohols or aldehydes makes the hydroxamic acid intermediates particularly useful [24].

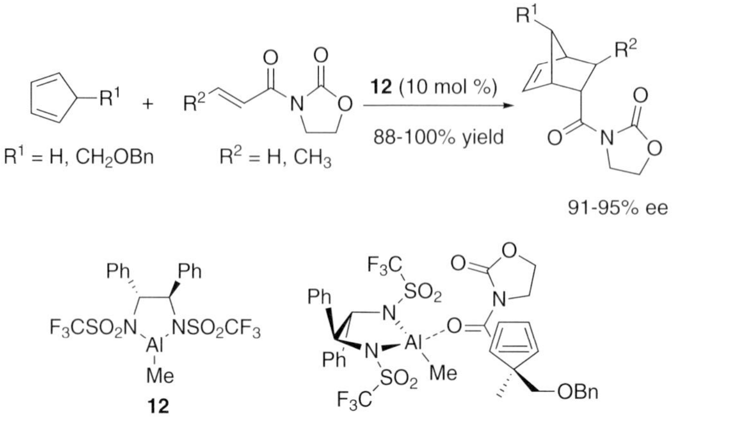

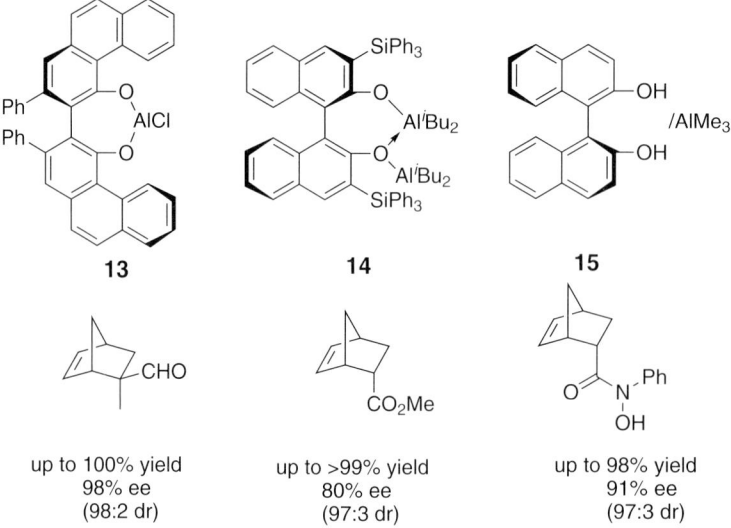

Scheme 1.7 Enantioselective DA reactions catalyzed by chiral aluminum complex **12**.

Scheme 1.8 Chiral aluminum complexes for DA reactions.

In 2005, Loh's group reported the first chiral indium complex **16** catalyzed asymmetric DA reactions [25a], in which allyltributylstannane was added to regenerate active chiral BINOL-In-allyl catalyst. The cycloaddition of a wide variety of cyclic or acyclic dienes with 2-methacrolein or 2-bromoacrolein gave the corresponding adducts in good yields and excellent enantioselectivities (Scheme 1.9).

In particular, the procedure is operationally simple and the catalyst can be easily prepared from commercially available chemicals at ambient temperature. This chiral indium catalyzed enantioselective DA reaction was also performed in ionic liquid, and catalysts could be recycled and reused for seven times without significant loss of reactivities and enantioselectivities [25b].

Scheme 1.9 Chiral indium catalyzed enantioselective DA reaction.

1.2.1.2 Chiral Copper, Magnesium, and Zinc Complexes

Chiral Copper Complexes Fruitful and excellent results for asymmetric DA reactions have been achieved under the catalysis of copper(II) complexes, and nitrogen-containing ligands are the most often used ligands [26]. Evans and coworkers demonstrated for the first time that bis(oxazoline)-copper(II) complex **17** could act as a Lewis acid to promote the enantioselective DA reaction of acryloyl-2-oxazolidinones and cyclopentadiene. Systematic investigation of the counterion effect, substrate scope, transition state model and synthetic applications was also carried out. It was disclosed that counterion structure and chelating dienophiles are critically important for this reaction [27]. The asymmetric DA reactions for substrates bearing other chelating templates (Scheme 1.10) catalyzed by **17** (1–16 mol%) have also been successfully achieved recently [28].

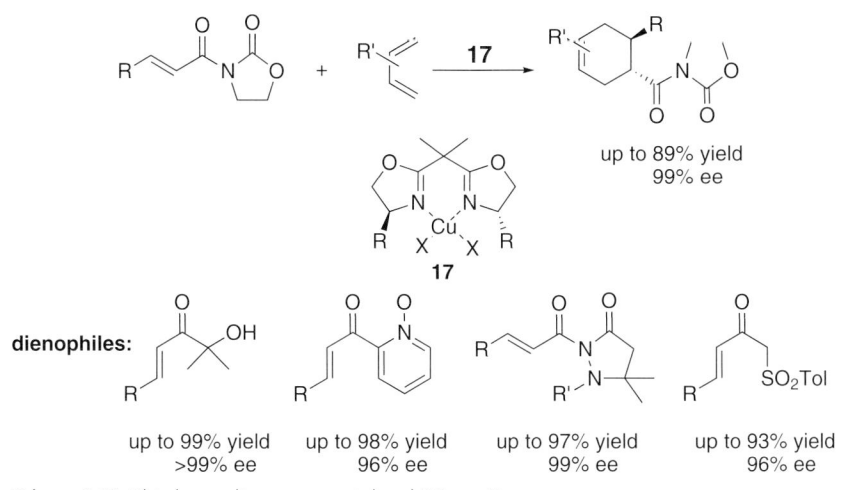

Scheme 1.10 Chiral oxazoline copper catalyzed DA reaction.

Bolm's group reported that C_2-symmetric bissulfoximines **18**, **19**/Cu(OTf)$_2$ complexes (10 mol%) catalyzed enantioselective DA reactions between cyclopentadiene and acryloyl-2-oxazolidinones, up to 93% ee of the adduct was obtained (Scheme 1.11) [29]. Further spectroscopic investigation of bissulfoximine copper (II) complexes indicated that the chiral ligand and the dienophile form a tetragonally distorted complex in CH$_2$Cl$_2$. The ligand binds to the Cu(II) center via the imine nitrogens, whereas the dieophile interacts via the carbonyl oxygen atoms [29b]. Ellman and coworkers developed analogous bis(sulfinyl)imidoamidine ligands **20** through a modular synthesis from readily available building blocks for Cu(II) catalyzed enantioselective DA reactions, providing a high level of asymmetric induction with 10 mol% of **20**/Cu(SbF$_6$)$_2$ (Scheme 1.11). Notably, the Cu(II) ligand complex exists as a unique M$_2$L$_4$ helicate in the solid state, and predominately coordinates via the sulfin oxygen in a solution of **20** [30a]. In 2004, Lassaletta *et al.* utilized chiral bis-hydrazones **21** as the ligands for Cu(OTf)$_2$-catalyzed DA reactions, the corresponding cycloadducts were obtained in good yields and enantioselectivities (Scheme 1.11) [30b].

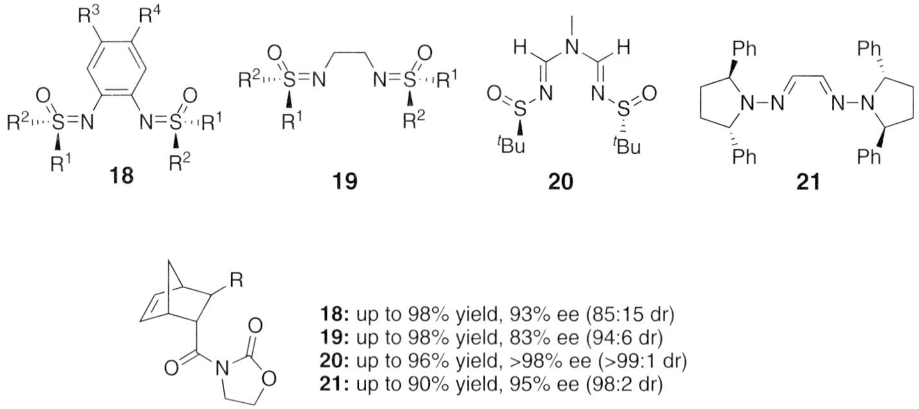

18: up to 98% yield, 93% ee (85:15 dr)
19: up to 98% yield, 83% ee (94:6 dr)
20: up to 96% yield, >98% ee (>99:1 dr)
21: up to 90% yield, 95% ee (98:2 dr)

Scheme 1.11 Enantioselective DA reaction catalyzed by **18–21**/Cu(II).

Ishihara's group designed a small-molecule artificial metalloenzyme which was prepared *in situ* from L-DOPA derived monopeptide **22** (2.2–11 mmol%) and Cu(OTf)$_2$ or Cu(NTf$_2$)$_2$ (2–10 mol%) for the enantioselective DA reactions between α,β-unsaturated 1-acyl-3,5-dimethylpyrazoles and cyclopentadienes. This biomimetic catalytic system provided the corresponding cycloadducts in good yields with excellent enantioselectivities. The authors suggested that the existence of intramolecular cation–π interaction in [**22a**·Cu(II)](OTf)$_2$ might be very important for controlling the conformation of sidearms in chiral ligands (Scheme 1.12) [31a]. Very recently, asymmetric DA reactions of propiolamide derivatives and cyclopentadienes catalyzed by **22b**/Cu(NTf$_2$)$_2$ (10 mol%) have been successfully realized in excellent yields and enantioselectivities [31b]. In the field of Cu(II)-catalyzed DA reactions, DNA-based Cu(II) catalysts (5–30 mol%) have also been successfully developed by Feringa et al., and excellent enantioselectivities (up to 98%) were attained [32]. The asymmetric induction in this catalytic system arises completely from the chirality of the DNA template.

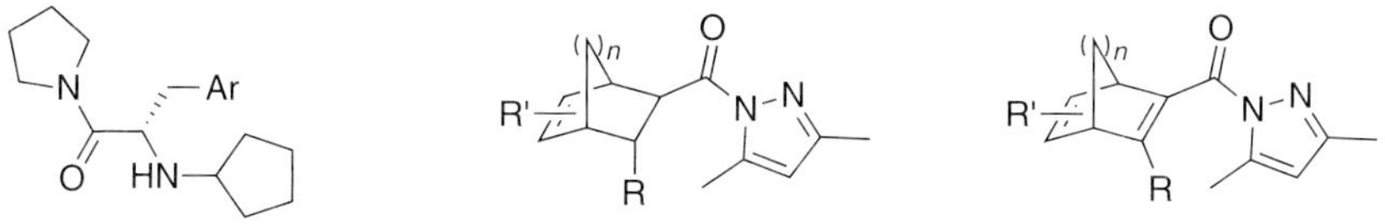

22a: Ar = 3,4-(MeO)$_2$C$_6$H$_3$,
22b: 2-naphthyl

22a/Cu(OTf)$_2$ 76->99%yield
or Cu(NTf$_2$)$_2$: (87-98% ee)
(88->99% endo)

22b/Cu(NTf$_2$)$_2$: 22-91% yield
(89-96% ee)

Scheme 1.12 Chiral monopeptide ligand for Cu(II)-catalyzed DA reactions.

Chiral Zinc and Magnesium Complexes In 1996, Evans and coworkers discovered that chiral bis(oxazoline)-Zn(II) complex was an effective catalyst for the enantioselective DA reaction and the counterion was found to be critical for this reaction, 92% ee of the adduct was achieved using **23** (10 mol%) as catalyst (Scheme 1.13) [27c,33a]. Chiral bis(oxazoline)-Mg(II) complex **24** (10 mol%) with 2 equiv of H$_2$O or tetramethylurea as auxiliary ligands has also proven to be an effective catalyst for the DA reaction with good yields and excellent enantioselectivities (Scheme 1.13) [33b,c,d]. In 2002, Renaud's group reported enantioselective DA reactions of cyclopentadiene and N-alkoxyacrylamides using a stoichiometric amount of BINOL-Zn complex prepared from BINOL and Et$_2$Zn, the corresponding adducts was obtained in up to 96% ee (Scheme 1.14) [33e].

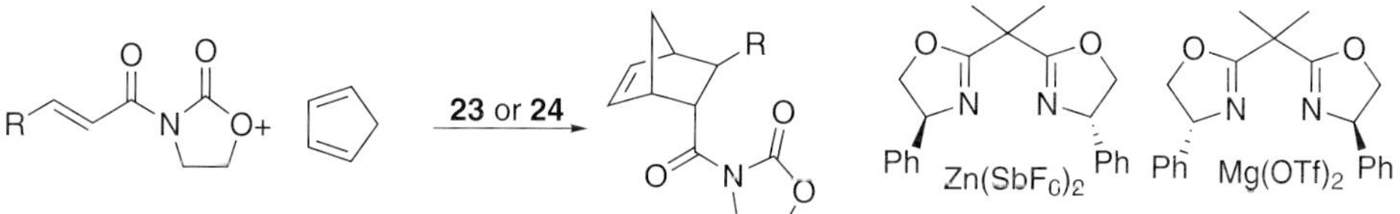

23
>90% yield
92% ee (98:2 dr)

24
100% yield
95% ee (94:6 dr)

Scheme 1.13 Chiral bis(oxazoline)/Cu(II)-catalyzed DA reaction.

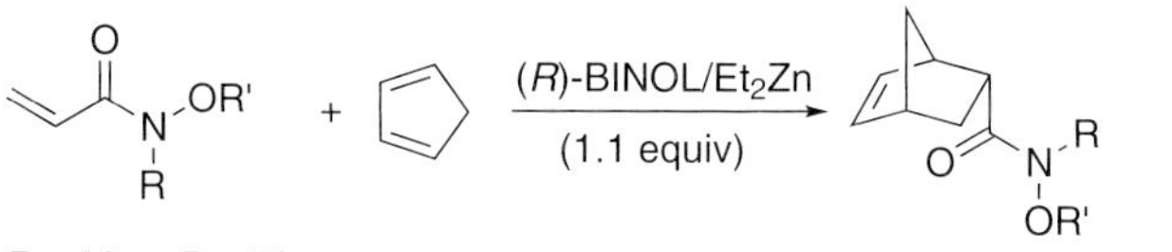

R = Me, t-Bu, Ph
R' = Me, Et, i-Pr

up to 100% conv.
96% ee
(54:1 dr)

Scheme 1.14 BINOLate zinc catalyzed DA reaction.

1.2.1.3 Chiral Transition Metal Complexes

Chiral Titanium Complexes Mikami and coworkers reported an asymmetric DA reaction between naphthoquinone and 1-methoxylbutadiene catalyzed by chiral titanium complexes **25** (R = H, X = Cl, 10 mol%) (Scheme 1.15) generated from BINOL and Ti(O^{i}Pr)$_2$Cl$_2$ in the presence of molecular sieves (MS), affording a single endo stereoisomer in 85% ee. This approach has provided a potential entry to the asymmetric synthesis of anthracycline antibiotics [34a]. Further studies showed the DA reactions of 5-hydroxynaphthoquinone (juglone) with butadienyl acetate using a MS-free catalyst **25** afforded the corresponding cycloadduct in up to 96% ee (Scheme 1.15). However, only 9% ee was obtained in the presence of MS [34b]. Employing the same catalysts for enantioselective cycloaddition of methacrolein to alkoxyadiene gave the corresponding products in good yields and ees (Scheme 1.15) [34b,c]. Yamamoto's group developed an alternative chiral helical titanium complex (**26**, Scheme 1.15) and applied this type of catalyst in enantioselective DA reactions of cyclopentadienes with methacrolein derivatives, affording the corresponding products in up to 86% yield with 98% ee (Scheme 1.15) [34d].

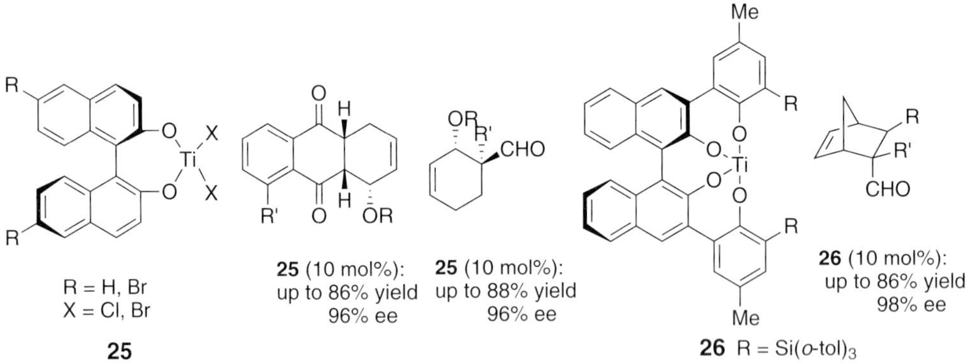

Scheme 1.15 Chiral titanium complex catalyzed DA reaction.

Chiral Chromium and Cobalt Complexes Rawal's group achieved a highly enantioselective DA reaction of substituted 1-amino-1,3-butadienes and methacrolein by the catalysis of Jacobsen's salen-Cr(III) complex **27** (5 mol%) in the presence of 4 Å MS in CH$_2$Cl$_2$, affording the corresponding highly functionalized cyclohexene derivatives having a quaternary chiral center with excellent yields and ees (Scheme 1.16). The high endo-selectivity of the DA reaction and the substituents at the 1-position of the diene were found to be very important for the success of this process [35]. Salen-Co(III) complexes **28** have also proven to be highly efficient for the same reaction, and as low as 0.05 mol% catalyst loading was enough to achieve high yield (93%) and enantioselectivity (98%) of the reaction (Scheme 1.16). Importantly, the reactions are conveniently carried out at room temperature, under an air atmosphere, and with minimal solvent. These conditions are highly desirable for industrial applications [36].

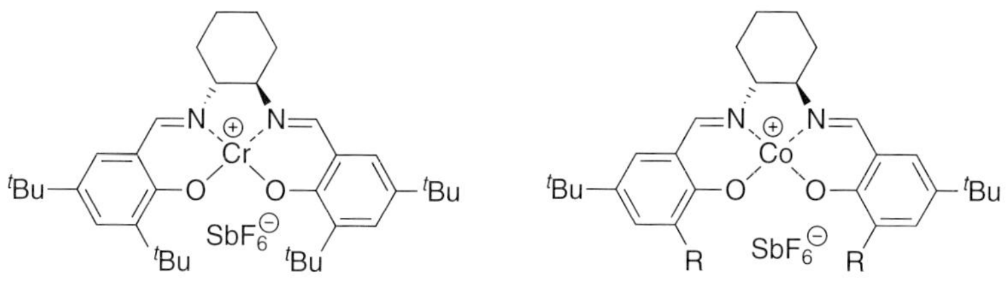

1.2 Asymmetric Diels–Alder Reaction

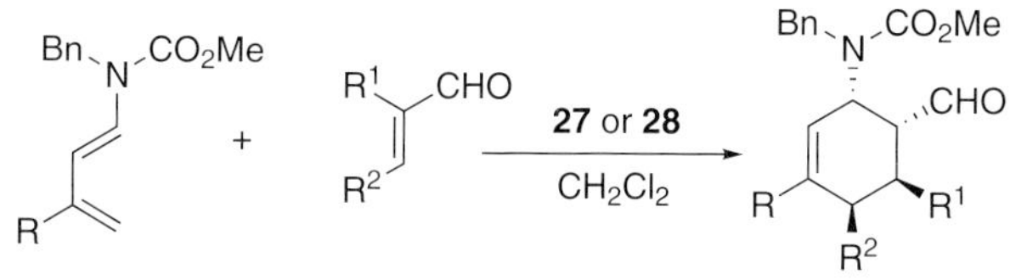

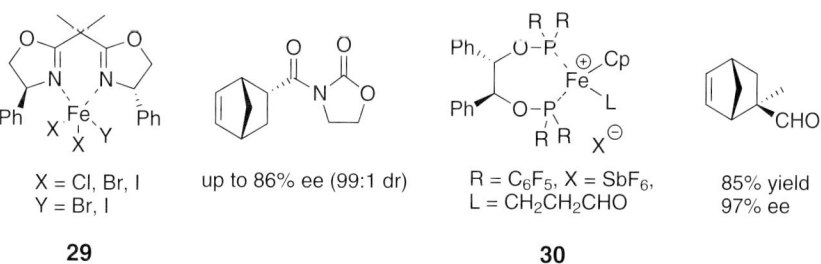

27: up to 85% yield (97% ee)
28: up to 93% yield (98% ee)

Scheme 1.16 Salen-chromium and Salen-cobalt catalyzed DA reaction.

Chiral Iron Complexes In 1991, Corey and coworkers employed chiral bis(oxazoline)/Iron(III) complexes **29** (10 mol%) (Scheme 1.17) as the catalysts for enantioselective DA reactions of cyclopentadiene and 3-acryloxyl-1,3-oxazolindin-2-one, the corresponding cycloadduct was obtained in 86% ee with 99:1 endo/exo selectivity. The chiral ligand was readily and efficiently recoverable from the reactions for reuse. The ready availablility of chiral ligand combined with the low cost of iron salts make this methodology potentially useful for practical application [37a]. Kündig's group developed chiral cyclopentadienyl-iron(III) complexes **30** bearing electron-poor phosphine ligands as catalysts (5 mol%) for cycloaddition of 2-methacrolein and cyclopentadiene in CH_2Cl_2, the corresponding product was afforded in 85% yield and 97% ee (Scheme 1.17) [37b].

X = Cl, Br, I
Y = Br, I

up to 86% ee (99:1 dr)

R = C_6F_5, X = SbF_6,
L = CH_2CH_2CHO

85% yield
97% ee

29

30

Scheme 1.17 Chiral iron(III) complex catalyzed DA reaction.

Chiral Ruthenium Complexes Davies and coworkers discovered that chiral arene ruthenium complexes **31** were highly efficient for the DA reaction between methacrolein and cyclopentadiene with only 0.5 mol% catalyst loading. The reaction proceeded rapidly at room temperature (0.25 h) to give the corresponding cycloadduct in high yield and ee value with good *exo* selectivity (Scheme 1.18) [38]. Faller's group prepared chiral bis(oxazoline) ligand modified ruthenium(II) complexes **32** and subjected this type of complex to the reaction of methacrolein and cyclopentadiene, up to 91% ee value and 98% *exo* selectivity were achieved (Scheme 1.18) [39a]. The same group also reported enantioselective DA reactions with the racemic Ru/BINAP-monoxide complex using proline or prolinamide as a chiral poisoning agent [39b]. Since 2001, Kündig's group has developed a series of chiral ruthenium catalysts (**33** and **34**) containing bidentate phosphine ligands and systematically investigated their application in enantioselective DA reactions of cyclopentadiene derivatives with methacrolein or α,β-unsaturated ketone. The corresponding adducts have been obtained in good yields with excellent diastereoselectivities and enantioselectivities (Scheme 1.18). Notably, this group reported the first example of a one-point binding transition metal Lewis acid catalyst capable of coordinating and activating an α,β-unsaturated ketone for enantioselective DA reactions [37b, 40]. Other transition metal complexes such as rhodium [41a], palladium [41b], and iridium [41c] have also been applied to asymmetric DA reactions.

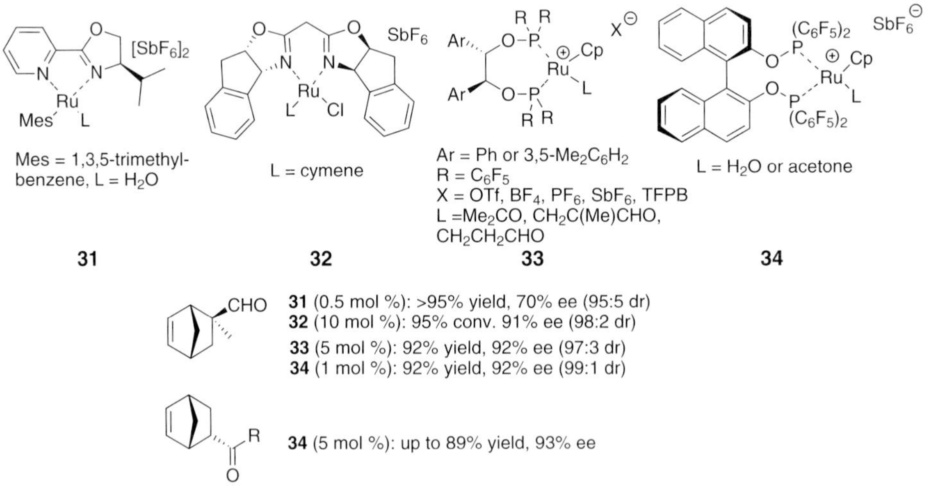

Scheme 1.18 Enantioselective DA reaction catalyzed by chiral ruthenium complexes.

1.2.1.4 Chiral Rare Earth Metal Complexes

In 1994, Kobayashi's group reported that enantioselective DA reaction of 3-acryl-1,3-oxazolindin-2-ones with cyclopentadienes using chiral ytterbium triflate catalyst (20 mol%) generated from BINOL, Yb(OTf)$_3$ and tertiary amine yielded endo adducts as the major isomers in up to 93% ee (Scheme 1.19). Other lanthanide (Lu, Tm, and Er) triflate complexes of BINOL were also found to be effective catalysts for the same reaction [42a]. Tertiary amines played a key role for both diastero and enantioselectivities, and *cis*-2,6-dimethyl-*N*-methylpiperidine proved to be the best additive. Most interestingly, both enantiomers are afforded with the same catalysts by addition or without addition of 1,3-diketones as achiral ligands [42a, b]. Markó

1.2 Asymmetric Diels–Alder Reaction

and coworkers successfully developed an asymmetric inverse electron-demand DA reaction of 3-carbomethoxy-2-pyrone derivatives and electron-rich dienophiles using **35** (20 mol%) as catalysts. The corresponding cycloadducts were obtained in up to 95% ee; they are key intermediates for the preparation of complex polycyclic compounds [43]. Very recently, Nishida and coworkers reported the first highly enantioselective DA reaction of electron-rich Danishefsky's diene with electron-deficient alkenes under the catalysis of complex **36**/Yb(OTf)$_3$ in the presence of DBU as the additive (Scheme 1.19). Remarkable (+)-nonlinear effects observed in this catalytic system suggested the possibility of the formation of a reservoir of nonreactive aggregates in the reaction system [44].

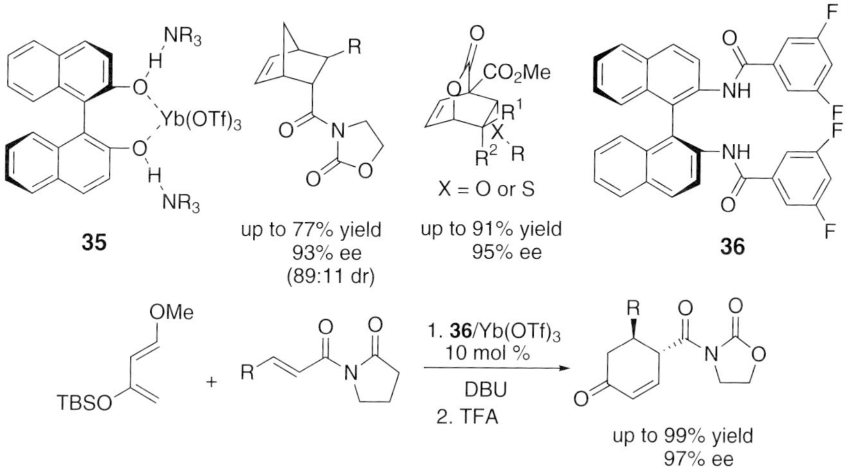

Scheme 1.19 Enantioselective DA reaction catalyzed by chiral lanthanide complexes.

Chiral 2,6-bis(oxazolidinyl) pyridines (pybox) were found to have a wide range application for lanthanide-catalyzed DA reactions (Scheme 1.20). In 2001, Furuzawa's group discovered that isopropyl-pybox (**37c**)/Sc(OTf)$_3$ complex (10 mol%) could catalyze the enantioselective DA reaction of acyl-1,3-oxazolidin-2-one with cyclic or acyclic dienes, yielding the corresponding products in good enantioselectivities (up to 90% ee). This reaction was also performed in less toxic benzotrifluoride and supercritical carbon dioxide, giving the products with good selectivity (65–75% ee) [45a]. Desimoni and coworkers also contributed a lot of effort on the pybox/lanthanide-catalyzed DA reactions [45b–f]. Under the catalysis of **38b**/Eu(OTf)$_3$ exo products were obtained for the reaction of acyl-1,3-oxazolidin-2-one with cyclopentadiene in 52% yield and >99% ee. The high enantioselectivity was attributed to the presence of a phenyl group at the 5-position of **38b** [45b]. Further studies showed that MS [45c], substitutents of pybox [45c–f], and lanthanide cations [45c–f] significantly affected both reactivities and enantioselectivities of the reaction. Xiao and coworkers investigated the impact of the electronic and steric properties of **38** on the enantioselectivity of the DA reaction of acyl-1,3-oxazolidin-2-one with cyclopentadiene, up to 96% ee of the corresponding cycloadduct was achieved under the catalysis of **39b**/Sc(OTf)$_3$

(5 mol%) [46]. In 2003, Evans and coworkers reported that chiral samarium and gadolinium complexes generated from pyridyl-bis(oxazoline) ligands (**37a** and **37b**) were able to catalyze enantioselective DA reaction of quinone with substituted 1,3-dienes, affording the corresponding products with excellent diastereoselectivities and enantioselectivities (Scheme 1.21). The absence of a nonlinear effect in this system suggested that neither catalyst aggregation nor dimer formation was occurring [47].

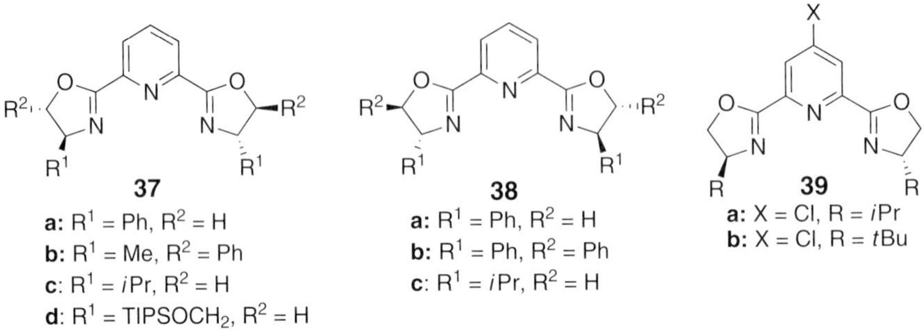

37
a: R^1 = Ph, R^2 = H
b: R^1 = Me, R^2 = Ph
c: R^1 = iPr, R^2 = H
d: R^1 = TIPSOCH$_2$, R^2 = H

38
a: R^1 = Ph, R^2 = H
b: R^1 = Ph, R^2 = Ph
c: R^1 = iPr, R^2 = H

39
a: X = Cl, R = iPr
b: X = Cl, R = tBu

Scheme 1.20 Representative pybox Ligands for lanthanides-catalyzed DA Reactions.

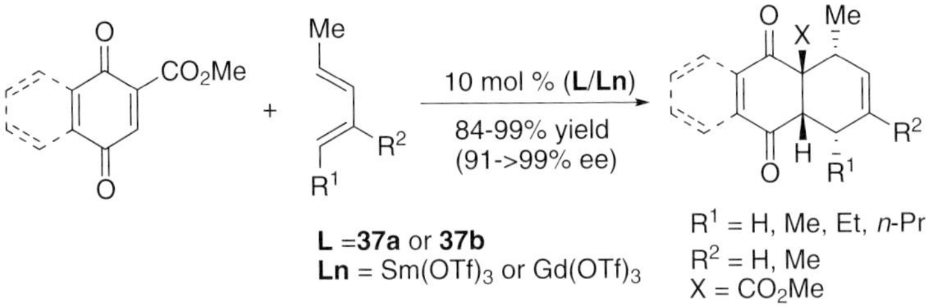

10 mol % (**L/Ln**)
84–99% yield
(91–>99% ee)

L = **37a** or **37b**
Ln = Sm(OTf)$_3$ or Gd(OTf)$_3$

R^1 = H, Me, Et, n-Pr
R^2 = H, Me
X = CO$_2$Me

Scheme 1.21 Enantioselective DA reaction catalyzed by chiral lanthanide complexes.

1.2.2
Organocatalysis of Asymmetric Diels–Alder Reaction

1.2.2.1 Chiral Secondary Amine Catalyzed Asymmetric Diels–Alder Reaction

Metal-based chiral Lewis acids have dominated the field of asymmetric catalysis for over 30 years, while organocatalysis emerging just in the last decade has been attracting extensive interest from chemists. In 2000, MacMillan's group reported the first highly enantioselective DA reaction of α,β-unsaturated aldehydes with 1,3-dienes under the catalysis of chiral secondary amine catalyst **40** (Scheme 1.22). The LUMO-lowering activation of aldehyde with amine to form reversible iminium ion intermediate (Scheme 1.22) provided great opportunities for exploring a novel asymmetric catalysis system [48]. Later, the chiral secondary amine catalyzed enantioselective DA reaction via iminium intermediate was thoroughly investigated by searching for new efficient catalysts and broadening the substrate scope. In 2002, MacMillan and coworkers successfully extended their work from α,β-unsaturated aldehydes to ketones with a modified catalyst **41** (20 mol%) (Scheme 1.23) [49]. The same group also developed the first organocatalytic intramolecular DA reaction under the catalysis of **42** (20 mol%), and completed the total synthesis of marine metabolite solanapyrone D in six steps based on the developed methodology (Scheme 1.23) [50].

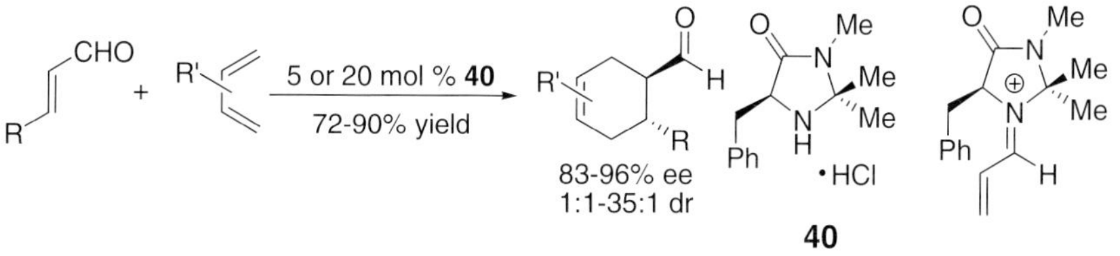

Scheme 1.22 The first organocatalysis of asymmetric DA reaction.

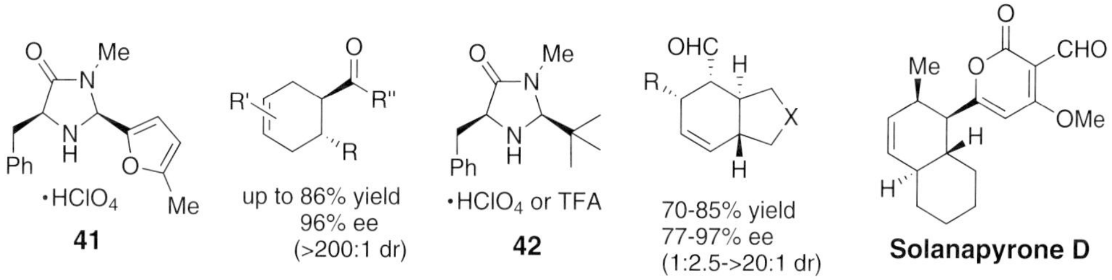

Scheme 1.23 Enantioselective DA reaction catalyzed by chiral secondary amine **41–42**.

In 2001, Bonni's group developed the HCl or $HClO_4$ salt of chiral aziridine **43** for catalysis of asymmetric DA reactions, moderate enantioselectivities with almost 1:1 diastereoselectivities were obtained (Scheme 1.24) [51]. Hayashi and coworkers found that **44a**, a trifluoroacetic acid (TFA) salt of diarylprolinol silyl ether, was an efficient organocatalyst for *exo*-selective DA reactions of α,β-unsaturated aldehydes with cyclopentadiene in toluene, up to 97% ee was achieved [52a]. Very recently, the same group successfully developed a practical procedure for enantioselective DA reaction using water as a solvent under the catalysis of **44a**. It was interesting to find that water could significantly accelerate the reaction and obviously improve the enantioselectivity of the catalysis [52b]. In 2008, Lee and coworkers discovered that camphor-derived sulfonyl hydrazines **45** (20 mol%) in the presence of trichloroacetic acid (10 mol%) could promote the same reaction at 0 °C or room temperature in brine without any added organic solvent to give the endo cycloadducts as the major isomers in 71–99% yields with good to excellent ees (93–96%) (Scheme 1.24) [53]. A binaphthyl-based diamine **46** was synthesized by Maruoka and coworkers in 2006 [54]. The application of diamine **46** (12 mol%) in combination with $TsOH \cdot H_2O$ (10 mol%) for the enantioselective DA reaction of α,β-unsaturated aldehydes with cyclopentadiene afforded the *exo* product (13 : 1 dr) with up to 92% ee (Scheme 1.24).

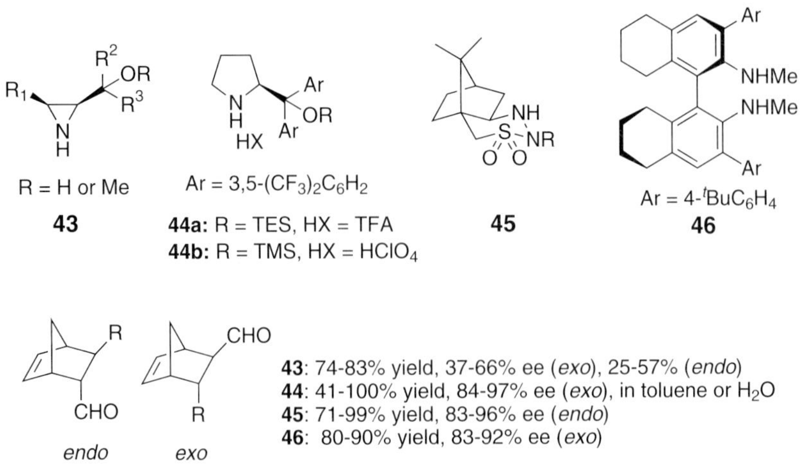
R = H or Me
43
Ar = 3,5-(CF3)2C6H2
44a: R = TES, HX = TFA
44b: R = TMS, HX = HClO4
45
Ar = 4-tBuC6H4
46
endo
exo
43: 74-83% yield, 37-66% ee (exo), 25-57% (endo)
44: 41-100% yield, 84-97% ee (exo), in toluene or H2O
45: 71-99% yield, 83-96% ee (endo)
46: 80-90% yield, 83-92% ee (exo)

Scheme 1.24 Enantioselective DA reaction catalyzed by chiral secondary amine **43–46**.

1.2.2.2 Chiral Primary Amine Catalyzed Asymmetric Diels–Alder Reaction

Chiral secondary amines have proven not to be effective catalysts for the DA reaction of sterically hindered α-substituted acroleins due to their bulkiness of chiral and the difficulty in the formation of iminium ions with α-substituted acroleins. In 2005, Ishihara and coworkers realized the first organocatalyzed DA reaction of α-substituted acroleins with a series of cyclic and acyclic dienes using less bulky primary amine **47** (Scheme 1.25), and the corresponding *exo* cycloadducts were attained in excellent yields and ees (Scheme 1.26) [55]. In an effort to improve the catalytic activity and enantioselectivity of the primary amine catalyzed DA reaction of α-substituted acroleins, Ishihara utilized a weak basic aromatic amine **48** (5 mol%) in combination with the strong Brønsted acid Tf_2NH as catalyst, up to 97% yield and 91% ee of the corresponding cycloadduct were afforded (Scheme 1.26) [56]. At the same time, Ha's group reported that the HCl salt of 1,2-diamino-1,2-diphenylethane **49** also promoted the asymmetric DA reaction of cyclopentadiene with crotonaldehyde, giving the corresponding products in high yields with good to excellent enantiomeric excesses (Scheme 1.26) [57].

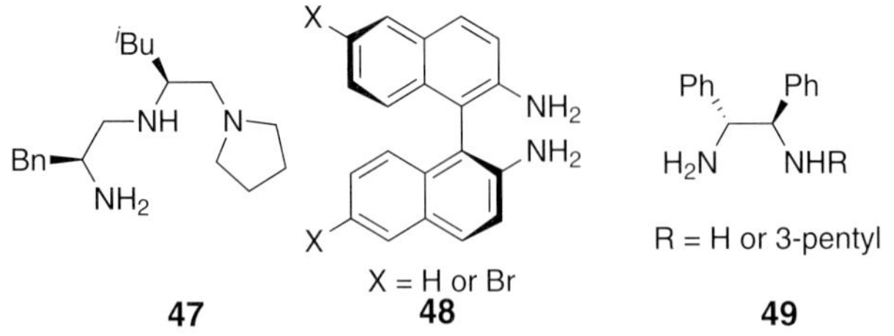
X = H or Br
R = H or 3-pentyl
47
48
49

Scheme 1.25 Chiral primary amine catalysts.

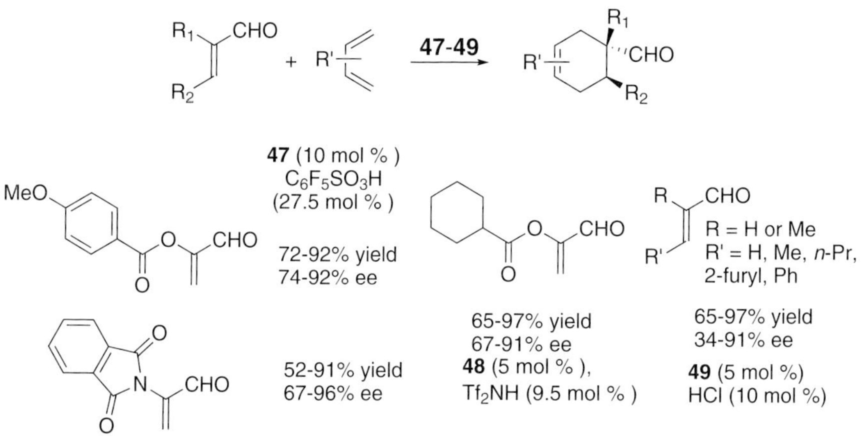

1.2 Asymmetric Diels–Alder Reaction | 17

Scheme 1.26 Enantioselective DA reaction catalyzed by chiral primary amine **47–49**.

Very recently, Deng and coworkers developed an efficient asymmetric DA reaction of 2-pyrone derivatives with a variety of α,β-unsaturated ketones, one type of more challenging substrates, under the catalysis of cinchona alkaloid **50**, the corresponding *exo* cycloadducts were obtained in moderate to excellent yields with 90–99% ees (Scheme 1.27) [58]. The corresponding cycloadduct could be transformed to one type of highly functionalized chiral intermediate by simple thermal decarboxylation without loss of enantioselectivity.

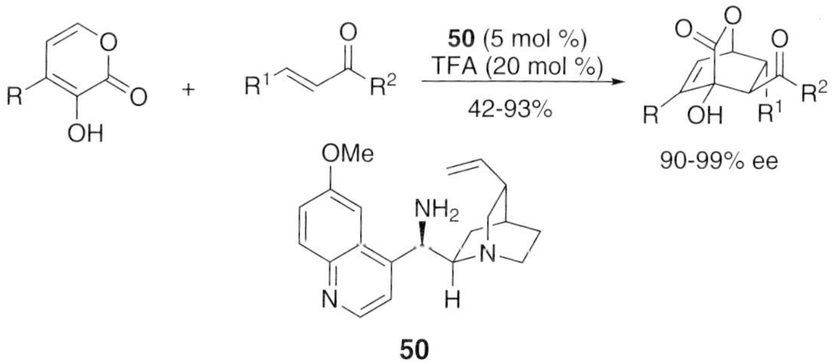

Scheme 1.27 Chiral primary amine **50** catalyzed DA reaction.

1.2.2.3 Brønsted Acid Catalyzed Asymmetric Diels–Alder Reaction

In the field of organocatalysis of DA reactions, the use of chiral Brønsted acid as catalyst [59a–c] has been less developed in comparison with chiral Lewis base. The first Brønsted acid catalyzed DA reactions were reported by Göbel's group by employing axially chiral amidinium ions **51** despite the fact that only 40% ee of the adduct was obtained with as high as 50 mol% catalyst loading (Scheme 1.28) [59d–e]. In this catalytic system, the diketone substrate was activated by chiral Brønsted acid via multiple hydrogen-bonding interaction. The first highly efficient and enantioselective DA reaction was realized by Rawal's group employing (α,α,α′,

1 Asymmetric Catalysis of Diels–Alder Reaction

α′-tetraaryl-1,3-dioxolane-4,5-dimethanol) TADDOL **52** as a Brønsted acid catalyst [60]. Up to 92% ee of the corresponding cycloadduct was obtained for the reaction of aminosiloxydienes and substituted acroleins (Scheme 1.29). A possible asymmetric induction model was proposed, in which the dienophile was supposed to be activated by an intermolecular hydrogen bonding between carbonyl and hydroxy groups, and the intramolecular hydrogen bonding between two hydroxy groups of TADDOL catalyst can facilitate the intermolecular hydrogen-bonding activation for the catalysis [60].

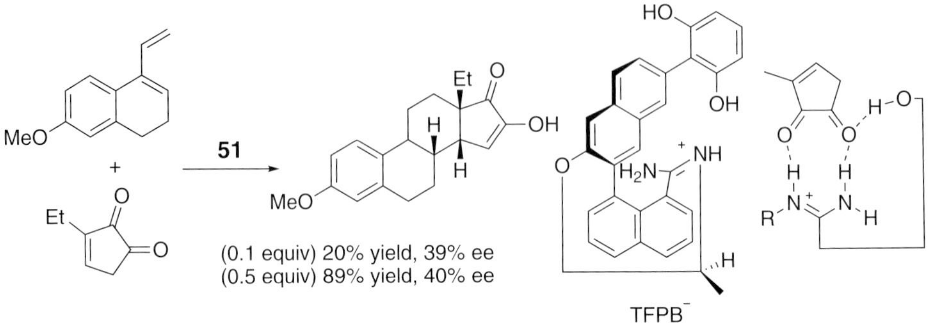

Scheme 1.28 Enantioselective DA reaction promoted by hydrogen bonding interaction.

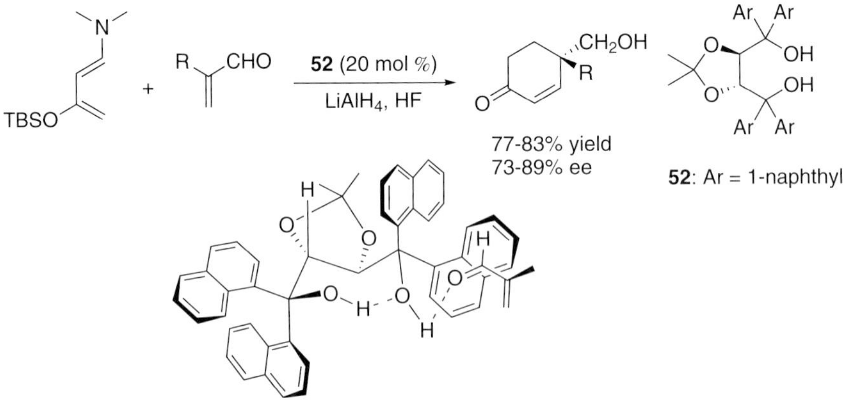

Scheme 1.29 Hydrogen bonding promoted DA reaction by TADDOL **52**.

On the basis of hydrogen-bonding activation strategy, Yamamoto's group developed an alternative chiral Brønsted acid, N-triflyl phosphoramide **53** derived from optically pure 3,3′-Ar$_2$BINOL, for the enantioselective DA reaction of ethyl vinyl ketone with electron-rich silyloxydiene, and the corresponding products were achieved in moderate to excellent yields with high enantioselectivities (Scheme 1.30). The success of this catalytic system was attributed to the strong Brønsted acidity of N-triflyl phosphoramide [61].

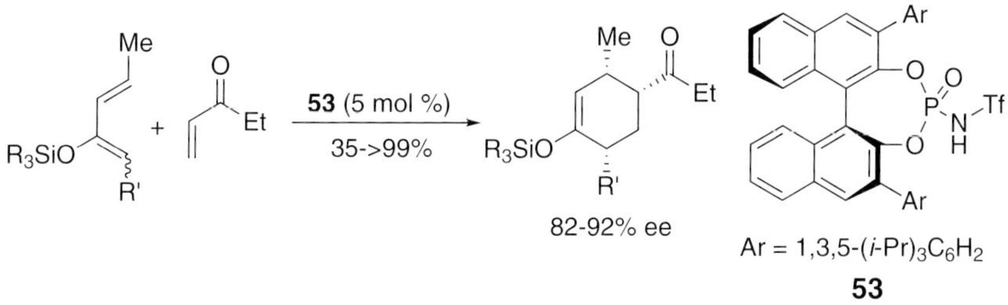

Scheme 1.30 Chiral phosphoramide catalyzed enantioselective DA reaction.

1.2.2.4 Bifunctional Organocatalysis of Asymmetric Diels–Alder Reaction via Hydrogen Bonding

In 2001, Yamamoto and coworkers realized the catalytic asymmetric DA reaction of anthrone and maleimide derivatives using chiral pyrrolidine **54** as a bifunctional catalyst, up to 97% yield and 87% ee of the corresponding adduct were obtained (Scheme 1.31) [62]. A possible transition state model for this [4 + 2] cycloaddition was proposed, in which two hydrogen bonds are believed to exist between the catalyst and the substrates. One is between the protonated pyrrolidine catalyst and anthrone enolate, and the other is between the hydroxy group of pyrrolidine and the carbonyl group of maleimide [62]. Such kinds of double hydrogen-bonding interactions allow the substrate activation and stereochemistry control to occur simultaneously. In 2006, Tan and coworkers successfully realized similar reactions using a chiral bicyclic guanidine **55** as catalyst, excellent yields and enantioselectivities of the reaction were afforded (Scheme 1.31) [63].

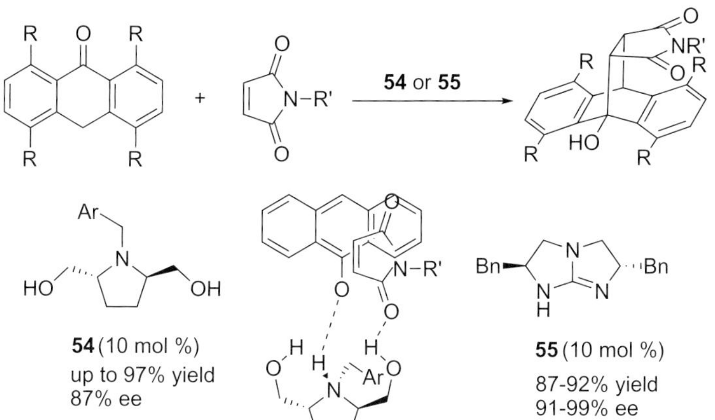

Scheme 1.31 Bifunctional organocatalysis of asymmetric DA reaction.

Deng's group developed an alternative asymmetric DA reaction of 2-pyrones with a variety of α,β-unsaturated ketones or esters using 5 mol% of bifunctional cinchona alkaloids catalysts **56**, the corresponding adducts were obtained in 60–100% yields with 82–91% ees (Scheme 1.32) [58]. Catalyst **56** has both hydrogen bond donor and acceptor motifs, and mechanistic studies suggest there are multiple hydrogen bonding interaction networks between catalyst and substrates which simultaneously raise the energy of the HOMO of the diene and lower the energy of the LUMO of the dienophile [64]. Deng and coworkers have also developed the asymmetric DA reaction between 2-pyrones and conjugated nitriles under the catalysis of cinchona-based thiourea **57** (5 mol%), good *exo* diastereoselectivity (up to 97:3) and enantioselectivity (up to 97% ee) of the catalysis were afforded (Scheme 1.32). In this reaction system, catalyst **57** was again considered as a bifunctional organocatalyst which could activate both dienes and dienophiles via hydrogen bonding interaction [64]. In 2008, Bernardi and coworker achieved the first asymmetric DA reactions between 3-vinylindoles and maleinides or quinones using a modified bifunctional thiourea catalyst **58**. The corresponding fused heterocycle could be readily obtained in good to excellent yields with very high ees (Scheme 1.33). It was suggested that the interaction between the basic moiety of the catalyst and the N–H group of the diene and the hydrogen bonding activation of the dienophile by the thiourea moiety might be responsible for the catalysis [65]. The resultant cycloadducts are potentially useful for the synthesis of a variety of biologically important natural or unnatural alkaloids.

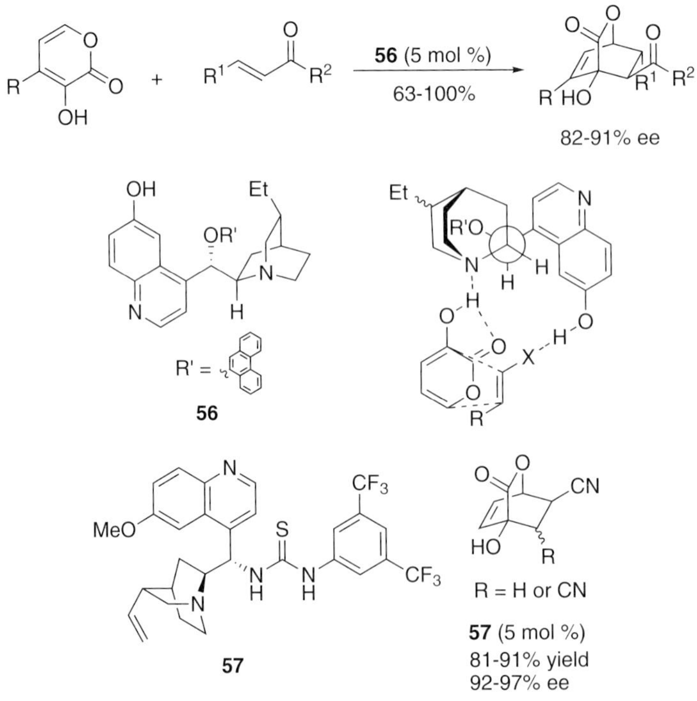

Scheme 1.32 Bifunctionalized catalysts promoted enantioselective DA reaction.

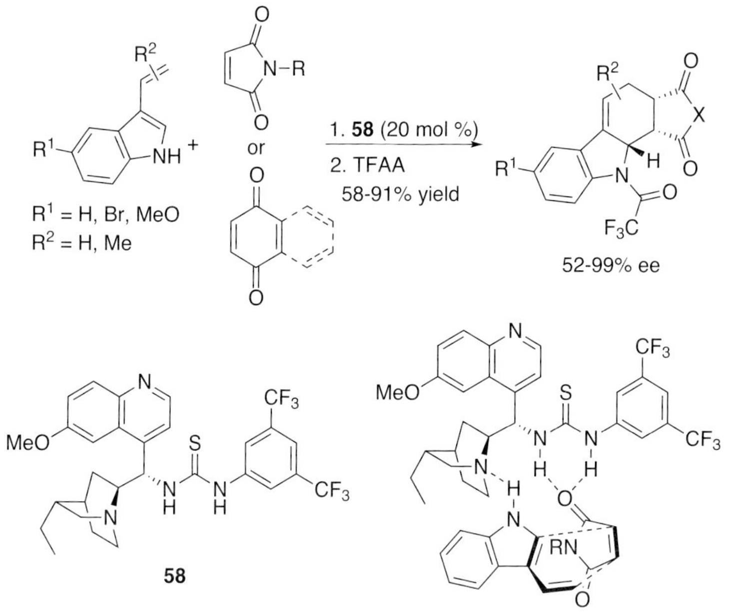

Scheme 1.33 Enantioselective DA reaction promoted by **58**.

1.3
Asymmetric Oxa-Diels–Alder Reaction

1.3.1
Lewis Acid Catalyzed Asymmetric Oxa-Diels-Alder Reaction

1.3.1.1 Chiral Aluminum, Boron, and Indium Complexes

Asymmetric hetero Diels Alder (HDA) reaction of carbonyl compounds with various dienes can afford dihydropyranone derivatives, one type of important synthons for natural or unnatural product synthesis. In 1988, Yamamoto's group reported the first HDA reactions of aldehydes and Danishefsky's dienes under the catalysis of 10 mol% of Al(III) complexes of **59** with good yields and up to 97% ee (Scheme 1.34). The steric hindrance on the 3,3′-position of binaphthyl was found to be crucial for the high reactivity and enantioselectivity. The chiral organoaluminum reagent derived from Me$_3$Al and 3,3′-dialkylbinaphthol (alkyl = H, Me or Ph) was employable only as a stoichiometric promoter and less satisfactory results were obtained in terms of both reactivity and enantioselectivity [66a]. Yamamoto and coworkers further applied an asymmetric poisoning strategy employing chiral ketone (3-bromocamphor) as an antagonist to enantioselectively deactivate one enantiomer of racemic BINOL-Al complex and the less deactivated enantiomer catalyzed the reaction to give the product in up to 82% ee [66b]. Jørgensen and Pu developed a type of chiral polymer aluminum catalyst **60**/Al for enantioselective HDA reaction of

2,3-dimethyl-1,3-butadiene with ethyl glyoxylate, and chiral polymer ligand **60** could be easily recycled and reused without lose of reactivity and enantioselectivity [67a]. In 2000, Jørgensen's group systematically investigated the effect of the steric and electronic environment of chiral Al complexes of **61** (10 mol%) for the asymmetric HDA reactions of benzaldehyde with Danishefsky's diene in *tert*-butyl methyl ether, up to 97% yield and 99.4% ee were achieved. It was found that bulkiness and hypercoordination of 3,3-substituents played a key role in this highly efficient and enantioselective reaction, and a possible hypercoordination model was also postulated to explain the asymmetric induction pathway. In this model, in addition to the coordination of benzaldehyde, one of the ether oxgen atoms of the chiral ligands was believed to coordinate to aluminum and form a trigonal-bipyrimidal structure at the aluminum center which accounted for the stereochemistry outcome of the reaction [67b].

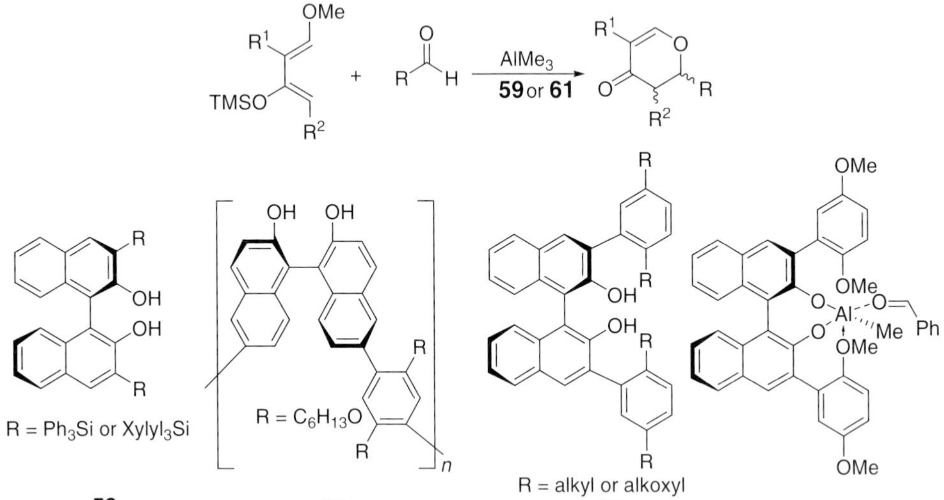

59
65-93% yield
67-97% ee

60
up to 80% yield
88% ee

61
up to 97% yield
99.4% ee

Scheme 1.34 Enantioselective HDA reaction catalyzed by chiral aluminum complexes.

Chiral boron complexes have also been successfully applied in asymmetric HDA reactions. In 1992, Yamamoto employed **1** (CAB) for the enantioselective HDA reaction of aldehydes with Danishefsky's diene, up to 95% ee of the corresponding adduct was obtained (Scheme 1.35) [68]. Complex **2** developed by Corey and coworkers was also an efficient catalyst, up to 82% ee of the product was obtained for the reaction of benzaldehyde with Danishefsky's diene via the Mukaiyama aldol reaction pathway [69]. Very recently, Feng's group developed a novel aromatic amide derived chiral *N,N*-dioxide/In(OTf)$_3$ complex **62** (Scheme 1.35) for the same type of HDA reaction. This catalyst system was applicable for a broad range of aromatic, aliphatic and heterocyclic aldehyde substrates [70]. This protocol was further employed to the sub-gram scale synthesis of triketide in 21% overall yield.

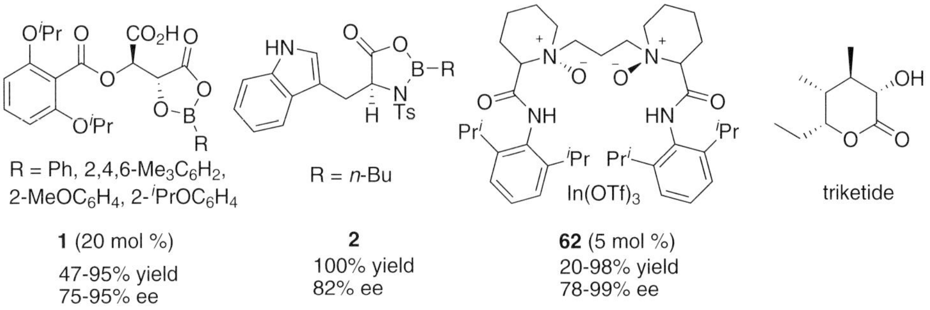

1.3 Asymmetric Oxa-Diels–Alder Reaction

Scheme 1.35 Enantioselective HDA reactions catalyzed by boron and indium complexes.

Compound **1** (20 mol %): R = Ph, 2,4,6-Me$_3$C$_6$H$_2$, 2-MeOC$_6$H$_4$, 2-^{i}PrOC$_6$H$_4$; 47–95% yield; 75–95% ee.

Compound **2**: R = n-Bu; 100% yield; 82% ee.

Compound **62** (5 mol %) In(OTf)$_3$: 20–98% yield; 78–99% ee.

triketide

1.3.1.2 Chiral Titanium and Zirconium Complexes

Chiral titanium (IV) complexes have been widely used as catalysts for asymmetric HDA reactions. In 1991, Mikami and coworkers reported an enantioselective HDA reaction between methyl glyoxylate and 1-methoxyldienes using BINOL-TiX$_2$ complexes **25** (10 mol%) as catalysts, cis cycloadducts were obtained as major products in high yields and excellent ees (Scheme 1.36). The observed cis-selectivity suggested the titanium catalyst **25** should be complexed in an *anti*-fashion and then the HDA reaction proceeded through an endo-orientation [71a, 34c]. Wada's group found that TADDOL-TiBr$_2$ (**63**, 5–10 mol%) could catalyze an inverse-electron demand HDA reaction of vinyl ether with α,β-unsaturated ketones, giving dihydropyranes in good to excellent yields and up to 97% ee (Scheme 1.36) [71b].

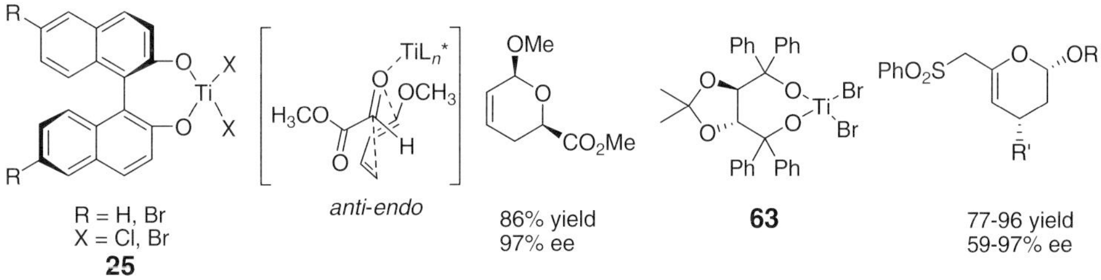

Scheme 1.36 Enantioselective HDA reactions catalyzed by chiral titanium complexes.

Compound **25**: R = H, Br; X = Cl, Br. *anti-endo*. 86% yield, 97% ee.

Compound **63**. 77–96 yield, 59–97% ee.

In 1995, Keck and coworkers reported the asymmetric HDA reaction of Danishefsky's diene with a variety of aldehydes using the catalyst (10 mol%) generated from BINOL **64** and Ti(O^{i}Pr)$_4$ in the ratio of 2:1 in the presence of 4 Å MS and TFA (0.3 mol%), the corresponding dihydropyrones were obtained in 75–97% ees [72]. Feng's group contributed a lot of effort on the chiral titanium-catalyzed HDA reaction of aldehydes and electron-rich dienes. A systematic investigation on the ligand screening, the effects of temperature, solvent, catalyst concentration and loading, ratio of ligand and Ti(O^{i}Pr)$_4$, as well as MS additive, disclosed that a titanium catalyst (20 mol%) prepared *in situ* by mixing a 1:1 ratio of H$_8$-BINOL **65** and Ti(O^{i}Pr)$_4$ could promote the HDA reaction between Danishefsky's diene and a wide variety of aldehydes efficiently via the Mukaiyama aldol pathway. High levels of enantioselectivities have been obtained in this catalytic system (Scheme 1.37) [73].

Scheme 1.37 Chiral BINOL-Ti complexes for HDA reaction.

Feng and coworkers have also expanded BINOL-Ti(IV) catalysts to the HDA reactions of substituted Danishefsky-type dienes (**67**, **68**) with aldehydes. Optically active 2,6-disubstituted, 2,2,6-trisubstituted, and 2,5-disubstituted dihydropyrones could be obtained in high yields with good to excellent enantioselectivities [74]. Yu and coworkers developed 3-diphenylhydroxymethyl-substituted BINOL (**66**)-titanium(IV) catalyst (20 mol%) for the same reaction via the DA pathway, affording 2,5-disubstituted dihydropyrones with up to 99% yield and 99% ee [75]. In 2008, an efficient asymmetric HDA reaction of Brassard's diene and aliphatic aldehydes was reported by Feng and coworkers using BINOL-Ti(IV) catalyst (10–20 mol%) in the presence of 4-picolyl hydrochloride as an additive, the corresponding α,β-unsaturated δ-lactones were afforded in 46–79% yields and 81–88% ees (Scheme 1.38) [76]. This methodology has also been successfully applied to a one-step synthesis of (R)-(+)-kavain and (S)-(+)-dihydrokavain.

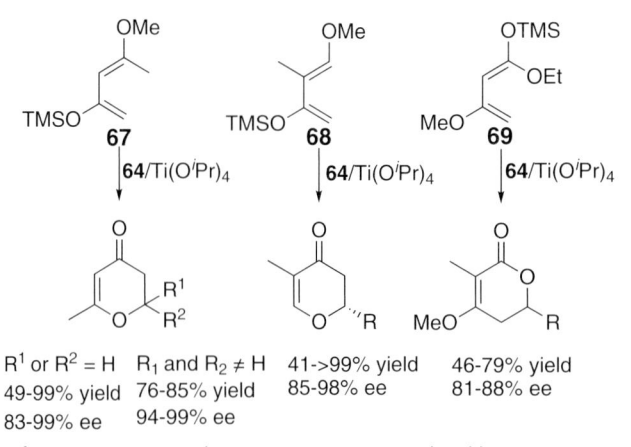

Scheme 1.38 Enantioselective HDA reactions catalyzed by BINOL-Ti complexes.

In 2002, Ding's group successfully discovered some exceptionally efficient enantioselective catalysts for the solvent-free HDA reactions of Danishefsky's diene with aldehydes using high-throughput screening of dynamic combinatorial catalyst libraries (104 members) of titanium complexes generated *in situ* from a series of chiral diols with Ti(O^{i}Pr)$_4$. It was found that catalysts **65**/Ti/**70** and **70**/Ti/**70** could promote the HDA reactions with only 0.5–0.05 mol% catalyst loading to give dihydropyranones in up to quantitative yield and >99% ee (Scheme 1.39). In particular, only 0.005 mol% of **65**/Ti/**70** was effective enough to promote the reaction between furfural and Danishefsky's diene, affording the corresponding cycloadducts in 63% yield and 96% ee, which represents one of the lowest catalyst loadings for Lewis acid catalysis in the HDA reaction [77].

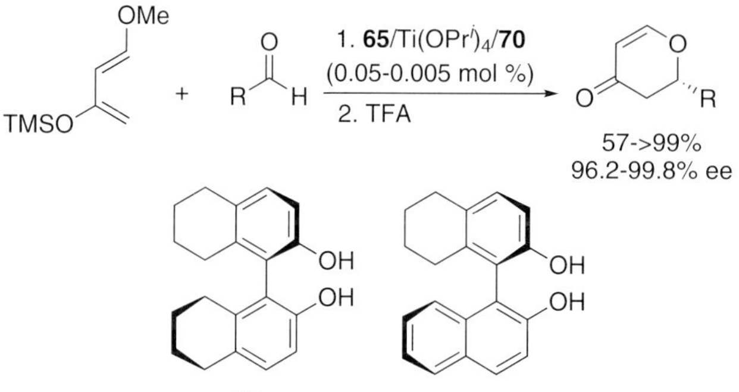

Scheme 1.39 Enantioselective HDA reaction catalyzed by complex **65**/Ti/**70**.

Ding and coworkers have also discovered a series of highly efficient chiral tridentate Schiff base **71** modified titanium catalysts for HDA reactions of Danishefsky's diene and aldehydes via a concerted [4 + 2] cycloaddition process (Scheme 1.40). A variety of 2-substituted dihydropyranones were afforded in up to 97% ee and >99% yield with 10 mol% of **71**/Ti/**71** as catalyst precursor in the presence of Naproxen **72** as an activator and 4 Å MS [78a, b]. Based on this type of catalyst system, some dendritic catalysts were designed and synthesized for the catalysis of the same reaction, affording the products in comparable enantiomeric excesses to those obtained with their homogeneous counterparts [78c, d]. The dendritic titanium catalyst could be recycled and reused at least three times without significant loss of activity and enantioselectivity. Very recently, Li and coworkers synthesized H$_4$-NOBIN derived tridentate Schiff base ligand **73**, which was then employed in the titanium-catalyzed HDA reaction with 2-naphthoic acid as additive. However, only moderate enantioselectivity was obtained [79].

BINOL-Zr(IV) complexes have also been applied to the catalytic asymmetric HDA reaction. In 2002, Kobayashi's group reported the first catalytic 2,3-trans-selective enantioselective HDA reaction between Danishefsky's dienes (R^1 = Me) and aldehydes in the presence of chiral zirconium complexes formed *in situ* with 3,3'-diiodobinaphthol derivatives **74**, Zr(O^{t}Bu)$_4$, a primary alcohol and a small

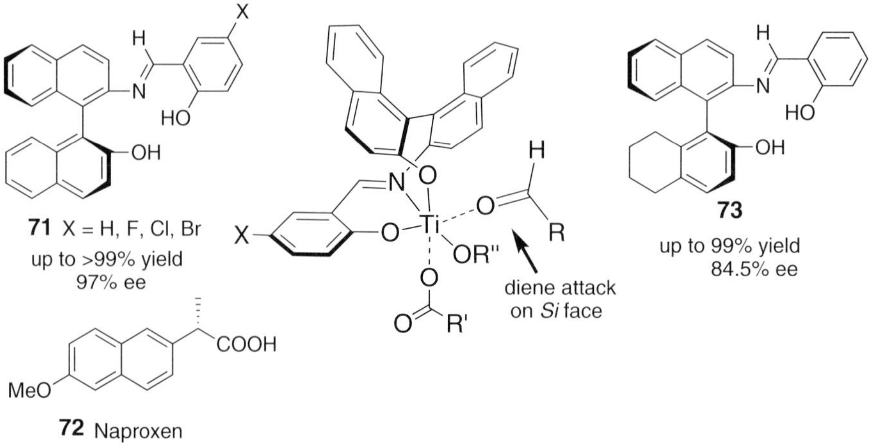

71 X = H, F, Cl, Br
up to >99% yield
97% ee

72 Naproxen

diene attack on *Si* face

73
up to 99% yield
84.5% ee

Scheme 1.40 Chiral Schiff base ligand for titanium catalyzed HDA reaction.

amount of water, the corresponding cycloadducts were afforded in up to quantitative yield and 98% ee (Scheme 1.41). This reaction was found to proceed in a stepwise cycloaddition process. For the diene substrates with $R^1 =$ BnO, 2,3-*cis*-dihydropyranones were achieved efficiently and enantioselectively (Scheme 1.41) [80]. (+)-Prelactone C and (+)-9-deoxygoniopypyrone have been concisely synthesized ultilizing this methodology.

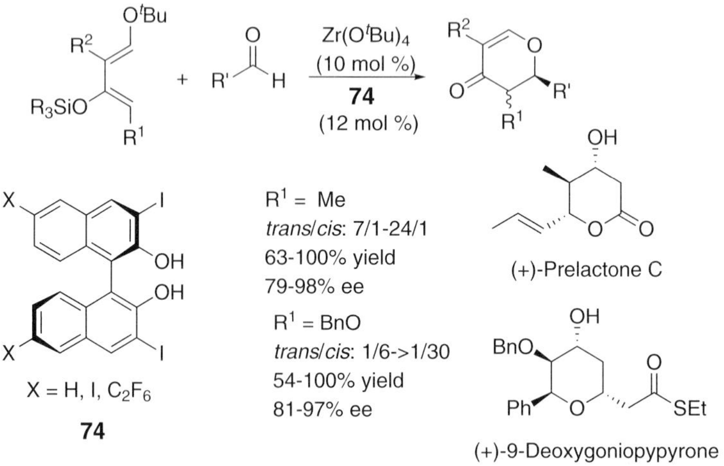

X = H, I, C₂F₆

74

R^1 = Me
trans/cis: 7/1-24/1
63-100% yield
79-98% ee

R^1 = BnO
trans/cis: 1/6->1/30
54-100% yield
81-97% ee

(+)-Prelactone C

(+)-9-Deoxygoniopypyrone

Scheme 1.41 Enantioselective HDA reaction catalyzed chiral zirconium complexes.

1.3.1.3 Chiral Chromium Complexes

Jacobsen and coworkers discovered that Salen-Cr(III) complex **75** (2 mol%) was an efficient catalyst for the enantioselective HDA of Danishefsky's diene and a variety of aldehydes in the presence of 4 Å MS, the corresponding cycloadducts were obtained in 65–98% yields and 62–93% ees (Scheme 1.42). The reaction was confirmed to proceed via a concerted [4 + 2] cycloaddition pathway [81]. In 2005, complex **75** was used as catalyst for a high-pressure HDA of 1,3-butadiene and glyoxylates by Jurczak's group. However, only moderate enantioselectivity was afforded (44–71% ee) [82].

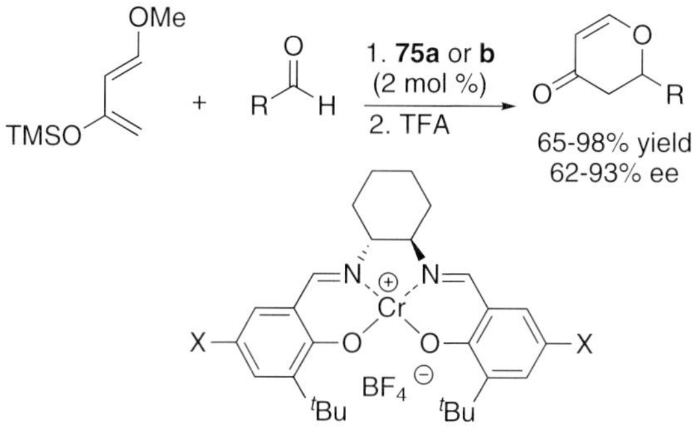

Scheme 1.42 Enantioselective HDA reaction catalyzed by Salen chromium complex.

Jacobsen and coworkers also developed chiral tridentate Schiff base chromium complexes **76** for enantioselective HDA reactions of trisilyloxy-2,4-hexadiene and a variety of aldehydes to afford *cis*-tetrahydropyranones in 28–97% yields and 90–99% ees (Scheme 1.43) [83a]. The catalyst loading could be reduced to 0.5 mol% for the reaction of 1-methoxyl-1,3-diene with $TBSOCH_2CHO$ without decrease in selectivity (91% yield, >99% ee). Complex **76a** was also found to be a highly efficient catalyst for inverse demand HDA reactions of α,β-unsaturated aldehydes with ethyl vinyl ether to give for an electron-inverse-demand HDA *cis*-dihydropyranes with a high level of enantioselectivities (Scheme 1.43) [83b]. The use of complex **76a** in the HDA reaction of Danishefsky's diene with chiral aldehydes showed very high diastereoselectivity (up to 1 : 33). This methodology provides selective access to any of four possible stereoisomers of the dihydropyranone products by judicious use of aldehyde and catalyst enantiomers, while achiral tridentated Schiff base chromium complex only led to low diastereoselectivities [83c].

In 2006, Berkessel's group developed two chiral chromium catalysts, **77** with chiral 2,5-diaminonorborane backbone Salen ligand and **78** bearing chiral porphyrin ligand, for enantioselective HDA reactions of Danishefsky's diene with aldehydes, up to 92–93% yields and 95–96% ees of the products have been obtained (Scheme 1.44) [84]. Moreover, chiral cobalt (**79**) [85a], vanadium (**80**) [85b], and manganese (**81**) [85c] complexes have also been successfully applied in the same HDA reaction (Scheme 1.45).

Scheme 1.43 Enantioselective HDA reactions catalyzed by chiral chromium complexes **76**.

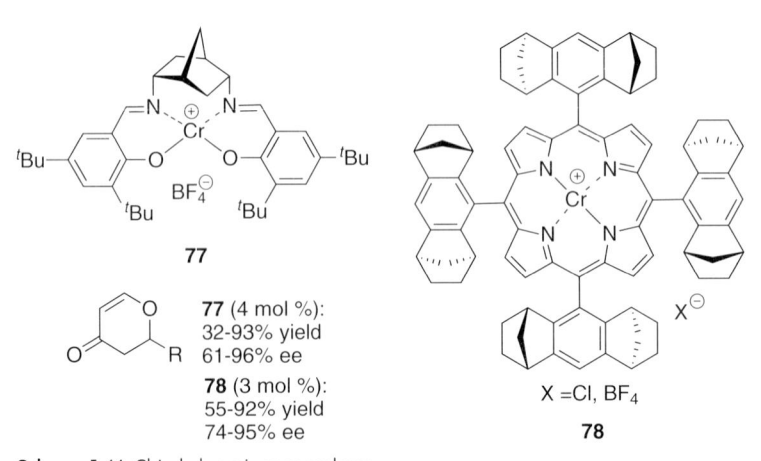

Scheme 1.44 Chiral chromium complexes.

1.3.1.4 Chiral Copper and Zinc Complexes

Chiral bis(oxazoline)-Cu(II) complexes (Scheme 1.46) represent one type of most efficient catalysts for asymmetric HDA reactions and have been intensively studied. In 1997, Jørgensen and coworkers reported the first highly enantioselective HDA reaction of Danishefsky's diene (Scheme 1.47, $R_3 = R_4 = H$) with various ketones including α-keto ester and α-diketone under the catalysis of **82a** (10 mol%), the corresponding cycloadducts have been achieved in 77–95% yields with up to 99% ee

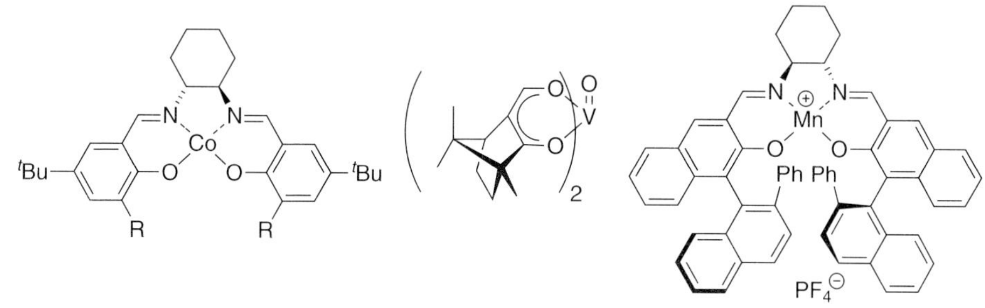

Scheme 1.45 Chiral cobalt, vanadium, and manganese complexes.

[86a]. Further research on this system led to a more general catalytic process for the HDA reaction of diketones with broad substrate scope, low catalyst loading, and excellent reactivity, diastereoselectivity, and enantioselectivity. For example, only 0.05 mol% of **82a** can catalyze the reaction of 2,3-pentanedinone with Danishefsky's diene (Scheme 1.47, $R^3 = R^4 = H$) efficiently, the reaction occurred at the methyl ketone fragment in 76% yield with 97.8% ee [86b].

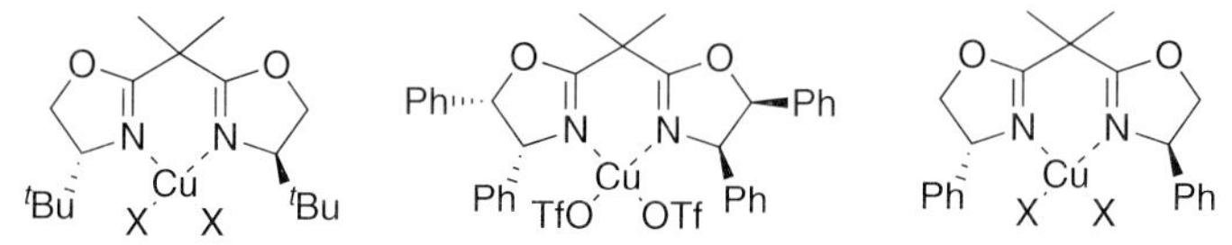

82a: X = OTf
82b: X = SbF$_5$

84a: X = OTf
84b: X = SbF$_5$

Scheme 1.46 Chiral copper complexes.

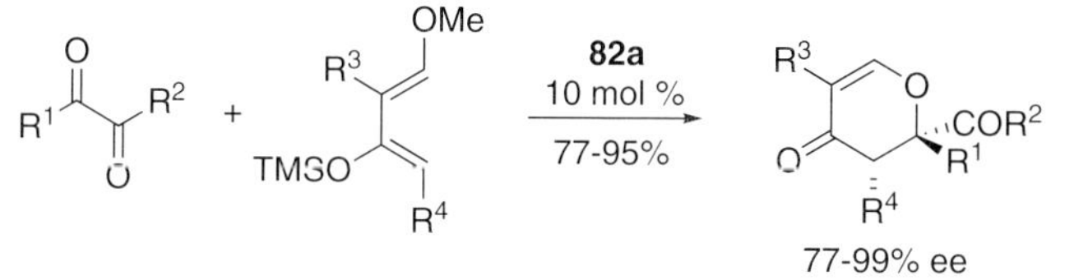

77–99% ee

Scheme 1.47 Enantioselective HDA reactions catalyzed by copper complex **82a**.

In 1998, Jørgensen's group developed a highly diastereo- (>95% de) and enantioselective inverse-electron-demand HDA reaction of β,γ-unsaturated α-keto esters bearing alkyl, ary and alkyoxy substitutents at the γ position with vinyl ether using **82a** as catalyst (Scheme 1.48), the corresponding cycloadducts were obtained in 51–96% yields with 90.4–99.5% ees [87a]. The substrate can be further expanded to γ-amino-protected β,γ-unsaturated α-keto esters and optically active amino sugar derivatives [87b,c]. Further experimental and theoretical investigations on the mechanism in chiral bis oxazoline Cu(II) catalyzed HDA reactions have also been carried out [87d]. In 2007, Jørgensen and coworkers realized an enantioselective HDA reaction of N-oxy-pyridine aldehydes and ketone derivatives with Danishefsky's diene under the catalysis of **83** (10 mol%), the corresponding cycloadducts were afforded in moderate yields and good enantioselectivities (Scheme 1.48). For the reaction of Brassard's diene, only the Mukaiyama aldol adduct was obtained in 84% yield and 99% ee [88].

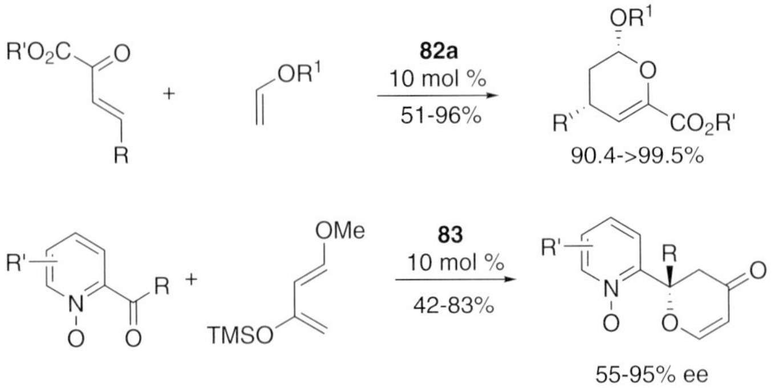

Scheme 1.48 Enantioselective HDA reactions catalyzed by copper complexes **82–83**.

Evans and coworkers reported a diastereo- and enantioselective HDA reaction of α,β-unsaturated acryl phosphonates with cyclic or acyclic enol ethers in the presence of chiral bis-oxazoline-Cu(II) (**82** or **84**) (2 mol%) to give the corresponding cyclic enol phosphonates in excellent yields (79–100%) and ees (88–99%) (Scheme 1.49). These cycloadducts are useful chiral synthons and can be conve-

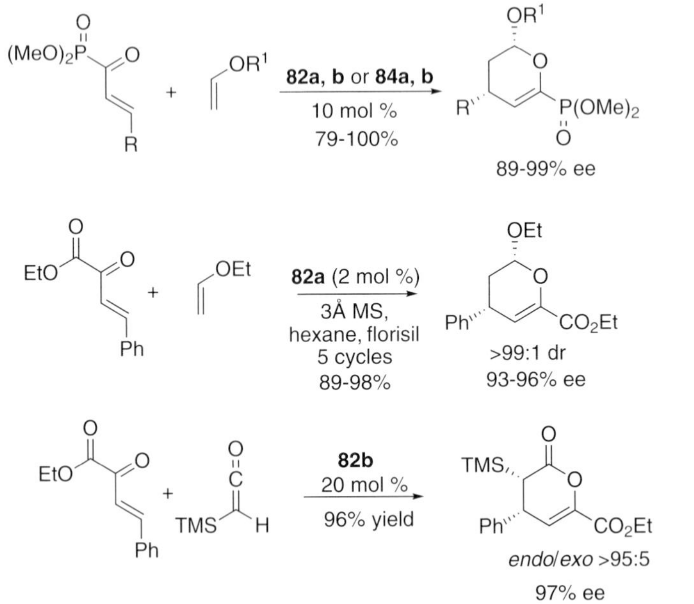

Scheme 1.49 Enantioselective HDA reactions catalyzed by copper complexes.

niently transformed to lactones, aldehydic ester and lactol methyl ether [89a]. Evans' group further extended these catalysts to the HDA reaction of β,γ-unsaturated α-keto esters and amides, affording the corresponding dihydropranes in excellent yields (87–99%), diastereo- (16:1 → 99:1), enantioselectivities (95–99%). Cycloadditions can be conducted with as low as 0.2 mol% of the chiral catalyst loading and are readily run on a multigram scale. When the reactions were conducted in hexane in the presence of absorbent (florisil), complexes **82** can be recycled and reused four times without loss of activity (Scheme 1.49) [89b,c]. In 2001, an enantioselective HDA reaction of β,γ-unsaturated α-keto esters with trimethylketene was realized to afford δ-lactone in 96% yield with 97% ee (Scheme 1.49) [89d]. Other chiral bis(oxazoline)-Cu(II)-catalyzed HDA reactions have also been reported recently [90].

Chiral sulfoximine ligands (Scheme 1.50) developed by Bolm's group are also effective for the Cu(II)-catalyzed HDA reaction [91]. In 2002, Bolm and coworkers reported enantioselective HDA reactions of 1,3-cyclohexadiene with ethyl glyoxylate or activated ketone under the catalysis of **85** (5 mol%), affording the corresponding cycloadducts in excellent yields and ees (Scheme 1.51) [91a]. Later, Bolm's group found that quinoline-based C_1-symmetric monosulfoximine-Cu(II) complexes **86** could promote the same HDA reaction efficiently, giving the products in up to 96% ee [91b]. Recently, ethylene-bridged chiral bissulfoximines have also been applied to Cu(II)-catalyzed HDA reaction. Up to 99% ee of the product was obtained in the presence of 5 mol% of catalyst **87** [91c].

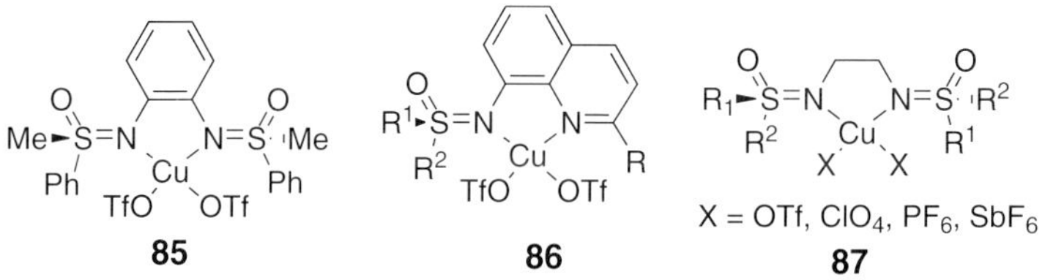

Scheme 1.50 Chiral sulfoximine-Cu(II) complexes.

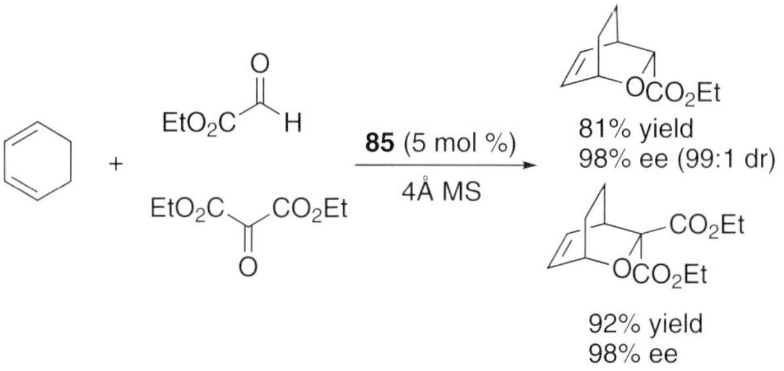

Scheme 1.51 Enantioselective HDA reactions catalyzed by chiral sulfoximine-Cu(II) complexes.

In comparison with chiral Cu(II) complexes, chiral bis(oxazoline)·Zn(II) complexes have received less successful application in the asymmetric HDA reaction. Jørgensen's group employed complexes **88** and **89** as catalysts for enantioselective

HDA reactions of ethyl glyoxylate with 2,3-dimethyl-1,3-butadiene or 1,3-cyclohexadiene, affording the corresponding adducts in only moderate ees (Scheme 1.52) [92]. For the HDA reaction of γ-amino-protected β,γ-unsaturated α-keto esters with ethyl vinyl ether, amino sugar derivative was obtained in 99% yield with 70% ee (Scheme 1.52) [87b].

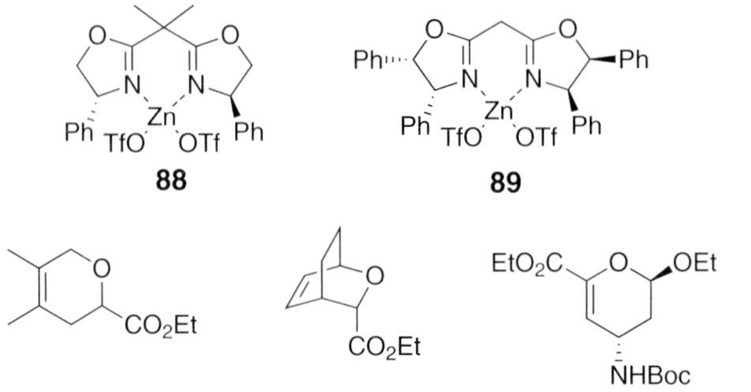

88: 26% yield, 87% ee   **88**: 84% yield, 27% ee   **88**: 99% yield
89: 42% yield, 79% ee   **89**: 84% yield, 65% ee   70% ee (70% de)

Scheme 1.52 Enantioselective HDA reaction catalyzed by chiral bis(oxazoline) copper complexes.

In 2002, Ding's group developed a highly efficient 3,3′-Br$_2$-BINOL-Zn(II) catalyst (Scheme 1.53) for the enantioselective HDA reaction of Danishefsky's diene with a variety of aldehydes [93]. Optically active 2-substituted dihydropyranones were afforded in up to 98% ee with excellent yields in the presence of 10 mol% of complex **89** which was generated *in situ* from the reaction of Et$_2$Zn with 3,3′-Br$_2$BINOL (Scheme 1.54) [93a]. Further studies showed that a series of chiral diimines activated BINOLate zinc complexes were also effective in promoting the HDA reaction in excellent enantioselectivities [93c]. In particular, two distinct asymmetric reactions including HDA of Danishefsky's diene and diethylzinc addition to aldehyde have been successfully integrated in one pot in the presence of a single catalyst (**90**, 10 mol%), up to 97.4% ee and 95.0% de of the products were obtained (Scheme 1.54) [93b,c]. In 2008, Ding and coworkers discovered that chiral BINOLate magnesium complexes were also highly efficient catalysts for the same HDA reaction to afford the products in up to 99% ees [94].

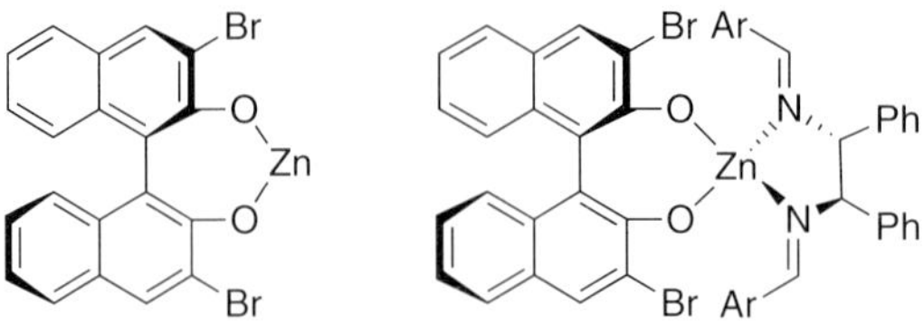

Ar = 2,4,6-Me$_3$C$_6$H$_2$

Scheme 1.53 Chiral zinc complexes.

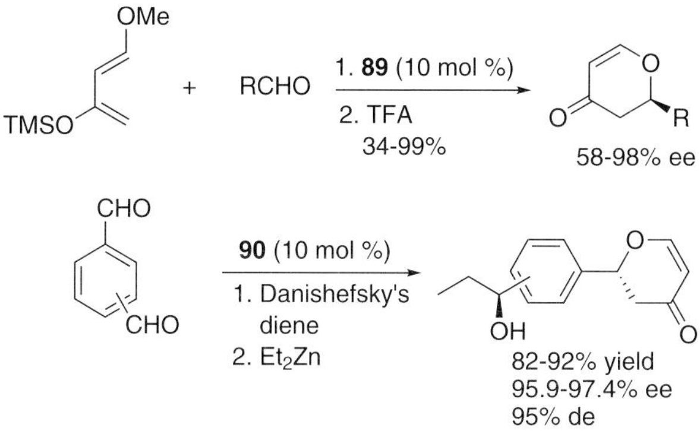

Scheme 1.54 Enantioselective HDA reactions catalyzed by chiral zinc complexes.

1.3.1.5 Chiral Rhodium, Palladium, and Platinum Complexes

Chiral rhodium complexes have been rapidly developed and successfully applied to asymmetric HDA reactions for the last decade. In 2001, Nishiyama and coworkers reported an enantioselective HDA reactions of Danishefsky's diene with n-butyl glyoxylate promoted by chiral 2,6-bis(oxazolinyl)RhCl$_2$(H$_2$O) complexes **91** as the catalysts (2 mol%), up to 84% ee of the corresponding adduct was obtained (Scheme 1.55). This reaction was confirmed to proceed via the concerted [4 + 2] cycloaddition pathway [95].

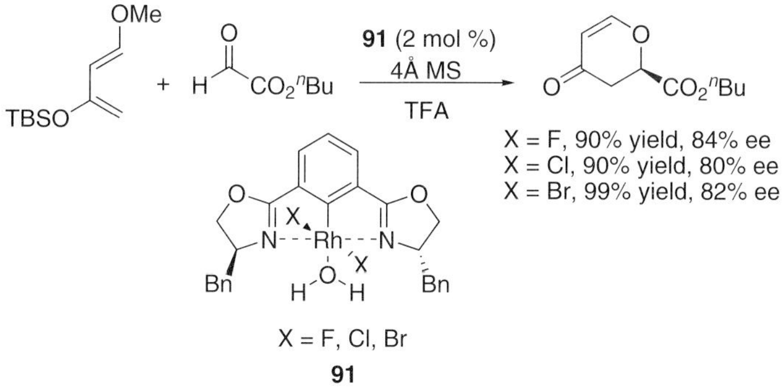

Scheme 1.55 Enantioselective HDA reaction catalyzed by chiral rhodium complexes **91**.

Chiral dirhodium(II) carboxamidate complexes (Scheme 1.56) have shown exceptional power for the catalysis of the asymmetric HDA reaction of diene with aldehydes. Doyle and coworkers have developed a variety of dirhodium complexes (**92–94** are the best) for promoting HDA reactions [96]. In 2001, Doyle's group discovered that complexes **92** and **93** are able to catalyze the addition of a variety of aldehydes to Danishefsky's diene (Scheme 1.57, $R^1 = R^2 = H$), affording the corresponding dihydropyranones in excellent yields and good ee values. When the catalyst loading was decreased to 0.01 mol% for the electron-poor aldehydes, cycloadducts could still be obtained in 71–81% yields with 61–73% ees, although a longer time was required (4–10 days) [96a, c, d]. Complex **93** (1 mol%) was also effective for the reaction of 2,3-dimethyl Danishefsky's diene with aldehydes to give cis-selective cycloadducts in up to 96% yield and 97% ee [96d]. In 2004, Hashimoto's group developed an alternative dirhodium(II) carboxamidate complex **95** which demonstrated exceptional efficiency in the catalysis of the HDA reaction between Danishefsky's diene and a wide range of aldehydes. Particularly, when the loading of catalyst **95** was reduced to as low as 0.002 mol%, up to 96% yield and 91% ee of cycloadduct could be obtained in the reaction of phenylacetylenyl aldehyde with Danishefsky's dine. The turnover number in this case reaches 48 000 which is among the highest values for Lewis acid catalyzed asymmetric HDA reaction [97].

Scheme 1.56 Chiral dirhodium complexes.

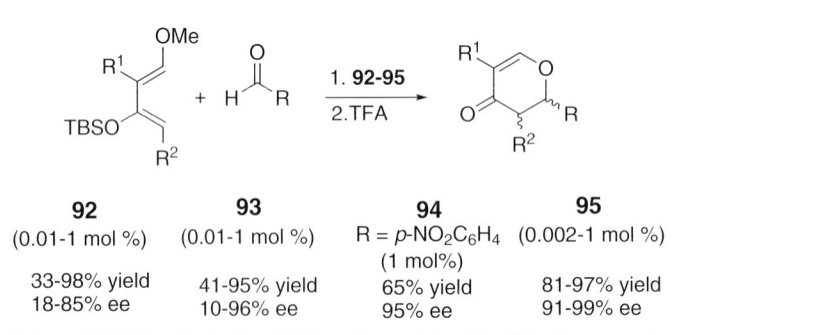

92	**93**	**94**	**95**
(0.01-1 mol %)	(0.01-1 mol %)	R = p-NO$_2$C$_6$H$_4$ (1 mol%)	(0.002-1 mol %)
33-98% yield	41-95% yield	65% yield	81-97% yield
18-85% ee	10-96% ee	95% ee	91-99% ee

Scheme 1.57 Enantioselective HDA reaction catalyzed by dirhodium complexes.

Cationic Pd(II) complexes (Scheme 1.58) can also act as one type of chiral Lewis acid catalyst to promote the enantioselective HDA reaction of diene with aldehydes. Oi and coworkers reported that BINAP-Pd(II) **96** complex is highly enantioselective for the HDA reaction of aryglyoxylals with cyclic or acyclic substituted 1,3-butadiene, giving the corresponding cycloadducts in 20–88% yields and 38–99% ees (Scheme 1.59) with 2 mol% catalyst loading. The molecular sieve (3 Å) additive was found to be critically important for the enantioselectivity of the reaction. When glyoxylate ester was employed as the substrate, the ene adducts were also obtained in addition to HDA adducts [98]. In 2002, Mikami's group developed an enantiopure biphenylphosphine(BIPHEP)-Pd complex (*R*)-**97** by asymmetric activation of the corresponding chirally flexible ligand-based racemic Pd complex (±)-**97** using enantiopure (R)-3,3′-dimethyl-2,2′-diamino-1,1′-binaphthyl (DM-DABN) as the chiral activator followed by *tropo*-inversion into the single Pd(II) diastereomer (R, R)-**98**) at 80 °C and protonation at 0 °C. HDA reaction of 1,3-cyclohexadiene with ethyl glyoxylate in the presence of (R,R)-**97** led to the corresponding cycloadducts in 60% yield with 82% ee (Scheme 1.59) [99a]. (R,R)-**98** was also an efficient catalyst for the reaction, giving the cycloadducts in 62% yield with 94% ee [99b]. Complex **99** developed by Gagné and coworkers catalyzed the same reaction efficiently to afford the product in 83% yield and 99% ee (Scheme 1.59) [100].

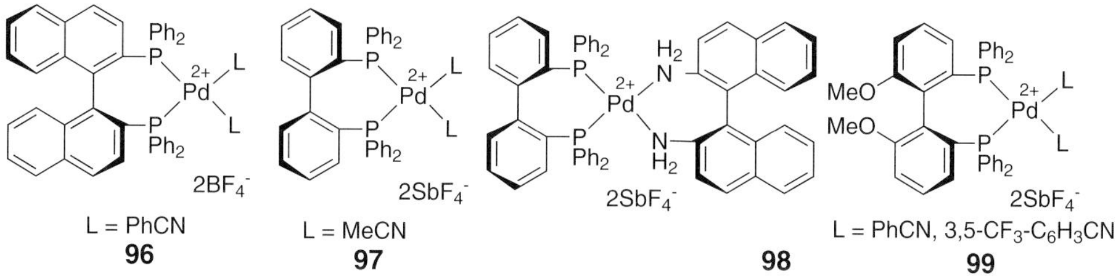

Scheme 1.58 Chiral palladium complexes.

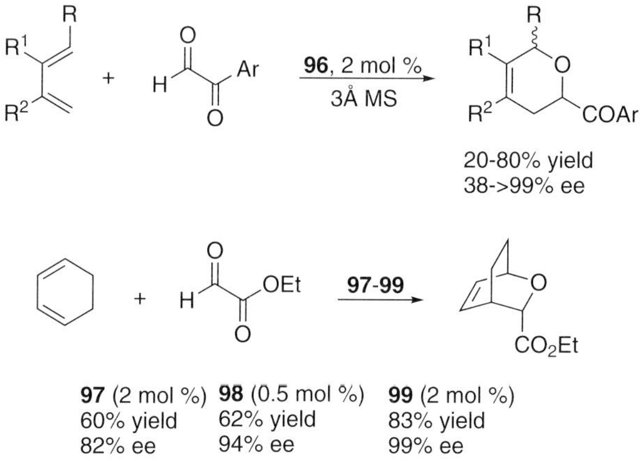

Scheme 1.59 Enantioselective HDA reaction catalyzed by chiral palladium complexes.

Oi's group also reported the use of BINAP-Pt(II) complex **100** as a chiral Lewis acid catalyst for the enantioselective HDA reaction of aryglyoxyal with acyclic or cyclic 1,3-dienes. The enantioselectivity of the catalysis could be up to 99% using 2 mol% of **100** (Scheme 1.60) [98]. Enantiopure flexible NUPHOS-Pt(II) complexes (**101** and **102**) were prepared by Doherty and coworkers using the similar strategy reported by Mikami [99a] using (S)-BINOL as the chiral template and trifluoromethanesulfonic acid as the protonation reagent to remove chiral auxiliary. The use of complexes **101** and **102** (2 mol%) in the catalysis of the HDA reaction of substituted 1,3-dienes with aryl glyoxals or glyoxylate esters showed that the enantioselectivity of the reaction could be as high as 99% with 60–99% of the substrate conversions (Scheme 1.61) [101].

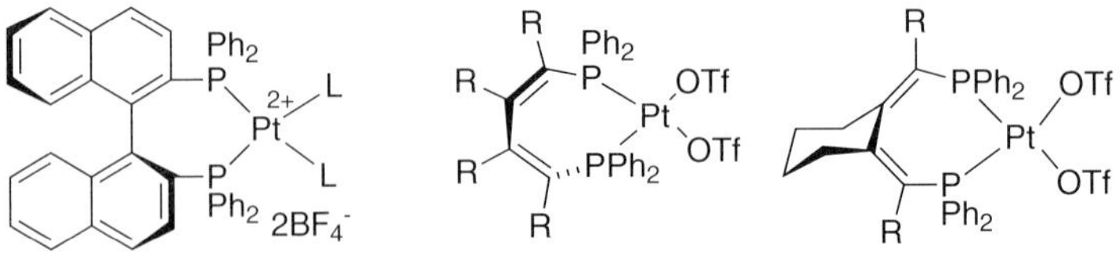

Scheme 1.60 Chiral platinum complexes.

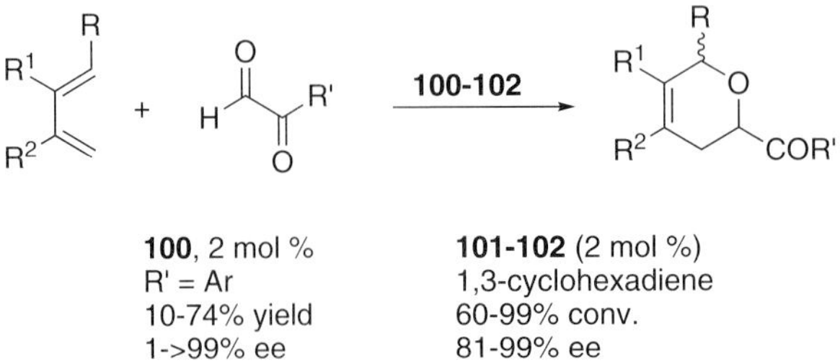

Scheme 1.61 Enantioselective HDA reaction catalyzed by chiral platinum complexes.

100, 2 mol %
R' = Ar
10–74% yield
1–>99% ee

101-102 (2 mol %)
1,3-cyclohexadiene
60–99% conv.
81–99% ee

1.3.1.6 Chiral Rare Earth Metal Complexes

Although chiral lanthanide complexes were among the earliest catalysts developed for the HDA reaction of Danishefsky's diene with aldehydes [102], only very few chiral lanthanide complexes have been reported for their catalytic asymmetric version. Inanaga's group developed a type of chiral lanthanide phosphate complexes **103** and **104** (Scheme 1.62) which were demonstrated to be applicable for the enantioselective HDA of Danishefsky's diene with a variety of aldehydes, affording the corresponding cycloadducts in up to 81% yield and 99% ee [103]. It was found that

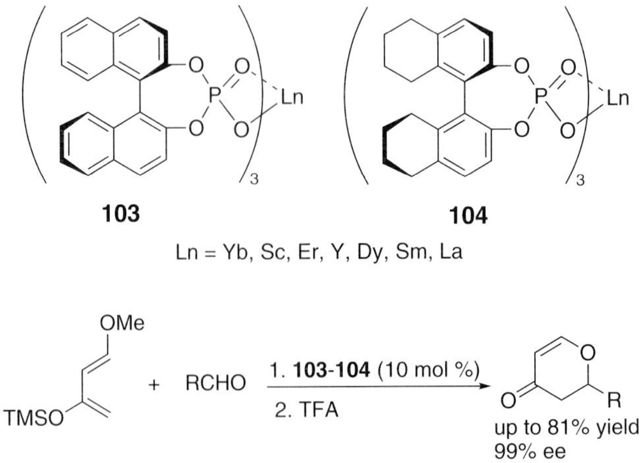

Ln = Yb, Sc, Er, Y, Dy, Sm, La

up to 81% yield
99% ee

Scheme 1.62 Enantioselective HDA reaction catalyzed by lanthanide complexes **103–104**.

the additive 2,6-lutidine was beneficial to enantioselectivity and the use of heavy lanthanides afforded higher ees than did the lighter ones [103b]. This type of catalyst is also effective for the reaction of Danishefsky's diene with phenylglyoxylates, affording the corresponding products in up to 99% ee [103e]. Very recently, Peters and coworkers discovered that the chiral complex generated *in situ* from norephedrine **105** (10–20 mol%) and Er(OTf)$_3$ (1.5 equiv) in the presence of diisopropylethyl amine can promote the HDA reaction of α,β-unsaturated acid chlorides with a variety of aromatic aldehydes to afford δ-lactone efficiently with up to 98% ee (Scheme 1.63) [104]. In this catalytic system, chiral Er complex acts as a bifunctional catalyst to bind both substrates with the same metal center. This formal HDA reaction is actually via an aldol reaction pathway followed by an intramolecular acylation [104].

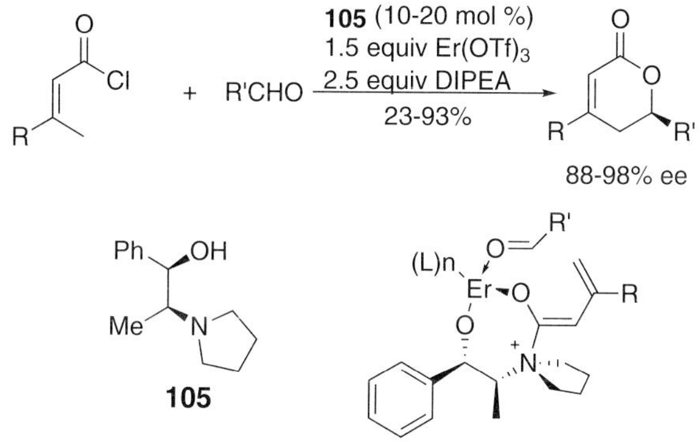

Scheme 1.63 Enantioselective HDA reaction catalyzed by Er complex.

1.3.2
Organocatalysis of Asymmetric Oxa-Diels–Alder Reaction

1.3.2.1 Hydrogen Bonding Promoted Asymmetric Oxa-Diels–Alder Reaction

Hydrogen bonding widely presents in nature, particularly in various large biomolecules such as proteins and nucleic acids, however, the use of this weak interaction with a small chiral molecule as hydrogen bonding donor (Brønsted acid) to promote enantioselective reactions was only a recent event [59a–c]. In 2003, Rawal's group reported the first highly efficient and enantioselective HDA reaction of aldehyde with electron-rich 1,3-diene through hydrogen bonding activation. The reaction of 1-amino-3-siloxy diene with a variety of aldehydes in the presence of a very simple hydrogen bonding donor TADDOL **52** proceeded smoothly to give the corresponding dihydropyranones after treatment with acetyl chloride in 52–97% yields and 92–98% ees (Scheme 1.64) [105a]. This pioneering work attracted a lot of interest in the area of asymmetric catalysis using hydrogen bonding activation [59a–c]. In 2005, Rawal and Yamamoto developed an alternative chiral Brønsted acid catalyst **106** bearing a partly reduced chiral binaphthyl backbone for the same HDA reaction, the corresponding products were obtained in 52–97% yields and 92–98% ee values (Scheme 1.64) [105b].

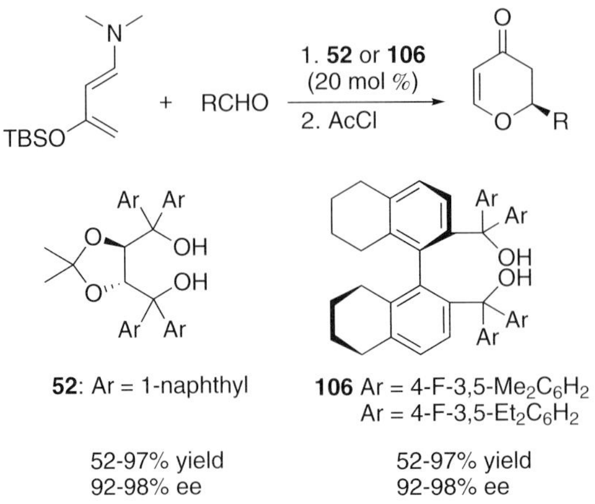

Scheme 1.64 Hydrogen bonding promoted HDA reaction.

In 2004, Ding and coworkers developed the first catalytic enantioselective HDA reaction of Brassard's diene with aldehydes through hydrogen bonding activation using **52** as the catalyst, the corresponding optically active δ-lactones were afforded in moderate-to-good yields with up to 91% ee (Scheme 1.65). Using this methodology, (S)-(+)-dihydrokawain has been synthesized in one step from 3-phenylpropionaldehyde in 50% isolated yield and 69% enantiomeric excess [106]. TADDOL **52** can also catalyze the asymmetric HDA reaction of the less active Danishefsky's diene with aldehydes to afford the corresponding dihydropyranones in moderate yields and ee values [107].

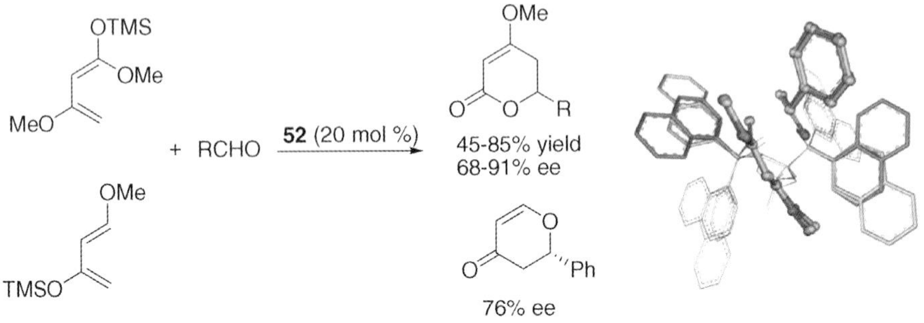

Scheme 1.65 Enantioselective HDA reactions catalyzed by TADDOL **52**.

The hydrogen bonding interaction model and asymmetric induction pathway in this type of chiral Brønsted acid catalyzed HDA reaction were systematically studied by Ding and Wu [107], on the basis of X-ray crystal structures of TADDOL-DMF and **106**-benzaldehyde complexes [105b, 106] and theoretical calculations. In agreement with the experimental findings, the calculation results indicate that this TADDOL-catalyzed HDA reaction proceeds via a concerted mechanism through an asynchronous and zwitterionic transition structure. The carbonyl group of benzaldehyde is activated by forming an intermolecular hydrogen bond with one of the hydroxy groups of TADDOL. Meanwhile, the intramolecular hydrogen bond between the two hydroxy groups in TADDOL is found to facilitate the intermolecular hydrogen bonding with benzaldehyde [107]. The computational studies on hydrogen bond promoted HDA reactions from other groups also supported the mechanism proposed above [108]. Frejd and coworkers reported a cleft organocatalyst bearing a single hydroxy group for the same type of enantioselective HDA reaction, only low to moderate ees have been obtained [109].

Chiral Brønsted acid catalyst **107** (Scheme 1.66) bearing a chiral oxazoline backbone was synthesized and applied to the enantioselective HDA reaction of 1-amino-3-siloxyl-1,3-butadiene with aldehydes in Sigman's group, up to 92% ee of the adduct was achieved (Scheme 1.67) [110]. It was found that both hydrogen bonds in **107** are critically important for effective catalysis. In 2005, Jørgensen and Mikami independently reported the HDA reaction of Danishefsky's diene with glyoxylate or glyoxal under the catalysis of chiral bis-sulfonamides catalyst **108a–b** (Scheme 1.66) through double hydrogen bonding activation, moderate to good yields of the products with up to 87% ee could be obtained [111, 112].

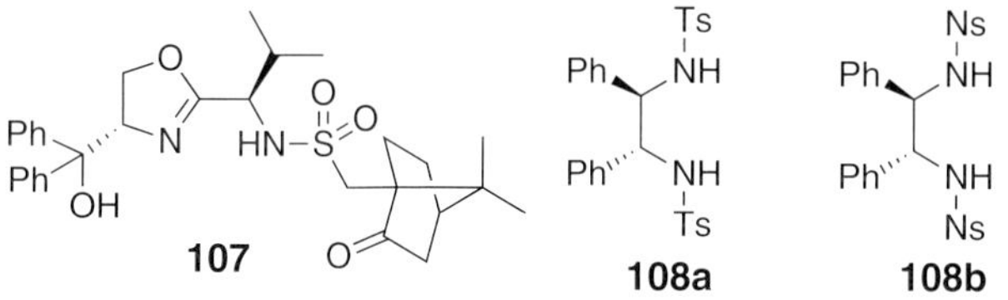

Scheme 1.66 Chiral Brønsted acid catalysts.

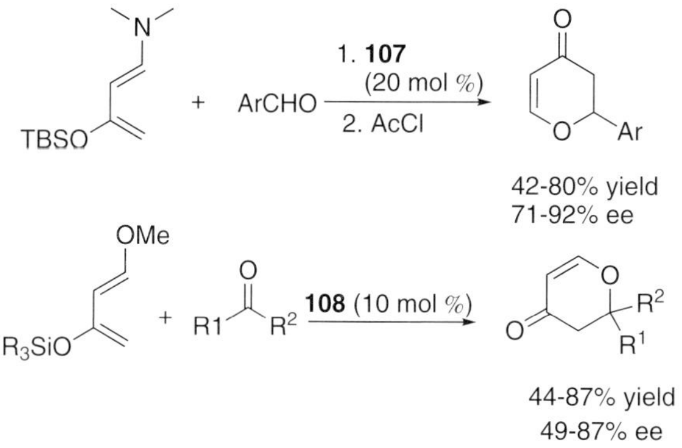

Scheme 1.67 Enantioselective HDA reactions through double hydrogen bonding activation.

1.3.2.2 Chiral Secondary Amine Catalyzed Asymmetric Oxa-Diels–Alder Reaction

The use of chiral amines as catalysts for the asymmetric HDA reaction has been rarely reported. A limited number of successful catalytic systems have been achieved in inverse-electron-demand HDA reactions recently by using secondary amines as the catalysts [113–116]. In 2003, Jørgensen and coworkers developed the first chiral amine **109** promoted HDA reactions of β,γ-unsaturated-α-ketones with aldehydes and *trans* lactones were afforded with good yields and up to 94% ee. The reaction proceeds through an active chiral enamine intermediate generated from chiral secondary amine and aldehyde, and silica gel is crucial for the regeneration of chiral amine catalyst (Scheme 1.68) [113]. Zhao and Liu reported the similar HDA reaction of β,γ-unsaturated-α-ketophosphonates and α,β-unsaturated trifluoromethyl ketones with catalysts **110** and **111**, respectively, and the corresponding cycloadducts were obtained in good yields and enantioselectivities (Scheme 1.69) [114, 115].

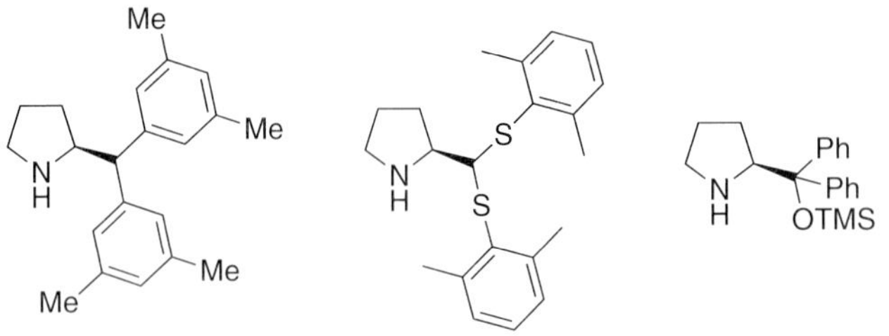

Scheme 1.68 Chiral secondary amine catalysts.

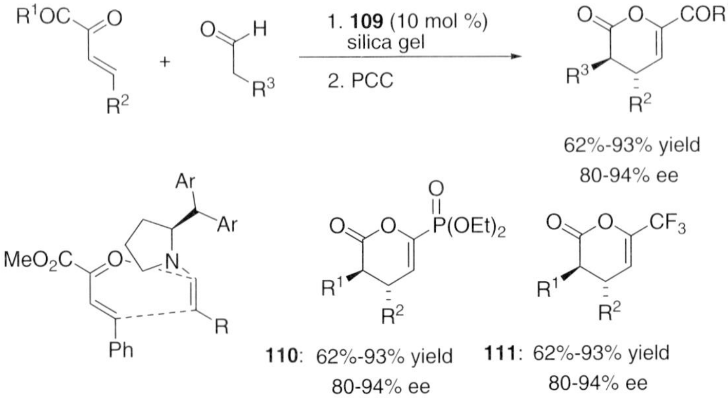

Scheme 1.69 Inverse-electron-demand HDA reactions catalyzed by **109–111**.

In 2007, an alternative enantioselective inverse-electron-demand HDA reaction of substituted quinones (X = H, Cl) and aldehydes was described by Dixon and coworkers using the secondary amine catalyst **111** (10 mol%) to afford the corresponding products with up to 81% ee. The resulting adduct could be further converted to optically active 2,3-dihydro-benzo[1, 4]dioxine compounds (Scheme 1.70) [116].

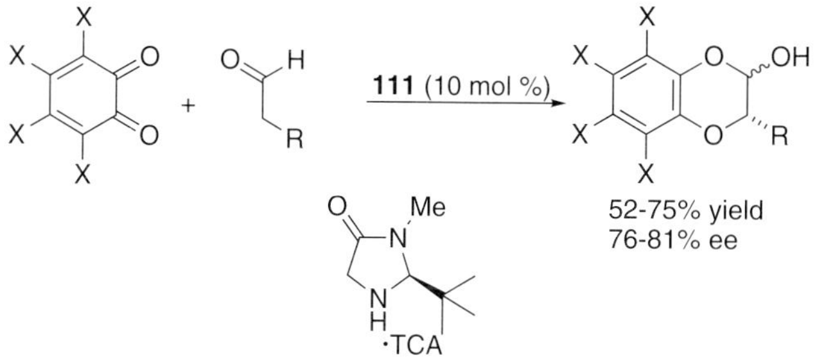

Scheme 1.70 Enantioselective HDA reaction catalyzed by chiral amine **112**.

1.4
Representative Applications in Total Synthesis

Chiral oxazaborolidine catalyzed enantioselective DA reactions have been applied to various organic syntheses of natural or unnatural products [7e]. A recent study reported by Corey et al. described the application of the catalytic asymmetric DA reaction as a key step for the synthesis of optically active intermediates of biologically important molecules, such as cortisone, dendrobine, or vitamin B12 and so on [117]. For example, the enantioselective DA reaction of benzoquinone **113** with diene **114** under the catalysis of **8b** (20 mol%), a chiral oxazaborolidine Lewis acid activated by Brønsted acid, in toluene at −78 °C to afford cycloadducts **115** in 95% yield with 90% ee and 100% de. The optically active **115** could be further transformed to **116** (>99% ee after recrystallization) in three steps, a key intermediate for the synthesis of cortisone (Scheme 1.71).

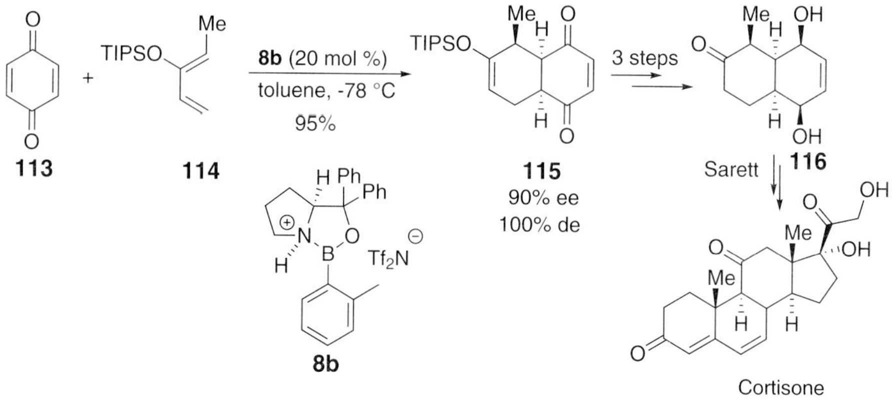

Scheme 1.71 Synthesis of optically active key intermediate of cortisone via catalytic asymmetric DA reaction.

In 2008, Danishefsky and coworkers successfully realized the total synthesis of Fluostatin C, a biologically important compound with antibiotic and antitumor activities, utilizing an enantioselective DA reaction as a key step. The cycloaddition reaction of diene **117** with methyl-substituted quinoneketal **118** under the catalysis of BINOL/Ti(IV) complex **25** afforded the corresponding adduct **119**, a key intermediate for the synthesis of Fluostatin C, in 93% yield with 65% ee (Scheme 1.72). Fluostatin C could be further converted to Fluostatin E after treatment with 1 M HCl [118]. The BINOL/Ti(IV) complex **25**-catalyzed enantioselective DA reaction was also used by Ward's group as a key step in the total synthesis of (−)-cyathin A_3 [119].

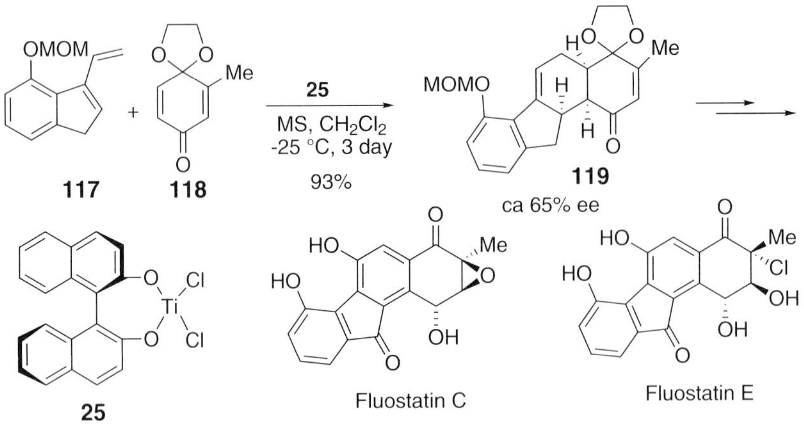

Scheme 1.72 Total synthesis of Fluostatin C and F.

A key intermediate **123** for the total synthesis of marine toxin immunogen (−)-gymnodimine has been obtained by Romo and coworkers using a catalytic asymmetric DA reaction as a key step in the presence of Evan's bis(oxazoline)/Cu(II) complex **82**. The highly functionalized cycloadduct **122** was constructed in 85% yield with 95% ee and 90% de (Scheme 1.73) using 20 mol% of **82** in the presence of MS at room temperature [120]. Moreover, Evan's bis(oxazoline)/Cu(II) complex-catalyzed HDA reaction has also been successfully applied in the total synthesis of (R)-dihydroactinidiolide and (R)-actinidiolide by Jørgensen and coworkers [121].

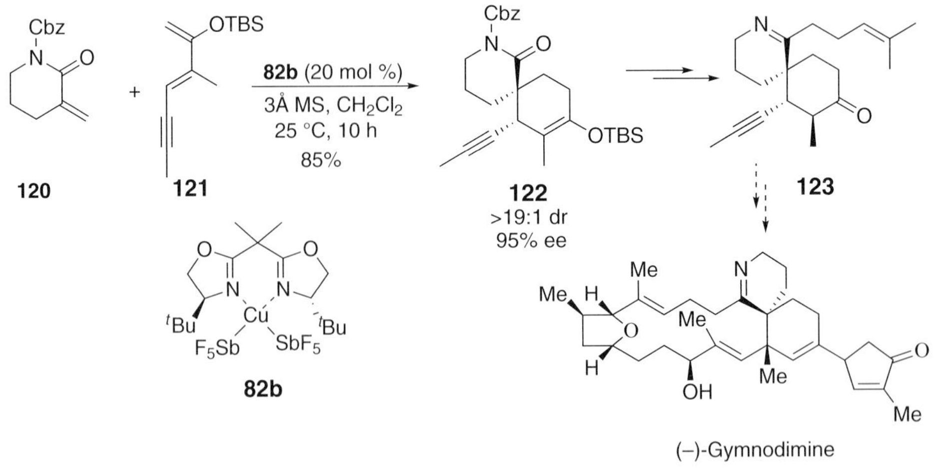

Scheme 1.73 Total synthesis of (−)-gymnodimine.

Tamiflu (oseltamivir phosphate) is a very important anti-influenza drug and several synthetic routes have been successfully developed with the approach for the drug production being from (−)-shikimic acid. In 2006, Corey and coworkers reported a facile approach to the total synthesis of Tamiflu employing the catalytic asymmetric DA reaction as a key step, avoiding the use of the relatively expensive and limited availability (−)-shikimic acid as starting material and the potentially hazardous and explosive azide-containing intermediates [122a]. The enantioselective DA reaction of 1,3-butadiene with trifluoroethyl acrylate in the presence of chiral boron complex **8b** (10 mol%) proceeded smoothly to give the corresponding cycloadduct **125** in 97% yield with >97%ee. Tamiflu was then efficiently synthesized in 11 steps from the cycloadduct intermediate **125** (Scheme 1.74) [122a].

In 2008, Shibasaki's group developed an alternative practical procedure for the synthesis of Tamiflu on the basis of the asymmetric DA reaction of 1-trimethylsiloxy-1,3-diene with dimethyl fumarate catalyzed by a chiral barium catalyst generated from chiral diol ligand **124** and $Ba(O^iPr)_2$ (Scheme 1.74) [123b]. The key intermediate **126** could be obtained in 91% yield and 95% ee on a scale of 58 g in the laboratory using 2.5 mol% of catalyst. CsF was found to be important for the removal of the TMS group of the diene to form the active barium dienolate via a transmetallation process. Starting from the key intermediate **126**, Shibashaki and coworkers have successfully developed a novel, efficient and easily handled route to the synthesis of Tamiflu [122b].

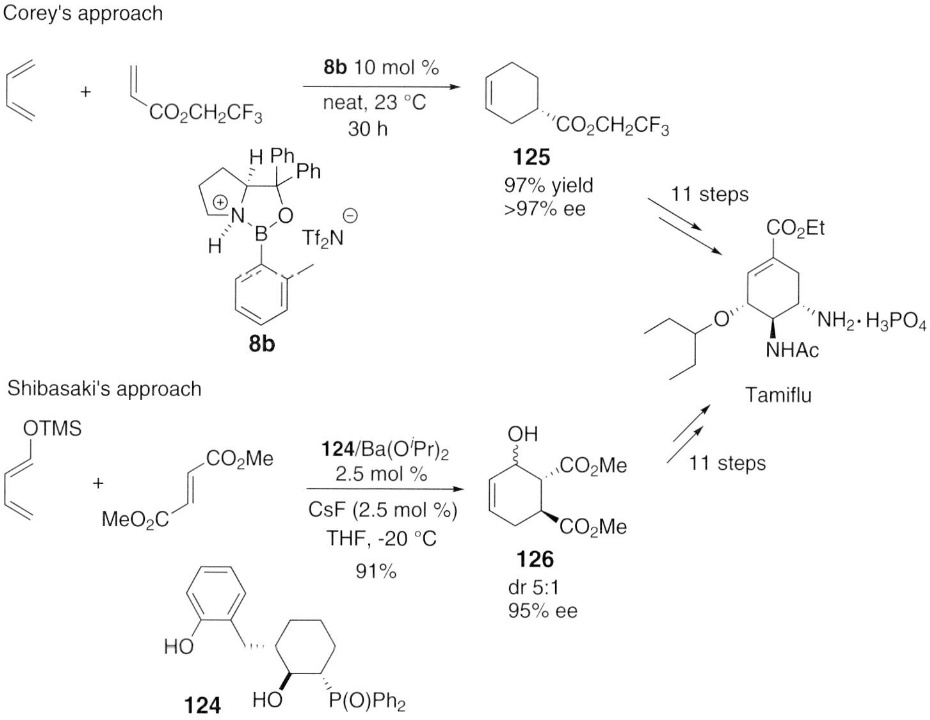

Scheme 1.74 Practical synthesis of Tamiflu based on catalytic asymmetric DA reactions.

Enantioselective HDA reactions promoted by the chiral chromium (III) complex (Jacobsen catalyst) are often used as the key step in the total syntheses of natural products, for example, FR901464, (−)-colombiasin A, and Elisapterosin B by Jacobsen's group [123], (−)-Lasonolide (A) by Ghosh's group [124], (+)-neopeltolide by Paterson's group [125], and Platencin by Nicolaou's group [126]. In 2008, Gademann's group reported the first total synthesis of the antitumor polyketide Anguinomycin C, in which the dihydropyran key fragment **127** was constructed quickly by a **76a**-catalyzed enantioselective HDA reaction of 1-trimethylsiloxy-1,3-diene with TES-protected propargylic aldehyde. In the presence of 2.3 mol% of **76a**, this reaction gave the intermediate **127** in 86% yield with 96% ee, which was then transformed into **128** by a seven-step reaction sequence including a Negishi cross-coupling process. Finally, the cross coupling of **128** and **129** followed by deprotection accomplished the total synthesis of anguinomycin C (Scheme 1.75) [127].

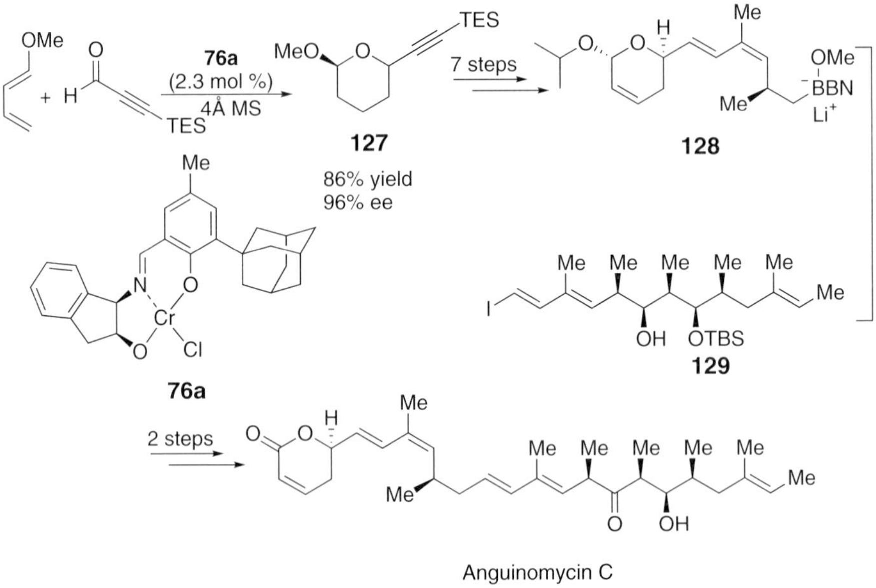

Scheme 1.75 Total synthesis of anguinomycin C on the basis of a catalytic asymmetric HDA reaction.

Organocatalysis of enantioselective DA reactions has also been used in the total synthesis of a variety of natural products. For example, the organocatalytic intramolecular DA reaction under the catalysis of **42** has been utilized in the total synthesis of marine metabolite solanapyrone D by the MacMillan group (Scheme 1.23) [50]. In 2003, the Kerr group achieved the first total synthesis of (+)-hapalindole Q with anti-algal and antimycotic activities, on the basis of an organocatalytic enantioselective DA reaction of diene **131** and dienophile **130** as the key step. A substoichiometric amount of **40** (40 mol%) promoted this reaction to give the key intermediate bearing four chiral centers including a quaternary center with good diastereoselectivity (85:15) and 92–93% ee in moderate yield (35%) (Scheme 1.76) [128].

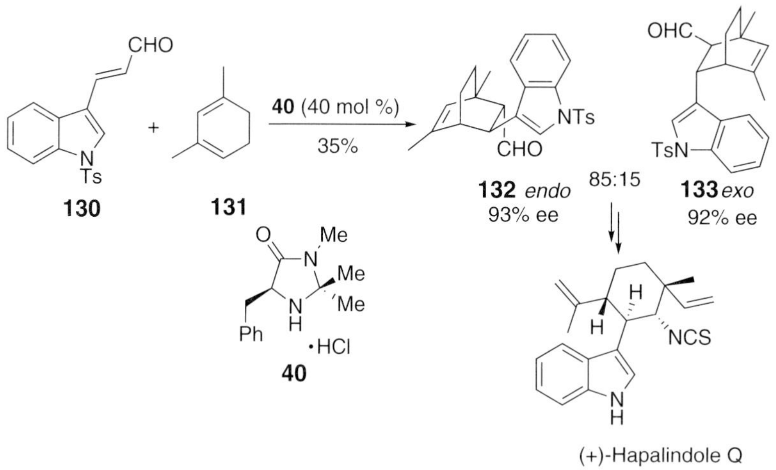

Scheme 1.76 Total synthesis of (+)-hapalindole on the basis of organocatalytic DA reaction.

In 2008, Bernardi and Ricci developed a hitherto elusive asymmetric DA reaction of 3-vinylindoles with maleinides or quinones using a bifunctional acid–base organocatalyst **58** (20 mol%), offering a direct approach to optically active tetrahydrocarbazole derivates [65]. The cycloadduct **133** could be obtained in 91% yield with 98% ee, and was further converted to the intermediates **134** through a 5-step reaction sequence with only a slightly drop in enantioselectivity (93% ee) (Scheme 1.77). Optically active **134** can serve as a very important synthon for the total synthesis of tubifolidine [129].

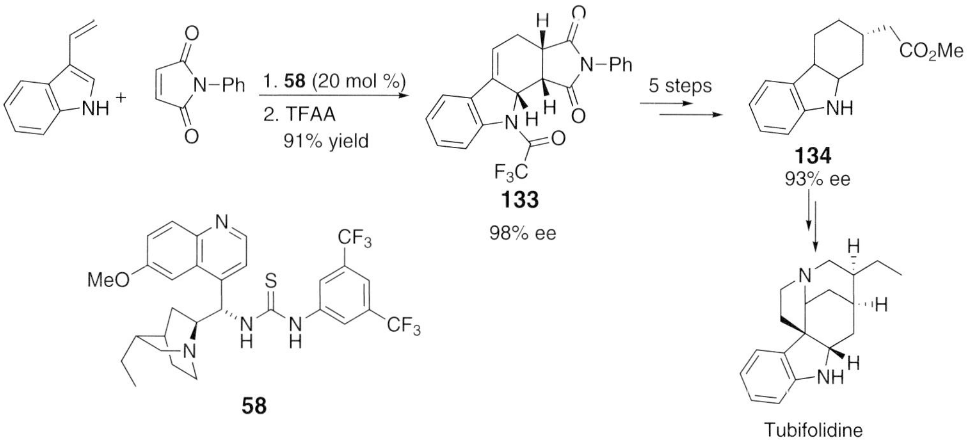

Scheme 1.77 Total synthesis of tubifolidine based on organocatalytic DA reaction.

1.5
Conclusion

As described in this chapter, there have been great successes for both metal and organo-based catalysts in enantioselective Diels–Alder and hetero-Diels–Alder reactions in the last 30 years, providing optically active six-membered carbon cyclic or oxa-cyclic compounds conveniently. Although this area has reached some degree of maturity, with numerous highly efficient chiral catalyst systems with diverse substrate types and very excellent enantioselectivities, as well as many successful applications in the total syntheses of natural products and biologically important compounds, the high catalyst loading (1–20 mol%) is still one of the drawbacks or bottlenecks in the catalysis, especially for the organocatalysis of DA reactions. The future effort in this area will be continuously directed to the development of new DA or HDA reaction systems and novel catalyst systems with high chemo-, regio-, and enantio-selectivities at a low catalyst loading. The last criterion is particularly important for the practical application of DA or HDA reactions on an industrial scale.

1.6
Experimental: Selected Procedures

1.6.1
Procedure for the Preparation of Chiral Boron Complex 8b (R = o-tol, Ar = Ph) (Scheme 1.4) and Its Application in Asymmetric Diels–Alder Reactions (Scheme 1.5) [18b]

To a 2-necked round-bottom flask (100 mL) equipped with a stir bar, a glass stopper and a 50-mL pressure-equalizing addition funnel (containing a cotton plug and about 10 g of 4 Å molecular sieves, and functioning as a Soxhlet extractor) fitted on top with a reflux condenser and a nitrogen inlet adaptor were added (S)-(−)-α,α-diphenyl-2-pyrolindinemethanol (0.082 g, 0.324 mmol), tri-o-tolylboroxine (0.038 g, 0.107 mmol) and toluene (25 mL). The resulting solution was then heated to reflux for 3 h (bath temperature ∼145 °C) before cooling to about 60 °C when the addition funnel and condenser were quickly replaced with a short-path distillation head. The mixture was concentrated to about 5 mL by distillation. This distillation protocol was repeated three times by re-charging with 3×5 mL of toluene. The solution was then allowed to cool to room temperature and the distillation head was quickly replaced with a vacuum adaptor. Concentration *in vacuo* (about 0.1 mmHg, 1 h) afforded the corresponding oxazaborolidine as a clear oil, which can then be dissolved in CH_2Cl_2 and used in two Diels–Alder experiments. To an aliquot of the oxazaborolidine precursor (0.160 mmol, theoretical) in 1.0 mL of CH_2Cl_2 at −25 °C was added trifluoromethanesulfonimide (0.20 M solution in CH_2Cl_2, freshly prepared, 0.667 mL, 0.133 mmol) dropwise. After 10 min at −25 °C, a colorless homogeneous catalyst solution was ready for use in the Diels–Alder reactions. To the resulting **8b** solution were successively charged 2,2,3-trimethyl-1,4-benzoquinone (0.665 mmol) in CH_2Cl_2 (0.7–1.0 mL) and 2-triisopropylsilyloxy-1,3-butadiene at −78 °C. The reaction mixture was stirred at −78 °C for 12 h (monitored by TLC or

^{1}H NMR) before it was concentrated by rotary evaporation. Water (5 mL) and hexanes (3 mL) were added to this residue. The aqueous layer was extracted with hexanes (4 × 5 mL). The combined organic layer was dried over Na$_2$SO$_4$, filtered, and concentrated to afford the crude product. Further purification by chromatography on silica gel gave adducts in 98% yield with 99% ee.

1.6.2
Procedure for Chiral Secondary Amine 42 Catalyzed Asymmetric Diels–Alder Reaction of Cyclopentadiene with (E)-Cinnamaldehyde (Scheme 1.22) [50]

To a solution of catalyst **42** (0.64 g, 2.5 mmol) in MeOH/H$_2$O (95/5 v/v, 2.5 mL) was added (*E*)-cinnamaldehyde (6.36 mL, 50.4 mmol). The solution was stirred for 1–2 min before addition of cyclopentadiene (12.5 mL, 151 mmol). The resulting reaction mixture was stirred at 23 °C for 8 h then diluted with diethyl ether and washed successively with water and brine. The organic layer was dried over Na$_2$SO$_4$, and concentrated. Hydrolysis of the product dimethyl acetal was performed by stirring the crude product mixture in TFA : H$_2$O : CHCl$_3$ (1 : 1 : 2) for 2 h at room temperature, followed by neutralization with saturated aq. NaHCO$_3$ and extraction with Et$_2$O. Further purification by chromatography on silica gel with 10% EtOAc/hexanes as eluent to give adducts as a colorless oil in 99% yield (12.2 g, endo/exo = 1/1.3, 93% ee for both endo and exo).

1.6.3
Procedure for 65/Ti/70 Catalyzed Asymmetric Hetero Diels–Alder Reaction of Benzaldehyde with Danishefsky's Diene (Scheme 1.39) [77]

To a dried Schlenk tube under argon atmosphere was charged the catalyst **65**/Ti/**70** (0.008 mmol) generated *in situ* by mixing **65**, **70** and Ti(O^{i}Pr)$_4$ in 1 : 1 : 1 molar ratio in toluene. Freshly distilled benzaldehyde (1.70 g, 16 mmol) and Danishefsky's diene (4.13 g, 24 mmol) were added sequentially to the catalyst. The reaction mixture was stirred for 48 h at 20 °C, diluted with diethyl ether (10 mL) and then treated with trifluoroacetic acid (3 mL). After the mixture was stirred for 0.5 h, saturated NaHCO$_3$ (20 mL) was added, the contents were stirred for 10 min, and the layers were separated. The aqueous layer was extracted with diethyl ether (3 × 50 mL), and the combined organic layer was dried over Na$_2$SO$_4$ and concentrated. The crude product was purified by flash chromatography on silica gel, eluting with hexanes/ethyl acetate (4 : 1), to afford 2.77 g (99.4% yield) of 2-phenyl-2,3-dihydro-4*H*-pyran-4-one as a colorless liquid with 99.4% ee.

1.6.4
Procedure for the Preparation of the Chiral Chromium Complex (1S, 2R)-76a and Its Application in the Asymmetric Hetero Diels–Alder Reaction of Ethyl Vinyl Ether with Crotonaldehyde (Scheme 1.43) [83b]

To a round-bottomed flask (200 mL) under a nitrogen atmosphere were added CrCl$_3$/THF (1 : 3) (2.80 g, 7.48 mmol), 3-(1-adamantyl)-2-hydroxy-5-methylbenzaldehyde-(1S, 2R)-1-aminoindanol imine (3.00 g, 7.48 mmol) and dry CH_2Cl_2 (60 mL) followed by dropwise addition of 2,6-lutidine (1.74 mL, 14.96 mmol). The solution was stirred for 3 h before dilution with CH_2Cl_2 (300 mL) and washing with water (3 × 180 mL) and brine (180 mL). The organic layer was dried over anhydrous Na_2SO_4, filtered and concentrated. The resulting solid was triturated with ice-cold acetone (10 mL), filtered, washed with an additional portion of cold acetone (10 mL), and air-dried to give the brown solid chromium complex (**76a**) as a dimer with a bridging water (2.3 g). To the filtrate (20 mL) was added water (2 mL) and the solution was allowed to stand, uncovered at 23 °C overnight. The resulting precipitate was filtered and washed with cold acetone to give an additional 0.6–0.8 g of the chromium complex ((1S, 2R)-**76a**) (combined yield 2.9–3.1 g, 75–80%). To an oven dried round-bottom flask (10 mL) were added freshly distilled ethyl vinyl ether (0.96 mL, 10.0 mmol) and (E)-crotonaldehyde (0.078 g, 1.0 mmol) followed by addition of freshly oven dried powdered 4 Å molecular sieves (0.150 g) and **76a** (0.024 g, 0.05 mmol). The resulting mixture was stirred at room temperature for 2 days then diluted with pentane and filtered through celite. The solvent was removed and the product was isolated by vacuum transfer to a flask cooled to −78 °C at 0.5 mmHg to give (2S,4S)-2-ethoxy-4-methyl-3,4-dihydro-2-H-pyran (0.106 g) as a clear oil in 75% yield with 94% ee.

1.6.5
Procedure for Chiral Copper Complex 82a (Scheme 1.46) Catalyzed Reaction of Asymmetric Hetero Ethyl Pyruvate with Danishefsky's Diene (Scheme 1.47) [86a]

82a was prepared by mixing $Cu(OTf)_2$ (0.036 g, 0.1 mmol) with 2,2'-isopropylidene-bis[(4S)-4-*tert*-butyl-2-oxazoline] (0.031 g, 0.105 mmol) in dry CH_2Cl_2 (2 mL) under an inert atmosphere with stirring for 2 h. The catalyst was cooled to −40 °C followed by addition of ethyl pyruvate (0.11 mL, 1 mmol) and Danishefsky's diene (0.25 mL, 1.2 mmol). The resulting solution was stirred for 30 h at −40 °C before quenched with TFA (0.1 mL dissolved in 20 mL of CH_2Cl_2). Then the mixture was stirred for an additional 1 h at 0 °C and neutralized with saturated aq. $NaHCO_3$. The solution was filtered through a plug of cotton, the organic phase was separated and the water phase was extracted twice with CH_2Cl_2. The combined organic phase was dried, filtered and concentrated. Purification of the crude product using flash chromatography (EtOAc/light petroleum, 1:4) gave the corresponding product in 78% yield with 99% ee.

1.6.6
Procedure for the Preparation of Chiral Dirhodium Carboxamidate Complex 95 and Its Application in the Asymmetric Hetero Diels–Alder Reaction of Phenylpropargyl Aldehyde with Danishefsky's Diene (Scheme 1.57) [97]

To a round-bottomed flask was added a mixture of $Rh_2(OAc)_4·2MeOH$ (0.76 g, 1.50 mmol) and (3S)-3-(1,3-dioxobenzo[f]isoindol-2-yl)-2-piperidinone (7.06 g, 24.0 mmol) in chlorobenzene (200 mL). The flask was fitted with a Soxhlet extraction apparatus and the thimble in the Soxhlet extraction apparatus was charged with an oven-dried mixture (3 g) of two parts sodium carbonate and one part sand. The mixture was heated to reflux with vigorous stirring for 30 h before cooling to room temperature, the unreacted excess of ligand was precipitated, filtered, and concentrated *in vacuo* to afford the crude product (3.91 g). Further purification by chromatography on silica gel (150 g, EtOAc) followed by recrystallization from CH_2Cl_2–MeCN (4:1, 20 mL) provided the acetonitrile adduct of **95** (1.12 g, 53%) as red fine needles, which turned into the tris(hydrate) as red purple fine needles after standing. To an ice-cooled solution of **95** (0.0002 mmol, 0.40 mL, 0.5×10^{-3} M in CH_2Cl_2) and phenylpropargyl aldehyde (1.30 g, 10.0 mmol) in CH_2Cl_2 (17 mL) was added a solution of Danishefsky's diene (2.57 g, 12.0 mmol) in CH_2Cl_2 (2.6 mL). The resulting mixture was stirred at this temperature for 64 h before quenching with a 10% solution of trifluoroacetic acid in CH_2Cl_2 (about 0.5 mL) and then stirred at 23 °C for an additional 0.5 h. The whole was partitioned between EtOAc (100 mL) and saturated aq. $NaHCO_3$ (10 mL), and the separated organic layer was successively washed with water and brine, and dried over Na_2SO_4, filtered and concentrated *in vacuo*. Column chromatography on silica gel (60 g, 2:1 hexane/EtOAc) afforded the corresponding product (1.46 g, 96%, 91% ee) as a pale yellow oil.

1.6.7
Procedure for TADDOL 52 Promoted Asymmetric Hetero Diels–Alder reaction of Benzaldehyde with 1-(N,N-Dimethylamino)-3-tert-butyldimethylsiloxy-1,3-Diene via Hydrogen Bonding Activation (Scheme 1.64) [105a]

To a solution of TADDOL **52** (0.1 mmol) and benzaldehyde (1.0 mmol) in toluene (0.5 mL) was added 1-(*N*,*N*-dimethylamino)-3-*tert*-butyldimethylsiloxy-1,3-diene (0.5 mmol) at −40 °C. The resulting solution was stirred for 24 h at this temperature before diluting with CH_2Cl_2 (2.0 mL). Acetyl chloride (1.0 mmol) was added dropwise at −78 °C, and the mixture was stirred for an additional 15 min then separated by chromatography on silica gel to afford 2-phenyl-2,3-dihydro-4*H*-pyran-4-one in 70% yield with 98% ee.

1.6.8
Procedure for Chiral Amine 109 Promoted Inverse Electron Demand Hetero Diels–Alder Reaction of Butyraldehyde and Enone (Scheme 1.69) [113]

To a solution of butyraldehyde (0.50 mmol) and enone (1.00 mmol) in CH_2Cl_2 (0.5 mL) was added catalyst **109** (0.05 mmol) followed by addition of silica gel (0.050 g) at −15 °C. The resulting mixture was allowed to warm to room temperature while stirring over night. The equilibrium mixture of intermediates was isolated by flash chromatography on silica gel (gradient CH_2Cl_2 to 15% Et_2O/CH_2Cl_2). Oxidation of the mixture of intermediates was performed in CH_2Cl_2 by addition of PCC (1.0 equiv) at room temperature. After 1 h, another equivalent of PCC was added, 5-ethyl-6-oxo-4-phenyl-5,6-dihydro-4*H*-pyran-2-carboxylic acid methyl ester was isolated in 69% yield with 84% ee by flash chromatography on silica gel with CH_2Cl_2 as the eluent.

Abbreviations

Ac	acetyl
Ar	aryl
BA	Brønsted acid
BINOL	1,1′-binaphth-2,2′-diol
BINAP	2,2′-bis(diphenylphosphino)-1,1′-binaphthyl
BIPHEP	2,2′-bis(diphenylphosphino)biphenyl
BLA	Brønsted acid assisted Lewis acid
Bn	benzyl
Bu	butyl
CAB	chiral (acyloxy) boron catalyst
Cp	cyclopentadienyl
DA reaction	Diels–Alder reaction
DBU	1,8-diazabicyclo[5.4.0]undec-7-ene
de	diastereomeric excess
DM-DABN	3,3′-dimethyl-1,1′-binaphthyl-2,2′-diamine
dr	diastereomeric ratio
ee	enantiomeric excess
ees	enantiomeric excesses
Et	ethyl
HDA reaction	hetero Diels–Alder reaction
HOMO	highest occupied molecular orbital
LA	Lewis acid
LUMO	lowest unoccupied molecular orbital
Me	methyl
MOM	methoxymethyl
MS	molecular sieve
NOBIN	2-amino-2′-hydroxy-1,1′-binaphthyl
NUPHOS	(1Z,3Z)-1,4-bis(diphenylphosphino)buta-1,3-diene
Ns	*p*-nitrobenzenesulfonyl
PCC	pyridinium chlorochromate
Ph	phenyl
Pr	propyl
pybox	pyridyl-bis(oxazoline)
Salen	a ligand prepared by condensation of salicylic aldehyde and ethylene diamine
TADDOL	α,α,α′,α′-tetraaryl-1,3-dioxolane-4,5-dimethanol
TBS	*tert*-butyldimethylsilyl
TCA	trichloroacetic acid
TES	triethylsilyl

Tf	trifluoromethanesulfonyl (triflate)
TFPB	tetrakis(3,5-bis(trifluoromethyl)phenyl)borate
TFA	trifluoroacetic acid
TFAA	trifluoroacetic acid anhydride
TIPS	triisopropylsilyl
TMS	trimethylsilyl
Ts	*p*-toluenesulfonyl
VAPOL	vaulted biphenanthrol

Acknowledgements

Financial support from the National Natural Science Foundation of China (No. 20532050, 20423001), the Chinese Academy of Sciences, the Major Basic Research Development Program of China (Grant No. 2006CB806106), the Science and Technology Commission of Shanghai Municipality, and Merck Research Laboratories is gratefully acknowledged.

References

1 Diels, O. and Alder, K. (1928) *Justus Liebigs Ann. Chem.*, **460**, 98–122.
2 For discussion on the discovery of the DA reaction see: Berson, J.A. (1992) *Tetrahedron*, **48**, 3–17.
3 For recent reviews on the application of DA reactions in total synthesis see: (a) Nicolaou, K.C., Snyder, S.A., Montagnon, T. and Vassilikogiannakis, G. (2002) *Angew. Chem. Int. Ed.*, **41**, 1668–1698; (b) Takao, K.-i., Munakata, R. and Tadano, K.-i. (2005) *Chem. Rev.*, **105**, 4779–4807.
4 For a representative review see: Engelbert, C. (1984) *Org. React.*, **32**, 1–374.
5 For a recent review see: Wessig, P. and Müller, G. (2008) *Chem. Rev.*, **108**, 2051–2063.
6 Oikawa, H. and Tokiwano, T. (2004) *Nat. Prod. Rep.*, **21**, 321–352.
7 (a) Kagan, H.B. and Riant, O. (1992) *Chem. Rev.*, **92**, 1007–1019; (b) Evans, D.A., Rovis, T. and Johnson, J.S. (1999) *Pure Appl. Chem.*, **71**, 1407–1415; (c) Jørgensen, K.A. (2000) *Angew. Chem. Int. Ed.*, **39**, 3558–3588; (d) Johnson, J.S. and Evans, D.A. (2000) *Acc. Chem. Res.*, **33**, 325–335; (e) Corey, E.J. (2002) *Angew. Chem. Int. Ed.*, **41**, 1650–1667; (f) Jørgensen, K.A. (2004) *Eur. J. Org. Chem.*, 2093–2102; (g) Yamamoto, H. and Futatsugi, K. (2005) *Angew. Chem. Int. Ed.*, **44**, 1924–1942; (h) Trost, B.M. and Jiang, C. (2006) *Synthesis*, 369–396; (i) Lin, L., Liu, X. and Feng, X. (2007) *Synlett*, 2147–2157; (j) Hansen, J. and Davies, H.M.L. (2008) *Coord. Chem. Rev.*, **252**, 545–555.
8 (a) Erkkil, A., Majander, I. and Pihko, P. (2007) *Chem. Rev.*, **107**, 5416–5470; (b) Shen, J. and Tan, C.-H. (2008) *Org. Biomol. Chem.*, **6**, 3229–3236; (c) Connon, S.J. (2008) *Chem. Commun.*, 2499–2510; (d) Denmark, S.E. and Beutner, G.L. (2008) *Angew. Chem. Int. Ed.*, **47**, 1560–1638; (e) Bartoli, G. and Melchiorre, P. (2008) *Synlett*, 1759–1772.
9 (a) Otto, S. and Engberts, J.B.F.N. (2000) *Pure Appl. Chem.*, **72**, 1365–1372; (b) Fringuelli, F., Piermatti, O., Pizzo, F. and Vaccaro, L. (2001) *Eur. J. Org. Chem.*, 439–455.

10 For reviews on boron complexes-catalyzed asymmetric DA reactions see: (a) Corey, E.J. (2002) *Angew. Chem. Int. Ed.*, **41**, 1650–1667; (b) Stemmler, R.T. (2007) *Synlett*, 997–998; (c) Harada, T. and Kusukawa, T. (2007) *Synlett*, 1823–1835.

11 (a) Furuta, K., Miwa, Y., Iwanaga, K. and Yamamoto, H. (1988) *J. Am. Chem. Soc.*, **110**, 6254–6255; (b) Furuta, K., Shimizu, S., Miwa, Y. and Yamamoto, H. (1989) *J. Org. Chem.*, **54**, 1481–1483; (c) Ishihara, K., Gao, Q. and Yamamoto, H. (1993) *J. Am. Chem. Soc.*, **115**, 10412–10413.

12 (a) Corey, E.J. and Loh, T.-P. (1991) *J. Am. Chem. Soc.*, **113**, 8966–8967; (b) Corey, E.J., Loh, T.-P., Roper, T.D., Azimioara, M.D. and Noe, M.C. (1992) *J. Am. Chem. Soc.*, **114**, 8290–8292.

13 (a) Singh, R.S. and Harada, T. (2005) *Eur. J. Org. Chem.*, 3433–3435; (b) Singh, R.S., Adachi, S., Tanaka, F., Yamauchi, T., Inui, C. and Harada, T. (2008) *J. Org. Chem.*, **73**, 212–218.

14 (a) Kobayashi, S., Murakami, M., Harada, T. and Mukaiyama, T. (1991) *Chem. Lett.*, 1341–1344; (b) Aggarwal, V.K., Anderson, E., Giles, R. and Zaparucha, A. (1995) *Tetrahedron: Asym.*, **6**, 1301–1306.

15 (a) Hayashi, Y., Rohde, J.J. and Corey, E.J. (1996) *J. Am. Chem. Soc.*, **118**, 5502–5503; (b) Sprott, K.T. and Corey, E.J. (2003) *Org. Lett.*, **5**, 2465–2467.

16 (a) Ishihara, K. and Yamamoto, H. (1994) *J. Am. Chem. Soc.*, **116**, 1561–1562; (b) Ishihara, K., Kurihara, H. and Yamamoto, H. (1996) *J. Am. Chem. Soc.*, **118**, 3049–3050; (c) Ishihara, K., Kurihara, H., Matsumoto, M. and Yamamoto, H. (1998) *J. Am. Chem. Soc.*, **120**, 6920–6930.

17 (a) Corey, E.J., Shibata, T. and Lee, T.W. (2002) *J. Am. Chem. Soc.*, **124**, 3808–3809; (b) Ryu, D.H., Lee, T.W. and Corey, E.J. (2002) *J. Am. Chem. Soc.*, **124**, 9992–9993.

18 (a) Ryu, D.H. and Corey, E.J. (2003) *J. Am. Chem. Soc.*, **125**, 6388–6390; (b) Ryu, D.H., Zhou, G. and Corey, E.J. (2004) *J. Am. Chem. Soc.*, **126**, 4800–4802; (c) Zhou, G., Hu, Q.-Y. and Corey, E.J. (2003) *Org. Lett.*, **5**, 3979–3982; (d) Canales, E. and Corey, E.J. (2008) *Org. Lett*, **10**, 3271–3273.

19 Payette, J.N. and Yamamoto, H. (2007) *J. Am. Chem. Soc.*, **129**, 9536–9537.

20 (a) Futatsugi, K. and Yamamoto, H. (2005) *Angew. Chem. Int. Ed.*, **44**, 1484–1487; (b) Liu, D., Canales, E. and Corey, E.J. (2007) *J. Am. Chem. Soc.*, **129**, 1498–1499; (c) Paddon-Row, M.N., Kwan, L.C.H., Willis, A.C. and Sherburn, M.S. (2008) *Angew. Chem. Int. Ed.*, **47**, 7013–7017.

21 Corey, E.J., Imwinkelried, R., Pikul, S. and Wang, Y.B. (1989) *J. Am. Chem. Soc.*, **111**, 5493–5495.

22 Bao, J., Rheingold, A.L. and Wulff, W.D. (1993) *J. Am. Chem. Soc.*, **115**, 3814–3815.

23 Ishihara, K., Kobayashi, J., Inanaga, K. and Yamamoto, H. (2001) *Synlett*, 394–396.

24 Corminboeuf, O. and Renaud, P. (2002) *Org. Lett.*, **4**, 1731–1733.

25 (a) Teo, Y.-C. and Loh, T.-P. (2005) *Org. Lett.*, **7**, 2539–2541; (b) Fu, F., Teo, Y.-P. and Loh, T.-P. (2006) *Org. Lett.*, **8**, 5999–6001.

26 For recent reviews on chiral bis(oxazoline) ligands see: (a) Schulz, E. (2005) *Top Organomet. Chem.*, **15**, 93–148; (b) Desimoni, G., Faita, G. and Jørgensen, K.A. (2006) *Chem. Rev.*, **106**, 3561–3651.

27 (a) Evans, D.A., Miller, S.J. and Lectka, T. (1993) *J. Am. Chem. Soc.*, **115**, 6460–6461; (b) Evans, D.A., Murry, J.A., von Matt, P. Norcross, R.D. and Miller, S.J. (1995) *Angew. Chem. Int. Ed.*, **34**, 798–800; (c) Evans, D.A., Miller, S.J., Lectka, T. and von Matt, F P. (1999) *J. Am. Chem. Soc.*, **121**, 7559–7573; (d) Evans, D.A., Barnes, D.M., Johnson, J.S., Lectka, T., von Matt P., Miller, S.J., Murry, J.A., Norcross, R.D., Shaughnessy, E.A. and Campos, K.R. (1999) *J. Am. Chem. Soc.*, **121**, 7582–7594.

28 (a) Palomo, C., Oiarbide, M., García, J.M., González, A. and Arceo, E. (2003) *J. Am. Chem. Soc.*, **125**, 13942–13943;

(b) Barroso, S., Blay, G. and Pedro, J.R. (2007) *Org. Lett.*, **9**, 1983–1986; (c) Sibi, M.P., Stanley, L.M., Nie, X., Venkatraman, L., Liu, M. and Jasperse, C.P. (2007) *J. Am. Chem. Soc.*, **129**, 395–405; (d) Barroso, S., Blay, G., Al-Midfa L. Muñoz, M.C. and Pedro, J.R. (2008) *J. Org. Chem.*, **73**, 6389–6392.

29 (a) Bolm, C., Martin, M., Simic, O. and Verrucci, M. (2003) *Org. Lett.*, **5**, 427–429; (b) Bolm, C., Martin, M., Gescheidt, G., Palivan, C., Neshchadin, D., Bertagnolli, H., Feth, M., Schweiger, A., Mitrikas, G. and Harmer, J. (2003) *J. Am. Chem. Soc.*, **125**, 6222–6227.

30 (a) Owens, T.D., Hollander, F.J., Oliver, A.G. and Ellman, J.A. (2001) *J. Am. Chem. Soc.*, **123**, 1539–1540; (b) Lassaletta, J.M., Alcarazo, M. and Fernández, R. (2004) *Chem. Commun.*, 298–299.

31 (a) Ishihara, K. and Fushimi, M. (2006) *Org. Lett.*, **8**, 1921–1924; (b) Ishihara, K. and Fushimi, M. (2008) *J. Am. Chem. Soc.*, **130**, 7532–7533.

32 (a) Boersma, A.J., Feringa, B.L. and Roelfes, G. (2007) *Org. Lett.*, **9**, 3647–3650; (b) Boersma, A.J., Klijn, J.E., Feringa, B.J. and Roelfes, G. (2008) *J. Am. Chem. Soc.*, **130**, 11783–11790.

33 (a) Evans, D.A., Kozlowski, M.C. and Tedrow, J.S. (1996) *Tetrahedron Lett.*, **37**, 7481–7484; (b) Takacs, J.M., Lawson, E.C., Reno, M.J., Youngman, M.A. and Quincy, D.A. (1997) *Tetrahedron: Asym.*, **8**, 3073–3078; (c) Honda, Y., Date, T. Hiramatsu, H. and Yamauchi, M. (1997) *Chem. Commun.*, 1411–1412; (d) Carbone, P., Desimoni, G., Faita, G., Filippone, S. and Righetti, P. (1998) *Tetrahedron*, **54**, 6099–6110; (e) Corminboeuf, O. and Renaud, P. (2002) *Org. Lett.*, **4**, 1735–1738.

34 (a) Mikami, K., Terada, M., Motoyama, Y. and Nakai, T. (1991) *Tetrahedron: Asym.*, **2**, 643–646; (b) Mikami, K., Motoyama, Y. and Terada, M. (1994) *J. Am. Chem. Soc.*, **116**, 2812–2820; (c) Motoyama, Y., Terada, M. and Mikami, K. (1995) *Synlett*, 967–968; (d) Maruoka, K., Murase, N. and Yamamoto, H. (1993) *J. Org. Chem.*, **58**, 2938–2939.

35 (a) Huang, Y., Iwama, T. and Rawal, V.H. (2000) *J. Am. Chem. Soc.*, **122**, 7843–7844; (b) Huang, Y., Iwama, T. and Rawal, V.H. (2002) *Org. Lett.*, **4**, 1146—1163.

36 (a) Huang, Y., Iwama, T. and Rawal, V.H. (2002) *J. Am. Chem. Soc.*, **124**, 5950–5951; (b) McGilvra, J.D. and Rawal, V.H. (2004) *Synlett*, 2440–2442.

37 (a) Corey, E.J., Imai, N. and Zhang, H.-Y. (1991) *J. Am. Chem. Soc.*, **113**, 728–729; (b) Kündig, E.P., Saudan, C.M. and Viton, F. (2001) *Adv. Synth. Catal.*, **343**, 51–56.

38 (a) Davies, D.L., Fawcett, J., Garratt, S.A. and Russell, D.R. (1997) *Chem. Commun.*, 1351–1352; (b) Davenport, A.J., Davies, D.L., Fawcett, J. and Russell, D.R. (2004) *Dalton Trans.*, 1481–1492.

39 (a) Faller, J.W. and Lavoie, A. (2001) *J. Organomet. Chem.*, **630**, 17–22; (b) Faller, J.W., Lavoie, A.R. and Grimmond, B.J. (2002) *Organometallics*, **21**, 1662–1666.

40 (a) Kündig, E.P., Saudan, C.M., Alezra, V., Viton, F. and Bernardinelli, G. (2001) *Angew. Chem. Int. Ed.*, **40**, 4481–4485; (b) Alezra, V., Bernardinelli, G., Corminboeuf, C., Frey, U., Kündig, E.P., Merbach, A.E., Saudan, C.M., Viton, F. and Weber, J. (2004) *J. Am. Chem. Soc.*, **126**, 4843–4853; (c) Rickerby, J., Vallet, M., Berardinelli, G., Viton, F. and Kündig, E.P. (2007) *Chem. Eur. J.*, **13**, 3354–3368.

41 For selected examples, see: (a) Davenport, A.J., Davies, D.L., Fawcett, J., Garratt, S.A., Lad, L. and Russell, D.R. (1997) *Chem. Commun.*, 2347–2348; (b) Nakano, H. Suzuki, Y. Kabuto, C. Fujita, R. and Hongo, H. *J. Org. Chem*, (2002) **67**, 5011–5014; (c) Carmona, D. Lamata, M.P. Viguri, F. Rodríguez, R. Lahoz, F.J. Dobrinovtch, I.T. and Oro, L.A. (2007) *Dalton Trans.*, 1911–1921.

42 (a) Kobayashi, S. and Ishitani, H. (1994) *J. Am. Chem. Soc.*, **116**, 4083–4084; (b) Kobayashi, S., Ishitani, H., Hachiya, I. and Akaki, M. (1994) *Tetrahedron*, **50**, 11623–11636.

43 (a) Markó, I.E., Evans, G.R., Seres, P., Chellé, I. and Janousek, Z. (1996) *Pure Appl. Chem.*, **68**, 113–122; (b) Markó, I.E., Chellé-Regnaut F I. Leroy, B. and Warriner, S.L. (1997) *Tetrahedron Lett.*, **38**, 4269–4272.

44 Sudo, Y., Shirasaki, D., Harada, S. and Nishida, A. (2008) *J. Am. Chem. Soc.*, **130**, 12588–12589.

45 (a) Fukuyawa, S.-i., Matsuzawa, H. and Metoki, K. (2001) *Synlett*, 709–711; (b) Desimoni, G., Faita, G., Guala, M. and Pratelli, C. (2002) *Tetrahedron*, **58**, 2929–2935; (c) Desimoni, G., Faita, G., Guala, M. and Pratelli, C. (2003) *J. Org. Chem.*, **68**, 7862–7866; (d) Desimoni, G., Faita, G., Guala, M. and Laurenti, A. (2004) *Eur. J. Org. Chem.*, 3057–3062; (e) Desimoni, G., Faita, G., Guala, M., Laurenti, A. and Mella, M. (2005) *Chem. Eur. J.*, **11**, 3816–3824; (f) Desimoni, G., Faita, G., Mella, M., Piccinini, F. and Toscanini, M. (2007) *Eur. J. Org. Chem.*, 1529–1534.

46 Wang, H., Wang, H., Liu, P., Yang, H., Xiao, J. and Li, C. (2008) *J. Mol. Catal. A: Chem.*, **285**, 128–131.

47 Evans, D.A. and Wu, J. (2003) *J. Am. Chem. Soc*, **125**, 10162–10163.

48 Ahrendt, K.A., Borths, C.J. and MacMillan, D.W.C. (2000) *J. Am. Chem. Soc*, **122**, 4243–4244.

49 Northrup, A.B. and MacMillan, D.W.C. (2002) *J. Am. Chem. Soc.*, **124**, 2458–2460.

50 Wilson, R.M., Jen, W.S. and MacMillan, D.W.C. (2005) *J Am. Chem. Soc.*, **127**, 11616–11617.

51 Bonini, B.F., Capitò, E., Comes-Franchini F M. Fochi, M., Ricci, A. and Zwanenburg, B. (2006) *Tetrahedron: Asym.*, **17**, 3135–3143.

52 (a) Gotoh, H. and Hayashi, Y. (2007) *Org. Lett.*, **9**, 2859–2862; (b) Hayashi, Y., Samanta, S., Gotoh, H. and Ishikawa, H. (2008) *Angew. Chem. Int. Ed.*, **47**, 6634–6637.

53 (a) He, H., Pei, B.-J., Chou, H.-H., Tian, T., Chan, W.-H. and Lee, A.W.M. (2008) *Org. Lett.*, **10**, 2421–2424; (b) Langlois, Y., Petit, A., Rémy, P., Scherrmann, M.-C. and Kouklovsky, C. (2008) *Tetrahedron Lett.*, **49**, 5576–5579.

54 Kano, T., Tanaka, Y. and Maruoka, K. (2006) *Org. Lett.*, **8**, 2687–2689.

55 (a) Ishihara, K. and Nakano, K. (2005) *J. Am. Chem. Soc.*, **127**, 10504–10505; (b) Ishihara, K., Nakano, K. and Akakura, M. (2008) *Org. Lett.*, **10**, 2893–2896.

56 (a) Sakakura, A., Suzuki, K., Nakano, K. and Ishihara, K. (2006) *Org. Lett.*, **8**, 2229–2232; (b) Sakakura, A., Suzuki, K. and Ishihara, K. (2006) *Adv. Synth. Catal.*, **348**, 2457–2465; (c) Kano, T., Tanaka, Y. and Maruoka, K. (2007) *Chem. Asian J.*, **2**, 1161–1165.

57 Kim, K.H., Lee, S., Lee, D.-W., Ko, D.-H. and Ha, D.-C. (2005) *Tetrahedron Lett.*, **46**, 5991–5994.

58 Singh, R.P., Bartelson, K., Wang, Y., Su, H., Lu, X. and Deng, L. (2008) *J. Am. Chem. Soc.*, **130**, 2422–2423.

59 For recent reviews on chiral Brønsted acid catalysis see: (a) Taylor, M.S. and Jacobsen, E.N. (2006), *Angew. Chem. Int. Ed.*, **45**, 1520–1543; (b) Doyle, A.G. and Jacobsen, E.N. (2007) *Chem. Rev.*, **107**, 5713–5743; (c) Akiyama, T. (2007) *Chem. Rev.*, **107**, 5744–5758. For examples, see: (d) Schuster, T., Bauch, M., Dürner, G. and Göbel, M.W. (2000) *Org. Lett.*, **2**, 179–181; (e) Tsogoeva, S.B., Dürner, G., Bolte, M. and Göbel, M.W. (2003) *Eur. J. Org. Chem.*, 1661–1664.

60 Thadani, A.N., Stankovic, A.R. and Rawal, V.H. (2004) *Proc. Natl. Acad. Sci. U.S.A.*, **101**, 5846–5850.

61 Nakashima, D. and Yamamoto, H. (2006) *J. Am. Chem. Soc.*, **128**, 9626–9627.

62 Uemae, K., Masuda, S. and Yamamoto, Y. (2001) *J. Chem. Soc., Perkin Trans. 1*, 1002–1006.

63 Shen, J., Nguyen, T.T., Goh, Y.-P., Ye, W., Fu, X., Xu, J. and Tan, C.-H. (2006) *J. Am. Chem. Soc.*, **128**, 13692–13693.

64 Wang, Y., Li, H., Wang, Y.-Q., Liu, Y., Foxman, B.M. and Deng, L. (2007) *J. Am. Chem. Soc.*, **129**, 6364–6365.

65. Gioia, C., Hauville, A., Bernardi, L., Fini, F. and Ricci, A. (2008) *Angew. Chem. Int. Ed.*, **47**, 9236–9239.
66. (a) Maruoka, K., Itoh, T., Shirasaka, T. and Yamamoto, H. (1988) *J. Am. Chem. Soc.*, **110**, 310–312; (b) Maruoka, K. and Yamamoto, H. (1989) *J. Am. Chem. Soc.*, **111**, 789–790.
67. (a) Johannsen, M., Jørgensen, K.A., Zheng, X.-F., Hu, Q.-S. and Pu, L. (1999) *J. Org. Chem.*, **64**, 299–301; (b) Simonsen, K.B., Svenstrup, N., Roberson, M. and Jørgensen, K.A. (2000) *Chem. Eur. J.*, **6**, 123–128.
68. Gao, Q., Maruyama, T., Mouri, M. and Yamamoto, H. (1992) *J. Org. Chem.*, **57**, 1951–1952.
69. Corey, E.J., Cywin, C.L. and Roper, T.D. (1992) *Tetrahedron Lett.*, **33**, 6907–6910.
70. Yu, Z., Liu, X., Dong, Z., Xie, M. and Feng, X. (2008) *Angew. Chem. Int. Ed.*, **47**, 1308–1311.
71. (a) Terada, M., Mikami, K. and Nakai, T. (1991) *Tetrahedron Lett.*, **32**, 935–938; (b) Wada, E., Pei, W., Yasuoka, H., Chin, U. and Kanamasa, S. (1996) *Tetrahedron*, **52**, 1205–1220.
72. Keck, G.E., Li, X.-Y. and Krishnamurthy, D. (1995) *J. Org. Chem.*, **60**, 5998–5999.
73. (a) Wang, B., Feng, X., Cui, X., Liu, H. and Jiang, Y. (2000) *Chem. Commun.*, 1605–1606; (b) Wang, B., Feng, X., Huang, Y., Liu, H., Cui, X. and Jiang, Y. (2002) *J. Org. Chem.*, **67**, 2175–2812.
74. (a) Huang, Y., Feng, X., Wang, B., Zhang, G. and Jiang, Y. (2002) *Synlett*, 2122–2124; (b) Fu, Z., Gao, B., Yu, Z., Yu, L., Huang, Y., Feng, X. and Zhang, G. (2004) *Synlett*, 1772–1775; (c) Yang, W., Shang, D., Liu, Y., Du, Y. and Feng, X. (2005) *J. Org. Chem.*, **70**, 8533–8537. For a chiral bis-titanium catalyzed HDA reaction, see: (d) Kii, S., Hashimoto, T. and Maruoka, K. (2002) *Synlett*, 931–932.
75. Yang, X.-B., Feng, J., Zhang, J., Wang, N., Wang, L., Liu, J.-L. and Yu, X.-Q. (2008) *Org. Lett.*, **10**, 1299–1302.
76. Lin, L., Chen, Z., Yang, X., Liu, X. and Feng, X. (2008) *Org. Lett.*, **10**, 1311–1314.
77. (a) Long, J., Hu, J., Shen, X., Ji, B. and Ding, K. (2002) *J. Am. Chem. Soc.*, **124**, 10–11; (b) Ding, K., Du, H., Yuan, Y. and Long, J. (2004) *Chem. Eur. J.*, **10**, 2872–2884.
78. (a) Yuan, Y., Long, J., Sun, J. and Ding, K. (2002) *Chem. Eur. J.*, **8**, 5033–5042; (b) Yuan, Y., Li, X., Sun, J. and Ding, K. (2002) *J. Am. Chem. Soc.*, **124**, 14866–14867; (c) Ji, B., Yuan, Y., Ding, K. and Meng, J. (2003) *Chem. Eur. J.*, **9**, 5989–5996; (d) Ji, B., Ding, K. and Meng, J. (2003) *Chinese J. Chem.*, **21**, 727–730.
79. Li, X., Meng, X., Su, H., Wu, X. and Xu, D. (2008) *Synlett*, 857–860.
80. (a) Yamashita, Y., Saito, S., Ishitani, H. and Kobayashi, S. (2002) *Org. Lett.*, **4**, 1221–1223; (b) Yamashita, Y., Saito, S., Ishitani, H. and Kobayashi, S. (2003) *J. Am. Chem. Soc.*, **125**, 3793–3798; (c) Kobayashi, S., Ueno, M., Saito, S., Mizuki, Y., Ishitani, H. and Yamashita, Y. (2004) *Proc. Natl. Acad. Sci. U.S.A.*, **101**, 5476–5481.
81. Schaus, S.E., Brånalt, J. and Jacobsen, E.N. (1998) *J. Org. Chem.*, **63**, 403–405.
82. Kosior, M., Kwiatkowski, P., Asztemborska, M. and Jurczak, J. (2005) *Tetrahedron: Asym.*, **16**, 2897–2900.
83. (a) Dossetter, A.G., Jamison, T.F. and Jacobsen, E.N. (1999) *Angew. Chem. Int. Ed.*, **38**, 2398–2400; (b) Gademann, K., Chavez, D.E. and Jacobsen, E.N. (2002) *Angew. Chem. Int. Ed.*, **41**, 3059–3061; (c) Joly, G.D. and Jacobsen, E.N. (2002) *Org. Lett.*, **4**, 1795–1798.
84. (a) Berkessel, A. and Vogl, N. (2006) *Eur. J. Org. Chem.*, 5029–5035; (b) Berkessel, A., Ertürk, E. and Laporte, C. (2006) *Adv. Synth. Catal.*, **348**, 223–228.
85. (a) Hu, Y.-J., Huang, X.-D., Yao, Z.-J. and Wu, Y.-L. (1998) *J. Org. Chem.*, **63**, 2456–2461; (b) Togni, A. (1990) *Organometallics*, **9**, 3106–3113; (c) Aikawa, K., Irie, R. and Katsuki, T. (2001) *Tetrahedron*, **57**, 845–851.
86. (a) Johannsen, M., Yao, S. and Jørgensen, K.A. (1997) *Chem. Commun.*, 2169–2170; (b) Yao, S., Johannsen, M., Audrain, H.,

Hazell, R.G. and Jørgensen, K.A. (1998) *J. Am. Chem. Soc.*, **120**, 8599–8605.

87 (a) Thorhauge, J., Johannsen, M. and Jørgensen, K.A. (1998) *Angew. Chem. Int. Ed.*, **37**, 2404–2406; (b) Zhang, W., Thorhauge, J. and Jørgensen, K.A. (2000) *Chem. Commun.*, 459–460; (c) Audrain, H., Thorhauge, J., Hazell, R.G. and Jørgensen, K.A. (2000) *J. Org. Chem.*, **65**, 4487–4497; (d) Thorhauge, J., Roberson, M., Hazell, R.G. and Jørgensen, K.A. (2002) *Chem. Eur. J.*, **8**, 1888–1898.

88 Landa, A., Richter, B., Johansen, R.L., Minkkila, A. and Jørgensen, K.A. (2007) *J. Org. Chem.*, **72**, 240–245.

89 (a) Evans, D.A. and Johnson, J.S. (1998) *J. Am. Chem. Soc.*, **120**, 4895–4896; (b) Evans, D.A., Olhava, E.J., Johnson, J.S. and Janey, J.M. (1998) *Angew. Chem. Int. Ed.*, **37**, 3372–3375; (c) Evans, D.A., Johnson, J.S. and Olhava, E.J. (2000) *J. Am. Chem. Soc.*, **122**, 1635–1649; (d) Evans, D.A. and Janey, J.M. (2001) *Org. Lett.*, **3**, 2125–2128.

90 (a) Ghosh, A.K. and Shirai, M. (2001) *Tetrahedron Lett.*, 6231–6233; (b) Wolf, C., Fadul, Z., Hawes, P.A. and Volpe, E.C. (2004) *Tetrahedron: Asym.*, **15**, 1987–1993; (c) Shin, Y.J., Yeom, C.-E., Kim, M.J. and Kim, B.M. (2008) *Synlett*, 89–93.

91 (a) Bolm, C. and Simić, O. (2001) *J. Am. Chem. Soc.*, **123**, 3830–3831; (b) Bolm, C., Verrucci, M., Simić, O., Cozzi, P.G., Raabe, G. and Okamura, H. (2003) *Chem. Commun.*, 2826–2827; (c) Bolm, C., Verrucci, M., Simić, O. and Hackenberger, C.P.R. (2005) *Adv. Synth. Catal.*, **347**, 1696–1700.

92 Yao, S., Johannsen, M. and Jørgensen, K.A. (1997) *J. Chem. Soc., Perkin Trans. 1*, 2345–2350.

93 (a) Du, H., Long, J., Hu, J., Li, X. and Ding, K. (2002) *Org. Lett*, **4**, 4349–4352; (b) Du, H. and Ding, K. (2003) *Org. Lett.*, **5**, 1091–1093; (c) Du, H., Zhang, X., Wang, Z. and Ding, K. (2005) *Tetrahedron*, **61**, 9465–9477.

94 Du, H., Zhang, X., Wang, Z., Bao, H., You, T. and Ding, K. (2008) *Eur. J. Org. Chem.*, 2248–2254.

95 Motoyama, Y., Koga, Y. and Nishiyama, H. (2001) *Tetrahedron*, **57**, 853–860.

96 (a) Doyle, M.P., Phillips, I.M. and Hu, W. (2001) *J. Am. Chem. Soc.*, **123**, 5366–5367; (b) Valenzuela, M., Doyle, M.P., Hedberg, C., Hu, W. and Holmstrom, A. (2004) *Synlett*, 2425–2428; (c) Doyle, M.P., Morgan, J.P., Fettinger, J.C., Zavalij, P.Y., Colyer, J.T., Timmons, D.J. and Carducci, M.D. (2005) *J. Org. Chem.*, **70**, 5291–5301; (d) Doyle, M.P., Valenzuela, M. and Huang, P. (2004) *Proc. Natl. Acad. Sci. U.S.A.*, **101**, 5391–5395.

97 Anada, M., Washio, T., Shimada, N., Kitagaki, S., Nakajima, M., Shiro, M. and Hashimoto, S. (2004) *Angew. Chem. Int. Ed.*, **43**, 2665–2668.

98 Oi, S., Terada, E., Ohuchi, K., Kato, T., Tachibana, Y. and Inoue, Y. (1999) *J. Org. Chem.*, **64**, 8660–8667.

99 (a) Mikami, K., Aikawa, K., Yusa, Y. and Hatano, M. (2002) *Org. Lett.*, **4**, 91–94. (b) Mikami, K., Aikawa, K. and Yusa, Y. (2002) *Org. Lett.*, **4**, 95–97.

100 Becker, J.J., Van Orden F L.J., White, P.S. and Gagné, M.R. (2002) *Org. Lett.*, **4**, 727–730.

101 Doherty, S., Knight, J.G., Hardacre, C., Lou, H., Newman, C.R., Rath, R.K., Campbell, S. and Nieuwenhuzen, M. (2004) *Organometallics*, **23**, 6127–6133.

102 (a) Bednarski, M., Maring, C. and Danishefsky, S. (1983) *Tetrahedron Lett.*, **24**, 3451–3454; (b) Bednarski, M. and Danishefsky, S. (1983) *J. Am. Chem. Soc.*, **105**, 6968–6969; (c) Bednarski, M. and Danishefsky, S. (1986) *J. Am. Chem. Soc.*, **108**, 7060–7067; (d) Qian, C. and Wang, L. (2000) *Tetrahedron Lett.*, **41**, 2203–2206.

103 (a) Inanaga, J., Sugimoto, Y. and Hanamoto, T. (1995) *New J. Chem.*, **19**, 707–712; (b) Hanamoto, T., Furuno, H., Sugimoto, Y. and Inanaga, J. (1997) *Synlett*, 79–80; (c) Furuno, H., Kambara,

T., Tanaka, Y., Hanamoto, T., Kagawa, T. and Inanaga, J. (2003) *Tetrahedron Lett.*, **44**, 6129–6132; (d) Furuno, H., Hayano, T., Kambara, T., Sugimoto, Y., Hanamoto, T., Tanaka, Y., Jin, Y.Z., Kagawa, T. and Inanaga, J. (2003) *Tetrahedron*, **59**, 10509–10523; (e) Fukuzawa, S.-i., Metoki, K. and Esumi, S.-i. (2003) *Tetrahedron*, **59**, 10445–10452.

104 Tiseni, P.S. and Peters, R. (2008) *Org. Lett.*, **10**, 2019–2022.

105 (a) Huang, Y., Unni, A.K., Thadani, A.N. and Rawal, V.H. (2003) *Nature*, **424**, 146; (b) Unni, A.K., Takenaka, N., Yamamoto, H. and Rawal, V.H. (2005) *J. Am. Chem. Soc.*, **127**, 1336–1337.

106 Du, H., Zhao, D. and Ding, K. (2004) *Chem. Eur. J.*, **10**, 5964–5970.

107 Zhang, X., Du, H., Wang, Z., Wu, Y.-D. and Ding, K. (2006) *J. Org. Chem.*, **71**, 2862–2869.

108 (a) Harriman, D.J., Lambropoulos, A. and Deslongchamps, G. (2007) *Tetrahedron Lett.*, **48**, 689–692; (b) Anderson, C.D., Dudding, T., Gordillo, R. and Houk, K.N. (2008) *Org. Lett.*, **10**, 2749–2752.

109 Friberg, A., Olsson, C., Ek, F., Berg, U. and Frejd, T. (2007) *Tetrahedron: Asymm*, **18**, 885–891.

110 Rajaram, S. and Sigman, M.S. (2005) *Org. Lett.*, **7**, 5473–5475.

111 Zhuang, W., Poulsen, T.B. and Jørgensen, K.A. (2005) *Org. Biomol. Chem.*, **3**, 3284–3289.

112 Tonoi, T. and Mikami, K. (2005) *Tetrahedron Lett.*, **46**, 6355–6358.

113 Juhl, K. and Jørgensen, K.A. (2003) *Angew. Chem. Int. Ed.*, **42**, 1498–1501.

114 Samanta, S., Krause, J., Mandal, T. and Zhao, C.-G. (2007) *Org. Lett.*, **9**, 2745–2748.

115 Zhao, Y. Wang, X.-J. and Liu, J.-T. (2008), *Synlett*, 1017–1021.

116 Hernandez-Juan, F.A., Cockfield, D.M. and Dixon, D.J. (2007) *Tetrahedron Lett.*, **48**, 1605–1608.

117 Hu, Q.-Y., Zhou, G. and Corey, E.J. (2004) *J. Am. Chem. Soc.*, **126**, 13708–13713, and references cited in.

118 Yu, M. and Danishefsky, S.J. (2008) *J. Am. Chem. Soc.*, **130**, 2783–2785.

119 Ward, D.E. and Shen, J. (2007) *Org. Lett.*, **9**, 2843–2846.

120 Kong, K., Moussa, A. and Romo, D. (2005) *Org. Lett.*, **7**, 5127–5130.

121 Yao, S., Johannsen, M., Hazell, R.G. and Jørgensen, K.A. (1998) *J. Org. Chem.*, **63**, 118–121.

122 (a) Hong, Y.S. and Corey, E.J. (2006) *J. Am. Chem. Soc.*, **128**, 6310–6311; (b) Yamatsugu, K., Yin, L., Kamijo, S., Kimura, Y., Kanai, M. and Shibasaki, M. (2009) *Angew. Chem. Int. Ed.*, **48**, 1070–1076.

123 (a) Thompson, C.F., Jamison, T.F. and Jacobsen, E.N. (2001) *J. Am. Chem. Soc.*, **123**, 9974–9983; (b) Boezio, A.A., Jarvo, E.R., Lawrence, B.M. and Jacobsen, E.N. (2005) *Angew. Chem. Int. Ed.*, **44**, 6046–6050.

124 Ghosh, A.K. and Gong, G. (2007) *Org. Lett.*, **9**, 1437–1440.

125 Paterson, I. and Miller, N.A. (2008) *Chem. Commun.*, 4708–4710.

126 Nicolaou, K.C., Tria, G.S. and Edmonds, D.J. (2008) *Angew. Chem. Int. Ed.*, **47**, 1780–1783.

127 Bonazzi, S., Güttinger, S., Zemp, I., Kutay, U. and Gademann, K. (2007) *Angew. Chem. Int. Ed.*, **46**, 8707–8710.

128 Kinsman, A.C. and Kerr, M.A. (2003) *J. Am. Chem. Soc.*, **125**, 14120–14125.

129 Shimizu, S., Ohori, K., Arai, T., Sasai, H. and Shibasaki, M. (1998) *J. Org. Chem.*, **63**, 7547–7551.

2
Catalytic Asymmetric Aza Diels–Alder Reactions
Yasuhiro Yamashita and Shū Kobayashi

2.1
Introduction

Asymmetric aza Diels–Alder reactions provide a useful route to optically active nitrogen-containing heterocyclic compounds [1]. Compared to the remarkable progress of catalytic asymmetric hetero Diels–Alder reactions to give oxygen-containing six-membered cycles, the development of catalytic asymmetric aza Diels–Alder reactions has been relatively slow [2]. This is because typical metal Lewis acids are deactivated by the nitrogen atoms of the products formed that are more basic than those of the starting imine compounds. However, accompanying recent progress with new types of metal Lewis acid catalysts which are tolerant to Lewis basic function groups, catalytic asymmetric reactions have been explored, and high enantioselectivities have been achieved. Furthermore, asymmetric organocatalysis and chiral Brønsted acid catalysis have also been applied to asymmetric aza Diels–Alder reactions. In this chapter, the recent progress in catalytic asymmetric aza Diels–Alder reactions is summarized.

Recent reports of catalytic asymmetric aza Diels–Alder reactions are classified into two types: those between a normal carbon diene and an imine of an azo dienophile ((i) and (ii) in Scheme 2.1) and those between an aza diene and a normal carbon dienophile ((iii) and (iv) in Scheme 2.1). While catalytic asymmetric processes have been investigated using some kinds of chiral catalysts in both types of reaction, the most well explored is the reaction of an imine as a dienophile and an activated diene, such as Danishefsky diene. Successful examples of the other type have been reported, however, in only a few cases.

2.2
Chiral Lewis Acid Catalysis

A pioneering study of enantioselective aza Diels–Alder reactions was reported in 1992 [3]. Chiral boron compounds were used in aza Diels–Alder reactions of benzyl

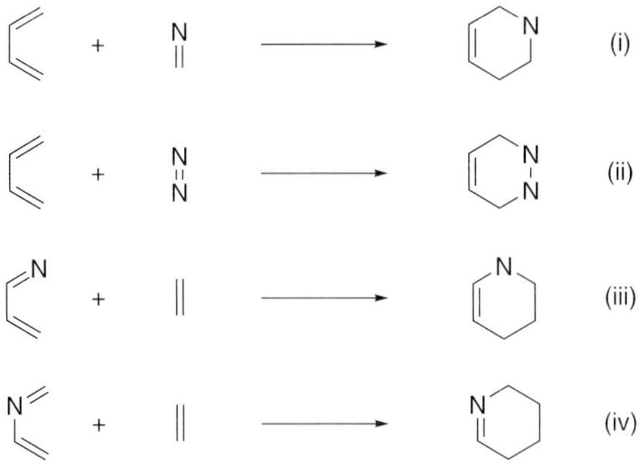

2 Catalytic Asymmetric Aza Diels–Alder Reactions

Scheme 2.1 Aza Diels–Alder reactions.

imines with 1-methoxy-3-silyloxy-1,3-diene (Danishefsky diene), and the desired piperidone derivatives were obtained in good yields with high enantioselectivities (Scheme 2.2). However stoichiometric amounts of the chiral compound were needed, and enantioselective aza Diels–Alder reactions using a catalytic amount of promoter were desired.

In 1996, the first example of catalytic enanioselective aza Diels–Alder reactions of azadienes was reported using a chiral lanthanide catalyst (Table 2.1) [4]. The choice of metal Lewis acids was crucial to realize significant catalytic turnover in the catalytic enantioselective aza Diels–Alder reactions and, finally, lanthanide triflate bearing a chiral ligand was selected. Aza Diels–Alder reactions of N-alkylidene-2-hydroxylanilines with cyclopentadiene or vinyl ethers proceeded in the presence of a chiral ytterbium complex prepared from Yb(OTf)$_3$, (R)-1,1′-binaphthol (R)-BINOL) and 1,8-diazabicyclo[5.4.0]undec-7-ene (DBU) to afford the desired optically active tetrahydroisoquinoline derivatives in good yields with good to high enantioselectivities.

Catalytic enantioselective aza Diels–Alder reactions of azadienophiles with 1-alkoxy-1,3-butadienes (Danishefsky diene) were also achieved using chiral zirconium catalysts in 1998 [5]. The active catalyst was prepared from zirconium tert-butoxide, (R)-6,6′-dibromo-1,1′-binaphthol (R)-6,6′-Br$_2$BINOL) and 2–3 equiv of N-methylimidazole [6], and the desired 4-piperidone derivatives were obtained in

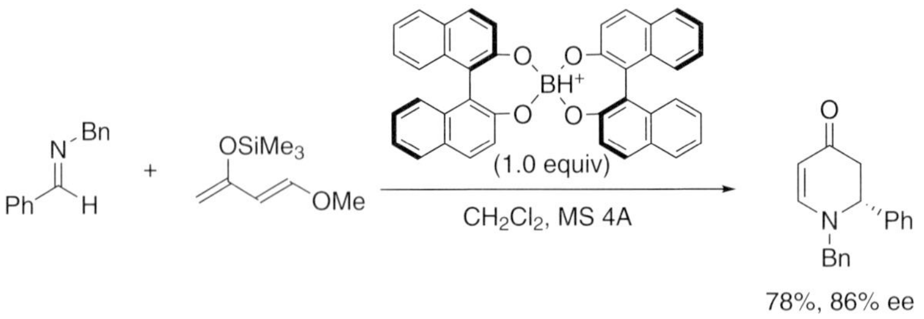

Scheme 2.2 Chiral boron-mediated aza Diels–Alder reaction.

Table 2.1 The first example of catalytic enantioselective aza Diels–Alder reaction of azadienes with olefins.

Entry	R¹	Olefin	Additive	Yield (%)	cis/trans	ee (cis, %)
1	Ph	cyclopentadiene	2,6-^{t}Bu$_2$-4-MePyridine	92	>99:1	71
2[a]	Ph	⟶OEt	2,6-^{t}Bu$_2$Pyridine	52	94:6	77
3	1-Naph	⟶OEt	2,6-^{t}Bu$_2$-4-MePyridine	74	>99:<1	91
4	1-Naph	⟶OEt (Et)	2,6-^{t}Ph$_2$Pyridine	67	93:7	86
5[b]	^{c}Hex	cyclopentadiene	2,6-^{t}Bu$_2$-4-MePyridine	58	>99:<1	73

[a] At −45 °C, 10 mol% catalyst.
[b] Sc(OTf)$_3$ was used instead of Yb(OTf)$_3$.

Chiral Yb catalyst

high yields with high enantioselectivities. The *ortho*-hydroxy group of the aromatic part on the imine nitrogen was important to construct rigid asymmetric environments. Aliphatic imines also reacted with the diene successfully (Table 2.2).

Synthesis of both enantiomers is a very important task, not only in organic chemistry but also in medicinal and bioorganic chemistry, because it is sometimes difficult to obtain both enantiomers from natural sources in the optically active form. A switch of enantioselectivity was observed in the current zirconium catalyst system using (R)-BINOL derivatives [7]. Interestingly, a chiral zirconium complex prepared from an (R)-BINOL derivative bearing phenyl groups at the 3,3′-position gave mainly in the aza Diels–Alder reactions the opposite enantiomers to those using Zr-(R)-6,6′-Br$_2$BINOL catalyst systems. The pyridone products were also obtained with high enantiomeric excesses (Table 2.3). The structure of the active catalyst was revealed, by NMR analysis in C$_6$D$_6$, to be a Zr-BINOL 1 : 1 complex, not a 1 : 2 complex as shown in

2 Catalytic Asymmetric Aza Diels–Alder Reactions

Table 2.2 Enantioselective aza Diels–Alder reactions using a chiral zirconium catalyst.

Entry	R^1	R^2	Yield (%)	ee (%)
1	1-Naph	H	96	88
2	1-Naph	Me	93	93
3[a]	5,6,7,8-tetrahydronaphthalen-1-yl	H	92	80
4	o-MeC$_6$H$_4$	H	83	82
5	Ph	Me	83	65
6[b]	^{c}Hex	Me	51	86

[a] 10 mol% catalyst,
[b] The imine was prepared from c-HexCHO and 2-amino-3-methylphenol.

Chiral Zr catalyst

Table 2.2. The sterically hindered substituents on the BINOL part may prevent formation of the Zr-BINOL 1:2 complex.

Preparation of several kinds of ligands is usually required to evaluate the asymmetric environment in the optimization of chiral catalyst structure. To simplify the investigation, synthesis and evaluation of chiral ligands in the solid phase are very effective. This method was adopted to optimize the zirconium catalyst systems, and several polymer-supported BINOL derivatives were synthesized and applied to the aza Diels–Alder reactions [8]. After successive final optimization in the liquid phase, a zirconium catalyst bearing 3,3′-bis(m-trifluoromethyl)BINOL and a CN group instead of an O^{t}Bu group was found to be the most effective. The desired cycloadducts were obtained in good yields with high enantioselectivities at lower catalyst loading (Table 2.4).

This kind of chiral zirconium catalyst has been successfully employed in the asymmetric aza Diels–Alder reaction of acyl hydrazones (Table 2.5) [9]. Acylhydrazones are stable imine equivalents, those derived from aromatic, α,β-unsaturated, and even aliphatic aldehydes, are stable crystals easy to handle at room temperature. Use of the hydrazones could expand the substrate scope of the reaction dramatically.

Table 2.3 Asymmetric aza Diels–Alder reaction using a chiral Zr catalyst prepared from 3,3'-disubstituted BINOL.

Entry	R¹	R²	Yield (%)	ee (%)
1	Ph	H	94	82
2	o-MeC₆H₄	H	93	91
3	1-Naph	H	88	84
4	(5,6,7,8-tetrahydronaphthyl)	Me	78	87
5	(benzo[d][1,3]dioxolyl)	H	74	86
6[a]	^{c}Hex	H	64	81

[a] The imine was prepared from c-HexCHO and 2-amino-3-methylphenol.

A chiral zirconium catalyst prepared from zirconium propoxide and 3,3',6,6'-tetraiodo-1,1'-binaphthol catalyzed the reactions of the hydrazones prepared from aliphatic aldehydes and Danishefsky diene in high enantioselectivities. This method was also applied for the asymmetric formal synthesis of (S)-(+)-coniine. The N—N bond of the piperidine ring system was successfully cleaved using SmI₂ (Scheme 2.3).

α-Imino esters are reactive dienophiles due to the strong electron-withdrawing effect of the ester part. Catalytic asymmetric aza Diels–Alder reactions of N-tosyl-α-imino ester with Danishefsky diene have been reported in the presence of a catalytic amount of chiral metal Lewis acid to afford the corresponding aza Diels–Alder adducts in a highly stereoselective manner [10]. Among combinations of late transition metal Lewis acids, copper (I), silver (I), palladium (II), zinc (II) with counter anions and chiral ligands, copper (I) perchlorate-tol-BINAP or the chiral

Table 2.4 Asymmetric aza Diels–Alder reactions using the optimized chiral zirconium catalyst.

Entry	R¹	R²	Catalyst (mol%)	Yield (%)	ee (%)
1	Ph	H	5	76	92
2	o-MeC$_6$H$_4$	H	2	68	94
3	1-Naph	H	5	80	92
4	(methylenedioxyphenyl)	Me	2	68	90
5	2-thienyl	H	2	61	83
6[a]	^{c}Hex	Me	5	75	84

[a] The imine was prepared from c-HexCHO and 2-amino-3-methylphenol.

Chiral Zr catalyst

phosphinooxazoline complex were found to form an effective asymmetric environment around the imino esters, and high diastereo- and enantioselectivities were attained (Table 2.6).

Furthermore, the method was expanded to aza Diels–Alder reactions with other dienes (Scheme 2.4). Not only cyclopentadiene, but also less reactive dienes such as cyclohexadiene and 2,3-dimethyl-1,3-butadiene, reacted with the imino ester in good yields with moderate to good enantioselectivities. N-(p-Methoxyphenyl)imino ester also reacted with Danishefsky diene in good enantioselectivity. The p-methoxyphenyl group could be removed under oxidative conditions. It is proposed that N-tosyl-α-imino ester coordinated in a tridentate fashion to copper (I)–tol-BINAP or copper

Table 2.5 Asymmetric aza Diels–Alder reactions of benzoylhydrazones.

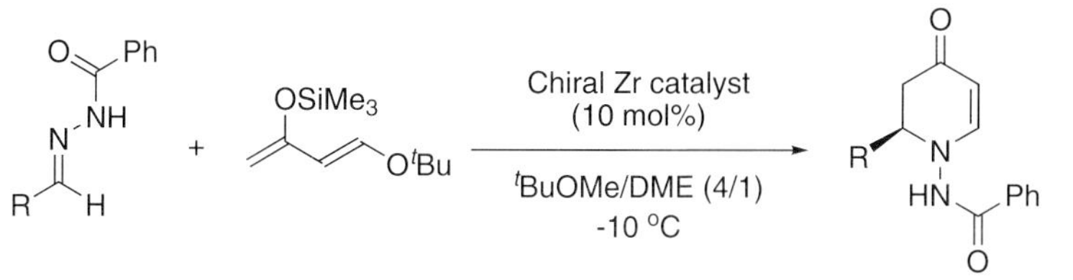

Entry	R	Yield (%)	ee (%)
1	PhCH$_2$CH$_2$	70	91
2	Pr	70	93
3	^{i}Bu	31	92
4[a]	^{i}Bu	50	91
5	^{c}Hex	44	89
6[a]	^{c}Hex	50	95

[a] In pure ^{i}BuOMe using 20 mol% of the Zr catalyst.

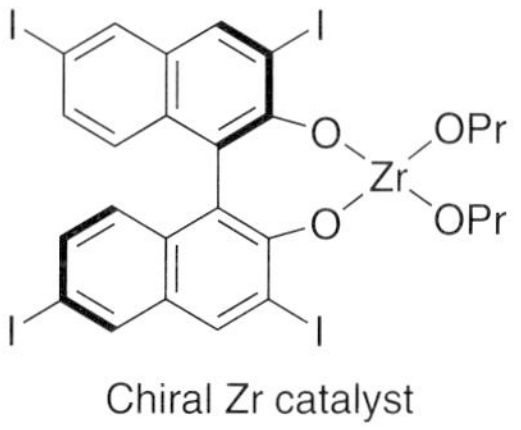

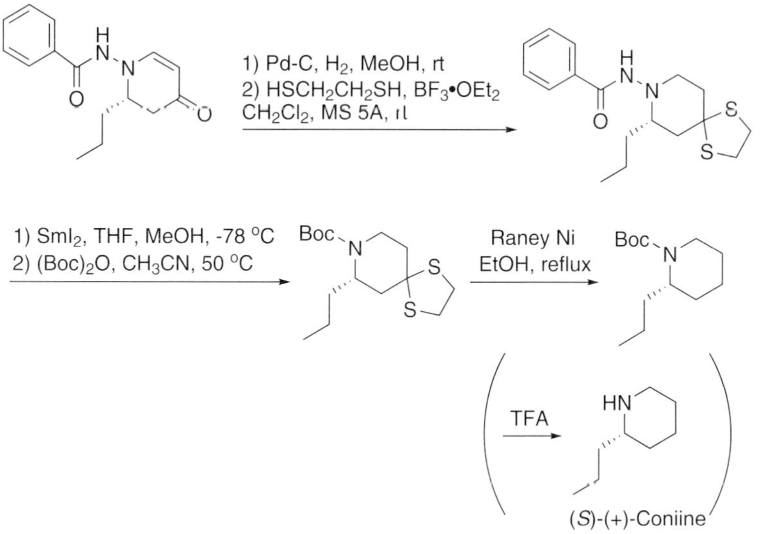

Scheme 2.3 Asymmetric synthesis of (S)-(+)-coniine.

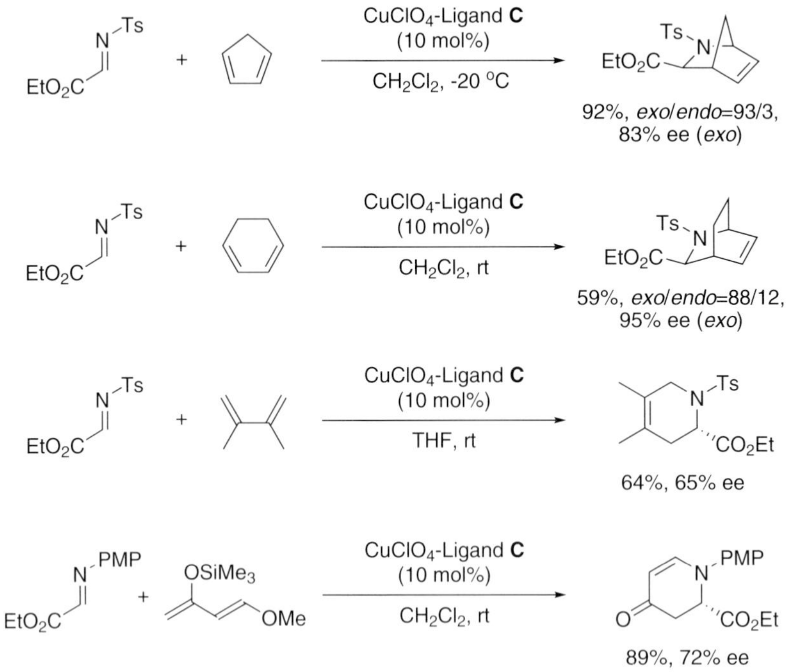

Scheme 2.4 Aza Diels–Alder reactions with other dienes and dienophile.

Table 2.6 Aza Diels–Alder reactions of N-tosyl-α-imino ester using a CuClO$_4$-chiral phosphine system.

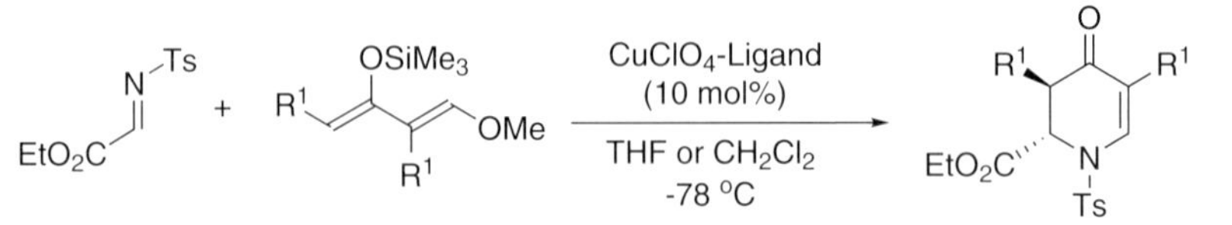

Entry	R^1	Ligand	Solvent	Yield (%)	trans/cis	ee (%, trans)
1	H	A	THF	82	–	87
2	H	B	CH$_2$Cl$_2$	77	–	86
3	Me	C	THF	91	91/9	94
4[a]	Me	C	THF	76	92/8	96

[a] 1 mol% catalyst was used.

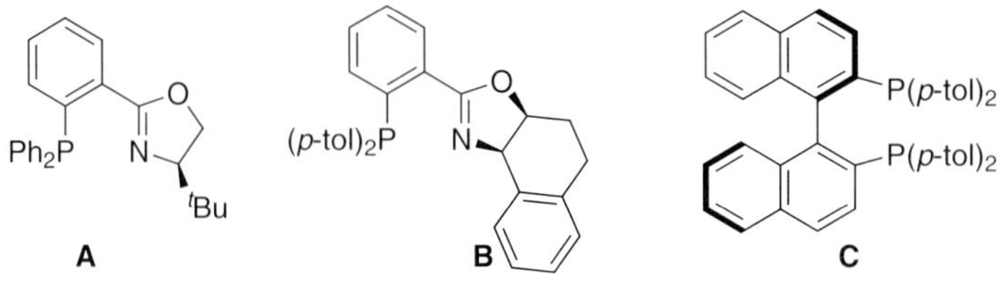

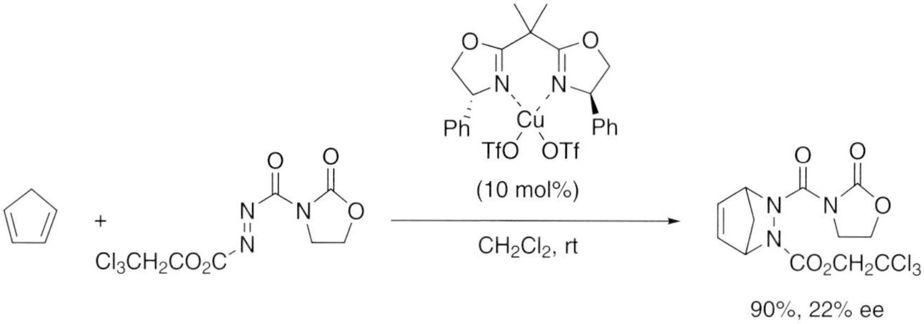

Scheme 2.5 Aza Diels–Alder reaction with azodicarboxylate.

(I)–phosphino-oxazoline catalyst, while the N-p-methoxyphenyl glyoxylate imine coordinated in a bidentate fashion.

An asymmetric aza Diels–Alder reaction of cyclopentadiene with azodicarboxylate as an azodienophile has also been reported; however, the ee was not high. Further improvement of the reaction system is required (Scheme 2.5) [11].

Parallel combinatorial examination to search for a good combination of a metal Lewis acid and a chiral ligand in the aza Diels–Alder reaction of an N-aryl-α-imino ester with Danishefsky diene was conducted. Combination of MgI_2, $Yb(OTf)_3$, $Cu(OTf)_2$, $FeCl_3$ with ligands **E–G** was screened, and finally high enantioselectivities were obtained under optimal conditions (Table 2.7, entries 1–4) [12] Furthermore, it was reported that the Zn-BINOL complex (**G**) was also found effective in affording the

Table 2.7 Aza Diels–Alder reactions of α-iminoester using some metal Lewis acids.

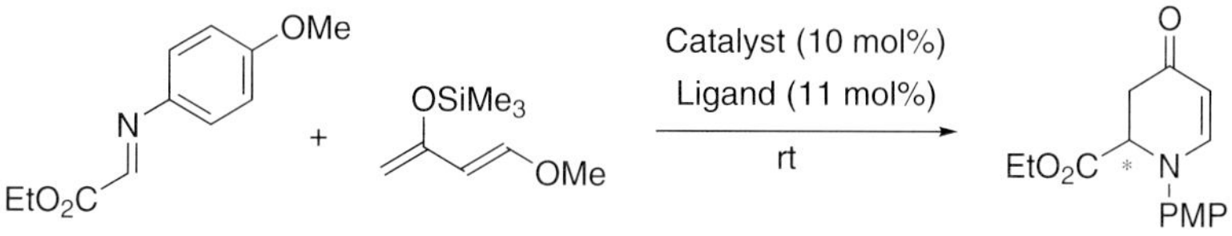

Entry	Catalyst	Ligand	Solvent	Additive	Yield (%)	ee (%)
1	MgI_2	E	MeCN	2,6-lutidine	64	97
2	$Yb(OTf)_3$	E	toluene	2,6-lutidine	60	87
3	$Cu(OTf)_2$	E	MeCN	none	58	86
4	$FeCl_3$	F	CH_2Cl_2	MS 4A	67	92
5	Et_2Zn	G	toluene	none	52	84(S)

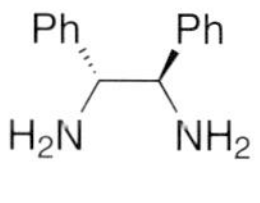

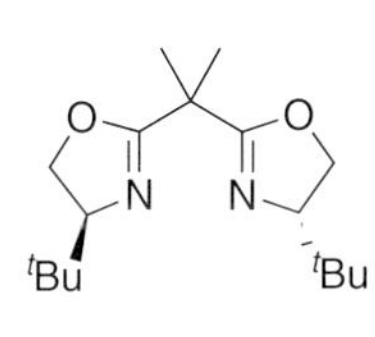

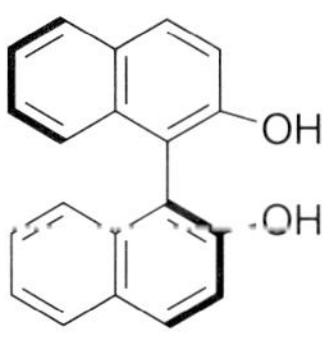

Table 2.8 Asymmetric aza Diels–Alder reaction of azadienes using a chiral Cu(II) complex.

Entry	R^1	R^2	R^3	Temp. (°C)	Yield (%)	ee (%)
1	Ph	Me	Me	−45	80	95
2	Ph	H	H	−45	83	98
3	Ph	Me	H	−45	96	98
4	Ph	H	Me	rt	80	93
5	(1-methyl-2-phenylvinyl)	Me	Me	−45	62	95

desired cyclic adduct in good enantioselectivity (entry 5) [13]. In this case, a less polar solvent, toluene, showed the best selectivity; however, a longer reaction time and lower temperature did not lead to further improvement in the enantioselectivity.

Asymmetric aza Diels–Alder reactions of electron-rich azadienes with olefins using a chiral copper complex have been reported (Table 2.8) [14]. The instability of these dienes in the presence of common Lewis acid systems has limited their applicability in synthesis. However, a combination of $Cu(OTf)_2$ and a chiral bisoxazoline ligand was found to promote the condensation of a range of 2-azadienes and α,β-unsaturated olefins bearing the oxazolidinone moiety in good to high yields and excellent stereoselectivities (exo/endo > 99 : 1). Further transformation of the product, including cleavage of the oxazolidine part, was demonstrated.

Asymmetric quinoline synthesis using a chiral titanium complex has been reported [15]. N-Benzylideneaniline was used as a diene component, and a titanium catalyst prepared from $TiCl_2(O^iPr)_2$ and a chiral amino diol promoted the reaction with vinyl ethers or cyclopentadiene to give the expected tricyclic adducts in moderate conversions with moderate to good diastereo- and enantioselectivities (Scheme 2.6). While the titanium Lewis acid could activate the N-benzylideneaniline catalytically, the turnover number was not high enough.

Silver acetate-chiral phosphine ligand was used for asymmetric aza Diels–Alder reactions of imines bearing an o-anisyl group on the nitrogen atom (Table 2.9) [16]. The chiral silver catalyst was prepared from a ligand derived from an inexpensive amino acid and other commercially available materials and worked well with one equivalent of proton source (e.g. ^iPrOH, H_2O etc.). The desired cycloadducts were

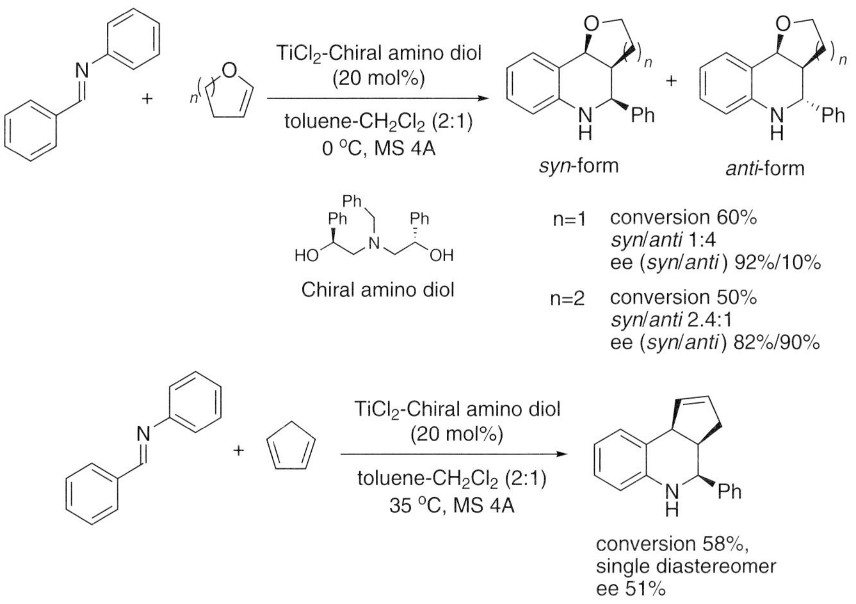

2.2 Chiral Lewis Acid Catalysis

Scheme 2.6 Chiral Ti complex-promoted asymmetric aza Diels–Alder reactions.

n=1 conversion 60%
syn/anti 1:4
ee (syn/anti) 92%/10%

n=2 conversion 50%
syn/anti 2.4:1
ee (syn/anti) 82%/90%

conversion 58%,
single diastereomer
ee 51%

Table 2.9 Chiral silver-catalyzed aza Diels–Alder reactions.

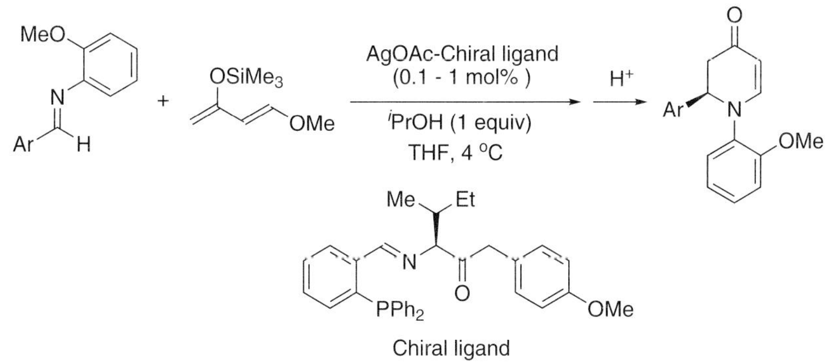

Entry	Ar	Cat. loading	Yield (%)	ee (%)
1	Ph	1	94	93
2	Ph	0.1	78	88
3	1-Naph	1	94	90
4	2-Naph	0.5	>98	95
5	p-MeOC$_6$H$_4$	1	86	91
6	p-ClC$_6$H$_4$	1	98	90
7	o-BrC$_6$H$_4$	1	91	89
8	m-NO$_2$C$_6$H$_4$	1	92	91
9	p-NO$_2$C$_6$H$_4$	1	>98	92
10	2-furyl	1	89	92

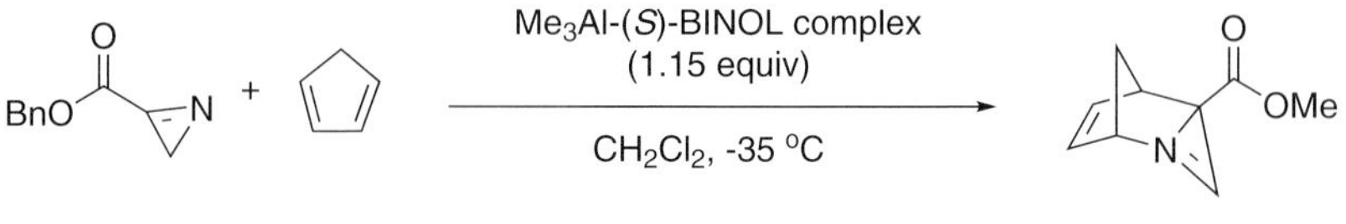

Scheme 2.7 Aza Diels–Alder reaction of an azirine mediated with a chiral Al complex.

obtained in high yields with high enantioselectivities in the presence of 0.1–1 mol% of the catalyst. The reaction also proceeded even in undistilled THF or under neat conditions with minor diminution in selectivity and yield.

A preliminary example of asymmetric aza Diels–Alder reactions using an azirine as a dienophile has been disclosed [17]. The use of azirines as imine electrophiles has some benefits in Lewis acid catalysis. Flexible E/Z conformations of a normal N-substituted imine which generate several possible reaction conformers with Lewis acids sometimes prevent high stereoselectivity of the corresponding product. However, in azirine cases, since the lone electron pairs of the nitrogen atoms are fixed in

Table 2.10 Aza Diels–Alder reactions using a chiral Cu catalyst.

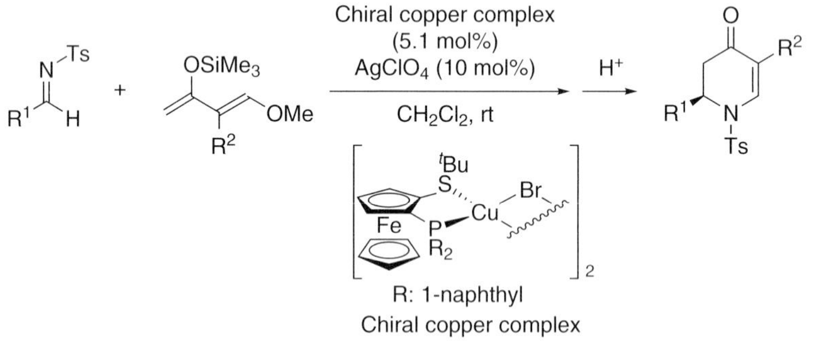

Entry	R^1	R^2	Yield (%)	ee (%)
1	Ph	H	90	93(97)[a]
2	o-MeC$_6$H$_4$	H	82	93
3	p-FC$_6$H$_4$	H	78	88(93)[a]
4	p-MeOC$_6$H$_4$	H	76	91
5	p-(Me$_2$N)C$_6$H$_4$	H	39	93
6	m,p-(MeO)$_2$C$_6$H$_3$	H	71[a]	94[a]
7	2-Naph	H	85	86(93)[a]
8	(E)-PhCH=CH	H	66	83(96)[a]
9	Pr	H	65[a]	73[a](82)[b]
10	Ph	Me	64	87(92)[c]
11	(E)-PhCH=CH	Me	57	88[a]

[a] −20 °C.
[b] −78 °C.
[c] 0 °C.

one position due to their cyclic structure, high stereoselectivities are expected. In addition, they have higher reactivity than typical N-aliphatic substituted imines. Indeed, asymmetric reactions of azirine bearing a chiral auxiliary with some dienes proceeded in high stereoselectivities. The Al–(S)-BINOL complex also promoted the reaction in moderate yield and selectivity, but a stoichiometric amount of the promoter was still required (Scheme 2.7).

Asymmetric aza Diels–alder reactions of tosyl imines with Danishefsky diene were catalyzed by a chiral Cu(I) complex (Table 2.10) [18]. The preformed dimeric CuBr-planar chiral phosphino sulfenyl ferrocene complex was activated by anion exchange using AgClO$_4$ to catalyze the desired reactions in good to high yields with high enantioselectivities. Not only tosyl imines prepared from aromatic aldehydes but also imines prepared from α,β-unsaturated and aliphatic aldehydes reacted with Danishefsky diene to afford the desired chiral 4-piperidone derivatives. The tosyl group on the product could be removed under reductive conditions (Zn-AcOH).

Table 2.11 Inverse-electron-demand asymmetric aza Diels–Alder reaction of N-sulfonyl-1-aza-1,3-diene.

Entry	Ar^1	Ar^2	endo/exo	Yield (%)	ee (%)
1	Ph	Ph	98/2	66	91
2	Ph	p-FC$_6$H$_4$	98/2	75	92
3	Ph	2-Naph	97/3	69	90
4	Ph	p-MeOC$_6$H$_4$	98/2	65	80
5	Ph	2-furyl	97/3	52	77
6	Ph	^{t}Bu	98/2	61	84
7	p-ClC$_6$H$_4$	Ph	97/3	73	90
8	p-CF$_3$C$_6$H$_4$	Ph	97/3	69	91
9	p-ClC$_6$H$_4$	(E)-PhCH=CH	98/2	63	92
10	p-NCC$_6$H$_4$	(E)-PhCH=CH	98/2	70	92

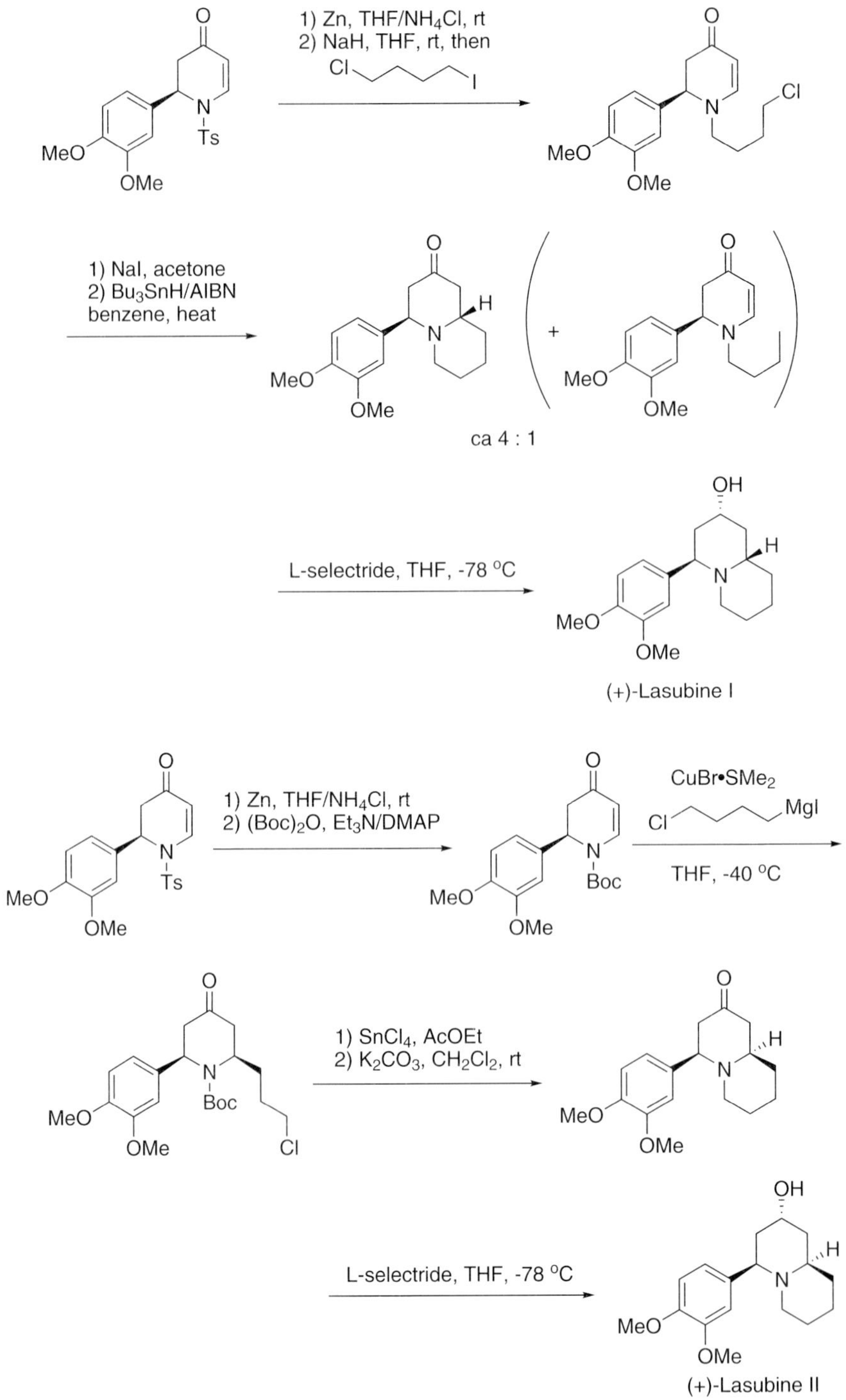

Scheme 2.8 Enantioselective synthesis of (+)-lasubine I and II.

The Cu-catalyzed asymmetric reaction was applied to total synthesis of (+)-lasubine I and II (Scheme 2.8) [19]. The key intermediate prepared by the asymmetric reaction was successfully converted to both (+)-lasubine I and II, diastereomeric isomers, via two different cyclization strategies.

Inverse-electron-demand asymmetric aza Diels–Alder reactions of N-sulfonyl-1-aza-1,3-dienes have been reported [20]. As N-arylsulfonyl groups, 2-pyridyl and 8-quinolinyl were found to be effective in improving yields and, especially, the azadiene bearing an 8-quinolinyl group showed good to high enantioselectivities in the presence of the Ni-DBFOX-Ph complex (Table 2.11). The effect of vinyl ethers was also examined, and propyl vinyl ether showed the best enantioselectivity while sterically hindered vinyl ethers such as *tert*-butyl vinyl ether did not work well. The substrate scope of the azadiene was then investigated, and the dienes containing a chalcone skeleton were found to be suitable for the current reaction.

Asymmetric aza Diels–Alder reactions of benzohydryl imines with Danishefsky diene using a combination of boron phenoxide and a chiral bisphenol derivative, a dual Lewis acid catalyst system, have been reported (Table 2.12) [21]. Chiral boron

Table 2.12 Asymmetric aza Diels–Alder reaction using dual boron Lewis acid catalyst system.

Entry	R	VAPOL (mol%)	Yield (%)	ee (%)
1	Ph	5	85	90
2	o-MeC$_6$H$_4$	5	83	93
3	1-Naph	10	78	90
4	p-BrC$_6$H$_4$	5	84	89
5	p-NO$_2$C$_6$H$_4$	10	69	73
6	p-MeOC$_6$H$_4$	10	71	90
7	p-F-o-MeC$_6$H$_3$	5	84	89
8	^{c}Hexenyl	10	45	93
9	^{c}Hex	10	90	93
10	^{i}Pr	10	64	90

The diene was added over 3 h via syringe pump.

Lewis acids are useful in asymmetric synthesis and are often used in the catalytic activation of carbonyl compounds. However, in the reactions of nitrogen-containing compounds, the binding interaction between a nitrogen atom and a boron Lewis acid is usually strong, and efficient catalytic turnover is difficult. In the current reaction, this problem was solved by using a stoichiometric amount of boron reagent and a catalytic amount of a chiral diol based on the difference in Lewis acidity between boron phenoxide and the boron-chiral ligand complex. The desired aza Diels–Alder reactions proceeded in moderate to good yields with good to high enantioselectivities.

Aza Diels–Alder reactions using a chiral scandium–N,N'-dioxide ligand complex have been reported recently. L-Proline-derived chiral ligands containing bis-N-oxide groups have been shown to form effective asymmetric environments around metal Lewis acids [22]. In the current investigation, scandium triflate was found to be

Table 2.13 Aza Diels–Alder reactions using a chiral scandium–bis-N-oxide ligand complex.

Entry	R	p-H$_2$NC$_6$H$_4$SO$_3$H (mol%)	Yield (%)	ee (%)
1	Ph	5	74	81
2	p-NCC$_6$H$_4$	15	84	90
3	p-MeOC$_6$H$_4$	15	66	86
4	1-Naph	15	80	87
5	2-Naph	15	85	84(99)[a]
6	(E)-PhCH=CH	15	92	84
7	p-MeC$_6$H$_4$	15	77	85(>99)[a]
8	p-O$_2$NC$_6$H$_4$	15	69	89(>99)[a]
9	p-ClC$_6$H$_4$	15	64	85(>99)[a]
10	2-thienyl	15	46	82
11	^{i}Pr	15	67	71

[a] After single recrystallization.

Table 2.14 Aza Diels–Alder reactions using a chiral niobium complex.

Entry	R	Yield (%)	ee (%)
1	Ph	81	96
2	1-Naph	89	92
3	p-MeC$_6$H$_4$	90	94
4	o-CF$_3$C$_6$H$_4$	94	99
5	o-ClC$_6$H$_4$	89	91
6	3-pyridyl	74	92
7[a,b]	^{c}Hex	67	90
8[a,c]	^{i}Pr	63	92
9[a,b,d]	^{i}Bu	47	90

[a]The imine prepared from o-amino-p-cresol was used.
[b]The reaction was carried out in CH$_2$Cl$_2$.
[c]The reaction was carried out in toluene at 0 °C.
[d]The imine was slowly added over 25 h. The reaction time was 96 h.

Chiral ligand

effective, and good enantioselectivity was achieved after fine optimization of the reaction conditions. p-Aminobenzenesulfonic acid worked well as an effective acid additive to improve the enantioselectivity. High enantioselectivities were obtained in the reactions of imines prepared from o-aminophenol (Table 2.13).

Chiral Nb-catalyzed aza Diels–Alder reactions were recently disclosed (Table 2.14) [23]. Chiral niobium Lewis acid catalysts have been less explored. In this report, novel chiral niobium Lewis acid catalyst systems using chiral multidentate BINOL derivatives were used successfully. The aza Diels–Alder reaction was catalyzed by the niobium complex with very high enantioselectivities. Interestingly, the imine containing the pyridyl component, strong Lewis basic moiety, reacted with the diene in high enantioselectivity. This method was also successfully applied to the synthesis of (+)-anabasine (Scheme 2.9).

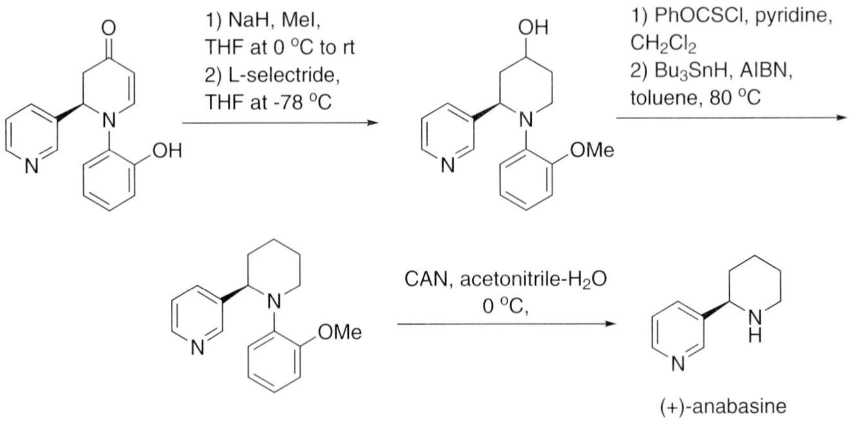

Scheme 2.9 Enantioselective synthesis of (+)-anabasine.

2.3
Chiral Organocatalysis

Asymmetric catalysis using chiral organic molecules without any metals is interesting in synthetic organic chemistry from the viewpoint of environmental friendliness.

Table 2.15 Proline-catalyzed three component aza Diels–Alder reaction.

Entry	Ketone	R^3	Conditions	Yield (%)	ee (%)
1	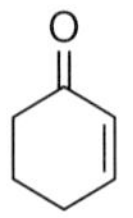 cyclohexanone	OMe	50 °C	82[a]	99
2	cyclohexanone	H	rt	54	>96
3	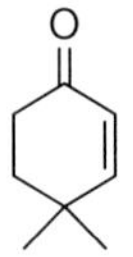 4-methylcyclohexanone	OMe	rt	70	>99
4	4-methylcyclohexanone	H	rt	69	98
5	4-methylcyclohexanone	Br	rt	20	>99
6	4-methylcyclohexanone	Cl	rt	32	>99
7	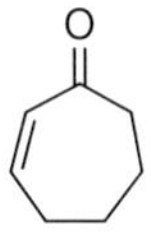 cycloheptanone	OMe	rt	75	98

[a] Yield of corresponding alcohol (1:1 cis/trans mixture).

While many types of chiral Lewis acids have been employed in aza Diels–Alder reactions, chiral organocatalysts in those reactions have also been developed.

Proline-catalyzed asymmetric aza Diels–Alder reactions of imines prepared from formaldehyde and p-anisidine with α,β-unsaturated cyclic ketones have been reported (Table 2.15) [24]. One-pot three-component direct-type catalytic enantioselective reactions proceeded in DMSO at rt to 50 °C in the presence of 30 mol% of the proline catalyst to afford the desired products in very high enantioselectivities. While six- and seven-membered cyclic ketones worked well, the five-membered cyclic ketone did not react. The reaction may proceed via a stepwise mechanism, Mannich addition and successive intramolecular 1,4-cyclization.

Chiral N-heterocyclic carbene (NHC)-catalyzed aza Diels–Alder reactions of azadiene with trans- α,β-unsaturated aldehydes bearing ester or ketone groups at the γ position were developed (Table 2.16) [25]. The carbene catalyst reacted with the

Table 2.16 Chiral N-heterocyclic carbine-catalyzed asymmetric aza Diels–Alder reaction.

Entry	R^1	R^2	Yield (%)	ee (%)
1	OEt	Ph	90	99
2	OEt	p-MeOC$_6$H$_4$	81	99
3	OEt	p-AcC$_6$H$_4$	55	99
4	OEt	2-furyl	71	99
5	OEt	^{n}Pr	58	99
6	O^{t}Bu	Ph	70	97
7	Me	Ph	51	99
8	Me	^{n}Pr	71	98
9	Ph	p-MeOC$_6$H$_4$	52	99

Assumed transition state

2 Catalytic Asymmetric Aza Diels–Alder Reactions

aldehyde to form an active enolate intermediate which further reacted with N-sulfonyl azadiene to afford the desired product as an almost single diastereomer in good yield and enantioselectivity.

Chiral Brønsted acid-catalyzed asymmetric reactions have become a rapidly growing area. Especially, chiral phosphoric acids bearing 3,3′-disubstituted-1,1′-binaphthol have often been employed for activation of several electrophiles. It was reported that efficient asymmetric activation of imines bearing a 2-hydroxyphenyl group on the nitrogen atom occurred using a chiral phosphonate in asymmetric Mannich-type reactions with ketene silyl acetals [26]. This reaction was expanded to aza Diels–Alder reactions with Danishefsky diene, and the desired 4-pyridone derivatives were obtained in high yields with good to high enantioselectivities (Table 2.17) [27]. Less polar aromatic solvents like toluene and mesitylene at lower temperature gave higher enanioselectivities.

Table 2.17 Aza Diels–Alder reaction using a chiral phosphoric acid as a catalyst.

Entry	Ar	Yield (%)	ee (%)
1	Ph	99	80
2	p-IC$_6$H$_4$	86	84
3	p-BrC$_6$H$_4$	100	84
4	p-ClC$_6$H$_4$	72	84
5	p-CF$_3$C$_6$H$_4$	77	81
6	o-BrC$_6$H$_4$	96	80
7	1-Naph	100	91

Chiral phosphoric acid

Table 2.18 Aza Diels–Alder reaction with Brassard's diene.

Entry	Ar	Yield (%)	ee (%)
1	Ph	87	94
2	p-BrC$_6$H$_4$	86	96
3	p-ClC$_6$H$_4$	90	97
4	p-FC$_6$H$_4$	76	98
5	p-CH$_3$C$_6$H$_4$	90	95
6	p-MeOC$_6$H$_4$	84	99
7	o-BrC$_6$H$_4$	83	98
8	o-ClC$_6$H$_4$	86	98
9	o-MeC$_6$H$_4$	76	96
10	1-Naph	79	98
11	2-Naph	91	97
12	2-furyl	63	97
13	(E)-PhCH=CH	76	98
14	^{c}Hex	69	99
15	^{i}Pr	65	93

Pyridinium salt of the chiral phosphoric acid

Aza Diels–Alder reactions with Brassard's diene using the chiral phosphate catalyst system have also been reported (Table 2.18) [28]. The chiral phosphate (3 mol%) effectively promoted the reaction in good yield with high enantioselectivity (92% ee) in mesitylene at −40 °C. The cyclic product was obtained after treatment with benzoic acid with heating. Furthermore, use of the pyridinium salt of the phosphate improved the yield significantly with comparable enantioselectivity.

2 Catalytic Asymmetric Aza Diels–Alder Reactions

Substrate scope was also investigated and it was found that not only aromatic but also α,β-unsaturated and aliphatic substrates gave the desired products in high enantioselectivities.

Furthermore, inverse electron-demand aza Diels–Alder reactions of azadienes with electron-rich olefins were investigated, and it was revealed that the desired reactions proceeded in good to high yields with high selectivities (Table 2.19) [29].

Table 2.19 Inverse electron-demand aza Diels–Alder reaction using a chiral phosphate.

Entry	Ar	R	Yield (%)	ee (%)
1	Ph	Et	89	94
2	Ph	^{n}Bu	82	96
3	Ph	Bn	76	91
4	Ph	(2,3-dihydrofuran)	86	90
5	Ph	(3,4-dihydro-2H-pyran)	95	97
6	p-BrC$_6$H$_4$	Et	77	90
7	p-ClC$_6$H$_4$	Et	79	88
8	p-CH$_3$C$_6$H$_4$	Et	59	91
9	2-Naph	Et	74	95

Pyridinium salt of the chiral phosphoric acid

Relatively wide scope of substrates was demonstrated, however the o-hydroxy group, which was always contained in the product structure, was necessary to achieve high enanitoselectivities.

Aza Diels–Alder reactions of a cyclic α,β-unsaturated ketone using a chiral phosphate have been reported [30]. Acetic acid was used as an additive to promote diene formation via enolization of cyclohexenone. *In situ* formation of diene via proton transfer followed by Mannich-type addition to imines and cyclization gave the corresponding products in moderate to good yields with good diastereo- and good to high enantioselectivities (Table 2.20).

Similarly chiral phosphoric acid also catalyzed asymmetric aza Diels–alder reactions of cyclohexenone with imines. In these reactions, acetic acid was not used, and

Table 2.20 Aza Diels–Alder reaction of 2-cyclohexenone using a chiral phosphoric acid as a chiral Brønsted catalyst and acetic acid as a co-catalyst.

Entry	Ar	Yield (%)	endo/exo	ee (%)[a]
1	Ph	71	4/1	86
2	p-CF$_3$C$_6$H$_4$	74	3/1	84
3	2-thienyl	70	4/1	88
4	p-ClC$_6$H$_4$	73	4/1	88
5	m,p-(MeO)$_2$C$_6$H$_3$	84	8/1	86
6	p-PhC$_6$H$_4$	54	4/1	84
7	o-FC$_6$H$_4$	74	4/1	88
8	m,m,p-(MeO)$_3$C$_6$H$_2$	71	4/1	84

[a] Ee of the endo adduct.

Table 2.21 Aza Diels–Alder reaction of 2-cyclohexenone.

Entry	Ar	Yield (%)	endo/exo	ee (%)
1	Ph	76	84/16	87
2	m-ClC$_6$H$_4$	74	81/19	83
3	p-ClC$_6$H$_4$	82	82/18	85
4	p-FC$_6$H$_4$	72	81/19	85
5	m-FC$_6$H$_4$	76	82/18	84
6	p-BrC$_6$H$_4$	81	82/18	85
7	m-BrC$_6$H$_4$	79	81/19	87
8	p-MeC$_6$H$_4$	81	83/18	83

the desired products were obtained in moderate to good yields with moderate to good diastereo- and enantioselectivities (Table 2.21) [31].

2.4
Conclusion

Recent progress in catalytic asymmetric aza Diels–Alder reactions has been surveyed in this chapter. Well-designed chiral Lewis acid catalysts promoted asymmetric aza Diels–Alder reactions efficiently to afford the piperidine derivatives in high yields with high enantioselectivities. Chiral organocatalysts have also been successfully employed; in particular, chiral Brønsted acids have shown high potential. The chiral piperidine derivatives obtained were well functionalized for further transformations, and useful intermediates for the synthesis of natural and biologically active compounds were obtained. Further improvement in catalytic turnover number as well as turnover frequency would be required in some cases for truly practical methods.

2.5 Experimental: Selected Procedures

2.5.1 Typical Experimental Procedure for Asymmetric aza Diels–Alder Reactions Using A Chiral Zr Complex (Table 2.4)

To a mixture of (R)-3,3'-bis(3-trifluoromethylphenyl)-2,2'-dihydroxy-1,1'-binaphthyl (22.5 mg, 0.039 mmol), N-methylimidazole (3.2 mg, 0.039 mmol), and MS 3A (160 mg) in benzene (0.75 mL) was added to a benzene (0.25 mL) solution of Zr $(O^tBu)_4$ (5.0 mg, 0.013 mmol), and the mixture was stirred at 80 °C for 2.5 h. To the mixture was added Me_3SiCN (2.7 mg, 0.027 mmol) in toluene (0.2 mL), and the mixture was stirred for 30 min at rt. A benzene solution (1.25 mL each) of the aldimine (0.65 mmol) and the diene (0.98 mmol) was added to this mixture over 1 h at the same temperature, and the mixture was further stirred for 1 h. Saturated $NaHCO_3$ was then added to quench the reaction. The aqueous layer was extracted with dichloromethane, and the crude adduct was treated with THF–1 M HCl (20:1) at 0 °C for 30 min. After a usual work up, the crude product was purified using preparative thin layer chromatography to afford the desired product. Enantiomeric excess of the product was determined after converting to a corresponding O-methylated compound.

2.5.2 Typical Experimental Procedure for Asymmetric Aza Diels–Alder Reactions Using a Chiral Ag Complex (Table 2.9)

(+)-2-[(2-Diphenylphosphanylbenzylidene)amino]-3-methylpentanoic acid (4-methoxyphenyl)-amide (6 mg, 12 µmg) and AgOAc (2.0 mg, 12 µmol) were weighed into a 16 × 150 mm test-tube in a N_2 atmosphere glovebox. The contents were dissolved in 0.5 mL of THF and allowed to stir for 15 min at 22 °C. The imine (1.20 mmol) was added, immediately followed by 1.5 mL of THF and 100 µL i-PrOH (1.25 mmol). The test-tube was capped with a septum, sealed with parafilm, removed from the glovebox, and allowed to stir in a cold room at 4 °C for 15 min. The diene (1.80 mmol) was added by syringe and the reaction mixture was allowed to stir at 4 °C for 12 h. The reaction was quenched with 2 mL of 10% aqueous HCl followed by vigorous stirring for 5 min. 5 mL of CH_2Cl_2 was added to the test-tube and the reaction was stirred vigorously for 1 min. The layers were allowed to separate, and the bottom layer (CH_2Cl_2) was removed by using a pipette and placed into a round-bottom flask. The aqueous layer was washed with 2 × 5 mL portions of CH_2Cl_2, as stated above. The combined organic layer was concentrated *in vacuo* (without drying) and purified by silica gel chromatography (EtOAc–hexanes) to deliver the desired product. The enantiomeric excesses of the products were determined by using chiral HPLC.

Abbreviations

AIBN	2,2′-azobisisobutyronitrile
Bn	benzyl
Boc	*tert*-butoxycarbonyl
DMAP	4-dimethylaminopyridine
^{c}Hex	cyclohexyl
L-selectride	lithium tris-*s*-butylborohydride
NMI	*N*-methylimidazole
PMP	*p*-methoxyphenyl
Ts	*p*-toluenesulfonyl

References

1 Kobayashi, S. and Jørgensen, K.A. (eds) (2002) *Cycloaddition Reactions in Organic Synthesis*, Wiley-VCH, Weinheim.

2 (a) Waldmann, H. (1994) *Synthesis*, 535–551; (b) Tietze, L.F. and Kettschau, G. (1997) *Top. Curr. Chem.*, **189**, 1–120; (c) Weinreb, S.M. (1991) *Comprehensive Organic Synthesis*, vol. 5 (eds B.M. Trost, I. Fleming and M.F. Semmelhock), Pergamon Press, Oxford, Chap. 4, pp. 401–449; (d) Boger, D.L. and Weinreb, S.M. (1987) *HeteroDiels-Alder Methodology in Organic Synthesis*, Academic Press, San Diego, Chap. 2 and 9; (e) Ooi, T. and Maruoka, K. (1999) *Comprehensive Asymmetric Catalysis*, vol. 3 (eds E.N. Jacobsen, A. Pfaltz and H. Yamamoto), Springer, Berlin, Germany, pp. 1237–1254; (f) Jørgensen, K.A. (2000) *Angew. Chem. Int. Ed.*, **39**, 3558–3588; (g) Nicolaou, K.C., Snyder, S.A., Montagnon, T. and Vassilikogiannakis, G. (2002) *Angew. Chem. Int. Ed.*, **41**, 1668–1698; (h) Jørgensen, K. A. (2004) *Eur. J. Org. Chem.*, 2093–2102.

3 (a) Hattori, K. and Yamamoto, H. (1992) *J. Org. Chem.*, **57**, 3264–3265; (b) Hattori, K. and Yamamoto, H. (1993) *Tetrahedron*, **49**, 1749–1760; (c) Ishihara, K., Miyata, M., Hattori, K., Tada, T. and Yamamoto, H. (1994) *J. Am. Chem. Soc.*, **116**, 10520–10524.

4 Ishitani, H. and Kobayashi, S. (1996) *Tetrahedron Lett.*, **37**, 7357–7360.

5 Kobayashi, S., Komiyama, S. and Ishitani, H. (1998) *Angew. Chem. Int. Ed.*, **37**, 979–981.

6 (a) Ishitani, H., Ueno, M. and Kobayashi, S. (1997) *J. Am. Chem. Soc.*, **119**, 7153–7154; (b) Kobayashi, S., Ishitani, H. and Ueno, M. (1998) *J. Am. Chem. Soc.*, **120**, 431–432; (c) Ishitani, H., Ueno, M. and Kobayashi, S. (2000) *J. Am Chem. Soc.*, **122**, 8180–8186.

7 Kobayashi, S., Kusakabe, K., Komiyama, S. and Ishitani, H. (1999) *J. Org. Chem.*, **64**, 4220–4221.

8 (a) Kobayashi, S., Kusakabe, K. and Ishitani, H. (2000) *Org. Lett.*, **2**, 1225–1227; (b) Kobayashi, S., Ueno, M., Saito, S., Mizuki, Y., Ishitani, H. and Yamashita, Y. (2004) *Proc. Nat. Acad. Sci. USA*, **101**, 5476–5481.

9 Yamashita, Y., Mizuki, Y. and Kobayashi, S. (2005) *Tetrahedron Lett.*, **46**, 1803–1806.

10 (a) Yao, S., Johannsen, M., Hazell, R.G. and Jørgensen, K.A. (1998) *Angew. Chem. Int. Ed.*, **37**, 3121–3124; (b) Yao, S., Saaby, S., Hazell, R.G. and Jørgensen, K.A. (2000) *Chem. Eur. J.*, **6**, 2435–2448.

11 Aburel, P.S., Zhuang, W., Hazell, R.G. and Jørgensen, K.A. (2005) *Org. Biomol. Chem.*, **3**, 2344–2349.

12 Bromidge, S., Wilson, P.C. and Whiting, A. (1998) *Tetrahedron Lett.*, **39**, 8905–8908.

13 Guillarme, S. and Whiting, A. (2004) *Synlett*, 711–713.

14 Jnoff, E. and Ghosez, L. (1999) *J. Am. Chem. Soc.*, **121**, 2617–2618.

15 Sundararajan, G., Prabagaran, N. and Varghese, B. (2001) *Org. Lett.*, **3**, 1973–1976.

16 Josephsohn, N.S., Snapper, M.L. and Hoveyda, A.H. (2003) *J. Am. Chem. Soc.*, **125**, 4018–4019.

17 Sjöholm, Å. and Somfai, P. (2003) *J. Org. Chem.*, **68**, 9958–9963.

18 Mancheño, O.G., Arrayás, R.G. and Carretero, J.C. (2004) *J. Am. Chem. Soc.*, **126**, 456–457.

19 Mancheño, O.G., Arrayás, R.G., Adrio, J. and Carretero, J.C. (2007) *J. Org. Chem.*, **72**, 10294–10297.

20 Esquivias, J., Arrayás, R.G. and Carretero, J.C. (2007) *J. Am. Chem. Soc.*, **129**, 1480–1481.

21 Newman, C.A., Antilla, J.C., Chen, P., Predeus, A.V., Fielding, L. and Wulff, W.D. (2007) *J. Am. Chem. Soc.*, **129**, 7216–7217.

22 Shang, D., Xin, J., Liu, Y., Zhou, X., Liu, X. and Feng, X. (2008) *J. Org. Chem.*, **73**, 630–637.

23 Jurcík, V., Arai, K., Salter, M.M., Yamashita, Y. and Kobayashi, S. (2008) *Adv. Synth. Catal.*, **350**, 647–651.

24 Sundén, H., Ibrahem, I., Eriksson, L. and Córdova, A. (2005) *Angew. Chem. Int. Ed.*, **44**, 4877–4880.

25 He, M., Struble, J.R. and Bode, J.W. (2006) *J. Am. Chem. Soc.*, **128**, 8418–8420.

26 (a) Akiyama, T., Itoh, J., Yokota, K. and Fuchibe, K. (2004) *Angew. Chem. Int. Ed.*, **43**, 1566–1568; (b) Akiyama, T., Saitoh, Y., Morita, H. and Fuchibe K. (2005) *Adv. Synth. Catal.*, **347**, 1523–1526.

27 Akiyama, T., Tamura, Y., Itoh, J., Morita, H. and Fuchibe, K. (2006) *Synlett*, 141–143.

28 Itoh, J., Fuchibe, K. and Akiyama, T. (2006) *Angew. Chem. Int. Ed.*, **45**, 4796–4798.

29 Akiyama, T., Morita, H. and Fuchibe, K. (2006) *J. Am. Chem. Soc.*, **128**, 13070–13071.

30 Rueping, M. and Azap, C. (2006) *Angew. Chem. Int. Ed.*, **45**, 7832–7835.

31 Liu, H., Cun, L.-F., Mi, A.-Q., Jiang, Y.-Z. and Gong, L.-Z. (2006) *Org. Lett.*, **8**, 6023–6026.

3
1,3-Dipolar Cycloaddition

Takuya Hashimoto and Keiji Maruoka

3.1
Introduction

In the comprehensive work of Rolf Huisgen and coworkers launched in the late 1950s [1], Huisgen defined 1,3-dipolar cycloadditions as "1,3-Dipole, *a-b-c*, must be defined, such that atom *a* possesses an electron sextet, that is, an incomplete valence shell combined with a positive formal charge, and that atom *c*, the negatively charged center, has an unshared electron pair. Combination of such a 1,3-dipole with a multiple bond system *d-e*, termed the dipolarophile, is referred to as a 1,3-dipolar cycloaddition." (Scheme 3.1).

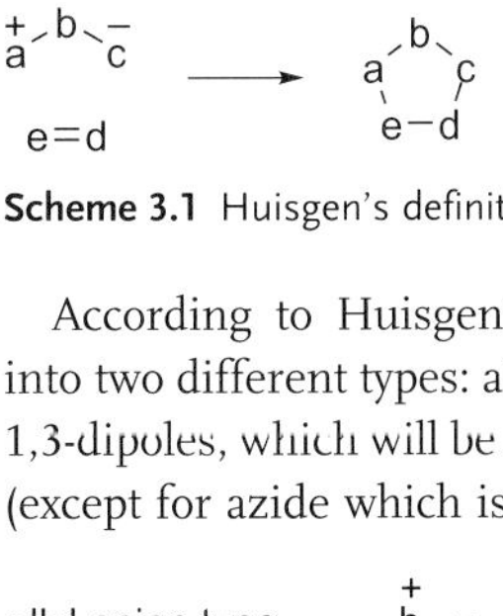

Scheme 3.1 Huisgen's definition of 1,3-dipolar cycloadditions.

According to Huisgen's classification of 1,3-dipoles, they can be categorized into two different types: allyl anion and propargyl anion. Important classes of these 1,3-dipoles, which will be reviewed in this article, are shown in Schemes 3.2 and 3.3 (except for azide which is reviewed in Chapter 18).

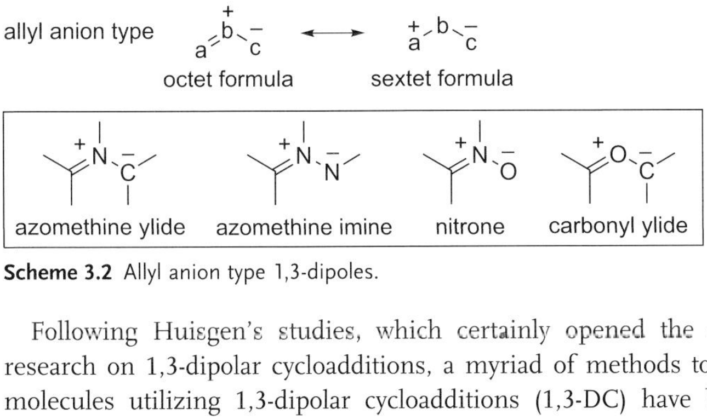

Scheme 3.2 Allyl anion type 1,3-dipoles.

Following Huisgen's studies, which certainly opened the door to intensive research on 1,3-dipolar cycloadditions, a myriad of methods to prepare complex molecules utilizing 1,3-dipolar cycloadditions (1,3-DC) have been disclosed to

Handbook of Cyclization Reactions. Volume 1.
Edited by Shengming Ma
Copyright © 2010 WILEY-VCH Verlag GmbH & Co. KGaA, Weinheim
ISBN: 978-3-527-32088-2

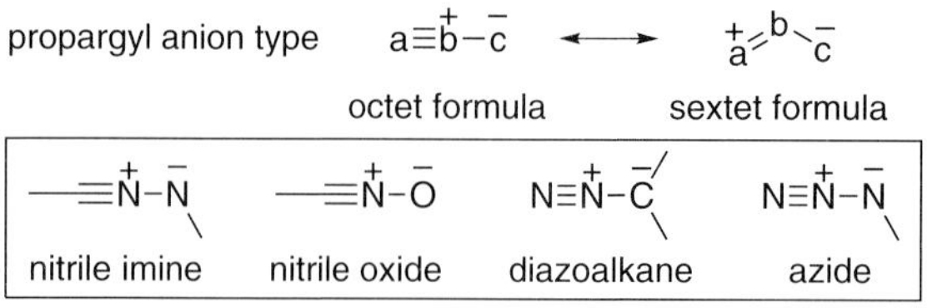

propargyl anion type a≡b⁻−c̄ ⇌ ⁺a−b=c̄

octet formula — sextet formula

−≡N⁺−N̄−	−≡N⁺−Ō	N≡N⁺−C̄⟨	N≡N⁺−N̄−
nitrile imine	nitrile oxide	diazoalkane	azide

Scheme 3.3 Propargyl anion type 1,3-dipoles.

date [2]. In this chapter, the focus is on the introduction of leading works of 1,3-dipolar cycloadditions, especially those reported in the past decade aiming at the realization of stereoselective syntheses by the intervention of the external mediator. At the same time, brand-new strategies for the generation of 1,3-dipoles will be described in each section, considering their importance in the further advance of 1,3-DCs. Although the theoretical aspects of 1,3-DCs have been extensively studied, they will not be covered in this article. Readers can refer to these works in the literature [3, 4].

3.2
Nitrones

Nitrones are the most commonly used 1,3-dipoles due to their ease of handling and attractiveness of the cycloadducts, isoxazolidines, as valuable precursors to β-amino acids, β-lactams or β-amino alcohols [5]. A general method for their preparation is the simple condensation of the carbonyl compound and N-monosubstituted hydroxylamine (Scheme 3.4). These nitrones are generally isolated as bench stable compounds.

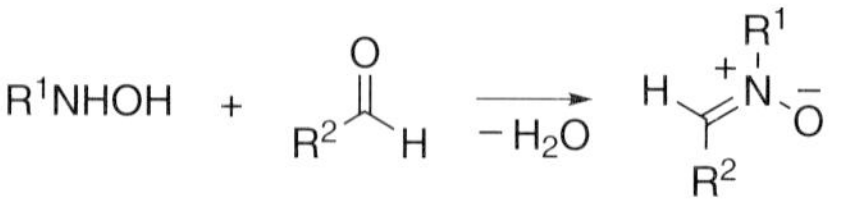

R^1NHOH + R^2CHO $\xrightarrow{-H_2O}$ nitrone (H, R^1, R^2, N⁺–O⁻)

Scheme 3.4 Synthesis of nitrones by the condensation of hydroxylamine.

Another reliable pathway to access nitrones is the tungstate-catalyzed oxidation of the corresponding secondary amine reported by Murahashi and coworkers [6]. Various secondary amines, such as aliphatic, benzylic and cyclic amines can be transformed into nitrones in practical yields (Scheme 3.5).

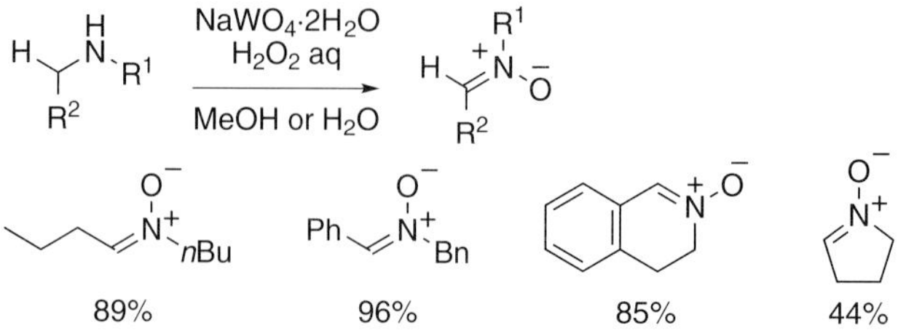

$R^2\text{-}NH\text{-}R^1$ $\xrightarrow[\text{MeOH or }H_2O]{NaWO_4\cdot 2H_2O,\ H_2O_2\ aq}$ nitrone

89%   96%   85%   44%

Scheme 3.5 Synthesis of nitrones by the tungsten-catalyzed oxidation of secondary amines.

Noteworthy is the recent literature which revealed new strategies for the in situ generation of electrophilic nitrones. One is the report from Denis' group in which the formation of N-Boc nitrone (Boc = tert-butoxycarbonyl) **3** by the base-promoted elimination of tert-butyl (phenylsulfonyl)alkyl-N-hydroxycarbamate **1** and cyclization of this labile intermediate with dimethyl acetylenedicarboxylate **2** was described (Scheme 3.6) [7].

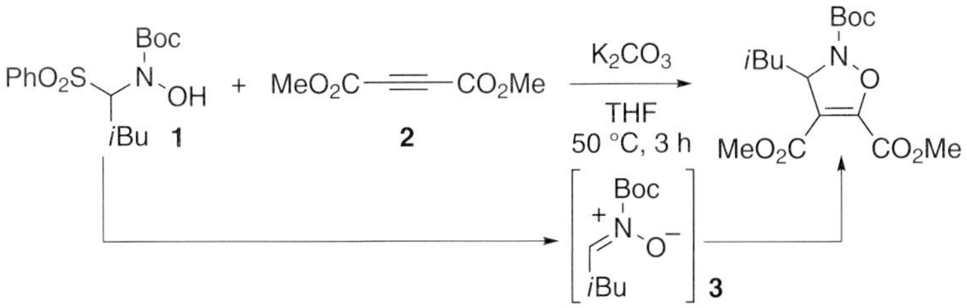

Scheme 3.6 Base-mediated in situ generation of nitrones.

Another method is the formation of N-sulfonyl nitrones by the Lewis acid catalyzed rearrangement of oxaziridines reported by Yoon and coworkers [8]. In the presence of $TiCl_4$, N-Ns oxaziridine (Ns = 4-nitrobenzenesulfonyl) **4** rearranges into N-Ns nitrone **5** and the intermediate reacts in situ with alkenes to give the isoxazolidines **6** with high syn-selectivity (Scheme 3.7).

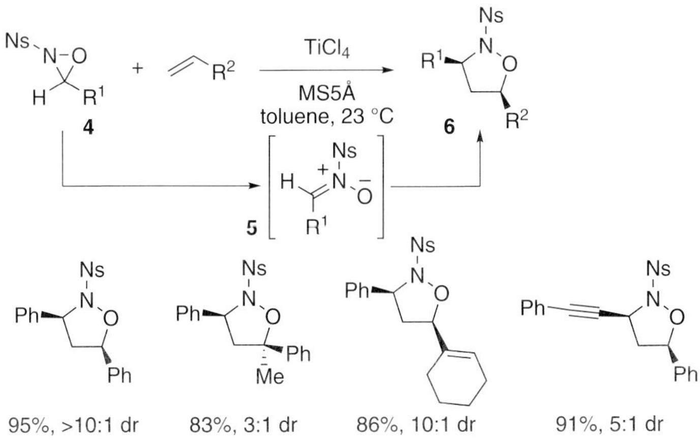

95%, >10:1 dr   83%, 3:1 dr   86%, 10:1 dr   91%, 5:1 dr

Scheme 3.7 Lewis acid-mediated in situ generation of nitrones.

3.2.1
Beginning of Lewis Acid Catalyzed 1,3-DC of Nitrones

The main obstacle faced in boosting the synthetic utility of 1,3-DC of nitrones is the necessity to control the regio-, diastereo- and enantioselectivities rigorously to obtain solely the desired cycloadduct, which is also a persistent problem in other 1,3-DCs (Scheme 3.8). Accordingly, intensive research has been undertaken to overcome this stereochemical issue by the action of external reagents.

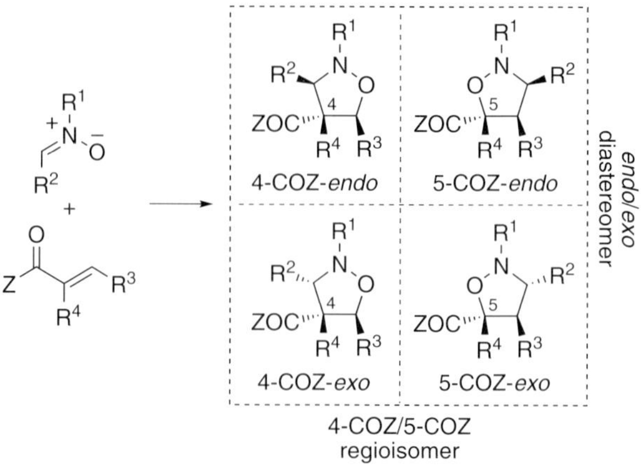

Scheme 3.8 Possible stereoisomers of 1,3-DC of nitrones.

1,3-DC of nitrones and an electron-deficient olefin are well-known to be accelerated by the use of Lewis acid catalyst. In the so-called normal electron demand 1,3-DC of nitrones, Lewis acid catalyst activates the alkene, typically the α,β-unsaturated carbonyl compound, leading to lowering of the LUMO of the double bond, and this lowered LUMO of the alkene then interacts with the HOMO of the nitrone to give the cycloadduct, isoxazolidine. Kanemasa et al. realized the first example of such Lewis acid catalyzed normal electron demand 1,3-dipolar cycloaddition between nitrones and α,β-unsaturated ketones containing a basic alkoxy group, which allowed the preferential chelation of alkene to Lewis acid in the presence of the rather basic and coordinative nitrone (Scheme 3.9) [9]. Importantly, the use of Lewis acid not only acts as an accelerator of the reaction but also as the stereocontroller, giving the endo-isomer exclusively. The thermal 1,3-DC gave the isoxzolidine as a mixture of endo- and exo-isomers.

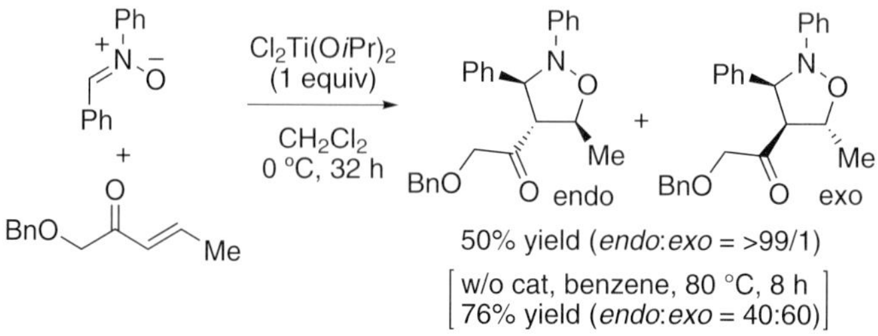

Scheme 3.9 Lewis acid-promoted 1,3-DC of nitrones and bidentate ketones.

3.2.2
Chiral Lewis Acid Catalyzed 1,3-DC of Nitrones and α,β-Unsaturated Carbonyl Compounds Bearing Ancillary Coordinating Group

Following Kanemasa's report, introduction of chiral Lewis acid catalyst to 1,3-DC of nitrone became of interest to organic chemists due to the importance of the thus-formed enantioenriched isoxazolidines as valuable intermediates for further elaboration. The first example of Lewis acid-catalyzed asymmetric normal electron demand 1,3-dipolar cycloadditions of nitrones was reported in 1994 by Jørgensen and coworkers [10]. Accomodation of titanium TADDOLate (TADDOL = 2,2-dimethyl-α,α,α',α'-tetraaryl-1,3-dioxolane-4,5-dimethanol) **7** as catalyst and the use of 3-N-(2-alkenoyl)oxazolidinones **8** as dipolarophile was found to be promising, giving the cycloadducts with moderate stereoselectivities (Scheme 3.10).

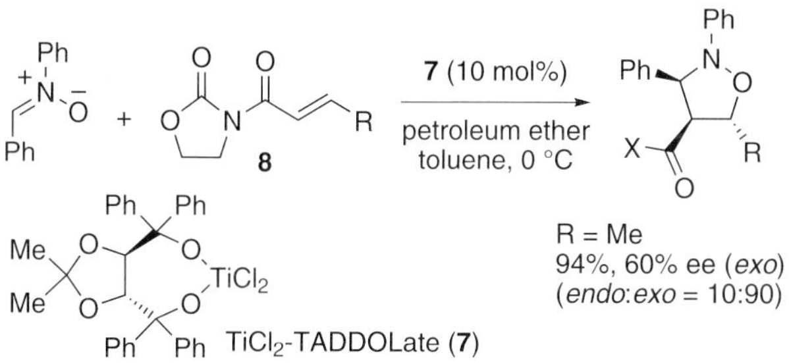

Scheme 3.10 Chiral titanium Lewis acid-catalyzed 1,3-DC of nitrones.

The oxazolidinone template attached to the α,β-unsaturated carbonyl was considered to be essential to secure the preferential coordination of dipolarophile to the Lewis acid by two point-binding, in the presence of the highly coordinative oxygen atom of the nitrone (Scheme 3.11).

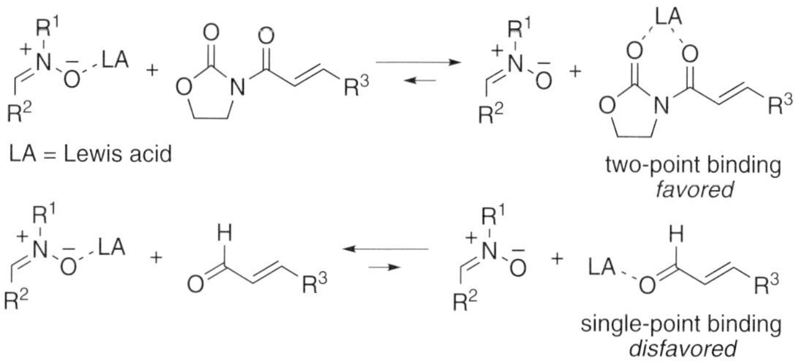

Scheme 3.11 Binding affinity of Lewis acid to α,β-unsaturated carbonyl compounds.

The last ten years have witnessed the development of highly efficient catalysts capable of achieving high stereoinduction in this typical substrate combination (Scheme 3.12) [11–17]. For example, in 1998, Kobayashi and coworkers reported that

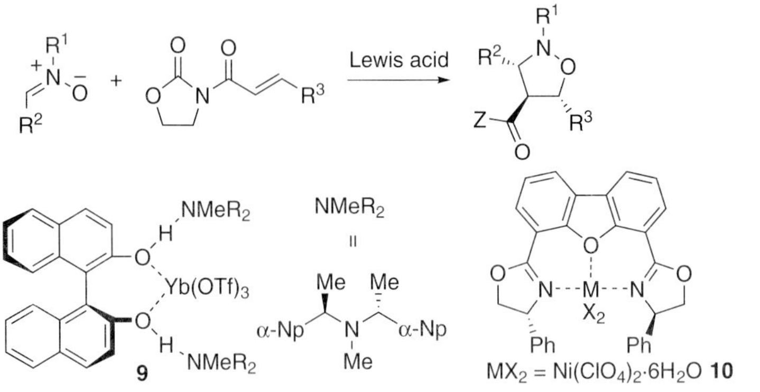

Scheme 3.12 Examples of highly enantioselective 1,3-DC of nitrones.

Catalyst 9 info:
R^1 = Bn, R^2 = aryl, R^3 = H or alkyl
up to 98% ee (endo)
endo:exo = >95:5

R^1 = Bn, R^2 = Et, R^3 = Me
88%, 96% ee (endo)
endo:exo = 53/47

Catalyst 10 info:
R^1 = Ph, Bn or Me, R^2 = aryl
R^3 = Me
up to >99% ee (endo)
endo:exo = >95:5

R^1 = Bn, R^2 = Et, R^3 = Me
92%, 97% ee (endo)
endo:exo = 94:6

MX_2 = Ni(ClO$_4$)$_2$·6H$_2$O **10**

a chiral ytterbium(III) catalyst **9** composed of BINOL (BINOL = 1,1′-bi(2-naphthol)) and chiral tertiary amine acts as a highly efficient catalyst to furnish the endo-cycloadduct exclusively with an excellent level of asymmetric induction [12]. In the same year, Kanemasa and coworkers discovered that the use of nickel(II)/DBFOX (DBFOX = 4,6-dibenzofurandiyl-2,2′-bis(4-phenyloxazoline)) complex **10** furnished the isoxazolidines in a highly endo- and enantioselective manner [13].

Attachment of other coordinative auxiliaries on α,β-unsaturated carbonyls has been developed by some groups. Palomo and coworkers employed α′-hydroxy enones **11** as the dipolarophile under the influence of the copper(II)/bisoxazoline **12** complex [18]. The distinct advantage of their method is the high efficiency in the reaction with β-unsubstituted enone (**11**, R^3 = H), giving the endo-cycloadduct with excellent enantiocontrol while minimizing the formation of the regioisomer to less than 10% (Scheme 3.13). The α′-hydroxy moiety of the cycloadduct could be

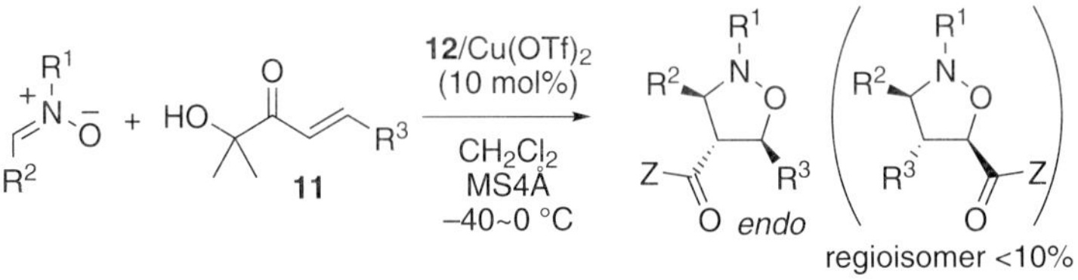

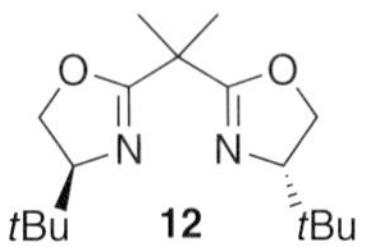

R^1	R^2	R^3	endo:exo	% yield	% ee (endo)
Bn	Ph	H	>98:2	85	94
Ph	Ph	H	76:24	98	>99
Ph	4-ClC$_6$H$_4$	Et	>98:2	79	>99

Scheme 3.13 Use of α′-hydroxy enones as the dipolarophile in 1,3-DC of nitrones.

easily removed by treatment with periodic acid to give the corresponding carboxylic acid.

Evans and coworkers employed 2-acyl imidazoles **13** in asymmetric 1,3-DC of nitrones, the synthetic utility of which had already been validated in asymmetric Friedel–Crafts alkylations [19]. Screening of various chiral lanthanides/pybox **14** complexes indicated the superiority of the early lanthanide complexes, exhibiting better overall yields and enantiofacial control, compared to late lanthanide complexes. By the use of cerium(IV)/**14** complex as catalyst, a broad range of substrates were converted to the endo-cycloadducts with excellent enantioselectivities, irrespective of the C-substituent of the nitrones or the β-substituent of the 2-acyl imidazoles (Scheme 3.14). In the case of C-alkyl nitrones, decrease in the endo/exo selectivity was observed. The cleavage of the imidazole moiety could be accomplished by consecutive N-methylation and hydrolysis under basic conditions.

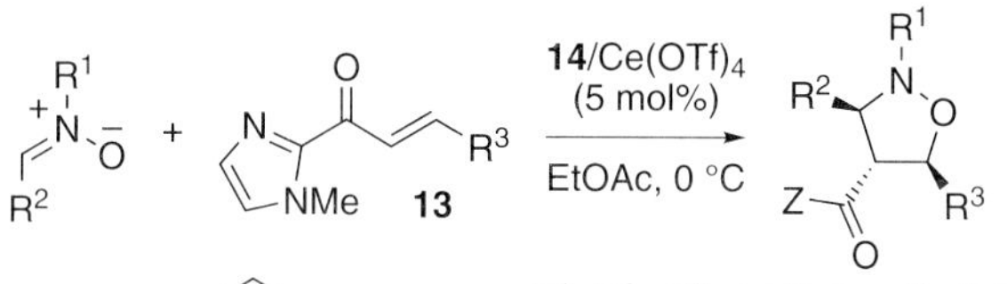

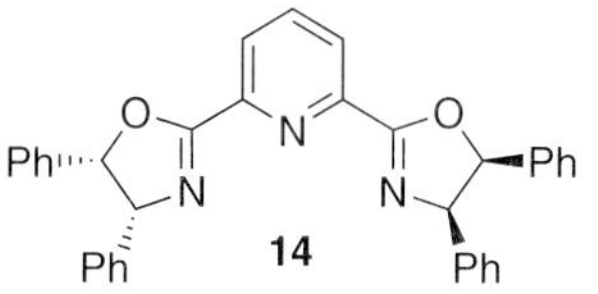

R^1	R^2	R^3	endo:exo	% yield	% ee (endo)
Bn	Ph	Me	>99:1	99	97
Ph	Bn	Et	Me	2.4:1	97	87
Bn	Ph	Ph	3:1	80	95
Ph	Ph	H	6:1	98	80

Scheme 3.14 Use of 2-acyl imidazoles as the dipolarophile in 1,3-DC of nitrones.

The possibility of using α′-phosphoric enones **15** in asymmetric 1,3-DC of nitrones was investigated by Kim et al. [20]. By the use of chiral copper(II)/bisoxazoline **16** complex, a wide range of nitrones could be converted to the isoxazolidines with high endo- and enantioselectivities (Scheme 3.15). Further elaboration of the cycloadduct by the Horner–Wadsworth–Emmons reaction was examined as its synthetic application.

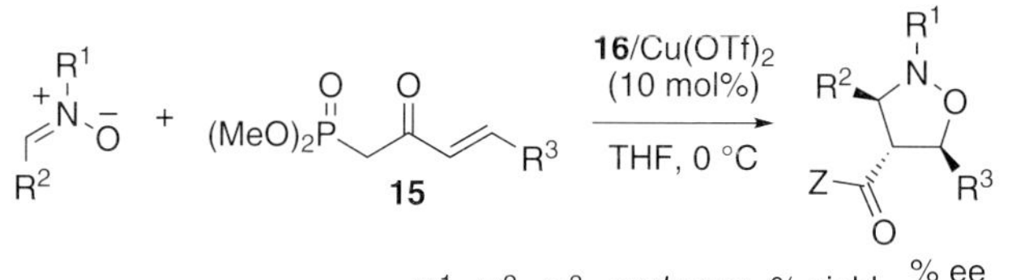

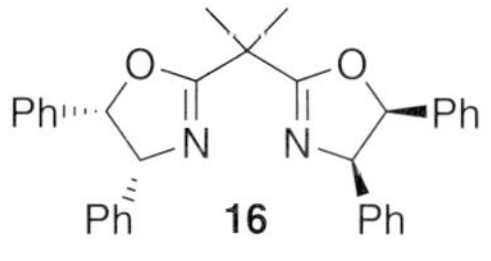

R^1	R^2	R^3	endo:exo	% yield	% ee (endo)
Bn	Ph	Me	>99:1	90	91
Bn	Et	Me	6:5	76	73
Bn	Cy	Me	>99:1	65	84
Bn	Ph	Ph	–	0	–

Scheme 3.15 Use of α′-phosphoric enones as the dipolarophile in 1,3-DC of nitrones.

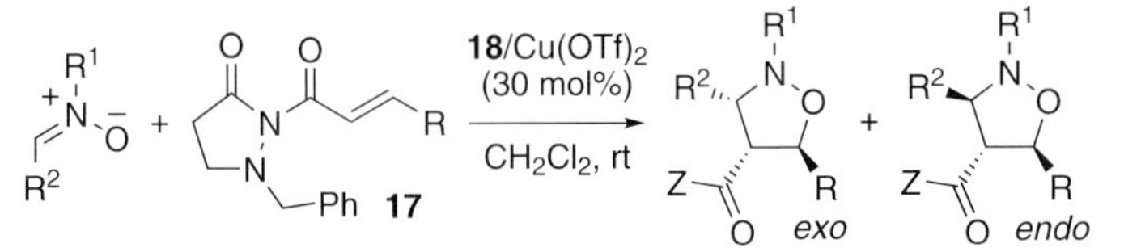

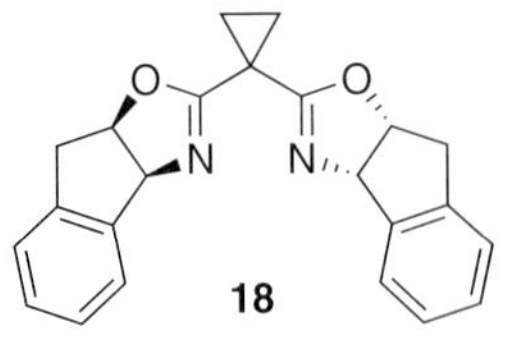

Scheme 3.16 Exo- and enantioselective 1,3-DC of nitrones.

R¹	R²	R	exo:endo	% yield	% ee (exo)
Me	Ph	H	66:34	85	99
Me	Ph	Me	96:4	94	98
Bn	4-ClC$_6$H$_4$	Me	85:15	52	98
Bn	4-MeOC$_6$H$_4$	Me	95:5	45	98
Ph	Ph	Me	52:48	85	99

Despite the fact that the first example of asymmetric 1,3-DC of nitrones developed by Jørgensen was the exo-selective cycloaddition, most of the following examples realized by various chiral Lewis acid catalysts provided the endo-cycloadducts selectively. In this context, Sibi et al. realized highly exo- and enantioselective 1,3-DC of nitrones by the use of α,β-unsaturated carbonyls bearing pyrazolidone template **17** in conjunction with a catalytic use of copper(II)/bisoxazoline **18** (Scheme 3.16) [21]. Uniquely, addition of molecular sieves led to the reversal or loss of the exo selectivity. Coordination of the substrate and ligand to copper(II) Lewis acid in a square-planar fashion is considered to be the crucial factor for exerting high exo selectivity.

Sibi et al. also succeeded in introducing rather unreactive α,β-disubstituted α,β-unsaturated carbonyl compounds in 1,3-DC of nitrones catalyzed by the same copper(II)/bisoxazoline **18** complex [22]. Use of acrylimides as α,β-disubstituted carbonyl compounds is the key factor in realizing the highly enantioselective process. Although a prolonged reaction time of five to ten days was required, highly substituted isoxazolidines could be produced in good yield with excellent enantioselectivity (Scheme 3.17).

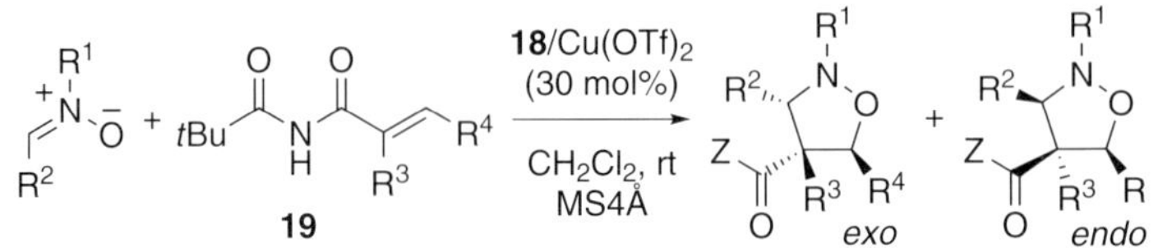

R¹	R²	R³	R⁴	exo:endo	% yield	% ee (exo)
Me	Ph	Me	Me	99:1	60	94
Me	Ph	Me	H	81:19	57	89
Bn	Ph	-(CH$_2$)$_4$-		99:1	89	94
Ph	Ph	-(CH$_2$)$_4$-		94:6	77	89

Scheme 3.17 Exo- and enantioselective 1,3-DC of nitrones with α,β-disubstituted carbonyl compounds.

Suga and coworkers developed exo-selective 1,3-DC of nitrones and 3-alkenoyl-2-thiazolidinethiones **20** catalyzed by chiral nickel(II)/binaphthyldiimine **21** complex (Scheme 3.18) [23]. This catalytic system could be applied to a broad range of nitrones and α,β-unsaturated carbonyls.

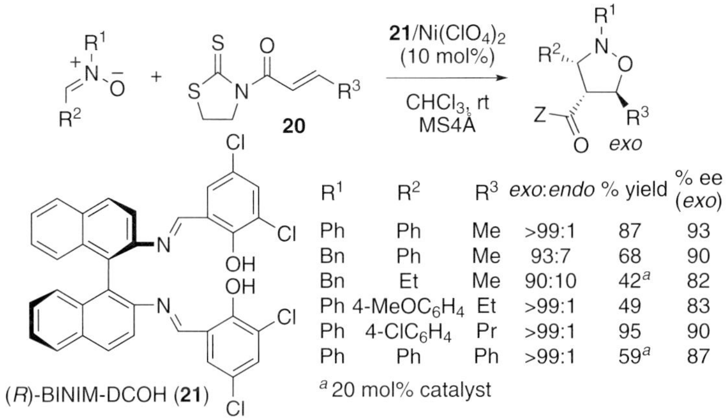

R^1	R^2	R^3	exo:endo	% yield	% ee (exo)
Ph	Ph	Me	>99:1	87	93
Bn	Ph	Me	93:7	68	90
Bn	Et	Me	90:10	42[a]	82
Ph	4-MeOC$_6$H$_4$	Et	>99:1	49	83
Ph	4-ClC$_6$H$_4$	Pr	>99:1	95	90
Ph	Ph	Ph	>99:1	59[a]	87

[a] 20 mol% catalyst

Scheme 3.18 Exo- and enantioselective 1,3-DC of nitrones with 3-alkenoyl-2-thiazolidinethiones.

3.2.3
Asymmetric 1,3-DC of Nitrones and α,β-Unsaturated Aldehydes

3.2.3.1 Organocatalysis

More recently, the use of α,β-unsaturated aldehydes lacking the ancillary basic template in asymmetric 1,3-DC of nitrones became the challenging goal, due to their easier accessibility and atom economic aspect, compared to substrates bearing a temporary template which must be removed in the further elaboration of the cycloadducts. In this respect, a novel strategy which does not rely on conventional Lewis acid catalysis appeared in 2000 [24, 25]. Namely, MacMillan and coworkers succeeded in applying their LUMO-lowering iminium catalysis to asymmetric 1,3-DC of nitrones and α,β-unsaturated aldehydes. The crucial factor for the successful implementation was the low affinity of the secondary amine catalyst to nitrones and the efficient activation of α,β-unsaturated aldehydes by the formation of iminium salts with the catalyst, whereas Lewis acid catalysts have been considered to coordinate to nitrones preferentially in the presence of poorly basic α,β-unsaturated aldehydes (see, Scheme 3.11). Under the influence of a catalytic amount of chiral imidazolidione catalyst **23**, highly enantioselective 1,3-DC of various N-alkyl nitrones with crotonaldehyde (**22**, $R^2 =$ Me) and acrolein (**22**, $R^2 =$ H) was accomplished, giving the endo-cycloadducts predominantly (Scheme 3.19).

Karlsson and Högberg extended this protocol to the reaction with cyclopent-1-enecarbaldehyde **24** [26]. Screening of various chiral pyrrolidinium salts revealed the superiority of the dihydrochloride salt of the diamine **25**, affording the bicyclic compound with high exo- and enantioselectivity (Scheme 3.20).

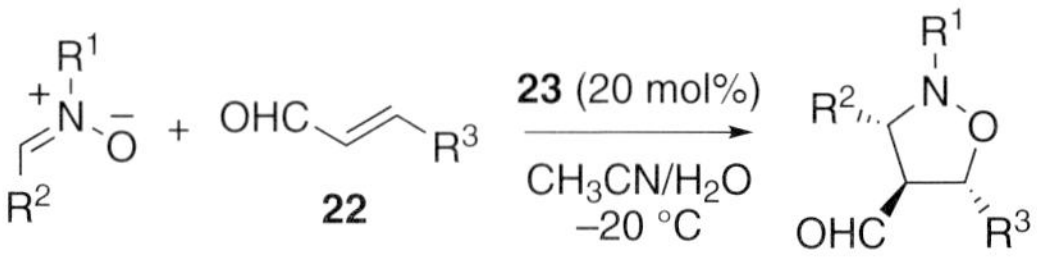

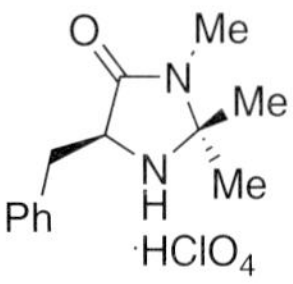

3 1,3-Dipolar Cycloaddition

Scheme 3.19 Pioneering work on organocatalytic enantioselective 1,3-DC of nitrones and α,β-unsaturated aldehydes.

R¹	R²	R³	endo:exo	% yield	% ee (endo)
Bn	Ph	Me	94:6	98	94
allyl	Ph	Me	93:7	73	98
Me	Ph	Me	95:5	66	99
Bn	Cy	Me	99:1	70	99
Bn	Ph	H	81:19	72	90
Bn	4-ClC₆H₄	H	80:20	80	91
Bn	4-MeC₆H₄	H	91:9	83	90

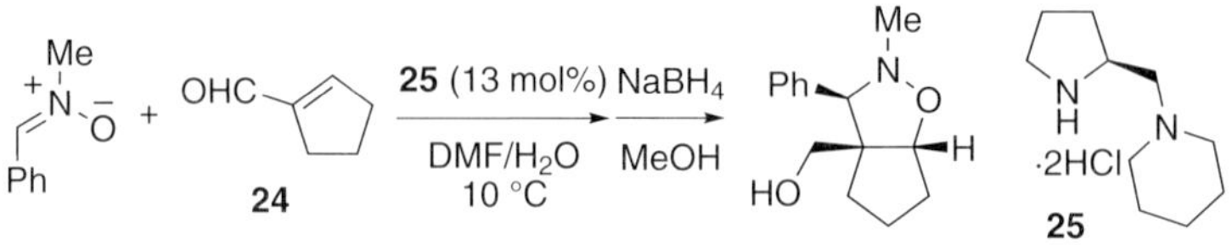

Scheme 3.20 Organocatalytic 1,3-DC of nitrones using cyclopent-1-enecarbaldehyde.

3.2.3.2 Lewis Acid Catalysis

Following the groundbreaking report of MacMillan, chiral Lewis acid catalyzed asymmetric 1,3-DC of nitrones regained momentum, aiming at the utilization of α,β-unsaturated aldehydes, and several catalysts have been developed, as summarized in Scheme 3.21.

The first successful report on this subject appeared in 2002, wherein Kündig and coworkers developed chiral iron **26a** and ruthenium **26b** Lewis acid catalyzed asymmetric 1,3-DC of nitrones [27, 28]. The outstanding feature of these Lewis acid catalysts is their aldehyde selective binding ability, allowing the preferential coordination of α,β-unsaturated aldehydes over nitrones. Their catalytic system using the iron complex **26a** was especially effective for the 1,3-DC of cyclic nitrones **31** and methacrolein (**32**, R^2 = Me, R^3 = H), in which the 5-*endo*-CHO adduct **34**, creating the C–O bond at the α-carbon of the aldehyde, was obtained exclusively with an excellent level of asymmetric induction (Table 3.1, entries 1, 4, 7). Complete inversion of the regioselectivity was observed in 1,3-DC with crotonaldehyde (**32**, R^2 = H, R^3 = Me) as dipolarophile, giving the 4-*endo*-CHO adduct **33** with 75% ee (Table 3.1, entry 5). The use of acyclic *N*-phenyl nitrones **35** as dipole provided the corresponding cycloadducts with the preferential formation of the 5-CHO-*endo* isomer **37** (60/40~100/0) and moderate to good enantioselectivities (e.g., Table 3.2, entries 1 and 2).

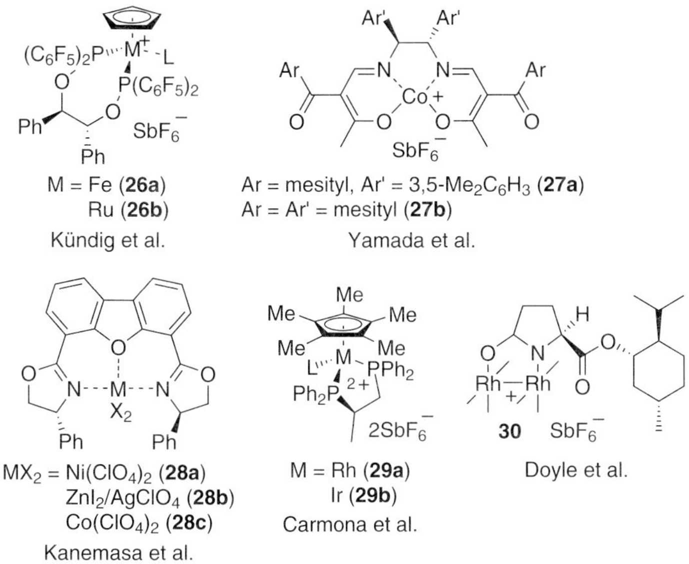

Scheme 3.21 Lewis acid catalysts utilized in enantioselective 1,3-DC of nitrones with α,β-unsaturated aldehydes.

For the 1,3-DC of acyclic N-phenyl nitrones and α,β-unsaturated aldehydes, the viability of several chiral Lewis acid catalysts having Mg, Ni, Zn, Co, Rh or Ir as metal sources was disclosed by some research groups, as showcased in Table 3.2. Yamada and coworkers developed the chiral β-ketoiminato cationic cobalt(III) complex **27** catalyzed asymmetric 1,3-DC [29], which is especially effective in the reaction of nitrones having o-halogenated aromatics as the C-substituent, providing cycloadducts with high enantiofacial discrimination (Table 3.2, entries 3, 10, 14, 16). The ratio of 4-CHO-*endo* and 5-CHO-*endo* highly depends on the nature of α,β-unsaturated aldehydes.

Kanemasa and coworkers utilized DBFOX/Ph ligand developed in their laboratory in conjunction with judiciously chosen metal sources, depending on the substituent pattern of α,β-unsaturated aldehydes [30]. For example, 1,3-DC of C,N-diphenyl nitrone (**35**, R^2 = Ph) and methacrolein (**32**, R^2 = Me, R^3 = H) was promoted efficiently by the use of the nickel(II)/DBFOX/Ph complex **28a**, giving the 5-CHO-*endo* cycloadduct **37** with 96% ee (Table 2.2, entry 4). On the other hand, the zinc(II)/DBFOX/Ph complex was the catalyst of choice in the 1,3-DC with α-bromoacrolein (**32**, R^2 = Br, R^3 = H), giving the 4-CHO-*endo* adduct **36** exclusively (Table 3.2, entry 18).

Carmona and coworkers investigated the utility of chiral rhodium and iridium/bisphosphine complexes **29** [31]. In their initial study, they discovered that 1,3-DC of C,N-diphenyl nitrone and methacrolein catalyzed by their rhodium complex **29a** furnishes the 4-CHO-*endo* adduct **37** preferentially, with high ee value (Table 3.2, entry 6), in contrast to other reports which showcased the 5-CHO-*endo* selective reactions (Table 3.2, entries 1–5,7). It should also be noted that, in the absence of catalyst, 1,3-DC of C,N-diphenyl nitrone and methacrolein gave the 5-CHO-*endo*

Table 3.1 Lewis acid-catalyzed enantioselective 1,3-DC of cyclic nitrones with α,β-unsaturated aldehydes.

Entry	Nitrone 31	R^2	R^3	Catalyst	4-:5-CHO	% yield	% ee 4-/5-CHO
1	31a	Me	H	26a	5-CHO selective	92	−/96
2		Me	H	29a	-:>99	100	−/86
3		Me	H	29b	-:100	100	−/86
4	31b	Me	H	26a	5-CHO selective	71	−/>96
5		H	Me	26a	4-CHO selective	75	75/−
6		Me	H	29b	-:>99	100	−/92
7	31c	Me	H	26a	5-CHO selective	71	−/94
8		Me	H	29b	-:100	100	−/93
9		H	H	29a	>90: 10	49	77/−
10		H	H	29b	>90: 10	69	66/−
11		H	Me	29a	>90: 10	100	74/−

adduct exclusively. In the 1,3-DC of cyclic nitrones **31**, use of methacrolein provided the 5-CHO-*endo* adduct (Table 3.1, entries 2, 3, 6 and 9) exclusively, whereas acrolein and crotonaldehyde were selectively transformed into 4-CHO-*endo* cycloadducts with moderate enantioselectivities (Table 3.1, entries 9–11). The structures of these rhodium and iridium Lewis acid-methacrolein complexes were intensively studied by X-ray crystallographic analysis. Carmona *et al.* also developed the chiral ruthenium/P,N-ligand complex which exerts moderate selectivities [32].

Recently, cationic chiral dirhodium carboxamidates were introduced in this sort of 1,3-DC by Doyle *et al.* [33]. Use of [Rh$_2${(5S,R)-menpy}$_4$]SbF$_6$**30** facilitated 1,3-DC of several C-aryl-N-phenyl nitrones and methacrolein with moderate to high regioselectivities and enantioselectivities. The preferential formation of the 4-CHO-*endo* adduct **36** was observed, when N-phenyl nitrones bearing electron-donating C-aryl groups were employed (Table 3.2, compare entries 7 and 8).

Despite the steady progresses of chiral Lewis acid catalyzed 1,3-DC of nitrones and α,β-unsaturated aldehydes, these approaches suffered from the generation of two regioisomeric products, especially in the reaction of acyclic nitrones, although the endo-selectivity was uniformly high with almost no exception, as exemplified above.

Table 3.2 Lewis acid-catalyzed enantioselective 1,3-DC of acyclic nitrones with α,β-unsaturated aldehydes.

$$\underset{35}{\overset{Ph}{\underset{R^1}{N}}\overset{+}{\underset{O^-}{\|}}} + \underset{32}{\overset{OHC}{\underset{R^2}{\searrow}}R^3} \xrightarrow{Lewis\ acid} \underset{4\text{-}CHO\text{-}endo\ (36)}{\overset{Ph}{\underset{OHC^{\prime\prime\prime}}{\overset{N}{\underset{R^2\ R^3}{\bigcirc}}}}R^1} + \underset{5\text{-}CHO\text{-}endo\ (37)}{\overset{Ph}{\underset{OHC^{\prime\prime\prime}}{\overset{N}{\underset{R^2\ R^3}{\bigcirc}}}}R^1}$$

Entry	R¹	R²	R³	Catalyst	4-:5-CHO	Yield (%)	ee (%) 4-/5-CHO
1	Ph	Me	H	26a	20: 80	85	91/87
2	Ph	Me	H	26b	40: 60	92	94/76
3	2,3-Cl₂C₆H₃	Me	H	27b	5: 95	91	–/82
4	Ph	Me	H	28a	>5: 95	73	–/96
5	Ph	Me	H	28b	45: 55	98	83/95
6	Ph	Me	H	29a	63: 37	100	90/75
7	Ph	Me	H	30	24: 76	95	95/53
8	4-Me₂N C₆H₄	Me	H	30	90: 10	50	90/45
9	Ph	H	H	28a	26: 74	75	91/98
10	2,3-Cl₂C₆H₃	H	H	27b	89: 11	90	79/–
11	Ph	H	H	29a	4-CHO selective	100	78/–
12	Ph	H	H	29b	4-CHO selective	100	90/–
13	Ph	H	Me	28b	4-CHO selective	78	ds = 80: 20 (49/84)
14	2,3-Cl₂C₆H₃	H	Me	27b	98: 2	94	63
15	Ph	-(CH₂)₃-		28c	4-CHO selective	46	92/–
16	2,3-Cl₂C₆H₃	-(CH₂)₃-		27a	>99: 1	quant	87/–
17	Ph	-(CH₂)₃-		27a	>99: 1	68	7/–
18	Ph	H	Br	28b	4-CHO selective	85	98/–
19	Ph	Me	Me	29b	4-CHO selective	96.5	92/–

In addition, use of N-phenyl nitrones hampered the elaboration of the cycloadducts due to the difficulty in removing the phenyl functionality by a conventional deprotection method.

One solution to these problems was reported by Maruoka and coworkers [34]. In their pursuit of the catalytic performance of bis-titanium chiral Lewis acids **39**, they found that these catalysts act as suitable promoters of the highly regio-, diastereo- and enantioselective 1,3-DC of N-benzyl nitrones **38** and α,β-unsaturated aldehydes, which had been unattainable by the use of other chiral Lewis acids. In the initial report, acrolein was utilized as a reactive α,β-unsaturated aldehyde under the influence of bis-titanium chiral Lewis acid **39a** to give the enantioenriched cycloadducts with the exclusive formation of the 4-CHO-endo adduct **40** (Scheme 3.22). For the C-substituent of nitrones, both electron-rich and electron-withdrawing groups were tolerated, as well as bulky alkyl groups such as *tert*-butyl.

The successful implementation of enantioselective 1,3-DC of N-benzyl nitrones and acrolein prompted the authors to apply other α,β-unsaturated aldehydes bearing a substituent at the α- or β-position of the aldehyde. Although subjection

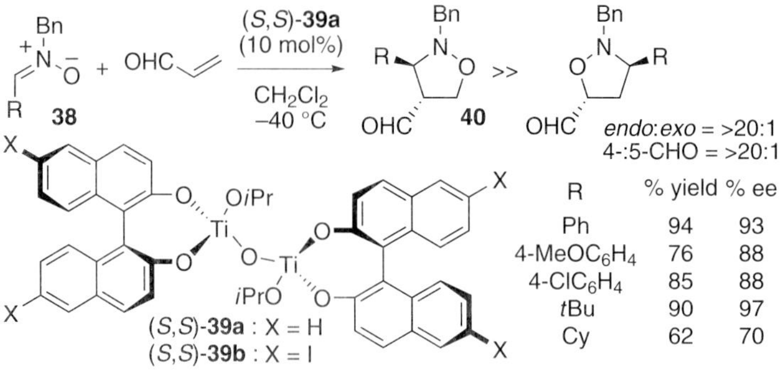
(S,S)-39a
(10 mol%)
CH2Cl2
−40 °C
38
40
endo:exo = >20:1
4-:5-CHO = >20:1
R % yield % ee
Ph 94 93
4-MeOC6H4 76 88
4-ClC6H4 85 88
tBu 90 97
Cy 62 70
(S,S)-39a : X = H
(S,S)-39b : X = I

Scheme 3.22 Bis-titanium Lewis acid-catalyzed enantioselective 1,3-DC of *N*-benzyl nitrones and acrolein.

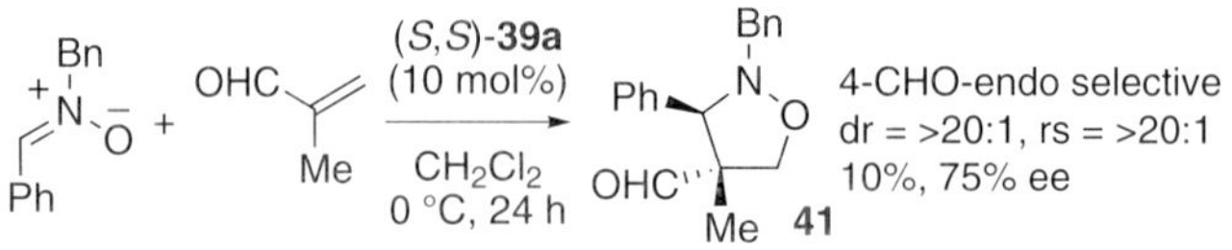
(S,S)-39a
(10 mol%)
CH2Cl2
0 °C, 24 h
41
4-CHO-endo selective
dr = >20:1, rs = >20:1
10%, 75% ee

Scheme 3.23 1,3-DC of *N*-benzyl-*C*-phenyl nitrone and methacrolein.

of *N*-benzyl-*C*-phenyl nitrone and methacrolein to the established reaction conditions led to poor conversion, the highly congested 4-CHO-adduct **41** was isolated as a single 4-CHO-*endo* isomer with moderate enantioselectivity (Scheme 3.23).

To address this poor reactivity problem, while keeping the unique regioselectivity, introduction of the sterically hindered functionality as the *N*-substituent of nitrone was tried [35]. This approach was supposed to minimize the undesired coordination of nitrone to Lewis acid, which is known to be the main cause of the deactivation of Lewis acid catalyst in 1,3-DC of nitrones, by the steric repulsion between the catalyst and *N*-substituent of nitrone. As such, substantial enhancement of reactivity may be acquired (Scheme 3.24).

After screening several *N*-substituents, *N*-diphenylmethyl nitrone **42** was found to be the *N*-substituent of choice, improving the reactivity dramatically. Additionally, a substantial increase in enantioselectivity was observed. Namely, starting from *N*-diphenylmethyl *C*-phenyl nitrone (**42**, $R^2 = Ph$) and methacrolein (**32**, $R^2 = Me$, $R^3 = H$), bis-titanium chiral Lewis acid **39a** catalyzed 1,3-DC afforded the *N*-diphenyl-

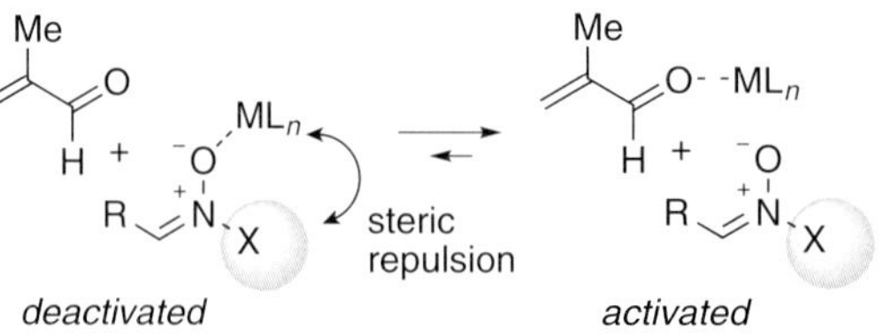
steric
repulsion
deactivated
activated

Scheme 3.24 Kinetic destabilization of Lewis acid-nitrone complex by steric repulsion.

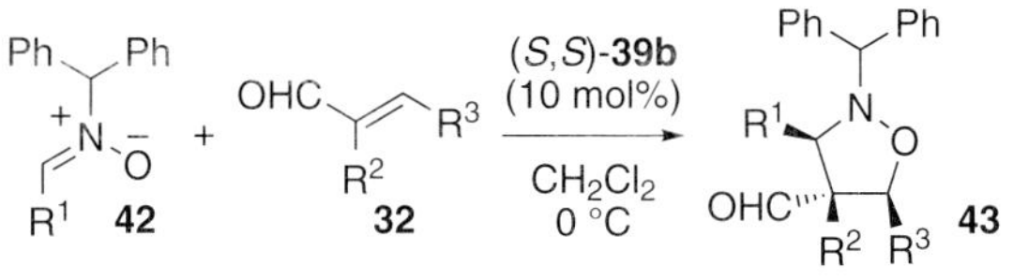

R^1 = aryl, alkenyl, R^2 = Me, R^3 = H
46~80%, 88~99% ee

R^1 = aryl, alkyl, R^2 = H, R^3 = Me
50~93%, 88~92% ee

Scheme 3.25 Bis-titanium Lewis acid-catalyzed enantioselective 1,3-DC of *N*-diphenylmethyl nitrones.

methyl isoxazolidine **43** in 58% yield with 90% ee. Further optimization identified the effectiveness of 6,6′-I_2-substituted BINOL as the chiral ligand, and with this catalyst **39b**, various *C*-aryl and alkenyl nitrones could be transformed into the corresponding isoxazolidines stereoselectively in moderate to high yields (Scheme 3.25). This catalytic system was also applicable to 1,3-DC with crotonaldehyde (**32**, R^2 = H, R^3 = Me). The *N*-diphenylmethyl moiety of the thus-formed cycloadduct could be easily removed in high yield by the use of oxidative or acidic conditions.

3.2.4
Asymmetric 1,3-DC of Nitrones and Other Electron-Deficient Olefins

Asymmetric 1,3-DC of *N*-aryl nitrones and dialkyl malonate was reported by Tang and coworkers [36], wherein cobalt(II)/TOX **45** (TOX = trioxazoline) complex was employed as catalyst (Scheme 3.26). A unique finding in their study is the switch of the diasteroselectivity depending on the reaction temperature. Namely, the catalyzed reaction of *C*,*N*-diphenyl nitrone and diethyl 2-benzylidenemalonate **44** conducted at −40 °C furnished the cis-adduct **46** with high enantioselectivity. On the other hand, the trans adduct **47** became the major product, when the reaction was conducted at 0 °C.

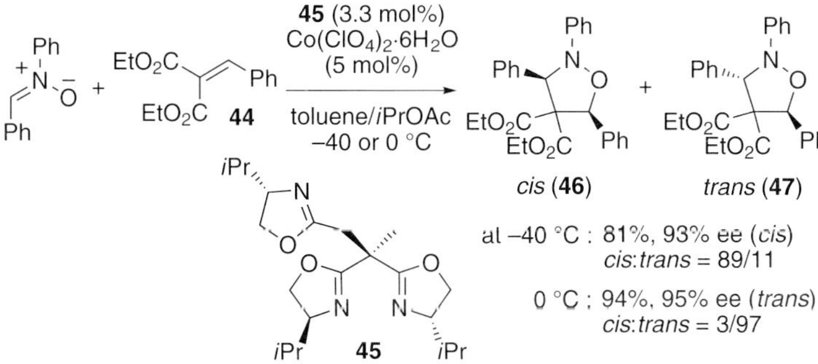

Scheme 3.26 Enantioselective 1,3-DC of nitrones and arylidenemalonate.

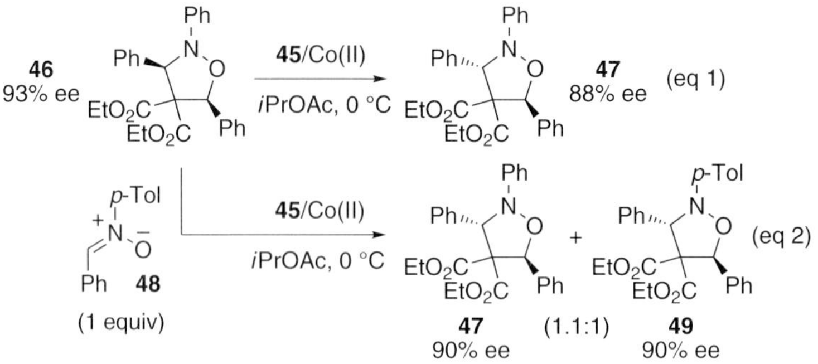

Scheme 3.27 Reversibility of cobalt(II)-catalyzed 1,3-DC of nitrones.

In their study on this intriguing reversal of diastereoselectivity, it was found that treatment of the cis-adduct **46** with the cobalt catalyst at 0 °C furnished the trans-product **47** with high asymmetric induction (Scheme 3.27, Eq. (1)). Additionally, when the cis-adduct **46** was subjected to the reaction with N-(p-tolyl) nitrone **48**, the crossover product **49** with the trans-configuration was observed (Eq. (2)). These observations could be explained by postulating the intervention of a retro 1,3-DC process. Thus, at low temperature, 1,3-DC provided the cis-ioxazolidine as a major product, wherein the kinetic control predominates. On the other hand, when the reaction was performed at 0 °C, the repetitive 1,3-DC and retro 1,3-DC finally led to the formation of the thermodynamically more favorable trans-adduct.

Carmona et al. applied their rhodium catalyst **29a** in 1,3-DC of 3,4-dyhydroisoquinoline N-oxide **31c** and methacrylonitrile **50** [37]. The existence of two isomeric catalyst–substrate complexes, (S_{Rh}, R_C)-**29a/50** and (R_{Rh}, R_C)-**29a/50**, arising from the chiral nature of the metal center was supposed to hamper the acquisition of high enantioselectivity (Scheme 3.28). Actually, use of the isolated (S_{Rh}, R_C)-**29a/50** complex under stoichiomeric reaction conditions furnished the cycloadduct **51** with higher selectivity.

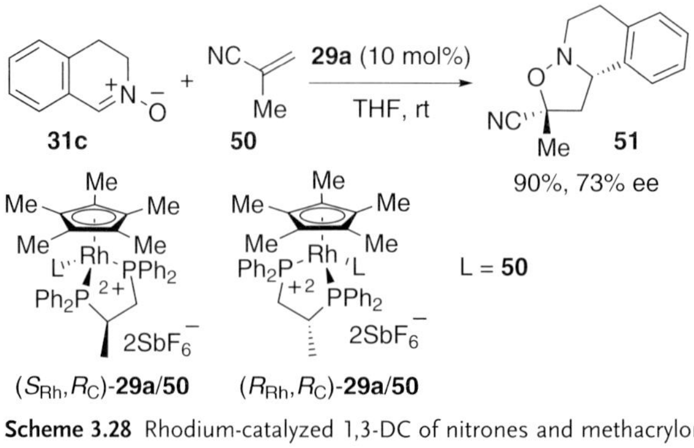

Scheme 3.28 Rhodium-catalyzed 1,3-DC of nitrones and methacrylonitrile.

3.2.5
Inverse Electron Demand 1,3-DC of Nitrones

Less studied is the Lewis acid catalyzed inverse electron demand 1,3-DC of nitrones, which proceeds via the LUMO lowering of nitrone by the coordination of Lewis acid, followed by the interaction of the HOMO of the electron-rich dipolarophile to this acid-nitrone complex. Allylic alcohol or alkyl vinyl ether are commonly employed as dipolarophiles. The ability of Lewis acid catalysts to promote inverse electron demand 1,3-DC of nitrone was revealed by Kanemasa and coworkers [38]. Use of magnesium, titanium or boron as the Lewis acid in the reaction of N-(benzoylmethylidene)aniline N-oxide **52** and crotyl alcohol **53** was found to give the *exo*-cycloadduct **54** with high regioselectivity (Scheme 3.29).

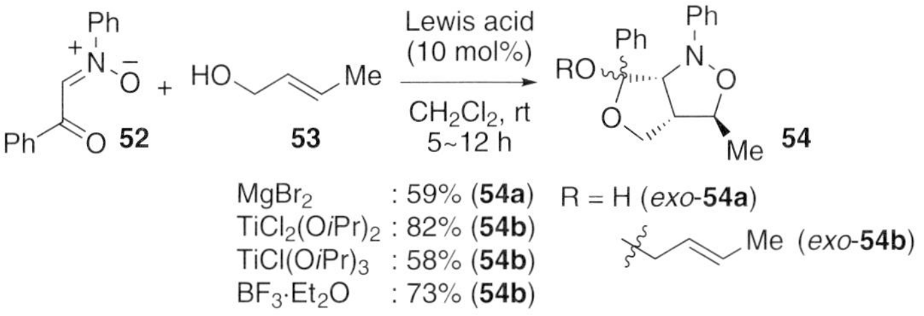

Scheme 3.29 Lewis acid-catalyzed inverse electron demand 1,3-DC of nitrones.

The first catalytic asymmetric inverse electron demand 1,3-DC was reported in 1994 by Scheeren and coworkers [39], wherein *C*,*N*-diphenyl nitrone and ketene acetals, such as **55**, were combined in the presence of the tyrosine-derived oxazaborolidine **56** as chiral Lewis acid catalyst (Scheme 3.30). Tuning of the catalyst structure and the reaction conditions led to the discovery of an optimal catalytic system, giving the cycloadducts with high regioselectivity and moderate enantioselectivity.

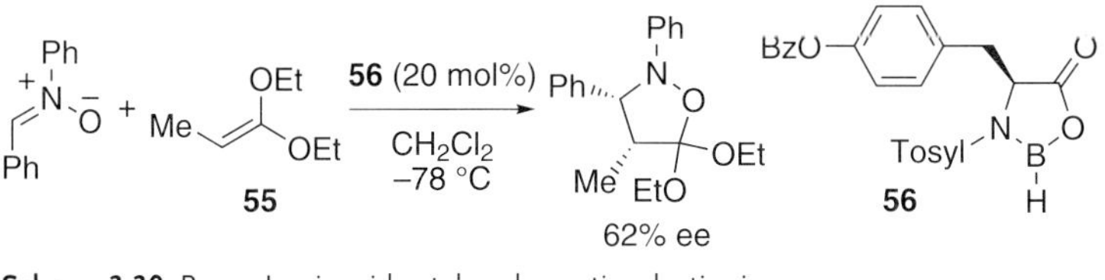

Scheme 3.30 Boron Lewis acid-catalyzed enantioselective inverse electron demand 1,3-DC of nitrones.

Following this report, Jørgensen and coworkers found that chiral aluminum Lewis acid prepared by the complexation of Me$_3$Al with 3,3′-diphenyl-BINOL **57a** functions as an effective catalyst for enantioselective inverse electron demand 1,3-DC of nitrones and vinyl ether [40]. Exo-cycloadducts could be obtained regioselectively in good yields with an excellent level of enantiocontrol (Scheme 3.31).

The use of dihydrofuran as the dipolarophile provided the bicyclic compound **58** in good yield with disappointingly low enantioselectivity. They also examined the

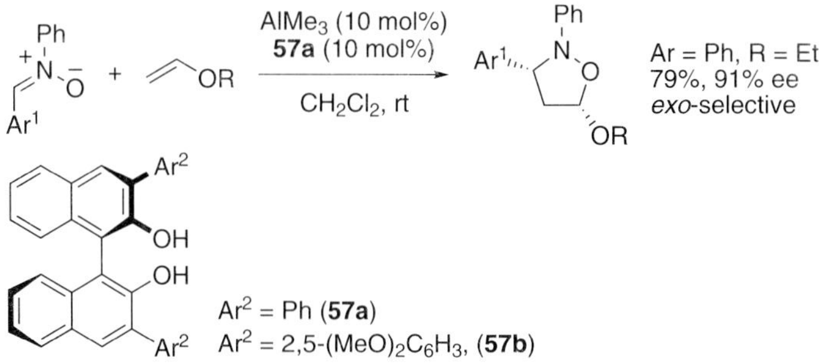
Ph
AlMe3 (10 mol%)
57a (10 mol%)
OR
CH2Cl2, rt
Ar1
Ph
Ar1
OR
Ar = Ph, R = Et
79%, 91% ee
exo-selective
Ar2
OH
OH
Ar2
Ar2 = Ph (57a)
Ar2 = 2,5-(MeO)2C6H3, (57b)

Scheme 3.31 Aluminum Lewis acid-catalyzed enantioselective inverse electron demand 1,3-DC of nitrones.

1,3-DC of cyclic nitrones and vinyl ether. By adjusting the 3,3′-substituent of BINOL to gain maximum asymmetric induction, *exo*-cycloadducts **59** could be obtained with high enantiomeric excesses (Scheme 3.32).

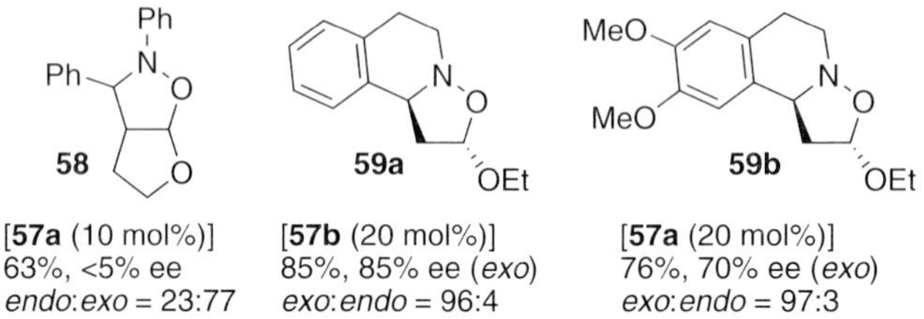
Ph
Ph
58
59a
OEt
MeO
MeO
59b
OEt
[57a (10 mol%)]
63%, <5% ee
endo:exo = 23:77
[57b (20 mol%)]
85%, 85% ee (exo)
exo:endo = 96:4
[57a (20 mol%)]
76%, 70% ee (exo)
exo:endo = 97:3

Scheme 3.32 Scope of aluminum Lewis acid-catalyzed enantioselective inverse electron demand 1,3-DC of nitrones.

To date, other chiral Lewis acid catalysts using copper(II)/bisoxazoline [41], titanium(IV)/diol [42, 43] and palladium(II)/tol-BINAP [44] have been applied to the inverse electron demand 1,3-DC of nitrones and vinyl ether.

Recently, the state of the art chiral Brønsted acid catalyst pushed forward this premature asymmetric transformation [45, 46]. Yamamoto and coworkers reported that chiral *N*-phosphoramide **60** facilitates the enantioselective inverse electron demand 1,3-DC of nitrones, especially nitrones bearing electron-withdrawing *N*-and *C*-substituents, and ethyl vinyl ether. Endo-cycloadducts were obtained predominantly with high enantioselectivities ranging from 70 to 93% ee (Scheme 3.33).

The endo-selectivity observed in Yamamoto's Brønsted acid-catalyzed reaction contrasts sharply with Jørgensen's chiral aluminum Lewis acid catalysis, which generally gave exo-cycloadducts. This observation was rationalized by considering the difference in the steric environment in the transition state structure (Scheme 3.34).

Use of allylic alcohol as dipolarophile in the inverse electron demand 1,3-DC of nitrones has been also developed. In 1997, Ukaji, Inomata and coworkers found that the zinc complex of (*R*,*R*)-DIPT (DIPT = diisopropyl tartrate) mediated the reaction of allyl alcohol and nitrones possessing an electron-withdrawing C-substituent (CN, CO_2tBu) to give the *trans*-isoxazolidine **62** with high enantioselectivity

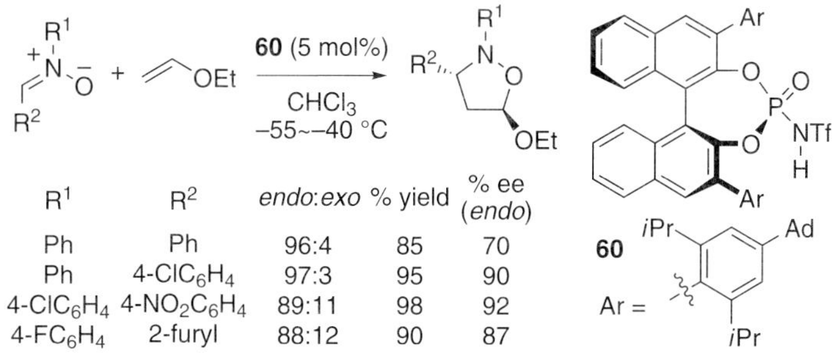

3.2 Nitrones

Scheme 3.33 Brønsted acid-catalyzed enantioselective inverse electron demand 1,3-DC of nitrones.

R^1	R^2	endo:exo	% yield	% ee (endo)
Ph	Ph	96:4	85	70
Ph	4-ClC$_6$H$_4$	97:3	95	90
4-ClC$_6$H$_4$	4-NO$_2$C$_6$H$_4$	89:11	98	92
4-FC$_6$H$_4$	2-furyl	88:12	90	87

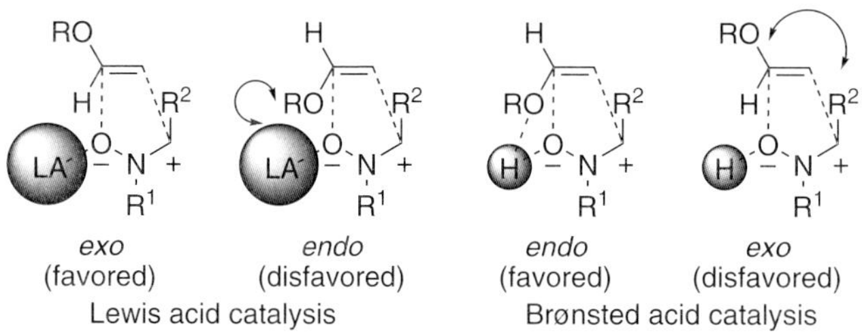

Scheme 3.34 Explanation of the stereochemical outcome of Brønsted acid-catalyzed inverse electron demand 1,3-DC of nitrones.

(Scheme 3.35) [47]. As a working model of this reaction, zinc chelated complex **61** was suggested as shown below. Additionally, scrutiny of the reaction conditions disclosed that only a catalytic amount of diisopropyl tartrate was required for successful implementation.

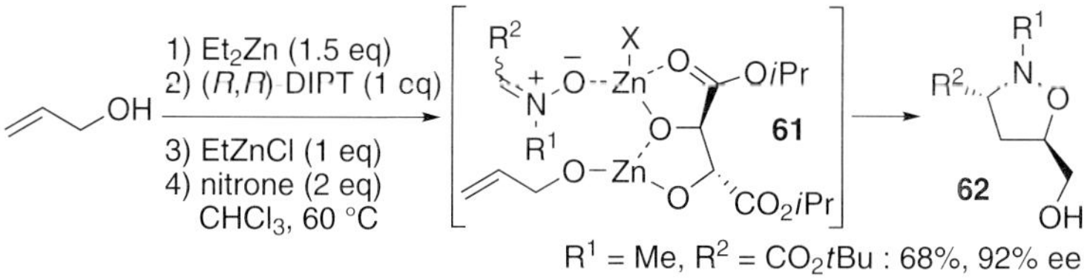

1) Et$_2$Zn (1.5 eq)
2) (R,R) DIPT (1 eq)
3) EtZnCl (1 eq)
4) nitrone (2 eq)
CHCl$_3$, 60 °C

R^1 = Me, R^2 = CO$_2$tBu : 68%, 92% ee

Scheme 3.35 Use of allylic alcohol in enantioselective inverse electron demand 1,3-DC of nitrones.

3.2.6
Kinugasa Reaction

In 1972, Kinugasa and Hashimoto reported a novel approach for the synthesis of β-lactams **65**, using nitrones and copper phenylacetylide **63** as substrates (Scheme 3.36) [48, 49]. The mechanistic perspective was later provided by Ding and Irwin in 1976 [50], wherein 1,3-DC of nitrone and copper acetylide and the

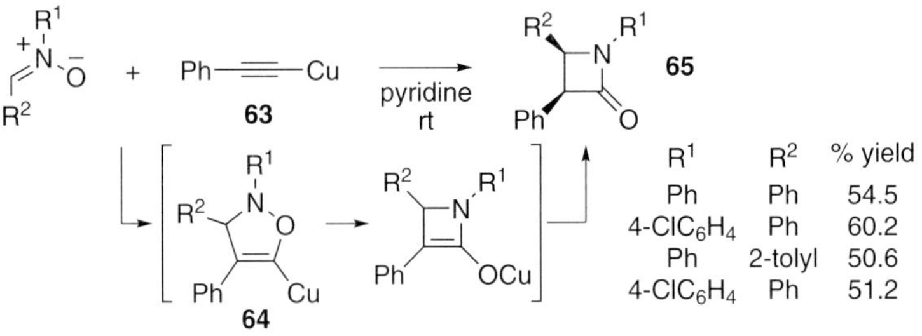

3 1,3-Dipolar Cycloaddition

Scheme 3.36 Kinugasa reaction.

R¹	R²	% yield
Ph	Ph	54.5
4-ClC$_6$H$_4$	Ph	60.2
Ph	2-tolyl	50.6
4-ClC$_6$H$_4$	Ph	51.2

successive rearrangement of the [3 + 2]-adduct **64** and protonation was suggested as a plausible pathway.

The pioneering work of Miura and coworkers revealed the catalytic and asymmetric Kinugasa reaction [51]. They demonstrated that a catalytic system consisting of copper(I) iodide/dppe (dppe = 1,2-bis(diphenylphosphino)ethane) complex and K$_2$CO$_3$ promoted the reaction of nitrones and terminal acetylenes, giving β-lactams. In their reaction conditions, copper acetylide was catalytically generated *in situ* from the terminal acetylene and copper complex. Use of a catalytic amount of copper(I)/chiral bisoxazoline or copper(I)/(−)-sparteine complex in this Kinugasa reaction was found to furnish β-lactams with moderate asymmetric induction.

A highly enantioselective catalytic system was reported in 2002 from Fu's laboratory [52]. By the accommodation of bisazaferrocene ligand **66**, originally developed in this group with copper(I) chloride, a series of *cis*-β-lactams could be obtained in moderate yields with high enantioselectivities (Scheme 3.37). Among the several N-substituents of the nitrones screened, the 4-anisyl group exerted the highest enantioselectivity. The broad functional group tolerance was revealed as for the C-substituent of nitrones and acetylenes.

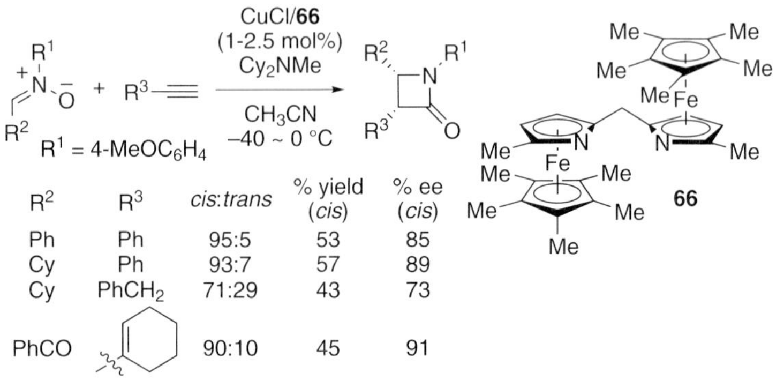

R¹ = 4-MeOC$_6$H$_4$

R²	R³	cis:trans	% yield (cis)	% ee (cis)
Ph	Ph	95:5	53	85
Cy	Ph	93:7	57	89
Cy	PhCH$_2$	71:29	43	73
PhCO	(cyclohexyl)	90:10	45	91

Scheme 3.37 Copper-catalyzed enantioselective Kinugasa reaction.

Intramolecular asymmetric Kinugasa reaction could be accomplished by the use of substrates (e.g., **67**) having terminal acetylene and nitrone moieties in an appropriate position, as reported by Shintani and Fu [53]. Under the influence of copper(I)/phosphaferrocene-oxazoline **68** catalyst, β-lactams with six or seven-membered

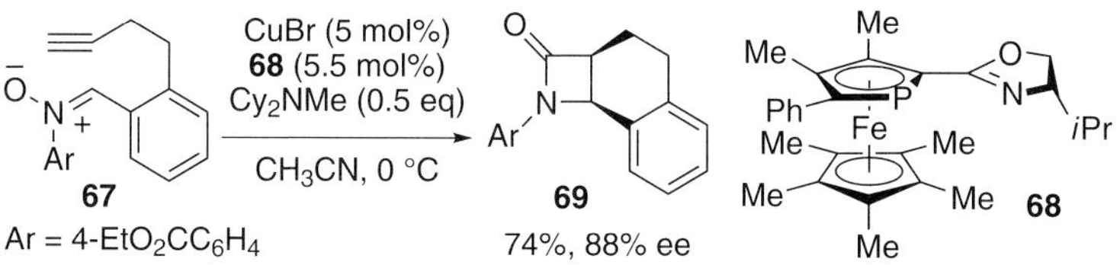

Scheme 3.38 Copper-catalyzed enantioselective intramolecular Kinugasa reaction.

carbocycles (e.g., **69**) were obtained in high yields with enantiomeric excesses ranging from 85 to 91% (Scheme 3.38).

Following Fu's report, Tang and coworkers applied their TOX ligand **45** in a copper (I)-catalyzed asymmetric Kinugasa reaction [54]. The judicious choice of an appropriate amine as base was important to give the β-lactams with high enantioselectivities (Scheme 3.39).

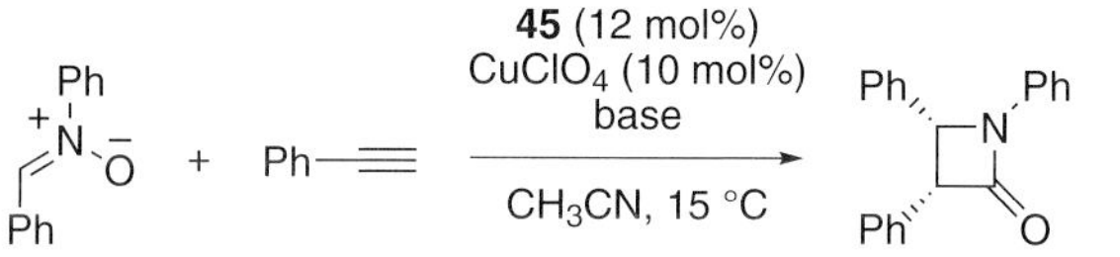

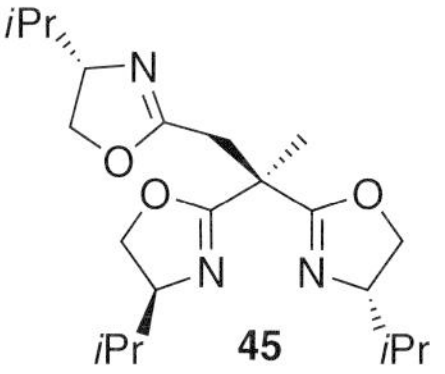

base	cis:trans	% yield (cis)	% ee (cis)
iPr$_2$NEt	>99:1	51	56
Cy$_2$NMe	96:4	45	58
iPr$_2$NH	90:10	61	72
Cy$_2$NH	93:7	63	80
iPrNH$_2$	83:17	39	59

Scheme 3.39 Effect of base in copper/TOX complex-catalyzed enantioselective Kinugasa reaction.

Recently, application of a chiral P,N-ligand in the asymmetric Kinugasa reaction was reported by Guiry *et al.* with limited success [55].

3.3
Azomethine Ylides

1,3-DC of azomethine ylides started from Huisgen and coworkers' investigation in which azomethine ylide **71** was generated by the treatment of *N*-alkyl-3,4-dihydroquinolinium bromide **70** with base and then scavenged by dimethyl fumarate (Scheme 3.40) [56]. Many methods for the generation of azomethine ylides as a transient species and successive 1,3-DC *in situ* with dipolarophiles have been

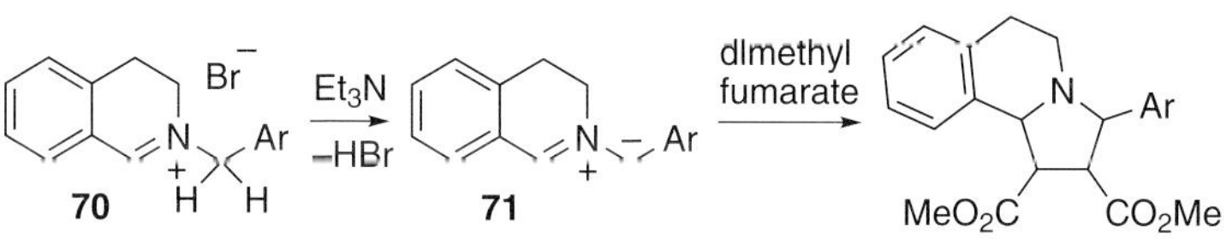

Scheme 3.40 Huisgen's report on 1,3-DC of azomethine ylide.

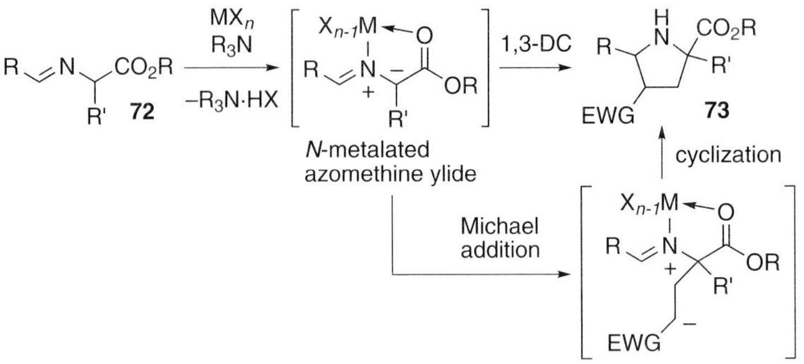

Scheme 3.41 Schematic explanation of 1,3-DC of N-metalated azomethine ylides.

developed to date, due to their synthetic value, especially as a means of furnishing highly functionalized pyrrolidine derivatives in a single operation [57].

Metal-mediated generation of azomethine ylides (N-metalated azomethine ylides) derived from the aldimine of α-amino acid esters and tandem 1,3-DC with electron-deficient olefins were extensively studied by Grigg and coworkers in the late 1980s as a facile means to provide pyrolidine derivatives in a regio- and stereodefined manner (Scheme 3.41) [58–61]. Mechanistically, the aldimine of α-amino acid ester **72** is considered to coordinate to the Lewis acidic metal salt first and then the α-hydrogen of the ester is deprotonated by the added base to generate N-metalated azomethine ylide. Subsequently, the thus-generated azomethine ylide undergoes 1,3-DC normally with electron-deficient olefin to give the pyrrolidine derivative **73**. In this regard, readers should be aware of the possibility that the 1,3-DC of azomethine ylides is not necessarily concerted. The stepwise Michael addition–cyclization sequence could be operative in some cases [62, 63].

In this section, the main focus is on the recent development of asymmetric 1,3-DC of N-metalated azomethine ylides derived from the imine of α-amino acid esters and electron-deficient olefins. The rise of cutting-edge organocatalysis in this area is also described. In the last part of this section, some novel approaches for the generation of azomethine ylides will be introduced.

3.3.1
Early Examples of Asymmetric 1,3-DC of N-Metalated Azomethine Ylides Derived from Aldimines of α-Amino Acid Esters

The use of a chiral metal complex in 1,3-DC of N-metalated azomethine ylides was disclosed in 1991 by Grigg's group [64]. Stoichiometric amounts of cobalt(II)/(1R,2S)-N-methylephedrine **76** complex facilitated the 1,3-DC of azomethine ylide derived from aldimines of glycine methyl ester **74** and methyl acrylate **75** to give the endo-cycloadducts **77** with high enantioselectivities (Scheme 3.42).

A stimulus work which became the basis of the current upsurge of catalytic asymmetric 1,3-DC of N-metalated azomethine ylides appeared in 1995 [65]. Grigg reported that the silver(I)/bisphosphine **80** complex catalyzes the enantioselective cyclization of N-(2-naphthylmethylidene)alanine methyl ester **78** with methyl vinyl

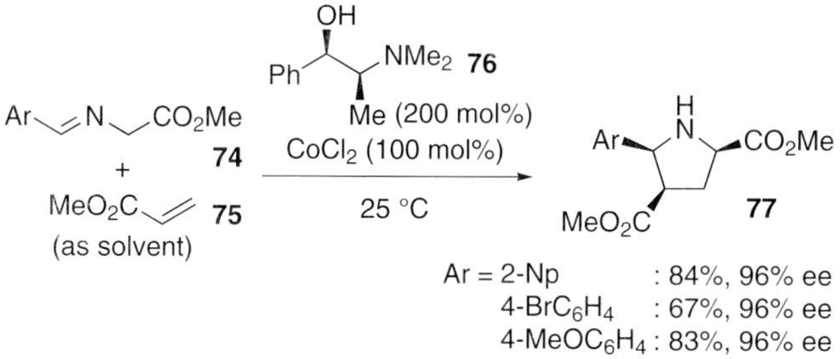

Scheme 3.42 Pioneering work on enantioselective 1,3-DC of N-metalated azomethine ylides mediated by chiral cobalt complex.

ketone **79a** or phenyl vinyl sulfone **79b**, although the enantiomeric excess of the cycloadduct **81** remained at a moderate level and no experimental detail was included in the article (Scheme 3.43).

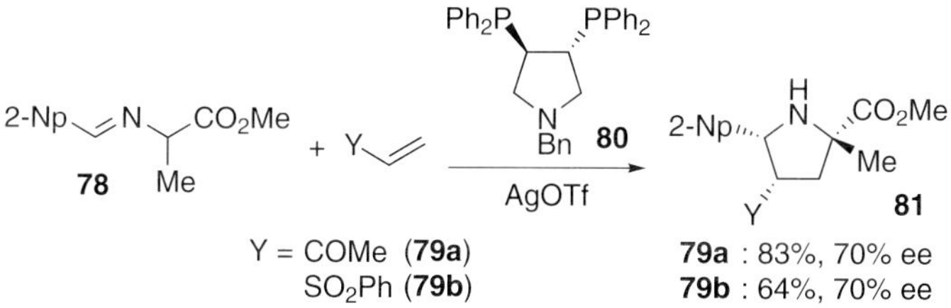

Scheme 3.43 Silver-catalyzed enantioselective 1,3-DC of N-metalated azomethine ylides.

3.3.2
Development of Asymmetric 1,3-DC of N-Metalated Azomethine Ylides with α,β-Unsaturated Carbonyl Compounds

Zhang's group and Jørgensen's group shed light on this premature chemistry concurrently in 2002. Zhang et al. re-examined Grigg's study, focusing especially on the finding of the appropriate bisphosphine ligand in 1,3-DC of N-arylidene glycine methyl ester (**74**, R = aryl) and dimethyl maleate **82** (Scheme 3.44) [66].

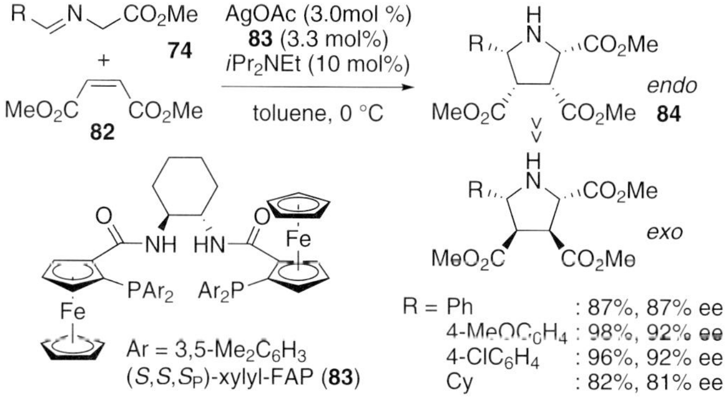

Scheme 3.44 Silver-catalyzed highly enantioselective 1,3-DC of N-metalated azomethine ylides.

Among a variety of ligands screened, they found the effectiveness of (S,S,S_p)-xylyl-FAP ligand **83**, originally designed by this group as a surrogate of Trost ligand having two planar chiral ferrocene moieties. With this optimal catalyst, a variety of N-arylidene glycine methyl esters (**74**, R = aryl) were converted to the corresponding pyrrolidines **84** as a single endo-isomer with high enantioselectivity. Use of N-alkylidne glycine methyl ester (e.g., **74**, R = Cy) neccesitated the longer reaction time, and the endo-cycloaddcut was obtained with slightly lower enantioselectivity.

The use of other electron-deficient olefins, such as dimethyl fumarate, methyl acrylate and N-methylmaleimide, as dipolarophiles was also explored to obtain the pyrrolidine derivatives **85a** to **85c**, respectively, as illustrated in Scheme 3.45.

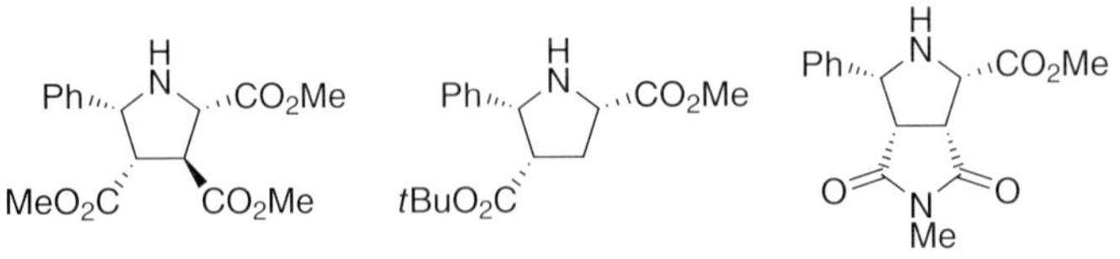

85a : 88%, 52% ee   **85b** : 85%, 93% ee   **85c** : 87%, 79% ee

Scheme 3.45 Scope of silver-catalyzed 1,3-DC of N-metalated azomethine ylides.

Jørgensen et al. focused on the use of the bisoxazoline ligand **12** for asymmetric 1,3-DC of azomethine ylide [67]. Examination of several metal sources revealed the inefficiency of the copper(II) salt which only led to the formation of racemate. On the other hand, the zinc(II) triflate/bisoxazoline **12** complex facilitated the reaction with a promising level of enantioselectivities (Scheme 3.46). Acrylate (**86**, R^1 = H) and crotonate (**86**, R^1 = Me) were employed as dipolarophiles, giving tri- or tetrasubstituted pyrolidines in an endo-selective manner.

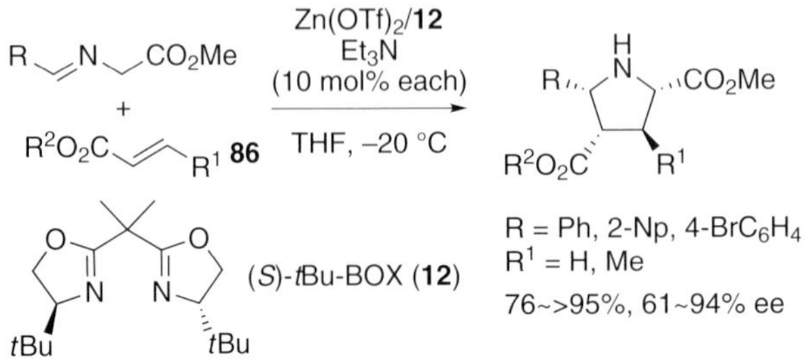

R = Ph, 2-Np, 4-BrC$_6$H$_4$
R^1 = H, Me
76~>95%, 61~94% ee

Scheme 3.46 Zinc-catalyzed highly enantioselective 1,3-DC of N-metalated azomethine ylides.

Due to the facile accessibility to various stereo-defined tri- or tetra-substituted pyrrolidines by this method, it is considered to be a proper choice of the reaction for the diversity-oriented synthesis. At this point, Schreiber and coworkers screened some commercially available mono- and bisphosphine ligands and P,N-ligand under the reaction conditions developed by Zhang to find a simple and practical

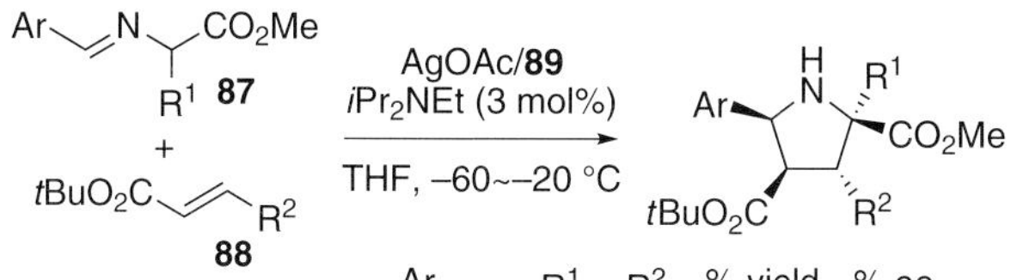

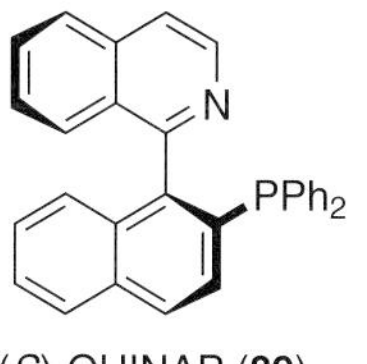

Scheme 3.47 Use of QUINAP ligand in silver-catalyzed 1,3-DC of N-metalated azomethine ylides.

Ar	R^1	R^2	% yield	% ee
Ph	H	H	84	91
4-MeC$_6$H$_4$	H	H	93	95
4-BrC$_6$H$_4$	H	H	89	95
Ph	H	Me	97	84
Ph	H	Ph	62	81 (endo:exo = 2:1)
Ph	Me	H	98	80
Ph	iBu	H	77	80
Ph	Bn	H	93	77

procedure [68]. The use of QUINAP **89** was found to be optimal to obtain endo-cycloadducts with high enantioselectivities in the 1,3-DC of N-arylidene glycine methyl ester (**87**, R^1 = H) and *tert*-butyl acrylate (**88**, R^2 = H) (Scheme 3.47). The use of crotonate (**88**, R^2 = Me) as dipolarophile also furnished the cycloadduct with high endo- and enantioselectivity, whereas the use of cinnamate (**88**, R^2 = Ph) led to the formation of the regioisomeric mixture. The catalytic system could also be applied to the reaction of the aldimine of α-alkyl-α-amino acid esters (**87**, R^1 = alkyl), thus enabling the formation of pyrrolidines with a chiral quaternary stereogenic center.

Carreira and coworkers designed a novel P,N-ligand **90** as an easily synthesized QUINAP surrogate and examined its viability briefly in the 1,3-DC of azomethine ylide [69]. The catalyst showed a comparable level of asymmetric induction to that reported by Schreiber (Scheme 3.48).

The examples of asymmetric 1,3-DC of azomethine ylide described above were limited to endo-selective cyclizations. In this context, Komatsu and coworkers discovered in their exploratory study that 1,3-DC of azomethine ylide derived from N-benzylidene glycine methyl ester and phenylmaleimide catalyzed by Cu(OTf)$_2$ produces the *exo*-cycloadduct *rac*-**93a** preferentially [70]. With this finding, they screened several chiral bisphosphine ligands and settled on the use of BINAP **91** and SEGPHOS **92** as an appropriate choice, giving pyrrolidines with high exo- and

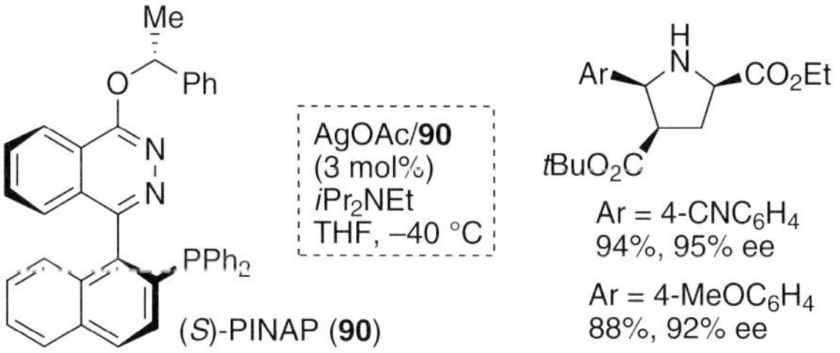

Scheme 3.48 Use of PINAP ligand in silver-catalyzed 1,3-DC of N-metalated azomethine ylides.

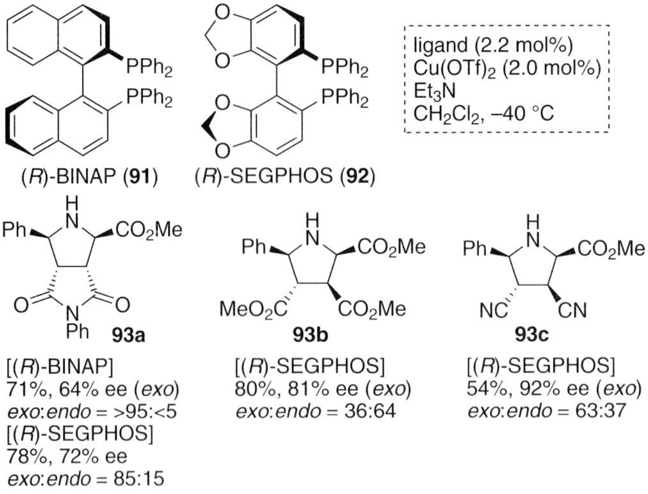

3 1,3-Dipolar Cycloaddition

Scheme 3.49 Copper/bisphosphine complex-catalyzed exo- and enantioselective 1,3-DC of N-metalated azomethine ylides.

enantioselectivities (Scheme 3.49). The optimized conditions could be applied to other dipolarophiles such as dimethyl fumarate and fumaronitrile, giving the cycloadducts **93b** and **93c** respectively with low to moderate exo-selectivities.

This exo-selectivity was explained as a result of the unfavorable steric repulsion between the catalyst and the phenyl moiety of the maleimide encountered in the case of the endo-transition state (Scheme 3.50).

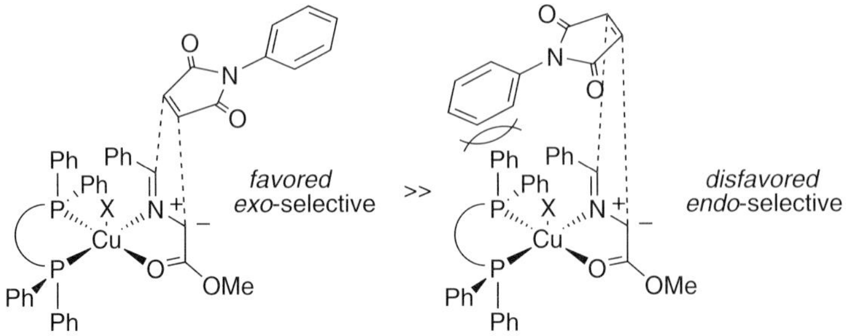

Scheme 3.50 Explanation of exo-selectivity.

Later, Zhang and coworkers utilized copper(I)/P,N-ligand **94** complex to realize the exo-selective 1,3-DC of azomethine ylide [71]. Under the influence of the planar chiral P,N ligand **94** and $CuClO_4$, the cycloaddition of various N-arylidene glycine methyl esters and *tert*-butyl acrylate gave the *exo*-adducts **95** with enantioselectivities ranging from 84 to 98% (Scheme 3.51). The reaction of dimethyl maleate also furnished the cycloadduct **96** with high exo-selectivity.

The transient N-metalated azomethine ylide of this copper(I)-catalyzed reaction is supposed to form a tetrahedral arrangement at the metal center. As such, attack of

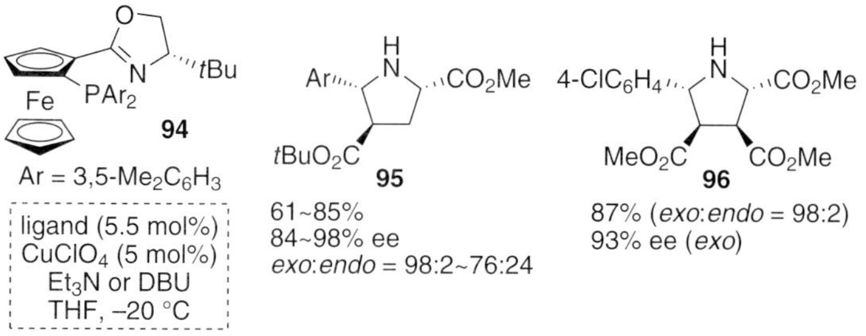

3.3 Azomethine Ylides

94: Fe-ferrocene with oxazoline (tBu) and PAr₂, Ar = 3,5-Me₂C₆H₃

ligand (5.5 mol%)
CuClO₄ (5 mol%)
Et₃N or DBU
THF, −20 °C

95: Ar....N(H)....CO₂Me, tBuO₂C-pyrrolidine
61~85%
84~98% ee
exo:endo = 98:2~76:24

96: 4-ClC₆H₄....N(H)....CO₂Me, MeO₂C, CO₂Me
87% (exo:endo = 98:2)
93% ee (exo)

Scheme 3.51 Copper/P,N-ligand complex-catalyzed exo- and enantioselective 1,3-DC of N-metalated azomethine ylides.

the diplarophile from the endo-approach is hampered by the bulk of the ligand (Scheme 3.52).

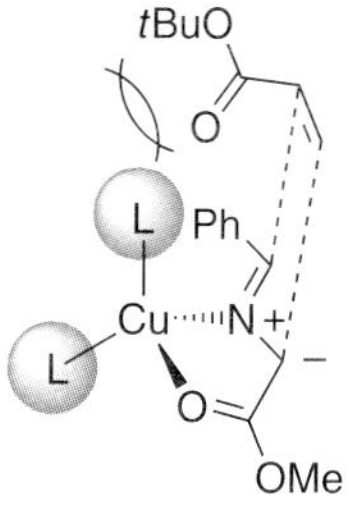

Scheme 3.52 Explanation of the observed exo-selectivity.

Carretero and coworkers accommodated planar chiral P,S-ligand Fesulphos **99** in combination with Cu(CH₃CN)₄ClO₄ to asymmetric 1,3-DC of azomethine ylides [72]. The clear-cut advantage of this metal–ligand combination is the broad scope, concerning both azomethine ylides and dipolarophiles. In the reaction with N-phenylmaleimide **98**, not only the aldimine of glycine ester (**97**, $R^1 = R^2 = H$) but also the aldimine of α-alkyl-α-amino acid ester (**97**, $R^1 = H$, $R^2 =$ alkyl), and the ketimine of glycine ester (**97**, $R^1 = $ Ph or Me, $R^2 = H$), could be successfully converted to the corresponding pyrrolidines in an endo-selective manner with high asymmetric induction (Scheme 3.53).

For dipolarophiles, the use of maleate, fumarate, fumaronitrile and acrylate all provided the corresponding pyrrolidines **100a** to **100d** with moderate to high endo- and enantioselectivities (Scheme 3.54). The inversion of endo/exo selectivity was observed when nitrostyrene was used as the dipolarophile, furnishing the almost single enantiomer of the *exo*-cycloadduct **100e** exclusively. Particularly noteworthy is the use of methacrolein as dipolarophile, which gave the endo-cycloadduct **100f** with an all-carbon quaternary center at C4, although the yield and ee remained at a moderate level.

In contrast to the intensive use of the aldimine of the glycine ester as the N-metalated azomethine ylide precursor, aldimines of other α-alkyl-α-amino acid esters have not been fully explored. In this regard, Nájera et al. succeeded in the highly

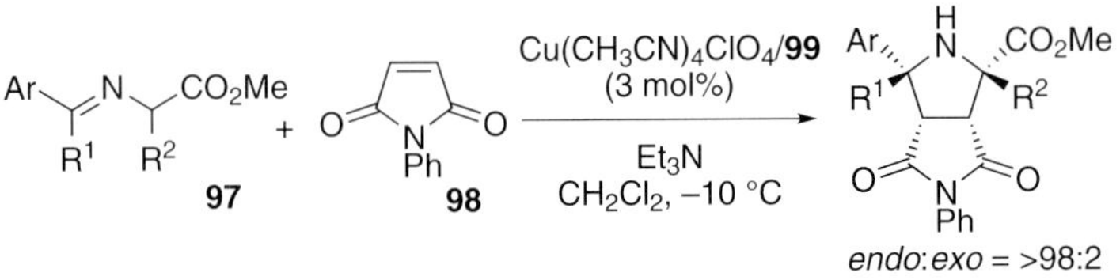

97
98
Cu(CH3CN)4ClO4/99
(3 mol%)
Et3N
CH2Cl2, –10 °C
endo:exo = >98:2

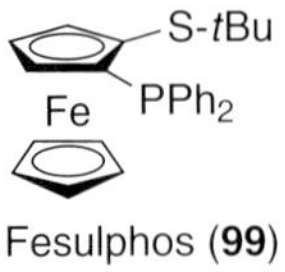

Fesulphos (**99**)

Ar	R^1	R^2	% yield	% ee
Ph	H	H	81	>99
4-FC_6H_4	H	H	82	>99
4-$MeOC_6H_4$	H	H	81	>99
Ph	H	Me	50	80
2-Np	H	Me	78	92
Ph	Ph	H	92	93
Ph	Me	H	80	94

Scheme 3.53 Use of copper/Fesulphos complex as a catalyst with a broad generality.

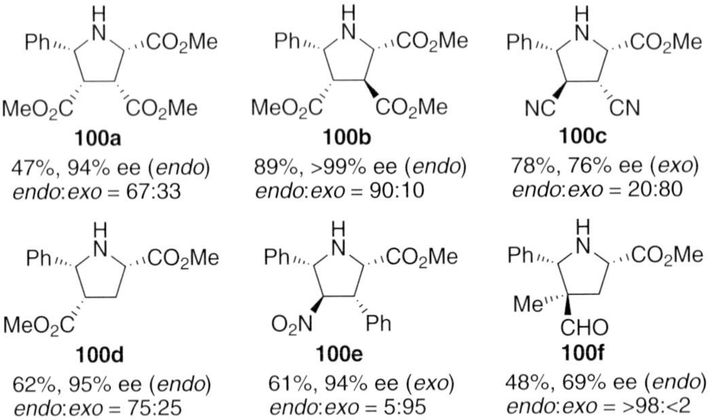

100a
47%, 94% ee (endo)
endo:exo = 67:33
100b
89%, >99% ee (endo)
endo:exo = 90:10
100c
78%, 76% ee (exo)
endo:exo = 20:80
100d
62%, 95% ee (endo)
endo:exo = 75:25
100e
61%, 94% ee (exo)
endo:exo = 5:95
100f
48%, 69% ee (endo)
endo:exo = >98:<2

Scheme 3.54 Scope of copper/Fesulphos complex-catalyzed enantioselective 1,3-DC of *N*-metalated azomethine ylides.

enantioselective 1,3-DC of azomethine ylide derived from aldimines of α-alkyl-α-amino acid esters by the use of the silver(I)/phosphoramidite **101** complex as catalyst (Scheme 3.55) [73].

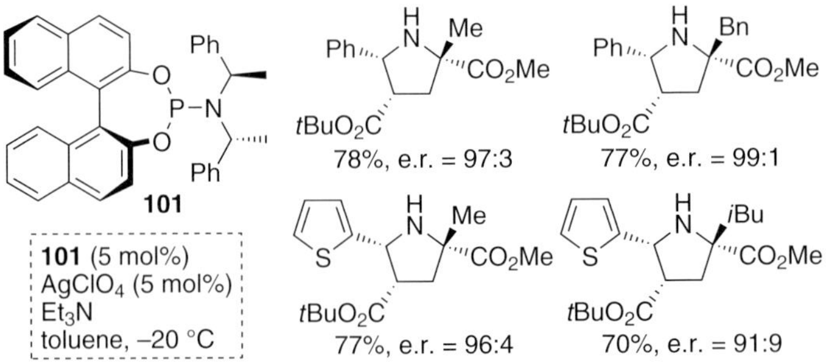

101
101 (5 mol%)
AgClO4 (5 mol%)
Et3N
toluene, –20 °C
78%, e.r. = 97:3
77%, e.r. = 99:1
77%, e.r. = 96:4
70%, e.r. = 91:9

Scheme 3.55 Silver/phosphoramidite complex-catalyzed 1,3-DC with aldimines of α-alkyl-α-amino acid esters.

A unique pair of chiral P,N-ligands **102** was discovered in 2007 by Zhou, Li and coworkers [74], which has the same absolute configuration at the basic domain of the ligand but exerts the opposite sense of asymmetric induction in 1,3-DC of azomethine ylide using AgOAc as a metal source. The only difference in these ligands is the nitrogen substituent; namely, one has the N,N-dimethyl group and another has the unsubstituted amino moiety (Scheme 3.56). The reaction of N-arylidene glycine methyl esters and dimethyl maleate catalyzed by AgOAc/**102a** complex provided the endo-cycloadducts with 88 to 91% ee. On the other hand, use of ligand **102b** led to the formation of the opposite enantiomer with 85 to 91% ee.

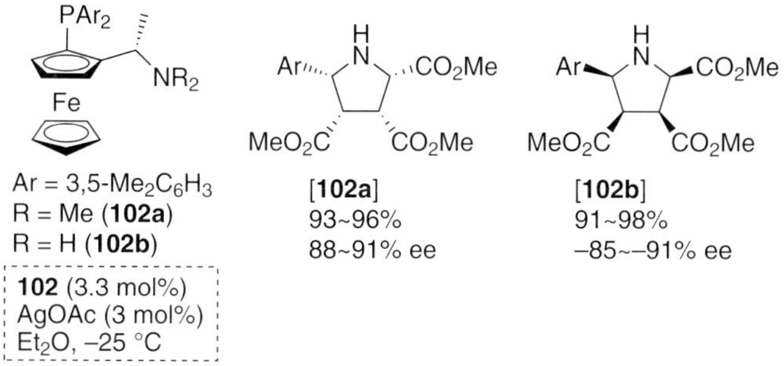

Scheme 3.56 Reversal of the asymmetric induction directed by the hydrogen-bonding.

In the case of catalysis with ligand **102a**, the N,N-dimethylamino group would shield one prochiral face of the glycinate (SM^1) and the maleate (SM^2) approaches from the bottom face of the transition state A, whereas the amino moiety of **102b** acts as a hydrogen-bonding donor, thereby rigidifying the direction of the approaching maleate from the top face of transition state B (Scheme 3.57).

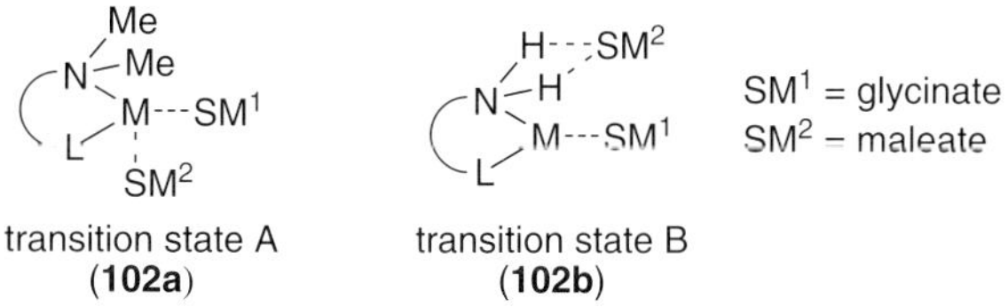

Scheme 3.57 Explanation of the reversal of the enantioselectivity.

In addition to the examples described above, there have been some other catalytic systems developed to date. Jørgensen *et al.* developed the catalytic system composed of AgF and hydrocinchonine, which can be manipulated without the requirement of dried, degassed solvent and inert atmosphere [75]. Zhou and coworkers discovered that the use of AgOAc as a metal source allowed 1,3-DC of azomethine ylide without external base [76]. Chiral P,N- and P,S-ferrocenyl ligands were accommodated in this catalytic system. Silver(I)/BINAP complex was introduced as a recyclable catalyst by Nájera and Sansano [77]. Shi and coworkers developed nickel(II)/binaphthyldiimine and copper(I)/thiophoshporamide complexes for the asymmetric 1,3-DC of

3 1,3-Dipolar Cycloaddition

azomethine ylide with maleimides [78]. Dogan, Garner and coworkers designed a chiral aziridino alcohol ligand and applied it as catalyst in combination with Zn (OTf)$_2$ [79].

An apparent example of an asymmetric Michael addition–cyclization sequence (see, Scheme 3.41) was developed by Kobayashi and coworkers for the preparation of pyrrolidine derivatives starting from aldimine and ketimine of amino acid esters [80]. This formal [3 + 2] cycloaddition was accomplished in a highly stereoselective manner by the use of Ca(OiPr)$_2$/bisoxazoline **106** complex (Scheme 3.58). For example, the catalytic reaction of benzophenoneimine of glycine methyl ester **103** and methyl crotonate **104** afforded the almost enantiopure pyrrolidine derivative **107**. Simple exchange of the Michael acceptor by methyl acrylate **105** switched the result of the reaction, giving the Michael adduct **108**.

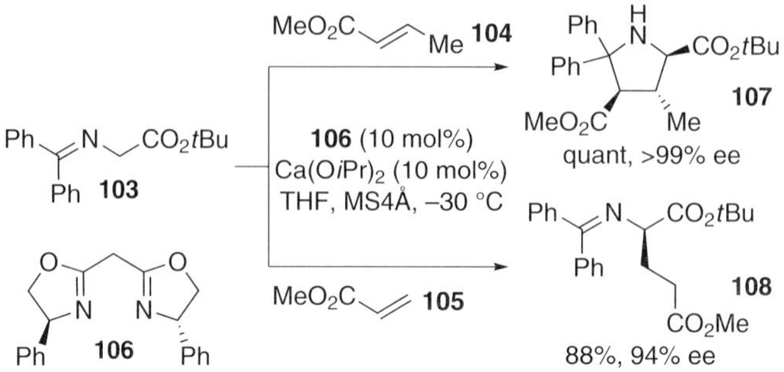

Scheme 3.58 Enantioselective Michael addition–cyclization sequence.

Intramolecular asymmetric 1,3-DC of azomethine ylide was reported by Pfaltz et al. using a catalytic system composed of AgOAc and PHOX **110** (PHOX = phosphinooxazoline) ligand [81]. Starting from aryl ether **109**, the tricyclic compounds **111** could be obtained as a single diastereomer with high enantioselectivities (Scheme 3.59). It seems to be of interest that 1,3-DC of such a harnessed molecule proceeds in a reversed sense of regioselectivity compared to that observed in the case of intermolecular reactions, opposing the electronic control.

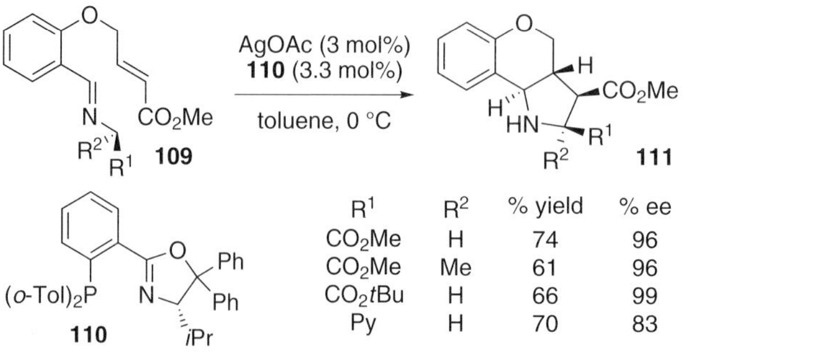

R^1	R^2	% yield	% ee
CO$_2$Me	H	74	96
CO$_2$Me	Me	61	96
CO$_2t$Bu	H	66	99
Py	H	70	83

Scheme 3.59 Intramolecular enantioselective 1,3-DC of N-metalated azomethine ylides.

3.3.3
Asymmetric 1,3-DC of N-Metalated Azomethine Ylides with Vinyl Sulfones and Nitroalkenes

Use of olefins bearing electron-withdrawing hetero-functionality, such as vinyl sulfones and nitroalkenes, in asymmetric 1,3-DC of *N*-metalated azomethine ylide has caught the attention of synthetic organic chemists. Phenyl vinyl sulfone **112** was first utilized in Grigg's pioneering research (see, Scheme 3.43), in which the pyrrolidine derivative having a sulfonyl moiety was obtained with 70% ee. Carretero and coworkers re-examined this reaction by screening various ligands and the metal sources, and identified the combination of Taniaphos **113** and Cu(CH$_3$CN)ClO$_4$ as an optimal catalyst [82]. With this catalyst, the exo-cycloadducts **114** could be obtained with ee values ranging from 41 to 85% (Scheme 3.60).

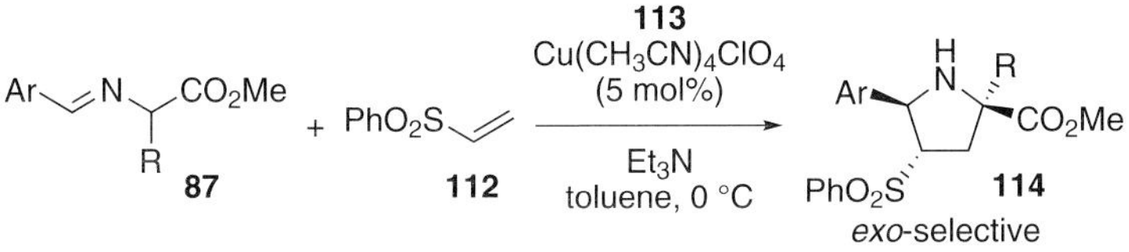

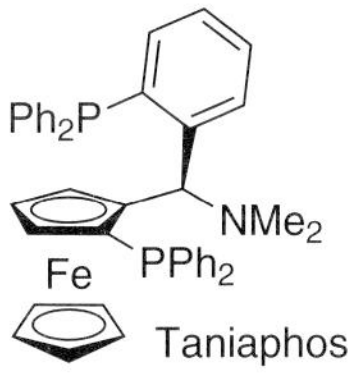

Ar	R	% yield	% ee
Ph	H	87	83
4-FC$_6$H$_4$	H	91	82
4-MeOC$_6$H$_4$	H	71	84
2-tolyl	H	92	41
Cy	H	50	69
Ph	Me	38	80

Scheme 3.60 Enantioselective 1,3-DC of *N*-metalated azomethine ylides with vinyl sulfone.

Following that report, Fukuzawa and Oki applied their ClickFerrophos ligand **115** to this reaction system, considering its structural similarity with Taniaphos [83]. Use of CuOAc allowed access to functionalized pyrrolidines with exceptionally high enantioselectivities (Scheme 3.61). The aldimine of alanine methyl ester could also be transformed into the pyrrolidine with a tertiary amine moiety.

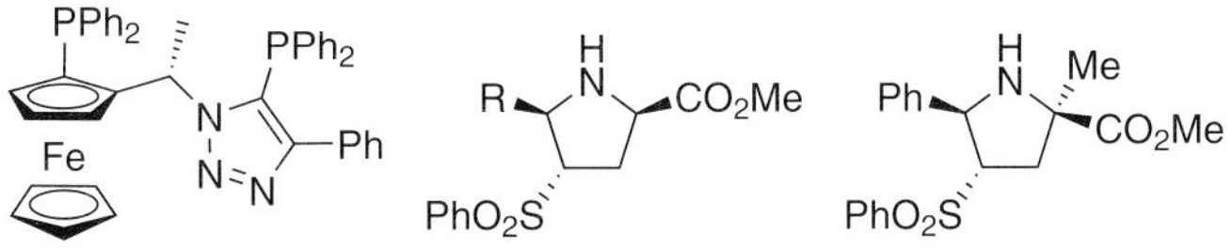

ClickFerrophos (**115**)

115 (5.5 mol%)
CuOAc (5 mol%)
ether, −40 °C

R = aromatic
80~91%, 94~96% ee

R = Cy
81%, 99% ee

83%, 93% ee

Scheme 3.61 Enantioselective 1,3-DC of *N*-metalated azomethine ylides with vinyl sulfone catalyzed by copper/ClickFerrophos complex.

More recently, Carrettero and coworkers introduced *trans*-bisphenylsulfonyl ethylene **116** as the dipolarophile in asymmetric 1,3-DC of azomethine ylides [84]. Aldimines of glycine methyl ester (**87**, R^2 = H) and alanine methyl ester (**87**, R^2 = Me) were converted to the bis-sulfonylated pyrrolidines **117** in high yields and enantioselectivities by the use of copper(I)/Fesulphos **99** catalyst (Scheme 3.62). Exo-cycloadducts were obtained exclusively in all cases. Use of the ketimine of glycine ester, instead of aldimine, was found to be unproductive. Since treatment of the bis-sulfonylated cycloadduct with Na-Hg amalgam led to introduction of the olefin into the pyrrolidine ring, bis-sulfonyl ethylene is regarded as a masked acetylene equivalent.

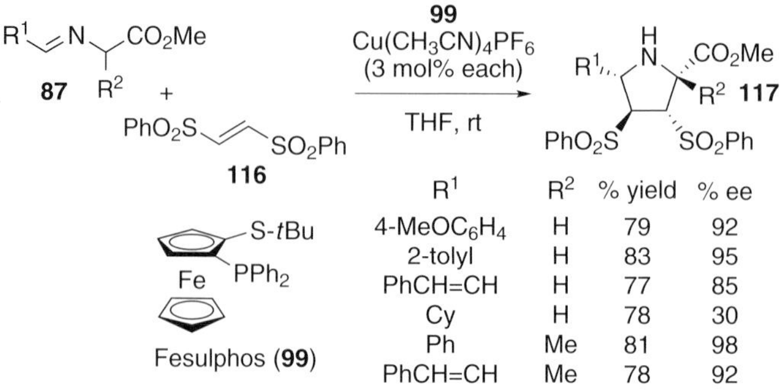

R^1	R^2	% yield	% ee
4-MeOC$_6$H$_4$	H	79	92
2-tolyl	H	83	95
PhCH=CH	H	77	85
Cy	H	78	30
Ph	Me	81	98
PhCH=CH	Me	78	92

Scheme 3.62 Use of *trans*-bisphenylsulfonyl ethylene as a dipolarophile.

Hou and coworkers developed 1,3-DC of azomethine ylides and nitroalkenes **118** catalyzed by the Cu(I)-P,N-ligand **119** complex (see also, Scheme 3.54), wherein the endo/exo selectivity can be switched by changing the electronic properties of the P-substituent of the ligand [85]. The reaction with P,N-chiral ferrocenyl ligand **119a** bearing the simple phenyl moiety furnished the pyrrolidines **120** with exo-selectivity, regardless of the β-substituent on the nitroalkenes (Scheme 3.63). On the other hand,

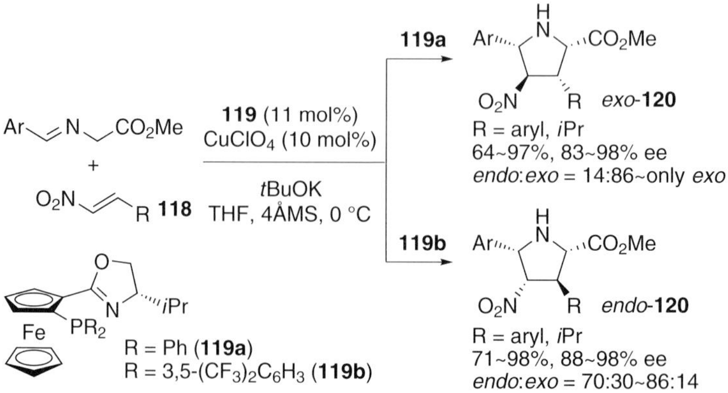

Scheme 3.63 Enantioselective 1,3-DC of N-metalated azomethine ylides with nitroolefins.

use of the ferrocenyl ligand **119b** with the electron-withdrawing aryl moiety led to the preferential formation of endo-cycloadducts **120**. This 1,3-DC is considered to proceed via a stepwise mechanism, since the non-cyclized Michael adduct was obtained as a side product. In this vein, use of strong base was indispensable in minimizing the formation of this undesired Michael adduct.

3.3.4
Organocatalytic Asymmetric 1,3-DC of Azomethine Ylides

As delineated in the above section, organocatalysis has emerged as a promising candidate to facilitate asymmetric 1,3-DC of nitrones and α,β-unsaturated aldehydes via iminium catalysis [24–26]. This ground-breaking report by MacMillan prompted the use of azomethine ylides as a 1,3-dipole in the iminium-catalyzed asymmetric 1,3-DC. In this context, Vicario et al. chose the aldimine of dialkyl 2-aminomalonate as a proper precursor of azomethine ylide, which does not require an external metal source for its generation [86]. Owing to the high acidity of the α-hydrogen atom, it is envisaged that the corresponding azomethine ylides can be spontaneously formed by thermal 1,2-prototropy under rather mild reaction conditions (Scheme 3.64) [87].

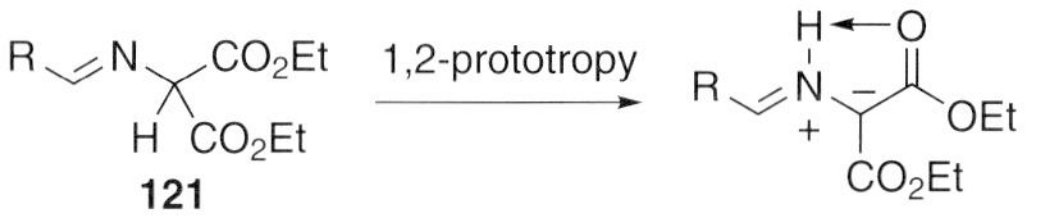

Scheme 3.64 Generation of azomethine ylide via 1,2-prototropy.

After screening several chiral secondary amine catalysts suitable for the LUMO-lowering activation of α,β-unsaturated aldehydes, they settled on the use of diphenylprolinol **123**, with which the enantioenriched pyrrolidines **124** were obtained in a highly endo- and enantioselective manner. A wide range of aldimines **121** and β-substituted enals **122** could be utilized as substrate (Scheme 3.65).

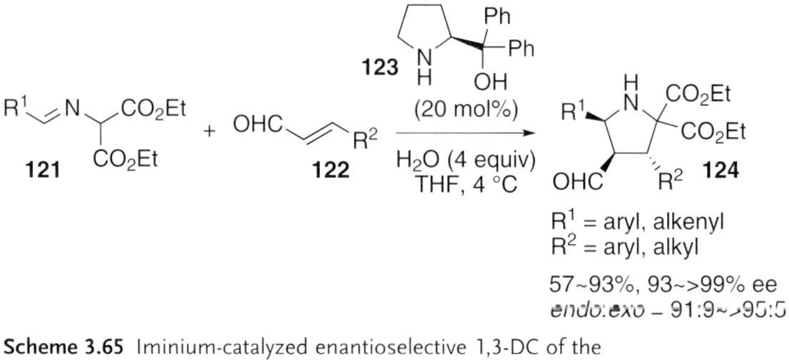

R^1 = aryl, alkenyl
R^2 = aryl, alkyl

57~93%, 93~>99% ee
endo:exo = 91:9~>95:5

Scheme 3.65 Iminium-catalyzed enantioselective 1,3-DC of the aldimine of dialkyl 2-aminomalonates.

A multicomponent version of this catalytic system was developed by Córdova et al. using O-TMS diphenylprolinol **126** as catalyst [88]. *In situ* generation of aldimine from the corresponding aldehyde and 2-aminomalonate **125** allowed one-pot access to the substituted pyrrolidines **124**, favoring the formation of the endo-isomer (Scheme 3.66). Addition of triethylamine led to improvement in the diastereoselectivity.

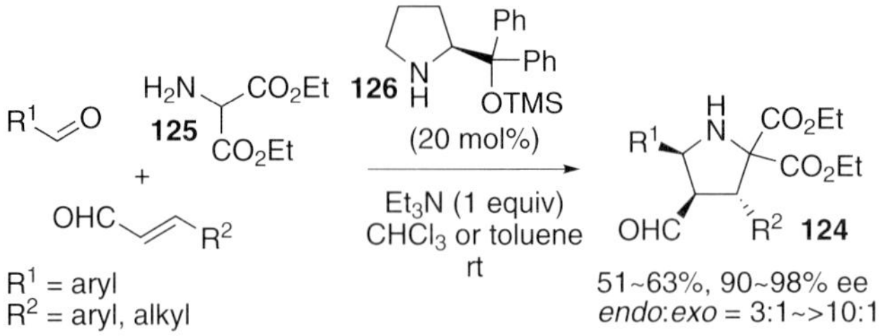

Scheme 3.66 Three-component iminium-catalyzed enantioselective 1,3-DC of the aldimine of dialkyl 2-aminomalonates.

Mechanistically distinctive organocatalytic 1,3-DC of azomethine ylides was reported by Gong et al. using chiral phosphoric acid as catalyst [89]. They believe that phosphoric acid complexes with the substrate to form an intermediary azomethine ylide species, as depicted in Scheme 3.67, and then this complex reacts with the dipolarophile.

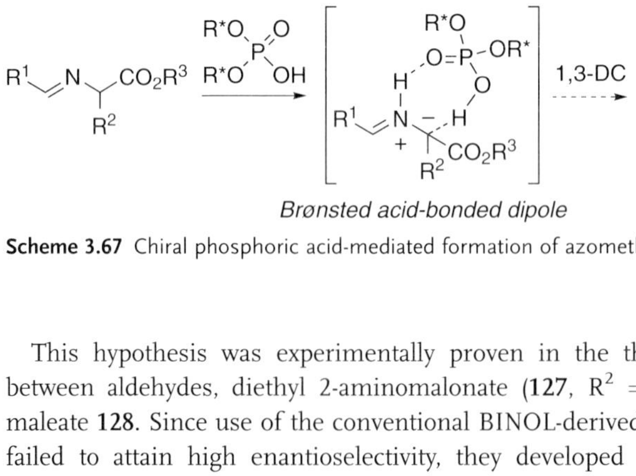

Brønsted acid-bonded dipole

Scheme 3.67 Chiral phosphoric acid-mediated formation of azomethine ylides.

This hypothesis was experimentally proven in the three-component 1,3-DC between aldehydes, diethyl 2-aminomalonate (**127**, $R^2 = CO_2Et$) and dimethyl maleate **128**. Since use of the conventional BINOL-derived monophosphoric acids failed to attain high enantioselectivity, they developed the bisphosphoric acid catalyst **129**, based on linked-BINOL, to overcome this deficiency (Scheme 3.68). With this catalyst, this three-component reaction proceeded highly stereoselectively, giving the endo-isomer **130**, regardless of the aldehydes employed. Not only 2-aminomalonate but also phenylglycine ester (**127**, $R^2 = Ph$) could be utilized as substrate, creating the pyrrolidine with one chiral quaternary stereocenter.

Zhang and coworkers developed asymmetric 1,3-DC between the benzophenone imine of glycine *tert*-butyl ester and nitroalkenes catalyzed by the cinchona alkaloid

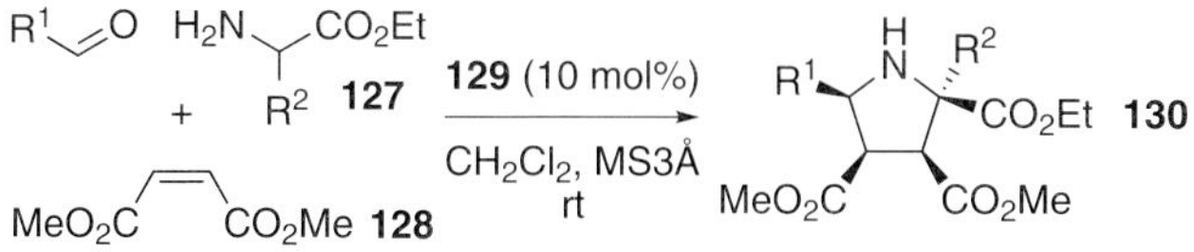

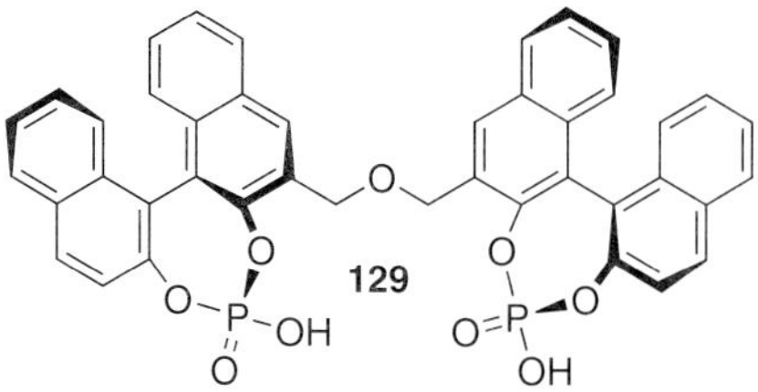

Scheme 3.68 Chiral phosphoric acid-catalyzed enantioselective 1,3-DC of the aldimine of dialkyl 2-aminomalonates.

R^1	R^2	% yield	% ee
2-$NO_2C_6H_4$	CO_2Et	97	96
Ph	CO_2Et	93	91
4-MeC_6H_4	CO_2Et	87	90
4-TolCH=CH	CO_2Et	70	92
Cy	CO_2Et	74	76
4-$NO_2C_6H_4$	Ph	92	97

derived acid–base organocatalyst **131** [90]. Exo-cyloadducts **132** were obtained as the major component with moderate enantioselectivity (Scheme 3.69).

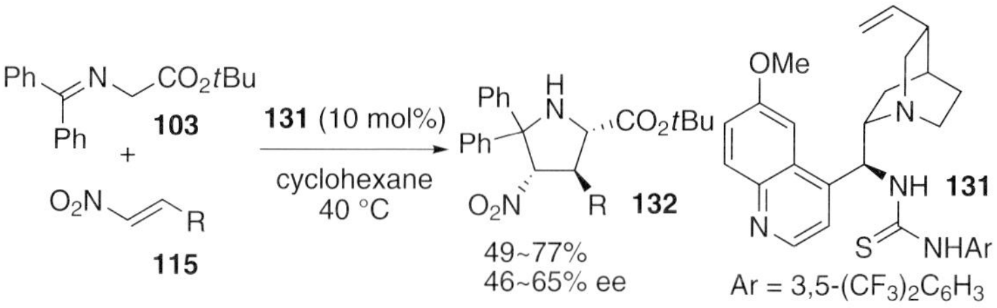

Scheme 3.69 Use of cinchona alkaloid derived acid–base organocatalyst in the reaction between the benzophenone imine of glycine tert-butyl ester and nitroalkenes.

3.3.5
Transition Metal Catalyzed 1,3-DC of Münchnones

Münchnone constitutes a class of mesoionic compounds which have an azomethine ylide moiety in their heterocyclic structure. In 2004, 1,3-DC of münchnone generated from azlactones **133** with olefinic compounds was reported by Tepe's group [91]. Its course of reaction is considered to involve N-metalation and hydrogen-abstraction leading to the N-metalated münchnone **134** as a first step, followed by 1,3-DC with electron-deficient olefin; the transient cycloadduct **135** is protonated to furnish Δ^1-pyrroline carboxylic acid **136**. This free-acid adduct is then esterified by TMSCHN$_2$ to facilitate the isolation of the final product **137** (Scheme 3.70).

In 2007, Toste and coworkers developed an enantioselective version of this 1,3-DC of münchnones in conjunction with their interest in gold catalysis [92]. Screening of various chiral bisphosphinegold(I) complexes led to the finding of the optimal gold catalyst **138** with the (S)-Cy-SEGPHOS ligand for the reaction of azlactones **133**

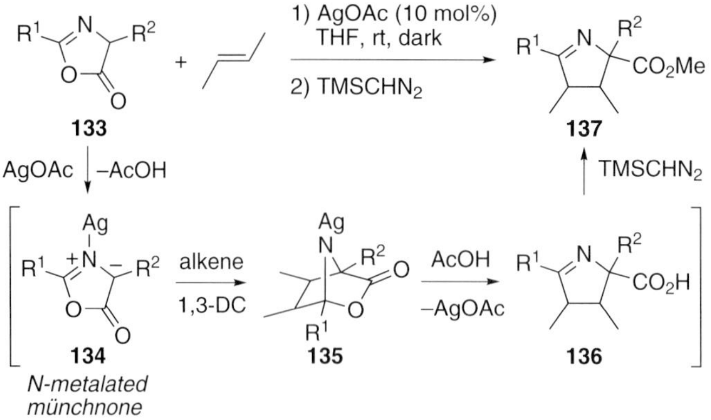

Scheme 3.70 Mechanistic pathway of transition metal-catalyzed 1,3-DC of münchnones.

and N-phenylmaleimide **98**, furnishing the exo isomer of the desired Δ^1-pyrrolines **139** with generally high enantioselectivity (Scheme 3.71).

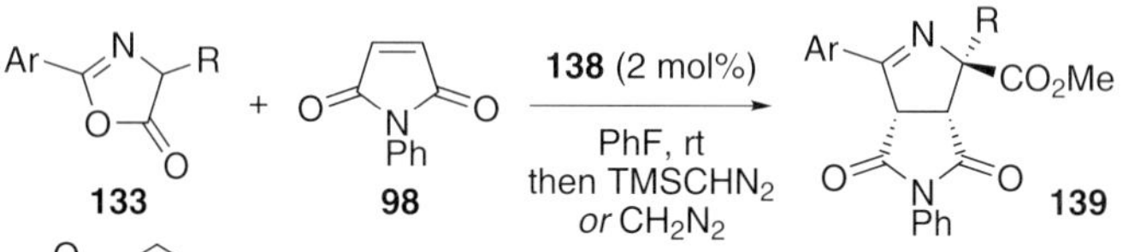

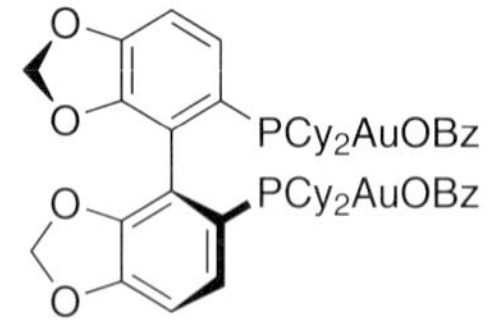

Ar	R	% yield	% ee
4-MeOC$_6$H$_4$	Me	77	95
4-BrC$_6$H$_4$	Me	75	93
2-tolyl	Me	73	86
Ph	allyl	86	87
Ph	Ph	35	78

Scheme 3.71 Gold-catalyzed enantioselective 1,3-DC of münchnones.

The catalytic performance was also examined with several dipolarophiles such as maleic anhydride, acrylate and acrylonitrile. In all cases, the *exo*-cycloadducts **140** were solely produced (Scheme 3.72).

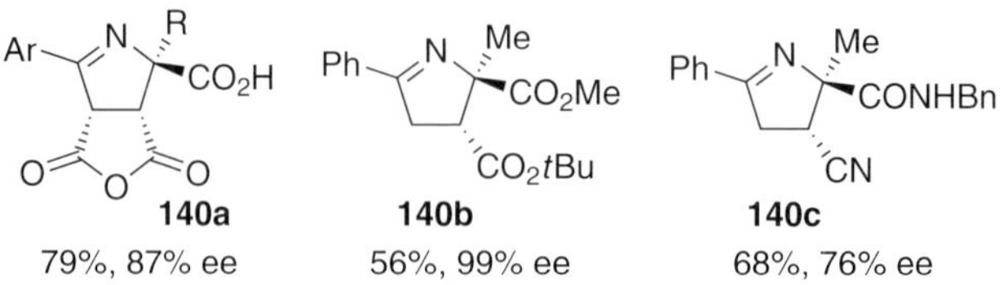

140a 79%, 87% ee
140b 56%, 99% ee
140c 68%, 76% ee

Scheme 3.72 Scope of the substrates.

3.3.6
Lewis Acid Catalyzed 1,3-DC of Azomethine Ylide Derived from the Imidate of 2-Aminomalonate

In 2004, Bowman and Johnson discovered that Lewis acid activation of the imidate of dimethyl 2-aminomalonate **141** leads to the generation of azomethine ylide **142** via 1,2-prototropy [93]. The thus-formed azomethine ylide can react with a broad range of electrophilic double bonds like C=O, C=N and C=C (Scheme 3.73), affording cyclic compounds **143** to **145**, respectively. Intervention of the magnesium(II)-chelated complex in the intermediate was confirmed by the experiment using the magnesium(II) salt having a chiral ligand, although the observed asymmetric induction was rather low.

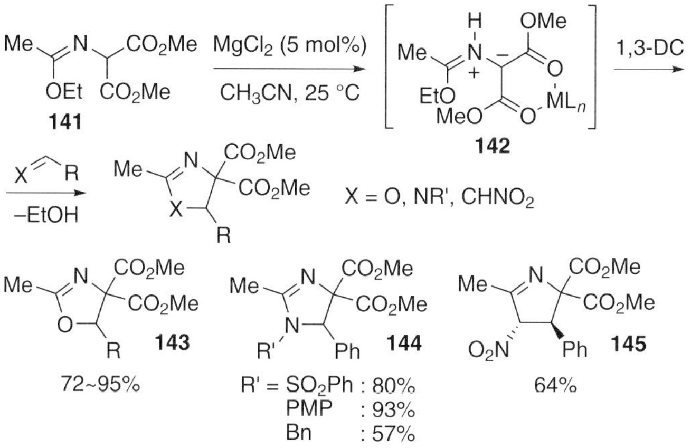

Scheme 3.73 Magnesium Lewis acid-catalyzed 1,3-DC of azomethine ylide derived from the imidate of 2-aminomalonate.

3.3.7
Lewis Acid Promoted Generation of Electrophilic Azomethine Ylides via C—C Bond Cleavage of Aziridines

Classically, thermal or photochemical ring opening of aziridines has been utilized as a way to generate azomethine ylides. One exception is the Lewis acid catalyzed ring opening of aziridine 2,2-dicarboxylate discovered by Vaultier and Carrie in 1978 [94]. Although the intermediacy of azomethine ylide was confirmed in their study, its inability to undergo 1,3-DC with electron-deficient olefin offset the importance of this finding. In 2004, Johnson and coworkers shed light on this unnoticed chemistry by assuming that this sort of azomethine ylide is highly electrophilc and would react with electron-rich alkenes preferentially [95]. As expected, subjection of acyclic vinyl ether as dipolarophile to the azomethine ylide **147** generated from the zinc(II)-promoted ring opening of the aziridine **146** furnished the desired cycloadducts with moderate yields and diastereoselectivities (Scheme 3.74). Mechanistically, both concerted and stepwise pathways are conceivable.

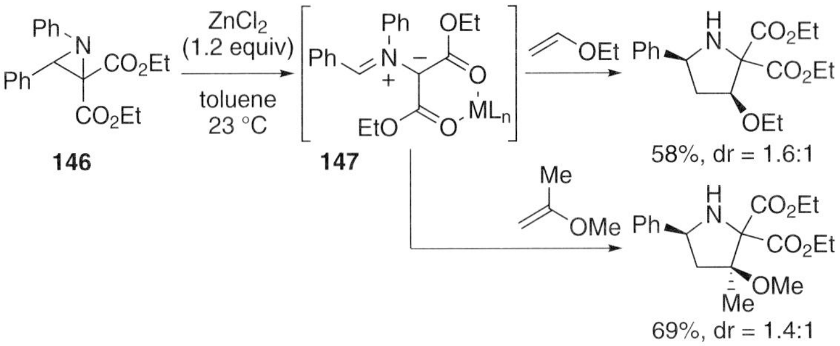

Scheme 3.74 1,3-DC of azomethine ylides generated via Lewis acid-promoted aziridine ring-opening.

Interestingly, use of cyclic vinyl ethers (e.g., **148**) or strained alkenes in this reaction system gave the [4 + 2]-adduct (e.g., **149**), as depicted in Scheme 2.75, presumably due to the steric difference of the vinyl ethers (Scheme 3.75).

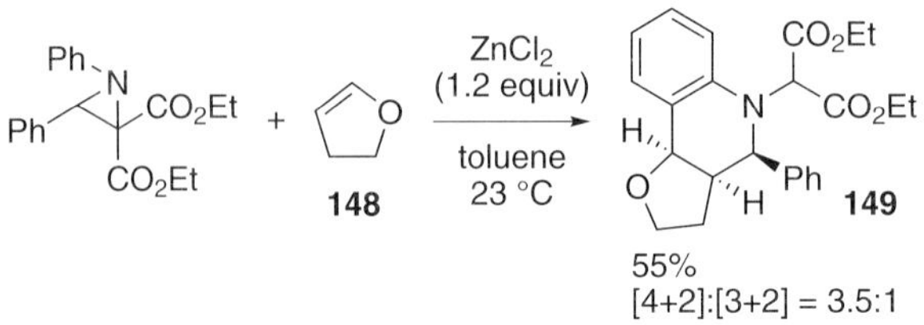

Scheme 3.75 Preference for the [4 + 2]-cycloaddition over [3 + 2]-cycloaddition.

3.3.8
Generation and 1,3-DC of Metal-Containing Azomethine Ylide via Electrophilic Activation of Alkynes

In 2002, Iwasawa and coworkers discovered a novel approach for the generation of azomethine ylides as a means of providing polycyclic indole derivatives [96]. Namely, photo-irradiation of the reactants composed of $W(CO)_6$, N-(o-alkynylphenyl)imine **150** and electron-rich alkene furnished the corresponding indole **152** with additional ring systems (Scheme 3.76). In the first stage of the reaction, $W(CO)_6$ dissociates carbon monoxide to generate $W(CO)_5$. Secondary, interaction of N-(o-alkynylphenyl)imine with $W(CO)_5$ leads to the electrophilic activation of the alkyne moiety. The nitrogen atom of the imine part then attacks this π-complex in a 5-endo fashion to give a tungsten-containing azomethine ylide **151**. The successive 1,3-DC with alkene and alkyl group migration gave the polycyclic indole. In contrast to the N-metalated azomethine ylide described above, wherein electron-deficient olefins are utilized as dipolarophile, electron-rich olefins react with this tungsten-containing azomethine ylide.

Electron-rich alkenes such as ketene silyl acetal, vinyl ether and enamine can be utilized as the dipolarophile. In the case of the reaction with terminal alkenes, a catalytic amount of tungsten generally provided the indole derivatives in fairly good yields, whereas the use of internal alkenes requires a stoichiometric amount of $W(CO)_6$ (Scheme 3.77).

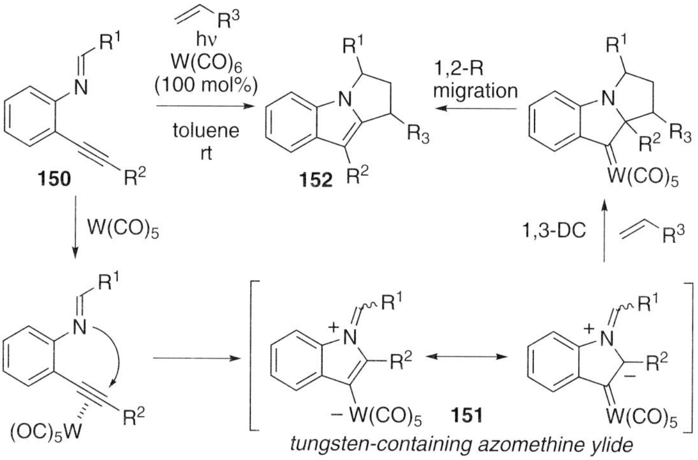

Scheme 3.76 Generation and reaction of tungsten-containing azomethine ylide.

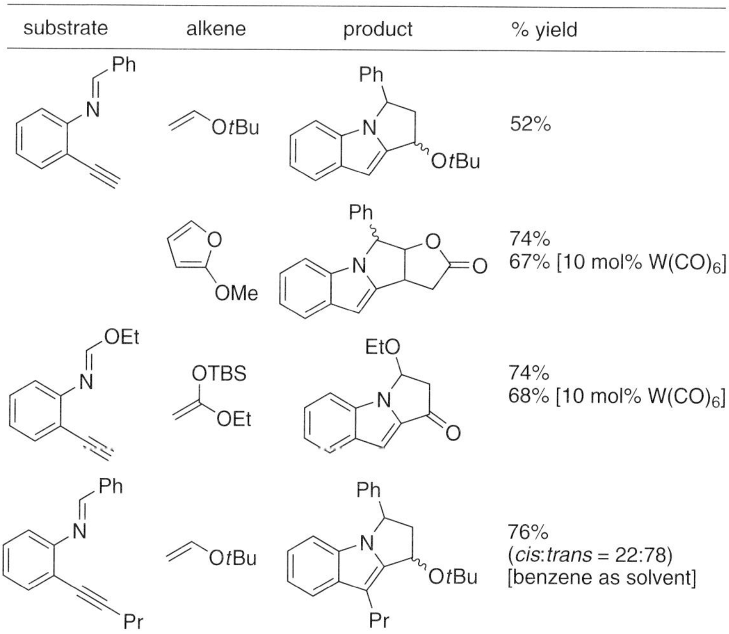

Scheme 3.77 Scope of the 1,3-DC of tungsten-containing azomethine ylide.

Extension of this 1,3-DC of metal-containing azomethine ylide was also accomplished by Iwasawa et al., which realized the catalytic use of metal sources in the reaction of N-(o-alkynylphenyl)imines **153** having internal alkynes with tert-butyl vinyl ether **154** [97]. From the investigation of various transition metals, they found the tendency that soft-metal complexes, such as third-row transition metals, act as efficient catalysts. Specifically, platinum(II) and gold(III) showed high catalytic activity for a broad range of substrates, providing tricyclic compounds **155** in yields

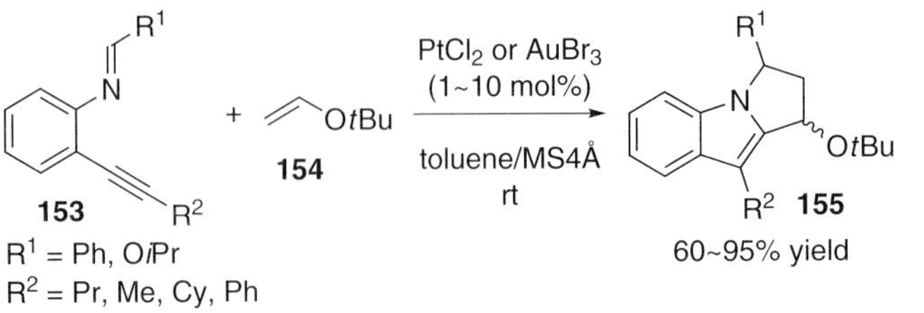

Scheme 3.78 Platinum- and gold-catalyzed 1,3-DC of metal-containing azomethine ylides.

of 60 to 95% (Scheme 3.78). Interestingly, this catalytic condition is not suitable for the reaction of terminal alkynes.

A mechanistically related, metal-mediated generation of azomethine ylides was developed by Su and Porco (Scheme 3.79) [98]. Treatment of o-alkynylarylidene glycinate **156** with AgOTf promotes 6-endo cyclization by the electrophilic activation of the alkyne and the successive nucleophilic attack of the imine nitrogen lone pair. After this cycloisomerization step, proton transfer occurs from the α-hydrogen atom of the glycinate to the C—Ag bond to give the transient azomethine ylide **157** with the recovery of AgOTf. 1,3-DC with electron-deficient alkene followed by an isomerization–oxidation sequence furnishes pyrrolo-isoquinoline **158**, which is regarded as the core structure of lamellarin alkaloids. This synthetic protocol can be applied to a broad range of o-alkynylarylidene glycinates, irrespective of the internal and terminal alkynes.

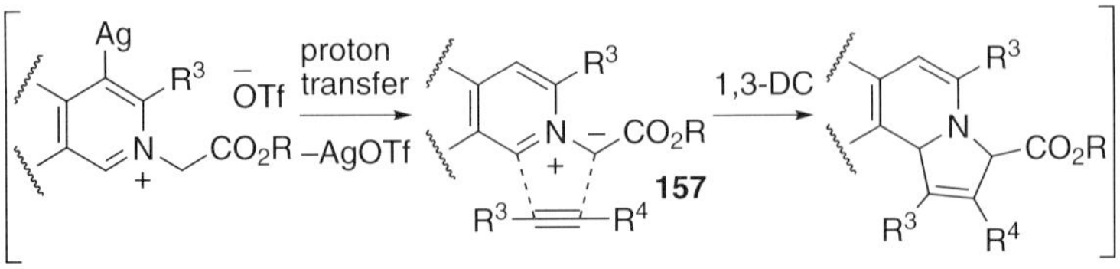

Scheme 3.79 Silver-mediated generation of azomethine ylides from o-alkynylarylidene glycinates.

3.4
Azomethine Imines

In 1960, azomethine imine was first defined and utilized as a 1,3-dipole by Huisgen and coworkers [99, 100]. In these early examples, azomethine imines were generated by the reaction of diazoalkane **159** and azo compound **160** or treatment of 1-aminopyridinium salt **161** with base (Scheme 3.80) [101].

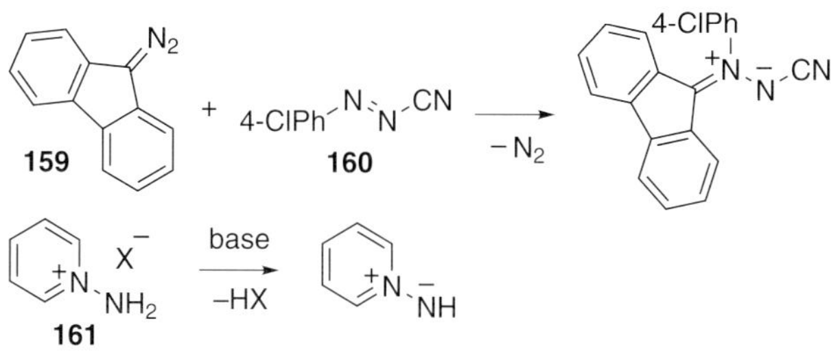

Scheme 3.80 Classical methods for the generation of azomethine imines.

Condensation of N-acyl-N'-alkylhydrazines **162** with aldehydes is employed as a simple and practical procedure to generate azomethine imines applicable in the 1,3-DC process, the foundation of which had been established by Oppolzer (Scheme 3.81) [102].

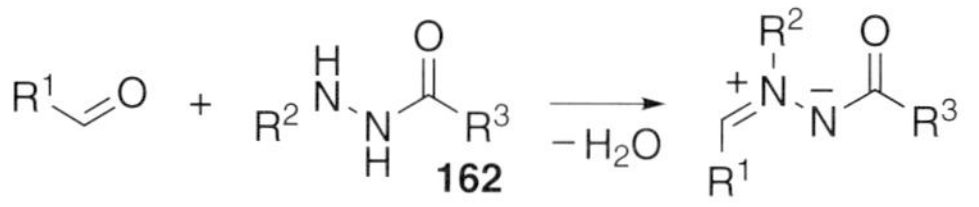

Scheme 3.81 Generation of azomethine imines by the condensation of N-acyl-N'-alkylhydrazines and aldehydes.

Although some applications of 1,3-DC of such transient azomethine imines in natural product syntheses clearly demonstrated their synthetic power [103], a journey to explore the catalytic asymmetric version has only recently started, utilizing 3-oxopyrazolidin-1-ium-2-ides (e.g., **163**), which were introduced as bench-stable azomethine imines and employed in 1,3-DC by Dorn and Otto as early as 1968 (Scheme 3.82) [104–106].

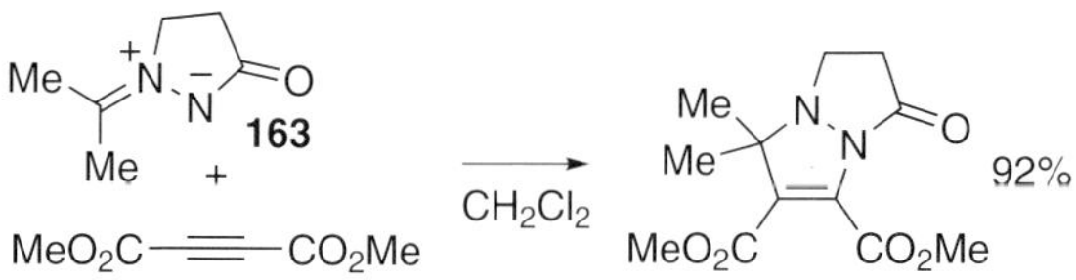

Scheme 3.82 Use of 3-oxopyrazolidin-1-ium-2-ide as a stable azomethine imine.

3.4.1
Asymmetric 1,3-DC of Azomethine Imines: Use of 3-Oxopyrazolidin-1-Ium-2-Ides as a 1,3-Dipole

In 2003, Shintani and Fu explored the enantioselective 1,3-DC of 3-oxopyrazolidin-1-ium-2-ides **164** with the catalytically generated copper(I)-acetylide complexed with chiral ligand [107]. On the basis of their finding in the asymmetric Kinugasa reaction employing a similar catalytic system [53, 54], preliminary experiments were done to identify the superiority of the phosphaferrocene-oxazoline ligand **166a** complexed with CuI, which regiospecifically gave the cycloadduct in 90% yield with 98% ee in the

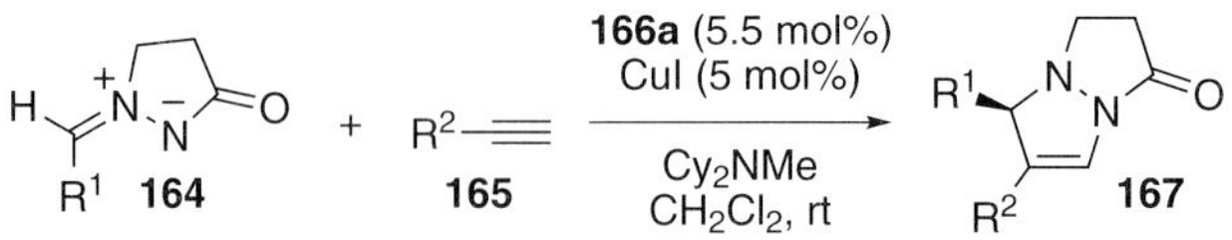

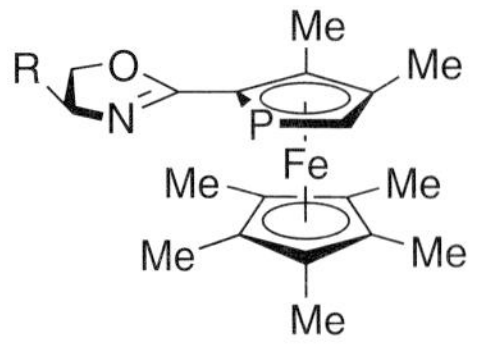

Scheme 3.83 Enantioselective 1,3-DC of 3-oxopyrazolidin-1-ium-2-ides with acetylenes.

R^1	R^2	% yield	% ee
Ph	CO_2Et	98	90
nC_5H_{11}	CO_2Et	92	82
Cy	CO_2Et	94	96
Ph	COMe	98	90
Ph	$4\text{-}CF_3C_6H_4$	90	86
Ph	Ph	73	88 (rs = 5.6:1)
Ph	nC_5H_{11}	63	74 (rs = 6.6:1)

R = iPr (**166a**)
R = Ph (**166b**)

reaction of azomethine imine (**164**, R^1 = Ph) and ethyl propiolate (**165**, R^2 = CO_2Et) (Scheme 3.83). Use of BINAP shut down the reaction, and use of a diasteromeric P,N-ligand proved to be less effective compared to **166a** with regard to the enantioselectivity. A broad range of azomethine imines and terminal acetylenes could be converted to the corresponding bicyclic pyrazolidinones **167** with high regio- and enantioselectivities.

Fu and coworkers further expanded this enantioselective 1,3-DC protocol to the kinetic resolution of azomethine imines **168** having a chiral stereogenic center at C-5 [108]. Treatment of azomethine imines bearing an alkyl or aryl substituent at C-5 with 0.5 to 0.6 equiv of terminal alkynes in the presence of copper(I)/phosphaferrocene-oxazoline **166b** complex generally provided the recovered azomethine imines **169** with high selectivity factor, concurrent with the formation of the cycloadduct **170** (Scheme 3.84). The synthetic utility of the thus-resolved enantioenriched azomethine imines was also demonstrated in that report.

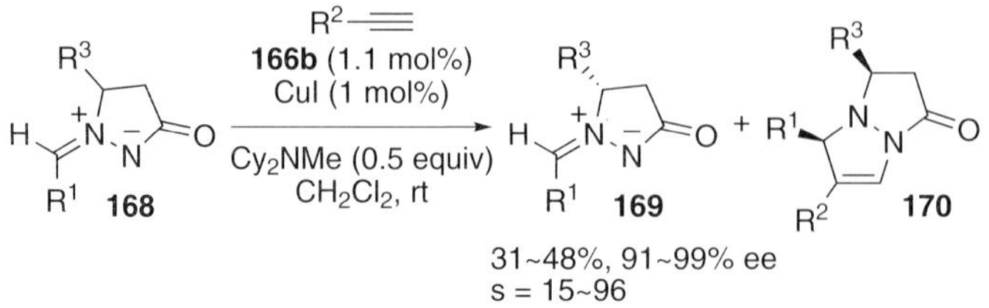

31~48%, 91~99% ee
s = 15~96

Scheme 3.84 Kinetic resolution of 3-oxopyrazolidin-1-ium-2-ides substituted at C-5 via catalytic 1,3-DC.

These works by Fu's group prompted further research, wherein electrophilically-activated α,β-unsaturated carbonyl compounds were introduced as the dipolarophile in 1,3-DC of azomethine imine employing 3-oxopyrazolidin-1-ium-2-ides. Activation of α,β-unsaturated aldehydes by iminium catalysis was undertaken by Chen and coworkers [109]. By the use of diarylprolinol **171** in combination with trifluoroacetic acid, exo-selective cyclization occurred smoothly to give **172** in a highly enantioselective manner (Scheme 3.85). As for the β-substituent of α,β-unsaturated aldehydes, only alkyl groups were tolerated.

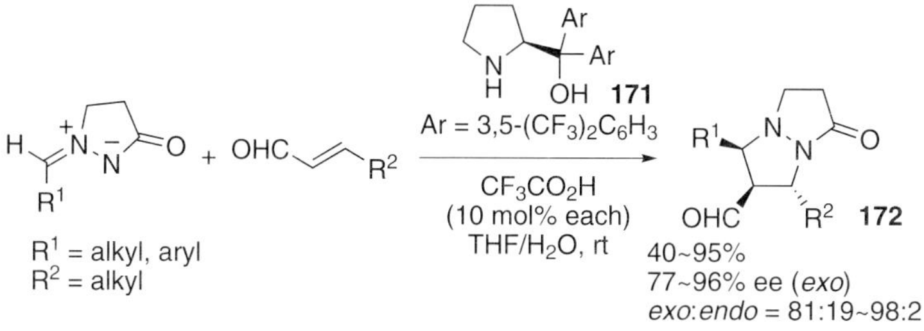

Scheme 3.85 Organocatalytic 1,3-DC of 3-oxopyrazolidin-1-ium-2-ides with α,β-unsaturated aldehydes.

To expand the scope of this enantioselective 1,3-DC of azomethine imines by iminium catalysis to include cyclic enones **173** as dipolarophile, Chen et al. introduced the quinine-based primary amine catalyst **174** complexed with 2 equiv arylsulfonic acid **175** as a suitable catalyst [110]. The reaction system is applicable to cyclohexenone, cyclopentenone and cycloheptanone. In these cases, exo-cycloadducts **176** were isolated as the only detectable isomer (Scheme 3.86). Additionally, the reaction promoted by pseudo-enantiomeric quinidine-derived catalyst could afford the cycloadduct having the opposite absolute configuration with a high degree of stereoinduction.

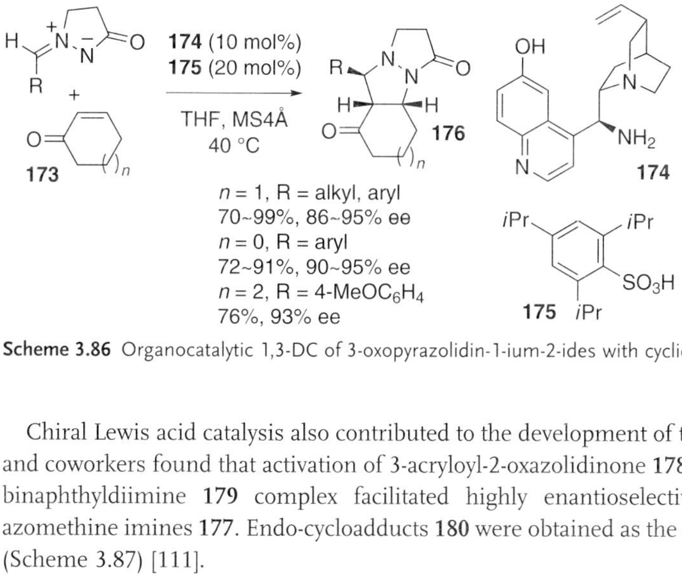

Scheme 3.86 Organocatalytic 1,3-DC of 3-oxopyrazolidin-1-ium-2-ides with cyclic enones.

Chiral Lewis acid catalysis also contributed to the development of this area. Suga and coworkers found that activation of 3-acryloyl-2-oxazolidinone **178** by nickel(II)/binaphthyldiimine **179** complex facilitated highly enantioselective 1,3-DC of azomethine imines **177**. Endo-cycloadducts **180** were obtained as the major product (Scheme 3.87) [111].

Aiming at the realization of the exo-selective 1,3-DC of azomethine imines, Sibi and coworkers adopted the reaction system developed for the exo-selective 1,3-DC of nitrones in which 2-acryloyl-3-pyrazolidinone **181** was utilized as the dipolarophile [21]. As in the case of 1,3-DC of nitrones, copper(II)/bisoxazoline **18** complex was found to be the suitable catalyst, giving the exo-cycloadduct **182** preferentially (Scheme 3.88) [112]. It was found that addition of molecular sieves had a detrimental effect on the exo-selectivity in some cases.

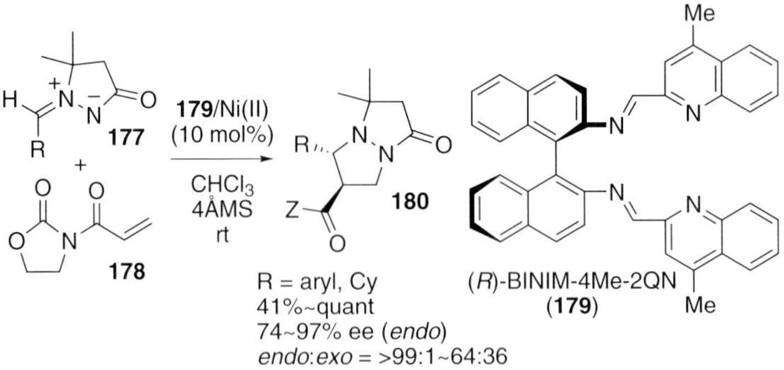

Me
177
179/Ni(II)
(10 mol%)
CHCl3
4ÅMS
rt
180
178
R = aryl, Cy
41%~quant
74~97% ee (endo)
endo:exo = >99:1~64:36
(R)-BINIM-4Me-2QN
(179)
Me

Scheme 3.87 Chiral Lewis acid-catalyzed enantioselective 1,3-DC of 3-oxopyrazolidin-1-ium-2-ide derivatives.

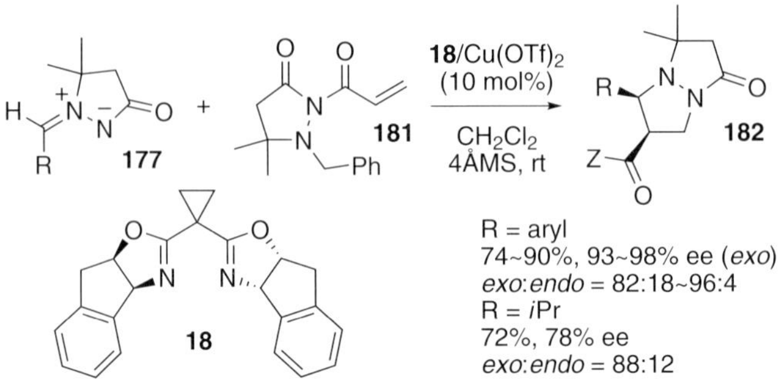

177
181
18/Cu(OTf)2
(10 mol%)
CH2Cl2
4ÅMS, rt
182
18
R = aryl
74~90%, 93~98% ee (exo)
exo:endo = 82:18~96:4
R = iPr
72%, 78% ee
exo:endo = 88:12

Scheme 3.88 Copper/bisoxazoline complex-catalyzed exo- and enantioselective 1,3-DC of 3-oxopyrazolidin-1-ium-2-ide derivatives.

They also attempted the use of 2-crotonoyl-3-pyrazolidinone **184** (Scheme 3.89). Although the azomethine imine derived from 5,5-dimethylpyrazolidin-3-one **183a** was found not to be a suitable reaction partner, use of unsubstituted azomethine imine **183b** and a stoichiometric amount of copper(II)/bisoxazoline **18** complex allowed the formation of the corresponding cycloadduct **185** with moderate enantioselectivity.

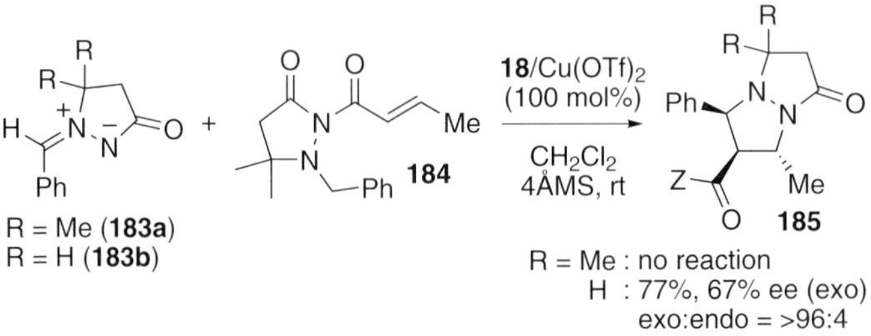

R = Me (183a)
R = H (183b)
184
18/Cu(OTf)2
(100 mol%)
CH2Cl2
4ÅMS, rt
185
R = Me : no reaction
H : 77%, 67% ee (exo)
exo:endo = >96:4

Scheme 3.89 Use of 2-crotonoyl-3-pyrazolidinone in Copper/bisoxazoline complex-catalyzed exo- and enantioselective 1,3-DC of azomethine imines.

Recently, asymmetric 1,3-DC of azomethine imines with allyl alcohol was reported by Ukaji, Inomata and co-workers, wherein the magnesium complex of diisopropyl (R,R)-tartrate was utilized as a stereocontroller (for the related work, see Scheme 3.35) [113].

3.4.2
Asymmetric [3 + 2] Cycloaddition of N-Acylhydrazones with Olefins

The other potentially useful method for the generation of azomethine imines is the direct activation of N-acylhydrazones. Brønsted acid activation and thermal 1,2-prototropy of acylhydrazones have been known for decades [114, 115], before Kobayashi and coworkers took notice of the possibility to promote this sort of reaction by the aid of Lewis acid catalyst [116]. They first discovered that the use of a Lewis acid, such as $Sc(OTf)_3$, $Zr(OTf)_4$, can be a promising tool for the cyclization of N-acylhydrazones with electron-rich olefins. With this preliminary research, they realized asymmetric intramolecular [3 + 2] cyclization catalyzed by chiral zirconium (IV) catalyst bearing modified BINOL ligand **187** [117]. The pyrazolidine derivatives **188** could generally beobtained with high trans-selectivity and enantioselectivity, starting from N-(4-nitrobenzoyl)hydrazones **186** (Scheme 3.90).

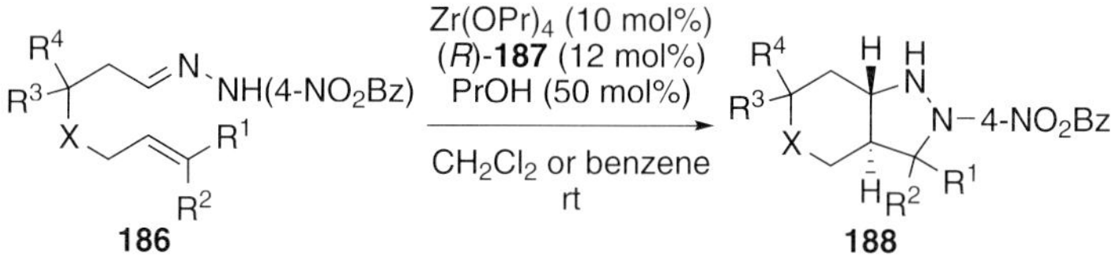

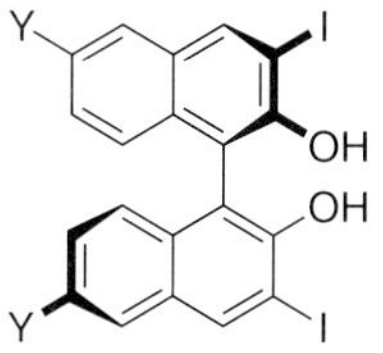

cat	R^1	R^2	R^3	R^4	X	% yield	% ee
(R)-**187a**	Me	Me	Me	Me	CH_2	99	96
(R)-**187b**	Me	Me	SCH_2CH_2S		CH_2	91	97
(R)-**187b**	SCH_2CH_2S		H	H	CH_2	57	72
(R)-**187a**	Me	Me	Me	Me	O	38	81

Y = I : (R)-**187a**
H : (R)-**187b**

Scheme 3.90 Zirconium-catalyzed intramolecular enantioselective [3 + 2]-cycloaddition of N-acylhydrazones.

Later, an intermolecular version of this asymmetric [3 + 2] cyclization was developed by the same group [118]. Use of N-benzoylhydrazones (**189**, Ar = Ph) was found to be beneficial in attaining higher enantioselectivity, when the reaction was conducted with ketene dimethyl dithioacetal **190** under similar catalytic conditions as above (Scheme 3.91). In the case of the cyclization with vinyl ethers, N-benzoylhydrazone was not suitable due to low reactivity, and N-(4-nitrobenzoyl) hydrazones (**189**, Ar = $4-NO_2C_6H_4$) were accommodated as dipolarophiles. Enantioenriched pyrazolidines thus-obtained **191** and **192** could be easily converted to 1,3-diamine or pyrazoline.

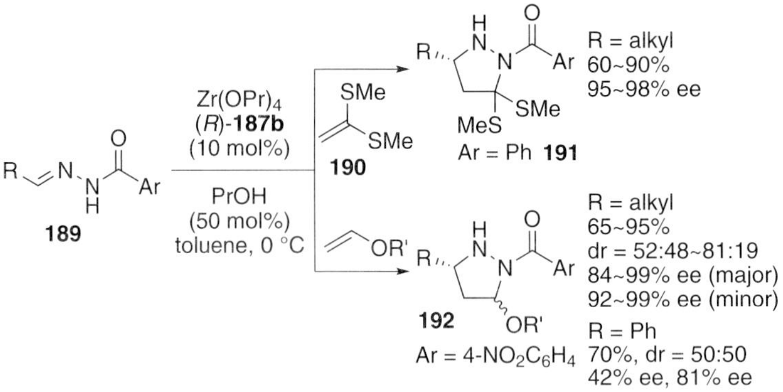

Scheme 3.91 Zirconium-catalyzed intermolecular enantioselective [3 + 2]-cycloaddition of N-acylhydrazones.

The substantial difference in the enantiomeric excesses of the major isomer and the minor isomer formed in the intermolecular reaction of acylhydrazones and vinyl ethers implied the intervention of N-metalated azomethine imine leading to the 1,3-DC pathway, rather than the stepwise nucleophilic addition of vinyl ether and successive cyclization (Scheme 3.92).

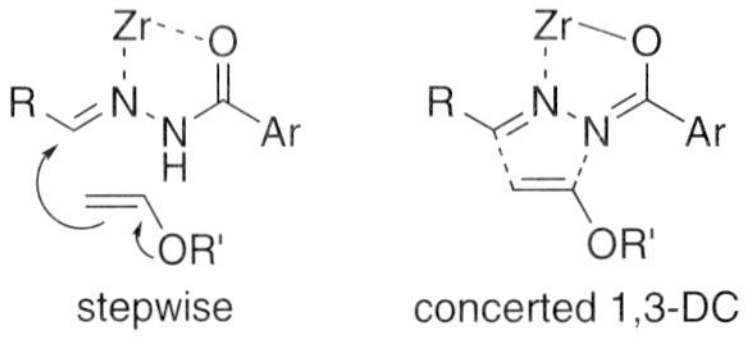

Scheme 3.92 Stepwise and concerted mechanism.

During the study on the use of strained silacycles as chiral Lewis acids, Leighton and coworkers found that chiral phenylsilane **193** promotes the cycloaddition of N-benzoylhydrazones and tert-butyl vinyl ether with high diastereo- and enantioselectivities [119]. With respect to N-acylhydrazones, various functionalities such as primary, secondary and tertiary alkyl groups, as well as the aromatic ring, were all tolerated (Scheme 3.93). The practicality of this process was demonstrated by

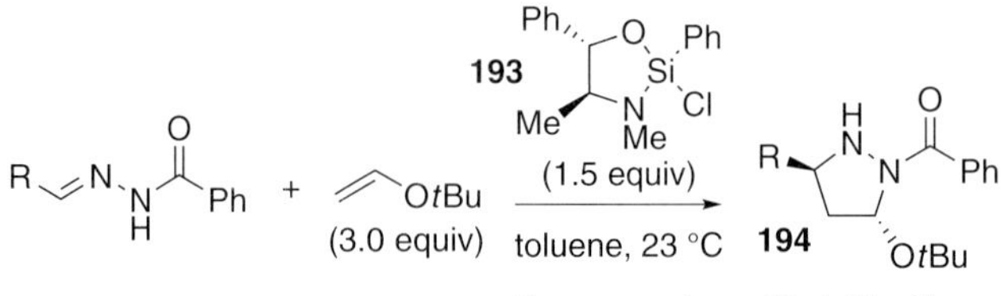

R	dr	% yield	% ee
PhCH$_2$CH$_2$	96:4	84	90
BnOCH$_2$	>97:3	85	90
Cy	>97:3	79	95
tBu	>97:3	76	98
Ph	>97:3	80	94

Scheme 3.93 Chiral silicon Lewis acid-mediated enantioselective [3 + 2]-cycloaddition of N-acylhydrazones.

Scheme 3.94 Chiral silicon Lewis acid-mediated enantioselective [3 + 2]-cycloaddition of N-acylhydrazones with cis-vinyl ether.

running the reaction with 5 g of the hydrazone, furnishing the cycloadduct **194** in high yield and ee by recrystallization without chromatography. Chiral ligand, pseudoephedrine, could be recovered nearly quantitatively by simple extraction.

Use of cis-vinyl ether **196** in the cycloaddition of N-acylhyrazone **195** led to the cycloadduct **197** projecting the methyl and tert-butoxy moieties in a trans-relationship (Scheme 3.94), and this fact strongly indicated the intervention of the stepwise mechanism (Scheme 3.95).

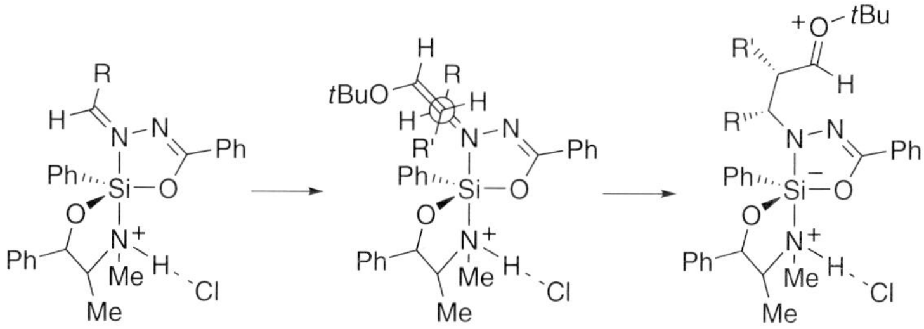

Scheme 3.95 Proposed mechanism of chiral silicon Lewis acid-mediated enantioselective [3 + 2]-cycloaddition of N-acylhydrazones.

They also succeeded in applying this strategy to the asymmetric synthesis of manzacidin C [120]. Use of protected methallyl alcohol **199** as olefin and proper adjustment of chiral silane **200** and N-acylhyrazone **198** to attain high asymmetric induction allowed facile access to the key precursor **201** of manzacidin C (Scheme 3.96).

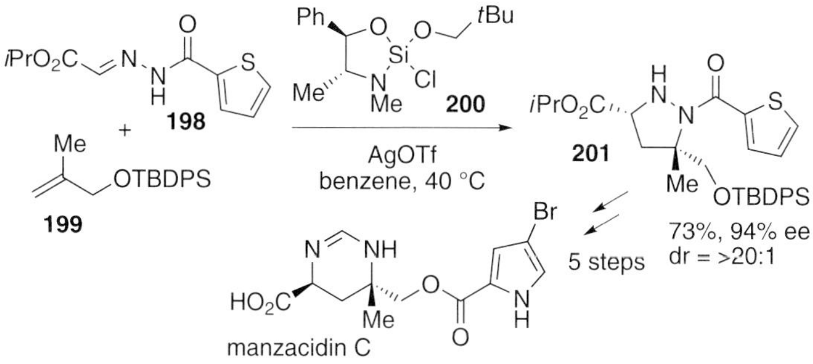

Scheme 3.96 Chiral silicon Lewis acid-mediated enantioselective [3 + 2]-cycloaddition of N-acylhydrazones and its application to the synthesis of manzacidin C.

3.5
Carbonyl Ylides

Like most of the other 1,3-dipoles, carbonyl ylides are normally generated as transient species *in situ*. The representative methods for the generation of carboyl ylides include the reaction of carbonyl compound and carbene, thermal or photochemical activation of epoxide, and thermal extrusion of nitrogen from 1,3,4-oxadiazoline. However, the required harsh reaction conditions obviously undermine their use in practical 1,3-DC (Scheme 3.97).

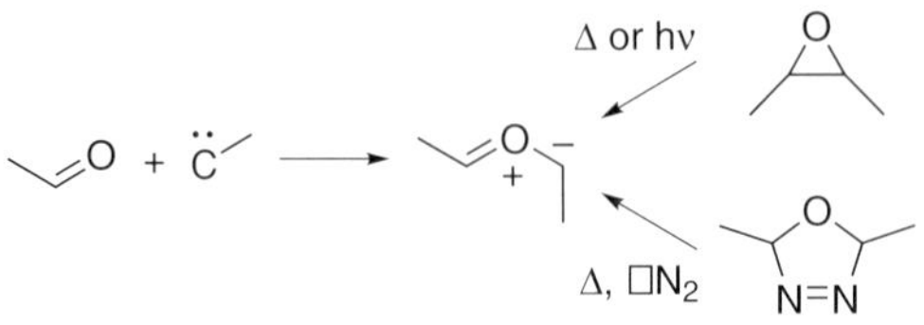

Scheme 3.97 Pathways for the generation of carbonyl ylides.

One exceptionally important strategy, which has been accommodated in a broad range of syntheses and also recently culminated in the catalytic asymmetric 1,3-DC of carbonyl ylides, is the metal carbenoid-mediated evolution of carbonyl ylides from diazocarbonyl compounds and carbonyl compounds as depicted in Scheme 3.98 [121]. Although the copper-mediated generation of carbonyl ylide has long been known since the report of Kharasch [122], it was the seminal study of Ibata, Takebayashi and coworkers which contributed to the early development of its use in 1,3-DC [123, 124]. Later, the synthetic power of this tandem carbonyl ylide formation–cycloaddition strategy has been further extended by the intensive research of Padwa and coworkers relying on the catalytic use of rhodium(II) carboxylate [125].

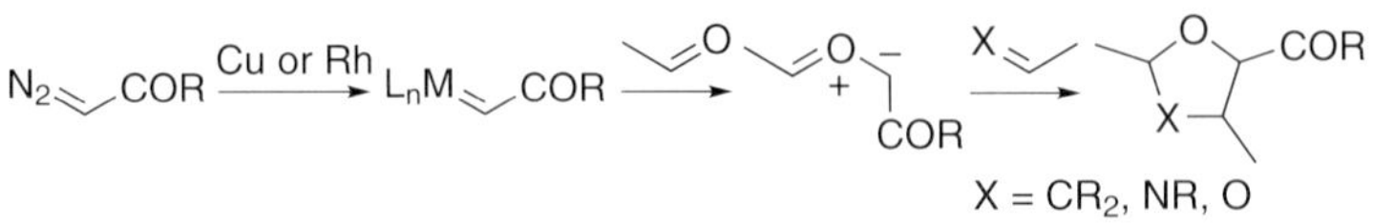

Scheme 3.98 Metal carbenoid-mediated evolution of carbonyl ylides.

In this section, the focus is on the beginning and advance of asymmetric 1,3-DC of carbonyl ylides generated by the rhodium(II)-catalyzed decomposition of diazo compounds. Additionally, a completely new method for the generation of carbonyl ylides developed in the past several years and their use in 1,3-DC will be introduced.

3.5.1
Dirhodium(II)-Catalyzed Asymmetric Tandem Carbonyl Ylide Formation-1,3-DC

According to the postulated mechanism of metal carbenoid-mediated generation of carbonyl ylides, it is considered to be unrealistic to use chiral rhodium or copper catalyst to achieve enantioselective 1,3-DC of carbonyl ylides, since the catalyst only associates with the substrate at the generation of the carbonyl ylide and does not interfere in the cyclization step which is crucial for asymmetric induction.

A hint came from two independent works of Padwa and Doyle, both of which observed the ligand-dependent regio- or diastereoselectivity in 1,3-DC of carbonyl ylides catalyzed by rhodium(II) complex [126, 127]. In 1997, Doyle and coworkers exemplified in the rhodium(II) catalyzed reaction of 4-nitrobenzaldehyde and ethyl diazoacetate that the diastereomeric ratio of thus-formed dioxolanes **202** varies depending on the ligand of the rhodium(II) catalyst (Scheme 3.99). This catalyst dependence can be explained by assuming the intervention of metal-associated carbonyl ylide **203** as the key intermediate of 1,3-DC step. Later, this proposal was further proved by the use of chiral rhodium catalyst $Rh_2(4S\text{-MEOX})_4$ (MEOX = methyl 2-oxooxazolidine-4-carboxylate), which gave the all-cis cycloadduct as a major product with 28% ee [121a].

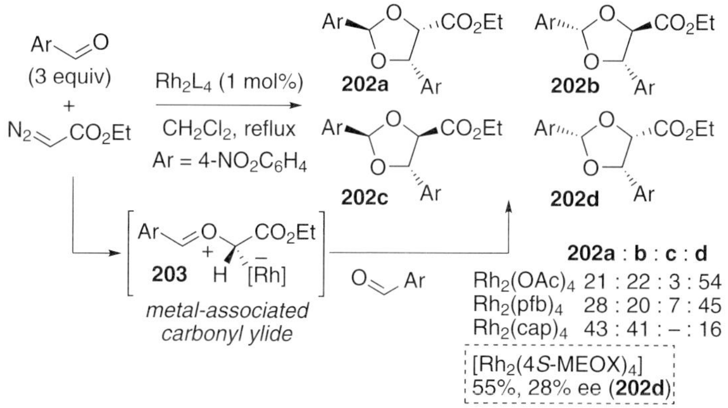

Scheme 3.99 The catalyst-dependence of the diastereoselectivity in tandem carbonyl ylide formation-1,3-DC.

The earliest example of asymmetric 1,3-DC of carbonyl ylides was reported in 1997 by Hodgson and coworkers [128]. They chose the substrate **204**, which was designed to undergo tandem carbonyl ylide formation-1,3-DC in an intramolecular fashion. By the use of $Rh_2(S\text{-DOSP})_4$ **205** as catalyst, the desired cycloadduct **207** was obtained with a moderate level of asymmetric induction (Scheme 3.100). In their following paper [129], they succeeded in increasing the enantioselectivity of this reaction by the use of $Rh_2(R\text{-DDBNP})_4$ **206**, wherein the cycloadduct was obtained in 66% yield with 90% ee.

The variation of substrates was also investigated in detail (Scheme 3.101) [130]. As dipolarophile, internal olefin and acetylene were both tolerated, giving the tricyclic compounds **208a** and **208b** with high enantioselectivities. Use of one-carbon homologated substrate provided the cycloadduct **208c** with a six-membered carbocycle. Replacement of one methylene unit by the oxygen atom led to lower reactivity and enantioselectivity (**208d**). Not only α-diazo-β-ketoesters but also aryldiazoketones can be utilized (**208e**) [131].

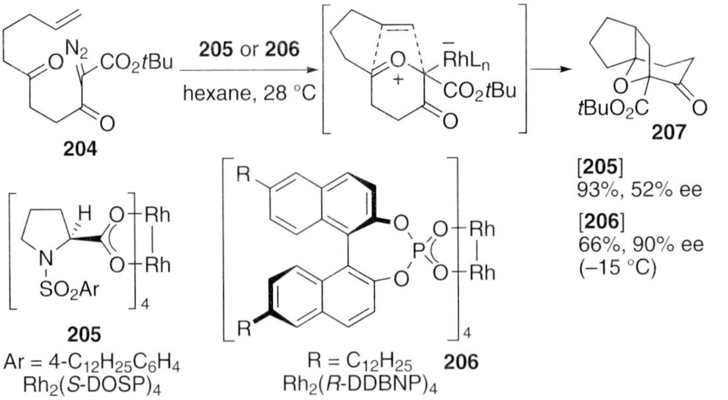

Scheme 3.100 Pioneering work on the enantioselective tandem carbonyl ylide formation-1,3-DC.

[205] Ar = 4-$C_{12}H_{25}C_6H_4$, Rh_2(S-DOSP)$_4$
[206] R = $C_{12}H_{25}$, Rh_2(R-DDBNP)$_4$

[205] 93%, 52% ee
[206] 66%, 90% ee (−15 °C)

208a 53%, 80% ee
208b 80%, 86% ee
208c 68%, 87% ee
208d 12%, 43% ee
208e quant, 51% ee Ar = 4-$NO_2C_6H_4$

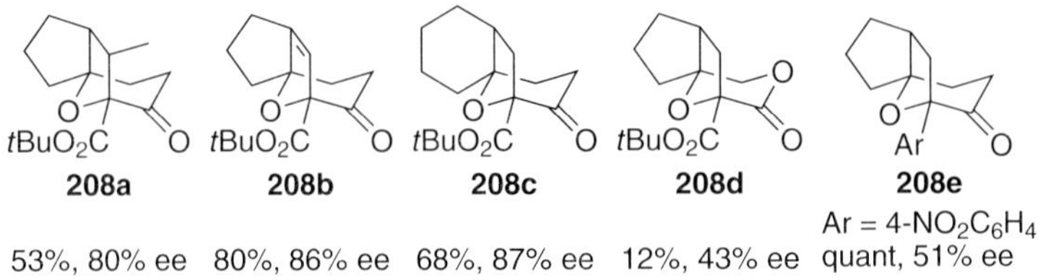

Scheme 3.101 Scope of Rh_2(R-DDBNP)$_4$-catalyzed intramolecular enantioselective tandem carbonyl ylide formation-1,3-DC.

Catalysis with Rh_2(S-DOSP)$_4$ **205** and Rh_2(R-DDBNP)$_4$ **206** was further extended to the intermolecular 1,3-DC [132]. For example, rhodium(II)-associated carbonyl ylide generated from 2-diazo-3,6-diketoester **209** reacted with styrene derivatives, arylacetylenes or strained alkenes to give the cycloadducts **210**, **211** or **212** respectively (Scheme 3.102).

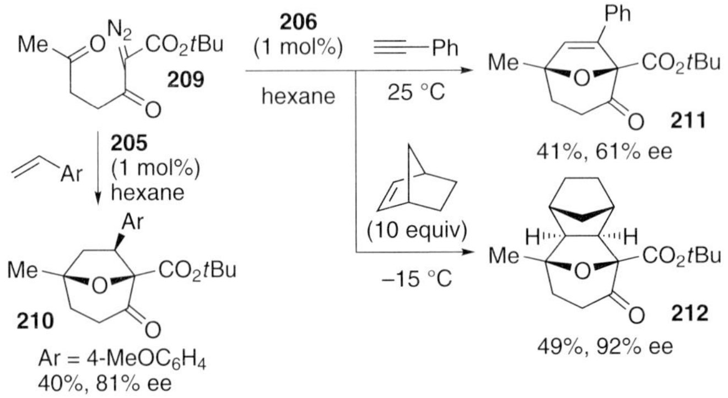

210 Ar = 4-MeOC_6H_4, 40%, 81% ee
211 41%, 61% ee
212 49%, 92% ee

Scheme 3.102 Rhodium-catalyzed intermolecular enantioselective tandem carbonyl ylide formation-1,3-DC.

3.5 Carbonyl Ylides

In 1998, Suga, Ibata et al. revealed that intermolecular 1,3-DC of carbonyl ylide generated from **213** and electron-deficient phthalimide furnishes the cycloadduct **214** with a detectable level of enantioselectivity [133], when the reaction was performed in the presence of $Rh_2(5S\text{-MEPY})_4$ (MEPY = methyl 2-pyrrolididone-5-carboxylate) (Scheme 3.103). They also observed slight asymmetric induction in the chiral copper(I)/bisoxazoline catalyzed reaction.

It should be noted that the example using α-diazoketones reported by Suga is considered to involve HOMO(dipole)–LUMO(dipolarophile) interaction. On the other hand, carbonyl ylides generated from the α-diazo-β-ketoester normally favor cycloaddition with an electron-rich π-bond via HOMO(dipolarophile)–LUMO(dipole) interaction, as in the case of Hodgson's work.

Scheme 3.103 Early example of enantioselective tandem carbonyl ylide formation-intermolecular 1,3-DC with electron-poor olefin.

A highly enantioselective tandem carbonyl ylide formation–intermolecular 1,3-DC sequence was realized by Hashimoto and coworkers in 1999 [134]. Cycloaddition of the rhodium-associated carbonyl ylide generated from α-diazoketone **215** with dimethyl acetylenedicarboxylate furnished the bicyclic compound **217** with high enantioselectivity, irrespective of the tether length and substituent pattern of the α-diazoketones, when $Rh_2(S\text{-BPTV})_4$ **216** having a chiral ligand derived from (S)-valine was employed as catalyst (Scheme 3.104). Other dirhodium(II) carboxylates

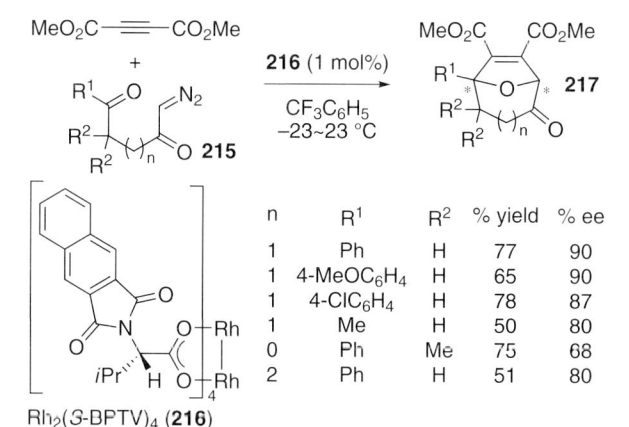

n	R^1	R^2	% yield	% ee
1	Ph	H	77	90
1	4-MeOC$_6$H$_4$	H	65	90
1	4-ClC$_6$H$_4$	H	78	87
1	Me	H	50	80
0	Ph	Me	75	68
2	Ph	H	51	80

Rh$_2$(S-BPTV)$_4$ (**216**)

Scheme 3.104 Highly enantioselective tandem carbonyl ylide formation-intermolecular 1,3-DC catalyzed by $Rh_2(S\text{-BPTV})_4$.

such as $Rh_2(S\text{-}DOSP)_4$ and $Rh_2(5S\text{-}MEPY)_4$ exhibited only poor enantioselectivity. Use of benzotrifluoride as solvent was found to be optimal to attain high yield and ee.

They extended the research to the reaction of benzene- and naphthalene-fused α-diazoketones [135]. In these cases, $Rh_2(S\text{-}PTTL)_4$ **220** derived from N-phthaloyl t-leucine gave better results than $Rh_2(S\text{-}BPTV)_4$, benzene-fused cycloadduct **218** and naphthalene-fused cycloadduct **219** could be obtained with 74% ee and 93% ee, respectively (Scheme 3.105).

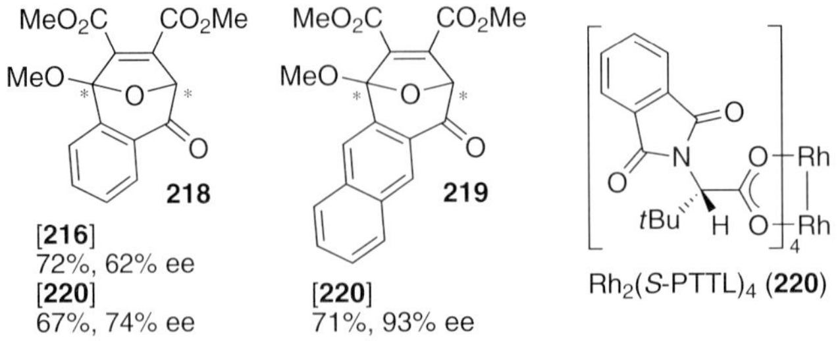

Scheme 3.105 Use of benzene- and naphthalene-fused α-diazoketones in enantioselective tandem carbonyl ylide formation-intermolecular 1,3-DC.

Hashimoto et al. also developed a suitable chiral catalyst which facilitates the tandem carbonyl ylilde formation-1,3-DC of α-diazo-β-ketoester with electron-rich acetylenes or alkenes [136]. Namely, the N-tetrachlorophthaloyl-t-leucine-derived $Rh_2(TCPLL)_4$**222** catalyzed reaction of α-diazo-β-ketoesters **221** with arylacetylenes furnished the cycloadducts **223** with rigorous control of enantiofacial discrimination and regioselectivity (Scheme 3.106). Aside from arylacetylenes, ethyl ethynyl ether can also be utilized as the dipolarophile with a similar level of enantioselectivities. Although use of styrene derivatives as olefinic dipolarophiles raised an additional stereochemical issue regarding the endo/exo selectivity in conjunction with the

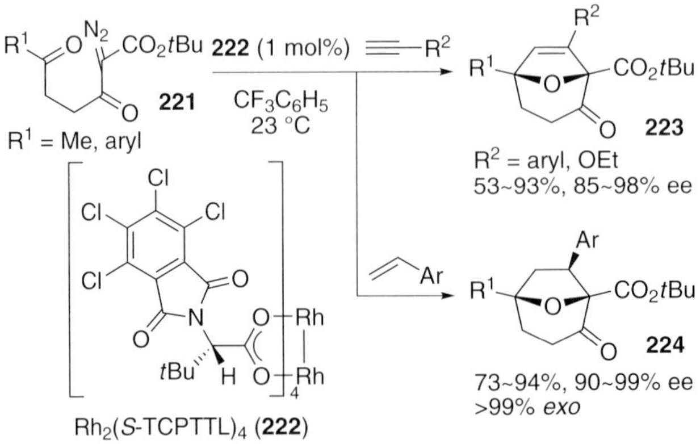

Scheme 3.106 Enantioselective tandem carbonyl ylilde formation-1,3-DC of α-diazo-β-ketoester with electron-rich acetylenes or alkenes catalyzed by $Rh_2(TCPLL)_4$.

regio- and enantioselectivity, this catalytic system cleanly provided exo-isomers **224** as a sole component with ee values ranging from 90 to 99%.

They further demonstrated the activity of their rhodium(II) catalysis in the reaction using aldehydes as dipolarophile [137]. Among the several para-substituted arylaldehyes examined, higher enantioselectivities were attained with the aldehydes having electron-withdrawing substituents (Scheme 3.107).

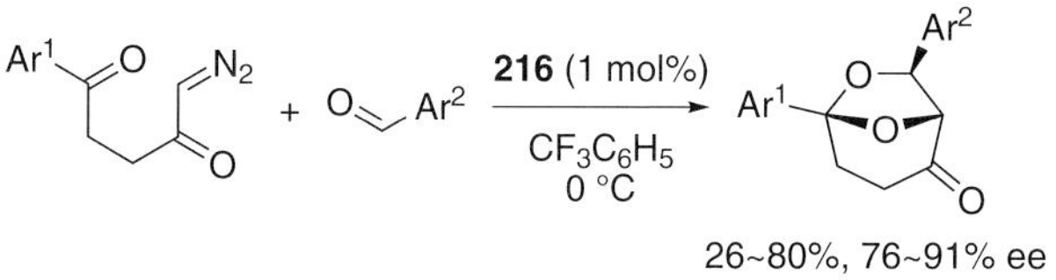

Scheme 3.107 Use of arylaldehydes as the dipolarophile in Rh_2-(S-BPTV)$_4$-catalyzed tandem carbonyl ylide formation-1,3-DC.

3.5.2
Three-Component Tandem Carbonyl Ylide Formation-1,3-DC

Generation of carbonyl ylides in an intermolecular fashion and successive 1,3-DC provided a fascinating means for the construction of complex polycyclic compounds. On the other hand, intermolecular generation of carbonyl ylides from diazo carbonyl compounds and aldehydes has been rarely examined in 1,3-DC with the third reaction partner. Normally, thus-generated carbonyl ylides further react with the second equivalent of the same aldehydes to give dioxolanes, as demonstrated in Doyle's study (Scheme 2.99). One unique exception is that reported by Hu and coworkers [138], wherein the electron-rich arylaldehyde selectively forms carbonyl ylide with phenyldiazoacetate **225** and successively cyclizes with electron-defficient arylaldehydes to give the dioxolane **226** composed of two different aldehydes and diazoacetate (Scheme 3.108).

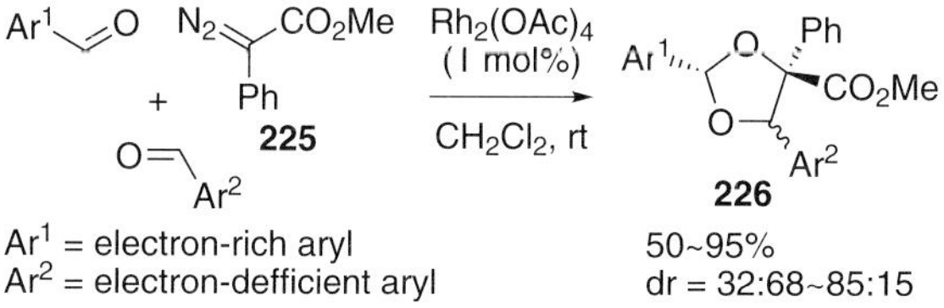

Ar^1 = electron-rich aryl
Ar^2 = electron-defficient aryl

50~95%
dr = 32:68~85:15

Scheme 3.108 Three-component tandem carbonyl ylide formation-1,3-DC using electron-rich and electron-defficient arylaldehydes.

Somfai and coworkers reported that the carbonyl ylide generated from benzaldehyde and ethyl diazoacetate **227** under the influence of rhodium(II) acetate cyclizes with various arylidenebenzylamines **228** to give oxazolidines **229** in high yields and diastereoselectivities (Scheme 3.109) [139]. Hydrolysis of the cycloadducts liberates *syn*-α-hydroxy-β-amino acid esters **230**. To examine the possibility of the

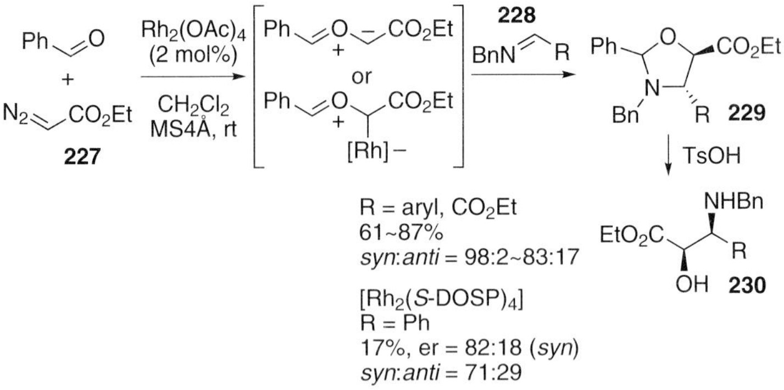

Scheme 3.109 Three-component tandem carbonyl ylide formation-1,3-DC using N-benzy imines as dipolarophile.

asymmetric version, they screened several chiral rhodium(II) complexes and observed a promising asymmetric induction of 64% ee by the use of $Rh_2(S\text{-}DOSP)_4$.

3.5.3
Chiral Lewis Acid Catalyzed 1,3-DC of Carbonyl Ylides Generated by Dirhodium(II) Co-Catalyst

The synthetic maneuver delineated above relies on the feasibility of chiral rhodium-associated carbonyl ylides to undergo 1,3-DC with a dipolarophile under mild conditions. However, dissociation of rhodium catalyst from the ylide, which leads to the immediate erosion of the enantioselectivity, remains as an inevitable threat faced in the process of broadening the scope. In this vein, Suga et al. developed a Lewis acid-rhodium(II) co-catalytic system, in which the rhodium catalyst promotes the generation of carbonyl ylide and the Lewis acid catalyzes the successive 1,3-DC by activating the dipolarophile independently (Scheme 3.110). In such a system, it is anticipated that the stereochemistry of cycloadducts can be controlled by the action of a chiral Lewis acid catalyst, whether or not rhodium is bound to the carbonyl ylide.

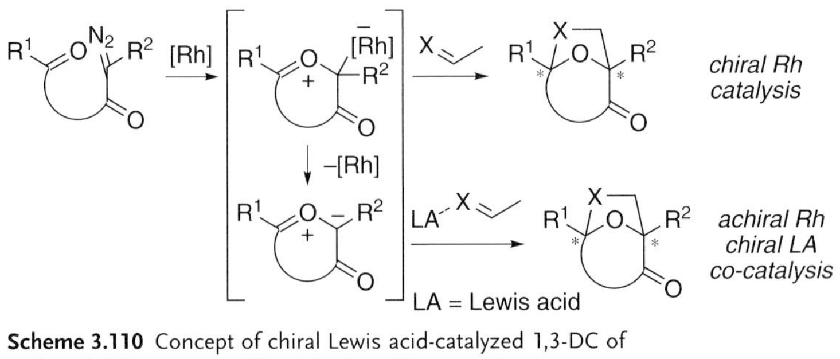

Scheme 3.110 Concept of chiral Lewis acid-catalyzed 1,3-DC of carbonyl ylides generated by dirhodium(II) co-catalyst.

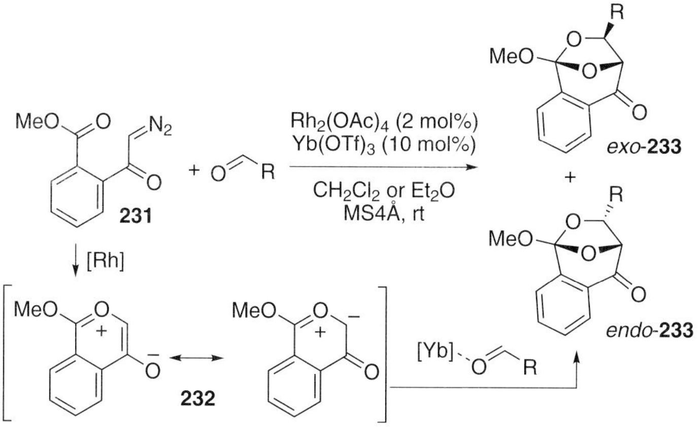

R = 4-MeOC$_6$H$_4$
77%, *exo*:*endo* = 87:13
[w/o Yb(OTf)$_3$]
68%, *exo*:*endo* = 46:54

R = BnOCH$_2$
97%, *exo*:*endo* = 93:7
[w/o Yb(OTf)$_3$]
96%, *exo*:*endo* = 69:31

Scheme 3.111 Effect of the Lewis acid catalyst in the tandem carbonyl ylide formation-1,3-DC.

In their early exploration of tandem cyclization-1,3-DC of carbonyl ylides using rhodium(II) and copper(I) for their generation, they noticed concomitant use of Lewis acid strongly affects the diastereoselectivity of the resultant cycloadducts [140]. As depicted in Scheme 3.111, *o*-(methoxycarbonyl)-α-diazoacetophenone **231** first decomposes by rhodium(II) catalysis, generating carbonyl ylide **232** (2-benzopyrylium-4-olate), and the ylide reacts with ytterbium triflate-activated electrophilic aldehydes to give the cycloadduct **233** with high exo-selectivity. In the absence of Lewis acid catalyst, almost no diastereoselectivity was observed. Other rare earth metal triflates such as Sc(OTf)$_3$, Lu(OTf)$_3$, Tm(OTf)$_3$, Ho(OTf)$_3$, Eu(OTf)$_3$ and Sm(OTf)$_3$ exhibited a similar tendency.

Importantly, the use of chiral ytterbium catalyst **234** clearly induced substantial asymmetric bias, although the exo-selectivity was cancelled out (Scheme 3.112).

Later, they discovered the proper Lewis acid/chiral ligand combination that realizes highly enantioselective 1,3-DC of 2-benzopyrylium-4-olate [141]. When benzyloxyacetaldehyde **236** and its analogs were utilized as the dipolarophile, Sc(OTf)$_3$/Pybox-*i*Pr **235a** complex exhibited an unsurpassed level of enantioselectivity

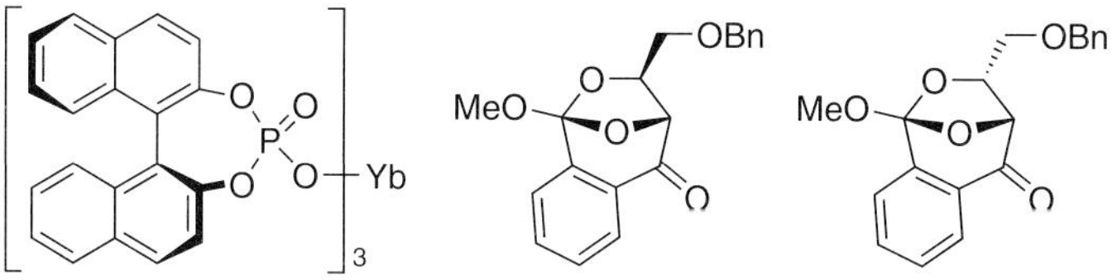

Yb[(*S*)-BNP]$_3$ (**234**)   14% ee   52% ee
exo:*endo* = 57:43

Scheme 3.112 Chiral Lewis acid-catalyzed 1,3-DC of carbonyl ylides generated by dirhodium(II) co-catalyst.

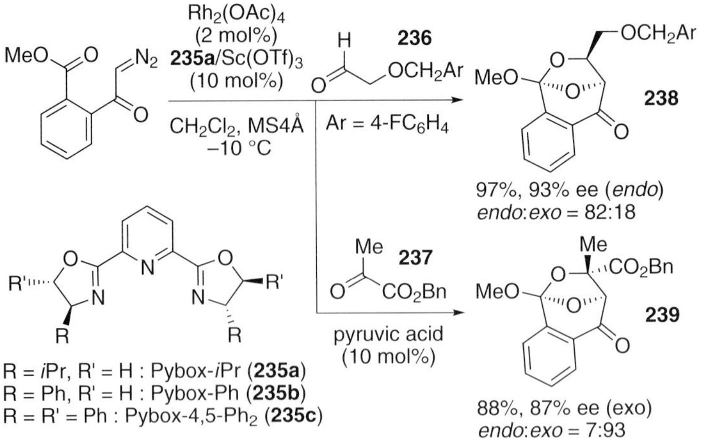

Scheme 3.113 Highly enantioselective chiral Lewis acid-catalyzed 1,3-DC of carbonyl ylides generated by dirhodium(II) co-catalyst.

compared to the other metal salts and chiral ligands, giving the endo-cycloadducts **238** preferentially (Scheme 3.113). Interestingly, the endo/exo selectivitity was easily switched by the choice of metal triflate. Pyruvate **237** can also be utilized as the dipolarophile, giving the *exo*-cycloadduct **239** with 87% ee, although the use of pyruvic acid as additive, which would associate with the catalyst, was necessary to accomplish high selectivity.

Employment of 3-acryloyl-2-oxazolidinone **178** as an olefinic dipolarophile was also attempted by using Sc(OTf)$_3$/Pybox-*i*Pr **235a**, with which only poor asymmetric induction was observed. However, slight modification of the catalytic system was found to be sufficient to gain high enantioselectivity. Namely, the combination of Yb(OTf)$_3$ and Pybox-Ph **235b** furnished the highly enantioenriched cycloadduct **240** in an exo-selective manner (Scheme 3.114). 1,3-DC with other 3-alkenoyl-2-oxazolidinones were also investigated in the following report [142].

In the above examples, Lewis acid lowers the LUMO energy of the dipolarophile, and promotes the reaction with the HOMO of the dipole (normal-electron demand 1,3-DC). On the other hand, the reaction of Lewis acid activated carbonyl ylide

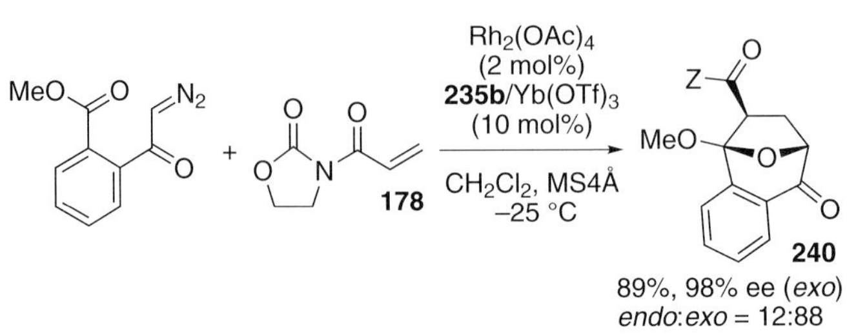

Scheme 3.114 Use of 3-acryloyl-2-oxazolidinone in chiral Lewis acid-catalyzed 1,3-DC of carbonyl ylides generated by dirhodium(II) co-catalyst.

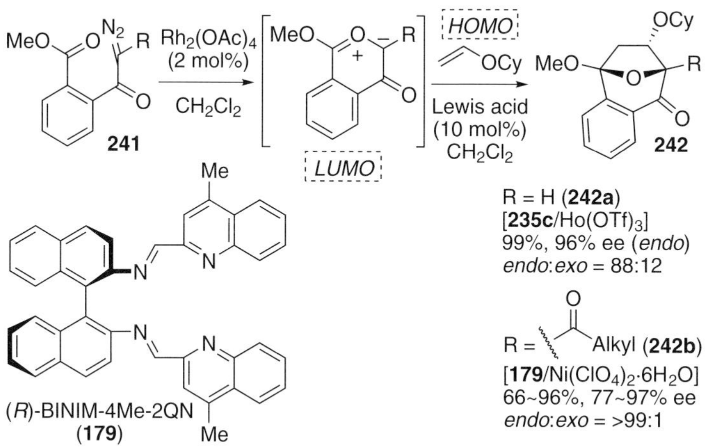

3.5 Carbonyl Ylides

Scheme 3.115 Use of electron-rich alkene as dipolarophile.

and electron-rich dipolarophile is also a conceivable approach to facilitate the so-called inverse-electron demand 1,3-DC. Screening of various chiral Lewis acids confirmed two appropriate catalysts for this purpose [143]. One is the rare earth metal triflate/Pybox-4,5-Ph$_2$ **235c** complex, which is especially suitable for the 1,3-DC of carbonyl ylides derived from α-diazoacetophenones (**241**, R = H) and vinyl ether (Scheme 3.115). In the reaction of α,α′-dicarbonyl diazo compounds (**241**, R = acyl), a highly enantioselective process could be realized by the use of the nickel(II)/(R)-BINIM-4Me-2QN **179** complex which gave endo-cycloadducts **242b** exclusively.

By tuning the metal sources, this inverse electron demand 1,3-DC of carbonyl ylides could be extended to the reaction of other diazocarbonyl compounds and vinyl ethers, as illustrated in Scheme 3.116.

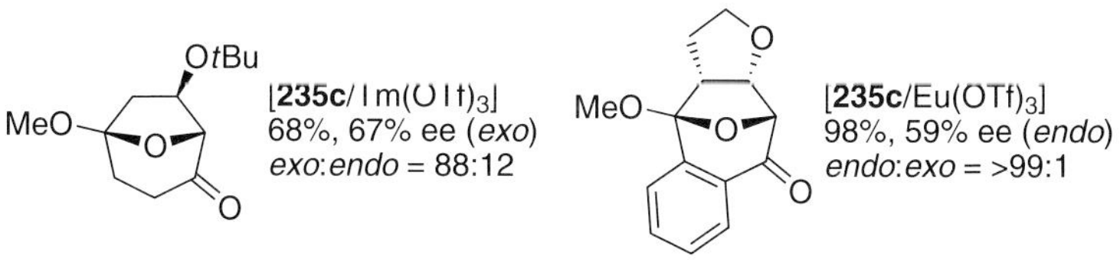

Scheme 3.116 Substrate scope of chiral Lewis acid-catalyzed 1,3-DC of carbonyl ylides generated by dirhodium(II) co-catalyst with electron-rich alkenes.

3.5.4
1,3-DC of Metal-Cotaining Carbonyl Ylides Generated via Electrophilic Activation of Alkynes

Electrophilic activation of an alkyne by a π-acidic transition metal and the successive nucleophilic addition of the internal carbonyl moiety to the alkyne were disclosed as a novel method for the generation of metal-containing carobonyl ylides by Iwasawa and

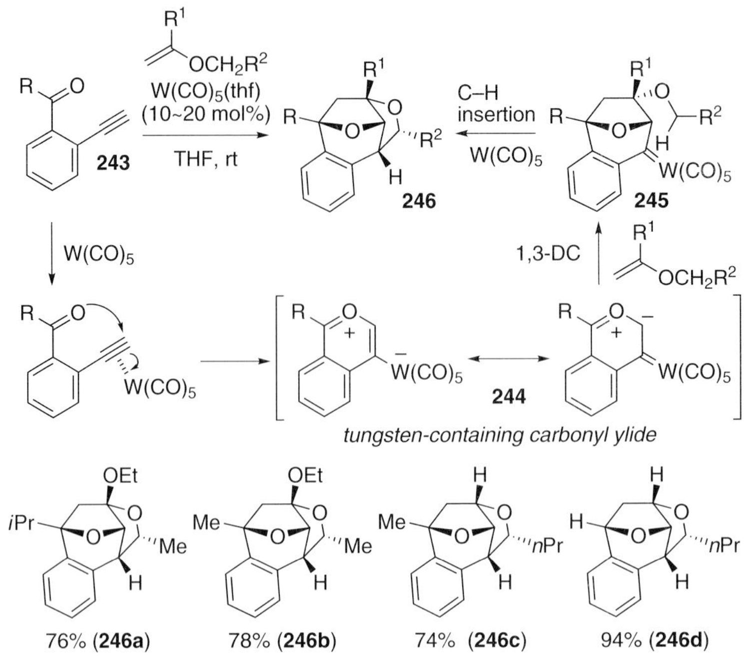

Scheme 3.117 Reaction pathway and substrate scope of 1,3-DC of tungsten-containing carbonyl ylides generated via electrophilic activation of alkynes.

coworkers [144]. In 2001, they discovered that treatment of o-ethynylphenyl ketones **243** and electron-rich alkenes with a catalytic amount of W(CO)$_5$(thf) gave rise to complex polycyclic compounds **246**, normally as a single isomer (Scheme 3.117) [145]. The mechanistic insight of this reaction was provided by the authors as follows. At first, 6-endo cyclization occurs via the nucleophilic addition of the carbonyl group to the W(CO)$_5$-π-complexed alkyne to give tungsten-containing carbonyl ylide **244**. The intermediacy of this transient species was directly confirmed by the NMR measurement. 1,3-DC then takes place with electron-rich olefins, such as vinyl ether and ketene acetal. Use of *E*- and *Z*-vinyl ether as the dipolarophile, respectively, provided the isomeric products, thus indicating the dominance of the concerted mechanism. They also pointed out the reversibility of this 1,3-DC step. Finally, the tungsten-carbene moiety of the initial cycloadduct **245** inserts into the C–H bond of the neighboring ethereal carbon to give the polycyclic compound **246**. As a precursor of carbonyl ylides, not only o-ethynylphenyl ketones but also o-ethynylbenzaldehyde can be utilized without difficulty, giving the polycycle **246d**.

In the following research investigating the possibility of other transition metal catalysts, they found that the reaction of o-ethynylbenzoate or thioate **247** with vinyl ether in the presence of platinum(II) leads to the unexpected formation of the naphthalene ring **249** [146, 147]. The course of the reaction is considered to involve cycloaddition of platinum(II)-containing ylide, 1,2-alkylmigration of thus-formed non-stabilized carbene intermediate **248** followed by aromatization (Scheme 3.118).

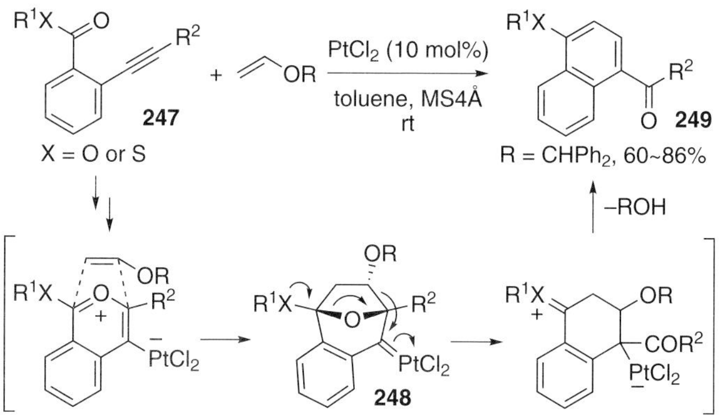

3.5 Carbonyl Ylides

Scheme 3.118 Generation and reaction of platinum-containing carbonyl ylides.

The successful examples presented above depend on the use of benzene-fused ynones and the application to acyclic γ,δ-ynones remained unsolved until 2008 [148]. Iwasawa and coworkers revealed that the generation of metal-containing carbonyl ylides from acyclic γ,δ-ynones **250** and subsequent 1,3-DC with electron-rich alkene can be realized by the use of PtCl$_2$ as catalyst (Scheme 3.119). The fate of the first 1,3-DC adducts depended on the nature of the substrates. In the case of γ,δ-ynones having a substituent at the propargylic position (**250**, R^4 = Me or OEt), the primary cycloadduct **251** underwent a 1,2-hydrogen shift and furnished the cyclic compound **254**. On the other hand, the primary cycloadduct **251** derived from γ,δ-ynones with no substituent at the propargylic position caused further ring-rearrangement to give 8-oxabicyclo[3.2.1]octane **255** as a major product, through the formation of the oxocarbenium intermediate **252** via 1,2-alkyl migration and

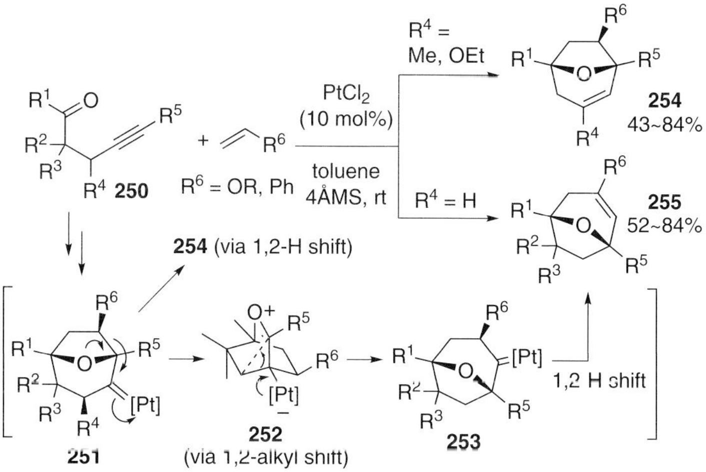

Scheme 3.119 Reaction sequence of metal-containing carbonyl ylides generated from acyclic γ,δ-ynones.

further 1,2-hydrogen shift of the nonstabilized carbene intermediate **253**. Other metal sources, such as W, Re, Ir, Au and Ag salts, were found to be ineffective.

Generation of metal-containing carbonyl ylides is not limited to 6-endo cyclization. Zhang et al. described in their gold(I)-catalyzed synthesis of bicycle[3.2.0]heptanes **258** that treatment of ethynyl ketones **256** with gold(I) species was assumed to afford the five-membered gold-containing carbonyl ylide **257** as a key intermediate (Scheme 3.120) [149]. 1,3-DC of this species with vinyl ether and subsequent skeletal rearrangement led to the formation of strained bicycles.

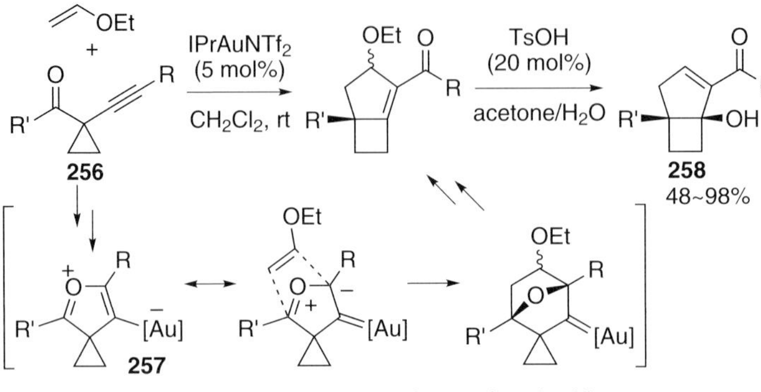

Scheme 3.120 Gold-catalyzed generation of 5-membered gold-containing carbonyl ylide and its reaction with vinyl ether.

Intramolecular 1,3-DC of metal-containing carbonyl ylides was reported by Oh and coworkers [150]. They first accommodated o-(1,6-enynyl)benzaldehydes **259** in rhodium(I) catalysis, aiming at the realization of Pauson–Khand-type cyclization. What they observed as a major product was the unexpected polycyclic compound **261**, presumably generated via 1,3-DC of rhodium-containing carbonyl ylide **260** (Scheme 3.121). Rhodium-bound carbene was assumed to undergo hydration and dehydrogenation to provide a keto moiety in the product at the any stage of the reaction. In the absence of water, extensive decomposition of the starting material

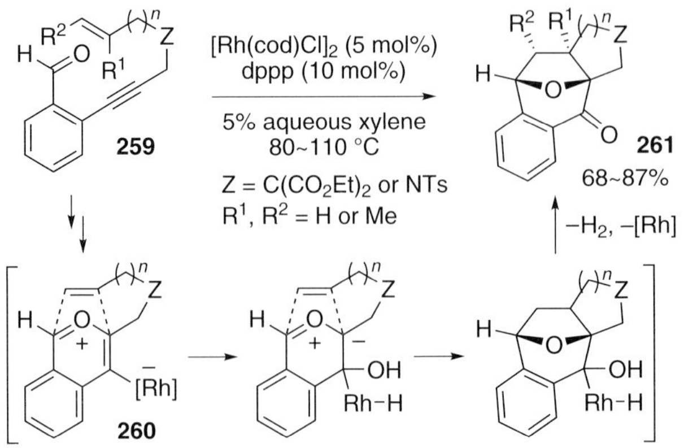

Scheme 3.121 Intramolecular 1,3-DC of rhodium-containing carbonyl ylides.

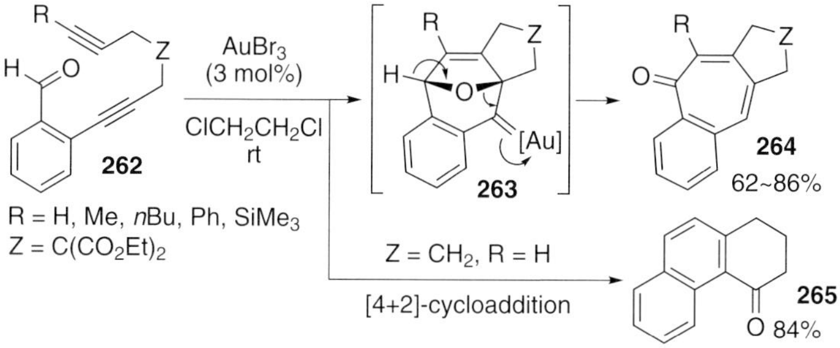

Scheme 3.122 Intramolecular 1,3-DC of gold-containing carbonyl ylides.

was observed. Under the optimized catalytic conditions composed of a catalytic amount of [Rh(cod)Cl]$_2$ and dppp, modifications of the tether length, the linker moiety and the substituent at the olefin were all tolerated.

Gold(III)-catalyzed reaction of o-(1,6-dienyl)benzaldehydes **262** furnished the tricyclic compounds **264** via 1,3-DC and subsequent fragmentation of gold-carbene species **263** (Scheme 3.122) [151]. In contrast, a naphthalene ring **265** was produced when the substrate having no substituent at the tether moiety was subjected to the reaction conditions. This discrepancy was explained by considering the involvement of a mechanistically distinctive [4 + 2]-cycloaddition.

Oh et al. also applied cyclic conjugated enynals **266** bearing an alkene moiety to this synthetic maneuver. The platinum(II)-carbene moiety generated after the key 1,3-DC was dissociated by 1,4-hydrogen-atom shift, while giving rise to [m,7,n]-tricycles **267** (Scheme 3.123).

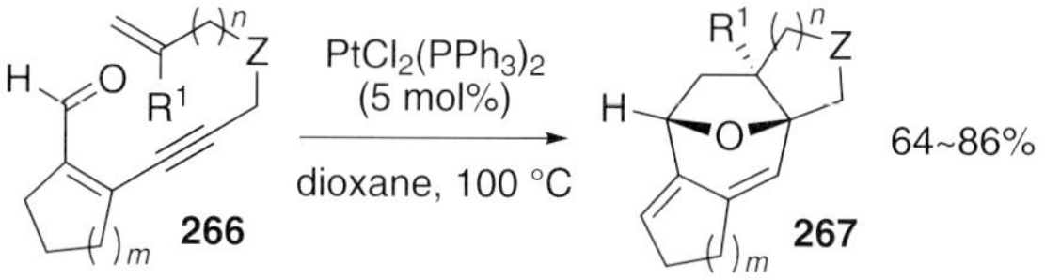

Scheme 3.123 Intramolecular 1,3-DC of platinum-containing carbonyl ylides generated from cyclic conjugated enynals.

3.6
Nitrile Oxides and Nitrile Imines

One classical and representative method for the generation of nitrile oxides [153] is the dehydrochlorination of hydroximoyl chlorides **269** under the influence of organic or inorganic base (Scheme 3.124). Hydroximoyl chlorides, in turn, can be prepared from the electrophilic chlorination of the corresponding oximes **268** by the use of reagents such as NCS or chlorine [154].

An additional practical option for the generation of nitrile oxides is the dehydration of nitroalkanes, reported by Mukaiyama and Hoshino in 1960 (Scheme 3.125) [155].

Due to the susceptibility of the thus-generated nitrile oxides to dimerize gradually giving furoxans **270**, freshly generated nitrile oxides are normally utilized without

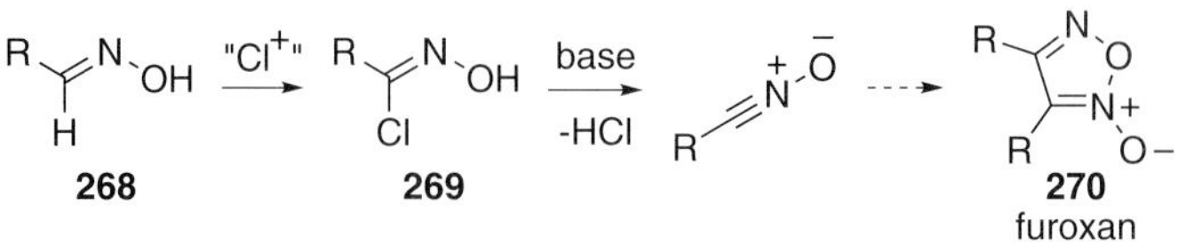

Scheme 3.124 Base-mediated generation of nitrile oxides from hydroximoyl chlorides.

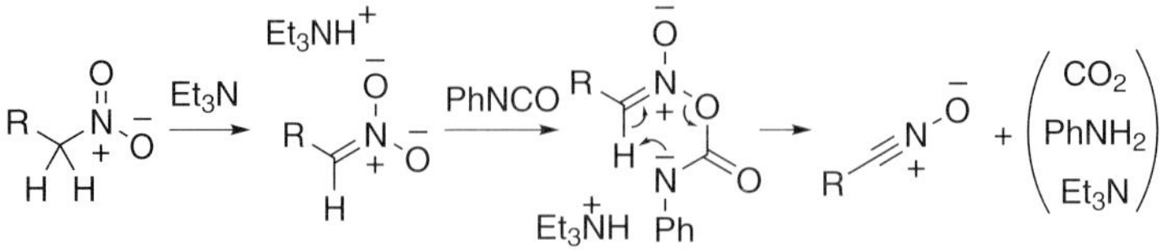

Scheme 3.125 Generation of nitrile oxides by the dehydration of nitroalkanes.

isolation. The only exception is the aromatic nitrile oxide having substituents at the 2- and 6-positions of the aryl moiety, such as mesityl nitrile oxide, which can be isolated as a bench-stable compound [156].

In the first part of this section, magnesium(II)-associated hydroxy-directed 1,3-DC of nitrile oxides, which was pioneered by Kanemasa's research and later developed as a synthetically useful method by Carreira and coworkers, will be described in detail. In the second part, asymmetric 1,3-DC of nitrile oxides accomplished by the use of an external chiral metal reagent is reviewed.

1,3-DC of nitrile imines is also included in this context, since basically the same strategy utilized in 1,3-DC of nitrile oxides can be applied to nitrile imines.

3.6.1
Magnesium(II)-Associated Hydroxy-Directed 1,3-DC of Nitrile Oxides

In search of a new method for the generation of nitrile oxides from hydroximoyl chlorides, Kanemasa and coworkers considered using an organometallic reagent as base [157]. Investigation of several organometallics led to the finding that treatment of benzohydroximoyl chloride 271 with nBuLi, EtMgBr or Et$_2$Zn and successive subjection of methyl acrylate successfully provided the cylcoadduct 273 in high yield (Scheme 3.126). One intriguing advantage of this generation method is the reluctance of the thus-generated nitrile oxides complexed with metal chlorides 272 for dimerization, compared to the naked nitrile oxides generated by the conventional triethylamine-mediated protocol. Another key finding is the induction of remarkably high diastereoselectivity in the 1,3-DC of the thus-generated nitrile oxides with allylic alcohol 274 having an α-chirality. Namely, 1,3-DC of benzonitrile oxide generated from the reaction of benzohydroximoyl chloride 271 and EtMgBr with allylic alcohol 274 (R = CO$_2$Me) provided the corresponding cycloadduct (275, R = CO$_2$Me) with high syn-selectivity, whereas the use of nitrile oxide generated by the treatment with triethylamine resulted in poor selectivity favoring the anti-isomer (syn:anti = 30 : 70). An allylic alcohol bearing no electron-withdrawing substituent (274, R = H) can

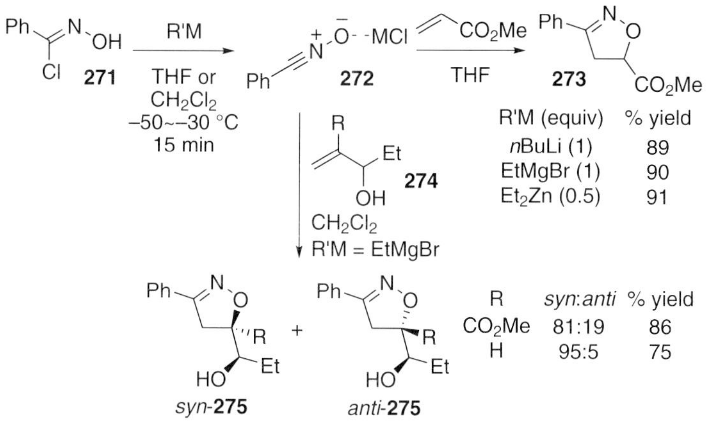

Scheme 3.126 1,3-DC of nitrile oxides generated by the treatment of hydroximoyl chloride with organometallic reagent.

R'M (equiv)	% yield
nBuLi (1)	89
EtMgBr (1)	90
Et$_2$Zn (0.5)	91

R	syn:anti	% yield
CO$_2$Me	81:19	86
H	95:5	75

also be used. Control of diastereoselectivity by the chelation of nitrile oxide and allylic alcohol with magnesium(II) is considered a crucial factor, since the use of other metal sources led to diminished selectivity.

In contrast to the terminal allylic alcohols, which are smoothly converted to 2,5-disubstituted isoxazoline as a single regioisomer, use of internal olefins as the dipolarophile raises a problem of regioselectivity, as well as poor reactivity. As a means to overcome these issues, Kanemasa et al. elaborated the diastereoselective magnesium(II)-chelation controlled 1,3-DC of nitrile oxides to be accommodated in the regioselective reaction with allylic alcohols having a 1,2-disubstituted alkene or a 1,1,2-trisubstituted alkene 276 (Scheme 3.127) [158]. During examination of the reaction conditions, they found that magnesium(II) allylic alkoxide 277 exhibits remarkably higher reactivity compared to the parent allylic alcohol. This reactive magnesium(II) allylic alkoxide underwent a regioselective 1,3-DC process with benzonitrile oxide to give isoxazolines 278 bearing a hydroxymethyl moiety at the 5-position exclusively. In the illustrated example, 1 equiv of magnesium(II) allylic alkoxide was consumed as base to generate nitrile oxide and it could be replaced by

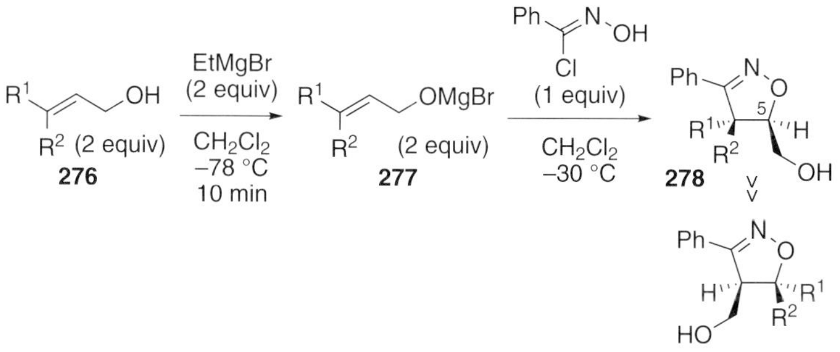

Scheme 3.127 Magnesium(II)-associated hydroxyl-directed regioselective 1,3-DC of nitrile oxides.

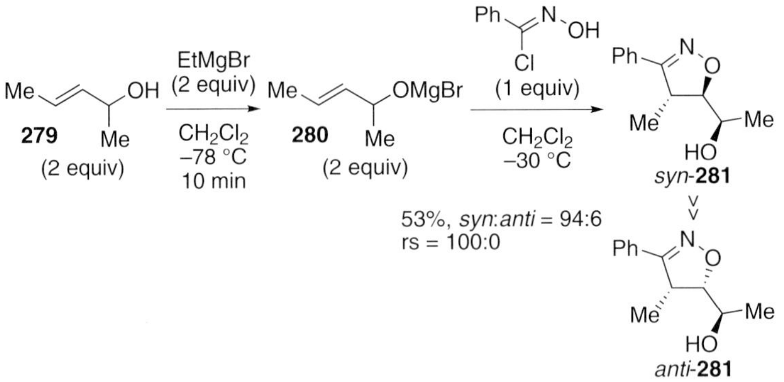

Scheme 3.128 Magnesium(II)-associated hydroxyl-directed regio- and syn-selective 1,3-DC of nitrile oxides.

other organometallic reagents, such as EtMgBr or iPrOLi. Direct use of allylic alcohols in combination with nBuOMgBr was also demonstrated as a practical method.

Subjection of an internal allylic alcohol **279** having an α-chirality to the newly developed reaction conditions afforded the corresponding cycloadduct **281** as a single regioisomer in a highly syn-selective manner (Scheme 3.128) [159].

The origin of this regio- and syn-selectivity can be explained by assuming the coordination of in situ generated nitrile oxide to the magnesium allylic alkoxide (Scheme 3.129). Diastereoselectivity arises from the preferential formation of the transient conformer with less 1,3-allylic strain.

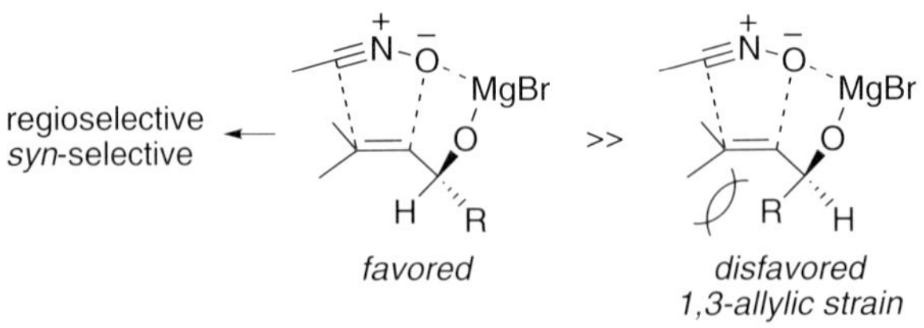

Scheme 3.129 Explanation of the observed stereoselectivity.

Despite the potential utility of Kanemasa's magnesium(II)-associated hydroxy-directed 1,3-DC of nitrile oxides, such as latent aldol adducts, synthetic application had been rarely examined, presumably due to the limitation regarding the use of aromatic nitrile oxides. In this context, a significant improvement was realized in 2001, wherein Carreira and coworkers established a brand new procedure allowing the use of aliphatic nitrile oxides [160]. Due to the lability of aliphatic nitrile oxides, they adopted the direct transfer of the hydroximinoyl chloride solution, prepared by the treatment of the corresponding oxime with tert-butyl hypochlorite at −78 °C, to the solution of magnesium allylic alkoxides (Scheme 3.130). Their protocol allowed facile access to all possible isomers of the masked polyketide building blocks starting from optically active allylic alcohols and oximes.

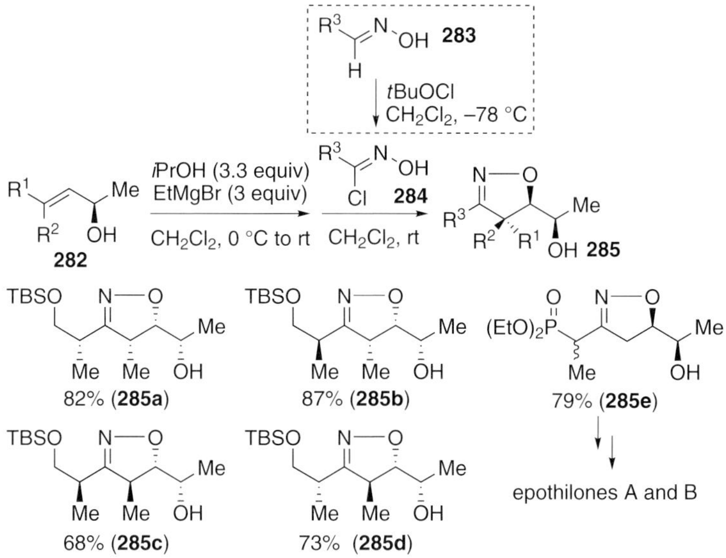

3.6 Nitrile Oxides and Nitrile Imines | 151

Scheme 3.130 Application of magnesium(II)-associated hydroxy-directed 1,3-DC of nitrile oxides to the polyketide building block synthesis.

Carreira's protocol can be extended to the cycloaddition of homoallylic alcohols (Scheme 3.131) [161], and cyclic allylic alcohols (Scheme 3.132) [162].

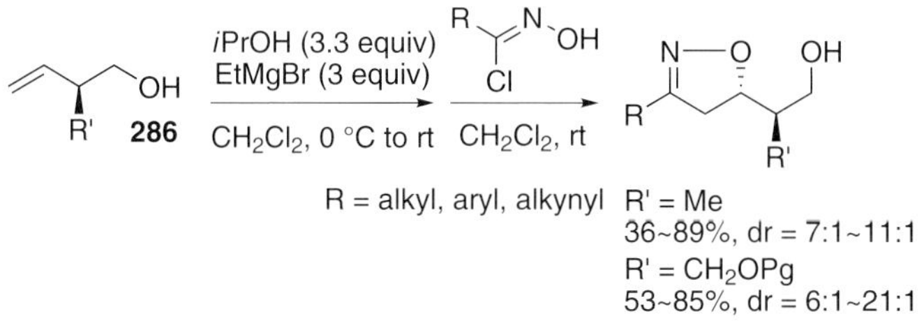

R = alkyl, aryl, alkynyl R' = Me
36~89%, dr = 7:1~11:1
R' = CH$_2$OPg
53~85%, dr = 6:1~21:1

Scheme 3.131 Use of homoallylic alcohols in magnesium(II)-associated hydroxy-directed 1,3-DC of nitrile oxides.

66%

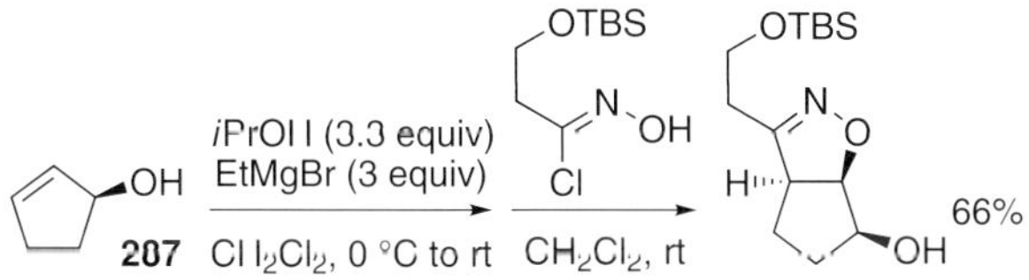

Scheme 3.132 Use of cyclic allylic alcohols in magnesium(II)-associated hydroxy-directed 1,3-DC of nitrile oxides.

3.6.2
Catalytic Asymmetric 1,3-DC of Nitrile Oxides

In the early study of Ukaji, Inomata and coworkers regarding asymmetric 1,3-DC of nitrile oxides by an external reagent, it was found that addition of hydroximoyl chloride to the solution of stoichiometric amounts of (R,R)-DIPT complexed with diethylzinc and allyl alcohol furnished the corresponding isoxazoline with a fairly good enantioselectivity. As an extension of this research, they developed the conditions that utilize only a catalytic amount of (R,R)-DIPT in 1996, thus realizing the first example of a catalytic asymmetric 1,3-DC of nitrile oxides (Scheme 3.133).

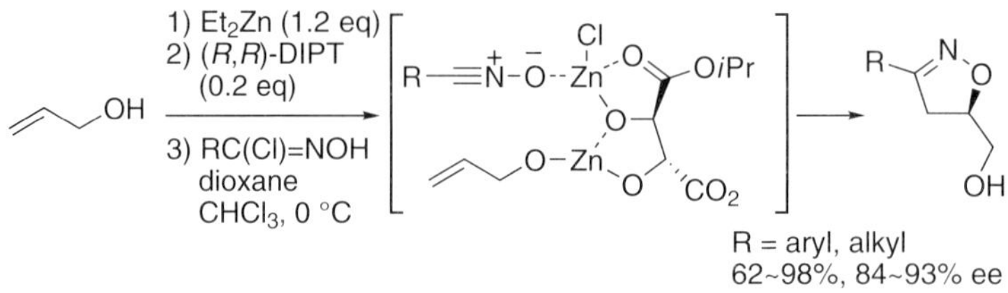

Scheme 3.133 Zinc-mediated enantioselective 1,3-DC of nitrile oxides.

Development of enantioselective 1,3-DC of nitrile oxides employing chiral Lewis acid-activated α,β-unsaturated carbonyl compounds had long been thwarted due to the highly coordinative character of nitrile oxides to Lewis acid catalyst, the feasibility of nitrile oxides to dimerize and the incompatibility of Lewis acid with the basic conditions required for the generation of nitrile oxides from hydroximoyl chlorides. In 2004, Sibi and coworkers reported that substoichiometric amounts of magnesium (II)/bisoxazoline **18** complex facilitates 1,3-DC of nitrile oxides with α,β-unsaturated carbonyl compounds having a coordinative template [164]. Two main concerns raised in their research were the difficulty of controlling the regioselectivity and the identification of a proper method for the generation of nitrile oxides. The former issue could be solved by using the pyrazolidinone template **17**, which would prevent the formation of one regioisomer **289** by the steric effect. To overcome the latter issue, they introduced a new method for the generation of nitrile oxides in another reaction flask that contains Amberlyst 21 as immobilized base. As such, asymmetric 1,3-DC of nitrile oxides with α,β-unsaturated carbonyl compounds could be realized, furnishing cycloadducts **288** with excellent levels of regio- and enantioselectivity (Scheme 3.134).

Accommodation of methacrolein, a single point-binding α,β-unsaturated carbonyl compound, in the asymmetric 1,3-DC of nitrile oxides was implemented by Kündig and coworkers using their ruthenium catalyst [165], which had also been utilized in 1,3-DC of nitrones [27]. By applying Sibi's method for the generation of nitrile oxides, enantioenriched isoxazolines **290** could be obtained with modest to high enantioselectivities (Scheme 3.135). The low catalyst loading required for the completion of the reaction clearly indicated the ability of the ruthenium catalyst to reversibly coordinate to both nitrile oxide and methacrolein.

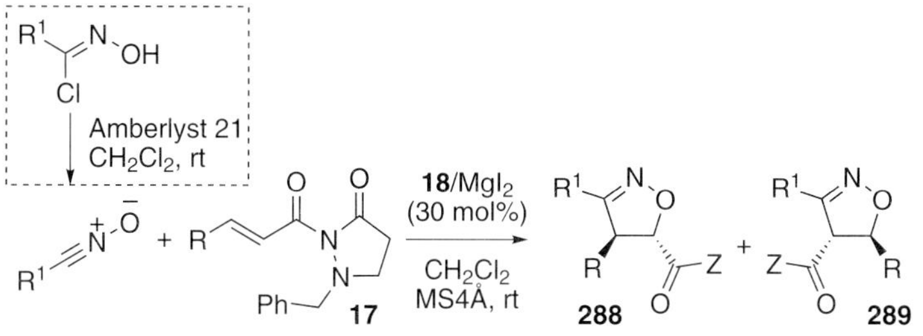

Scheme 3.134 Magnesium/bisoxazoline complex-catalyzed enantioselective 1,3-DC of nitrile oxides.

R¹	R	% yield	288:289	% ee (288)
mesityl	Me	84	99:1	99
mesityl	Ph	85	99:1	99
mesityl	CO_2Et	75	99:1	99
4-ClC$_6$H$_4$	Me	70	99:1	96
4-MeC$_6$H$_4$	Me	61	10:1	99
tBu	Me	44	99:1	92

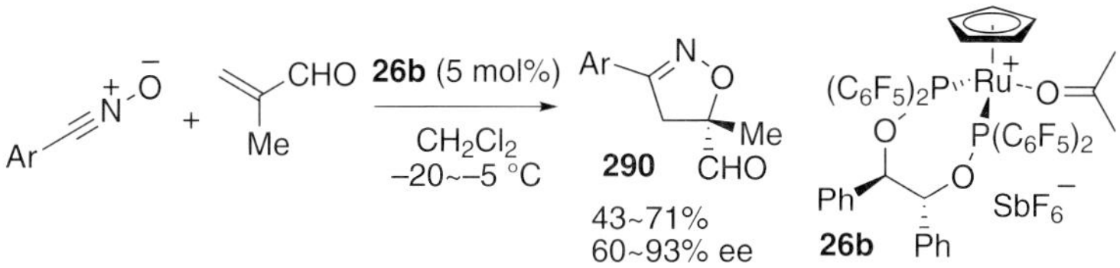

Scheme 3.135 Enantioselective 1,3-DC of nitrile oxides with methacrolein catalyzed by ruthenium complex.

Other examples of asymmetric 1,3-DC of nitrile oxides include magnesium(II)/pybox and ytterbium(III)/pybox-mediated reactions with 2-acryloyl-3-oxazolidinone and its analogs [166], and ytterbium(III)/(−)-sparteine complex-mediated reaction with allylic alcohol as well as methyl crotonate [167].

3.6.3
1,3-DC of Nitrile Imines

Kanemasa *et al.* applied their hydroxy-directed approach to the 1,3-DC of nitrile imines [157]. 1,3-DC of nitrile imine generated by the treatment of hydrazonyl chloride **291** with magnesium alkoxide or amide and allylic alchol furnished the 2-pyrazoline **292** with high syn-selectivity (Scheme 3.136).

The only successful example of catalytic asymmetric 1,3-DC of nitrile imines to date was reported by Sibi and coworkers in 2005 [168]. One key feature of their methodology is the mild generation of nitrile imines from hydrazonyl bromides **293** *in situ*, which could be realized by the use of diisopropylethylamine as a Lewis acid compatible base (Scheme 3.137). In the presence of magnesium(II)/bisoxazoline

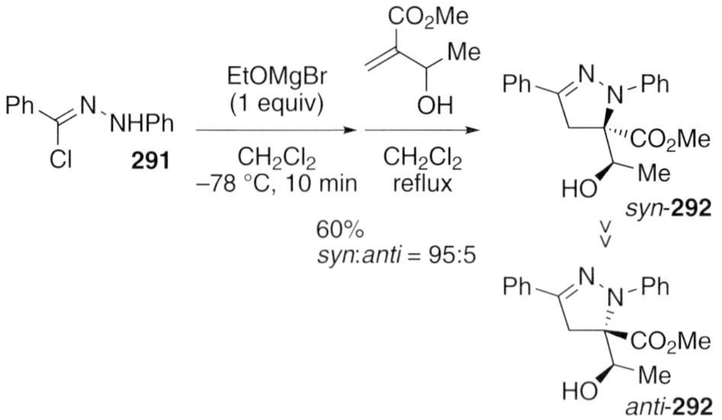

Scheme 3.136 Magnesium(II)-associated hydroxyl-directed 1,3-DC of nitrile imines.

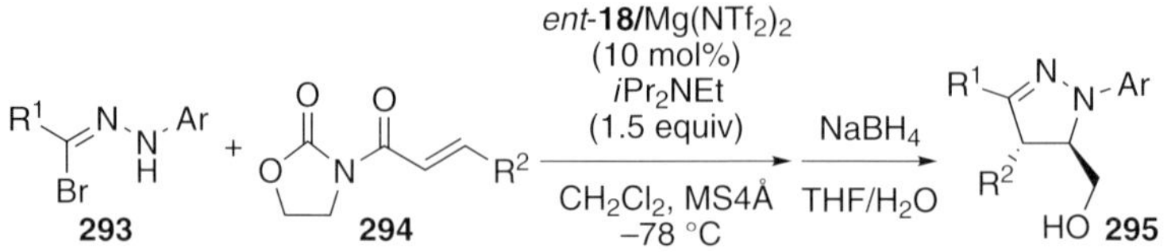

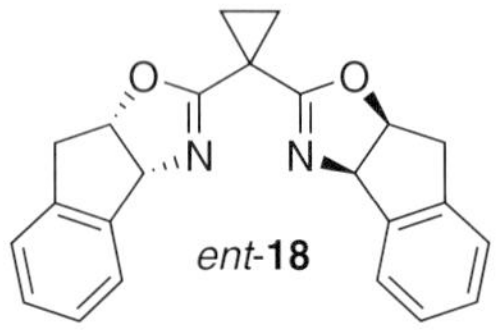

Scheme 3.137 Magnesium/bisoxazoline complex-catalyzed enantioselective 1,3-DC of nitrile imines.

R¹	Ar	R²	% yield	% ee
Ph	4-BrC$_6$H$_4$	Me	91	99
Ph	4-BrC$_6$H$_4$	Ph	95	97
Ph	4-BrC$_6$H$_4$	OBz	97	96
Ph	4-BrC$_6$H$_4$	H	95	84
iPr	4-BrC$_6$H$_4$	Me	98	99
Ph	Ph	Me	92	95

ent-**18** complex, 1,3-DC of the thus-generated nitrile oxides with various 3-alkenoyl-2-oxazolidiones **294** afforded enantioenriched 2-pyrazolines **295** with high enantioselectivity in a regiospecific manner. Use of α-substituted and α,β-disubstituted α,β-unsaturated carbonyl compounds in this asymmetric 1,3-DC can also be accomplished by the utilization of the pyrazolidinone template instead of oxazolidinone [169].

3.7
Diazo Compounds

Despite the long history of 1,3-DC of diazo compounds which started in 1888, when E. Buchner carried out the reaction of methyl diazoacetate **296** and methyl acrylate to obtain Δ²-pyrazoline **297** (Scheme 3.138) [170], the synthetic value of this cycloaddition had been rarely paid attention until recently.

The use of diazo compounds (diazoalkanes and diazo carbonyls) spreads over a range of synthetic methods, such as transition-metal catalyzed cyclopropanation, C−H insertion, and Wolff rearrangement [171]. Accordingly, the practical proce-

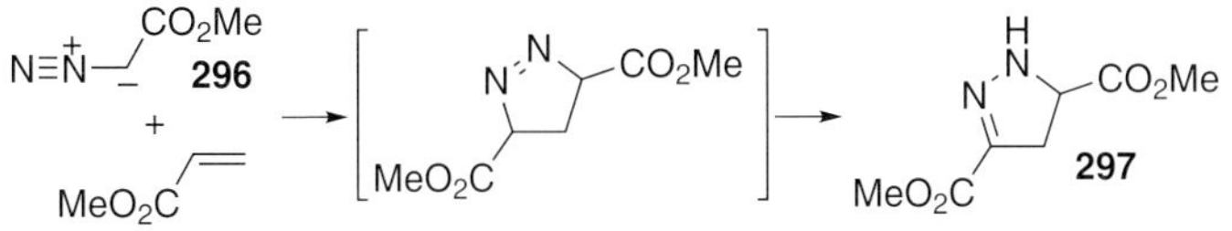

Scheme 3.138 Early example of 1,3-DC of the diazo compound.

dures to access diazo compounds have also been extensively developed synergistically.

One classical method for the preparation of alkyl diazoacetate is the diazotization of alkyl glycinate, although this is not a reliable method due to the competitive decomposition of the thus-formed diazoacetate under the acidic reaction condition [172].

Diazo transfer from sulfonyl azide is commonly applied for the synthesis of diazoacetate, which is known as Regitz synthesis (Scheme 3.139) [173].

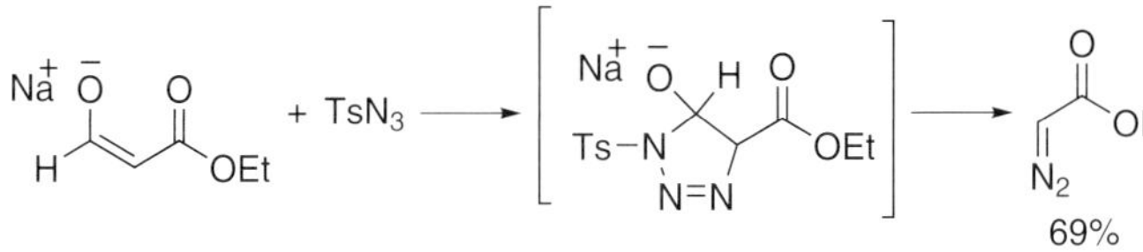

Scheme 3.139 Regitz deformylating synthesis of ethyl diazoacetate.

Another important diazo compound used in 1,3-DC is (trimethylsilyl)diazomethane, which can be prepared by Shioiri's method (Scheme 3.140) [174].

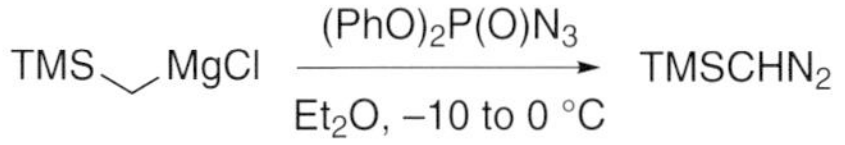

Scheme 3.140 Shioiri's procedure for the synthesis of (trimethylsilyl)diazomethane.

3.7.1
Lewis Acid Catalyzed Asymmetric 1,3-DC of Diazo Compounds with α,β-Unsaturated Carbonyl Compounds

In 1997, Carreira and coworkers demonstrated the synthetic utility of camphor sultam-derived dipolarophiles **298** in 1,3-DC using TMSCHN$_2$ as the 1,3-dipole, which certainly encouraged the development of Lewis acid catalyzed 1,3-DC of diazo compounds [175]. Optically enriched Δ^2-pyrazolines **300** could be obtained in excellent yields and high diastereoselectivities after the acidic desilylation of Δ^1-pyrazolines **299** (Scheme 3.141). Noteworthy is the fact that they were aware of promoting the reaction by the action of a Lewis acid catalyst.

The first example of a chiral Lewis acid catalyzed asymmetric 1,3-DC of TMSCHN$_2$ and α,β-unsaturated carbonyl compounds was realized by Kanemasa and coworkers in 2000 [176]. They demonstrated that magnesium(II)/DBFOX/Ph **302** complex

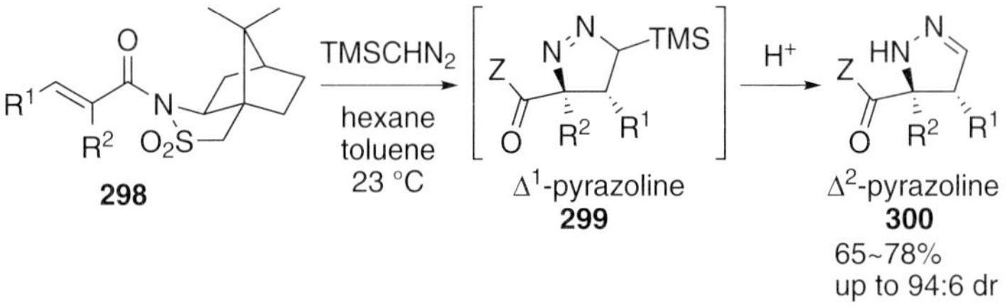

Scheme 3.141 Asymmetric 1,3-DC of (trimethylsilyl) diazomethane with camphor sultam-derived dipolarophiles.

effectively catalyzes the 1,3-DC of TMSCHN$_2$ and 3-alkenoyl-4,4-dimethyl-2-oxazolidinones **301**. The initially formed Δ^1-pyrazoline was intercepted *in situ* by desilylative acetylation to provide *N*-acetyl-Δ^2-pyrazoline **303** in high yield and excellent enantioselectivity (Scheme 3.142).

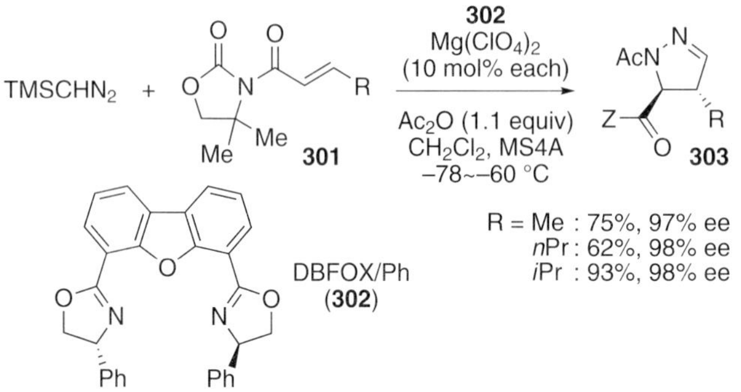

Scheme 3.142 Chiral Lewis acid-catalyzed enantioselective 1,3-DC of (trimethylsilyl)diazomethane and α,β-unsaturated carbonyl compounds.

Use of alkyl diazoacetate as the 1,3-dipole in catalytic asymmetric synthesis was first described briefly by Jørgensen and Gothelf in their review article in 1998 [2b], no experimental detail was included (Scheme 3.143).

In 2006, Maruoka and coworkers discovered the utility of α-alkyl-α,β-unsaturated aldehydes **304** as the dipolarophile in enantioselective 1,3-DC of alkyl diazoacetate catalyzed by titanium(IV)-BINOLate, such as 1:2 mixture of Ti(O*i*Pr)$_4$ and

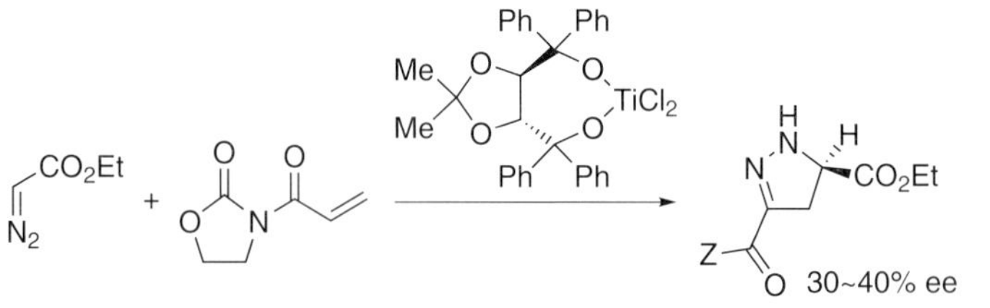

Scheme 3.143 Use of ethyl diazoacetate as 1,3-dipole in chiral Lewis acid-catalyzed enantioselective 1,3-DC.

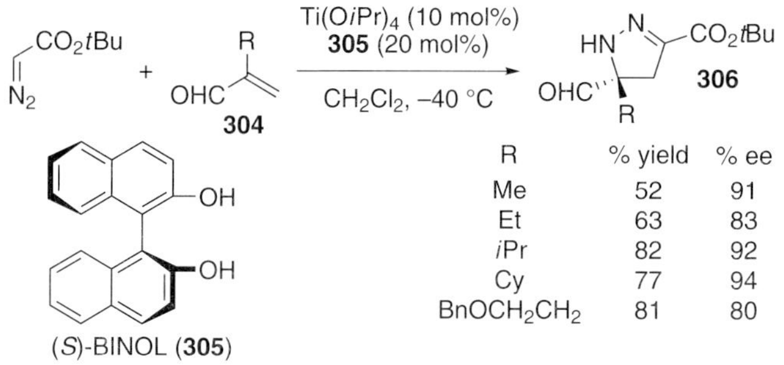

3.7 Diazo Compounds

Scheme 3.144 Titanium-BINOLate-catalyzed enantioselective 1,3-DC of diazoacetate and α-substituted enals.

R	% yield	% ee
Me	52	91
Et	63	83
iPr	82	92
Cy	77	94
BnOCH$_2$CH$_2$	81	80

BINOL **305** complex (Scheme 3.144) [177]. The synthetic utility of the thus-formed optically enriched Δ^2-pyrazoline **306** with a quaternary stereogenic center could be demonstrated by the four-step synthesis of bromopyrrole alkaloid manzacidin A (Scheme 3.145).

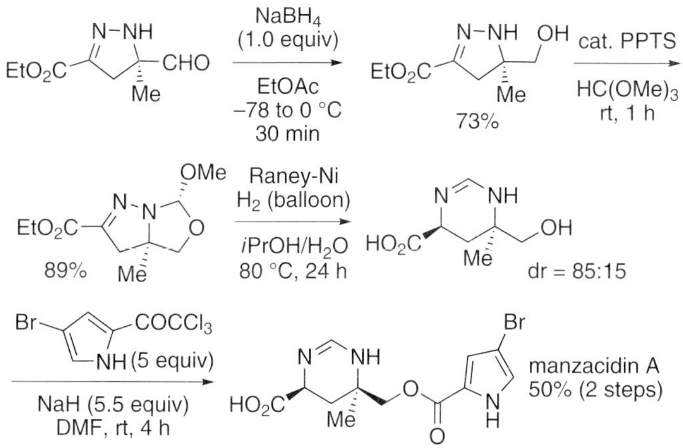

Scheme 3.145 Total synthesis of manzacidin A from the enantiomerically enriched pyrazoline.

Following this report, Sibi and coworkers have developed the enantioselective 1,3-DC of diazoacetates and α,β-unsaturated pyrazolidinone imides **307** using Mg(NTf$_2$)$_2$/bisoxazoline *ent*-**18** as chiral Lewis acid catalyst (Scheme 3.146) [178]. The notable advantage of this protocol is the applicability to various α- and/or β-substituted unsaturated carbonyls providing the cycloadducts **308** in good yields with an excellent level of enantiocontrol.

3.7.2
Lewis acid Catalyzed 1,3-DC of Diazo Compounds with Alkynes

Li *et al.* discovered that InCl$_3$ could act as a Lewis acid catalyst which facilitates the 1,3-DC of diazoacetates and propiolate to give the pyrazoles in water [179]. 1,3-DC

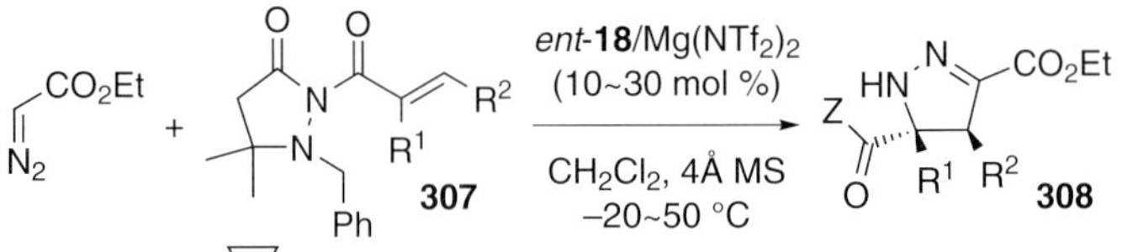

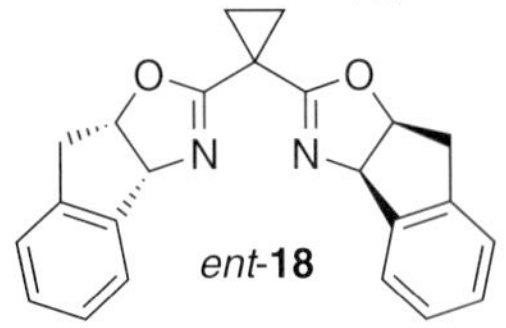

Scheme 3.146 Magnesium/bisoxazoline complex-catalyzed 1,3-DC of ethyl diazoacetate.

R^1	R^2	% yield	% ee
H	Me	72	99
H	Ph	54	90
Me	Me	52	98
Me	H	99	97

of simple α-unsubstituted diazoacetate and propiolate led to the pyrazoline **310** via the 1,3-hydrogen shift of the primary cycloadduct **309** (Scheme 3.147). On the other hand, use of aryldiazoacetate as the 1,3-dipole led to the sequential 1,3-DC, 1,5-aryl shift and 1,3-hydrogen shift via intermediate **311**, giving the rearranged arylpyrazole **312**.

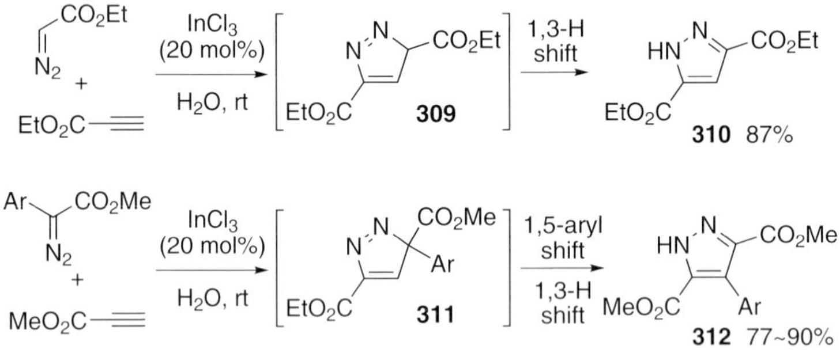

Scheme 3.147 Indium-catalyzed 1,3-DC of diazoacetates and propiolate.

3.8
Conclusions

In the realm of ever-evolving 1,3-dipolar cycloadditions, the last decade especially has witnessed the upsurge of enantioselective 1,3-DCs catalyzed by well-tuned chiral Lewis acids. The recent advent of chiral organocatalysis has further widened the synthetic options in catalytic enantioselective 1,3-DCs and this area is expected to grow rapidly. Additionally, novel methods for the generation of transient 1,3-dipoles have been emerging continuously, offering facile access to complex structures. The application of these strategies in target-oriented synthesis will be of particular interest in the future, as the conventional 1,3-DCs led to the various elegant syntheses of valuable complex molecules in the past.

3.9
Experimental: Selected Experimental

3.9.1
Preparation of N-Benzylidenebenzylamine N-oxide [6]

$$Ph\diagdown N(H)\diagdown Ph \xrightarrow[\text{MeOH} \\ 0\,°C\,2h\,\text{then rt, 18 h}]{\text{NaWO}_4\cdot 2H_2O\,(0.4\,\text{mol\%})\\ H_2O_2\,\text{aq (3.3 equiv)}} Ph\diagdown N^+(O^-)\diagdown Bn \quad 96\%$$

In a three-necked flask equipped with a mechanical stirrer, a thermometer, and a dropping funnel were placed dibenzylamine (182 g, 0.923 mol), $Na_2WO_4\cdot 2H_2O$ (12.2 g, 3.69 mmol), and methanol (1.8 L). To the stirred solution was added 30% aqueous hydrogen peroxide (314 g, 2.77 mol) dropwise with ice cooling over a period of 2 h. After the addition was complete, the reaction mixture was stirred at room temperature for 18 h. Methanol was removed under reduced pressure, and to the residue were added CH_2Cl_2 (500 mL × 3) and water (500 mL). The organic layer was separated, washed with saturated aqueous sodium bisulfite (500 mL), and dried over anhydrous sodium sulfate. Removal of the solvent gave N-benzylidenebenzylamine N-oxide (186 g, 96%) as crystals. Further purification was achieved by recrystallization from petroleum ether/CH_2Cl_2.

3.9.2
General Procedure for the Preparation of the Aldimine of N-Arylidene Amino Acid Ester [72b]

R^1	R^2	R^3	Yield (%)
Ph	H	Me	89
4-MeOPh	H	Mc	72
2-tolyl	H	Me	80
Ph	H	tBu	80
Ph	Me	Me	87

To a suspension of the corresponding amino acid ester hydrochloride (23.9 mmol) and $MgSO_4$ (25.0 mmol) in CH_2Cl_2 (25 mL) was added Et_3N (3.4 mL, 23.9 mmol). The mixture was stirred at room temperature for 1 h, and then the corresponding aldehyde (20.0 mmol) was added. The reaction was stirred at room temperature overnight, and the resulting precipitate was removed by filtration. The filtrate was washed with water (15 mL), the aqueous phase was extracted with CH_2Cl_2 (3 × 10 mL), and the combined organic phases were washed with brine, dried over $MgSO_4$, and concentrated. The resulting imino esters were obtained pure and used in 1,3-dipolar cycloadditions without further purification.

3.9.3
Preparation of 1-Benzylidene-3-oxopyrazolidin-1-ium-2-ide [107]

Ph–CHO + pyrazolidin-3-one → (ylide) 72%
MeOH, rt, 10 min

Benzaldehyde (1.07 mL, 10.5 mmol) was added to a solution of pyrazolidin-3-one (861 mg, 10.0 mmol) in methanol (1.0 mL). The mixture was stirred for 10 min at room temperature and then diluted with Et_2O (5.0 mL). The precipitate was collected by filtration, washed with Et_2O, and dried under vacuum to afford 1-benzylidene-3-oxopyrazolidin-1-ium-2-ide as a white solid (1.26 g, 7.24 mmol; 72%).

3.9.4
Preparation of *tert*-Butyl Diazoacetate [180]

Caution: Diazo compounds are potentially explosive and should be handled with special care.

tert-butyl acetoacetate + TsN_3 → tert-butyl diazoacetate 75%
TBAB (5 mol%), pentane/3N NaOH, 25 °C, 15 h

To a solution of *tert*-butyl acetoacetate (7.13 mL, 43 mmol), TsN_3 (8.40 g, 43 mmol) and TBAB (tetrabutylammonium bromide) (708 mg, 2.2 mmol) in pentane (100 mL) was added 3 N NaOH (30 mL, 90 mmol) dropwise at room temperature. After vigorous stirring for 15 h at the same temperature, the organic phase was separated, dried over Na_2SO_4 and carefully evaporated. The residue was purified by column chromatography on silica gel (eluting with pentane/ether = 20: 1) to give *tert*-butyl diazoacetate (4.58 g, 32.2 mmol; 75%).

References

1. (a) Huisgen, R. (1963) *Angew. Chem., Int. Ed.*, **2**, 565–598; (b) Huisgen, R. (1963) *Angew. Chem., Int. Ed.*, **2**, 633–645.
2. (a) For general reviews, see: Padwa, A. and Pearson, W.H.(eds) (2003) *Synthetic Applications of 1,3-Dipolar Cycloaddition Chemistry Toward Heterocycles and Natural Products*, Wiley and Sons, Hoboken, NJ; (b) Gothelf, K.V. and Jørgensen, K.A. (1998) *Chem. Rev.*, **98**, 863–911; (c) Gothelf, K.V. (2002) *Cycloaddition Reactions in Organic Synthesis* (eds S. Kobayashi and K.A. Jørgensen), Wiley-VCH, Weinheim, Chapter 6, pp. 211–247; (d) Kanemasa, S. (2002) *Cycloaddition Reactions in Organic Synthesis* (eds S. Kobayashi and K.A. Jørgensen), Wiley-VCH, Weinheim, Chapter 7, pp. 249–300; (e) Kanemasa, S. (2002) *Synlett*, 1371–1387; (f) Pellissier, H. (2007) *Tetrahedron*, **63**, 3235–3285; (g) Stanley, L.M. and Sibi, M.P. (2008) *Chem. Rev.*, **108**, 2887–2902.
3. Jørgensen, K.A. (2002) *Cycloaddition Reactions in Organic Synthesis* (eds S. Kobayashi and K.A. Jørgensen),

Wiley-VCH, Weinheim, Chapter 8, pp. 301–327.

4 Ess, D.H. and Houk, K.N. (2008) *J. Am. Chem. Soc.*, **130**, 10187–10198, and references therein.

5 (a) Frederickson, M. (1997) *Tetrahedron*, **53**, 403–425; (b) Gothelf, K.V. and Jørgensen, K.A. (2000) *Chem. Commun.*, 1449–1458.

6 Murahashi, S.-I., Mitsui, H., Shiota, T., Tsuda, T. and Watanabe, S. (1990) *J. Org. Chem.*, **55**, 1736–1744.

7 Guinchard, X., Vallée, Y. and Denis, J.-N. (2005) *Org. Lett.*, **7**, 5147–5150.

8 Partridge, K.M., Anzovino, M.E. and Yoon, T.P. (2008) *J. Am. Chem. Soc.*, **130**, 2920–2921.

9 Kanemasa, S., Uemura, T. and Wada, E. (1992) *Tetrahedron Lett.*, **33**, 7889–7892.

10 (a) Gothelf, K.V. and Jørgensen, K.A. (1994) *J. Org. Chem.*, **59**, 5687–5691; (b) Gothelf, K.V., Hazell, R.G. and Jørgensen, K.A. (1995) *J. Am. Chem. Soc.*, **117**, 4435–4436; (c) Gothelf, K.V., Thomsen, I. and Jørgensen, K.A. (1996) *J. Am. Chem. Soc.*, **118**, 59–64; (d) Gothelf, K.V., Hazell, R.G. and Jørgensen, K.A. (1996) *J. Org. Chem.*, **61**, 346–355; (e) Jensen, K.B., Gothelf, K.V. and Jørgensen, K.A. (1997) *Helv. Chim. Acta*, **80**, 2039–2046; (f) Jensen, K.B., Gothelf, K.V., Hazell, R.G. and Jørgensen, K.A. (1997) *J. Org. Chem.*, **62**, 2471–2477; (g) Sanchez-Blanco, A.I., Gothelf, K.V. and Jorgensen, K.A. (1997) *Tetrahedron Lett.*, **38**, 7923–7926; (h) Gothelf, K.V. and Jørgensen, K.A. (1997) *J. Chem. Soc., Perkin Trans. 2*, 111–116; (i) Gothelf, K.V., Hazell, R.G. and Jørgensen, K.A. (1998) *J. Org. Chem.*, **63**, 5483–5488.

11 (a) Hori, K., Kodama, H., Ohta, T. and Furukawa, I. (1996) *Tetrahedron Lett.*, **37**, 5947–5950; (b) Hori, K., Kodama, H., Ohta, T. and Furukawa, I. (1999) *J. Org. Chem.*, **64**, 5017–5023; (c) Kodama, H., Ito, J., Hori, K., Ohta, T. and Furukawa, I. (2000) *J. Organomet. Chem.*, **603**, 6–12.

12 Kobayashi, S. and Kawamura, M. (1998) *J. Am. Chem. Soc.*, **120**, 5840–5841.

13 Kanemasa, S., Oderaotoshi, Y., Tanaka, J. and Wada, E. (1998) *J. Am. Chem. Soc.*, **120**, 12355–12356.

14 (a) Desimoni, G., Faita, G., Mortoni, A. and Righetti, P. (1999) *Tetrahedron Lett.*, **40**, 2001–2004; (b) Crosignani, S., Desimoni, G., Faita, G., Filippone, S., Mortoni, A., Righetti, P. and Zema, M. (1999) *Tetrahedron Lett.*, **40**, 7007–7010; (c) Desimoni, G., Faita, G., Mella, M. and Boiocchi, M. (2005) *Eur. J. Org. Chem.*, 1020–1027.

15 (a) Iwasa, S., Tsushima, S., Shimada, T. and Nishiyama, H. (2001) *Tetrahedron Lett.*, **42**, 6715–6717; (b) Iwasa, S., Tsushima, S., Shimada, T. and Nishiyama, H. (2002) *Tetrahedron*, **58**, 227–232; (c) Iwasa, S., Maeda, H., Nishiyama, K., Tsushima, S., Tsukamoto, Y. and Nishiyama, H. (2002) *Tetrahedron*, **58**, 8281–8287; (d) Iwasa, S., Ishima, Y., Widagdo, H.S., Aoki, K. and Nishiyama, H. (2004) *Tetrahedron Lett.*, **45**, 2121–2124; (e) Phomkeona, K., Takemoto, T., Ishima, Y., Shibatomi, K., Iwasa, S. and Nishiyama, H. (2008) *Tetrahedron*, **64**, 1813–1822.

16 Suga, H., Kakehi, A., Ito, S. and Sugimoto, H. (2003) *Bull. Chem. Soc. Jpn.*, **76**, 327–334.

17 (a) Saito, T., Yamada, T., Miyazaki, S. and Otani, T. (2004) *Tetrahedron Lett.*, **45**, 9581–9584; (b) Saito, T., Yamada, T., Miyazaki, S. and Otani, T. (2004) *Tetrahedron Lett.*, **45**, 9585–9587.

18 Palomo, C., Oiarbide, M., Arceo, E., García, J.M., López, R., González, A. and Linden, A. (2005) *Angew. Chem., Int. Ed.*, **44**, 6187–6190.

19 Evans, D.A., Song, H.-J. and Fandrick, K.R. (2006) *Org. Lett.*, **8**, 3351–3354.

20 Lim, K.-C., Hong, Y.-T. and Kim, S. (2008) *Adv. Synth. Catal.*, **350**, 380–384.

21 Sibi, M.P., Ma, Z. and Jasperse, C.P. (2004) *J. Am. Chem. Soc.*, **126**, 718–719.

22 Sibi, M.P., Ma, Z., Itoh, K., Prabagaran, N. and Jasperse, C.P. (2005) *Org. Lett.*, **7**, 2349–2352.

23 Suga, H., Nakajima, T., Itoh, K. and Kakehi, A. (2005) *Org. Lett.*, **7**, 1431–1434.
24 Jen, W.S., Wiener, J.J.M. and MacMillan, D.W.C. (2000) *J. Am. Chem. Soc*, **122**, 9874–9875.
25 (a) Puglisi, A., Benaglia, M., Cinquini, M., Cozzi, F. and Celentano, G. (2004) *Eur. J. Org. Chem.*, 567–573; (b) Chow, S.S., Nevalainen, M., Evans, C.A. and Johannes, C.W. (2007) *Tetrahedron Lett.*, **48**, 277–280; (c) Rios, R., Ibrahem, I., Vesely, J., Zhao, G.-L. and Córdova, A. (2007) *Tetrahedron Lett.*, **48**, 5701–5705; (d) Lemay, M., Trant, J. and Ogilvie, W.W. (2007) *Tetrahedron*, **63**, 11644–11655.
26 (a) Karlsson, S. and Högberg, H.-E. (2002) *Tetrahedron: Asymmetry*, **13**, 923–926; (b) Karlsson, S. and Högberg, H.-E. (2003) *Eur. J. Org. Chem.*, 2782–2791.
27 (a) Viton, F., Bernardinelli, G. and Kündig, E.P. (2002) *J. Am. Chem. Soc.*, **124**, 4968–4969; (b) Badoiu, A., Bernardinelli, G., Mareda, J., Kündig, E.P. and Viton, F. (2008) *Chem. Asian J.*, **3**, 1298–1311.
28 (a) Achiral bulky aluminum Lewis acid-catalyzed reaction appeared concurrently: Kanemasa, S., Ueno, N. and Shirahase, M. (2002) *Tetrahedron Lett.*, **43**, 657–660; (b) For ab initio study see: Tanaka, J. and Kanemasa, S. (2001) *Tetrahedron*, **57**, 899–905.
29 (a) Mita, T., Ohtsuki, N., Ikeno, T. and Yamada, T. (2002) *Org. Lett.*, **4**, 2457–2460; (b) Ohtsuki, N., Kezuka, S., Kogami, Y., Mita, T., Ashizawa, T., Ikeno, T. and Yamada, T. (2003) *Synthesis*, 1462–1466; (c) Kezuka, S., Ohtsuki, N., Mita, T., Kogami, Y., Ashizawa, T., Ikeno, T. and Yamada, T. (2003) *Bull. Chem. Soc. Jpn.*, **76**, 2197–2207.
30 (a) Shirahase, M., Kanemasa, S. and Oderaotoshi, Y. (2004) *Org. Lett.*, **6**, 675–678; (b) Shirahase, M., Kanemasa, S. and Hasegawa, M. (2004) *Tetrahedron Lett.*, **45**, 4061–4063.
31 (a) Carmona, D., Lamata, M.P., Viguri, F., Rodríguez, R., Oro, L.A., Balana, A.I., Lahoz, F.J., Tejero, T., Merino, P., Franco, S. and Montesa, I. (2004) *J. Am. Chem. Soc.*, **126**, 2716–2717; (b) Carmona, D., Lamata, M.P., Viguri, F., Rodríguez, R., Oro, L.A., Lahoz, F.J., Balana, A.I., Tejero, T. and Merino, P. (2005) *J. Am. Chem. Soc.*, **127**, 13386–13398; (c) Carmona, D., Lamata, M.P., Viguri, F., Rodríguez, R., Fischer, T., Lahoz, F.J., Dobrinovitch, I.T. and Oro, L.A. (2007) *Adv. Synth. Catal.*, **349**, 1751–1758.
32 Carmona, D., Lamata, M.P., Viguri, F., Ferrer, J., García, N., Lahoz, F.J., Martín, M.L. and Oro, L.A. (2006) *Eur. J. Inorg. Chem.*, 3155–3166.
33 Wang, Y., Wolf, J., Zavalij, P. and Doyle, M.P. (2008) *Angew. Chem., Int. Ed.*, **47**, 1439–1442.
34 Kano, T., Hashimoto, T. and Maruoka, K. (2005) *J. Am. Chem. Soc.*, **127**, 11926–11927.
35 (a) Hashimoto, T., Omote, M., Kano, T. and Maruoka, K. (2007) *Org. Lett.*, **9**, 4805–4808; (b) Hashimoto, T., Omote, M., Hato, Y., Kano, T. and Maruoka, K. (2008) *Chem. Asian J.*, **3**, 407–412; (c) Hashimoto, T., Omote, M. and Maruoka, K. (2008) *Org. Biomol. Chem.*, **6**, 2263–2265.
36 Huang, Z.-Z., Kang, Y.-B., Zhou, J., Ye, M.-C. and Tang, Y. (2004) *Org. Lett.*, **6**, 1677–1679.
37 Carmona, D., Lamata, M.P., Viguri, F., Rodríguez, R., Lahoz, F.J. and Oro, L.A. (2007) *Chem. Eur. J.*, **13**, 9746–9756.
38 Kanemasa, S., Tsuruoka, T. and Wada, E. (1993) *Tetrahedron Lett.*, **34**, 87–90.
39 (a) Seerden, J.-P.G., Scholte op Reimer, A.W.A. and Scheeren, H.W. (1994) *Tetrahedron Lett.*, **35**, 4419–4422; (b) Seerden, J.-P.G., Scholte op Reimer, A.W.A. and Scheeren, H.W. (1995) *Tetrahedron: Asymmetry*, **6**, 1441–1450; (c) Seerden, J.-P.G., Boeren, M.M.M. and Scheeren, H.W. (1997) *Tetrahedron*, **53**, 11843–11852.
40 (a) Simonsen, K.B., Bayón, P., Hazell, R.G., Gothelf, K.V. and Jørgensen, K.A. (1999) *J. Am. Chem. Soc.*, **121**, 3845–3853;

(b) Jensen, K.B., Roberson, M. and Jørgensen, K.A. (2000) *J. Org. Chem.*, **65**, 9080–9084; (c) Simonsen, K.B., Jørgensen, K.A., Hu, Q.-S. and Pu, L. (1999) *Chem. Commun.*, 811–812.

41 Jensen, K.B., Hazell, R.G. and Jørgensen, K.A. (1999) *J. Org. Chem.*, **64**, 2353–2360.

42 (a) Bayón, P., de March, P., Espinosa, M., Figueredo, M. and Font, J. (2000) *Tetrahedron: Asym.*, **11**, 1757–1765; (b) Bayón, P., de March, P., Figueredo, M., Font, J. and Medrano, J. (2000) *Tetrahedron: Asym.*, **11**, 4269–4278.

43 Mikami, K., Ueki, M., Matsumoto, Y. and Terada, M. (2001) *Chirality*, **13**, 541–544.

44 Hori, K., Itoh, J., Ohta, T. and Furukawa, I. (1998) *Tetrahedron*, **54**, 12737–12744.

45 Jiao, P., Nakashima, D. and Yamamoto, H. (2008) *Angew. Chem., Int. Ed.*, **47**, 2411–2413.

46 Achiral thioureas have been noticed as an accelerator of inverse electron 1,3-DC of nitrone: Wittkopp, A. and Schreiner, P.R. (2003) *Chem. Eur. J.*, **9**, 407–414.

47 Ukaji, Y., Taniguchi, K., Sada, K. and Inomata, K. (1997) *Chem. Lett.*, **26**, 547–548; (b) Ding, X., Taniguchi, K., Ukaji, Y. and Inomata, K. (2001) *Chem. Lett.*, **30**, 468–469; (c) Ding, X., Ukaji, Y., Fujinami, S. and Inomata, K. (2002) *Chem. Lett.*, **31**, 302–303; (d) Ding, X., Taniguchi, K., Hamamoto, Y., Sada, K., Fujinami, S., Ukaji, Y. and Inomata, K. (2006) *Bull. Chem. Soc. Jpn.*, **79**, 1069–1083.

48 For a review see: Marco-Contelles, J. (2004) *Angew. Chem., Int. Ed.*, **43**, 2198–2200.

49 Kinugasa, M. and Hashimoto, S. (1972) *J. Chem. Soc., Chem. Commun.*, 466–467.

50 Ding, L.K. and Irwin, W.J. (1976) *J. Chem. Soc., Perkin Trans. 1*, 2382–2386.

51 (a) Okuro, K., Enna, M., Miura, M. and Nomura, M. (1993) *J. Chem. Soc., Chem. Commun.*, 1107–1108; (b) Miura, M., Enna, M., Okuro, K. and Nomura, M. (1995) *J. Org. Chem.*, **60**, 4999–5004.

52 Lo, M.M.-C. and Fu, G.C. (2002) *J. Am. Chem. Soc.*, **124**, 4572–4573.

53 Shintani, R. and Fu, G.C. (2003) *Angew. Chem., Int. Ed.*, **42**, 4082–4085.

54 (a) Ye, M.-C., Zhou, J., Huang, Z.-Z. and Tang, Y. (2003) *Chem. Commun.*, 2554–2555; (b) Ye, M.-C., Zhou, J. and Tang, Y. (2006) *J. Org. Chem.*, **71**, 3576–3582.

55 Coyne, A.G., Müller-Bunz, H. and Guiry, P.J. (2007) *Tetrahedron: Asym.*, **18**, 199–207.

56 Huisgen, R., Grashey, R. and Steingruber, E. (1963) *Tetrahedron Lett.*, **4**, 1441–1445.

57 For reviews see: (a) Nájera, C. and Sansano, J. (2003) *Curr. Org. Chem.*, **7**, 1105–1150; (b) Nájera, C. and Sansano, J.M. (2005) *Angew. Chem., Int. Ed.*, **44**, 6272–6276; (c) Coldham, I. and Hufton, R. (2005) *Chem. Rev.*, **105**, 2765–2810; (d) Husinec, S. and Savic, V. (2005) *Tetrahedron: Asym.*, **16**, 2047–2061; (e) Pandey, G., Banerjee, P. and Gadre, S.R. (2006) *Chem. Rev.*, **106**, 4484–4517.

58 Grigg, R., Kemp, J., Malone, J. and Tangthongkum, A. (1980) *J. Chem. Soc., Chem. Commun.*, 648–650.

59 (a) Grigg, R. and Gunaratne, H.Q.N. (1982) *J. Chem. Soc., Chem. Commun.*, 384–386; (b) Grigg, R., Gunaratne, H.Q.N. and Sridharan, V. (1987) *Tetrahedron*, **43**, 5887–5898.

60 (a) Barr, D.A., Grigg, R., Gunaratne, H.Q.N., Kemp, J., McMeekin, P. and Sridharan, V. (1988) *Tetrahedron*, **44**, 557–570; (b) Amornraksa, K., Barr, D., Donegan, G., Grigg, R., Ratanankul, P. and Sridharan, V. (1989) *Tetrahedron*, **45**, 4649–4668; (c) Barr, D.A., Grigg, R. and Sridharan, V. (1989) *Tetrahedron Lett.*, **30**, 4727–4730; (d) Barr, D.A., Dorrity, M.J., Grigg, R., Malone, J.F., Montgomery, J., Rajviroongit, S. and Stevenson, P. (1990) *Tetrahedron Lett.*, **31**, 6569–6572; (e) Grigg, R., Montgomery, J. and Somasunderam, A. (1992) *Tetrahedron*, **48**, 10431–10442.

61 (a) Tsuge, O., Kanemasa, S. and Yoshioka, M. (1988) *J. Org. Chem.*, **53**, 1384–1391; (b) Kanemasa, S., Yoshioka, M. and Tsuge, O. (1989) *Bull. Chem. Soc. Jpn.*, **62**,

869–874; (c) Kanemasa, S., Uchida, O. and Wada, E. (1990) *J. Org. Chem.*, **55**, 4411–4417.
62. (a) Ayerbe, M., Arrieta, A., Cossío, F.P. and Linden, A. (1998) *J. Org. Chem.*, **63**, 1795–1805; (b) Vivanco, S., Lecea, B., Arrieta, A., Prieto, P., Morao, I., Linden, A. and Cossío, F.P. (2000) *J. Am. Chem. Soc.*, **122**, 6078–6092.
63. See also: ref. [58], [61b] and [61c].
64. Allway, P. and Grigg, R. (1991) *Tetrahedron Lett.*, **32**, 5817–5820.
65. Grigg, R. (1995) *Tetrahedron: Asym.*, **6**, 2475–2486.
66. Longmire, J.M., Wang, B. and Zhang, X. (2002) *J. Am. Chem. Soc.*, **124**, 13400–13401.
67. Gothelf, A.S., Gothelf, K.V., Hazell, R.G. and Jørgensen, K.A. (2002) *Angew. Chem., Int. Ed.*, **41**, 4236–4238.
68. Chen, C., Li, X. and Schreiber, S.L. (2003) *J. Am. Chem. Soc.*, **125**, 10174–10175.
69. Knöpfel, T.F., Aschwanden, P., Ichikawa, T., Watanabe, T. and Carreira, E.M. (2004) *Angew. Chem., Int. Ed.*, **43**, 5971–5973.
70. Gao, W., Zhang, X. and Raghunath, M. (2005) *Org. Lett.*, **7**, 4241–4244.
71. Oderaotoshi, Y., Cheng, W., Fujitomi, S., Kasano, Y., Minakata, S. and Komatsu, M. (2003) *Org. Lett.*, **5**, 5043–5046.
72. (a) Cabrera, S., Arrayás, R.G. and Carretero, J.C. (2005) *J. Am. Chem. Soc.*, **127**, 16394–16395; (b) Cabrera, S., Arrayás, R.G., Martín-Matute, B., Cossíno, F.P. and Carretero, J.C. (2007) *Tetrahedron*, **63**, 6587–6602.
73. Nájera, C., Retamosa, M.d.G. and Sansano, J.M. (2008) *Angew. Chem., Int. Ed.*, **47**, 6055–6058.
74. Zeng, W., Chen, G.-Y., Zhou, Y.-G. and Li, Y.-X. (2007) *J. Am. Chem. Soc.*, **129**, 750–751.
75. Alemparte, C., Blay, G. and Jørgensen, K.A. (2005) *Org. Lett.*, **7**, 4569–4572.
76. (a) Zeng, W. and Zhou, Y.-G. (2005) *Org. Lett.*, **7**, 5055–5058; (b) Zeng, W. and Zhou, Y.-G. (2007) *Tetrahedron Lett.*, **48**, 4619–4622.
77. Nájera, C., Retamosa, M.d.G. and Sansano, J.M. (2007) *Org. Lett.*, **9**, 4025–4028.
78. Shi, J.-W., Zhao, M.-X., Lei, Z.-Y. and Shi, M. (2008) *J. Org. Chem.*, **73**, 305–308.
79. Dogan, Ö., Koyuncu, H., Garner, P., Bulut, A., Youngs, W.J. and Panzner, M. (2006) *Org. Lett.*, **8**, 4687–4690.
80. Saito, S., Tsubogo, T. and Kobayashi, S. (2007) *J. Am. Chem. Soc.*, **129**, 5364–5365.
81. Stohler, R., Wahl, F. and Pfaltz, A. (2005) *Synthesis*, 1431–1436.
82. (a) Llamas, T., Arrayás, R.G. and Carretero, J.C. (2006) *Org. Lett.*, **8**, 1795–1798; (b) Llamas, T., Arrayás, R.G. and Carretero, J.C. (2007) *Synthesis*, 950–956.
83. López-Pérez, A., Adrio, J. and Carretero, J.C. (2008) *J. Am. Chem. Soc.*, **130**, 10084–10085.
84. Fukuzawa, S.-i. and Oki, H. (2008) *Org. Lett.*, **10**, 1747–1750.
85. Yan, X.-X., Peng, Q., Zhang, Y., Zhang, K., Hong, W., Hou, X.-L. and Wu, Y.-D. (2006) *Angew. Chem., Int. Ed.*, **45**, 1979–1983.
86. Vicario, J.L., Reboredo, S., Badía, D. and Carrillo, L. (2007) *Angew. Chem., Int. Ed.*, **46**, 5168–5170.
87. Grigg, R. (1987) *Chem. Soc. Rev.*, **16**, 89–121.
88. Ibrahem, I., Rios, R., Vesely, J. and Córdova, A. (2007) *Tetrahedron Lett.*, **48**, 6252–6257.
89. (a) Chen, X.-H., Zhang, W.-Q. and Gong, L.-Z. (2008) *J. Am. Chem. Soc.*, **130**, 5652–5653; (b) For the acid-mediated generation of azomethine ylides, see Ref. [59].
90. Xue, M.-X., Zhang, X.-M. and Gong, L.-Z. (2008) *Synlett*, 691–694.
91. Peddibhotla, S. and Tepe, J.J. (2004) *J. Am. Chem. Soc.*, **126**, 12776–12777.
92. Melhado, A.D., Luparia, M. and Toste, F.D. (2007) *J. Am. Chem. Soc.*, **129**, 12638–12639.
93. Bowman, R.K. and Johnson, J.S. (2004) *J. Org. Chem.*, **69**, 8537–8540.
94. Vaultier, M. and Carrié, R. (1978) *Tetrahedron Lett.*, **19**, 1195–1198.

95 Pohlhaus, P.D., Bowman, R.K. and Johnson, J.S. (2004) *J. Am. Chem. Soc.*, **126**, 2294–2295.

96 (a) Kusama, H., Takaya, J. and Iwasawa, N. (2002) *J. Am. Chem. Soc.*, **124**, 11592–11593; (b) Takaya, J., Kusama, H. and Iwasawa, N. (2004) *Chem. Lett.*, **33**, 16–17.

97 Kusama, H., Miyashita, Y., Takaya, J. and Iwasawa, N. (2006) *Org. Lett.*, **8**, 289–292.

98 Su, S. and Porco, J.A. Jr (2007) *J. Am. Chem. Soc.*, **129**, 7744–7745.

99 Schantl, J.G. (2004) *Sci. Synth.*, **27**, 731–824.

100 (a) Huisgen, R., Fleischmann, R. and Eckell, A. (1960) *Tetrahedron Lett.*, **1**, (33), 1–4; (b) Huisgen, R. and Eckell, A. (1960) *Tetrahedron Lett.*, **1**, (33), 5–8.

101 Huisgen, R., Grashey, R. and Krischke, R. (1962) *Tetrahedron Lett.*, **3**, 387–391.

102 (a) Oppolzer, W. (1970) *Tetrahedron Lett.*, **11**, 2199–2204; (b) Oppolzer, W. (1970) *Tetrahedron Lett.*, **11**, 3091–3094; (c) Oppolzer, W. (1977) *Angew. Chem., Int. Ed.*, **16**, 10–23.

103 (a) Jacobi, P.A., Brownstein, A., Martinelli, M. and Grozinger, K. (1981) *J. Am. Chem. Soc.*, **103**, 239–241; (b) Jacobi, P.A., Martinelli, M.J. and Solvenko, P. (1984) *J. Am. Chem. Soc.*, **106**, 5594–5598; (c) Overman, L.E., Rogers, B.N., Tellew, J.E. and Trenkle, W.C. (1997) *J. Am. Chem. Soc.*, **119**, 7159–7160; (d) Nilsson, B.L., Overman, L.E., Read de Alaniz, J. and Rohde, J.M. (2008) *J. Am. Chem. Soc.*, **130**, 11297–11299.

104 Godtfresden, W.O. and Vangedal, S. (1955) *Acta Chem. Scand.*, **9**, 1498–1509.

105 Howard, J.C., Gever, G. and Wei, P. (1963) *J. Org. Chem.*, **28**, 868–870.

106 (a) Dorn, H. and Otto, A. (1968) *Angew. Chem., Int. Ed.*, **7**, 214–215; (b) Dorn, H. and Otto, A. (1968) *Chem. Ber.*, **101**, 3287–3301.

107 Shintani, R. and Fu, G.C. (2003) *J. Am. Chem. Soc.*, **125**, 10778–10779.

108 Suárez, A., Downey, C.W. and Fu, G.C. (2005) *J. Am. Chem. Soc.*, **127**, 11244–11245.

109 Chen, W., Yuan, X.-H., Li, R., Du, W., Wu, Y., Ding, L.-S. and Chen, Y.-C. (2006) *Adv. Synth. Catal.*, **348**, 1818–1822.

110 Chen, W., Du, W., Duan, Y.-Z., Wu, Y., Yang, S.-Y. and Chen, Y.-C. (2007) *Angew. Chem., Int. Ed.*, **46**, 7667–7670.

111 Suga, H., Funyu, A. and Kakehi, A. (2007) *Org. Lett.*, **9**, 97–100.

112 Sibi, M., Rane, D., Stanley, L.M. and Soeta, T. (2008) *Org. Lett.*, **10**, 2971–2974.

113 Kato, T., Fujinami, S., Ukaji, Y. and Inomata, K. (2008) *Chem. Lett.*, **37**, 342–343.

114 Hesse, K.-D. (1971) *Liebigs Ann. Chem.*, **743**, 50–56.

115 (a) Grigg, R., Kemp, J. and Thompson, N. (1978) *Tetrahedron Lett.*, **19**, 2827–2830; (b) Grigg, R., Jordan, M. and Malone, J.F. (1979) *Tetrahedron Lett.*, **20**, 3877–3878; (c) Grigg, R., Dowling, M., Jordan, M.W. and Sridharan, V. (1987) *Tetrahedron*, **43**, 5873–5886.

116 Kobayashi, S., Hirabayashi, R., Shimizu, H., Ishitani, H. and Yamashita, Y. (2003) *Tetrahedron Lett.*, **44**, 3351–3354.

117 Kobayashi, S., Shimizu, H., Yamashita, Y., Ishitani, H. and Kobayashi, J. (2002) *J. Am. Chem. Soc.*, **124**, 13678–13679.

118 Yamashita, Y. and Kobayashi, S. (2004) *J. Am. Chem. Soc.*, **126**, 11279–11282.

119 (a) Shirakawa, S., Lombardi, P.J. and Leighton, J.L. (2005) *J. Am. Chem. Soc.*, **127**, 9974–9975; (b) Tran, K. and Leighton, J.L. (2006) *Adv. Synth. Catal.*, **348**, 2431–2436.

120 Tran, K., Lombardi, P.J. and Leighton, J.L. (2008) *Org. Lett.*, **10**, 3165–3167.

121 (a) Doyle, M.P. and Forbes, D.C. (1998) *Chem. Rev.*, **98**, 911–936; (b) Hodgson, D.M., Pierard, F.Y.T.M. and Stupple, P.A. (2001) *Chem. Soc. Rev.*, **30**, 50–61; (c) Mehta, G. and Muthusamy, S. (2002) *Tetrahedron*, **58**, 9477–9504; (d) Chiu, P. (2005) *Pure Appl. Chem.*, **77**, 1183–1189; (e) Padwa, A. (2005) *J. Organomet. Chem.*, **690**, 5533–5540; (f) Singh, V., Murali Krishna, U., Vikrant, and Trivedi, G.K. (2008) *Tetrahedron*, **64**, 3405–3428.

122 Kharasch, M.S., Rudy, T., Nudenberg, W. and Büchi, G. (1953) *J. Org. Chem.*, **18**, 1030–1044.

123 (a) Takebayashi, M., Ibata, T. and Ueda, K. (1970) *Bull. Chem. Soc. Jpn.*, **43**, 1500–1505; (b) Takebayashi, M., Ibata, T., Ueda, K. and Ohashi, T. (1970) *Bull. Chem. Soc. Jpn.*, **43**, 3964; (c) Ueda, K., Ibata, T. and Takebayashi, M. (1972) *Bull. Chem. Soc. Jpn.*, **45**, 2779–2782; (d) Ibata, T., Motoyama, T. and Hamaguchi, M. (1976) *Bull. Chem. Soc. Jpn.*, **49**, 2298–2301; (e) Ibata, T. and Jitsuhiro, K. (1979) *Bull. Chem. Soc. Jpn.*, **52**, 3582–3585; (f) Ibata, T., Jitsuhiro, K. and Tsubokura, Y. (1981) *Bull. Chem. Soc. Jpn.*, **54**, 240–244; (g) Ibata, T. and Toyoda, J. (1985) *Bull. Chem. Soc. Jpn.*, **58**, 1787–1792; (h) Ibata, T., Toyoda, J., Sawada, M. and Tanaka, T. (1986) *J. Chem. Soc., Chem. Commun.*, 1266–1267; (i) Ibata, T. and Toyoda, J. (1986) *Bull. Chem. Soc. Jpn.*, **59**, 2489–2493.

124 (a) de March, P. and Huisgen, R. (1982) *J. Am. Chem. Soc.*, **104**, 4952; (b) Huisgen, R. and de March, P. (1982) *J. Am. Chem. Soc.*, **104**, 4953–4954.

125 (a) Padwa, A. (1991) *Acc. Chem. Res.*, **24**, 22–28; (b) Padwa, A. and Hornbuckle, S.F. (1991) *Chem. Rev.*, **91**, 263–309; (c) Padwa, A. and Weingarten, M.D. (1996) *Chem. Rev.*, **96**, 223–270; (d) Padwa, A. (1998) *Chem. Commun.*, 1417–1424; (e) Padwa, A. (2005) *Helv. Chim. Acta*, **88**, 1357–1374.

126 (a) Padwa, A., Austin, D.J., Hornbuckle, S.F. and Price, A.T. (1992) *Tetrahedron Lett.*, **33**, 6427–6430; (b) Padwa, A., Austin, D.J. and Hornbuckle, S.F. (1996) *J. Org. Chem.*, **61**, 63–72.

127 Doyle, M.P., Forbes, D.C., Protopopova, M.N., Stanley, S.A., Vasbinder, M.M. and Xavier, K.R. (1997) *J. Org. Chem.*, **62**, 7210–7215.

128 Hodgson, D.M., Stupple, P.A. and Johnstone, C. (1997) *Tetrahedron Lett.*, **38**, 6471–6472.

129 (a) Hodgson, D.M., Stupple, P.A. and Johnstone, C. (1999) *Chem. Commun.*, 2185–2486; (b) Hodgson, D.M., Stupple, P.A., Pierard, F.Y.T.M., Labande, A.H. and Johnstone, C. (2001) *Chem. Eur. J.*, **7**, 4465–4476; (c) Hodgson, D.M., Selden, D.A. and Dossetter, A.G. (2003) *Tetrahedron: Asym.*, **14**, 3841–3849; (d) Hodgson, D.M., Stupple, P.A. and Johnstone, C. (2003) *ARKIVOC*, **vii**, 49–58.

130 (a) Hodgson, D.M., Labande, A.H. and Pierard, F.Y.T.M. (2003) *Synlett*, 59–62; (b) Hodgson, D.M., Labande, A.H., Pierard, F.Y.T.M. and Expósito Castro, M.Á. (2003) *J. Org. Chem.*, **68**, 6153–6159.

131 Hodgson, D.M., Glen, R., Grant, G.H. and Redgrave, A.J. (2003) *J. Org. Chem.*, **68**, 581–586.

132 (a) Hodgson, D.M., Glen, R. and Redgrave, A.J. (2002) *Tetrahedron Lett.*, **43**, 3927–3930; (b) Hodgson, D.M., Labande, A.H., Glen, R. and Redgrave, A.J. (2003) *Tetrahedron: Asym.*, **14**, 921–924; (c) Hodgson, D.M., Brückl, T., Glen, R., Labande, A.H., Selden, D.A., Dossetter, A.G. and Redgrave, A.J. (2004) *Proc. Nat. Acad. Sci. U.S.A.*, **101**, 5450–5454.

133 Suga, H., Ishida, H. and Ibata, T. (1998) *Tetrahedron Lett.*, **39**, 3165–3166.

134 Kitagaki, S., Anada, M., Kataoka, O., Matsuno, K., Umeda, C., Watanabe, N. and Hashimoto, S. (1999) *J. Am. Chem. Soc.*, **121**, 1417–1418.

135 Kitagaki, S., Yasugahira, M., Anada, M., Nakajima, M. and Hashimoto, S. (2000) *Tetrahedron Lett.*, **41**, 5931–5935.

136 Shimada, N., Anada, M., Nakamura, S., Nambu, H., Tsutsui, H. and Hashimoto, S. (2008) *Org. Lett.*, **10**, 3603–3606.

137 Tsutsui, H., Shimada, N., Abe, T., Anada, M., Nakajima, M., Nakamura, S., Nambu, H. and Hashimoto, S. (2007) *Adv. Synth. Catal.*, **349**, 521–526.

138 Lu, C.-D., Chen, Z.-Y., Liu, H., Hu, W.-H. and Mi, A.-Q. (2004) *Org. Lett.*, **6**, 3071–3074.

139 (a) Torssell, S., Kienle, M. and Somfai, P. (2005) *Angew. Chem. Int. Ed.*, **44**, 3096–3099; (b) Torssell, S. and Somfai, P. (2006) *Adv. Synth. Catal.*, **348**, 2421–2430.

140 (a) Suga, H., Kakehi, A., Ito, S., Inoue, K., Ishida, H. and Ibata, T. (2000) *Org. Lett.*, **2**, 3145–3148; (b) Suga, H., Kakehi, A., Ito, S., Inoue, K., Ishida, H. and Ibata, T. (2001) *Bull. Chem. Soc. Jpn.*, **74**, 1115–1121.

141 (a) Suga, H., Inoue, K., Inoue, S. and Kakehi, A. (2002) *J. Am. Chem. Soc.*, **124**, 14836–14837; (b) Suga, H., Inoue, K., Inoue, S., Kakehi, A. and Shiro, M. (2005) *J. Org. Chem.*, **70**, 47–56.

142 Suga, H., Suzuki, T., Inoue, K. and Kakehi, A. (2006) *Tetrahedron*, **62**, 9218–9225.

143 Suga, H., Ishimoto, D., Higuchi, S., Ohtsuka, M., Arikawa, T., Tsuchida, T., Kakehi, A. and Baba, T. (2007) *Org. Lett.*, **9**, 4359–4362.

144 Kusama, H. and Iwasawa, N. (2006) *Chem. Lett.*, **35**, 1082–1087.

145 Iwasawa, N., Shido, M. and Kusama, H. (2001) *J. Am. Chem. Soc.*, **123**, 5814–5815; (b) Kusama, H., Funami, H., Shido, M., Hara, Y., Takaya, J. and Iwasawa, N. (2001) *J. Am. Chem. Soc.*, **127**, 2709–2716.

146 (a) Kusama, H., Funami, H., Takaya, J. and Iwasawa, N. (2004) *Org. Lett.*, **6**, 605–608; (b) Kusama, H., Funami, H. and Iwasawa, N. (2007) *Synthesis*, 2014–2024.

147 For related reactions which were proposed to proceed via [4 + 2]-cycloaddition, see: (a) Asao, N., Takahashi, K., Lee, S., Kasahara, T. and Yamamoto, Y. (2002) *J. Am. Chem. Soc.*, **124**, 12650–12651; (b) Asao, N., Nogami, T., Lee, S. and Yamamoto, Y. (2003) *J. Am. Chem. Soc.*, **125**, 10921–10925; (c) Theoretical calculations suggest the involvement of [3 + 2] in these examples, see: Straub, B.F. (2004) *Chem. Commun.*, 1726–1728.

148 Kusama, H., Ishida, K., Funami, H. and Iwasawa, N. (2008) *Angew. Chem., Int. Ed.*, **47**, 4903–4905.

149 Li, G., Huang, X. and Zhang, L. (2008) *J. Am. Chem. Soc.*, **130**, 6944–6945.

150 (a) Shin, S., Gupta, A.K., Rhim, C.Y. and Oh, C.H. (2005) *Chem. Commun.*, 4429–4431; (b) Gupta, A.K., Rhim, C.Y., Oh, C.H., Mane, R.S. and Han, S.-H. (2006) *Green Chem.*, **8**, 25–28.

151 Kim, N., Kim, Y., Park, W., Sung, D., Gupta, A.K. and Oh, C.H. (2005) *Org. Lett.*, **7**, 5289–5291.

152 Oh, C.H., Ryu, J.H. and Lee, H.I. (2007) *Synlett*, 2337–2342.

153 For a review see: Kanemasa, S. (2004) *Sci. Synth.*, **19**, 17–40.

154 Liu, K.-C., Shelton, B.R. and Howe, R.K. (1980) *J. Org. Chem.*, **45**, 3916–3918.

155 Mukaiyama, T. and Hoshino, T. (1960) *J. Am. Chem. Soc.*, **82**, 5339–5342.

156 Grundmann, C. and Dean, J.M. (1965) *J. Org. Chem.*, **30**, 2809–2812.

157 (a) Kanemasa, S., Kobayashi, S., Nishiuchi, M., Yamamoto, H. and Wada, E. (1991) *Tetrahedron Lett.*, **32**, 6367–6370; (b) Kanemasa, S. and Kobayashi, S. (1993) *Bull. Chem. Soc. Jpn.*, **66**, 2685–2693.

158 Kanemasa, S., Nishiuchi, M. and Wada, E. (1992) *Tetrahedron Lett.*, **33**, 1357–1360.

159 (a) Kanemasa, S., Nishiuchi, M., Kamimura, A. and Hori, K. (1994) *J. Am. Chem. Soc.*, **116**, 2324–2339; (b) Kanemasa, S., Okuda, K., Yamamoto, H. and Kaga, S. (1997) *Tetrahedron Lett.*, **38**, 4095–4098; (c) Kamimura, A., Kaneko, Y., Ohta, A., Kakehi, A., Matsuda, H. and Kanemasa, S. (1999) *Tetrahedron Lett.*, **40**, 4349–4352.

160 (a) Bode, J.W. and Carreira, E.M. (2001) *J. Am. Chem. Soc.*, **123**, 3611–3612; (b) Bode, J.W. and Carreira, E.M. (2001) *J. Org. Chem.*, **66**, 6410–6424; (c) Bode, J.W., Fraefel, N., Muri, D. and Carreira, E.M. (2001) *Angew., Chem. Int. Ed.*, **40**, 2082–2085; (d) Fader, L.D. and Carreira, E.M. (2004) *Org. Lett.*, **6**, 2485–2488; (e) Muri, D., Lohse-Fraefel, N. and Carreira, E.M. (2005) *Angew. Chem. Int. Ed.*, **44**, 4036–4038.

161 Lohse-Fraefel, N. and Carreira, E.M. (2005) *Org. Lett.*, **7**, 2011–2014.

162 Becker, N. and Carreira, E.M. (2007) *Org. Lett.*, **9**, 3857–3858.

163 (a) Ukaji, Y., Sada, K. and Inomata, K. (1993) *Chem. Lett.*, **22**, 1847–1850; (b) Shimizu, M., Ukaji, Y. and Inomata, K. (1996) *Chem. Lett.*, **25**, 455–456; (c) Yoshida, Y., Ukaji, Y., Fujinami, S. and Inomata, K. (1998) *Chem. Lett.*, **27**, 1023–1024; (d) Tsuji, M., Ukaji, Y. and Inomata, K. (2002) *Chem. Lett.*, **31**, 1112–1113.

164 Sibi, M.P., Itoh, K. and Jasperse, C.P. (2004) *J. Am. Chem. Soc.*, **126**, 5366–5367.

165 Brinkmann, Y., Madhushaw, R.J., Jazzar, R., Bernardinelli, G. and Kündig, E.P. (2007) *Tetrahedron*, **63**, 8413–8419.

166 Yamamoto, H., Hayashi, S., Kubo, M., Harada, M., Hasegawa, M., Noguchi, M., Sumimoto, M. and Hori, K. (2007) *Eur. J. Org. Chem.*, 2859–2864.

167 Gucma, M. and Golebiewski, W.M. (2008) *J. Heterocyclic Chem.*, **45**, 241–245.

168 Sibi, M.P., Stanley, L.M. and Jasperse, C.P. (2005) *J. Am. Chem. Soc.*, **127**, 8276–8277.

169 Sibi, M.P., Stanley, L.M. and Soeta, T. (2006) *Adv. Synth. Catal.*, **348**, 2371–2375.

170 Buchner, E. (1888) *Ber. Dtsch. Chem. Ges.*, **21**, 2637–2647.

171 (a) Doyle, M.P., McKervey, M.A. and Ye, T. (1998) *Modern Catalytic Methods for Organic Synthesis with Diazo Compounds*, John Wiley and Sons; (b) Ye, T. and McKervey, M.A. (1994) *Chem. Rev.*, **94**, 1091–1160.

172 Searle, N.E. (1956) *Org. Synth.*, **36**, 25–28.

173 (a) Regitz, M. (1967) *Angew. Chem. Int. Ed.*, **6**, 733–749; (b) Regitz, M., Menz, F. and Rüter, J. (1967) *Tetrahedron Lett.*, **8**, 739–742.

174 Shioiri, T., Aoyoama, T. and Mori, S. (1990) *Org. Synth.*, **68**, 1–7.

175 (a) Mish, M.R., Guerra, F.M. and Carreira, E.M. (1997) *J. Am. Chem. Soc.*, **119**, 8379–8380; (b) Whitlock, G.A. and Carreira, E.M. (1997) *J. Org. Chem.*, **62**, 7916–7917; (c) Guerra, F.M., Mish, M.R. and Carreira, E.M. (2000) *Org. Lett.*, **2**, 4265–4267; (d) Sasaki, H. and Carreira, E.M. (2000) *Synthesis*, 135–138; (e) Whitlock, G.A. and Carreira, E.M. (2000) *Helv. Chim. Acta*, **83**, 2007–2022.

176 Kanemasa, S. and Kanai, T. (2000) *J. Am. Chem. Soc.*, **122**, 10710–10711.

177 (a) Kano, T., Hashimoto, T. and Maruoka, K. (2006) *J. Am. Chem. Soc.*, **128**, 2174–2175; (b) Hashimoto, T. and Maruoka, K. (2008) *Org. Biomol. Chem.*, **6**, 829–835.

178 Sibi, M.P., Stanley, L.M. and Soeta, T. (2007) *Org. Lett.*, **9**, 1553–1556.

179 Jiang, N. and Li, C.-J. (2004) *Chem. Commun.*, 394–395.

180 (a) Ledon, H. (1974) *Synthesis*, 347–348; (b) Ledon, H.J. (1979) *Org. Synth.*, **59**, 66–69. The procedure provided herein is the modified method of Ledon's synthesis accommodated in the author's laboratory.

4
Intramolecular 1,2-Addition and 1,4-Addition Reactions
Xiyan Lu and Xiuling Han

4.1
Introduction

This chapter concerns cyclization reactions involving intramolecular 1,2-addition or 1,4-addition of carbon–heteroatom multiple bonds, including aldehydes or ketones, imines and nitriles. In these carbon–heteroatom multiple bonds, there is a very large difference in electronegativity between the carbon and the heteroatoms. Since these bonds are strongly polar, with the carbon always the positive end (except for isonitriles), nucleophilic species will go to the carbon, and electrophilic species to the heteroatom. The nucleophilic species can be oxygen, nitrogen, sulfur and carbon nucleophiles. If the carbon–heteroatom multiple bond and the nucleophiles exist in the same substrate molecule, intramolecular 1,2-addition reaction will occur and a cyclization reaction can be realized. A similar situation occurs in 1,4-addition.

Stereochemistry is important in these reactions. Taking aldehyde as an example, the two faces of the carbonyl group are prochiral (except formaldehyde). The two sites above and below the plane of the carbonyl group of an aldehyde can be described as enantiotopic, which is the important factor in forming stereoisomers.

4.2
Cyclization via Intramolecular 1,2-Addition Reactions

4.2.1
1,2-Addition of Carbonyl Compounds

4.2.1.1 1,2-Addition of Dicarbonyl Compounds (Aldol Reaction)
When an enolizable carbonyl compound is added to an aldehyde (or ketone) leading to an aldol product, the reaction is known as the aldol reaction which was first described by Kane in 1848, and is a classical transformation in organic synthesis. A new C–C bond is formed and two new stereocenters are generated in this reaction. The initial product of the reaction is a β-hydroxycarbonyl compound. Under certain

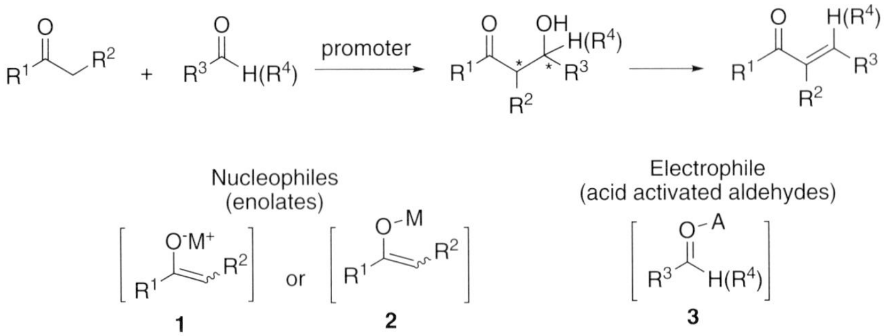

4 Intramolecular 1,2-Addition and 1,4-Addition Reactions

Scheme 4.1 Intermolecular aldol reaction of two carbonyl compounds.

conditions, dehydration will occur to generate the corresponding α,β-unsaturated carbonyl compound (Scheme 4.1) [1].

The aldol addition reaction has been one of the most widely used synthetic operations for the construction of stereochemically complex natural and non-natural products. The intramolecular aldol reaction is a powerful method for the synthesis of cyclic products [2].

Originally, the aldol reaction was carried out with Brönsted acid or base catalysis. An early and rather prominent example of the intramolecular aldol condensation was found in which two aldehyde groups were used (Scheme 4.2) [3].

Other examples using one aldehyde and one ketone group or two ketone groups are also known (Scheme 4.3) [4].

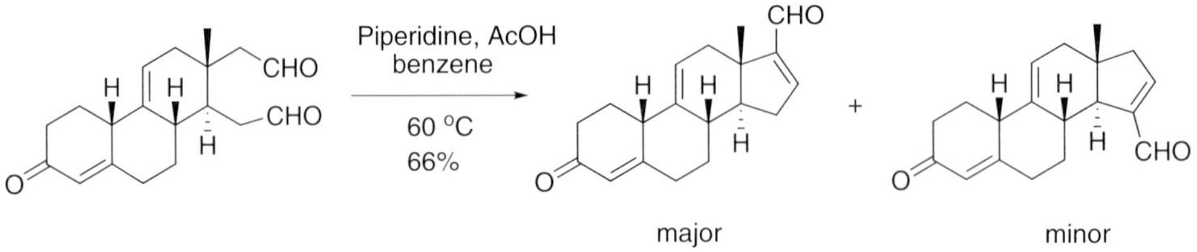

Scheme 4.2 Woodward's synthesis of cholesterol.

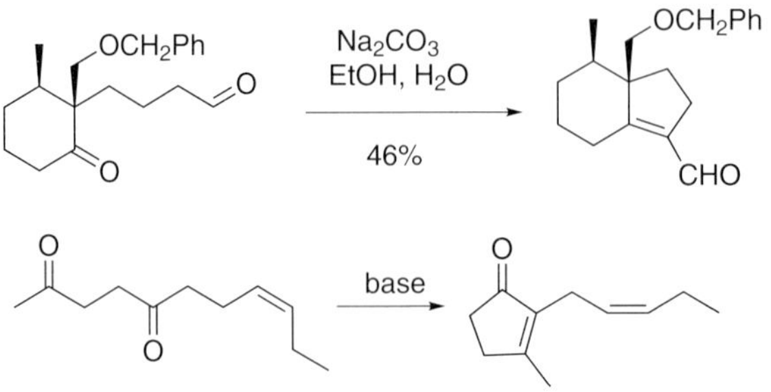

Scheme 4.3 Early examples of intramolecular aldol reaction.

In all these reactions, the dehydrated products are generally obtained. By careful control of the reaction conditions, products with multiple chiral centers can also be obtained. Now, the asymmetric aldol reaction is becoming one of the most important topics in modern catalytic synthesis [5].

In the aldol methodology, development of methods for the formation and application of preformed enolates was a key step in the early stage. Over the past decade, rapid progress has been made in the development of enantioselective aldol reactions under organocatalysis [6].

The reaction of heptanedial with a catalytic amount of (S)-proline in dichloromethane gave aldol in high yield and diastereoselectivity, and with excellent enantioselectivity [7a]. A high enantioselectivity and a high catalytic efficiency have been exhibited by (4R,2S)-tetrabutylammonium 4-TBDPSoxy-prolinate in the aldolization of 3-(4-oxocyclohexyl)propionaldehyde to give highly enantiomerically enriched (1S,5R,8R)-8-hydroxybicyclo[3.3.1]nonan-2-one [7b,7c]. The diastereo- and enantioselective synthesis of 3-hydroxy 2,3-dihydrobenzofurans via an intramolecular cis-selective aldol reaction employing proline-type organocatalysts was developed enabling a new entry to coumarin natural products [7d] (Scheme 4.4).

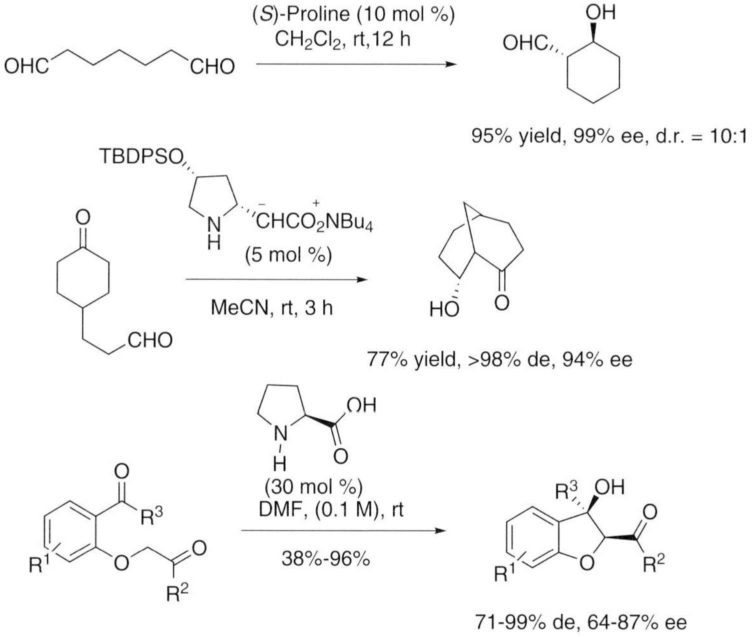

Scheme 4.4 Amino acids catalyzed enantioselective intramolecular aldol reaction.

Monosalts of N-substituted bimorpholine derivatives are efficient organocatalysts in intramolecular aldol reactions; isopropyl-substituted bimorpholine is the most stereoselective catalyst, affording products in high yield with enantioselectivities up to 95% ee [8a]. The trifluoroacetic acid salt of 2-(pyrrolidinylmethyl)pyrrolidine was found to be an effective organocatalyst of an asymmetric intramolecular aldol

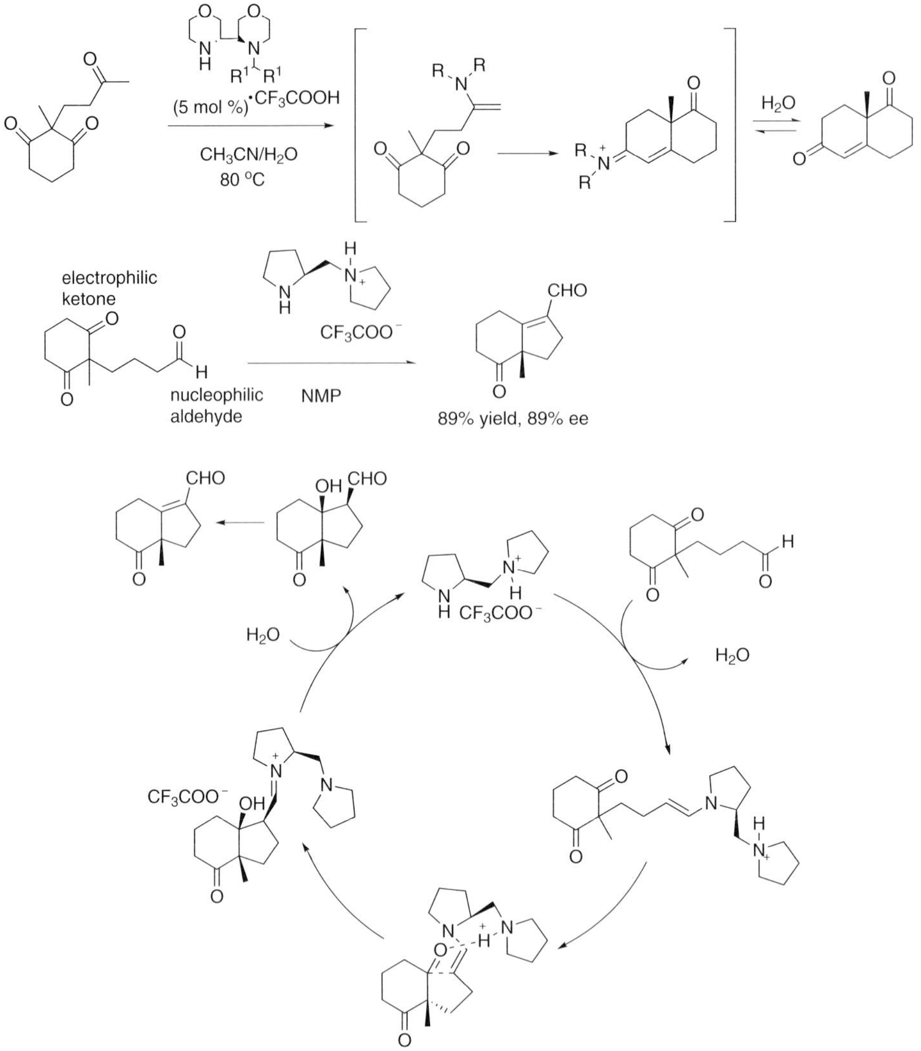

Scheme 4.5 Two different modes of the reactivity of dicarbonyl compounds.

reaction, affording bicyclo[4.3.0]nonane derivatives with a high enantioselectivity, in which the rare combination of aldehyde as a nucleophile and ketone as an electrophile was realized [8b]. The reaction is thought to proceed as follows: organocatalyst reacts with the aldehyde moiety of tricarbonyl starting compound to generate the enamine, which reacts with one of the carbonyl groups intramolecularly to generate the iminium salt. Hydrolysis of the iminium ion generates the aldol product with regeneration of the organocatalyst (Scheme 4.5).

Organocatalytic methods for the intramolecular aldol reaction have become an important tool for stereoselective synthesis over the past years, for further details see Chapter 25.

4.2.1.2 Vinylogous Aldol Reaction

The aldol reaction can be extended in vinylogous terms by employing an α,β-unsaturated carbonyl compound either as the electrophilic component or as the dienol (dienolate) source (Scheme 4.6)[9].

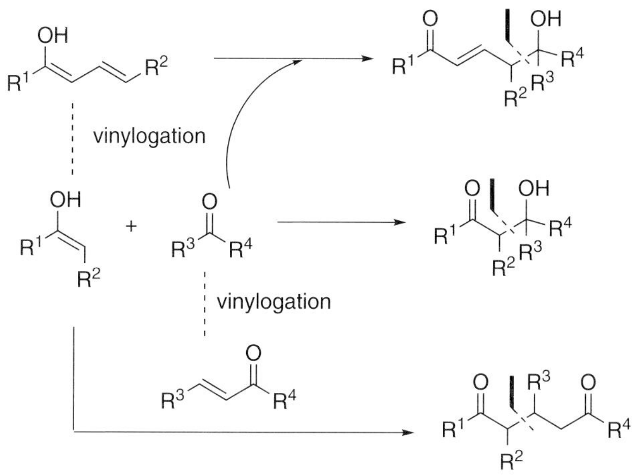

Scheme 4.6 Vinylogous aldol reaction.

The stereoselective tandem conjugate addition–cyclization reactions have been extensively explored [10]. In particular, intramolecular aldol and Michael cyclizations are useful strategies [11]. Oshima et al. have recently developed a highly stereoselective intramolecular reaction between the α,β-unsaturated ketone unit and the aldehyde functionality via a sequential Michael addition–aldol reaction with the use of halides as nucleophiles (Scheme 4.7) [12].

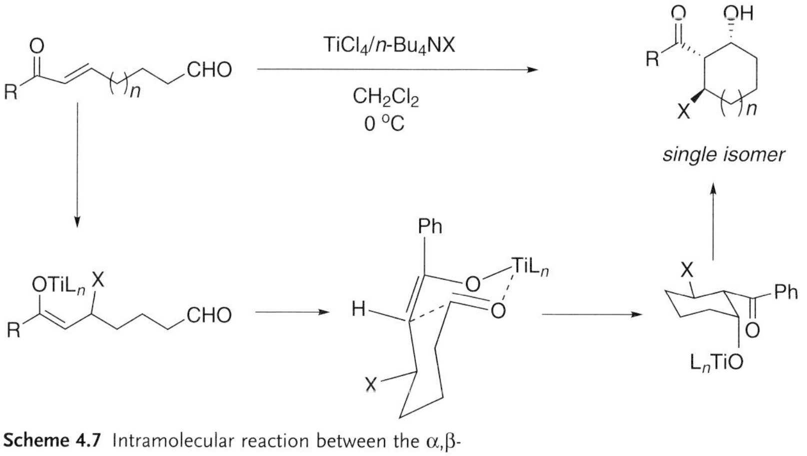

Scheme 4.7 Intramolecular reaction between the α,β-unsaturated ketone unit and the aldehyde functionality.

4 Intramolecular 1,2-Addition and 1,4-Addition Reactions

A similar reaction can also take place between a ketone and 2-alkenoate through an intramolecular iodo-aldol cyclization to afford hetero- or carbocycles containing quaternary centers in good yields and high trans selectivity having the hydroxy and iodomethyl groups on opposite faces of the ring system (Scheme 4.8) [13].

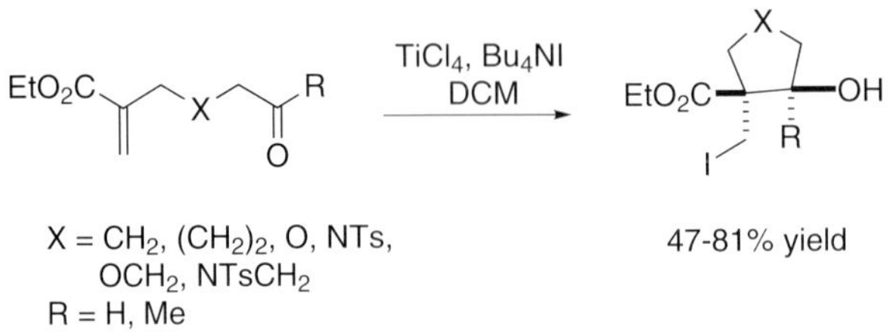

X = CH_2, $(CH_2)_2$, O, NTs, OCH_2, $NTsCH_2$
R = H, Me

47–81% yield

Scheme 4.8 Intramolecular reaction between a ketone and 2-alkenoate.

An intramolecular version of the Yamamoto vinylogous aldol reaction employing the bulky Lewis acid ATPH (aluminum tris-(2,6-diphenylphenoxide)) to control the site of aldolization has been described (Scheme 4.9) [14]. This macrocyclization process is effective for the construction of 10-, 12-, and 14-membered macrolides.

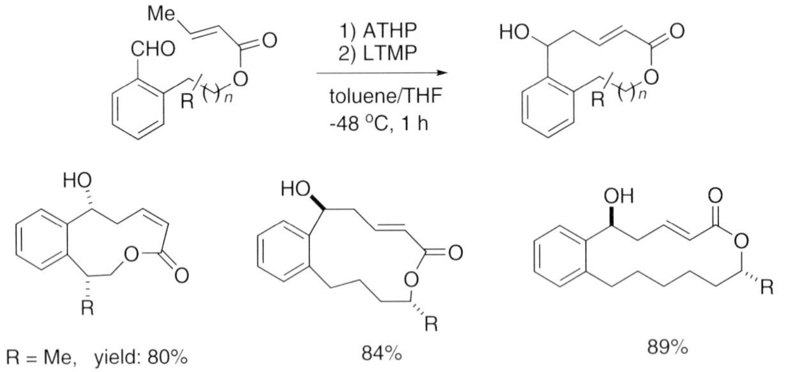

R = Me, yield: 80%   84%   89%

Scheme 4.9 Intramolecular Yamamoto vinylogous aldol reaction.

A rhodium(I)-catalyzed tandem hydrosilylation–intramolecular aldol reaction sequence of the α,β-unsaturated ester unit and the aldehyde functionality can be used to prepare α-ethoxycarbonyl cycloalkanols in moderate to excellent yields under mild reaction conditions (Scheme 4.10) [15].

Chiu used the commercially available Stryker's reagent, [$(Ph_3P)CuH]_6$, to react with the enones to form stable copper enolates, which was followed by intramolecular aldol cyclization to generate five- and six-membered carbocycles in one pot (Scheme 4.11). This tandem reaction is generally diastereoselective and provides good yields of β-hydroxyketones without any dehydration at low temperature [16].

Catalytic enone hydrometallation represents a promising strategy for enolate generation, circumventing the utilization of preformed enol or enolate derivatives. The metal-catalyzed reductive condensation of α,β-unsaturated carbonyl compounds

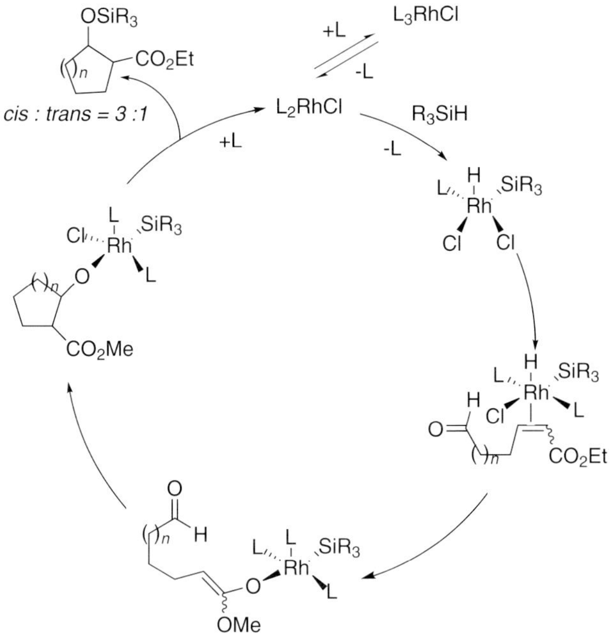

Scheme 4.10 Rhodium(I)-catalyzed tandem hydrosilylation-intramolecular aldol reaction.

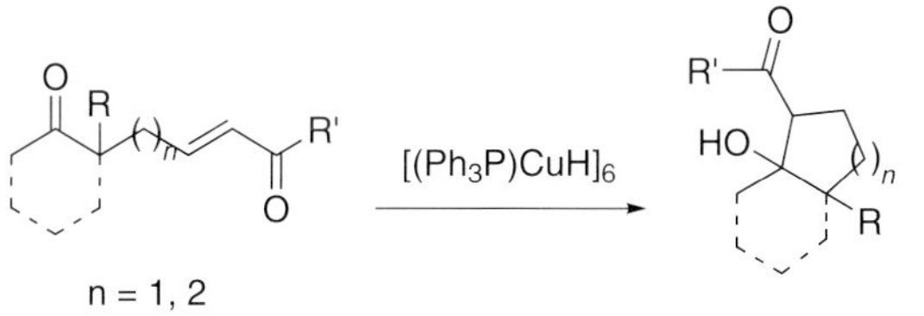

Scheme 4.11 Intramolecular cyclization of enone and ketone functionality using Stryker's reagent.

with aldehydes in the presence of a hydride donor, that is, a "reductive aldol" reaction, has been described (Scheme 4.12) [11b,17].

In this reaction, exposure of Co(dpm)$_2$ to phenylsilane generates the hydrido-cobalt species **4** which, upon hydrometallation of the enone, yields cobalt enolate **5**. Subsequent 1,2-addition to the appendant aldehyde results in the formation of cobalt-alkoxide **6**. σ-Bond metathesis liberates the product to regenerate the hydrido-cobalt species **4** and complete the catalytic cycle. Enolates generated by the catalysis of Rh under hydrogen in a similar way participate in the addition with ketone partners. The use of appendant dione partners enables diastereoselective formation of cycloaldol products possessing 3-stereogenic centers, including 2-contiguous quaternary centers (Scheme 4.13) [18].

Similar methods for the diastereoselective synthesis of lactones or lactams by the intramolecular reductive aldol cyclization of the corresponding α,β-unsaturated esters or amides in the presence of TMDS using Cu(I)-bisphosphine catalysts have

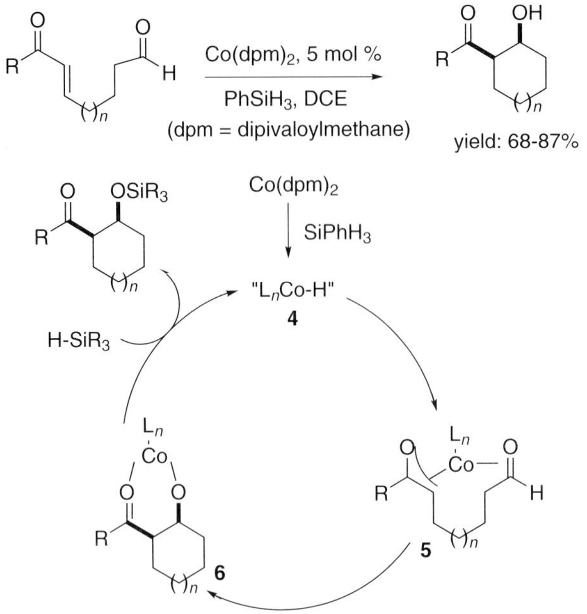

Scheme 4.12 A "reductive aldol" reaction.

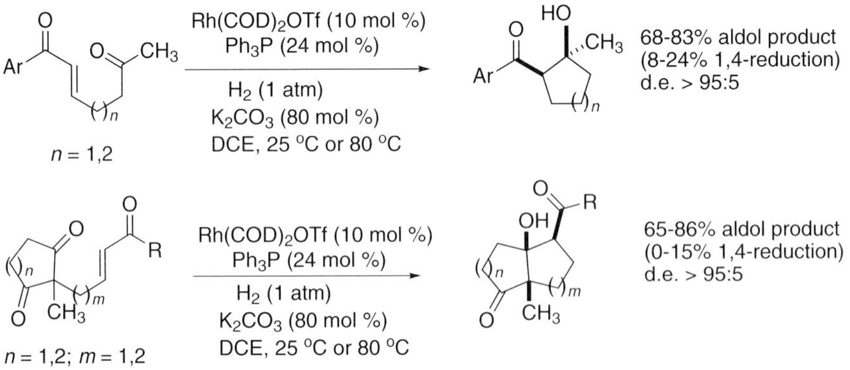

Scheme 4.13 Catalytic aldol cycloreduction of keto-enones under hydrogen.

been reported [19]. Utilization of chiral nonracemic bisphosphines renders the cyclizations enantioselective [19a].

The proline-catalyzed intramolecular aldol cyclization of triketones is recognized today as one of the first contributions to enantioselective organocatalysis. In the early 1970s, two groups, Hajos and Parrish at Hoffmann La Roche, and Eder, Sauer, and Wiechert at Schering AG, published a series of papers and patents involving these transformations [20]. Very recently, the Baylis–Hillman reaction has been added to the list of these useful carbon–carbon bond-forming reactions [21]. Frater reported the first intramolecular Baylis–Hillman reaction [22] and Krishna reported the first

4.2.1.3 1,2-Addition of Alkene-, Alkyne- or Arene-Substituted Carbonyl Compounds

The ene reaction is a powerful synthetic tool for C—C bond formation that allows construction of two vicinal chiral centers (Scheme 4.14). Like the Diels–Alder addition, it proceeds thermally and can be accelerated by Lewis acids. Its intramolecular version has been extensively studied and developed into a reliable methodology for the synthesis of cyclic products [24].

X=Y: $\mathrm{C{=}O}$, $\mathrm{C{=}N{-}}$, $\mathrm{C{=}S}$, $\mathrm{C{=}C}$, $-\mathrm{C{\equiv}C}-$ etc.

Scheme 4.14 General scope of the ene reaction.

When carbonyl compounds are used as enophiles, alcohols, rather than ethers, are formed exclusively (Scheme 4.15). From the synthetic point of view, the carbonyl-ene reaction should, in principle, constitute a more efficient alternative to the carbonyl addition reaction of allyl metals, which has now become one of the most useful methods for carbon skeleton construction with stereocontrol featuring acyclic stereocontrol.

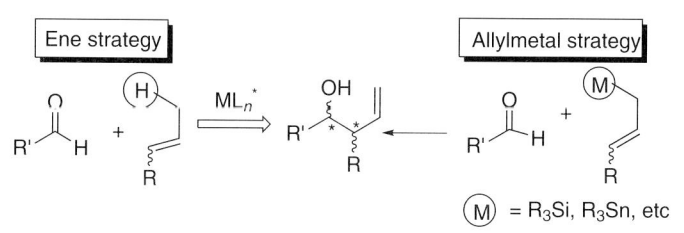

M = R_3Si, R_3Sn, etc

Scheme 4.15 Carbonyl-ene reaction.

Intramolecular carbonyl ene reactions present an attractive method for ring closure, forming two contiguous stereocenters with an often high degree of stereocontrol. One of the best studied carbonyl ene reactions is the cyclization of citronellal **7** affording a mixture of isopulegol **8** and neoisopulegol **9** (Scheme 4.16). Cyclization can be effected under Lewis acidic conditions with a diastereomeric ratio of up to 95 : 5 in favor of the trans diastereoisomer **8**. Cyclization in aqueous acid results in formation of the menthane diols **10** and **11** [24a]. Reversal of the stereoselectivity of cyclization of citronellal was also reported. Using $PhCH_2(Et)_3N^+[Mo(CO)_4ClBr_2]^-$

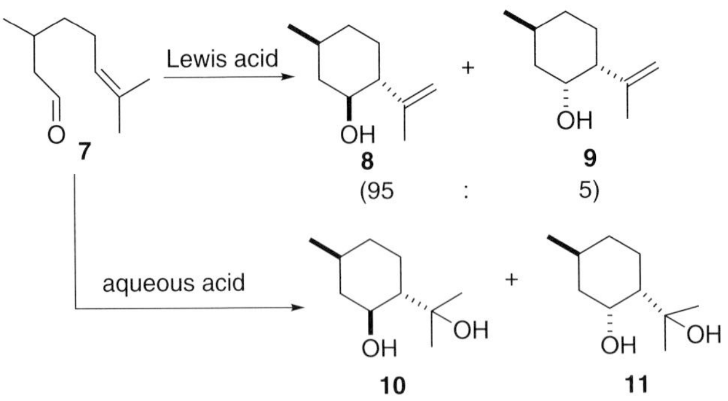

Scheme 4.16 Carbonyl-ene reaction of citronellal.

as the catalyst, the cyclization of (±)-**7** in dry DME at room temperature affording a mixture of **8** and **9** in a 25: 75 ratio [25].

This reaction has been applied for the synthesis of 3,4-disubstituted piperidines (Scheme 4.17). Carbonyl ene cyclization of aldehydes **12** catalyzed by the Lewis acid methyl aluminum dichloride in refluxing chloroform affords trans-piperidines **14** with diastereomeric ratios of up to 93: 7. By contrast, Prins cyclization of **12** catalyzed by hydrochloric acid at low temperatures affords cis-products **13** with diastereomeric ratios of up to 98: 2 [26].

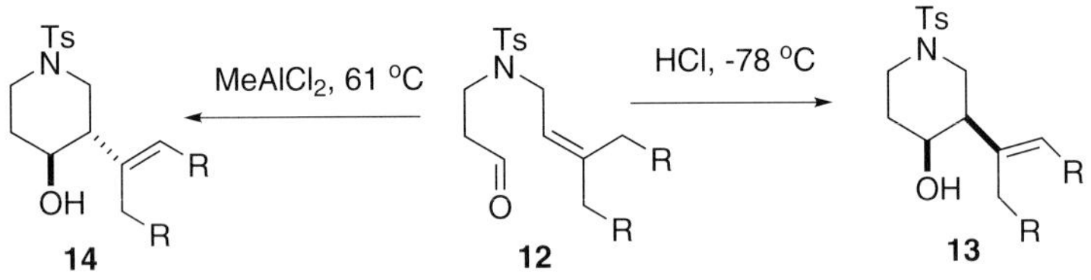

R = H, CH$_3$, -(CH$_2$)$_2$-, -(CH$_2$)$_3$-, -(CH$_2$)$_4$-

Scheme 4.17 Synthesis of 3,4-disubstituted piperidines.

R=Ph, 80%
R=4-BrC$_6$H$_4$, 84%
R=CMe$_3$, 90%

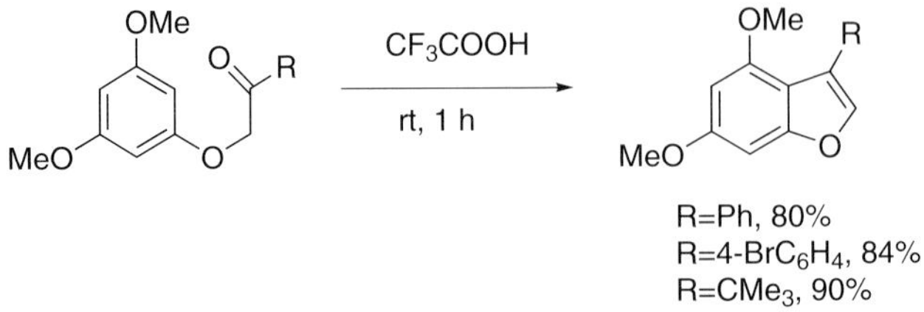

Scheme 4.18 Formation of benzofurans.

For substituted electron-rich arenes, the aryl carbon atom can easily attack the intramolecular carbonyl carbon atom to form the benzofurans (Scheme 4.18) [27].

Larock *et al.* reported a reaction which was first initiated by the interaction of an electrophile with the alkyne, then the formed carbonium ion was attacked by the

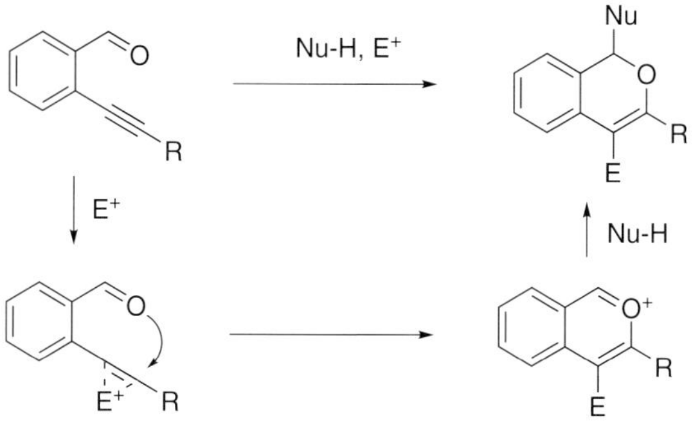

Scheme 4.19 Electrophilic cyclization of acetylenic aldehydes and ketones.

carbonyl oxygen intramolecularly, which was followed by the reaction with a nucleophile to afford the benzocyclic products (Scheme 4.19). For example, highly substituted oxygen heterocycles can be prepared in good yields at room temperature by reacting o-(1-alkynyl)-substituted arene carbonyl compounds with NBS, I_2, ICl, p-$O_2NC_6H_4$SCl, or PhSeBr and various alcohols or carbon-based nucleophiles [28].

The use of ketones, esters, and nitriles as terminal electrophiles in Cu-catalyzed tandem conjugate addition–electrophilic trapping has been developed (Scheme 4.20) [29]. Cobalt can also be used as the catalyst [11b,30].

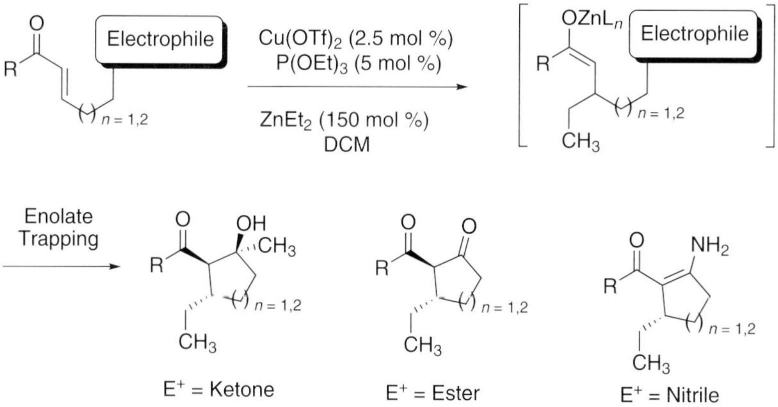

Scheme 4.20 Cu-catalyzed tandem conjugate addition-electrophilic trapping cyclization.

Important catalytic processes have been developed based on the reactivity of metal oxametallacycles derived from conjugated enals or enones and alkenes or alkynes (Scheme 4.21 and 4.22). Montgomery's group reported the nickel catalyzed intramolecular coupling of an enone with an alkyne in the presence of organozinc compounds to yield the cyclic products in a chemoselective and stereoselective fashion (Scheme 4.21). They proposed the mechanism involving the formation of the metallacycle **16** from the oxidative cyclization of a nickel (0) π-complex of **15** first, followed by the transmetalation with the organozinc to form **17**. Finally the product was formed by reductive elimination [31].

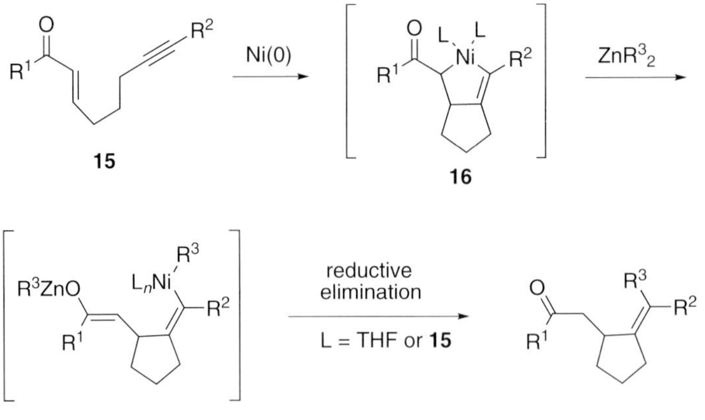

Scheme 4.21 General scope of the intramolecular coupling of an enone-alkyne in the presence of orgnozinc compounds.

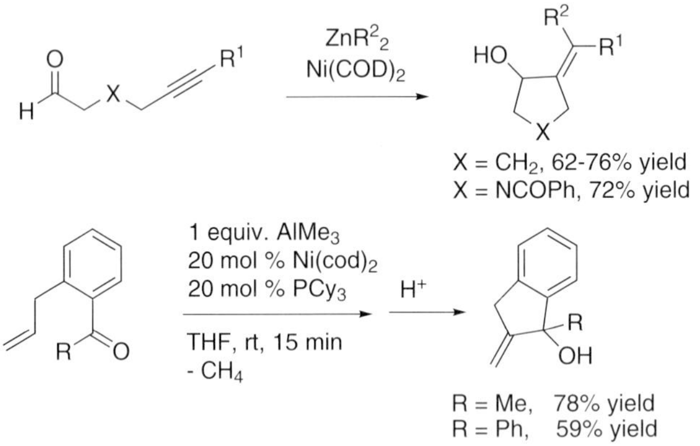

$X = CH_2$, 62–76% yield
$X = NCOPh$, 72% yield

$R = Me$, 78% yield
$R = Ph$, 59% yield

Scheme 4.22 Nickel-catalyzed cyclization of alkene or alkyne substituted carbonyl compounds.

Other transition metals (e.g., rhodium [32] and palladium [33]) also can be used as catalysts. Enantioselective cyclization was also reported in the case of the reductive coupling of acetylenic aldehydes using rhodium catalyst in the presence of hydrogen with a chiral phosphine ligand (Scheme 4.23) [32b].

Similar reaction occurs in the case of dienyl aldehydes [34], alkynyl aldehydes [32b,35] and allenyl aldehydes [36].

Another type of cyclization of acetylenic aldehydes was developed by Hayashi's group. They found that the addition/cyclization of arylboronic acids to acetylenic aldehydes proceeds smoothly with the use of a rhodium/diene catalyst, leading to cyclic allylic alcohols with a tetrasubstituted olefin (Scheme 4.24). At the same time, they achieved effective asymmetric catalysis by the use of chiral diene ligands [37].

For the arene substituted carbonyl substrates, either the halosubstituted arenes or the arylboronic acids were used to generate first the arylpalladium species, followed by insertion of alkynes or dienes, and the subsequent nucleophilic addition of the

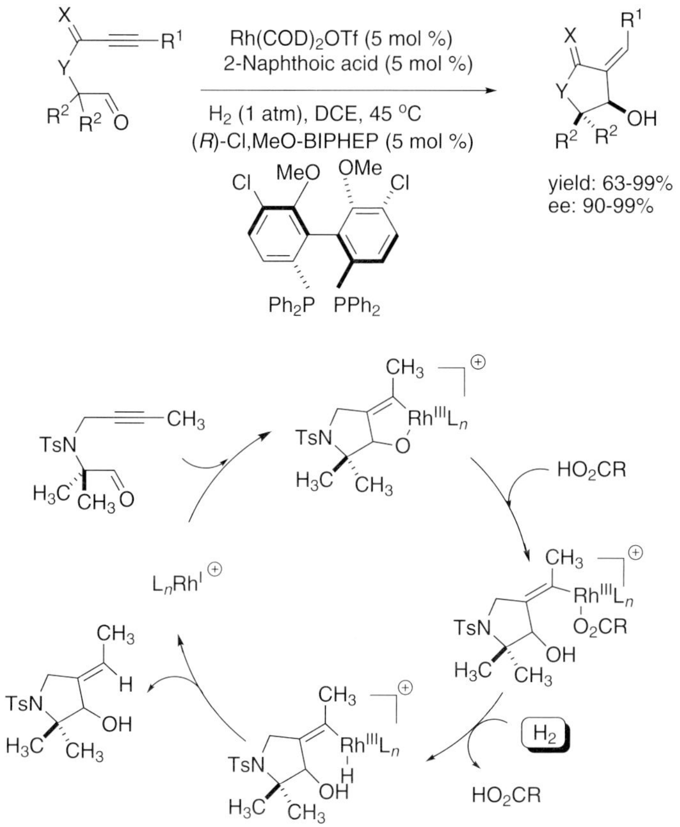

Scheme 4.23 Reductive cyclization of acetylenic aldehydes in the presence of hydrogen.

formed metal species to the carbonyl groups. Rhodium was the most commonly used metal. Palladium and iridium complexes were also used as catalysts [38]. Recently, cationic palladium complexes have been used as catalysts for the 1,2-addition of ketones with high enantioselectivity in the presence of chiral ligands [39]. Intramolecular tandem reactions can be realized by the arylboronic acids, alkynes and carbonyl groups to obtain cyclic compounds of different ring size (Scheme 4.25) [40].

In the presence of a Pd catalyst, arynes, generated *in situ* from 2-(trimethylsilyl)aryl triflates and CsF, undergo annulation with *o*-haloarenecarboxaldehydes providing a useful new method for the synthesis of fluoren-9-ones in good yields (Scheme 4.26) [41].

4.2.2
1,2-Addition of Imines

Imino groups are less electrophilic than the corresponding carbonyl groups in aldehydes and ketones so that many fewer reactions of imines have been reported. Due to the importance of forming heterocycles and amino acids from imino compounds, the chemistry is now receiving more attention.

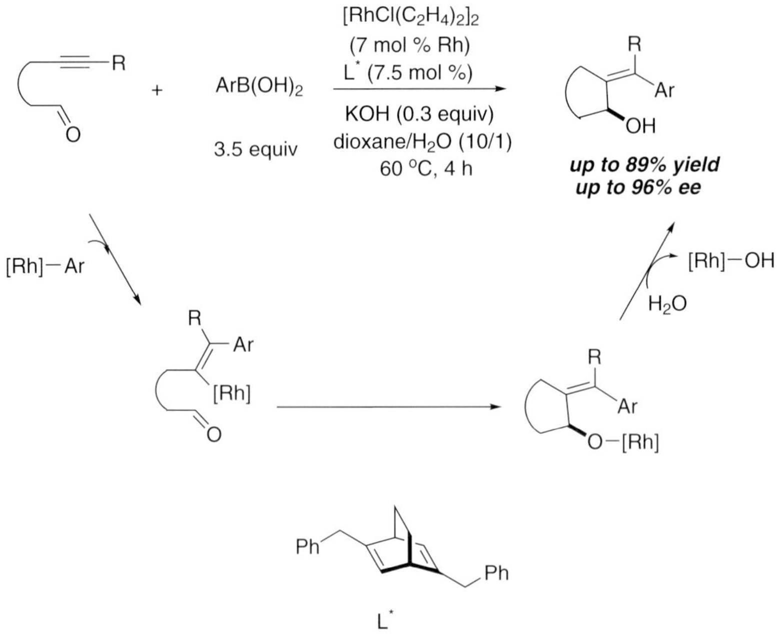

Scheme 4.24 Catalytic asymmetric arylative acyclization of alkynals.

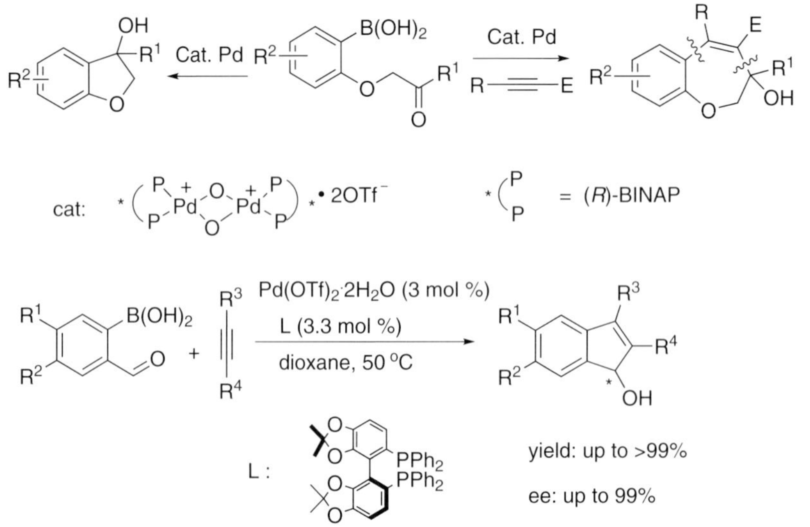

Scheme 4.25 Tandem reactions of arylboronic acids, alkynes and carbonyl groups to obtain different benzocyclic compounds.

4.2.2.1 1,2-Addition of Alkene or Alkyne Substituted Imino Compounds (Imino Ene Reactions)

Two pathways are possible for imino ene reactions (Scheme 4.27): (i) carbon–carbon bond formation could occur with hydrogen transfer to the nitrogen to give primary

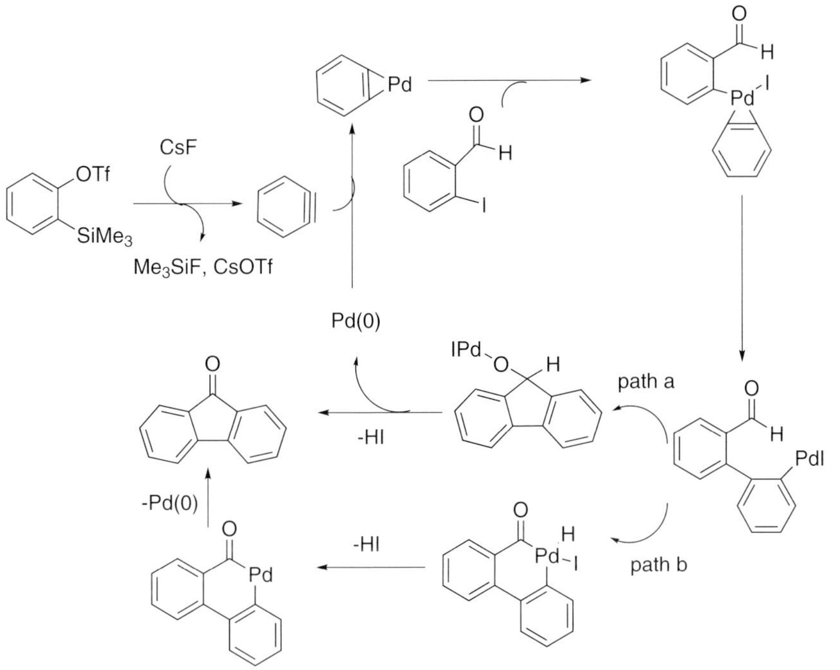

Scheme 4.26 Palladium-catalyzed annulation of arynes with o-haloarenecarboxaldehydes.

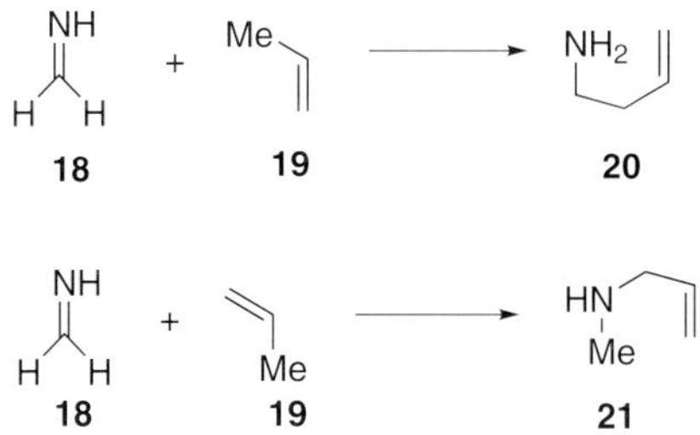

Scheme 4.27 General scope of the imino ene reactions.

homoallyl amine **20**, (ii) carbon–nitrogen bond formation and hydrogen transfer to the methylene group of imine **18** would lead to the formation of secondary allyl amine **21**. The first type of imino ene reaction is the more common [42].

In general, the imino group is not so active as the carbonyl groups. Direct ene reactions of simple unactivated imines are rare. Lewis acid is necessary to activate the imino group, for example, stannic chloride is used to activate the benzylimino group as shown in Scheme 4.28 [43].

Many reports deal with the activation of the imino groups by the introduction of electron withdrawing substituents on the nitrogen or carbon atoms. Tietze et al. reported the cyclization of doubly activated alkenylimines in the presence of either Lewis acid or trialkyl triflates. The reactivity of monoactivated imines is lower than that of the doubly activated alkenylimines, as expected (Scheme 4.29) [44].

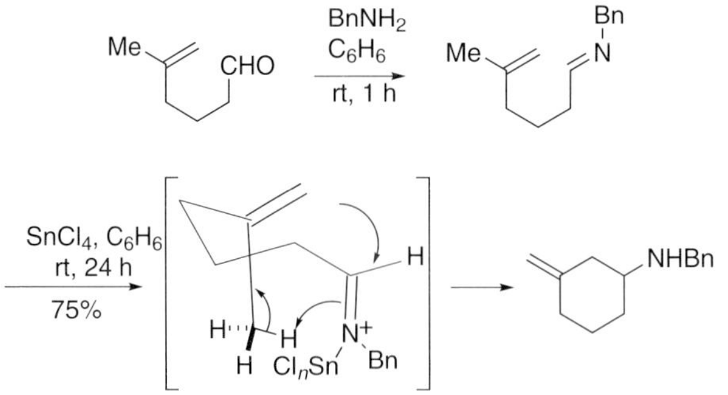

Scheme 4.28 Lewis acid activated imino ene reaction.

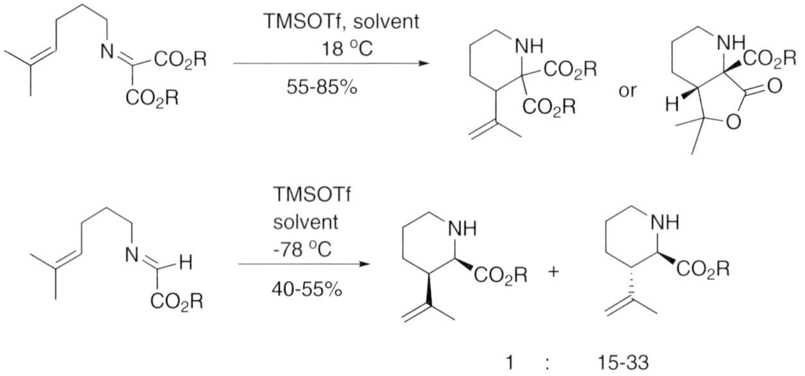

Scheme 4.29 Cyclization of doubly activated alkenylimines compared with monoactivated alkenylimines.

N-Sulfonyl imine derivatives were often used in imino ene reactions. Compound **22** was heated in o-dichlorobenzene to afford a single lactone product **23** (Scheme 4.30) [42a,45].

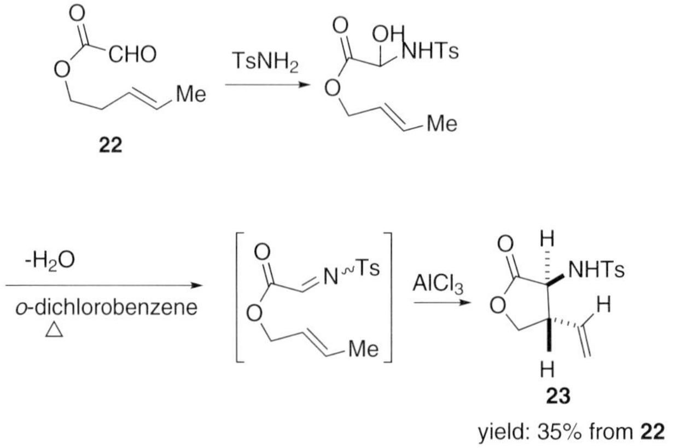

yield: 35% from **22**

Scheme 4.30 The use of N-sulfonyl imine derivatives in an imino ene reaction.

Scheme 4.31 Retrosynthesis of substituted piperidines from amino acids.

Alkenyl imines prepared from amino acid derivatives can be cyclized diastereo-selectively in the presence of Lewis acids to give 3-amino-2,4-dialkyl-substituted piperidines (Scheme 4.31). The product distribution and diastereoselectivity depends on the type of Lewis acid used and the nitrogen-protecting group (Scheme 4.32) [46].

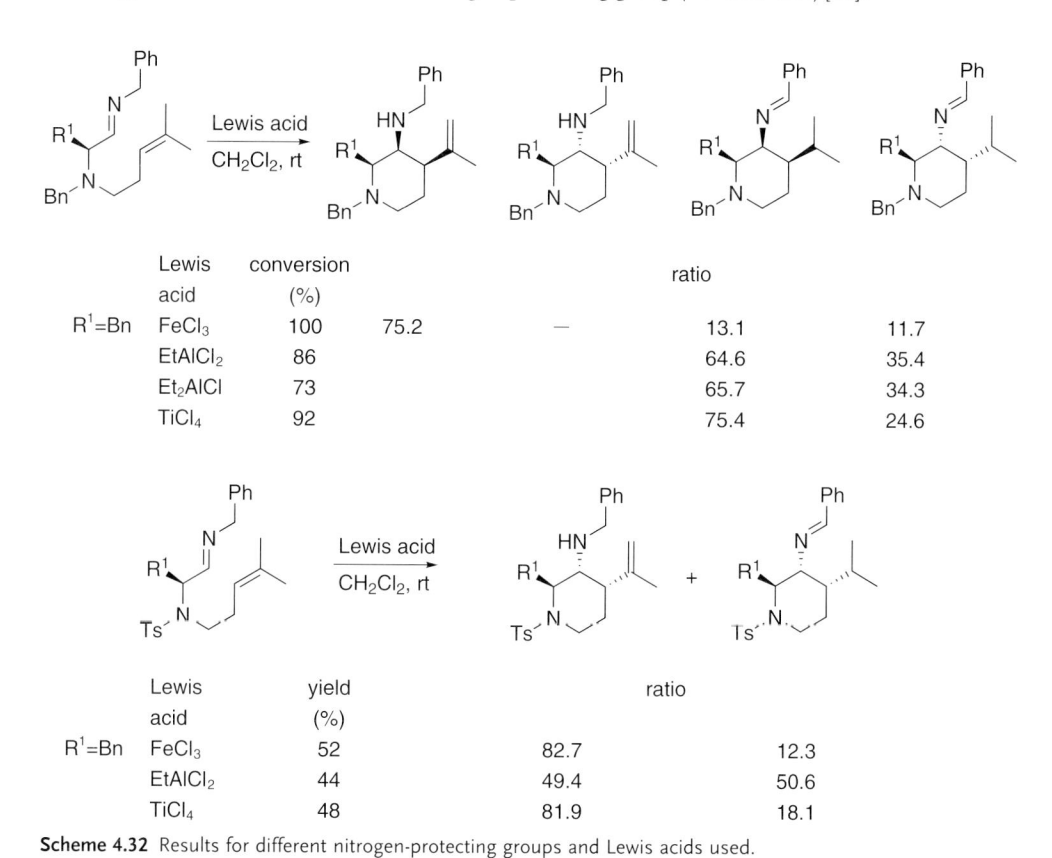

	Lewis acid	conversion (%)			ratio	
R^1=Bn	$FeCl_3$	100	75.2	—	13.1	11.7
	$EtAlCl_2$	86			64.6	35.4
	Et_2AlCl	73			65.7	34.3
	$TiCl_4$	92			75.4	24.6

	Lewis acid	yield (%)		ratio	
R^1=Bn	$FeCl_3$	52	82.7		12.3
	$EtAlCl_2$	44	49.4		50.6
	$TiCl_4$	48	81.9		18.1

Scheme 4.32 Results for different nitrogen-protecting groups and Lewis acids used.

Recently, the intramolecular coupling reaction of diyne, diene, and enyne using Cp_2ZrBu_2 has been shown to be very effective for the formation of cyclic compounds [47]. As multiple bonds, nitrile, isocyanate, and hydrazone can also react with Cp_2ZrBu_2, forming various heterocycles and nitrogen-containing cyclic compounds (Scheme 4.33). The coupling reaction of ene-imine generated from ene-aldehyde and the reactivity of the formed azazirconacycle were investigated [48].

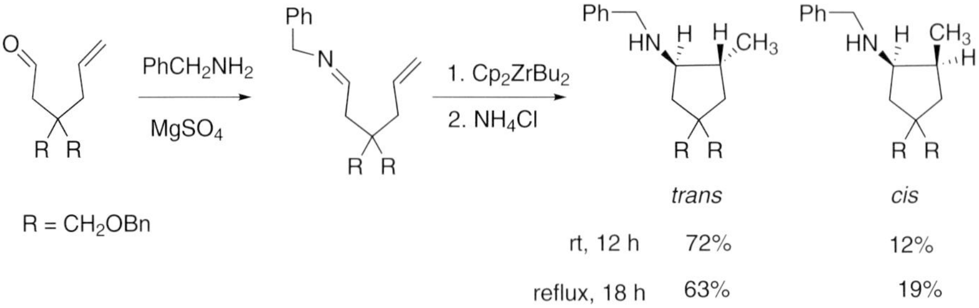

4 Intramolecular 1,2-Addition and 1,4-Addition Reactions

Scheme 4.33 Reaction of Cp$_2$ZrBu$_2$ with ene-imines.

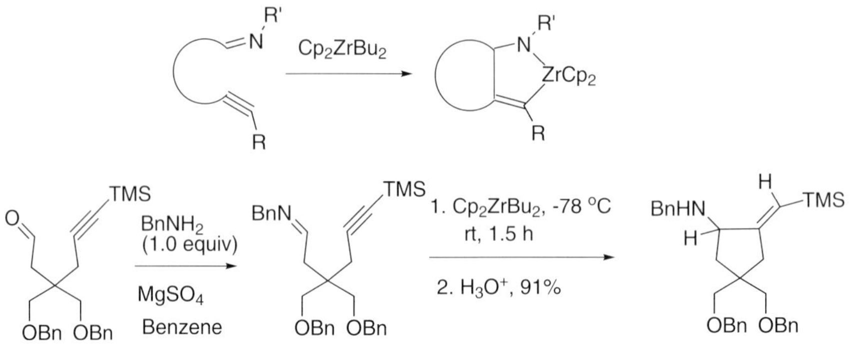

Scheme 4.34 Cyclization reaction of yne-imines.

The cyclization reaction of yne-imines generated from yne-aldehydes has also been reported (Scheme 4.34) [49].

The oxidative cyclization of an imine and an alkyne with nickel(0) gives a nickelapyrroline and the subsequent insertion of a second alkyne gives a nickeladihydroazepine. This complex then undergoes reductive elimination to give a 1,2-dihydropyridine. This sequential reaction process was developed into a nickel-catalyzed [2 + 2 + 2] cycloaddition of two alkynes and an imine that yields 1,2-dihydropyridines (Scheme 4.35) [50].

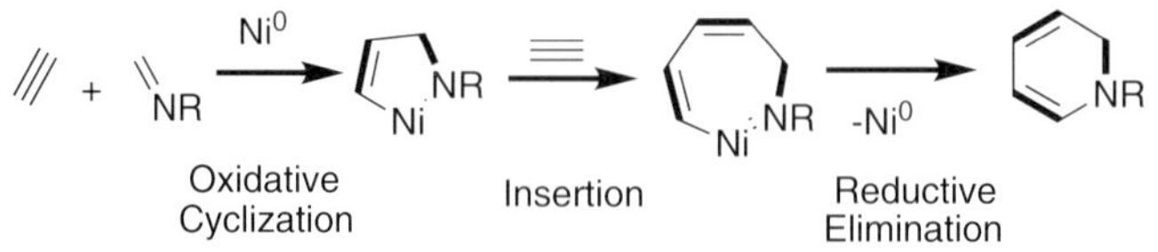

Scheme 4.35 Nickel-catalyzed [2 + 2 + 2] cycloaddition.

4.2.2.2 Nucleophilic Addition to Imino Groups

It was found that reaction of **24** (X = COOEt, R = C$_6$H$_5$) under a variety of base-catalyzed conditions afforded the indolines **25** (cis/trans mixture) in about 80% yield (Scheme 4.36) [51].

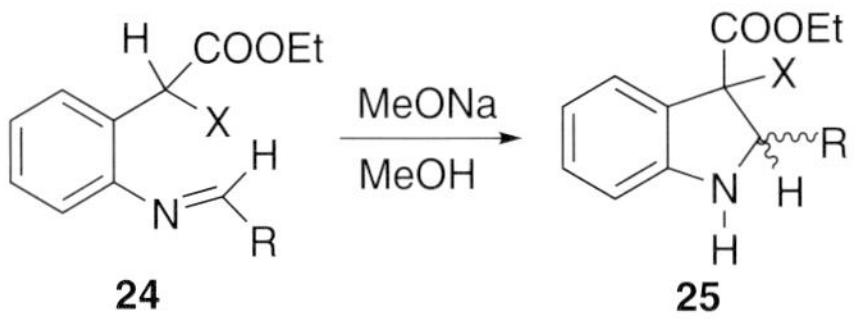

Scheme 4.36 Cyclization involving nucleophilic addition to imino group.

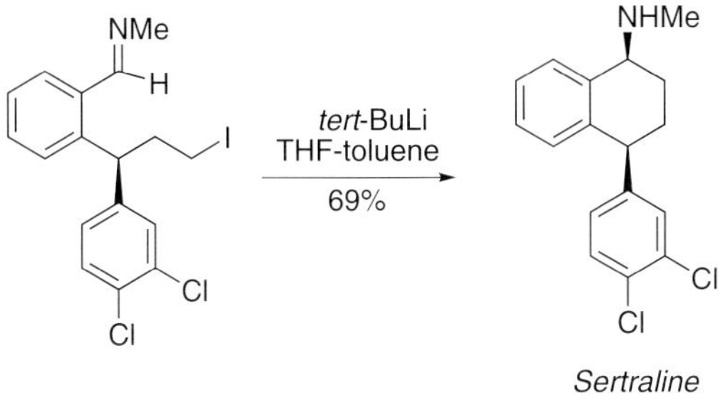

Scheme 4.37 Synthesis of sertraline.

Using similar methodology, a synthesis of sertraline, an antidepressant, as a single diastereoisomer was developed. The nucleophilic lithium reagent was first formed via Li-I exchange (Scheme 4.37) [52].

Yamamoto et al. reported the stereoselective synthesis of β-aminotetrahydro-pyran and -furan via the Lewis acid-mediated cyclic allylation of oxygen-tethered hydrazone-allylstannanes. The author also reported the asymmetric synthesis of β-amino cyclic ethers via the cyclic allylation of imine-allylstannanes with the (R)-(+)-l-phenylethylamine as the chiral auxiliary (Scheme 4.38) [53, 54].

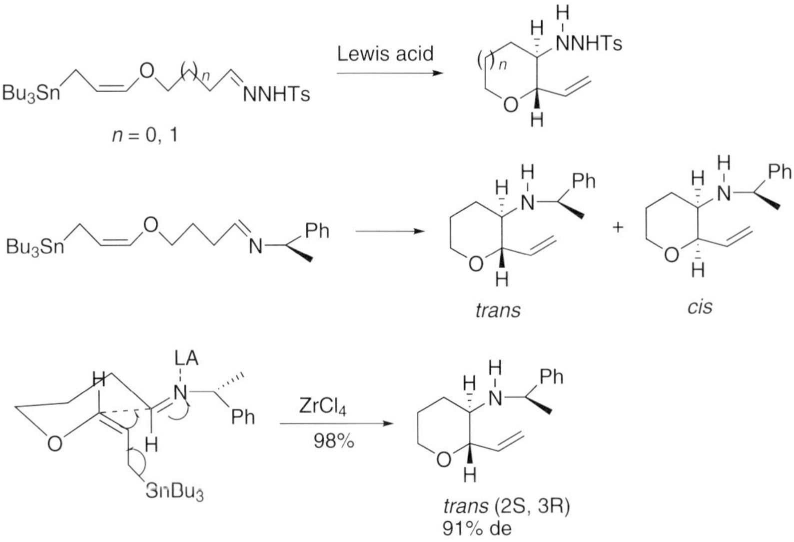

Scheme 4.38 Asymmetric synthesis of β-amino cyclic ethers via the cyclization of imine-allylstannanes using chiral auxiliary.

A tandem reaction using an ortho-alkynylarylimine as the starting material, which was designed as the nucleophilic addition of an allyl indium reagent (from transmetalation of the allyl tin compound with the indium salt) to an imine moiety and the subsequent intramolecular addition of the resulting amine to the activated carbon–carbon triple bond by using an indium salt formed *in situ*, has been developed [55]. After careful study, it is believed that the plausible mechanism is nucleophilic attack of the imine nitrogen atom at the indium-coordinated alkyne **26** to form the iminium ion **27**, followed by the trapping of **27** by allylindium to form **28**, and subsequent protonation produces 1,2-dihydroisoquinoline as the product (Scheme 4.39).

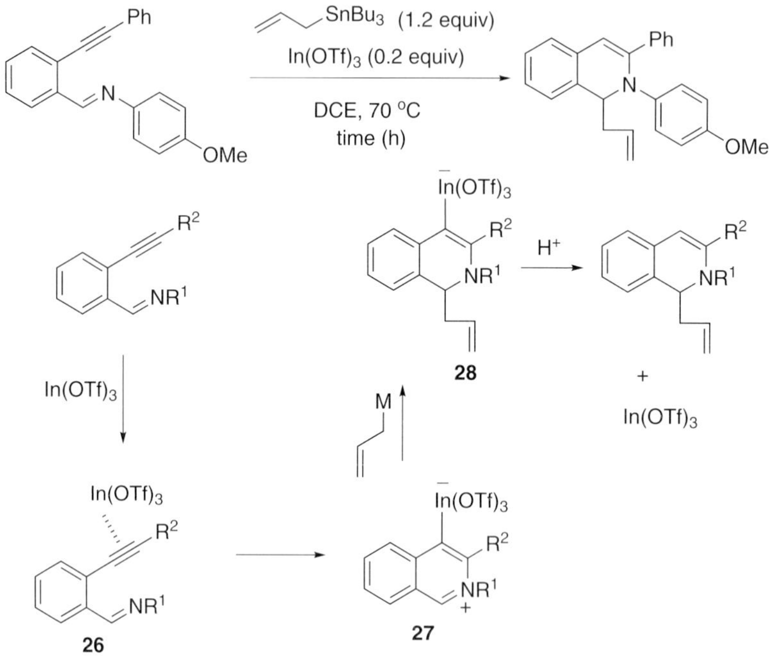

Scheme 4.39 Cyclization of ortho-alkynylarylimine catalyzed by In(OTf)$_3$.

A new diastereoselective Pd/In bimetallic intramolecular cascade reaction employing allenyl-sulfinimines and aryl iodides has been developed. The reaction affords optically active highly substituted *cis*-pyrrolidines and piperidines. The *tert*-butanesulfinyl group was found to be a useful chiral auxiliary (Scheme 4.40) [56].

Similarly, a diastereoselective synthesis of pyrrolidines was achieved from a carbocyclization of allene-hydrazones by the Pd-catalyzed distannylation of an allene moiety, followed by TiCl$_4$ mediated cyclic allylation to afford the pyrrolidines (Scheme 4.41) [57].

Palladium-catalyzed intramolecular cyclization of the alkyne and imine functional groups of 2-(1′-alkynyl)-N-alkylideneanilines gives functionalized 3-alkenylindoles in good yields (Scheme 4.42) [58].

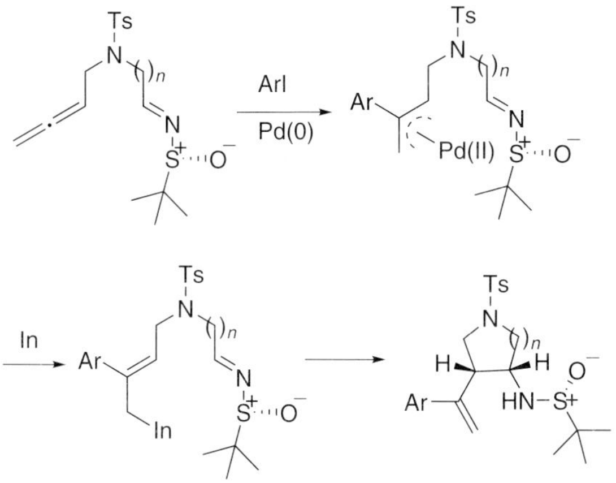

Scheme 4.40 Synthesis of optically active highly substituted *cis*-pyrrolidines and piperidines.

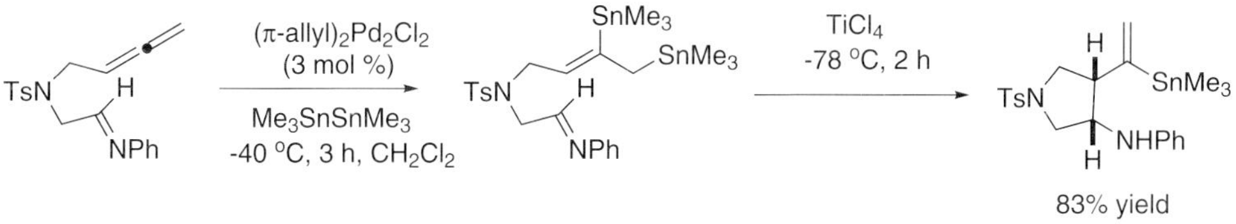

Scheme 4.41 Carbocyclization of allene-hydrazones.

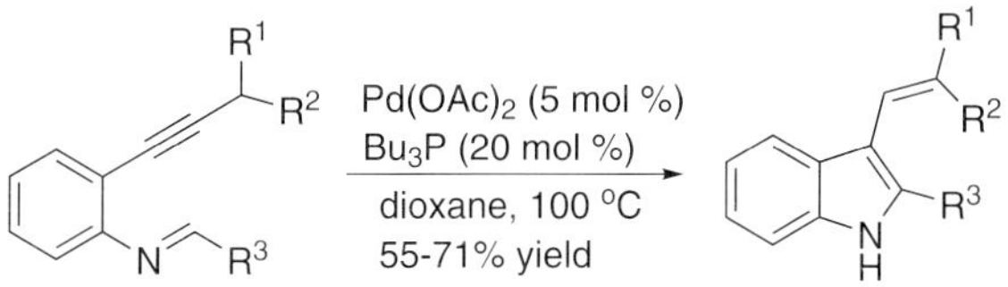

Scheme 4.42 Palladium-catalyzed cyclization of 2-(1′-alkynyl)-N-alkylideneanilines.

In addition to the above-described catalysts, by using carbophilic Lewis acids, In(OTf)$_3$, NiCl$_2$, and AuCl(PPh$_3$)/AgNTf$_2$, a concise and efficient synthesis of 1,3-disubstituted 1,2-dihydroisoquinolines has been achieved via tandem nucleophilic addition and cyclization of 2-(1′-alkynyl)arylaldimines [59–61]. Takemoto et al. summarized three different mechanisms for different kinds of catalysts, as shown in Scheme 4.43 [59].

The intramolecular nucleophilic addition of organometallic species, derived from C–H bond activation followed by sequential insertion of an alkyne and an imine moiety to yield indene derivatives has been reported. The following rhenium catalyzed reaction was shown here as a typical example (Scheme 4.44) [62].

Due to the recent rapid development of transition metal chemistry, some new reactions have been developed, for example, a new Ru$_3$(CO)$_{12}$-catalyzed cyclocarbonylation of yne-imines leading to α,β-unsaturated lactams (Scheme 4.45) [63].

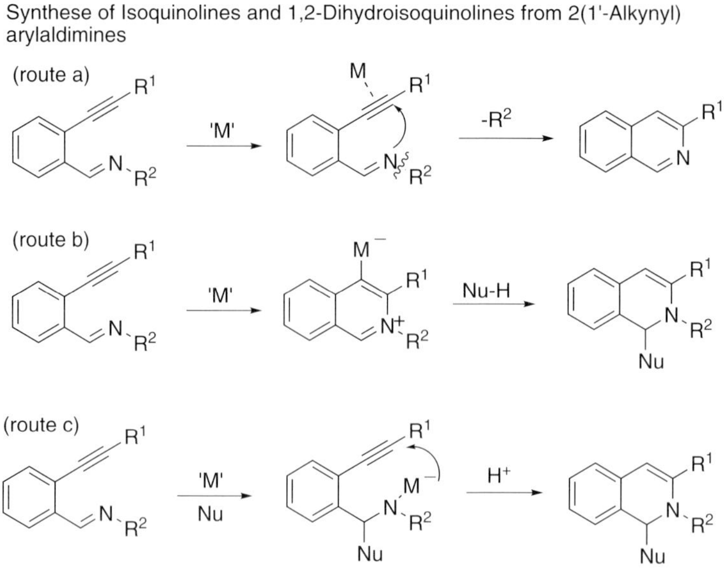
Synthese of Isoquinolines and 1,2-Dihydroisoquinolines from 2(1'-Alkynyl) arylaldimines
(route a)
'M'
-R2
(route b)
'M'
Nu-H
(route c)
'M'
Nu
H+

Scheme 4.43 Three different types of cyclizations of 2-(1′-alkynyl)arylaldimines.

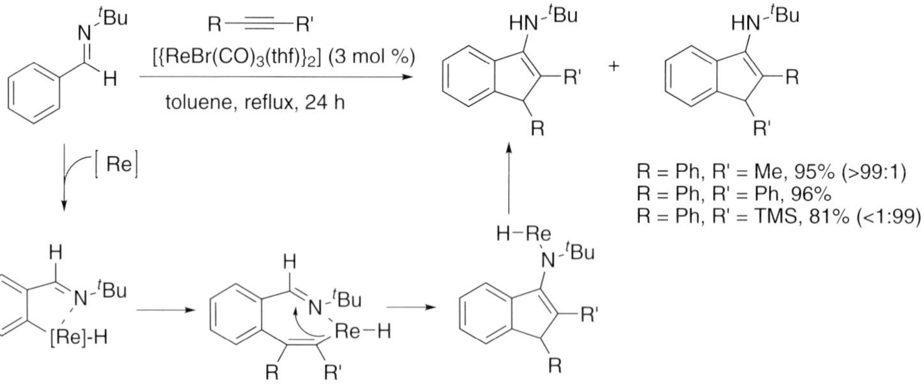
[{ReBr(CO)3(thf)}2] (3 mol %)
toluene, reflux, 24 h
[Re]
R = Ph, R' = Me, 95% (>99:1)
R = Ph, R' = Ph, 96%
R = Ph, R' = TMS, 81% (<1:99)

Scheme 4.44 Rhenium catalyzed synthesis of indenes initiated by C–H activation.

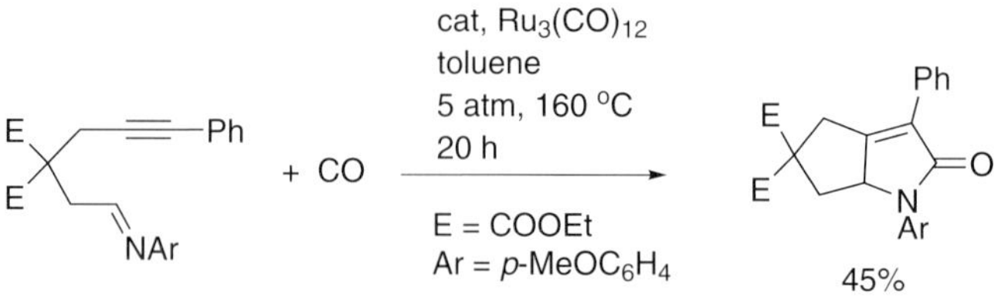
cat, Ru3(CO)12
toluene
5 atm, 160 °C
20 h
+ CO
E = COOEt
Ar = p-MeOC6H4
45%

Scheme 4.45 $Ru_3(CO)_{12}$-catalyzed cyclocarbonylation of yne–imines.

Hayashi developed a new palladium-catalyzed [3 + 3] cycloaddition of trimethylenemethane (TMM) with azomethine imines to produce highly functionalized hexahydropyridazine derivatives under simple and mild conditions (Scheme 4.46) [64].

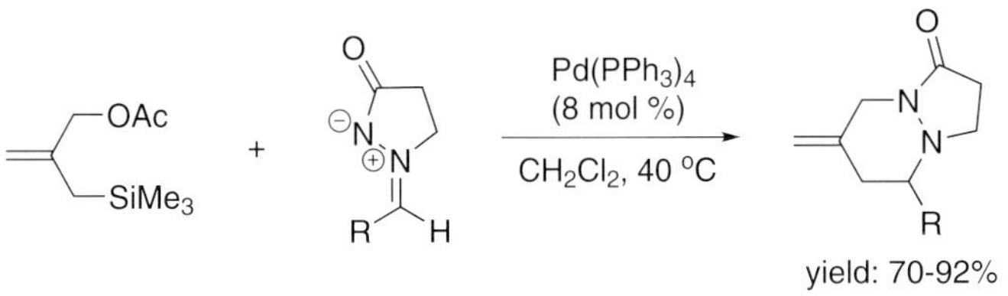

Scheme 4.46 Palladium-catalyzed [3 + 3] cycloaddition of trimethylenemethane (TMM) with azomethine imines.

4.2.3
1,2-Addition of Nitriles

Despite the long histroy of nitrile chemistry, the exact nature of this area remains elusive, reflecting a complex interplay between solvation, charge delocalization and inductive stabilization. Conceptually, the powerful electron-withdrawing effect of the nitrile is most consistent with an sp^3 hybridized carbanion **29**, as either a contact or separated ion pair, whereas the greater electronegativity of nitrogen over carbon suggests an sp^2 hybridized "keteniminate" **30** as the preferred structure (Scheme 4.47). In the extreme, these species might be envisaged as C- or N-metallated nitriles **31** and **32**, respectively. The unusual charge stabilization of nitriles correlates with the excellent nucleophilicity of nitrile anions [65].

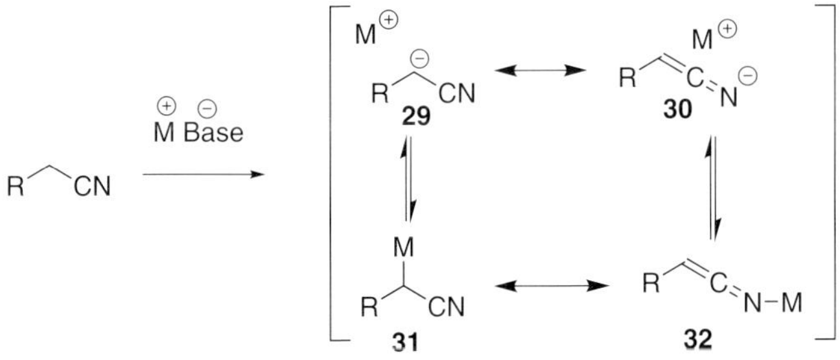

Scheme 4.47 Property of nitrile group.

4.2.3.1 Acid-Catalyzed Reactions

The acid-catalyzed intramolecular cyclization of 4-phenylbutyronitrile may occur to yield 1-tetralone imine (isolated as 1-tetralone) in good yield (Scheme 4.48) [66].

Another example is the intramolecular cyclizations of δ-hydroxynitriles promoted by triflic anhydride. In this reaction, triflic anhydride first reacts with the hydroxy

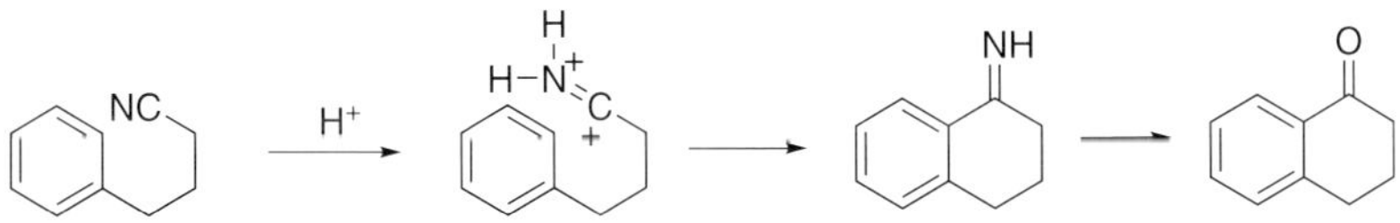

Scheme 4.48 1-Tetralone by the acid-catalyzed cyclization of 4-phenylbutyronitrile.

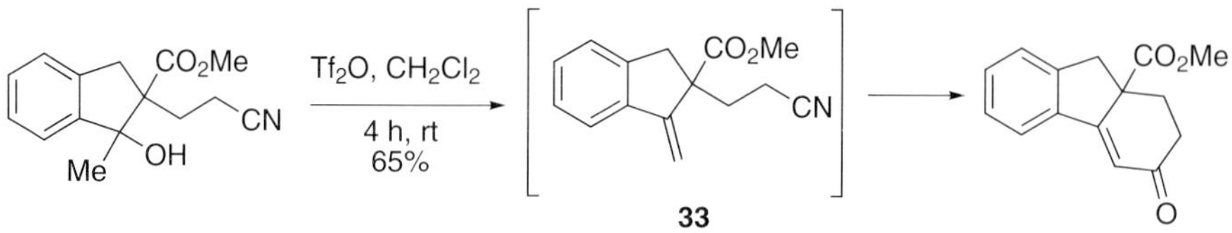

Scheme 4.49 Acid-catalyzed cyclization reactions of nitriles.

group and eliminates to form the double bond in **33**, which is added to the nitrile group activated by triflic acid [66] (Scheme 4.49).

4.2.3.2 Base-Catalyzed Reactions

2-Amino-4-cyanomethyl-6-dialkylamino-3,5-pyridinedicarbonitriles were found to react with substituted oxiranes yielding **34**. The oxirane ring was shown to be opened selectively from the unsubstituted side and with the pyridylacetonitrile afforded 5-amino-2,3-dihydrofurans. The latter further reacted with the adjacent nitrile to form the final product [67] (Scheme 4.50).

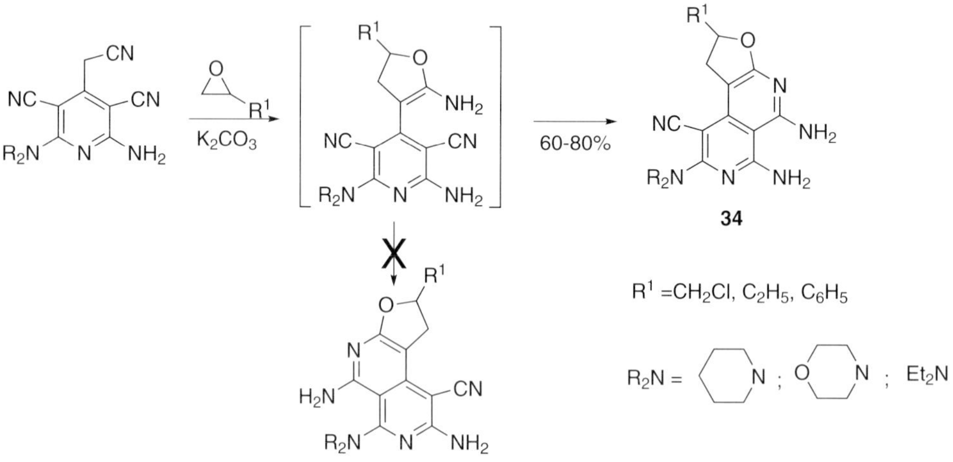

$R^1 = CH_2Cl, C_2H_5, C_6H_5$

Scheme 4.50 Base-catalyzed cyclization of pyridylacetonitrile with oxiranes.

o-Cyano-β,β-difluorostyrenes react with organolithiums selectively at the cyano carbon to generate the corresponding sp^2 nitrogen anions, which in turn undergo intramolecular replacement of the vinylic fluorine to afford 3-fluoroisoquinolines. Similarly, the reaction of β,β-difluoro-o-isocyanostyrenes with organomagnesiums or -lithiums generates the corresponding sp^2 carbanions on the isocyano carbon. Subsequent cyclization via substitution of the fluorine leads to 3-fluoroquinolines (Scheme 4.51) [68].

Recently, a synthesis of RHPS4, a potent telomerase inhibitor, via an anionic ring closing reaction was reported. The treatment of **35** with 2.2 equiv of *tert*-butyllithium at −78 °C produced the pentacyclic core of RHSP4 **36** in 66% yield after the

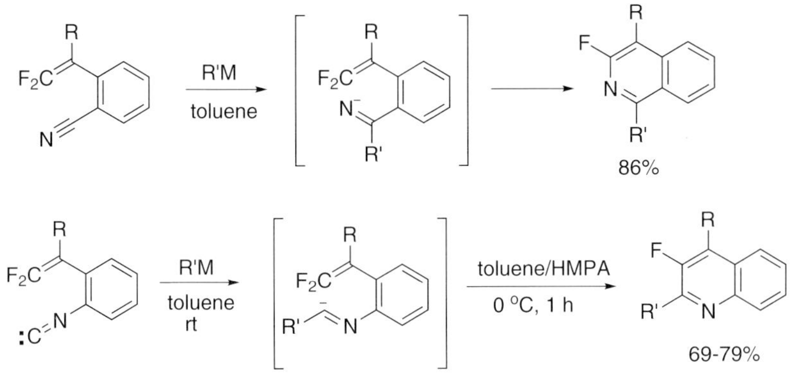

Scheme 4.51 Cyclization reactions initiated by the nucleophilic attack of the nitrile and isonitrile groups.

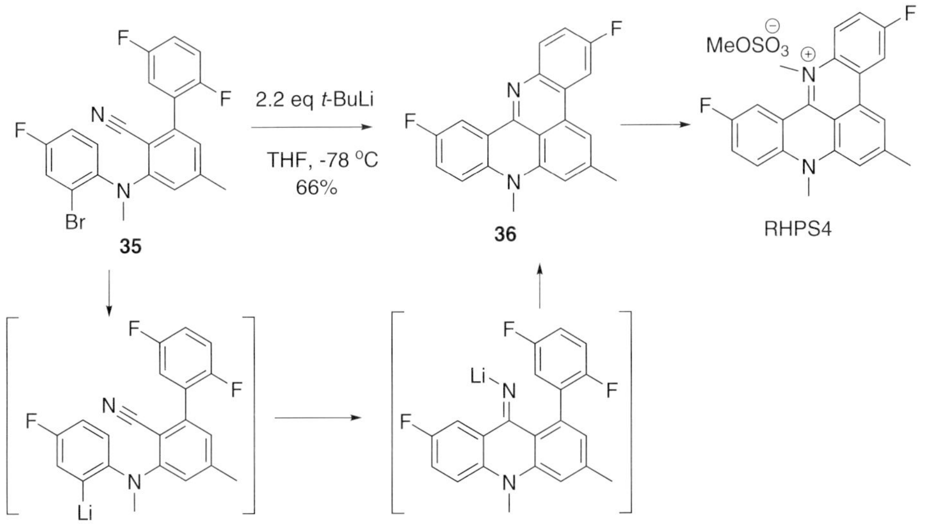

R = n-Bu
R = sec-Bu

Scheme 4.52 Synthesis of RHPS4.

recrystallization of the crude product. Pentacycle **36** could subsequently be converted to RHPS4 (Scheme 4.52) [69].

Alkanesulfonates of cyanohydrins or sulfonamides of aminonitriles are treated with bases such as DBU or sodium hydride to produce the 4-amino-5H-1,2-oxathiole 2,2-dioxide and 4-amino-2,3-dihydroisothiazole 1,1-dioxide ring systems, respectively (Scheme 4.53). The mechanism involves the removal of a proton from the methylene neighboring the SO_2 group, the *in situ* generated carbanion attacks the cyano group to give eventually the cyclic enamine system. This reaction is both rapid and clean. It was shown that not only complex chiral polyoxygenated derivatives from sugars or nucleosides but also simple ketones (acetone, benzyl methyl ketone and acetol

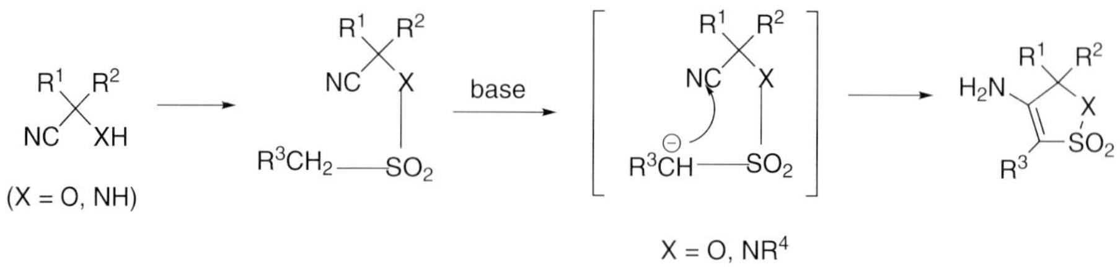

Scheme 4.53 Cyclization of alkanesufonates or sulfonamides of nitrile derivatives.

derivatives) and aliphatic aldehydes may be successfully used as the starting materials in this reaction [70].

Cyclization of dinitriles is an important process in constructing heterocyclic natural products. Using sodium hydride and N-methylaniline as the base in highly dilute conditions, the dimeric intermolecular reaction of symmetrical dinitrile **37** gave the large membered ring product **38** (Scheme 4.54) [71].

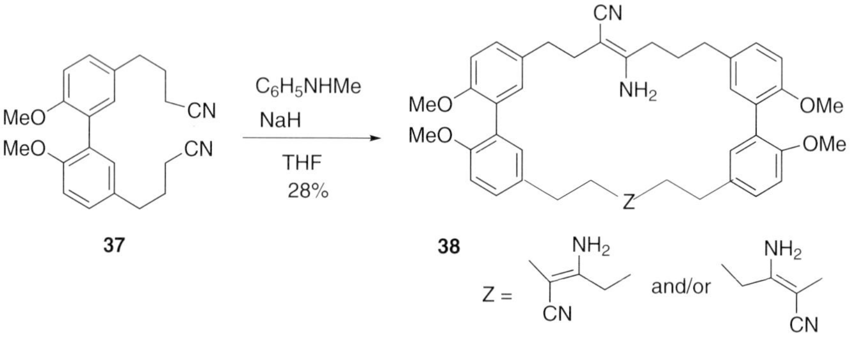

Scheme 4.54 Base-promoted dimeric intermolecular reaction of symmetrical dinitrile.

4.2.3.3 Nucleophilic Addition to Nitriles

The nucleophilic addition of main group organometallics such as organolithium and organomagnesium compounds to carbon–nitrogen triple bonds provides a facile method for forming carbon–carbon bonds. On the other hand, the organometallic compounds of late transition metals are typically less polar, and hence less nucleophilic, than other organometallics. Therefore, their synthetic potential as nucleophiles toward polar electrophilic multiple bonds has been overshadowed compared to their addition to non-polar carbon–carbon multiple bonds.

There are also a number of examples of simple ortho- and para-substituted aromatic nitriles readily undergoing palladium-catalyzed processes without involvement of the nitrile functionality. In fact, palladium chloride bis-acetonitrile and bis-benzonitrile are widely used reagents or catalysts and acetonitrile is one of the most commonly employed solvents in organopalladium chemistry. Larock and others independently reported the participation of the carbon–nitrogen triple bond of simple aryl nitriles in organopalladium catalyzed annulation reactions.

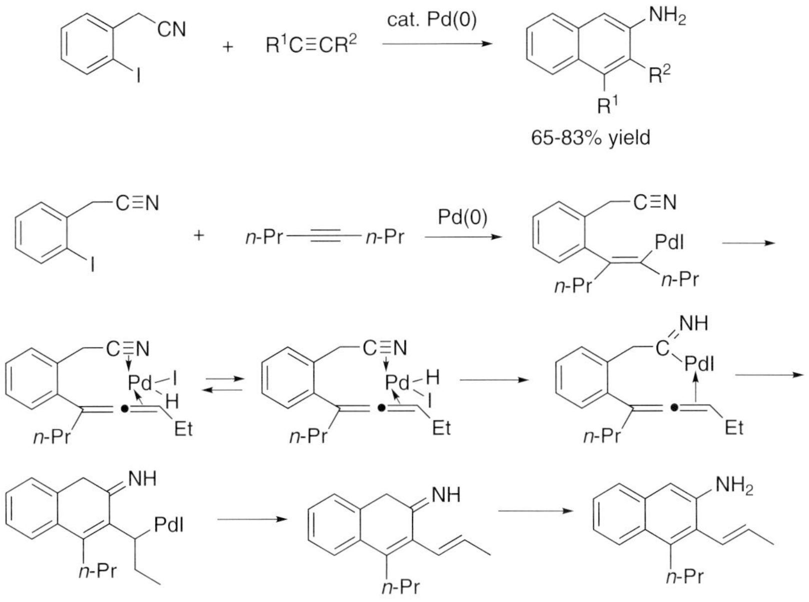

Scheme 4.55 Palladium-catalyzed tandem cyclization reaction of 2-iodophenylacetonitriles and alkynes.

Different styles of addition reactions to nitriles were reported [33b], for example, 2-iodobenzonitrile and 2-iodophenylacetonitriles react with internal alkynes and bicyclic alkenes to form indenones, naphthenones, 2-aminonaphthalenes, and bicyclic aryl ketones in good yields (Schemes 4.55 and 4.56) [72–74].

Recently, it has been reported that organorhodium(I) species undergo nucleophilic addition to aldehydes, ketones, esters, acid anhydrides and imines. For example, the

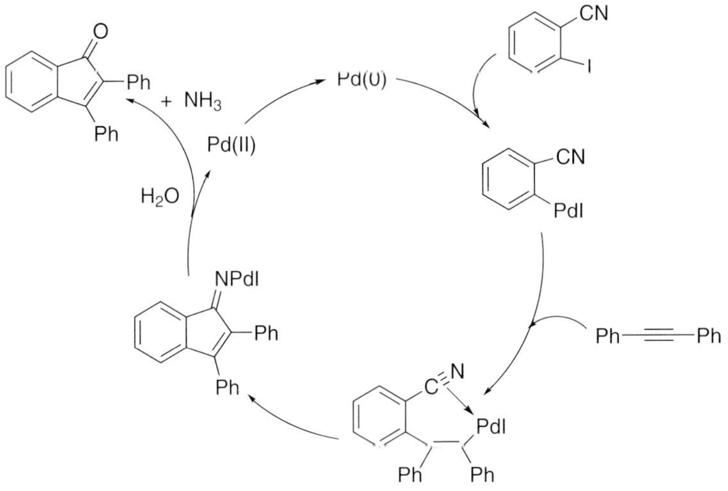

Scheme 4.56 Palladium-catalyzed tandem cyclization reaction of 2-iodobenzonitrile and alkynes.

rhodium-catalyzed reaction of 4-cyanobenzaldehyde with phenylboronic acid produces the corresponding secondary alcohol, in which the cyano group remains intact [75]. This result suggests that a cyano group is significantly less reactive than an aldehydic carbonyl group. The first example of a new cyclization reaction of cyano-substituted alkynes – a nucleophilic addition of an organorhodium(I) species to a cyano group in which the insertion of a carbon–carbon triple bond into an arylrhodium(I) species is followed by intramolecular addition to the carbon–nitrogen triple bond of the cyano group was developed recently (Scheme 4.57) [76].

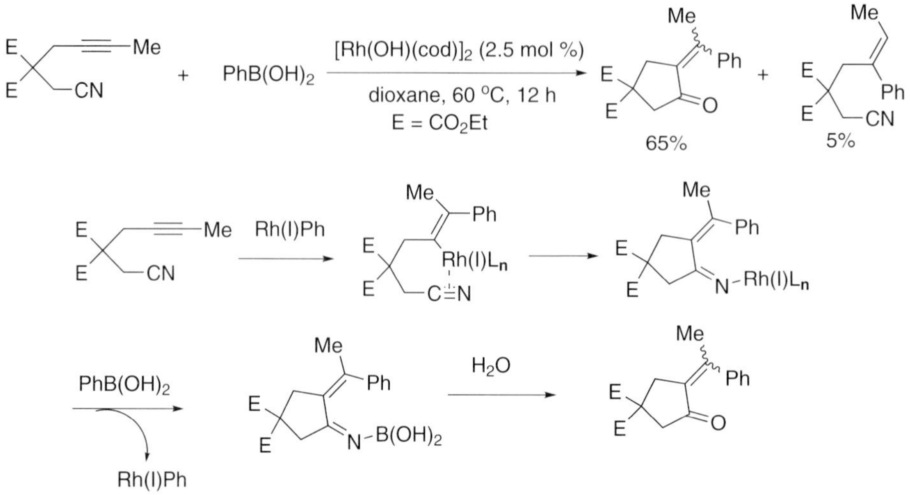

Scheme 4.57 Rhodium-catalyzed cyclization reaction of cyano-substituted alkynes.

A new rhodium-catalyzed [3 + 2] annulation reaction in which 2-cyanophenylboronic acid reacted as a three-carbon component with alkynes or alkenes to afford substituted indenones or indanones has also been reported (Scheme 4.58) [77].

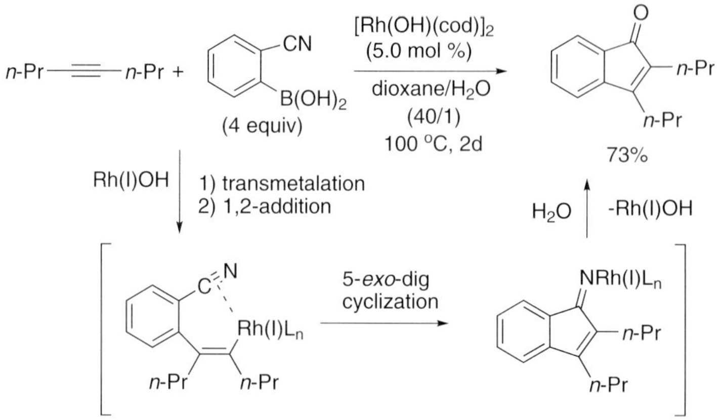

Scheme 4.58 Rhodium-catalyzed [3 + 2] annulation reaction of 2-cyanophenylboronic acid with alkynes.

A new cascade reaction triggered by the rhodium-catalyzed conjugate addition of organoboron species, followed by the addition of (oxa-π-allyl)rhodium(I) intermediates to the cyano group, allowing the formation of five- and six-membered ring was possible (Scheme 4.59) [78].

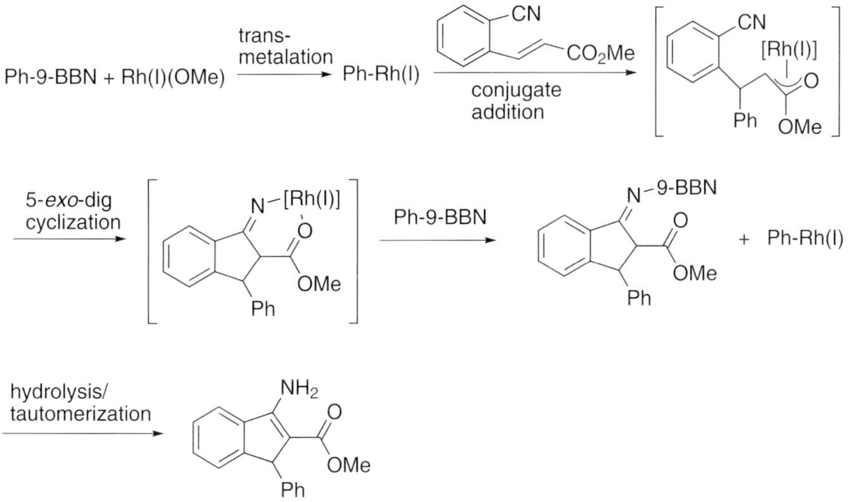

Scheme 4.59 Cascade reaction of conjugate addition and 1,2-addition to nitrile catalyzed by Rh species.

Krische et al. explored conjugate addition–electrophilic trapping as a modular platform for catalytic reaction development. Nitriles were also used as the terminal electrophiles similar to ketones and esters (see Scheme 4.20) [29].

4.2.3.4 Cycloaddition Mediated by a Stoichiometric Amount of Transition Metals

Cycloaddition of two alkynes and a nitrile using transition metals is a straightforward and attractive method for the preparation of pyridines [79]. Cobalt complexes have been extensively used for this process, and the mechanistic aspects of the reaction have been intensively investigated [79b, 79c]. It is generally accepted that metallacyclopentadienes formed by coupling of two alkynes are intermediates for the formation of pyridines in catalytic and stoichiometric reactions. However, reaction of metallacyclopentadienes prepared from two different alkynes with a nitrile affords a mixture of two regioisomers due to the orientation of the nitrile molecule (Scheme 4.60). Takahashi developed a novel coupling reaction of azazirconacyclopentadienes, which were prepared in situ from an alkyne and a nitrile, with a different alkyne to afford only a single isomer of the pyridines as shown in Eq. (2) (Scheme 4.60) [80].

A catalytic asymmetric [2 + 2 + 2] cycloaddition of two alkynes with nitriles was developed in the presence of a chiral cobalt(I) complex for the synthesis of enantiomerically enriched atropoisomers of 2-arylpyridines [81]. Bulky substituents R′ and R″ attached at the ortho positions of the 2-arylpyridine derivatives should prevent internal rotation, thus leading to the formation of two stable atropoisomers.

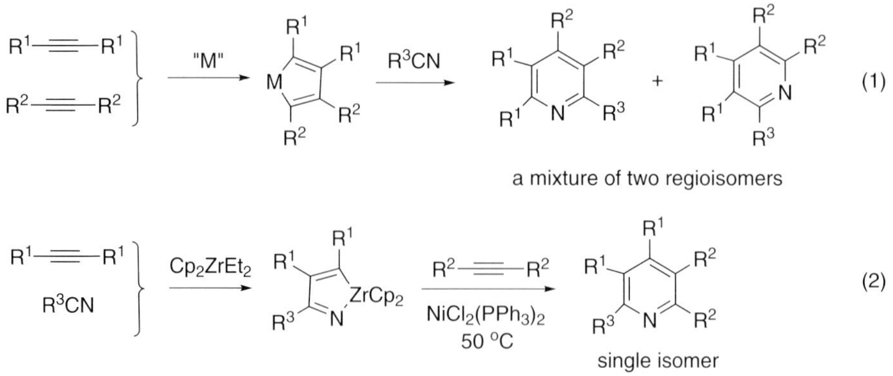

Scheme 4.60 A novel method to prepare a single isomer of pyridine.

A chiral ligand in the cobalt catalyst would be responsible for preferential formation of only one of them. Thus, when 2-methoxy-1-(1,7-octadiynyl)naphthalene reacted with nitriles in the presence of chiral Co catalyst, a product with high yield and high enantioselectivity was obtained (Scheme 4.61).

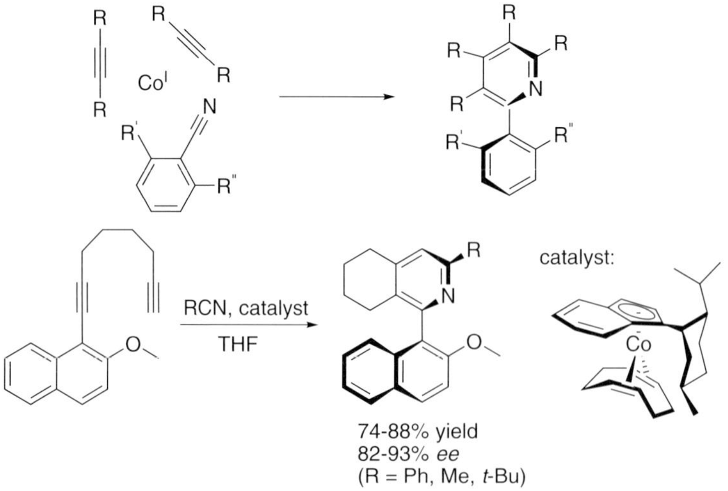

Scheme 4.61 [2 + 2 + 2] Cycloaddition giving axially chiral 2-arylpyridines.

The reaction of α,β-acetylenic γ-hydroxy nitriles with thiosemicarbazide, under mild conditions (rt, no catalyst, in 1:1 aqueous ethanol, 4–14 h), proceeded chemo-, regio- and stereoselectively to give tri-functionalized (amino, hydroxy and thioamide groups) pyrazoles in 53–91% yields (Scheme 4.62) [82].

The reactions of aryl isocyanates are similar to the corresponding aryl nitriles. In the reaction of Scheme 4.63, products with E/Z ratio of 99/1 were obtained in most cases [83].

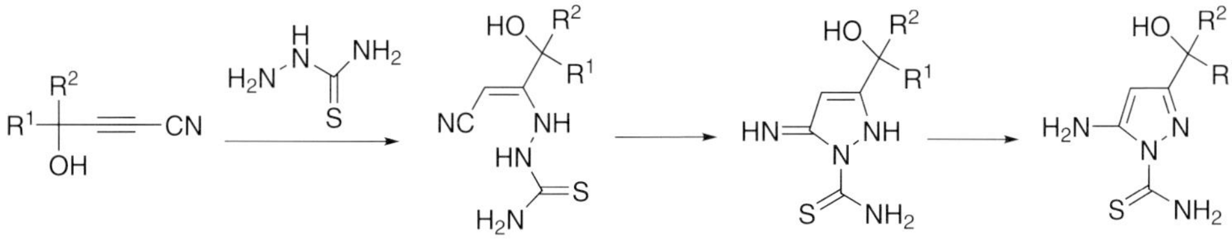

Scheme 4.62 Reaction of α,β-acetylenic γ-hydroxy nitriles with thiosemicarbazide.

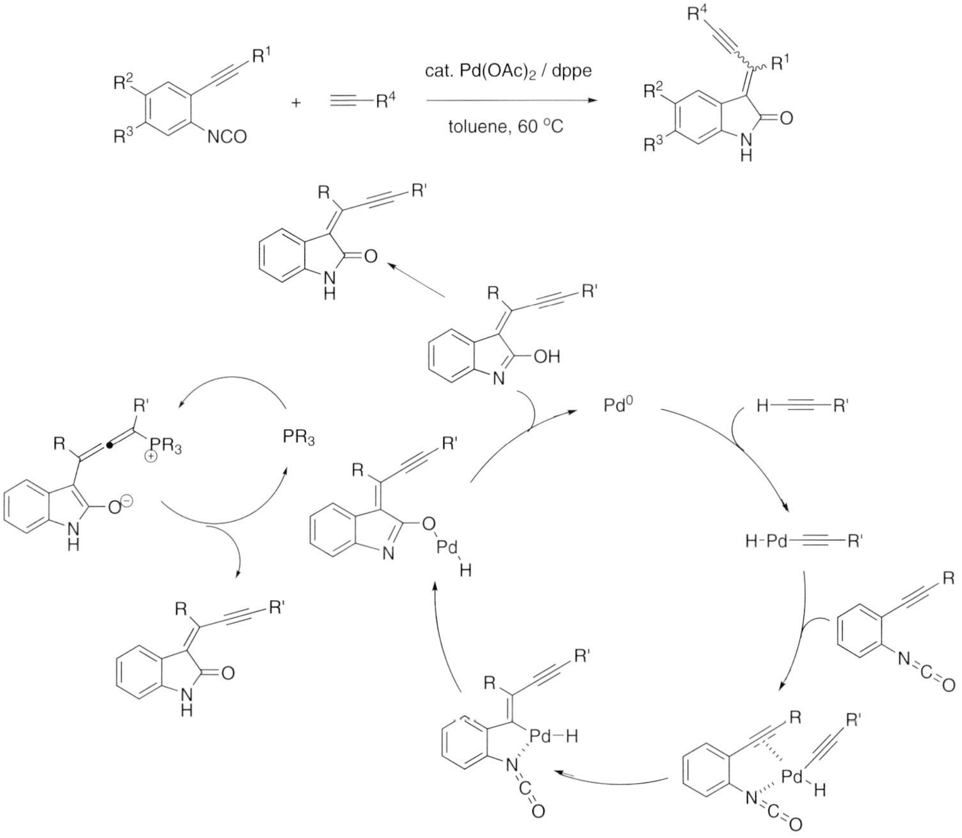

Scheme 4.63 Palladium-catalyzed reaction of o-alkynyl arylisocyanates with alkynes.

4.3
Cyclization via Intramolecular 1,4-Addition Reactions

4.3.1
General

1,4-Addition (or conjugate addition) refers to the addition of any class of nucleophile to an unsaturated system in conjugation with an activating group, usually an electron-withdrawing group. The Michael reaction refers to the addition of stabilized carbanions to an unsaturated system in conjugation with a carbonyl group. 1,4-Additions have a relatively long history since the first publication of a paper in 1883 by Komnenos [84]. However, the chemistry of 1,4-addition really began soon after the work of Arthur Michael in 1887 [85]. Many new types of conjugate acceptors have been used. Some of these consist of alkenes or alkynes activated by groups other than carbonyls or related functional groups. Many nucleophiles other than the carbanions such as nitrogen, oxygen, and sulfur nuclephiles have also been used, especially for the synthesis of heterocycles. Recently, much attention has been paid to catalytic 1,4-addition reactions and various intramolecular and tandem reactions for the synthesis of natural products have been developed [86].

Over the last 30 years the asymmetric 1,4-addition of nucleophiles to α,β-unsaturated carbonyl compounds has been a hot topic in asymmetric reactions [86c]. More recently, the tandem reaction resulting from the 1,4-addition of a nucleophile to a Michael acceptor followed by trapping of the anionic intermediate with an electrophile forming two contiguous stereocenters has been realized. A variety of compatible electrophiles have been demonstrated, including aldehydes, ketones, esters, and nitriles, π-allyl palladium species, halides and tosylates, oxocarbenium ions and, less frequently, a proton source. Stoichiometric methods include the use of chiral auxiliaries, chiral nucleophiles, and chiral radical sources.

4.3.2
1,4-Addition of Carbanions (Michael Addition)

4.3.2.1 1,4-Additions Mediated by Bases

The syntheses of the racemic axane-4 family of sesquiterpenoids were completed using a stereoselective intramolecular 1,4-addition in which the carbanion was stabilized by the carbonyl group (Scheme 4.64) [87].

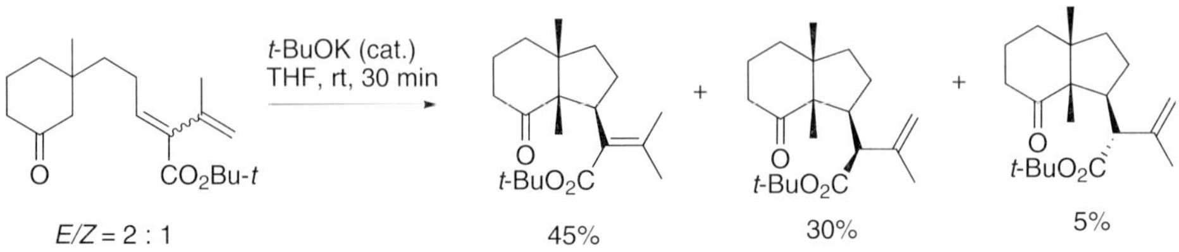

Scheme 4.64 The synthesis of the axane-4 family of sesquiterpenoids.

This method was further improved by treatment of **39** with pyrrolidine affording the bicyclic product **40** in 86% yield, the pyrrolidine having been first reacted with the ketone to form the imino group (Scheme 4.65) [88].

Azocycles with contiguous quaternary and tertiary stereocenters have been prepared in high enantiomeric purity by intramolecular 1,4-addition of enolates generated from α-amino acid derivatives *via* memory of chirality (Scheme 4.66) [89].

Similarly, the cyclization of phenylthioether substituted esters is another example of this type of reaction (Scheme 4.67) [90].

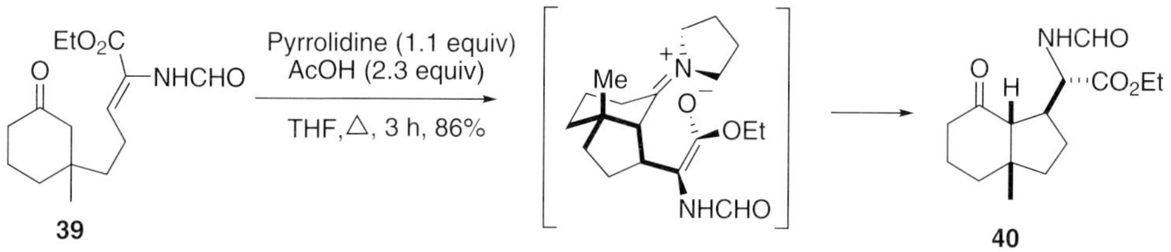

4.3 Cyclization via Intramolecular 1,4-Addition Reactions

Scheme 4.65 Improvement of the reaction by the addition of pyrrolidine.

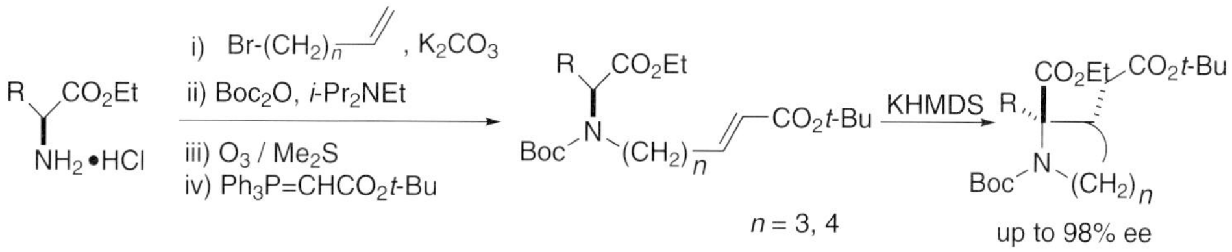

Scheme 4.66 Highly enantioselective synthesis of nitrogen heterocycles.

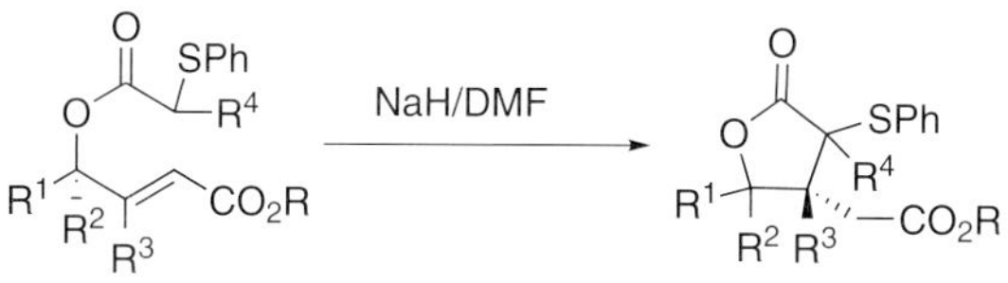

Scheme 4.67 Cyclization of phenylthioether substituted esters.

Synthesis of heterocyclic compounds via intramolecular conjugate addition of heteronucleophiles (X = N, O, S $\cdots$) is a well known process. However, intramolecular Michael addition of carbanions (X$^-$ = $-$NC$^-$, $-$OC$^-$, $-$SC$^-$) often causes undesired side reactions especially when an anion stabilizing heteratom Y (Y = O, N, $\cdots$) is substituted at the γ-position, since the use of a significantly strong base is needed to generate the carbanions from the corresponding hetroalkyls ($-$NCH$_3$, $-$OCH$_3$, $-$SCH$_3$) and the base abstracts the γ-hydrogen atom competitively. Using the higher order cyano copper (or silver) amide (2LDA-CuCN or 2LDA-AgCN) at $-78\,^{\circ}$C enables one to transform γ-alkylsulfonyloxy-α,β-unsaturated ester **41** into γ-sulfones **42** in good to high yields (Scheme 4.68) [91].

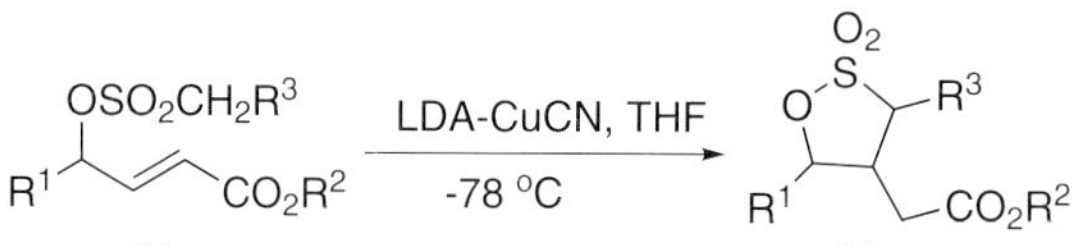

Scheme 4.68 Transformation of γ-alkylsulfonyloxy-α,β-unsaturated esters into γ-sulfones.

In the synthesis of taxol, an efficient method for the construction of the eight-membered ring (B-ring) has been a major problem. Fukumoto et al. developed the intramolecular Michael addition of the sulfonyl carbanion to construct the eight-membered ring (Scheme 4.69) [92].

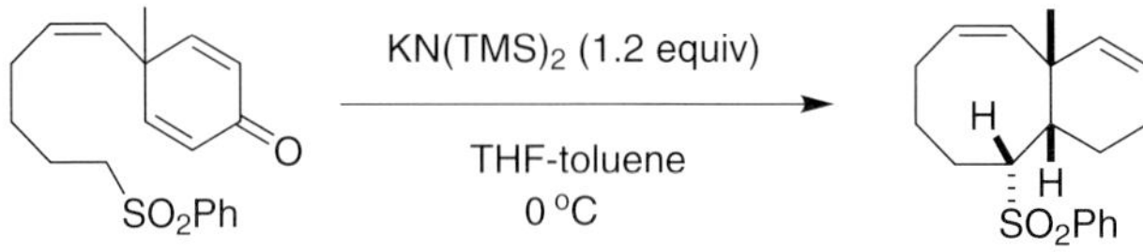

Scheme 4.69 Construction of the eight-membered ring.

A general and convergent method for the synthesis of tricyclic and tetracyclic intermediates related to sterols was developed in which 1,4-addition was used as the main step (Scheme 4.70). Such compounds are interesting intermediates for the synthesis of polycyclic natural and unnatural products [93].

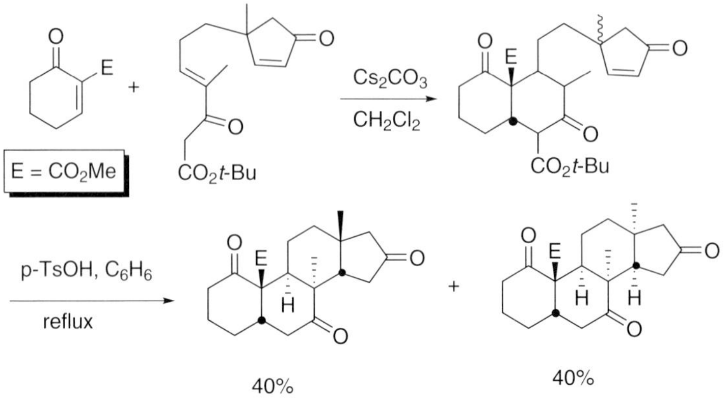

Scheme 4.70 Synthesis of tricyclic and tetracyclic intermediates related to sterols.

A significantly highly asymmetric tandem 1,4-addition of certain metal amide reagents to nona-2,7-diene-1,9-dioic acid ester was realized in which even a simple auxiliary such as the menthyl group produces a high diastereo- and enantioselectivity (Scheme 4.71) [94].

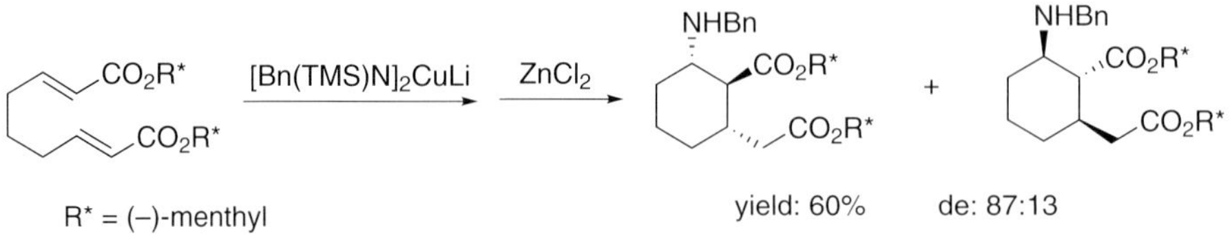

R* = (−)-menthyl   yield: 60%   de: 87:13

Scheme 4.71 Asymmetric tandem 1,4-addition of metal amide reagents to nona-2,7-diene-1,9-dioic acid ester.

An asymmetric induction was observed in the intramolecular Michael addition reaction of (R)-lithiated vinylic sulfoxide to (Z)-enoates. The Michael addition proceeds with high diastereoselectivity to give the adducts with (R)-configuration at the β-position of the ester in the five-membered ring (Scheme 4.72). The selectivity was reversed in the six-membered ring formation. On the other hand, the corresponding (E)-enoates provided Michael adducts with poor diastereoselectivity [95].

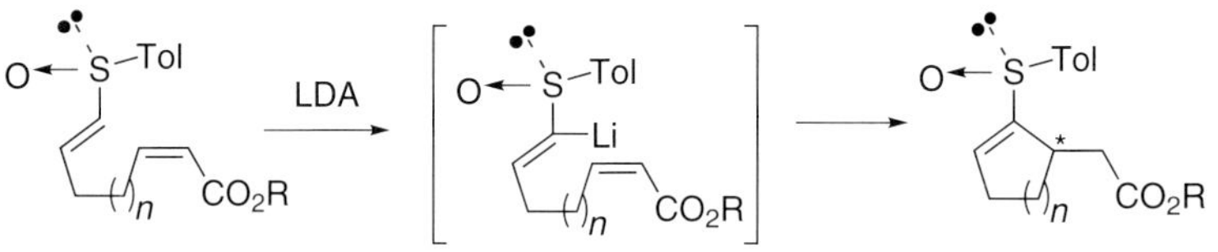

Scheme 4.72 Asymmetric intramolecular Michael addition induced by sulfoxide group.

A metal–halogen exchange reaction may be used to introduce internal nucleophilic centers (a carbanion) which may then undergo efficient intramolecular conjugate addition reactions leading to three-, four-, five-, and six-membered rings as shown in Scheme 4.73 [96].

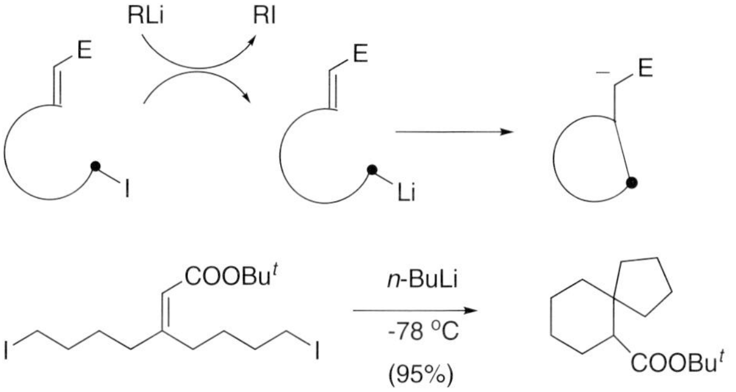

Scheme 4.73 A halide-base induced intramolecular conjugate addition.

4.3.2.2 1,4-Additions Mediated by Lewis Acids

Novel spirocyclization based on intramolecular 1,4-addition has been developed using a combination of Lewis acid and 1,2-diol. Treatment of five- and six-membered α,β-unsaturated cyclic ketones having a 4-oxopentyl group at the β-position with Lewis acid and ethylene glycol gave spiro[4.5]decane-2,7-dione and spiro[5.5]undecane-2,8-dione, respectively (Scheme 4.74). The asymmetric version of this reaction has been developed by using optically active 1,2-diol such as cyclohexane-l,2-diol to afford the spirocyclic products of up to 85% ee [97].

1,3,4,9-Tetrahydropyrano[3,4-b]indole derivatives have been prepared from commercially available and inexpensive indole-2-carboxylic acid using intramolecular InBr$_3$-catalyzed Michael addition of indole with α,β-unsaturated ketones (Scheme 4.75) [98].

The reaction of diethyl 2-[(N-methyl-N-phenylcarbamoyl)methylene]malonate **43** in the presence of ZnCl$_2$ (1.2 equiv) in CH$_2$Cl$_2$ at room temperature for 16 h gave diethyl 2-(1-methyl-2-oxoindolin-3-yl)malonate **44** in 98% yield (Scheme 4.76). The reaction also proceeded with catalytic amounts (0.2 equiv) of Lewis acids such as AlCl$_3$, ZnCl$_2$, ZnBr$_2$, Sc(OTf)$_3$, and InBr$_3$ [99].

A new five-membered ring formation in which the key step involves a CuCl-mediated intramolecular conjugate addition of a vinyltrimethylstannane to a cyclic enone system was realized. It is concluded that a vinylcopper species is formed as

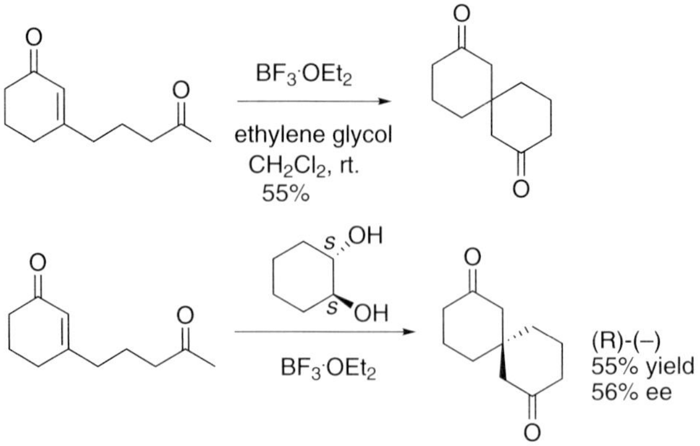

Scheme 4.74 Spirocyclization based on intramolecular 1,4-addition.

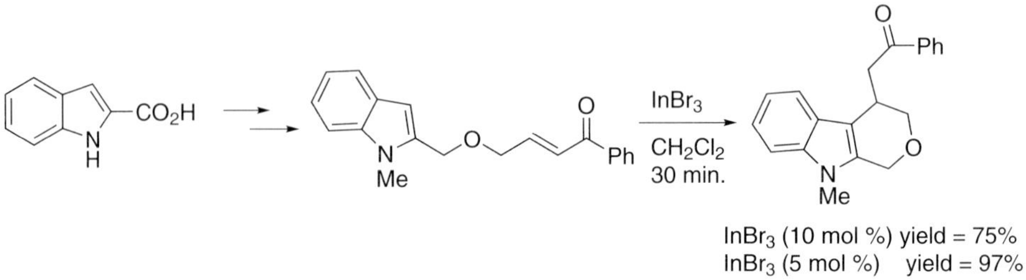

Scheme 4.75 Intramolecular cyclization catalyzed by InBr$_3$ for the synthesis of tetrahydro-pyranylindole.

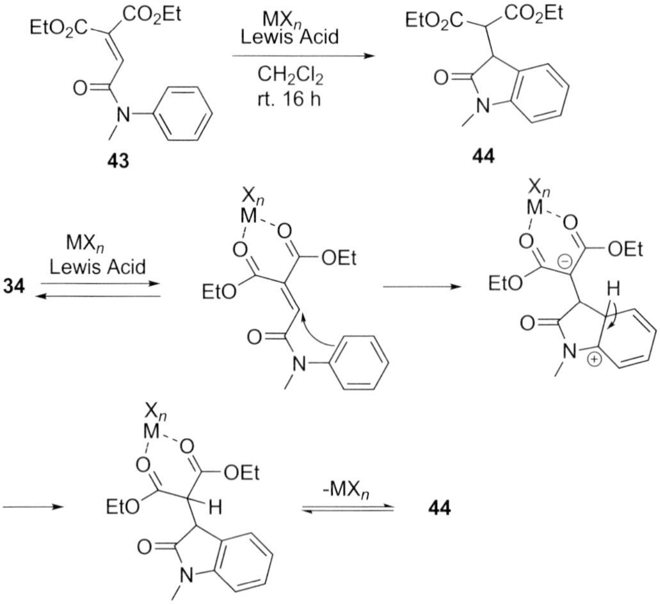

InBr$_3$ (10 mol %) yield = 75%
InBr$_3$ (5 mol %) yield = 97%

Scheme 4.76 Lewis acid-catalyzed intramolecular 1,4-addition.

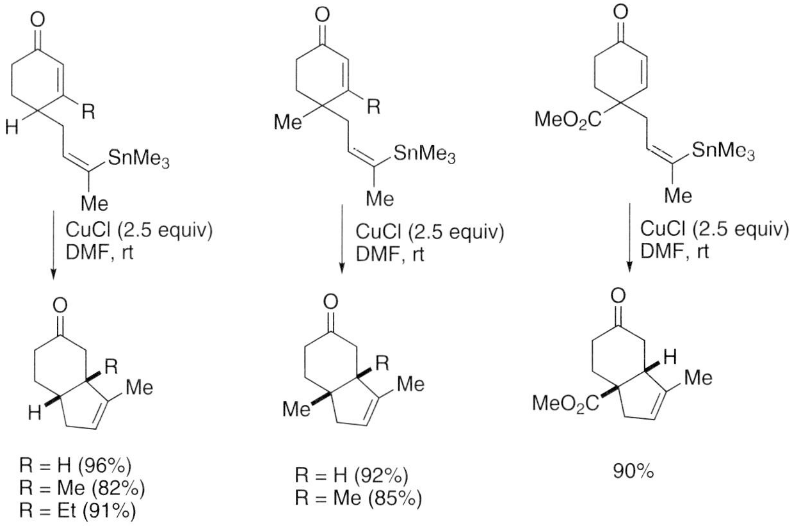

Scheme 4.77 Intramolecular conjugate addition of a vinyltrimethylstannane to a cyclic enone.

an intermediate in this coupling process (Scheme 4.77) [100]. A rhodium catalyst may also be used in similar reactions [101].

The reaction of Me$_2$CuLi with 1,1-dibromoalkene leads to a (Z)-vinylcopper species which in turn undergoes an intramolecular conjugate addition reaction with the α,β-unsaturated ester moiety (Scheme 4.78). Five-, six-, and seven-membered carbocycles were constructed by this method [102].

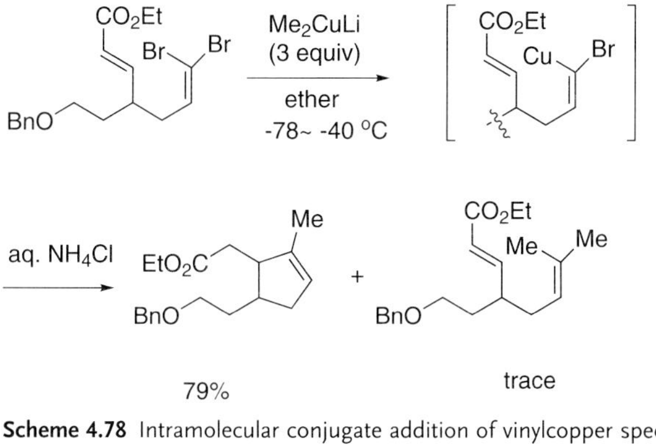

Scheme 4.78 Intramolecular conjugate addition of vinylcopper species with enoates.

4.3.2.3 Catalytic 1,4-Addition Reactions

It was reported that tetrabutylammonium triphenyldifluorosilicate (TBAT) is an effective catalyst in the cyclic allenylation of α,β-unsaturated esters, nitriles and amides, a reaction, which cannot be achieved by the TiCl$_4$-promoted procedure (Scheme 4.79) [103].

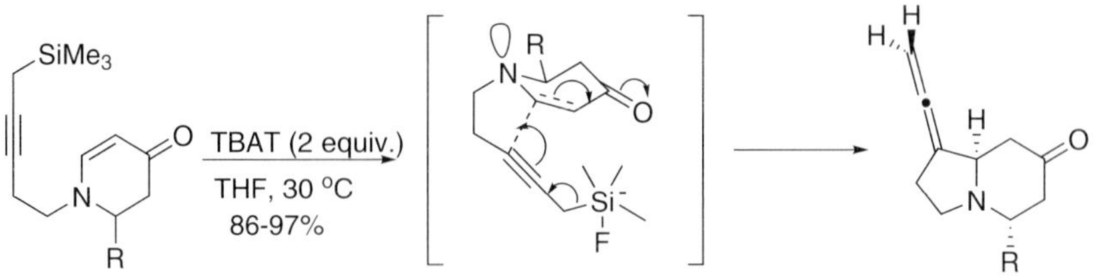

Scheme 4.79 TBAT-initiated intramolecular conjugate addition of propargylsilane to dihydropyridones.

A copper-catalyzed intermolecular 1,4-additon of dialkyl zinc or trialkyl aluminum reagents with enones and subsequent cyclic conjugate addition with the enoate moiety in the presence of homochiral phosphoramidite ligands can yield the trisubstituted cyclohexanes highly diastereoselectively (Scheme 4.80) [104].

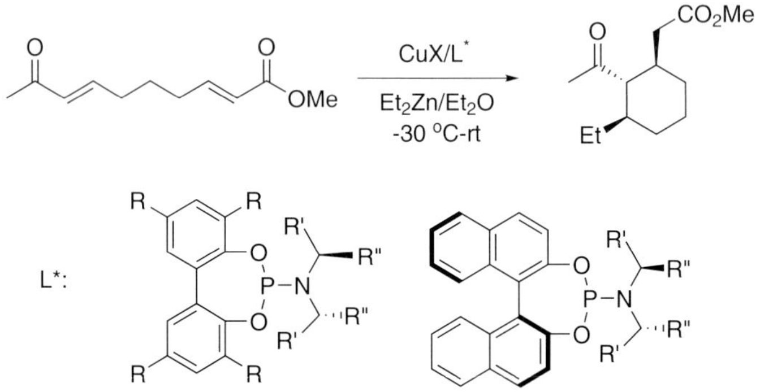

Scheme 4.80 CuX-catalyzed tandem asymmetric conjugate addition–cyclization reaction.

For the nickel-catalyzed intramolecular cyclization of alkynyl enones, a different mechanism involving the formation of metallacycle first from the oxidative cyclization of a nickel (0) π-complex with the enynes instead of the 1,4-addition of dialkylzinc as the first step was proposed, as shown in Scheme 4.81 [31a].

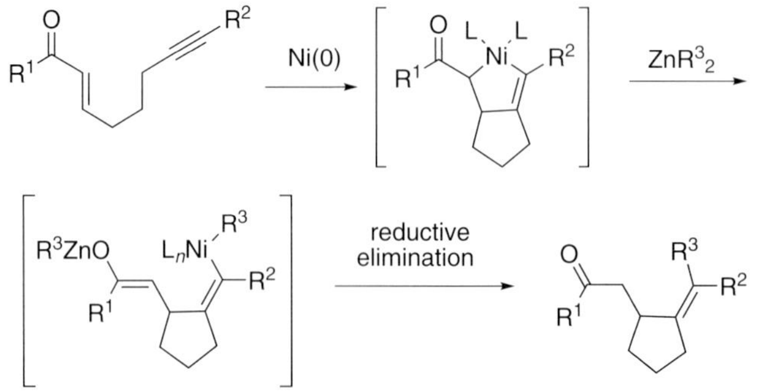

Scheme 4.81 Mechanism of nickel-catalyzed intramolecular cyclization of alkynyl enones.

4.3 Cyclization via Intramolecular 1,4-Addition Reactions

A tandem Pd-mediated intramolecular cyclization followed by intermolecular aryl–alkenyl cross coupling has been used successfully in a synthesis of a carbacyclin. In this reaction, the formed arylpalladium species reacts with the enone via conjugate addition and the palladium species is finally quenched by protonolysis or β-hydride elimination (Scheme 4.82) [105].

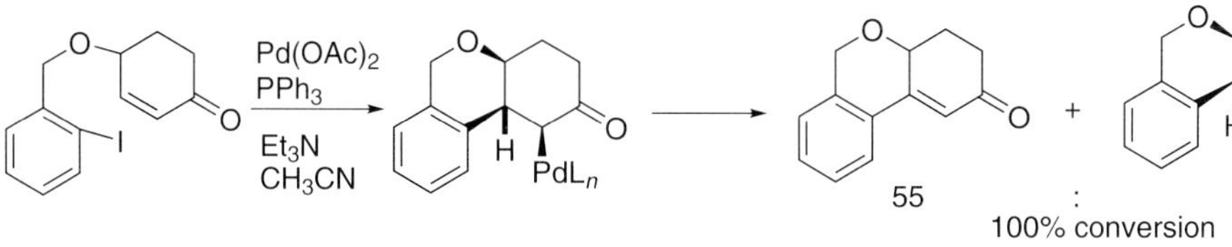

Scheme 4.82 Palladium-catalyzed intramolecular conjugation addition to cyclohexenone.

Using methyl 6-acetoxymethylhepta-2,6-dienoate as the starting material, conjugate addition of the enoates will occur with the active methylene compounds in the presence of a base. In the presence of a palladium(0) catalyst, the π-allyl palladium complex will form first, followed by the addition of active methylene compounds and finally the six-membered adducts resulting from 1,4-addition sequence were obtained with moderate to good yields (Scheme 4.83) [106].

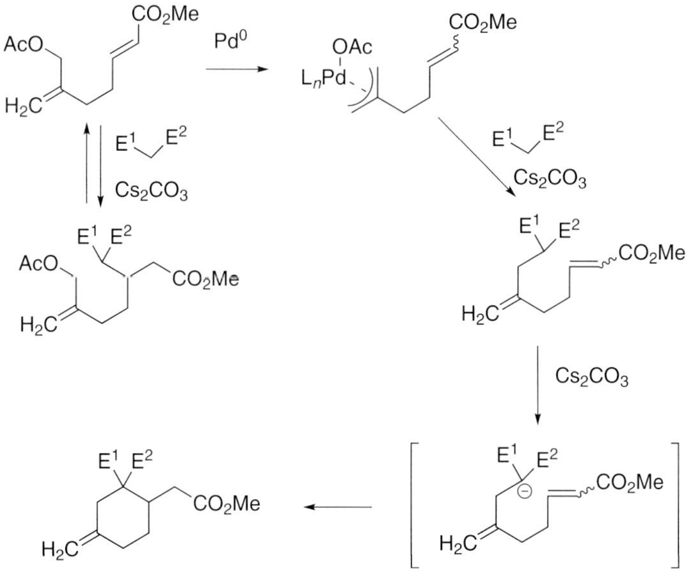

Scheme 4.83 Different products obtained in the presence or absence of Pd catalyst.

A rhodium-catalyzed addition-cyclization of 2,6-enynones or esters with aryl- or alkenylboronic acids occurring under mild conditions has been developed (Scheme 4.84) [107].

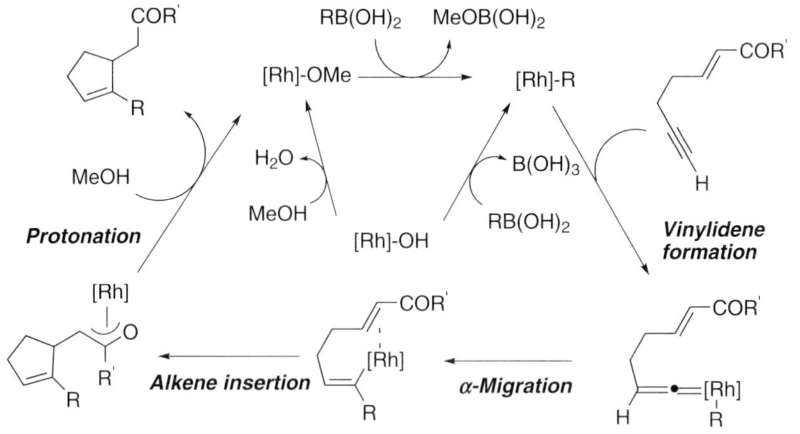

Scheme 4.84 Rhodium-catalyzed addition–cyclization of 2,6-enynones with aryl- or alkenylboronic acids.

Arylboronates bearing a pendant Michael-type acceptor olefin or acetylene linkage at the ortho-position undergo transmetalation with a rhodium-based catalytic complex to generate a functionalized organorhodium intermediate which can add to nonterminal acetylenes followed by conjugate addition to yield the indenes in good to excellent yields (Scheme 4.85) [108].

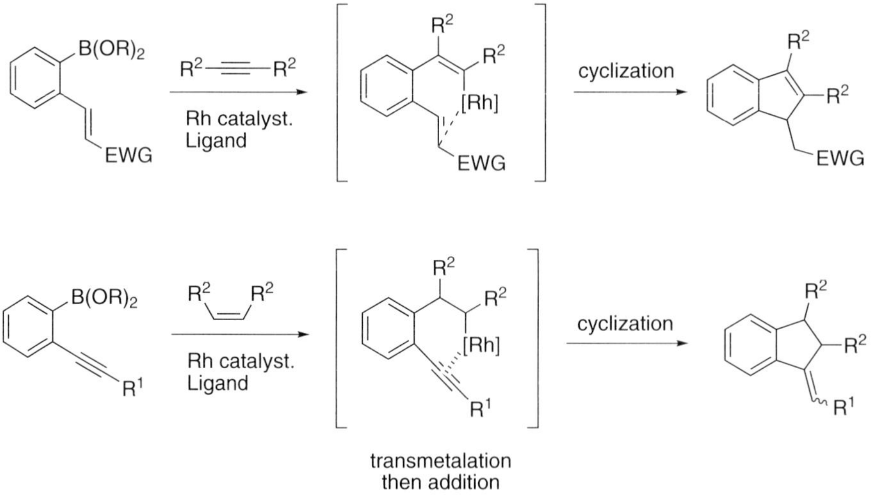

Scheme 4.85 Rhodium-catalyzed tandem cyclization reactions of transmetalation, insertion, and 1,4-addition.

The intramolecular Stetter reaction is a useful method for construction of cyclic compounds. According to the proposed mechanism, intermediate **45** would result from nucleophilic attack of the carbene on the aldehyde. Subsequent proton transfer would afford the acyl-anion equivalent **46**. Carbon–carbon bond formation resulted from nucleophilic attack of acyl-anion equivalent **46** into a Michael acceptor, generating enolate **47**. Enolate protonation followed by catalyst turnover generated the desired product (Scheme 4.86). For the intramolecular cyclization reaction, products with high enantioselectivity were obtained using the homochiral NHC carbene as catalyst (Scheme 4.87) [109].

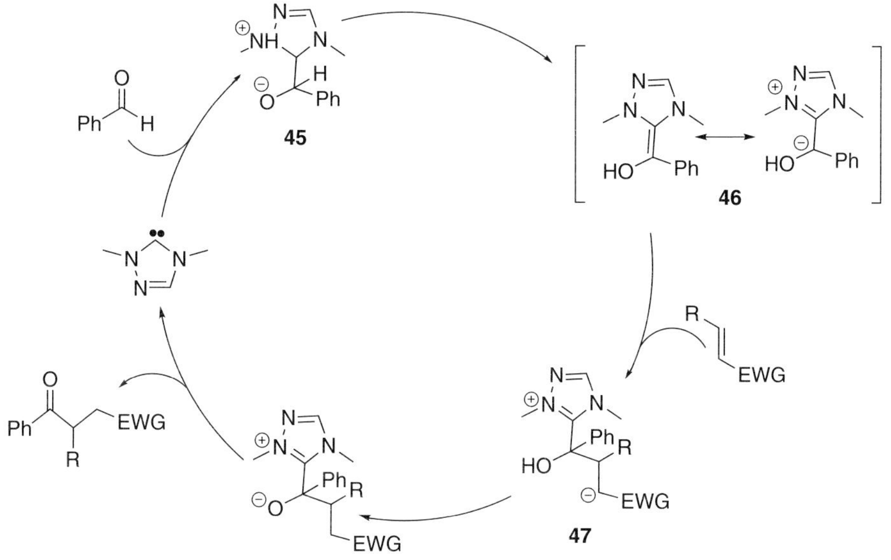

Scheme 4.86 Carbene-catalyzed Stetter reaction.

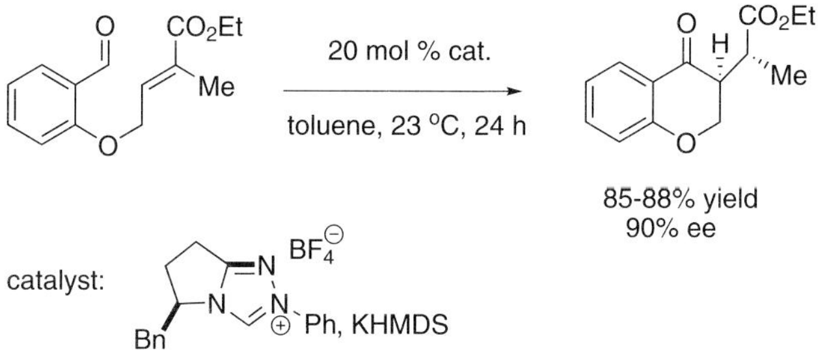

Scheme 4.87 Enantioselective cyclization catalyzed by homochiral carbene.

A new variant of the Morita–Baylis–Hillman reaction has been developed for the synthesis of substituted cyclopentenes or cyclohexenes by trialkylphosphine-catalyzed cyclizations of 1,5-hexadienes or 1,6-heptadienes in which the alkenes are activated by electron-withdrawing groups [110]. Conjugate addition of tributylphosphine to the bis(enone) **48** provides enolate **49**. Subsequent intramolecular conjugate addition to the appendant enone yields zwitterionic intermediate **50**. Finally, proton-transfer enables elimination of tributylphosphine to complete the catalytic cycle and give the cycloalkenes product **51** (Scheme 4.88).

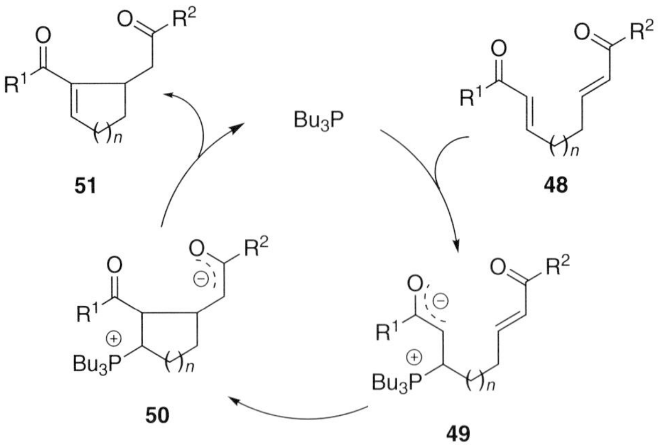

Scheme 4.88 Tertiary phosphine-catalyzed synthesis of cycloalkenes from bis-enones.

A highly efficient method for constructing heterocycles was achieved by tertiary phosphine-catalyzed tandem umpolung addition and intramolecular conjugate addition (Scheme 4.89). This strategy offers a simple and promising method for constructing synthetically useful heterocycles under neutral conditions with high atom economy [111].

4.3.3
1,4-Addition of Heteronucleophiles

Tetrabutylammonium fluoride can also be used as the catalyst to promote the intramolecular hetero-nucleophilic attack of a carbodiimide group on an α,β-unsaturated enoate for the construction of nitrogen heterocycles (Scheme 4.90) [112, 113].

A viable strategy for the construction of heterocyclic ring systems lies in the tandem conjugate additions of an amine functionality bearing a Michael acceptor **52** (Scheme 4.91). The intermediate anion **53** undergoes a further conjugate addition to give the final heterocycle **54**. Nitroalkenes are ideal candidates as electrophilic olefins for this procedure. Benzyloxy nitroalkene **55** can be considered as a synthetic equivalent of nitrodiene **56** and its reaction with hydroxyester **57** in ethanol affords pyrrolidine quantitatively as a sole diastereomer **58**. The formation of nitrodiene is presumably promoted by the amino group. Simple functional group transformations allow the total synthesis of (−)-α-kainic acid [114].

An intramolecular amino-Michael reaction of 6-aminohex-4-en-2-ynoates followed by aromatization proceeded smoothly to afford pyrrole frameworks (Scheme 4.92) [115].

A sequential nucleophilic 1,4-addition of primary amines to the enamides involving a chiral auxiliary and subsequent hydrogenation enabled the synthesis of the enantiopure piperidines in very good overall yields and high stereoselectivity from the bifunctional substrate **59**. The amine initially attacks the aldehyde to form the imine which then adds intramolecularly to the conjugate double bond (Scheme 4.93) [116].

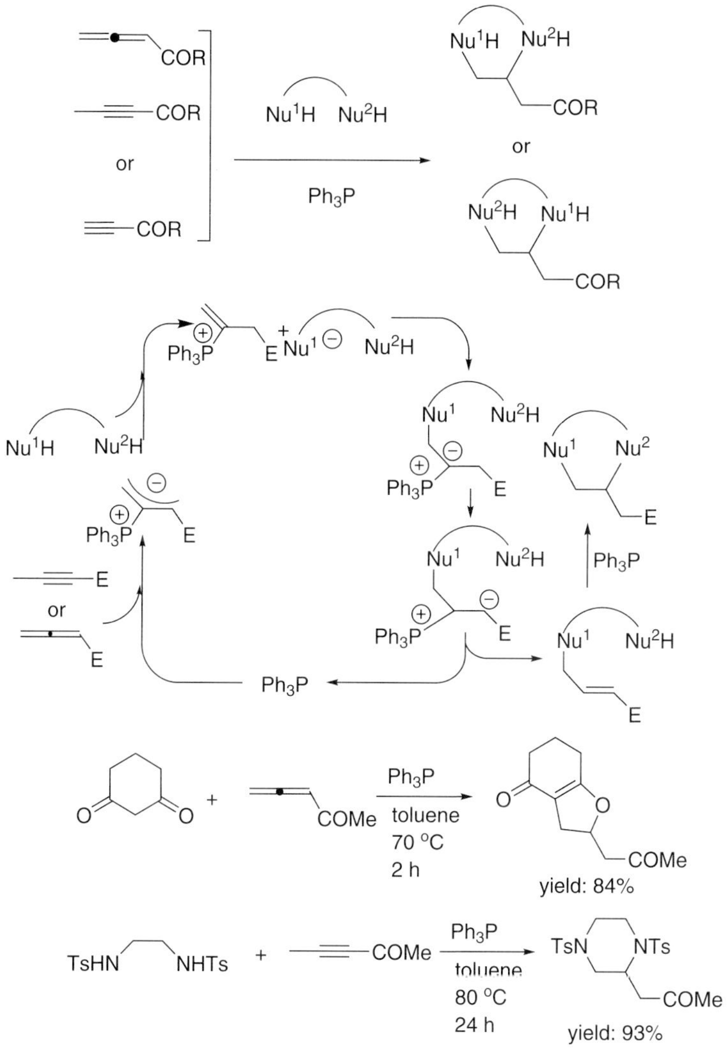

Scheme 4.89 Tertiary phosphine-catalyzed tandem umpolung addition and conjugate addition for constructing heterocycles.

The α-lithiumaziridines **60** can add to a conjugate system of pyrrole without change of the aziridine skeleton to form the saturated five-membered ring **61** (Scheme 4.94) [117].

The oxyanion derived from hydroxyacrylate *E*-**62** undergoes smooth intramolecular Michael addition to give the *trans*-2,6-disubstituted tetrahydropyran **63** as the major product. In contrast, the oxyanion obtained from isomer *Z*-**62** cyclizes to give the *cis*-2,6-disubstituted tetrahydropyran **64** as the major product (Scheme 4.95). Such chemistry has been extended to the enantioselective synthesis of these tetrahydropyran derivatives [118].

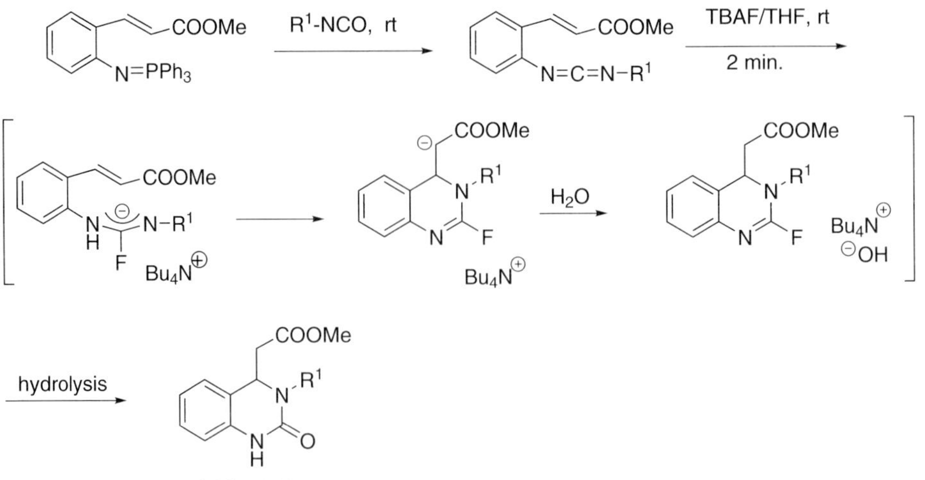

yield: 40-88%

Scheme 4.90 Fluoride ion-catalyzed conjugate addition of carbodiimide on enoates.

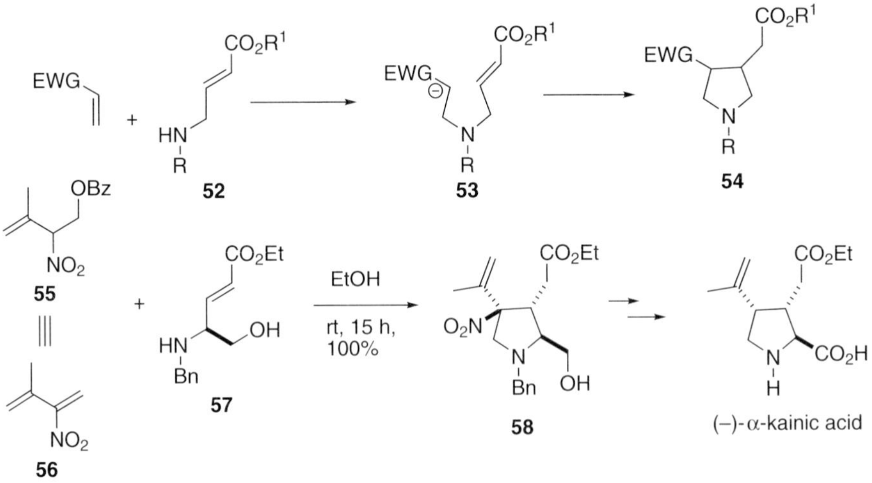

Scheme 4.91 Construction of heterocyclic ring systems via tandem conjugate additions.

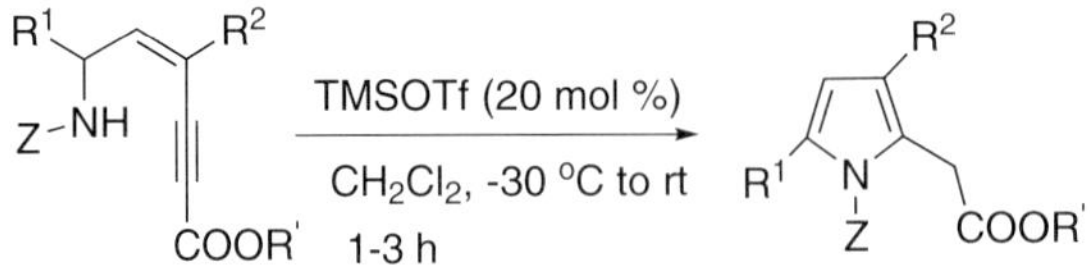

yield: 31-96%

Scheme 4.92 An intramolecular amino-Michael reaction.

Bismuth(III) salts provide a convenient, inexpensive, and environmentally benign source of the corresponding Brønsted acid. For example, the hydrolysis of bismuth (III) bromide is known to afford two equivalents of hydrogen bromide and insoluble bismuth oxybromide, of which the former is responsible for the observed catalysis.

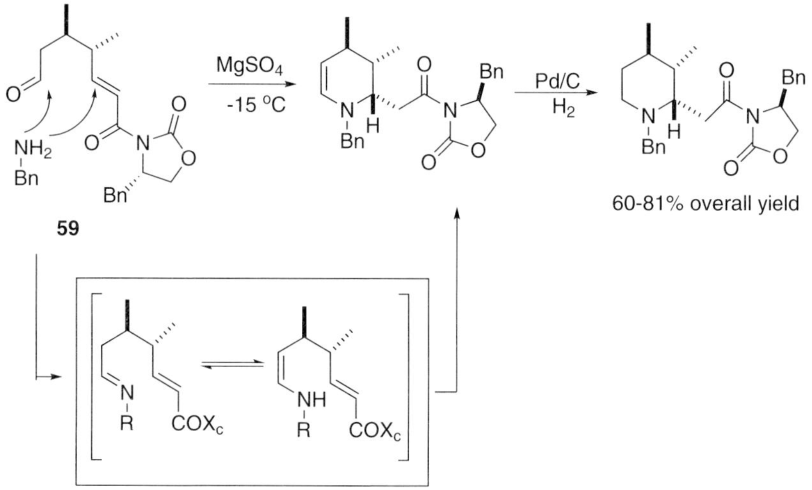

59

Scheme 4.93 A sequential nucleophilic 1,4-addition of primary amines to the enamides.

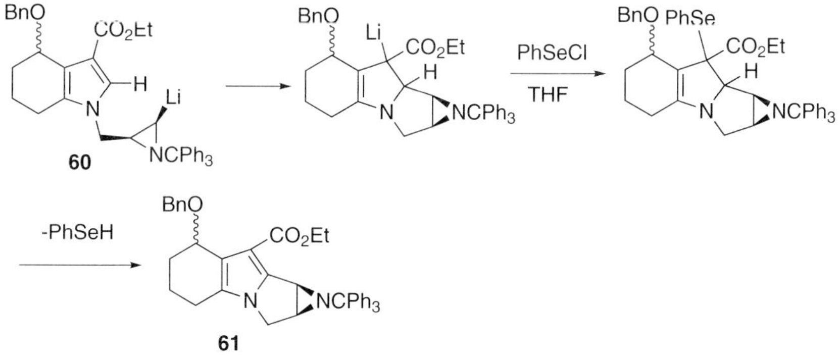

60

61

20% overall yield

Scheme 4.94 1,4-Addition of lithium aziridine to pyrrole.

This reagent promotes the desilylation and intramolecular etherification for the construction of *cis*- and *trans*-2,6-disubstituted tetrahydropyrans, respectively (Scheme 4.96) [119].

An approach to highly functionalized tetrahydrofuran derivatives based upon a novel oxa-Michael/Michael dimerization of *cis*-γ-hydroxyenones is presented. The reaction begins with either 1,2-dioxines or *trans*-γ-hydroxyenones and proceeds by addition of one molecule of *trans*-γ-hydroxyenone to another molecule of *cis*- or *trans*-γ-hydroxyenone catalyzed by an alkoxide or hydroxide base. Subsequent intramolecular Michael addition of the keto-enolate gives the observed tetrahydrofurans (Scheme 4.97). Substitution at both the 2- and 4-positions of the γ-hydroxyenone is tolerated [120].

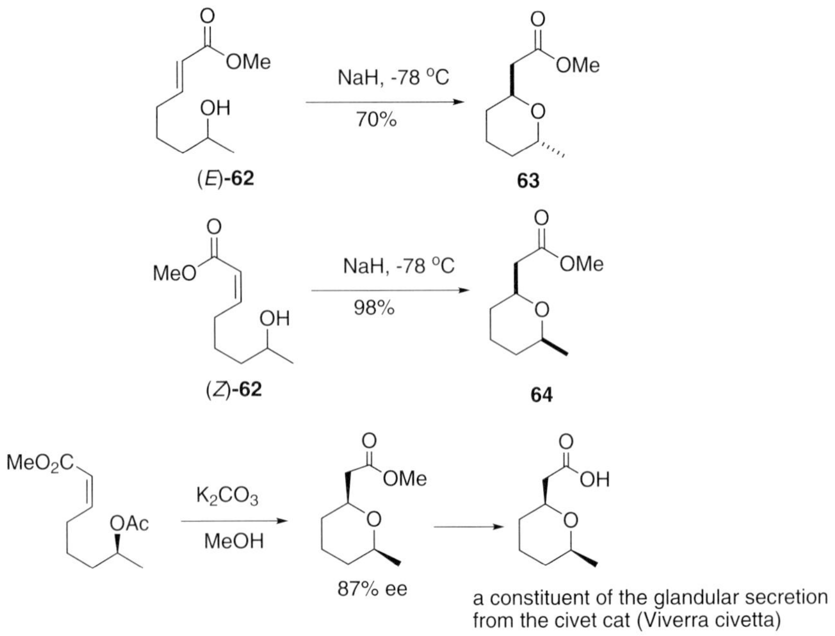

Scheme 4.95 Intramolecular 1,4-addition by oxygen nucleophile.

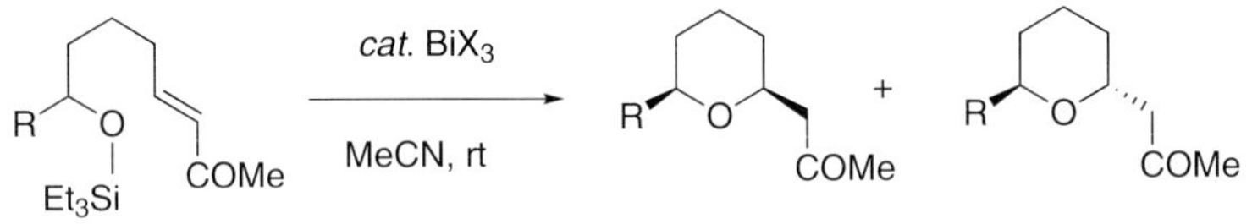

Scheme 4.96 BiX$_3$-promoted desilylation and intramolecular etherification for the construction of tetrahydrofurans.

Similarly, oxazine ring is easily synthesized by the use of the *N*-hydroxymethyl group after conjugate addition (Scheme 4.98) [121].

A rapid, high-yielding synthesis of benzothiazines in which a precursor aryl thiourea is prepared on a solid phase has been developed. Addition of trifluoroacetic acid catalyzes a conjugate addition reaction to form the desired heterocycle and releases it from the resin (Scheme 4.99) [122].

4.3.4
Other Acceptors for 1,4-Addition Reactions

Numerous anionic nucleophiles react conjugately with unsaturated carbonyl derivatives, but many analogous reactions involving unsaturated nitriles are problematic. Compound **65** first reacted with butyllithium to form the dithiane anion **66**. The cyclization did indeed proceed favorably when catalyst 12-crown-4 was added to the preformed dithiane anion, to provide a 4:1 ratio of indolizidines **67** and **68** in 90%

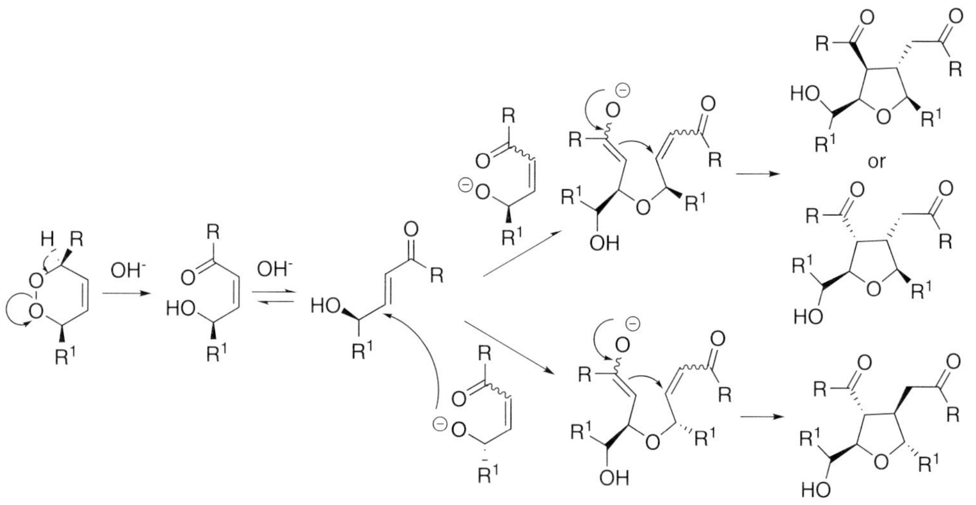

Scheme 4.97 Synthesis of functionalized tetrahydrofurans by oxa-Michael/Michael cyclization of γ-hydroxyenones.

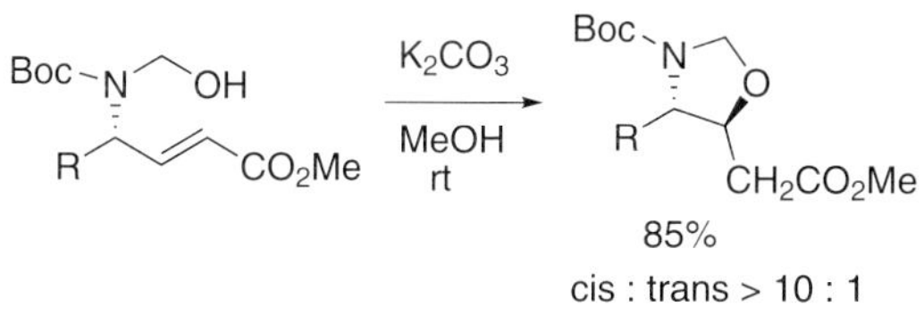

Scheme 4.98 The use of N-hydroxymethyl group for stereoselective conjugate addition.

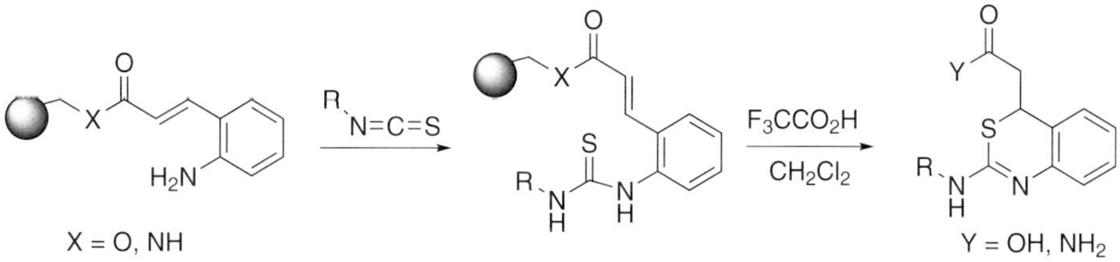

Scheme 4.99 A solid phase synthesis of benzothiazines.

yield (Scheme 4.100). The requirement for 12-crown-4 may reflect the need for a free carbanion with an orbital that can overlap with the π-system of the unsaturated nitrile, as depicted in **53** [123].

A highly stereospecific 1,4-addition in the ω-hydroxy α,β-unsaturated nitrile **70**, obtained from 'diacetone glucose **69**, leading to annulated furanose **71** has been reported (Scheme 4.101) [124].

A vinylsulfone may also act as a Michael acceptor (Scheme 4.102) [125].

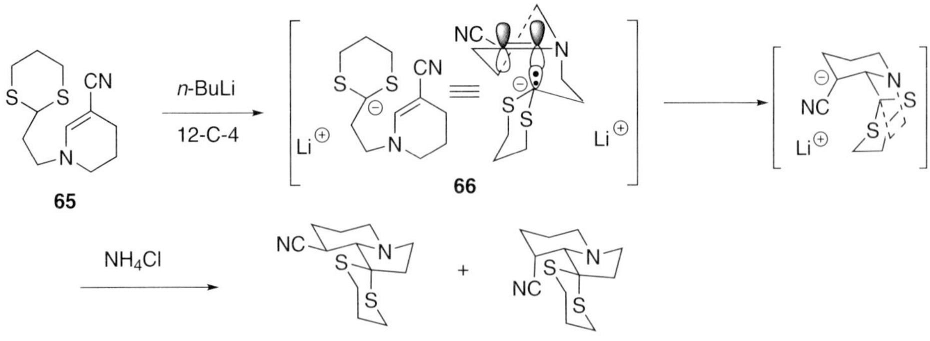

Scheme 4.100 Stereoselective 1,4-addition reactions of α,β-unsaturated nitriles.

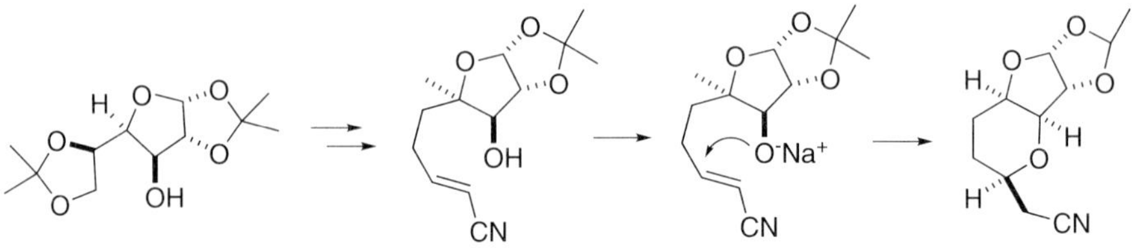

Scheme 4.101 1,4-Addition in the ω-hydroxy α,β-unsaturated nitrile.

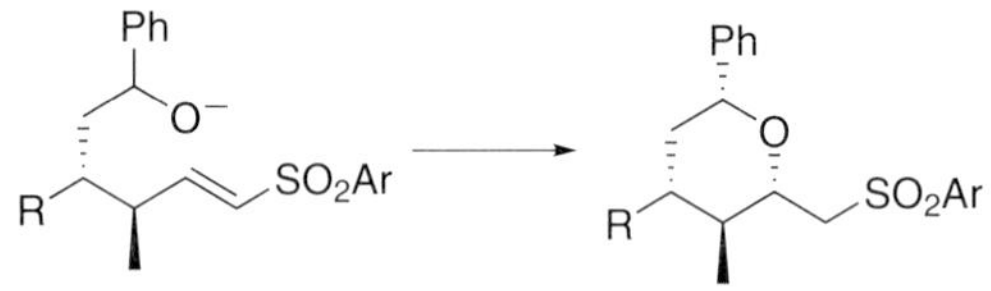

Scheme 4.102 Cyclic 1,4-addition of vinylsulfone.

4.3.5
Catalytic Asymmetric Intramolecular Cyclization via 1,4-Addition

Many intramolecular cyclization reactions via 1,4-addition are stereoselective. It is common to use the homochiral substrates to obtain the homochiral products. Recently, with the developments in chiral technology, the catalytic asymmetric intramolecular cyclization via 1-4-addition is becoming a topic of interest. The enantioselective 1,4-addition of arylboronic acids to β-(o-hydroxyaryl)enones to give β-diaryl ketones which was transformed into hydroxydihydropyrans in the presence of a dicationic palladium(II) catalyst, [Pd(S,S-chiraphos)(PhCN)$_2$](SbF$_6$)$_2$, by intramolecular 1,2-addition to the ketone group has been studied (Scheme 4.103). The product was further dehydrated yielding optically active chromenes with up to 99% ee [126].

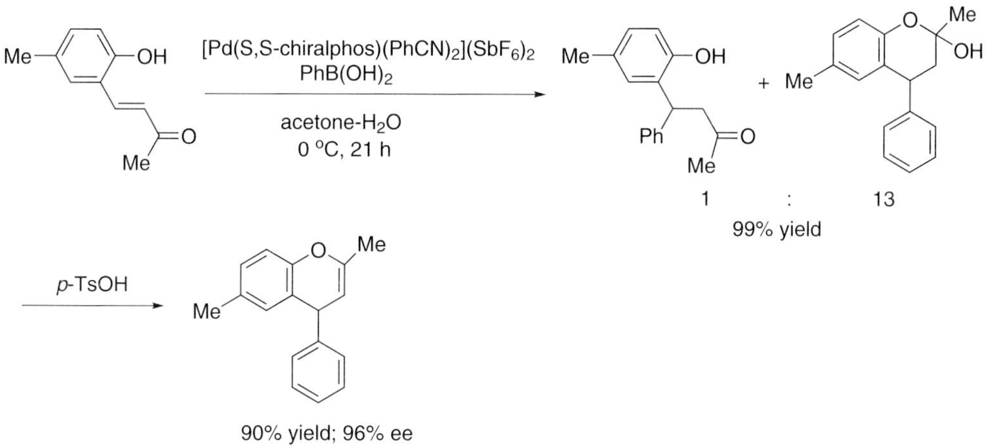

Scheme 4.103 Enantioselective cyclization with tandem 1,4- and 1,2-additions.

4.4
Conclusion and Perspectives

Cyclization reactions involving intramolecular 1,2 addition or 1,4-addition of carbon–heteroatom multiple bonds have a long history in the field of synthetic chemistry. Many useful reactions have been developed for the synthesis of natural products or bioactive target molecules. The chemistry of intramolecular 1,2- and 1,4-addition of carbon–heteroatom bonds has also been studied at a high level with the recent development of synthetic methods.

In recent decades, the chemistry of transition-metal catalyzed reactions has become well developed. Different kinds of transition-metal catalyzed reactions have been investigated with the use of different substrates and the formation of new cyclization products. Especially, with the development of catalytic asymmetric reactions in the presence of homochiral ligands, many asymmetric syntheses of optically active products will become possible in the future.

4.5
Experimental: Selected Procedures

4.5.1
1,2-Addition of Alkyne Substituted Carbonyl Compounds (See Section 4.2.1.3)

4.5.1.1 General Procedure for Cyclization of 5-Yn-1-ones [32c]
To an oven-dried, N_2-purged flask were added [Rh(OH)(cod)]$_2$ (2.28 mg, 0.5 μmol, 0.05 equiv of Rh), arylboronic acid (1.0 mmol, 5.0 equiv), 1,4-dioxane (1 mL), and H_2O (20 μL). A solution of 5-yn-1-one (0.2 mmol, 1.0 equiv) in 1,4-dioxane (1.0 mL) was added to the reaction mixture at r.t. After complete consumption of substrate, H_2O was added. The aqueous layer was extracted with EtOAc three times.

The combined extracts were washed with brine and dried over MgSO$_4$. The solvent was removed under reduced pressure and the residue was purified by preparative thin-layer chromatography (hexane–EtOAc) to give the product.

4.5.2
1,2-Addition of Arene Substituted Carbonyl Compounds (See Section 4.2.1.3)

4.5.2.1 Representative Procedure for Asymmetric Cyclization of o-(Acylmethyloxy)-arylboronic Acid to Furnish the Optically Active Cycloalkanols [39c]

Under nitrogen, Amberlite IRA(OH) (1.5 equiv) was added to a solution of o-(acylmethyloxy)-arylboronic acid (51.2 mg, 0.2 mmol) and catalyst (9.0 mg, 2.5 mol%) in toluene (2 mL). The reaction mixture was stirred at 40 °C for 12 h. After the reaction was complete, as monitored by TLC, the reaction mixture was quenched with 1 N NaOH (8 mL) and the aqueous layer was extracted with EtOAc (3 × 15 mL). The organic layers were dried over Na$_2$SO$_4$ and concentrated under reduced pressure. The residue was purified by flash column chromatography (petroleum ether/EtOAc, 8: 1) to obtain the product cycloalkanol (36 mg, 85%). The ee value was determined by chiral HPLC using a Chiralcel OD-H column with hexane/isopropanol (90: 10), flow: 0.7 mL min^{-1}; ee: 92%; $[\alpha]_D^{20}$ −114.5 (c 1.00, CHCl$_3$).

4.5.3
Nucleophilic Addition to Imino Groups (See Section 4.2.2.2)

4.5.3.1 Typical Procedure for the Palladium-Catalyzed Indole Synthesis [60]

n-Bu$_3$P (50 μL, 0.2 mmol) was added to a solution of Pd(OAc)$_2$ (11.2 mg, 0.05 mmol) dissolved in 1,4-dioxane (2.0 mL) under an argon atmosphere. The color of the solution turned from pale reddish brown to pale yellow immediately. This palladium solution was added to imine (247 mg, 1.0 mmol) placed in the reaction vessel sealed with a screw-cap, and the mixture was heated at 100 °C for 25 h under an argon atmosphere. The reaction mixture was cooled to room temperature and filtered to remove palladium catalyst. The concentrated crude mixture was purified by column chromatography (Al$_2$O$_3$, hexane–AcOEt 20∼10: 1) to afford 143 mg of the product, Alternatively, the crude product was hydrogenated with Pd-C (10%) in EtOH to give 3-butyl-2-phenylindole in 60% isolated yield.

4.5.4
1,2-Addition of Nitriles (See Section 4.2.4)

4.5.4.1 Procedure for the 1.2-Addition of Nitrile [68]

To a solution of o-(1-butyl-2,2-difluorovinyl)benzonitrile (93 mg, 0.42 mmol) in Et$_2$O (3 mL) was added butyllithium (0.34 mL, 1.49 M in hexane, 0.50 mmol) at −78 °C under nitrogen. After the mixture had been stirred for 0.5 h at −78 °C, the reaction was quenched with phosphate buffer (pH 7). Organic materials were extracted with AcOEt three times. The combined extracts were washed with brine and dried over Na$_2$SO$_4$. After removal of the solvent under reduced pressure, the residue was purified by PTLC on silica gel (hexane–AcOEt 5: 1) to give 1,4-dibutyl-3-fluoroisoquinoline (94 mg, 86%) as a pale yellow liquid.

4.5.5
1,2-Addition of Nitriles (See Section 4.2.4)

4.5.5.1 Synthesis of Enantiomerically Enriched Atropoisomers of 2- Arylpyridines [81]

A thermostatted (3 °C) reaction vessel was loaded with a diyne (524 mg, 2 mmol), catalyst (8.4 mg, 0.02 mmol), THF (20 mL), and benzonitrile (412 mL, 4 mmol) under argon atmosphere. The mixture was stirred and irradiated by two 460-W lamps ($\lambda \sim 420$ nm) for 24 h. The reaction was quenched by switching off the lamps and simultaneously letting in air. The reaction mixture was filtered through a thin pad of silica gel, eluting with THF. The solvent was removed in vacuo to give an oily residue, which was further dissolved in Et_2O (10 mL). Colorless crystals of the product (413 mg, 57% yield) were filtered off and washed with Et_2O. The optical purity was determined to be >98%ee by HPLC.

4.5.6
Catalyzed 1,4-Addition Reactions (See Section 4.3.2.3)

4.5.6.1 Typical Procedure for the Phosphine-Catalyzed Tandem Nucleophilic Additions [111]

To a solution of cyclohexa-1,3-dione (56 mg, 0.5 mmol) and triphenylphosphine (6.5 mg, 0.025 mmol) in toluene (1.5 mL) at 70 °C under nitrogen was added a solution of penta-3,4-dien-2-one (41 mg, 0.5 mmol) in toluene (1 mL). After the reaction was complete, the solvent was removed under reduced pressure and the residue was purified by flash chromatography on silica gel (ethyl acetate:petroleum ether = 2: 1) to give the addotion product as a yellow oil (81 mg, 84%).

4.5.7
Catalytic Asymmetric Intramolecular Cyclization via 1,4- Addition (See Section 4.3.5)

4.5.7.1 General Procedure for Asymmetric 1,4-Addition [126]

$ArB(OH)_2$ (1.2 mmol), acetone (2 mL), $AgBF_4$ (0.1 mmol, if used), enone (1 mmol) and water (0.2 mL) were added to a flask under nitrogen. [Pd(S,S-chiraphos)(PhCN)$_2$] $(SbF_6)_2$ (0.001 mmol) was then added at 0 °C. After stirring for 21 h at 0 °C or 25 °C, chromatography on silica gel gave the corresponding 1,4-adduct, chromanol. Next, chromanol (0.5 mmol), 4 Å molecular sieves (1 g), *p*-toluenesulfonic acid (0.05 mmol) and toluene (4 mL) were added to a flask under nitrogen. The mixture was then heated to reflux. After being stirred for 3 h, chromatography on silica gel gave a corresponding chromene product. The enantiomer excess was determined by chiral HPLC analysis.

Abbreviations

AcOH	acetic acid
ATPH	aluminum tris-(2,6-diphenylphenoxide)
Cl-MeOBIPHEP	chloro-methyoxy-biphenylphosphine
DCM	dichloromethane
DCE	dichloroethane
DME	dimethoxyethane
dpm	dipivaloylmethane
Ph-9-BBN	9-phenyl-9-borabicyclo[3.3.1]-nonane
TBAT	tetrabutylammonium triphenyldifluorosilicate
TBDPSO	*tert*-butyldiphenylsilyloxy
TMDS	1,1,3,3-tetramethylhydrosiloxane
TMS	trimethylsilyl

Acknowledgments

The authors would like to thank Dr Miao Yang, Mr Feng Zhou, and Miss Xufen Yu for their help in the literature search.

References

1 (a) Trost, B.M. Fleming, I. and Heathcock, C.H. (eds) (1991) *Comprehensive Organic Synthesis*, vol. 2, Pergamon, Oxford, pp. 133–319; (b) Palomo, C., Oiarbide, M. and García, J.M. (2002) *Chem. Eur. J.*, 8, 37–44; (c) Mahrwald, R. (2004) *Modern Aldol Reactions*, Wiley-VCH, Weinheim, vols. 1–2; (d) Saito, S. and Yamamoto, H. (2004) *Acc. Chem. Res*, 37, 570–579.

2 For an overview of total syntheses involving the aldol addition reaction, see: (a) Mukaiyama, T. (1999) *Tetrahedron*, 55, 8609–8670; (b) Nicolaou, K.C., Vourloumis, D., Winssinger, N. and Baran, P.S. (2000) *Angew. Chem.*, 112, 46–126; (2000) *Angew. Chem. Int. Ed.*, 39, 44–122.

3 Woodward, R.B., Sondheimer, F., Taub, D., Heusler, K. and Mclamore, W.I. (1952) *J. Am. Chem. Soc.*, 74, 4223–4251.

4 (a) Marshall, J.A. and Greene, A. (1972) *J. Org. Chem.*, 37, 982–985; (b) Ho, T.-L. (1974) *Synth. Commun.*, 265–287.

5 (a) Bach, T. (1994) *Angew. Chem*, 106, 433–435; (1994) *Angew. Chem. Int. Ed.*

Engl., **33**, 417–419; (b) Machajewski, T.D. and Wong, C.-H. (2000) *Angew. Chem.*, **112**, 1406–1430; (2000) *Angew. Chem. Int. Ed.*, **39**, 1352–1374; (c) Denmark, S.E. and Stavenger, R.A. (2000) *Acc. Chem. Res.*, **33**, 432–440.

6 (a) Guillena, G., Nájera, C. and Ramón, D.J. (2007) *Tetrahedron: Asym.*, **18**, 2249–2293; (b) Migneco, L.M., Leonelli, F. and Bettolo, R.M. (2004) *Arkivoc*, 253–265.

7 (a) Pidathala, C., Hoang, L., Vignola, N. and List, B. (2003) *Angew. Chem.*, **115**, 2891–2894; (b) Itagaki, N., Sugahara, T. and Iwabuchi, Y. (2005) *Org. Lett.*, **7**, 4181–4185; (c) Itagaki, N., Kimura, M., Sugahara, T. and Iwabuchi, Y. (2005) *Org. Lett.*, **7**, 4185–4188; (d) Enders, D., Niemeier, O. and Straver, L. (2006) *Synlett*, 3399–3402.

8 (a) Kanger, T., Kriis, K., Laars, M., Kailas, T., Müürisepp, A.-M., Pehk, T. and Lopp, M. (2007) *J. Org. Chem.*, **72**, 5168–5173; (b) Hayashi, Y., Sekizawa, H., Yamaguchi, J. and Gotoh, H. (2007) *J. Org. Chem.*, **72**, 6493–6499.

9 (a) Casiraghi, G. and Zanardi, F. (2000) *Chem. Rev.*, **100**, 1929–1972; (b) Denmark, S.E., Heemstra, J.R. Jr. and Beutner, G.L. (2005) *Angew. Chem.*, **117**, 4760–4777; (2005) *Angew. Chem. Int. Ed.*, **44**, 4682–4698.

10 (a) Little, R.D., Masjedizadeh, M.R., Wallquist, O. and McLoughlin, J. I. (1995) *Org. React.*, **47**, 315–552; (b) Ho, T.L. (1992) *Tandem Organic Reactions*, Wiley, New York; (c) Tietze, L.F. and Beifuss, U. (1993) *Angew. Chem.*, **105**, 137–170; (1993) *Angew. Chem., Int. Ed. Engl.*, **32**, 131–163; (d) Ihara, M. and Fukumoto, K. (1993) *Angew. Chem.*, **105**, 1059–1971; (1993) *Angew. Chem. Int. Ed. Engl*, **32**, 1010–1022.

11 For recent examples of aldol and Michael cyclizations, see: (a) Enholm, E.J., Xie, Y. and Abboud, K.A. (1995) *J. Org. Chem.*, **60**, 1112–1113; (b) Baik, T.-G., Luis, A.L., Wang, L.C. and Krische, M.J. (2001) *J. Am. Chem. Soc.*, **123**, 5112–5113; (c) Nagaoka, Y. and Tomioka, K. (1999) *Org. Lett.*, **1**, 1467–1469; (d) Ono, M., Nishimura, K., Nagaoka, Y. and Tomioka, K. (1999) *Tetrahedron Lett.*, **40**, 6979–6982; (e) Ono, M., Nishimura, K., Tsubouchi, H., Nagaoka, Y. and Tomioka, K. (2001) *J. Org. Chem.*, **66**, 8199–8203; (f) Kamenecka, T.M., Overman, L.E. and Ly Sakata, S.K. (2002) *Org. Lett.*, **4**, 79–82; (g) Suwa, T., Nishino, K., Miyatake, M., Shibata, I. and Baba, A. (2000) *Tetrahedron Lett.*, **41**, 3403–3406.

12 Yagi, K., Turitani, T., Shinokubo, H. and Oshima, K. (2002) *Org. Lett.*, **4**, 3111–3114.

13 (a) Douelle, F., Capes, A.S. and Greaney, M.F. (2007) *Org. Lett.*, **9**, 1931–1934; (b) Basavaiah, D. Rao, A.J. and Satyanarayana, T. (2003) *Chem. Rev.*, **103**, 811–891.

14 Abramite, J.A. and Sammakia, T. (2007) *Org. Lett*, **9**, 2103–2106.

15 (a) Emiabata-Smith, D. McKillop, A., Mills, C., Motherwell, W.B. and Whitehead, A.J. (2001) *Synlett*, 1302–1304; (b) Freiria, M., Whitehead, A.J., Tochera, D.A. and Motherwell, W.B. (2004) *Tetrahedron*, **60**, 2673–2692.

16 (a) Chiu, P., Szeto, C.-P., Geng, Z. and Cheng, K.-F. (2001) *Org. Lett.*, **3**, 1901–1903; (b) Chiu, P., Chen, B. and Cheng, K.F. (1998) *Tetrahedon Lett.*, **39**, 9229–9232; (c) Chiu, P., Szeto, C.P., Geng, Z. and Cheng, K.F. (2001) *Tetrahedron Lett.*, **42**, 4091–4093.

17 Huddleston, R.R. and Krische, M.J. (2003) *Synlett*, 12–21.

18 (a) Huddleston, R.R. and Krische, M.J. (2003) *Org. Lett.*, **5**, 1143–1146; (b) Koech, P.K. and Krische, M.J. (2004) *Org. Lett.*, **6**, 691–694.

19 (a) Lam, H.W. and Joensuu, P.M. (2005) *Org. Lett.*, **7**, 4225–4228; (b) Lam, H.W., Murray, G.J. and Firth, J.D. (2005) *Org. Lett.*, **7**, 5743–5746.

20 (a) Hajos, Z.G. and Parrish, D.R. (1974) *J. Org. Chem.*, **3**, 1615–1621; (b) Eder, U.

Sauer, G. and Wiechert, R. (1971) *Angew. Chem.*, **83**, 492–493; (1971) *Angew. Chem. Int. Ed.*, **10**, 496–497; (c) Ruppert, J., Eder, U. and Wiechert, R. (1973) *Chem. Ber.*, **106**, 3636–3644.

21 (a) Basavaiah, D., Rao, A.J. and Satyanarayana, T. (2003) *Chem. Rev.*, **103**, 811–891; (b) Yeo, J.E., Yang, X., Kim, H.J. and Koo, S. (2004) *Chem. Commun.*, 236–237.

22 Roth, F., Gygax, P. and Frater, G. (1992) *Tetrahedron Lett.*, **33**, 1045–1048.

23 Krishna, P.R., Kannan, V. and Sharma, G.V.M. (2004) *J. Org. Chem.*, **69**, 6467–6469.

24 (a) Snider, B.B. (1991) in *Comprehensive Organic Synthesis* (eds B.M. Trost and I. Fleming), Pergamon, Oxford, U.K, vol. **2**, pp. 527–567; (b) Mikami, K. and Shimizu, M. (1992) *Chem. Rev.*, **92**, 1021–1050.

25 Kočovský, P., Ahmed, G., Sÿrogl, J., Malkov, A.V. and Steele, J. (1999) *J. Org. Chem.*, **64**, 2765–2775.

26 Williams, J.T., Bahia, P.S. and Snaith, J.S. (2002) *Org. Lett.*, **4**, 3727–3730.

27 Black, D.St.C., Craig, D.C., Kumar, N. and Rezaie, R. (1999) *Tetrahedron*, **55**, 4803–4814.

28 Yue, D., Della Cà, N. and Larock, R.C. (2004) *Org. Lett.*, **6**, 1581–1584.

29 Agapiou, K., Cauble, D.F. and Krische, M.J. (2004) *J. Am. Chem. Soc.*, **126**, 4528–4529.

30 Wang, L.-C., Jang, H.-Y., Roh, Y., Lynch, V., Schultz, A.J., Wang, X. and Krische, M.J. (2002) *J. Am. Chem. Soc.*, **124**, 9448–9453.

31 (a) Montgomery, J. (2000) *Acc. Chem. Res.*, **33**, 467–473; (b) Tang, X.-Q. and Montgomery, J. (1999) *J. Am. Chem. Soc.*, **121**, 6098–6099; (c) Shibata, K., Kimura, M., Shimizu, M. and Tamaru, Y. (2001) *Org. Lett.*, **3**, 2181–2183; (d) Ogoshi, S., Ueta, M., Arai, T. and Kurosawa, H. (2005) *J. Am. Chem. Soc.*, **127**, 12810–12811; (e) Herath, A. and Montgomery, J. (2006) *J. Am. Chem. Soc.*, **128**, 14030–14031; (f) Tekevac, T.N. and Louie, J. (2005) *Org. Lett.*, **7**, 4037–4039; (g) Oblinger, E. and Montgomery, J. (1997) *J. Am. Chem. Soc.*, **119**, 9065–9066.

32 (a) Cauble, D.F., Gipson, J.D. and Krische, M.J. (2003) *J. Am. Chem. Soc.*, **125**, 1110–1111; (b) Rhee, J.U. and Krische, M.J. (2006) *J. Am. Chem. Soc.*, **128**, 10674–10675; (c) Miura, T., Shimada, M. and Murakami, M. (2005) *Synlett*, 667–669.

33 (a) Gevorgyan, V., Quan, L.G. and Yamamoto, Y. (1999) *Tetrahedron Lett.*, **40**, 4089–4092; (b) Zhao, L. and Lu, X. (2002) *Angew. Chem.*, **114**, 4519–4521; (2002) *Angew. Chem. Int. Ed.*, **41**, 4343–4345; (c) Tsukamoto, H., Ueno, T. and Kondo, Y. (2007) *Org. Lett.*, **9**, 3033–3036.

34 (a) Sato, Y., Takimoto, M., Hayashi, K., Katsuhara, T., Takagi, K. and Mori, M. (1994) *J. Am. Chem. Soc.*, **116**, 9771–9772; (b) Sato, Y., Saito, N. and Mori, M. (2000) *J. Am. Chem. Soc.*, **122**, 2371–2372; (c) Sato, Y., Takimoto, M. and Mori, M. (2000) *J. Am. Chem. Soc.*, **122**, 1624–1634; (d) Sato, Y., Takanashi, T., Hoshiba, M. and Moil, M. (1998) *Tetrahedron Lett.*, **39**, 5579–5582.

35 (a) Tsukamoto, H., Ueno, T. and Kondo, Y. (2006) *J. Am. Chem. Soc.*, **128**, 1406–1407; (b) Shintani, R., Okamoto, K., Otomaru, Y., Ueyama, K. and Hayashi, T. (2005) *J. Am. Chem. Soc.*, **127**, 54–55; (c) Miura, T., Shimada, M. and Murakami, M. (2007) *Tetrahedron.*, **63**, 6131–6140.

36 (a) Montgomery, J. and Song, M. (2002) *Org. Lett.*, **4**, 4009–4011; (b) Song, M. and Montgomery, J. (2005) *Tetrahedron.*, **61**, 1440–1448; (c) Gai, X., Grigg, R., Collard, S. and Muir, J.E. *Chem. Commun.*, **2000**, 1765–1766; (d) Kang, S.-K. and Yoon, S.-K. (2002) *Chem. Commun.*, 2634–2635; (e) Ha, Y.-H. and Kang, S.-K. (2002) *Org. Lett.*, **4**, 1143–1146; (f) Tsukamoto, H., Matsumoto, T. and Kondo, Y. (2008) *J. Am. Chem. Soc.*, **130**, 388–389; (g) Ha, Y.-H. and Kang, S.-K. (2002) *Org. Lett.*, **4**, 1143–1146.

37 Shintani, R., Okamoto, K., Otomaru, Y., Ueyama, K. and Hayashi, T. (2005) *J. Am. Chem. Soc.*, **127**, 54–55.

38 (a) Larock, R.C. and Doty, M.J. (1993) *J. Org. Chem.*, **58**, 4579–4583; (b) Quan, L.G., Gevorgyan, V. and Yamamoto, Y. (1999) *J. Am. Chem. Soc.*, **121**, 3545–3546; (c) Quan, L.G., Lamrani, M. and Yamamoto, Y. (2000) *J. Am. Chem. Soc.*, **122**, 4827–4828; (d) Solé, D., Vallverdú, L., Peidró, E. and Bonjoch, J. (2001) *Chem. Commun.*, 1888–1889; (e) Takezawa, A., Yamaguchi, K., Ohmura, T., Yamamoto, Y. and Miyaura, N. (2002) *Synlett.*, 1733–1736; (f) Solé, D., Vallverdú, L., Solans, X., Font-Bardía, M. and Bonjoch, J. (2003) *J. Am. Chem. Soc.*, **125**, 1587–1594; (g) Miura, T. and Murakami, M. (2007) *Chem. Commun.*, 217–224; (h) Nishimura, T., Yasuhara, Y. and Hayashi, T. (2007) *J. Am. Chem. Soc.*, **129**, 7506–7507.

39 (a) Liu, G. and Lu, X. (2006) *J. Am. Chem. Soc.*, **128**, 16504–16505; (b) Song, J., Shen, Q., Xu, F. and Lu, X. (2007) *Org. Lett*, **9**, 2947–2950; (c) Liu, G. and Lu, X. (2008) *Tetrahedron*, **64**, 7324–7330.

40 (a) Liu, G. and Lu, X. (2007) *Adv. Synth. Catal.*, **349**, 2247–2252; (b) Yang, M., Zhang, X. and Lu, X. (2007) *Org. Lett.*, **9**, 5131–5133.

41 Zhang, X. and Larock, R.C. (2005) *Org. Lett.*, **7**, 3973–3976.

42 (a) Borzilleri, R.M. and Weinreb, S.M. (2005), *Synthesis*, 347–360; (b) Oppolzer, W. and Snieckus, V. (1978) *Angew. Chem.*, **90**, 506–516; (1978) *Angew. Chem. Int. Ed. Engl.*, **17**, 476–486.

43 Demailly, G. and Solladie, G. (1981) *J. Org. Chem.*, **46**, 3102–3108.

44 (a) Tietze, L.F. and Bratz, M. (1989) *Chem. Ber.*, **122**, 997–1002; (b) Tietze, L.F. and Bratz, M. (1989) *Synthesis*, 439–442.

45 Tschaen, D.M., Turos, E. and Weinreb, S.M. (1984) *J. Org. Chem.*, **49**, 5058–5064.

46 Laschat, S., Fröhlich, R. and Wibbeling, B. (1996) *J. Org. Chem.*, **61**, 2829–2838.

47 Negishi, E., Cederbaum, F.E. and Takahashi, T. (1986) *Tetrahedron Lett.*, **27**, 2829–2832.

48 Makabe, M., Sato, Y. and Mori, M. (2004) *J. Org. Chem.*, **69**, 6238–6243.

49 Makabe, M., Sato, Y. and Mori, M. (2004) *Synthesis*, 1369–1374.

50 Ogoshi, S., Ikeda, H. and Kurosawa, H. (2007) *Angew. Chem.*, **119**, 5018–5020; (2007) *Angew. Chem. Int. Ed.*, **46**, 4930–4932.

51 Dijkink, J., Zonjee, J.N., de Jong, B.S. and Speckamp, W.N. (1983) *Heterocycles*, **20**, 1255–1258.

52 Chen, C. and Reamer, R.A. (1999) *Org. Lett.*, **1**, 293–294.

53 Kadota, L., Park, J.-Y. and Yamamoto, Y. (1996) *J. Chem. Soc., Chem. Commun.*, 841–842.

54 Park, J.-Y., Park, C.-H., Kadota, I. and Yamamoto, Y. (1998) *Tetrahedron Lett.*, **39**, 1791–1794.

55 Yanada, R., Obika, S., Kono, H. and Takemoto, Y. (2006) *Angew. Chem.*, **118**, 3906–3909; (2006) *Angew. Chem. Int. Ed.*, **45**, 3822–3825.

56 Cooper, I.R., Grigg, R., Hardie, M.J., MacLachlan, W.S., Sridharana, V. and Thomasa, W.A. (2003) *Tetrahedron Lett.*, **44**, 2283–2285.

57 Kim, S.-H., Oh, S.-J., Kim, Y. and Yu, C.-M. (2007) *Chem. Commun.*, 5025–5027.

58 Takeda, A., Kamijo, S. and Yamamoto, Y. (2000) *J. Am. Chem. Soc.*, **122**, 5662–5663.

59 Obika, S., Kono, H., Yasui, Y., Yanada, R. and Takemoto, Y. (2007) *J. Org. Chem.*, **72**, 4462–4468.

60 Asao, N. Yudha, S. Nogami, T. and Yamamoto, Y. (2005), *Angew. Chem.*, **117**, 5662–5664; (2005) *Angew. Chem. Int. Ed.*, **44**, 5526–5528.

61 Su, S. and Porco, J.A. Jr. (2007) *J. Am. Chem. Soc.*, **129**, 7744–7745.

62 Kuninobu, Y., Kawata, A. and Takai, K. (2005) *J. Am. Chem. Soc.*, **127**, 13498–13499.

63 Chatani, N., Morimoto, T., Kamitani, A., Fukumoto, Y. and Murai, S. (1999) *J. Organomet. Chem.*, **579**, 177–181.

64 Shintani, R. and Hayashi, T. (2006) *J. Am. Chem. Soc.*, **128**, 6330–6331.

65 (a) Fleming, F.F. and Shook, B.C. (2002) *Tetrahedron*, **58**, 1–23; (b) Friedrich, K. and Wallenfels, K. (1970) in *The Chemistry of the Cyano Group* (ed. Z. Rappoport), The Chemistry of Functional Groups (ed. S. Patai), Interscience Publishers, London, pp. 67–122; (c) Pichat, L. *ibid.*, pp. 743–790.

66 (a) Sato, Y., Yato, M., Ohwada, T., Saito, S. and Shudo, K. (1995) *J. Am. Chem. Soc.*, **117**, 3037–3043; (b) Justribó, V., Pellegrinet, S.C. and Colombo, M. (2007) *J. Org. Chem.*, **72**, 3702–3712.

67 Tverdokhlebov, A.V., Zavada, A.V., Tolmachev, A.A., Kostyuk, A.N., Chernegad, A.N. and Rusan, E.B. (2005) *Tetrahedron*, **61**, 9618–9623.

68 Ichikawa, J., Wada, Y., Miyazaki, H., Mori, T. and Kuroki, H. (2003) *Org. Lett*, **5**, 1455–1458.

69 Kristensen, J.L. (2008) *Tetrahedron Lett.*, **49**, 2351–2354.

70 Marco, J.L., Ingate, S.T. and Chinchón, P.M. (1999) *Tetrahedron.*, **55**, 7625–7644.

71 (a) Dansou, B., Pichon, C., Dhal, R., Brown, E. and Mille, S. (2000) *Eur. J. Org. Chem.*, 1527–1533; (b) Malassene, R., Vanquelef, E., Toupet, L., Hurvois, J.-P. and Moinet, C. (2003) *Org. Biomol. Chem.*, **1**, 547–551; (c) Marchaín, Š., Baumlová, B., Baran, P., Oulyadi, H. and Daïch, A. (2006) *J. Org. Chem.*, **71**, 9114–9127.

72 Yang, C.-C., Sun, P.-J. and Fang, J.-M. (1994) *J. Chem. Soc., Chem. Commun.*, 2629–2630.

73 Larock, R.C., Tian, Q. and Pletnev, A.A. (1999) *J. Am. Chem. Soc.*, **121**, 3238–3239.

74 (a) Wei, L.-M., Lin, C.-F. and Wu, M.-J. (2000) *Tetrahedron Lett.*, **41**, 1215–1218; (b) Pletnev, A.A. and Larock, R.C. (2002) *Tetrahedron Lett.*, **43**, 2133–2136; (c) Pletnev, A.A., Tian, Q. and Larock, R.C. (2002) *J. Org. Chem.*, **67**, 9276–9287; (d) Pletnev, A.A. and Larock, R.C. (2002) *J. Org. Chem.*, **67**, 9428–9438; (e) Tian, Q., Pletnev, A.A. and Larock, R.C. (2003) *J. Org. Chem.*, **68**, 339–347; (f) Zhao, B. and Lu, X. (2006) *Org. Lett.*, **8**, 5987–5990.

75 Sakai, M., Ueda, M. and Miyaura, N. (1998) *Angew. Chem.*, **110**, 3475–3477; (1998) *Angew. Chem. Int. Ed.*, **37**, 3279–3281.

76 Miura, T., Nakazawa, H. and Murakami, M. (2005) *Chem. Commun.*, 2855–2856.

77 Miura, T. and Murakami, M. (2005) *Org. Lett.*, **7**, 3339–3341.

78 Miura, T., Harumashi, T. and Murakami, M. (2007) *Org. Lett.*, **9**, 741–743.

79 (a) Schore, N.E. (1991) *Comprehensive Organic Synthesis*, vol. 5 (eds Trost, B.M. and Fleming, I.), Pergamon Press Ltd, Oxford, pp. 1129–1162; (b) Schore, N.E. (1988), *Chem. Rev.*, **88**, 1081–1119; (c) Bönnemann, H. (1978), *Angew. Chem.*, **90**, 517–526; (1978) *Angew. Chem. Int. Ed. Engl.*, **17**, 505–515; (d) Vollhardt, K.P.C. (1984) *Angew. Chem.*, **96**, 525–541; (1984) *Angew. Chem. Int. Ed.*, **23**, 539–556; (e) Bönnemann, H. (1985) *Angew. Chem.*, **97**, 264–279. (1985) *Angew. Chem. Int. Ed. Engl.*, **24**, 248–262; (f) Tanaka, R., Yuza, A., Watai, Y., Suzuki, D., Takayama, Y., Sato, F. and Urabe, H. (2005) *J. Am. Chem. Soc.*, **127**, 7774–7780.

80 (a) Takahashi, T., Tsai, F.-Y. and Kotora, M. (2000) *J. Am. Chem. Soc.*, **122**, 4994–4995; (b) Takahashi, T., Tsai, F.-Y., Li, Y., Wang, H., Kondo, Y., Yamanaka, M., Nakajima, K. and Kotora, M. (2002) *J. Am. Chem. Soc.*, **124**, 5059–5067.

81 Gutnov, A., Heller, B., Fischer, C., Drexler, H.-J., Spannenberg, A., Sundermann, B. and Sundermann, C. (2004), *Angew. Chem.*, **116**, 3825–3827; (2004) *Angew. Chem. Int. Ed.*, **43**, 3795–3797.

82 Trofimov, B.A., Mal'kina, A.G. Borisova, A.P., Nosyreva, V.V., Shemyakina, O.A., Kazheva, O.N., Shilov, G.V. and Dyachenko, O.A. (2008) *Tetrahedron Lett.*, **49**, 3104–3107.

83 Kamijo, S., Sasaki, Y., Kanazawa, C., Schüßeler, T. and Yamamoto, Y. (2005) *Angew. Chem.*, **117**, 7896–7899; (2005) *Angew. Chem. Int. Ed.*, **44**, 7718–7721.

84 Komnenos, T. (1883) *Liebigs Ann. Chem.*, **218**, 145–169.

85 Michael, A. (1887) *J. Prakt. Chem.*, **3**, 349–351.

86 (a) Perlmutter, P. (1992) *Conjugate Addition Reactions in Organic Synthesis*, Pergamon Press, Oxford; (b) Tietze, L.-F. and Beifuss, U. (1993), *Angew. Chem.*, **105**, 137–170; (1993) *Angew. Chem. Int. Ed. Engl.*, **32**, 131–163; (c) Sibi, M.P. and Manyem, S. (2000), *Tetrahedron.*, **56**, 8033–8061; (d) Hayashi, T. and Yamasaki, K. (2003) *Chem. Rev.*, **103**, 2829–2844; (e) Balme, G., Bouyssi, D. and Monteiro, N. (2006), *Top. Organomet. Chem.*, **19**, 115–148; (f) Koripelly, G., Rosiak, A. and Rössle, M. (2007) *Synthesis.*, 1279–1300.

87 (a) Hart, D.J. and Lai, C. (1989) *Synlett*, 49–51; (b) Mahboobi, S. Kuhr, S. and Koller, M. (1996) *Tetrahedron.*, **52**, 6373–6382; (c) Nara, S., Toshlma, H. and Ichihara, A. (1996) *Tetrahedron Lett.*, **37**, 6745–6748.

88 Guével, A.-C. and Hart, D.J. (1994) *Synlett.*, 169–170.

89 Kawabata, T., Majumdar, S., Tsubaki, K. and Monguchi, D. (2005) *Org. Biomol. Chem.*, **3**, 1609–1611.

90 Rodriguez, C.M., Martin, T., Ramirez, M.A. and Martin, V.S. (1994) *J. Org. Chem.*, **59**, 4461–4472.

91 Guével, A.-C. and Hart, D.J. (1994) *Synlett*, 185–186.

92 (a) Ihara, M., Suzuki, S., Tokunaga, Y., Takeshita, H. and Fukumoto, K. (1996) *Chem. Commun.*, 1801–1802; (b) Fujishima, H., Takeshita, H., Suzuki, S., Toyota, M. and Ihara, M. (1999) *J. Chem. Soc., Perkin Trans. 1*, 2609–2616; (c) Bachi, M.D., Bilokin, Y.V. and Melman, A. (1998) *Tetrahedron Lett.*, **39**, 3035–3038.

93 Guay, B. and Deslongchamps, P. (2003) *J. Org. Chem.*, **68**, 6140–6148.

94 (a) Shida, N., Uyehara, T. and Yamamoto, Y. (1992) *J. Org. Chem.*, **57**, 5049–5051; (b) Davies, S.G., Díez, D. Dominguez, S.H., Garrido, N.M., Kruchinin, D., Price, P.D. and Smith, A.D. (2005) *Org. Biomol. Chem.*, **3**, 1284–1301; (c) Yasuhara, T., Nishimura, K., Yamashita, M., Fukuyama, N., Yamada, K., Muraoka, O. and Tomioka, K. (2003) *Org. Lett.*, **5**, 1123–1126; (d) Yasuhara, T., Osafune, E., Nishimura, K., Yamashita, M., Yamada, K., Muraoka, O. and Tomioka, K. (2004) *Tetrahedron Lett.*, **45**, 3043–3045.

95 (a) Maezaki, N., Yuyama, S., Sawamoto, H., Suzuki, T., Izumi, M. and Tanaka, T. (2001) *Org. Lett.*, **3**, 29–31; (b) Maezaki, N., Sawamoto, H., Yuyama, S., Yoshigami, R., Suzuki, T., Izumi, M., Ohishi, H. and Tanaka, T. (2004) *J. Org. Chem.*, **69**, 6335–6340; (c) Yoshizaki, H., Tanaka, T., Yoshii, E., Koizumi, T. and Takeda, K. (1998) *Tetrahedron Lett.*, **39**, 47–50.

96 (a) Cooke, M.P. Jr. (1993) *J. Org. Chem.*, **58**, 2910–2912; (b) Cooke, M.P. Jr. and Gopal, D. (1994) *J. Org. Chem.*, **59**, 260–263; (c) Piers, E., Harrison, C.L. and Zetina-Rocha, C. (2001) *Org. Lett.*, **3**, 3245–3247.

97 Yamada, S., Karasawa, S., Takahashi, Y., Aso, M. and Suemune, H. (1998) *Tetrahedron*, **54**, 15555–15566.

98 Agnusdei, M., Bandini, M., Melloni, A. and Umani-Ronchi, A. (2003) *J. Org. Chem.*, **68**, 7126–7129.

99 Yamazaki, S., Morikawa, S., Iwata, Y., Yamamoto, M. and Kuramoto, K. (2004) *Org. Biomol. Chem.*, **2**, 3134–3138.

100 (a) Piers, E., McEachern, E.J. and Burns, P.A. (1995) *J. Org. Chem.*, **60**, 2322–2323; (b) Piers, E., Skupinska, K.A. and Wallace, D.J. *Synlett.*, **1999**, 1867–1870; (c) Piers, E., McEachern, E.J. and Burns, P.A. (2000) *Tetrahedron.*, **56**, 2753–2765.

101 Dziedzic, M., Maøecka, M. and Furman, B. (2005) *Org. Lett.*, **7**, 1725–1727.

102 Tanino, K., Arakawa, K., Satoh, M., Iwata, Y. and Miyashita, M. (2006) *Tetrahedron Lett.*, **47**, 861–864.

103 (a) Furman, B. and Dziedzic, M. (2003) *Tetrahedron Lett.*, **44**, 6629–6631; (b) Dziedzic, M., Lipner, G., Illangua, J.M. and Furman, B. (2005) *Tetrahedron.*, **61**, 8641–8647.

104 Li, K. and Alexakis, A. (2005) *Tetrahedron Lett.*, **46**, 8019–8022.

105 (a) Riestad, G.K. and Branchaud, B.P. (1995) *Tetrahedron Lett.*, **36**, 7047–7050;

(b) Lindahl, K.-F., Carroll, A., Quinn, R.J. and Ripper, J.A. (2006) *Tetrahedron Lett.*, **47**, 7493–7495.

106 Jousse-Karinthi, C. Zouhiri, F., Mahuteau, J. and Desmaële, D. (2003) *Tetrahedron*, **59**, 2093–2099.

107 Chen, Y. and Lee, C. (2006) *J. Am. Chem. Soc.*, **128**, 15598–15599.

108 Lautens, M. and Marquardt, T. (2004) *J. Org. Chem.*, **69**, 4607–4614.

109 (a) de Alaniz, J.R. and Rovis, T. (2005) *J. Am. Chem. Soc.*, **127**, 6284–6289; (b) Phillips, E.M. Wadamoto, M. Chan, A. and Scheidt, K.A. (2007) *Angew. Chem.*, **119**, 3167–3170; (2007) *Angew. Chem. Int. Ed.*, **46**, 3107–3110.

110 (a) Wang, L.-C., Luis, A.L., Agapiou, K., Jang, H.-Y. and Krische, M.J. (2002) *J. Am. Chem. Soc.*, **124**, 2402–2403; (b) Frank, S.A., Mergott, D.J. and Roush, W.R. (2002) *J. Am. Chem. Soc.*, **124**, 2404–2405; (c) Mergott, D.J., Frank, S.A. and Roush, W.R. (2002) *Org. Lett.*, **4**, 3157–3160; (d) Agapiou, K. and Krische, M.J. (2003) *Org. Lett.*, **5**, 1737–1740; (e) Seidel, F. and Gladysz, J.A. (2007) *Synlett*, 986–988; (f) Brown, P.M., Käppel, N., Murphy, P.J., Coles, S.J. and Hursthouse, M.B. (2007) *Tetrahedron*, **63**, 1100–1106.

111 (a) Lu, C. and Lu, X. (2002) *Org. Lett.*, **4**, 4677–4679; (b) Henry, C.E. and Kwon, O. (2007) *Org. Lett.*, **9**, 3069–3072.

112 (a) Molina, P., Aller, E. and Lorenzo, A. (1998) *Synthesis*, 283–287; (b) Molina, P., Aller, E., Lorenzo, A., Foces-Foces, C. and Saiz, A.L.L. (1996) *Tetrahedron*, **52**, 13671–13680; (c) Xin, Z., Pei, Z., von Geldem, T. and Jirousek, M. (2000) *Tetrahedron Lett.*, **41**, 1147–1150.

113 (a) Kim, S.H. and Fuchs, P.L. (1996) *Tetrahedron Lett.*, **37**, 2545–2548. (b) Bland, D., Hart, D.J. and Lacoutiere, S. (1997) *Tetrahedron.*, **53**, 8871–8880; (c) Matsushima, Y. and Kino, J. (2006) *Tetrahedron Lett.*, **47**, 8777–8780.

114 (a) Ballini, R., Bosica, G., Fiorini, D., Palmieri, A. and Petrini, M. (2005) *Chem. Rev.*, **105**, 933–971; (b) Bland, D., Chambournier, G., r Dragan, V. and Hart, D.J. (1999) *Tetrahedron.*, **55**, 8953–8966.

115 Ishikawa, T., Aikawa, T., Watanabe, S. and Saito, S. (2006) *Org. Lett.*, **8**, 3881–3884.

116 Schneider, C., Börner, C. and Schuffenhauer, A. (1999) *Eur. J. Org. Chem.*, 3353–3362.

117 (a) Vedejs, E., Little, J.D. and Seaney, L.M. (2004) *J. Org. Chem.*, **69**, 1788–1793; (b) Vedejs, E. and Little, J.D. (2004) *J. Org. Chem.*, **69**, 1794–1799.

118 (a) Banwell, M.G., Bissett, B.D., Bui, C.T., Pham, H.T.T. and Simpson, G.W. (1998) *Aust. J. Chem.*, **51**, 9–18; (b) Ishikawa, T., Oku, Y., Kotake, K.-I. and Ishii, H. (1996) *J. Org. Chem.*, **61**, 6484–6485; (c) Ballini, R., Fiorini, D., Gil, M.V., Palmieri, A., Románb, E. and Serrano, J.A. (2003) *Tetrahedron Lett.*, **44**, 2795–2797; (d) Gao, B., Yu, Z., Fu, Z. and Feng, X. (2006) *Tetrahedron Lett.*, **47**, 1537–1539.

119 Evans, P.A. and Andrews, W.J. (2005) *Tetrahedron Lett.*, **46**, 5625–5627.

120 Greatrex, B.W., Kimber, M.C., Taylor, D.K. and Tiekink, E.R.T. (2003) *J. Org. Chem.*, **68**, 4239–4246.

121 Yoo, D., Oh, J.S. and Kim, Y.G. (2002) *Org. Lett.*, **4**, 1213–1215.

122 Hari, A. and Miller, B.L. (2000) *Org. Lett.*, **2**, 3667–3670.

123 Fleming, F.F., Hussain, Z., Weaver, D. and Norman, R.E. (1997) *J. Org. Chem.*, **62**, 1305–1309.

124 Marco, J.L. and Fernández, S. (1999) *J. Chem. Res. (S)*, 544–545.

125 Grimaud, L., Rotulo, D., Ros-Perez, R. Guitry-Azam, L. and Prunet, J. (2002) *Tetrahedron Lett.*, **43**, 7477–7479.

126 Nishikata, T., Yamamoto, Y. and Miyaura, N. (2007) *Adv. Synth. Catal.*, **349**, 1759–1764.

5
Cyclic Carbometallation of Alkenes, Arenes, Alkynes and Allenes
Ron Grigg and Martyn Inman

5.1
Introduction

The development and application of transition metal catalysts has revolutionized organic synthesis, facilitating rapid access to complex targets which would otherwise require laborious multistep syntheses. Cyclic carbometallation has proved a particularly useful and versatile method for the synthesis of a wide variety of ring systems; this chapter surveys both catalytic and stoichiometric reactions in which carbometallation is a key step.

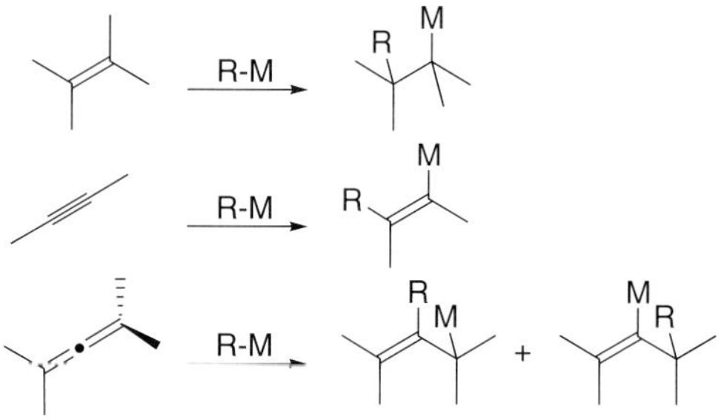

Scheme 5.1 Carbometallation of alkenes, alkynes and allenes.

Carbometallation can be defined as the addition, usually syn, of an organometallic species to an unsaturated carbon–carbon bond with formation of two new bonds (Scheme 5.1). Typically, intramolecular carbometallation occurs exo, although a number of endo-selective reactions have been reported. A variety of 3- to 20-membered carbocycles and heterocycles have been synthesized using carbometallation as a key step, highlighting its synthetic importance.

When M is a Group I or Group II metal, there is usually no way of restoring the active R–M species, and the process is stoichiometric in terms of metal. Where transition metals are used, it is often possible to combine carbometallation with other steps which regenerate an active metal species resulting in catalytic processes. For example, in the Heck reaction, carbopalladation of an alkene is followed by

Handbook of Cyclization Reactions. Volume 1.
Edited by Shengming Ma
Copyright © 2010 WILEY-VCH Verlag GmbH & Co. KGaA, Weinheim
ISBN: 978-3-527-32088-2

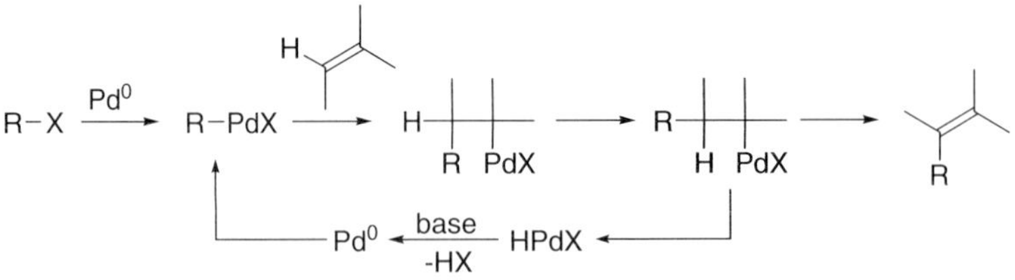

5 Cyclic Carbometallation of Alkenes, Arenes, Alkynes and Allenes

Scheme 5.2 Catalytic cycle for Heck reaction.

β-hydride elimination and reductive elimination of HPd(II)X, which recycles the catalytically active Pd(0) catalyst via reductive elimination in the presence of base (Scheme 5.2).

Catalytic processes are of particular interest, as they provide significant economic and environmental benefits. By far the most common catalyst for carbometallation reactions is palladium; however, other metals have found application in this area, including nickel, cobalt and rhodium.

5.2
Carbometallation of Alkenes

5.2.1
Stoichiometric Intramolecular Carbometallation of Alkenes

Methods for stoichiometric carbometallation have focused primarily on lithium species [1]. Several approaches to the synthesis of organolithiums have been employed (Scheme 5.3).

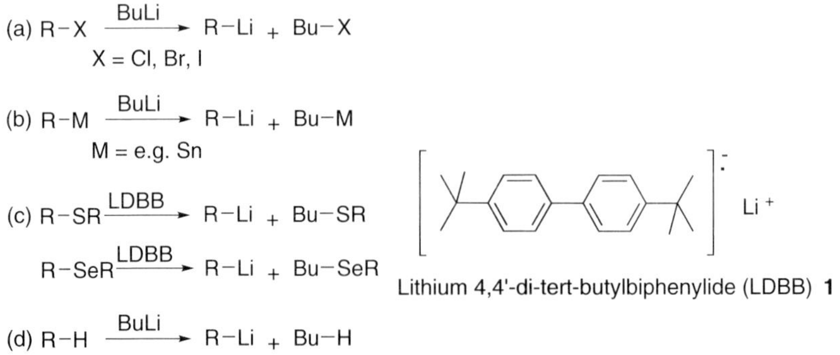

(a) R−X $\xrightarrow{\text{BuLi}}$ R−Li + Bu−X
    X = Cl, Br, I

(b) R−M $\xrightarrow{\text{BuLi}}$ R−Li + Bu−M
    M = e.g. Sn

(c) R−SR $\xrightarrow{\text{LDBB}}$ R−Li + Bu−SR
    R−SeR $\xrightarrow{\text{LDBB}}$ R−Li + Bu−SeR

Lithium 4,4'-di-tert-butylbiphenylide (LDBB) **1**

(d) R−H $\xrightarrow{\text{BuLi}}$ R−Li + Bu−H

Scheme 5.3 Generation of organolithiums.

Halide-lithium exchange (Scheme 5.3a) is a convenient and regioselective method for generation of organolithiums. The 5-*exo-trig* cyclization of 6-iodohex-1-enes, for example, **2** → **4** has been studied in depth, and was shown to give greatly improved yields in the presence of TMEDA (Scheme 5.4) [2].

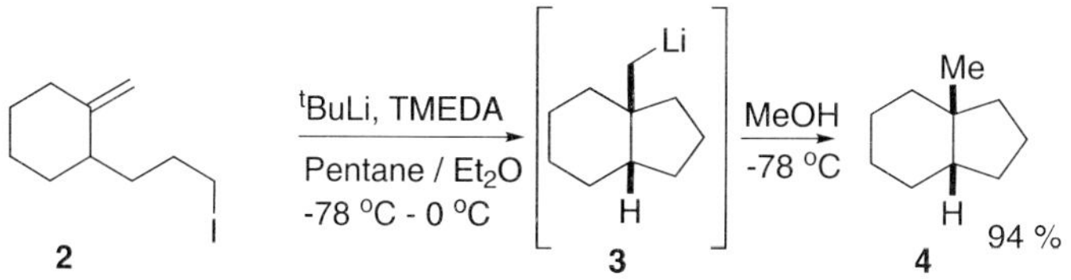

Scheme 5.4 Intramolecular carbolithiation.

The cyclization of alkenyl aryl lithiums has also been studied. Ortho-lithiated aryl allyl ethers undergo a rearrangement to give cyclopropyl phenols, for example, 7, through carbolithiation of the alkene and C—O bond cleavage (Scheme 5.5) [3]. Chiral carbolithiations of substituted aryl allyl amines, using sparteine and other chiral ligands, have also been investigated, giving ees of over 80% [4].

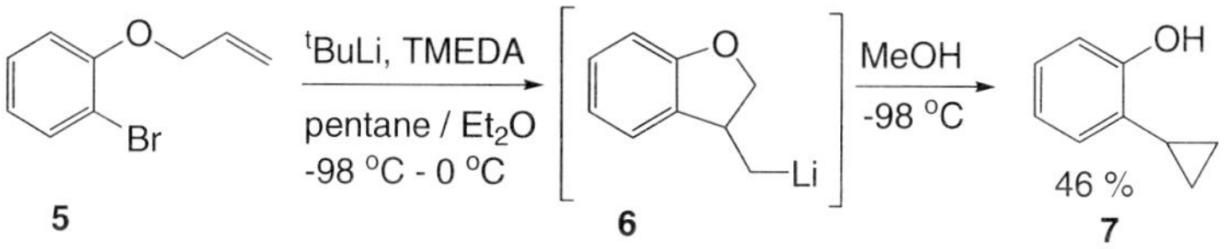

Scheme 5.5 Cyclic carbolithiation-C—O bond cleavage.

Transmetallation of lithiated species to metals such as zinc and copper, or direct synthesis of organozincs from organohalides [5, 6], permit other electrophilic reactions such as stoichiometric metallo-ene reactions [7] and cross-coupling, for example, 8 → 11 (Scheme 5.6) [8].

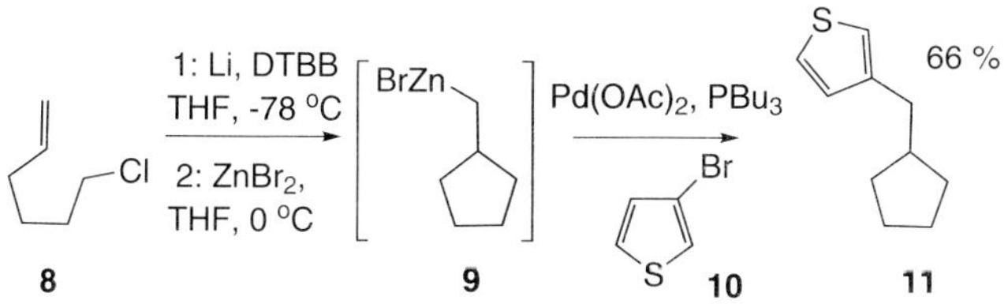

Scheme 5.6 Cyclic carbolithiation–Negishi coupling.

Lithium transmetallation of organometallic species (Scheme 5.3b) represents another facile, regioselective route to organolithiums, and one which can be applied with stereoretention. Chiral tributylstannanes 12 adjacent to nitrogen have been employed in a diastereoselective manner in the synthesis of pyrrolizidine alkaloids 14 (Scheme 5.7) [9]. Further examples of similar chiral processes have been reported [10].

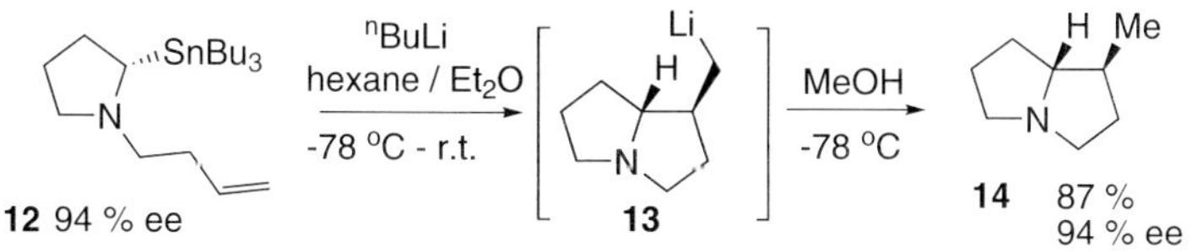

Scheme 5.7 Asymmetric carbolithiation.

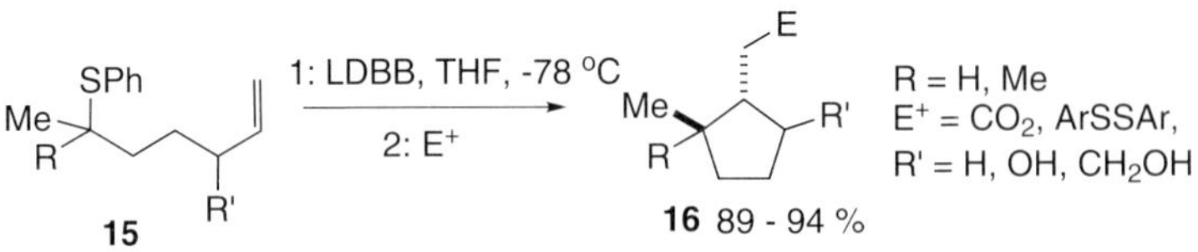

Scheme 5.8 Carbolithiation followed by trapping with E$^+$.

Reductive lithiation, usually using aromatic radical anions to displace thioethers **15** or selenoethers **17**, has been used to cyclize secondary and tertiary alkyl thioethers onto alkenes (Scheme 5.8) [11, 12]. Primary thioethers were found only to work when the alkene was allylic or homoallylic, that is when R′ = OH or CH$_2$OH. Interestingly, Krief et al. found that when using achiral methyl selenoethers, the diastereoselectivity could be switched by changing the reaction conditions from ^{n}BuLi in THF to ^{t}BuLi in pentane (Scheme 5.9) [13]. Related alternatives to sulfur and selenium reductions include the displacement of nitriles using lithium 4,4′-di-^{t}butyl biphenylide (LDBB) **1** [14] and the use of the Shapiro reaction to generate vinyl anions [15].

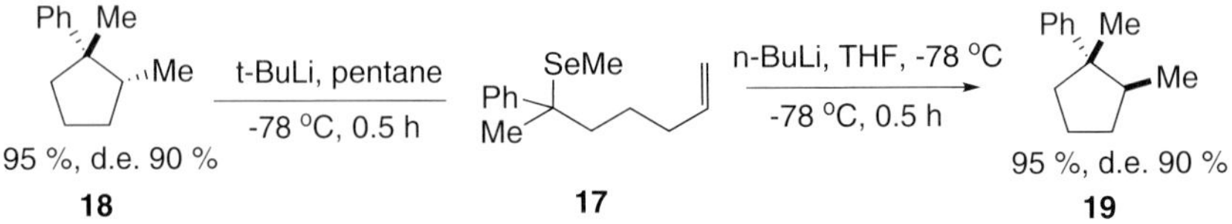

Scheme 5.9 Switch of diastereoselectivity in carbolithiation.

5.2.2
Intramolecular Heck Reactions

The Heck reaction (Scheme 5.2) is the archetypal catalytic reaction of alkenes. Since the first intramolecular examples were described in the late seventies [16], the intramolecular Heck reaction has become a fundamental tool for ring formation. It is outside the scope of this chapter to discuss this literature in detail, but a few representative examples will be given. For a more detailed account of this area, readers are directed to reviews of palladium-catalyzed intramolecular Heck reactions [17] and their application in natural product synthesis [18].

Significant attention has recently focused on asymmetric intramolecular Heck reactions. Where the substrate already contains chirality, this can be exploited to generate diastereoselective Heck reactions, for example, the use of chiral arylsulfinyl auxiliaries in the enantioselective synthesis of carbocycles **20** → **21** (Scheme 5.10) [19].

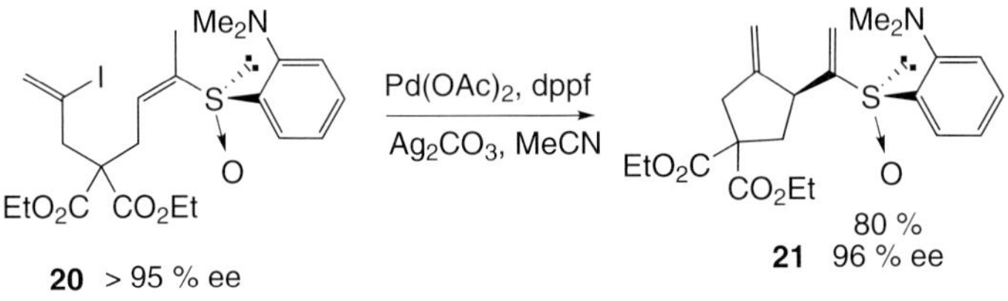

Scheme 5.10 Cyclic carbopalladation of chiral 1-alkenyl sulfoxide.

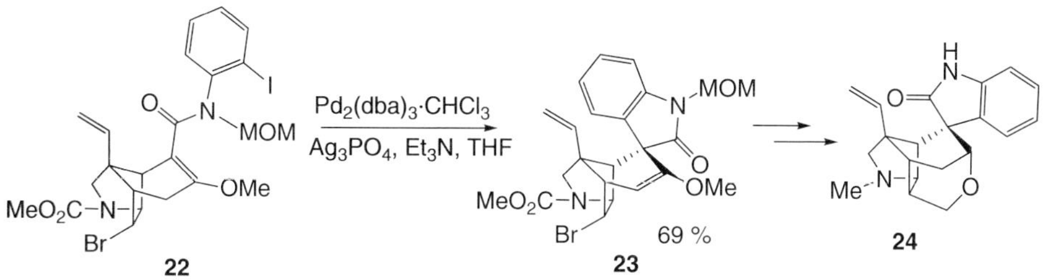

Scheme 5.11 Cyclic carbopalladation for lactam.

Diastereoselective Heck reactions have found substantial applications in total synthesis, as the intermediates are often rich in chiral information. Examples include the syntheses of gelsimine **24** (Scheme 5.11) [20] and morphine **27** (Scheme 5.12) [21].

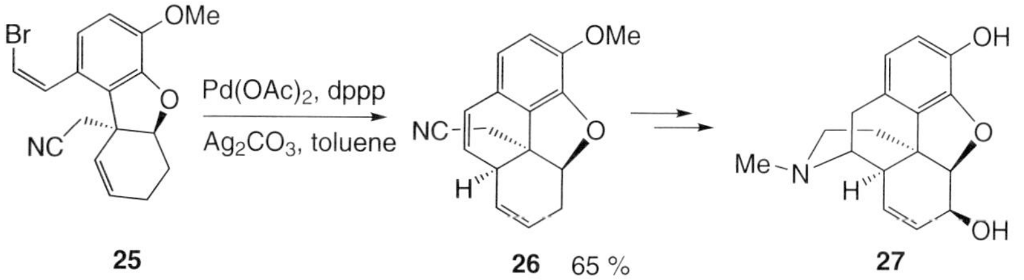

Scheme 5.12 Carbopalladation in polycyclic skeleton.

Chiral catalysts employing bidentate ligands, particularly biphosphines such as BINAP, have been widely used in Heck reactions [22]. Cationic Heck reactions generally give better enantioselectivity, so triflate starting materials or silver additives are often beneficial. This subject has been comprehensively reviewed [18a] and will not be discussed here, although the development of monodentate phosphoramadite ligands able to induce high enantioselectivity is noteworthy [23].

Oxidative C–H bond activation is an emerging field in palladium catalysis, and a number of applications to intramolecular Heck reactions have been reported. These can proceed through stoichiometric Pd(II) mediation [24], but it is usually much more efficient to include an oxidizing agent to return the Pd(0) species to an active Pd(II) form. Activated alkenes such as p-benzoquinone and electron-rich arenes are particularly effective coupling partners in such reactions; the synthesis of indoloquinones **29** has been reported using either 2.5 equiv. Cu(OAc)$_2$ [25] or an oxygen atmosphere [26] as the oxidant (Scheme 5.13a). The O$_2$ system has also been applied to the synthesis of carbazoles **30** (Scheme 5.13b) [26].

Pd(II) catalyzed syntheses of benzofurans **32** and dihydrobenzofurans **33** have been devised from electron-rich aryl allyl ethers (Scheme 5.14 a) using benzoquinone (BQ) as an oxidant. The stereochemistry of more complex examples **34** → **36** (Scheme 5.14b) seems to confirm that syn-palladation occurs, rather than Pd(II) activation of the alkene to nucleophilic attack [27]. This supports related earlier work by the same group using indoles as the arene [28].

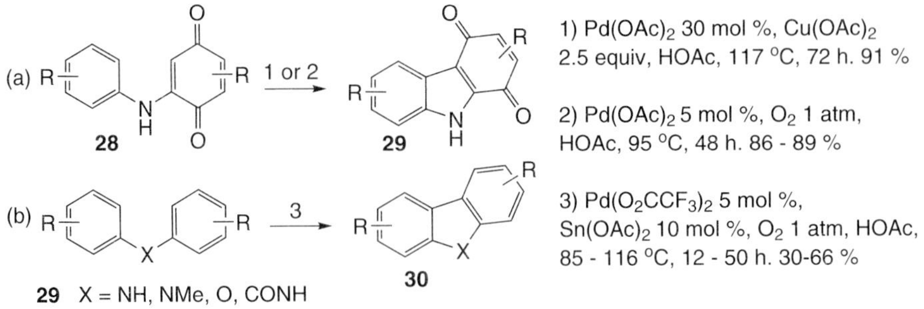

Scheme 5.13 Pd(II)-Catalyzed C–H activation-cyclic carbopalladation of arenes.

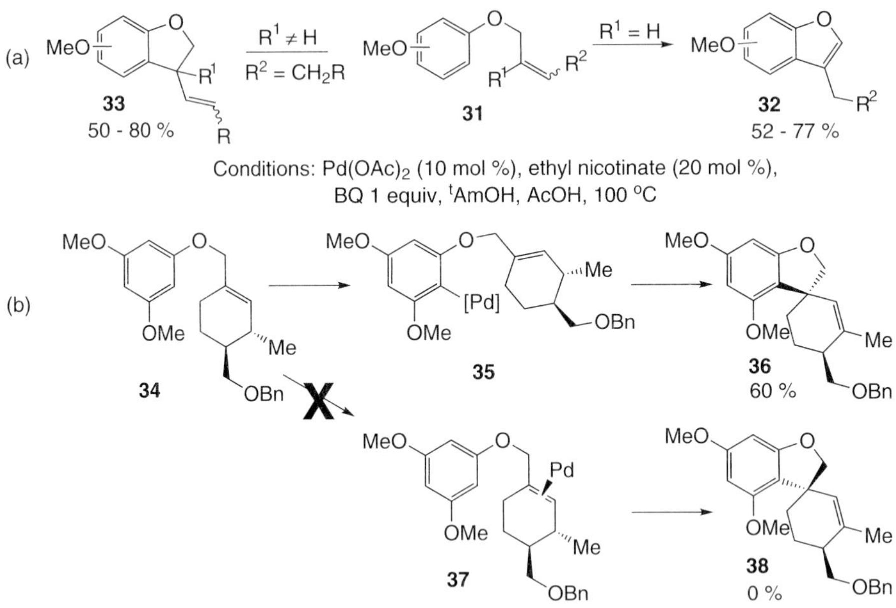

Scheme 5.14 Cyclic carbopalladation of arenes based on Pd(II)-catalyzed C–H activation.

The most atom efficient variation of the Heck reaction is the enyne cycloisomerization, whereby a vinylpalladium species is generated by hydropalladation of an alkyne before undergoing a normal Heck reaction [29]. A common hydridopalladium species is HPdOAc, generated from Pd^0 and HOAc. This method has found widespread use recently, including applications in a number of total syntheses [30]. A hydropalladation–Heck–Diels–Alder cascade has been devised by the De Meijere group to give various polycyclic compounds, for example, **39** → **41** (Scheme 5.15) [31].

When allylic alcohols are used, β-hydride elimination can give an enol which tautomerizes to the corresponding aldehyde. Derivatization of this by reductive amination using H_2 and racemic N-acetylphenylalanine (NAP) **44** as a Brönsted acid gives the corresponding aminoethyl heterocycle **43** (Scheme 5.16). Wittig reaction of the aldehyde intermediate is also described [32].

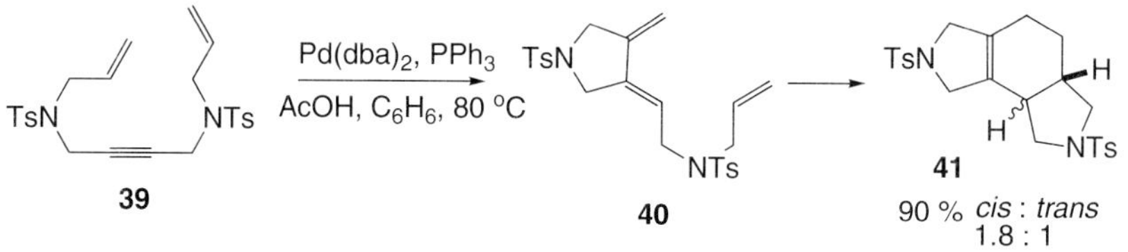

Scheme 5.15 Polycycle via hydropalladation–cyclic carbopalladation–Diels–Alder sequence.

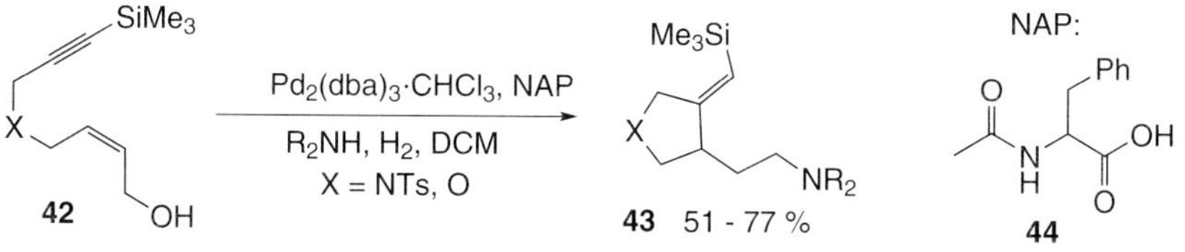

Scheme 5.16 Cyclic carbopalladation of allylic alcohol.

Although palladium is an effective catalyst, its expense may cause problems when large-scale Heck reactions are required. Intermolecular Heck reactions using rhodium and iridium [33] and intramolecular examples using copper [34] and cobalt [33, 35] have been reported, but it is nickel which currently provides the best alternative to palladium (Scheme 5.17) [36]. Recent developments include the use of nickel carbene complexes as efficient intermolecular Heck reaction catalysts [37].

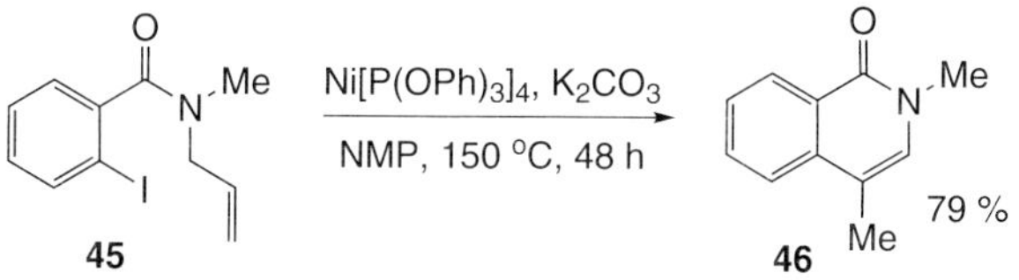

Scheme 5.17 Ni-catalyzed synthesis of benzolactam.

5.2.3
Intramolecular Catalytic Alkene Carbometallation: Other Termination Steps

Termination steps in palladium-catalyzed Heck reactions are not limited to β-hydride elimination. The alkylpalladium intermediate may, for example, be cross-coupled with an organometallic reagent, substituted with a nucleophile, carbonylated to the acylpalladium, or may undergo further intermolecular or intramolecular carbopalladations (Scheme 5.18). In order to avoid so-called "shunt" reactions, where steps

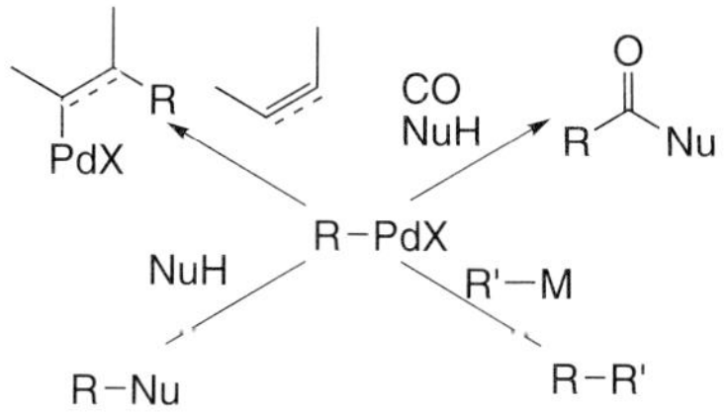

Scheme 5.18 Reactivity of R-PdX.

Scheme 5.19 Cyclic carbopalladation of olefin followed by Stille coupling.

R	Yield	48 : 49
phenylethynyl	38	1 : 1
2-styryl	46	8 : 1
vinyl	63	1 : 0
2-pyridyl	38	1 : 0
2-thiazolyl	40	1 : 0

occur in the wrong order and give rise to side-products, it is important to consider the relative rate of each step, and to bear in mind that the intramolecular carbopalladation is likely to be much faster than an intermolecular equivalent and, in the case of carbonylation, sensitive to pressure.

Many organometallic reagents can be used to capture organopalladium species, the most commonly used being organostannanes and boronic acids. Both have been successfully applied to the capture of carbopalladated alkenes; the first published case was the coupling of norbornene amide **47** with various stannanes, in a diastereoselective manner (Scheme 5.19) [38]. The use of zinc reagents in the same reaction was found to give poor yields, possibly due to the greater reactivity of organozinc compounds [39]. This process is part of a general and wide-ranging cascade cyclization–anion capture protocol developed by the Grigg group, outlined in Table 5.1. The permutations of each starter species, relay species and anion transfer agent, in both intra- and inter-molecular processes, gives rise to vast potential for molecular diversity, much of which remains to be realized [40]. Additionally, the relay stage can be extended by the use of several relay species in sequence, in either an intramolecular (see Section 5.6) or intermolecular fashion. The latter is particularly interesting, as it provides the possibility of multicomponent reactions and the diversity-oriented syntheses associated with them. Reaction components are hence

Table 5.1 Potential combinations for (poly) cyclization–anion capture processes.

Starter species	Relay species	Terminating species	Transfer agent
Alkyl	Alkene	Alkene	Anionic (H$^-$, AcO$^-$, $^-$CN, N$_3^-$, TsNR$^-$, RSO$_2^-$, (EWG)$_2$CH$^-$)
Allyl	Alkyne	Alkyne	Neutral (R$_2$NH, CO/MeOH, allenes, acrylates)
Aryl	Allene	Allene	Organometallics (RM, M = Sn(IV), B(III), Zn(II), Zr(IV), Al(III), Mg(II) etc.)
Vinyl	1,3-diene	1,3-diene	
Allenyl	CO		
Acyl	Aryne		
Carbamyl			
Oxycarbonyl			

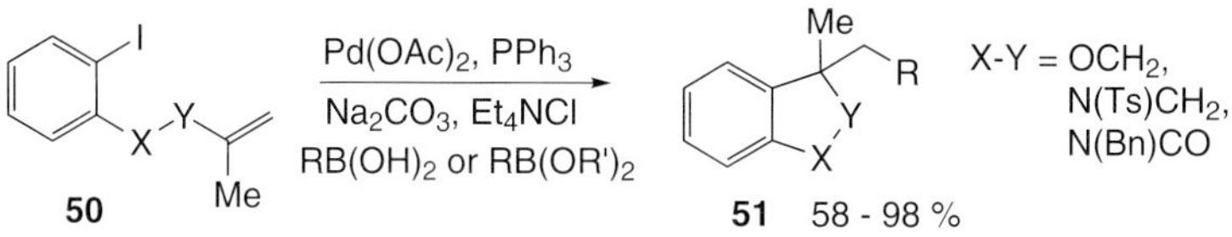

Scheme 5.20 Cyclic carbopalladation followed by Suzuki coupling.

required which will switch the reaction pathway from intra- to inter-molecular at the correct point in the sequence; such components have been termed relay switches, and include most prominently allenes, carbon monoxide and arynes.

Boronic acids were used in the synthesis of a number of heterocyclic compounds from allyl or acryloyl iodobenzene derivatives **50** in good to excellent yield (Scheme 5.20) [41]. Further cyclization–anion transfer reactions of boronic acids with vinyl, aryl and benzyl halides have been reported [42].

The palladium catalyst employed in these reactions can also be used to generate organometallic coupling reagents *in situ* as part of a temperature-controlled extended cascade, as is demonstrated by the synthesis of up to 17-membered spiromacrocycles **54** from terminal alkynes tethered through alkenes to aryl halides (Scheme 5.21) [43].

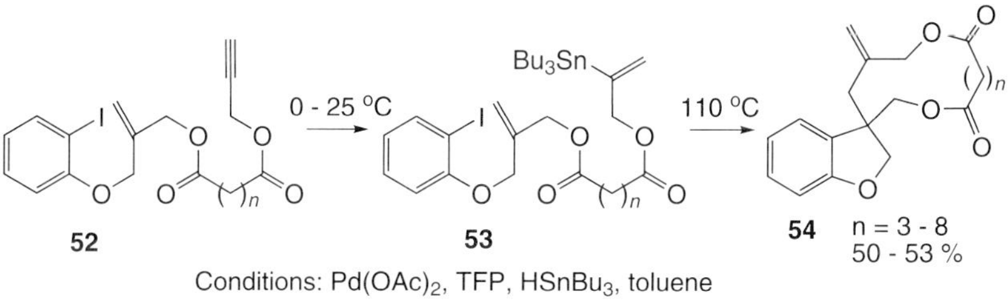

Scheme 5.21 Cyclic carbopalladation–cyclic Stille coupling.

By addition of ZnI_2 to the reaction mixture, alkene carbopalladations can be followed by transmetallation to give the corresponding organozinc iodide, which is in turn reactive towards alkyl halides [44]. Additionally, Stille cross-coupling has been demonstrated to be an effective means of terminating enyne cycloisomerization reactions [45].

In addition to organometallic anion transfer agents, the terminating species may also be a formal nucleophile. Reactions analogous to those in Schemes 5.19 and 5.20 have been performed using HCO_2Na as a source of hydride ions for reductive cyclizations (R = H) [46]; cyanide sources such as KCN and $K_4[Fe(CN)_6]$ have also been employed to give the corresponding nitriles (R = CN) [47]. Stoichiometric nickel-promoted carbometallation–cyanation sequences have also been reported [48].

An interesting variation is the carbopalladation of 1,3-dienes, which gives an electrophilic π allyl palladium species that undergoes regioselective nucleophilic attack at the less hindered terminal position in good yield for example, **55** → **57** (Scheme 5.22) [49].

Scheme 5.22 Cyclic carbopalladation of 1,3-diene.

Conditions: Pd(OAc)$_2$, PPh$_3$, NaCH(CN)$_2$, MeCN, 80 °C, 1 h

Carbonylation provides a means of introducing ketone, ester and amide moieties into cyclized products via a highly electrophilic acylmetal intermediate. Palladium again overwhelmingly dominates as the catalyst, with a variety of terminating agents such as hydride [50], stannanes, boranes [51] and carbon-, oxygen- and nitrogen-centered nucleophiles (Scheme 5.23) [52]. Several intermolecular examples were published over thirty years ago [53], whilst the first reported intramolecular cyclization-carbonylations used stoichiometric amounts of Ni(CO)$_4$ both as a catalyst and CO source [48c-d, 54].

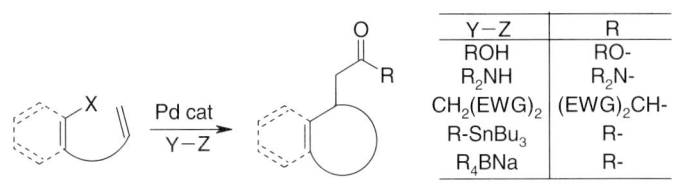

Y–Z	R
ROH	RO-
R$_2$NH	R$_2$N-
CH$_2$(EWG)$_2$	(EWG)$_2$CH-
R-SnBu$_3$	R-
R$_4$BNa	R-

Scheme 5.23 Cyclic carbopalladation–carbonylation.

In the presence of AcOH and CO, enynes such as **58** undergo cyclization and carbonylation to the mixed anhydride, which gives the corresponding carboxylic acid **59** in good yield upon aqueous work-up; *in situ* hydrolysis of the acylpalladium formed from vinyl bromide **60** gave the same products in generally lower yield (Scheme 5.24) [55].

Scheme 5.24 Pd-catalyzed cyclic carbopalladation–carboxylation.

Several recent total syntheses exemplify the versatility of the cyclization–carbonylation cascade [56]. Oxidative carbopalladation of unhalogenated alkenyl indoles with a carbonylative termination has also been reported [57].

Intermolecular carbopalladation of allenes occurs smoothly after intramolecular alkene carbopalladation, and the resulting π-allylpalladium(II) intermediate can

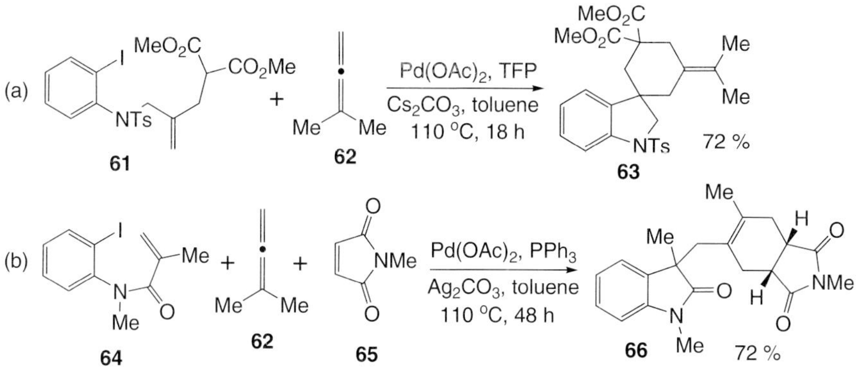

Scheme 5.25 Cyclic carbopalladation–intermolecular carbopalladation of allenes.

undergo either inter- or intra-molecular nucleophilic attack, for example, **61** → **63** (Scheme 5.25a) [58] or β-hydride elimination and Diels–Alder reaction, in cases where the allene contains appropriate hydrogen atoms, for example, **64** → **66** (Scheme 5.24b) [59].

Metallo-ene reactions represent an interesting form of carbometallation, proceeding through a 6π-electron transition state with a metal ion replacing the normal migrating proton (Scheme 5.26) [60]. Many catalytic and stoichiometric methods for metallo-ene reactions have been developed using a diverse range of metals including lithium, magnesium, zinc, titanium, indium, tin, rhodium, platinum, palladium and nickel. Almost all examples terminate in protolysis of the C–M bond; other terminating steps have been developed, such as cross-coupling with an organostannane (Scheme 5.27). The reaction proceeds with complete diastereoselectivity [61].

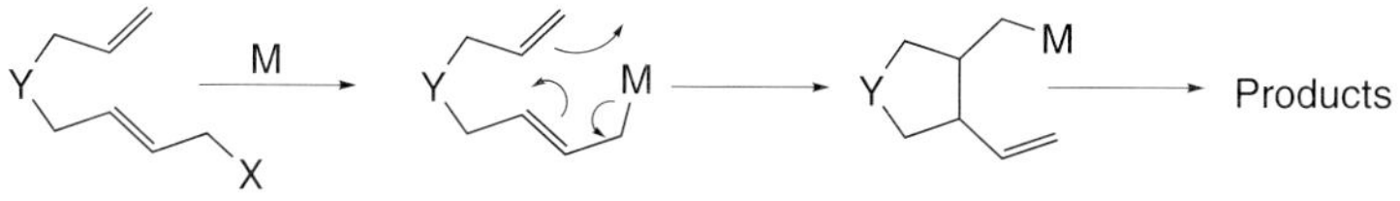

Scheme 5.26 Metallo-ene reaction.

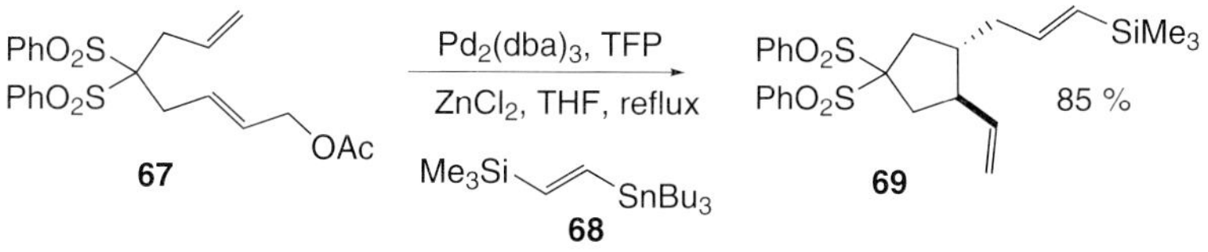

Scheme 5.27 Pd-catalyzed metallo-ene reaction and Stille coupling.

Rhodium is rapidly emerging as a versatile catalyst, as its reactivity becomes better understood [62]. Organorhodium species, generated by transmetallation of boronic acids, will readily undergo carborhodation onto alkynes, alkenes and, unlike organopalladium species, carbonyl moieties. Intramolecular racemic and chiral Michael-like addition to activated alkenes has been reported, for example, **70** → **71**

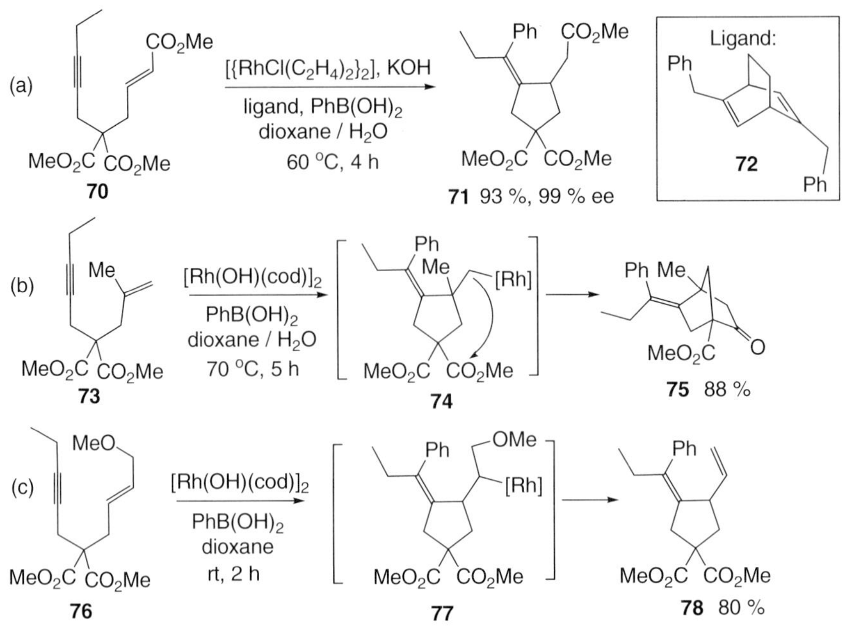

Scheme 5.28 Rh-catalyzed cyclic carbometallation.

(Scheme 5.28a) [63]. A further cyclization onto the ester has also been reported when unactivated alkenes are used, for example, **73** → **75** (Scheme 5.28b) [64]. Addition to allylic ethers **76**, followed by deoxyrhodation, gives vinylcyclopentanes **78** (Scheme 5.28c) [65]. An interesting reaction proceeding through aldehyde C–H insertion, hydrorhodation and decarborhodation has recently been reported, for example, **79** → **82** (Scheme 5.29) [66].

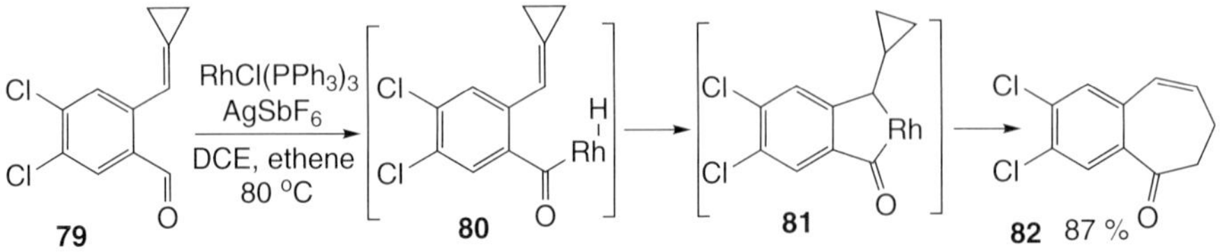

Scheme 5.29 Rh-catalyzed formation benzocycloheptenone.

5.3
Alkyne Carbometallation

5.3.1
Stoichiometric Methods

As with alkenes, the stoichiometric carbometallation of alkynes is predominantly mediated by lithium. The basic principles are the same as those described in

Section 5.2.1, usually proceeding via syn addition to give the more stable vinyl lithium species. Recent literature has focused on selective syn and anti carbolithiations, and includes the stereoselective synthesis of five- and six-membered carbocycles using the Cby directing group. Cby-protected alcohols **83** undergo diastereoselective syn addition to alkynes in the presence of sparteine (Scheme 5.30a) [67]. By contrast, using the Cb group on both the alcohol and the alkyne gives anti addition, for example, **85** → **86** (Scheme 5.30b) [68].

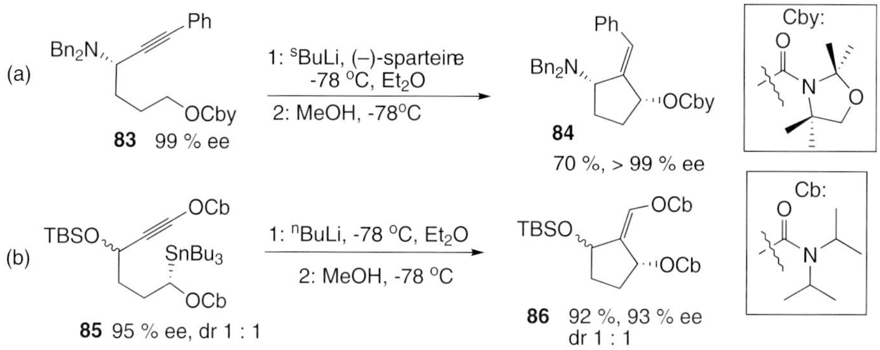

Scheme 5.30 Cyclic carbolithiation of alkynes.

A tandem intermolecular–intramolecular carbolithiation sequence has been reported by the Taylor group. Switching the solvent from Et_2O to 4 : 1 v/v Et_2O : THF resulted in a switch from *syn* **89** to *anti* **90** carbolithiation of the alkyne (Scheme 5.31) [69].

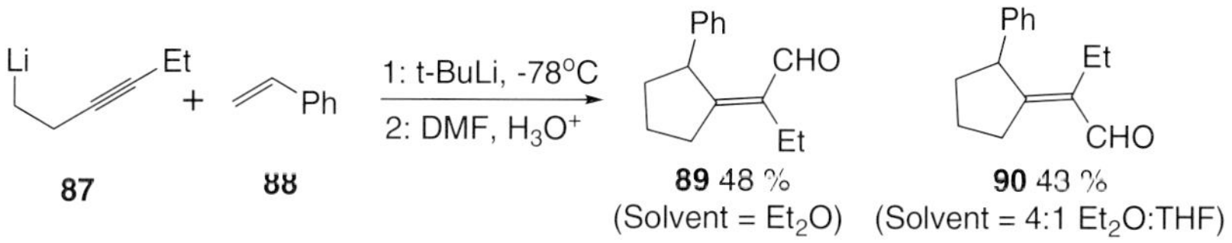

Scheme 5.31 Intermolecular carbolithiation of styrene followed by cyclic alkyne-carbolithiation.

5.3.2
Catalytic Alkyne Carbometallations

Almost all examples of catalytic intramolecular alkyne cyclizations are based on palladium chemistry, and most of the concepts discussed in Section 5.2.3 apply. The major difference is that β-hydride elimination does not occur on vinylpalladium species, and so Heck-type termination cannot occur. The terminating steps which are available to vinylpalladium species are broadly the same as those summarized in Scheme 5.18. A host of alkyne cyclization–anion capture sequences have been reported; the first example of such a reaction **91** → **92** used propargyl 2-iodoaniline derivatives and a variety of aryl, vinyl, alkynyl and allyl stannanes (Scheme 5.32) [38].

Scheme 5.32 Cyclic carbopalladation of alkyne–Stille coupling.

Boronic acids and organozinc reagents – including alkyl Reformatsky reagents – were found to react in a similar fashion. Yields from alkyne starting materials were generally better than for the corresponding alkenes [39, 70]. In a comprehensive study of various cross-coupling partners, it was found that vinyl zinc chlorides or stannanes, and aryl alanes, gave the best yield from aryl iodide starting materials **93** (Scheme 5.33). A similar study indicated that zirconocenes (M = ZrCp$_2$Cl) worked best with vinyl iodide starting materials [71]. Boronic acids have been the organometallic reagent of choice in recent years, being both reactive and non-toxic [72].

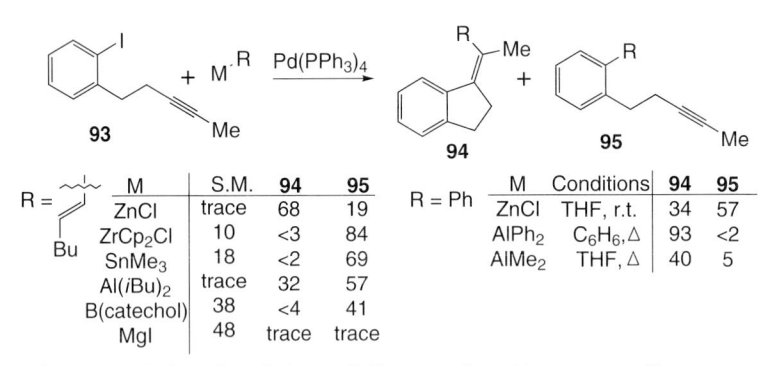

R =	M	S.M.	94	95
	ZnCl	trace	68	19
	ZrCp$_2$Cl	10	<3	84
Bu	SnMe$_3$	18	<2	69
	Al(iBu)$_2$	trace	32	57
	B(catechol)	38	<4	41
	MgI	48	trace	trace

R = Ph	M	Conditions	94	95
	ZnCl	THF, r.t.	34	57
	AlPh$_2$	C$_6$H$_6$, Δ	93	<2
	AlMe$_2$	THF, Δ	40	5

Scheme 5.33 Cyclic carbopalladation of alkyne-coupling with organometallics.

Alkynyl Stille couplings have been applied to the synthesis of the neocarzinostatin core **98** (Scheme 5.34). The insertion of palladium into the correct C–Br bond of the starting material **96** was attributed to coordination of either the pendant alkyne or alcohol to palladium prior to insertion [73].

Scheme 5.34 Cyclic carbopalladation starting from selective oxidative addition.

A tandem palladation–anion capture–electrocyclization has been reported, converting alkynyl propargylic carbonates **99** into naphthalene derivatives **102**

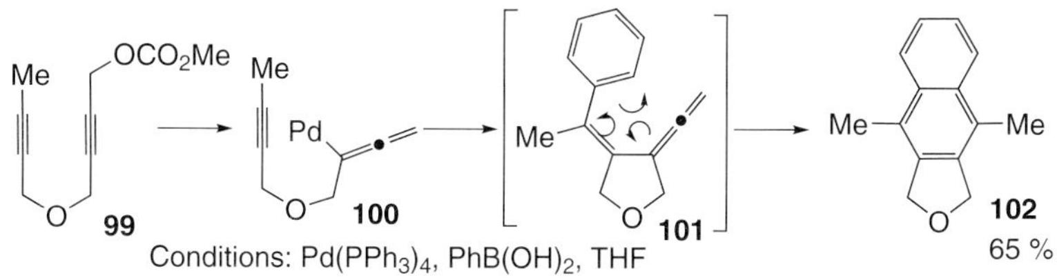

5.3 Alkyne Carbometallation

Scheme 5.35 Cyclic carbopalladation of alkyne based on the oxidative addition of propargylic carbonates.

(Scheme 5.35) [74]. Similar compounds have been synthesized by carbopalladation and oxidative insertion into an aryl C–H bond [75].

An interesting synthesis of cyclobutanes has been reported which proceeds through a 4-*exo*-trig cyclization onto an alkyne and terminates with a Stille coupling. When a vinylstannane such as **106** was used, 6π-electrocyclization also occurred (Scheme 5.36) [76].

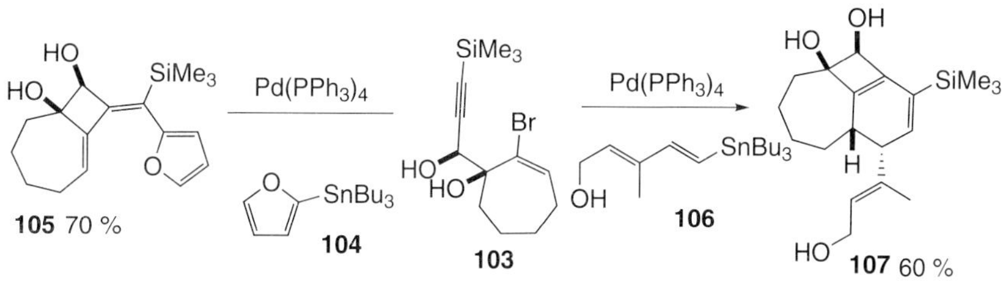

Scheme 5.36 Cyclobutane synthesis based on cyclic carbopalladation.

Carbonylation of vinylpalladium intermediates has been explored in some depth. The use of methanol gives α,β-unsaturated methyl esters [77], while intramolecular capture by a pendant amine or alcohol gives the corresponding lactam, for example, **109**, or lactone, for example, **111** (Scheme 5.37) [78].

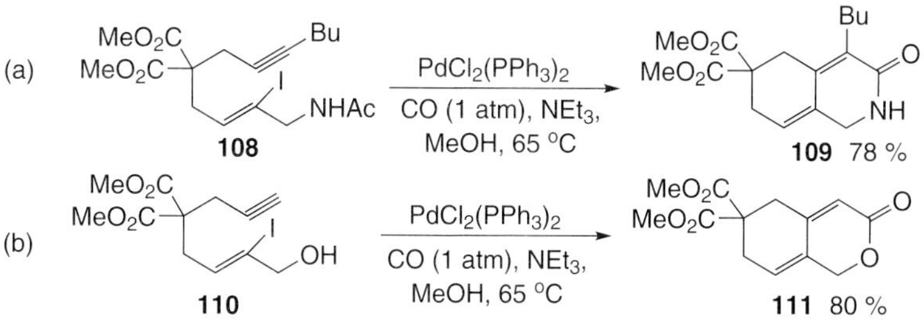

Scheme 5.37 Alkyne carbopalladation trapped with amide or alcohol.

Reductive cyclization of diynes and halogenated enynes has been effected using various hydride sources such as HCO_2Na, HCO_2H + piperidine and Et_3SiH (Scheme 5.38) [79].

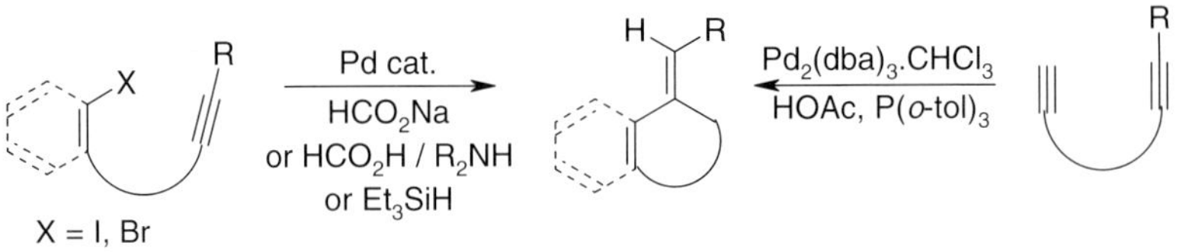

Scheme 5.38 Cyclic carbopalladation followed by reduction with H⁻.

In addition, vinylpalladium intermediates derived from alkynes have been captured by allene insertion and nucleophilic substitution [80], or by intermolecular Heck reaction [81].

An interesting series of intramolecular palladium and nickel-catalyzed arylcyanations (Scheme 5.39a) and borocyanations (Scheme 5.39b) have recently been reported; both proceed through insertion into the C−CN or B−CN bond, carbometallation and reductive elimination [82]. Nickel- and rhodium-catalyzed silylative cyclizations of diynes have been studied in depth, particularly by Ojima and Matsuda [83]. Recent developments in this field include the cyclization of allenynes **116** to vinylpyrrolidines **117** (Scheme 5.40), catalyzed by a rhodium carbonyl complex under CO (1 atm). The role of the CO is not commented on but it would seem that the increased π-Lewis acidity of (CO)Rh(II) complexes promotes alkyne coordination [83c].

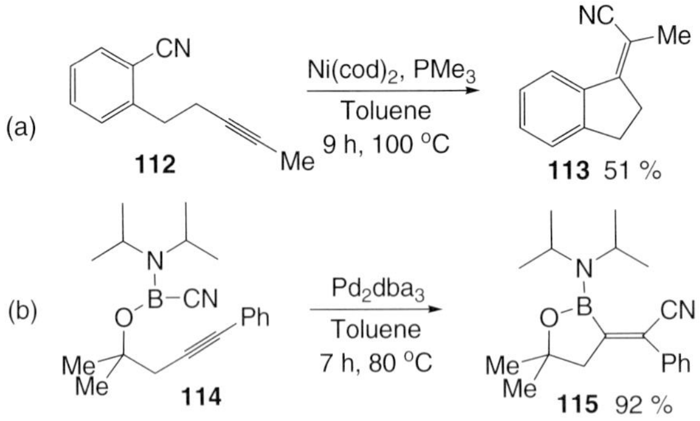

Scheme 5.39 Cyclic carbometallation and cyanation based on the oxidative addition of nitriles.

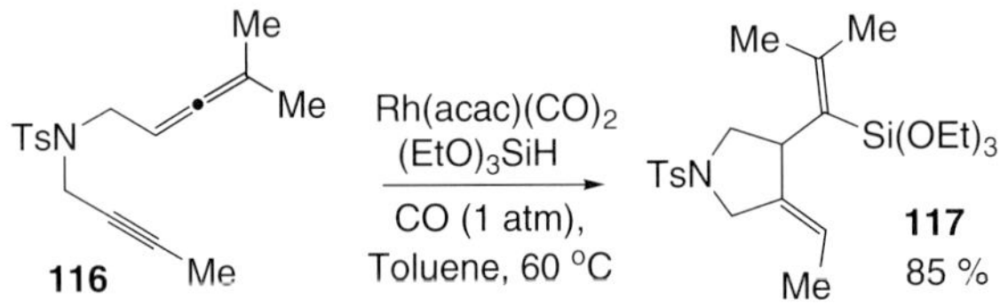

Scheme 5.40 Rh-catalyzed cyclization of allenyne.

Alkynes can also undergo metallo-ene type reactions, and the resulting vinylpalladium species can undergo anion capture with boronic acids, for example, **118 → 119** (Scheme 5.41) [84].

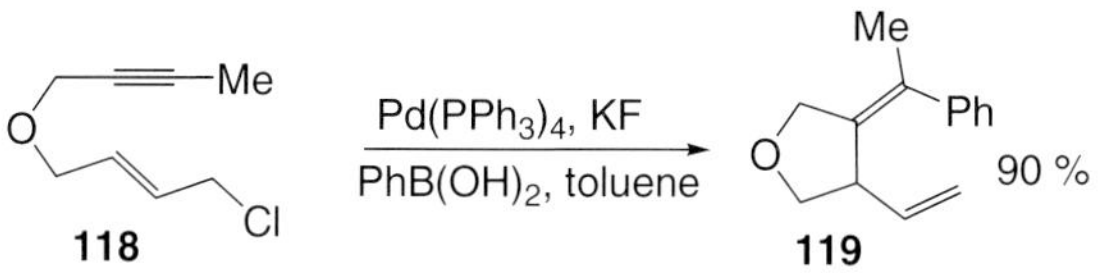

Scheme 5.41 Metallo-ene reaction of alkyne.

5.4
Carbometallations of Arenes

The addition of an organometallic species to an arene is inhibited by the significant energy barrier associated with loss of aromaticity. Nonetheless, a number of stoichiometric examples, involving organolithium species, have been reported but are outside the scope of this chapter. Readers are directed to a very comprehensive recent review by Ortiz et al. of nucleophilic dearomatizations, which includes a number of cyclative examples [85].

5.5
Carbometallations of Allenes

Allenes provide a particularly interesting class of substrates for carbometallation reactions. Ignoring possible regioisomers, there are two possible structurally isomeric outcomes when an allene is carbometallated (Scheme 5.42). Structure (ii) is usually the preferred product when transition metals such as palladium are used. This can be attributed to the fact that the sp hybridized central carbon is a harder electrophile than the sp^2 hybridized terminal carbon, and the latter is a better match for the organopalladium(II) species. A significant change in reactivity occurs upon carbopalladation, as the resulting π-allyl palladium species are very electrophilic, making allenes a particularly interesting relay switch in catalytic cyclization–anion capture reactions.

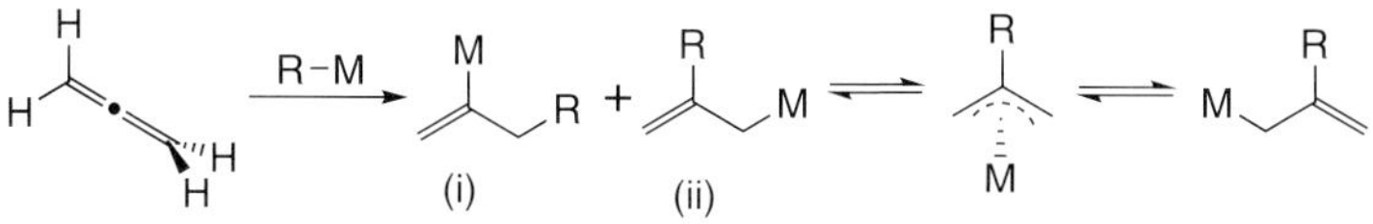

Scheme 5.42 Carbometallation of allene.

Intramolecular carbopalladation of allenes can be followed by intermolecular nucleophilic attack (Scheme 5.43) [86]. Intermolecular nucleophiles attack preferentially at the exo position to give **121** when K_2CO_3 is used as the base, and the endo position to give **122** when Ag_2CO_3 is used. Sodium azide can be used as a nucleophile, giving an alkyl azide which is reactive towards dipolarophiles [87]. Intramolecular nucleophilic attack is also possible; in such cases, the regiochemistry of the attack is governed by the position of the nucleophile relative to the allylic system

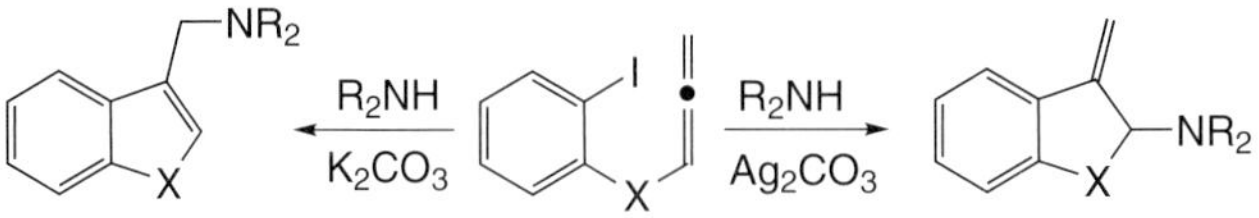

Scheme 5.43 Pd-catalyzed cyclization of 2-iodoaryl allene in the presence of amines.

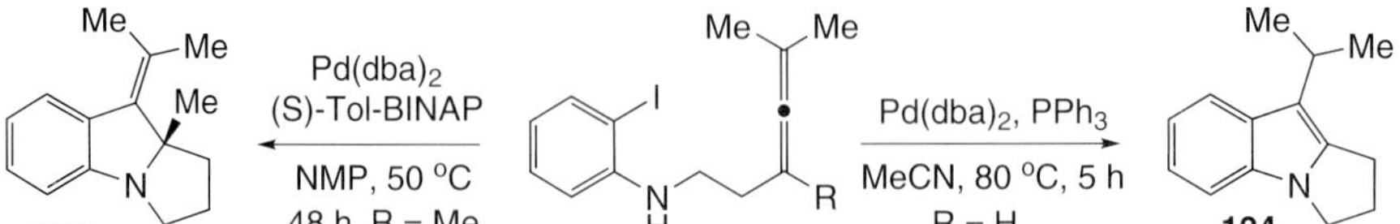

Compound **121**: 88–91 %
Compound **120**: X = O, CON(Me), CH$_2$O, CH$_2$N(Ts)
Conditions: Pd(OAc)$_2$, PPh$_3$, MeCN, 80 °C, 28 h
Compound **122**: 36–88 %

Scheme 5.44 Cyclic allene carbopalladation.

Compound **125**: 90 % brsm, 88 % ee
Compound **124**: 100 %

(Scheme 5.44) [58]. When 2-iodoaniline derivatives are used, the double bond migrates into the endocyclic position, to give an annulated indole **124** in quantitative yield, when R = H [88]. Chiral ligands have also been screened in intramolecular reactions, with (S)-Tol-BINAP giving an ee of 88% [89].

A rearrangement resulting in formation of a nucleophilic species *in situ* has been reported (Scheme 5.45) [90].

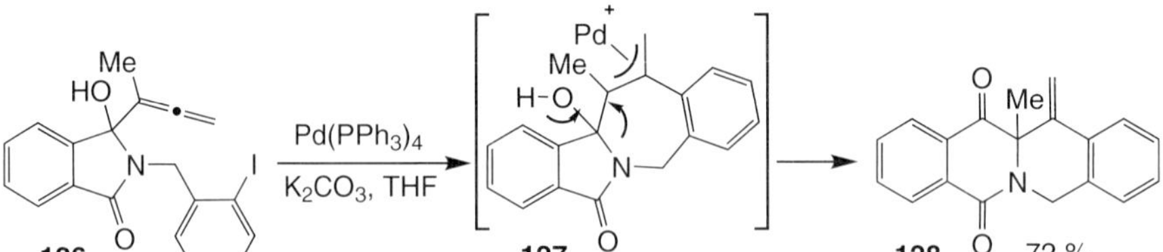

Scheme 5.45 Cyclic allene carbopalladation followed by ring expansion.

Compound **128**: 72 %

Where alkyl allenes are used, β-hydride elimination to a diene is possible. In intramolecular cases, exo-dig addition to the allene is favored, followed by β-hydride elimination (if present) to give **130**; in their absence, *endo*-β-hydride elimination occurs to give **131** (Scheme 5.46). A range of 5- to 20-membered rings was made in

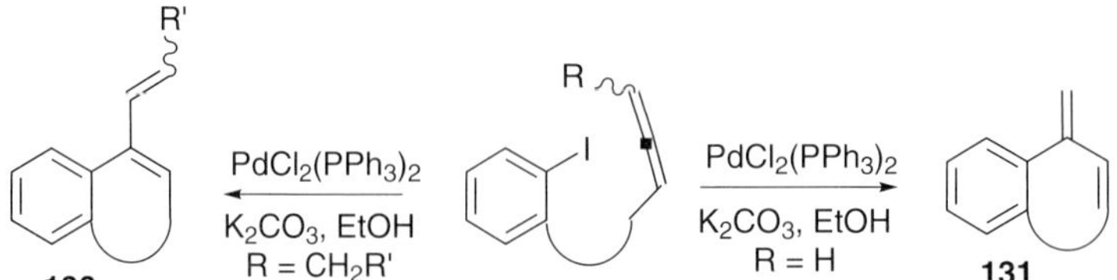

Scheme 5.46 Cyclic carbopalladation of allene and β-H elimination.

Compound **130**: 35–61 %
Compound **131**: 52–86 %

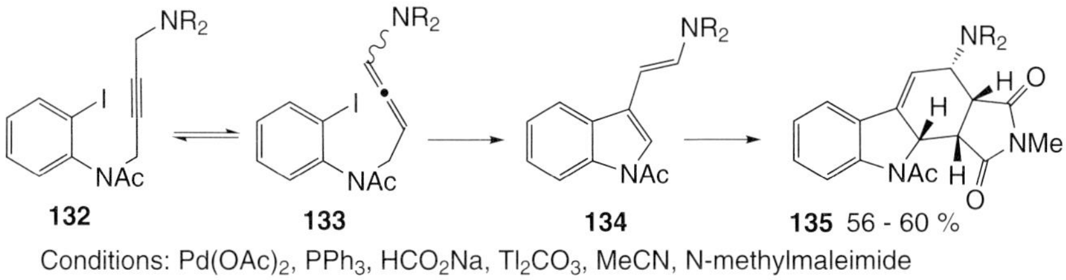

Scheme 5.47 Cyclic carbopalladation for indole derivative.

Conditions: Pd(OAc)$_2$, PPh$_3$, HCO$_2$Na, Tl$_2$CO$_3$, MeCN, N-methylmaleimide

this way [91]. Dienes **134** and corresponding Diels–Alder products **135** can be generated from alkynyl aryl iodides **132**, probably through alkyne–allene equilibration under the reaction conditions (Scheme 5.47) [79b].

Norbornadiene **137** has been used as a relay switch leading to intramolecular allene carbopalladation followed by β-hydride elimination and retro-Diels–Alder reaction giving vinyl naphthalenes **139** in good yield (Scheme 5.48) [92].

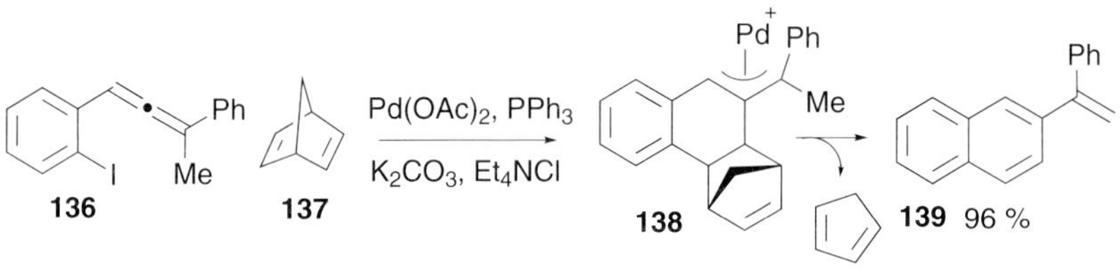

Scheme 5.48 Intermolecular olefin-cyclic allene carbopalladations–β-H elimination for the synthesis of naphthylene.

5.6
Multiple Cyclization Reactions

In addition to monocyclization, it is possible to design reactions in which a number of rings are formed through sequential carbometallation of various unsaturated systems. These processes are broadly termed "living" organometallic processes. It is especially important in such cases to plan the desired reaction pathway carefully, with consideration given to the relative rates and likely regiochemistry of each step. When planned and executed correctly, such processes are very powerful; a variety of complex fused, bridged and spiro polycyclic systems can be synthesized efficiently from linear or branched starting materials in a single step. Multiple carbopalladations can broadly be categorized into the five classes outlined in Scheme 5.48. Linear polyunsaturated compounds, in which the metal starts in the middle of the chain, give transannular or "zipper" reactions (Scheme 5.49a); where it starts at the end of a linear chain, a "circular" or "dumbbell" sequence occurs (Scheme 5.49b). Branched chains with a series of exo double bonds give spiro products (Scheme 5.49c); if the branches are homologated, linear annulation occurs (Scheme 5.49d); finally,

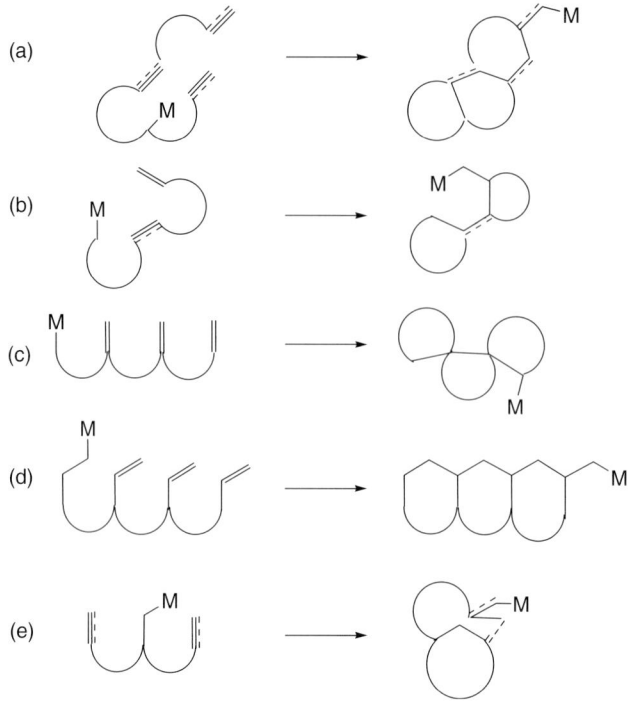

Scheme 5.49 Modes for polycyclization.

bridged products can be accessed by attaching a number of alkenes or alkynes to an initiating central position (Scheme 5.49e).

In addition to the above patterns of "living" processes, multiple cyclizations may be devised in which each cyclization is independent of the others. Examples of this include the intramolecular double-Heck reaction of geminal dibromoalkenes (Scheme 5.50a) [93], and the triple-Heck reaction of 1,3,5-tribromobenzene derivatives (Scheme 5.50b) [94].

Scheme 5.50 Double and triple Heck reactions.

5.6.1
Stoichiometric Polycyclization Methods

Relatively few stoichiometric methods exist for polycyclization, partly because the required intermediates are highly reactive and undergo numerous reaction pathways besides the desired one. However, some examples of tandem carbolithiations have been documented. A number of diene carbolithiation cascades have been reported, including the synthesis of the norbornane skeleton **145** (Scheme 5.51) [95] and tandem alkyne–alkene and diyne carbolithiation, for example, **146** → **147** (Scheme 5.52) [96].

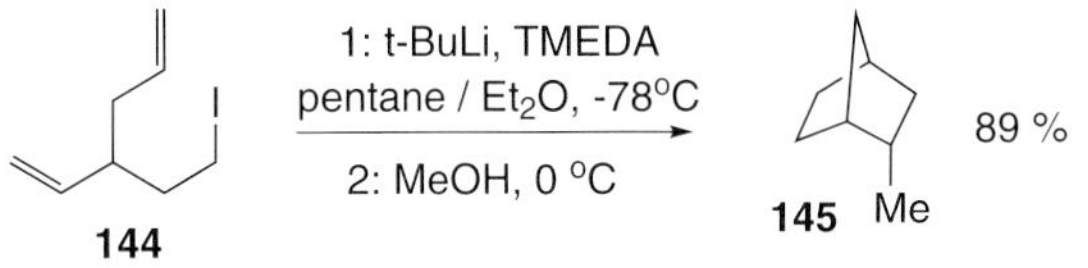

Scheme 5.51 Sequential carbolithiations of diene.

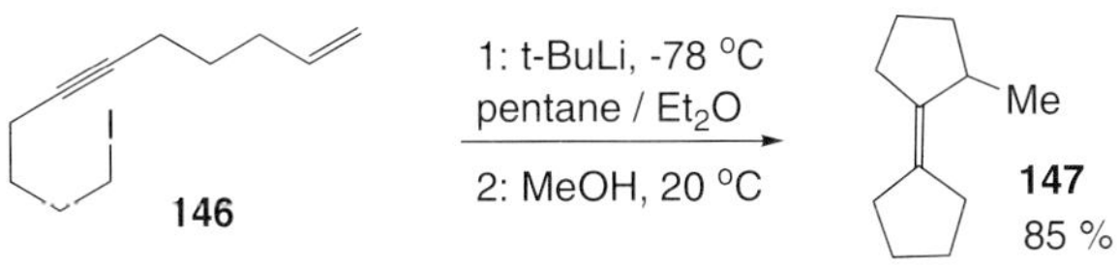

Scheme 5.52 Sequential carbolithiations of enyne.

Catalytic polycyclizations are generally more viable than stoichiometric ones, and have accordingly received much greater attention; again, palladium is the catalyst of choice due to its predictable and malleable behavior.

5.6.2
Fused- and Bridged-Ring Forming Catalytic Cyclizations

The first examples of living palladium reactions were the intramolecular spiro, fused and bridged double-Heck reactions described by Overman et al., applied to the synthesis of the scopadulcic acid skeleton **152** (Scheme 5.53) [97]. An asymmetric

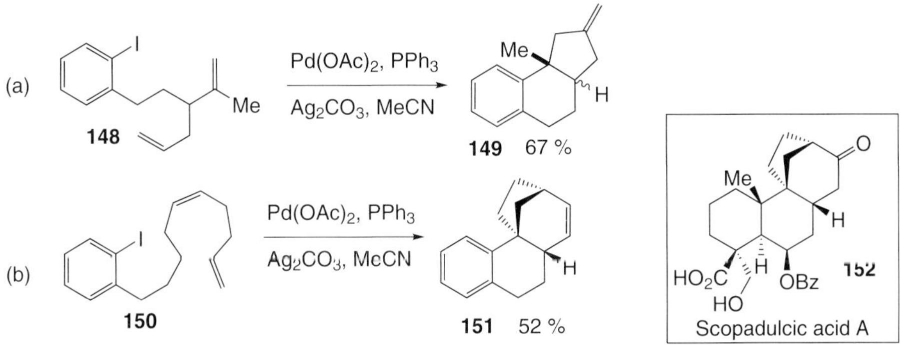

Scheme 5.53 Double carbopalladations.

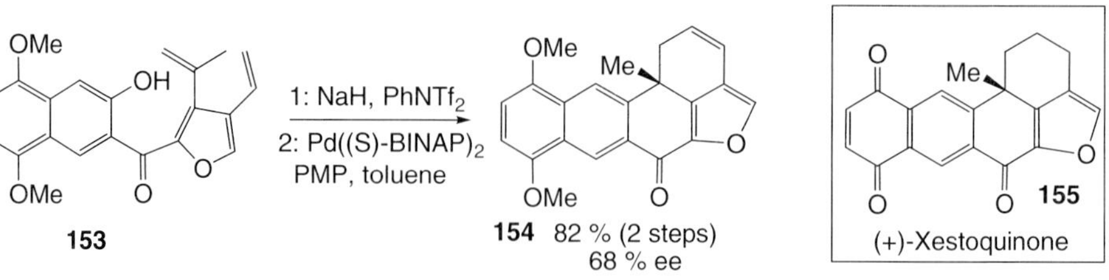

Scheme 5.54 Enantioselective sequential double carbopallations.

version of the fused 6-*exo*, 6-*endo* double-Heck reaction was developed with a view to total synthesis, and was applied to xestoquinone **155** and halenaquinone (Scheme 5.54) [22b, 98].

Variations on fused-mode cyclizations include termination by hydride or phenyl anion transfer, for example, **156** → **157** (Scheme 5.55a) [99] and by carbopalladation onto indoles, pyrroles and pyridones, for example, **158** → **159** (Scheme 5.55b); the latter case requires a formal anti-hydride elimination but may proceed via an oxyallylpalladium (II) intermediate [100].

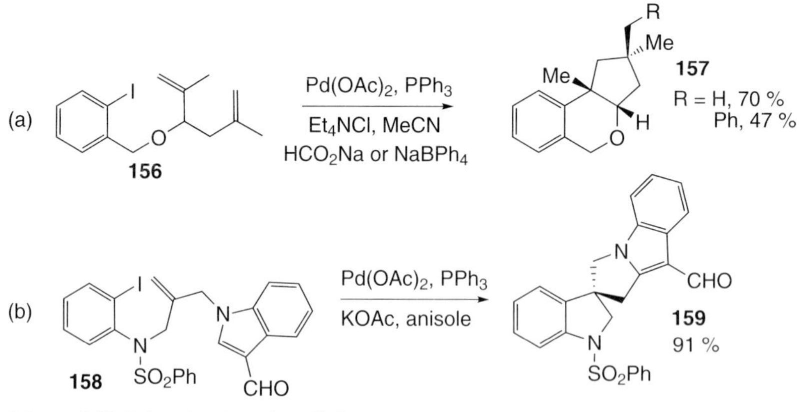

Scheme 5.55 Polycycles via carbopalladations.

A synthesis of propellanes **161** using fused-mode polycyclization methodology has been reported (Scheme 5.56) [101] and recent applications of bridged polycyclizations include the synthesis of bridged benzazepines **163** (Scheme 5.57) [102].

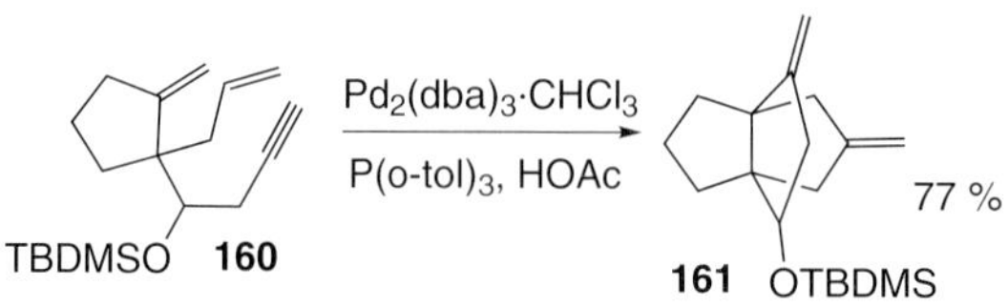

Scheme 5.56 Hydropalladation initiated cycloisomerization of ene-ene-yne.

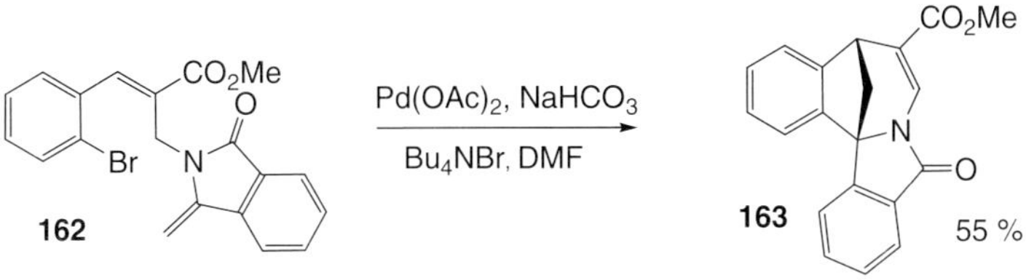

Scheme 5.57 Pd-catalyzed polycyclizations.

5.6.3
Spiro-Mode Polycyclizations

Palladium catalyzed spiro-polycyclizations were first described alongside fused and bridged processes (Scheme 5.58). It was found that the stereochemistry of the spiro center of **165** was determined by the presence of groups on the linker chain [97a]. Soon after, reports of similar reactions from benzyl chlorides **166** and enol triflates **168** were published (Scheme 5.59) [103].

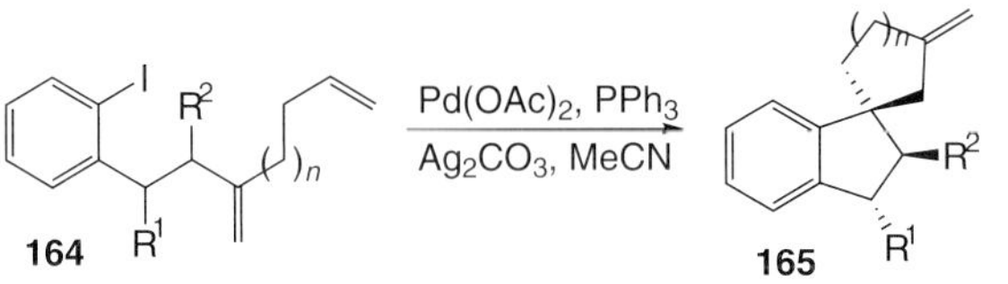

n	R^1	R^2	Yield %
1	H	H	86
2	H	H	85
1	Me	H	81
1	H	Me	73

Scheme 5.58 Double cyclic carbopalladations for spiro compounds.

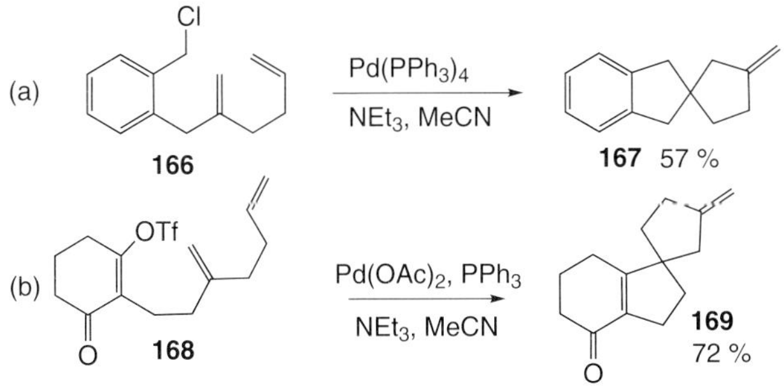

Scheme 5.59 Pd-catalyzed spiro-mode cyclization.

The spirocyclization can be iterated many times in sequence; cascades of seven spirocyclizations have been achieved from polyenynes using $Pd_2(dba)_3$/HOAc and terminating in β-hydride elimination to give the heptaspirocycle **171**. This process is 100% atom efficient, and proceeds in good yield (Scheme 5.60) [104].

As with previous palladium-catalyzed processes, other terminating steps are available. Hydride and phenyl anion capture, from HCO_2Na and $NaBPh_4$ respectively, have been described in diastereoselective bicyclizations, for example, **172** → **173** (Scheme 5.61) [99].

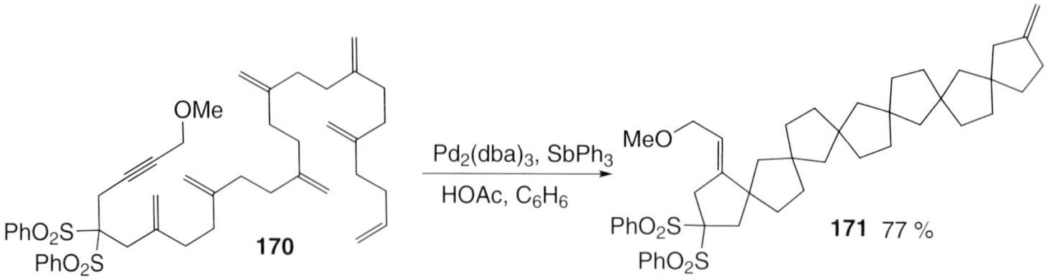

Scheme 5.60 Hydropalladation initiated spiro-mode polycyclization.

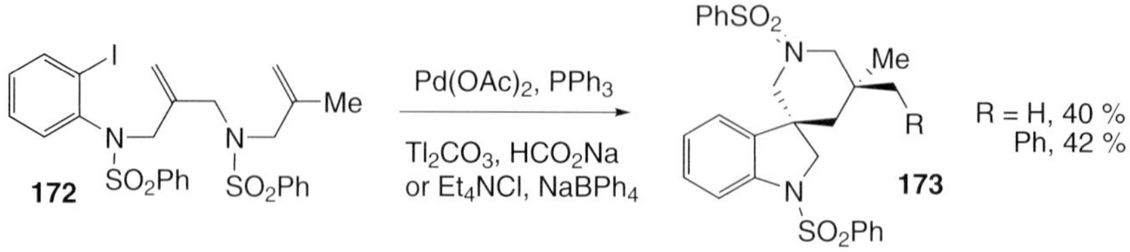

Scheme 5.61 Double cyclization-nucleophilic trapping.

More recently, spiro-mode bicyclization has been used as a key step in the synthesis of gloiosiphone A **176** (Scheme 5.62) [105]. Intermolecular carbometallation, carbonylation and direct nucleophilic substitution have not yet been applied as termination steps in spirocyclizations. It can be speculated that these should also be compatible, as intermolecular steps can usually be delayed until the end of the living cascade by judicious substrate design and selection of reaction conditions.

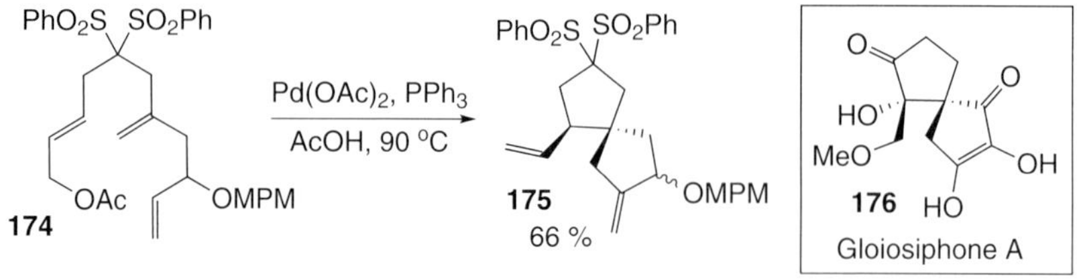

Scheme 5.62 Cyclic metallo-ene-carbopalladation.

5.6.4
Transannular or "Zipper" Mode Polycyclizations

Alkynes are particularly suitable for transannular polycyclization reactions, as β-hydride elimination is suppressed and each carbopalladation directs the catalytic center towards the next substrate. The first example, a bicyclization terminating in β-hydride elimination to give **179**, illustrates this (Scheme 5.63) [77].

The scope of this type of cyclization was quickly extended to tri-, tetra- and pentacyclizations, including the synthesis of a steroid-like skeleton **181** (Scheme 5.64) [106].

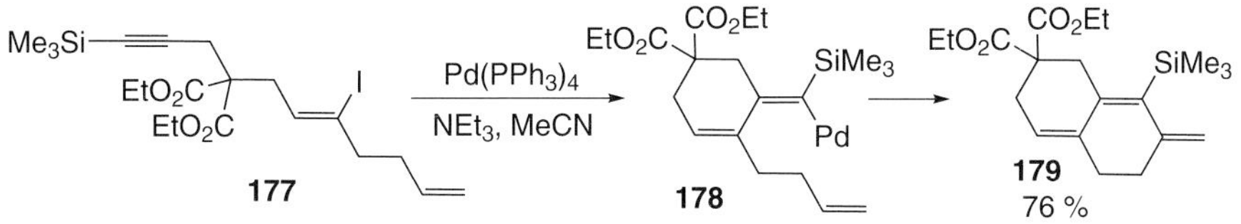

Scheme 5.63 Bicyclic compound via cyclic carbopalladation of alkyne and alkene.

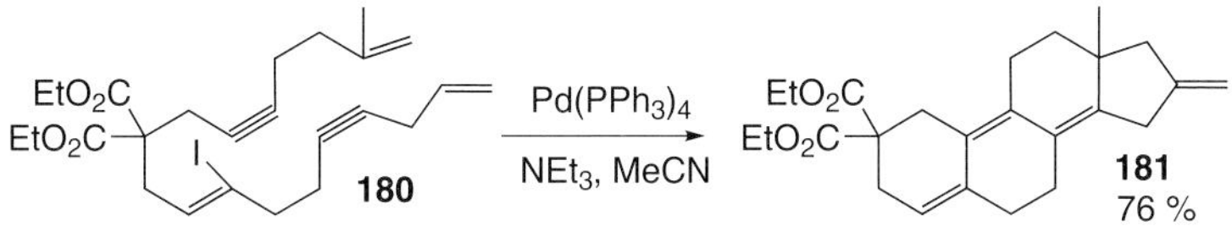

Scheme 5.64 Zipper reaction for steroid-like skeleton.

Carbonylation coupled with nucleophilic substitution with alcohols and amines, both intermolecular and intramolecular, has been studied in depth as a termination step in transannular polycyclizations (Scheme 5.65) [78].

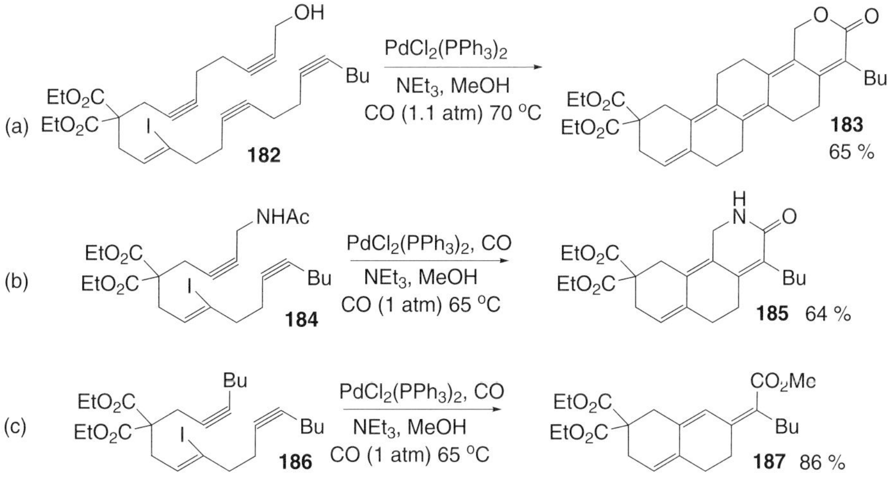

Scheme 5.65 Polycyclization terminated by carbonylation.

Intermolecular carbopalladation of allenes, followed by nucleophilic substitution, has been combined with transannular tricyclizations, using amines and sulfinic acid salts as nucleophiles, for example, **188 → 190** and **191 → 193** (Scheme 5.66) [107].

5.6.5
Circular or "Dumbbell" Polycyclizations

5.6.5.1 2 + 2 + 2 Cyclization Reactions
Intramolecular versions of the Pd(II)-catalyzed cyclotrimerization of alkynes (Scheme 5.67) [108] provide an interesting route to fused arenes, and offer a solution

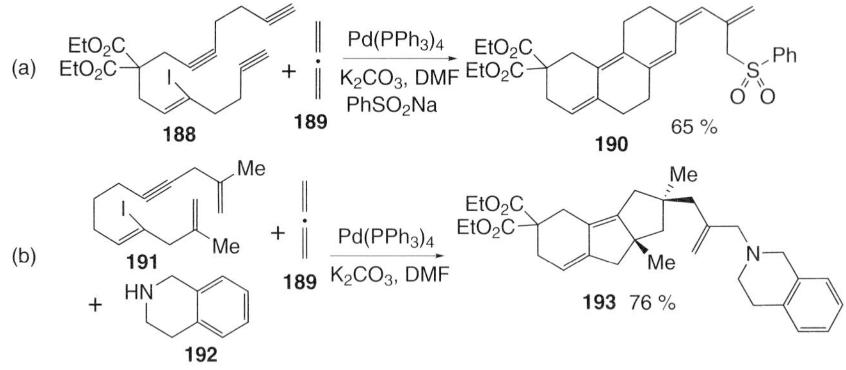

Scheme 5.66 Polycyclization terminated by intermolecular allene carbopalladation.

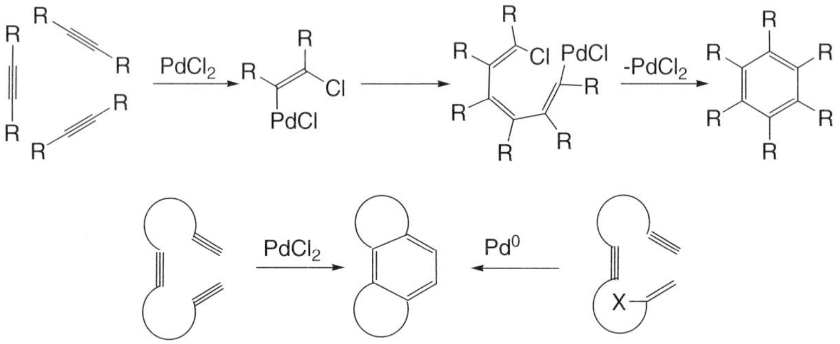

Scheme 5.67 Pd-catalyzed [2 + 2 + 2] cyclization.

to the absence of regioselectivity which is inherent in intermolecular reactions. In such reactions, the palladium species completes one circuit of the new ring, carbopalladating each alkene in turn, before the reaction is complete; hence, such reactions can be described as "circular" cascade reactions. Such reactions have been studied in great depth using a variety of metals including rhodium, iridium, ruthenium, nickel and cobalt, the majority of which proceed via intermediate metallocycles [109]. Only a few representative examples using palladium catalysis will be discussed here.

Syntheses of benzene derivatives by intramolecular trimerization may start either from haloenediynes such as **196** or triynes, for example, **198**. The latter requires a Pd (0) catalyst, although this can be generated *in situ* from Pd(II) salts (Scheme 5.68)

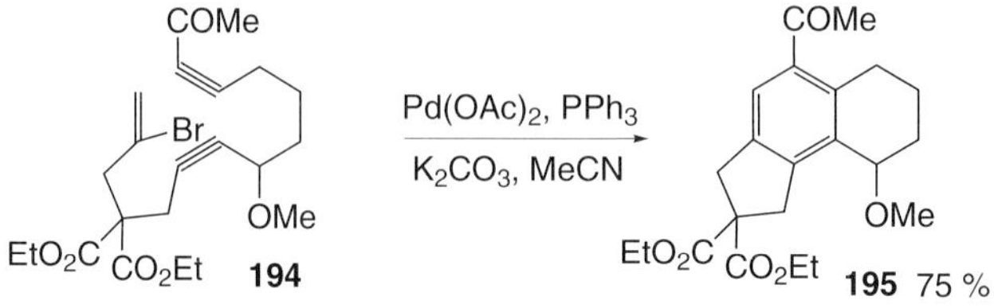

Scheme 5.68 Pd-catalyzed cyclization of vinylic bromide-diyne.

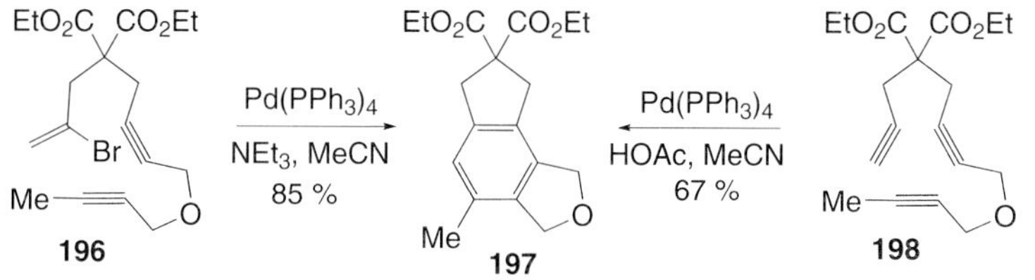

Scheme 5.69 Pd-catalyzed [2 + 2 + 2] cyclization of 2-bromo-6,11-diynene.

[110]. A direct comparison of the two approaches – using Pd(PPh$_3$)$_4$ and HOAc as a source of HPd(II)OAc for the triyne method – indicated that the halide approach gave higher yields (Scheme 5.69) [111]. It suffers only from its lower atom efficiency, as HBr is lost at the end of the catalytic cycle.

Where a propargylic carbonate forms part of a triyne system, an allenylpalladium species is the first intermediate, and cyclization occurs via a 6-exo-trig carbopalladation to give a benzylpalladium species **201**. This can conceptually be trapped with any terminating step, but only β-hydride elimination to give **202** and cross-coupling to **203** have so far been reported (Scheme 5.70) [107].

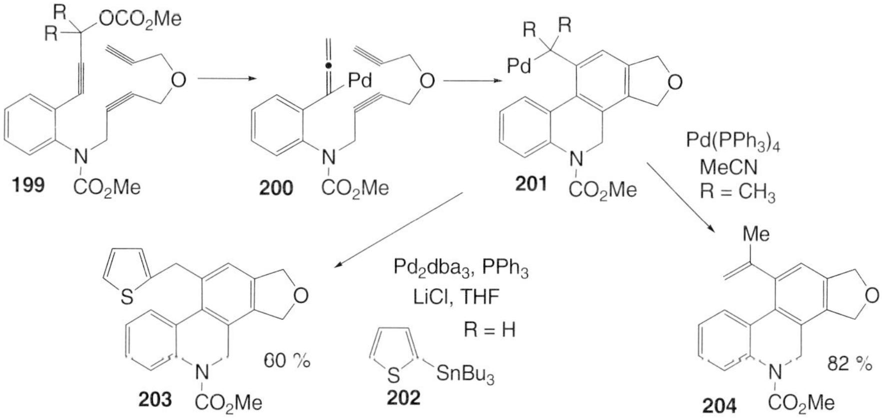

Scheme 5.70 Pd-catalyzed cyclization of progargylic carbonate-diyne.

Aryl diynes **205** have also been reported to undergo 2 + 2 + 2 cyclization, possibly proceeding through an oxidative C–H insertion generating a Pd^{IV} intermediate **206** (Scheme 5.71); an alternative mechanism proceeding through carbopalladation of the arene cannot be ruled out [79b].

Replacement of an alkyne moiety by an alkene yields cyclohexadienes such as **210**, although a circular palladation is not thought to be involved. Rather, a hexatriene **209** is formed, which undergoes 6π-electrocyclization (Scheme 5.72) [112].

This methodology was applied to the synthesis of steroid-like skeletons **212** and **214** using cyclohexene or benzene derivatives as starting materials (Scheme 5.73) [113]. If an *exo* β-hydride can be eliminated, 6π-electrocyclization is forestalled and

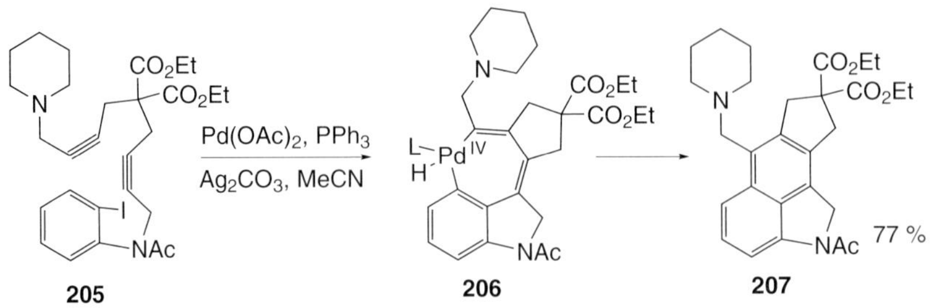

Scheme 5.71 Pd-catalyzed synthesis of tetracycle from aryl iodide-1,1-diyne.

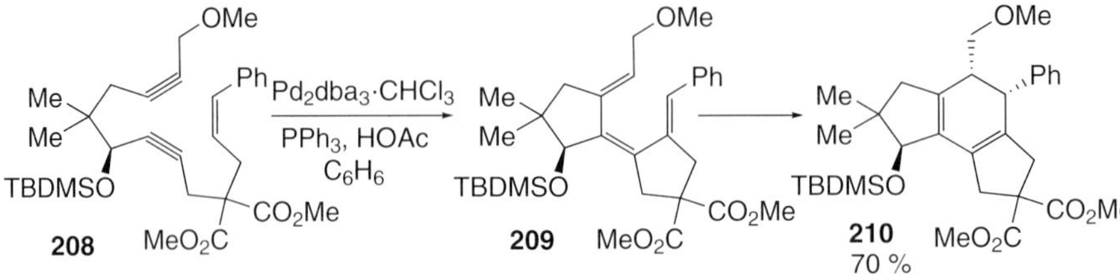

Scheme 5.72 6,11-Diynene cyclization–6π-electrocyclization.

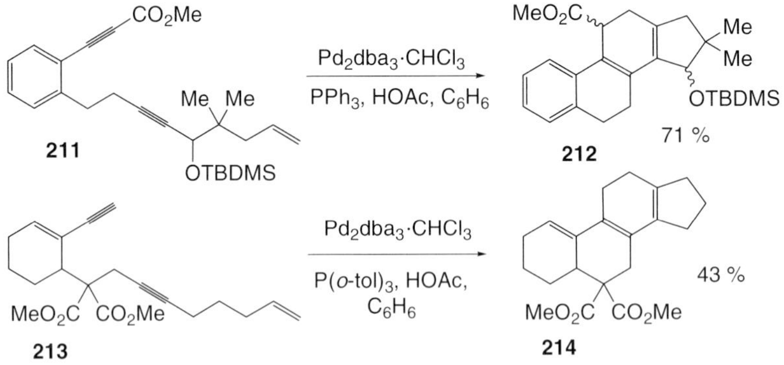

Scheme 5.73 Synthesis of steroid-like skeletons via Pd-catalyzed cyclization of 6,12-diynenes.

replaced by a Diels–Alder reaction to give a highly crowded polycyclic system **217** (Scheme 5.74) [114].

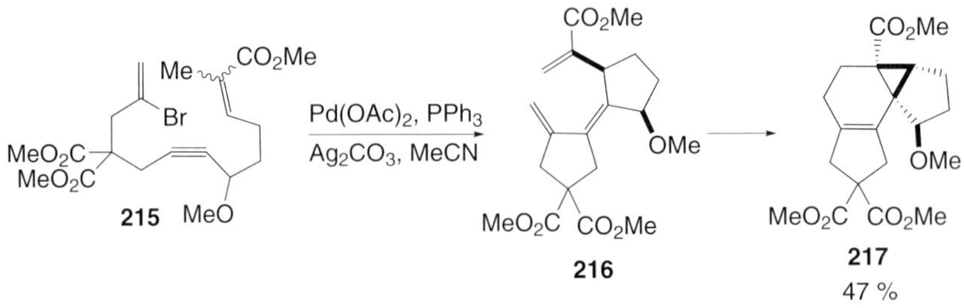

Scheme 5.74 Combination of cyclization of 2-bromo-6,11-diynene and Diels–Alder reaction for polycycle.

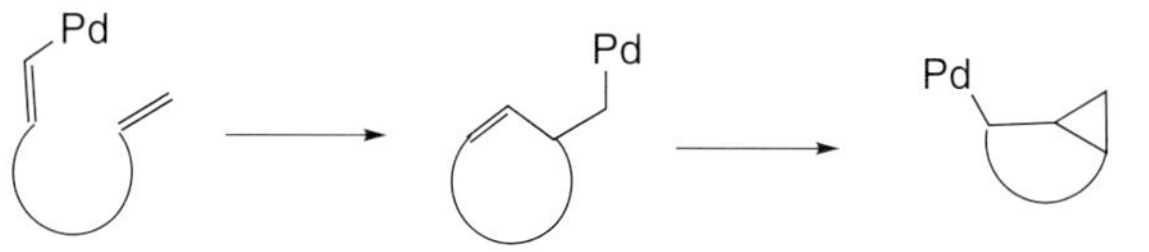

Scheme 5.75 Pd-catalyzed cyclopropane formation.

5.6.5.2 **Cyclopropanations**

In the absence of any alternative further step, homoallylic organopalladium species undergo 3-exo-trig cyclization to give cyclopropyl derivatives (Scheme 5.75). The first documented example of this was a regioselective 3-exo-trig cyclization which was favored over the potentially competitive 5-exo-trig process, giving cyclopropane **223** (Scheme 5.76) [77]. The regioselectivity was explained by the more stable π-allyl palladium intermediate afforded by cyclopropanation, compared to the homoallylic product of cyclopentanation. The latter is also disadvantaged by the strain inherent in the double bond at the ring junctions. Cyclopropanation can also be delayed until a circular cascade has been completed, for example, **224** → **226** (Scheme 5.77) [115]. The De Meijere group has been particularly active in cyclopropanation cascades, and readers are directed to an account of their work in this field [116].

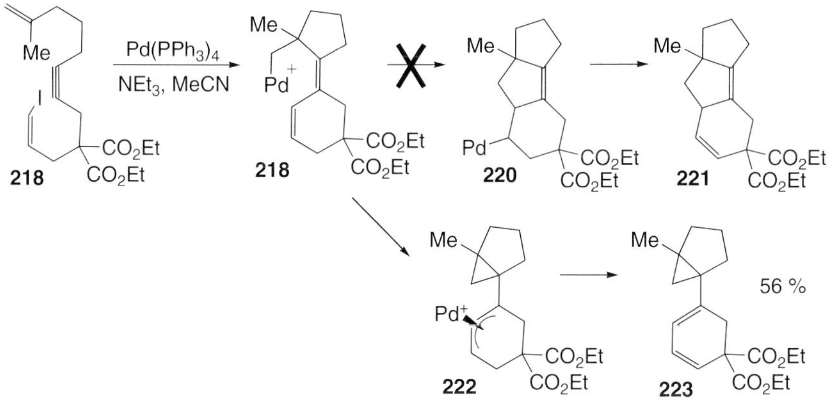

Scheme 5.76 Cyclic carbopalladation of 1,3-diene for the formation of three-membered ring.

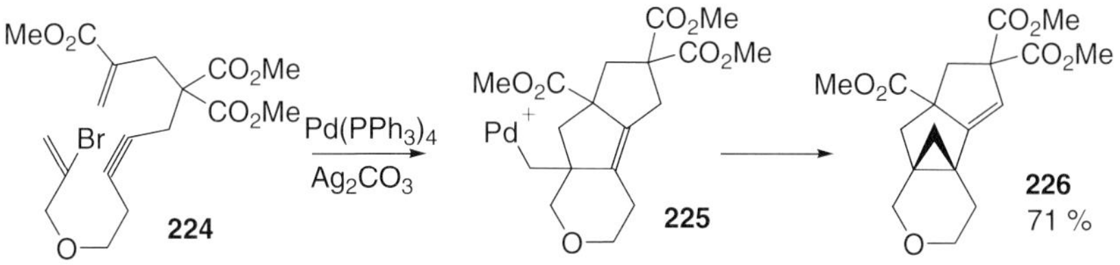

Scheme 5.77 Pd-catalyzed cyclization of 2-bromo-1,12-dien-7-yne.

Vinyl palladium species react with norbornenes to give cyclopropyl-fused norbornane derivatives **230** in a stereoselective manner (Scheme 5.78) [117].

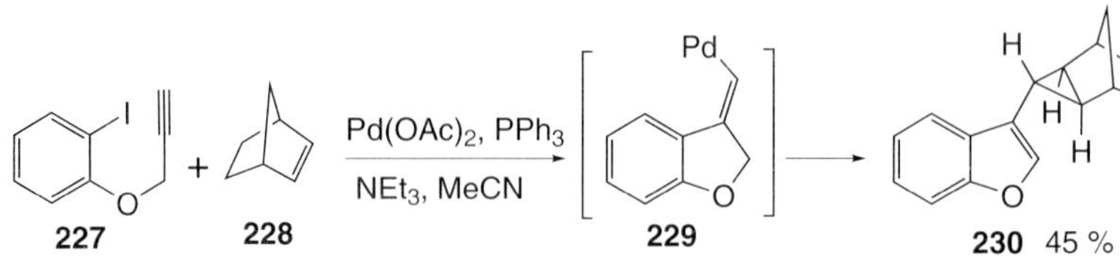

256 | *5 Cyclic Carbometallation of Alkenes, Arenes, Alkynes and Allenes*

Scheme 5.78 Intermolecular carbopalladation for cyclopropane.

Allenyl palladium species readily undergo cyclopropanation reactions with olefins, with a highly diastereoselective 3-exo-trig cyclization giving a vinyl palladium species **233** which can be trapped in various ways (Scheme 5.79) [118]. Similarly, two examples of cyclobutanations, both terminating in β-hydride elimination, have been reported (Scheme 5.80) [103b, 119].

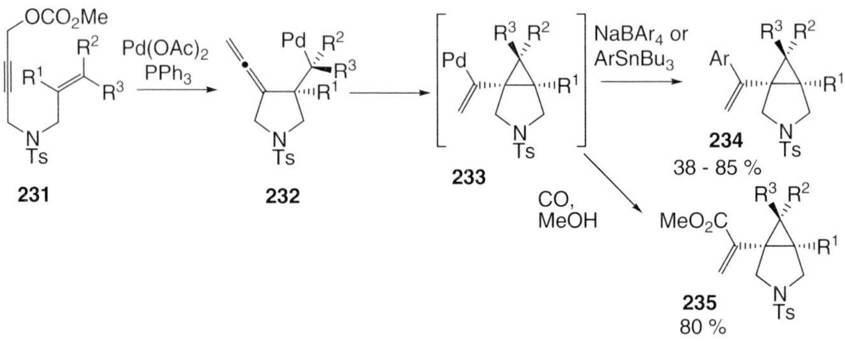

Scheme 5.79 Cyclopropane formation via cyclic carbopalladation of allene.

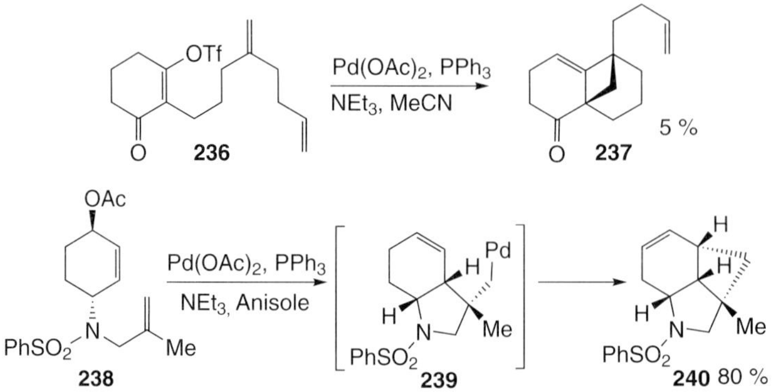

Scheme 5.80 Pd-catalyzed formation of cyclobutanes.

Rhodium catalysis has also been reported to effect cyclopropanation by a different mechanism to the above examples, for example, **241** → **243** and **244** → **247** (Scheme 5.81) [120].

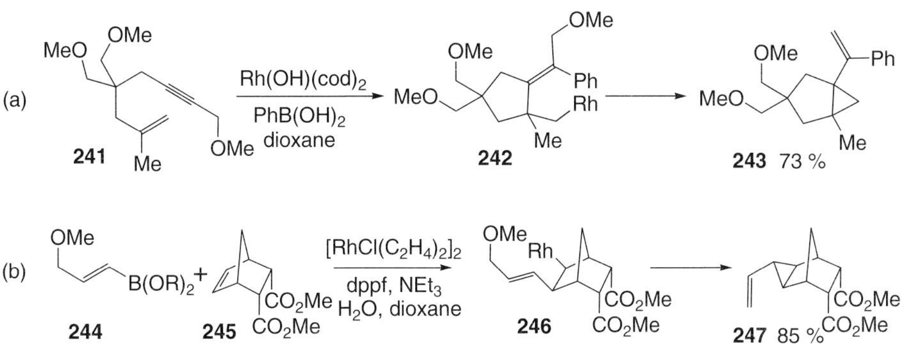

Scheme 5.81 Rh-catalyzed formation of cyclopropanes.

5.7
Acylpalladations

Carbonylation has, in the preceding sections, been described as a terminating step, with the highly electrophilic acylpalladium species reacting with a C, N, O or H-centered nucleophile to yield a carbonyl compound and regenerate the catalytic species. Carbonylation of organopalladium compounds is a reversible process (Scheme 5.82), which could compete at each step with the desired pathway. Usually, intramolecular carbopalladations are rapid enough that, with 1 atm pressure of CO and 5–7-membered ring formation, carbonylation does not compete effectively. In such cases, successive carbopalladations occur until the end of the cascade, when CO is inserted, followed by nucleophilic substitution. It is not unusual to see side products as a result of carbonylation competing too early in the reaction sequence (Scheme 5.83), but in most cases this can be minimized by optimizing the reaction conditions. The relative rate of steps 1 and 2 in Scheme 5.71 is crucial; step 1 is controlled principally by the ring size, and step 2 by CO pressure, temperature and solvent. Carbopalladation to give 5- and 6-membered rings is usually extremely fast, and carbonylation does not interfere, but 3-exo and 4-exo cyclizations are much slower and CO will usually insert beforehand; the resulting increase in ring size usually makes cyclization much more viable, and step 4 is able to compete with step 3. By using conditions that favor steps 2 and 4, an interesting synthesis of cyclic ketones through acylpalladation becomes viable.

Cyclopentenones are excellent targets for acylpalladation, as the cyclization is greatly accelerated by carbonylation. Early reports utilized stoichiometric Pd(PPh$_3$)$_4$ [121] but catalytic methods were soon established. A number of examples are shown

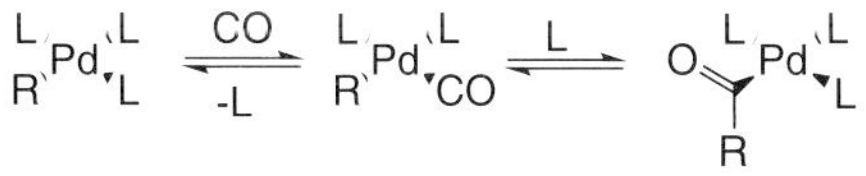

Scheme 5.82 Carbonylation of C–Pd Bond.

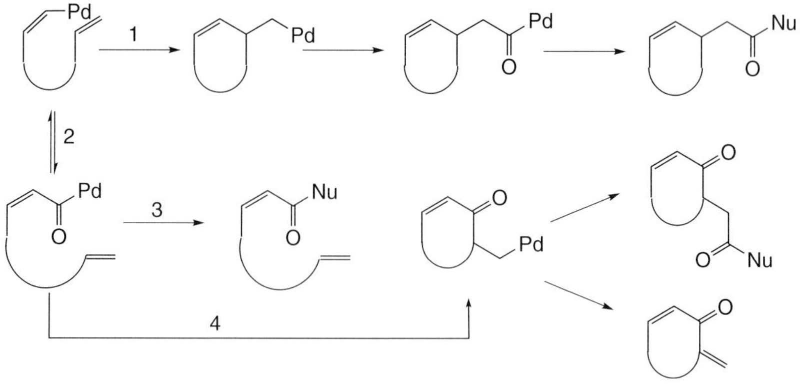

Scheme 5.83 Different mode of acylpalladations.

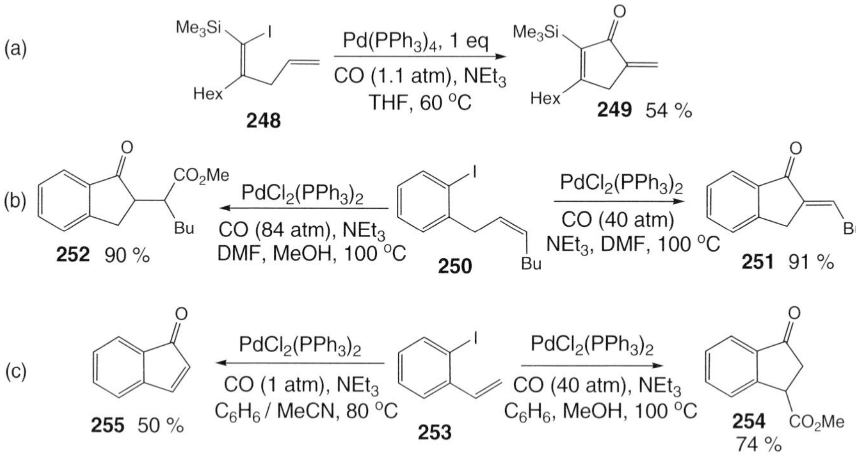

Scheme 5.84 Cyclic acylpalladation for cyclic ketones.

in Scheme 5.84, indicating the scope of mono-acylpalladation [54c, 122]. It is worth noting that β-hydride elimination usually gives the conjugated enone, either directly by regioselective elimination, or by elimination "away" from the carbonyl, followed by double bond migration under the reaction conditions. Although *exo* cyclization is still favored, formal endo cyclization occurs in some cases such as the synthesis of indenones **255** from 2-iodostyrene [122]. A synthesis of the martinellene alkaloid skeleton **258** has recently been reported, using cyclic acylpalladation as a key step (Scheme 5.85) [123].

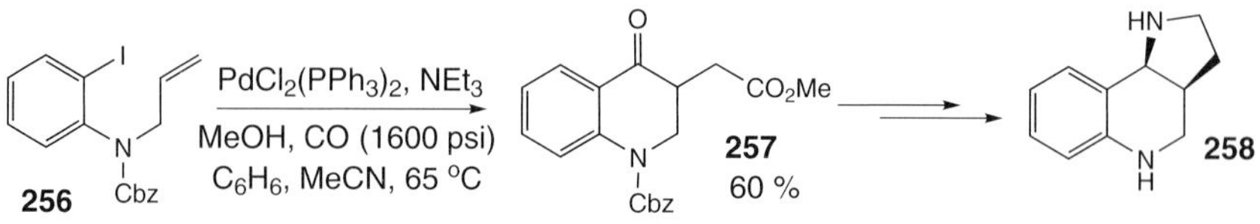

Scheme 5.85 Cyclic acylpalladation–esterification.

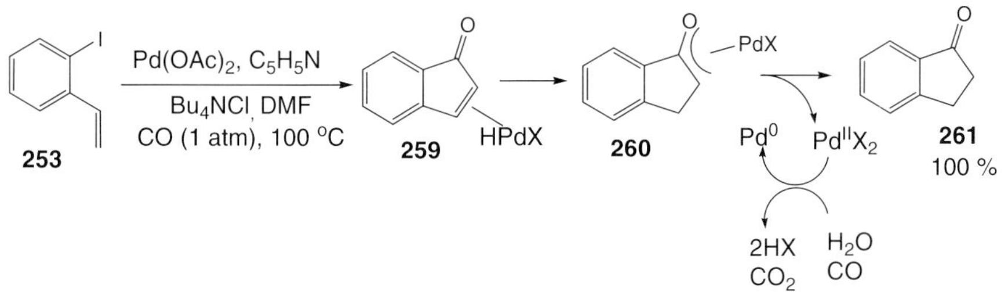

Scheme 5.86 Indanones from Pd-catalyzed cyclization of 2-vinylphenyl iodide in the presence of CO.

A reductive synthesis of indanones **261** and other cyclopentenones has been reported which is thought to proceed through elimination of Pd(II) followed by reduction back to Pd(0) with CO as reductant (Scheme 5.86) [124].

Where carbonylation gives a ketone with α-protons, the enol form can act as a nucleophile, yielding enol lactones. The use of benzyl halides **262** as starter species exemplifies this sequence (Scheme 5.87) [125].

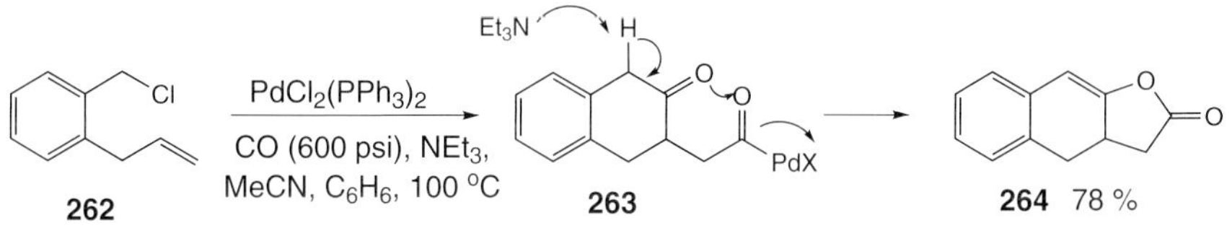

Scheme 5.87 Tricyclic lactone via double carbonylation.

Alternative terminations reported in acylpalladation reactions include cross-coupling with stannanes, for example, **265** → **267** (Scheme 5.88a) [126]. and allene acylpalladation-nucleophilic quench, for example, **268** → **270** (Scheme 5.88b) [127].

Where allylic species are used as starting materials, a metallo-ene reaction is followed by carbonylation and acylpalladation of the allylic double bond

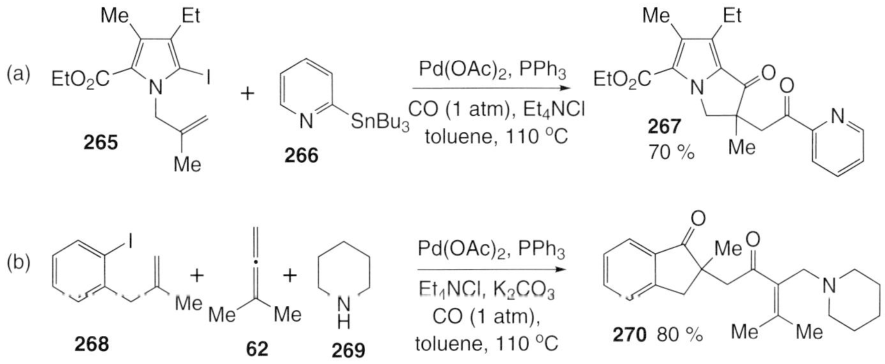

Scheme 5.88 Pd-catalyzed double carbonylation involving allene and amine.

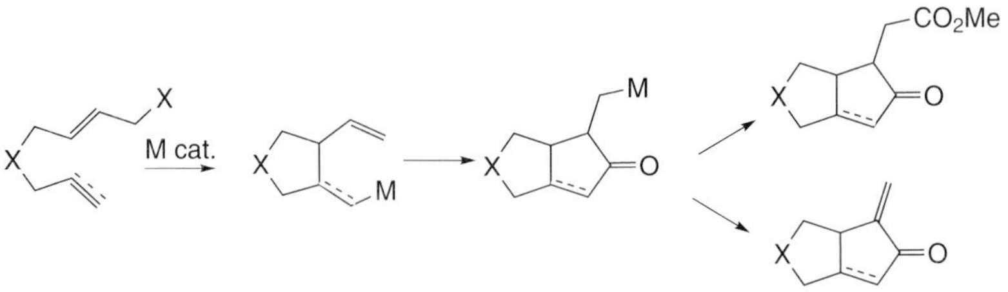

Scheme 5.89 Metallo-ene followed by carbonylation.

(Scheme 5.89). This methodology works with pendant alkenyl or alkynyl groups, and also functions with a nickel catalyst [128]. A number of total syntheses, including hirsutene **273** [129], pentalenolactone A methyl ester **276** [130], isorauniticine **279** [131] and fenestranes **281** [132] have included such reactions as key steps (Scheme 5.90).

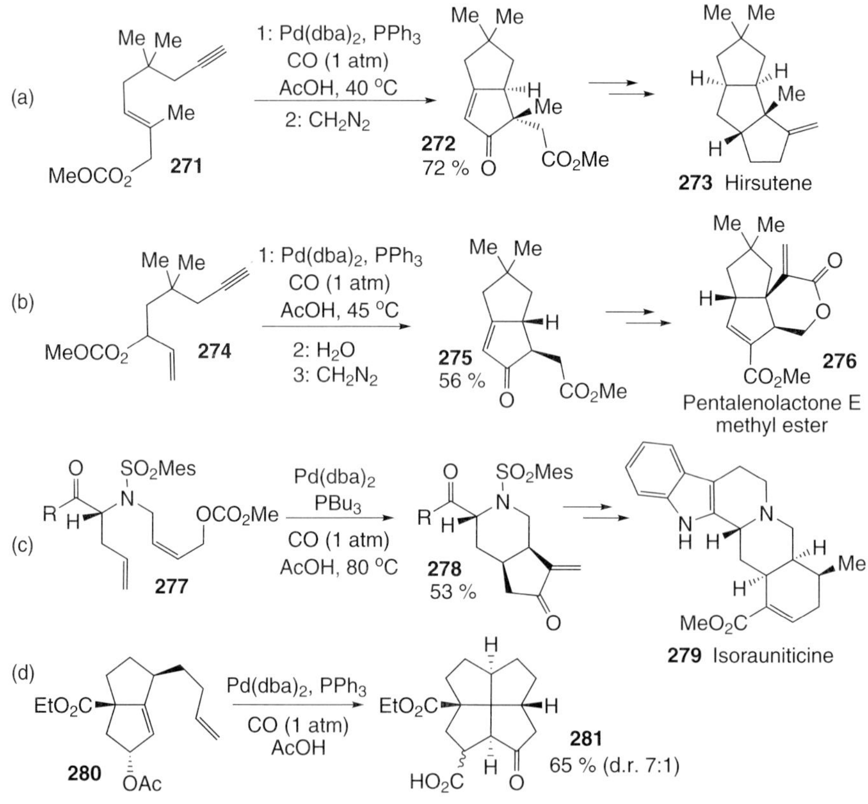

Scheme 5.90 Pd-catalyzed synthesis of cyclic ketones via acylpalladation.

Multiple cyclizations involving acylpalladation can be more complex than the corresponding carbopalladation processes, due to the greater reactivity of acylpalla-

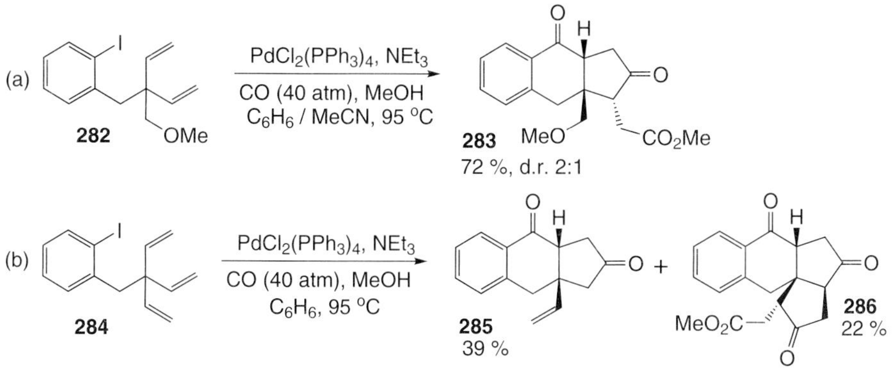

Scheme 5.91 Double or triple carbonylation for multiple cyclic ketones.

dium species. However, a number of examples have been reported. Fused bi-and tricyclizations have been performed on polyenes, although the tricyclization gave only a 22% yield of **286** (Scheme 5.91) [133]. The complex bridged tricyclic system **291** was also synthesized from an allylic acetate with pendant allenyl and alkenyl functionalities, as shown in Scheme 5.92. Again, the yield is modest, although the increase in molecular complexity is significant and includes a bridged five-membered ring and four new contiguous stereocentres [134].

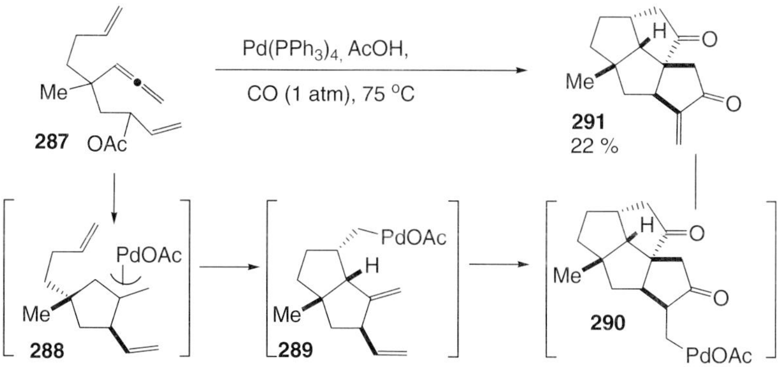

Scheme 5.92 Pd-Catalyzed double carbonylation of allyl acetate–allene-ene.

5.8 Conclusion

Cyclic carbometallation, employing both main group and transition metals, has proved an exceptionally fertile general protocol for generating new methods for the synthesis of complex ring systems. Alkenes, arenes, alkynes and allenes provide highly reactive carbometallation "triggers" allowing access to highly diverse and complex ring systems. Palladium catalysis has provided a particularly effective and versatile approach to ring construction with excellent region- and stereoselectivity. Growing areas such as rhodium catalysis and stoichiometric carbolithiation look set to provide new complementary methodologies in the future.

5.9
Experimental: Selected Procedures

5.9.1
Synthesis of 11 [8]

6-Chlorohex-1-ene 8 (0.138 mL, 1.0 mmol) was added at $-30\,°C$ under an argon atmosphere to a stirred green suspension of lithium powder (40 mg, 5.8 mmol) and DTBB (13.3 mg, 0.05 mmol) in THF (4 mL). The color disappeared after the substrate addition was complete, the reaction mixture was stirred until the green color was regenerated (40 min), and the excess lithium was filtered off under inert conditions. The resulting solution was added at room temperature to a solution of zinc bromide (260 mg, 1.1 mmol) in THF (5 mL), and the color of the mixture changed to brown. A solution of $Pd(OAc)_2(PtBu_3)_2$ [prepared by dissolving palladium(II) acetate (11.3 mg, 0.05 mmol) and tri-*tert*-butylphosphane (22 mg, 0.1 mmol) in THF (7 mL)] was added to the resulting mixture, followed by 3-bromothiophene 10 (1.1 mmol). The mixture was stirred at reflux for 6 h, hydrolyzed with water (10 mL), acidified (2 M HCl, 10 mL) and extracted with ethyl acetate (3×20 mL). The organic layer was washed with saturated NH_4Cl solution (2×15 mL), dried ($MgSO_4$), filtered and the filtrate evaporated (15 Torr). The residue was purified by column chromatography (silica gel, hexane/ethyl acetate) to give 11 (66%).

5.9.2
Synthesis of 41 [31]

A round-bottomed flask with reflux condenser was charged with *N,N'*-diallyl-*N,N'*-ditosylbut-2-yne-1,4-diamine 39 (0.473 g, 1.00 mmol), $Pd(dba)_2$ (0.026 g, 0.045 mmol), PPh_3 (0.026 g, 0.099 mmol), and benzene (4 mL) and the mixture degassed and put under nitrogen. Acetic acid (0.012 g, 0.2 mmol) was then added and the mixture was stirred and heated under reflux for 2 h and then concentrated in vacuo to 1 mL. This residue was purified by column chromatography (1.5×13 cm, 15 g of silica gel, CH_2Cl_2, Rf = 0.12), yielding 0.426 g (90%) of 41 as a colorless solid.

5.9.3
Synthesis of 185 [78b]

A mixture of 184 (169 mg, 0.33 mmol), Et_3N (0.21 mL, 153 mg, 1.5 mmol), $Cl_2Pd(PPh_3)_2$ (13 mg, 5 mol%) in MeOH (1.3 mL) was stirred for 24 h at $65\,°C$ under 1 atm of CO. The resultant reaction mixture was concentrated, diluted with Et_2O, washed with H$_2$O, dried (MgSO$_4$), filtered and evaporated. Chromatography on silica gel (20/80 n-pentane-Et$_2$O) afforded 125 mg (64%) of **185**.

5.9.4
Synthesis of 207 [79b]

A mixture of **205** (1 mmol), palladium acetate (0.1 mmol), triphenylphosphine (0.2 mmol), sodium formate (1.5 mmol) and silver carbonate (1.0 mmol) in dry acetonitrile (2 mL) was boiled under reflux for 1 h. After completion of the reaction the mixture was filtered and the filtrate evaporated under reduced pressure. The residue was extracted with chloroform (3 × 25 mL), the chloroform extracts washed with water (50 ml) and the combined organic extracts dried (MgSO$_4$) and concentrated. The residue was purified by flash column chromatography to afford **207** (77%).

Abbreviations

Ac	acetyl
Acac	acetylacetonate
^{t}AmOH	2-methyl-2-butanol
BINAP	2,2'-bis(diphenylphosphino)-1,1'-binaphthalene
Bn	benzyl
BQ	1,4-benzoquinone
brsm	based on recovered starting material
BuLi	butyl lithium
Cb	N,N-diisopropylaminocarbonyl
Cby	2,2,4,4-tetramethyloxazolidine-3-carbonyl
cod	1,4-cyclooctadiene
dba	dibenzylideneacetone
DCE	1,2-dichloroethane
DMF	N,N-dimethylformamide
dppf	1,1'-bis(diphenylphosphino)ferrocene
dppp	1,3-bis(diphenylphosphino)propane
LDBB	lithium 4,4-di-*tert*-butyl-biphenylide
Mes	mesityl (Scheme 5.90c)
NAP	N-acetylphenylalanine
PMP	1,2,2,6,6-pentamethylpiperidine
P(o-tol)$_3$	tri(2-tolyl)phosphine
TBDMS	*tert*-butyldimethylsilyl
Tf	trifluoromethanesulfonyl
TFA	trifluoroacetate
TFP	tri(2-furyl)phosphine
THF	tetrahydrofuran
TMEDA	N,N,N',N'-tetramethylethylenediamine
Tol-BINAP	2,2'-bis(di(2-tolyl)phosphino)-1,1'-binaphthalene

References

1 Mealy, M.J. and Bailey, W.F. (2002) *J. Organomet. Chem.*, **646**, 59–67.

2 Bailey, W.F., Nurmi, T.T., Patricia, J.J. and Wang, W. (1987) *J. Am. Chem. Soc.*, **109**, 2442–2448.

3 Bailey, W.F. and Punzalan, E.R. (1996) *Tetrahedron Lett.*, **37**, 5435–5436.

4 (a) Bailey, W.F. and Mealy, M.J. (2000) *J. Am. Chem. Soc.*, **122**, 6787–6788; (b) Mealy, M.J., Luderer, M.R., Bailey, W.F. and Sommer, M.B. (2004) *J. Org. Chem.*, **69**, 6042–6049.

5 Meyer, C., Marek, I., Courtemanche, G. and Normant, J.-F. (1994) *Tetrahedron*, **50**, 11665–11692.

6 Karoyan, P., Quancard, J., Vaissermann, J. and Chassaing, G. (2003) *J. Org. Chem.*, **68**, 2256–2265.

7 Meyer, C., Marek, I., Courtemanche, G. and Normant, J.-F. (1995) *J. Org. Chem.*, **60**, 863–871.

8 Yus, M. and Ortiz, R. (2004) *Eur. J. Org. Chem.*, 3833–3841.

9 Coldham, I., Hufton, R. and Snowden, D.J. (1996) *J. Am. Chem. Soc.*, **118**, 5322–5323.

10 (a) Christoph, G. and Hoppe, D. (2002) *Org. Lett.*, **4**, 2189–2192; (b) Broka, C.A., Lee, W.J. and Shen, T. (1988) *J. Org. Chem.*, **53**, 1336–1338.

11 Deng, K., Bensari, A. and Cohen, T. (2002) *J. Am. Chem. Soc.*, **124**, 12106–12107.

12 Deng, K., Bensari-Bouguerra, A., Whetstone, J. and Cohen, T. (2006) *J. Org. Chem.*, **71**, 2360–2372.

13 Krief, A. and Bousbaa, J. (1996) *Synlett*, 1007–1009.

14 (a) Rychnovsky, S.D., Hata, T., Kim, A.I. and Buckmelter, A.J. (2001) *Org. Lett.*, **3**, 807–810; (b) Rychnovsky, S.D. and Takaoka, L.R. (2003) *Angew. Chem. Int. Ed.*, **42**, 818–820.

15 Chamberlin, A.R., Bloom, S.H., Cervini, L.A. and Fotsch, C.H. (1988) *J. Am. Chem. Soc.*, **110**, 4788–4796.

16 Mori, M., Chiba, K. and Ban, Y. (1977) *Tetrahedron Lett.*, **18**, 1037–1040.

17 (a) Gibson, S.E. and Middleton, R.J. (1996) *Contemp. Org. Synth.*, **3**, 447–471; (b) Zeni, G. and Larock, R.C. (2006) *Chem. Rev.*, **106**, 4644–4680; (c) Tietze, L.F., Ila, H. and Bell, H.P. (2004) *Chem. Rev.*, **104**, 3453–3516.

18 (a) Dounay, A.B. and Overman, L.E. (2003) *Chem. Rev.*, **103**, 2945–2963; (b) Overman, L.E. (1994) *Pure Appl. Chem.*, **66**, 1423–1430.

19 Buezo, N.D., Mancheño, O.G. and Carretero, J.C. (2000) *Org. Lett.*, **2**, 1451–1454.

20 Madin, A., O'Donnell, C.J., Oh, T., Old, D.W., Overman, L.E. and Sharp, M.J. (2005) *J. Am. Chem. Soc.*, **127**, 18054–18065.

21 Trost, B.M., Tang, W. and Toste, F.D. (2005) *J. Am. Chem. Soc.*, **127**, 14785–14803.

22 (a) Fitzpatrick, M.O., Coyne, A.G. and Buiry, P.J. (2006) *Synlett*, **18**, 3150–3154; (b) Hopkins, J.M., Gorobets, E., Wheatley, B.M.M., Parvez, M. and Keay, B.M. (2006) *Synlett*, **18**, 3120–3124; (c) McDermott, M.C., Stephenson, G.R., Hughes, D.L. and Walkington, A.J. (2006) *Org. Lett.*, **8**, 2917–2920.

23 (a) Imbos, R., Minnaard, A.J. and Feringa, B.L. (2002) *J. Am. Chem. Soc.*, **124**, 184–185; (b) Imbos, R., Minnaard, A.J. and Feringa, B.L. (2003) *Dalton Trans.*, 2017–2023.

24 (a) Baran, P.S., Guerrero, C.A. and Corey, E.J. (2003) *J. Am. Chem. Soc.*, **125**, 5628–5629; (b) Garg, N.K., Caspi, D.D. and Stoltz, B.M. (2006) *Synlett*, **18**, 3081–3087.

25 (a) Knölker, H.-J., Reddy, K.R. and Wagner, A. (1998) *Tetrahedron Lett.*, **39**, 8267–8270; (b) Knölker, H.-J., Fröhner, W. and Reddy, K.R. (2002) *Synthesis*, 557–564.

26 Hagelin, H., Oslob, J.D. and Åkermark, B. (1999) *Chem. Eur. J.*, **5**, 2413–2416.

27 Zhang, H., Ferreira, E.M. and Stoltz, B.M. (2004) *Angew. Chem. Int. Ed.*, **43**, 6144–6148.

28 Ferreira, E.M. and Stoltz, B.M. (2003) *J. Am. Chem. Soc.*, **125**, 9578–9579.
29 Trost, B.M. (1990) *Acc. Chem. Res.*, **23**, 34–42.
30 (a) Wender, P.A., D'Angelo, N., Elitzin, V.I., Ernst, M., Jackson-Ugueto, E.E., Kowalski, J.A., McKendry, S., Rehfeuter, M., Sun, R. and Voigtlaender, D. (2007) *Org. Lett.*, **9**, 1829–1832; (b) Trost, B.M., Corte, J.R. and Gudiksen, M.S. (1999) *Angew. Chem. Int. Ed.*, **38**, 3662–3664.
31 van Boxtel, L.J., Körbe, S., Noltemeyer, M. and de Meijere, A. (2001) *Eur. J. Org. Chem.*, 2283–2292.
32 (a) Kressierer, C.J. and Müller, T.J.J. (2005) *Org. Lett.*, **7**, 2237–2240; (b) Kressierer, C.J. and Müller, T.J.J. (2004) *Angew. Chem. Int. Ed.*, **43**, 5997–6000.
33 Iyer, S. (1995) *J. Organomet. Chem.*, **490**, C27–C28.
34 Iyer, S., Ramesh, C., Sarkar, A. and Wadgaonkar, P.P. (1997) *Tetrahedron Lett.*, **38**, 8113–8116.
35 (a) Fujioka, T., Nakamura, T., Yorimitsu, H. and Oshima, K. (2002) *Org. Lett.*, **4**, 2257–2259; (b) Ikeda, Y., Nakamura, T., Yorimitsu, H. and Oshima, K. (2002) *J. Am. Chem. Soc.*, **124**, 6514–6515.
36 (a) Iyer, S., Ramesh, C. and Ramani, A. (1997) *Tetrahedron Lett.*, **38**, 8533–8536; (b) Boldrini, G.P., Savola, D., Tagliavini, E., Trombini, C. and Ronchi, A.U. (1986) *J. Organomet. Chem.*, **301**, C62–C64; (c) Trost, B.M. and Tour, J.M. (1988) *J. Am. Chem. Soc.*, **110**, 5231–5233; (d) Trost, B.M. and Matsuda, K. (1988) *J. Am. Chem. Soc.*, **110**, 5233–5235.
37 Inamoto, K., Kuroda, J.-I., Danjo, T. and Sakamoto, T. (2005) *Synlett*, **10**, 1624–1626.
38 Burns, B., Grigg, R., Ratananukul, P., Sridharan, V., Stevenson, P., Sukirthalingam, S. and Worakun, T. (1988) *Tetrahedron Lett.*, **29**, 5565–5568.
39 Burns, B., Grigg, R., Sridharan, V., Stevenson, P., Sukirthalingam, S. and Worakun, T. (1989) *Tetrahedron Lett.*, **29**, 1135–1138.
40 (a) Grigg, R. and Sridharan, V. (1999) *J. Organomet. Chem.*, **576**, 65–87; (b) Anwar, U., Fielding, M.R., Grigg, R., Sridharan, V. and Urch, C.J. (2006) *J. Organomet. Chem.*, **691**, 1476–1487.
41 Grigg, R., Sansano, J.M., Santhakumar, V., Sridharan, V., Thangavelanthum, R., Thornton-Pett, M. and Wilson, D. (1997) *Tetrahedron*, **53**, 11803–11826.
42 (a) Lee, C.-W., Oh, K.S., Kim, K.S. and Ahn, K.H. (2000) *Org. Lett.*, **2**, 1213–1216; (b) Grigg, R., Kilner, C., Mariani, E. and Sridharan, V. (2006) *Synlett*, **18**, 3021–3024; (c) Grigg, R., Mariani, E. and Sridharan, V. (2001) *Tetrahedron Lett.*, **42**, 8677–8680; (d) Grigg, R., Sukirthalingam, S. and Sridharan, V. (1997) *Tetrahedron Lett.*, **32**, 2545–2548; (e) Grigg, R. and Sansano, J.M. (1996) *Tetrahedron*, **52**, 13441–13454.
43 (a) Casaschi, A., Grigg, R., Sansano, J.M., Wilson, D. and Redpath, J. (1996) *Tetrahedron Lett.*, **37**, 4413–4416; (b) Casaschi, A., Grigg, R. and Sansano, J.M. (2001) *Tetrahedron*, **57**, 607–615; (c) Casaschi, A., Grigg, R. and Sansano, J.M. (2000) *Tetrahedron*, **56**, 7541–7551; (d) Casaschi, A., Grigg, R. and Sansano, J.M. (2000) *Tetrahedron*, **56**, 7553–7560.
44 (a) Stadtmüller, H., Vaupel, A., Tucker, C.E., Stüdermann, T. and Knochel, P. (1996) *Chem. Eur. J.*, **2**, 1204–1220; (b) Vaupel, A. and Knochel, P. (1995) *Tetrahedron Lett.*, **36**, 231–232.
45 Yamada, H., Aoyagi, S. and Kibayashi, C. (1997) *Tetrahedron Lett.*, **38**, 3027–3030.
46 Burns, B., Grigg, R., Ratananukul, P., Sridharan, V., Stevenson, P. and Worakun, T. (1988) *Tetrahedron Lett.*, **29**, 4329–4332.
47 (a) Grigg, R., Santhakumar, V. and Sridharan, V. (1993) *Tetrahedron Lett.*, **34**, 3163–3164; (b) Pinto, A., Jia, Y., Neuville, L. and Zhu, J. (2007) *Chem. Eur. J.*, **13**, 961–967.
48 (a) Hinojosa, S., Delgado, A. and Llebaria, A. (1999) *Tetrahedron Lett.*, **40**, 1057–1060; (b) Cancho, Y., Martín, J.M., Martínez, M., Llebaria, A., Moretó, J.M. and Delgado, A. (1998) *Tetrahedron*, **54**, 1221–1232; (c)

Solé, D., Cancho, Y., Llebaria, A., Moréto, J.M. and Delgado, A. (1994) *J. Am. Chem. Soc.*, **116**, 12133–12134; (d) Solé, D., Cancho, Y., Llebaria, A., Moréto, J.M. and Delgado, A. (1996) *J. Org. Chem.*, **61**, 5895–5904.

49 Grigg, R., Sridharan, V., Sukirthalingam, S. and Worakum, T. (1989) *Tetrahedron Lett.*, **30**, 1139–1142.

50 Brown, S., Clarkson, S., Grigg, R. and Sridharan, V. (1995) *J. Chem. Soc., Chem. Commun.*, 1135–1136.

51 Grigg, R., Redpath, J., Sridharan, V. and Wilson, D. (1994) *Tetrahedron Lett.*, **35**, 4429–4432.

52 (a) Grigg, R. and Sridharan, V. (1993) *Tetrahedron Lett.*, **34**, 7471–7474; (b) Copéret, C. and Negishi, E. (1999) *Org. Lett.*, **1**, 165–167; (c) Negishi, E., Ma, S., Amanfu, J., Copéret, C., Miller, J.A. and Tour, J.M. (1996) *J. Am. Chem. Soc.*, **118**, 5919–5931.

53 (a) Schoenberg, A., Bartoletti, I. and Heck, R.F. (1974) *J. Org. Chem.*, **39**, 3318–3326; (b) Schoenberg, A. and Heck, R.F. (1974) *J. Org. Chem.*, **39**, 3327–3331.

54 Llebaria, A., Camps, F. and Moretó, J.M. (1992) *Tetrahedron Lett.*, **33**, 3683–3686.

55 (a) Aggarwal, V.K., Butters, M. and Davies, P.W. (2003) *Chem. Commun.*, 1046–1047; (b) Aggarwal, V.K., Davies, P.W. and Moss, W.O. (2002) *Chem. Commun.*, 972–973.

56 (a) Clique, B., Fabritius, C.-H., Couturier, C., Monteiro, N. and Balme, G. (2003) *Chem. Commun.*, 272–273; (b) Aggarwal, V.K., Davies, P.W. and Schmidt, A.T. (2004) *Chem. Commun.*, 1232–1233; (c) Artman, G.D. and Weinreb, S.M. (2003) *Org. Lett.*, **5**, 1523–1526.

57 Liu, C. and Widenhoefer, R.A. (2004) *J. Am. Chem. Soc.*, **126**, 10250–10251.

58 (a) Grigg, R., Köppen, I., Rasparini, M. and Sridharan, V. (2001) *Chem. Commun.*, 964–965; (b) Grigg, R., Sridharan, V. and Terrier, C. (1996) *Tetrahedron Lett.*, **37**, 4221–4224.

59 Grigg, R., Brown, S., Sridharan, V. and Uttley, M.D. (1998) *Tetrahedron Lett.*, **39**, 3247–3250.

60 (a) Oppolzer, W. (1989) *Angew. Chem. Int. Ed. Engl.*, **28**, 38–52;(b) Davies, A.D. (1995) Hydrogen-ene and Metallo-ene Reactions, in *Organic Reactivity. Physical and Biological Aspects* (eds B.T. Golding, R.J. Griffin and H. Maskell), Royal Society of Chemistry, Cambridge, pp. 264–277.

61 Oppolzer, W. and Ruiz-Montes, J. (1993) *Helv. Chim. Acta*, **76**, 1266–1274.

62 (a) Fagnou, K. and Lautens, M. (2003) *Chem. Rev.*, **103**, 169–196; (b) Miura, T. and Murakami, M. (2007) *Chem. Commun.*, 217–224.

63 (a) Shintani, R., Tsurusaki, A., Okamoto, K. and Hayashi, T. (2005) *Angew. Chem. Int. Ed.*, **44**, 3909–3912; (b) Chen, Y. and Lee, C. (2006) *J. Am. Chem. Soc.*, **128**, 15598–15599.

64 Shintani, R., Okamoto, K., Otomaru, Y., Ueyama, K. and Hayashi, T. (2005) *J. Am. Chem. Soc.*, **127**, 54–55.

65 Miura, T., Shimada, M. and Murakami, M. (2005) *J. Am. Chem. Soc.*, **127**, 1094–1095.

66 Aïssa, C. and Fürstner, A. (2007) *J. Am. Chem. Soc.*, **129**, 14836–14837.

67 Oestreich, M., Fröhlich, R. and Hoppe, D. (1999) *J. Org. Chem.*, **64**, 8616–8626.

68 (a) Gralla, G., Wibbeling, B. and Hoppe, D. (2002) *Org. Lett.*, **4**, 2193–2195; (b) Gralla, G., Wibbeling, B. and Hoppe, D. (2003) *Tetrahedron Lett.*, **44**, 8979–8982.

69 Wei, X. and Taylor, R.J.K. (2000) *Angew. Chem. Int. Ed.*, **39**, 409–412.

70 Wang, R.-T., Chou, F.-L. and Luo, F.-T. (1990) *J. Org. Chem.*, **55**, 4846–4849.

71 Negishi, E., Noda, Y., Lamaty, F. and Vawter, E.J. (1990) *Tetrahedron Lett.*, **31**, 4393–4396.

72 (a) Yu, H., Richey, R.N., Carson, M.W. and Coghlan, M.J. (2006) *Org. Lett.*, **8**, 1685–1688; (b) Yanada, R., Obika, S., Inokuma, T., Yanada, K., Yamashita, M., Ohta, S. and Takemoto, Y. (2005) *J. Org. Chem.*, **70**, 6972–6975; (c) Oh, C.H. and Lim, Y.K. (2003) *Tetrahedron Lett.*, **44**, 267–270.

73 (a) Nuss, J.M., Levine, B.H., Rennels, R.A. and Heravi, M.M. (1991) *Tetrahedron Lett.*,

32, 5243–5246; (b) Nuss, J.M., Rennels, R.A. and Levine, B.H. (1993) *J. Am. Chem. Soc.*, **115**, 6991–6992; (c) Torii, S., Okumoto, H., Tadakoro, T., Nishimura, A. and Rashid, M.A. (1993) *Tetrahedron Lett.*, **34**, 2139–2142; (d) Madec, D. and Férézou, J.-P. (2006) *Eur. J. Org. Chem.*, 92–104.
74 Wang, F., Tong, X., Cheng, J. and Zhang, Z. (2004) *Eur. J. Org. Chem.*, **10**, 5338–5344.
75 Ohno, H., Yamamoto, M., Iuchi, M. and Tanaka, T. (2005) *Angew. Chem. Int. Ed.*, **44**, 5103–5106.
76 Salem, B., Klotz, P. and Suffert, J. (2003) *Org. Lett.*, **5**, 845–848.
77 Zhang, Y. and Negishi, E. (1989) *J. Am. Chem. Soc.*, **111**, 3454–3456.
78 (a) Copéret, C., Sugihara, T. and Negishi, E. (1995) *Tetrahedron Lett.*, **36**, 1771–1774; (b) Copéret, C., Ma, S., Sugihara, T. and Negishi, E. (1996) *Tetrahedron*, **52**, 11529–11544; (c) Sugihara, T., Copéret, C., Owczarczyk, Z., Harring, L.S. and Negishi, E. (1994) *J. Am. Chem. Soc.*, **116**, 7923–7924.
79 (a) Burns, B., Grigg, R., Sridharan, V. and Worakun, T. (1988) *Tetrahedron Lett.*, 4325–4328; (b) Grigg, R., Loganathan, V., Sridharan, V., Stevenson, P., Sukirthalingam, S. and Worakun, T. (1996) *Tetrahedron*, **52**, 11479–11502; (c) Trost, B.M. and Lee, D.C. (1988) *J. Am. Chem. Soc.*, **110**, 7255–7258; (d) Finch, H., Pegg, N.A. and Evans, B. (1993) *Tetrahedron Lett.*, **34**, 8353–8356; (e) Oh, C.H. and Park, S.J. (2000) *Tetrahedron Lett.*, **44**, 3785–3787; (f) Donets, P.A. and Van der Eycken, E.V. (2007) *Org. Lett.*, **9**, 3017–3020.
80 Schelper, M. and de Meijere, A. (2005) *Eur. J. Org. Chem.*, 582–592.
81 Grigg, R. and Savic, V. (1996) *Tetrahedron Lett.*, **37**, 6565–6568.
82 (a) Suginome, M., Yamamoto, A. and Murakami, M. (2003) *J. Am. Chem. Soc.*, **125**, 6358–6359; (b) Suginome, M., Yamamoto, A. and Murakami, M. (2005) *J. Organomet. Chem.*, **690**, 5300–5308; (c) Nakao, Y., Oda, S., Yada, A. and Hiyama, T. (2006) *Tetrahedron*, **62**, 7567–7576.
83 (a) Tamao, K., Kobayashi, K. and Ito, Y. (1989) *J. Am. Chem. Soc.*, **111**, 6478–6480; (b) Muraoka, T., Matsuda, I. and Itoh, K. (1988) *Tetrahedron Lett.*, **39**, 7324–7328; (c) Shibata, T., Kadowaki, S. and Takagi, K. (2004) *Organometallics*, **23**, 4116–4120.
84 (a) Zhu, G. and Zhang, Z. (2003) *Org. Lett.*, **5**, 3645–3648; (b) Zhu, G., Tong, X., Cheng, J., Sun, Y., Li, D. and Zhang, Z. (2005) *J. Org. Chem.*, **70**, 1712–1717.
85 Ortiz, F.L., Iglesias, M.J., Fernández, I., Sánchez, C.M.A. and Gómez, G.R. (2007) *Chem. Rev.*, **107**, 1580–1691.
86 Grigg, R., Sridharan, V. and Xu, L.-H. (1995) *J. Chem. Soc., Chem. Commun.*, 1903–1904.
87 Gardiner, M., Grigg, R., Sridharan, V. and Vicker, N. (1998) *Tetrahedron Lett.*, **39**, 435–438.
88 Hiroi, K., Hiratsuka, Y., Watanabe, K., Abe, I., Kato, F. and Hiroi, M. (2001) *Synlett*, **2**, 263–265.
89 Hiroi, K., Hiratsuka, Y., Watanabe, K., Abe, I., Kato, F. and Hiroi, M. (2002) *Tetrahedron: Asymm.*, **13**, 1351–1353.
90 Jeong, I.-Y. and Nagao, Y. (1998) *Tetrahedron Lett.*, **39**, 8677–8680.
91 (a) Ma, S. and Negishi, E. (1994) *J. Org. Chem.*, **59**, 4730–4732; (b) Ma, S. and Negishi, E. (1995) *J. Am. Chem. Soc.*, **117**, 6345–6357.
92 Grigg, R. and Xu, L.-H. (1996) *Tetrahedron Lett.*, **37**, 4251–4254.
93 (a) Ma, S. and Xu, B. (1998) *J. Org. Chem.*, **63**, 9156–9157; (b) Ma, S., Xu, B. and Ni, B. (2000) *J. Org. Chem.*, **65**, 8532–8543.
94 Ma, S. and Ni, B. (2002) *J. Org. Chem.*, **67**, 8280–8283.
95 (a) Bailey, W.F. and Rossi, K. (1988) *J. Am. Chem. Soc.*, **111**, 765–766; (b) Bailey, W.F., Khanolkar, A.D. and Gavaskar, K.V. (1992) *J. Am. Chem. Soc.*, **114**, 8053–8060; (c) Krief, A. and Barbeaux, P. (1991) *Tetrahedron Lett.*, **32**, 417–420.
96 Bailey, W.F. and Ovaska, T.V. (1993) *Chem. Lett.*, 819–820.
97 (a) Abelman, M.M. and Overman, L.E. (1988) *J. Am. Chem. Soc.*, **110**, 2328–2329; (b) Overman, L.E., Abelman, M.M., Kucera, D.J., Tran, V.D. and Ricca, D.J.

(1992) *Pure App. Chem.*, **64**, 1813–1819; (c) Overman, L.E., Ricca, D.J. and Tran, V.D. (1993) *J. Am. Chem. Soc.*, **115**, 2042–2044.

98 (a) Maddaford, S.P., Andersen, N.G., Cristofoli, W.A. and Keay, B.A. (1996) *J. Am. Chem. Soc.*, **118**, 10766–10773; (b) Lau, S.Y.W. and Keay, B.A. (1999) *Synlett*, **5**, 605–607; (c) Lau, S.Y.W., Andersen, N.G. and Keay, B.A. (2001) *Org. Lett.*, **3**, 181–184.

99 (a) Grigg, R., Dorrity, M.J., Malone, J.F., Sridharan, V. and Sukirthalingam, S. (1990) *Tetrahedron Lett.*, **31**, 1343–1346; (b) Burns, B., Grigg, R., Santhakumar, V., Sridharan, V., Stevenson, P. and Worakun, T. (1992) *Tetrahedron*, **48**, 7297–7320; (c) Grigg, R., Loganathan, V., Sukirthalingam, S. and Sridharan, V. (1990) *Tetrahedron Lett.*, **31**, 6573–6576.

100 Grigg, R., Fretwell, P., Meerholtz, C. and Sridharan, V. (1994) *Tetrahedron*, **50**, 359–370.

101 Trost, B.M. and Shi, Y. (1991) *J. Am. Chem. Soc.*, **113**, 701–703.

102 Gowrisankar, S., Lee, H.S., Lee, K.Y., Lee, J.-E. and Kim, J.N. (2007) *Tetrahedron Lett.*, **48**, 8619–8622.

103 (a) Wu, G.-Z., Lamaty, F. and Negishi, E. (1989) *J. Org. Chem.*, **54**, 2507–2508; (b) Carpenter, N.E., Kucera, D.J. and Overman, L.E. (1989) *J. Org. Chem.*, **54**, 5846–5848.

104 Trost, B.M. and Shi, Y. (1993) *J. Am. Chem. Soc.*, **115**, 9421–9438.

105 Doi, T., Iijima, Y., Takasaki, M. and Takahashi, T. (2007) *J. Org. Chem.*, **72**, 3667–3671.

106 Zhang, Y., Wu, G.-Z., Agnel, G. and Negishi, E. (1990) *J. Am. Chem. Soc.*, **112**, 8590–8592.

107 Grigg, R., Rasul, R. and Savic, V. (1997) *Tetrahedron Lett.*, **38**, 1825–1828.

108 Maitlis, P.M. (1976) *Acc. Chem. Res.*, **9**, 93–99.

109 (a) Gandon, V., Aubert, C. and Malacria, M. (2006) *Chem. Commun.*, 2209–2217; (b) Chopade, P.R. and Louie, J. (2006) *Adv. Synth. Catal.*, **348**, 2307–3227; (c) Kotha, S., Brahmachary, E. and Lahiri, K. (2005) *Eur. J. Org. Chem.*, 4741–4767; (d) Grigg, R., Kilner, C., Senthilnanthanan, M., Seabourne, C.R., Sridharan, V. and Murrer, B.A. (2007) *Arkivoc*, **xi**, 145–160.

110 Meyer, F.E. and de Meijere, A. (1991) *Synlett*, 777–778.

111 Negishi, E., Harring, L.S., Owczarczyk, Z., Mohamud, M.M. and Ay, M. (1992) *Tetrahedron Lett.*, **33**, 3253–3256.

112 Trost, B.M. and Shi, Y. (1993) *J. Am. Chem. Soc.*, **115**, 12491–12509.

113 Trost, B.M. and Shi, Y. (1992) *J. Am. Chem. Soc.*, **114**, 791–792.

114 Henniges, H., Meyer, F.E., Schick, U., Funke, F., Parsons, P.J. and de Meijere, A. (1996) *Tetrahedron*, **52**, 11545–11578.

115 Meyer, F.E., Parsons, P.J. and de Meijere, A. (1991) *J. Org. Chem.*, **56**, 6488–6491.

116 de Meijere, A., von Zezschwitz, P. and Bräse, S. (2004) *Acc. Chem. Res.*, **38**, 413–422.

117 (a) Catellani, M., Chiusoli, G.P., Giroldini, W. and Salerno, G. (1980) *J. Organomet. Chem.*, **199**, C21–C23; (b) Brown, D., Grigg, R., Sridharan, V., Tambyrajah, V. and Thornton-Pett, M. (1998) *Tetrahedron*, **54**, 2595–2606.

118 (a) Grigg, R., Rasul, R., Redpath, J. and Wilson, D. (1996) *Tetrahedron Lett.*, **37**, 4609–4612; (b) Böhmer, J., Grigg, R. and Marchbank, J.D. (2002) *Chem. Commun.*, 768–769.

119 Grigg, R., Sridharan, V. and Sukirthalingam, S. (1991) *Tetrahedron Lett.*, **32**, 3855–3858.

120 Miura, T., Sasaki, T., Harumashi, T. and Murakami, M. (2006) *J. Am. Chem. Soc.*, **128**, 2516–2517.

121 Negishi, E. and Miller, J.A. (1983) *J. Am. Chem. Soc.*, **105**, 6761–6763.

122 Negishi, E., Copéret, C., Ma, S., Mita, T., Sugihara, T. and Tour, J.M. (1996) *J. Am. Chem. Soc.*, **118**, 5904–5918.

123 (a) Nieman, J.A. and Ennis, M.D. (2000) *Org. Lett.*, **2**, 1395–1397; (b) Nieman, J.A. and Ennis, M.D. (2001) *J. Org. Chem.*, **66**, 2175–2177.

124 Gagnier, S.V. and Larock, R.C. (2003) *J. Am. Chem. Soc.*, **125**, 4804–4807.

125 Wu, G., Shimoyama, I. and Negishi, E. (1991) *J. Org. Chem.*, **56**, 6506–6507.

126 Grigg, R., Redpath, J., Sridharan, V. and Wilson, D. (1994) *Tetrahedron Lett.*, **35**, 7661–7664.

127 Grigg, R. and Pratt, R. (1997) *Tetrahedron Lett.*, **38**, 4489–4492.

128 (a) Oppolzer, W., Bedoya-Zurita, M. and Switzer, C.Y. (1988) *Tetrahedron Lett.*, **29**, 6433–6436; (b) Oppolzer, W., Keller, T.H., Bedoya-Zurita, M. and Stone, C. (1989) *Tetrahedron Lett.*, **30**, 5883–5886; (c) Camps, F., Coll, J., Moréto, J.M. and Torras, J. (1988) *Tetrahedron Lett.*, **28**, 4745–4748.

129 Oppolzer, W. and Robyr, C. (1994) *Tetrahedron*, **50**, 415–424.

130 Oppolzer, W., Xu, J.-Z. and Stone, C. (1991) *Helv. Chim. Acta*, **74**, 465–468.

131 Oppolzer, W., Bienayme, H. and Genevois-Borella, A. (1991) *J. Am. Chem. Soc.*, **113**, 9660–9661.

132 Keese, R., Guidetti-Grept, R. and Herzog, B. (1992) *Tetrahedron Lett.*, **33**, 1207–1210.

133 Copéret, C., Ma, S. and Negishi, E. (1996) *Angew. Chem. Int. Ed.*, **35**, 2125–2126.

134 Doi, T., Yanagisawa, A., Nakanishi, S., Yamamoto, K. and Takahashi, T. (1996) *J. Org. Chem.*, **61**, 2602–2603.

6
Transition Metal-Catalyzed Intramolecular Allylation Reactions
Zhan Lu and Shengming Ma

6.1
Introduction

In transition metal-catalyzed intramolecular allylation reactions, *in situ* generated π-allylic transition metal species would be easily attacked by soft nucleophiles to afford the cyclic compounds [1]. Due to facile formation of C−C, −N, −O, −S bonds, and the tolerance of a wide variety of different functional groups, this reaction is a powerful tool for the synthesis of cyclic products and its potential has been demonstrated in the total synthesis of many natural products. In this chapter, we will show some of the most typical recent advances in this area.

6.2
Palladium-Catalyzed Intramolecular Allylation Reactions

6.2.1
Cyclic Allylations of Carbonucleophiles

Under the catalysis of $PdCl_2(PhCN)_2$ and $Yb(OTf)_3$, the stereocontrolled highly regiostereoselective intramolecular cyclization reaction of Meldrum's ester-allylic acetate **1** afforded the vinylic *trans*-cyclopropanes **2** in 68–71% yields exclusively (Eq. (6.1)) [2].

$$\text{1} \quad (Ar = Ph, 3\text{-}ClC_6H_4) \xrightarrow[\text{THF, RT, 24 h}]{\substack{PdCl_2(PhCN)_2 \ (10\ mol\%) \\ Yb(OTf)_3\ (20\ mol\%)}} \text{2} \quad (Y: 68\text{-}71\%) \quad (6.1)$$

Optically active 2-cyclohexyl E-vinylsilane S-4 can be obtained via the palladium-catalyzed intramolecular allylic alkylation of optically active 8,8-bis(methoxycarbonyl)-2(E)-octenyl-1-acetoxy-(*tert*-butyldimethyl)silane R-3 (Eq. (6.2)) [3].

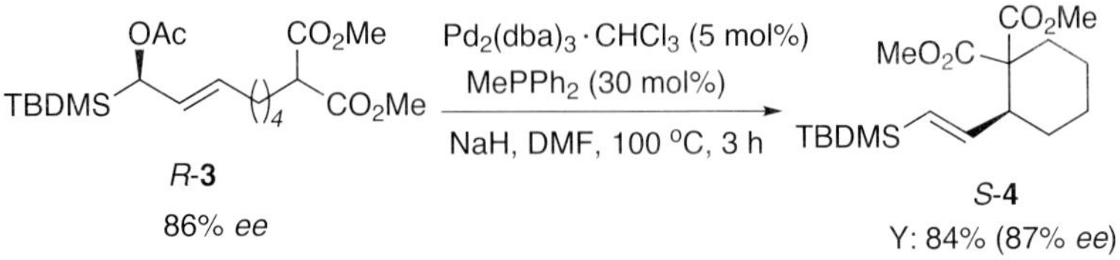

$$\text{R-3} \xrightarrow[\text{NaH, DMF, 100 °C, 3 h}]{\text{Pd}_2(\text{dba})_3 \cdot \text{CHCl}_3 \text{ (5 mol\%)}\atop \text{MePPh}_2 \text{ (30 mol\%)}} \text{S-4} \quad (6.2)$$

R-3: 86% ee
S-4: Y: 84% (87% ee)

Enantioselective intramolecular allylation of 7,7′-bis(methoxycarbonyl)hept-2-enyl carbonate **5** using S-**7a** as a chiral ligand produced vinylcyclopentane derivative **6** in 60% yield with 87% ee (Scheme 6.1) [4]. A five-membered 1-alkylidene-2-vinyl cyclopentane derivative **9** was also facilely constructed in 100% conversion with 95% ee via an intramolecular reaction of hexa-3,5-dien-2-yl ester **8** bearing a malonate moiety with chiral ligand S-**7b** (Scheme 6.1) [5].

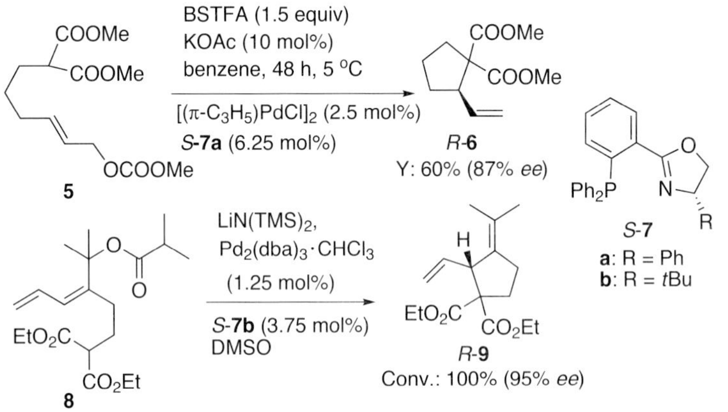

Scheme 6.1 Enantioselective synthesis of carbocycles via enantioselective intramolecular allylations.

An asymmetric preparation of (+)-methyl pederate **12**, a key intermediate for the total synthesis of mycalamides A and B **13**, based on a palladium-catalyzed intramolecular allylic alkylation reaction was developed by Toyota and Ihara *et al.* Under the catalysis of Pd(OAc)$_2$ with PPh$_3$ as the ligand, the intramolecular cyclization reaction of the optically active malonate-allylic methyl carbonate **10** in DMF afforded the δ-lactone (+)-(4R,5R)-**11** as a mixture of four diastereomers in 87% yield (Scheme 6.2) [6].

In an aqueous ethyl acetate biphasic system, the palladium-catalyzed intramolecular reaction of the allylic acetate moiety with the acetyl acetate part in **14** or **18** afforded medium- and large-sized lactones in high yields using a water-soluble ligand **17**. These lactones retain the original configuration of the olefin moiety

Scheme 6.2 Asymmetric synthesis of methyl pederate **12**.

(Scheme 6.3) [7]. However, when $n = 2$, the reaction afforded δ-lactone **16** in 51% yield by 6-exo cyclization and the formation of the eight-membered lactone was not observed.

The Pd(0)-catalyzed intramolecular allylation reaction of α-sulfonyl δ-lactone in **20** in CH_2Cl_2 constructed the vinyl-substituted bicyclic lactones **21** and **22** in 88% yield with a dr ratio of 30/1. Based on this reaction, bacillariolide III **23** was synthesized in 15 steps in 14.6% overall yield (Scheme 6.4) [8].

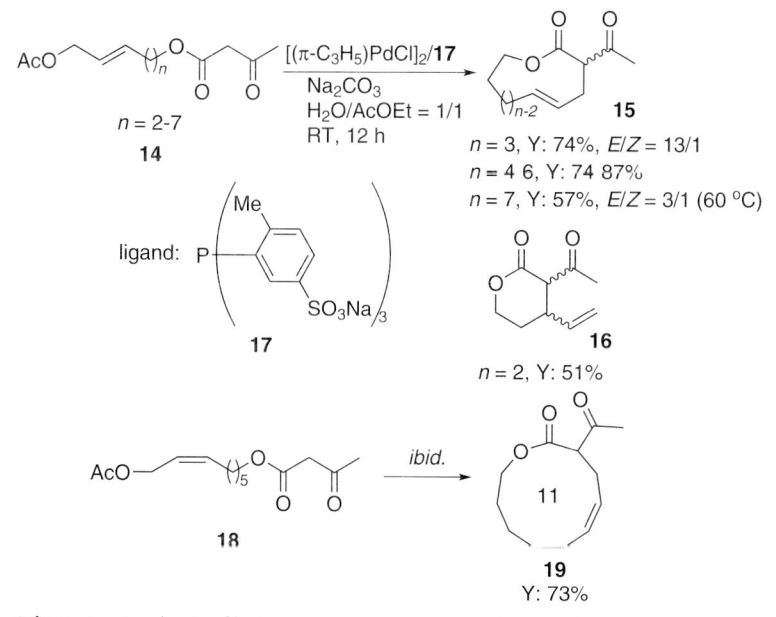

Scheme 6.3 Synthesis of lactones via an aqueous–organic biphasic system.

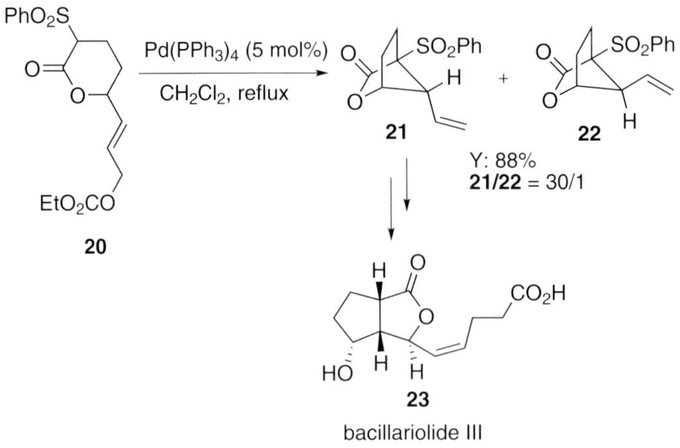

Scheme 6.4 Synthesis of bacillariolide III.

The palladium-catalyzed cyclization of N-(4-acetoxy-2(Z)-butenyl) amides Z-24 with an electron-withdrawing group at the α-position may afford trans-pyrrolidin-2-ones 25 highly diastereoselectively. The reaction showed a great tolerance towards the acetamide anion stabilizers such as Me(O)C, MeO$_2$C, NC, PhSO$_2$, and (EtO)$_2$(O)P. Not only Z-24a (EWG = CO$_2$Me) but also E-24a (EWG = CO$_2$Me) may react under the same condition to afford the same cyclization product trans-25a (EWG = CO$_2$Me) [9a]. Ti(OiPr)$_4$ could be added to promote this reaction [10]. Under phase-transfer conditions A or in the presence of a crown ether (conditions B), the same products could also be furnished (Scheme 6.5) [11].

The Pd-catalyzed intramolecular allylation of similarly structured cis-26a,b,d with a cyclic allylic acetate moiety, in which EWG is CO$_2$Me or SO$_2$Ph, afforded the hexahydroindol-2-one 27a,b,d in high yields and dr values under the conditions A (Scheme 6.6). With EWG being CN, the similar reaction of cis-26c afforded the corresponding product 27c with lower dr value (60/40). The same products can also be synthesized by the Pd-catalyzed intramolecular allylation of trans-26a-b under the conditions B (Scheme 6.6) [12].

The intramolecular cyclization of N-(3-triethylsilyl-2(E)-butenyl) amide 28 afforded the trans-disubstituted γ-butyrolactone 29 in 90% yield with exclusive trans diastereoselectivity. However, the similar 5-exo-trig cyclization of the 2-triethylsilyl substituted analogue 30 proceeded with lower yield. The possible explanation is that intramolecular C=O−SiEt$_3$ coordination might be responsible for the exclusive formation of the isomer featuring a cis-(CO$_2$Me/Et$_3$Si) relationship in 31 (Scheme 6.7) [13]. Cyclization has also been observed between an allylic sulfone and a phosphoroacetamide in 32 leading to the formation of the cyclic product trans-33 in quantitative yield with high dr value (>95/5) (Scheme 6.7) [14].

Craig's group developed a diastereoselective palladium-catalyzed intramolecular allylation of chiral N-(1-substituted-4-(methoxycarbonyloxy)-2(E)-butenyl) α-tosylacetamides 34 using TTMPP (TTMPP = tris(2,4,6-trimethoxyphenyl)phosphine) as the ligand to provide 35 in 78–90% yields with a cis/trans ratio of 67/33-90/10

6.2 Palladium-Catalyzed Intramolecular Allylation Reactions | 275

Conditions A: [(π-C$_3$H$_5$)PdCl]$_2$ (5 mol%), dppe (12.5 mol%), n-Bu$_4$NBr (10 mol%)
aq. KOH (2.0 equiv), CH$_2$Cl$_2$/H$_2$O = 1/1, RT, 0.5–72 h
Y: 72–96%, dr = 75/25–>99/1

Conditions B: Pd(OAc)$_2$ (5 mol%), dppe (12.5 mol%), 15-crown-5 (1.2 equiv)
NaH (1.1 equiv), DMF, RT, 2 h, Y: 76% (**25a**)

Scheme 6.5 Intramolecular allylation of N-(4-acetoxy-2-butenyl) amides.

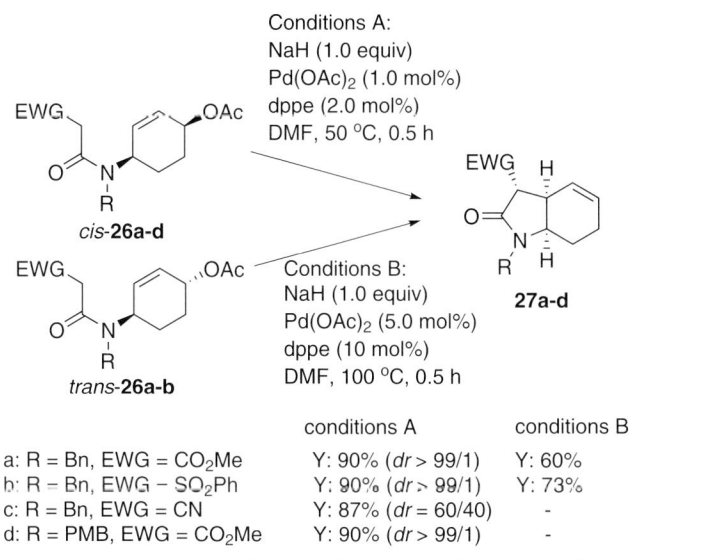

	conditions A	conditions B
a: R = Bn, EWG = CO$_2$Me	Y: 90% (dr > 99/1)	Y: 60%
b: R = Bn, EWG = SO$_2$Ph	Y: 90% (dr > 99/1)	Y: 73%
c: R = Bn, EWG = CN	Y: 87% (dr = 60/40)	–
d: R = PMB, EWG = CO$_2$Me	Y: 90% (dr > 99/1)	–

Scheme 6.6 Intramolecular allylation of N-(4-acetoxy-2-cyclohexenyl) amides.

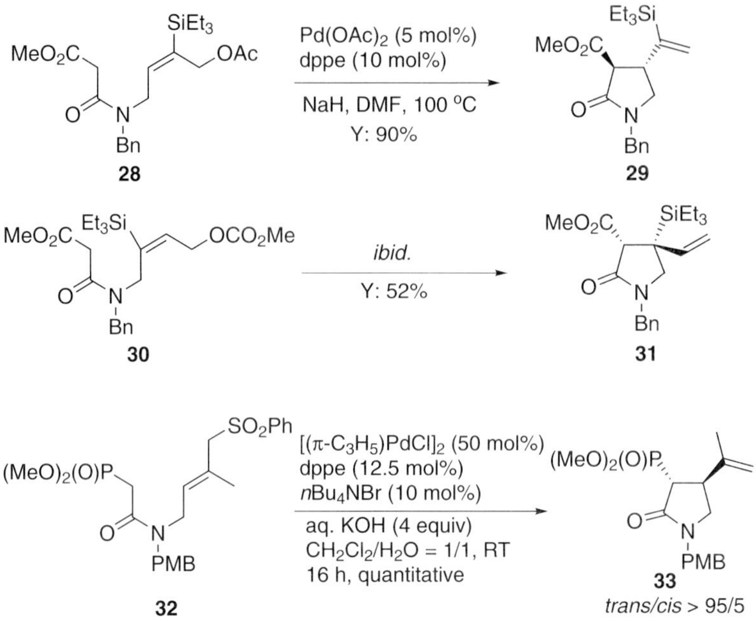

Scheme 6.7 Intramolecular allylation of N-2- or 3-substituted-4-acetoxy(or sulfonyl)-2-butenyl amides.

(Scheme 6.8) [15]. Tricyclic product **37** was also prepared similarly with high diastereoselectivity (Scheme 6.8) [16].

3-Vinylquinuclidine azabicyclo[2.2.2]octane can be constructed by a ring closure reaction of **38** using an intramolecular palladium-catalyzed allylation of enol

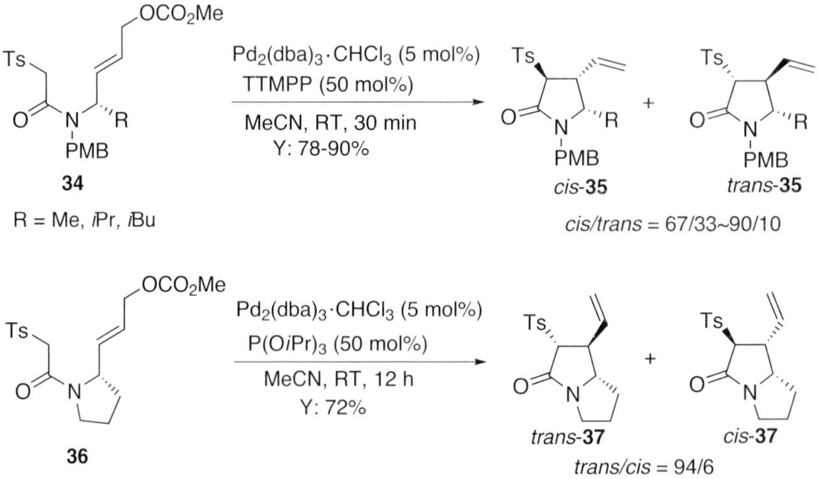

Scheme 6.8 Intramolecular highly diastereoselective allylation of optically active α-tosylacetamides-allylic carbonates.

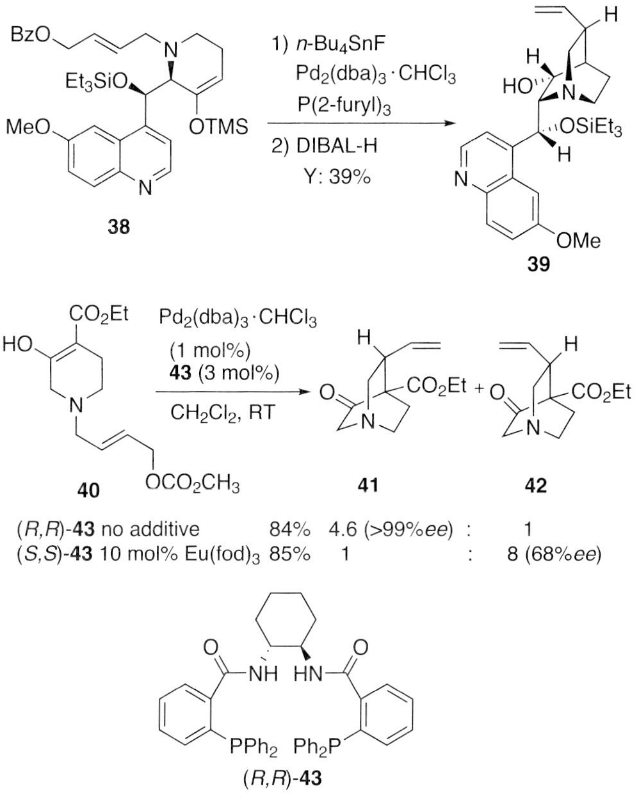

Scheme 6.9 Intramolecular allylation using enol or enol silyl ether as carbonucleophiles.

silyl ether with excellent regio- and diastereoselectivity (Scheme 6.9) [17]. The enantioselective intramolecular reaction of α-keto ester-allylic carbonate **40** using chiral ligand (R,R)-**43** afforded a mixture of ethyl 3-oxo-8-vinylquinuclidine-4-carboxylates **41** and **42** in 84% yield with the ratio of **41**(>99% ee)/**42** being 4.6/1 or 85% yield with the ratio of **41/42** (68% ee) being 1/8 in the presence of Eu(fod)$_3$ (Scheme 6.9) [18].

Saicic et al. demonstrated a synergic combination of organotransition metal catalysis and organocatalysis to allow the Tsuji–Trost-type α-allylation of aldehydes. In the reaction, the organocatalyst pyrrolidine reacted with aldehyde to form the nucleophilic enamine (Scheme 6.10) [19]. Using R-BINAP as a chiral ligand, a catalytic asymmetric variant of the reaction is also possible to afford the cyclopentane derivative **47** in 40% yield and 91% ee (Scheme 6.10) [19].

Recently, Shintani and Hayashi et al. developed a highly regio- and stereoselective palladium-catalyzed synthesis of spiro[2,4]heptanes **49** by the reaction of α-(methoxycarbonyl)-γ-methylene-δ-valerolactones **48** with electron-deficient olefins through a sequential allylic cleavage, elimination of CO_2, Michael addition, and intramolecular nucleophilic trapping of the π-allyl palladium intermediate at the center carbon atom, and a reductive elimination process (Scheme 6.11) [20].

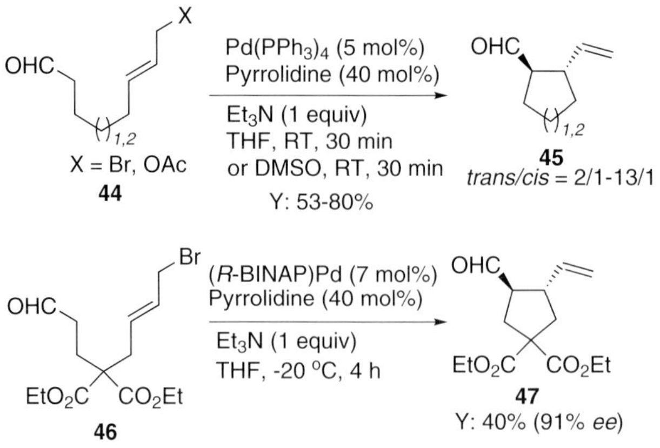

Scheme 6.10 Palladium and pyrrolidine co-catalyzed intramolecular α-allylation of aldehydes.

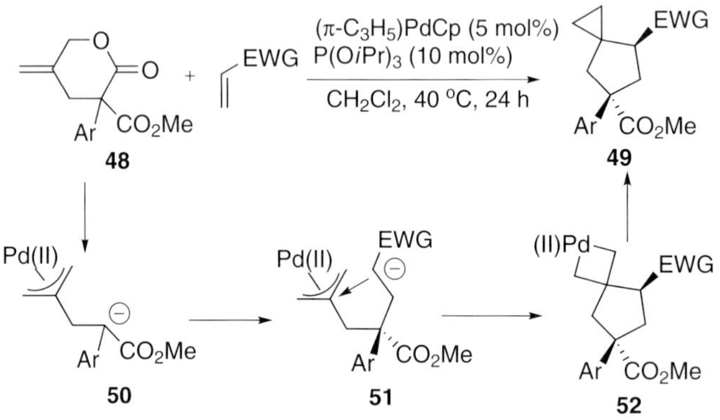

Scheme 6.11 Palladium-catalyzed synthesis of spiro[2,4] heptanes **49** by the reaction of γ-methylene-δ-lactones **48** with electron-deficient olefins.

Krische et al. demonstrated the palladium-catalyzed intramolecular allylation reaction of allylic carbonate moiety with the nucleophilic carbon center generated by the nucleophilic conjugate addition reaction of the electron-deficient olefin moiety in **53** with PBu$_3$ to form the carbocycle **54** in 64–92% yield (Eq. (6.3)) [21].

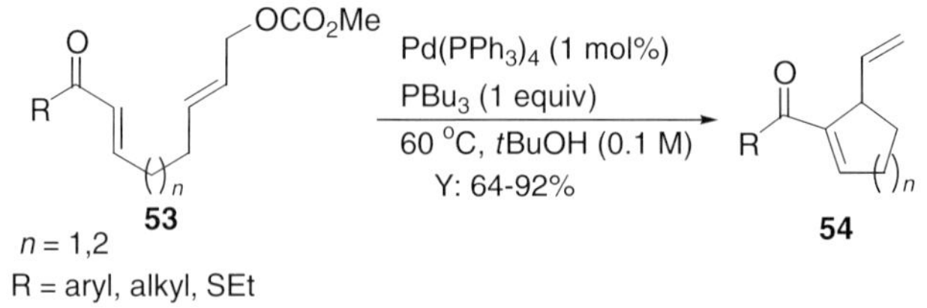

(6.3)

A palladium-catalyzed two-component cyclization of electron-deficient olefin **55** with ω-hydroxyallylic carbonates **56** afforded the cyclic ethers **57** in 31% to approximately quantitative yield with trans/cis ratios of 73/27-1/>99 [22] via the intermolecular Michael addition of the hydroxyl group in **56** with the electron-deficient carbon–carbon double bond in **55** followed by cyclic allylation (Scheme 6.12). In the formation of the five-membered product, a bidentate ligand, such as dppe, gave the best result whereas the monodentate bulky ligand, (o-tolyl)$_3$P, produced the best result for the formation of the six-membered ring. The enantioselective allylation reaction using chiral bidentate ligands (R,R)-**43** or **58** afforded the corresponding cyclic compounds in 29–75% yield with 46–92% ee (Scheme 6.12) [22].

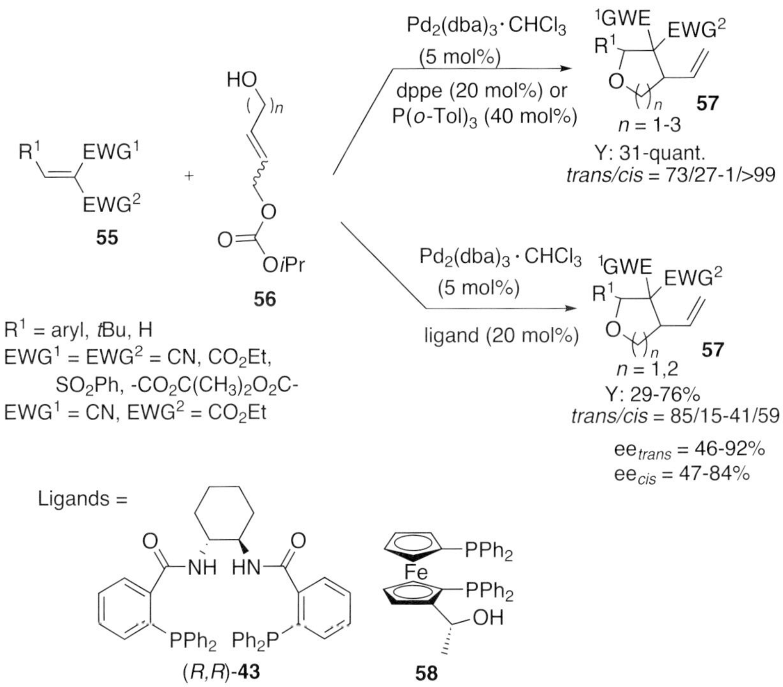

Scheme 6.12 Palladium-catalyzed cyclization of electron-deficient olefin **55** with ω-hydroxylallylic carbonates.

Lee and coworkers demonstrated an intramolecular palladium-catalyzed cross-coupling reaction of allyl indium generated *in situ* by the reaction of allyl acetates **59** with indium and indium trichloride in the presence of Pd(0)-catalyst with intramolecular aryl halide in DMF (Scheme 6.13) [23].

The coupling reaction of the bis-π-allylpalladium complexes formed from allylic chloride-allylic tin **64a** ($n = 1$) with 2-aryl-1,1-dicyanoethene afforded medium-sized carbocycles **66**, **68**, and **69** (Scheme 6.14) [24]. The similar reaction of **64b,c** ($n = 2$ and 3) afforded the corresponding carbocycles **66** and **67**.

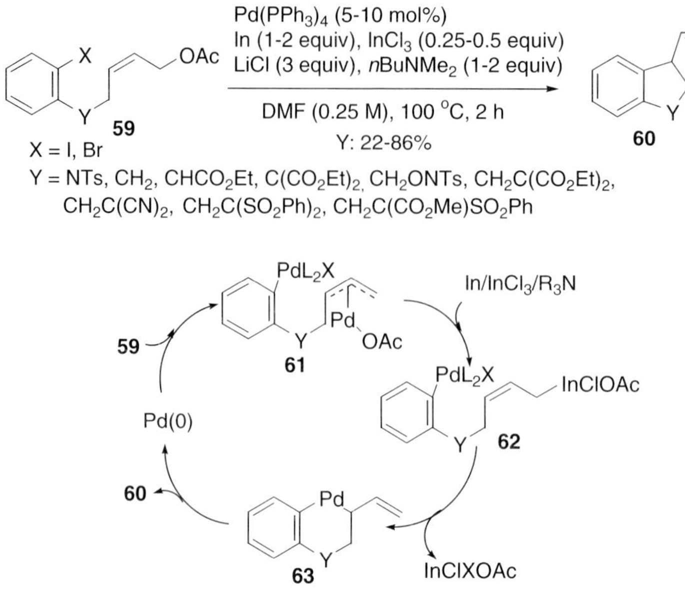

6 Transition Metal-Catalyzed Intramolecular Allylation Reactions

Scheme 6.13 Pd-catalyzed intramolecular cyclization of aryl halide moiety with allylic acetate moiety in the presence of indium.

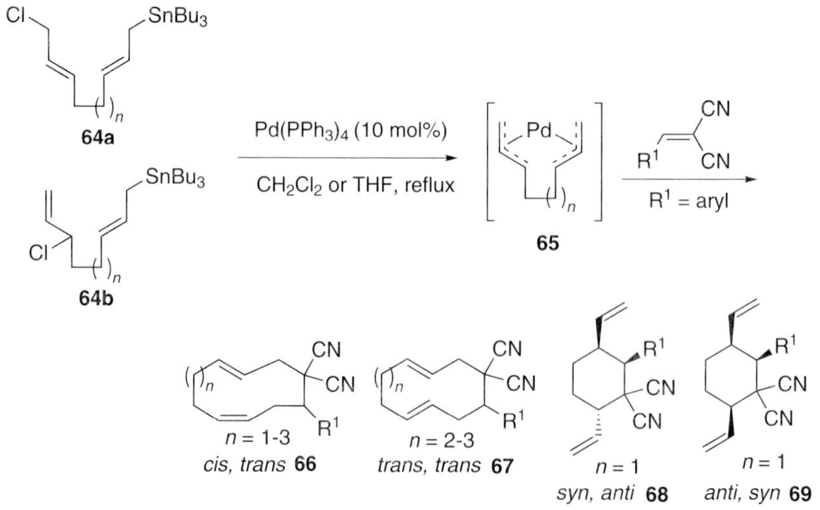

Reaction conditions: Pd(PPh$_3$)$_4$ (5-10 mol%), In (1-2 equiv), InCl$_3$ (0.25-0.5 equiv), LiCl (3 equiv), nBuNMe$_2$ (1-2 equiv), DMF (0.25 M), 100 °C, 2 h, Y: 22-86%.

59: X = I, Br; Y = NTs, CH$_2$, CHCO$_2$Et, C(CO$_2$Et)$_2$, CH$_2$ONTs, CH$_2$C(CO$_2$Et)$_2$, CH$_2$C(CN)$_2$, CH$_2$C(SO$_2$Ph)$_2$, CH$_2$C(CO$_2$Me)SO$_2$Ph

Catalytic cycle intermediates: **61** (PdL$_2$X), **62** (PdL$_2$X with InClOAc), **63** (Pd with InClXOAc), **60**.

Scheme 6.14 Coupling reaction of allylic chloride–allylic tins **64** in the presence of 2-aryl-1,1-dicyanoethenes.

64a, **64b** + Pd(PPh$_3$)$_4$ (10 mol%), CH$_2$Cl$_2$ or THF, reflux → **65** → products **66** (n = 1-3, cis, trans), **67** (n = 2-3, trans, trans), **68** (n = 1, syn, anti), **69** (n = 1, anti, syn); R^1 = aryl.

The carbon atom at the 3-position of the indole moiety in **70** may also readily act as the nucleophile to react with allylic carbonate moiety at the 2-position of the indole ring to form cyclic indole alkaloids **72** (Scheme 6.15) [25].

The enantioseletive version of this reaction was realized by using **76** as the chiral ligand (Scheme 6.16) [26].

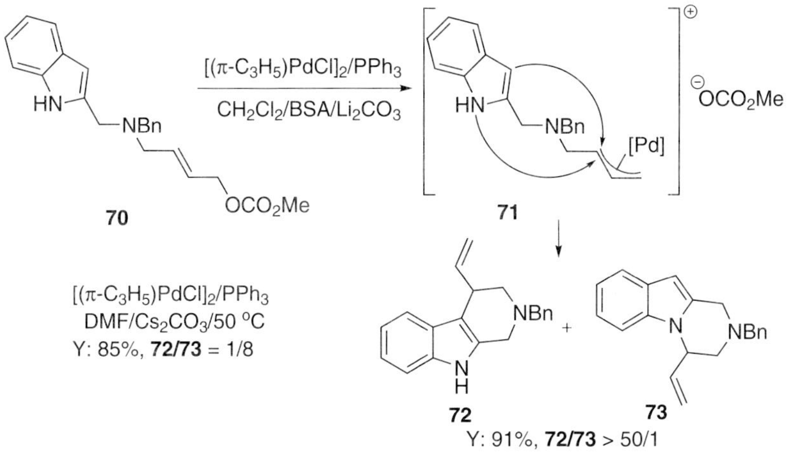

Scheme 6.15 Intramolecular allylation of the carbon atom at the 3-position of indole.

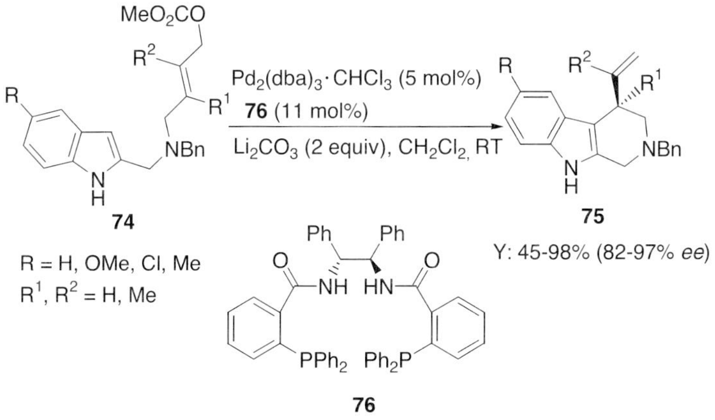

Scheme 6.16 Enantioselective cyclic allylation of indoles.

Using [(π-C₃H₅)PdCl]₂ and ferrocene-based chiral *P,P*-ligand **80**, the bicyclic α-keto ester **77** may react with 2-acetoxymethylallyl acetate **78** to afford the product **79** in 82% yield with 90.3% ee, which is a key intermediate in the total synthesis of (−)-huperzine A **81** (Scheme 6.17) [27].

Palladium-catalyzed [3 + 2] cycloaddition [28] of 2-(trimethylsilylmethyl)allyl acetate **83** with the optically active electron-deficient cyclic olefin **82** afforded 5-methylene-3-(2-naphthoxy)tetrahydrocyclopenta[c]furan-1-one **84** in 90% yield [29]. Trost and coworkers developed such enantioselective palladium-catalyzed [3 + 2] cycloaddition reactions by the use of chiral phosphorus amidites **87** [30], **93**, and **94** [31] to provide methylenecyclopentane derivatives in high enantiopurity (Scheme 6.18).

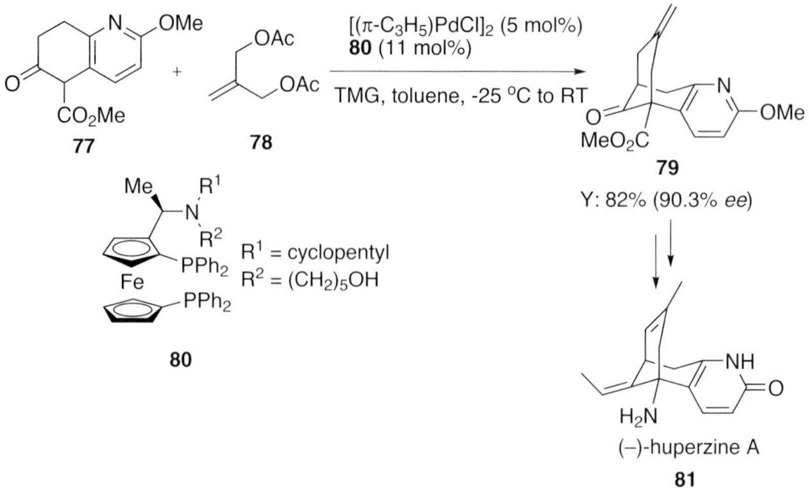

Scheme 6.17 Enantioselective synthesis of a key intermediate for the total synthesis of (−)-huperzine A.

The oxidative addition of vinylaziridine **95** with Pd(0) may also generate a π-allyl palladium intermediate bearing a nucleophilic tosyamidio group, which may undergo a conjugate addition with electron-deficient olefins to generate an intramolecular carbon nucleophile to trap the π-allyl palladium moiety to form the five-membered product **98A** or **98B** (Scheme 6.19) [32].

Larock's group reported that (2′-iodophenyl)malonate **100** reacted with 1,2-nonadiene or 4,5-undecadiene in the presence of a palladium catalyst and the chiral bisoxazoline ligand to construct five- and six-membered carbocycles [33]. This group developed a controllable regioselective palladium-catalyzed coupling–cyclization reaction of 2-(2′-substituted-2′,3′-butadienyl)malonates **103** with phenyl iodide to afford the three- or five-membered products under different reaction conditions (Scheme 6.20) [34].

6.2.2
Cyclic Allylic Aminations

The palladium-catalyzed intramolecular allylamination reaction of cyclic allylic acetate **107** tethered with a primary amine generated *in situ* from reduction of the azide functionality in **106** in CH_3CN in the presence of Et_3N afforded the spiro bicyclic compounds **108** (Scheme 6.21) [35]. The similar reaction of cyclic allylic acetate **109** tethered with secondary amine can be realized to synthesize a pyrrolidine derivative **110** in 74% yield (Scheme 6.21) [35].

The palladium-catalyzed intramolecular allylation reaction of 4-(benzylamino)-2-butenyl carbonate afforded the aziridine **111** in 79% yield highly regioselectively (Eq. (6.4)) [36].

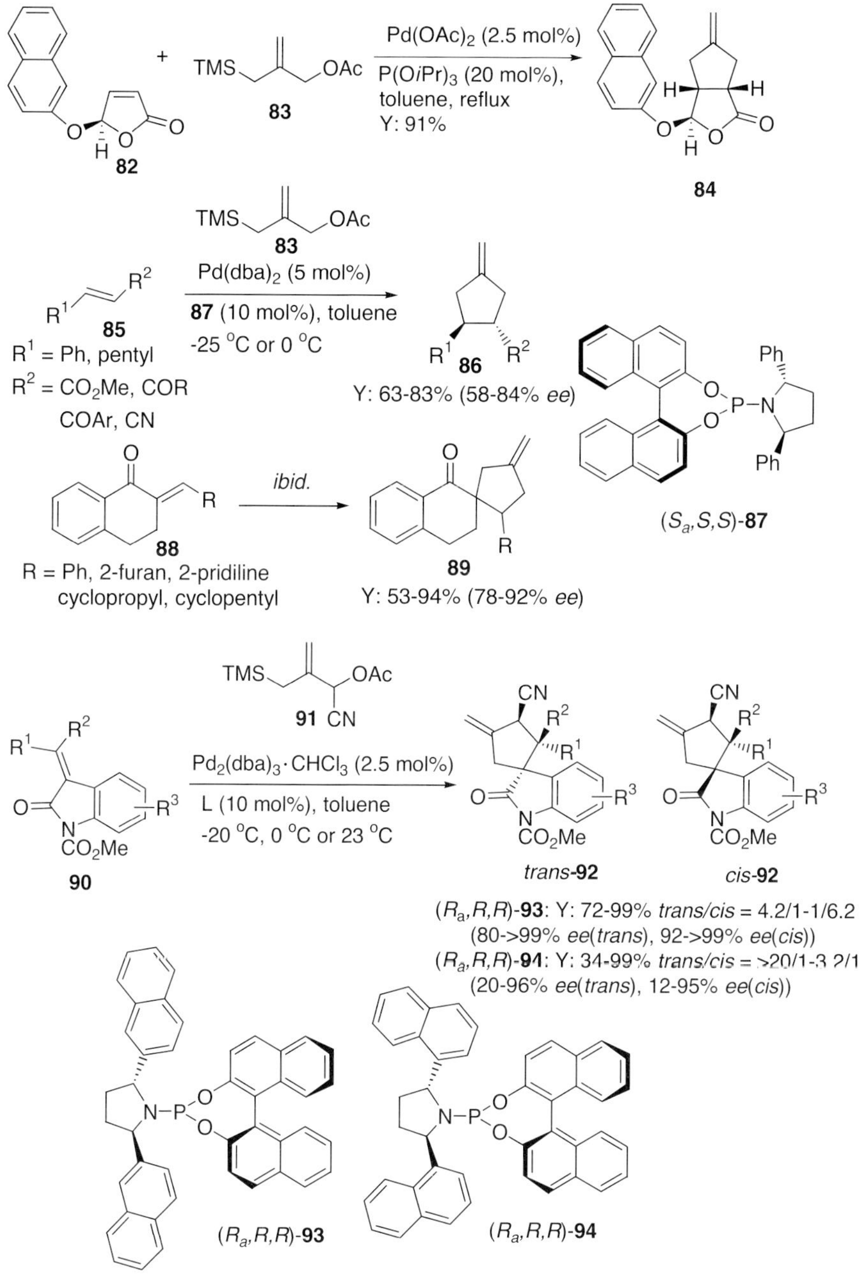

Scheme 6.18 Palladium-catalyzed [3 + 2] cycloaddition of 2-(trimethylsilylmethyl)allyl acetates.

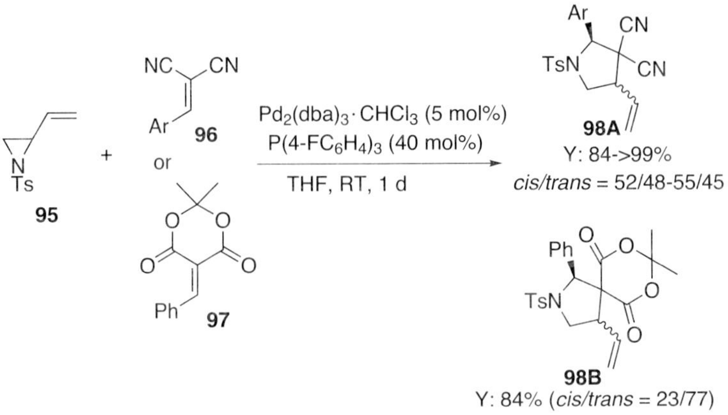

Scheme 6.19 Palladium-catalyzed ring-opening cycloaddition of N-tosyl vinylaziridine with electron-deficient olefins.

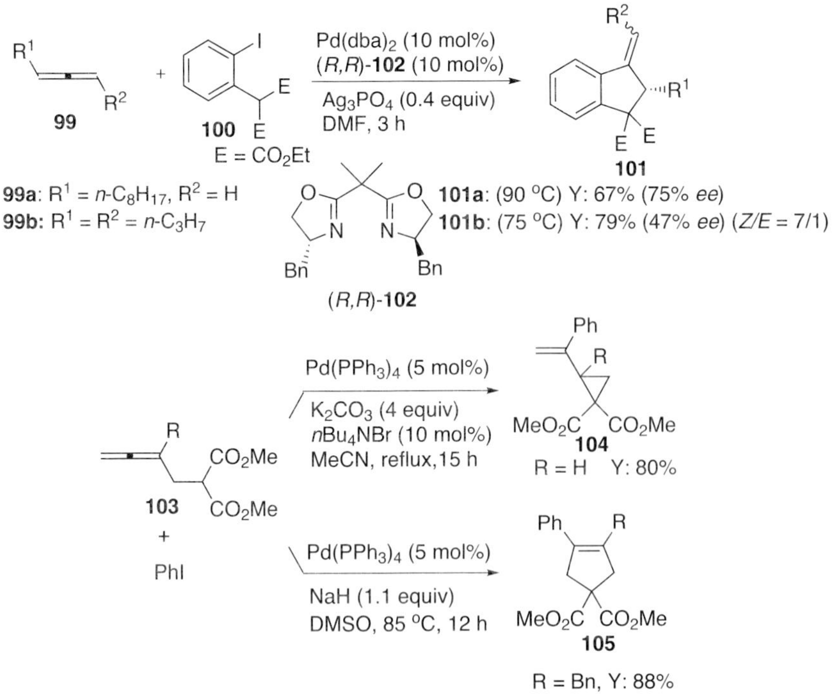

Scheme 6.20 Intermolecular carbopalladation of allenes followed by intramolecular allylation.

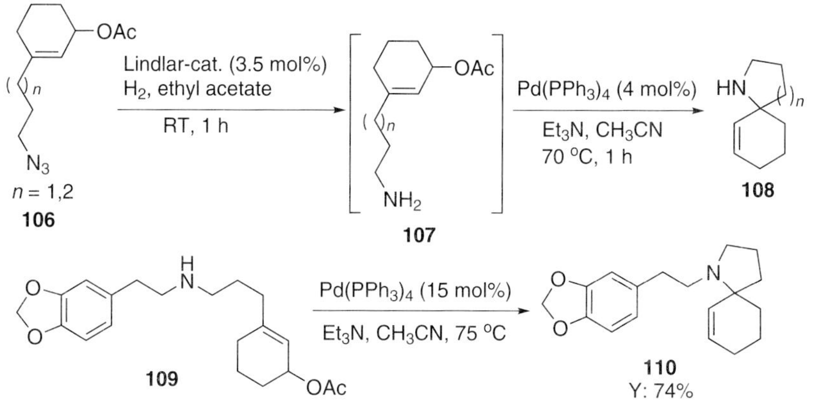

Scheme 6.21 Synthesis of spirobicyclic compounds via intramolecular allylamination reactions of cyclic allylic acetate.

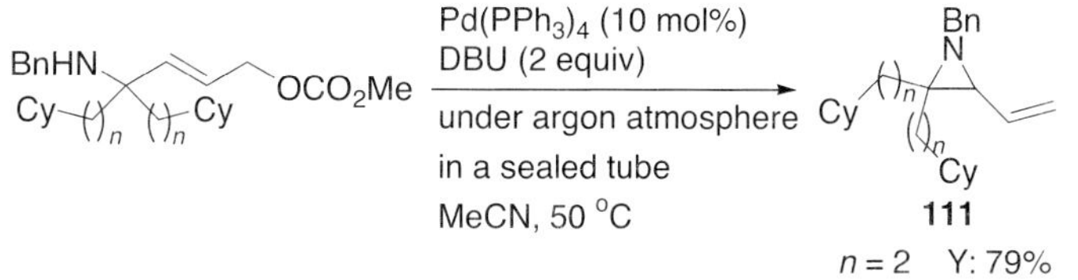

$$(6.4)$$

Using benzotriazole as a leaving group in the presence of PPh_3 and K_2CO_3, palladium-catalyzed intramolecular allylic amination reactions of 6- or 7-amino allylic benzotriazolate **113**, generated *in situ* via the nucleophilic substitution reaction of the chloride with amines, have been applied to the synthesis of five- and six-membered azacycles (Scheme 6.22) [37].

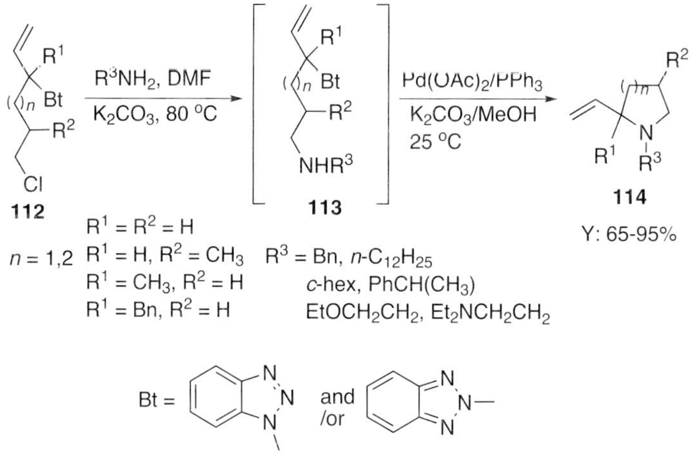

Scheme 6.22 Synthesis of substituted pyrrolidines and piperidines via intramolecular amination of *in-situ* generated 6 or 7-amino allylic benzotriazolates.

Palladium-catalyzed asymmetric intramolecular allylic amination of 8-(*p*-methoxybenzylamino)octen-3-yl acetate **115** with chiral ligand **43** afforded the seven-membered cyclic amine **116** in 84% yield with 92% *ee* (Eq. (6.5)) [38].

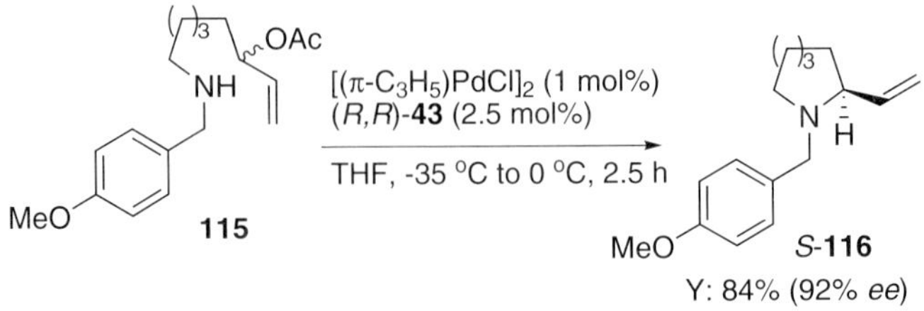
OAc
NH
MeO
115
[(π-C3H5)PdCl]2 (1 mol%)
(R,R)-43 (2.5 mol%)
THF, -35 °C to 0 °C, 2.5 h
N
H
MeO
S-116
Y: 84% (92% ee)

(6.5)

The tetrahydroquinoline **118** can be constructed in high yield through the intramolecular allylic amination of **117**. The diastereoselectivity of the reaction depends on the configuration of the double bond in the substrates (Scheme 6.23) [39].

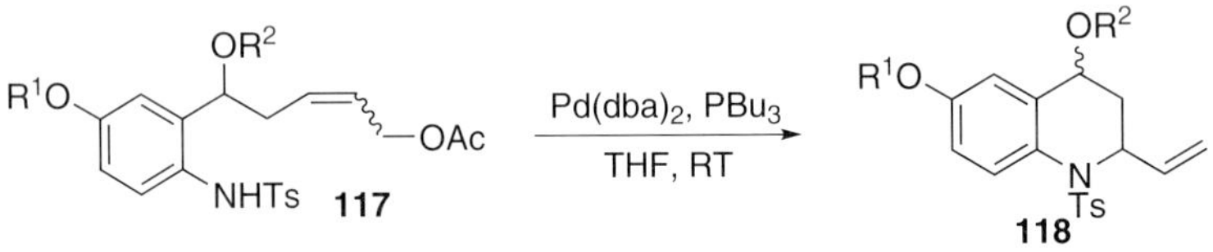
OR2
R1O
OAc
NHTs
117
Pd(dba)2, PBu3
THF, RT
OR2
R1O
N
Ts
118

$R^1 = R^2$ = TBS,	*E* only	20 h	Y: 92%, only *cis*
R^1 = Bn, R^2 = H,	mixture of *E* and *Z*	17 h	Y: 96%, *cis*/*trans* = 3/1
R^1 = Bn, R^2 = Ac,	*Z* only	44 h	Y: 96%, *cis*/*trans* = 3/1
R^1 = Bn, R^2 = TBS,	*E* only	16 h	Y: 96%, only *cis*

Scheme 6.23 Synthesis of 2-vinyl tetrahydroquinoline via intramolecular allylic amination.

Optically active benzocyclic amide **120** and bridged bicyclic tosylamide **123** could be obtained in 100% conversion (95% *ee*) and 90% yield (88% *ee*), respectively, via palladium-catalyzed asymmetric intramolecular allylic amination of trifluroacetylamide **119** [40] with chiral ligand **121** or the eight-membered cyclic tosylated amino allylic carbonate **122** with 1,2-diaminohexane-based phosphine-pyridine ligand **124** [41] (Scheme 6.24).

Under the catalysis of $Pd(OAc)_2$ and $P(OEt)_3$, the cyclization reaction of **125** was conducted in DMF in the presence of nBu_4NCl at 100 °C to provide the *cis*-bicyclic product **126** in 77% yield (Eq. (6.6)) [42]. The inversion of the configuration of the C=C bond is surprising.

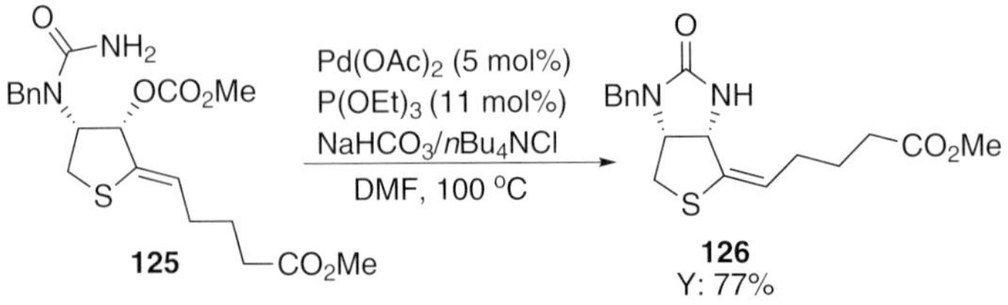
O
NH2
BnN
OCO2Me
S
125
CO2Me
Pd(OAc)2 (5 mol%)
P(OEt)3 (11 mol%)
NaHCO3/nBu4NCl
DMF, 100 °C
O
BnN
NH
S
CO2Me
126
Y: 77%

(6.6)

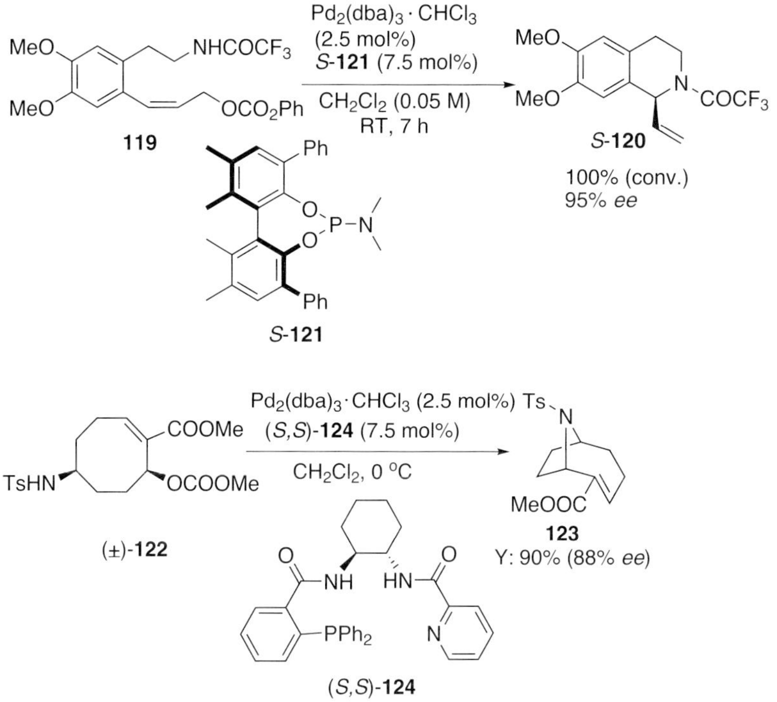

Scheme 6.24 Enantioselective synthesis of benzocyclic amide and bridged bicyclic tosylamide.

Using chiral bisphosphorus-amide ligands **43** and **129–132** [43], palladium-catalyzed enantioselective intramolecular desymmetrizative allylic amination of cis-1,4-dicarbamate-2-cyclopentene or cyclohexene **127** affords the optically active cis-fused bicyclic products **128** with high enantiopurity (Scheme 6.25).

In the presence of Cs_2CO_3 and DMF, the cyclic indole alkaloid **73** can be synthesized as a major product via the regioselectivity-controlled palladium-catalyzed allylation reaction of indole-tethered with allylic carbonates (Scheme 6.15) [25].

The palladium-catalyzed intramolecular allylic amination reactions via an allylic C–H activation of N-allyl (N'-tosylanthraniloyl)amides **133** in DMSO using NaOAc as a base at 100 °C formed 2-vinylquinazolin-4-ones **134** in 78–96% yields. With N-(2'-methylallyl) (N'-tosylanthraniloyl)amides **135**, the competition of activation of a C–H bond from a methylene group or a methyl group led to the production of the six-membered ring 2-(2'-methylvinyl)quinazolin-4-one **136** and the eight-membered ring 2-methylene-1,4-benzodiazepin-5-one **137** in 62% and 16% yields via the intermediacy of **138** and **139**, respectively (Scheme 6.26) [44].

Yamamoto and coworkers developed a method for the synthesis of azacycles **141** via the Pd(0)-catalyzed intramolecular hydroamination of amine-alkynes **140**. The enantioselective Pd(0)-catalyzed intramolecular reaction of **140** can be conducted to afford the optically active products: the higher the catalyst loading, the higher the enantioselectivity (Scheme 6.27) [45]. This reaction may proceed via the oxidative

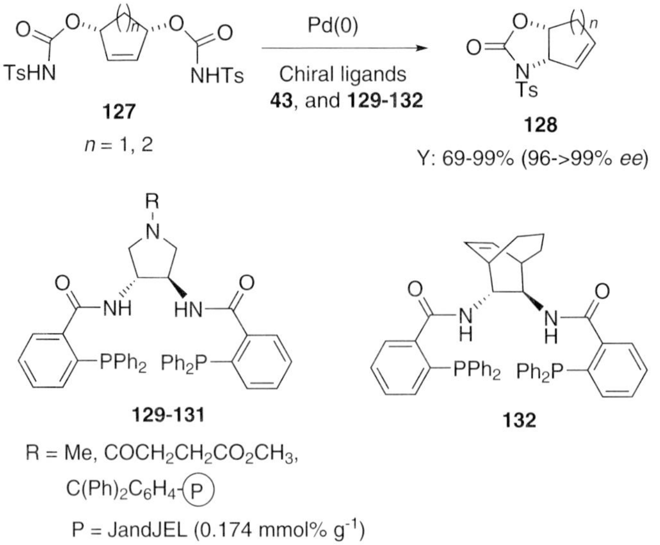

Scheme 6.25 Enantioselective intramolecular desymmetrizative allylic amination.

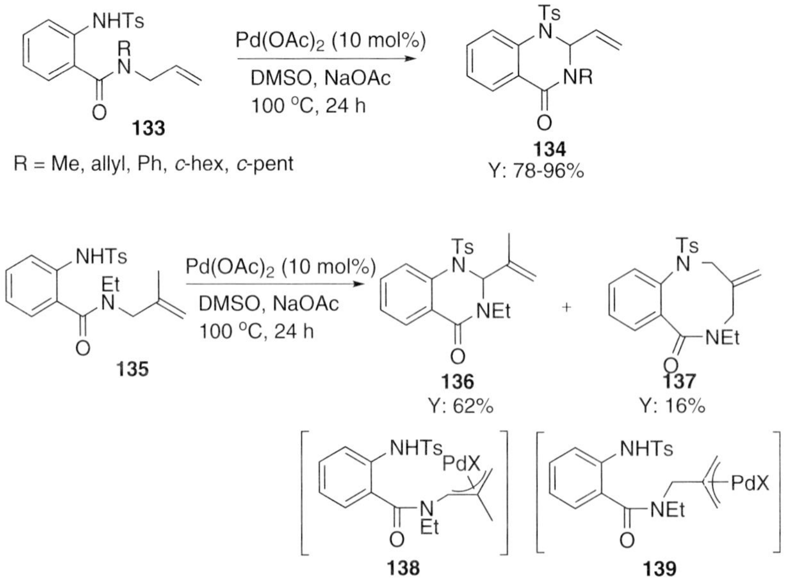

129-131
R = Me, COCH$_2$CH$_2$CO$_2$CH$_3$,
C(Ph)$_2$C$_6$H$_4$-(P)

P = JandJEL (0.174 mmol% g^{-1})

Scheme 6.26 Intramolecular allylamination reactions via an allylic C–H activation.

addition, insertion, β-H elimination, hydropalladation of 1,2-diene forming the π-allylic palladium intermediate, and an allylic amination sequence.

Larock and coworkers reported the reaction of amino-iodide with terminal allenes to furnish the azacycles [33]. 2,3-Dihydropyrroles **149**, 1,2,3,6-tetrahydropyrridines **151**, and azetidines **150** could be synthesized via the intermolecular

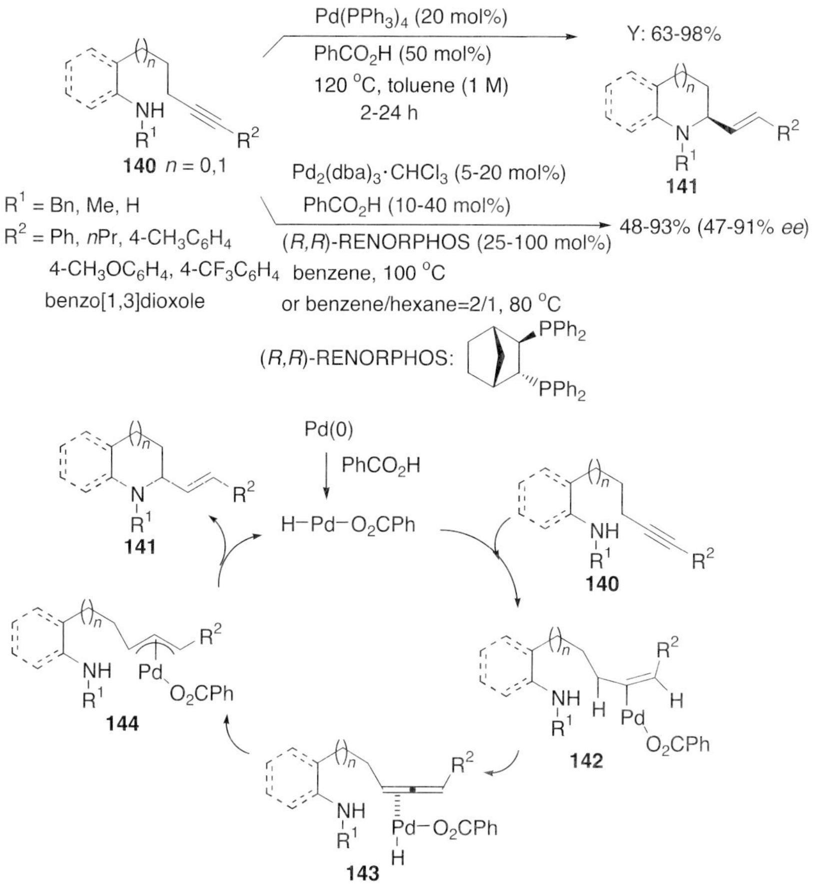

Scheme 6.27 Pd(0)-catalyzed intramolecular hydroamination of amine-alkynes.

carbopalladation of N-3,4-alkadienyl toluenesulfonamides 148 with organic halides followed by regioselective intramolecular allylic aminations [33b,c,46]. 5-Hydroxypyrolin-2(5H)-one 153 was obtained via the coupling reaction of 2,3-allenamide with phenyl iodide, followed by intramolecular allylic aminations and γ-hydroxylation [47a]. Optically active trans-1,2-diazetidines 155 were synthesized by the palladium-catalyzed reaction of chiral 2,3-allenyl hydrazines 154 with aryl halides [47b]. Optically active tricyclic products 157 may be efficiently constructed via the intramolecular carbopalladation of allene and the subsequent asymmetric allylic amination of 2-(N-allenyl)aminophenyl iodides 156 (Scheme 6.28) [48].

The tandem Pd(OAc)$_2$-catalyzed carbopalladation, β-H elimination, hydropalladation, allylic substitution reaction of ω-olefinic N-tosyl amides 159 with vinylic bromides 160 using P(o-tol)$_3$ as the ligand in the presence of Na$_2$CO$_3$ and n-Bu$_4$NCl afforded the substituted pyrrolidones and piperidones in 49–82% isolated yield (Scheme 6.29) [49].

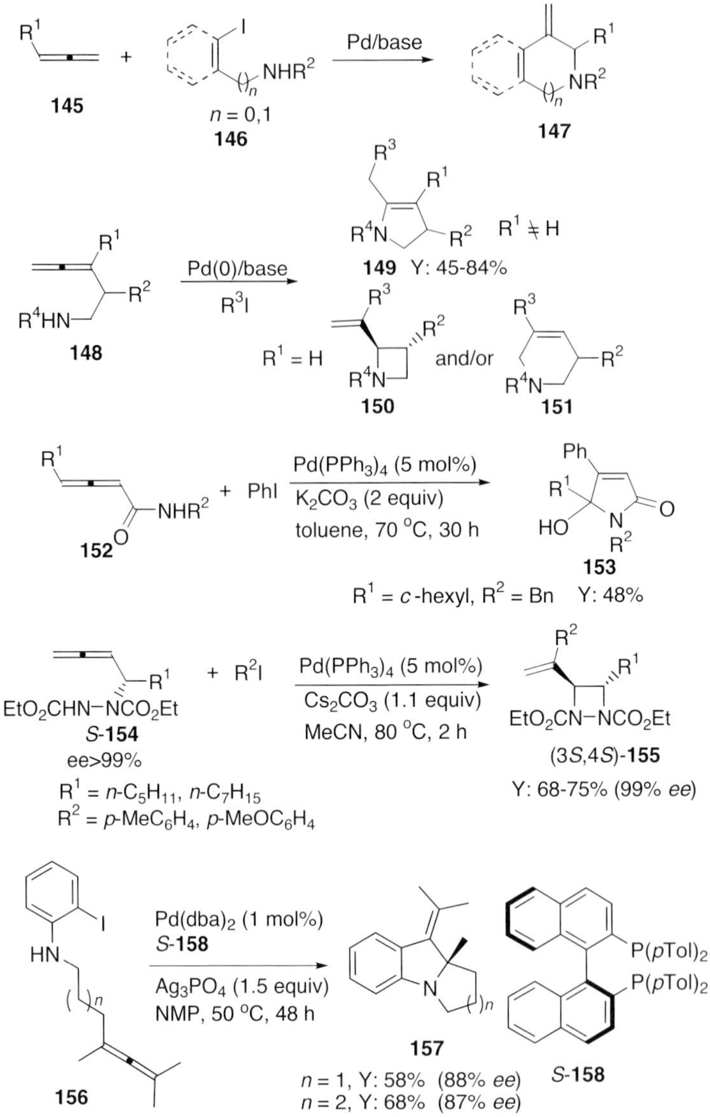

Scheme 6.28 Intermolecular carbopalladation of allenes followed by intramolecular amination of functionalized allenes with aryl iodides or allenes with amino iodides.

Trost et al. observed that the amine functionality could even react intramolecularly with an allylic p-nitrophenyl ether moiety in **169**, which was formed from the Ru-catalyzed cyclometallation of proparylic tosylamide **165** with 3-butenyl p-nitrophenyl ether **166**, affording cyclic N-tosylamide **167** in 92% yield with 94% ee (Scheme 6.30) [50].

Scheme 6.29 Tandem Pd(OAc)$_2$-catalyzed reaction of 1-alkenyl bromides with alkenamides.

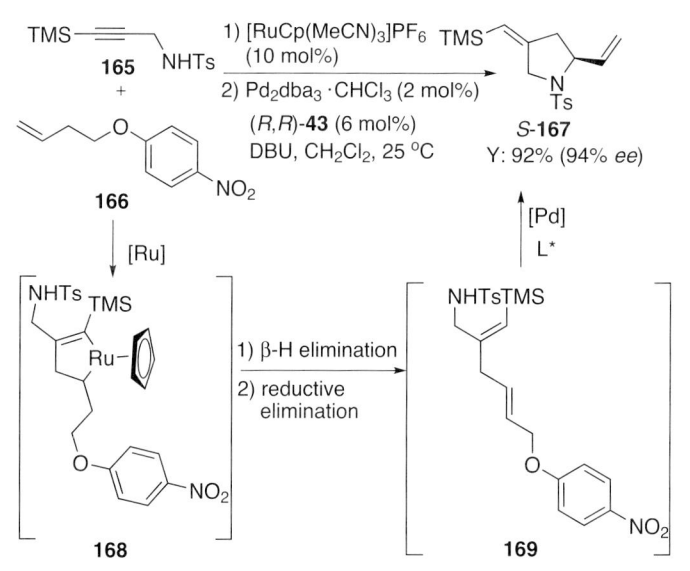

Scheme 6.30 Intramolecular allylic cyclization of *in situ* generated allylic amide-allyl *p*-nitrophenyl ether.

The regio- and stereo-selective formation of five-membered heterocycles by the palladium-catalyzed ring-opening cycloaddition reactions of vinyloxiranes [51], vinylaziridines [52], and 2-vinylthiiranes [53] with heterocumulenes could be easily realized (Scheme 6.31). Using chiral ligands S-TolBINAP, R-BINAP, or (R,R)-**180**, the enantioselective version of this reaction of vinyloxiranes [51a] or 2-vinylthiiranes [53] with heterocumulenes afforded the optically active heterocycles (Scheme 6.31).

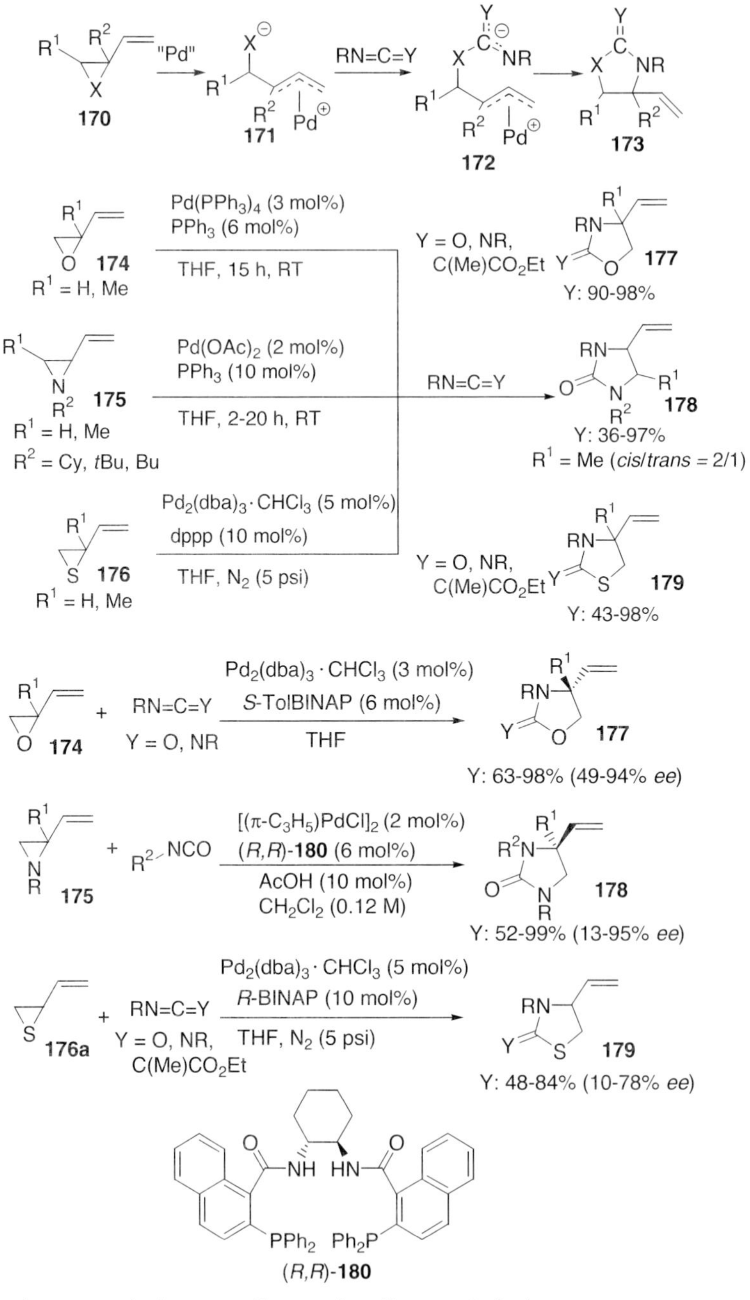

Scheme 6.31 The formation of five-membered heterocycles by the palladium-catalyzed ring-opening cycloaddition reactions of vinyloxiranes, vinylaziridines, and 2-vinylthiiranes with heterocumulenes.

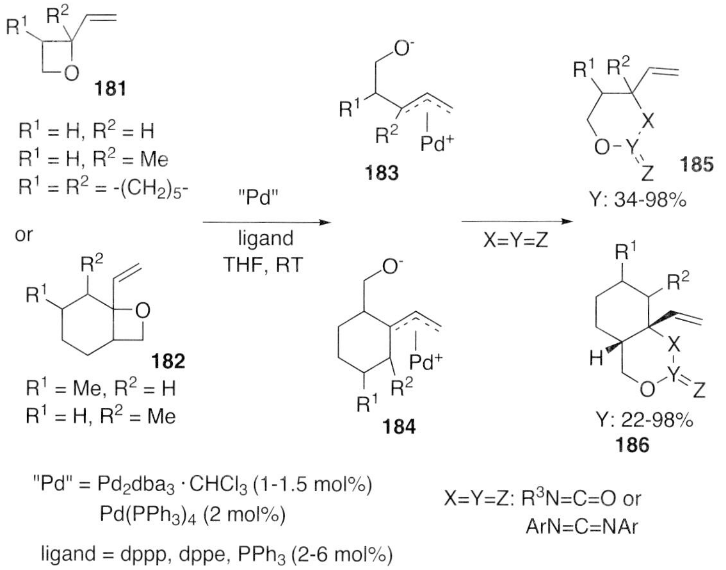

6.2 Palladium-Catalyzed Intramolecular Allylation Reactions

Scheme 6.32 Palladium-catalyzed cycloaddition reactions of vinyloxetanes with heterocumulenes.

Alper and Larksarp et al. also developed the palladium-catalyzed cycloaddition reactions of vinyloxetanes with heterocumulenes to prepare 4-vinyl-1,3-oxazin-2-imines (**185**, Z = NAr), 4-vinyl-1,3-oxazin-2-ones (**185**, Z = O), cis-3-aza-1-oxo-9-vinyl[4.4.0]decane derivatives **186** (Scheme 6.32) [54].

Alper et al. even observed a palladium-catalyzed ring-opening cyclization of 2-vinylpyrrolidines **187** with aryl isocyanates to synthesize the seven-membered diazepin-2-ones **188** (Scheme 6.33) [55].

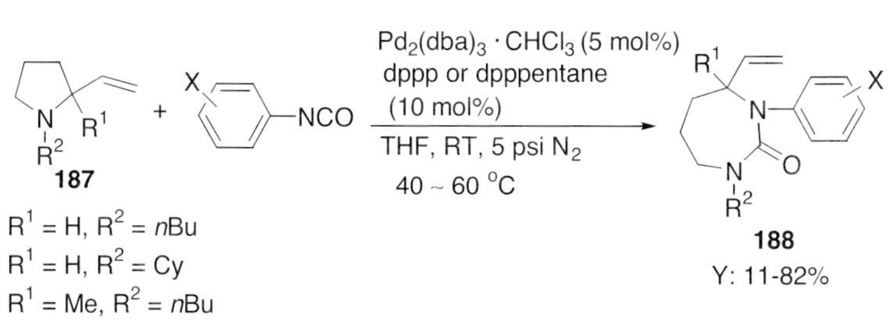

Scheme 6.33 Palladium-catalyzed ring-opening cyclization of 2-vinylpyrrolidines with aryl isocyanates.

The N-Boc-protected optically active pyrrolidines **190** were obtained via the palladium-catalyzed enantioselective [3 + 2] cycloaddition of 2-trimethylsilylmethylallyl acetate with imines **189** (Eq. (6.7)) [56].

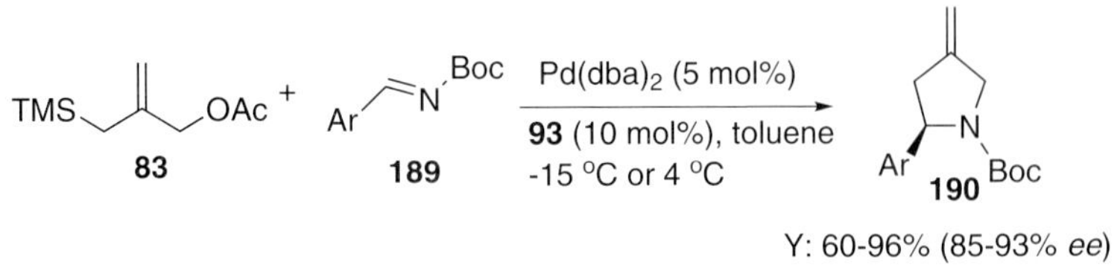

6.2.3
Cyclic O- or S-Allylation

A palladium-catalyzed ligand-controlled cyclization reaction of C2-symmetric 1,9-diacetoxy-2,7-nonadien-4,6-diol **191** yielded tetrahydrofuran **192** with three chiral centers highly stereoselectively. Simple functional group manipulation led to the desired F-ring of halichondrin B **193**. Desymmetrization of the *meso*-**191** enantioselectively provided the diastereomer **194** (Scheme 6.34) [57].

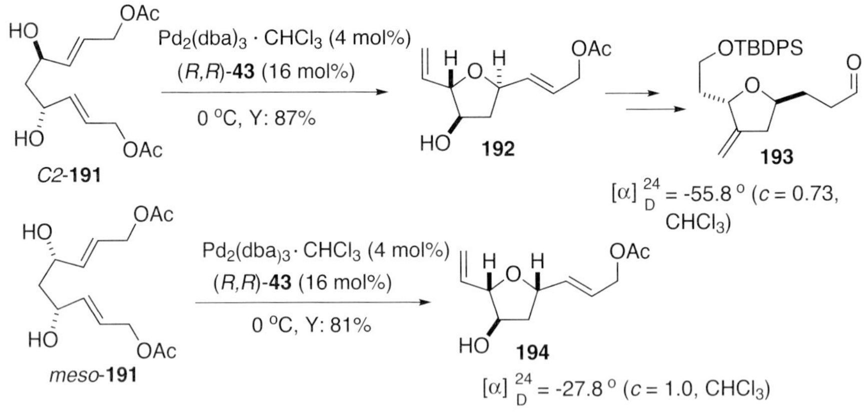

Scheme 6.34 Palladium-catalyzed allylic etherification reaction of 1,9-diacetoxy-2,7-nonadien-4,6-diols.

A palladium-catalyzed double cyclization allowed the selective formation of two stereocenters in *trans/threo/trans* bis-THF core structures in a single step by the use of the chiral ligand (R,R)-**43** (Scheme 6.35) [58].

The enantioselective synthesis of 2-isopropenyl-2,3-dihydrobenzofuran from 4-(2-hydroxyphenyl)-2(E)-methyl-2-butenyl methyl carbonate was realized via the palladium-catalyzed intramolecular allylation reaction in the presence of the chiral Trost ligand (R,R)-**43** (Eq. (6.8)) [59].

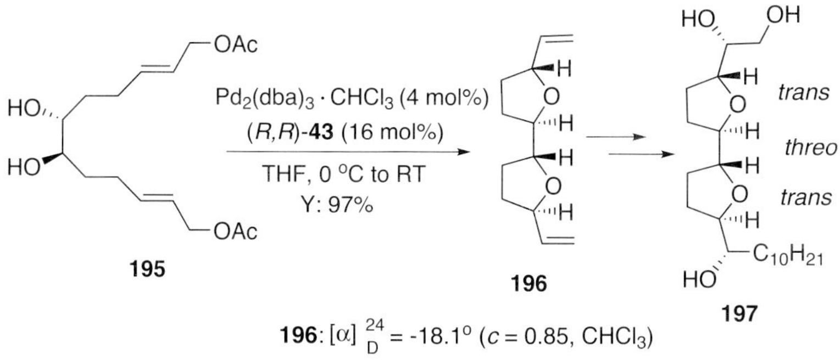

Scheme 6.35 Palladium-catalyzed double cyclization of 1,12-diacetoxy-2,10-dodecadien-6,7-diol **195**.

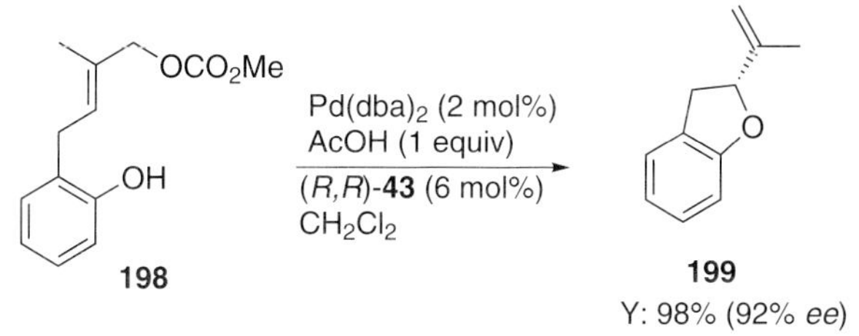

196: $[\alpha]_D^{24}$ = -18.1° (c = 0.85, CHCl$_3$)

(6.8)

199
Y: 98% (92% ee)

An intramolecular Pd-catalyzed allylation of hydroxyamine-allylic acetate cis- and trans-**200** has been reported to preferentially afford trans- or cis-3-substituted-5-vinyl isoxazolidines trans- or cis-**201**, respectively (Scheme 6.36) [60].

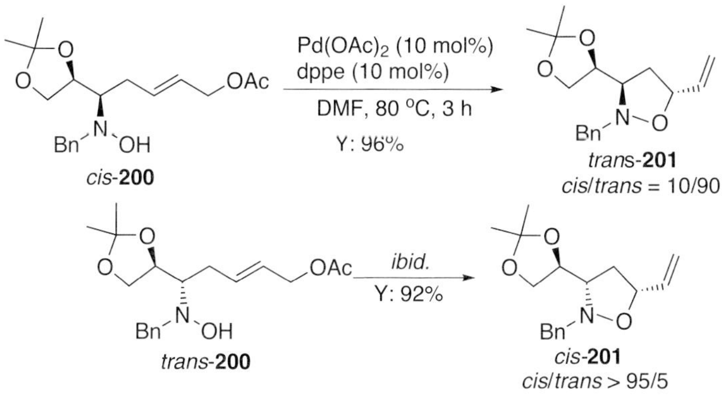

Scheme 6.36 Intramolecular Pd-catalyzed allylic alkylation of hydroxyamine-allylic acetate.

Different from intramolecular allylation of **1** with the carbonucleophile, an efficient palladium-catalyzed intramolecular O-allylaltion reaction of **1b** with alcohol or water as cosolvent was also developed to afford the α-alkoxycarbonyl-β-phenyl-γ-vinyl γ-lactones **202** in 73–87% yield with excellent diastereoselectivity via the reaction of

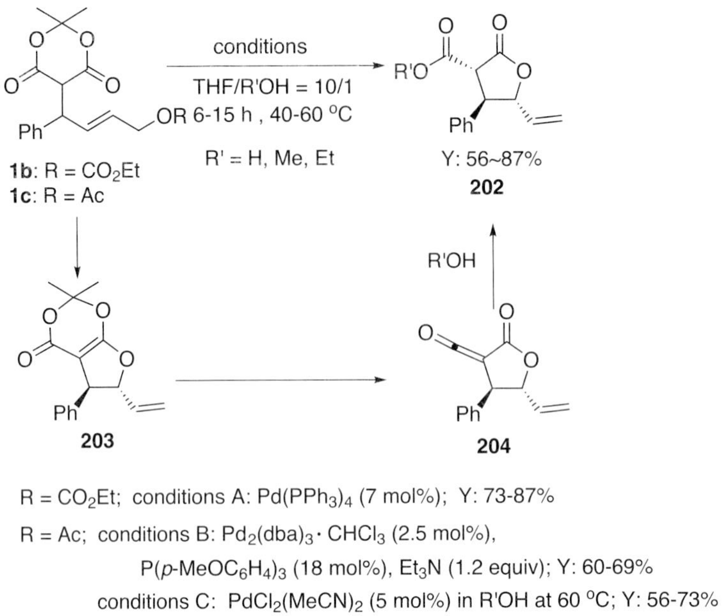

Scheme 6.37 Synthesis of α-alkoxycarbonyl-β-phenyl-γ-vinyl γ-lactones.

R = CO₂Et; conditions A: Pd(PPh₃)₄ (7 mol%); Y: 73–87%
R = Ac; conditions B: Pd₂(dba)₃·CHCl₃ (2.5 mol%),
P(p-MeOC₆H₄)₃ (18 mol%), Et₃N (1.2 equiv); Y: 60–69%
conditions C: PdCl₂(MeCN)₂ (5 mol%) in R'OH at 60 °C; Y: 56–73%

the ketene intermediate **204** with the alcohol in the solvent. Under the modified conditions B or C, similar reaction of acetate **1c** could also be achieved (Scheme 6.37) [2].

The intramolecular O-allylation reaction of the allylic carbonate moiety with the enolic oxygen of the 1,3-diketone in *S*-**205** under the catalysis of Pd₂(dba)₃·CHCl₃ and dppf afforded 5-(propen-1(*E*)-yl)-4,5-dihydrofuran *R*-(*E*)-**206** and 5-(propen-1(*Z*)-yl)-4,5-dihydrofuran *S*-(*Z*)-**206** in a ratio of 3/1 with high enantioselectivity (97% ee) (Scheme 6.38) [61].

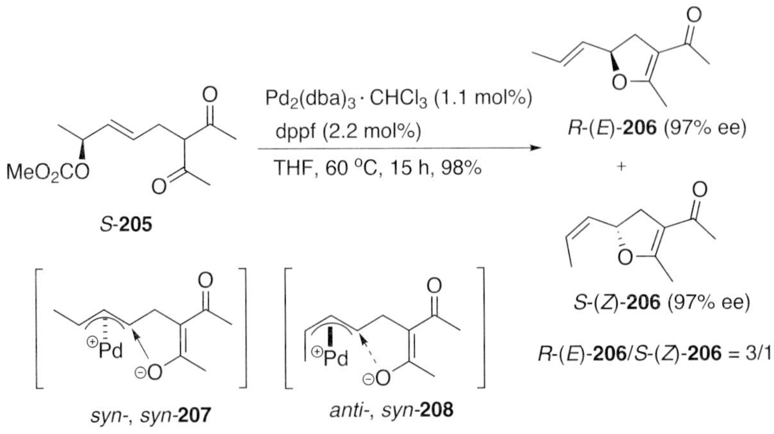

Scheme 6.38 Pd-catalyzed intramolecular O-allylation reaction with 1,3-diketone moiety-1,3-chirality transfer.

6.2 Palladium-Catalyzed Intramolecular Allylation Reactions

The palladium-catalyzed intramolecular reaction of allyl chloride moiety with the oxygen atom at the carbonyl benzamide in the presence of NaH and n-Bu$_4$NI in THF of 5-benzamidio-4-silyloxy-2(E)-alkenyl chloride **209** or 6-(4′-methoxyphenyl)-5-benzamidio-4-silyloxy-2-hexenyl chloride **212** afforded the kinetically or thermodynamically favored products, depending on the reaction temperature. Among the compounds prepared **213** has been used as a key intermediate for the enantio-selective total synthesis of anisomycin (Scheme 6.39) [62].

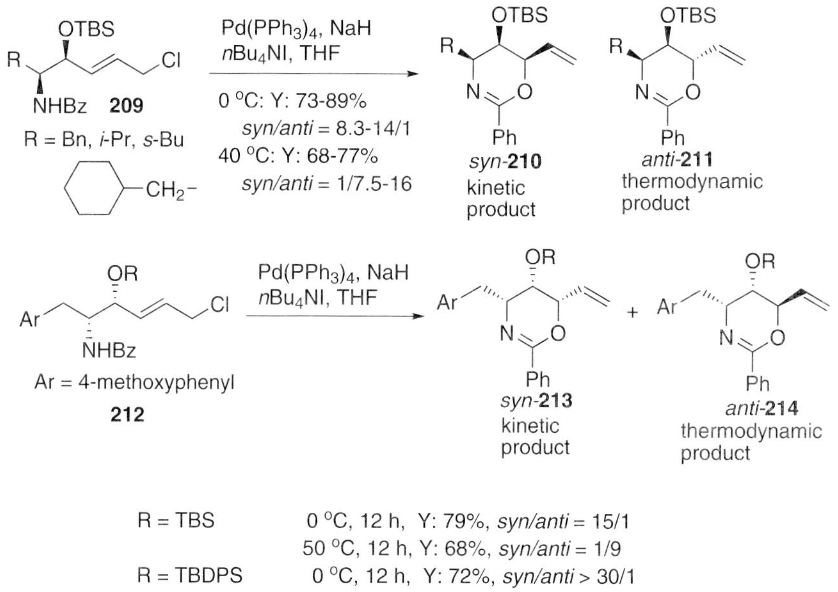

Scheme 6.39 Stereoselective synthesis of oxazines.

The asymmetric tandem double allylic substitution of catechol, 1,2-diol, diamine, or hydroxylamine **215** with 2(Z)-butenol diacetate or dicarbonate **216** using chiral ligands led to the formation of vinyl-substituted six-membered heterocycles **217** in 48–87% yield with 44–94% ee (Scheme 6.40) [63].

Yoshida and Ihara et al. developed a palladium-catalyzed CO$_2$-recycling reaction of 4-hydroxy-2-alkenylic carbonates **223** to afford cyclic carbonates **224** in 60–79% yield (Scheme 6.41) [64]. A highly stereoselective reaction of optically active allylic carbonate S-**225** (95% ee) occurred smoothly to produce chiral heterocycle S-**226** in 85% yield with 93% ee (Scheme 6.41) [64a]. In the presence of CO$_2$ (1 atm) and DBU, the similar reaction of 4-amino allylic carbonates, acetate, or benzoate **228** was reported to afford the vinyloxazolidinones **229** in 53–73% yield (Scheme 6.41) [36]. It is proven that the presence of DBU is necessary for the efficient fixation of CO$_2$.

A palladium-catalyzed domino reaction of 4-hydroxy-2-butynyl carbonates **230** with phenols was also reported to synthesize cyclic carbonates **231** via a CO$_2$

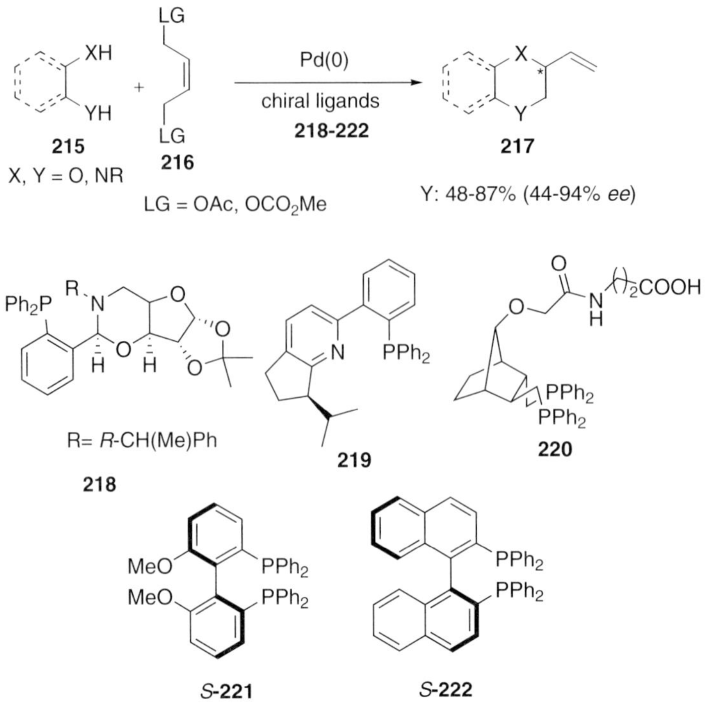

6 Transition Metal-Catalyzed Intramolecular Allylation Reactions

Scheme 6.40 Asymmetric tandem double allylic substitution of catechol, 1,2-diol, diamine, or hydroxylamine **215** with 2(Z)-butenol diacetate or dicarbonate **216**.

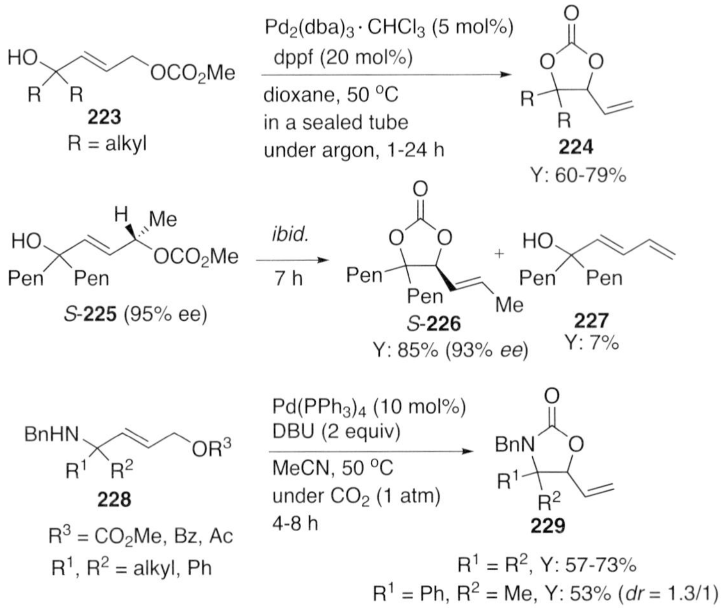

Scheme 6.41 Palladium-catalyzed CO_2-recycling reaction of allylic carbonates, or benzoates, or acetates.

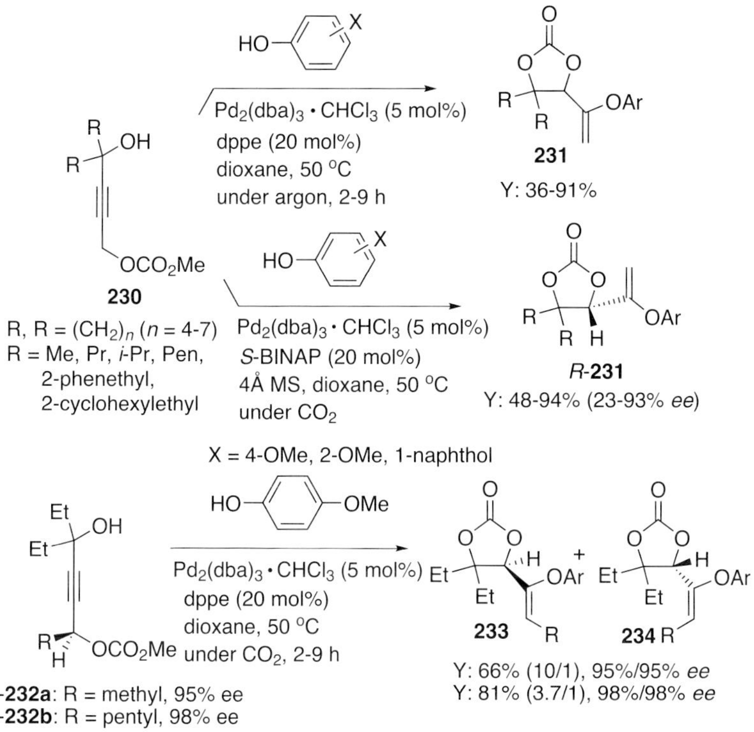

Scheme 6.42 Palladium-catalyzed CO_2-recycling reaction of 4-hydroxy-2-butynyl carbonates with phenols.

elimination-fixation process (Scheme 6.42) [65]. Using chiral ligands, optically active products could be yielded in 48–94% yield with 23–93% ee [66]. When optically active substrates were used, a cascade 1,3-chirality transfer process was achieved under similar reaction conditions (Scheme 6.42) [67].

Sinou's group developed a palladium-catalyzed stereoselective annulation reaction of 2-hydroxyphenol with propargylic carbonates **235** in the presence of a chiral bidentate P,P-ligand to afford 2-alkylidene-1,4-benzodioxanes **236** and **237** (Scheme 6.43) [68]. Reynolds and coworkers reported that the palladium-catalyzed cyclization reaction of the dihydroxypyrrole **241** with methyl propargylic carbonate afforded the cyclized product 3-methyl-2-methylidene-1,4-pyrrolo[c]dioxane **242** in 95–98% yield (Scheme 6.43) [69].

The palladium-catalyzed ring-opening cyclization reaction of 2-vinylthiirane with diphenylketene **247** afforded the oxathiolane derivative **248** in 59% yield (Scheme 6.44) [53]. The similar reaction with isothiocyanates **249** afforded the corresponding 1,3-dithiocyanates **250** in 72–98% yields. Azothiocyanate **251** was obtained via the reaction of 2-vinylaziridine **175a** with phenylisothiocyanate **249a** (Scheme 6.44) [52a].

3-Methylene-2,3-dihydrobenzofurans, 6-methylenebenzolactones, and 3-methyleneisochromans **252** were obtained by the reaction of terminal allenes with

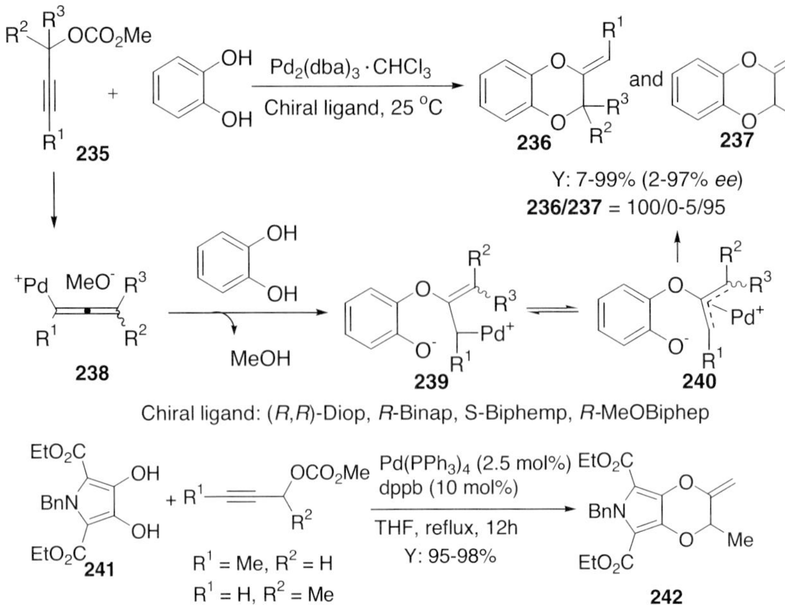

Scheme 6.43 Palladium-catalyzed cyclization reaction of catechol or dihydroxypyrrole with methyl propargylic carbonates.

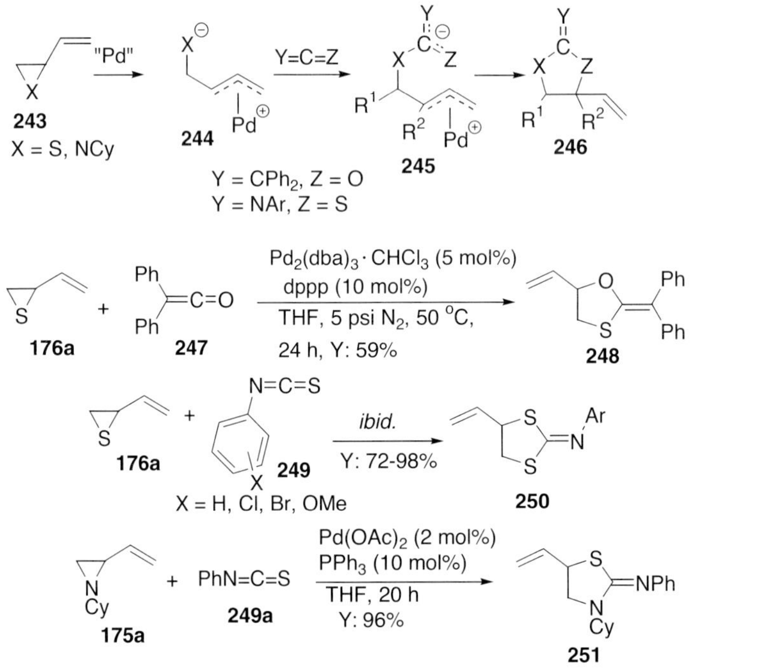

Scheme 6.44 Synthesis of oxathiolane derivatives, 1,3-dithiocyanates, and azothiocyanate.

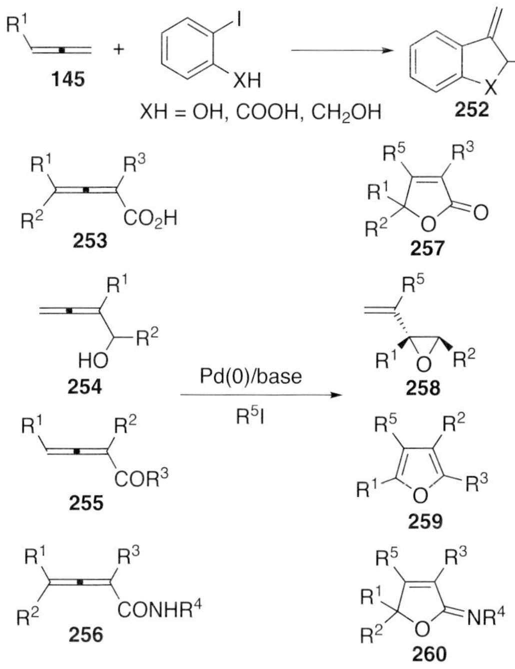

Scheme 6.45 Intermolecular carbopalladation of allenes followed by intramolecular allylation of functionalized allenes **253–256** or simple allenes **145** with aryl iodides.

2-iodophenol, 2-iodobenzoic acid, or 2-iodophenylmethanol (Scheme 6.45) [33]. In the presence of the Pd(0)-catalyst, the reaction of organic halides with 2,3-allenoic acid [70], allenols [71], 1,2-allenyl ketone [72], 2,3-allenamide [47] afforded the corresponding butenolides, *trans*-vinylic oxiranes, polysubstituted furans, and iminolactones (Scheme 6.45).

6.3
Iridium-Catalyzed Intramolecular Allylation Reactions

Using (S,S,S_a)-(+)-(3,5-dioxa-4-phosphacyclohepta[2,1-a;3,4-a']dinaphthalen-4-yl) bis(1-(2'-methoxyphenyl)ethyl)amine **263**, Helmchen's group developed the enantioselective iridium-catalyzed intramolecular allylic alkylation of 7,7'- or 8,8'-bis (methoxycarbonyl)alk-2(E)-enyl carbonates **261** for the formation of the optically active carbocycles **262** with 96–97% ee (Scheme 6.46) [73]. The enantioselectivity of the Ir-catalyzed reaction is better than that of the Pd-catalyzed reaction (compare Scheme 6.46 with Scheme 6.1). Cyclic amines **266** may be prepared by the reaction of amine-allylic acetate **264** or amine-allylic carbonate **265** by using chiral phosphane-oxazoline *R*-**7c** or chiral phosphorus amidite **267** as the ligand (Scheme 6.46) [74].

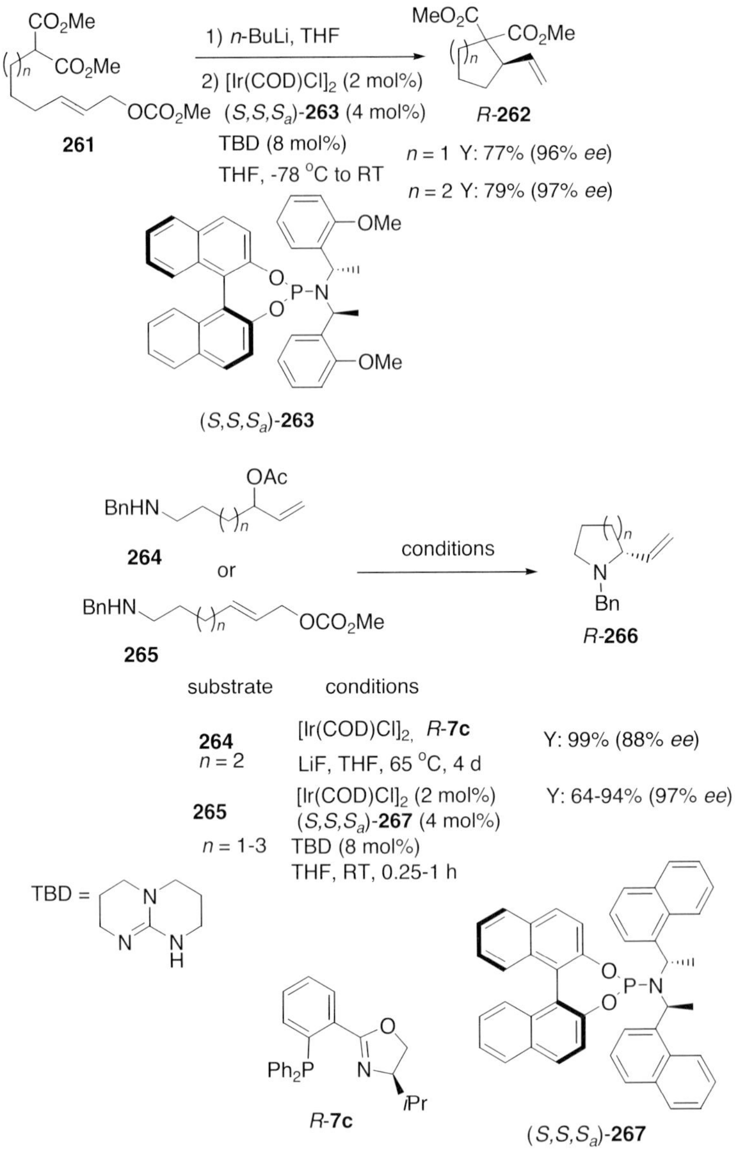

6 Transition Metal-Catalyzed Intramolecular Allylation Reactions

Scheme 6.46 Synthesis of optically active carbocycles and azacycles via Ir-catalyzed intramolecular allylation.

Takemoto's group reported an iridium-catalyzed sequential double allylic amination of **268** or **269** having two allylic carbonate moieties with primary amine to afford the azacycles **270** in 42–84% yield with low diastereoselecitivity (Scheme 6.47) [75].

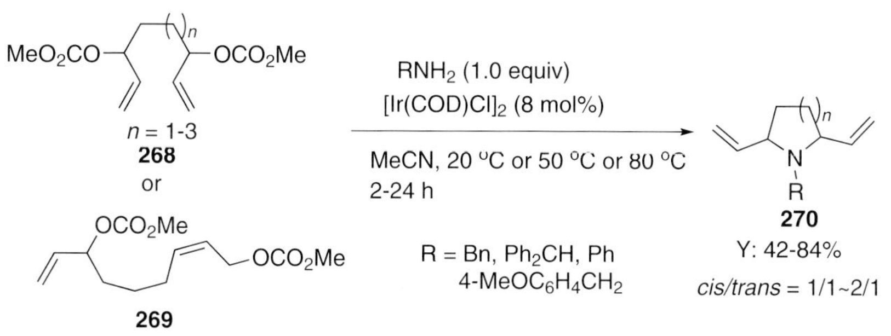

Scheme 6.47 Synthesis of divinylsubstituted 5-, 6-, and 7-membered azacycles via Ir-catalyzed intermolecular and intramolecular allylic amination.

By using chiral ligand (S,S_a)-**272**, Helmchen's group developed an iridium-catalyzed enantioselective double allylic amination of 8-(methoxycarbonyl)octa-2(E),6(E)-dienyl carbonate or 9-(methoxycarbonyl)nona-2(E),7(E)-dienyl carbonate **271** with benzylamine, which afforded the optically active azacycle **270a** in 59–77% yield and >99% ee with high diastereoselecitivity (Scheme 6.48) [74b].

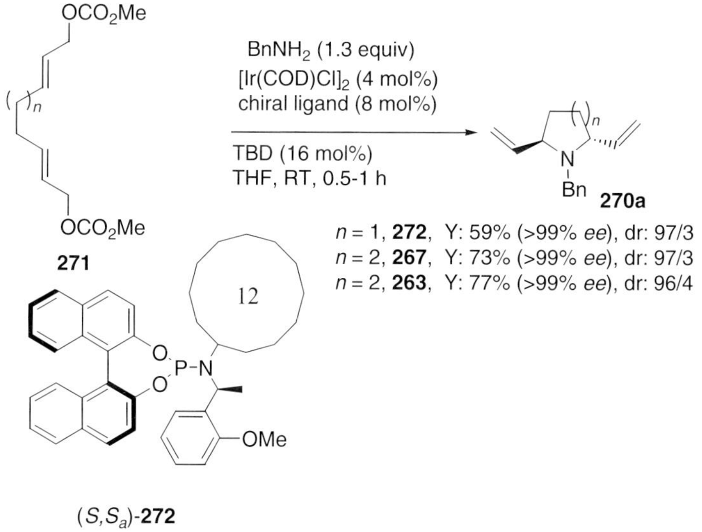

Scheme 6.48 Enantioselective synthesis of 2,5- or 2,6-divinyl-substituted five- or six-membered azacycles.

By using phosphorus amidite **267** as a chiral ligand, an Ir-catalyzed intramolecular allylic etherification reaction of 5-(2'-hydroxyphenyl)pent-2(E)-enyl carbonate **273** was realized to afford the chromane derivative R-(−)-**274** in 60% yield with 95% ee (Eq. (6.9)) [74b].

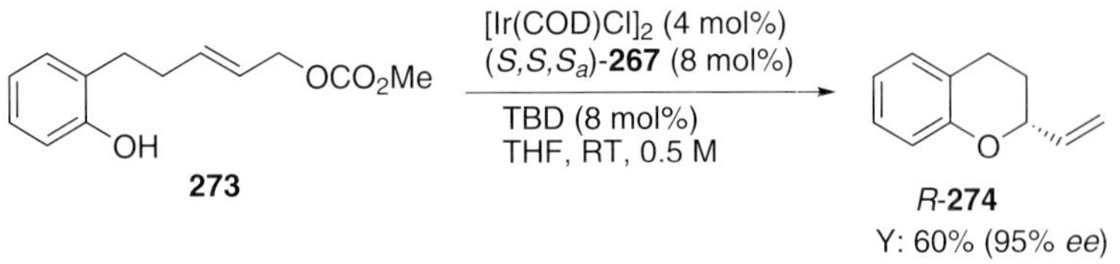

6.4
Ni-Catalyzed Intramolecular Allylation Reactions

With *R*-MeOBiphep being the best chiral ligand among the many chiral bidentate P, P-ligands screened, the asymmetric Ni(COD)$_2$-catalyzed intramolecular allylic amination of ethyl 4-[(4-methoxyanilino)carbonyloxy]-2(Z)-butenyl carbonate **275** afforded N-(p-methoxyphenyl)-4-vinyl-2-oxazolidinone **276** with 75% ee in 88% yield. The ee of the product could be improved from 75% to 97% with 64% overall yield after recrystallization (Scheme 6.49) [76].

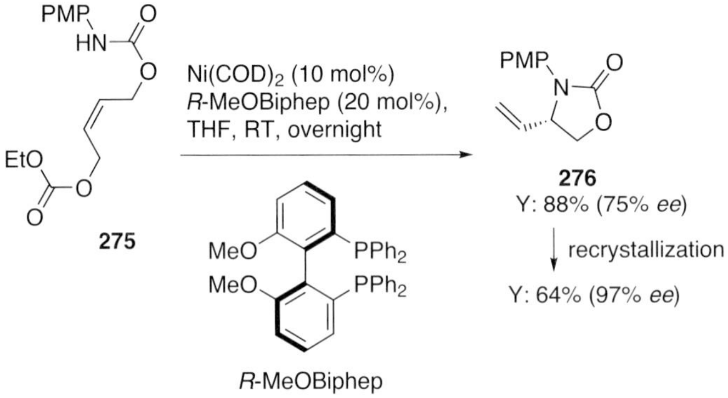

Scheme 6.49 Synthesis of optically active 4-vinyl-2-oxazolidinone based on enantioselective Ni(COD)$_2$-catalyzed intramolecular allylic amination.

6.5
Rh-Catalyzed Intramolecular Allylation Reactions

Martin *et al.* reported that the Rh-catalyzed intramolecular allylation reaction of 5-methoxycarbonyloxy-3(Z)-pentenyl acetoacetate **277** afforded the not-readily-available eight-membered 2-acetyl-4(Z)-ζ-heptenlactone **278**. Compared with the catalysis of Rh, the same reaction under the catalysis of Pd(PPh$_3$)$_4$ afforded a mixture of the eight-membered ring and the six-membered ring with a ratio of 60/40 (Eq. (6.10)) [77].

$$\text{277} \xrightarrow[\text{K}t\text{OBu, DMF, 0 °C} \atop \text{Y: 68\%}]{[\text{Rh(CO)}_2\text{Cl}]_2 \text{ (5 mol\%)}} \text{278} \tag{6.10}$$

Using NaH as the base, the [Rh(CO)₂Cl]₂-catalyzed cyclization of methyl 3-oxo-8-(methoxycarbonyloxy)-6(Z)-octenoate **279** in DMF afforded 2(E)-methoxycarbonyl-methylene-5-vinyltetrahydrofuran **280** exclusively in 71% yield. The formation of the corresponding carbocycle was not observed (Eq. (6.11)) [78].

$$\text{279} \xrightarrow[\text{DMF, 0 °C} \atop \text{Y: 71\%}]{[\text{Rh(CO)}_2\text{Cl}]_2, \text{ NaH}} \text{280} \tag{6.11}$$

6.6
Conclusion and Perspectives

In the last 30 years, transition-metal catalyzed allylation has been developed systematically and widely applied in organic synthesis due to its broad functional group tolerance and capability. In particular, the transition metal-catalyzed intramolecular allylation reaction has become one of the most efficient methods to construct very differently substituted carbocycles and heterocycles, which are widely existent in natural products and pharmaceutically interesting products. Acetoxyl and alkoxycarbonyloxyl are usually used as the leaving groups. When chiral ligands are used, the optically active cyclic compounds may be produced with high enantioselectivity. Palladium was the main metal studied to cope with different allylic substrates in the intramolecular allylation reaction, and the use of Ir, Ni, Rh complexes for this type of reaction is just at the very beginning. It is reasonable to believe that more efficient and stereoselective methods will be developed in this area in the near future.

6.7
Experimental: Selected Procedures

6.7.1
Synthesis of Optically Active Carbocycle 9 with Chiral Ligand S-7b (Scheme 6.1) [5]

To a flask were added, under N_2, $Pd_2(dba)_3 \cdot CHCl_3$ (28.5 mg, 0.031 mmol, 2.5 mol% Pd), phosphinooxazoline **S-7b** (36.1 mg, 0.093 mmol), and then degassed

DMSO (15 mL). Meanwhile, LiHMDS (2.74 mL of a 1 M solution in hexanes, 2.74 mmol, 0.92 equiv to malonate **8**) was added to a solution of diene **8** (1.00 g of 92:8 mass:mass with diethyl malonate, 2.49 mmol) in degassed DMSO (17 mL). After stirring at room temperature for 30 min, the reaction solution was sparged with N_2 for 2 min following which the Pd solution was transferred via cannula to the enolate solution. After stirring for 12 h, the reaction solution was diluted with EtOAc (200 mL) and water (200 mL). The organic layer was separated and the aqueous layer was extracted with EtOAc (2×200 mL). The combined organic layers were washed with brine, dried over $MgSO_4$, filtered and concentrated. The crude product was dissolved in Et_2O/petroleum ether $= 1/3$ (100 mL) and extracted with water (2×100 mL) to remove the remaining DMSO. The organic layer was dried over $MgSO_4$, filtered and concentrated to provide 0.675 g crude product **9**. ^1H NMR analysis indicated complete conversion of the starting diene, with no other products formed in >2% yield. The analytical sample of compound **9** was prepared by flash chromatography (eluent: hexane/EtOAc $= 98/2$ to 96/4).

6.7.2
Synthesis of Lactone 15 (n = 4) in an Aqueous–Organic Biphasic System (Scheme 6.3) [7]

To a 20-mL flask, $[(\pi\text{-}C_3H_5)PdCl]_2$ (4.6 mg, 0.0125 mmol) and the trisodium salt of tris(5-sulfonato-2-tolyl)phosphine **17** (67.5 mg, 0.11 mmol) were added under argon atmosphere. After degassed water (5 mL) was added, the mixture was stirred vigorously at 75 °C for 15 min. A homogeneous yellow clear solution was obtained. The solution was cooled to room temperature and then Na_2CO_3 (127.2 mg, 1.2 mmol) and ethyl acetate (3 mL) were added. A solution of 7-acetoxyhept-5-enyl 3-oxobutyrate (**14**(n = 4), 128.1 mg, 0.5 mmol) in ethyl acetate (2 mL) was added via a syringe to the ethyl acetate phase. After stirring for 12 h, the mixture was extracted with ethyl acetate. The organic extracts were dried through a short pad of silica gel on Na_2SO_3 layer. Concentration followed by purification of the residual oil afforded 3-acetyl-3,4,7,8,9,10-hexahydrooxecin-2-one (**15**(n = 4), 85.7 mg, 0.44 mmol) in 87% yield.

6.7.3
Synthesis of Lactone 25a in an Aqueous–Organic Biphasic system (Scheme 6.5) [9a]

Bis(trimethylsilyl)acetamide (0.3 mL, 1.2 mmol) and AcOK (10 mg, 0.1 mmol) are added to a solution of the acyclic substrate (1.0 mmol) in THF (20 mL) sequentially with stirring, under a nitrogen atmosphere. In a separate flask $Pd_2(dba)_3 \cdot CHCl_3$ (45 mg, 0.05 mmol) and PPh_3 (130 mg, 0.5 mmol) are added to the reaction vessel. After the resulting mixture is refluxed for 12 h, a saturated aqueous solution of NH_4Cl is added and the organic phase is extracted with Et_2O. The collected organic phases are dried, and the solvent is removed *in vacuo*. Flash chromatography (hexanes/AcOEt) gives the pure compounds as oils (**25a**, 69% from Z-**24a**, 73% from E-**24a**).

6.7.4
Synthesis of (4S,5S)-3-oxo-5-vinyl-1-aza-bicyclo[2.2.2]octane-4-carboxylic Acid Ethyl Ester (Scheme 6.9) [18]

To a dry flask carbonate **40** (720 mg, 2.4 mmol), Pd$_2$(dba)$_3$·CHCl$_3$ (24.7 mg, 0.024 mmol), and (R,R)-**43** (50.2 mg, 0.073 mmol) were added. The flask was sealed and flushed with argon. Degassed CH$_2$Cl$_2$ (25 mL) was added and the reaction mixture was stirred at room temperature for 43 h. After concentration *in vacuo*, the crude mixture was purified by flash chromatography on silica gel (50–100% ethyl acetate/petroleum ether) to give 409 mg (76%) of esters **41** and **42**.

6.7.5
Synthesis of trans-2-vinylcyclopentanecarbaldehyde (Scheme 6.10) [19]

A solution of Pd(PPh$_3$)$_4$ (8.9 mg, 7.8 µmol, 5 mol%), aldehyde **44** (X = Br, n = 1) (32 mg, 0.156 mmol), pyrrolidine (4.4 mg, 5.16 µL, 6.24 µmol, 40 mol%) and Et$_3$N (15.7 mg, 21.6 µL, 0.156 mmol) was stirred at room temperature, under an argon atmosphere. The reaction was monitored by TLC (eluent: petroleum ether/AcOEt 4/1). After 30 min., the reaction mixture was quenched with methanol (2 mL) and then cooled at 0 °C and NaBH$_4$ (58 mg, 10 equiv) was added in several portions. The reaction mixture was diluted with EtOAc (60 mL), washed with water, dried over MgSO$_4$ and concentrated under reduced pressure. Purification of the residue by dry-flash chromatography (eluent: hexanes/AcOEt = 4/1) afforded 14.1 mg (72%) of 2-hydroxymethyl-1-vinylcyclopentane, as a colorless oil. The product was obtained as a mixture of isomers in a ratio trans : cis = 11 : 1, as determined by GC.

6.7.6
Synthesis of 3-vinylindan-1,1-dicarboxylic Acid Diethyl Ester (Scheme 6.13) [23]

A solution of N,N-dimethylbutylamine (101.2 mg, 2.0 mmol) and diethyl 2-iodophenyl *cis*-but-2-en-4-ethoxycarbonyl-1-yl malonate **59** (X = I, Y = C(CO$_2$Et)$_2$) (237.1 mg, 0.5 mmol) in DMF (1 mL) was added to a suspension of indium (114.8 mg, 1.0 mmol), indium (III) chloride (55.29 mg, 0.25 mmol), lithium chloride (63.5 mg, 1.5 mmol), and [Pd(PPh$_3$)$_4$] (10 mol%, 57.3 mg) in DMF (1 mL). After the reaction mixture had been stirred at 100 °C for 1 h, it was quenched with saturated aqueous Na$_2$S$_2$O$_3$. The aqueous layer was extracted with ether (3 × 20 mL) and the combined organics were washed with water and brine, dried with MgSO$_4$, filtered, and concentrated under reduced pressure. The residue was purified by silica gel column chromatography (eluent: n-hexane) to give 3-vinylindan-1,1-dicarboxylic acid diethyl ester **60a** (155.3 mg, 80%).

6.7.7
Synthesis of (3S,3Sa,6Ra)-5-methylene-3-(2-naphthoxy)tetrahydrocyclopenta[c]furan-1-one (Scheme 6.18) [29]

To a solution of *ent*-**82** (2.26 g, 10.0 mmol), 2-(trimethylsliylmethyl)allyl acetate **83** (2.79 g, 15.0 mmol), and palladium(II) acetate (56.1 mg, 0.250 mmol) in toluene

(100 mL) was added triisopropyl phosphite (491 µL, 416 mg, 2.00 mmol). The reaction mixture was stirred under reflux for 12 h. After concentration *in vacuo* to 10 mL, flash chromatography (eluent: petroleum ether/Et$_2$O = 3/1) gave product ent-**84** as a white solid (2.53 g, 90%).

6.7.8
Synthesis of 1-Ethenyl-2-trifluoroacetyl-6,7-dimethoxy-1,2,3,4-tetrahydro-isoquinoline (Scheme 6.24) [40b]

A solution of ligand S-**121** (3.5 mg, 0.0075 mmol) and Pd$_2$(dba)$_3$ (1.2 mg, 0.00125 mmol) in CH$_2$Cl$_2$ (0.5 mL) was added under N$_2$ to a 5 mL round-bottomed flask fitted with a stirring bar. The solution was stirred at room temperature until the color of the solution turned to light yellow from purple. Then, **119** (23 mg, 0.05 mmol) in CH$_2$Cl$_2$ (0.5 mL) was added to the catalyst solution via a syringe. The mixture was stirred at room temperature until TLC indicated completion of the reaction. The reaction mixture was passed through a short pad of silica gel (eluent: hexanes/AcOEt = 10/1–6/1). The filtrate was then concentrated and subjected to HPLC analysis, using a Chiralcel OD-H column (hexanes/iPrOH = 98/2, 0.5 mL min^{-1}), which indicated that the enantiopurity of the product S-**120** was 95% ee with 100% conversion and product selectivity. The product **120** was isolated as a colorless oil as a mixture of two rotamers in 75/25 ratio (determined by ^{1}H NMR at 25 °C).

6.7.9
Synthesis of (R)-2-Isopropenyl-2,3-dihydrobenzofuran (eq. 6.8) [59a]

To a two-neck round-bottom flask was added CH$_2$Cl$_2$ (10 mL) followed by Pd(dba)$_2$ (5.0 mg, 0.0087 mmol). The solution was thoroughly degassed by bubbling Ar into the solution for 5 min. The solution was stirred for 30 min after adding the (R,R)-Trost ligand **43** (18.0 mg, 0.0261 mmol). Degassed acetic acid (26 mg, 0.44 mmol, 25 µL) was added to the solution at rt, and then after 5 min stirring, the allyl carbonate **198** (100 mg, 0.423 mmol) was added. After 18 h, analysis of the reaction mixture by TLC indicated that a new product had formed at a higher R$_f$ and that a small amount of starting material still persisted in the solution. Evaporation of the solvent and purification by column chromatography (eluent: hexane/AcOEt = 9/1) afforded the desired benzofuran R-**199** (66.6 mg, 98%) and 92% ee, as determined by chiral HPLC.

6.7.10
Synthesis of Optically Active Cyclic Carbonate S-226 (Scheme 6.41) [64a]

Pd$_2$(dba)$_3$·CHCl$_3$ (4.9 mg, 4.7 µmol) and dppe (7.5 mg, 18.8 µmol) were added to a stirred solution of allylic carbonate S-**225** (28.2 mg, 94.0 µmol) in dioxane (2 mL) in a sealed tube at rt. After stirring for 7 h at 50 °C, the reaction mixture was concentrated and the residue was chromatographed on silica gel (eluent: hexane/AcOEt = 9/1) to give cyclic carbonate S-**226** (21.8 mg, 85%) and diene **227** (1.5 mg, 7%) as colorless oils.

6.7.11
Synthesis of 2,3-dihydro-1,4-benzodioxine (Scheme 6.43) [68b]

A mixture of $Pd_2(dba)_3 \cdot CHCl_3$ (20.8 mg, 0.022 mmol), and diphosphine (0.091 mmol) in THF (7 mL) was stirred under a nitrogen atmosphere at rt for 30 min. This catalyst solution was added to a mixture of catechol (100 mg, 0.9 mmol) and the corresponding propargylic carbonate **235** (1.1 mmol) in the presence of triethylamine (269 mg, 2.6 mmol). The resulting solution was stirred at rt for 24 h. The solvent was evaporated and the residue was chromatographed over silica (eluent: petroleum ether/ethyl acetate) to afford the corresponding 2,3-dihydro-1,4-benzodioxine **236**.

6.7.12
Synthesis of Optically Active Carbocycles and Azacycles via Ir-Catalyzed Intramolecular Allylation (Scheme 6.46) [74b]

Under an atmosphere of argon in a dry Schlenk tube, a mixture of $[Ir(COD)Cl]_2$ (13.4 mg, 0.02 mmol), ligand **267** (0.04 mmol), TBD (11.2 mg, 0.08 mmol) and dry [D8]-THF (1 mL) was stirred at room temperature for 2 h to give a red solution. The catalyst solution was transferred to a dry NMR tube. Then the substrate **265** (1.0 mmol) was added and the mixture was reacted at room temperature until NMR monitoring showed complete conversion. The solvent was removed under reduced pressure and the crude product was purified by bulb-to-bulb distillation.

6.7.13
Synthesis of Compound 270 (n = 2) (Scheme 6.47) [75]

A mixture of **268** (n = 2) (100 mg, 0.387 mmol), benzylamine (41.5 mg, 0.387 mmol), and $[Ir(COD)Cl]_2$ (10.4 mg, 0.0155 mmol) in MeCN (1.0 mL) was stirred under argon atmosphere at 20 °C for 2 h. The reaction mixture was concentrated at reduced pressure. Purification of the residue by preparative TLC (eluent: hexane/EtOAc = 10/1) afforded cis-**270** (n = 2) (43 mg, 52%) and trans-**270** (n = 2) (26 mg, 32%).

6.7.14
Synthesis of N-(p-methoxyphenyl)-4-vinyl-2-oxazolidinone (Scheme 6.49) [76a]

The reaction was performed with **275** (400 mg, 1.30 mmol), $Ni(COD)_2$ (36 mg, 0.13 mmol) and R-MeOBIPHEP (152 mg, 0.26 mmol). After 6 h, the reaction afforded cyclization product **276** (250 mg, 88%, 75% ee). This compound was recrystallized from ether–hexane with a few drops of CH_2Cl_2, followed by addition of a seed crystal to give **276** (181 mg, 64% overall, 97% ee). A second recrystallization from the same solvent system afforded S-**276** (127 mg, 45%, >99% ee).

Abbreviations

Ac	acetyl
dba	dibenzylideneacetone
Bn	benzyl
BSA	N,O-bis(trimethylsilyl)acetamide
BSTFA	N,O-bis(trimethylsilyl)trifluoroacetamide
Bz	benzoyl
COD	*cis,cis*-1,5-cyclooctadiene
Cp	cyclopentadienyl
Cy	cyclohexyl
DBU	1,8-diazabicyclo[5,4,0]undec-7-ene
dppb	1,4-bis(diphenylphosphino)butane
dppe	1,2-bis(diphenylphosphino)ethane
dppf	1,1′-bis(diphenylphosphino)ferrocene
dppp	1,3-bis(diphenylphosphino)propane
dpppentane	1,5-bis(diphenylphosphino)pentane
EWG	electron-withdrawing group
fod	6,6,7,7,8,8-heptafluoro-2,2-dimethyl 3,5-octanedionate
Pen	pentyl
PMB	*p*-methoxybenzyl
TBD	1,5,7-triazabicyclo[4.4.0]dec-5-ene
TBDMS	*tert*-butyldimethylsilyl
TBS	*tert*-butyldimethylsilyl
Tf	*p*-trifluoromethansulfonyl
TMG	N,N,N′,N′-tetramethylguanidine
TMS	trimethylsilyl
Tol	tolyl
Ts	*p*-toluenesulfonyl
TTMPP	tris(2,4,6-trimethoxyphenyl)phosphine

References

1 For general reviews, see: (a) Trost, B.M. and van Vranken, D.L. (1996) *Chem. Rev.*, **96**, 395–422; (b) Trost, B.M. (1996) *Acc. Chem. Res.*, **29**, 355–364; (c) Trost, B.M. and Crawley, M.L. (2003) *Chem. Rev.*, **103**, 2921–2943; (d) Trost, B.M. (2004) *J. Org. Chem.*, **69**, 5813–5837; (e) Lu, Z. and Ma, S. (2008) *Angew. Chem.*, **120**, 264–303; (2008) *Angew. Chem. Int. Ed.*, **47**, 258–297.

2 Fillion, E., Carret, S., Mercier, L.G. and Trépanier, V.É. (2008) *Org. Lett.*, **10**, 437–440.

3 Sakaguchi, K., Yamada, T. and Ohfune, Y. (2005) *Tetrehedron Lett.*, **46**, 5009–5012.

4 Koch, G. and Pfaltz, A. (1996) *Tetrahedron: Asym.*, **7**, 2213–2216.

5 Bian, J., Van Wingerden, M. and Ready, J.M. (2006) *J. Am. Chem. Soc.*, **128**, 7428–7429.

6 Toyota, M., Hirota, M., Nishikawa, Y., Fukumoto, K. and Ihara, M. (1998) *J. Org. Chem.*, **63**, 5895–5902.

7 Kinoshita, H., Shinokubo, H. and Oshima, K. (2005) *Angew. Chem.*, **117**, 2449–2452;

(2005) *Angew. Chem. Int. Ed.*, **44**, 2397–2400.

8 Seo, S.-Y., Jung, J.-K., Paek, S.-M., Lee, Y.-S., Kim, S.-H., Lee, K.-O. and Suh, Y.-G. (2004) *Org. Lett.*, **6**, 429–432.

9 (a) Giambastiani, G., Pacini, B., Porcelloni, M. and Poli, G. (1998) *J. Org. Chem.*, **63**, 804–807; (b) Lemaire, S., Prestat, G., Giambastiani, G., Madec, D., Pacini, B. and Poli, G. (2003) *J. Organomet. Chem.*, **687**, 291–300; (c) Poli, G. and Giambastiani, G. (2002) *J. Org. Chem.*, **67**, 9456–9459.

10 Poli, G., Giambastiani, G. and Mordini, A. (1999) *J. Org. Chem.*, **64**, 2962–2965.

11 Madec, D., Prestat, G., Martini, E., Fristrup, P., Poli, G. and Norrby, P.-O. (2005) *Org. Lett.*, **7**, 995–998.

12 Lemaire, S., Giambastiani, G., Prestat, G. and Poli, G. (2004) *Eur. J. Org. Chem.*, 2840–2847.

13 Poli, G., Giambastiani, G., Malacria, M. and Thorimbert, S. (2001) *Tetrehedron Lett.*, **42**, 6287–6289.

14 Thuong, M.B.T., Sottocornola, S., Prestat, G., Broggini, G., Madec, D. and Poli, G. (2007) *Synlett*, 1521–1524.

15 Craig, D., Hyland, C.J.T. and Ward, S.E. (2005) *Chem. Commun.*, 3439–3441.

16 Craig, D., Hyland, C.J.T. and Ward, S.E. (2006) *Synlett*, 2142–2144.

17 Johns, D.M., Mori, M. and Williams, R.M. (2006) *Org. Lett.*, **8**, 4051–4054.

18 Trost, B.M., Sacchi, K.L., Schroeder, G.M. and Asakawa, N. (2002) *Org. Lett.*, **4**, 3427–3430.

19 Bihelovic, F., Matovic, R., Vulovic, B. and Saicic, R.N. (2007) *Org. Lett.*, **9**, 5063–5066.

20 Shintani, R., Park, S. and Hayashi, T. (2007) *J. Am. Chem. Soc.*, **129**, 14866–14867.

21 Jellerichs, B.G., Kong, J.-R. and Krische, M.J. (2003) *J. Am. Chem. Soc.*, **125**, 7758–7759.

22 Sekido, M., Aoyagi, K., Nakamura, H., Kabuto, C. and Yamamoto, Y. (2001) *J. Org. Chem.*, **66**, 7142–7147.

23 Seomoon, D., Lee, K., Kim, H. and Lee, P.H. (2007) *Chem. Eur. J.*, **13**, 5197–5206.

24 Nakamura, H., Aoyagi, K., Shim, J.-G. and Yamamoto, Y. (2001) *J. Am. Chem. Soc.*, **123**, 372–377.

25 Bandini, M., Melloni, A. and Umani-Ronchi, A. (2004) *Org. Lett.*, **6**, 3199–3202.

26 (a) Bandini, M., Melloni, A., Piccinelli, F., Sinisi, R., Tommasi, S. and Umani-Ronchi, A. (2006) *J. Am. Chem. Soc.*, **128**, 1424–1425; (b) Bandini, M., Melloni, A., Tommasi, S. and Umani-Ronchi, A. (2005) *Synlett*, 1199–1222.

27 (a) Kaneko, S. Yoshino, T. Katoh, T. and Terashima, S. (1998) *Tetrahedron*, **54**, 5471–5484; (b) Kaneko, S., Yoshino, T., Katoh, T. and Terashima, S. (1997) *Tetrahedron: Asym.*, **8**, 829–832; (c) He, X.-C., Wang, B., Yu, G. and Bai, D. (2001) *Tetrahedron: Asym.*, **12**, 3213–3216; (d) He, X.-C., Wang, B. and Bai, D. (1998) *Tetrahedron Lett.*, **39**, 411–414.

28 (a) Lautens, M., Klute, W. and Tam, W. (1996) *Chem. Rev.*, **96**, 49–92; (b) Trost, B.M. (1986) *Angew. Chem.*, **98**, 1–20; (1986) *Angew. Chem. Int. Ed.*, **25**, 1–20; (c) Trost, B.M. (1988) *Pure Appl. Chem.*, **60**, 1615–1626; (d) Marquand, P.L. and Tam, W. (2008) *Angew. Chem.*, **120**, 2968–2970; (2008) *Angew. Chem. Int. Ed.*, **47**, 2926–2928.

29 Trost, B.M. and Crawley, M.L. (2004) *Chem. Eur. J.*, **10**, 2237–2252.

30 Trost, B.M., Stambuli, J.P., Silverman, S.M. and Schwörer, U. (2006) *J. Am. Chem. Soc.*, **128**, 13328–13329.

31 (a) Trost, B.M., Cramer, N. and Bernsmann, H. (2007) *J. Am. Chem. Soc.*, **129**, 3086–3087; (b) Trost, B.M., Cramer, N. and Silverman, S.M. (2007) *J. Am. Chem. Soc.*, **129**, 12396–12397.

32 Aoyagi, K., Nakamura, H. and Yamamoto, Y. (2002) *J. Org. Chem.*, **67**, 5977–5980.

33 (a) Zenner, J.M. and Larock, R.C. (1999) *J. Org. Chem.*, **64**, 7312–7322; (b) Ma, S. (2003) *Handbook of Organopalladium Chemistry for Organic Synthesis* (ed E.-i. Negishi), Wiley-VCH, New York, pp. 1491–1521; (c) Mandai, T. (2004) *Modern Allene Chemistry*, vol. 2

34 (a) Ma, S. and Zhao, S. (2000) *Org. Lett.*, **2**, 2495–2497; (b) Ma, S., Jiao, N., Zhao, S. and Hou, H. (2002) *J. Org. Chem.*, **67**, 2837–2847.

35 Tietze, L.F., Schirok, H., Wöhrmann, M. and Schrader, K. (2000) *Eur. J. Org. Chem.*, 2433–2444.

36 Yoshida, M., Ohsawa, Y., Sugimoto, K., Tokuyama, H. and Ihara, M. (2007) *Tetrahedron Lett.*, **48**, 8678–8682.

37 Katritzky, A.R., Yao, J. and Yang, B. (1999) *J. Org. Chem.*, **64**, 6066–6070.

38 Trost, B.M., Krische, M.J., Radinov, R. and Zanoni, G. (1996) *J. Am. Chem. Soc.*, **118**, 6297–6298.

39 Hara, O., Sugimoto, K. and Hamada, Y. (2004) *Tetrahedron*, **60**, 9381–9390.

40 (a) Ito, K., Akashi, S., Saito, B. and Katsuki, T. (2003) *Synlett*, 1809–1812; (b) Shi, C. and Ojima, I. (2007) *Tetrahedron*, **63**, 8563–8570.

41 Trost, B.M. and Oslob, J.D. (1999) *J. Am. Chem. Soc.*, **121**, 3057–3064.

42 Seki, M., Hatsuda, M., Mori, Y. and Yamada, S.-i. (2002) *Tetrahedron Lett.*, **43**, 3269–3272.

43 (a) Trost, B.M. and Patterson, D.E. (1998) *J. Org. Chem.*, **63**, 1339–1341; (b) Trost, B.M. Pan, Z. Zambrano, J. and Kujat, C. (2002) *Angew. Chem.*, **114**, 4885–4887; (2002) *Angew. Chem. Int. Ed.*, **41**, 4691–4693; (c) Trost, B.M. Zambrano, J.L. and Richter, W. (2001) *Synlett*, 907–909; (d) Lee, S.-g. Lim, C.W. Song, C.E. Kim, K.M. and Jun, C.H. (1999) *J. Org. Chem.*, **64**, 4445–4451; (e) Song, C.E. Yang, J.W. Roh, E.J. Lee, S.-g. Ahn, J.H. and Han, H. (2002) *Angew. Chem.*, **114**, 4008–4010; (2002) *Angew. Chem. Int. Ed.*, **41**, 3852–3854.

44 Beccalli, E.M., Broggini, G., Paladino, G., Penoni, A. and Zoni, C. (2004) *J. Org. Chem.*, **69**, 5627–5630.

45 (a) Patil, N.T., Wu, H. and Yamamoto, Y. (2007) *J. Org. Chem.*, **72**, 6577–6579; (b) Lutete, L.M., Kadota, I. and Yamamoto, Y. (2004) *J. Am. Chem. Soc.*, **126**, 1622–1623; (c) Patil, N.T., Lutete, L.M., Wu, H., Pahadi, N.K., Gridnev, I.D. and Yamamoto, Y. (2006) *J. Org. Chem.*, **71**, 4270–4279.

46 (a) Ma, S., Yu, F., Li, J. and Gao, W. (2007) *Chem. Eur. J.*, **13**, 247–254; (b) Ma, S. and Gao, W. (2002) *Org. Lett.*, **4**, 2989–2992.

47 (a) Ma, S. and Xie, H. (2002) *J. Org. Chem.*, **67**, 6575–6578; (b) Cheng, X. and Ma, S. (2008) *Angew. Chem.*, **120**, 4657–4659; (2008) *Angew. Chem. Int. Ed.*, **47**, 4581–4583.

48 Hiroi, K., Hiratsuka, Y., Watanabe, K., Abe, I., Kato, F. and Hiroi, M. (2002) *Tetrahedron: Asym.*, **13**, 1351–1353.

49 Pinho, P., Minnaard, A.J. and Feringa, B.L. (2003) *Org. Lett.*, **5**, 259–261.

50 Trost, B.M., Machacek, M.R. and Faulk, B.D. (2006) *J. Am. Chem. Soc.*, **128**, 6745–6754.

51 (a) Larksarp, C. and Alper, H. (1997) *J. Am. Chem. Soc.*, **119**, 3709–3715; (b) Larksarp, C. and Alper, H. (1998) *J. Org. Chem.*, **63**, 6229–6233; (c) Shim, J.-G. and Yamamoto, Y. (1998) *J. Org. Chem.*, **63**, 3067–3071.

52 (a) Butler, D.C.D., Inman, G.A. and Alper, H. (2000) *J. Org. Chem.*, **65**, 5887–5890; (b) Trost, B.M. and Fandrick, D.R. (2003) *J. Am. Chem. Soc.*, **125**, 11836–11837.

53 Larksarp, C., Sellier, O. and Alper, H. (2001) *J. Org. Chem.*, **66**, 3502–3506.

54 Larksarp, C. and Alper, H. (1999) *J. Org. Chem.*, **64**, 4152–4158.

55 Zhou, H.-B. and Alper, H. (2003) *J. Org. Chem.*, **68**, 3439–3445.

56 Trost, B.M., Silverman, S.M. and Stambuli, J.P. (2007) *J. Am. Chem. Soc.*, **129**, 12398–12399.

57 Jiang, L. and Burke, S.D. (2002) *Org. Lett.*, **4**, 3411–3414.

58 Burke, S.D. and Jiang, L. (2001) *Org. Lett.*, **3**, 1953–1955.

59 (a) Pelly, S.C. Govender, S. Fernandes, M.A. Schmalz, H.-G. and de Koning, C.B. (2007) *J. Org. Chem.*, **72**, 2857–2864. Some reports about enantioselecitve intramolecular allylations: (b) Labrosse, J.-R. Poncet, C. Lhoste, P. and Sinou, D.

(1999) *Tetrahedron: Asym.*, **10**, 1069–1078; (c) Trost, B.M. Shen, H.C. Dong, L. and Surivet, J.-P. (2003) *J. Am. Chem. Soc.*, **125**, 9276–9277; (d) Trost, B.M. Shen, H.C. and Surivet, J.-P. (2004) *J. Am. Chem. Soc.*, **126**, 12565–12579; (e) Trost, B.M. Shen, H.C. Dong, L. Surivet, J.-P. and Sylvain, C. (2004) *J. Am. Chem. Soc.*, **126**, 11966–11983; (f) Trost, B.M. and Machacek, M.R. (2002) *Angew. Chem.*, **114**, 4887–4891; (2002) *Angew. Chem. Int. Ed.*, **41**, 4693–4697.

60 Merino, P., Tejero, T., Mannucci, V., Prestat, G., Madec, D. and Poli, G. (2007) *Synlett*, 944–948.

61 Hayashi, T., Yamane, M. and Ohno, A. (1997) *J. Org. Chem.*, **62**, 204–207.

62 Joo, J.-E., Lee, K.-Y., Pham, V.-T., Tian, Y.-S. and Ham, W.-H. (2007) *Org. Lett.*, **9**, 3627–3630.

63 (a) Nakano, H., Yokoyama, J.-i., Fujita, R. and Hongo, H. (2002) *Tetrahedron Lett.*, **43**, 7761–7764; (b) Wilkinson, M.C. (2005) *Tetrahedron Lett.*, **46**, 4773–4775; (c) Ito, K., Imahayashi, Y., Kuroda, T., Eno, S., Saito, B. and Katsuki, T. (2004) *Tetrahedron Lett.*, **45**, 7277–7281; (d) Yamazaki, A., Achiwa, I. and Achiwa, K. (1996) *Tetrahedron: Asym.*, **7**, 403–406; (e) Massacret, M., Lakhmiri, R., Lhoste, P., Nguefack, C., Abdelouahab, F.B.B., Fadel, R. and Sinou, D. (2000) *Tetrahedron: Asym.*, **11**, 3561–3568; (f) Massacret, M., Lhoste, P., Lakhmiri, R., Parella, T. and Sinou, D. (1999) *Eur. J. Org. Chem.*, 2665–2673.

64 (a) Yoshida, M., Ohsawa, Y. and Ihara, M. (2004) *J. Org. Chem.*, **69**, 1590–1597; (b) Yoshida, M. and Ihara, M. (2004) *Chem. Eur. J.*, **10**, 2886–2893.

65 Yoshida, M. and Ihara, M. (2001) *Angew. Chem.*, **113**, 636–639; (2001) *Angew. Chem. Int. Ed.*, **40**, 616–619.

66 Yoshida, M., Fujita, M., Ishii, T. and Ihara, M. (2003) *J. Am. Chem. Soc.*, **125**, 4874–4881.

67 Yoshida, M., Fujita, M. and Ihara, M. (2003) *Org. Lett.*, **5**, 3325–3327.

68 (a) Labrosse, J.-R., Lhoste, P. and Sinou, D. (2000) *Org. Lett.*, **2**, 527–529; (b) Dominczak, N., Damez, C., Rhers, B., Labrosse, J.-R., Lhoste, P., Kryczka, B. and Sinou, D. (2005) *Tetrahedron*, **61**, 2589–2599; (c) Labrosse, J.-R., Lhoste, P. and Sinou, D. (2002) *Eur. J. Org. Chem.*, 1966–1971; (d) Damez, C., Labrosse, J.-R., Lhoste, P. and Sinou, D. (2003) *Tetrahedron Lett.*, **44**, 557–560; (e) Labrosse, J.-R., Lhoste, P. and Sinou, D. (2001) *J. Org. Chem.*, **66**, 6634–6642; (f) Labrosse, J.-R., Lhoste, P., Delbecq, F. and Sinou, D. (2003) *Eur. J. Org. Chem.*, 2813–2822; (g) Labrosse, J.-R., Lhoste, P. and Sinou, D. (1999) *Tetrahedron Lett.*, **40**, 9025–9028.

69 Zong, K., Abboud, K.A. and Reynolds, J.R. (2004) *Tetrahedron Lett.*, **45**, 4973–4975.

70 (a) Ma, S. and Shi, Z. (1998) *J. Org. Chem.*, **63**, 6387–6389; (b) Ma, S., Shi, Z. and Wu, S. (2001) *Tetrahedron: Asym.*, **12**, 193–195.

71 (a) Ma, S. and Zhao, S. (1999) *J. Am. Chem. Soc.*, **121**, 7943–7944; (b) Xu, D., Li, Z. and Ma, S. (2002) *Chem. Eur. J.*, **8**, 5012–5018.

72 Ma, S. and Zhang, J. (2000) *Chem. Commun.*, 117–118.

73 Streiff, S. Welter, C. Schelwies, M. Lipowsky, G., Miller, N. and Helmchen, G. (2005) *Chem. Commun.*, 2957–2959.

74 (a) Welter, C., Koch, O., Lipowsky, G. and Helmchen, G. (2004) *Chem. Commun.*, 896–897; (b) Welter, C., Dahnz, A., Brunner, B., Streiff, S., Dübon, P. and Helmchen, G. (2005) *Org. Lett.*, **7**, 1239–1242.

75 Miyabe, S.H. Yoshida, K. Kobayashi, Y., Matsumura, A. and Takemoto, Y. (2003) *Synlett*, 1031–1033.

76 (a) Berkowitz, D.B. and Maiti, G. (2004) *Org. Lett.*, **6**, 2661–2664; (b) Berkowitz, D.B., Shen, W. and Maiti, G. (2004) *Tetrohedron: Asym.*, **15**, 2845–2851.

77 Ashfeld, B.L., Miller, K.A. and Martin, S.F. (2004) *Org. Lett.*, **6**, 1321–1324.

78 Ashfeld, B.L., Miller, K.A., Smith, A.J., Tran, K. and Martin, S.F. (2007) *J. Org. Chem.*, **72**, 9018–9031.

7
Cyclic Coupling Reactions
Xuefeng Jiang and Shengming Ma

7.1
Introduction

Coupling reactions have been shown to be very powerful for the formation of carbon–carbon, carbon–oxygen or –nitrogen bonds, and so on [1]. During the last 10 years intramolecular coupling reactions have been applied to the synthesis of many cyclic compounds. This chapter reviews the most representative reports on the preparation of cyclic products via intramolecular cross-coupling reactions from 1997 to 2008.

7.2
Cyclic Coupling Involving Organoboron

In the Pd-catalyzed Suzuki coupling reaction [2], the base played an essential role in this reaction. Because organoboron compounds may be easily synthesized by hydroboration [3] and are compatible with many functional groups, synthetic organic chemists have demonstrated the potential of the intramolecular Suzuki coupling reaction for the synthesis of different cyclic compounds.

In 1997, by using Suzuki aryl–aryl cross-coupling reaction Schlüter *et al.* prepared macrocyclic oligophenylenes **2** [4], which are useful for studies on aromaticity, host–guest chemistry, aggregation behavior, and molecular constructions. It is a feasible method for the gram scale synthesis of this type of cyclic compounds in yields as high as 85% (Scheme 7.1).

During the total synthesis of complestatin **5** [5], Zhu *et al.* utilized the palladium-catalyzed intramolecular Suzuki reaction of an aryl boronate and an aryl iodide in linear tripeptide **3** affording the 16-membered ring of complestatin **4** in good yield with excellent atropdiastereoselectivity (Scheme 7.2).

Sulikowski *et al.* constructed a 20-membered macrolactone **7** also by intramolecular Suzuki–Miyaura alkenyl–alkenyl cross-coupling during their total synthesis of apoptolidinone **8** (Scheme 7.3) [6].

Handbook of Cyclization Reactions. Volume 1.
Edited by Shengming Ma
Copyright © 2010 WILEY-VCH Verlag GmbH & Co. KGaA, Weinheim
ISBN: 978-3-527-32088-2

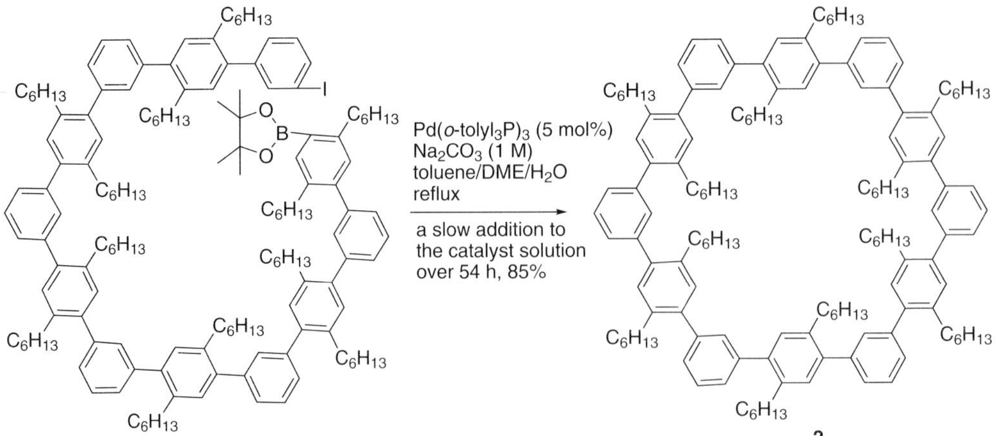

1 2

Scheme 7.1 Suzuki coupling for macrocyclic oligophenylene.

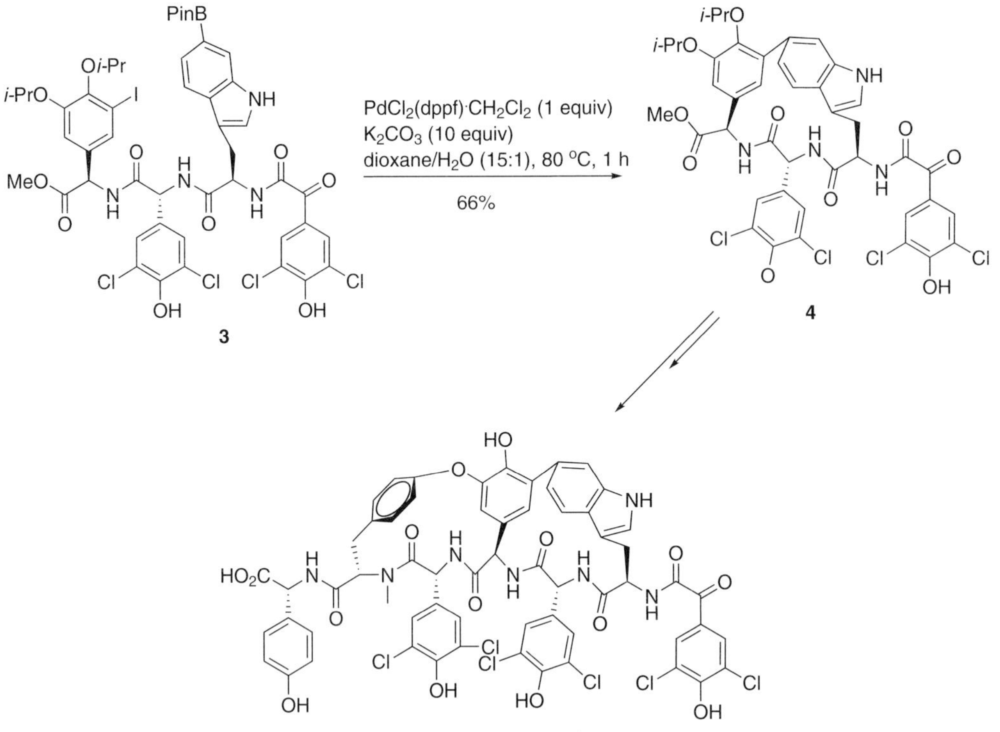

3 4

complestatin (**5**)

Scheme 7.2 Suzuki coupling of dipeptide.

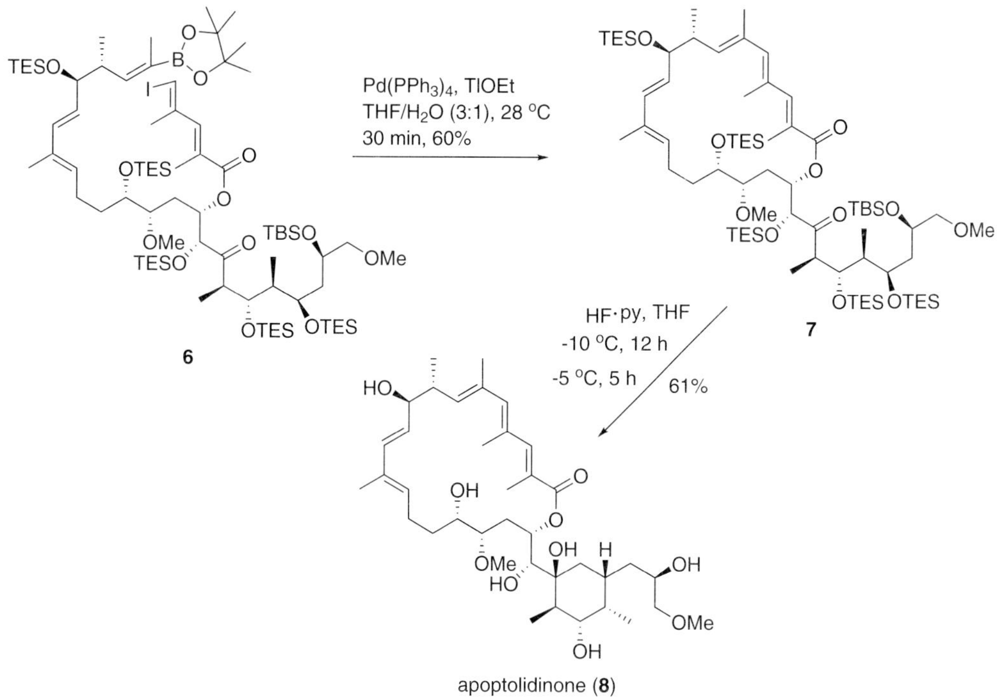

Scheme 7.3 Alkenyl–alkenyl coupling for lactone.

Recently, Roush et al. established a 24-membered ring in the total synthesis of (+)-superstolide A **11** [7]. Treatment of a 0.001 M solution of **9** in wet THF with TlOEt and a catalytic amount of Pd(PPh$_3$)$_4$ effected an intramolecular Suzuki reaction of the 1,3-dialkadienyl boronate moiety with alkenyl iodide providing the macrocyclic octaene **10** in 35–40% yields (Scheme 7.4).

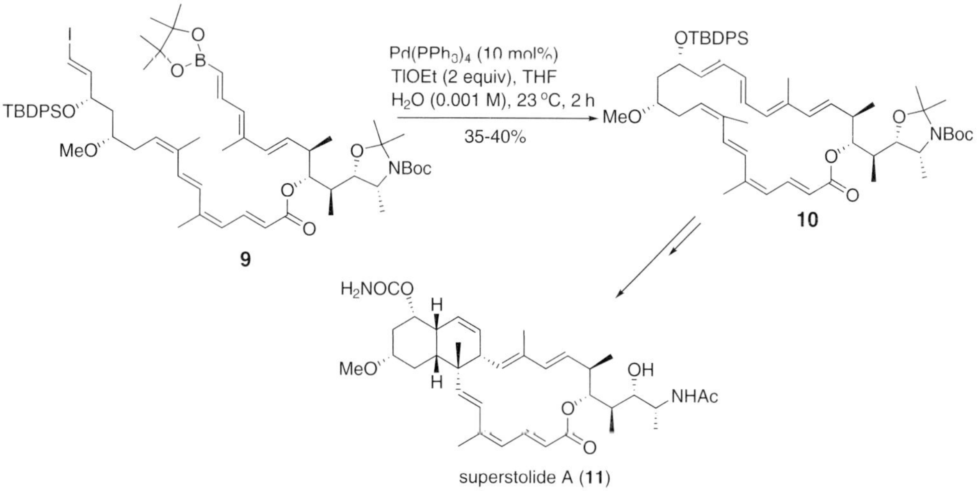

Scheme 7.4 Coupling for the total synthesis of (+)-superstolide A.

Molander et al. used an intramolecular Suzuki cross-coupling between an alkenyl potassium trifluoroborate and 1,3-alkadienyl bromide to construct the polyunsaturated 12-membered macrolactone core **13** of oximidine II **14** (Scheme 7.5) [8]. Other boranes, such as catecholborane, are ineffective.

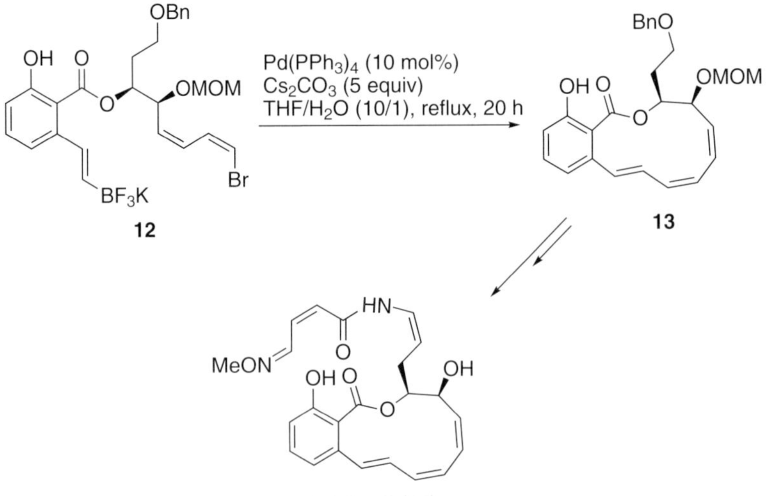

Scheme 7.5 Cyclic coupling for benzocyclic lactone.

In 2007, Nicolaou et al. reported the total synthesis of marinomycins A, B, and C [9]. Stille and Heck coupling reactions led to monomeric marinomycin A, but failed to deliver useful amounts of marinomycin A **17**. Then with slight modifications, they tested the synthetic strategies based on the Suzuki, Stille, and Heck coupling reactions as a potential tool for the ring closure. The Suzuki coupling reaction turned out to be the most expedient, and both Stille and Heck approaches delivered the monomeric marinomycin A only (Scheme 7.6).

A cyclic coupling between alkenyl iodide and alkyl borane was realized by the hydroboration of the terminal of the alkene in **18** with 9-BBN (5 equiv) and the cross-coupling sequence to afford **20**, then a stereospecific synthesis of salicylihalamide A was realized under highly dilute conditions by adding the borane **19** solution via a syringe pump to a mixture of a palladium catalyst and NaOH in benzene/H$_2$O at 80 °C (Scheme 7.7) [10]. The advantage of intramolecular alkyl borane–alkenyl iodide coupling over the complementary ring-closing metathesis reaction is the higher degree of control of the olefin geometry among the resulting macrocyclic adducts.

This group developed a double (**21**) or triple (**23**) intramolecular Suzuki coupling reaction for the efficient synthesis of tri- or tetracyclic products with a benzene core (Scheme 7.8) [11].

In 2000, Zhu's group prepared the 15-membered *m,m*-cyclophane **27** via aryl–aryl coupling with two distinct cross-coupling manifolds [12a], Miyaura's arylboronic ester synthesis [12b] and an intramolecular Suzuki reaction, proceeding in an

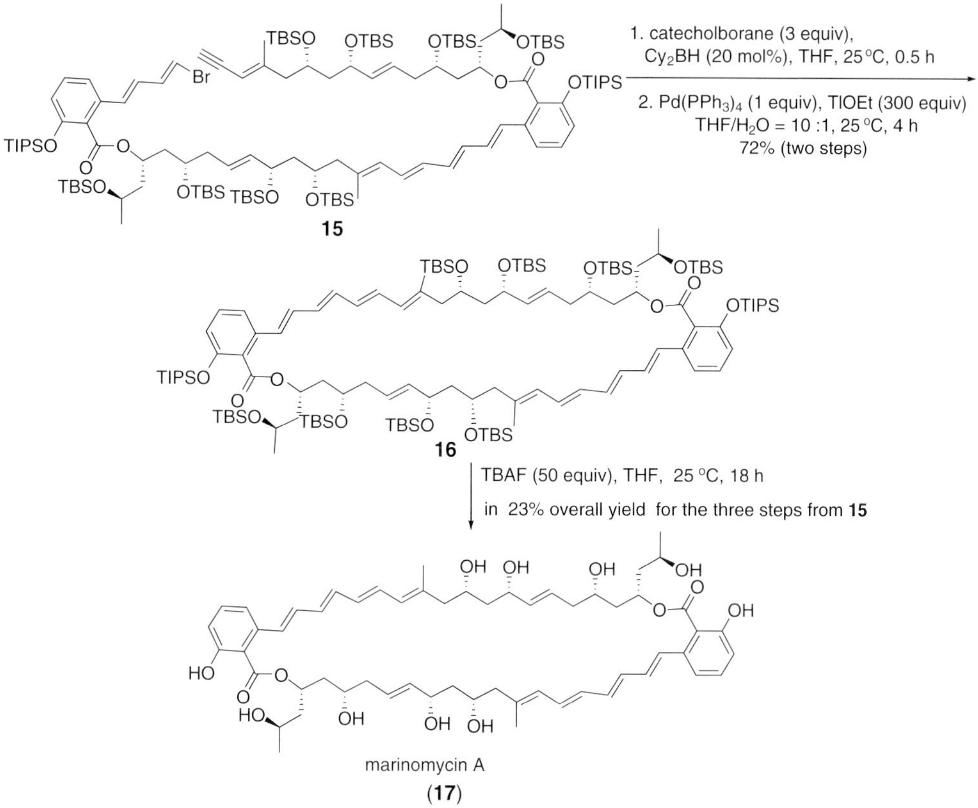

Scheme 7.6 Hydroborylation – cyclic Suzuki coupling for marinomycin A.

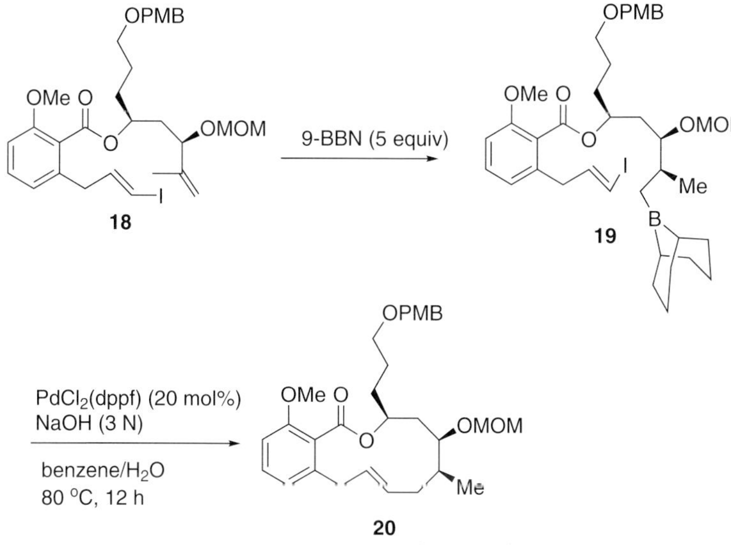

Scheme 7.7 Benzocyclic lactone via sp^2- C–sp^3-C Suzuki coupling.

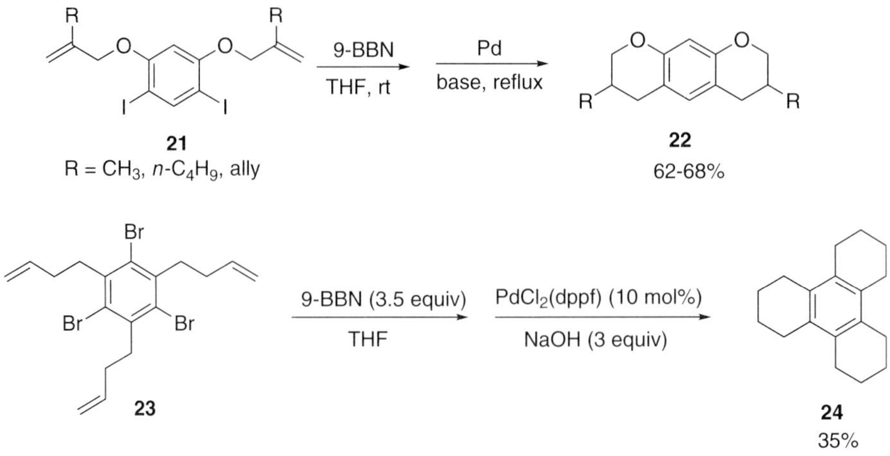

Scheme 7.8 Double or triple cyclic Suzuki coupling.

ordered fashion (Eq. (7.1)). Concentration is an important factor for the success of this process.

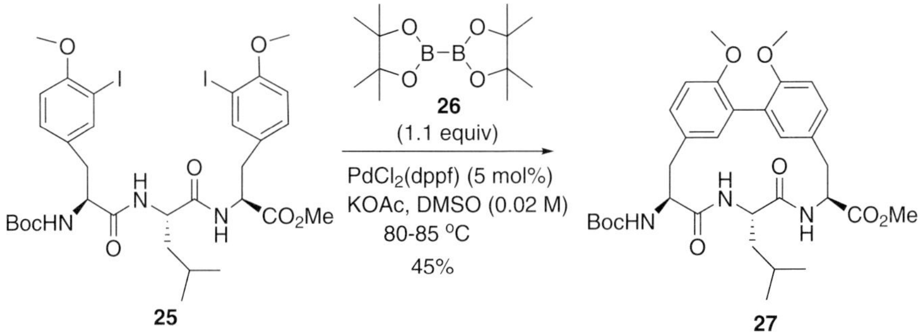

(7.1)

The intramolecular copper-mediated O-arylation furnishes macrocyclic inhibitors via the coupling between phenol and boronic acid in **28** [13]. The mild conditions using a weak base (Et$_3$N) at room temperature allow the synthesis of such base-sensitive amino acids without racemization (Eq. (7.2)). This methodology provided a general route to macrocyclic biphenyl ethers.

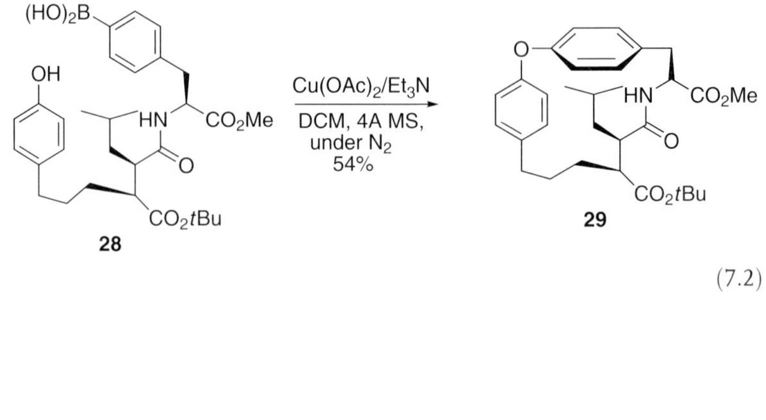

(7.2)

7.3
Cyclic Coupling Involving Organotin

Stille of Colorado State University developed the Pd-catalyzed coupling between organic halides and tin compounds [14]. Although organotin compounds are not so "green" for the environment, the compatibility, stereospecificity, and regioselectivity of the Stille coupling reaction are interesting, thus attracting many synthetic organic chemists to use the method within their synthetic processes.

Piers et al. have shown that it is possible to use Stille coupling between the cycloalkenyl triflate and cycloalkenyl tin in **30** to form a five-membered ring which converts the starting bicyclic compound **30** to a fused tricyclic system **31** (Scheme 7.9) [15].

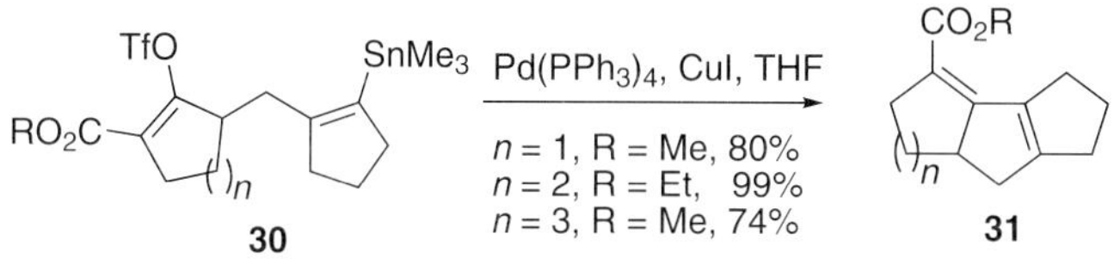

Scheme 7.9 Stille coupling for fused tricyclic compounds.

De Lucchi et al. have reported the triple Stille reaction of a bifunctional substrate **32** that leads to the formation of polycyclic products **33** (Eq. (7.3)) [16]. The mechanism by which trimers **33** are formed could be attractively envisaged as a 'head to tail' coupling between the tin and bromine termini of the double bonds.

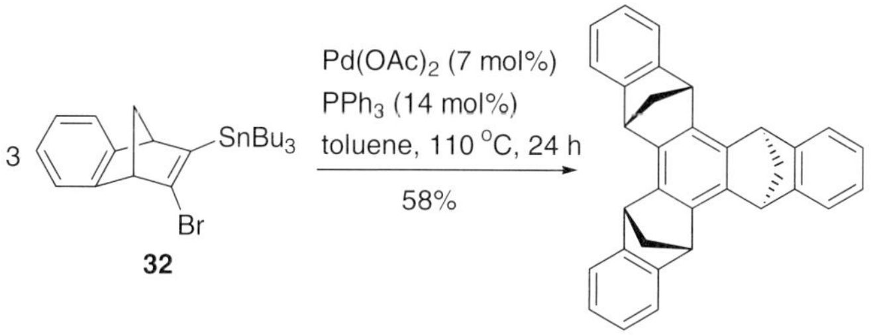

In 2000, Grigg et al. developed an interesting cascade reaction. Starting from **34** and Bu$_3$SnH, the hydrostannylation was followed by oxidative addition of Pd(0), 5-exo-trig cyclic carbopalladation, and an intramolecular Stille coupling to afford a wide range of bicyclic spiro- and bridged heterocycles (Scheme 7.10) [17].

Stereoselective construction of (E)-γ-tributylstannylmethylene butenolides **39** was achieved through the palladium-catalyzed tandem cross coupling/cyclization

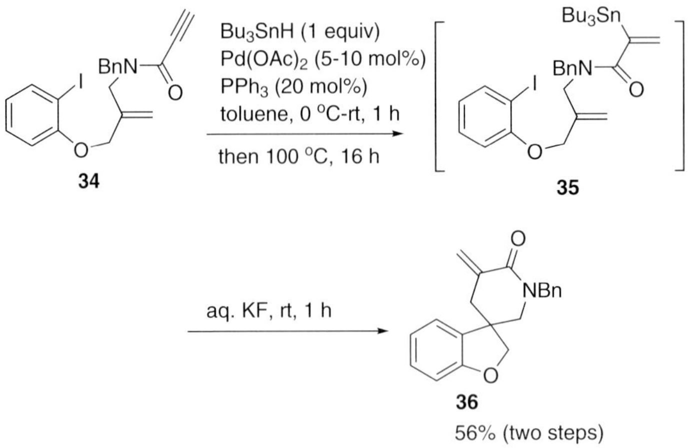

Scheme 7.10 Hydrostannylation – cyclic carbopalladation – cyclic Stille coupling.

reactions of tributylstannyl 3-iodopropenoate derivatives **37** with tributyltinacetylene **38** by Duchêne and Parrain et al. (Eq. (7.4)) [18].

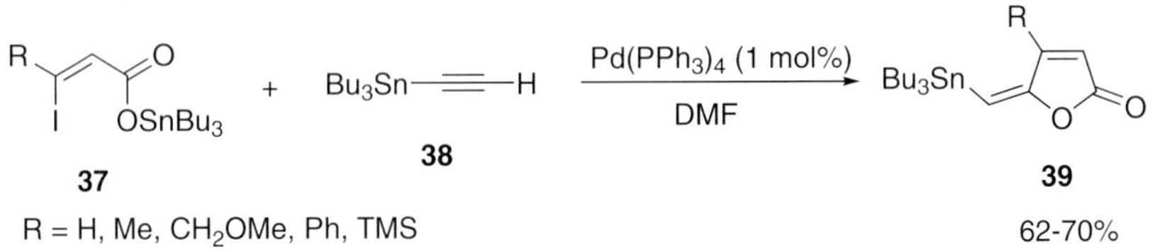

R = H, Me, CH$_2$OMe, Ph, TMS

62-70%

(7.4)

The reaction between acyl chlorides **42** and an O-stannyl ester-functionalized vinyltin **41** also involves a subsequent cyclization to give 1-pyran-2-ones **44** in good yields [19]. After radical hydrostannation of **40**, Stille coupling yields **43** by oxidative addition, transmetalation, and reductive elimination. Cyclization then occurs via a lactonization reaction on the dienol derived from **43** to afford **44** (Scheme 7.11).

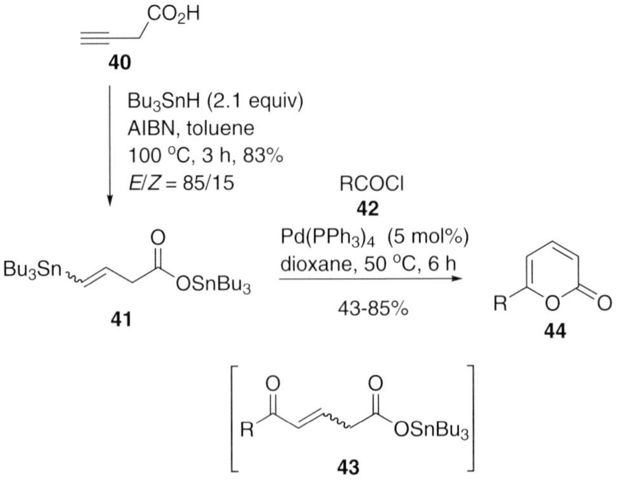

Scheme 7.11 Intermolecular coupling – lactonization for 1-pyran-2-ones.

The potent naturally occurring immunomodulators (−)-rapamycin **48** and (−)-27-demethoxyrapamycin **49** were synthesized by Smith, III in 1997 [20]. The macrocyclic framework was assembled by the effective Stille macrocyclization mediated by [(2-furyl)$_3$P]$_2$PdCl$_2$. He also achieved the total synthesis of the potent antiviral agent (−)-macrolactin A and two related family members, (+)-macrolactin E **52** and (−)-macrolactinic acid, via the Pd-catalyzed Stille cross-coupling reaction for the rapid stereospecific construction of the triene forming the macrocycles (Scheme 7.12) [21].

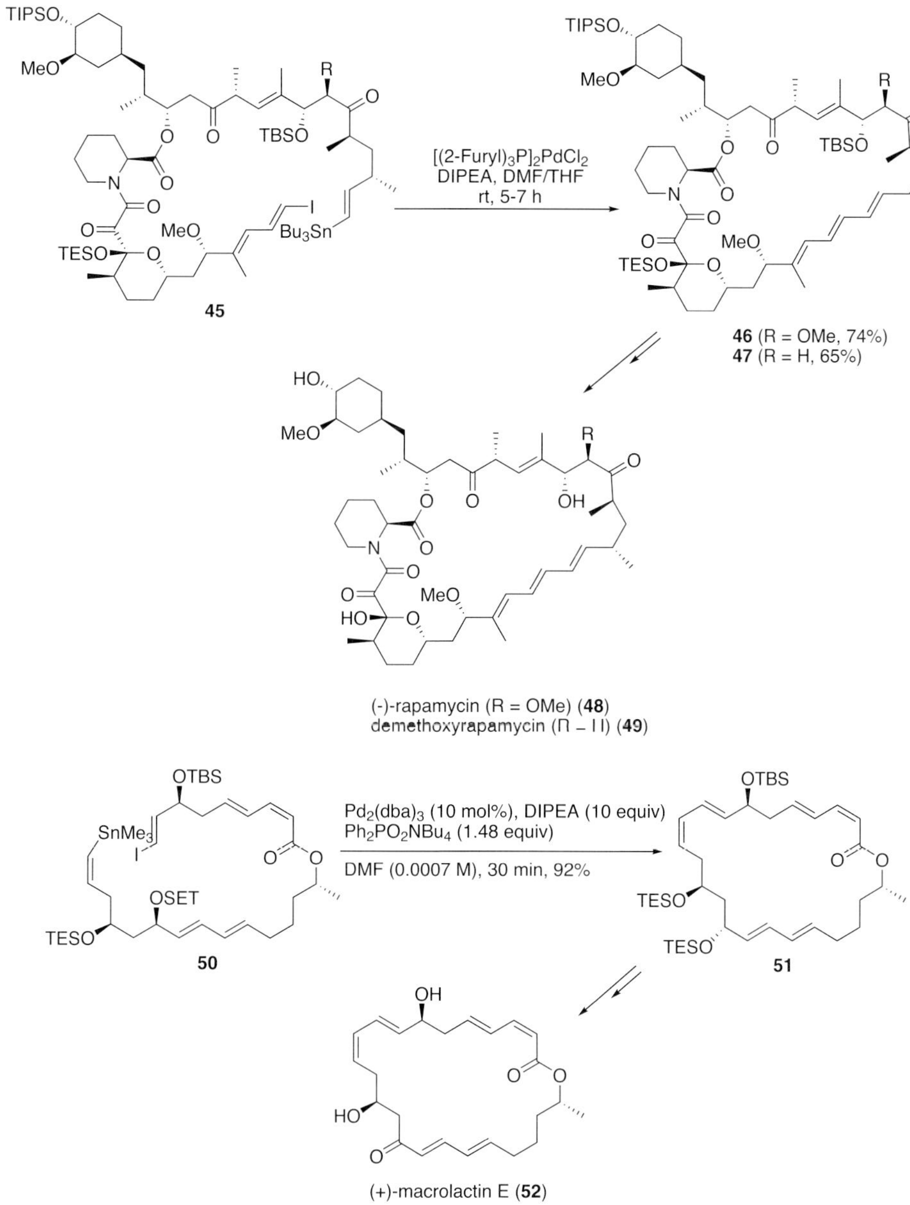

Scheme 7.12 Formation of macrolactones via cyclic Stille coupling.

In 1998, Nicolaou et al. described a solid-phase construction of macrocycles by employing Stille coupling in the total synthesis of (S)-zearalenone **55** by releasing the polymer-supported tin (Scheme 7.13) [22].

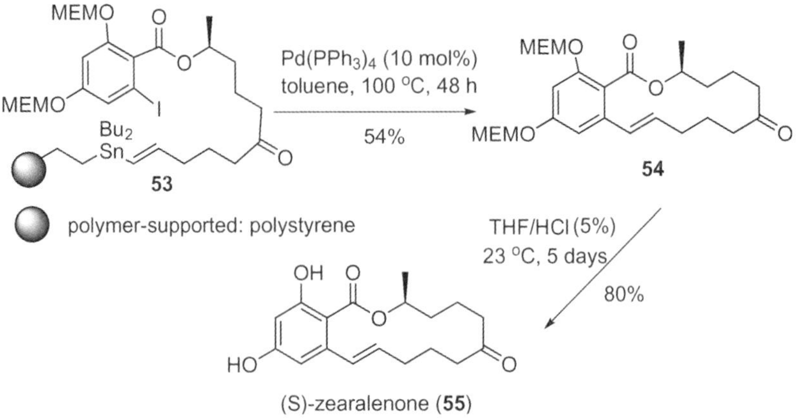

Scheme 7.13 Polymer-supported coupling for (S)-zearalenone.

Sanglifehrin A **62** is a newly discovered immunosuppressive agent containing a 22-membered macrolactone ring [23]. Nicolaou and his coworkers again developed a regioselective intramolecular Stille coupling to establish the main structure of sanglifehrin A **62** [24]. Formation of an amide bond between amine **56** and carboxylic acid **57** was effected by HATU/i-Pr$_2$NEt furnishing the cyclization precursor **58**. Ring closure of **58** took place regioselectively in a dilute solution of DMF (0.01 mM) in the presence of [Pd$_2$(dba)$_3$]·CHCl$_3$/AsPh$_3$ at ambient temperature to give the desired macrocycle **59** in 40% yield through an intramolecular Stille coupling reaction. A second Stille reaction was carried out intermolecularly between vinyl stannane **60** and vinyl iodide **59** ([Pd$_2$(dba)$_3$]·CHCl$_3$/AsPh$_3$, DMF, 35 °C) to afford the protected form of the target molecule **61** (Scheme 7.14).

A copper-mediated cross coupling between two molecules of **63** was established by Paterson et al. in the synthesis of elaiolide **66**, in which treatment of the ester **63** with 10 equiv of copper(I) thiophene-2-carboxylate **64** in NMP led to the formation of the 16-membered C2-symmetric macrolide **65** in 80% yield (Scheme 7.15) [25]. This process follows intermolecular and intramolecular coupling, suggesting the occurrence of a rapid Cu(I)-mediated cyclization without competing oligomerization.

The "Stille–Kelly" reaction, the Pd-catalyzed intramolecular coupling of an aryl dihalide in the presence of a ditin reagent offers a means for highly regioselective cyclization of unsymmetrical phenyl groups, which is always useful in building cyclic compounds in the total synthesis of natural products [26].

Sakamoto's group synthesized N-methylsulfonyl carbolines **68** by Stille–Kelly coupling of ortho-bromo-substituted anilinopyridines **67** (Eq. (7.5)) [27].

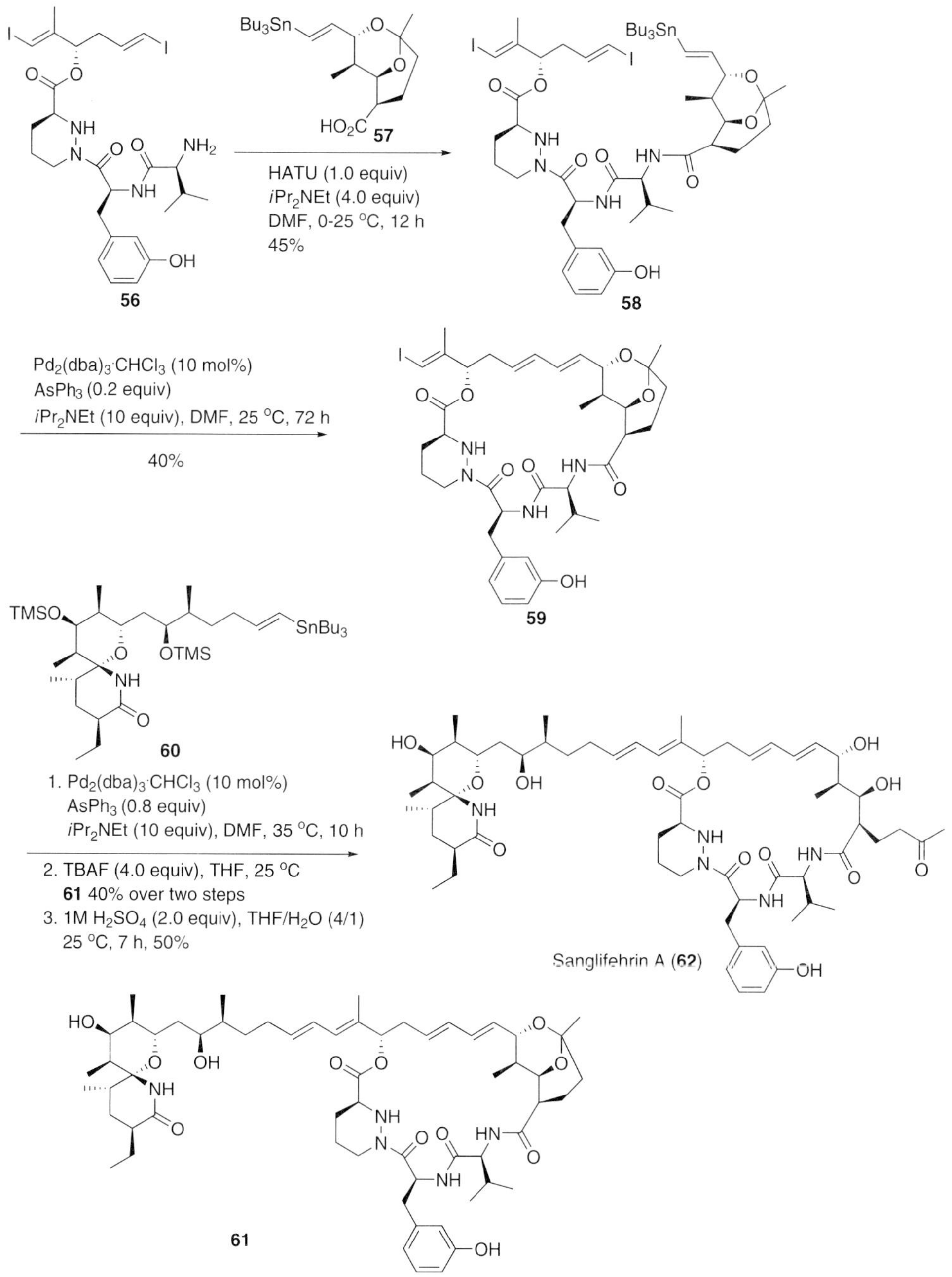

Scheme 7.14 Stille coupling for the synthesis of sanglifehrin A.

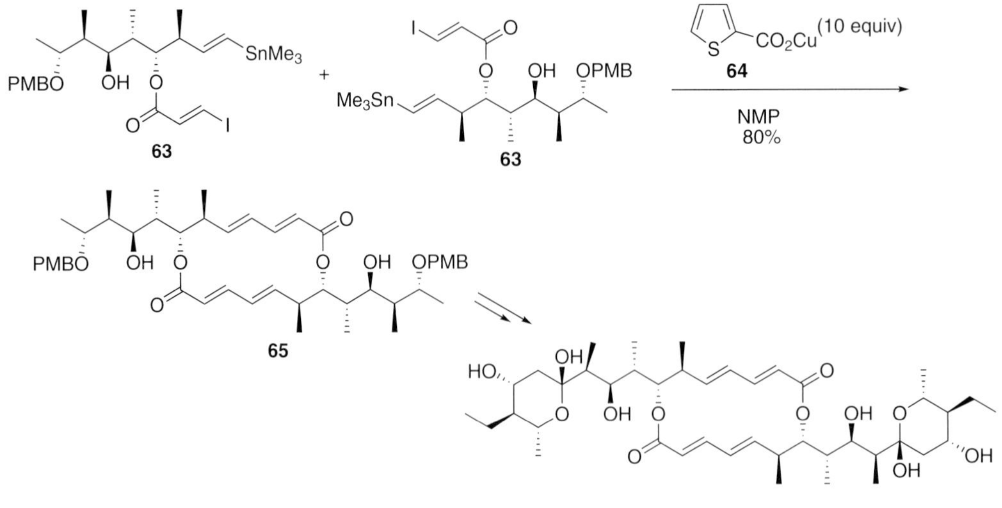

elaiolide (**66**)

Scheme 7.15 Cu-mediated inter– intramolecular alkenyl tin – alkenyl iodide homo-coupling.

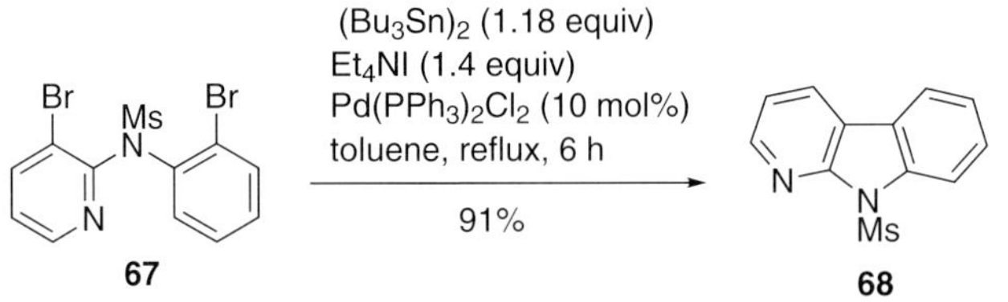

(7.5)

Total syntheses of plagiochins A **70** and D **71** with neurotrophic activity [28] have also been accomplished by Fukuyama *et al.* using Pd(0)-catalyzed intramolecular Stille-Kelly reaction of the dibromoperrottetin derivative **69** (Eq. (7.6)) [28].

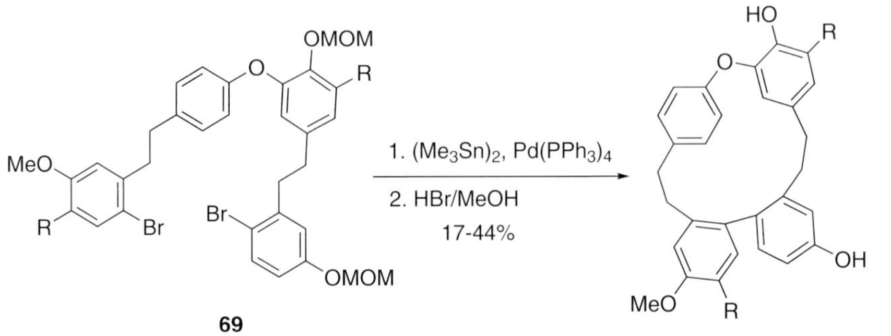

plagiocin A: R = OH (**70**)
plagiocin D: R = H (**71**)

(7.6)

Li *et al.* synthesized benzo[4,5]furo[3,2-c]pyridine **73** by employing the Stille–Kelly conditions, in which diiodide **72** was treated with hexamethylditin in the presence of catalytic PdCl$_2$(PPh$_3$)$_2$ in refluxing xylene to give benzo[4,5]furo[3,2-c]pyridine **73** in 92% yield (Eq. (7.7)) [29].

$$\text{72} \xrightarrow[\text{92\%}]{\substack{(\text{Me}_3\text{Sn})_2,\ \text{Pd}(\text{PPh}_3)_2\text{Cl}_2 \\ \text{xylene, reflux}}} \text{73} \qquad (7.7)$$

Bach et al. synthesized the tetrameric N-benzyl-protected purine **75** by the Stille–Kelly coupling reaction [30]. The final ring closure was achieved within the tetrameric 6′′′-chloro-8-iodoquaterpurine **74** by a reductive intramolecular cross-coupling in the presence of hexamethylditin using Pd$_2$(dba)$_3$ and P(2-furyl)$_3$ as the catalyst (Eq. (7.8)).

$$\text{74} \xrightarrow[\text{quant.}]{\substack{(\text{Me}_3\text{Sn})_2 \\ \text{Pd}_2(\text{dba})_3,\ \text{TFP} \\ \text{dioxane, 85 °C}}} \text{75} \qquad (7.8)$$

The preparation of the western hemisphere subtarget **77** for the construction of the heptacyclic cores of (+)-nodulisporic acids A and B and (−)-nodulisporic acid D was also demonstrated by Smith, III utilizing the Stille–Kelly coupling conditions (Eq. (7.9)) [31].

$$\text{76} \xrightarrow[\substack{\text{2. H}_2\text{NNH}_2\cdot\text{H}_2\text{O/EtOH} \\ \text{25 °C, 40 h} \\ \text{59\% over 2 steps}}]{\substack{\text{1. Me}_3\text{SiSnBu}_3\ (2.1\ \text{equiv}) \\ \text{Pd(PPh}_3)_4\ (10\ \text{mol\%}) \\ \text{Bu}_4\text{NBr}\ (3\ \text{equiv}) \\ \text{Li}_2\text{CO}_3\ (1.01\ \text{equiv}) \\ \text{xylene, reflux, 40 min}}} \text{77} \qquad (7.9)$$

In addition, Nicolaou et al. have established the 29-membered ring of rapamycin **80** based on the coupling reaction of diiodide **78** with 1,2-distannane in the presence of Pd(CH$_3$CN)$_2$Cl$_2$ and Hünig's base in DMF/THF in 28% yield together with unreacted starting material **78** (30% yield) and an iodo-stannane intermediate **79** (30% yield) (Scheme 7.16) [32].

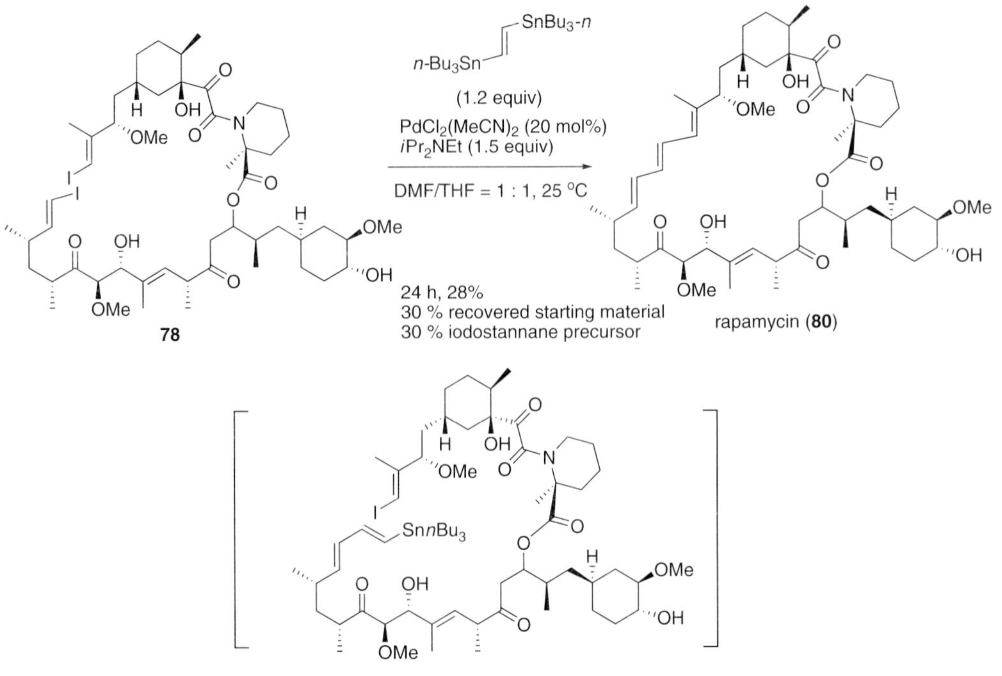

Scheme 7.16 Double Stille coupling of ditin–diiodides.

This Stille-type "double bond stitching coupling" was used again in the total synthesis of (+)-mycotrienin I **83** by Panek et al. to furnish the required conjugated triene system concurrently with the 21-membered ring **82** of the target molecule **83** (Scheme 7.17) [33].

The Danishefsky laboratory also features a double Stille-type coupling in its assembly of the enediyne unit by using bis(alkynyl iodide)/Z-1,2-distannylethylene coupling transformation (Scheme 7.18) [34].

7.4
Cyclic Coupling Involving Organosilicon

The intermolecular Hiyama coupling between organosilicon compounds like vinyl-, ethynyl-, and allyl-silanes and organic halides such as aryl, vinyl, and allyl halides gave the corresponding coupled products in a stereospecific and chemoselective manner [35]. However, cyclic coupling involving organosilicon is not so common in the literature. Denmark et al. describe a series of cyclic silyl ether substrates bearing a tethered alkenyl halide which undergo the cross-coupling reaction to produce medium-sized 9–12-membered rings **89** (Eq. (7.10)) [36].

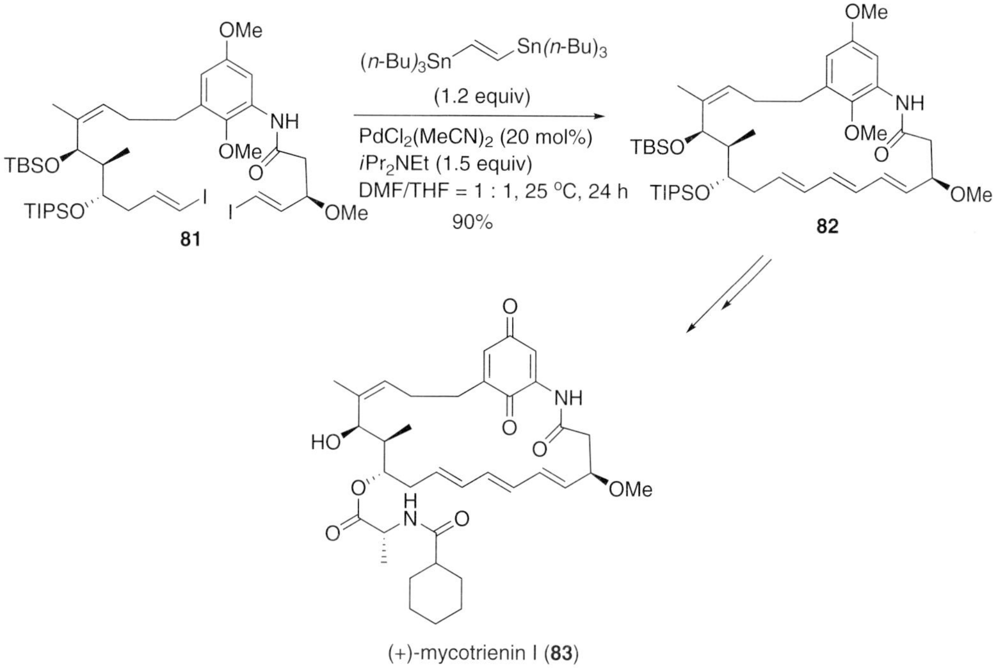

Scheme 7.17 Double bond stitching coupling for (+)-mycotrienin I.

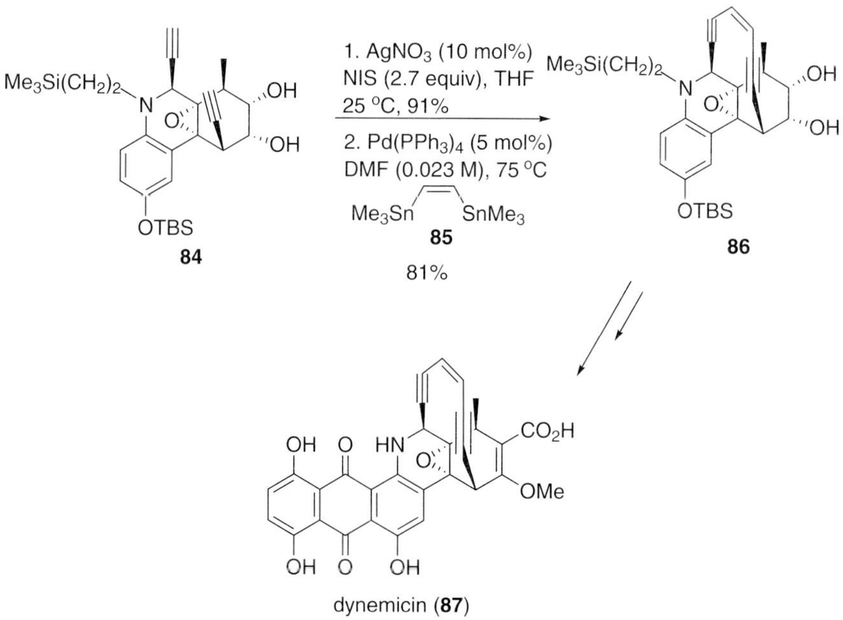

Scheme 7.18 Double coupling for cyclic conjugated diynene.

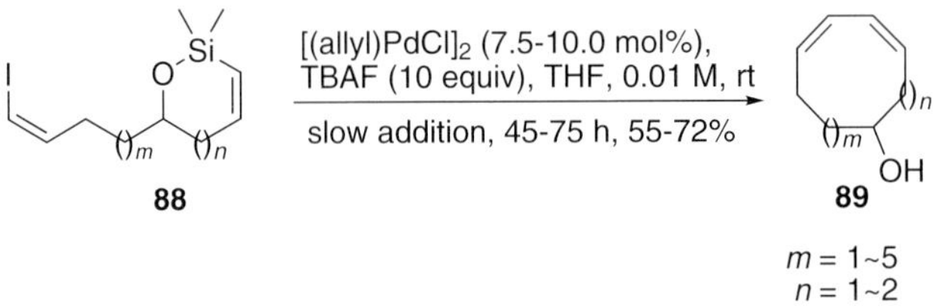
[(allyl)PdCl]2 (7.5-10.0 mol%), TBAF (10 equiv), THF, 0.01 M, rt
slow addition, 45-75 h, 55-72%
88
89
m = 1~5
n = 1~2

(7.10)

Denmark and coworkers also achieved the first enantioselective total synthesis of (+)-brasilenyne **92** in 19 linear steps with 5.1% overall yield from L-(*S*)-malic acid [37]. The construction of the nine-membered oxonin core **91** containing a 1,3-*cis*, *cis*-diene unit was accomplished with an intramolecular cross-coupling reaction, in which 7.5 mol% of $[allylPdCl]_2$ was used as the catalyst and 10 equiv of 1.0 M TBAF solution were introduced as the activator by a syringe-pump (Scheme 7.19).

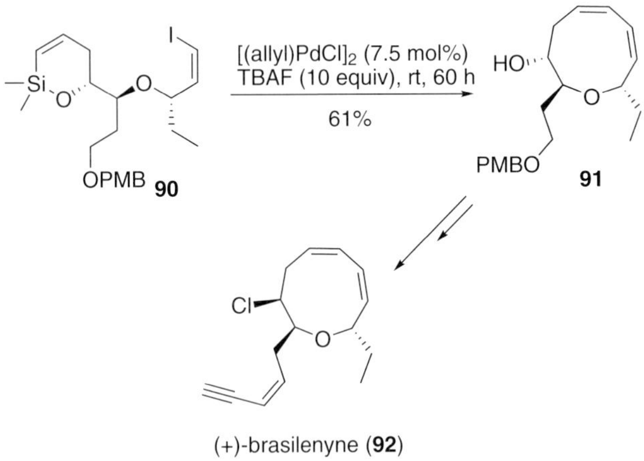
[(allyl)PdCl]2 (7.5 mol%)
TBAF (10 equiv), rt, 60 h
61%
OPMB 90
PMBO
91
(+)-brasilenyne (92)

Scheme 7.19 Cyclic coupling of cyclic silyloxane with 1-alkenyl iodide.

Recently, Suginome's group established a new synthetic access to 2,4-disubstituted siloles **95** via the Pd-catalyzed cyclization reaction of terminal alkynes **94** with (dialkylaminosilyl)pinacolboranes **93**, which serve as new silylene equivalents [38]. The elimination of (dialkylamino)borane seems to be the key driving force for the reaction (Scheme 7.20).

7.5
Cyclic Coupling Involving Amines or Alcohols

Buchwald [39] and Hartwig [40] both developed the palladium-catalyzed arylation of amines, alcohols and carbon nucleophiles with aryl halides or sulfonates, which has proven to be a powerful tool in organic synthesis. Using this method to establish heterocycles is becoming more and more popular.

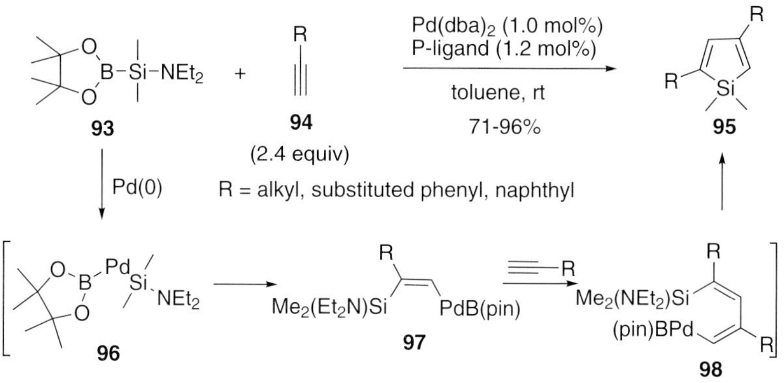

Scheme 7.20 Siloles via coupling.

In 1999, Buchwald et al. developed efficient palladium-catalyzed intramolecular cyclization reactions of secondary amides and carbamates [41]. With the proper choice of palladium catalyst, ligand, and base, five-, six-, and seven-membered rings **100** are formed efficiently from secondary amide or secondary carbamate precursors **99**, offering significant improvements to cross-coupling chemistry (Eq. (7.11)).

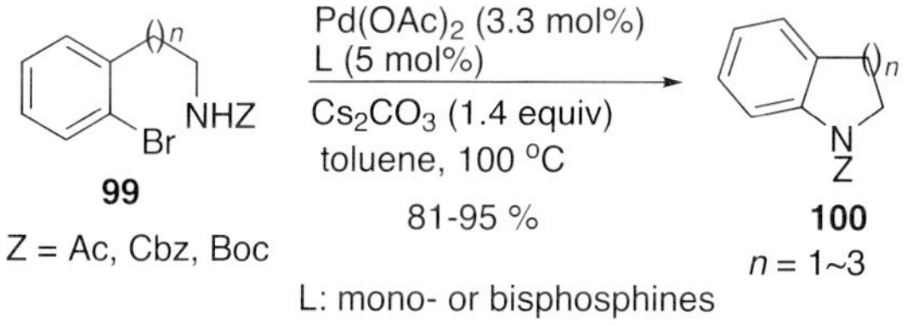

$$\text{(7.11)}$$

In 1998, Abouabdellah and Dodd et al. reported the preparation of a series of pharmaceutically interesting α-carbolines **102** employing Pd-catalyzed intramolecular amination of **101** as a key step (Eq. (7.12)) [42].

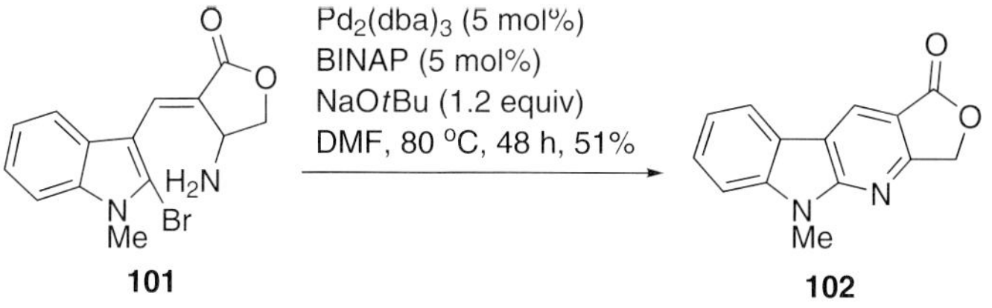

$$\text{(7.12)}$$

In Snider's total synthesis of fumiquinazolines **105**, the imidazoindolone unit **104** was formed via a Pd-catalyzed cyclization of an iodoindole carbamate **103** (Scheme 7.21) [43].

A novel stereoselective approach to the ring system of the mitomycins was described by Coleman in 2001 [44]. The synthesis was based on a convergent strategy involving a stereocontrolled addition of a α-phenyl silyl enol ether to a pyrroline N-acyliminium ion followed by an intramolecular palladium-catalyzed aryl triflate

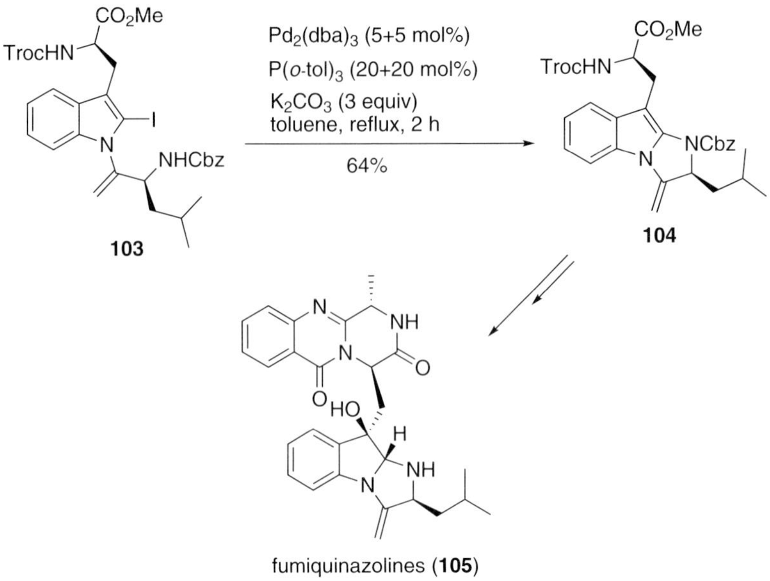

Scheme 7.21 Cyclic C–N bond formation of indole.

amination to afford the (9R*,9aR*)-tetrahydropyrrolo[1,2-a]indole ring system **107** (Eq. (7.13)).

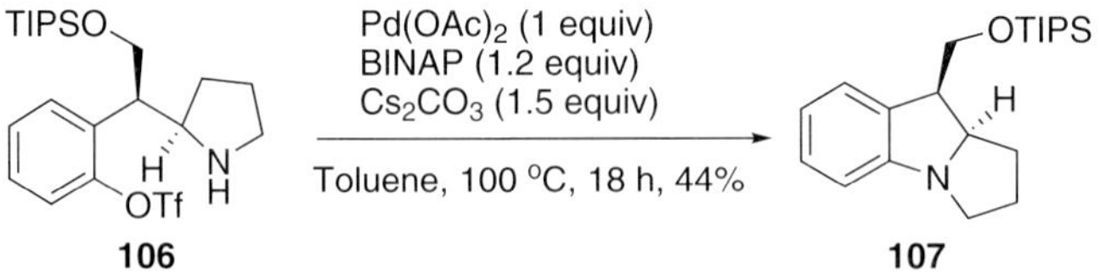

$$\text{(7.13)}$$

Cyclic peptidic RGD models **109** were efficiently synthesized by Pd(P(t-Bu)$_3$)$_2$-catalyzed carbonylative macrolactamization of **108** in the presence of 4 Å molecular sieves under 10 atm of carbon monoxide (Scheme 7.22) [45].

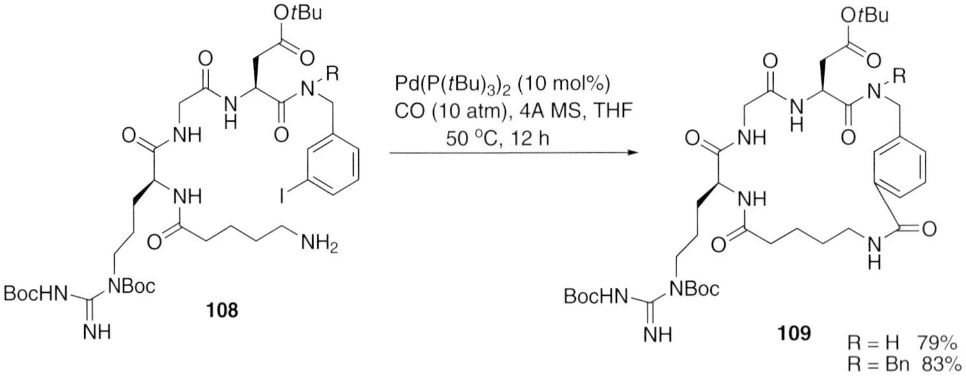

Scheme 7.22 Macrolactone via C–N coupling.

7.5 Cyclic Coupling Involving Amines or Alcohols | 333

Phenazines **111** were prepared by the palladium(II)-catalyzed intramolecular amination of aryl bromides **110**, using BINAP as ligand and Cs_2CO_3 as base (Eq. (7.14)) [46].

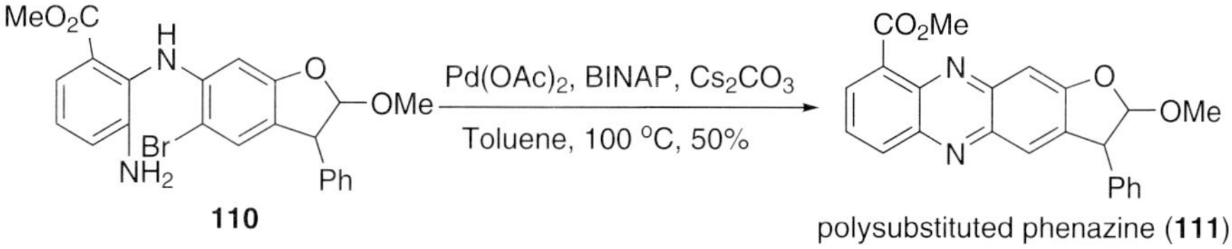

$$\text{(7.14)}$$

Roger *et al.* reported an improved route for the construction of the heterobenzazepine ring system **113** which involved a Pd-catalyzed intramolecular C–N bond forming reaction [47]. Both oxazepine and thiazepine were prepared (Scheme 7.23).

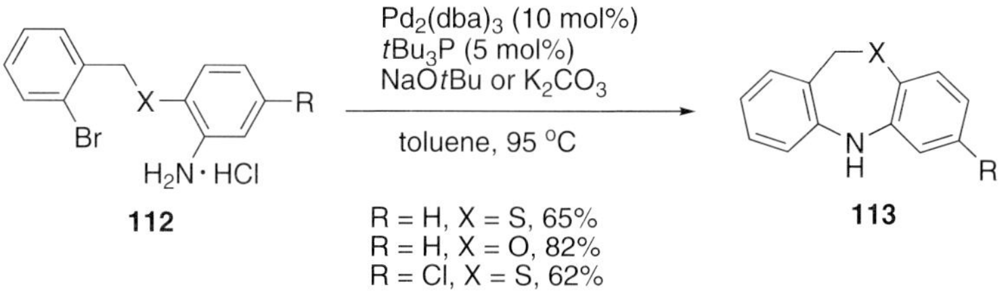

R = H, X = S, 65%
R = H, X = O, 82%
R = Cl, X = S, 62%

Scheme 7.23 Heterobenzazepine via C–N bond formation.

A divergent and regiocontrolled Pd-catalyzed domino sequence involving an intramolecular *N*-arylation and an intermolecular Heck reaction has been developed by Neuville and Zhu, providing rapid access to functionalized benzodiazepine-2,5-diones **115** (Scheme 7.24) [48].

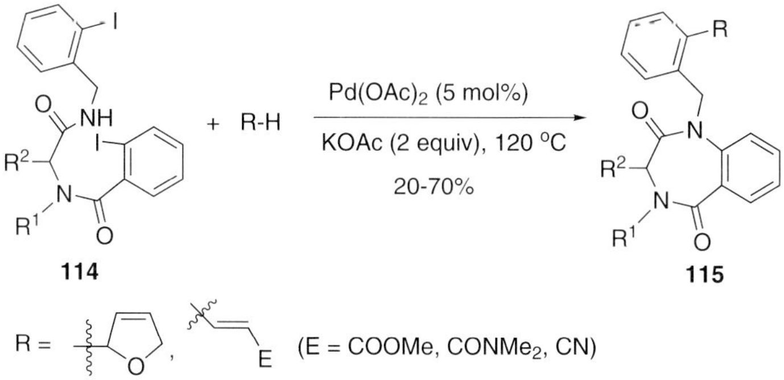

(E = COOMe, CONMe$_2$, CN)

Scheme 7.24 Benzodiazepine-2,5-diones via C–N coupling.

Smith, III *et al.* achieved the assembly of the E-ring of the highly strained parent tetracycle **117** [49], whose structural motif was found in the potent ectoparasiticidal agents (+)-nodulisporic acids A and B and related congeners (Eq. (7.15)).

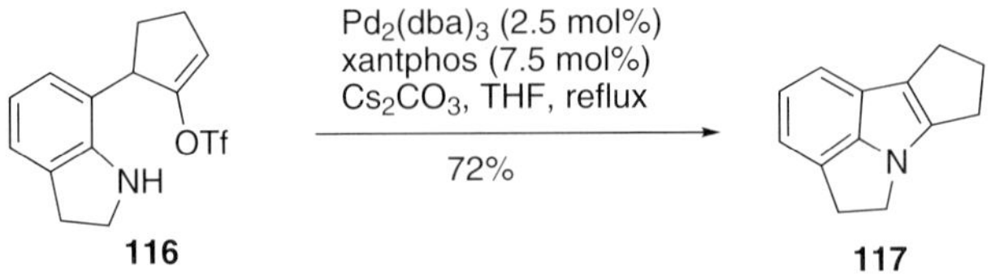

$$\text{116} \xrightarrow[\text{72\%}]{\substack{\text{Pd}_2(\text{dba})_3 \ (2.5 \ \text{mol\%}) \\ \text{xantphos} \ (7.5 \ \text{mol\%}) \\ \text{Cs}_2\text{CO}_3, \ \text{THF}, \ \text{reflux}}} \text{117} \quad (7.15)$$

A highly efficient method of indole synthesis using gem-dihalovinylaniline substrates **118** and organoboron reagents was developed via a Pd-catalyzed tandem intramolecular amination and an intermolecular Suzuki coupling by Lautens' group [50]. Aryl, alkenyl, and alkyl boron reagents may all be successfully employed, making it a versatile modular approach. The reaction tolerates a variety of substitution patterns on the aniline, leading to indoles **119** with the group at C2–C7 (Scheme 7.25).

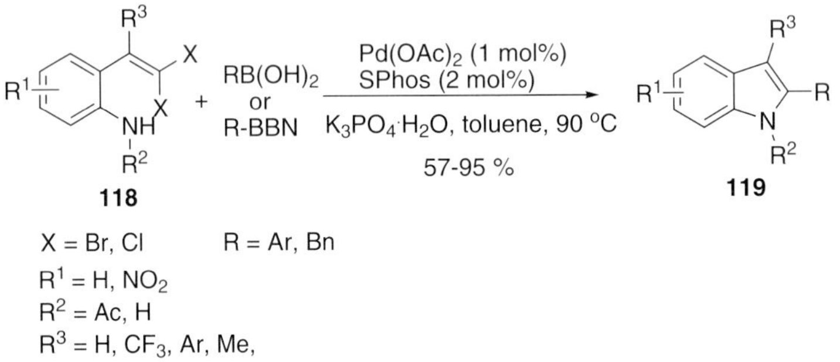

118 + RB(OH)$_2$ or R-BBN $\xrightarrow[\text{57-95 \%}]{\substack{\text{Pd(OAc)}_2 \ (1 \ \text{mol\%}) \\ \text{SPhos} \ (2 \ \text{mol\%}) \\ \text{K}_3\text{PO}_4 \cdot \text{H}_2\text{O}, \ \text{toluene}, \ 90 \ °\text{C}}}$ **119**

X = Br, Cl   R = Ar, Bn
R^1 = H, NO$_2$
R^2 = Ac, H
R^3 = H, CF$_3$, Ar, Me,

Scheme 7.25 Cyclic C–N coupling – intermolecular coupling for substituted indoles.

The multicomponent assembly of an alkyne, an amine, and 2-iodobenzoic acid gave indole derivatives **122** in high regioselectivity by a one-pot Curtius rearrangement/palladium-catalyzed indolization process (Scheme 7.26) [51].

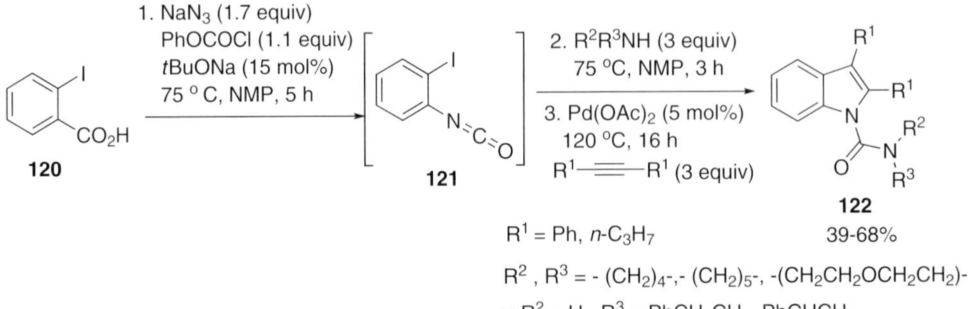

120 $\xrightarrow{\substack{\text{1. NaN}_3 \ (1.7 \ \text{equiv}) \\ \text{PhOCOCl} \ (1.1 \ \text{equiv}) \\ t\text{BuONa} \ (15 \ \text{mol\%}) \\ 75 \ °\text{C}, \ \text{NMP}, \ 5 \ \text{h}}}$ **121** $\xrightarrow{\substack{\text{2. R}^2\text{R}^3\text{NH} \ (3 \ \text{equiv}) \\ 75 \ °\text{C}, \ \text{NMP}, \ 3 \ \text{h} \\ \text{3. Pd(OAc)}_2 \ (5 \ \text{mol\%}) \\ 120 \ °\text{C}, \ 16 \ \text{h} \\ \text{R}^1\text{≡\!=\!≡R}^1 \ (3 \ \text{equiv})}}$ **122**   39-68%

R^1 = Ph, n-C$_3$H$_7$

R^2, R^3 = - (CH$_2$)$_4$-,- (CH$_2$)$_5$-, -(CH$_2$CH$_2$OCH$_2$CH$_2$)-
or R^2 = H , R^3 = PhCH$_2$CH$_2$, PhCHCH$_3$

Scheme 7.26 Indoles from 2-iodobenzoic acid.

The intramolecular coupling between a hydrazide and an aryl bromide was achieved by Katayama's group to access a variety of indolo[1,2-b]indazole precursors **124** in good yields (Eq. (7.16)) [52].

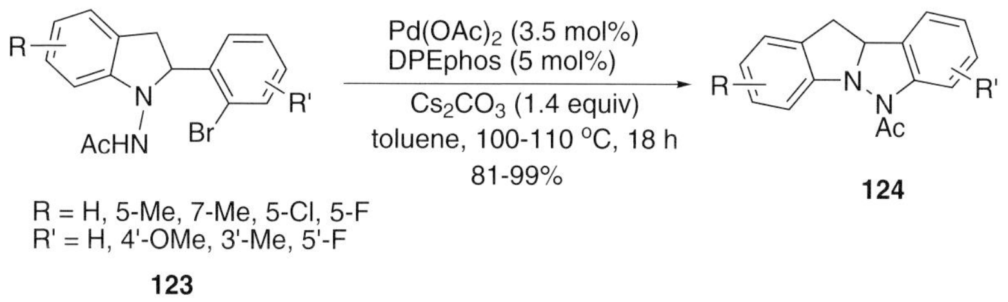

7.5 Cyclic Coupling Involving Amines or Alcohols

[Structure 123]

Pd(OAc)$_2$ (3.5 mol%)
DPEphos (5 mol%)
Cs$_2$CO$_3$ (1.4 equiv)
toluene, 100–110 °C, 18 h
81–99%

[Structure 124]

R = H, 5-Me, 7-Me, 5-Cl, 5-F
R' = H, 4'-OMe, 3'-Me, 5'-F

123

(7.16)

The use of guanidines as nucleophiles in intramolecular coupling processes was achieved with either a Pd or a Cu catalyst (Scheme 7.27) [53].

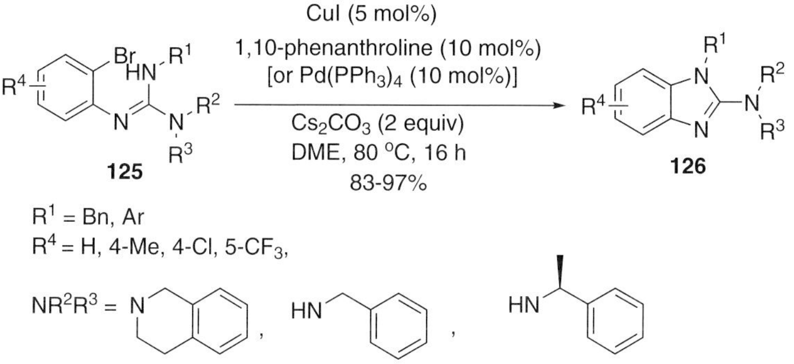

[Structure 125]

CuI (5 mol%)
1,10-phenanthroline (10 mol%)
[or Pd(PPh$_3$)$_4$ (10 mol%)]
Cs$_2$CO$_3$ (2 equiv)
DME, 80 °C, 16 h
83–97%

[Structure 126]

R^1 = Bn, Ar
R^4 = H, 4-Me, 4-Cl, 5-CF$_3$,

NR^{2}R^3 = [tetrahydroisoquinoline], HN-CH$_2$-Ph, HN-CH(Me)-Ph

Scheme 7.27 Cyclic coupling of 2-bromophenyl guanidines.

In the synthesis of carbapenems, a Pd-catalyzed intramolecular amination of the alkenyl bromide with lactamic nitrogen was observed (Eq. (7.17)) [54].

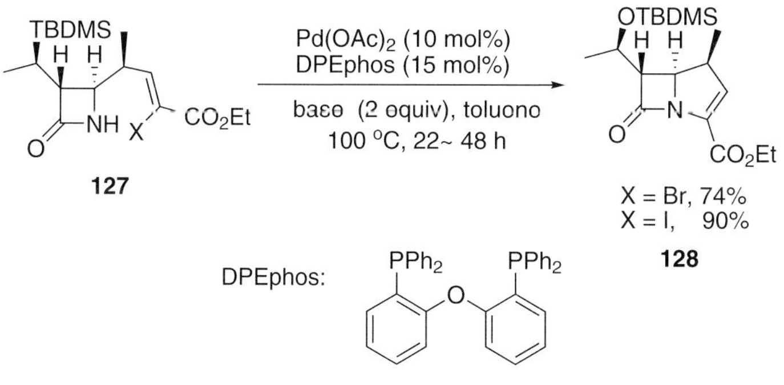

[Structure 127]

Pd(OAc)$_2$ (10 mol%)
DPEphos (15 mol%)
base (2 equiv), toluene
100 °C, 22~48 h

[Structure 128]

X = Br, 74%
X = I, 90%

128

(7.17)

DPEphos: [structure with two PPh$_2$ groups on diphenyl ether]

Fukuyama's group make great efforts on CuI-mediated intermolecular aryl amination reaction [56a] and the intramolecular amination of aryl halides [56b]. These methods were successfully used in the total synthesis of the duocarmycins **133** [56c] and (+)-yatakemycin **142** [56d,e], but in some cases stoichiometric (or greater) quantities of CuI were required (Scheme 7.28).

Ma et al. developed the amino acid-promoted Ullmann reaction [57a], which enables the coupling of aryl halides with β-amino acids and ester in the synthesis of

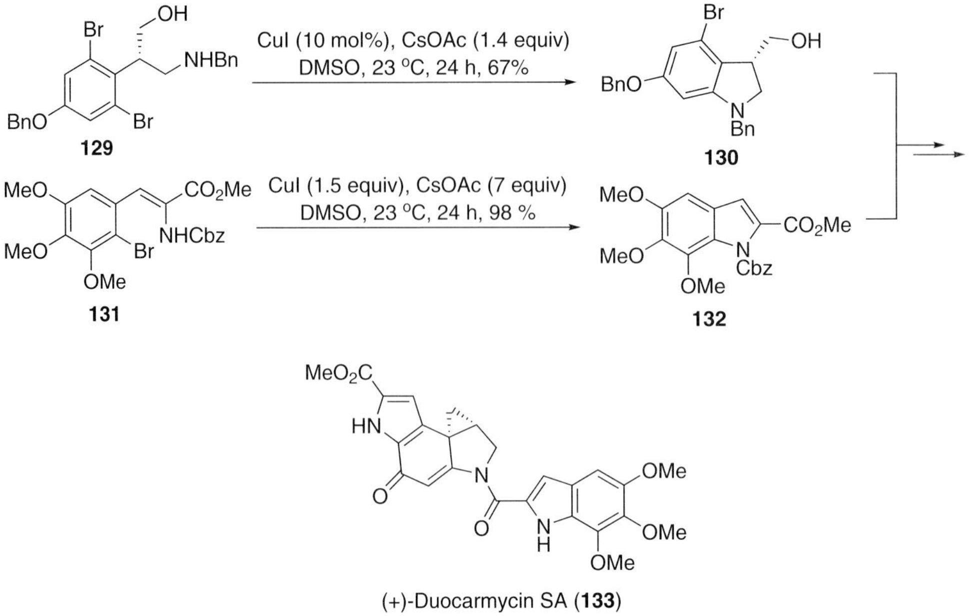

Scheme 7.28 Cu-catalyzed C−N bond formation for the synthesis of duocarmycin SA and yatakemycin.

SB-214857 **145** [57]. In 2007, he also used the CuI/N,N-dimethylglycine-catalyzed intermolecular coupling reaction to form an enamide **148**, which was followed by cyclization to afford Ziziphine N **149** [58]. For complicated cases, a large amount of CuI is required (Scheme 7.29).

Compared with these two protocols, in 2003, Buchwald's group developed a mild and efficient protocol for the coupling of amides **150** and carbamates with vinyl bromides and iodides **151**, just using a combination of 5 mol% CuI and 20 mol% N,N'-dimethyl ethylenediamine to afford the enamide moiety **152** [59]. During the total synthesis of paliurine F **155**, the Evano group made a brief survey of various catalytic systems for the amidation of vinyl iodides and revealed dramatic differences in reactivity [60]. While initial investigations of the intramolecular amidation using combinations of copper iodide and triphenylphosphine [61] or N,N'-dimethylglycine [62] failed to give any cyclization product; a combination of [Cu(CH$_3$CN)$_4$]PF$_6$ with 3,4,7,8-tetramethyl-1,10-phenanthroline as ligand afforded the desired product, but only in trace amounts. They eventually found that subjecting the acyclic skeleton to copper(I) thiophene-2-carboxylate in NMP at 90 °C smoothly provided the desired macrocyclic enamide in 60% yield. In 2000, Porco *et al.* developed the cross-coupling reaction of 1-alkenyl iodides **157** with amides **156** using 30 mol% of CuTC [61]. During the synthesis of CJ-15, 801 **163**, 10 mol% of Cu(MeCN)$_4$PF$_6$, 20 mol% of 3,4,7,8-tetramethyl-1,10-phenanthroline **161** and 3 equiv of Rb$_2$CO$_3$ were used to afford the coupling product **162**, in which the CuTC/Cs$_2$CO$_3$ catalyzed system was not effective [61]. Switching to the CuI/N,N'-dimethylethylenediamine catalytic

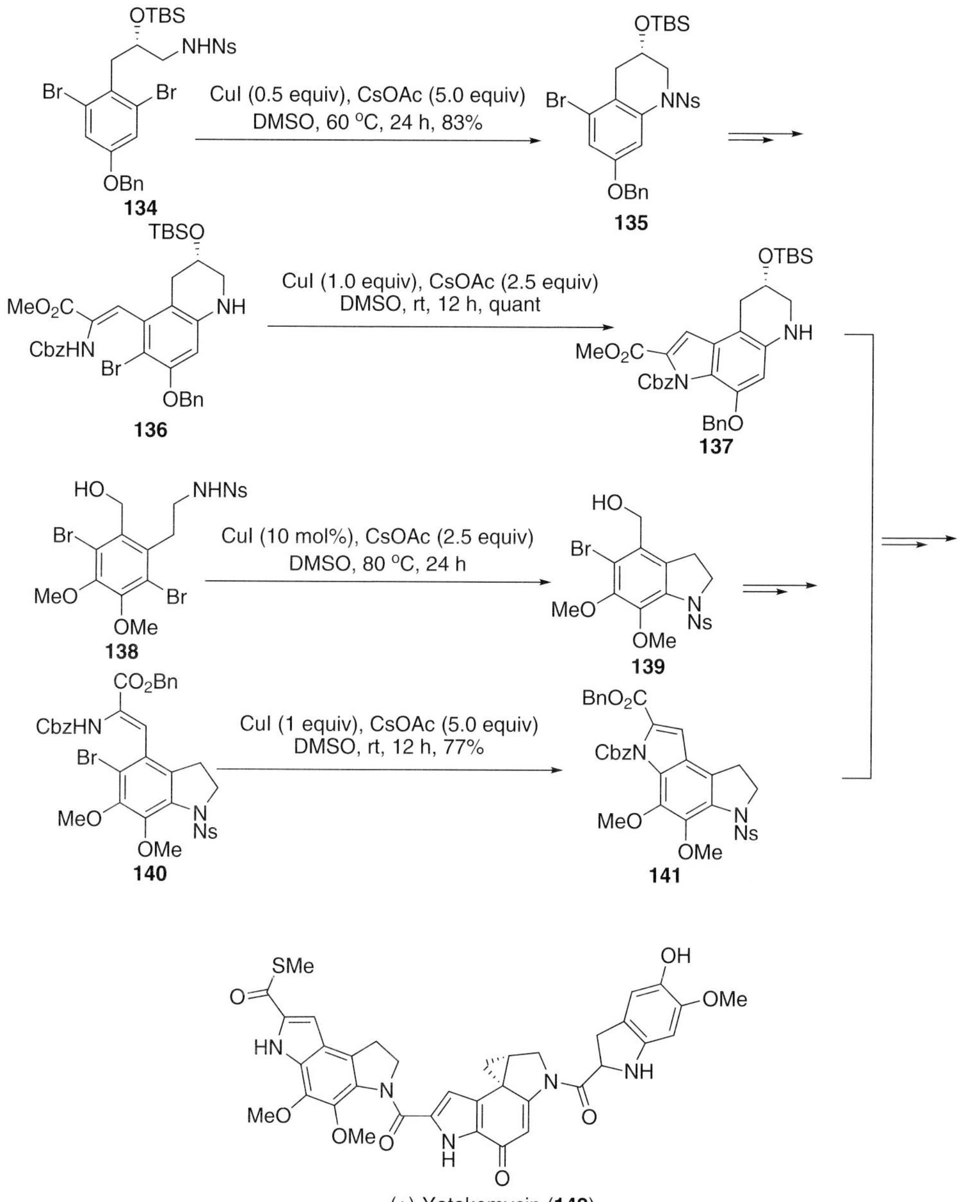

7.5 Cyclic Coupling Involving Amines or Alcohols | 337

Scheme 7.28 (*Continued*).

system (Buchwald's system) improved the yield slightly, thus providing product in 70% yield, along with 20% of recovered starting material (Scheme 7.30).

In 2008, Evano finished the total synthesis of the cyclopeptide alkaloid mucronine E **166** again by subjection of iodoamide **164** to catalytic copper(I) iodide and *N,N'*-

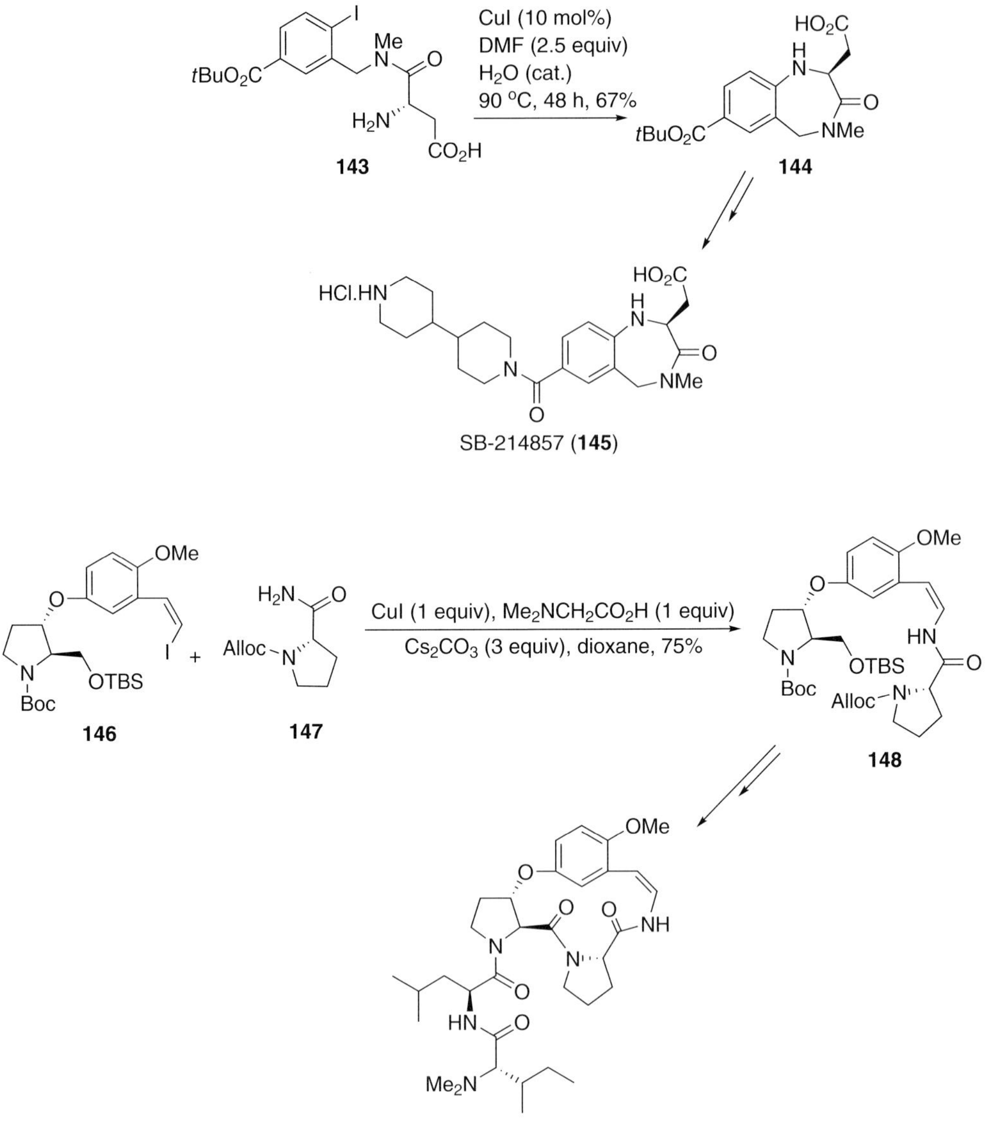

Scheme 7.29 Cu-catalyzed C–N bond formation for SB-214857 and ziziphine N.

dimethylethylenediamine in THF under high dilution conditions at 63 °C providing the 15-membered ring **165** in 84% yield. The most striking feature of this approach is that this mild intramolecular amidation protocol proceeds with complete regioselectivity for the terminal amide (Scheme 7.31) [63].

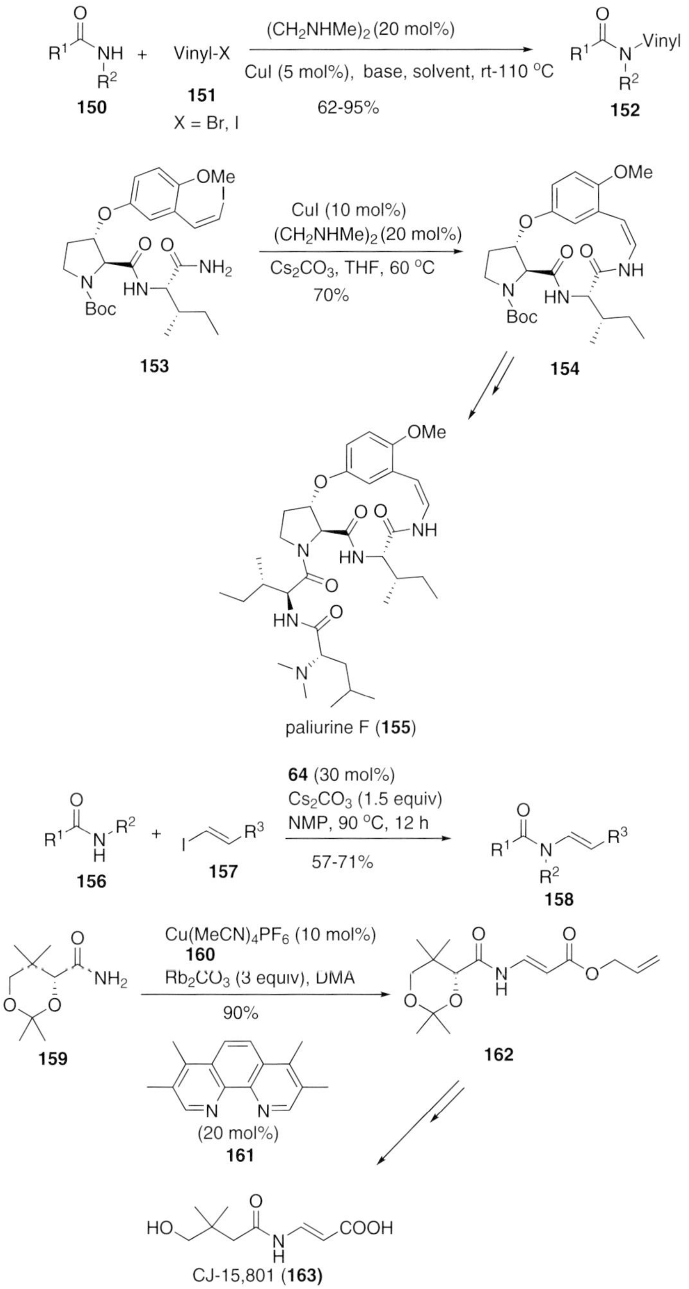

7.5 Cyclic Coupling Involving Amines or Alcohols | **339**

Scheme 7.30 Cyclic amide–1-alkenyl iodide coupling.

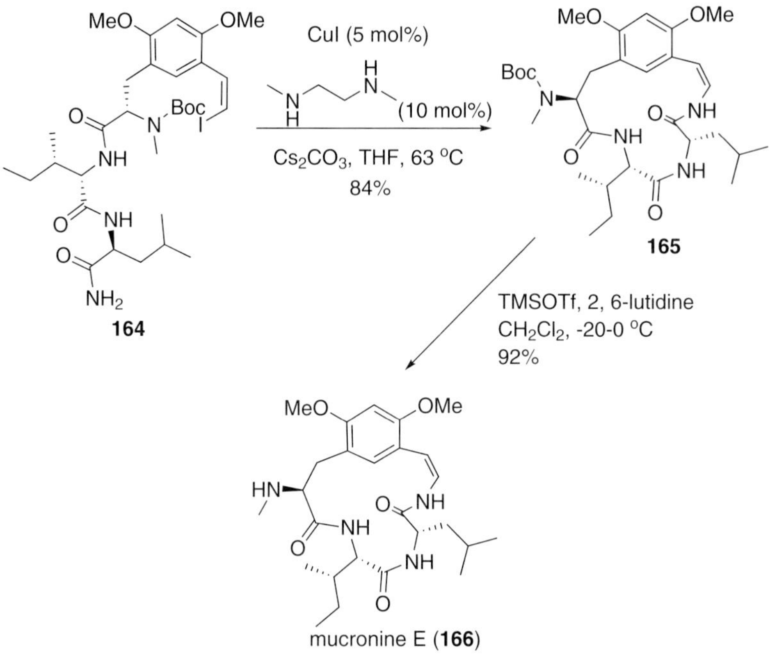

Scheme 7.31 Mucronine E via CuI–TMEDA-catalyzed C−N coupling.

The Ullman reaction catalyzed by copper has had a long history in forming new bonds in organic synthesis [55]. Some of the recent advances in this area are of particular interest. Li *et al.* developed a general and highly efficient synthesis of β-lactams **168** via copper-catalyzed intramolecular N-vinylation, with preference for 4-exo ring closure (Eq. (7.18)) [64].

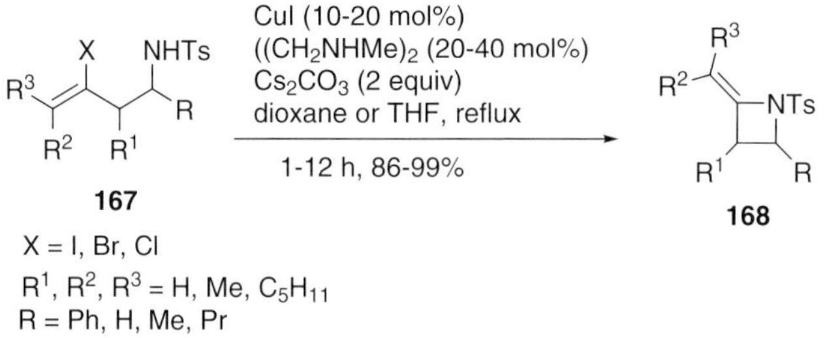

X = I, Br, Cl
R^1, R^2, R^3 = H, Me, C_5H_{11}
R = Ph, H, Me, Pr

Also in the same year, Urabe's group reported a copper-catalyzed 1,2-double amination reaction of 1-halo-1-alkynes **169** to afford protected tetrahydropyrazine compounds **171** (Scheme 7.32) [65].

The use of a Pd/dppf catalyst gives access to the tricyclic phenothiazine scaffold **176** starting from 2-bromothiophenols **173**, aliphatic or aromatic amines **174**, and 1-bromo-2-iodobenzenes **175** in a single reaction flask [66]. This transformation involves thioether formation and subsequent intermolecular and intramolecular aryl

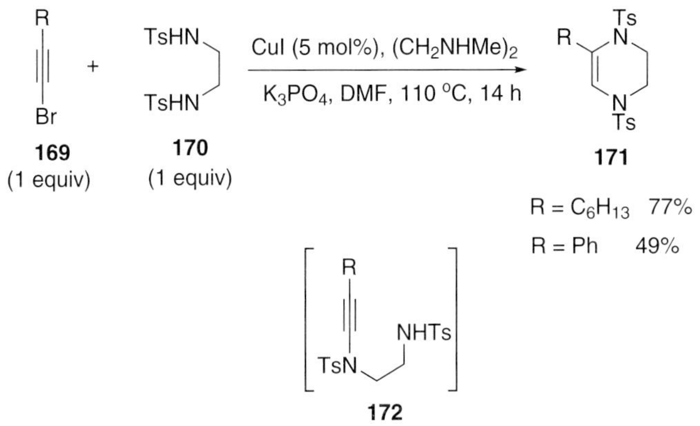

7.5 Cyclic Coupling Involving Amines or Alcohols | 341

Scheme 7.32 Cu-catalyzed 1,2-double amination of 1-alkynyl bromides.

amination reactions (Eq. (7.19)).

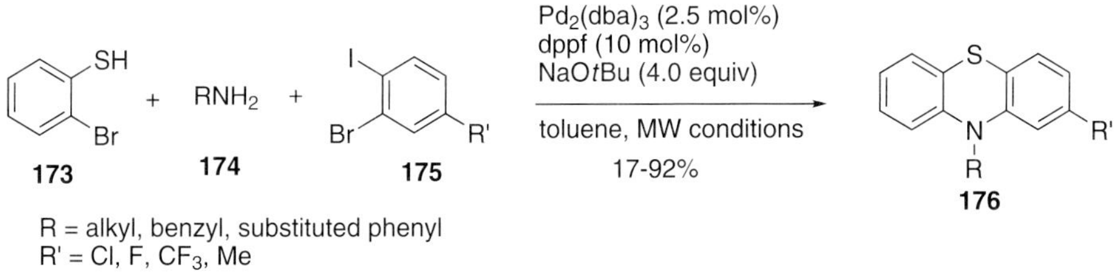

(7.19)

The intramolecular Ullmann reaction for C–O bond formation was also used to establish the ring structures in natural product synthesis. In 1997, Nicolaou et al. used the method to form the 16-membered macrocyclic biaryl ethers systems **178** of vancomycin **179** (Scheme 7.33) [67].

Ma's group also established the 13-membered ring **181** by intramolecular coupling of phenol and phenyl bromide [68]. However, 3 equiv of CuI were used (Eq. (7.20)).

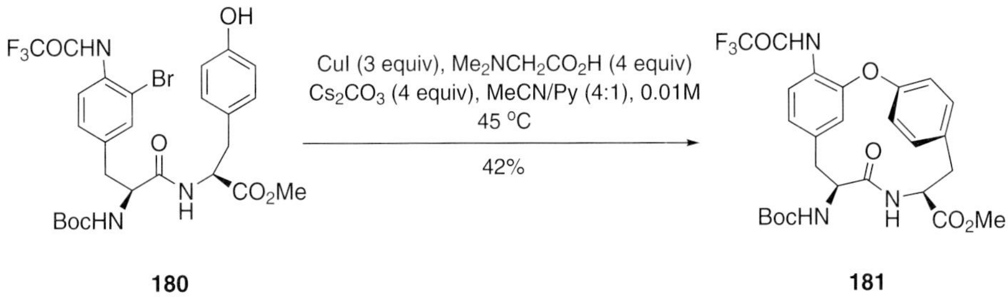

(7.20)

Another palladium-free cyclic coupling reaction was found during the synthesis of vancomycin aglycon **186** by Boger et al. in 1999 [69]. Two bisaryl ether macrocycles were formed by two nucleophilic aromatic substitution reactions promoted by o-nitro groups (Scheme 7.34).

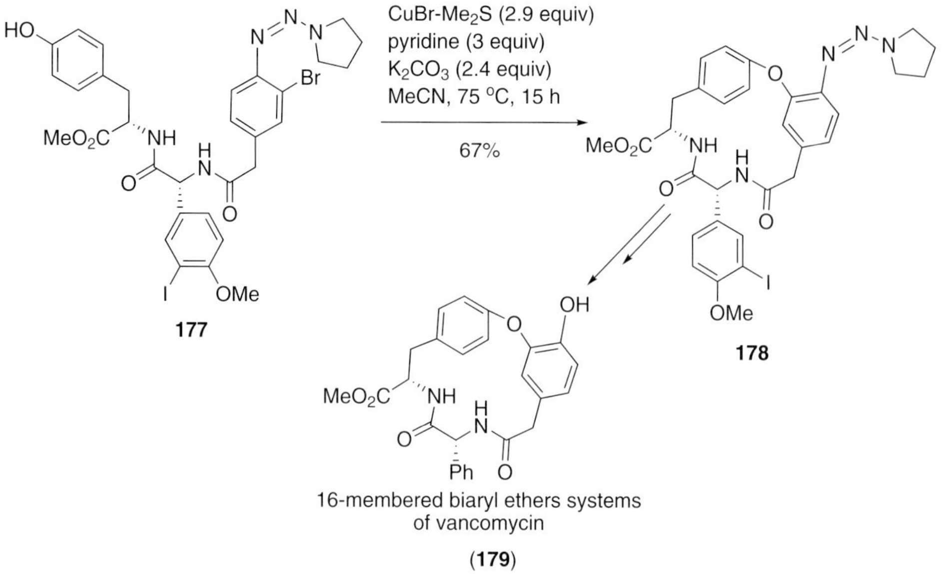

Scheme 7.33 Approach to vancomycin via Cu-catalyzed C−O coupling.

Overman's group constructed the final ring of (+)-minfiensine **189** by an intramolecular palladium-catalyzed ketone enolate-vinyl iodide coupling of **187** (Scheme 7.35) [70].

Solé et al. have also developed Pd(0)-catalyzed intramolecular coupling of ketone enolate formed from **190** and β-(2-iodoanilino)esters **192** for the synthesis of bridged azacycles **191** and indole-3-carboxylic acid esters **193** (Scheme 7.36) [71].

The palladium-catalyzed asymmetric intramolecular α-arylation of amides afforded oxindoles **196** in high yield and excellent enantiomeric purity by using the C2-symmetric bulky N-heterocyclic carbene ligand **195** (Scheme 7.37) [72].

Liebeskind et al. demonstrated that CuTC could be efficiently used to mediate the Ullmann reaction at room temperature under very mild conditions tolerating a wide variety of functional groups to form the carbon–carbon bond (Eq. (7.21)) [73].

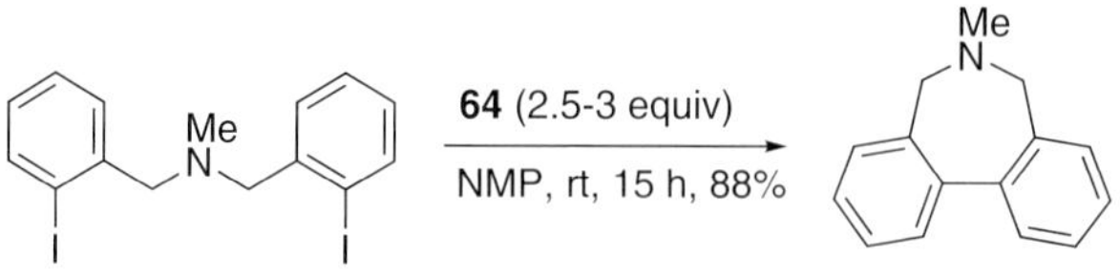

Formation of a carbon–carbon bond was reported by Lin's group using asymmetrical intramolecular oxidative coupling between two aryl bromide units in **199** to achieve the asymmetrical synthesis of the naturally occurring (+)-kotanin **201** [74a]. Fan and Chan also used this asymmetric intramolecular Ullmann coupling with central-to-axial chirality transfer. A bridged C2-symmetric biphenyl phosphine ligand **204** possessing additional chiral centers on the linking unit of the biphenyl groups was synthesized [74b] (Scheme 7.38).

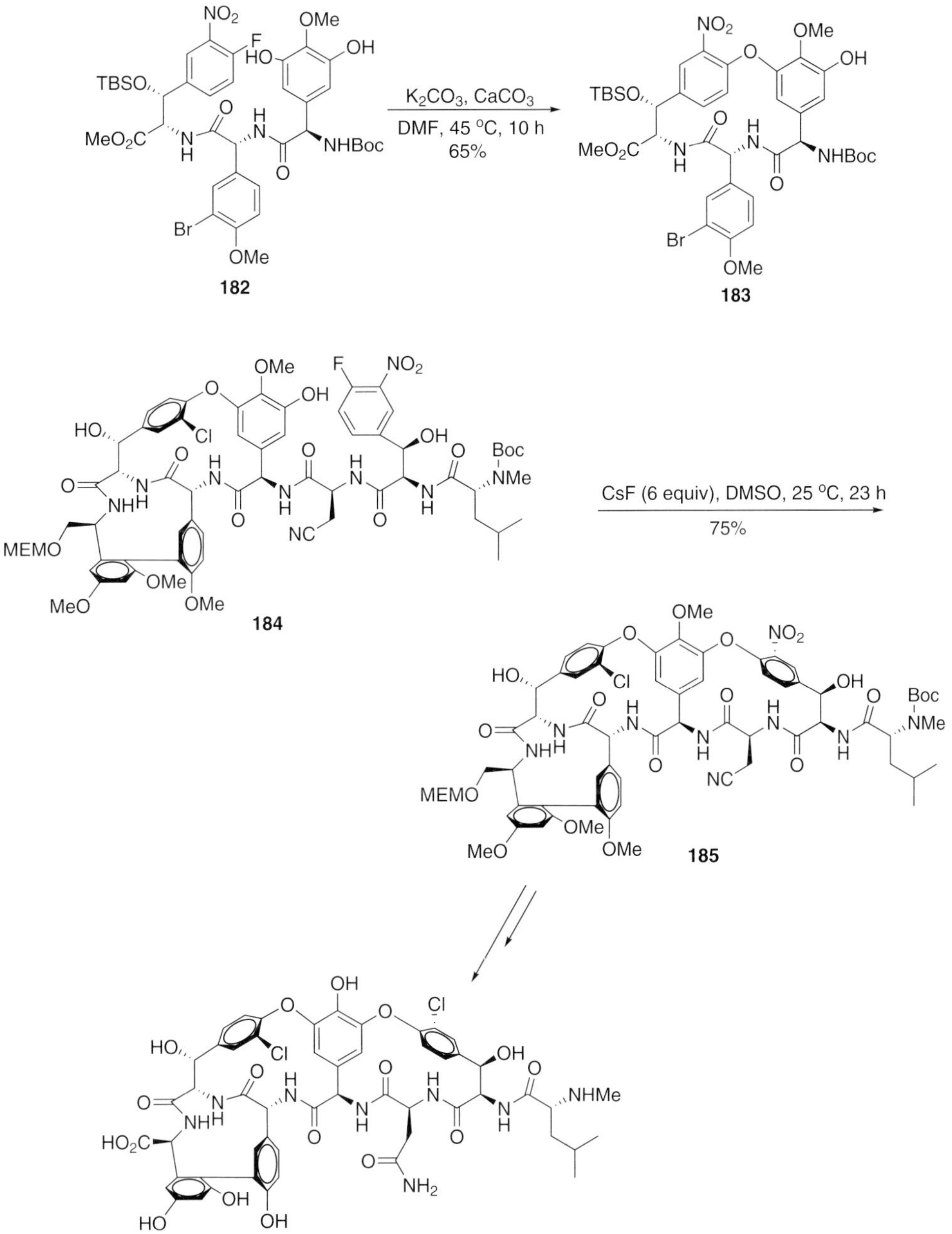

7.5 Cyclic Coupling Involving Amines or Alcohols | 343

Scheme 7.34 Non-catalyzed cyclic C—O bond formation.

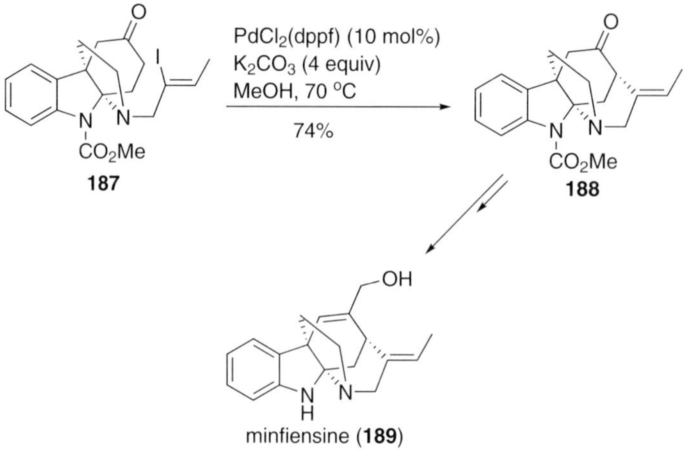

Scheme 7.35 Cyclic α-alkenylation of ketone for minfiensine.

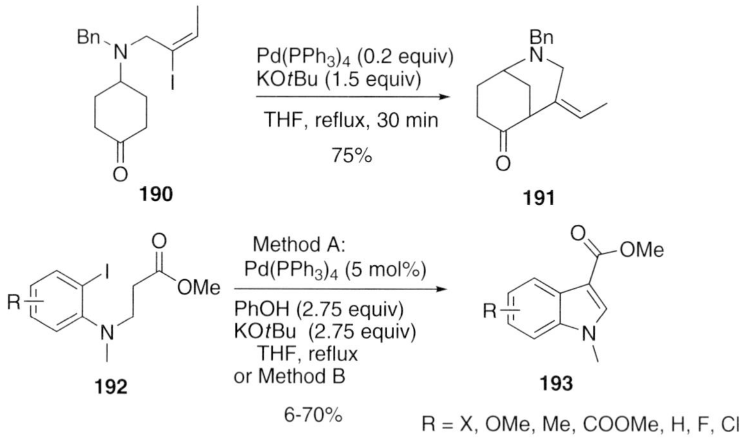

Scheme 7.36 Pd-catalyzed cyclic α-alkenylation or arylation of ketones.

Method B: Pd(PPh$_3$)$_4$ (10 mol%), PhOH (0.3 equiv), K$_3$PO$_4$ (3 equiv), DMF, 90 °C

Harrowven *et al.* described a synthesis of the pyrrolophenanthridone alkaloid hippadine **207** via the Ullman-type coupling reaction mediated by CuI to effect the construction of the pentacyclic skeleton **206** (Scheme 7.39) [75].

7.6
Cyclic Coupling via the Heck Reaction

The Heck reaction has been developed for nearly 40 years [76], proving to be one of the most powerful tools for the formation of carbon–carbon bonds. Overman [77], Beletskaya [78] and others [79] have comprehensively summarized the advance in

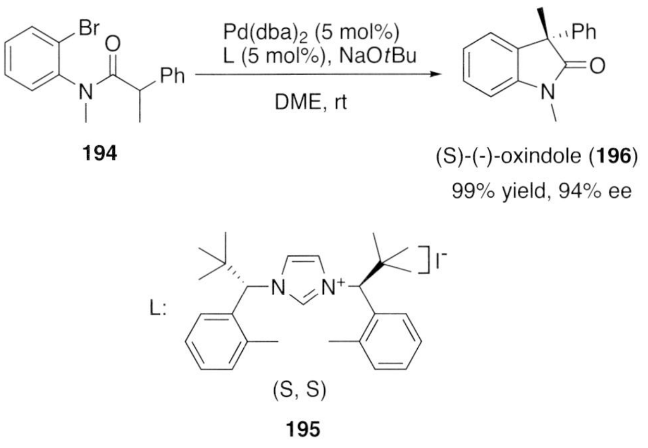

7.6 Cyclic Coupling via the Heck Reaction

Scheme 7.37 Enantioselective α-arylation of amide.

the Heck reaction before 2002. Grigg also summarized some of the recent advances in Chapter 4 in this book. We review the developments since 2002 of cyclic coupling via the Heck reaction which are not included in Chapter 4.

Fuwa and Sasaki designed a strategy for the synthesis of 2-substituted indoles **209** starting from N-(o-bromophenyl)enecarbamates **208** by a 5-endo Heck-type cyclization (Eq. (7.22)) [80].

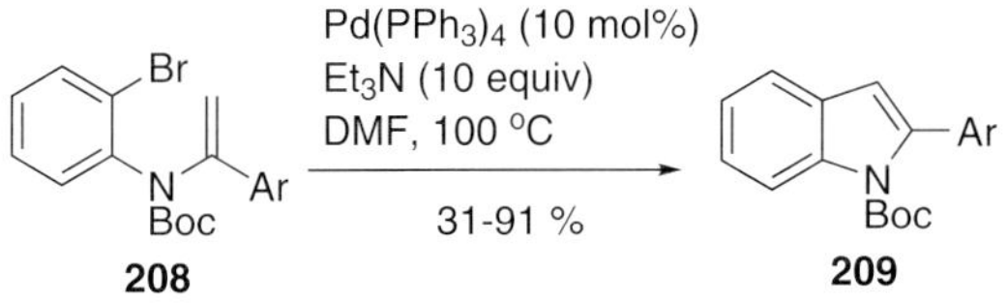

(7.22)

Fu's group from MIT reported the first palladium-catalyzed intramolecular Heck reactions of unactivated, β-hydrogen-containing alkyl bromides and chlorides **210**, in which the catalyst $Pd_2(OMe\text{-}dba)_3$/SIMes-HBF_4 suppressed the β-hydrogen elimination with palladium oxidative addition of sp^3 carbon halide bond (Scheme 7.40) [81].

Yorimitsu and Oshima found that the combination of [$CoCl_2$(dppb)] catalyst and Me_3SiCH_2MgCl induces intramolecular 5-exo Heck-type cyclization reactions of **212**, possibly via a radical process, to afford **213** (Scheme 7.41) [82].

Several bicyclic compounds **215** bearing a 1,2-cyclopentenediol unit have been prepared from the 5-exo cyclic carbopalladation-intermolecular Stille coupling reaction by Suffert et al. [83]. They also developed an intramolecular carbopalladation-Stille cross-coupling or Pd-shift-Stille coupling reaction of benzocycloheptenyl bromide **216** (Scheme 7.42) [84].

A Pd/dppf-based catalyst provides access to 3-substituted indoles **223** from 1,2-dihalogenated aromatic compounds **221** and allylic amines via sequential intermolecular aryl amination, 5-exo carbopalladation, and C=C bond isomerization (Eq. (7.23)) [85].

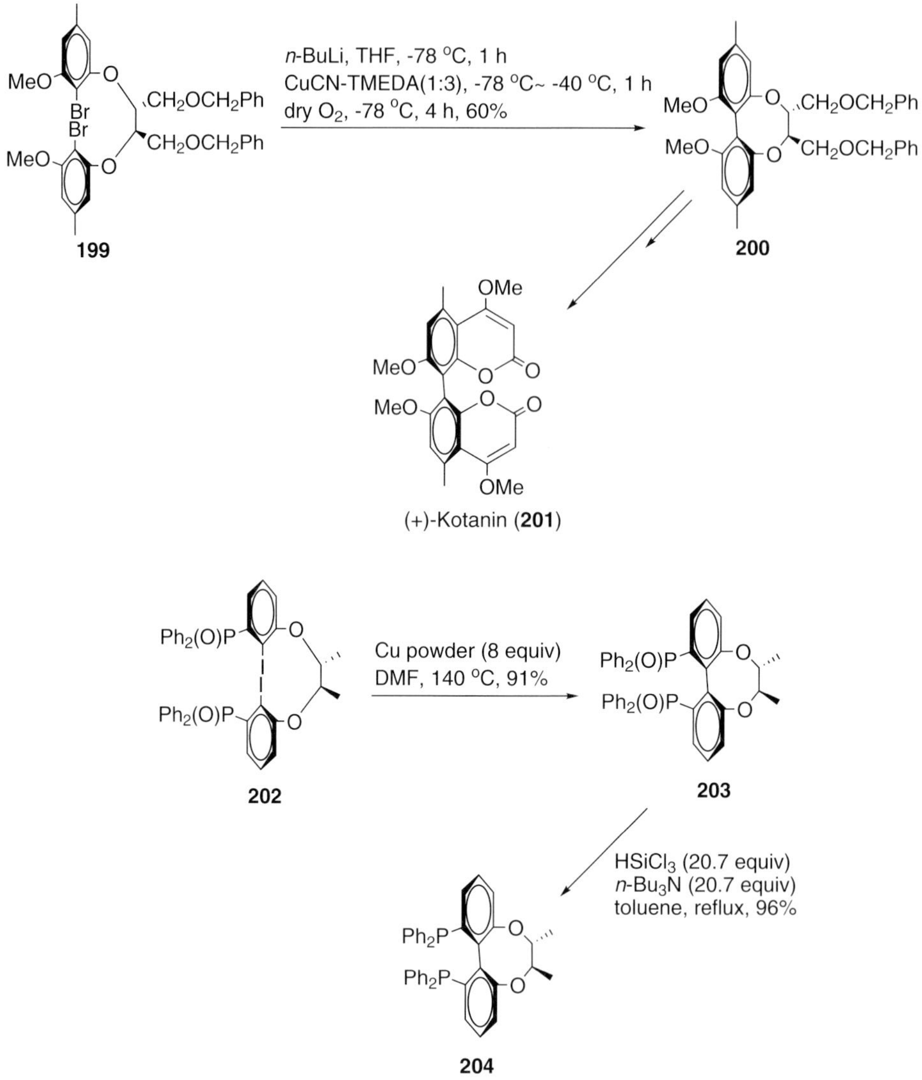

Scheme 7.38 Cyclic biaryl coupling.

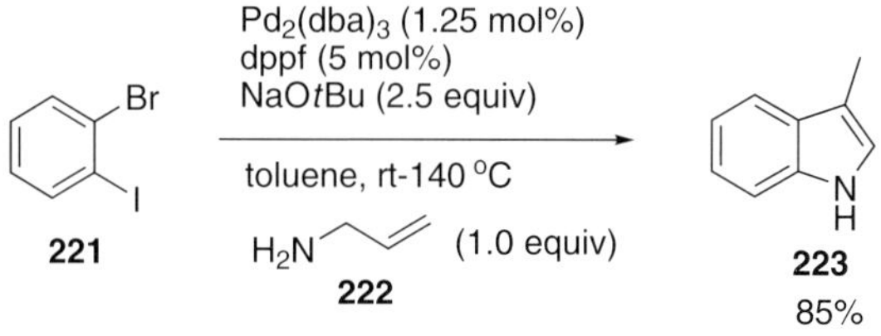

(+)-Kotanin (201)

$$\text{221} + \text{222} \xrightarrow[\text{toluene, rt-140 °C}]{\substack{\text{Pd}_2(\text{dba})_3 \text{ (1.25 mol\%)} \\ \text{dppf (5 mol\%)} \\ \text{NaO}t\text{Bu (2.5 equiv)}}} \text{223} \quad 85\% \tag{7.23}$$

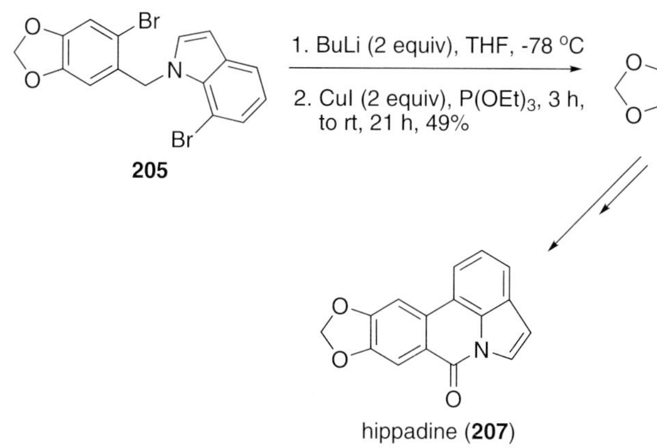

7.6 Cyclic Coupling via the Heck Reaction

Scheme 7.39 Cyclic biaryl coupling for hippadine.

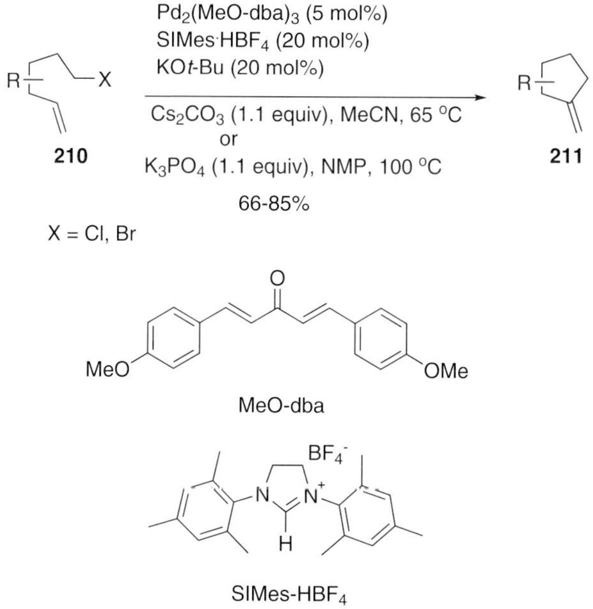

Scheme 7.40 Cyclic heck reaction of sp³ C—X bonds.

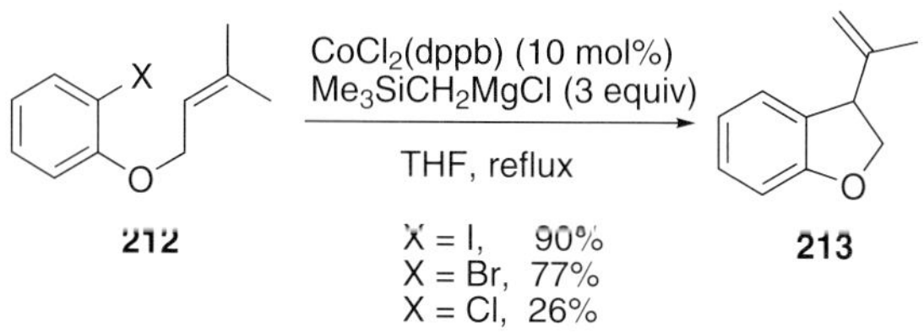

Scheme 7.41 Co-catalyzed Heck–type reaction.

X = I, 90%
X = Br, 77%
X = Cl, 26%

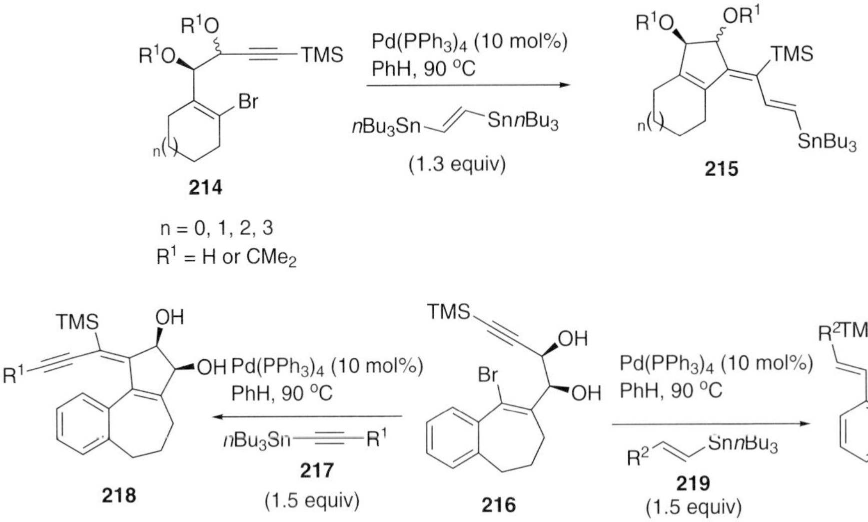

Scheme 7.42 Cyclic carbopallation of alkynes – coupling for bi- or tricyclic compounds.

Overman et al. used a diastereoselective intramolecular 5-exo Heck reaction to build the C3 quaternary center during the total synthesis of asperazine **226** [86]. The precursor **224** was efficiently coupled with the tethered aryl iodide moiety obtaining product **225** (Scheme 7.43).

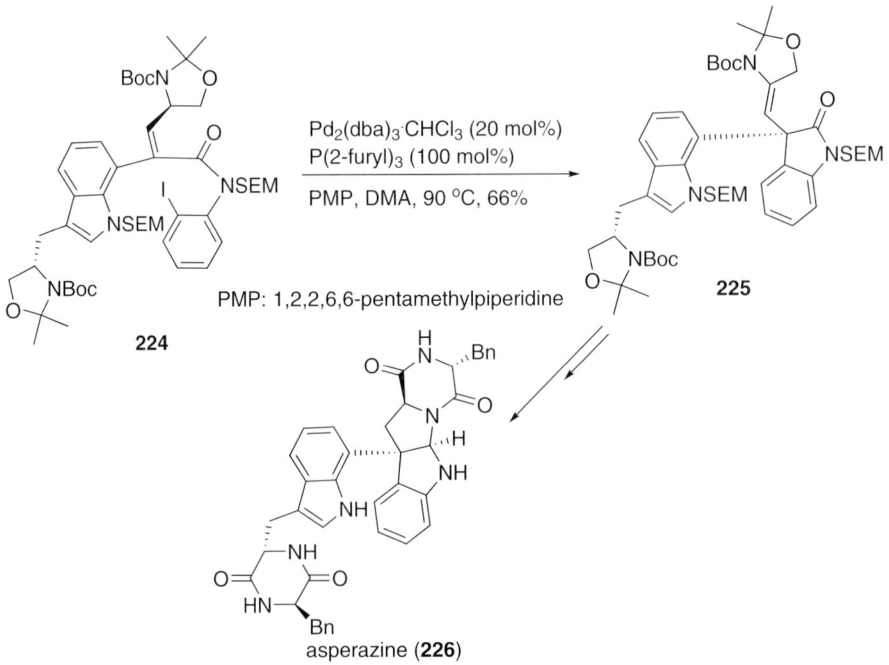

Scheme 7.43 Asperazine via cyclic Heck reaction.

7.6 Cyclic Coupling via the Heck Reaction

A practical synthesis involving asymmetric 5-exo Heck cyclization has also been developed by Overman's group for establishing oxindoles **228** containing an all-carbon quaternary carbon center in high enantiopurity (71–98% ee) with the Pd-BINAP catalyst (Scheme 7.44) [87].

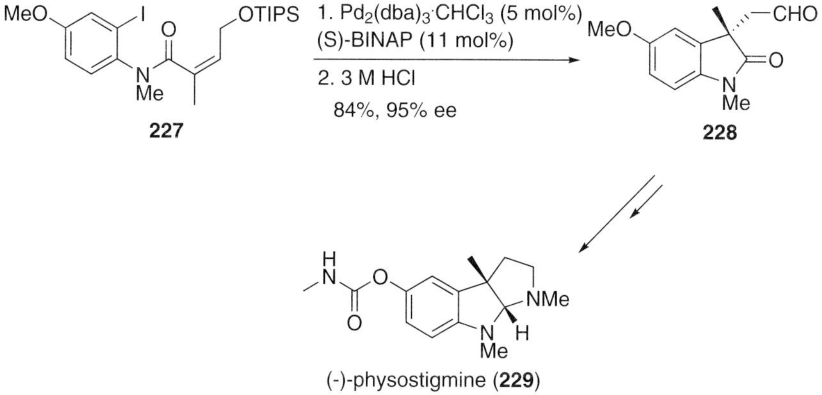

Scheme 7.44 Enantioselective cyclic Heck reaction.

Fukuyama *et al.* applied the intramoecular 6-exo Heck reaction as the key step in the assembly of the central bicyclo[3.3.1] ring system **231** in the total synthesis of ecteinascidin 743 **232** (Scheme 7.45) [88].

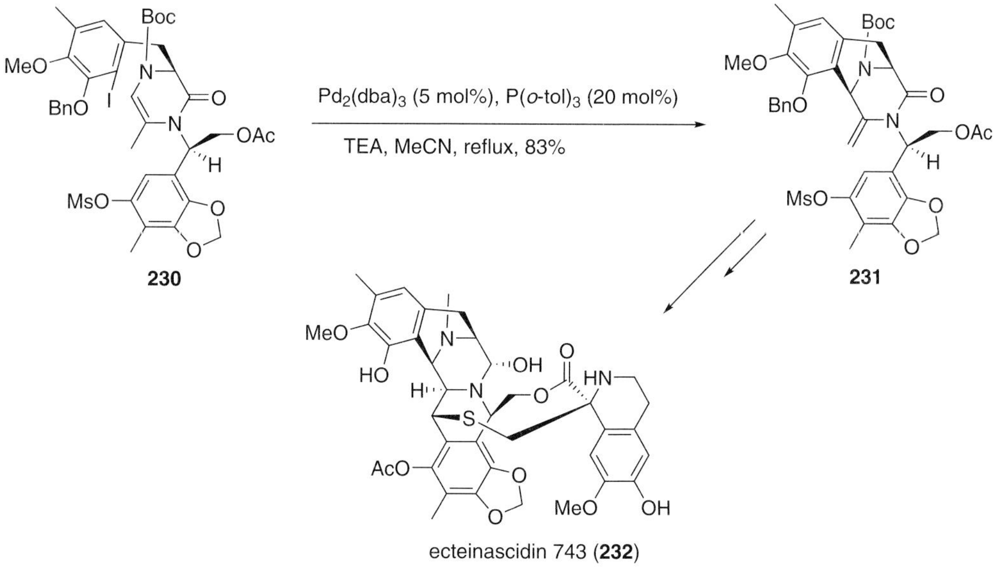

Scheme 7.45 Ecteinascidin 743 via cyclic Heck reaction.

Hexahydronaphthacene **234** can be formed by a palladium-catalyzed domino Tsuji–Trost-allylation–Heck reaction of aryl iodide **233** that contains an allyl acetate moiety and a undephilic unit (Eq. (7.24)) [89].

$$(7.24)$$

In 2002, this group developed an intramolecular triple 6-exo Heck reaction for synthesis of fused tetracycles **236** with a benzene core (Eq. (7.25)) [90].

$$(7.25)$$

Tietze's group reported an enantioselective total synthesis of desogestrel **242** using two consecutive Heck reactions as the key steps [91]. For the first intermolecular Heck coupling of **237** with **238**, the use of Pd(OAc)$_2$ in the presence of PPh$_3$ and Ag$_2$CO$_3$ in DMF is most appropriate due to the insertion into the double bond of **238** taking place exclusively from the α-face anti to the angular ethyl group, establishing the stereochemistry needed for the synthesis of **239**. Construction of the tetracyclic steroid core from **239** was then successfully performed using the palladacene **240** at 135°C to afford **241** in 94% yield (Scheme 7.46).

Keay et al. finished the total synthesis of (+)-xestoquinone **249** in 68% ee by a palladium(0)-catalyzed polyene cyclization of naphthyl triflate **243** using (S)-(+)-BINAP as the chiral ligand (Scheme 7.47) [92].

In 1997, Negishi et al. established the seven-membered ring **251** via an intramolecular 7-endo Heck reaction of **250** during the synthesis of (±)-7-*epi*-β-bulnesene **252** (Scheme 7.48) [93].

Regioselective 7-endo Heck cyclization of **253** gave rise to a tricyclic intermediate, which may be converted to guanacastepene N **255** (Scheme 7.49) [94].

Stoltz's group also applied the 7-exo Heck reaction to build the seven-membered ring **258** of the carbocyclic skeleton of the natural product cephalotaxine **259** in a racemic form (Scheme 7.50) [95].

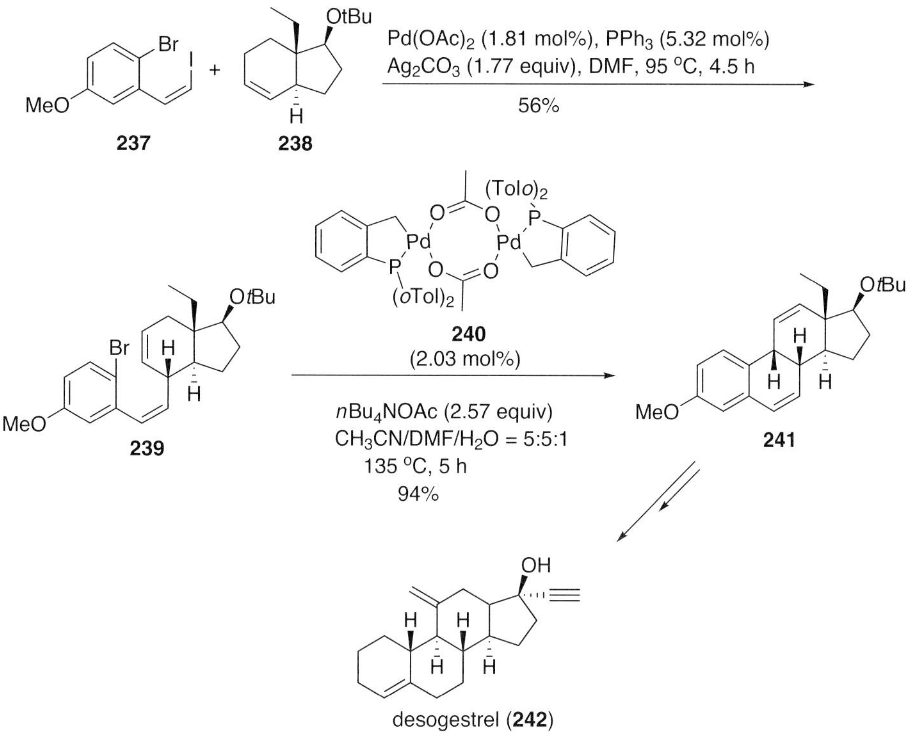

Scheme 7.46 Desogestrel via cyclic arylation of cyclohexene.

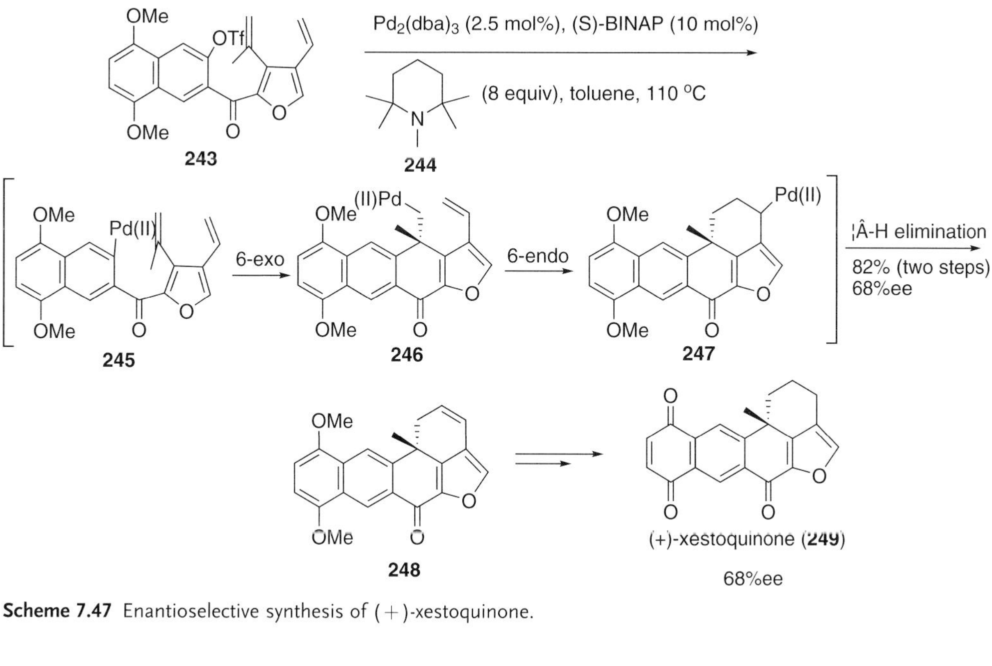

Scheme 7.47 Enantioselective synthesis of (+)-xestoquinone.

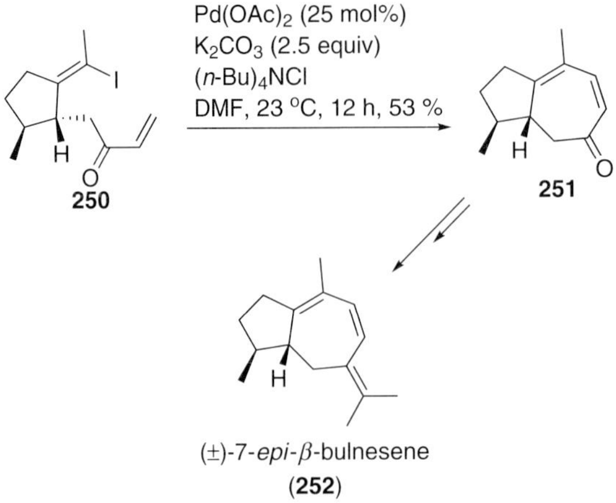

Scheme 7.48 Endo-mode cyclic Heck reaction of 2-enone for 7-epi-β-bulnesene.

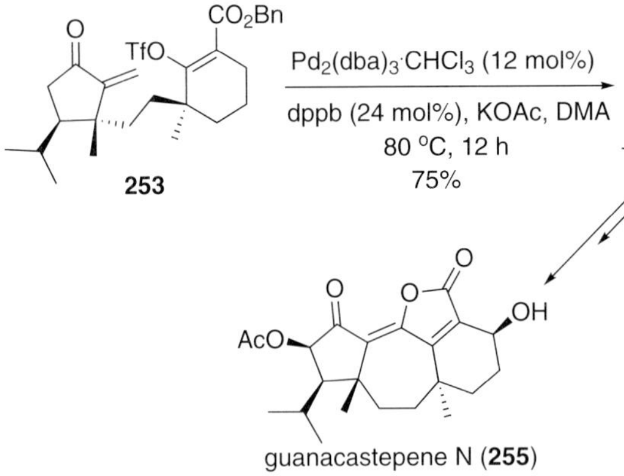

Scheme 7.49 Synthesis of guanacastepene N.

A synthetic sequence was developed by Overman's group for advancing 3,4-dihydro-9a,4a-(iminoethano)-9H-carbazole to (+)-minfiensine **262** by a reductive 6-exo Heck cyclization of **260** to form the fifth ring of (+)-minfiensine **262** (Scheme 7.51) [96].

A type of palladium-catalyzed cascade cyclization–coupling reaction that proceeds with suppressed β-hydride elimination, due to coordinative stabilization of an alkylpalladium intermediate by one of the N-sulfonyl oxygens, was found by Kim's group (Eq. (7.26)) [97].

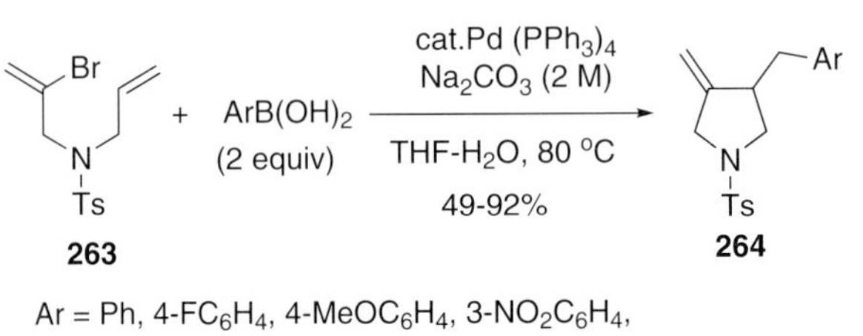

$$Ar = Ph, 4\text{-}FC_6H_4, 4\text{-}MeOC_6H_4, 3\text{-}NO_2C_6H_4,$$

(7.26)

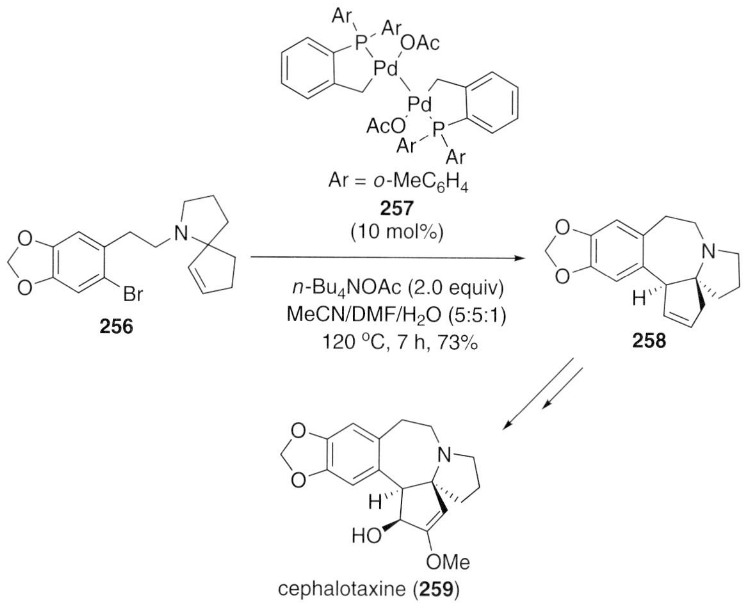

Scheme 7.50 7-exo cyclic Heck reaction for cephalotaxine.

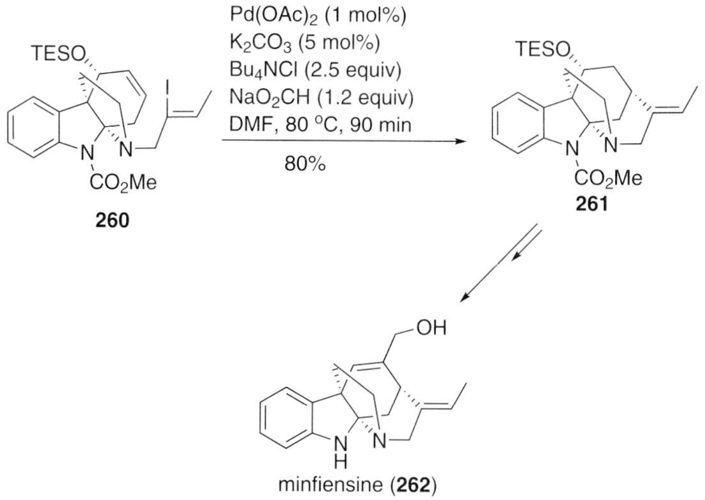

Scheme 7.51 Synthesis of minfiensine via cyclic carbopalladation–reduction.

In 2007, Baran et al. finished the synthesis of hapalindole U **267** and ambiguine H **268** using the same key structure **266**, which was also afforded by a reductive Heck-type reaction [98]. Compound **265** proceeded by 6-exo-trig cyclization in the presence of HCO$_2$Na to afford **266** (Scheme 7.52).

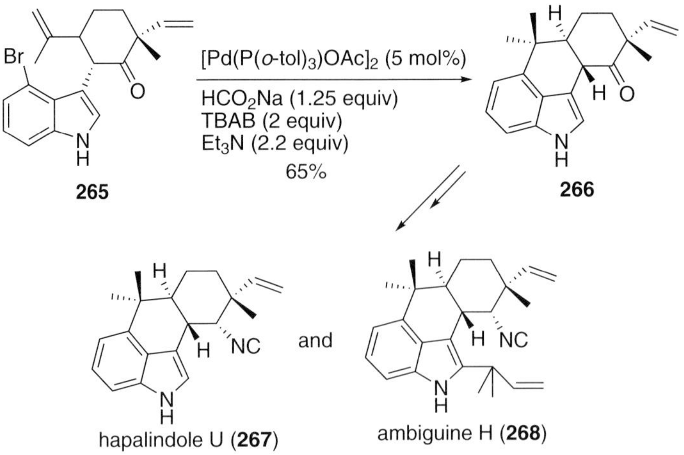

Scheme 7.52 Cyclic carbopalladation–reduction for hapalindole U and ambiguine H.

7.7
Cyclic Coupling Involving Terminal Alkynes

Sonogashira coupling involving coupling between terminal alkynes with aryl or vinyl halides under the catalysis of palladium and copper(I) was discovered in 1975 [99]. This reaction was reviewed last year [100]. We focus on the most recent developments in this area for the efficient synthesis of cyclic compounds.

The Iyoda group carried out a simple and efficient synthesis of tribenzohexadehydro[12]annulene **270** and its derivatives using the coupling reaction of 2-iodoaryl acetylenes **269** in the presence of catalytic amounts of CuI and PPh$_3$, together with 3 equiv of K$_2$CO$_3$ in DMF (Eq. (7.27)) [101]. This synthetic procedure was applied to the synthesis of a large annulenoannulene derivative.

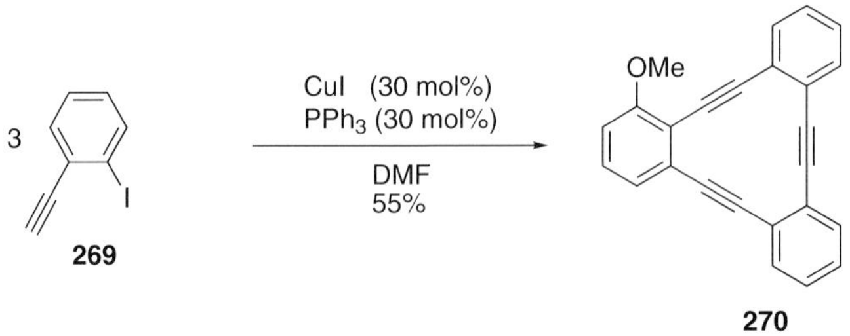

(7.27)

Recently, Fallis *et al.* developed a general synthetic strategy for the rapid assembly of both acetylenic and allenic macrocycles from optically active propargyl alcohols [102]. The sequential application of Sonogashira cross-coupling reactions generated the cyclic alkynes in **272** and **275**, which were converted to tri-meta-allenophane **273** and tetra-meta-allenophane **276** (Scheme 7.53).

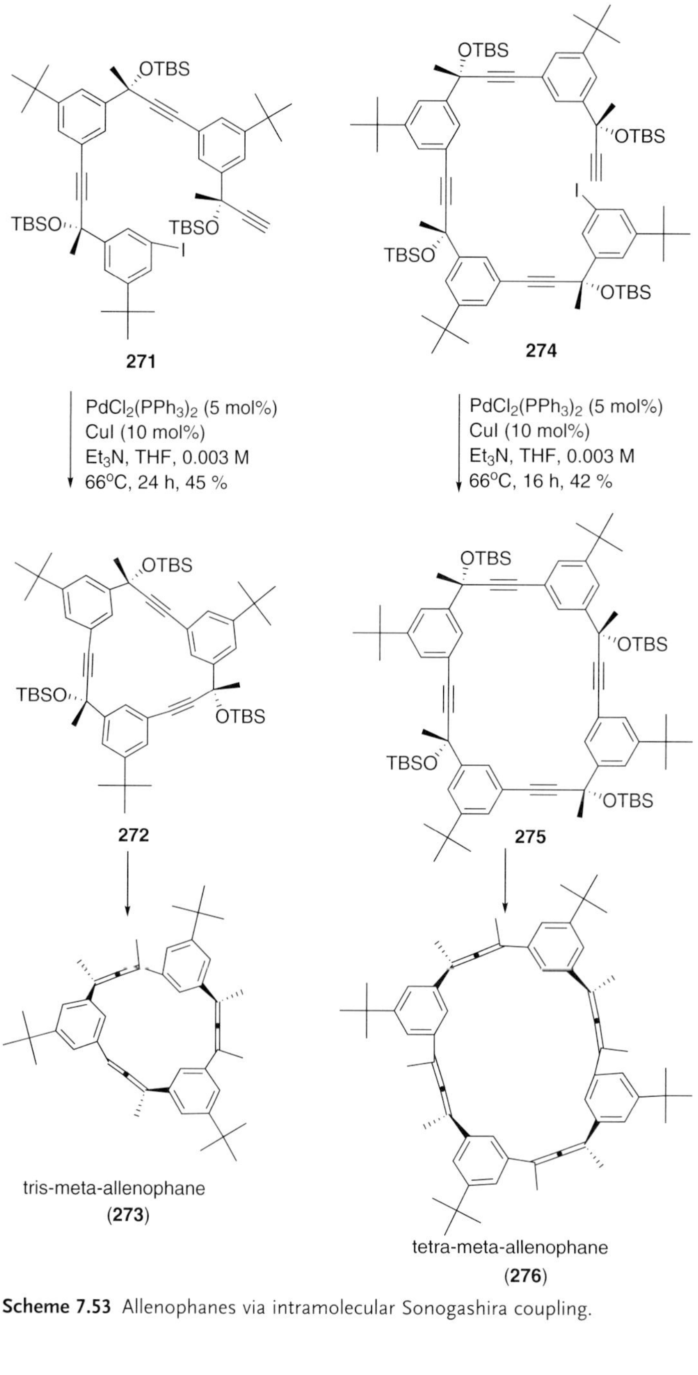

7.7 Cyclic Coupling Involving Terminal Alkynes | 355

271

PdCl$_2$(PPh$_3$)$_2$ (5 mol%)
CuI (10 mol%)
Et$_3$N, THF, 0.003 M
66°C, 24 h, 45 %

274

PdCl$_2$(PPh$_3$)$_2$ (5 mol%)
CuI (10 mol%)
Et$_3$N, THF, 0.003 M
66°C, 16 h, 42 %

272

275

tris-meta-allenophane
(**273**)

tetra-meta-allenophane
(**276**)

Scheme 7.53 Allenophanes via intramolecular Sonogashira coupling.

A CuI-catalyzed Sonogashira-type intramolecular coupling reaction between terminal alkyne and vinyl iodide was also reported by Coleman et al. for stereocontrolled construction of the 12-membered cyclic enyne in **278**. Stereoselective reduction of the cyclic alkyne **278** introduced the *cis*-olefin in target **279** (Scheme 7.54) [103].

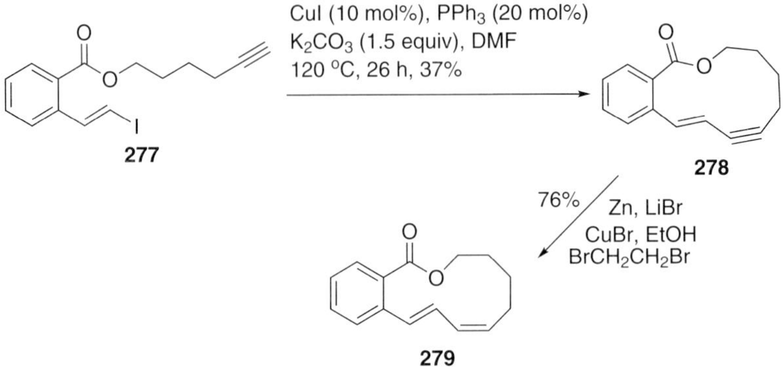

Scheme 7.54 Sonogashira coupling for benzolactone.

The Glasser reaction is an oxidative coupling by two terminal alkynes [104]. Myers' group established the key structure by Glasser coupling in the total synthesis of N1999A2 **292**, in which two intramolecular terminal alkynes were coupled with $Cu(OAc)_2$ as catalyst (Scheme 7.55) [105].

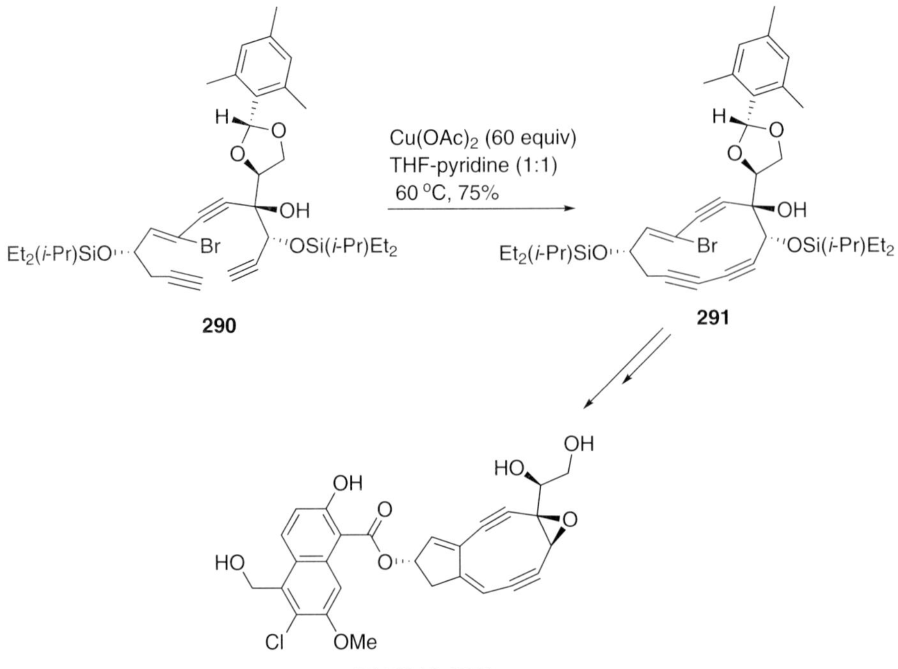

Scheme 7.55 $Cu(OAc)_2$-mediated cyclic coupling of two terminal alkynes.

7.8
Conclusion and Perspectives

Cross coupling has been extensively demonstrated as one of most important tools in the syntheses of drugs and advanced materials, especially those with cyclic structural units. High catalyst turnover, wide scope, selectivity, mild conditions and application in muti-component reactions are targets in this field. Cyclic coupling will achieve sustainable development both in academia and in industry.

7.9
Experimental: Selected Procedures

7.9.1
Synthesis of Compound 4 (Scheme 7.2) [5]

A solution of tripeptide **3** (46 mg, 0.04 mmol), and K_2CO_3 (55 mg, 0.4 mmol) in 1,4-dioxane/H_2O (40:3, 86 mL) was degassed for 30 min. $PdCl_2$(dppf)·CH_2Cl_2 (32 mg, 0.04 mmol) was added to the solution and the resulting mixture was stirred at 90 °C for 1 h. After cooling to room temperature, 1% HCl was added to pH 6 and the solvent was passed through celite to remove palladium. The filtrate was evaporated and the residue was partitioned with EtOAc and 1% HCl. The organic phase was washed with H_2O and brine and dried over Na_2SO_4. The solvent was evaporated under reduced pressure and the residue was purified by flash column chromatography (silica gel, from 4 : 5 : 1 heptane:EtOAc:MeOH to 10% MeOH in CH_2Cl_2, gradient elution) to provide the desired product **4** (24 mg, 66%).

7.9.2
Synthesis of Compound 17 (Scheme 7.6) [9]

To a 25 °C solution of Suzuki cyclization precursor **15** (120.0 mg, 0.052 mmol, 1.0 equiv) in degassed THF (0.10 mL) was added a solution of catecholborane (160.0 μL, 1.0 M in THF, 0.16 mmol, 3.0 equiv) and the reaction mixture was stirred at 25 °C for 1 h in the dark. Then, a freshly prepared solution of dicyclohexylborane (100.0 μL, 0.1 M in THF, 0.010 mmol, 0.2 equiv) was added at 25 °C and the reaction mixture was stirred for an additional 30 min in the dark (until the starting material was consumed, as determined by TLC). The reaction mixture was then carefully quenched with degassed H_2O (0.30 mL) and stirred for 1 h in the dark. The resulting boronic acid in the THF/H_2O solution was diluted with degassed THF (56 mL) and degassed H_2O (5.3 mL), and Pd(PPh$_3$)$_4$ (60.0 mg, 0.052 mmol, 1.0 equiv) was added, followed by TlOEt (1.1 mL, 15.6 mmol, 300 equiv). The reaction mixture was stirred at 25 °C for 4 h in the dark, then quenched with H_2O (50 mL) and extracted with EtOAc (3 × 100 mL). The combined organic layers were washed with brine (3 × 100 mL), dried (MgSO$_4$), and concentrated. The resulting oil was dissolved in THF (40 mL) and TBAF (2.6 mL, 2.6 mmol, 50 equiv) was added at 25 °C. The reaction mixture was stirred at 25 °C for 18 h, at which time the volume of the reaction was reduced to about 5 mL. The solution was purified by HPLC (C8-Luna 5 μ column, 100 Å, 250 mm × 10 mm, 60% MeCN in H$_2$O) in 1 mL increments and provided marinomycin A **17** (11.9 mg, 0.012 mmol, 23% yield in 3 steps).

7.9.3
Synthesis of Compound 44 (R = Ph) (Scheme 7.11) [19]

Under N$_2$ atmosphere, a toluene solution (20 mL) of but-3-ynoic acid **40** (1 g, 0.012 mol), tributyltin hydride (8.6 g, 0.029 mol), and azobis(isobutyronitrile) (10 mg) was stirred at 100 °C. After 3 h, toluene was removed under vacuum and then carbon tetrachloride (15 mL) was added to react with the excess tributyltin hydride. After the solution was stirred for 1 h, potassium fluoride solution (0.5 M, 15 mL) and acetone (15 mL) were added. The solution was extracted with diethyl ether after filtration and then dried over magnesium sulfate. After removal of the solvents, tributylstannyl-4-(tributylstannyl)but-3-enoate **41** (6.56 g) was obtained as a 85/15 mixture of Z and E (83% yield). Then in a Schlenk tube, to a degassed solution of 2.11 g (15 mmol) of benzoyl chloride **42** in 50 mL of distilled dioxane was added dropwise 6.64 g (10 mmol) of tributylstannyl-4-(tributylstannyl)but-3-enoate **41** (E/Z = 91/9), and then 570 mg (5 mol%) of tetrakis(triphenylphosphine)palladium(0) was added. After stirring for 6 h at room temperature, the mixture was hydrolyzed with a saturated solution of ammonium chloride and extracted with diethyl ether (3 × 50 mL). The organic phase was washed with a saturated solution of sodium chloride and dried with magnesium sulfate. A 1.24 g (7.20 mmol) yield of 6-phenylpyran-2-one **44** was obtained after purification by column chromatography on silica gel (95/5 hexane/diethyl ether).

7.9.4
Synthesis of Compound 86 (Scheme 7.18) [34]

The epoxide **84** (560 mg, 1.005 mmol, 1.0 equiv) was dissolved in THF (20 mL). To this solution was added AgNO$_3$ (18 mg, 0.10 mmol, 0.1 equiv) followed by N-iodosuccinimide (610 mg, 2.71 mmol, 2.7 equiv), and the reaction was stirred in the dark for 3.5 h. After which time, the reaction was diluted with H$_2$O (100 mL) and EtOAc (300 mL). The layers were separated, and the organic layer was washed saturated brine (100 mL), dried (MgSO$_4$), filtered, and concentrated. Purification by column chromatography (SiO$_2$, 2:1 hexane/EtOAc) provided the bis-iodoalkyne 740 mg (91%) as a very pale yellow glass. The diiodide (480 mg, 0.592 mmol, 1.0 equiv) was dissolved in DMF (56 mL) and dry argon was passed through the solution for 20 min. The reaction was then heated to 75 °C, at which temperature Pd(PPh$_3$)$_4$ (34 mg, 0.0296 mmol, 0.05 equiv) was added under a stream of argon. A 0.023 M solution of cis-1,2-bis-(trimethylstannyl)ethylene **85** (33 mL, 0.769 mmol, 1.3 equiv) in degassed DMF was then added dropwise to the reaction via syringe-pump over 1.2 h. After complete addition, the yellow–tan reaction mixture was allowed to warm to room temperature and then poured into H$_2$O (100 mL) and Et$_2$O (250 mL).

The layers were separated, and the organic layer was washed with H_2O (5 × 60 mL). The organic layer was then washed with saturated brine, dried ($MgSO_4$), filtered, and concentrated *in vacuo*. Purification by column chromatography (SiO_2, 3 : 1 EtOAc/hexane) gave 284 mg (81%) of the enediyne **86** as an off-white foam.

7.9.5
Synthesis of Compound 91 (Scheme 7.19) [37]

In a three-necked, 50-mL, round-bottomed flask fitted with a N_2 inlet adapter, a thermocouple (inserted through a septum), rubber septa and a magnetic stir bar was placed a solution of TBAF·$3H_2O$ (9.47 g, 30.0 mmol, 10.0 equiv) in THF (30 mL). π-Allylpalladium chloride dimer (APC) (82.4 mg, 0.225 mmol, 0.075 equiv, purchased from ACROS) was added. A solution of **90** (1.55 g, 3.0 mmol) in THF (10 mL) was then added by syringe-pump for 60 h at room temperature. After complete addition of **90**, the mixture was stirred for 1 h and then concentrated. The residue was filtered through a short silica gel column, which was eluted with Et_2O (600 mL). The solvent was removed by rotary evaporation to give a brown residue, which was purified by chromatography (silica gel, hexane/EtOAc, 9/1 to 4/1) to afford 607 mg (61%) of **91** as a colorless oil.

7.9.6
Synthesis of Compound 100 (Z = Boc, n = 2) (Eq. 7.11) [41]

A dry 25-mL sealable Schlenk tube was charged with Pd(OAc)$_2$ (3.2 mg, 0.014 mmol) and (±)-BINAP (13.4 mg, 0.022 mmol). The reaction vessel was evacuated and flushed with argon. A solution of **99** (135 mg, 0.43 mmol) in toluene (1 mL) was added via cannula. The mixture was heated under argon at 100 °C for 2 min to dissolve the solids. The reaction vessel was removed from the oil bath, charged quickly with Cs_2CO_3 (196 mg, 0.60 mmol) and toluene (0.3 mL), sealed with a Teflon screw cap, and heated at 100 °C until the aryl bromide was consumed (24 h). The reaction mixture was cooled to rt, filtered through a short plug of SiO_2, and concentrated. The residue was purified by flash column chromatography (9% EtOAc-hexanes) to afford **100** (Z = Boc, n = 2) (81 mg, 81% yield) as a colorless oil.

7.9.7
Synthesis of Compound 135 (Scheme 7.28)) [56d,e]

A flame-dried 100-mL round-bottomed flask equipped with a magnetic stirrer bar was charged with **134** (3.13 g, 4.39 mmol), copper iodide (417.1 mg, 2.19 mmol), cesium acetate (4.21 g, 21.9 mmol), and dry DMSO (14.6 mL) under argon atmosphere. The resulting pale-green solution was stirred at 60 °C for 24 h, after which TLC (hexane/dichloromethane = 3 : 2) indicated complete consumption of **134**. After the flask was cooled to room temperature, the reaction mixture was poured into 5% aqueous sodium chloride in 10% aqueous ammonium hydroxide and extracted with ethyl acetate. The combined organic extracts were washed with brine, dried over anhydrous sodium sulfate, and filtered. The filtrate was concentrated under reduced pressure. The residue was purified by silica-gel chromatography (dichloromethane/hexane = 1:1–3:1, gradient) to afford pure (−)-(3S)-7-benzyloxy-5-bromo-1-(o-nitrobenzenesulfonyl)-1,2,3,4-tetrahydro-3-(*tert*-butyldimethylsilyloxy)quinoline **135** (2.32 g, 3.66 mmol, 83%) as a white solid, which was recrystallized from hexane/dichloromethane to afford fine colorless needles.

7.9.8
Synthesis of Compound 261 (Scheme 7.51) [96]

A suspension of vinyl iodide **260** (1.20 g, 2.07 mmol), K_2CO_3 (1.43 g, 10.4 mmol), $Bu_4NCl·H_2O$ (1.43 g, 5.18 mmol), NaO_2CH (170 mg, 2.48 mmol) and DMF (30 mL) was degassed with Ar for 15 min. To this suspension was added $Pd(OAc)_2$ (5 mg, 0.02 mmol)), and degassing was continued for an additional 10 min. The sealed reaction vial was then heated at 80 °C for 90 min at which time TLC analysis indicated complete consumption of starting material. After cooling to rt, the reaction was diluted with diethyl ether and washed with brine. The organic layer was then dried (Na_2SO_4), concentrated, and the residue was chromatographed over silica gel (5–10% EtOAc/hexanes) to afford **261** (750 mg, 1.66 mmol, 80%) as a clear, colorless oil.

7.9.9
Synthesis of Compound 278 (Scheme 7.54) [103]

Alkyne **277** (1.0 mmol) was added to a mixture of CuI (0.1 mmol), Ph_3P (0.2 mmol), K_2CO_3 (1.5 mmol) in DMF (25 mL) under nitrogen. The reaction mixture was stirred at 110 °C for 26 h, when it was poured into water (20 mL), extracted with ether (20 mL × 3). The combined organic extracts were dried ($MgSO_4$), filtered and concentrated *in vacuo*. The residue was purified by flash chromatography (5 × 20 cm silica gel, 0–15% EtOAc/hexane gradient) to afford cycloalkyne **278** (37%).

7.9.10
Synthesis of Compound 291 (Scheme 7.53) [105]

A 100-mL round-bottom flask was charged under an argon atmosphere with $Cu(OAc)_2$ (918 mg, 5.06 mmol, 60 equiv), tetrahydrofuran (15 mL), and pyridine (15 mL), in that order. The mixture was stirred at 23 °C for 5 min, then was warmed to 60 °C. After 10 min, a solution of the bromoenetriyne **290** (61.5 mg, 84.3 μmol, 1 equiv) in a 1:1 mixture of THF and pyridine (30 mL) was added via cannula over 12 min. After 6 min, the reaction mixture was allowed to cool to ambient temperature and the cooled solution was partitioned between a 1:1 mixture of ether and hexanes (120 mL, cooled in an ice-water bath) and 1.0 M aqueous copper (II) sulfate solution (80 mL). The layers were separated and the organic layer was washed with 1.0 M aqueous copper (II) sulfate solution (2 × 80 mL). The aqueous layers were combined and the combined solution was extracted with a 1:1 mixture of ether and hexanes (80 mL). The organic extracts were combined and the combined solution was washed with brine (30 mL). The washed solution was dried over anhydrous sodium sulfate and toluene (3 mL) was added. The dried solution was filtered and the filtrate was concentrated to a volume of ∼3 mL. The residue was purified by flash-column chromatography (13% ether–pentane). To determine the yield of the reaction, fractions containing pure **291** were combined and the combined solution (∼50 mL) was diluted with C_6D_6 (50 mL). The solution was concentrated to a volume of about 1 mL. A ^{1}H NMR spectrum was taken to confirm purity and that no toluene was present, then an internal standard (toluene, 5.0 µL, 47 µmol) was added. The yield of **291** was determined by comparing integrated peak areas for **291** relative to toluene (calculated: 46.0 mg, 75%).

Abbreviations

Ac	acetyl
AIBN	azobisisobutyronitrile
9-BBN	9-borabicyclo[3.3.1]nonane
BINAP	2,2′-bis(diphenylphosphino)-1,1′-binaphthyl
Bn	benzyl
Boc	*tert*-butyloxycarbonyl
Cbz	carboxybenzyl
Cy	cyclohexyl
CuTC	copper(I) thiophene-2-carboxylate
dba	dibenzylideneacetone
DCM	dichloromethane
DIPEA	*N,N*-diisopropylethylamine
DMA	dimethylacetamide
DME	1,2-dimethoxyethane
DMF	dimethylformamide
DMSO	dimethyl sulfoxide
DPEphos	bis(2-diphenylphosphinophenyl)ether
dppb	1-4-bis(diphenylphosphino)butane
dppe	1-4-bis(diphenylphosphino)ethane
dppf	1,1′-bis(diphenylphosphino)ferrocene
HATU	o-(7-azabenzotriazol-1-yl)-*N,N,N′,N′*-tetramethyluronium hexafluorophosphate
MEM	2-methoxyethoxymethyl
MOM	methoxymethyl
MS	molecular sieve
Ms	methanesulfonyl
NIS	*N*-iodosuccinimide
NMP	*N*-methylpyrrolidone
Ns	2-nitrobenzene-1-sulfonyl
PMB	*p*-methoxybenzyl

PMP	1,2,2,6,6-pentamethylpiperidine
Py	pyridine
SEM	2-(trimethylsilyl)ethoxymethyl
Sphos	2-dicyclohexylphosphino-2′,6′-dimethoxy-1,1′-biphenyl
TBAB	tetra-n-butylammonium bromide
TBAF	tetra-n-butylammonium fluoride
TBDPS	*tert*-butyldiphenylsilyl
TBS	*tert*-butyldimethylsilyl
TES	triethylsilyl
Tf	trifluoromethylsulfonyl
TFP	trifuran-2-ylphosphine
THF	tetrahydrofuran
TIPS	triisopropylsilyl
TMEDA	tetramethylethylenediamine
TMS	trimethylsilyl
Troc	2,2,2-trichloroethoxycarbonyl
Ts	tosyl
Xantphos	4,5-bis(diphenylphosphino)-9,9-dimethylxanthene

References

1 (a) de Meijere, A. and Diederich, F. (eds) (2004) *Metal-Catalyzed Cross-Coupling Reactions*, Wiley-VCH, Weinheim; (b) Corbet, J.-P. and Mignani, G. (2006) *Chem. Rev.*, **106**, 2651–2710.

2 (a) Suzuki, A. (1985) *Pure Appl. Chem.*, **57**, 1749–1758; (b) Suzuki, A. (1991) *Pure Appl. Chem.*, **63**, 419–422; (c) Suzuki, A. (1994) *Pure Appl. Chem.*, **66**, 213–222.

3 Brown, H.C. (1975) *Organic Synthesis via Boranes*, John Wiley & Sons, New York, pp. 38–44.

4 Hensel, V., Lützow, K., Jacob, J., Gessler, K., Saenger, W. and Schluter, A.D. (1997) *Angew. Chem. Int. Ed. Engl.*, **36**, 2654–2656.

5 Jia, Y., Bois-Choussy, M. and Zhu, J. (2007) *Org. Lett.*, **9**, 2401–2404.

6 Wu, B., Liu, Q. and Sulikowski, G.A. (2004) *Angew. Chem. Int. Ed.*, **43**, 6673–6675.

7 Tortosa, M., Yakelis, N.A. and Roush, W.R. (2008) *J. Am. Chem. Soc.*, **130**, 2722–2723.

8 Molander, G.A. and Dehmel, F. (2004) *J. Am. Chem. Soc.*, **126**, 10313–10318.

9 Nicolaou, K.C., Nold, A.L., Milburn, R.R., Schindler, C.S., Cole, K.P. and Yamaguchi, J. (2007) *J. Am. Chem. Soc.*, **129**, 1760–1768.

10 Bauer, M. and Maier, M.E. (2002) *Org. Lett.*, **4**, 2205–2208.

11 Ma, S., Ni, B., Lin, S. and Liang, Z. (2005) *J. Organomet. Chem.*, **690**, 5389–5395.

12 (a) Carbonnelle, A.-C. and Zhu, J. (2000) *Org. Lett.*, **2**, 3477–3480; (b) Ishiyama, T., Murata, M. and Miyaura, N. (1995) *J. Org. Chem.*, **60**, 7508–7510.

13 Decicco, C.P., Song, Y. and Evans, D.A. (2001) *Org. Lett.*, **3**, 1029–1032.

14 (a) Stille, J.K. (1986) *Angew. Chem. Int. Ed. Engl.*, **25**, 508–524; (b) Mitchell, T.N. (1992) *Synthesis*, 803–815.

15 Piers, E., Romero, M.A. and Walker, S.D. (1999) *Synlett*, 1082–1084.

16 Paulon, A., Cossu, S., De Lucchi, O. and Zonta, C. (2000) *Chem. Commun.*, 1837–1838.

17 Casaschi, A., Grigg, R., Sansano, J.M., Wilson, D. and Redpath, J. (2000) *Tetrahedron*, **56**, 7541–7551.

18 Rousset, S., Abarbri, M., Thibonnet, J., Duchêne, A. and Parrain, J.-L. (1999) *Org. Lett.*, **1**, 701–703.

19 Thibonnet, J., Abarbri, M., Parrain, J.-L. and Duchêne, A. (2002) *J. Org. Chem.*, **67**, 3941–3944.

20 Smith, A.B. III, Condon, S.M., McCauley, J.A., Leazer, J.L. Jr, Leahy, J.W. and Maleczka, R.E. Jr, (1997) *J. Am. Chem. Soc.*, **119**, 962–973.

21 Smith, A.B. III, and Ott, G.R. (1998) *J. Am. Chem. Soc.*, **120**, 3935–3948.

22 Nicolaou, K.C., Winssinger, N., Pastor, J. and Murphy, F. (1998) *Angew. Chem. Int. Ed. Engl.*, **37**, 2534–2537.

23 (a) Sanglier, J.-J., Quesniaux, V., Fehr, T., Hofmann, H., Mahnke, M., Memmert, K., Schuler, W., Zenke, G., Gschwind, L., Mauer, C. and Schilling, W. (1999) *J. Antibiot.*, **52**, 466; (b) Fehr, T., Kallen, J., Oberer, L., Sanglier, J.-J. and Schilling, W. (1999) *J. Antibiot.*, **52**, 474.

24 Nicolaou, K.C., Xu, J., Murphy, F., Barluenga, S., Baudoin, O., Wei, H.-X., Gray, D.L.F. and Ohshima, T. (1999) *Angew. Chem. Int. Ed.*, **38**, 2447–2451.

25 (a) Paterson, I., Lombart, H.-G. and Allerton, C. (1999) *Org. Lett.*, **1**, 19–22; (b) Paterson, I. and Man, J. (1997) *Tetrahedron Lett.*, **38**, 695–698.

26 (a) Kelly, T.R., Li, Q. and Bushan, V. (1990) *Tetrahedron Lett.*, **31**, 161–164; (b) Kelly, T.R., Xu, W., Ma, Z., Li, Q. and Bushan, V. (1993) *J. Am. Chem. Soc.*, **115**, 5843–5844; (c) Fukuyama, Y., Yaso, K., Nakamura, K. and Kodama, M. (1999) *Tetrahedron Lett.*, **40**, 105–108.

27 Iwaki, T., Yasuhara, A. and Sakamoto, T. (1999) *J. Chem. Soc., Perkin Trans. 1*, 1505–1510.

28 Fukuyama, Y., Yaso, H., Mori, T., Takahashi, H., Minami, H. and Kodama, M. (2001) *Heterocycles*, **54**, 259–274.

29 Yue, W.S. and Li, J.J. (2002) *Org. Lett.*, **4**, 2201–2203.

30 Guthmann, H., Könemann, M. and Bach, T. (2007) *Eur. J. Org. Chem.*, 632–638.

31 Smith, A.B. III, Davulcu, A.H., Cho, Y.S., Ohmoto, K., Kürti, L. and Ishiyama, H. (2007) *J. Org. Chem.*, **72**, 4596–4610.

32 (a) Nicolaou, K.C., Chakraborty, T.K., Piscopio, A.D., Minowa, N. and Bertinato, P. (1993) *J. Am. Chem. Soc.*, **115**, 4419–4420; (b) Piscopio, A.D., Minowa, N., Chakraborty, T.K., Koide, K., Bertinato, P. and Nicolaou, K.C. (1993) *J. Chem. Soc. Chem. Commun.*, 617–618; (c) Nicolaou, K.C., Bertinato, P., Piscopio, A.D., Chakraborty, T.K. and Minowa, N. (1993) *J. Chem. Soc. Chem. Commun.*, 619–621; (d) Nicolaou, K.C., Piscopio, A.D., Bertinato, P., Chakraborty, T.K., Minowa, N. and Koide, K. (1995) *Chem. Eur. J.*, **1**, 318–333.

33 (a) Panek, J.S. and Masse, C.E. (1997) *J. Org. Chem.*, **62**, 8290–8291; (b) Masse, C.E., Yang, M., Solomon, J. and Panek, J.S. (1998) *J. Am. Chem. Soc.*, **120**, 4123–4134.

34 Shair, M.D., Yoon, T.Y., Mosny, K.K., Chou, T.C. and Danishefsky, S.J. (1996) *J. Am. Chem. Soc.*, **118**, 9509–9525.

35 (a) Hatanaka, Y. and Hiyama, T. (1988) *J. Org. Chem.*, **53**, 918–920; (b) Hatanaka, Y. and Hiyama, T. (1989) *J. Org. Chem.*, **54**, 268–270; (c) Hatanaka, Y., Fukushima, S. and Hiyama, T. (1989) *Chem. Lett.*, 1711; (d) Hatanaka, Y. and Hiyama, T. (1989) *Tetrahedron Lett.*, **30**, 6051–6054; (e) Hatanaka, Y. and Hiyama, T. (1990) *J. Am. Chem. Soc.*, **112**, 7793–7794; (f) Hatanaka, Y., Ebina, Y. and Hiyama, T. (1991) *J. Am. Chem. Soc.*, **113**, 7075–7076; (g) Gouda, K., Hagiwara, E., Hatanaka, Y. and Hiyama, T. (1996) *J. Org. Chem.*, **61**, 7232–7233.

36 Denmark, S.E. and Yang, S.-M. (2002) *J. Am. Chem. Soc.*, **124**, 2102–2103.

37 Denmark, S.E. and Yang, S.M. (2004) *J. Am. Chem. Soc.*, **126**, 12432–12440.

38 Ohmura, T., Masuda, K. and Suginome, M. (2008) *J. Am. Chem. Soc.*, **130**, 1526–1527.

39 For seminal papers, see: (a) Guram, A.S. and Buchwald, S.L. (1994) *J. Am. Chem. Soc.*, **116**, 7901–7902; (b) Guram, A.S., Rennels, R.A. and Buchwald, S.L. (1995) *Angew. Chem., Int. Ed. Engl.*, **34**, 1348–1350. For review, see: (c) Wolfe, J.P., Wagaw, S., Marcoux, J.F. and Buchwald, S.L. (1998) *Acc. Chem. Res.*, **31**, 805–818.

40 For seminal papers, see: (a) Louie, J. and Hartwig, J.F. (1995) *Tetrahedron Lett.*, **36**, 3609–3612; (b) Driver, M.S. and Hartwig, J.F. (1996) *J. Am. Chem. Soc.*, **118**, 7217–7218. For review, see: (c) Hartwig, J.F. (1998) *Acc. Chem. Res.*, **31**, 852–860.

41 (a) Yang, B.H. and Buchwald, S.L. (1999) *Org. Lett.*, **1**, 35–37; (b) Yang, B.H. and Buchwald, S.L. (1999) *J. Organomet. Chem.*, **576**, 121; (c) Klapars, A., Antilla, J.C., Huang, X. and Buchwald, S.L. (2001) *J. Am. Chem. Soc.*, **123**, 7727.

42 Abouabdellah, A. and Dodd, R.H. (1998) *Tetrahedron Lett.*, **39**, 2119–2122.

43 Snider, B.B. and Zeng, H. (2000) *Org. Lett.*, **2**, 4103–4106.

44 Coleman, R.S. and Chen, W. (2001) *Org. Lett.*, **3**, 1141–1144.

45 Doi, T., Kamioka, S., Shimazu, S. and Takahashi, T. (2008) *Org. Lett.*, **10**, 817–819.

46 Emoto, T., Kubosaki, N., Yamagiwa, Y. and Kamikawa, T. (2000) *Tetrahedron Lett.*, **41**, 355–358.

47 Margolis, B.J., Swidorski, J.J. and Rogers, B.N. (2003) *J. Org. Chem.*, **68**, 644–647.

48 Salcedo, A., Neuville, L., Rondot, C., Retailleau, P. and Zhu, J. (2008) *Org. Lett.*, **10**, 857–860.

49 (a) Smith, A.B. III, Kürti, L. and Davulcu, A.H. (2006) *Org. Lett.*, **8**, 2167–2170; (b) Smith, A.B. III, Kürti, L., Davulcu, A.H., Cho, Y.S. and Ohmoto, K. (2007) *J. Org. Chem.*, **72**, 4611–4620.

50 Fang, Y.-Q. and Lautens, M. (2008) *J. Org. Chem.*, **73**, 538–549.

51 Leogane, O. and Lebel, H. (2008) *Angew. Chem. Int. Ed.*, **47**, 350–352.

52 Zhu, Y.M., Kiryu, Y. and Katayama, H. (2002) *Tetrahedron Lett.*, **43**, 3577–3580.

53 Evindar, G. and Batey, R.A. (2003) *Org. Lett.*, **5**, 133–136.

54 Kozawa, Y. and Mori, M. (2002) *Tetrahedron Lett.*, **43**, 111–114.

55 (a) Ullmann, F. (1903) *Ber. Dtsch. Chem. Ges.*, **36**, 2382; (b) Fanta, P.E. (1946) *Chem. Rev.*, **38**, 139–196; (c) Fanta, P.E. (1964) *Chem. Rev.*, **64**, 613–632; (d) Hassan, J., Sevignon, M., Gozzi, C., Schulz, E. and Lemaire, M. (2002) *Chem. Rev.*, **102**, 1359–1470.

56 (a) Yamada, K., Kubo, T., Tokuyama, H. and Fukuyama, T. (2002) *Synlett*, 231–234; (b) Okano, K., Tokuyama, H. and Fukuyama, T. (2003) *Org. Lett.*, **5**, 4987–4990; (c) Yamada, K., Kurokawa, T., Tokuyama, H. and Fukuyama, T. (2003) *J. Am. Chem. Soc.*, **125**, 6630–6631; (d) Okano, K., Tokuyama, H. and Fukuyama, T. (2006) *J. Am. Chem. Soc.*, **128**, 7136–7137; (e) Okano, K., Tokuyama, H. and Fukuyama, T. (2008) *Chem. Asian J.*, **3**, 296–309.

57 (a) Ma, D., Zhang, Y., Yao, J., Wu, S. and Tao, F. (1998) *J. Am. Chem. Soc.*, **120**, 12459–12467; (b) Ma, D. and Xia, C. (2001) *Org. Lett.*, **3**, 2583–2586.

58 He, G., Wang, J. and Ma, D. (2007) *Org. Lett.*, **9**, 1367–1369.

59 Jiang, L., Job, G.E., Klapars, A. and Buchwald, S.L. (2003) *Org. Lett.*, **5**, 3667–3669.

60 (a) Toumi, M., Couty, F. and Evano, G. (2007) *Angew. Chem. Int. Ed.*, **46**, 572–575. (b) Toumi, M., Couty, F. and Evano, G. (2007) *J. Org. Chem.*, **72**, 9003–9009.

61 (a) Shen, R. and Porco, J.A. Jr (2000) *Org. Lett.*, **2**, 1333–1336; (b) Han, C., Shen, R., Su, S. and Porco, J.A. Jr (2004) *Org. Lett.*, **6**, 27–30.

62 Pan, X., Cai, Q. and Ma, D. (2004) *Org. Lett.*, **6**, 1809–1812.

63 Toumi, M., Couty, F. and Evano, G. (2008) *Synlett*, 29–32.

64 (a) Fang, Y. and Li, C. (2007) *J. Am. Chem. Soc.*, **129**, 8092–8093; (b) Lu, H. and Li, C. (2006) *Org. Lett.*, **8**, 5365–5367.

65 Fukudome, Y., Naito, H., Hata, T. and Urabe, H. (2008) *J. Am. Chem. Soc.*, **130**, 1820–1821.

66 Dahl, T., Tornøe, C.W., Bang-Andersen, B., Nielsen, P. and Jørgensen, M. (2008) *Angew. Chem. Int. Ed.*, **47**, 1726–1728.

67 Nicolaou, K.C., Ramanjulu, J.M., Natarajan, S., Bräse, S., Li, H., Boddy, C.N.C. and Rübsam, F. (1997) *Chem. Commun.*, 1899–1990.

68 Cai, Q., Zou, B. and Ma, D. (2006) *Angew. Chem., Int. Ed.*, **45**, 1276–1279.

69 (a) Boger, D.L., Miyazaki, S., Kim, S.H., Wu, J.H., Loiseleur, O. and Castle, S.L. (1999) *J. Am. Chem. Soc.*, **121**, 3226–3227; (b) Crowley, B.M., Mori, Y., McComas, C.C., Tang, D. and Boger, D.L. (2004) *J. Am. Chem. Soc.*, **126**, 4310–4317.

70 Dounay, A.B., Humphreys, P.G., Overman, L.E. and Wrobleski, A.D. (2008) *J. Am. Chem. Soc.*, **130**, 5368–5377.

71 (a) Solé, D., Diaba, F. and Bonjoch, J. (2003) *J. Org. Chem.*, **68**, 5746–5749; (b) Solé, D. and Serrano, O. (2008) *J. Org. Chem.*, **73**, 2476–2479.

72 Kündig, E.P., Seidel, T.M., Jia, Y.-X. and Bernardinelli, G. (2008) *Angew. Chem. Int. Ed.*, **46**, 8484–8487.

73 Zhang, S., Zhang, D. and Liebeskind, L.S. (1997) *J. Org. Chem.*, **62**, 2312–2313.

74 (a) Lin, G.-Q. and Zhong, M. (1997) *Tetrahedron: Asym.*, **8**, 1369–1372; (b) Qiu, L., Wu, J., Chan, S., Au-Yeung, T.T.-L., Ji, J.-X., Guo, R., Pai, C.-C., Zhou, Z., Li, X., Fan, Q.-H. and Chan, A.S.C. (2004) *Proc. Natl. Acad. Sci.*, **101**, 5815–5820.

75 Harrowven, D.C., Lai, D. and Lucas, M.C. (1999) *Synthesis*, **8**, 1300–1302.

76 (a) Mizoroki, T., Mori, K. and Ozaki, A. (1971) *Bull. Chem. Soc. Jpn.*, **44**, 581; (b) Heck, R.F. and Nolley, J.P. (1972) *J. Org. Chem.*, **37**, 2320–2322; (c) Heck, R.F. (1982) *Org. React.*, **27**, 345; (d) Heck, R.F. (1985) *Palladium Reagent in Organic Syntheses*, Academic Press, New York.

77 Dounay, A.B. and Overman, L.E. (2003) *Chem. Rev.*, **103**, 2945–2964.

78 Beletskaya, I.P. and Cheprakov, A.V. (2000) *Chem. Rev.*, **100**, 3009–3066.

79 Selected reviews include: de Meijere, A. and Meyer, F.E., (1994) *Angew. Chem.*, **33**, 2473–2506; (b) Stang, P.J. and Diederick, F. (eds) (1998) *Metal-Catalyzed Cross Coupling Reactions*, Wiley-VCH, Weinheim. For recent reviews of the asymmetric Heck reaction, see: (c) Shibasaki, M. and Vogl, E.M. (1999) *J. Organomet. Chem.*, **576**, 1–15. For recent reviews on mechanistic studies of the Heck reaction, see: (d) Amatore, C. and Jutand, A. (1999) *J. Organomet. Chem.*, **576**, 254–278.

80 Fuwa, H. and Sasaki, M. (2007) *Org. Lett.*, **9**, 3347–3350.

81 Firmansjah, L. and Fu, G.C. (2007) *J. Am. Chem. Soc.*, **129**, 11340–11341.

82 Affo, W., Ohmiya, H., Fujioka, T., Ikeda, Y., Nakamura, T., Yorimitsu, H., Oshima, K., Imamura, Y., Mizuta, T. and Miyoshi, K. (2006) *J. Am. Chem. Soc.*, **128**, 8068–8077.

83 Salem, B., Delort, E., Klotz, P. and Suffert, J. (2003) *Org. Lett.*, **5**, 2307–2310.

84 Bour, C. and Suffert, J. (2005) *Org. Lett.*, **7**, 653–656.

85 Jensen, T., Pedersen, H., Bang-Andersen, B., Madsen, R. and Jørgensen, M. (2008) *Angew. Chem. Int. Ed.*, **47**, 888–890.

86 Govek, S.P. and Overman, L.E. (2001) *J. Am. Chem. Soc.*, **123**, 9468–9469.

87 (a) Dounay, A.B., Hatanaka, K., Kodanko, J.J., Oestreich, M., Overman, L.E., Pfeifer, L.A. and Weiss, M.M. (2003) *J. Am. Chem. Soc.*, **125**, 6261–6271; (b) Matsuura, T., Overman, L.E. and Poon, D.J. (1998) *J. Am. Chem. Soc.*, **120**, 6500–6503.

88 Endo, A., Yanagisawa, A., Abe, M., Tohma, S., Kan, T. and Fukuyama, T. (2002) *J. Am. Chem. Soc.*, **124**, 6552–6554.

89 Tietze, L.F., Redert, T., Bell, H.P., Hellkamp, S. and Levy, L.M. (2008) *Chem. Eur. J.*, **14**, 2527–2535.

90 (a) Ma, S., Xu, B. and Ni, B. (2000) *J. Org. Chem.*, **65**, 8532; (b) Ma, S. and Ni, B. (2002) *J. Org. Chem.*, **67**, 8280.

91 Tietze, L.F. and Krimmelbein, I.K. (2008) *Chem. Eur. J.*, **14**, 1541–1551.
92 (a) Maddaford, S.P., Andersen, N.G., Cristofoli, W.A. and Keay, B.A. (1996) *J. Am. Chem. Soc.*, **118**, 10766–10773; (b) Cristofoli, W.A. and Keay, B.A. (1994) *Synlett*, 625–626.
93 Negishi, E., Ma, S., Sugihara, T. and Noda, Y. (1997) *J. Org. Chem.*, **62**, 1922–1923.
94 Iimura, S., Overman, L.E., Paulini, R. and Zakarian, A. (2006) *J. Am. Chem. Soc.*, **128**, 13095–13101.
95 Liu, Q., Ferreira, E.M. and Stoltz, B.M. (2007) *J. Org. Chem.*, **72**, 7352–7358.
96 Dounay, A.B., Overman, L.E. and Wrobleski, A.D. (2005) *J. Am. Chem. Soc.*, **127**, 10186–10187.
97 Lee, C.-W., Oh, K.S., Kim, K.S. and Ahn, K.H. (2000) *Org. Lett.*, **2**, 1213–1216.
98 Baran, P.S., Maimone, T.J. and Richter, J.M. (2007) *Nature*, **446**, 404–408.
99 (a) Sonogashira, K., Tohda, Y. and Hagihara, N. (1975) *Tetrahedron Lett.*, **16**, 4467–4470; (b) Sonogashira, K. (2004) *Metal-Catalyzed Cross-Coupling Reactions*, vol. 1 (eds F. Diederich and A. de Meijera), Wiley-VCH, Weinheim, p. 319; (c) Sonogashira, K. (2002) *Handbook of Organopalladium Chemistry for Organic Synthesis* (eds E. Negishi and A. de Meijere), Wiley-Interscience, New York, p. 493; (d) Sonogashira, K. (2002) *J. Organomet. Chem.*, **653**, 46–49; (e) Rossi, R., Carpita, A. and Bellina, F. (1995) *Org. Prep. Proced. Int.*, **27**, 127; (f) Sonogashira, K. (1991) *Comprehensive Organic Synthesis*, vol. 3 (eds B.M. Trost and I. Fleming), Pergamon, Oxford, p. 521.
100 (a) Chinchilla, R. and Nájer, C. (2007) *Chem. Rev.*, **107**, 874–922; (b) Doucet, H. and Hierso, J.-C. (2007) *Angew. Chem. Int. Ed.*, **46**, 834–871.
101 Iyoda, M., Sirinintasak, S., Nishiyama, Y., Vorasingha, A., Sultana, F., Nakao, K., Kuwatani, Y., Matsuyama, H., Yoshida, M. and Miyake, Y. (2004) *Synthesis*, **9**, 1527–1531.
102 Leclère, M. and Fallis, A.G. (2008) *Angew. Chem. Int. Ed.*, **47**, 568–572.
103 Coleman, R.S. and Garg, R. (2001) *Org. Lett.*, **3**, 3487–3490.
104 (a) Glaser, C. (1869) *Berichte*, **2**, 422; (b) Siemsen, P., Livingston, R.C. and Diederich, F. (2000) *Angew. Chem. Int. Ed. Engl.*, **39**, 2632–2657.
105 Ji, N., O'Dowd, H., Rosen, B.M. and Myers, A.G. (2006) *J. Am. Chem. Soc.*, **128**, 14825–14827.

8
Transition Metal-Mediated [2 + 2 + 2] Cycloadditions
David Lebœuf, Vincent Gandon, and Max Malacria

8.1
Introduction

Since the discovery of benzene formation by thermal cyclization of three molecules of acetylene by Berthelot [1] and the pioneering work of Reppe [2] in transition metal-mediated cyclizations, many catalysts have been developed for the cyclotrimerization of acetylenic compounds. In addition to alkynes, a large variety of other unsaturated partners such as alkenes, allenes, nitriles, aldehydes and ketones, imines, isocyanates, isothiocyanates, and so on, can take part in [2 + 2 + 2] cyclizations. Many of these reactions proceed with good chemo-, regio- and stereoselectivities and have found applications in the synthesis of complex polycyclic compounds. In the last three decades, this reaction has been extensively investigated and the topic has been thoroughly reviewed [3–9]. The formation of pyridines by [2 + 2 + 2] cycloadditons of two alkynes to one nitrile has been recently reviewed by Heller and Hapke [9], therefore this topic will not be treated in this chapter. The co-trimerization of acetylenic compounds to give benzene derivatives has also been covered comprehensively from 1980 up to the middle of 2004 [8]. Therefore, the present chapter is devoted to the most recent applications of the title reaction, from 2004 to 2008, emphasizing the modern aspects of this well-established strategy for constructing six-membered rings. We have divided this chapter into three sections covering the still very hot topic of [2 + 2 + 2] co-trimerization of three alkynes to give benzenes, the no less hot topic of cyclohexadiene formation by [2 + 2 + 2] co-cyclization of two alkynes to one alkene, and the new and very promising area of cyclohexene formation by [2 + 2 + 2] co-cyclization of two alkenes to one alkyne. Over the last four years, these three aspects have received a strong resurgence of interest, notably due to the use of new cyclization partners, new catalytic systems, and the blooming of long-awaited enantioselective transformations. We will not focus on the mechanistic aspects of the [2 + 2 + 2] cycloaddition reactions. Those wishing to find useful information on that matter can refer to the recent literature [10 15].

8.2
Benzene Derivatives

8.2.1
Rapid Construction of Functionalized Benzenes Using Heteroatom-Substituted Alkynes

Alkynes bearing heteroatoms are of particular interest because they give rise to products able to undergo subsequent transformations after the cyclization step. In that respect, alkynylsilanes [16] and alkynylstannanes [17, 18] have been used to produce "metalated" aromatic derivatives which were transformed by electrophilic substitution or metal-halogen exchange. In this context, new unsaturated partners exhibiting heteroatoms at the triple bond have been used recently in cyclotrimerization reactions, either to improve the chemo- and the regioselectivity, to allow functionalization of the compounds obtained, or simply to build new functionalized compounds. Another application of these substrates in [2 + 2 + 2] cyclizations lies in the temporary connection of two alkynes with a cleavable tether, such as boron or silyl groups. This strategy was implemented to allow highly chemo- and regioselective cyclotrimerization of three unsymmetrical alkynes (see Sections 8.2.1.1 and 8.2.1.2).

8.2.1.1 Alkynylsilanes

Malacria and coworkers achieved the formal intermolecular [2 + 2 + 2] cycloaddition of three different alkynes via the use of a temporary silylated tether. Under refluxing conditions, irradiation, and in the presence of 5 mol% of CpCo(CO)$_2$, the triyne **1** led to the corresponding cycloadduct **2** after cleavage of the silicon groups using TBAF (Scheme 8.1) [19, 20]. In sharp contrast, attempts to generate **2** in an intermolecular fashion from the three different untethered alkynes led to a very complex mixture. Thus, this approach is a useful tool for controlling chemo- and regioselectivity.

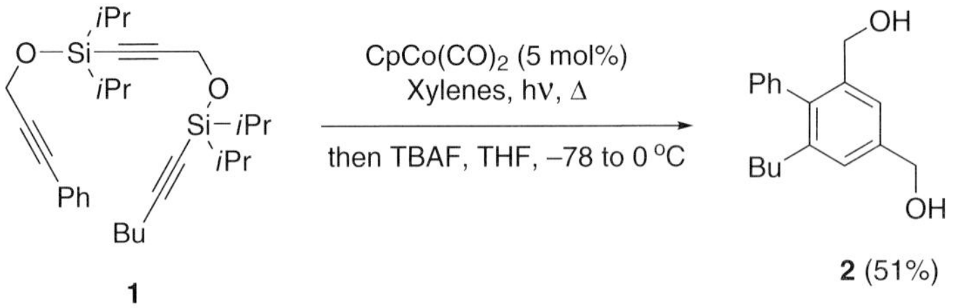

Scheme 8.1 Co-catalyzed [2 + 2 + 2] cycloaddition of the silylated tethered triyne **1** and fluorodesilylation.

Moreover, this methodology has been extended to grant access to the taxane framework. The polyunsaturated compound **3** undergoes cobalt-catalyzed [2 + 2 + 2] cycloaddition and the ether moiety of the resulting product **4** is oxidized into an enone prior to [4 + 2] cycloaddition (Scheme 8.2) [21]. The use of a temporary silicon tether allowed the selective formation of a polysubstituted arene after cleavage of the silicon group and the possibility of easily functionalizing the aromatic ring.

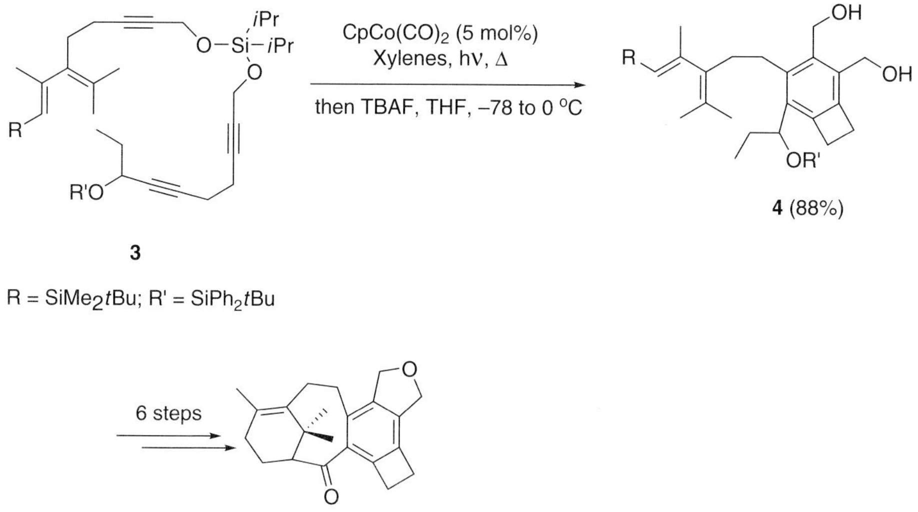

3

R = $SiMe_2tBu$; R' = $SiPh_2tBu$

4 (88%)

5 (27%)

Scheme 8.2 Access to the taxane framework using the temporary silicon-tether strategy.

8.2.1.2 **Alkynylboronates**

Yamamoto and coworkers reported the Ru(II)-catalyzed chemo- and regioselective cyclotrimerization of three unsymmetrical alkynes carried out in the presence of a temporary boron linkage [22–24]. The crude arylboronates were subjected to Suzuki–Miyaura couplings with aryl iodides to afford biaryls. As a result, a four-component approach to highly substituted biaryls was successfully accomplished by combining these two reactions in a one-pot process (Scheme 8.3).

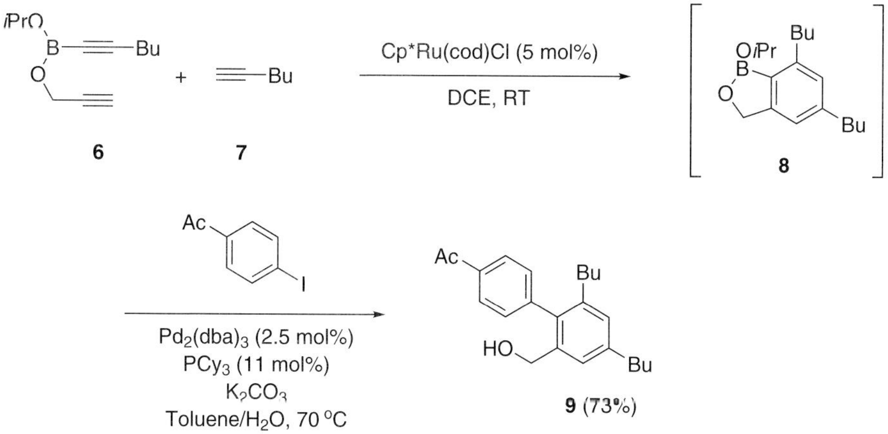

6 7 8

9 (73%)

Scheme 8.3 One-pot four-component coupling via Ru-catalyzed [2 + 2 + 2] cycloaddition/Suzuki–Miyaura coupling.

8.2.1.3 Ynamides

Ynamides bearing an electron-withdrawing group at the nitrogen atom are versatile reagents in organic chemistry. In that respect, Witulski and coworkers reported their use as partners in [2 + 2 + 2] cycloaddition for the synthesis of indoles or carbazoles [25–27]. For example, Witulski and Alayrac achieved the intramolecular [2 + 2 + 2] cycloaddition of the triyne **10** in the presence of 2 mol% of Wilkinson's catalyst to obtain the carbazole **11** in 99% yield (Scheme 8.4, Eq. (1)) [26]. Very recently, Hsung and coworkers investigated the [2 + 2 + 2] cycloaddition of diynes to chiral cyclic ynamides. New biarylic amides were prepared in excellent yields. For instance, the cycloaddition of 1,6-heptadiyne **12** with the aryl ynamide **13** incorporating Evans' auxiliary by using Wilkinson's catalyst in combination with $AgSbF_6$, led to the biaryls **14** and **15** with a low diastereoselectivity but in 96% yield (Scheme 8.4, Eq. (2)) [28]. Thereafter, they developed an asymmetric version of this reaction using a cationic rhodium complex in combination with a chiral ligand [29]. At the same time, Tanaka and coworkers synthesized axially chiral anilides such as **18** by enantioselective [2 + 2 + 2] cycloaddition of diynes with trimethylsilylynamides (Scheme 8.4, Eq. (3)) [30, 31].

8.2.1.4 Alkynylphosphines

Bulky phosphines play an important role in metal-catalyzed reactions. Nevertheless, methods aimed at their preparation are limited. Therefore, Oshima and coworkers developed a new approach using alkynylphosphine sulfides as precursors for [2 + 2 + 2] cycloadditions with diynes by using a rhodium catalyst and BINAP [32]. For that purpose, they used a neutral rhodium complex, $[RhCl(cod)]_2$, in combination with silver tetrafluoroborate to generate an active cationic rhodium species. The reaction of the diyne **19** with the alkynylphosphine sulfide **20** was carried out in dichloromethane at 25 °C to afford the corresponding phosphine sulfide **21** in good yield (Scheme 8.5). Afterwards, **21** was reacted with Chatgilialoglu reagent TTMSS under radical conditions to give the corresponding phosphine **22** in high yield. To show the interest of the products, one of the synthesized phosphines was used as ligand for Pd-catalyzed amination of an aryl chloride with morpholine.

8.2.1.5 Alkynylhalides

After having achieved the synthesis of arylboronates by ruthenium-catalyzed [2 + 2 + 2] cycloaddition [33, 34], Yamamoto and coworkers became interested in the synthesis of iodobenzenes, which are useful intermediates for the preparation of a wide range of important organic molecules. In particular, they can be subjected to Suzuki–Miyaura, Mirozoki–Heck or Sonogashira couplings. However, the direct iodination of arenes is often difficult because of the low electrophilicity of iodine, therefore they worked on the cyclotrimerization of diiododiynes with acetylene to afford diiodobenzenes in good yields [35, 36]. They next extended this methodology to the synthesis of *p*-phenylenes. Thus, the reaction of the tetrayne **23** and the monoborylated diyne **24** with acetylene in the presence of Cp*Ru(cod)Cl in dichloroethane at room temperature provided the corresponding

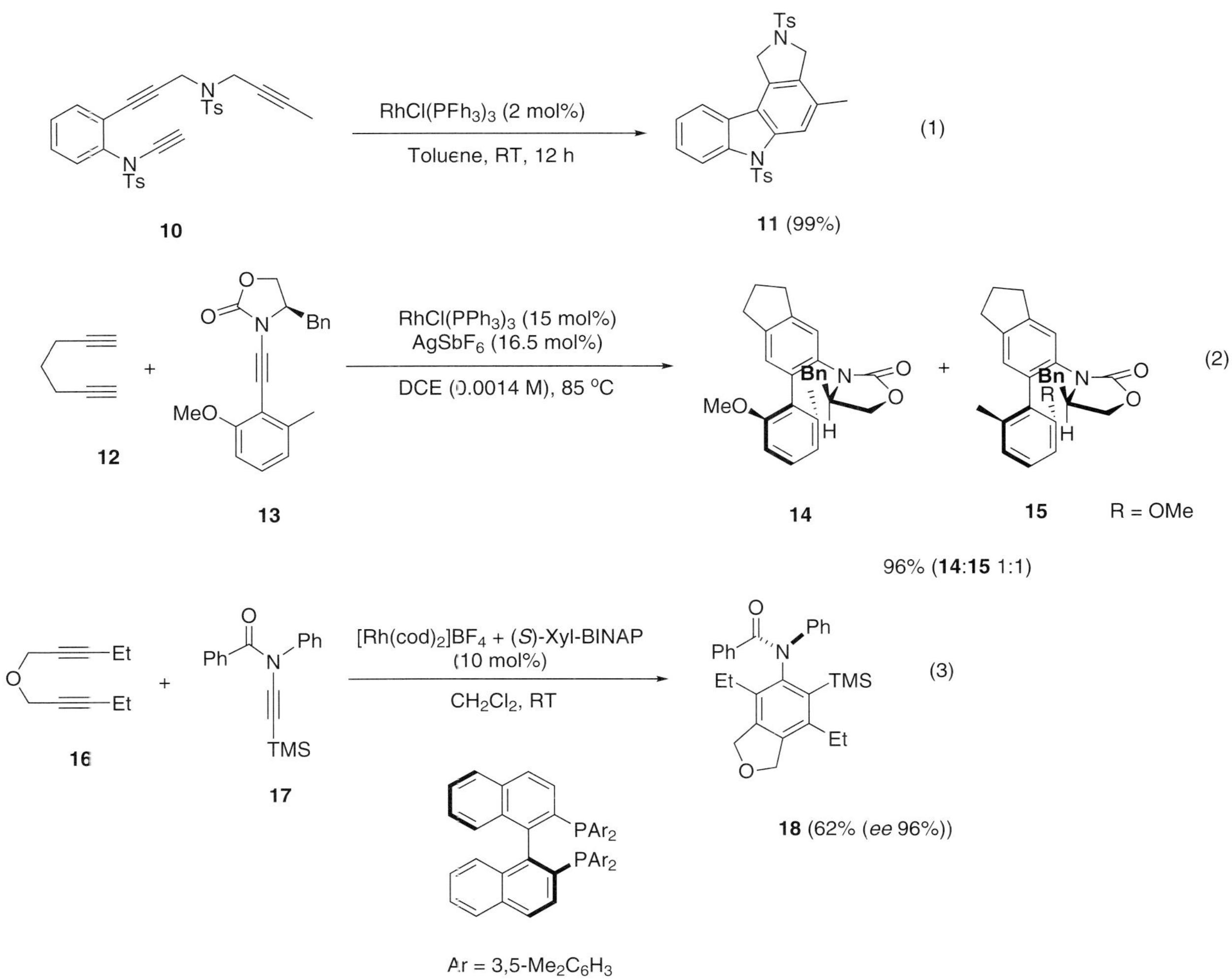

Scheme 8.4 Ynamides as partners for Rh-catalyzed [2 + 2 + 2] cycloadditions.

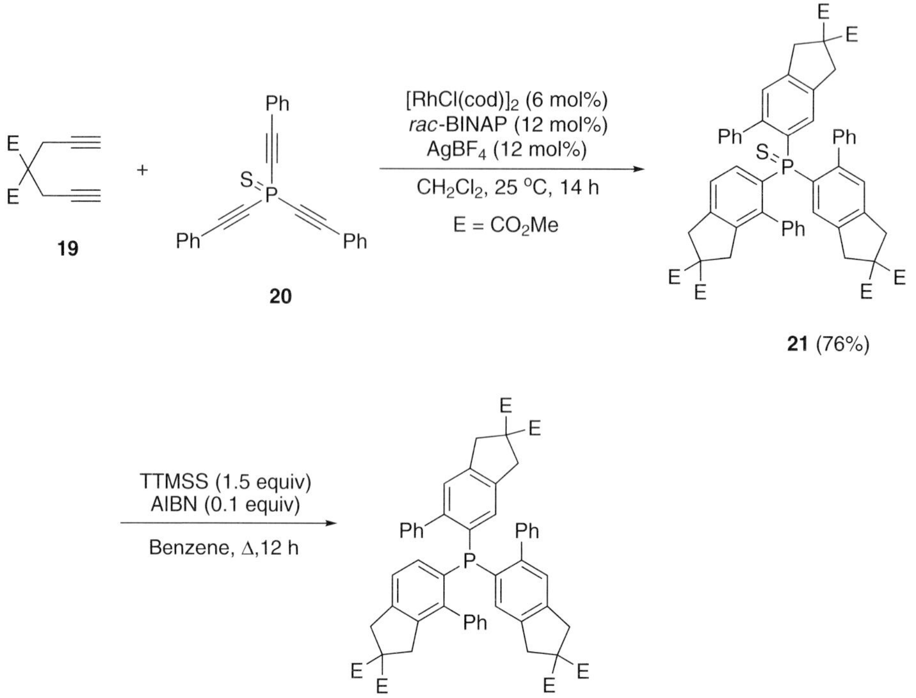

Scheme 8.5 Rh-catalyzed [2 + 2 + 2] cycloaddition of diyne **19** with 1-alkynylphosphine sulfide **20** and desulfidation with TTMSS.

cycloadducts **25** and **26**, in 87% and 73% yields, respectively (Scheme 8.6). After a double Suzuki–Miyaura coupling, the hexa-*p*-phenylene **27** was isolated in 36% yield.

8.2.2
Rapid Construction of the Naphthalene Core from Transient Benzynes

In recent years, the introduction of transition-metal catalysis to generate arynes under smooth reaction conditions has significantly broadened their usefulness in organic synthesis. The palladium and nickel-catalyzed [2 + 2 + 2] cycloadditions of arynes [37] have emerged as powerful methods for the synthesis of polyunsaturated compounds (e.g., helicenes [38, 39], naphthalenes [40–42], [N]-phenylenes [43, 44]). Sato *et al.* applied this strategy to the synthesis of arylnaphthalenes, such as Taiwanin C, by [2 + 2 + 2] co-cyclization of a diyne bearing an electron-deficient group with an aryne [45, 46]. The aryne **30** was generated *in situ* from compound **28** using $Pd_2(dba)_3$, $P(o\text{-tol})_3$ and CsF at room temperature. The total synthesis of Taiwanin C was finally achieved by standard reactions from compound **32** (Scheme 8.7).

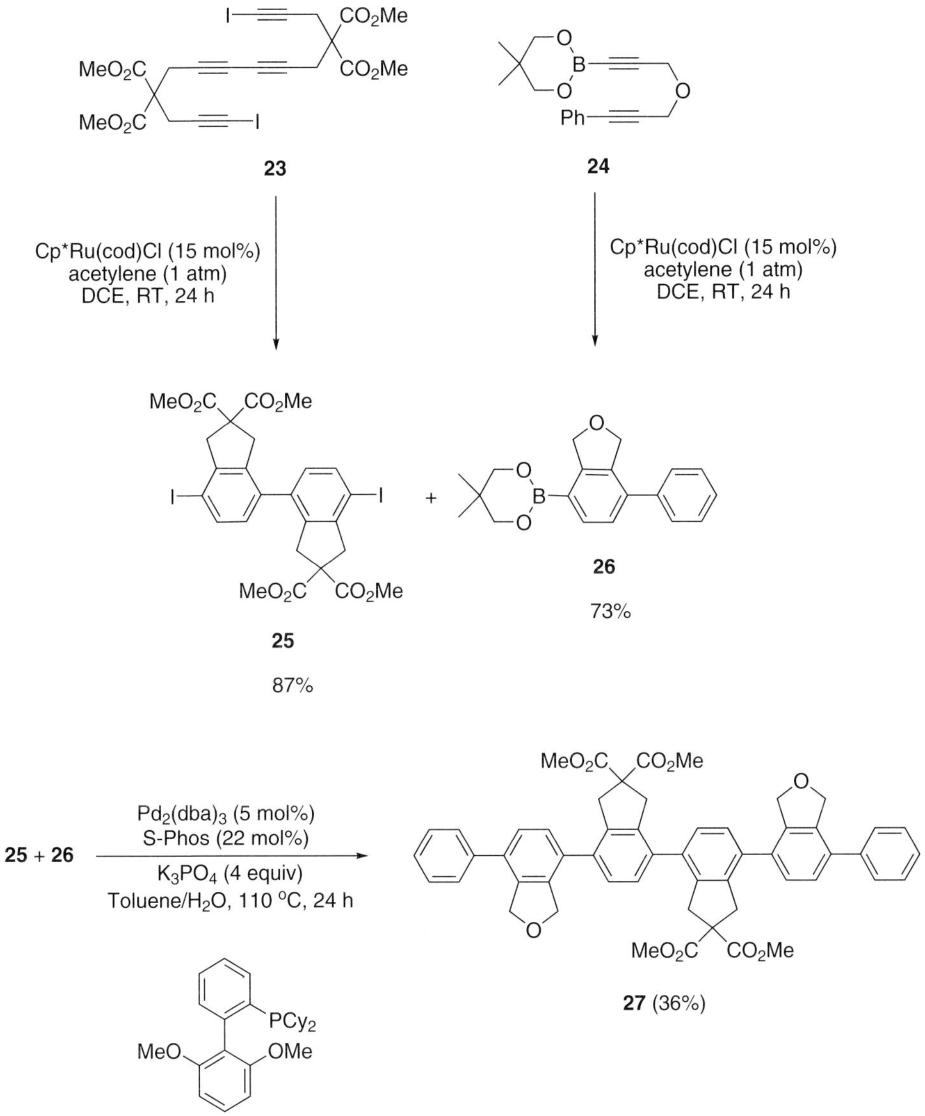

Scheme 8.6 Ru-catalyzed [2 + 2 + 2] cycloadditions aimed at the synthesis of hexa-p-phenylene **27**.

8.2.3
Rapid Construction of Polyphenylenes

8.2.3.1 Helicenes
Helicenes are fascinating compounds with a wide range of potential applications. The [2 + 2 + 2] cycloaddition is an expedient method that allows the synthesis of angularly fused polycyclic molecules like helicenes and related compounds [47–50],

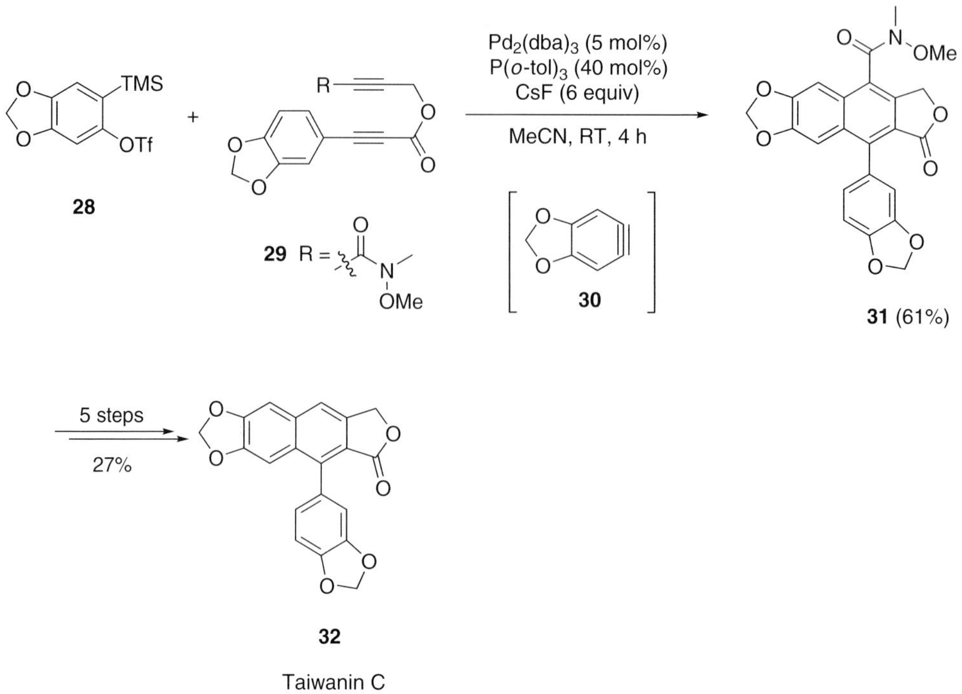

Scheme 8.7 Total synthesis of Taiwanin C.

including heliphenes [51, 52]. However, there are a few examples that give a satisfying stereocontrol of this reaction. Recently, Tanaka and coworkers reported the synthesis of helically chiral compounds via a Rh-catalyzed [2 + 2 + 2] cycloaddition (Scheme 8.8, Eq. (4)) [53], whereas Stara et al. realized the total diastereoselective cyclotrimerization of the triyne **35** exhibiting *p*-tolyl groups at the alkyne termini. A stoichiometric amount of CpCo(CO)$_2$ and 2 equiv of triphenylphosphine were used to prevent rapid decomposition of the catalyst (Scheme 8.8, Eq. (5)) [54, 55].

8.2.3.2 Silafluorenes

Silafluorenes are interesting compounds that contain a highly emissive silole unit. These compounds have been studied since the 1950s, but their synthesis and that of their derivatives, such as spirosilabifluorenes, proved very difficult via conventional methods and was mostly limited to biaryls. Murakami and coworkers reported a new strategy for the synthesis of 9-silafluorenes based on iridium-catalyzed [2 + 2 + 2] cycloaddition of silicon-bridged 1,6-diynes with alkynes [56]. For example, they achieved the synthesis of **39** from the tetrayne **37** and the internal alkyne **38** in the presence of 5 mol% of [IrCl(cod)]$_2$ and 20 mol% of triphenylphosphine in good yield (Scheme 8.9). The silafluorene **39** exhibits an absorption wavelength of 362 nm and a high fluorescence efficiency of 91%.

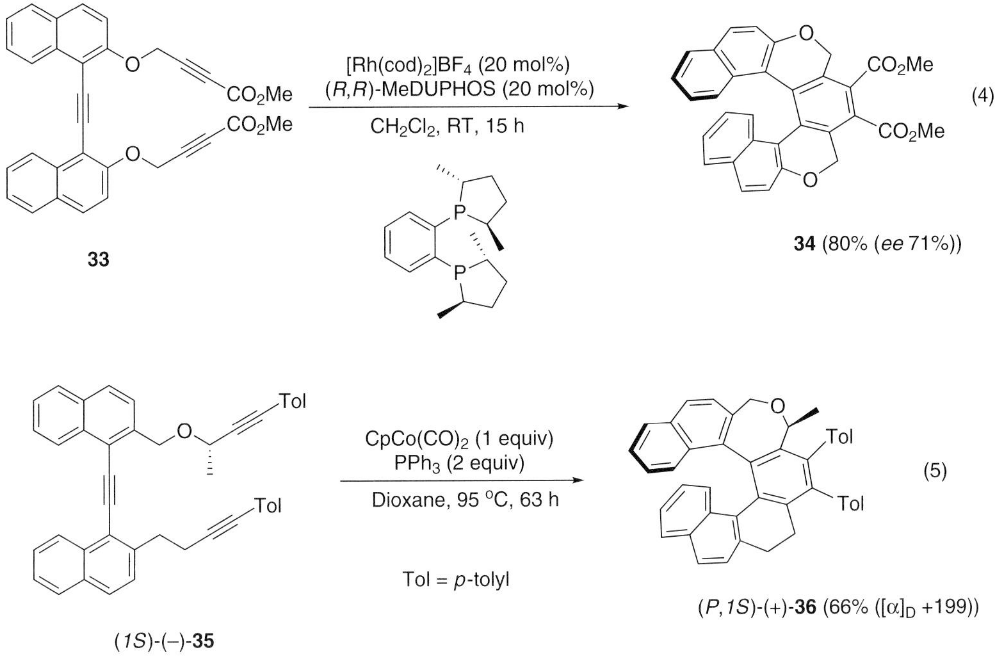

Scheme 8.8 Synthesis of [7]helicene-like compounds via [2 + 2 + 2] cyclotrimerization.

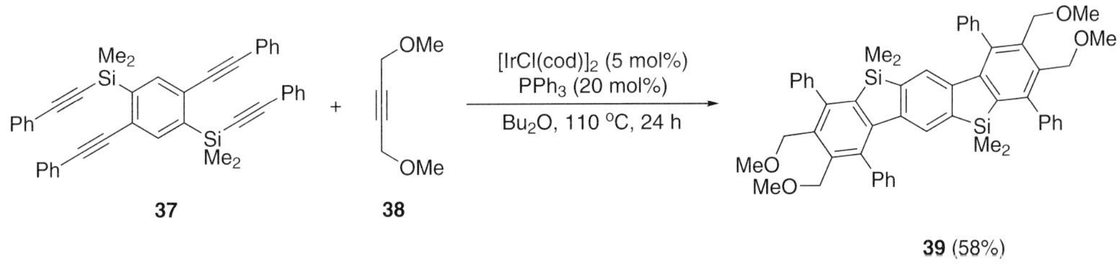

Scheme 8.9 Synthesis of ladder-type silafluorene **39**.

8.2.3.3 Angular [N]-Phenylenes

Phenylenes are the object of intense study by both theoretical and synthetic chemists because of their unique combination of aromatic and anti-aromatic properties and their potential as molecular magnets and conducting materials. Numerous angular [N]-phenylenes are available from intramolecular cyclotrimerization of alkynes [57, 58]. For instance, Vollhardt and coworkers synthesized the C_{3h}-symmetric [7]-phenylene **41** via cobalt-catalyzed triple [2 + 2 + 2] cycloaddition of the nonayne **40**, albeit in poor yield (Scheme 8.10, Eq. (7)). This compound, Archimedene, is yet a promising precursor of C_{120}-fullerene [59]. They also assembled *syn* and *anti* doublebent [5]-phenylenes **43** and **45** by double cobalt-catalyzed [2 + 2 + 2] cycloaddition of hexaynes (Scheme 8.10, Eqs. (8) and (9)) [60].

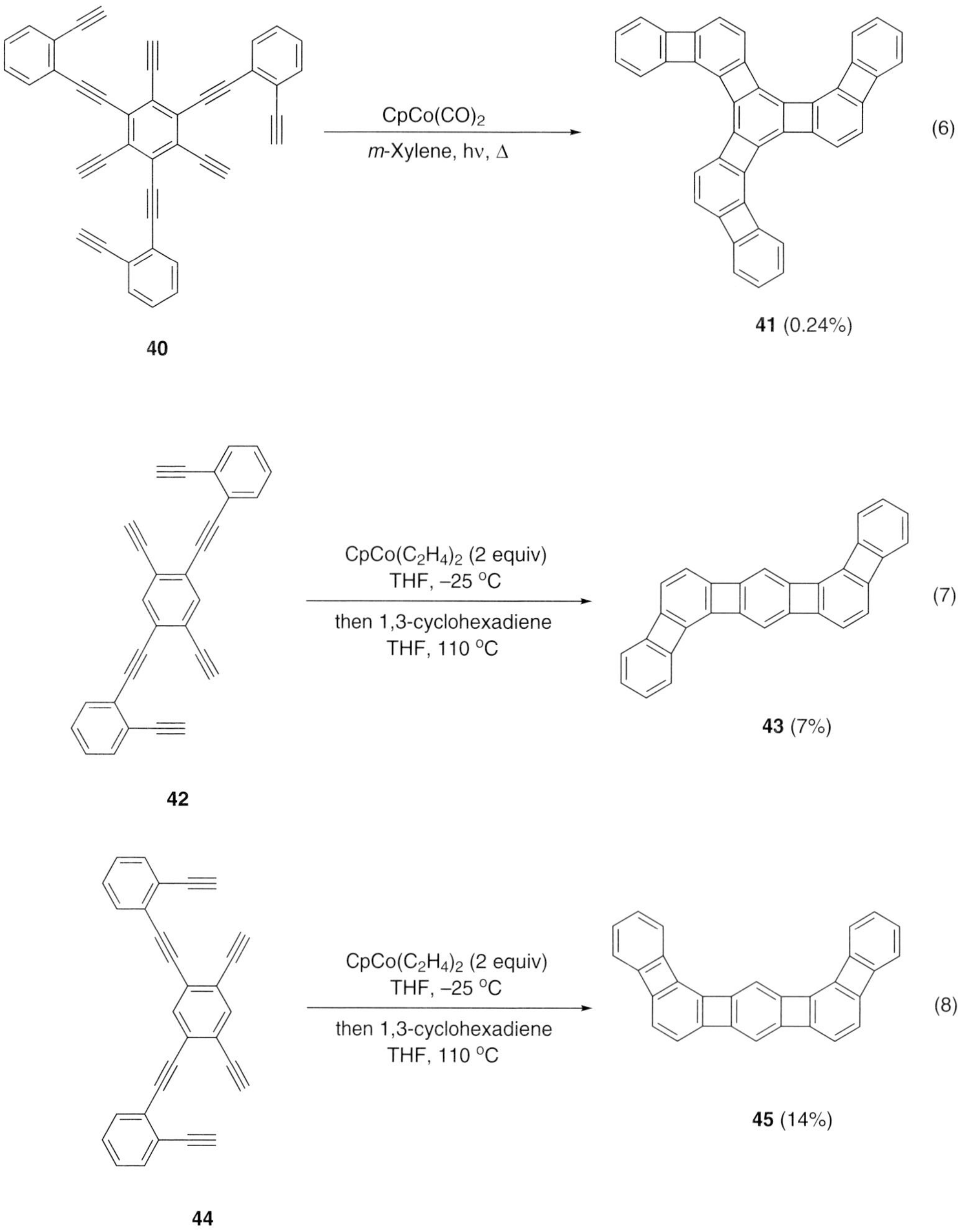

376 | *8 Transition Metal-Mediated [2 + 2 + 2] Cycloadditions*

(6)

40 → **41** (0.24%)

CpCo(CO)$_2$
m-Xylene, hν, Δ

(7)

42 → **43** (7%)

CpCo(C$_2$H$_4$)$_2$ (2 equiv)
THF, −25 °C

then 1,3-cyclohexadiene
THF, 110 °C

(8)

44 → **45** (14%)

CpCo(C$_2$H$_4$)$_2$ (2 equiv)
THF, −25 °C

then 1,3-cyclohexadiene
THF, 110 °C

Scheme 8.10 Synthesis of angular [N]-phenylenes via Co-catalyzed [2 + 2 + 2] cyclotrimerization.

8.2.4
Development of New Catalytic Systems for the [2 + 2 + 2] Cycloaddition

8.2.4.1 Rhodium Cationic Complexes

Catalysts based on many different metals have been developed to catalyze inter- and intramolecular [2 + 2 + 2] cyclotrimerization of terminal alkynes. In the last decade, we have witnessed a dramatic increase in the development of new catalytic systems. Although the neutral rhodium complex [RhCl(PPh$_3$)$_3$] has been used successfully to catalyze cyclotrimerization of diynes or triynes, it often proved inefficient for the cyclotrimerization of terminal alkynes and led to the formation of linear dimers [61]. The pioneering work of Tanaka and coworkers on the use of cationic rhodium complexes for the [2 + 2 + 2] cycloaddition of alkynes opened new horizons in this domain. They showed that the rhodium catalyst [Rh(cod)$_2$]BF$_4$ in combination with DTBM-Segphos was able to catalyze the cyclotrimerization of terminal alkynes in high yields with very good regioselectivities in favor of the 1,2,4-trisubstituted arenes (Table 8.1) [62, 63].

Applications of this catalytic system to more complex transformations will be provided throughout the rest of this chapter.

Table 8.1 Rh-catalyzed intermolecular [2 + 2 + 2] cyclotrimerization of terminal alkynes.

R	Yield [%] (a : b)
nC$_{10}$H$_{21}$ (46)	91 (83 : 17)
nC$_6$H$_{13}$ (47)	81 (82 : 18)
Bn (48)	94 (82 : 18)
(CH$_2$)$_3$Cl (49)	90 (86 : 14)
1-cyclohexenyl (50)	89 (97 : 3)
Ph (51)	93 (83 : 17)
TMS (52)	74 (70 : 30)

8.2.4.2 Inexpensive Catalytic Systems

In recent years, several groups have demonstrated the relevance of low-valent metal species (e.g., Rh [62, 63], Co [64, 65], or Ru [66]) for the cyclotrimerization of alkynes.

8 Transition Metal-Mediated [2 + 2 + 2] Cycloadditions

In that vein, Okamoto and coworkers reported that stable and inexpensive sources of iron and cobalt, $FeCl_3 \cdot 6H_2O$ and $CoCl_2 \cdot 6H_2O$, in the presence of zinc powder as reducing agent and an N-heterocyclic carbene or a diimine as ligand, could also generate a low-valent metal fragment able to catalyze intramolecular cyclotrimerization of triynes [67–69]. The application of the iron-based catalytic system to hexayne **53** afforded the biaryl **54** quantitatively with total chemoselectivity (Scheme 8.11), whereas the use of the cobalt-based catalytic system with different triynes led to the corresponding cycloadducts in good yields (Table 8.2).

Scheme 8.11 Fe-catalyzed intramolecular [2 + 2 + 2] cyclotrimerization of triyne **53**.

8.2.4.3 Microwave and Solid-Supported Cycloadditions [70–72]

Since the first report of a microwave-promoted organic synthesis in 1986, this process has been greatly developed and became a common alternative which helps to save

Table 8.2 Co-catalyzed intramolecular [2 + 2 + 2] cyclotrimerization of triynes.

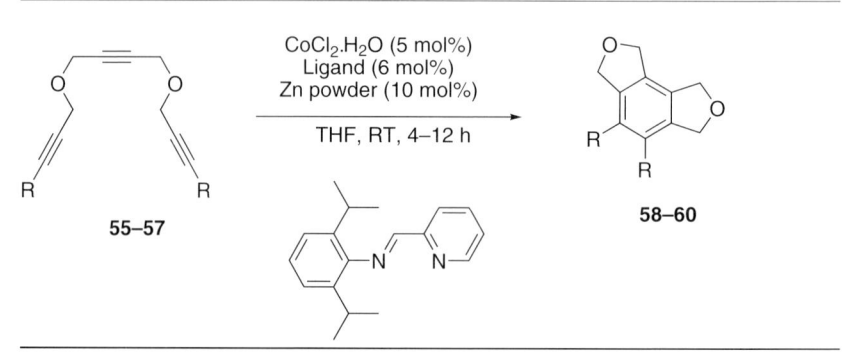

R	Yield [%]
H (**55**)	62
TMS (**56**)	97
Ph (**57**)	82

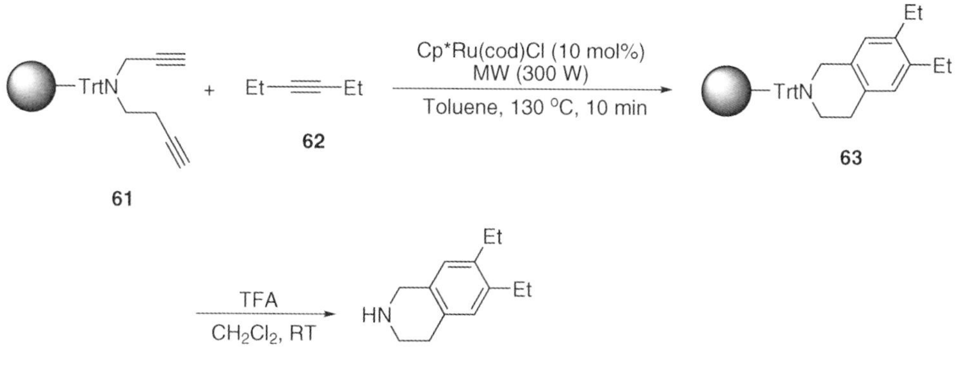

Scheme 8.12 Microwave-assisted [2 + 2 + 2] cycloaddition aimed at the synthesis of tetrahydroisoquinoline **64**.

time and costs [73]. Deiters and coworkers, as well as other groups, reported cobalt- [74–76], nickel- [77], rhodium- [78] and ruthenium-catalyzed cyclotrimerizations performed under microwave irradiation, sometimes in combination with solid supports [78]. The implementation of microwave irradiation together with a solid support allowed a substantial increase in yields and shortened reaction times dramatically. Moreover, activating additives, excessive heating or light irradiation are no longer necessary to promote the cyclization. For example, the synthesis of the tetrahydroisoquinoline **64** was achieved after a microwave- and ruthenium-mediated cyclotrimerization of the diyne **61** to 3-hexyne. The diyne **61** was immobilized on a standard polystyrene resin using a trityl linker. Finally, after cleavage from the resin, the compound **64** was obtained in 78% yield (Scheme 8.12) [79].

8.2.5
Enantioselective Reactions

Although transition metal-catalyzed [2 + 2 + 2] is a well-established process for the synthesis of polycyclic benzenes, only a few examples of enantioselective transformations had been reported before 2003: Mori and coworkers described the asymmetric synthesis of isoindoline and isoquinoline derivatives using nickel-catalyzed [2 + 2 + 2] cyclotrimerization of triynes in 1994, whereas Stara and coworkers achieved the synthesis of a [6]helicene-like molecule by nickel-catalyzed [2 + 2 + 2] cycloaddition of triynes in 1999 [48]. Since 2003, the development of enantioselective [2 + 2 + 2] cycloadditions is essentially due to the use of cationic rhodium complexes, even if other metals like iridium and cobalt were occasionally employed successfully as well [80, 81].

Historically, the first efforts for achieving enantioselective [2 + 2 + 2] cycloadditions were directed toward the synthesis of axially chiral biaryls. In this context, Shibata and coworkers reported the first asymmetric synthesis of axially chiral teraryls by iridium-catalyzed [2 + 2 + 2] cycloaddition of 1,6-diynes disubstituted

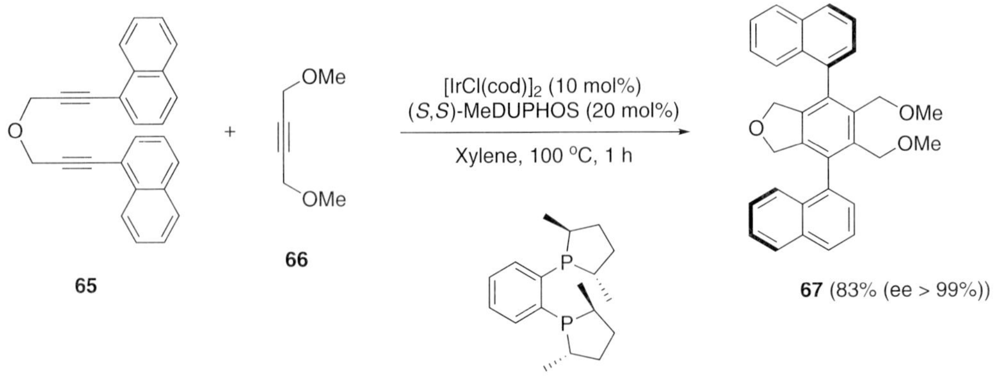

Scheme 8.13 Synthesis of the axially chiral teraryl **67** by Ir-catalyzed [2 + 2 + 2] cycloaddition.

by aryl or naphthyl groups with internal alkynes in the presence of (*S*,*S*)-MeDUPHOS. For example, the teraryl **67** was obtained in high yield with >99% ee (Scheme 8.13) [80, 82–85].

Tanaka and coworkers also became interested in the synthesis of axially chiral biaryls using rhodium cationic complexes: they achieved the synthesis of phthalides by [2 + 2 + 2] cycloaddition of unsymmetrical 1,4-diynes with alkynes [86], and the synthesis of biaryls through double [2 + 2 + 2] cycloaddition of electron-deficient 1,6-diynes with 1,2-diynes (or tetraynes with terminal alkynes) [87], or by intermolecular cotrimerization of three internal alkynes [88]. Recently, they applied this methodology to the construction of axially chiral phosphorus-containing biaryl compounds through [2 + 2 + 2] cycloaddition of diynes with alkynylphosphonates [89]. For example, the diyne **68** and the arylethynyl phosphonate **69** gave the phosphonate **70** in the presence of [Rh(cod)$_2$]BF$_4$/(*R*)-H$_8$-BINAP in quantitative yield and excellent enantiomeric excess (Scheme 8.14).

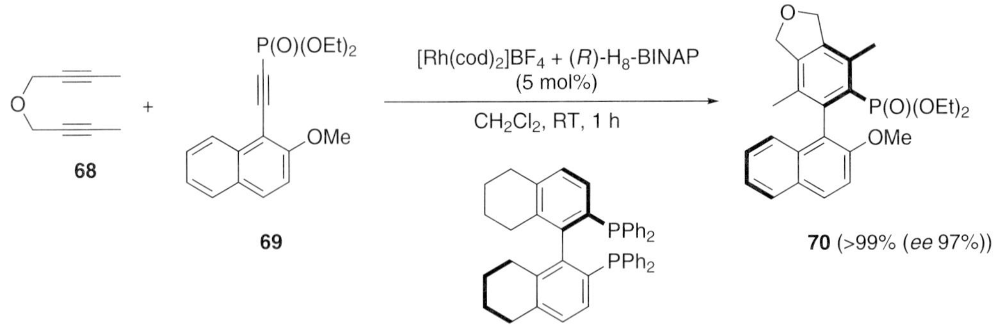

Scheme 8.14 Synthesis of the axially chiral biaryl phosphonate **70** by Rh-catalyzed [2 + 2 + 2] cycloaddition.

Cyclophanes have received widespread attention in organic chemistry as they are challenging targets for synthesis and have interesting physical properties. They were also used as intermediates in the total synthesis of natural products. In the last decades, much of the research on cyclophanes has been focused on the inclusion of guest

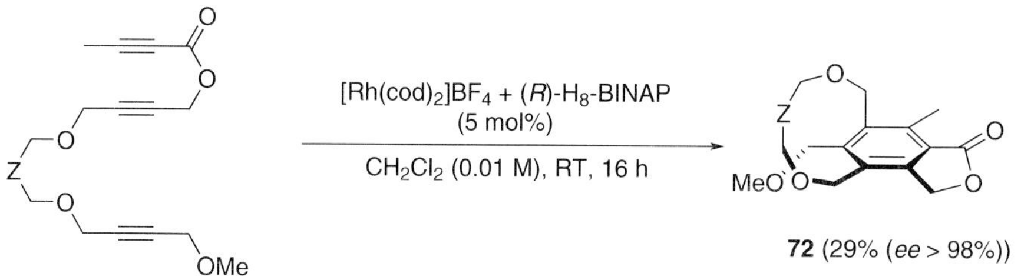

Scheme 8.15 Rh-catalyzed enantioselective synthesis of m-cyclophane **72**.

molecules into their cavity. However, there were no enantioselective syntheses of cyclophanes reported in the literature apart from the resolution of racemates [62, 63, 90, 91]. Tanaka and coworkers described the formation of the m-cyclophane **72**, which results from the cyclotrimerization of the triyne **71** using 5 mol% of Rh(I)/(R)-H$_8$-BINAP at room temperature in dichloromethane. The compound **72** was also obtained with high enantiomeric excess (Scheme 8.15) [92, 93].

8.3
1,3-Cyclohexadienes and their Heterocyclic Counterparts

8.3.1
Emergence (or Re-emergence) of Peculiar Unsaturated Partners

8.3.1.1 C≡C Bonds
Alkynylboronates Siebert et al. showed that alkynyl catechol boronic esters easily undergo catalytic cyclotrimerization with Co$_2$(CO)$_8$ or CpCo(CO)$_2$ giving 1,2,4- and 1,3,5-triborylbenzene derivatives [94–97]. On the other hand, under the same experimental conditions, Aubert et al. found that alkynyl pinacol boronic esters did not give the corresponding cycloadducts [98]. These substrates proved to be much better candidates for co-cyclization reactions with diynes or alkenes. They achieved the synthesis of 1,3- and 1,4-diboryl-1,3-cyclohexadienes resulting from the [2 + 2 + 2] co-cyclization of alkynylboronates to alkenes using CpCo(C$_2$H$_4$)$_2$, previously exploited as an active source of CpCo for the cocyclization of alkynes to alkenes. For instance, with indene, the reaction proceeded with complete control of the regioselectivity in very good yield (Scheme 8.16). The free diene could be liberated through rapid oxidative demetallation using iron(III) chloride in acetonitrile [99, 100]. With p-iodoanisole, the cyclohexadiene **75** was converted into the compound **78** in good yield (77%). The use of the [Pd(tBu$_3$P)$_2$]/NaOH system for the coupling reaction allowed the cyclohexadiene ring to be kept intact (no aromatization). However, only the less hindered borylated site proved reactive.

Arynes Several unsaturated partners like alkynes (see Section 8.2.2), allenes [101] or allyl chlorides [101] have been used in reactions with arynes to generate products

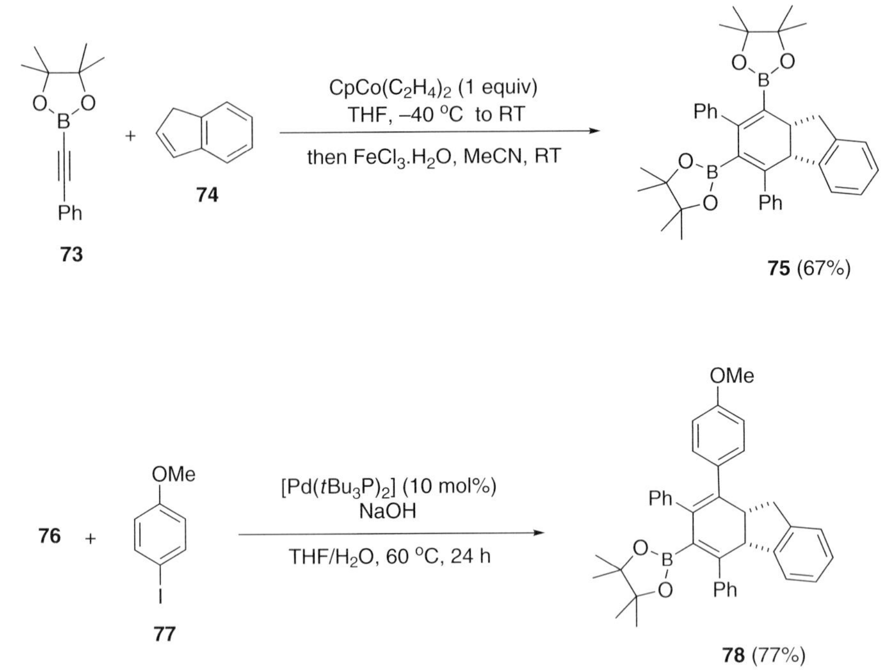

Scheme 8.16 CpCo-mediated co-cyclization of alkynylboronate **73** with indene and Suzuki-Miyaura coupling with aryliodide **77**.

arising from formal [2 + 2 + 2] cycloadditions. However, the use of alkenes in such reactions remained unexplored until Cheng and coworkers reported the palladium-catalyzed [2 + 2 + 2] cycloaddition of bicyclic alkenes with benzynes to provide annellated 9,10-dihydrophenanthrenes in good yields (Scheme 8.17, Eq. (9)) [102]. More recently, Perez, Pena and coworkers showed that acyclic alkenes substituted with an electron-withdrawing group could also promote this reaction and furnish both the 9,10-dihydrophenanthrene cycloadducts and the products resulting from the β-hydride elimination. For example, the benzyne derived from **82** and methyl acrylate afforded the corresponding dihydrophenanthrene **84** and the ortho-olefinated biaryls **85** and **86** in good yields via palladium-catalyzed co-cyclization (Scheme 8.17, Eq. (10)) [103].

8.3.1.2 C=C Bonds

Biscyclopropylidenes Although methylenecyclopropanes and biscyclopropylidenes are very useful intermediates in organic synthesis and have been employed as partners for the Pauson–Khand reaction, their potential as [2 + 2 + 2] partners was evaluated only recently [104]. de Meijere, Malacria and coworkers described how these compounds behave as simple alkenes without participation of the cyclopropyl units [105, 106]. For instance, they achieved the [2 + 2 + 2] cycloaddition of the enediyne **87** with an electron-withdrawing group at the terminal alkyne in the

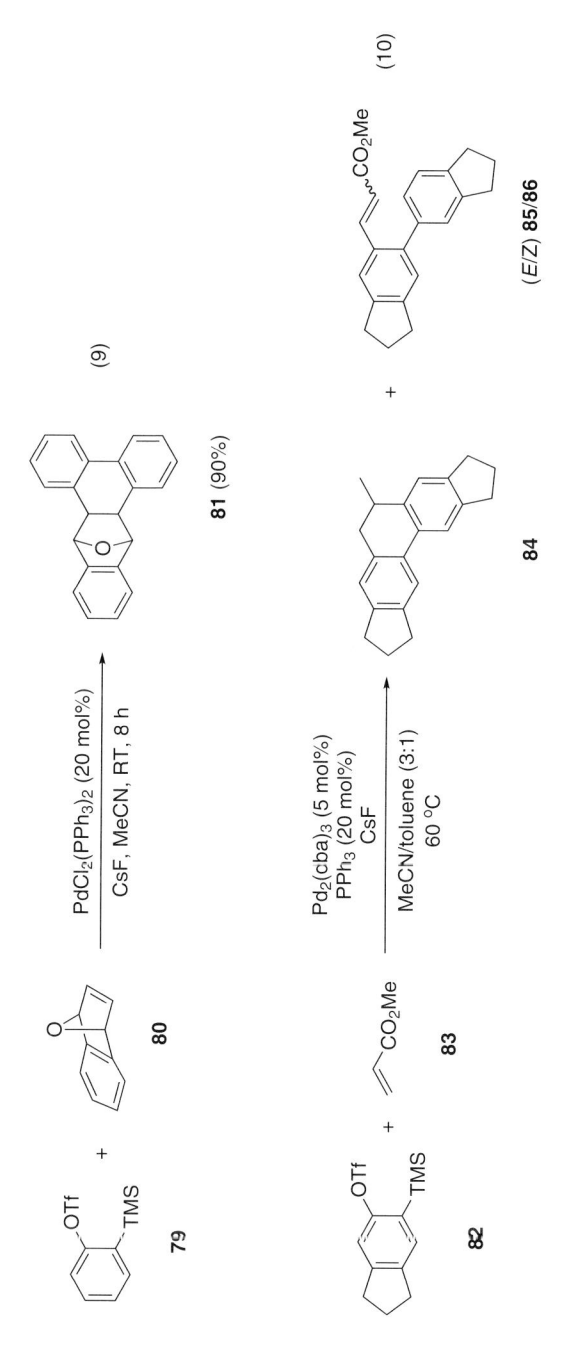

8.3 1,3-Cyclohexadienes and their Heterocyclic Counterparts

Scheme 8.17 Examples of Pd-catalyzed co-trimerization of benzynes with alkenes.

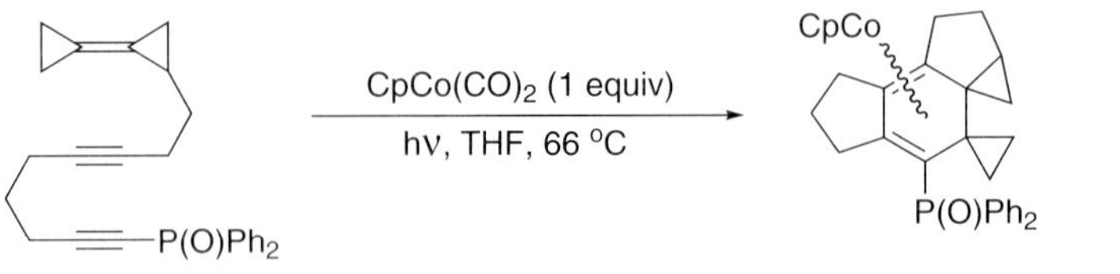

Scheme 8.18 Co-mediated intramolecular co-cyclization of the enediyne **87** with a methylenecyclopropane moiety.

presence of a stoichiometric amount of CpCo(CO)$_2$ under irradiation leading to an inseparable mixture of the diastereoisomeric spirocyclopropanated compounds **88** and **89** in excellent yield (Scheme 8.18).

Allenes Allenes are versatile reagents in organic synthesis [107]. Although allenes are good ligands, only a few examples employing this unsaturated partner in [2 + 2 + 2] cycloaddition have been reported [101, 108–114]. Malacria and coworkers showed that allenes could be relevant partners for intramolecular [2 + 2 + 2] cycloaddition [110, 111]. For instance, upon treatment with CpCo(CO)$_2$, allene-diynes furnished the corresponding cycloadducts in high yields with complete chemo-, regio- and diastereoselectivities. Moreover, with optically active allenes, the process could be performed with total transfer of chirality [112]. Afterwards, they extended this strategy to the synthesis of the 11β-aryl steroid frameworks using an intramolecular cobalt-mediated [2 + 2 + 2] cycloaddition of allenediynes, even if this reaction showed often moderate to low diastereoselectivities. The cyclization and decomplexation sequence of the allenediyne **90** could be carried out in the same pot, which allowed the formation of the 11β-aryl steroid skeleton in 48% overall yield (Scheme 8.19) [113, 114].

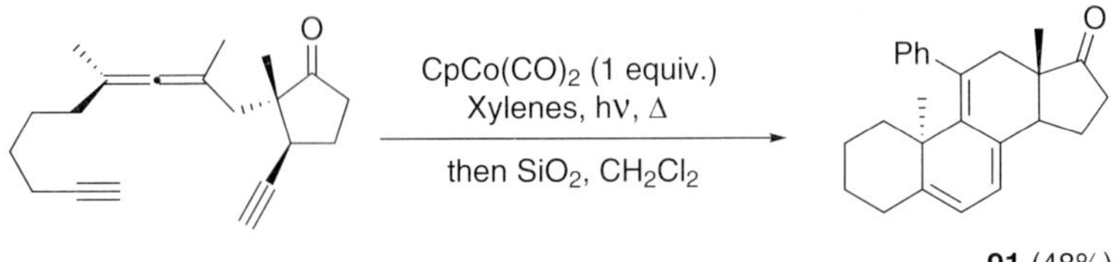

Scheme 8.19 Synthesis of the 11β-aryl steroid framework via Co-mediated [2 + 2 + 2] cycloaddition of allenediyne **90**.

Heteroaromatics Vollhardt and coworkers demonstrated that double bonds of heteroaromatic compounds could be used in cobalt-mediated [2 + 2 + 2] cycloadditions (furans [115], benzofurans [116], thiophenes [115], indoles [117, 118], pyrroles [119], imidazoles [120], and pyrimidines [121, 122]). The success of these reactions led them to consider pyridones and pyrazinones as potential cyclization partners [123]. For example in the presence of CpCo(C$_2$H$_4$)$_2$, the compound **92** led to the tetracycle **93** via a chemo-, regio- and diastereo-selective [2 + 2 + 2] cycloaddition (Scheme 8.20, Eq. (11)). The cocyclization of the N-3-butynylpyridinone **94** to

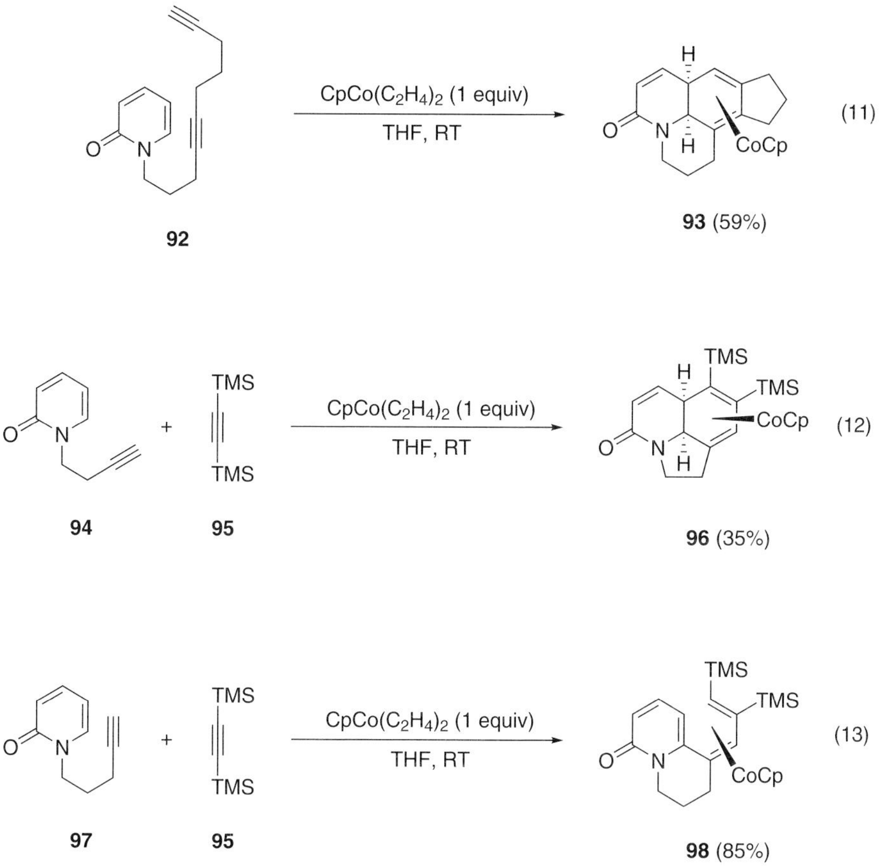

Scheme 8.20 Behavior of alkynyl-pyridinones in the presence of CpCo(C$_2$H$_4$)$_2$.

BTMSA gave the cycloadduct **96** in moderate yield (Scheme 8.20, Eq. (12)). Surprisingly, the substrate **97**, which differs from the former by the length of the tether, furnished the pyridinone **98** resulting from a C–H bond activation (Scheme 8.20, Eq. (13)). These results were rationalized by means of DFT computations [124].

This methodology inspired a new approach for the synthesis of the galanthan alkaloid core, as examplified by anhydrolycorinone. The cobalt-catalyzed [2 + 2 + 2] cycloaddition of **99** with BTMSA furnished the pentacycles **100** and **101** in a 2 : 1 ratio in 57% yield (Scheme 8.21). Removal of the two silyl groups with fluoride, followed by oxidative demetallation afforded anhydrolycorinone in 41% yield.

8.3.1.3 C=X Bonds

Carbon Dioxide, Aldehydes and Ketones Carbon dioxide represents an interesting starting material in organic synthesis because of its abundance, low-cost, safety and relative non-toxicity. However, its use has been limited because of the harsh conditions often necessary to activate it. In the last decades, some catalysts proved

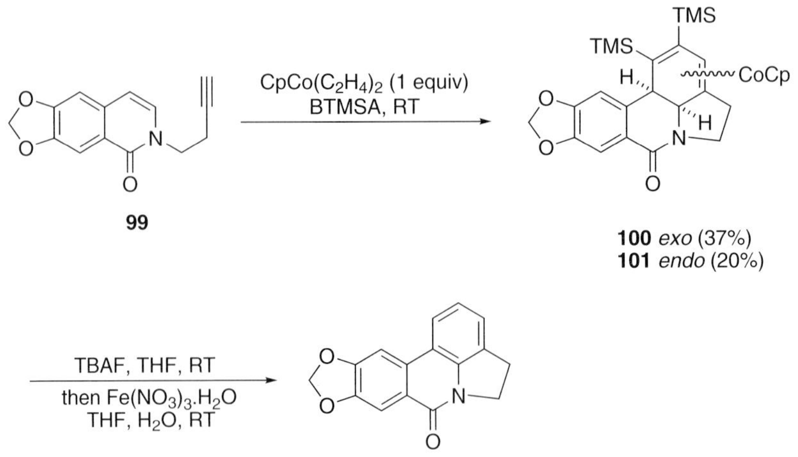

102 (41%)
Anhydrolycorinone

Scheme 8.21 Total synthesis of anhydrolycorinone.

able to activate carbon dioxide toward [2 + 2 + 2] cycloadditions [125, 126]. Louie and coworkers developed a new catalytic system, a nickel complex [Ni(cod)$_2$] in combination with an *N*-heterocyclic carbene, which allowed [2 + 2 + 2] cycloadditions of diynes with carbon dioxide under mild conditions to furnish pyrones [127, 128]. Under these conditions ([Ni(cod)$_2$] in combination with 1,3-*bis*(2,6-diisopropylphenyl)-imidazol-2-ylidene (IPr)), the cycloaddition of the various diynes **103–107** under one atmosphere of carbon dioxide led to the corresponding pyrones **108–112** in excellent yields (Table 8.3).

Table 8.3 Ni-catalyzed [2 + 2 + 2] cycloaddition of diynes with carbon dioxide.

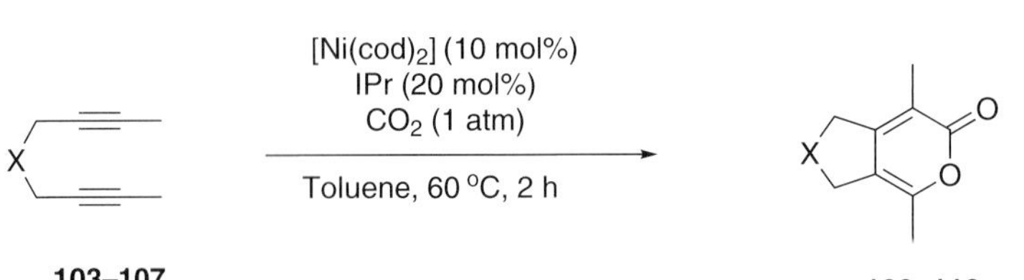

103–107 → **108–112**

X	Yield [%]
C(CO_2Me)$_2$ (**103**)	93
C(CH_2OBn)$_2$ (**104**)	93
C(CH_2OTBDMS)$_2$ (**105**)	92
CH(CO_2Me) (**106**)	82
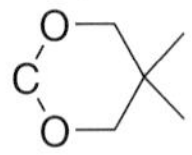 (**107**)	96

The first example of [2 + 2 + 2] cycloaddition of alkynes to carbonyl compounds was described by Vollhardt in 1989, but it required a stoichiometric amount of cobalt [129]. In 2005, Tekavec and Louie reported the Ni-catalyzed cycloaddition of diynes to aldehydes [130]. This reaction required mild conditions in comparison with different methods described by Tsuda and Saegusa (Ni) or Yamamoto and Itoh (Ru) [131, 132]. However, the furan **115** obtained after cyclization proved unstable and afforded the dienone **116** via an electrocyclic ring opening (Scheme 8.22, Eq. (14)). More recently, Tanaka and Shibata demonstrated that this reaction could also be catalyzed by [Rh(cod)$_2$]BF$_4$ under mild conditions (Scheme 8.22, Eq. (15)) [133, 134].

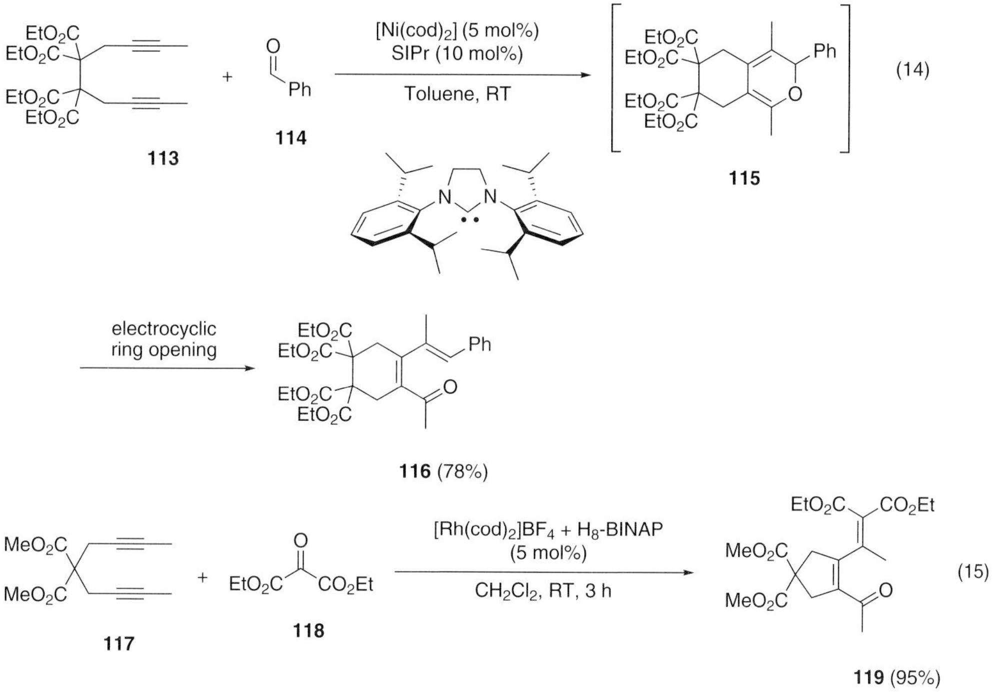

Scheme 8.22 Synthesis of dienones **116** and **119** via Ni- and Rh-catalyzed [2 + 2 + 2] cycladdition/electrocyclic ring opening.

Isothiocyanates and Carbon Disulfide The pioneering work directed toward the [2 + 2 + 2] cycloaddition of alkynes with isothiocyanates or carbon disulfide was achieved by Wakatsuki and Yamazaki using a stoichiometric amount of cobaltacyclopentadiene [135, 136]. Yamamoto and coworkers reported the first catalytic version of this reaction using Cp*Ru(cod)Cl as catalyst. They carried out [2 + 2 + 2] cycloaddition of various 1,6-diynes with benzoyl isothiocyanate to generate the corresponding thiopyran-2-imines in good yields (Table 8.4) [137]. The mechanism of the cycloaddition of two acetylene units with isothiocyanate catalyzed by CpRuCl was studied by theoretical calculations by Kirchner and Schmid [138].

In 2006, Tanaka and coworkers developed a new methodology to achieve the cycloaddition of diynes with carbon disulfide. The corresponding dithiopyrones were

Table 8.4 Ru-catalyzed [2 + 2 + 2] cycloaddition of 1,6-diynes with phenylisothiocyanate.

X	Yield [%]
C(CO$_2$Me)$_2$ (19)	88
C(Ac)$_2$ (120)	74
C(CN)$_2$ (121)	58
(122)	67

obtained in good yields in the presence of 5 mol% of [RhCl(cod)]$_2$ in combination with BINAP (Table 8.5) [139].

Isocyanates and Imines Transition-metal catalyzed [2 + 2 + 2] cycloaddition of two alkynes to one isocyanate to afford pyridinones has been accomplished by using Co [140–142], Ni [143–145], Ru [137, 146, 147] and Rh [148]. The formation of pyridinone via [2 + 2 + 2] cycloaddition was first reported by Yamazaki and Hong by using a stoichiometric amount of cobalt [140, 141], before Hoberg and Oster achieved it with Ni catalysts [143–145]. Takahashi *et al.* developed another method to synthesize pyridinones through the formation of an azazirconacyclopentenone followed by transmetallation with Ni(PPh$_3$)$_2$Cl$_2$ [149]. However, this reaction required stoichiometric amounts of both Ni and Zr. More recently, Yamamoto and coworkers investigated the cycloaddition of diynes with isocyanates using [CpRuCl(cod)] [146, 147].

Table 8.5 Rh-catalyzed [2 + 2 + 2] cycloaddition of 1,6-diynes with carbon disulfide.

R	Yield [%]
CO$_2$Me (19)	88
Ac (120)	74
CH$_2$(OMe) (127)	75

8.3 1,3-Cyclohexadienes and their Heterocyclic Counterparts

After having examined the rhodium-catalyzed [2 + 2 + 2] cycloaddition of terminal alkynes (or diynes) with isocyanates [150], Tanaka and coworkers extended this strategy to the synthesis of axially chiral 2-pyridinones by using a cationic rhodium complex. Previously, the reaction had always been carried out in the presence of a neutral rhodium complex, but it led to pyridinones in low yields [148]. In contrast, the [2 + 2 + 2] cycloaddition of the unsymmetrical diyne **132** with the isocyanate **133**, in the presence of the catalytic system [Rh(cod)$_2$]BF$_4$/(R)-DTBM-Segphos, furnished the 2-pyridinone **134** in 89% yield with total chemo-, regio- and enantio-selectivity (Scheme 8.23).

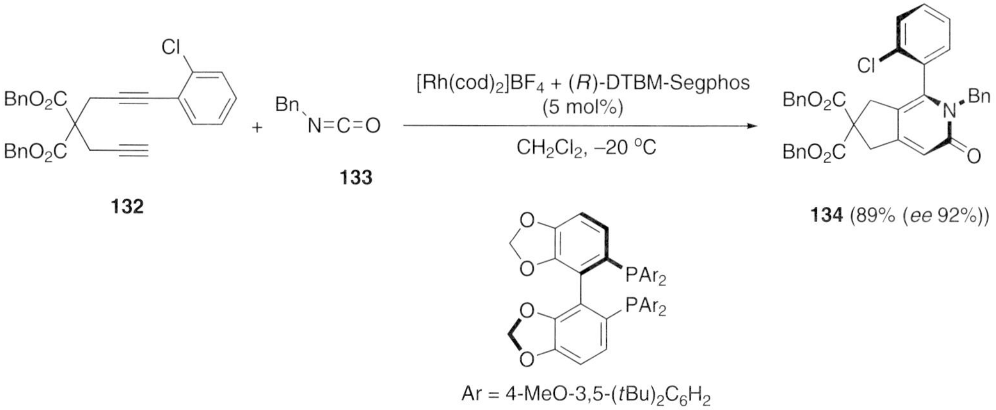

Scheme 8.23 Rh-catalyzed regio- and enantioselective [2 + 2 + 2] cycloaddition of diyne **132** with isocyanate **133**.

After having achieved the cobalt-catalyzed [2 + 2 + 2] cycloaddition of diynes and nitriles to construct pyridine-macrocycles [151], Maryanoff and coworkers reported the corresponding reaction with isocyanates. For example, reaction of the diyne **135** with the isocyanate **136** in the presence of 30 mol% of CpCo(CO)$_2$ furnished a mixture of 2-oxopyridinophanes **137** and **138** in 68% yield (Scheme 8.24). Among all possible regioisomeric products, only two cyclophanes, the *m*- and *p*-oxopyridinophanes, were obtained in a 1:2 ratio [152–154].

On the other hand, Louie and coworkers demonstrated that the catalytic system based on [Ni(cod)$_2$] and a N-heterocyclic carbene was not only able to catalyze cycloaddition of diynes with isocyanates [155], but also of one alkyne with two isocyanates to afford pyrimidine-diones [156], and finally of three isocyanates to furnish isocyanurates under mild conditions [157].

As mentioned above, there are several examples of the formation of 1,2-dihydropyrans by [2 + 2 + 2] cycloaddition of alkynes with carbonyl compounds. Several groups also reported examples with insertion of C=N bonds (isocyanates (see above) and carbodiimides [74, 140, 149, 158] (see Section 8.4)) into metallacycles to prepare pyridinones or iminopyridines. However, no such reactions had been reported with imines until, recently, Ogoshi and coworkers described the first example of [2 + 2 + 2] cycloaddition of two alkynes to one imine to build 1,2-dihydropyridines

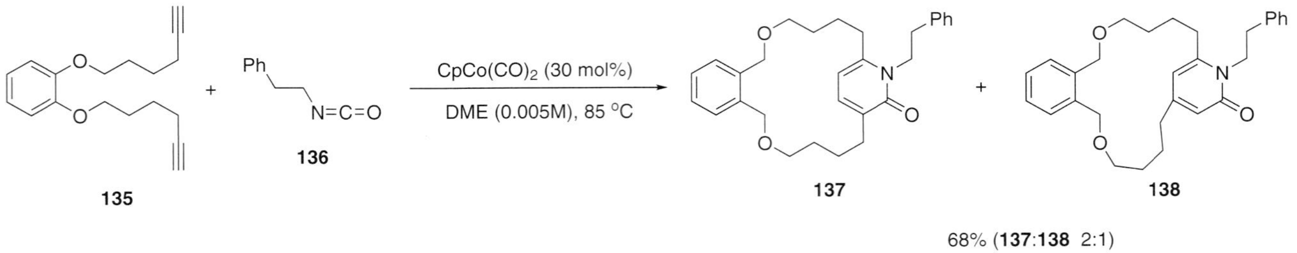

Scheme 8.24 Co-catalyzed [2 + 2 + 2] cycloaddition aimed at the synthesis of macrocyclic pyridinophanes.

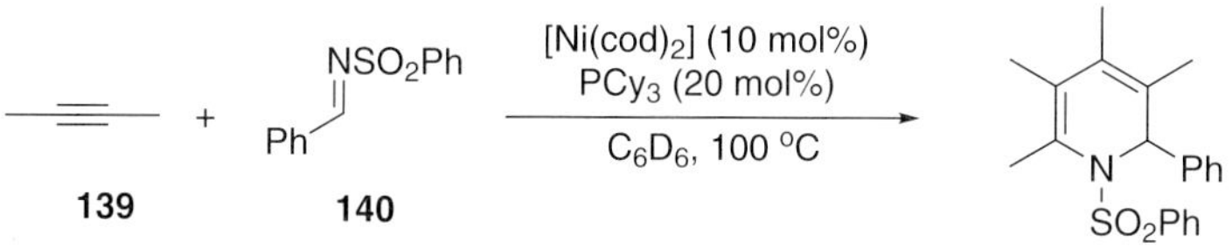

Scheme 8.25 Ni-catalyzed [2 + 2 + 2] cycloaddition of 2-butyne with imine **140**.

using a nickel catalyst. The reaction of the N-(benzene sulfonyl)benzaldimine with 2-butyne in the presence of [Ni(cod)$_2$] and PCy$_3$ at 100 °C gave the dihydropyridine **141** in 64% yield (Scheme 8.25) [159].

8.3.2
Enantioselective [2 + 2 + 2] Cycloaddition

The first enantioselective [2 + 2 + 2] cycloadditions of enynes with alkynes were reported by Evans et al. [160] and Shibata et al. in 2005 [161]. Unlike Shibata's group (Scheme 8.26, Eq. (17)), Evans and coworkers did not directly use a rhodium cationic complex but the neutral complex [RhCl(cod)]$_2$, in combination with silver tetrafluoroborate and the chiral phosphine (S)-Xyl-P-PHOS. For instance, the 1,3-cyclohexadiene **145** was obtained in 98% yield and 97% ee (Scheme 8.26, Eq. (16)).

Afterwards, Shibata and coworkers developed a new approach for the enantioselective synthesis of 1,3-cyclohexadienes by Rh-catalyzed [2 + 2 + 2] cycloaddition of diynes and exo-methylene cyclic compounds to furnish chiral spirocyclic products (Scheme 8.27, Eq. (18)), or unfunctionalized alkenes such as norbornene [162]. Finally, the same team described the enantioselective intramolecular [2 + 2 + 2] cycloaddition of (E)-enediynes which gave 1,3-cyclohexadienes with good yields and an excellent enantioselectivity (Scheme 8.27, Eq. (19)) [163].

8.4
Cyclohexenes and their Heterocyclic Counterparts

As we have seen before, several versions of transition metal-catalyzed [2 + 2 + 2] cycloaddition have been developed to afford arenes and cyclohexadienes. However, examples of [2 + 2 + 2] cycloaddition of one alkyne unit to two alkene units remained very scarce until recently. Montgomery and coworkers reported an unexpected dimerization of an alkynyl enone involving a formal [2 + 2 + 2] cycloaddition in the presence of a nickel catalyst in 1999 [164]. In 2006, after having examined different enantioselective [2 + 2 + 2] cycloadditions of alkynes, Shibata and coworkers described the intramolecular cycloaddition of 1,4-dienynes to build bridged compounds with the creation of a quaternary carbon stereocenter by using Rh(I)$^+$/Tol-BINAP as the catalytic system. For example, the intramolecular [2 + 2 + 2] cycloaddition of the dienyne **154**, displaying a nitrogen-containing tether, led to the cycloadduct **155** in dichloroethane at 60 °C in 83% yield with high enantiomeric

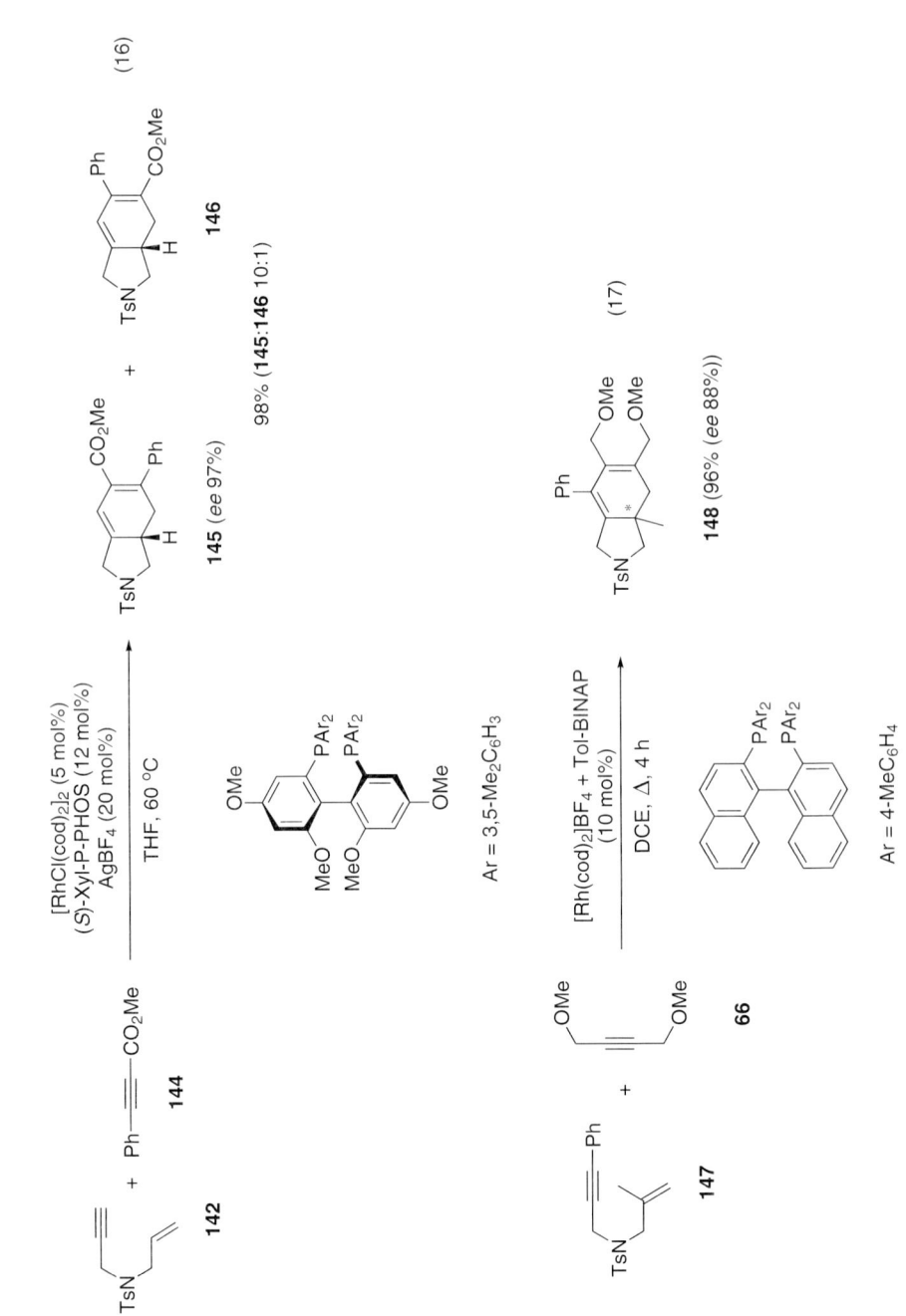

Scheme 8.26 Rh-catalyzed enantioselective [2 + 2 + 2] carbocyclization reactions.

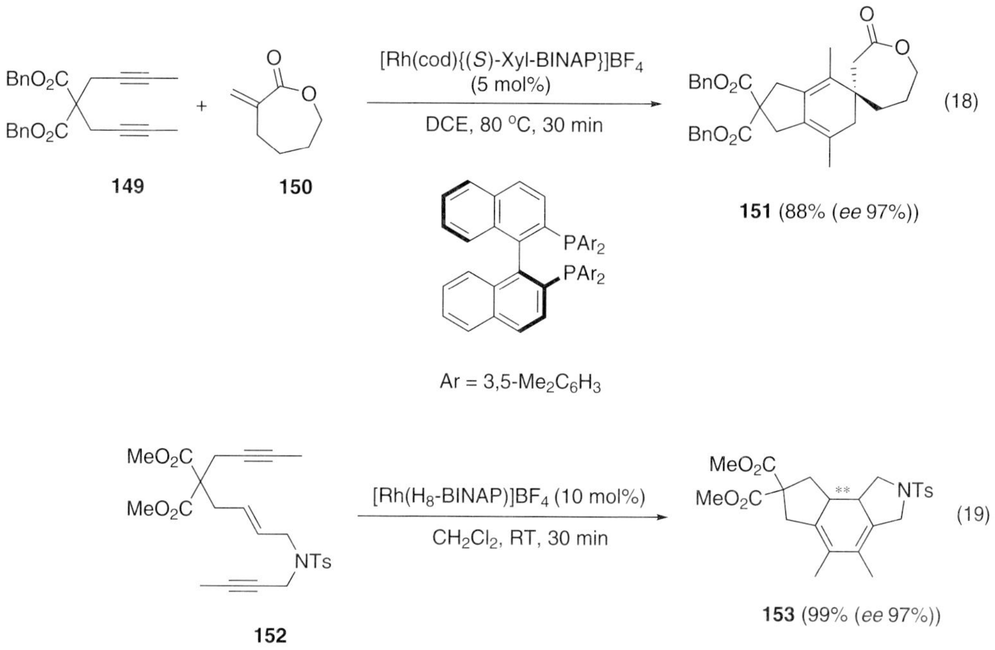

Scheme 8.27 Rh-catalyzed enantioselective [2 + 2 + 2] co-cyclization of diyne **149** with the *exo*-methylene cyclic compound **150**, and Rh-catalyzed enantioselective [2 + 2 + 2] cycloaddition of enediyne **152**.

excess (Scheme 8.28, Eq. (20)) [165]. In 2008, they worked on other 1,*n*-dienynes to obtain bicyclic, tricyclic or spirocyclic compounds, depending on the nature of the substrate [166]. Afterwards, Sato's and Tanaka's groups reported independently the intramolecular [2 + 2 + 2] cycloaddition of ene-yne-enes to furnish the corresponding cyclohexenes by using ruthenium and rhodium catalysts (Scheme 8.28, Eqs. (21) and (22)) [167, 168].

Other unsaturated partners such as ketones, isocyanates or diimides have been used instead of alkenes to achieve [2 + 2 + 2] cycloaddition of two double bonds to one alkyne. Rovis and coworkers have been interested successively in alkenyl isocyanates and alkenyl diimides as cyclization partners with alkynes. Rovis and Yu reported the regio- and enantio-selective Rh-catalyzed [2 + 2 + 2] cycloaddition of alkenyl isocyanates with alkynes substituted by aryl groups to afford indolizidines and/or vinylogous amides using chiral phosphoramidites as ligands [169, 170]. The formation of vinylogous amides is the result of a migration of the carbonyl group and is favored over that of the indolizidine (Scheme 8.29).

Furthermore, this methodology was applied to the total synthesis of (+)-Lasubine II. The reaction of the alkenyl isocyanate **162** with the terminal alkyne **165** in the presence of the rhodium complex [Rh(C$_2$H$_4$)$_2$Cl]$_2$ and a ligand derived from TADDOL furnished the vinylogous amide **166** in an excellent enantiomeric excess of 98% (Scheme 8.30). After the diastereoselective hydrogenation of **166** followed

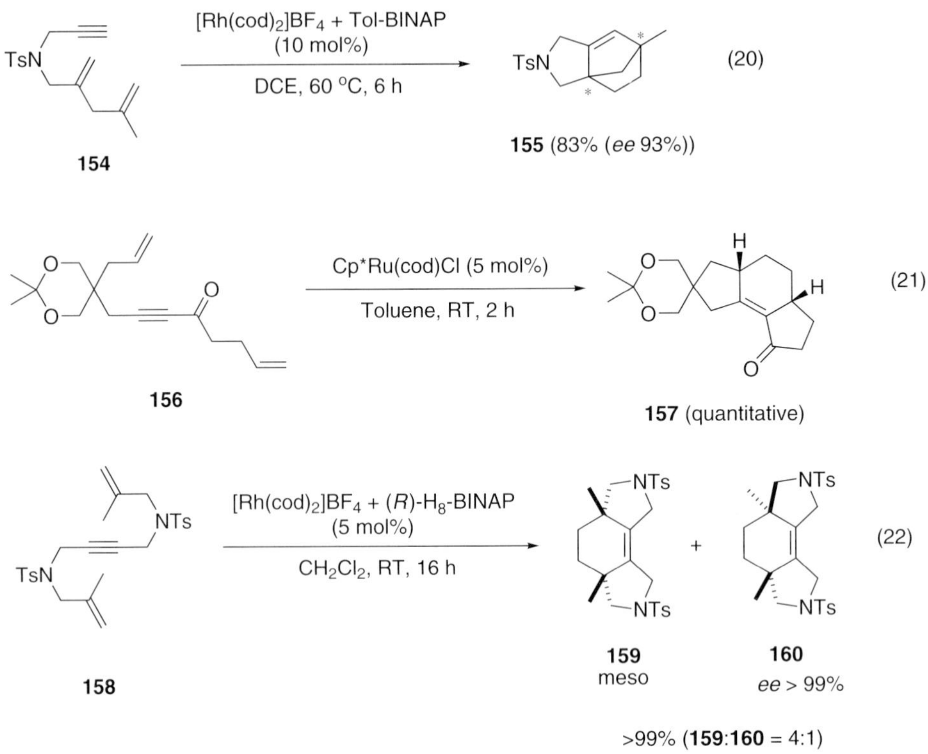

Scheme 8.28 Rh- and Ru-catalyzed enantioselective [2 + 2 + 2] cycloadditions of dienynes.

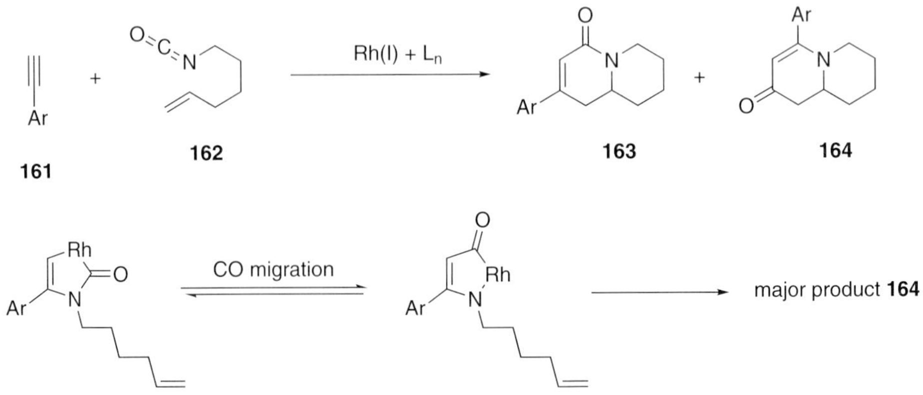

Scheme 8.29 Mechanistic pathway for the formation of vinylogous amide.

by a Mitsunobu reaction, the synthesis of the target compound was achieved in only three steps. Very recently, Rovis and Lee showed that bulkier alkenes could reverse the ratio between indolizidines and vinylogous amides in favor of the former [171].

The [2 + 2 + 2] cycloadditions involving carbodiimides as unsaturated partners were very scarce in the literature and concerned only the reaction of alkynes with carbodiimides [140, 149, 158]. Recently, Deiters and coworkers described the

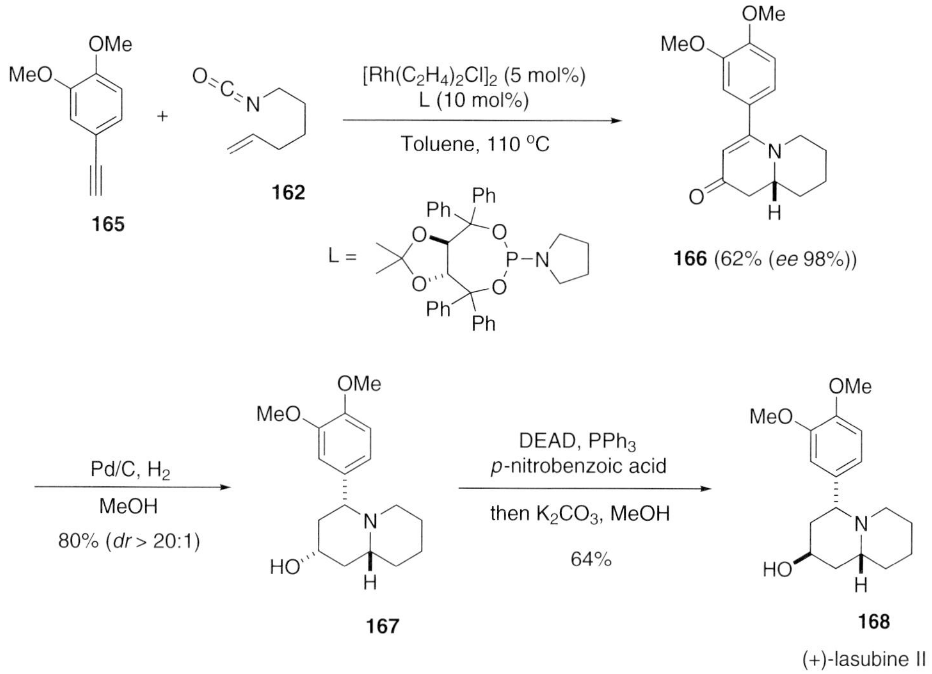

Scheme 8.30 Total synthesis of (+)-Lasubine II.

cycloaddition of solid-supported diynes with carbodiimides in high yield [74]. Under these conditions, Rovis and Yu achieved the first enantioselective cycloaddition employing alkenyl carbodiimides: they achieved the [2 + 2 + 2] cycloaddition of phenylacetylene with the alkenyl carbodiimide **169**, which led to the bicyclic amidine **170** as major product in good yield (70%) and excellent enantiomeric excess (97%), but admixed with the product resulting from the migration of the isocyanide (very scarce in the literature) (Scheme 8.31) [172].

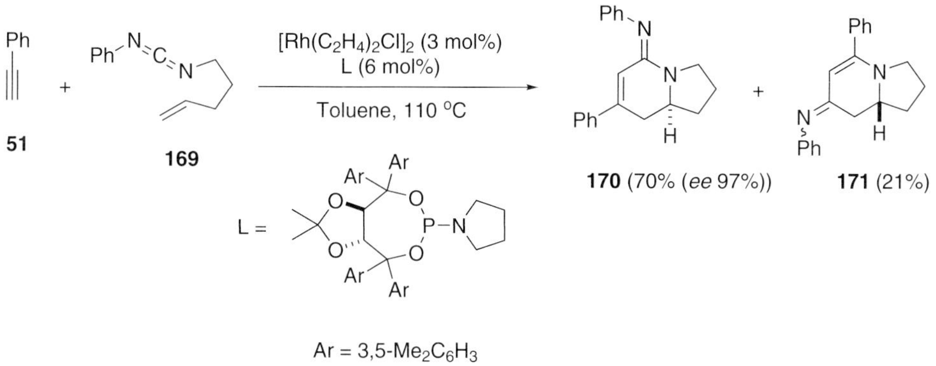

Scheme 8.31 Rh-catalyzed enantioselective [2 + 2 + 2] cycloaddition aimed at the preparation of the bicyclic amidine **170**.

As mentioned above, several groups studied the [2 + 2 + 2] cycloaddition of diynes with carbonyl compounds involving various transition metals (e.g., Co, Ni, Ru or Rh). However, the pyran cycloadducts resulting from this reaction often proved unstable and gave dienones after electrocyclic ring opening. In contrast, a transition metal-catalyzed [2 + 2 + 2] cycloaddition of enynes with ketones would furnish dihydropyrans with creation of two quaternary centers, which eliminates the possibility of an electrocyclic ring opening. Louie's and Tanaka's groups also investigated the [2 + 2 + 2] cycloaddition of enynes with carbonyl compounds. Tanaka and coworkers worked on the enantioselective [2 + 2 + 2] cycloaddition of enynes with ketones using a rhodium cationic complex. For example the enyne **172** transformed into the expected dihydropyran **174** with ethyl pyruvate with good regio-, diastereo- and enantio-selectivity in high yield (Scheme 8.32) [173]. On the other hand, Louie and coworkers obtained similar results with ketones, but, in the case of aldehydes, they also observed enones arising from a β-H elimination (Scheme 8.33) [174].

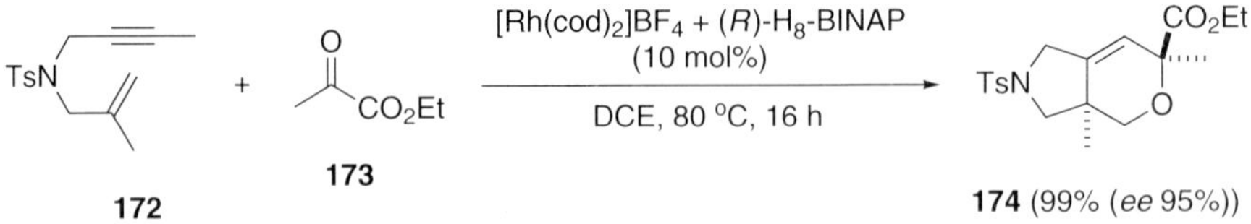

Scheme 8.32 Rh-catalyzed regio-, diastereo-, and enantio-selective [2 + 2 + 2] cycloaddition of 1,6-enyne **172** with ketone **173**.

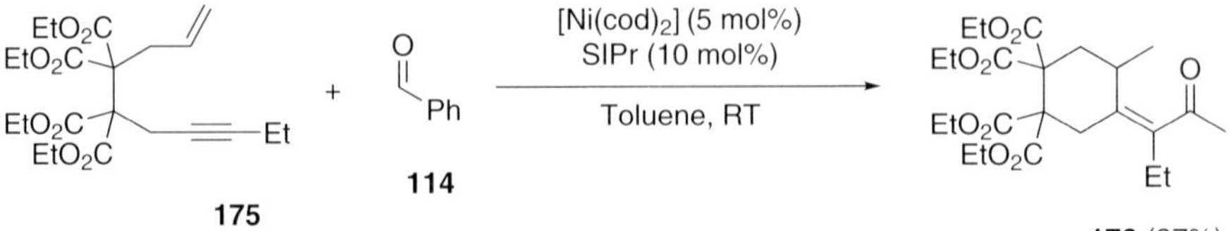

Scheme 8.33 Ni-catalyzed [2 + 2 + 2] cycloaddition of enyne **175** with benzaldehyde.

8.5
Conclusion and Perspectives

Although it is a mature reaction, the transition metal-mediated [2 + 2 + 2] cycloaddition still attracts significant interest by many research groups throughout the world. The three-component coupling of three unsaturated partners to generate six-membered rings is indeed very elegant, and allows the formation of benzenes, cyclohexadienes, cyclohexene, and many types of heterocycles. Over the last four years, catalytic and enantioselective reactions have been flourishing, and it seems clear that many new developments of this reaction will be discovered in the near future. Efforts are still needed to make this reaction industrially viable, notably by the development of recyclable supported catalysts.

8.6
Experimental: Selected Procedures

8.6.1
Synthesis of Compound 9 [22]

To a solution of Cp*Ru(cod)Cl (9.5 mg, 0.025 mmol), 1-hexynylboronate (105.1 mg, 0.50 mmol), and 1-hexyne **7** (104.3 mg, 2.0 mmol) in dry degassed 1,2-dichloroethane (1.0 mL) was added a solution of propargyl alcohol (30.8 mg, 0.55 mmol) in dry degassed 1,2-dichloroethane (3.0 mL) within 15 min at room temperature under an atmosphere of argon. The solution was stirred at room temperature for 5 h, and then, the solvent was removed under reduced pressure. To the residue were added $Pd_2(dba)_3 \cdot CHCl_3$ (13 mg, 0.025 mmol), PCy_3 (15.9 mg, 0.055 mmol), p-iodoacetophenone (184.5 mg, 0.75 mmol), toluene (3.5 mL), and 2 M aqueous K_2CO_3 (1.5 mL), and the reaction mixture was degassed at $-78\,°C$ under reduced pressure. After heating at $70\,°C$ under an atmosphere of argon for 3 h, the resulting solution was extracted with AcOEt (3 mL × 3) and the organic layer was dried over $MgSO_4$. The solvent was evaporated and the crude product was purified by silica gel flash chromatography (hexane/AcOEt = 6/1) to give biaryl **9** (123.4 mg, yield 73%) as pale yellow oil.

8.6.2
Synthesis of Compound 31 [45]

$Pd_2(dba)_3 \cdot CHCl_3$ (26 mg, 0.025 mmol) and $P(o\text{-tol})_3$ (61 mg, 0.20 mmol) were dissolved in acetonitrile (1.2 mL), and the mixture was stirred at room temperature for 15 min. The catalyst solution was added via cannula (using acetonitrile (1.0 mL) for rinsing) to a solution of **28** (160 mg, 0.51 mmol), **29** (530 mg, 1.6 mmol) and CsF (472 mg, 3.1 mmol) in acetonitrile (1.8 mmol) at $0\,°C$, and the mixture was stirred at room temperature for 4 h. The reaction mixture was quenched with saturated aqueous NH_4Cl solution and extracted with AcOEt. The organic layer was washed with brine and dried over Na_2SO_4. After removal of the solvent, the residue was purified by column chromatography on silica gel (hexane/AcOEt = 3/2) to give **31** (134 mg, yield 61%) as a yellowish solid along with the recovery of unchanged **29** (257 mg, 48% recovery).

8.6.3
Synthesis of Compound 67 [80]

(S,S)-MeDUPHOS (6.4 mg, 0.021 mmol) and $[IrCl(cod)]_2$ (7.1 mg, 0.0105 mmol) were stirred in degassed xylene (1.0 mL) at room temperature to give a reddish yellow solution. After the addition of a xylene solution (1.5 mL) of 1,4-dimethoxy-2-butyne (36.0 mg, 0.315 mmol) and a xylene solution (1.5 mL) of diyne **65** (36.5 mg, 0.105 mmol), the resulting mixture was further stirred under reflux for 1 h. The solvent was removed under reduced pressure, and purification of the crude products by thin layer chromatography (toluene/AcOEt = 15/1) gave pure **67** (40.3 mg, 83% yield).

8.6.4
Synthesis of Compounds 100 and 101 [123]

A solution of isoquinolone **99** (121 mg, 0.5 mmol) in dry degassed THF (4.0 mL) and a solution of CpCo(C$_2$H$_4$)$_2$ (90 mg, 0.5 mmol) in dry degassed THF (4.0 mL) were added simultaneously over 1 h to a mixture of degassed BTMSA (6.0 mL). The mixture was stirred for another 1 h, THF and excess BTMSA removed *in vacuo*, and the residue was purified by chromatography (hexane/AcOEt = 1/2) to recover some **99** (16.4 mg, 14%), and afford the *exo* complex **100** (98.4 mg, 37% yield, 43% based on converted starting material) and the *endo* complex **101** (54.6 mg, 20% yield, 24% based on converted starting material).

8.6.5
Synthesis of Compound 116 [130]

In a glove box, the diyne **113** (165.6 mg, 0.39 mmol) and benzaldehyde (52 mg, 0.49 mmol) were introduced in an oven dried scintillation vial equipped with a magnetic stir bar and dissolved in toluene (3.9 mL). To the stirring solution were added [Ni(cod)$_2$] (5.4 mg, 0.020 mmol) and SIPr (15.3 mg, 0.039 mmol), previously equilibrated for at least 6 h. The reaction was stirred at room temperature for 2 h over which time the color changed from orange to brown. The mixture was concentrated *in vacuo* and purified by flash chromatography (12% AcOEt/hexanes then 15% AcOEt/hexanes) to yield dienone **116** (166.3 mg, yield 80%) as a sticky pale yellow oil.

8.6.6
Synthesis of Compound 145 [160]

[RhCl(cod)]$_2$ (6.2 mg, 5 mol%) and AgBF$_4$ (9.7 mg, 20 mol%) were suspended in anhydrous THF (1.0 mL) and stirred at room temperature under an atmosphere of argon for 10 min. (S)-Xyl-P-PHOS (22.7 mg, 12 mol%) in anhydrous THF (3.0 mL) was then added to the yellow suspension, and the mixture was stirred at room temperature for an additional 30 min. Methyl phenylpropiolate (120.1 mg, 0.75 mmol) was added in one portion, followed by addition of 1,6-enyne **142** (62.3 mg, 0.25 mmol) in anhydrous THF (2.0 mL) via syringe pump over 2 h at 60 °C, followed by an additional 30 min (TLC control). The reaction mixture was allowed to cool to room temperature, and the resultant mixture was filtered through a short pad of silica gel (eluting with 50% AcOEt/hexanes) and concentrated *in vacuo* to afford the crude product. Purification by flash chromatography (silica gel, eluting with a 10–30% AcOEt/hexanes gradient) afforded the bicyclohexadienes **145/146** (94.1 mg, yield 98%) as a white solid, with 10:1 regioselectivity and 97% ee.

8.6.7
Synthesis of Compound 174 [173]

(R)-H$_8$-BINAP (18.9 mg, 0.030 mmol) and [Rh(cod)$_2$]BF$_4$ (12.2 mg, 0.030 mmol) were dissolved in CH$_2$Cl$_2$ (2.0 mL), and the mixture was stirred at room temperature for 5 min. H$_2$ was introduced to the resulting solution in a Schlenk tube. After stirring at room temperature for 0.5 h, the resulting solution was concentrated to dryness and dissolved in (CH$_2$Cl)$_2$ (0.5 mL). To this solution was added a (CH$_2$Cl)$_2$ (1.0 mL) solution of **172** (83.2 mg, 0.30 mmol) and **173** (69.6 mg, 0.60 mmol) at room temperature. The mixture was stirred at 80 °C for 16 h. The resulting solution was concentrated and purified by preparative TLC (hexane/AcOEt = 5/1), which furnished **174** (113.5 mg, 99% yield) as a pale yellow oil.

Abbreviations

AIBN	azo*bis*isobutyronitrile
Ac	acetyl
Ar	aryl
BINAP	2,2′-*bis*(diphenylphosphino)-1,1′-binaphthyl
BINOL	1,1′-binaphthol
Bn	benzyl
BTMSA	*bis*(trimethylsilyl)acetylene
Bu	butyl
Cod	1,5-cyclooctadiene
Cp	η^5-cyclopentadienyl
Cp*	η^5-pentamethylcyclopentadienyl
Cy	cyclohexyl
Dba	dibenzylideneacetone
DCE	1,2-dichloroethane
DEAD	diethyl azodicarboxylate
DME	1,2-dimethoxyethane
Et	ethyl
*i*Pr	isopropyl
IPr	N,N′-*bis*(2,6-diisopropylphenyl)-imidazol-2-ylidene
Me	methyl
Ph	phenyl
RT	room temperature
SIPr	N,N′-*bis*(2,6-diisopropylphenyl)-4,5-dihydroimidazol-2-ylidene
*t*Bu	*tert*-butyl
TADDOL	α,α,α′,α′-tetraphenyl-1,3-dioxolane-4,5-dimethanol
TBAF	tetra-*n*-butylammonium fluoride
TBDMS	*tert*-butyldimethylsilyl
Tf	triflate
TFA	trifluoroacetic acid

THF	tetrahydrofuran
TLC	thin layer chromatography
TMS	trimethylsilyl
Tol	tolyl
Trt	trityl
Ts	tosyl
TTMSS	tris(trimethylsilyl)silane

References

1 Berthelot, M. (1866) *C. R. Hebd. Seances Acad. Sci.*, **62**, 905–909.
2 Reppe, W. and Schweckendiek, W.J. (1948) *Justus Liebigs Ann. Chem.*, **560**, 104–115.
3 Yamamoto, Y. (2005) *Curr. Org. Chem.*, **9**, 503–519.
4 Gandon, V., Aubert, C. and Malacria, M. (2005) *Curr. Org. Chem.*, **9**, 1699–1712.
5 Kotha, S., Brahmachary, E. and Lahiri, K. (2005) *Eur. J. Org. Chem.*, 4741–4767.
6 Gandon, V., Aubert, C. and Malacria, M. (2006) *Chem. Commun.*, 2209–2217.
7 Chopade, P.R. and Louie, J. (2006) *Adv. Synth. Catal.*, **348**, 2307–2327.
8 Agenet, N., Gandon, V., Buisine, O., Slowinski, F. and Malacria, M. (2007) *Organic Reactions*, vol. 68 (ed. T.V. RajanBabu), John Wiley and Sons, Hoboken, pp. 1–302.
9 Heller, B. and Hapke, M. (2007) *Chem. Soc. Rev.*, **36**, 1085–1094.
10 Kirchner, K., Calhorda, M.J., Schmid, R. and Veiros, L.F. (2003) *J. Am. Chem. Soc.*, **125**, 11721–11729.
11 Yamamoto, Y., Arakawa, T., Ogawa, R. and Itoh, K. (2003) *J. Am. Chem. Soc.*, **125**, 12143–12160.
12 Dahy, A.A. and Koga, N. (2005) *Bull. Chem. Soc Jpn.*, **78**, 781–791.
13 Dahy, A.A. and Koga, N. (2005) *Bull. Chem. Soc Jpn.*, **78**, 792–803.
14 Gandon, V., Agenet, N., Vollhardt, K.P.C., Malacria, M. and Aubert, C. (2006) *J. Am. Chem. Soc.*, **128**, 8509–8520.
15 Agenet, N., Gandon, V., Vollhardt, K.P.C., Malacria, M. and Aubert, C. (2007) *J. Am. Chem. Soc.*, **129**, 8860–8871.
16 Hillard, R.L. III and Vollhardt, K.P.C. (1977) *J. Am. Chem. Soc.*, **99**, 4058–4069.
17 Parnell, C.A. and Vollhardt, K.P.C. (1985) *Tetrahedron*, **41**, 5791–5796.
18 Hirthammer, M. and Vollhardt, K.P.C. (1986) *J. Am. Chem. Soc.*, **108**, 2481–2482.
19 Petit, M., Chouraqui, G., Aubert, C. and Malacria, M. (2003) *Org. Lett.*, **5**, 2037–2040.
20 Chouraqui, G., Petit, M., Aubert, C. and Malacria, M. (2004) *Org. Lett.*, **6**, 1519–1521.
21 Chouraqui, G., Petit, M., Phansavath, P., Aubert, C. and Malacria, M. (2006) *Eur. J. Org. Chem.*, 1413–1421.
22 Yamamoto, Y., Ishii, J.-i., Nishiyama, H. and Itoh, K. (2004) *J. Am. Chem. Soc.*, **126**, 3712–3713.
23 Yamamoto, Y., Ishii, J.-i., Nishiyama, H. and Itoh, K. (2005) *J. Am. Chem. Soc.*, **127**, 9625–9631.
24 Yamamoto, Y., Ishii, J.-i., Nishiyama, H. and Itoh, K. (2005) *Tetrahedron*, **61**, 11501–11510.
25 Witulski, B. and Stengel, T. (1999) *Angew. Chem. Int. Ed.*, **38**, 2426–2430.
26 Witulski, B. and Alayrac, C. (2002) *Angew. Chem. Int. Ed.*, **41**, 3281–3284.
27 Witulski, B., Stengel, T. and Fernandez-Hernandez, J. M. (2000) *Chem. Commun.*, 1965–1966.
28 Tracey, M.R., Oppenheimer, J. and Hsung, R.P. (2006) *J. Org. Chem.*, **71**, 8629–8632.

29 Oppenheimer, J., Hsung, R.P., Figueroa, R. and Johnson, W.L. (2007) *Org. Lett.*, **9**, 3969–3972.

30 Tanaka, K., Takeishi, K. and Noguchi, K. (2006) *J. Am. Chem. Soc.*, **128**, 4586–4587.

31 Tanaka, K. and Takeishi, K. (2007) *Synthesis*, 2920–2923.

32 Kondoh, A., Yorimitsu, H. and Oshima, K. (2007) *J. Am. Chem. Soc.*, **129**, 6996–6997.

33 Yamamoto, Y., Hattori, K., Ishii, J.-i., Nishiyama, H. and Itoh, K. (2005) *Chem. Commun.*, 4438–4440.

34 Yamamoto, Y., Hattori, K., Ishii, J.-i. and Nishiyama, H. (2006) *Tetrahedron*, **62**, 4294–4305.

35 Yamamoto, Y., Hattori, K. and Nishiyama, H. (2006) *J. Am. Chem. Soc.*, **128**, 8336–8340.

36 Yamamoto, Y. and Hattori, K. (2008) *Tetrahedron*, **64**, 847–855.

37 Guitian, E., Perez, D. and Pena, D. (2005) *Top. Organomet. Chem.*, **14**, 109–146.

38 Pena, D., Cobas, A., Perez, D., Guitian, E. and Castedo, L. (2003) *Org. Lett.*, **5**, 1863–1866.

39 Caeiro, J., Pena, D., Cobas, A., Perez, D. and Guitian, E. (2006) *Adv. Synth. Catal.*, **348**, 2466–2474.

40 Deaton, K.R. and Gin, M.S. (2003) *Org. Lett.*, **5**, 2477–2480.

41 Hsieh, J.-C. and Cheng, C.-H. (2005) *Chem. Commun.*, 2459–2461.

42 Deaton, K.R., Strouse, C.S. and Gin, M.S. (2004) *Synthesis*, 3084–3088.

43 Iglesias, B., Cobas, A., Perez, D., Guitian, E. and Vollhardt, K.P.C. (2004) *Org. Lett.*, **6**, 3557–3560.

44 Romero, C., Pena, D., Perez, D. and Guitian, E. (2006) *Chem. Eur. J.*, **12**, 5677–5684.

45 Sato, Y., Tamura, T. and Mori, M. (2004) *Angew. Chem. Int. Ed.*, **43**, 2436–2440.

46 Sato, Y., Tamura, T., Kinbara, A. and Mori, M. (2007) *Org. Lett.*, **349**, 647–661.

47 Stara, I.G., Stary, I., Kollarovic, A., Teply, F., Saman, D. and Tichy, M. (1998) *J. Org. Chem.*, **63**, 4046–4050.

48 Stara, I.G., Stary, I., Kollarovic, A., Teply, F., Vyskocil, S. and Saman, D. (1999) *Tetrahedron Lett.*, **40**, 1993–1996.

49 Teply, F., Stara, I.G., Stary, I., Kollarovic, A., Saman, D., Rulisek, L. and Fiedler, P. (2002) *J. Am. Chem. Soc.*, **124**, 9175–9180.

50 Teply, F., Stara, I.G., Stary, I., Kollarovic, A., Saman, D., Vyskocil, S. and Fiedler, P. (2003) *J. Org. Chem.*, **68**, 5193–5197.

51 Han, S., Bond, A.D., Disch, R.L., Holmes, D., Schulman, J.M., Teat, S.J., Vollhardt, K.P.C. and Whitener, G.D. (2002) *Angew. Chem. Int. Ed.*, **41**, 3223–3227.

52 Han, S., Anderson, D.R., Bond, A.D., Chu, H.V., Disch, R.L., Holmes, D., Schulman, J.M., Teat, S.J., Vollhardt, K.P.C. and Whitener, G.D. (2002) *Angew. Chem. Int. Ed.*, **41**, 3227–3230.

53 Tanaka, K., Kamisawa, A., Suda, T., Noguchi, K. and Hirano, M. (2007) *J. Am. Chem. Soc.*, **129**, 12078–12079.

54 Stara, I.G., Alexandrova, Z., Teply, F., Sehnal, P., Stary, I., Saman, D., Budesinsky, M. and Cvacka, J. (2005) *Org. Lett.*, **7**, 2547–2550.

55 Sehnal, P., Krausova, Z., Teply, F., Stara, I.G., Stary, I., Rulisek, L., Saman, D. and Cisarova, I. (2008) *J. Org. Chem.*, **73**, 2074–2082.

56 Matsuda, T., Kadowaki, S., Goya, T. and Murakami, M. (2007) *Org. Lett.*, **9**, 133–136.

57 Vollhardt, K.P.C. and Mohler, D.L. (1996) *Advances in Strain in Organic Chemistry* (ed. B. Halton), JAI, London, pp. 121–160.

58 Miljanic, O. and Vollhardt, K.P.C. (2006) *Carbon-Rich Compounds: From Molecules to Materials* (eds M.M. Haley and R.R. Tykwinski), Wiley-VCH, Weinheim, pp. 140–197.

59 Bruns, D., Miura, H., Vollhardt, K.P.C. and Stanger, A. (2003) *Org. Lett.*, **5**, 549–552.

60 Bong, D.T.Y., Chan, E.W.L., Diercks, R., Dosa, P.I., Haley, M.M., Matzger, A.J., Miljanic, O.S. and Vollhardt, K.P.C. (2004) *Org. Lett.*, **6**, 2249–2252.

61 Ohshita, J., Furumori, K., Matsuguchi, A. and Ishikawa, M. (1990) *J. Org. Chem.*, **55**, 3277–3280.
62 Tanaka, K. and Shirasaka, K. (2003) *Org. Lett.*, **5**, 4697–4699.
63 Tanaka, K., Toyoda, K., Wada, A., Shirasaka, K. and Hirano, M. (2005) *Chem. Eur. J.*, **11**, 1145–1156.
64 Hilt, G., Hess, W., Vogler, T. and Hengst, C. (2005) *J. Organomet. Chem.*, **690**, 5170–5181.
65 Hilt, G., Vogler, T., Hess, W. and Galbiati, F. (2005) *Chem. Commun.*, 1474–1475.
66 Rüba, E., Schmid, R., Kirchner, K. and Calhorda, M.J. (2003) *J. Organomet. Chem.*, **682**, 204–211.
67 Saino, N., Kogure, D. and Okamoto, S. (2005) *Org. Lett.*, **7**, 3065–3067.
68 Saino, N., Kogure, D., Kase, K. and Okamoto, S. (2006) *J. Organomet. Chem.*, **691**, 3129–3136.
69 Saino, N., Amemiya, F., Tanabe, E., Kase, K. and Okamoto, S. (2006) *Org. Lett.*, **8**, 1439–1442.
70 Senaiar, R.S., Young, D.D. and Deiters, A. (2006) *Chem. Commun.*, 1313–1315.
71 Young, D.D., Senaiar, R.S. and Deiters, A. (2006) *Chem. Eur. J.*, **12**, 5563–5568.
72 Senaiar, R.S., Teske, J.A., Young, D.D. and Deiters, A. (2007) *J. Org. Chem.*, **72**, 7801–7804.
73 Loupy, A. (2006) *Microwaves in Organic Synthesis*, 2nd edn, Wiley-VCH, Weinheim.
74 Young, D.D. and Deiters, A. (2007) *Angew. Chem. Int. Ed.*, **46**, 5187–5190.
75 Zhou, Y., Porco, J.A. Jr and Snyder, J.K. (2007) *Org. Lett.*, **9**, 393–396.
76 Hrdina, R., Kadlcikova, A., Valterova, I., Hodacova, J. and Kotora, M. (2006) *Tetrahedron: Asym.*, **17**, 3185–3191.
77 Teske, J.A. and Deiters, A. (2008) *J. Org. Chem.*, **73**, 342–345.
78 Sripada, L., Teske, J.A. and Deiters, A. (2008) *Org. Biomol. Chem.*, **6**, 263–265.
79 Young, D.D., Sripada, L. and Deiters, A. (2007) *J. Comb. Chem.*, **9**, 735–738.
80 Shibata, T., Fujimoto, T., Yokota, K. and Takagishi, K. (2004) *J. Am. Chem. Soc.*, **126**, 8382–8383.
81 Gutnov, A., Heller, B., Fischer, C., Drexler, H.-J., Spannenberg, A., Sundermann, B. and Sundermann, C. (2004) *Angew. Chem. Int. Ed.*, **43**, 3795–3797.
82 Shibata, T. and Tsuchikama, K. (2005) *Chem. Commun.*, 6017–6019.
83 Shibata, T., Tsuchikama, K. and Otsuka, M. (2006) *Tetrahedron: Asymmetry*, **17**, 614–619.
84 Shibata, T., Arai, Y., Takami, K., Tsuchikama, K., Fujimoto, T., Takebayashi, S. and Takagi, K. (2006) *Adv. Synth. Catal.*, **348**, 2475–2483.
85 Shibata, T., Yoshida, S., Arai, Y., Otsuka, M. and Endo, K. (2008) *Tetrahedron*, **64**, 821–830.
86 Tanaka, K., Nishida, G., Wada, A. and Noguchi, K. (2004) *Angew. Chem. Int. Ed.*, **43**, 6510–6512.
87 Nishida, G., Suzuki, N., Noguchi, K. and Tanaka, K. (2006) *Org. Lett.*, **8**, 3489–3492.
88 Tanaka, K., Nishida, G., Ogino, M., Hirano, M. and Noguchi, K. (2005) *Org. Lett.*, **7**, 3119–3121.
89 Nishida, G., Noguchi, K., Hirano, M. and Tanaka, K. (2007) *Angew. Chem. Int. Ed.*, **46**, 3951–3954.
90 Ueda, T., Kanomata, N. and Machida, H. (2005) *Org. Lett.*, **7**, 2365–2368.
91 Tanaka, K., Sagae, H., Toyoda, K. and Noguchi, K. (2006) *Eur. J. Org. Chem.*, 3575–3581.
92 Tanaka, K., Sagae, H., Toyoda, K., Noguchi, K. and Hirano, M. (2007) *J. Am. Chem. Soc.*, **129**, 1522–1523.
93 Tanaka, K., Sagae, H., Toyoda, K. and Hirano, M. (2008) *Tetrahedron*, **64**, 831–846.
94 Maderna, A., Pritzkow, H. and Siebert, W. (1996) *Angew. Chem., Int. Ed. Engl.*, **35**, 1501–1503.
95 Ester, C., Maderna, A., Pritzkow, H. and Siebert, W. (2000) *Eur. J. Inorg. Chem.*, 1177–1184.
96 Gu, Y., Pritzkow, H. and Siebert, W. (2001) *Eur. J. Inorg. Chem.*, 373–379.

97 Goswami, A., Maier, C.-J., Pritzkow, H. and Siebert, W. (2004) *Eur. J. Inorg. Chem.*, 2635–2645.

98 Gandon, V., Leca, D., Aechtner, T., Vollhardt, K.P.C., Malacria, M. and Aubert, C. (2004) *Org. Lett.*, **6**, 3405–3407.

99 Gandon, V., Leboeuf, D., Amslinger, S., Vollhardt, K.P.C., Malacria, M. and Aubert, C. (2005) *Angew. Chem. Int. Ed.*, **44**, 7114–7118.

100 Geny, A., Leboeuf, D., Rouquié, G., Vollhardt, K.P.C., Malacria, M., Gandon, V. and Aubert, C. (2007) *Chem. Eur. J.*, **13**, 5408–5425.

101 Hsieh, J.-C., Rayabarapu, D.K. and Cheng, C.-H. (2004) *Chem. Commun.*, 532–533.

102 Jayanth, T.T., Jeganmohan, M. and Cheng, C.-H. (2004) *J. Org. Chem.*, **69**, 8445–8450.

103 Quintana, I., Boersma, A.J., Pena, D., Perez, D. and Guitian, E. (2006) *Org. Lett.*, **8**, 3347–3349.

104 de Meijere, A., Becker, H., Stolle, A., Kozhushkov, S., Bes, M.T., Salaün, J. and Noltemeyer, M. (2005) *Chem. Eur. J.*, **11**, 2417–2482.

105 Schelper, M., Buisine, O., Kozhushkov, S., Aubert, C., de Meijere, A. and Malacria, M. (2005) *Eur. J. Org. Chem.*, 3000–3007.

106 Zhao, L. and de Meijere, A. (2006) *Adv. Synth. Catal.*, **348**, 2484–2492.

107 Ma, S. (2005) *Chem. Rev.*, **105**, 2829–2871.

108 Shanmugasundaram, M., Wu, M.-S. and Cheng, C.-H. (2001) *Org. Lett.*, **3**, 4233–4236.

109 Shanmugasundaram, M., Wu, M.-S., Jeganmohan, M., Huang, C.-W. and Cheng, C.-H. (2002) *J. Org. Chem.*, **67**, 7724–7729.

110 Aubert, C., Llerena, D. and Malacria, M. (1994) *Tetrahedron Lett.*, **35**, 2341–2344.

111 Llerena, D., Buisine, O., Aubert, C. and Malacria, M. (1998) *Tetrahedron*, **54**, 9373–9392.

112 Buisine, O., Aubert, C. and Malacria, M. (2000) *Synthesis*, 985–989.

113 Petit, M., Aubert, C. and Malacria, M. (2004) *Org. Lett.*, **6**, 3937–3940.

114 Petit, M., Aubert, C. and Malacria, M. (2006) *Tetrahedron*, **62**, 10582–10593.

115 Boese, R., Harvey, D.F., Malaska, M.J. and Vollhardt, K.P.C. (1994) *J. Am. Chem. Soc.*, **116**, 11153–11154.

116 Perez, D., Siesel, B.A., Malaska, M.J., David, E., Peter, K. and Vollhardt, C. (2000) *Synlett*, 306–310.

117 Grotjahn, D.B. and Vollhardt, K.P.C. (1986) *J. Am. Chem. Soc.*, **108**, 2091–2093.

118 Boese, R., Van Sickle, A.P. and Vollhardt, K.P.C. (1994) *Synthesis*, 1374–1382.

119 Sheppard, G.S. and Vollhardt, K.P.C. (1986) *J. Org. Chem.*, **51**, 5496–5498.

120 Boese, R., Knölker, H.-J. and Vollhardt, K.P.C. (1987) *Angew. Chem., Int. Ed. Engl.*, **26**, 1035–1037.

121 Boese, R., Rodriguez, J. and Vollhardt, K.P.C. (1991) *Angew. Chem., Int. Ed. Engl.*, **30**, 993–994.

122 Pellissier, H., Rodriguez, J. and Vollhardt, K.P.C. (1999) *Chem. Eur. J.*, **5**, 3945–3961.

123 Aubert, C., Betschmann, P., Eichberg, M.J., Gandon, V., Heckrodt, T.J., Lehmann, J., Malacria, M., Masjost, B., Paredes, E., Vollhardt, K.P.C. and Whitener, G.D. (2007) *Chem. Eur. J.*, **13**, 7443–7465.

124 Aubert, C., Gandon, V., Geny, A., Heckrodt, T.J., Malacria, M., Paredes, E. and Vollhardt, K.P.C. (2007) *Chem. Eur. J.*, **13**, 7466–7478.

125 Tsuda, T., Morikawa, S., Sumiya, R. and Saegusa, T. (1988) *J. Org. Chem.*, **53**, 3140–3145.

126 Tsuda, T., Morikawa, S., Hasegawa, N. and Saegusa, T. (1990) *J. Org. Chem.*, **55**, 2978–2981.

127 Louie, J., Gibby, J.E., Farnworth, M.V. and Tekavec, T.N. (2002) *J. Am. Chem. Soc.*, **124**, 15188–15189.

128 Tekavec, T.N., Arif, A.M. and Louie, J. (2004) *Tetrahedron*, **60**, 7431–7437.

129 Harvey, D.F., Johnson, B.M., Ung, C.S. and Vollhardt, K.P.C. (1989) *Synlett*, 15–18.

130 Tekavec, T.N. and Louie, J. (2005) *Org. Lett.*, **7**, 4037–4039.

131 Tsuda, T., Kiyoi, T., Miyane, T. and Saegusa, T. (1988) *J. Am. Chem. Soc.*, **110**, 8570–8572.

132 Yamamoto, Y., Takagishi, H. and Itoh, K. (2002) *J. Am. Chem. Soc.*, **124**, 6844–6845.

133 Tanaka, K., Otake, Y., Wada, A., Noguchi, K. and Hirano, M. (2007) *Org. Lett.*, **9**, 2203–2206.

134 Tsuchikama, K., Yoshinami, Y. and Shibata, T. (2007) *Synlett*, 1395–1398.

135 Wakatsuki, Y. and Yamazaki, H. (1973) *J. Chem. Soc., Chem. Commun.*, 280.

136 Yamazaki, H. (1987) *J. Synth. Org. Chem. Jpn.*, **45**, 244–257.

137 Yamamoto, Y., Takagishi, H. and Itoh, K. (2002) *J. Am. Chem. Soc.*, **124**, 28–29.

138 Schmid, R. and Kirchner, K. (2003) *J. Org. Chem.*, **68**, 8339–8344.

139 Tanaka, K., Wada, A. and Noguchi, K. (2006) *Org. Lett.*, **8**, 907–909.

140 Hong, P. and Yamazaki, H. (1977) *Tetrahedron Lett.*, **18**, 1333–1336.

141 Hong, P. and Yamazaki, H. (1977) *Synthesis*, 50–52.

142 Earl, R.A. and Vollhardt, K.P.C. (1984) *J. Org. Chem.*, **49**, 4786–4800.

143 Hoberg, H. and Oster, B.W. (1982) *J. Organomet. Chem.*, **234**, C35–C38.

144 Hoberg, H. and Oster, B.W. (1982) *Synthesis*, 324–325.

145 Hoberg, H. and Oster, B.W. (1983) *J. Organomet. Chem.*, **252**, 359–364.

146 Yamamoto, Y., Takagishi, H. and Itoh, K. (2001) *Org. Lett.*, **3**, 2117–2119.

147 Yamamoto, Y., Kinpara, K., Saigoku, T., Takagishi, H., Okuda, S., Nishiyama, H. and Itoh, K. (2005) *J. Am. Chem. Soc.*, **127**, 605–613.

148 Flynn, S.T., Hasso-Henderson, S.E. and Parkins, A.W. (1985) *J. Mol. Catal.*, **32**, 101–105.

149 Takahashi, T., Tsai, F.-Y., Li, Y., Wang, H., Kondo, Y., Yamanaka, M., Nakajima, K. and Kotora, M. (2002) *J. Am. Chem. Soc.*, **124**, 5059–5067.

150 Tanaka, K., Wada, A. and Noguchi, K. (2005) *Org. Lett.*, **7**, 4737–4739.

151 Maryanoff, B.E. and Zhang, H.-C. (2007) *Arkivoc*, **12**, 7–35.

152 Bonaga, L.V.R., Zhang, H.-C., Gauthier, D.A., Reddy, I. and Maryanoff, B.E. (2003) *Org. Lett.*, **5**, 4537–4540.

153 Bonaga, L.V.R., Zhang, H.-C. and Maryanoff, B.E. (2004) *Chem. Commun.*, 2394–2395.

154 Bonaga, L.V.R., Zhang, H.-C., Moretto, A.F., Ye, H., Gauthier, D.A., Li, J., Leo, G.C. and Maryanoff, B.E. (2005) *J. Am. Chem. Soc.*, **127**, 3473–3485.

155 Duong, H.A., Cross, M.J. and Louie, J. (2004) *J. Am. Chem. Soc.*, **126**, 11438–11439.

156 Duong, H.A. and Louie, J. (2006) *Tetrahedron*, **62**, 7552–7559.

157 Duong, H.A., Cross, M.J. and Louie, J. (2004) *Org. Lett.*, **6**, 4679–4681.

158 Hoberg, H. and Burkhart, G. (1979) *Synthesis*, 525.

159 Ogoshi, S., Ikeda, H. and Kurosawa, H. (2007) *Angew. Chem. Int. Ed.*, **46**, 4930–4932.

160 Evans, P.A., Lai, K.W. and Sawyer, J.R. (2005) *J. Am. Chem. Soc.*, **127**, 12466–12467.

161 Shibata, T., Arai, Y. and Tahara, Y.-k. (2005) *Org. Lett.*, **7**, 4955–4957.

162 Shibata, T., Kawachi, A., Ogawa, M., Kuwata, Y., Tsuchikama, K. and Endo, K. (2007) *Tetrahedron*, **63**, 12853–12860.

163 Shibata, T., Kurokawa, H. and Kanda, K. (2007) *J. Org. Chem.*, **72**, 6521–6525.

164 Seo, J., Chui, H.M.P., Heeg, M.J. and Montgomery, J. (1999) *J. Am. Chem. Soc.*, **121**, 476–477.

165 Shibata, T. and Tahara, Y.-k. (2006) *J. Am. Chem. Soc.*, **128**, 11766–11767.

166 Shibata, T., Tahara, Y.-k., Tamura, K. and Endo, K. (2008) *J. Am. Chem. Soc.*, **130**, 3451–3457.

167 Tanaka, D., Sato, Y. and Mori, M. (2007) *J. Am. Chem. Soc.*, **129**, 7730–7731.

168 Tanaka, K., Nishida, G., Sagae, H. and Hirano, M. (2007) *Synlett*, 1426–1430.

169 Yu, R.T. and Rovis, T. (2006) *J. Am. Chem. Soc.*, **128**, 2782–2783.

170 Yu, R.T. and Rovis, T. (2006) *J. Am. Chem. Soc.*, **128**, 12370–12371.

171 Lee, E.E. and Rovis, T. (2008) *Org. Lett.*, **10**, 1231–1234.

172 Yu, R.T. and Rovis, T. (2008) *J. Am. Chem. Soc.*, **130**, 3262–3263.

173 Tanaka, K., Otake, Y., Sagae, H., Noguchi, K. and Hirano, M. (2008) *Angew. Chem. Int. Ed.*, **47**, 1312–1316.

174 Tekavec, T.N. and Louie, J. (2008) *J. Org. Chem.*, **73**, 2641–2648.

9
Cyclizations Based on Cyclometallation
Hao Guo and Shengming Ma

9.1
Introduction

Cyclizations based on cyclometallation are one of most powerful synthetic methods for constructing mono- or multi-ring systems. Usually, in this type of reaction, metal in a low valence state coordinates with two unsaturated bonds to generate a five-membered metallacyclic intermediate. This intermediate can readily undergo β-H elimination, reductive elimination, insertion, or isomerization to give the cyclized products. It may also react further with electrophilic or nucleophilic reagents, resulting in a series of new products. Coupling products can also be afforded by treatment with organometallic reagents. In this chapter, we will present a selection of the most recent advances in this area. Related reviews are cited in each section.

9.2
Titanium-Catalyzed Cyclizations [1]

In the presence of *c*-hexMgCl, $(2,6\text{-Me}_2\text{C}_6\text{H}_3\text{O})_4\text{Ti}$ catalyzes the cyclization reaction of 1,6- or 1,7-dienes to afford the corresponding 1-methylene-2-methylcyclopentane or cyclohexane derivatives as the major products (Scheme 9.1) [2].

On introduction of an acetoxymethyl group into one of the two alkene moieties, the $\text{Ti(OPr}^i)_4$-promoted cyclization afforded the 1-methyl-2-vinylcyclopentane derivative in the presence of *i*PrMgBr (Scheme 9.2). The reaction is believed to proceed via cyclometallation of the 1,6-diene substrate with titanium(II) which is reduced *in situ* from $\text{Ti(OPr}^i)_4$ with *i*PrMgBr, β-OAc elimination, and hydrolysis [3].

In the presence of *i*PrMgCl, $\text{Ti(OPr}^i)_3\text{Cl}$ promotes the cyclization of ethyl 3-(2-allylphenyl)-2(*E*)-propenoate. The titanium(II) generated *in situ* reacts with the diene substrate to afford a bicyclotitanapentane intermediate. The C–Ti bond α to the ester moiety reacts preferentially with H$^+$ or EtCHO. Subsequent intramolecular reaction of the remaining C–Ti bond with the ester group results in the formation of the five-membered cyclic ketone ring (Scheme 9.3) [4].

Handbook of Cyclization Reactions. Volume 1.
Edited by Shengming Ma
Copyright © 2010 WILEY-VCH Verlag GmbH & Co. KGaA, Weinheim
ISBN: 978-3-527-32088-2

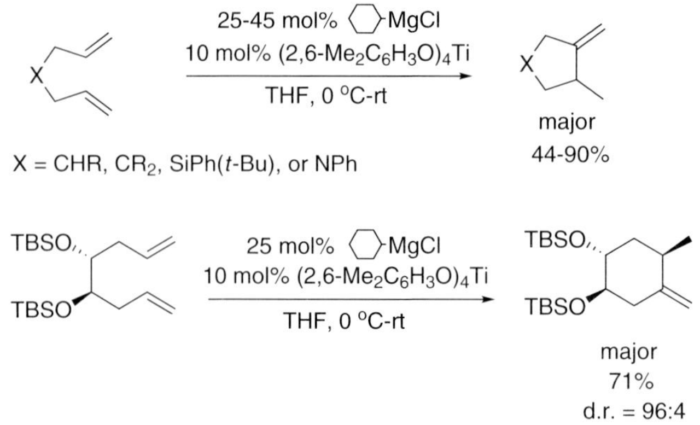

Scheme 9.1 (2,6-Me$_2$C$_6$H$_3$O)$_4$Ti-catalyzed cyclization of 1,6-or 1,7-dienes.

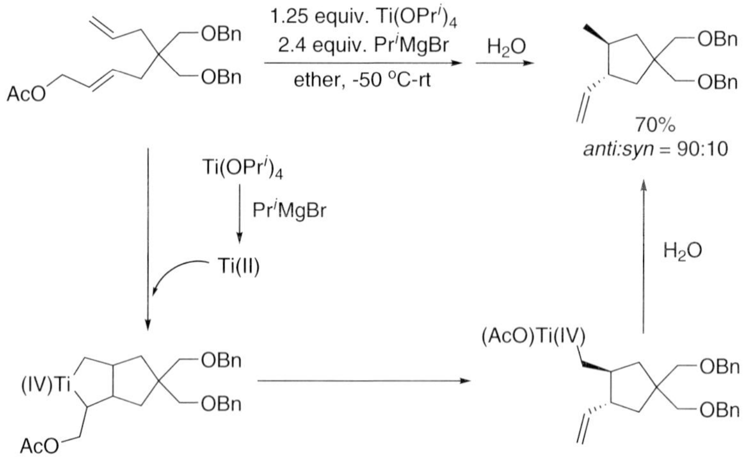

Scheme 9.2 Ti(OPr^i)$_4$-promoted cyclization of 1-acetoxy-2,7-dienes.

For the non-functionalized 1,6- or 1,7-enyne substrates, the titanabicyclic species generated *in situ* reacts directly with water to form alkylidenecyclopentane or cyclohexane derivatives. It should be noted that with aldehyde or ketone the sp^3-C–Ti bond reacts preferentially to afford the alcohol while the sp^2-C–Ti bond is protonated (Scheme 9.4) [4].

When a carbonate functionality is introduced into the allylic position of the alkene moiety in (E)-oct-2-en-7-yne or (E)-non-2-en-8-yne derivatives, the enyne undergoes cyclometallation with titanium(II) to afford the metallabicyclic intermediate. Subsequent β-elimination of EtOCO$_2$– forms the 1-alkenyltitanium(IV) species, which could react further with H$^+$, I$_2$, or aldehydes (Scheme 9.5) [3].

A very similar cyclization to that shown in Scheme 9.3 was observed with alk-7- or -8-yn-2-enoates (Scheme 9.6) [4].

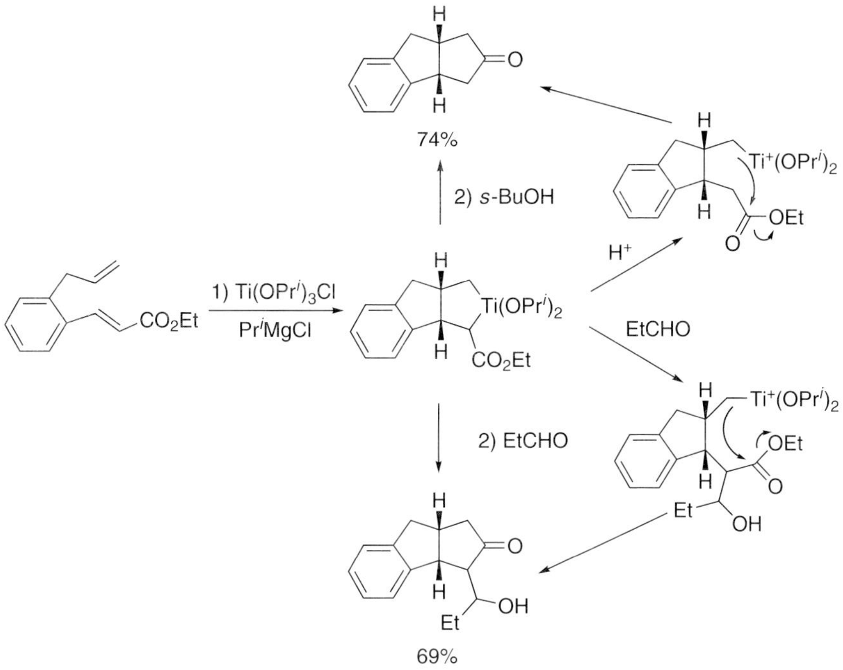

Scheme 9.3 Ti(OPr^i)$_3$Cl-promoted cyclization of 2,7-alkadienoate.

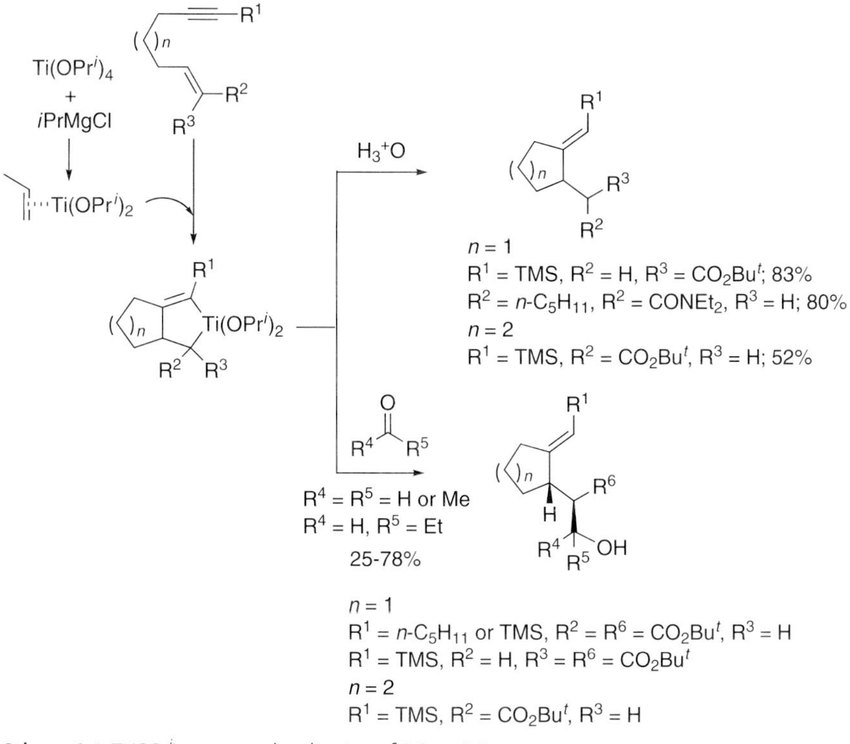

Scheme 9.4 Ti(OPr^i)$_4$-promoted cyclization of 1,6- or 1,7-enynes.

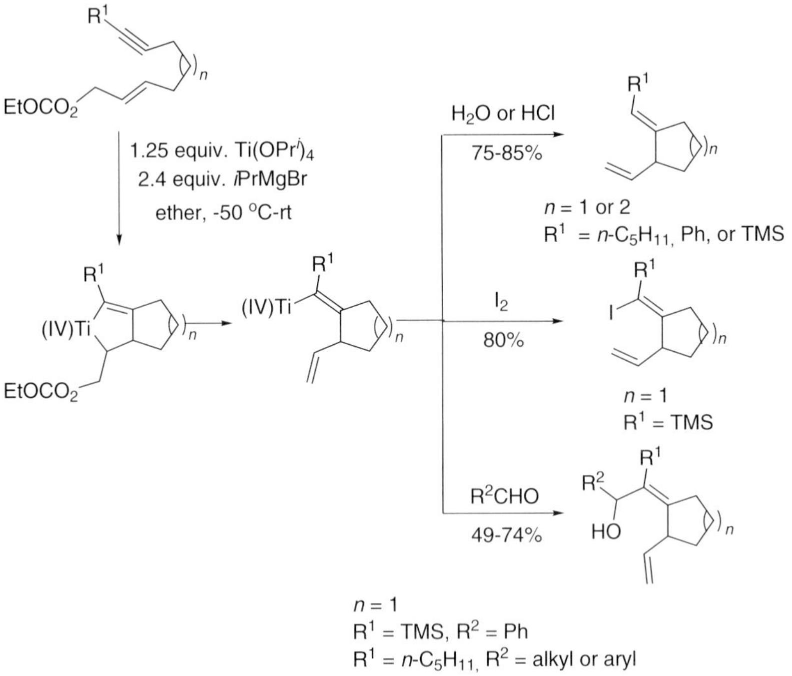

Scheme 9.5 Ti(OPr^i)$_4$-promoted cyclization of 2,7- or 2,8-alkenynyl carbonates.

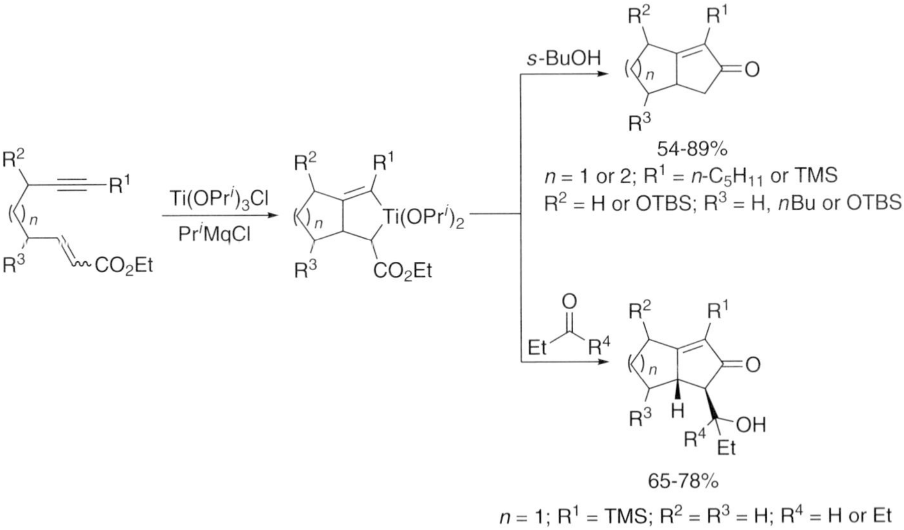

Scheme 9.6 Ti(OPr^i)$_3$Cl-promoted cyclization of 1-ethoxycarbonyl-1-en-6- or 7-ynes.

9.2 Titanium-Catalyzed Cyclizations

If the ethoxycarbonyl group is introduced into the terminal position of the alkyne moiety in 1,6-enynes, the titanabicycle undergoes rearrangement to afford a titanium carbene intermediate at 0 °C. Subsequent treatment with sec-BuOH or pentan-3-one formed different bicyclo[3.1.0]hexane derivatives (Scheme 9.7) [4].

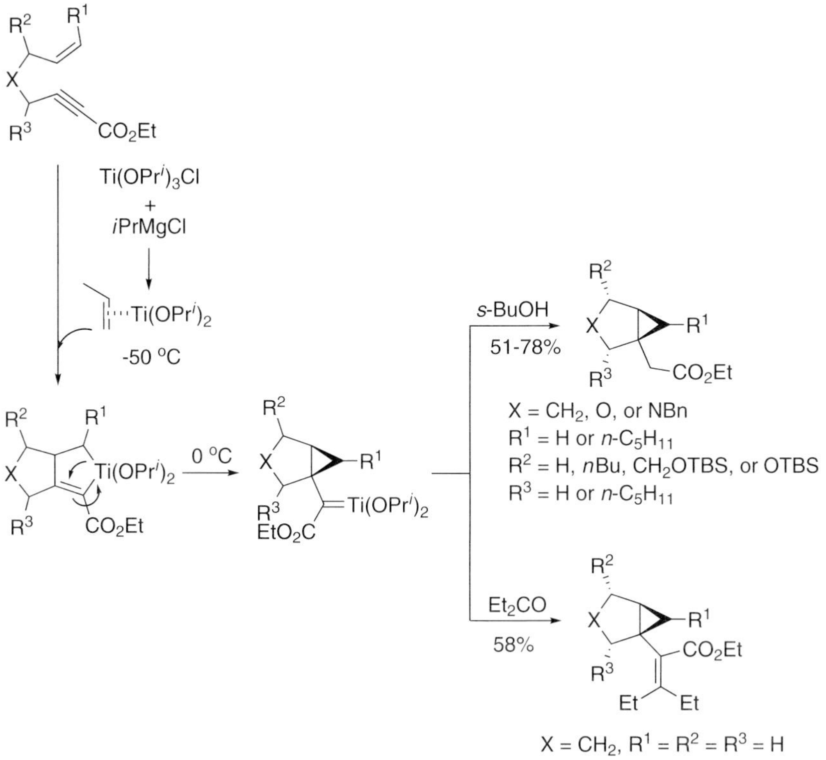

Scheme 9.7 Ti(OPr^i)$_3$Cl-promoted cyclization of 7-en-2-ynoates.

In the absence of electrophiles, the Cp$_2$Ti(CO)$_2$-catalyzed cyclization reaction of 1,6-enynes afforded the corresponding 1-alkylidene-2-alkenylcyclopentane derivatives (Scheme 9.8) via cyclometallation, β-H elimination, and reductive elimination [5].

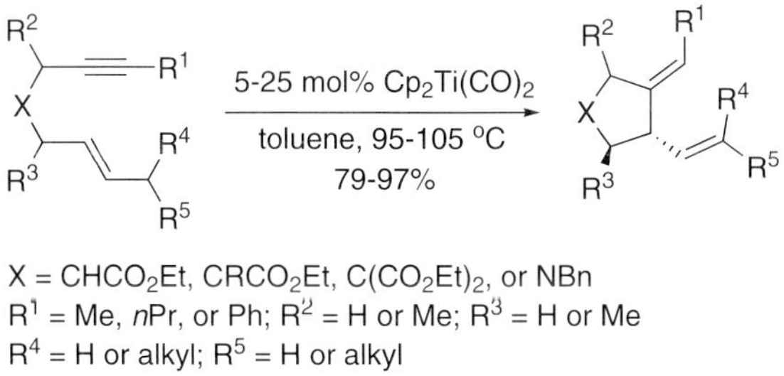

X = CHCO$_2$Et, CRCO$_2$Et, C(CO$_2$Et)$_2$, or NBn
R^1 = Me, nPr, or Ph; R^2 = H or Me; R^3 = H or Me
R^4 = H or alkyl; R^5 = H or alkyl

Scheme 9.8 Cp$_2$Ti(CO)$_2$-catalyzed cyclization of 1,6-enynes.

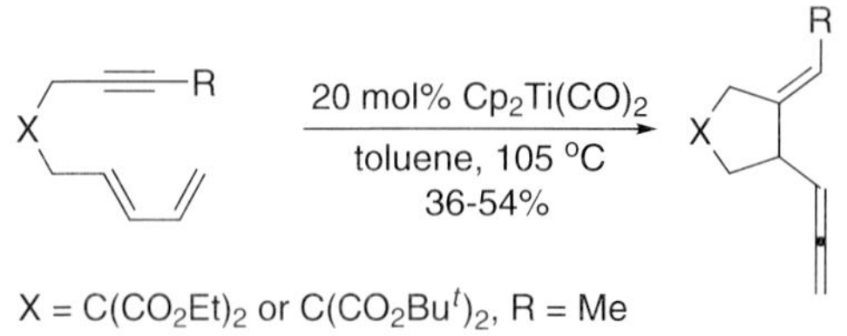

9 Cyclizations Based on Cyclometallation

Scheme 9.9 Cp$_2$Ti(CO)$_2$-catalyzed cyclization of alka-1,3(E)-dien-8-ynes.

Reaction conditions: 20 mol% Cp$_2$Ti(CO)$_2$, toluene, 105 °C, 36–54%.

X = C(CO$_2$Et)$_2$ or C(CO$_2$Bu^t)$_2$, R = Me
X = NPh, R = n-C$_5$H$_{11}$

When alka-1,3(E)-dien-8-ynes were applied in the above reaction, interestingly, the allene moiety was formed via β-H elimination (Scheme 9.9) [5].

The C–C double bond closer to the tether X and the C–C triple bond in 1,2-octadien-7-yne derivatives undergo highly regioselective cyclometallation with the in situ generated (η^2-propene)Ti(OPr^i)$_2$ to afford 1-alkylidene-2-alkenylcyclopentane derivatives after treatment with HCl. When the optically active reactant with an axial chirality in the allene moiety was applied in this reaction, the axial chirality of the allene moiety was nicely transferred into the center chirality of the product (Scheme 9.10) [6].

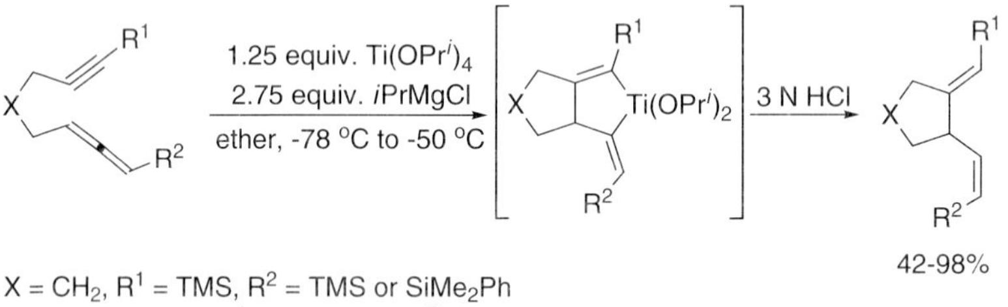

Reaction conditions: 1.25 equiv. Ti(OPr^i)$_4$, 2.75 equiv. iPrMgCl, ether, −78 °C to −50 °C; then 3 N HCl. 42–98%.

X = CH$_2$, R^1 = TMS, R^2 = TMS or SiMe$_2$Ph
X = C(CH$_2$OBn)$_2$, R^1 = H or TMS, R^2 = H or n-C$_5$H$_{11}$
When R^2 = n-C$_5$H$_{11}$, the E-isomer was also formed.

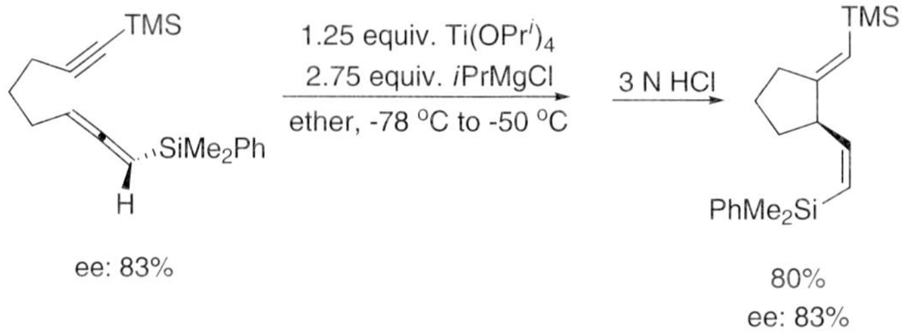

Starting material: TMS, SiMe$_2$Ph, H; ee: 83%

Reaction: 1.25 equiv. Ti(OPr^i)$_4$, 2.75 equiv. iPrMgCl, ether, −78 °C to −50 °C; then 3 N HCl

Product: TMS, PhMe$_2$Si; 80%, ee: 83%

Scheme 9.10 Ti(OPr^i)$_4$-promoted cyclization of 1,2-octadien-7-yne derivatives.

When 1,2-heptadien-6-yne derivatives were applied in the above reaction, 1,2-bisalkylidenecyclopentane derivatives were formed upon further reaction with H$^+$ or aldehydes/ketones. An efficient chirality transfer was also observed in the reaction of chiral reactants with aldehyde, ketone, or amine. In all these cases, the reaction with electrophiles occurred at the γ-position of the allylic titanium moiety in the metallabicyclic intermediate while the sp^2-C–Ti bond was protonated (Scheme 9.11) [6].

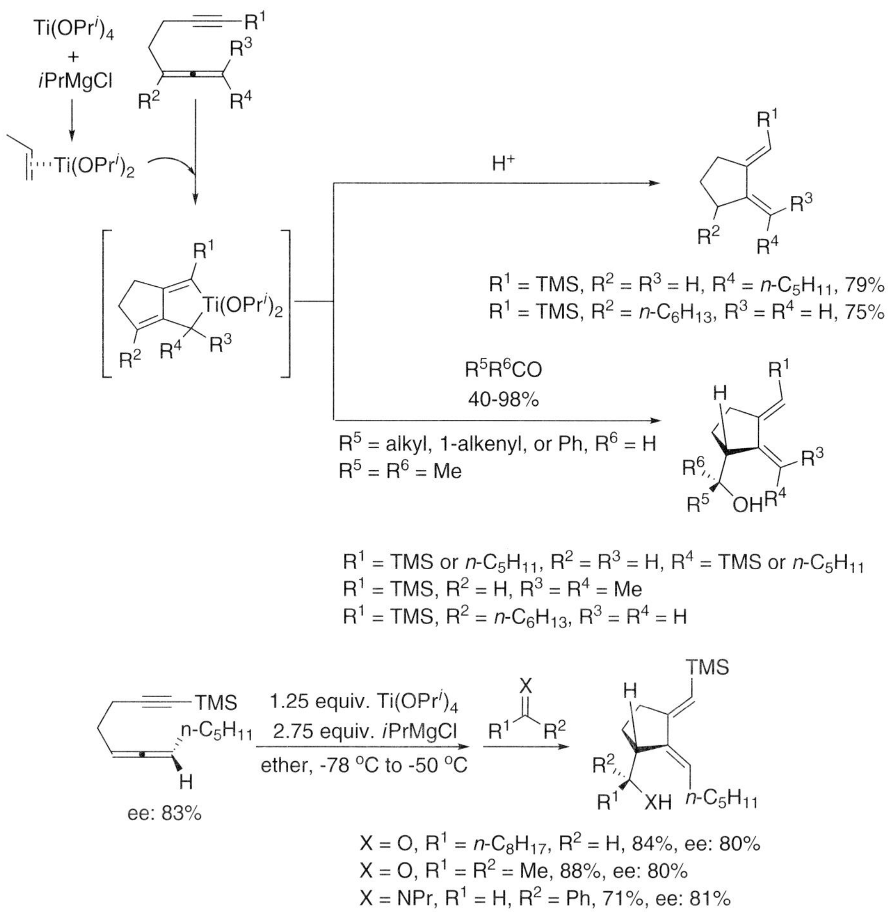

Scheme 9.11 Ti(OPr^i)$_4$-promoted cyclization of 1,2-heptadien-6-yne derivatives.

(η^2-Propene)Ti(OPr^i)$_2$ also promotes the reaction of trideca-2,7-diynyl acetate to afford 1-vinylidene-2-((E)-alkylidene)cyclopentane via cyclometallation, β-OAc elimination, and protonolysis (Eq. 9.1) [3].

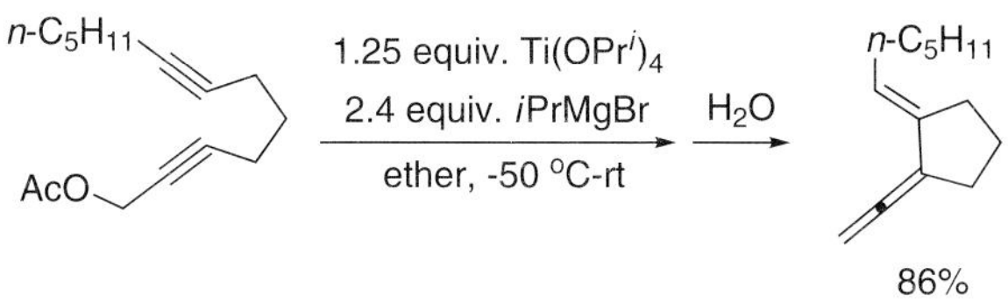

In the presence of iPrMgCl, the Ti(OPr^i)$_4$-promoted reaction of 1,6- or 1,7-diynes afforded a titanabicyclic intermediate which further reacts with PhPCl$_2$ to produce 2-phenyl-2,4,5,6-tetrahydrocyclopenta[c]phosphole, 5-phenyl-3,5-dihydro-1H-phospholo[3,4-c]furan, or 2-phenyl-4,5,6,7-tetrahydro-2H-isophosphindole derivatives (Scheme 9.12) [7].

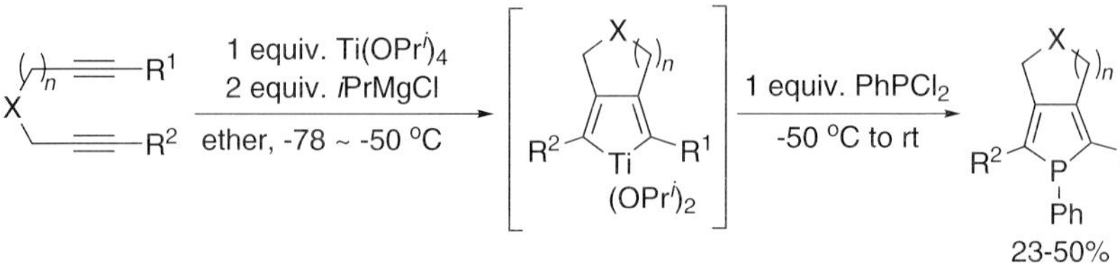

n = 1; X = CH$_2$ or O; R^1 = CO$_2$Et; R^2 = CO$_2$Me, CO$_2$Et, Ph, 2-pyridyl, or 2-thienyl;
n = 2; X = CH$_2$; R^1 = R^2 = CO$_2$Et

Scheme 9.12 Ti(OPr^i)$_4$-promoted reaction of 1,6- or 1,7-diynes with PhPCl$_2$.

9.3
Zirconium-Catalyzed Cyclizations [8]

The Zr-catalyzed reaction of hepta-1,6-diene derivatives in the presence of 1.5 equiv. of nBu$_2$Mg or 3.0 equiv. of nBuMgCl in ether afforded a mixture of trans- and cis-1,2-dimethylcyclopentane derivatives with the trans-isomer being the major product. The reaction of octa-1,7-diene in toluene gave similar stereoselectivity, however, when it was carried out in ether, the cis-isomer became the major product (Scheme 9.13) [9].

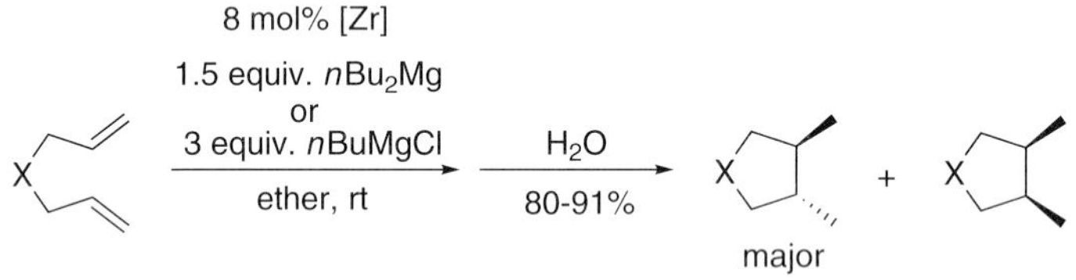

[Zr] = 2,2'-biphenyl-bis(3,4-dimethylcyclopentadienyl)zirconium dichloride

X = CH$_2$, CH(TBDMSO), NBu^t, SiMe$_2$, or

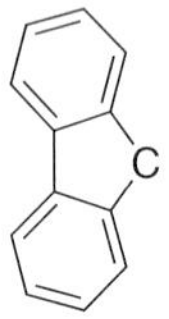

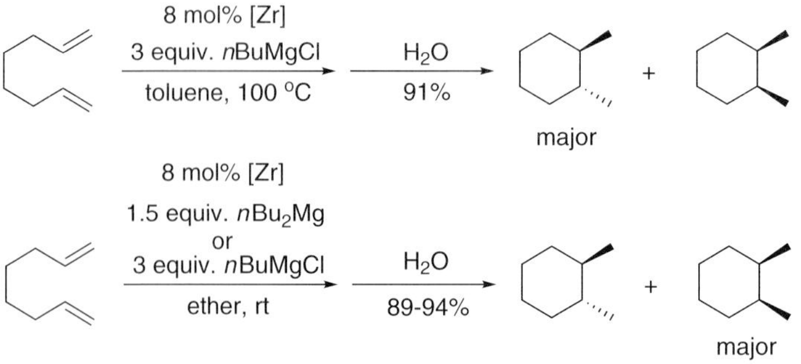

[Zr] = 2,2'-biphenyl-bis(3,4-dimethylcyclopentadienyl)zirconium dichloride

Scheme 9.13 Zr-catalyzed cyclization of 1,6- or 1,7-dienes.

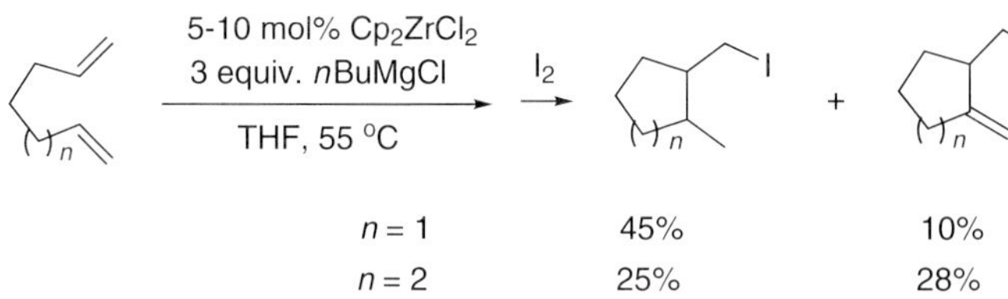
5-10 mol% Cp2ZrCl2
3 equiv. nBuMgCl
THF, 55 °C
I2
n = 1 45% 10%
n = 2 25% 28%

Scheme 9.14 Cp_2ZrCl_2-catalyzed reaction of 1,6- or 1,7-dienes with I_2.

When I_2 was applied instead of water in the above reaction, a mixture of 1-(iodomethyl)-2-methyl or methylene cyclopentane or cyclohexane was afforded (Scheme 9.14) [9a].

(*S*)-(EBTHI)Zr(BINOL) catalyzes the reaction of nitrogen-tethered 1,6- or 1,7-dienes in the presence of nBuMgCl to afford pyrrolidine or piperidine derivatives with high ee values (Scheme 9.15). However, a mixture of diastereomeric products is formed in most cases [10].

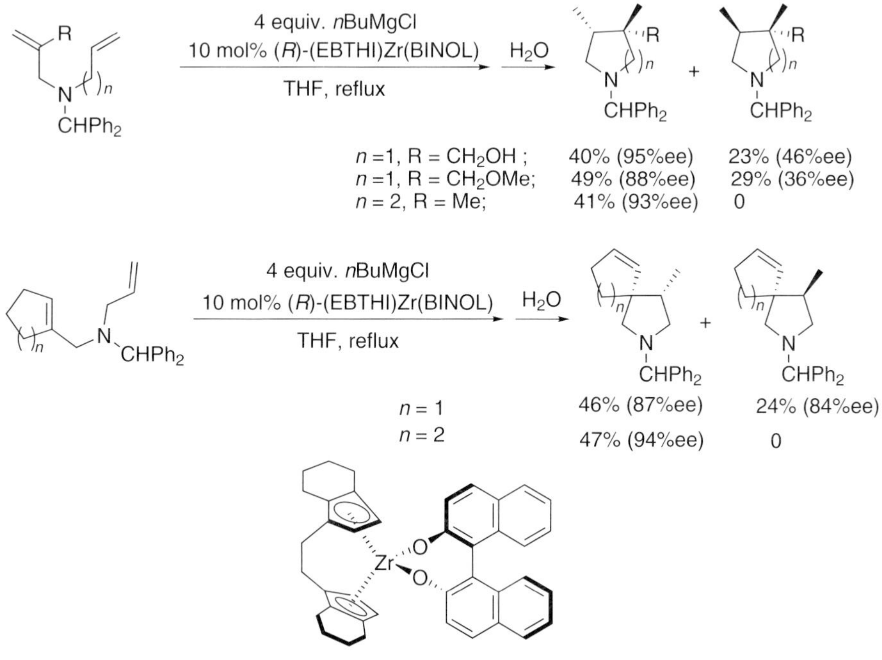
4 equiv. nBuMgCl
10 mol% (R)-(EBTHI)Zr(BINOL)
THF, reflux
H2O
CHPh2
n =1, R = CH2OH ; 40% (95%ee) 23% (46%ee)
n =1, R = CH2OMe; 49% (88%ee) 29% (36%ee)
n = 2, R = Me; 41% (93%ee) 0
n = 1 46% (87%ee) 24% (84%ee)
n = 2 47% (94%ee) 0

(*S*)-(EBTHI)Zr(BINOL)

Scheme 9.15 (*S*)-(EBTHI)Zr(BINOL)-catalyzed enantioselective cyclization of *N*-tethered 1,6- or 1,7-dienes.

The zirconocene-1-butene complex generated *in situ* reacts stoichiometrically with 2-bromohepta-1,6-diene derivatives to yield the 1-methyl-2-methylenecyclopentane derivatives via cyclometallation, β-Br elimination, and protonolysis (Scheme 9.16) [11].

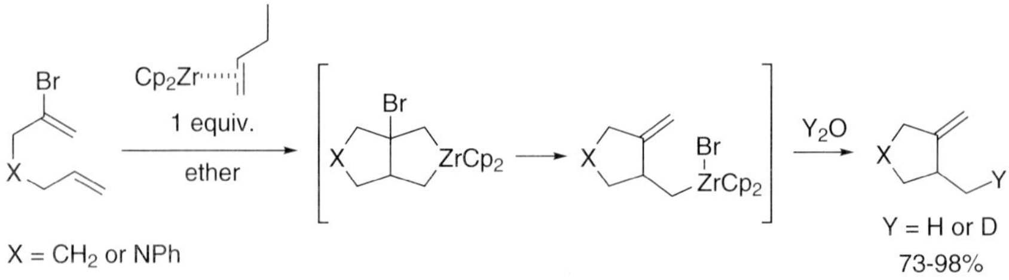

Scheme 9.16 Zr-promoted cyclization of 2-bromohepta-1,6-diene derivatives.

Zr-mediated reaction of 1,6- or 1,7-enynes may also afford bicyclic zirconacyclopentene species. This intermediate may undergo further transformation such as iodination, protonalysis, or transmetallation/oxidation to afford different products (Scheme 9.17) [12].

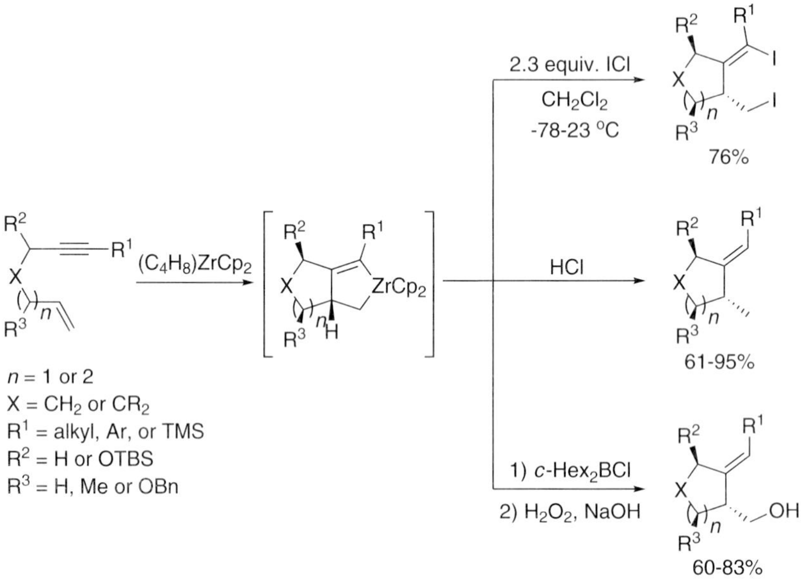

Scheme 9.17 Zr-mediated reaction of 1,6- or 1,7-enynes.

These *in situ* generated bi- or tri-cyclic zirconacycles from 1,6-dienes, 1,6-enynes, or 1,6-diynes with a stoichiometric amount of zirconocene-1-butene complex react with α-lithiated allylic chlorides to afford -ate complexes. Subsequent alkylative or alkenylative migration–LiCl elimination affords σ,η³-zirconacycles, which can be hydrolyzed by water to afford cyclopentane, pyrrolidine, or octahydro-1*H*-indole derivatives. If PhCHO is added in the presence of $BF_3 \cdot Et_2O$ before hydrolysis, an alcohol is generated from the η³-allyl Zr moiety (Scheme 9.18) [13].

The (zirconocene-1-butene)-mediated reaction of 1-bromooct-1(*E*)-en-7-yne derivatives afforded 1-vinylic cyclohexene derivatives via cyclometallation, rearrangement via the arrow shown in Scheme 9.20, and protonolysis (Scheme 9.19) [14].

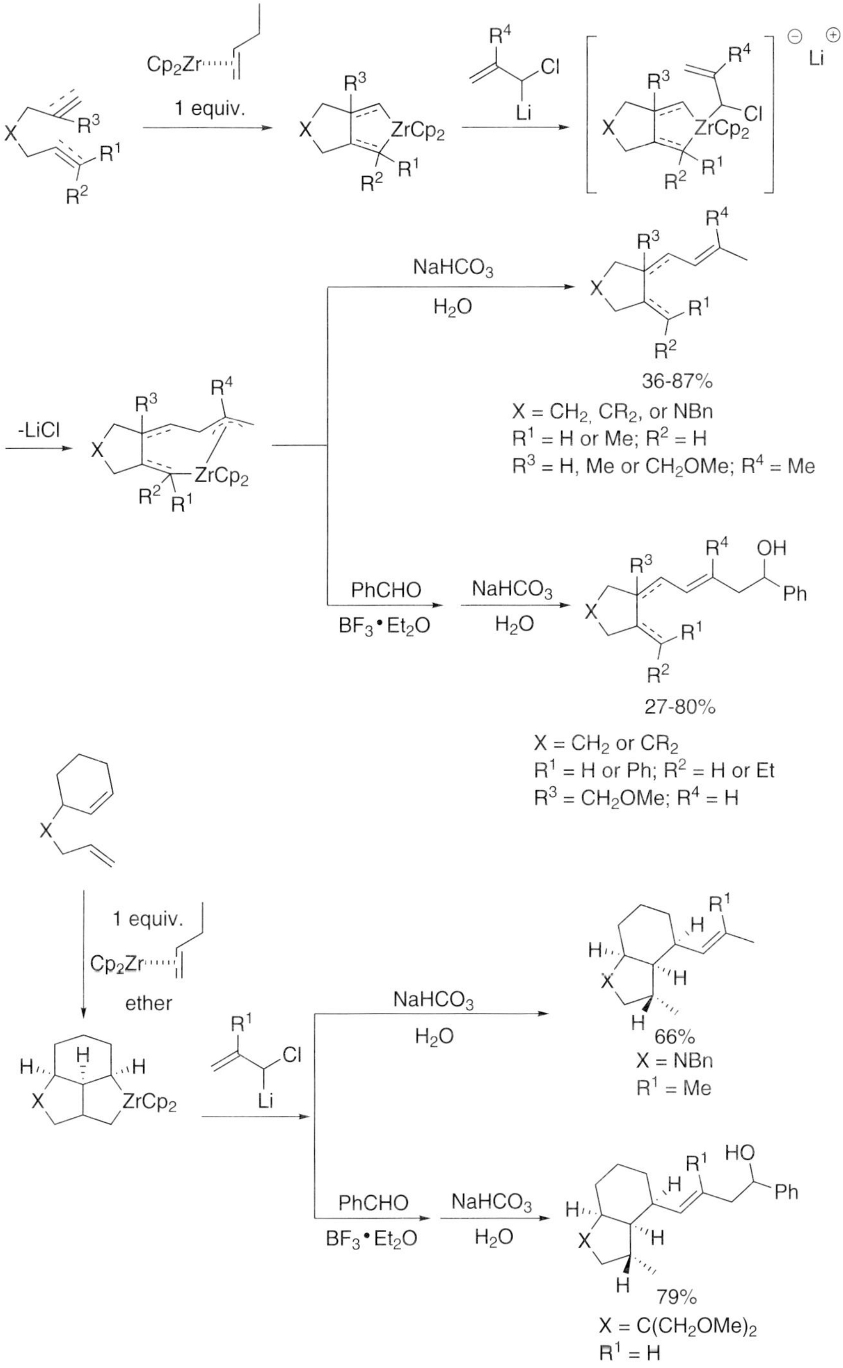

Scheme 9.18 Zr-mediated reaction of 1,6-dienes, 1,6-enynes, or 1,6-diynes.

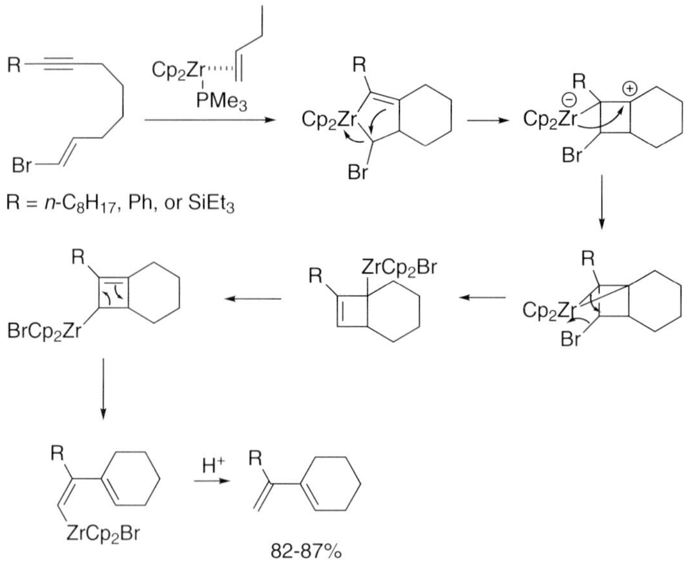

Scheme 9.19 Zr-mediated reaction of 1-bromooct-1(E)-en-7-yne derivatives.

Zirconocene catalyzes cyclobutene formation from 1-alkynyl chlorides and an excess of EtMgBr. In the presence of a conjugated C=C bond, only the alkyne moiety reacts. 1-Bromo- or 1-iodoalkynes do not give cyclobutenes under catalytic conditions, however, they undergo this reaction under the promotion of a stoichiometric zirconocene reagent. Furthermore, 1-hexyl-2-iodocyclobutene is afforded when the reaction of 1-chlorooct-1-yne with 1.25 equiv. of Cp_2ZrEt_2 is quenched with I_2 (Scheme 9.20) [15].

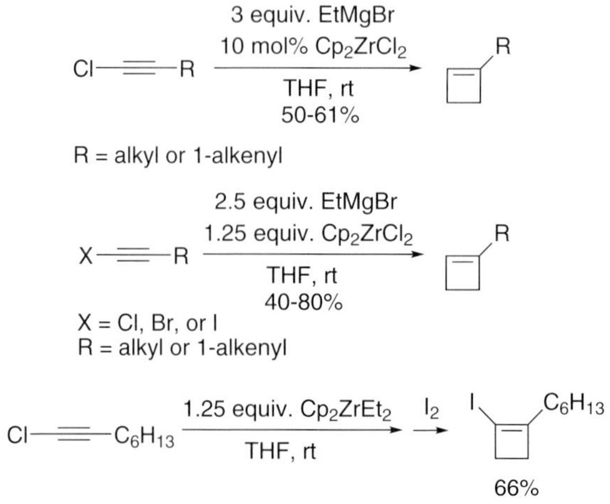

Scheme 9.20 Zr-catalyzed or promoted cyclobutene formation from 1-haloalkynes.

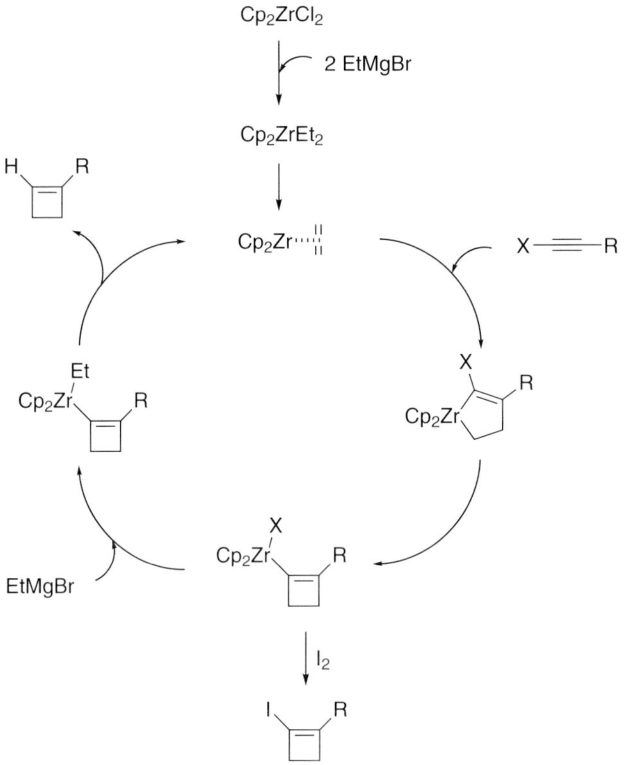

Scheme 9.21 The mechanism for Zr-catalyzed or promoted cyclobutene formation from 1-haloalkynes.

The mechanism is shown in Scheme 9.21. The zirconocene–ethylene complex generated *in situ* reacts with haloalkyne to form the α-halozirconacyclopentene intermediate, which undergoes a ring-closing reaction to give the cyclobutenyl zirconocene halide species via reductive elimination of the zirconacyclopentene intermediate and the subsequent oxidative addition of the cyclobutenyl halide moiety to zirconocene. The cyclobutenyl zirconocene halides formed react with EtMgBr to afford the cyclobutenyl ethyl zirconocene, which undergoes β-H abstraction and reductive elimination to give the cyclobutene and regenerates the zirconocene–ethylene complex. Upon treatment with I_2, cyclobutenyl iodide was formed [15].

9.4
Ruthenium-Catalyzed Cyclizations [1d, 8b, 16]

Ruthenium-catalyzed intramolecular cycloisomerization of 1,6-dienes afforded 1-methyl-2-methylenecyclopentanes efficiently (Scheme 9.22) via a cyclometallation, β-H elimination, and reductive elimination process [17].

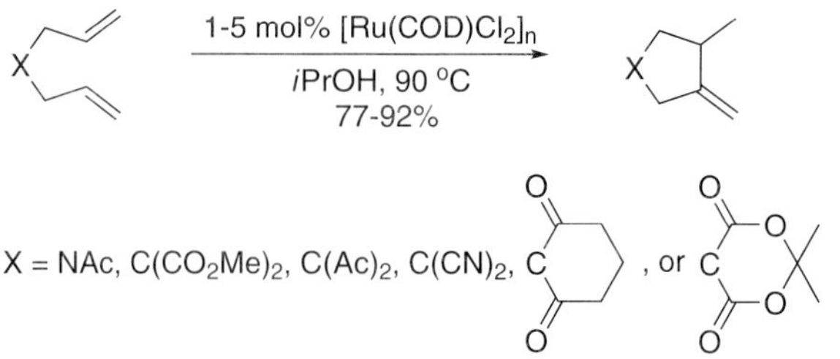

Scheme 9.22 [Ru(COD)Cl$_2$]$_2$-catalyzed cyclization of 1,6-dienes.

X = NAc, C(CO$_2$Me)$_2$, C(Ac)$_2$, C(CN)$_2$, C(cyclohexanedione), or C(dioxane-dimethyl)

In a similar mechanistic manner, the cationic ruthenium catalyst [CpRu(MeCN)$_3$]$^+$[PF$_6$]$^-$ catalyzed the cycloisomerization of 1,6- or 1,7-enynes to form the corresponding 1-alkylidene-2-alkenylcyclopentane or cyclohexane derivatives (Scheme 9.23) [18].

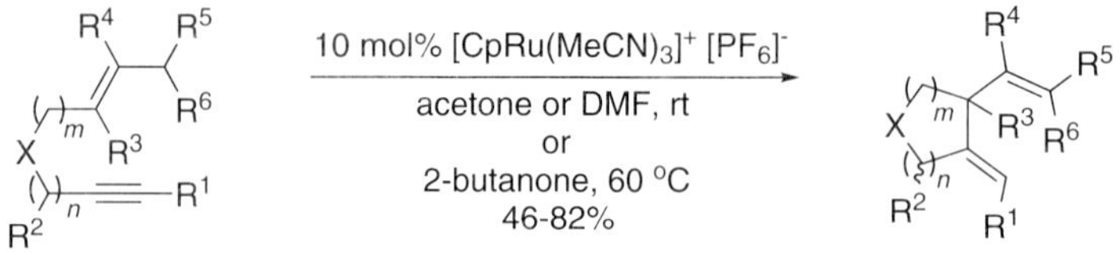

$n = m = 1$
X = O, NTs, CH$_2$, C(CO$_2$Me)$_2$, or C(SO$_2$Ph)$_2$
R^1 = H, CO$_2$Me, or CO$_2$Et
R^2 = H, Ph, PMBO, or TBSO
R^3 = H or Me
R^4 = H or Me
R^5 = H or alkyl
R^6 = H or alkyl

$n = 2, m = 1$
X = NTs or C(CO$_2$Et)$_2$
R^1 = R^2 = R^3 = R^5 = R^6 = H
R^4 = H or Me
$n = 1, m = 2$
X = C(SO$_2$Ph)$_2$
R^1 = R^2 = R^3 = R^5 = R^6 = H
R^4 = Me

Scheme 9.23 [CpRu(MeCN)$_3$]$^+$[PF$_6$]$^-$-catalyzed cyclization of 1,6- or 1,7-enynes.

Interestingly, a Cp*Ru(COD)Cl-catalyzed intramolecular cyclization of 1,6-enynes in the presence of ethylene afforded allylidenecyclopentane derivatives (Scheme 9.24) via a sequential cyclometallation, intermolecular insertion into the C–C double bond of ethylene, β-H elimination, and reductive elimination [19].

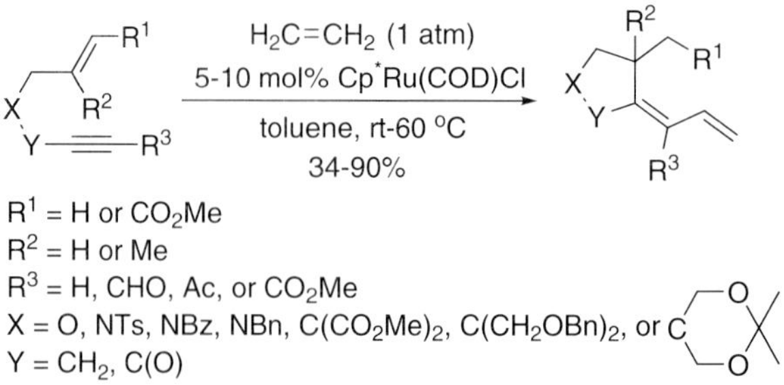

R^1 = H or CO$_2$Me
R^2 = H or Me
R^3 = H, CHO, Ac, or CO$_2$Me
X = O, NTs, NBz, NBn, C(CO$_2$Me)$_2$, C(CH$_2$OBn)$_2$, or C(dioxane-dimethyl)
Y = CH$_2$, C(O)

Scheme 9.24 Cp*Ru(COD)Cl-catalyzed cyclization of 1,6-enynes.

Sato and Mori *et al.* observed a new type of ruthenium-catalyzed cyclization of 1,6-enynes having a keto-carbonyl group on the alkyne moiety in the presence of ethylene affording 1-(1-(2-methylcyclopent-1-enyl)cyclopropyl) ketone derivatives (Scheme 9.25) [19b].

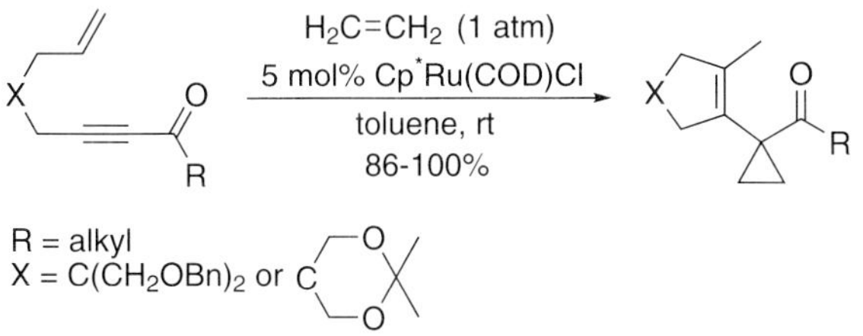

Scheme 9.25 Cp*Ru(COD)Cl-catalyzed cyclization of 6-en-1-ynyl ketones in the presence of ethylene.

Mechanistically, coordination of the carbonyl oxygen to the ruthenium metal in the ruthenacyclopentene intermediate formed by cyclometallation causes the formation of a ruthenium carbene intermediate, which is followed by [2 + 2]-cycloaddition of the [Ru=C] and ethylene and reductive elimination to form the cyclopropane ring (Scheme 9.26) [19b].

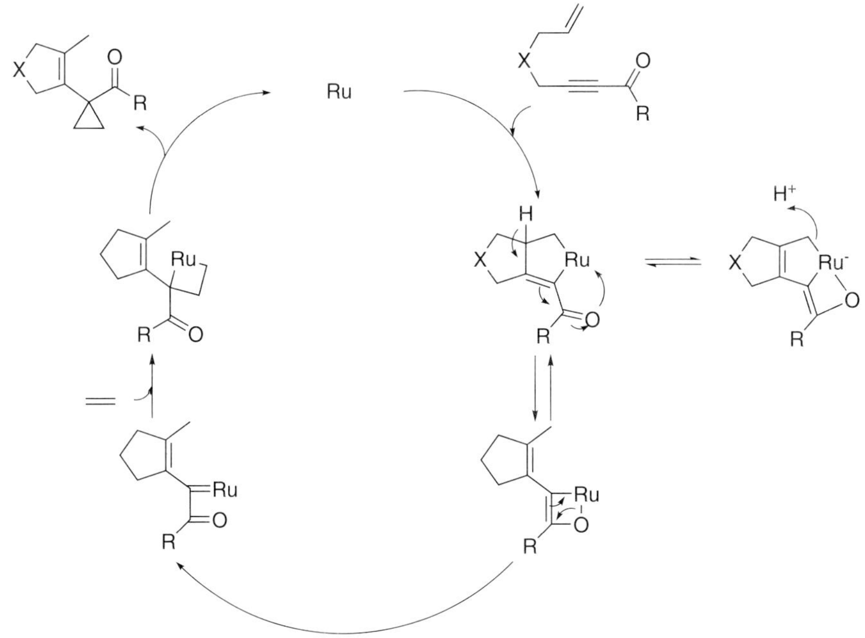

Scheme 9.26 The mechanism for Cp*Ru(COD)Cl-catalyzed cyclization of 6-en-1-ynyl ketones in the presence of ethylene.

Trost et al. reported that the cationic ruthenium catalyst [CpRu(MeCN)$_3$]$^+$ [PF$_6$]$^-$ could also catalyze the cyclization of 1,6-diynes in the presence of water to afford the cyclopent-1-enyl ketone derivatives. The ketone functionality was formed at the more sterically hindered side. If the steric differences of the two alkyne moieties are not significant, a mixture of regioisomers is formed. This reaction could also be extended to 1,7-diynes (Scheme 9.27) [20].

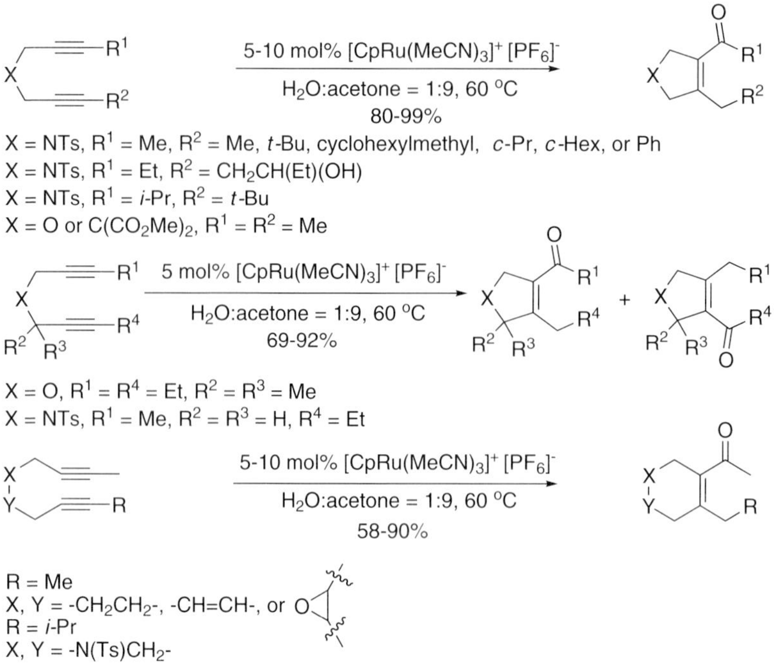

X−C≡C−R^1, X−C≡C−R^2 →[5-10 mol% [CpRu(MeCN)$_3$]$^+$ [PF$_6$]$^-$][H$_2$O:acetone = 1:9, 60 °C; 80-99%] product

X = NTs, R^1 = Me, R^2 = Me, t-Bu, cyclohexylmethyl, c-Pr, c-Hex, or Ph
X = NTs, R^1 = Et, R^2 = CH$_2$CH(Et)(OH)
X = NTs, R^1 = i-Pr, R^2 = t-Bu
X = O or C(CO$_2$Me)$_2$, R^1 = R^2 = Me

→[5 mol% [CpRu(MeCN)$_3$]$^+$ [PF$_6$]$^-$][H$_2$O:acetone = 1:9, 60 °C; 69-92%]

X = O, R^1 = R^4 = Et, R^2 = R^3 = Me
X = NTs, R^1 = Me, R^2 = R^3 = H, R^4 = Et

→[5-10 mol% [CpRu(MeCN)$_3$]$^+$ [PF$_6$]$^-$][H$_2$O:acetone = 1:9, 60 °C; 58-90%]

R = Me
X, Y = -CH$_2$CH$_2$-, -CH=CH-, or O⟨⟩
R = i-Pr
X, Y = -N(Ts)CH$_2$-

Scheme 9.27 [CpRu(MeCN)$_3$]$^+$ [PF$_6$]$^-$-catalyzed cyclization of 1,6- or 1,7-diynes in the presence of water.

This reaction may proceed via cyclometallation, nucleophilic attack by the hydroxyl group in water, and protonation (Scheme 9.28) [20, 21].

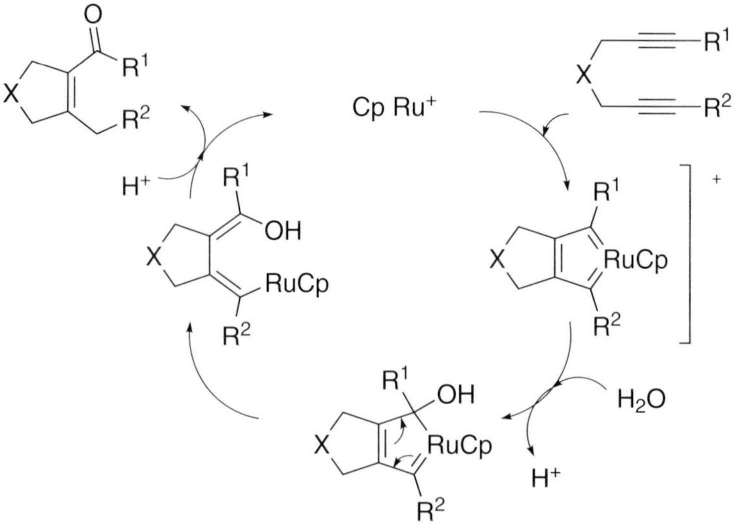

Scheme 9.28 The mechanism for [CpRu(MeCN)$_3$]$^+$ [PF$_6$]$^-$-catalyzed cyclization of 1,6-diynes in the presence of water.

9.4 Ruthenium-Catalyzed Cyclizations

When the above reaction was carried out in the presence of MeOH, the 1,3-dienol ethers were afforded as the single products in excellent yields highly stereoselectively. When N-(but-2-ynyl)-N-(6-hydroxyhex-2-ynyl)-4-methylbenzenesulfonamide was applied in this reaction in acetone, the hydroxyl group in the substrate behaved in a manner similar to that of MeOH to give the cyclic ether as the only product (Scheme 9.29) [20].

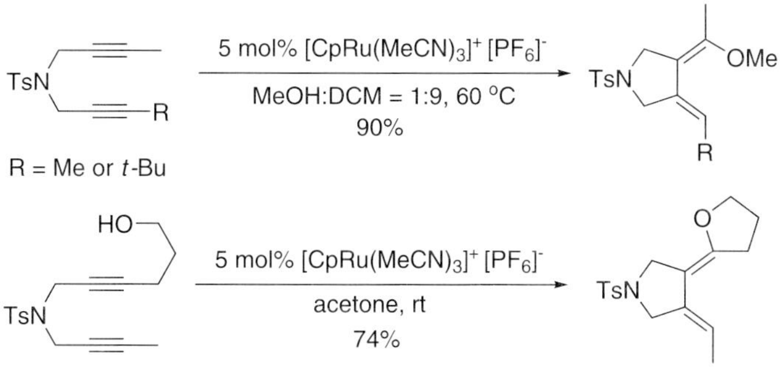

Scheme 9.29 $[CpRu(MeCN)_3]^+ [PF_6]^-$-catalyzed cyclization of 1,6-diynes in the presence of MeOH or an intramolecular hydroxyl group.

When octa-2,7-diyn-1-ol or nona-2,8-diyn-1-ol derivatives were applied in the above reaction, 1-acyl-2-alkenylcyclopentene or cyclohexene derivatives were formed (Scheme 9.30) via cyclometallation, dehydration forming the required water, nucleophilic attack by the water generated *in situ*, β-H elimination, and reductive elimination (Scheme 9.31) [22].

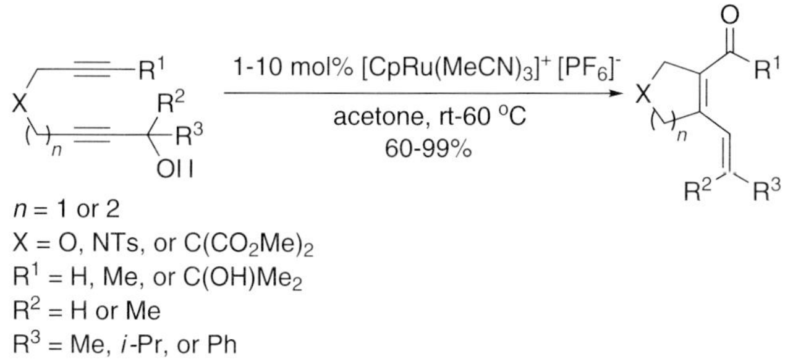

$n = 1$ or 2
$X = O$, NTs, or $C(CO_2Me)_2$
$R^1 = H$, Me, or $C(OH)Me_2$
$R^2 = H$ or Me
$R^3 = Me$, *i*-Pr, or Ph

Scheme 9.30 $[CpRu(MeCN)_3]^+ [PF_6]^-$-catalyzed cyclization of octa-2,7-diyn-1-ol or nona-2,8-diyn-1-ol derivatives.

When 9-hydroxy-9-methyldeca-2,7-diyn-4-yl acetate derivatives were applied in the above reaction, the acetoxy group was also eliminated to afford 1-alkenyl-2-acylcyclopenta-1,3-diene derivatives (Scheme 9.32) [22].

Hepta- or octa-1,6-diyn-3-ol derivatives could also undergo the above cyclization in the presence of malonic acid to give 2-formyl-3-alkylidenecyclopent-1-ene derivatives (Scheme 9.33) [22].

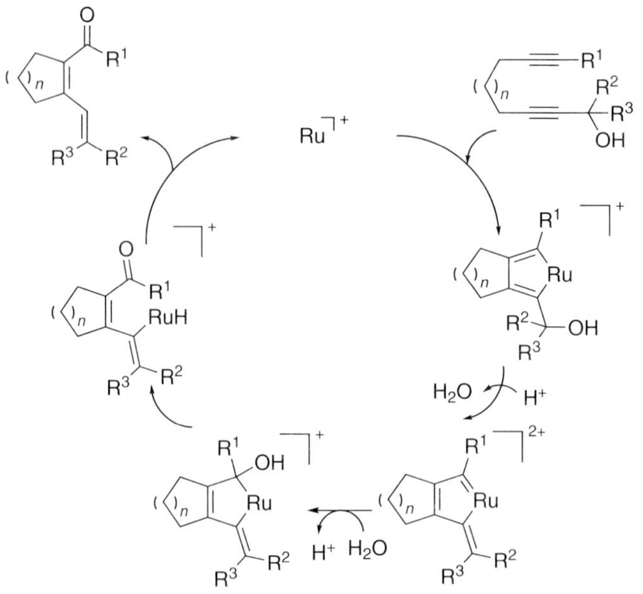

Scheme 9.31 The mechanism for $[CpRu(MeCN)_3]^+ [PF_6]^-$-catalyzed cyclization of octa-2,7-diyn-1-ol or nona-2,8-diyn-1-ol derivatives.

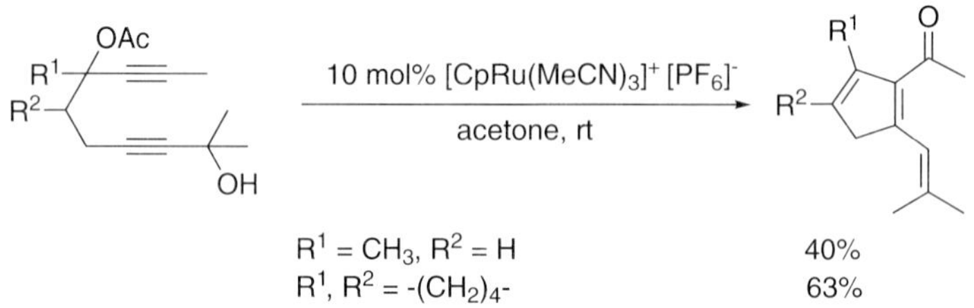

$R^1 = CH_3, R^2 = H$ — 40%
$R^1, R^2 = -(CH_2)_4-$ — 63%

Scheme 9.32 $[CpRu(MeCN)_3]^+ [PF_6]^-$-catalyzed cyclization of 9-hydroxy-9-methyldeca-2,7-diyn-4-yl acetate derivatives.

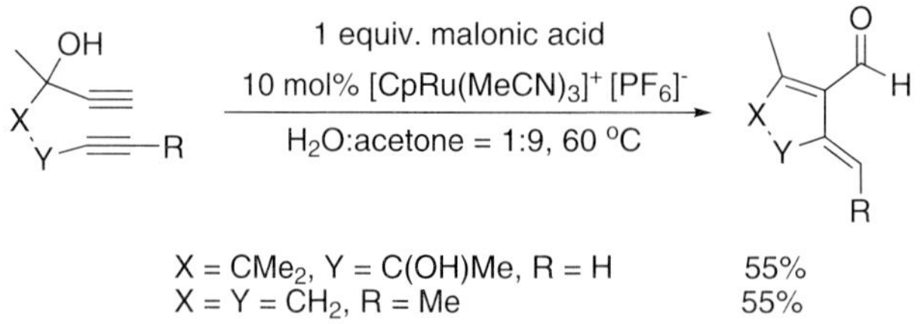

$X = CMe_2, Y = C(OH)Me, R = H$ — 55%
$X = Y = CH_2, R = Me$ — 55%

Scheme 9.33 $[CpRu(MeCN)_3]^+ [PF_6]^-$-catalyzed cyclization of hepta- or octa-1,6-diyn-3-ol derivatives in the presence of malonic acid.

9.5
Cobalt-Catalyzed Cyclizations [1d]

Cobalt-catalyzed diastereoselective intermolecular carbometallation of allenes and α,β-unsaturated enones formed the cis-3-alkylidenecyclopentanol derivatives with excellent chemo-, regio-, and stereo-selectivities (Scheme 9.34) [23].

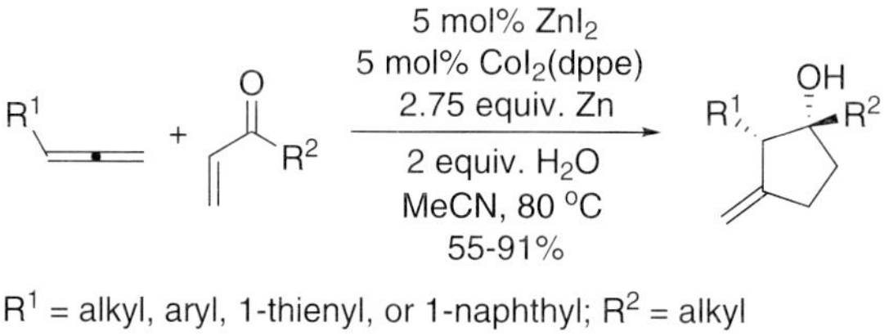

R^1 = alkyl, aryl, 1-thienyl, or 1-naphthyl; R^2 = alkyl

Scheme 9.34 Cobalt-catalyzed diastereoselective intermolecular carbometallation of allenes and enones.

The catalytic cycle is likely initiated by the reduction of Co(II) to Co(I) by zinc, followed by the highly regioselective cyclometallation of Co(I) with the relatively more electron-rich non-terminal C=C bond in the allene and the C=C bond in the enone to form the cobaltacyclopentane intermediate which isomerizes to the cyclic enolate species. Selective protonation generates the carbonyl coordinated Co(III) intermediate which subsequently undergoes highly diastereoselective carbonyl insertion into the cobalt–carbon bond to form the cobalt alkoxide. Subsequent protonation leads to the observed product and the resulting Co(III) is reduced to Co(I) again by zinc dust (Scheme 9.35) [23].

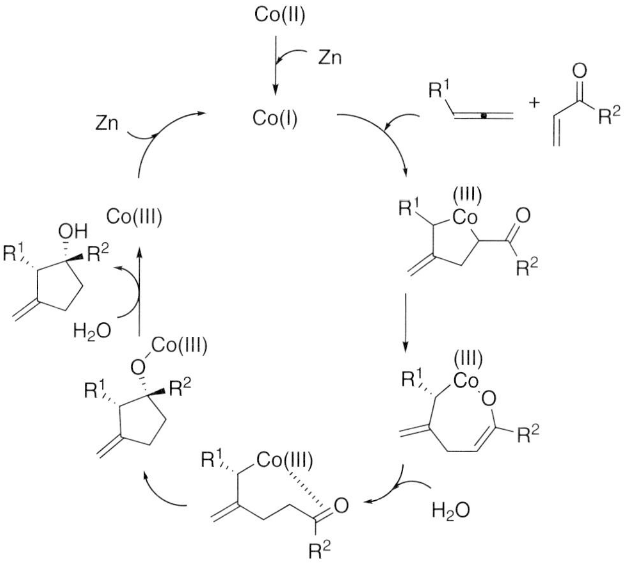

Scheme 9.35 The mechanism for cobalt-catalyzed diastereoselective intermolecular carbometallation of allenes and conjugated enones.

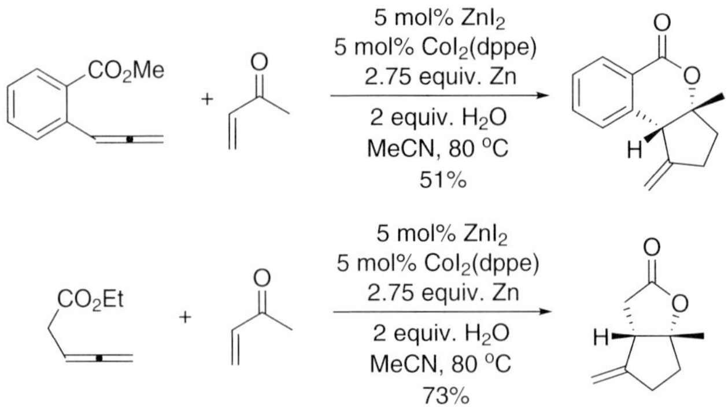

Scheme 9.36 Cobalt-catalyzed diastereoselective intermolecular carbometallation of 2-propadienyl benzoate or β-allenoate with but-3-en-2-one.

When the above conditions were applied to the reaction of 2-propadienyl benzoate or β-allenoate with but-3-en-2-one, the tricyclic benzopyranone derivative or bicyclic γ-lactone was afforded (Scheme 9.36) [23].

Co(CO)$_6$ promoted the cyclization reaction of 1,6- or 1,7-enynes to afford the corresponding 1-alkylidene-2-((E)-alkylidene)cyclopentane or cyclohexane derivatives (Scheme 9.37) [24].

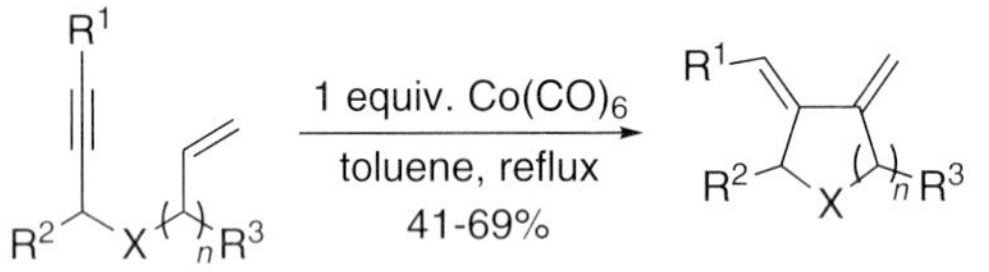

n = 1
X = CH$_2$ or CHR
R^1 = alkyl or TMS
R^2 = H or OH
R^3 = H, alkyl, or OTBS

n = 2
X = NTs
R^1 = Me
R^2 = R^3 = H

Scheme 9.37 Co(CO)$_6$-promoted cyclization of 1,6- or 1,7-enynes.

Stoichiometric CpCo(CO)$_2$ also promoted the cyclization of 1,6-enynes. However, this reaction was highly substrate-dependent. If dimethyl 2-(but-2(E)-enyl)-2-(3-phenylprop-2-ynyl)malonate was used, only 1(E)-propylidene-2(E)-benzylidenecyclopentane could be afforded via cyclometallation, β-H elimination, and reductive elimination. However, the reaction of dimethyl 2-(but-2(E)-enyl)-2-(4,4-dimethylpent-2-ynyl)malonate gave the cyclopentadiene derivative via C−C double bond isomerizations. Furthermore, when dimethyl 2-(but-2(E)-enyl)-2-(3-(trimethylsilyl)prop-2-ynyl)malonate was applied, a mixture of these two types of products was formed (Scheme 9.38) [25].

For the reaction of dimethyl 2-(2-methylallyl)-2-(3-phenylprop-2-ynyl)malonate, product **1** was formed as the only product, possibly via cyclometallation and subsequent reductive elimination (Scheme 9.39) [25].

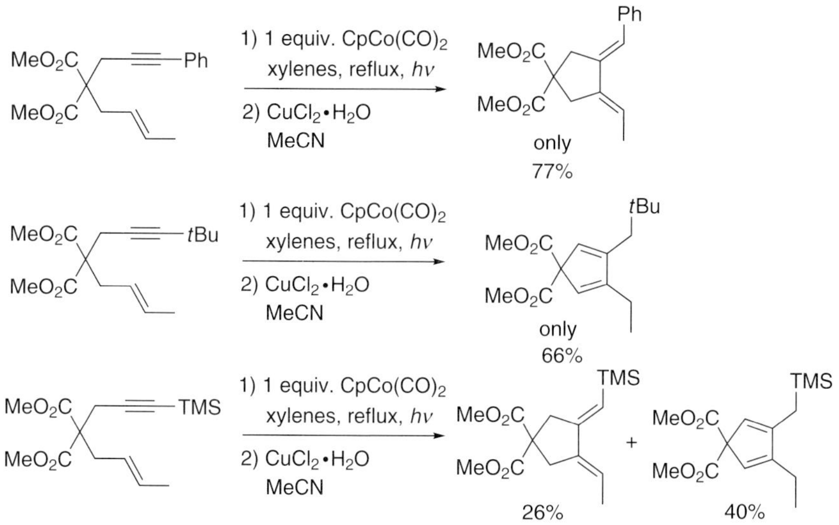

Scheme 9.38 CpCo(CO)$_2$-promoted cyclization of 1,6-enynes.

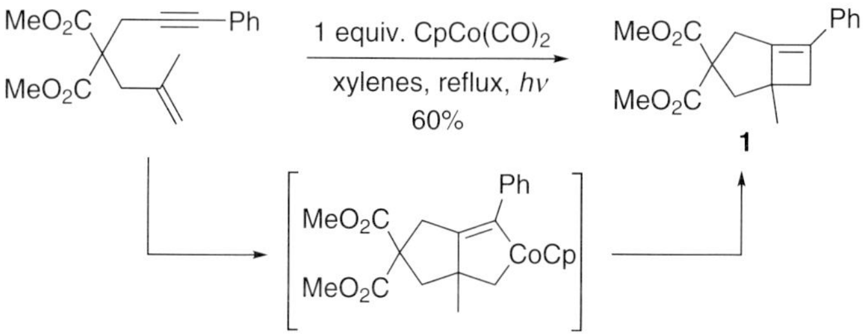

Scheme 9.39 CpCo(CO)$_2$-promoted cyclization of dimethyl 2-(2-methylallyl)-2-(3-phenylprop-2-ynyl)malonate.

When 1,7-enynes with a terminal C=C bond were applied under the above conditions, a mixture of 1-methylene-2-alkylidenecyclohexanes and 1 vinylcyclohex-1-enes was formed. The bicyclic cobaltacyclopentene intermediate undergoes β-H elimination and subsequent reductive elimination to give the 1-alkylidene-2-alkylidenecyclohexanes as the major products. Reductive elimination of the cyclometalled species resulted in the formation of intermediates, **2** followed by ring-opening isomerization to produce the 1-vinylcyclohex-1-enes as the minor products (Scheme 9.40) [25].

9.6
Rhodium-Catalyzed Cyclizations [1d, 16a, 16c, 26]

Rhodium catalyzed the asymmetric cyclodimerization of oxa- or aza-benzonorbornadienes to afford a tetrahydrofuran or pyrrolidine ring with excellent ee value with (R)-BINAP as the ligand (Scheme 9.41) [27].

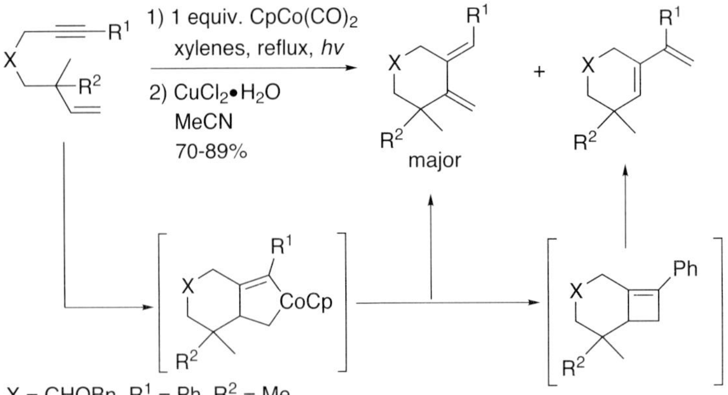

Scheme 9.40 CpCo(CO)$_2$-promoted cyclization of 1,7-enynes with a terminal C=C bond.

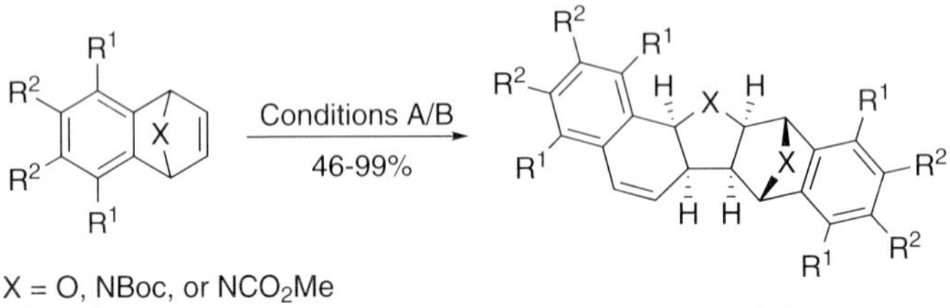

X = CHOBn, R^1 = Ph, R^2 = Me
X = C(CO$_2$Me), R^1 = Ph or TMS, R^2 = H

X = O, NBoc, or NCO$_2$Me
R^1 = R^2 = H, F, or -(CH$_2$)$_4$-
R^1 = H, R^2 = Me, OMe, F, or Br
R^1 = Me, R^2 = H or Br
Conditions A: 0.5 mol% [[RhCl((R)-BINAP)]$_2$ / 4 NaBAr$_4$],
  DCE, 40-80 °C or toluene 100 °C
Conditions B: 2 mol% [[RhCl(COD)]$_2$ / 2 (R)-BINAP / 2 AgBF$_4$],
  DCE, 60 °C

ee: 88-99%

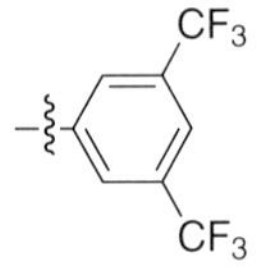

Scheme 9.41 Rh-catalyzed asymmetric cyclodimerization of oxa- or aza-benzonorbornadienes with (R)-BINAP as the ligand.

The proposed catalytic cycle begins with the Rh(I)-catalyzed cyclometallation of two molecules of reactants, which yields a polycyclic rhodacyclopentane intermediate. Subsequent β-oxygen elimination followed by reductive elimination to form the C—O bond yields the final products and regenerates the catalytic Rh(I) species (Scheme 9.42) [27].

Rhodium-catalyzed intramolecular cycloisomerization of allenenes formed 1-alkylidene-2-alkenylcyclopentane or cyclohexane derivatives when R^3 or R^4 does not contain a hydrogen atom available for β-H elimination. However, when 2-butenyl 4-methyl-2,3-pentadienyl malonate was used, 1,1-bis(methoxycarbonyl)-3-(2-methyl-

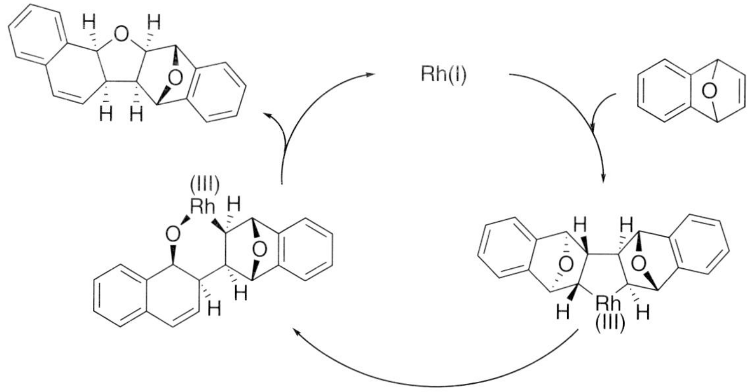

Scheme 9.42 The mechanism for Rh-catalyzed asymmetric cyclodimerization of oxabenzonorbornadienes.

propenyl)-4-vinylcyclopentane was formed indicating the preferred β-H elimination from the methyl group connected to the olefin moiety (Scheme 9.43) [28].

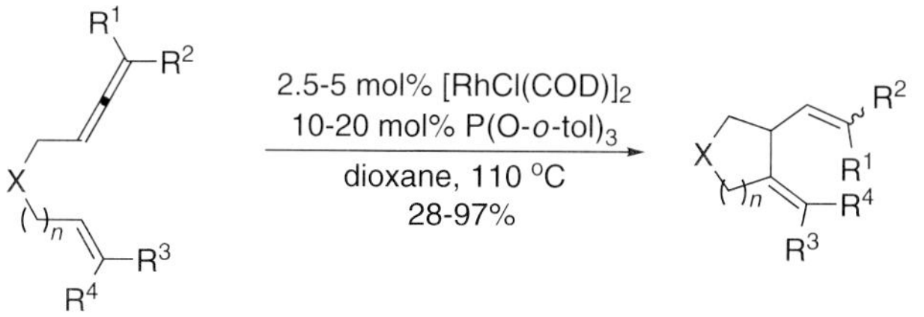

$n = 1$
$R^1 = R^2 = R^4 = H$, Me, or $-(CH_2)_5-$, $R^3 = H$, Ph, or TMS
$R^1 = R^3 = R^4 = H$, $R^2 = n-C_5H_{11}$ or t-Bu
$R^1 = R^2 = $ Me, $R^3 = H$, $R^4 = $ Ph

X = O, NTs, NBn, C(CO$_2$Me)$_2$, or [barbiturate structure]

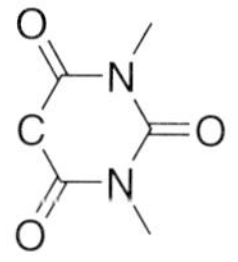

$n = 2$
$R^1 = R^2 = $ Me, $R^3 = R^4 = H$
X = C(CO$_2$Me)$_2$

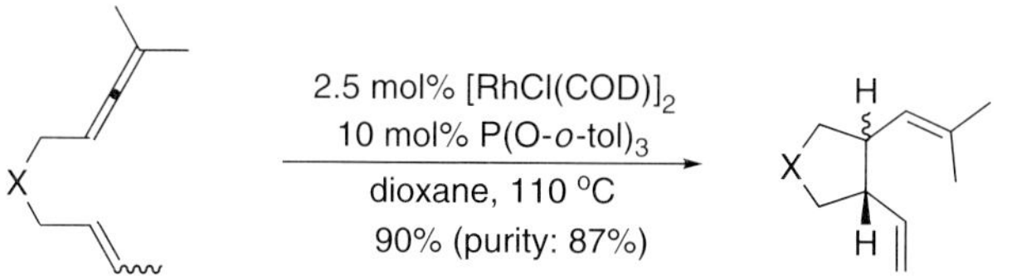

$E/Z = 3:1$
X = C(CO$_2$Me)$_2$

cis/trans = 2.8:1

Scheme 9.43 Rh-catalyzed intramolecular cycloisomerization of allenenes.

A *cis*-5,7-fused bicyclic system **3** was prepared by the Rh-catalyzed cyclometallation, β-decarbopalladation, and reductive elimination of 1-cyclopropyl-1,6-dienes. However, when 1-cyclopropanyl-1,7-diene was applied, a *trans*-fused bicyclo[5.4.0] product **4** was formed. The related reaction of 1-cyclopropanyl-1,6-enynes afforded bicyclo[5.3.0]octadienes (Scheme 9.44) [29].

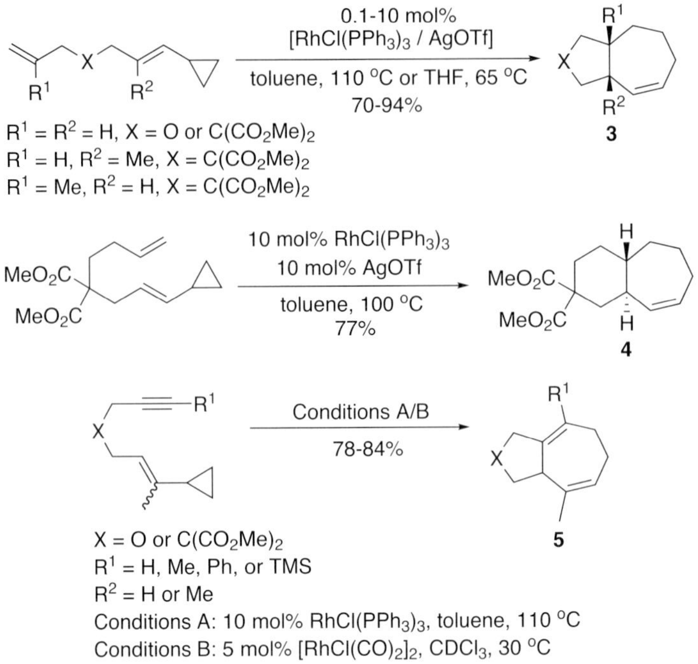
0.1-10 mol%
[RhCl(PPh3)3 / AgOTf]
toluene, 110 °C or THF, 65 °C
70-94%
3
R1 = R2 = H, X = O or C(CO2Me)2
R1 = H, R2 = Me, X = C(CO2Me)2
R1 = Me, R2 = H, X = C(CO2Me)2
10 mol% RhCl(PPh3)3
10 mol% AgOTf
toluene, 100 °C
77%
MeO2C
MeO2C
4
Conditions A/B
78-84%
5
X = O or C(CO2Me)2
R1 = H, Me, Ph, or TMS
R2 = H or Me
Conditions A: 10 mol% RhCl(PPh3)3, toluene, 110 °C
Conditions B: 5 mol% [RhCl(CO)2]2, CDCl3, 30 °C

Scheme 9.44 Rh-catalyzed cyclization of 1-cyclopropyl-1,6- or 1,7-dienes or 1-cyclopropyl-1,6-enynes.

Rhodium also catalyzed the asymmetric cross-cyclodimerization of oxabenzonorbornadienes and DMAD to yield the corresponding dihydronaphthofuran derivatives (Scheme 9.45) [27].

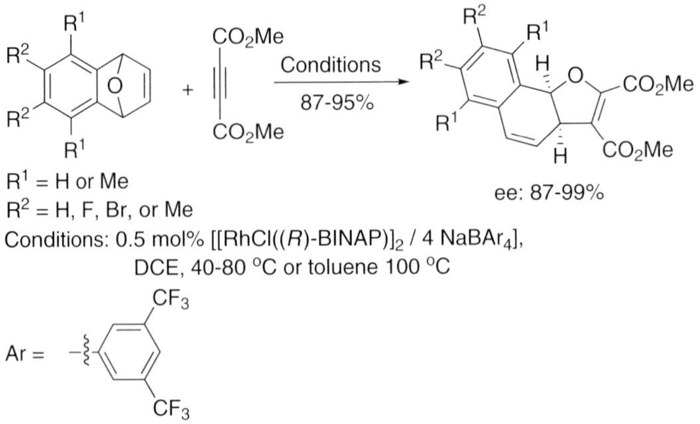
CO2Me
CO2Me
Conditions
87-95%
ee: 87-99%
R1 = H or Me
R2 = H, F, Br, or Me
Conditions: 0.5 mol% [[RhCl((R)-BINAP)]2 / 4 NaBAr4],
DCE, 40-80 °C or toluene 100 °C
Ar =
CF3
CF3

Scheme 9.45 Rh-catalyzed asymmetric cross-cyclodimerization of oxabenzonorbornadienes and DMAD.

9.6 Rhodium-Catalyzed Cyclizations

This reaction could also be extended to the intermolecular cyclization of 7-alkylidene-4,5-*trans*-bis(ethoxycarbonyl)spiro[2.4]heptane **6** with but-2-yne, yielding 1,2-*trans*-bis(ethoxycarbonyl)-substituted product **7** as the major product (Eq. 9.2) [30].

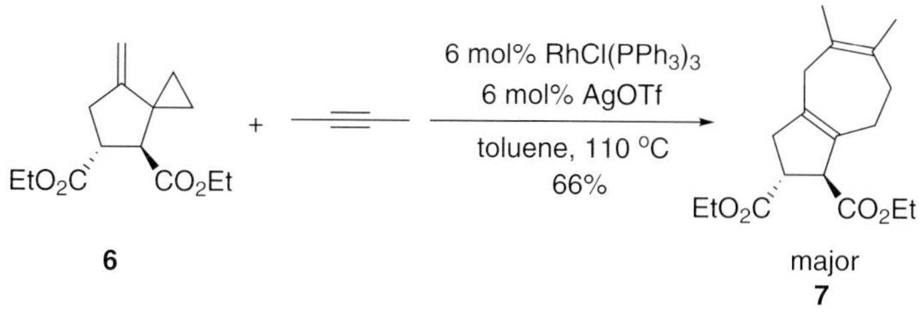

$$\text{6 mol\% RhCl(PPh}_3)_3,\ \text{6 mol\% AgOTf, toluene, 110 °C, 66\%}$$

6 + but-2-yne → **7** (major)

(9.2)

The reductive cyclization of 1,6-enynes with a terminal C=C bond was realized with H_2 (1 atm) under the catalysis of $Rh(COD)_2OTf$ to afford the corresponding alkylidene cyclopentanes highly stereoselectively (Scheme 9.46) via the sequential ligand exchange, cyclometallation, reductive elimination, oxidative addition with H_2, and reductive elimination (Scheme 9.47) [31].

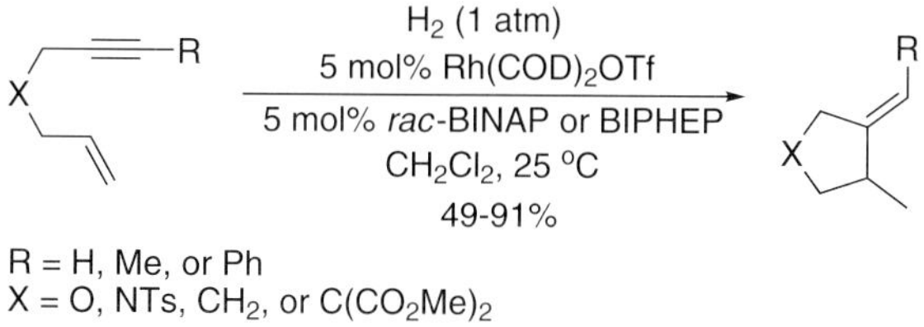

R = H, Me, or Ph
X = O, NTs, CH_2, or $C(CO_2Me)_2$

Scheme 9.46 $Rh(COD)_2OTf$-catalyzed reductive cyclization of 1,6-enynes with a terminal C=C bond under 1 atm of H_2.

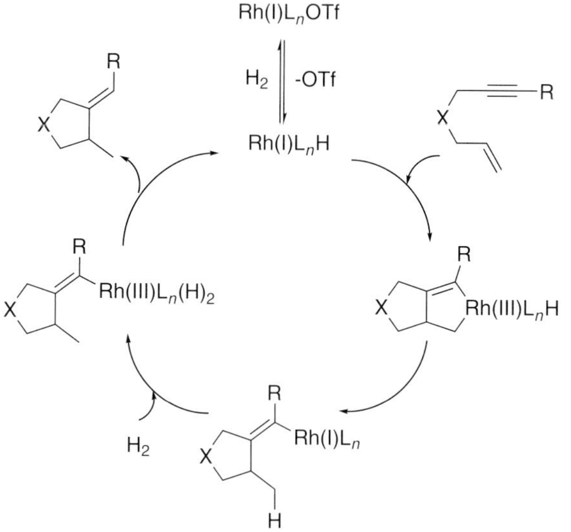

Scheme 9.47 The mechanism for $Rh(COD)_2OTf$-catalyzed reductive cyclization of 1,6-enynes with a terminal C=C bond under 1 atm of H_2.

The cationic Rh-catalyst [Rh(dppb)]$^+$, prepared *in situ* from the reaction of [Rh(dppb)Cl]$_2$ with AgSbF$_6$, catalyzes the cyclization of non-terminal 1(Z),6-enynes to afford 1-alkylidene-2-(1′E)-alkenyl cyclic products. However, the reaction of a 1(E),6-enyne, such as (E)-N-(but-2-ynyl)-N-(pent-2-enyl)benzenesulfonamide, afforded not only the 1-alkylidene-2-(E)-alkenyl pyrrolidine derivative as the major product, but also bisalkylidene tetrahydropyrrole as a by-product (Scheme 9.48) [32].

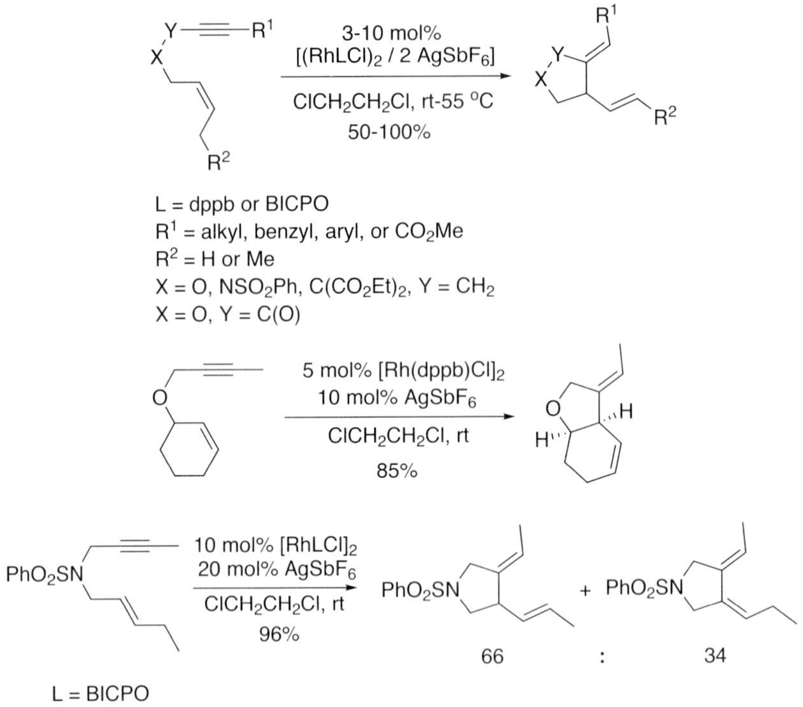

L = dppb or BICPO
R^1 = alkyl, benzyl, aryl, or CO$_2$Me
R^2 = H or Me
X = O, NSO$_2$Ph, C(CO$_2$Et)$_2$, Y = CH$_2$
X = O, Y = C(O)

L = BICPO

Scheme 9.48 Cationic Rh-catalyzed cyclization of non-terminal 1,6-enynes.

In 2000, Zhang et al. reported the enantioselective Rh-catalyzed cyclization of enyne with β-H connected to the carbon atom linked to R^3, affording 1-alkylidene-2-vinyl cyclic products. It is interesting to note that the reaction of 4-hydroxy-2(Z)-butenyl 2-alkynoates afforded α-(Z)-alkylidene-β-formylmethylbutyrolactones in >99% ee. The aldehyde functionality was produced via an enol intermediate formed by β-H elimination. In all cases, the R-ligand resulted in R-products and the S-ligand led to S-products (Scheme 9.49) [33].

Rhodium also catalyzes the kinetic resolution of 5-hydroxy-3(Z)-hexen-2-yl 2-butynyl ether affording the ketone and remaining starting material in >99% ee. All four diastereoisomers may be prepared by applying (R)- and (S)-BINAP (Scheme 9.50) [34].

In 2004, Mikami et al. reported the cationic [Rh((S,S)-skewphos)]$_2$(SbF$_6$)$_2$-catalyzed enantioselective cyclization of nitrogen tethered-1,6-enynes affording the similar

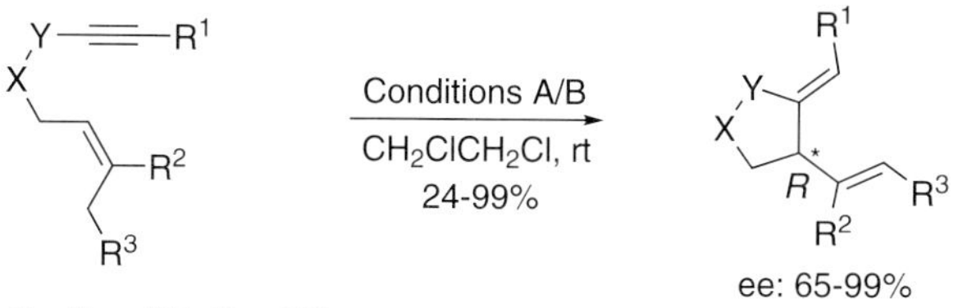

X = O or NR, Y = CH$_2$
X = O, Y = C(O)
R^1 = alkyl, aryl, C(O)R, or CO$_2$Et
R^2 = H or Me
R^3 = H, alkyl, alkoxy, OAc, or NRR

Conditions A: 5-10 mol% [(RhClL*)$_2$ / AgSbF$_6$], L* = (R,R)-Me-Duphos,
(R,R,R,R)-BICP, or (R,R,R,R)-BICPO;
Conditions B: 5 mol% [[Rh(COD)Cl]$_2$ / 2 (R)-BINAP / 4 AgSbF$_6$],
When (S)-BINAP was applied, the S-products were afforded.

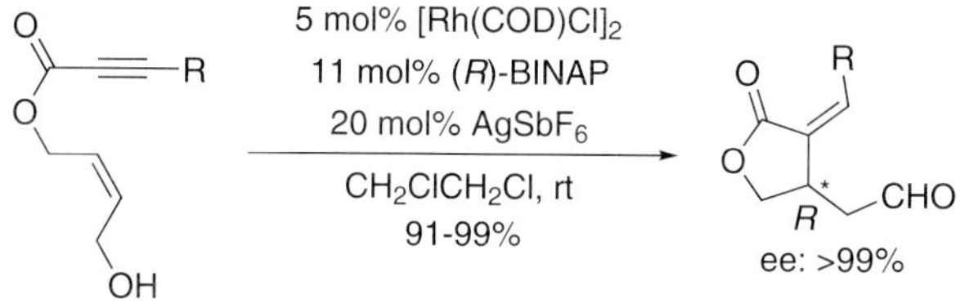

R = alkyl or phenyl.
When (S)-BINAP was applied, the S-products were afforded.

Scheme 9.49 Rh-catalyzed enantioselective cyclization of 1,6-enynes.

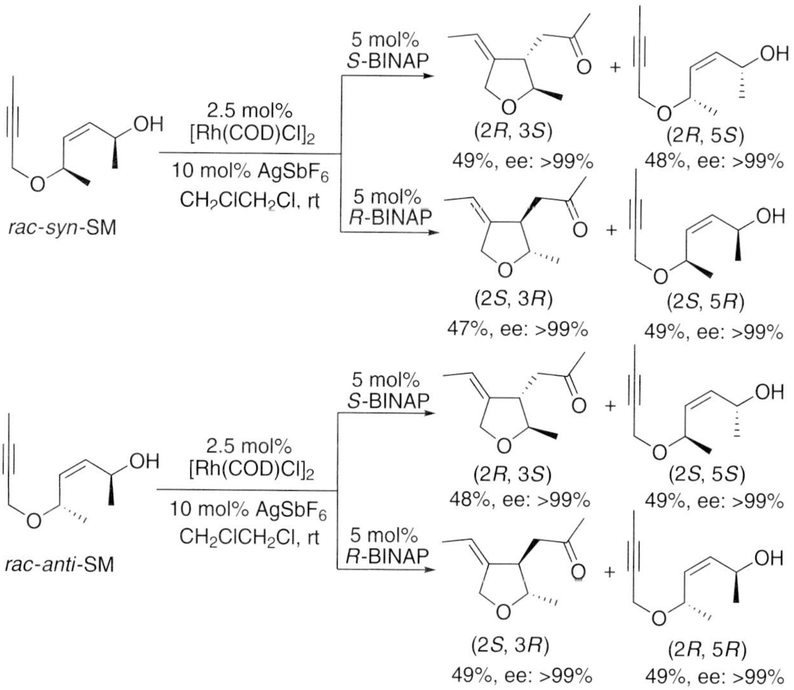

Scheme 9.50 Rh-catalyzed kinetic resolution of 5-hydroxy-3(Z)-hexen-2-yl 2-butynyl ether.

optically active cyclization products. However, the reaction of oxygen-tether substrates afforded a mixture of regioisomers depending on the relative position of the carbon–carbon double bonds (Scheme 9.51) [35].

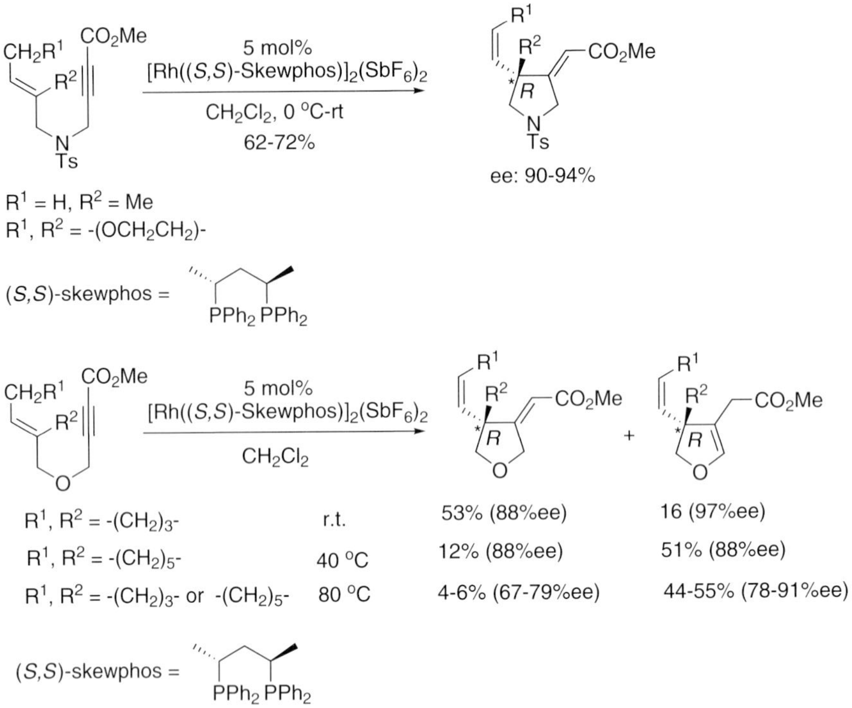

R^1 = H, R^2 = Me
R^1, R^2 = -(OCH$_2$CH$_2$)-

ee: 90-94%

(S,S)-skewphos =

R^1, R^2	conditions	product A	product B
-(CH$_2$)$_3$-	r.t.	53% (88%ee)	16 (97%ee)
-(CH$_2$)$_5$-	40 °C	12% (88%ee)	51% (88%ee)
-(CH$_2$)$_3$- or -(CH$_2$)$_5$-	80 °C	4-6% (67-79%ee)	44-55% (78-91%ee)

(S,S)-skewphos =

Scheme 9.51 [Rh((S,S)-skewphos)]$_2$(SbF$_6$)$_2$-catalyzed enantioselective cyclization of 1,6-enynes.

A very different type of reaction was observed in the transition metal-catalyzed intramolecular cycloisomerization of dodeca-6(E),11-dien-1-yne derivatives **8** affording the tetracyclic product **9**. Several transition metal complexes such as [Rh(OOCCF$_3$)$_2$]$_2$, [RuCl$_2$(CO)$_3$]$_2$, PtCl$_2$, [IrCl(CO)$_3$]$_n$, and ReCl(CO)$_5$ catalyze this reaction (Scheme 9.52) [36].

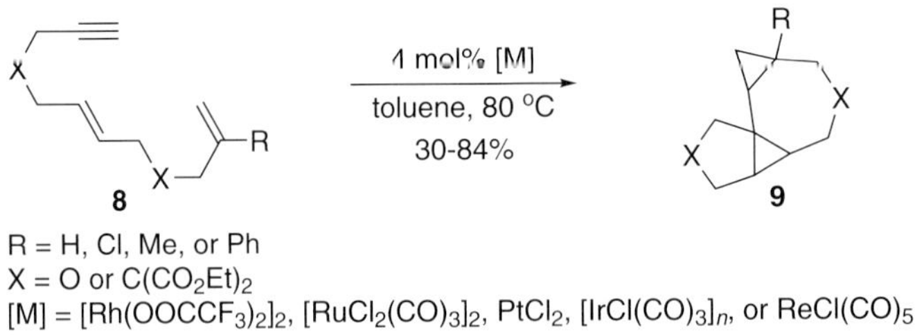

R = H, Cl, Me, or Ph
X = O or C(CO$_2$Et)$_2$
[M] = [Rh(OOCCF$_3$)$_2$]$_2$, [RuCl$_2$(CO)$_3$]$_2$, PtCl$_2$, [IrCl(CO)$_3$]$_n$, or ReCl(CO)$_5$

Scheme 9.52 Transition metal-catalyzed cycloisomerization of alka-6(E),11-dien-1-yne derivatives.

The cyclometallation results in a metallacyclopentene intermediate which undergoes a rearrangement to yield a carbenoid intermediate. Intramolecular cyclopropanation of the C–C double bond forms the final product and regenerates the catalytically active species (Scheme 9.53) [36].

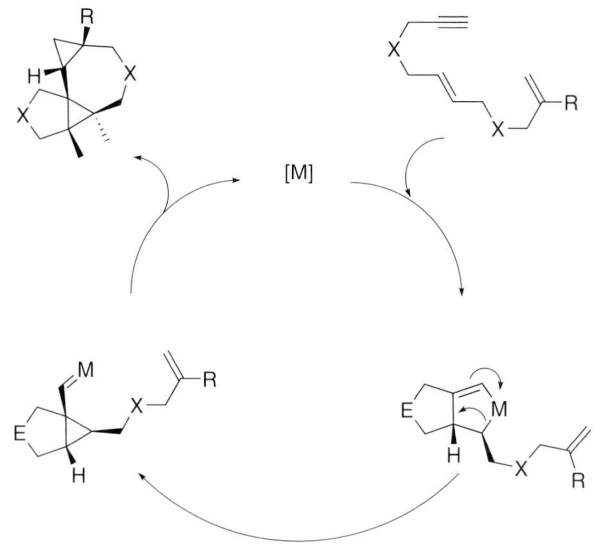

Scheme 9.53 The mechanism for transition metal-catalyzed cycloisomerization of alka-6(E),11-dien-1-yne derivatives.

Rh-catalyzed reaction of tetraethyl 6-methyleneundec-1-en-10-yne-4,4,8,8-tetracarboxylate afforded the product **10** via a similar process to that mentioned above, however, it had to compete with the skeletal reorganization reaction which formed **11** as a by-product. When [RuCl$_2$(CO)$_3$]$_2$ or PtCl$_2$ was used as the catalyst, only the skeletal reorganization product was formed (Scheme 9.54) [36].

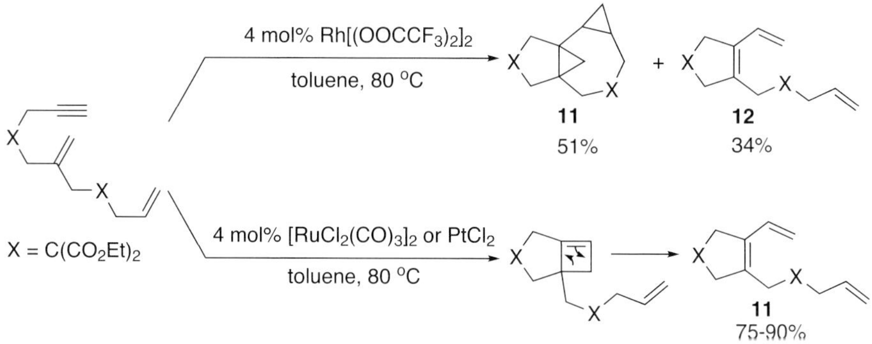

Scheme 9.54 Transition metal-catalyzed reaction of tetraethyl 6-methyleneundec-1-en-10-yne-4,4,8,8-tetracarboxylate.

The Rh-catalyzed reaction of allenynes afforded six- or seven-membered 2-alkenyl-3-alkylidenecycloalkene derivatives (Scheme 9.55) [37].

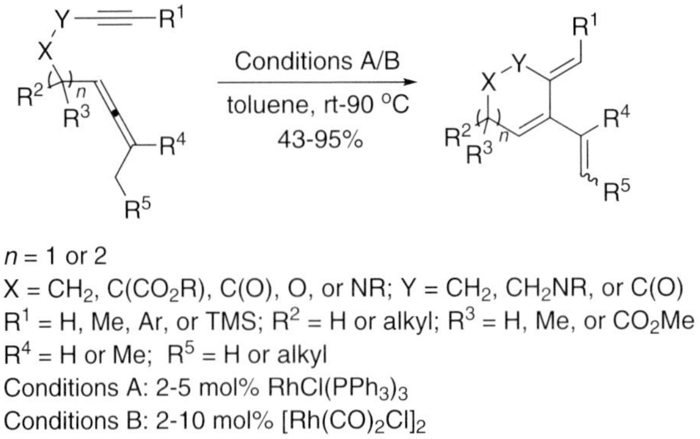
Conditions A/B
toluene, rt-90 °C
43-95%
n = 1 or 2
X = CH2, C(CO2R), C(O), O, or NR; Y = CH2, CH2NR, or C(O)
R1 = H, Me, Ar, or TMS; R2 = H or alkyl; R3 = H, Me, or CO2Me
R4 = H or Me; R5 = H or alkyl
Conditions A: 2-5 mol% RhCl(PPh3)3
Conditions B: 2-10 mol% [Rh(CO)2Cl]2

Scheme 9.55 Rh-catalyzed reaction of allenynes.

However, it was observed in our group that with the 2-alkynyl 2,3-allenoates, *trans*-$RhCl(CO)(PPh_3)_2$ should be used. The catalyzed monomeric and dimeric cycloisomerization of propargylic 2,3-dienoates yielded α,β-unsaturated δ-lactone derivatives (Scheme 9.56) [38].

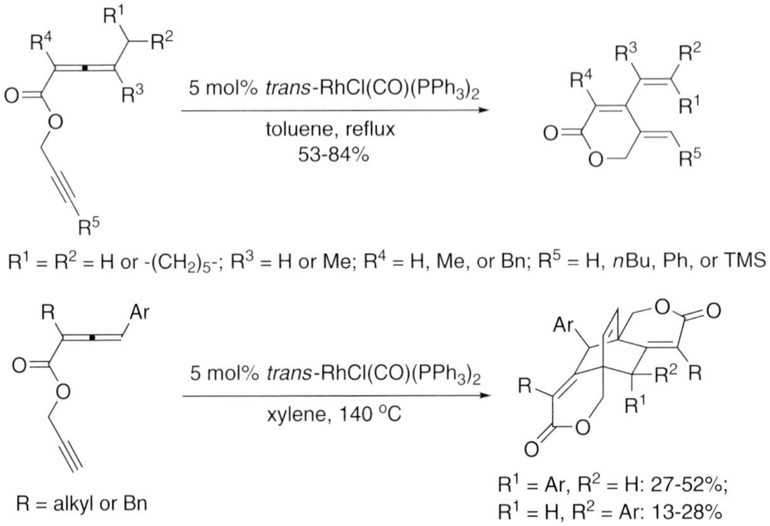
5 mol% trans-RhCl(CO)(PPh3)2
toluene, reflux
53-84%
R1 = R2 = H or -(CH2)5-; R3 = H or Me; R4 = H, Me, or Bn; R5 = H, nBu, Ph, or TMS
5 mol% trans-RhCl(CO)(PPh3)2
xylene, 140 °C
R = alkyl or Bn
R1 = Ar, R2 = H: 27-52%;
R1 = H, R2 = Ar: 13-28%

Scheme 9.56 *Trans*-$RhCl(CO)(PPh_3)_2$-catalyzed monomeric and dimeric cycloisomerization of propargylic 2,3-allenoates.

In the presence of H_2, the cyclometallation of 1,6-diynes followed by reductive hydrogenation produced the corresponding 1,2-dialkylidenecyclopentanes (Scheme 9.57) [31].

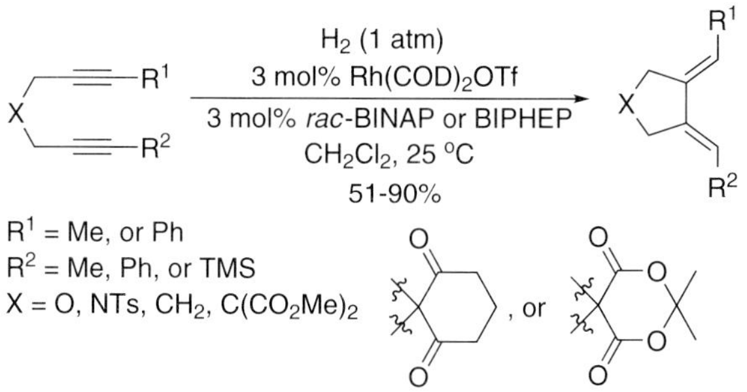

9.6 Rhodium-Catalyzed Cyclizations

Scheme 9.57 Rh-catalyzed reductive cyclization of 1,6-diynes under 1 atm of H_2.

Reaction conditions:
- H_2 (1 atm)
- 3 mol% Rh(COD)$_2$OTf
- 3 mol% rac-BINAP or BIPHEP
- CH_2Cl_2, 25 °C
- 51–90%

R^1 = Me, or Ph
R^2 = Me, Ph, or TMS
X = O, NTs, CH_2, $C(CO_2Me)_2$, cyclohexanedione, or dioxo-dimethyl-dioxane

The cationic Rh(I)/Segphos-catalyzed cycloisomerization of 1,6- or 1,7-diynes in the presence of 1,2-cyclohexandione forming 2-vinyl-3-alkylidenecylcoalkenes has also been reported (Scheme 9.58). Cyclometallation of the Rh(I) complex with the diyne unit results in a rhodabicyclic intermediate. Subsequent β-H elimination forming an allene moiety is followed by reductive elimination and double-bond isomerization to afford the final products [39].

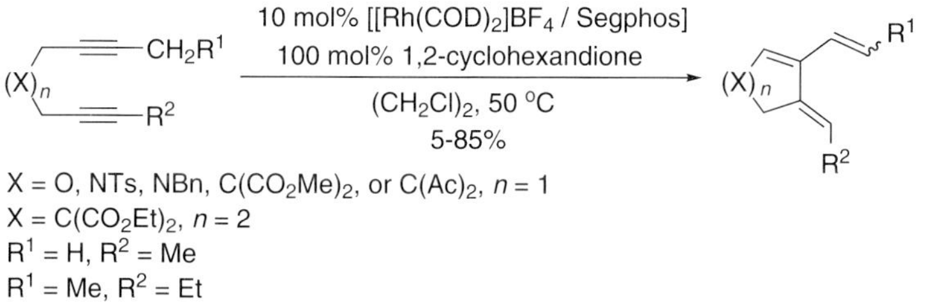

Reaction conditions:
- 10 mol% [[Rh(COD)$_2$]BF$_4$ / Segphos]
- 100 mol% 1,2-cyclohexandione
- $(CH_2Cl)_2$, 50 °C
- 5–85%

X = O, NTs, NBn, $C(CO_2Me)_2$, or $C(Ac)_2$, n = 1
X = $C(CO_2Et)_2$, n = 2
R^1 = H, R^2 = Me
R^1 = Me, R^2 = Et

Scheme 9.58 Cationic Rh(I)/Segphos-catalyzed cycloisomerization of 1,6- or 1,7-diynes in the presence of 1,2-cyclohexandione.

A new cycloisomerization pattern of 1,5-bisallenes catalyzed by Rh(I) has been reported to afford the 3,4-bis(alkylidene) seven-membered cycloheptene derivatives, in which the cyclic double bond was formed with the more substituted side (Scheme 9.59) [40].

18,19-Norsteroid derivates can be prepared by the Rh(I)-catalyzed bimolecular cyclization of unsubstituted 1,5-bisallenes. *trans*-RhCl(CO)(PPh$_3$)$_2$ could also catalyze the bimolecular cyclization between two different 1,5-bisallenes, which provided a combinatorial one-step approach to heterosteroids (Scheme 9.60) [41].

This reaction begins with the cyclometallation of Rh(I) with one terminal carbon–carbon double bond and one internal carbon–carbon double bond of the first bisallene to form a metallabicyclic intermediate, which undergoes carbometallation with one of the two allene moieties in the second molecule of bisallene to afford a

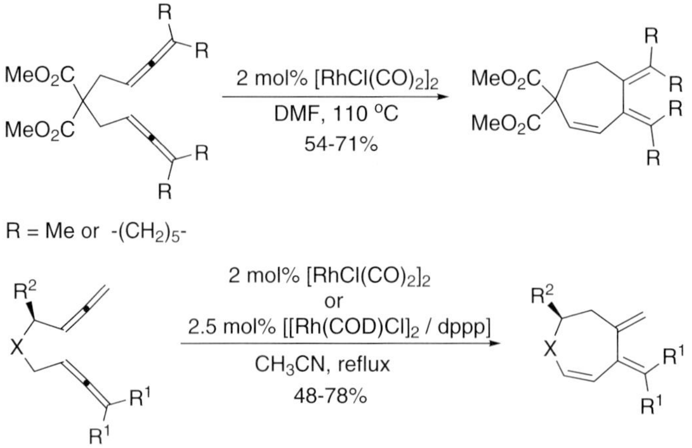

X = O, NTs, $C(CO_2Me)_2$, or $C(CO_2Me)(Ac)$
R^1 = Me, Et, or $-(CH_2)_5-$; R^2 = H or Me

Scheme 9.59 Rh-catalyzed cyclization of 1,5-bisallenes.

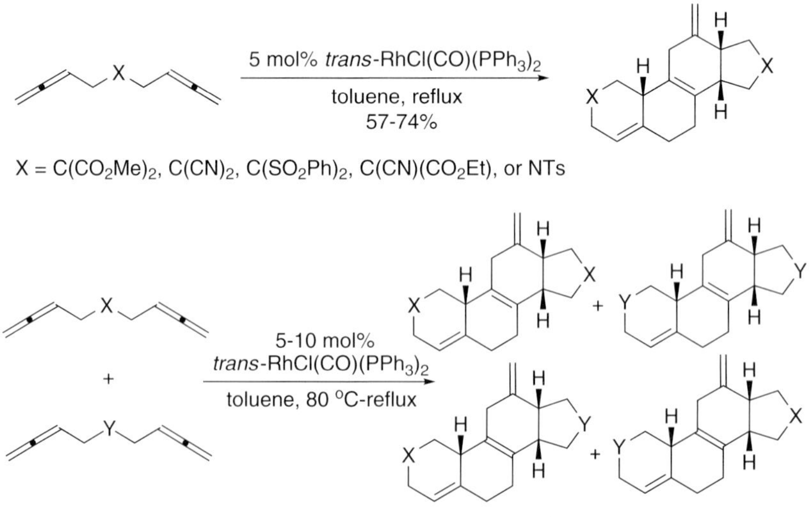

X = O, NTs, or SO_2; Y = O, SO_2, $C(CO_2Me)$, $C(CN)_2$, or $C(SO_2Ph)_2$

Scheme 9.60 Rh(I)-catalyzed bimolecular cyclization of unsubstituted 1,5-bisallenes.

π-allyl rhodium species. Subsequent reductive elimination and intramolecular Diels–Alder reaction gives the final product (Scheme 9.61) [41].

Recently, the bicyclo[4.4.0]decene with a conjugated exocyclic diene unit was constructed by *trans*-$RhCl(CO)(PPh_3)_2$-catalyzed cross-cyclization of 1,5-bisallenes in the presence of monoallenes (Scheme 9.62) [42].

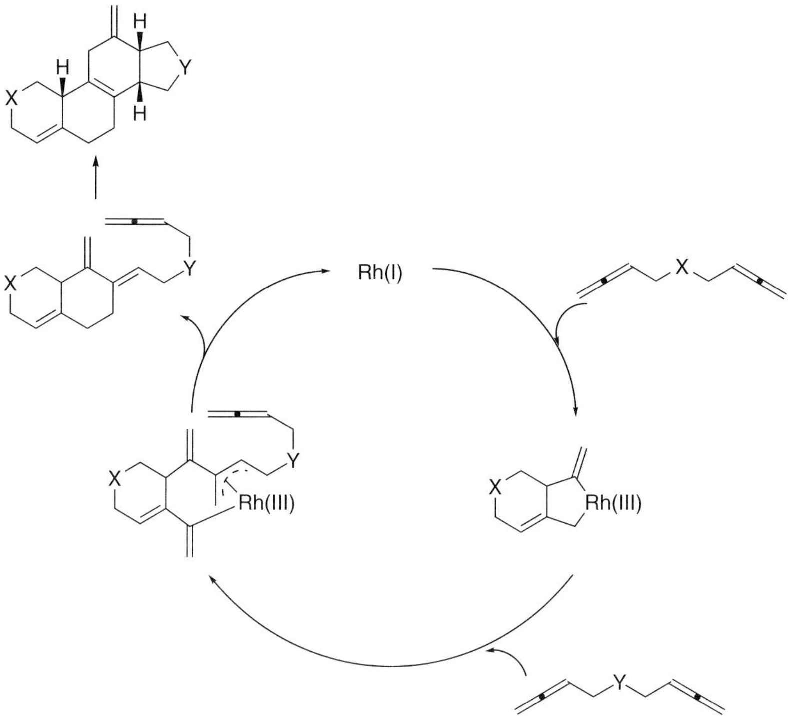

Scheme 9.61 The mechanism for Rh(I)-catalyzed bimolecular cyclization of unsubstituted 1,5-bisallenes.

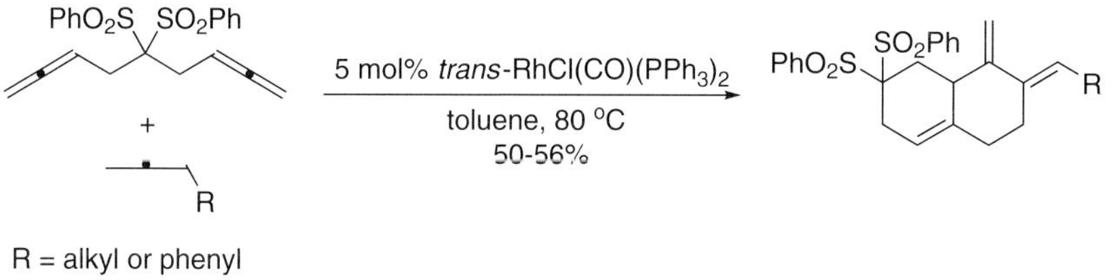

R = alkyl or phenyl

Scheme 9.62 *Trans*-RhCl(CO)(PPh₃)₂-catalyzed cross-cyclization of 1,5-bisallenes with monoallenes.

9.7
Iridium-Catalyzed Cyclizations [1d]

[IrCl(COD)]$_2$-dppf also catalyzed the cyclization of 1,6-enynes with a terminal C=C bond, affording 1-methylene-2(*E*)-alkylidenecyclopentanes in toluene or the (*Z*)-isomer in benzene (Scheme 9.63) [43].

Cyclometallation of Ir(I) and 1,6-enynes gives the bicyclic iridacyclopentene intermediate, which affords the *E*-vinyl Ir-H species via β-H elimination. Direct

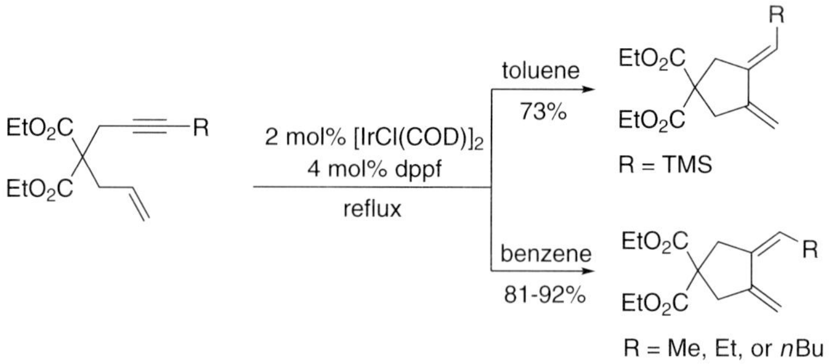

Scheme 9.63 [IrCl(COD)]$_2$-dppf-catalyzed cyclization of 1,6-enynes.

reductive elimination forms the *E*-products. Isomerization of the *E*-vinyl iridium hydride species to its *Z*-isomer via the zwitterionic carbene intermediate and the subsequent reductive elimination produces the *Z*-products (Scheme 9.64) [43].

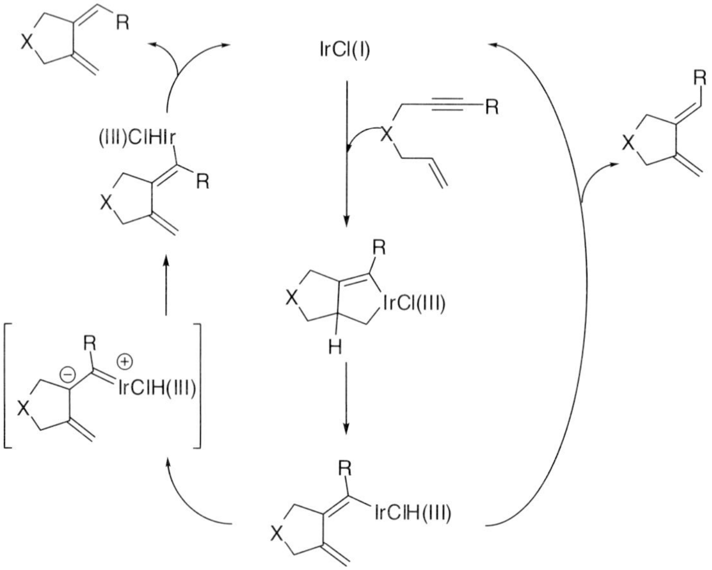

Scheme 9.64 The mechanism for [IrCl(COD)]$_2$-dppf-catalyzed cyclization of 1,6-enynes.

9.8
Nickel-Catalyzed Cyclizations [1d, 8b, 16a, 16c, 44]

Ni(COD)$_2$ catalyzes the dimerization of ethyl cyclopropylideneacetate to give (*E*)-ethyl 7-(2-ethoxycarbonylmethylene)spiro[2.4]heptane-4-carboxylate in the presence of

PPh$_3$ (Scheme 9.65). As shown in Scheme 9.65, cyclometallation of ethyl cyclopropylideneacetate with the Ni(0) catalyst results in the formation of a nickelacyclopentane, which undergoes β-decarbometallation to open one of the two three-membered rings in the intermediate to afford a nickelacyclohexane intermediate. Subsequent reductive elimination provides the product and regenerates the catalyst [45].

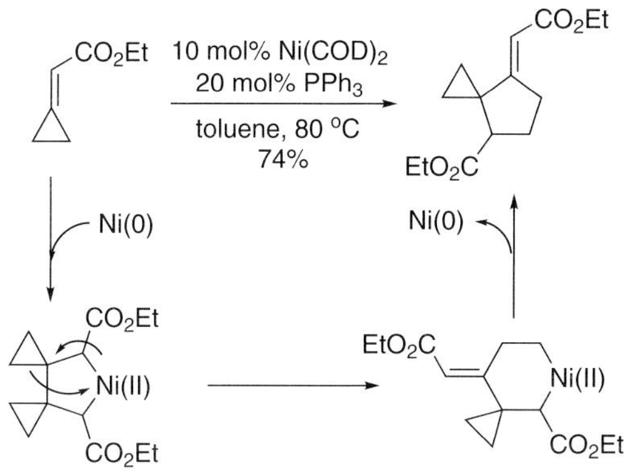

Scheme 9.65 Ni(COD)$_2$-catalyzed dimerization of ethyl cyclopropylideneacetate.

Ni(COD)$_2$ catalyzes the cyclization of 1-phenyldeca-2(E),7(E),9-trien-1-one to afford *trans*-1-(2-oxo-2-phenylethyl)-2-(1′(E)-propenyl)cyclopentane in the presence of PPh$_3$ and Et$_2$Zn (Eq. 9.3) [46].

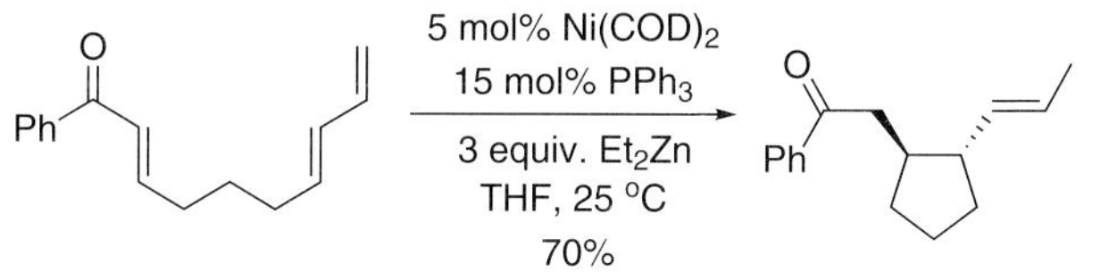

(9.3)

Coumarine derivatives are formed by the nickel-catalyzed highly regio- and stereoselective cyclization of oxanorbornenes and 2-alkynoates (Scheme 9.66). The reduction of Ni(II) by zinc likely initiates this catalytic reaction. Cyclometallation of the reactants with Ni(0) results in a nickelacyclopentene intermediate, which undergoes β-oxy elimination to form the nickelacyclohexene species. Subsequent β-H elimination forming the ketone functionality affords the nickel hydride which gives the bicyclic ketone product and regenerates the catalyst via reductive elimination. Z → E isomerization of this product and the subsequent lactonization produces the final tricyclic product (Scheme 9.66) [47].

Nickel also catalyzes the intramolecular reaction of the electron-deficient alkene with an alkyne moiety of alk-1,6-enynyl carboxylates and ketones. When MeLi/ZnCl$_2$ was applied in this reaction, 1-(alkylidene)-2-alkylcyclopentane derivatives were afforded via cyclometallation, transmetallation, and reductive elimination. When

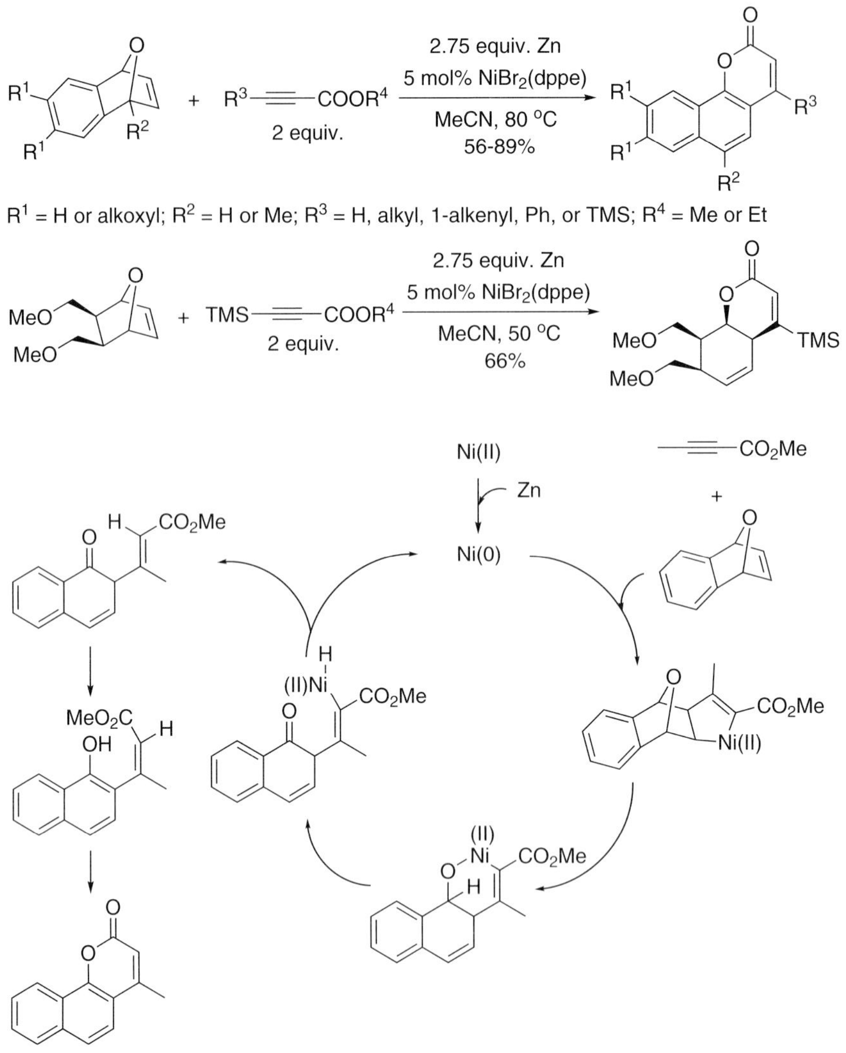

Scheme 9.66 Ni-catalyzed cyclization of oxanorbornenes and 2-alkynoates.

Et$_2$Zn was used, 1-alkylidene-2-alkylcyclopentane derivatives and ethene were formed via the sequential cyclometallation, transmetallation, β-H elimination, and reductive elimination process (Scheme 9.67) [46, 48].

Nickel-mediated carboxylative cyclization of enynes in the presence of DBU under 1 atm of CO$_2$ afforded the cyclized carboxylation products (Scheme 9.68) [49].

Nickel catalyzes the regio- and stereo-selective cyclometallation and CO$_2$ insertion reaction of 1,3,8,10- or 1,3,9,11-tetraenes and an excess amount of organozinc reagent to afford the *trans*-cyclopentane or cyclohexane derivatives (Scheme 9.69) [50].

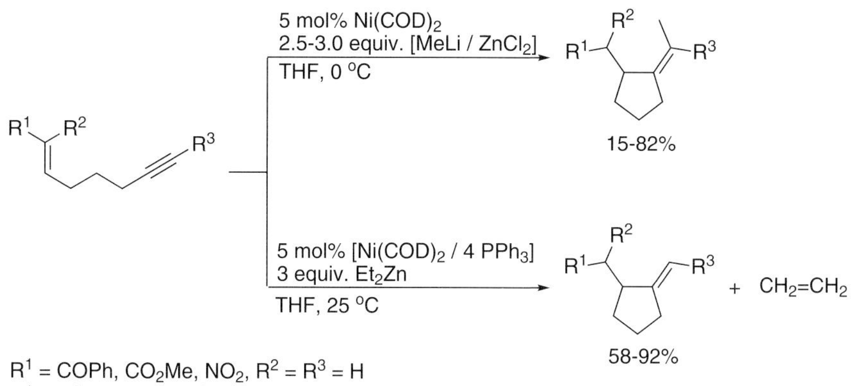

R^1 = COPh, CO_2Me, NO_2, R^2 = R^3 = H
R^1 = R^2 = CO_2Me, R^3 = H, Me, or TMS
R^1 = COPh, R^2 = H, R^3 = n-Bu or Ph

Scheme 9.67 Ni-catalyzed intramolecular reaction of alk-1,6-enynyl carboxylates and ketones.

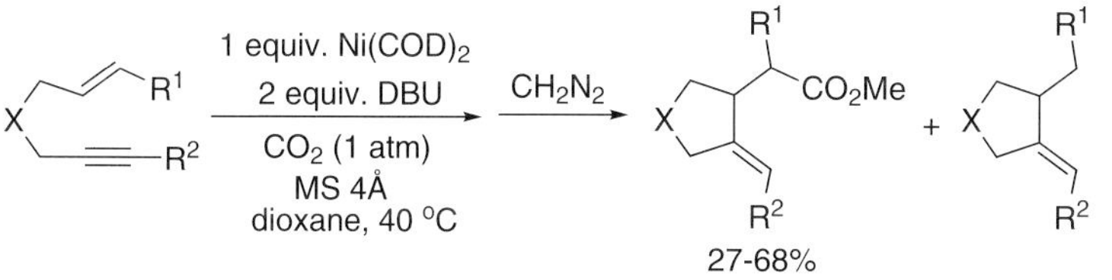

X = $C(CO_2Me)_2$, R^1 = CN or CO_2Me, R^2 = Me, CH_2OMe, Ph or CO_2Me
X = CH_2, NTs, or O, R^1 = CO_2Me, R^2 = Me
X = NTs, R^1 = CO_2Me, R^2 = CH_2OMe

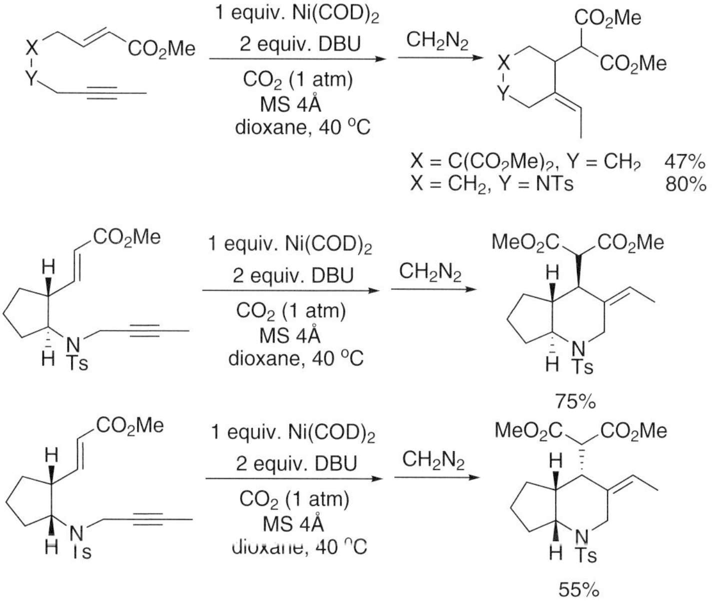

X = $C(CO_2Me)_2$, Y = CH_2 47%
X = CH_2, Y = NTs   80%

75%

55%

Scheme 9.68 Ni-mediated carboxylative cyclization of 1,6- or 1,7-enynes.

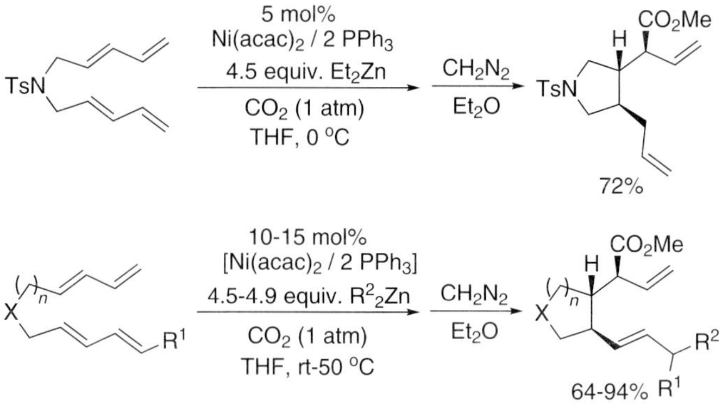

Scheme 9.69 Ni-catalyzed reaction of 1,3,8,10- or 1,3,9,11-tetraenes with CO_2 and organozinc reagent.

$n = 1$, $X = C(CO_2Me)_2$ or NTs, $R^1 = H$ or Me, $R^2 = Me$ or Ph
$n = 2$, $X = NTs$, $R^1 = R^2 = Me$

This reaction seems to start with cyclometallation of the bis(1,3-diene) with the Ni(0) complex generated *in situ* to produce the bis(π-allyl) nickel intermediate. The subsequent insertion of CO_2 into one of the two nickel–carbon bonds affords the π-allylic nickel carboxylate species which transmetallates with the organozinc reagent. Subsequent transmetallation, β-H elimination, and reductive elimination afford the products. However, in the presence of Me_2Zn or Ph_2Zn, a coupling reaction followed to yield the corresponding coupling products (Scheme 9.70) [50].

When (S)-MeO-MOP was applied as the chiral ligand in the above reaction, excellent enantioselectivities were observed (Scheme 9.71) [51].

2-Alkylidenecyclopentanol derivatives were formed by the $Ni(COD)_2$-catalyzed alkylative cyclization of 5-hexynal derivatives with organozincs via cyclometallation, transmetallation, and reductive elimination. Notably, when Et_2Zn was applied, ethene was formed as the by-product via a β-H elimination process (Scheme 9.72) [52].

Alk-7-yn-2-enal reacts with a stoichiometric amount of $Ni(COD)_2$ to afford **12**. However, the reaction of 1-phenylnon-2(E)-en-7-yn-1-one or 1,8-diphenyloct-2(E)-en-7-yn-1-one formed **13**. When alk-7-yn-2-enone, with a terminal carbon–carbon triple bond, was applied in this reaction, only the monocyclic 2-(2-methylenecyclopentyl)-1-phenylethanone **14** was produced in low yield (Scheme 9.73) [53].

In addition, the intermediate formed by the cyclometallation of 8-phenyloct-2(E)-en-7-ynal with a stoichiometric amount of $Ni(COD)_2$ in the presence of tmeda reacted with dry O_2 to afford **15**. The α-nickelated aldehyde moiety in the formed nickel metallabicyclic intermediate could isomerize to the enolate to form a nickelabicyclo[5.3.0]-intermediate. When this was treated with Me_2Zn, 2-(E)-(1-phenylethylidene)cyclopentyl)acetaldehyde **16** was formed via transmetallation and reductive elimination. When it was treated with electrophiles such as water,

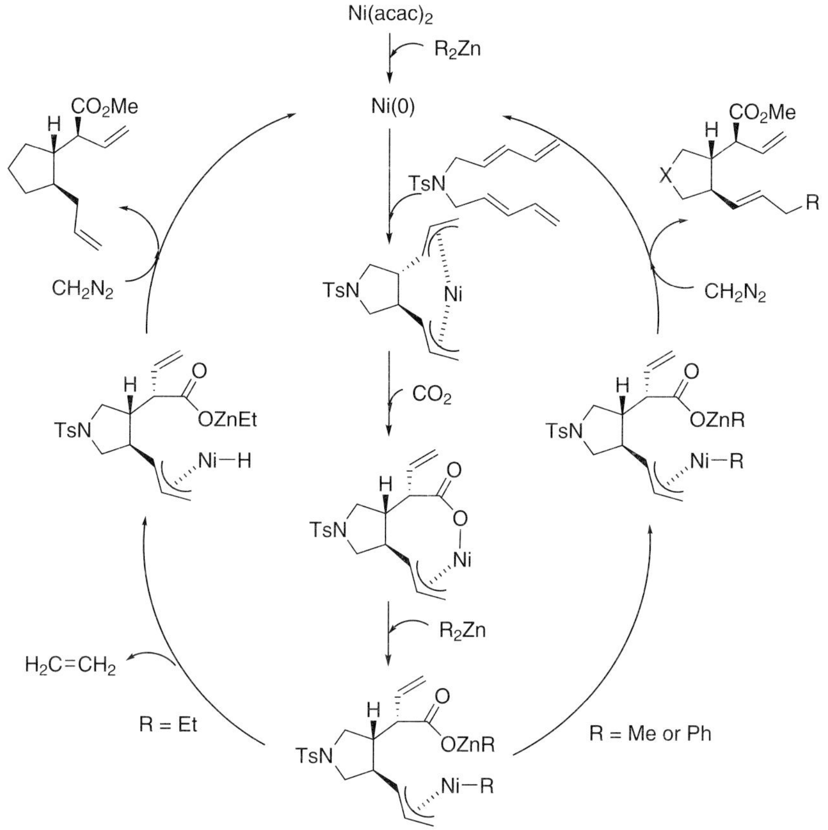

Scheme 9.70 The mechanism for Ni-catalyzed reaction of 1,3,8,10- or 1,3,9,11-tetraenes with CO_2 and organozinc reagent.

X = NTs; R^1 = H, R^2 = Me or Ph
X = NTs, $C(CO_2Me)_2$, $C(CH_2OBn)_2$, or $-(CH_2OCOOCH_2)-$; R^1 = H; R^2 = Me or Ph

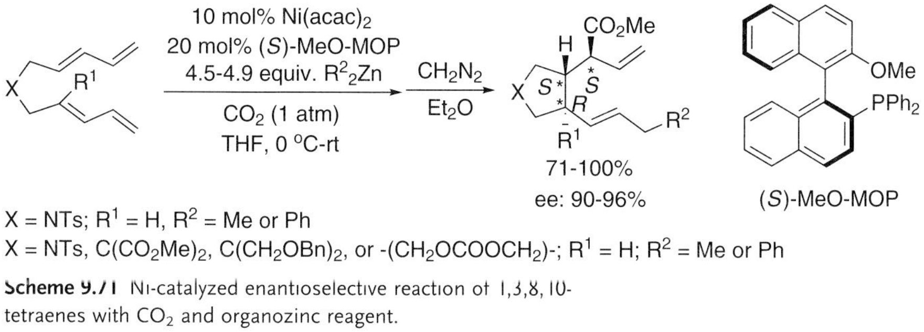

Scheme 9.71 Ni-catalyzed enantioselective reaction of 1,3,8,10-tetraenes with CO_2 and organozinc reagent.

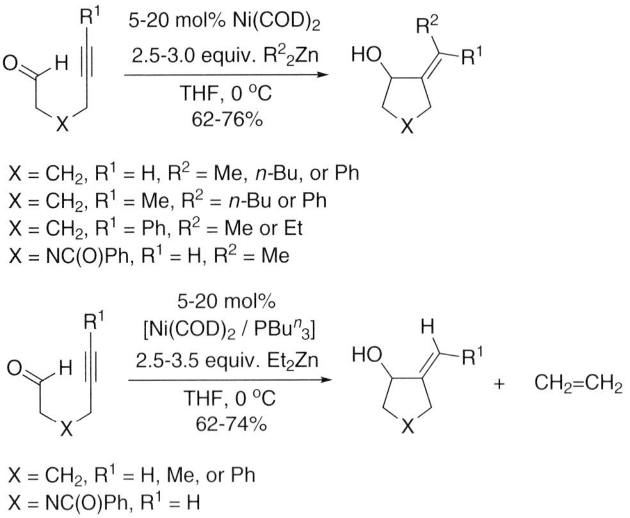

Scheme 9.72 Ni(COD)$_2$-catalyzed alkylative cyclization of 5-hexyanl derivatives with organozincs.

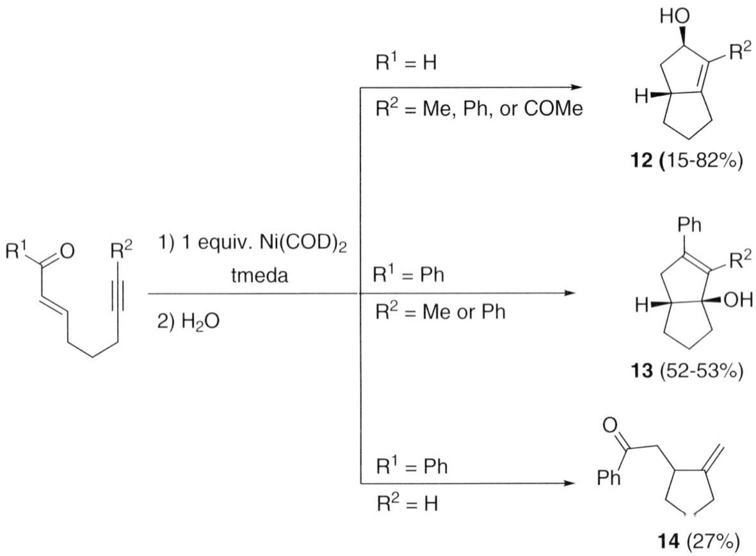

Scheme 9.73 Ni(COD)$_2$-promoted reaction of alk-7-ynyl-2-enals or ketones.

alkyl iodides, benzoyl chloride, or aldehydes, the corresponding products **17–20** were afforded. If acrylaldehyde was applied in this reaction, product **21** produced (Scheme 9.74) [53, 54].

Ni(COD)$_2$ could also catalyze the organozinc-promoted intramolecular carbocyclization of benzoyl substituted C–C double bonds with an aldehyde functionality to form the corresponding cyclopentanol or cyclohexanol (Scheme 9.75) [46].

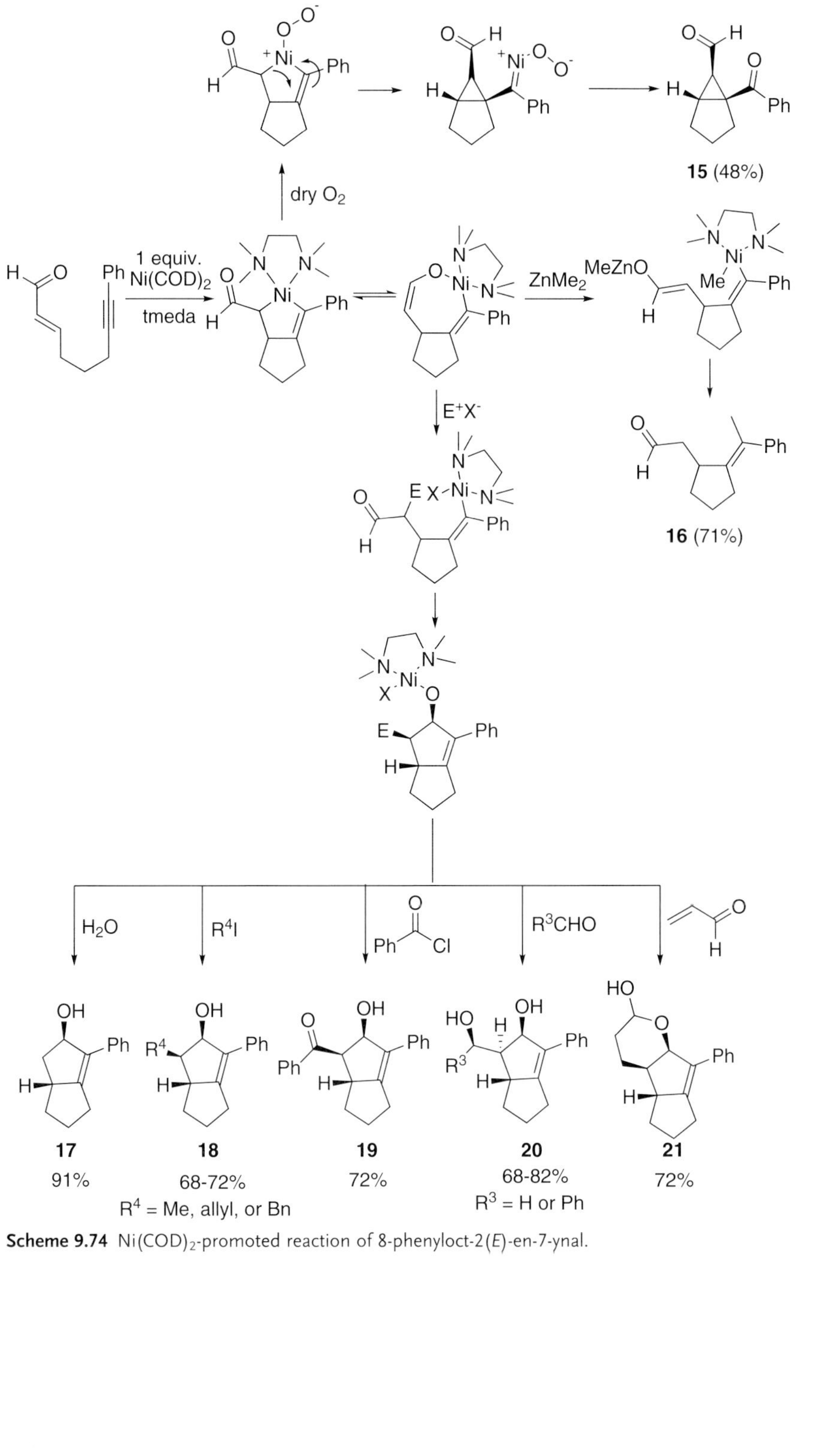

9.8 Nickel-Catalyzed Cyclizations | 447

Scheme 9.74 Ni(COD)$_2$-promoted reaction of 8-phenyloct-2(E)-en-7-ynal.

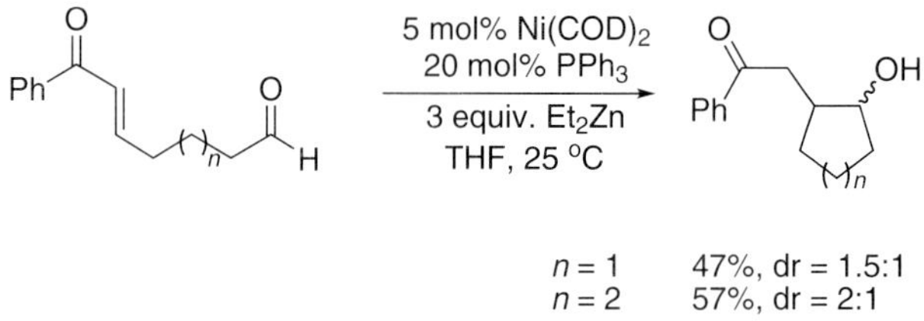
Ph
O
O
H
n
5 mol% Ni(COD)2
20 mol% PPh3
3 equiv. Et2Zn
THF, 25 °C
O
Ph
OH
n
n = 1 47%, dr = 1.5:1
n = 2 57%, dr = 2:1

Scheme 9.75 Ni(COD)$_2$-catalyzed organozinc-promoted intramolecular carbocyclization of enals.

9.9
Palladium-Catalyzed Cyclizations [1d, 8b, 16c, 16d, 26]

The palladium(0)-catalyzed cyclization of *N*-(3-(methoxycarbonyl)-2-propynyl)-*N*-(3-methylbut-2-enyl)-4-methylbenzenesulfonamide **22** afforded 3-(methoxycarbonyl) methylene-4-(propen-2-yl)-1-tosylpyrrolidine *Z*-**23** via the cyclometallation, β-H elimination, and reductive elimination. However, when *N*-(3-(methoxycarbonyl)prop-2-ynyl)-*N*-(3-phenylpropa-2(*E*)-enyl)-4-methylbenzenesulfonamide **24** was applied, (*Z*)-3-benzylidene-4-(methoxycarbonyl)methylene-1-tosylpyrrolidine **25** was formed via the different regioselective β-H elimination (Scheme 9.76) [55].

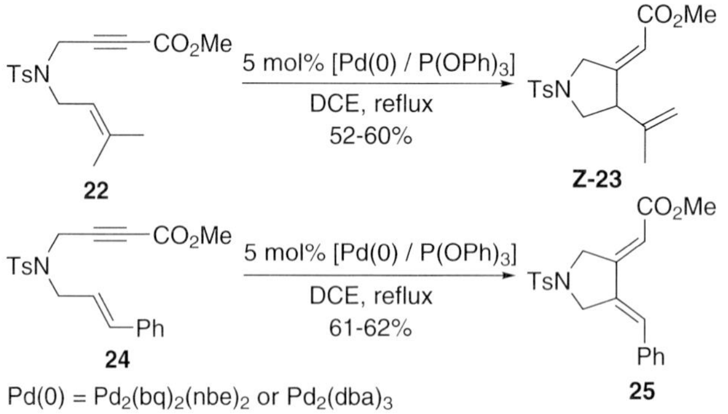
TsN
CO2Me
5 mol% [Pd(0) / P(OPh)3]
DCE, reflux
52-60%
CO2Me
TsN
22
Z-23
TsN
CO2Me
Ph
5 mol% [Pd(0) / P(OPh)3]
DCE, reflux
61-62%
CO2Me
TsN
Ph
24
25
Pd(0) = Pd2(bq)2(nbe)2 or Pd2(dba)3

Scheme 9.76 Pd(0)-catalyzed cyclization of 1,6-enynes.

When methyl 7-substituted oct-7-en-2-ynoates **26** were applied under the above reaction conditions, 3-alkylidenecyclohex-1-ene derivatives **27** were afforded. However, in some cases, 4-(methoxycarbonylmethylene)cyclohex-1-ene derivatives **28** were formed in small amounts as by-products (Scheme 9.77) [55].

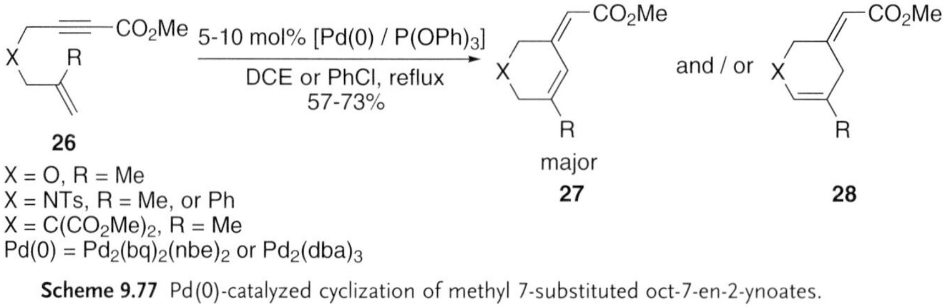
X
R
CO2Me
5-10 mol% [Pd(0) / P(OPh)3]
DCE or PhCl, reflux
57-73%
CO2Me
X
R
major
and / or
CO2Me
X
R
26
27
28
X = O, R = Me
X = NTs, R = Me, or Ph
X = C(CO2Me)2, R = Me
Pd(0) = Pd2(bq)2(nbe)2 or Pd2(dba)3

Scheme 9.77 Pd(0)-catalyzed cyclization of methyl 7-substituted oct-7-en-2-ynoates.

The mechanism for the formation of the two isomers is shown in Scheme 9.78. It starts with cyclometallation of the enyne esters leading to the palladabicyclopentene intermediate, which subsequently undergoes isomerization to afford the bicycloalkyl carbenoid species. The ring opening of the cyclopropane unit results in the formation of a zwitterionic intermediate. Its proton transfer gives two regioisomers of alkenyl palladium hydride, which produces the corresponding isomers via reductive elimination [55].

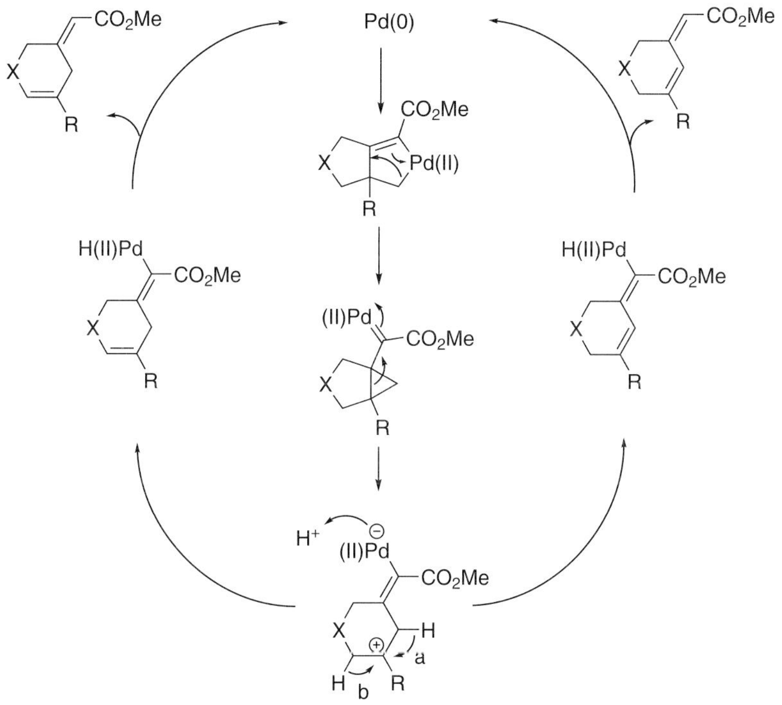

Scheme 9.78 The mechanism for Pd(0)-catalyzed cyclization of methyl 7-substituted oct-7-en-2-ynoates.

9.10
Conclusion and Perspectives

Cyclizations based on cyclometallation have been extensively explored during the past decades and a series of transition metal complexes have been proven to be efficient catalysts for this reaction. The efficiency in constructing cyclic structures has led to this type of reaction being widely applied in synthetic chemistry. However, in several cases, the catalytic reaction is still not realized. Discovering a means of developing the stoichiometric reactions into catalytic ones is obviously an important task in this field. Furthermore, highly enantioselective reactions are still desired.

9.11
Experimental: Selected Procedures

9.11.1
Typical Procedure for the Titanium-Catalyzed Cyclizations Shown in Scheme 9.4 [4]

A solution of *tert*-butyl 8-(trimethylsilyl)oct-2(*E*)-en-7-ynoate (53.3 mg, 0.20 mmol) in 1 mL of Et$_2$O and *i*PrMgCl (0.30 mL, 1.60 M solution in Et$_2$O, 0.48 mmol) were added to a stirred solution of Ti(OPr^i)$_4$ (0.070 mL, 0.24 mmol) in Et$_2$O (2 mL) at $-78\,°$C under an argon atmosphere. The reaction mixture was then gradually allowed to warm to $-20\,°$C over 1 h and stirred for an additional 2 h. Then 1 N HCl and Et$_2$O were added, and the organic layer was separated, washed with aqueous NaHCO$_3$ solution, and dried over MgSO$_4$. Evaporation of the solvent and flash chromatography on silica gel (eluent: hexane–ether) afforded *tert*-butyl (2-((*E*)-(trimethylsilyl) methylene)cyclopent-1-yl)acetate (44.5 mg, 83%).

9.11.2
Typical Procedure for the Zirconium-Catalyzed Cyclizations Shown in Scheme 9.16 [11]

Zirconocene dichloride (146 mg, 0.5 mmol) and *N*-allyl-*N*-(2-bromoallyl)aniline (126 mg, 0.5 mmol) were added separately to two 25-mL Schlenk flasks under a nitrogen atmosphere. After both Schlenk flasks were removed to a vacuum manifold, 3 mL of diethyl ether was added to each flask. The zirconocene solution was cooled to $-78\,°$C and 0.4 mL of 2.5 M n-butyllithium was added and the mixture was stirred for 1 h. The aniline solution was then added to the zirconocene solution, and the resulting solution was warmed to room temperature over 1 h. The reaction was monitored by GC. Then 1.0 mL of D$_2$O was added and the mixture was stirred for 2 h. It was then poured into 30 mL of mixed water/ether (1: 1) in a separatory funnel, and the organic layer was separated. The crude product was filtered through neutral alumina and then washed with 10 mL of saturated aqueous NaHCO$_3$. After separation, the organic layer was dried over MgSO$_4$. The solvent was removed on a rotary evaporator to afford 3-deuteromethyl-4-methylene-1-phenylpyrrolidine (85 mg, 98%).

9.11.3
Typical Procedure for the Ruthenium-Catalyzed Cyclizations Shown in Scheme 9.22 [17]

Under nitrogen, [Ru(COD)Cl$_2$]$_n$ (2.8 mg, 1 mol%) was added to a solution of dimethyl 2,2-diallylmalonate (212 mg, 1.0 mmol) in degassed *i*PrOH (4 mL). The reaction mixture was then stirred at 90 °C for 24 h. Removal of the solvent and purification by flash column chromatography on silica-gel (eluent: hexane:ethyl acetate $= 20:1$) afforded dimethyl 3-methyl-4-methylenecyclopentane-1,1-dicarboxylate (195 mg, 92%).

9.11.4
Typical Procedure for the Cobalt-Catalyzed Cyclizations Shown in Scheme 9.34 [23]

CoI_2(dppe) (0.0500 mmol), zinc powder (2.75 mmol), and zinc iodide (0.0500 mmol) were added to a Schlenk tube with a screw cap. After evacuation and purging three times with nitrogen gas, freshly distilled dry CH_3CN (2.0 mL), 1-phenylallene (1.00 mmol), methyl vinyl ketone (1.20 mmol), and H_2O (2.00 mmol) were added to this tube. The reaction mixture was then allowed to react at 80 °C for 6 h. It was then cooled, diluted with dichloromethane, and stirred in air for 10 min. The resulting mixture was filtered through a Celite and silica gel pad and washed with dichloromethane. Evaporation of the solvent and purification via chromatography on silica gel (eluent: hexane–ethyl acetate) afforded 1-methyl-3-methylene-2-phenylcyclopentanol (88%).

9.11.5
Typical Procedure for the Rhodium-Catalyzed Cyclizations Shown in Scheme 9.47 [30]

A solution of *trans*-diethyl 7-alkylidenespiro[2.4]heptane-4,5-dicarboxylate (1.22 g (93%), 4.50 mmol) and 2-butyne (0.27 g, 5 mmol) in toluene (10 mL) was added to a solution of $RhCl(PPh_3)_3$ (0.25 g, 0.27 mol) and AgOTf (70 mg, 0.27 mmol) in toluene (20 mL) in a 100-mL Schlenk flask. The mixture was heated at 110 °C for 3 h. After complete consumption of the starting material, as monitored by GC analysis, the mixture was filtered through a silica gel pad and evaporated to dryness. The residue weighed 1.14 g (83%) and according to GC it consisted of one major (66%) and three minor components (20, 5 and 9%, respectively). The crude product was purified by flash chromatography on silica gel (eluent: hexane:ether $= 1:1$). The major product was identified as diethyl 3,4-dimethylbicyclo[5.3.0]deca-1(7),3-diene-*trans*-8,9-dicarboxylate.

9.11.6
Typical Procedure for the Iridium-Catalyzed Cyclizations Shown in Scheme 9.63 [43]

A solution of $[Ir(COD)Cl]_2$ (14.7 mg, 0.022 mmol), dppf (25.3 mg, 0.046 mmol), and diethyl 2-allyl-2-(but-2-ynyl)malonate (252.3 mg, 1.0 mmol) in benzene (5 mL) was stirred under reflux for 2 h under an argon atmosphere. After the reaction was complete, as monitored by GLC, the solution was evaporated *in vacuo*. The residue was purified by flash column chromatography on silica gel (eluent: hexane:ethyl acetate $= 98:2$) to give 3-ethylidene-4-methylenecyclopentane-1,1-dicarboxylate (230.9 mg, 92%).

9.11.7
Typical Procedure for the Nickel-Catalyzed Cyclizations Shown in eq 9.3 [46]

A solution of PPh_3 (20 mg, 0.08 mmol) and $Ni(COD)_2$ (7 mg, 0.03 mmol) in THF was stirred at 25 °C for 5 min. It was then added to a solution of Et_2Zn (1.46 mmol) in THF at 0 °C, and the resulting mixture was immediately transferred by cannula to a solution of 1-phenyldeca-2(E),7(E),9-trien-1-one (115 mg, 0.51 mmol) in THF at 0 °C. After the reaction was complete, as monitored by TLC, it was subjected to an extractive work-up (NaHCO$_3$/EtOAc) followed by flash chromatography on SiO$_2$ (eluent: hexanes:ethyl acetate = 9 : 1) to give trans-1-(2-oxo-2-phenylethyl)-2-(E)-(1-propenyl)cyclopentane (81 mg, 70%).

9.11.8
Typical Procedure for the Palladium-Catalyzed Cyclizations Shown in Scheme 9.77 [55]

A solution of **26** (X = O; R = Me) (84 mg, 0.50 mmol), Pd(dq)$_2$(nbe)$_2$ (8 mg, 0.013 mmol) and P(OPh)$_3$ (8 mg, 0.026 mmol) in dry degassed (CH$_2$Cl)$_2$ (5 mL) was refluxed under Ar for 12 h. After evaporation of the solvent and purification by flash column chromatography on silica gel (eluent: hexane:ethyl acetate = 10 : 1), product **27** (48 mg, 57%) was afforded.

Abbreviations

Ac	acetyl
acac	2,4-pentanedionato
BICP	2,2-bis(diphenylphosphanyl)-1,1'-dicyclopentane
BICPO	2,2'-bis(diphenylphosphinooxy)bi(cyclopentane)
BINAP	2,2'-bis(diphenylphosphosino)-1,1'-binaphthyl
BIPHEP	2,2'-bis(diphenylphosphino)-1,1'-biphenyl
Bn	benzyl
Boc	tert-butoxycarbonyl
bq	p-benzoquinone
Bz	benzoyl
COD	cyclooctadiene
Cp	cyclopentadienyl
Cp∗	1,2,3,4,5-pentamethylcyclopenta-1,3-dienyl
dba	1,5-diphenylpenta-1(E),4(E)-dien-3-one
DBU	1,8-diazabicyclo[5.4.0]undec-7-ene
DCE	1,2-dichloroethane
DCM	dichloromethane
DMAD	dimethyl 2-butynedioate
DMF	dimethyl formamide
dppb	1,4-bis(diphenylphosphino)butane
dppe	bis(diphenylphosphino)ethane
Duphos	1,2-bis(phospholano)benzene
OTf	trifluoromethanesulfonate
PMB	p-methoxybenzyl
Segphos	5,5'-bis(diphenylphosphino)-4,4'-bi-1,3-benzodioxole
TBDMS	tert-butyldimethylsilyl

TBS	*tert*-butyldimethylsilyl
THF	tetrahydrofuran
tmeda	N^1,N^1,N^2,N^2-tetramethylethane-1,2-diamine
TMS	trimethylsilyl
tol	tolyl
Ts	tosyl

References

1 (a) Sato, F., Urabe, H. and Okamoto, S. (2000) *Chem. Rev.*, **100**, 2835–2886; (b) Sato, F., Urabe, H. and Okamoto, S. (2000) *Synlett*, 753–775; (c) Sato, F. and Okamoto, S. (2001) *Adv. Synth. Catal.*, **343**, 759–784; (d) Aubert, C., Buisine, O. and Malacria, M. (2002) *Chem. Rev.*, **102**, 813–834.

2 Okamoto, S. and Livinghouse, T. (2000) *J. Am. Chem. Soc.*, **122**, 1223–1224.

3 Takayama, Y., Gao, Y. and Sato, F. (1997) *Angew. Chem. Int. Ed.*, **36**, 851–853.

4 Urabe, H., Suzuki, K. and Sato, F. (1997) *J. Am. Chem. Soc.*, **119**, 10014–10027.

5 Sturla, S.J., Kablaoui, N.M. and Buchwald, S.L. (1999) *J. Am. Chem. Soc.*, **121**, 1976–1977.

6 Urabe, H., Takeda, T., Hideura, D. and Sato, F. (1997) *J. Am. Chem. Soc.*, **119**, 11295–11305.

7 Matano, Y., Miyajima, T., Nakabuchi, T., Matsutani, Y. and Imahori, H. (2006) *J. Org. Chem.*, **71**, 5792–5795.

8 (a) Dzhernilev, U.M., Sultanov, R.M. and Gaimaldinov, R.G. (1995) *J. Organomet. Chem.*, **491**, 1–10; (b) Ojima, I., Tzamarioudaki, M., Li, Z. and Donovan, R.J. (1996) *Chem. Rev.*, **96**, 635–662; (c) Mori, M. (2005) *Chem. Pharm. Bull.*, **53**, 457–470.

9 (a) Negishi, E., Rousset, C.J., Choueiry, D., Maye, J.P., Suzuki, N. and Takahashi, T. (1998) *Inorg. Chim. Acta*, **280**, 8–20; (b) Martín, S. and Brintzinger, H. (1998) *Inorg. Chim. Acta*, **280**, 189–192.

10 Yamaura, Y., Hyakutake, M. and Mori, M. (1997) *J. Am. Chem. Soc.*, **119**, 7615–7616.

11 Millward, D.B. and Waymouth, R.M. (1997) *Organometallics*, **16**, 1153–1158.

12 (a) Negishi, E., Ma, S., Sugihara, T. and Noda, Y. (1997) *J. Org. Chem.*, **62**, 1922–1923; (b) Gorman, J.S.T., Iacono, S.T. and Pagenkopf, B.L. (2004) *Org. Lett.*, **6**, 67–70.

13 Gordon, G.J., Luker, T., Tuckett, M.W. and Whitby, R.J. (2000) *Tetrahedron*, **56**, 2113–2129.

14 Takahashi, T., Xi, Z., Fischer, R., Huo, S., Xi, C. and Nakajima, K. (1997) *J. Am. Chem. Soc.*, **119**, 4561–4562.

15 Kasai, K., Liu, Y., Harab, R. and Takahashi, T. (1998) *Chem. Commun.*, 1989–1990.

16 (a) Nakamura, I. and Yamamoto, Y. (2004) *Chem. Rev.*, **104**, 2127–2198; (b) Trost, B.M., Frederiksen, M.U. and Rudd, M.T. (2005) *Angew. Chem. Int. Ed.*, **44**, 6630–6666; (c) Trost, B.M. and Krische, M.J. (1998) *Synlett*, 1; (d) Diver, S.T. and Giessert, A.J. (2004) *Chem. Rev.*, **104**, 1317–1382.

17 Yamamoto, Y., Ohkoshi, N., Kameda, M. and Itoh, K. (1999) *J. Org. Chem.*, **64**, 2178–2179.

18 (a) Trost, B.M. and Toste, F.D. (2000) *J. Am. Chem. Soc.*, **122**, 714–715; (b) Trost, B.M. and Toste, F.D. (2002) *J. Am. Chem. Soc.*, **124**, 5025–5036.

19 (a) Mori, M., Saito, N., Tanaka, D., Takimoto, M. and Sato, Y. (2003) *J. Am. Chem. Soc.*, **125**, 5606–5607; (b) Tanaka, D., Sato, Y. and Mori, M. (2006) *Organometallics.*, **25**, 799–801.

20 Trost, B.M. and Rudd, M.T. (2003) *J. Am. Chem. Soc.*, **125**, 11516–11517.

21 (a) Becker, E., Rüba, E., Mereiter, K., Schmid, R. and Kirchner, K. (2001)

22 Trost, B.M. and Rudd, M.T. (2002) *J. Am. Chem. Soc.*, **124**, 4178–4179.
23 Chang, H., Jayanth, T.T. and Cheng, C. (2007) *J. Am. Chem. Soc.*, **129**, 4166–4167.
24 Krafft, M.E., Wilson, A.M., Dasse, O.A., Bonaga, L.V.R., Cheung, Y.Y., Fu, Z., Shao, B. and Scott, I.L. (1998) *Tetrahedron Lett.*, **39**, 5911–5914.
25 Buisine, O., Aubert, C. and Malacria, M. (2001) *Chem. Eur. J*, **7**, 3517–3525.
26 Lautens, M., Klute, W. and Tam, W. (1996) *Chem. Rev.*, **96**, 49–92.
27 (a) Nishimura, T., Kawamoto, T., Sasaki, K., Tsurumaki, E. and Hayashi, T. (2007) *J. Am. Chem. Soc.*, **129**, 1492–1493; (b) Allen, A., Marquand, P.L., Burton, R., Villeneuve, K. and Tam, W. (2007) *J. Org. Chem.*, **72**, 7849–7857.
28 (a) Makino, T. and Itoh, K. (2003) *Tetrahedron Lett.*, **44**, 6335–6338; (b) Makino, T. and Itoh, K. (2004) *J. Org. Chem.*, **69**, 395–405.
29 (a) Wender, P.A., Husfeld, C.O., Langkopf, E. and Love, J.A. (1998) *J. Am. Chem. Soc.*, **120**, 1940–1941; (b) Wender, P.A. and Sperandio, D. (1998) *J. Org. Chem.*, **63**, 4164–4165.
30 Binger, P., Wedemanna, P., Kozhushkovc, S.I. and de Meijere, A. (1998) *Eur. J. Org. Chem.*, 113–119.
31 Jang, H. and Krische, M.J. (2004) *J. Am. Chem. Soc.*, **126**, 7875–7880.
32 Cao, P., Wang, B. and Zhang, X. (2000) *J. Am. Chem. Soc.*, **122**, 6490–6491.
33 (a) Cao, P. and Zhang, X. (2000) *Angew. Chem. Int. Ed.*, **39**, 4104–4106; (b) Lei, A., He, M. and Zhang, X. (2002) *J. Am. Chem. Soc.*, **124**, 8198–8199; (c) Lei, A., Waldkirch, J.P., He, M. and Zhang, X. (2002) *Angew. Chem. Int. Ed.*, **41**, 4526–4529; (d) Lei, A., He, M., Wu, S. and Zhang, X. (2002) *Angew. Chem. Int. Ed.*, **41**, 3457–3460.
34 Lei, A., He, M. and Zhang, X. (2003) *J. Am. Chem. Soc.*, **125**, 11472–11473.
35 (a) Mikami, K., Yusa, Y., Hatano, M., Wakabayashi, K. and Aikawa, K. (2004) *Chem. Commun.*, 98–99; (b) Mikami, K., Yusa, Y., Hatano, M., Wakabayashi, K. and Aikawa, K. (2004) *Tetrahedron*, **60**, 4475–4480.
36 Chatani, N., Kataoka, K. and Murai, S. (1998) *J. Am. Chem. Soc.*, **120**, 9104–9105.
37 (a) Brummond, K.M., Chen, H., Sill, P. and You, L. (2002) *J. Am. Chem. Soc.*, **124**, 15186–15187; (b) Shibata, T., Takesue, Y., Kadowaki, S. and Takagi, K. (2003) *Synlett*, 268–270; (c) Brummond, K.M. and Mitasev, B. (2004) *Org. Lett.*, **6**, 2245–2248; (d) Brummond, K.M., Painter, T.O., Probst, D.A. and Mitasev, B. (2007) *Org. Lett.*, **9**, 347–349.
38 Jiang, X. and Ma, S. (2007) *J. Am. Chem. Soc.*, **129**, 11600–11607.
39 Tanaka, K., Otake, Y. and Hirano, M. (2007) *Org. Lett.*, **9**, 3953–3956.
40 Lu, P. and Ma, S. (2007) *Org. Lett.*, **9**, 2095–2097.
41 (a) Ma, S., Lu, P., Lu, L., Hou, H., Wei, J., He, Q., Gu, Z., Jiang, X. and Jin, X. (2005) *Angew. Chem. Int. Ed.*, **44**, 5275–5278; (b) Ma, S. and Lu, L. (2007) *Chem. Asian. J.*, **2**, 199–204.
42 Lu, P. and Ma, S. (2007) *Org. Lett.*, **9**, 5319–5321.
43 Kezuka, S., Okado, T., Niou, E. and Takeuchi, R. (2005) *Org. Lett.*, **7**, 1711–1714.
44 (a) Montgomery, J. (2000) *Acc. Chem. Res.*, **33**, 467–473; (b) Montgomery, J. (2004) *Angew. Chem. Int. Ed.*, **43**, 3890–3908.
45 Kawasaki, T., Saito, S. and Yamamoto, Y. (2002) *J. Org. Chem.*, **67**, 4911–4915.
46 Montgomery, J., Oblinger, E. and Savchenko, A.V. (1997) *J. Am. Chem. Soc.*, **119**, 4911–4920.
47 (a) Rayabarapu, D.K., Sambaiah, T. and Cheng, C. (2001) *Angew. Chem. Int. Ed.*, **40**, 1286–1288; (b) Rayabarapu, D.K. and Cheng, C. (2002) *Pure. Appl. Chem.*, **74**, 69–75.
48 Montgomery, J. and Seo, J. (1998) *Tetrahedron*, **54**, 1131–1144.

49 (a) Takimoto, M., Mizuno, T., Satoa, Y. and Mori, M. (2005) *Tetrahedron Lett.*, **46**, 5173–5176; (b) Takimoto, M., Mizuno, T., Morib, M. and Sato, Y. (2006) *Tetrahedron*, **62**, 7589–7597.

50 Takimoto, M. and Mori, M. (2002) *J. Am. Chem. Soc.*, **124**, 10008–10009.

51 Takimoto, M., Nakamura, Y., Kimura, K. and Mori, M. (2004) *J. Am. Chem. Soc.*, **126**, 5956–5957.

52 Oblinger, E. and Montgomery, J. (1997) *J. Am. Chem. Soc.*, **119**, 9065–9066.

53 (a) Chowdhury, S.K., Amarasinghe, K.K.D., Heeg, M.J. and Montgomery, J. (2000) *J. Am. Chem. Soc.*, **122**, 6775–6776; (b) Mahandru, G.M., Skauge, A.R.L., Chowdhury, S.K., Amarasinghe, K.K.D., Heeg, M.J. and Montgomery, J. (2003) *J. Am. Chem. Soc.*, **125**, 13481–13485.

54 (a) Amarasinghe, K.K.D., Chowdhury, S.K., Heeg, M.J. and Montgomery, J. (2001) *Organometallics*, **20**, 370–372; (b) Hratchian, H.P., Chowdhury, S.K., Gutiérrez-García, V.M., Amarasinghe, K.K.D., Heeg, M.J., Schlegel, H.B. and Montgomery, J. (2004) *Organometallics*, **23**, 4636–4646.

55 Yamamoto, Y., Kuwabara, S., Ando, Y., Nagata, H., Nishiyama, H. and Itoh, K. (2004) *J. Org. Chem.*, **69**, 6697–6705.

10
Transition Metal Catalyzed Cyclization Reactions of Functionalized Alkenes, Alkynes, and Allenes
Nitin T. Patil and Yoshinori Yamamoto

10.1
Introduction

In studying the evolution of organic chemistry one can quickly judge that cyclization reactions are very important in shaping the domain of organic chemistry. In the last two decades many carbon–carbon and carbon–heteroatom bond-forming cyclization reactions have evolved as powerful tools in synthetic organic chemistry. Among them, transition metal catalyzed cyclization reactions are most valuable. A variety of multiply substituted carbocycles and heterocycles can be obtained by using cyclization processes. These reactions are very interesting due not only to their ability to construct complex molecules from readily accessible starting materials but also to the fact that they can be carried out under mild conditions and, in some cases, with high atom economy. There are many different types of catalytic cyclization processes reported in the literature and these will be discussed in the chapters by other experts. In this chapter, the transition metal catalyzed cyclization reaction of functionalized alkenes, allenes, and alkynes, for the synthesis of carbocycles and heterocycles, is described. Since there are innumerable reports on this topic [1], it is difficult for us to describe them all and so this chapter provides a personal selection among recent examples of this category.

10.2
Cyclization Reactions of Alkenes

10.2.1
Cyclization with Tethered Nucleophiles

The activation of the C–C double bond of alkenes by a transition metal is generally considered difficult compared to that of alkynes and allenes. Out of several metal catalysts [2] platinum, gold and silver are known to be the best for this purpose. They

Handbook of Cyclization Reactions. Volume 1.
Edited by Shengming Ma
Copyright © 2010 WILEY-VCH Verlag GmbH & Co. KGaA, Weinheim
ISBN: 978-3-527-32088-2

have an ability to coordinate with π-bonds of the alkene, thereby allowing the nucleophiles to attack. One of the valuable transformations in this category was described by He and coworkers who have reported silver(I) triflate mediated intramolecular addition of a hydroxy or carboxyl group into inert olefins, to form cyclic ethers **2** and lactones **4** (Scheme 10.1) [3]. In general, good to excellent yields were obtained for substrates **1** and **2** under the experimentally simple reaction conditions.

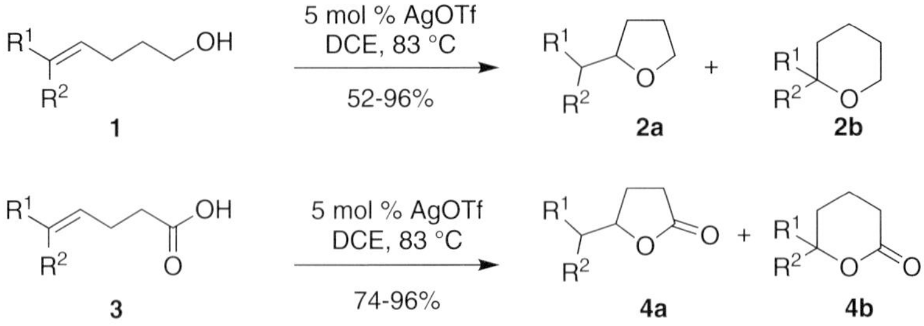

Scheme 10.1 Silver-catalyzed hydroalkoxylation and hydrocarboxylation reactions of alkenes.

The gold-catalyzed intramolecular cyclization of alkenol **5** was reported by Yang and He [4]. They observed that the formation of cyclic ether **6a/6b** from **5** using 5 mol% Ph$_3$PAuOTf in toluene at 85 °C took place smoothly (Scheme 10.2). Although, other metal catalyzed reactions for hydroalkoxylation are known [5], this is the first example in the history of gold catalysis.

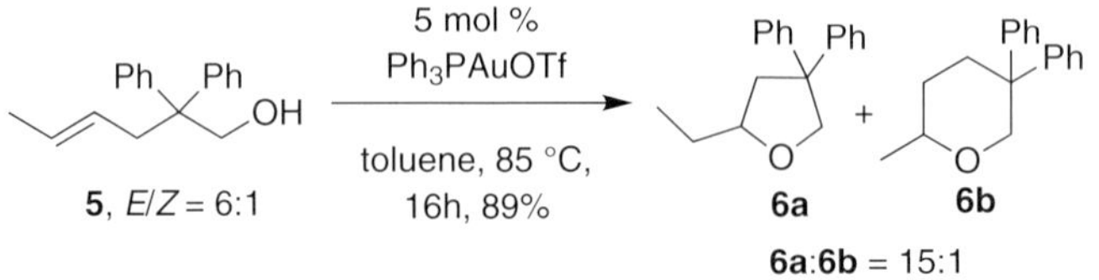

Scheme 10.2 Gold-catalyzed hydroalkoxylation of alkenes.

The He group also reported the Au(I)-catalyzed synthesis of dihydrobenzofurans **8** from aryl allyl ethers **7** (Scheme 10.3) [6]. They explored the mechanism of the reaction and revealed that it proceeded by a Claisen rearrangement (cf. **9**) followed by an intramolecular addition of the resulting phenol to the C—C bond of the allyl group. The same research group reported gold(I)-catalyzed intramolecular hydroamination of unactivated olefins **10** which led to a variety of nitrogen heterocycles **11** (Scheme 10.4) [7]. They further examined the hydroamination of 1,5-dienes **12** with TsNH$_2$ (Scheme 10.5). The first intermolecular hydroamination of a 1,5-diene by TsNH$_2$ followed by a second intramolecular hydroamination produces pyrrolidines **13** in one pot. After the publication of the above paper, a similar report on intramolecular hydroamination has appeared [8].

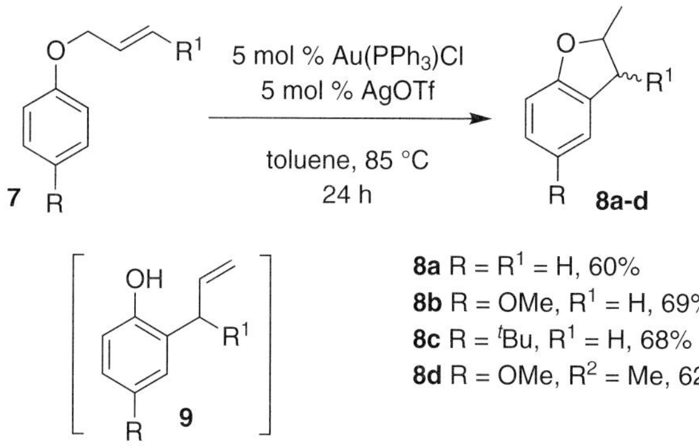

10.2 Cyclization Reactions of Alkenes

Scheme 10.3 Gold-catalyzed tandem Claisen rearrangement/hydroalkoxylation.

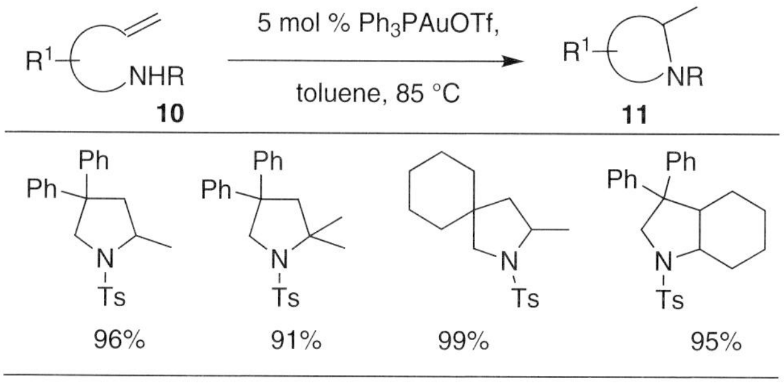

8a R = R^1 = H, 60%
8b R = OMe, R^1 = H, 69%
8c R = ^{t}Bu, R^1 = H, 68%
8d R = OMe, R^2 = Me, 62%

Scheme 10.4 Gold-catalyzed hydroamination of alkenes.

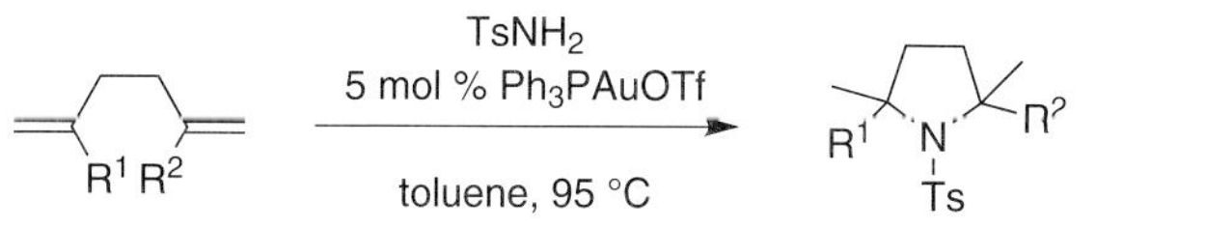

96%   91%   99%   95%

Scheme 10.5 Synthesis of pyrrolidines via double hydroamination of 1,5-dienes.

12a R^1 = H, R^2 = H
12b R^1 = H, R^2 = Me
12c R^1 = Me, R^2 = Me

13a R^1 = H, R^2 = H,
64% (cis/trans: 37/63)
13b R^1 = H, R^2 = Me, 90%
13c R^1 = Me, R^2 = Me, 80%

The addition of carbon nucleophiles to unactivated alkenes under gold catalysis was reported by Che and Zhou. For example, the substrates **14** underwent cyclization in the presence of Au[P(t-Bu)$_2$(o-biphenyl)]Cl **16** (5 mol%) and AgOTf (5 mol%) under mild conditions to produce highly substituted lactams **15** in high yields and regioselectivities (Scheme 10.6) [9].

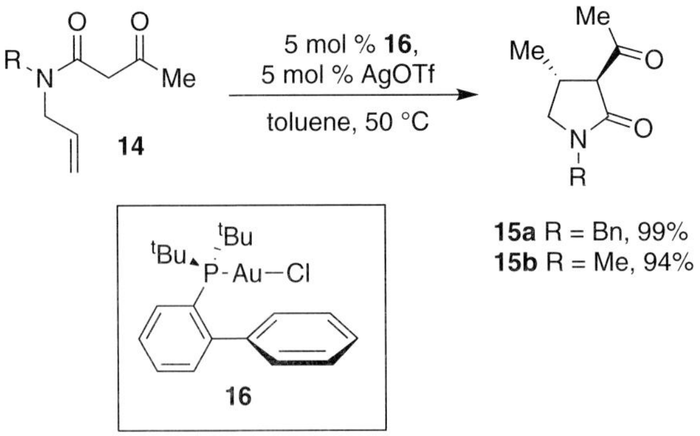

Scheme 10.6 Gold-catalyzed hydroamination of alkenes.

10.2.2
Hydroarylation Reactions

Intramolecular hydroarylation of alkenes via activation of the C—C double bond provides a direct route to heterocycles and carbocycles. Since the reactivity of alkenes is generally considered low compared to that of allenes and alkynes, only a few examples are found in the literature concerning this topic. The Sames group developed a bimetallic system which consists of a combination of RuCl$_3$ and AgOTf for an intramolecular hydroarylation of arene-enes **17** (Scheme 10.7) [10]. A variety of annulated products are accessible under mild conditions using this method.

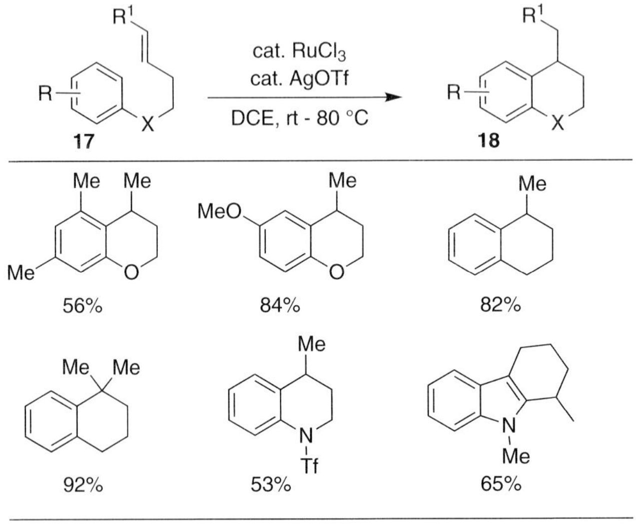

Scheme 10.7 Ruthenium-catalyzed hydroarylation of alkenes.

The Widenhoefer group reported a mild platinum-catalyzed protocol for the intramolecular alkylation of indoles with unactivated olefins. For example, the reaction of **19** with 2 mol% $PtCl_2$ in 1,4-dioxane, that contained a trace of HCl (5 mol%), at 60 °C led to the formation of **20** (Scheme 10.8) [11]. An asymmetric variant of this process was reported by the same research group (Scheme 10.9) [12].

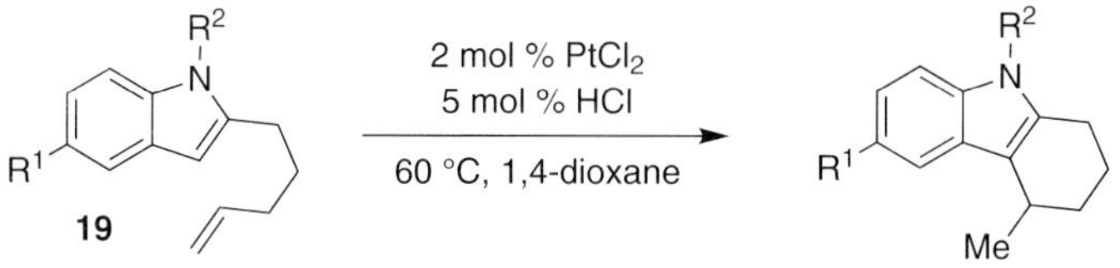

20a R^1 = H, R^2 = Me, 92%
20b R^1 = H, R^2 = Bn, 85%
20c R^1 = OMe, R^2 = Me, 94%

Scheme 10.8 Platinum-catalyzed hydroarylation of alkenes.

20aa R^1 = H, R^2 = Bn, 93%, 88%ee
20ab R^1 = OMe, R^2 = Me, 96%, 87%ee

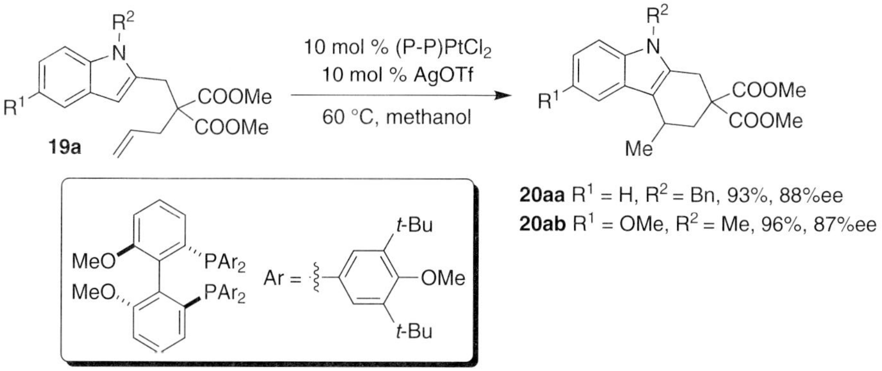

Scheme 10.9 Platinum-catalyzed asymmetric hydroarylation of alkenes.

Li and coworkers reported the annulation of phenols with cyclohexadiene to form the benzofurans **21** by using a combination of $AuCl_3$/AgOTf catalyst (Scheme 10.10) [13]. It was found that the presence of electron-donating groups on the aromatic ring promoted the reaction. A Friedel–Craft type reaction (cf. **22**) and intramolecular hydroalkoxylation (cf. **23**) are the key features of the mechanism. Youn and Eom reported the silver-catalyzed annulation of phenols and naphthols with cyclic as well as acyclic dienes to form dihydrobenzopyran and dihydrobenzofuran [14]. Recently, the air and moisture stable $Cu(OTf)_2$-bipy catalyst was reported for this purpose [15].

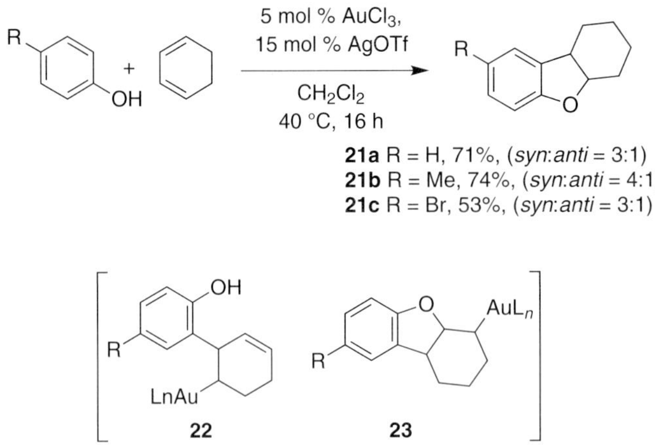

462 | 10 Transition Metal Catalyzed Cyclization Reactions of Functionalized Alkenes, Alkynes, and Allenes

21a R = H, 71%, (syn:anti = 3:1)
21b R = Me, 74%, (syn:anti = 4:1)
21c R = Br, 53%, (syn:anti = 3:1)

Scheme 10.10 Gold-catalyzed reactions of phenols with dienes.

10.2.3
Other Cascade Processes

Chemler and coworkers reported the oxidative cyclization of tosyl-o-allylaniline **24** to produce tetracycle **25** (Scheme 10.11) [16]. The oxidative cyclization of **24** occurred when treated with 3 equiv. Cu(OAc)$_2$ and 1 equiv. Cs$_2$CO$_3$ in CH$_3$CN or DMF at 120 °C. On the other hand, when **24** was treated with Pd/Cu bimetallic catalyst the indole product **26** was obtained through an aminopalladation–β-hydride elimination sequence. A plausible mechanism for the former case is depicted in Scheme 10.12. The formation of a nitrogen–copper(II) bond takes place (cf. **27**) at the beginning, followed by intramolecular migratory insertion to form the indole nucleus **28**. Intramolecular cyclization of **28** via a radical pathway affords **29**, which on loss of a hydrogen radical gives **25**.

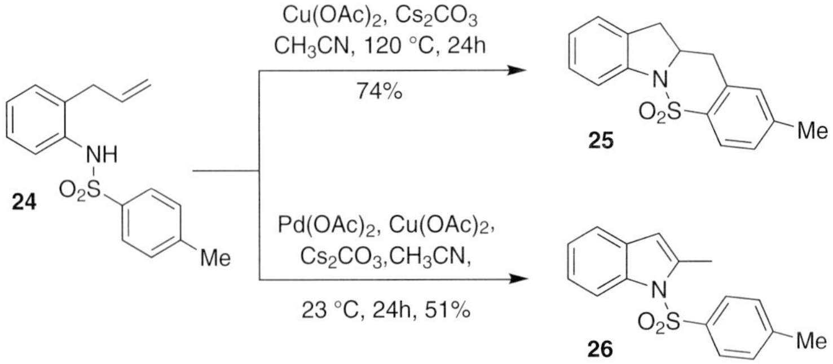

Scheme 10.11 Copper and palladium-catalyzed cyclization of aminoalkenes.

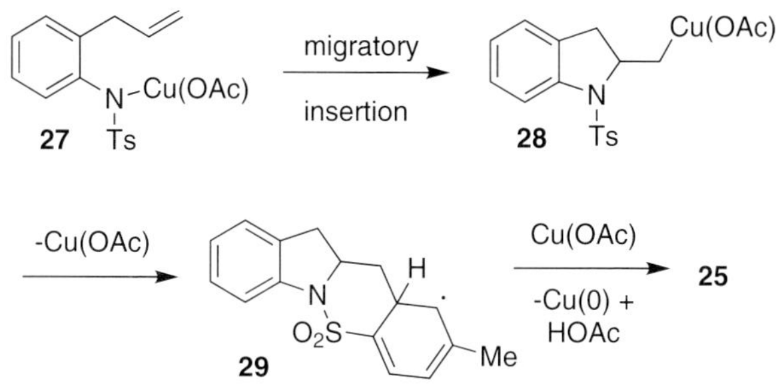

Scheme 10.12 Mechanism of copper-catalyzed cyclization of aminoalkenes.

They also reported the intramolecular 1,2-diamination of unactivated olefin **30** under Cu(OAc)$_2$ catalysis which gave the nitrogen-containing heterocycle **31** (Scheme 10.13) [17]. The *in situ* generated intermediate **32** on migratory insertion gives the organocopper species **33**, which on ligand exchange with a nitrogen followed by reductive elimination affords product **31**. The synthesis of bicyclic guanidines through palladium-catalyzed diamination of olefins is also known [18].

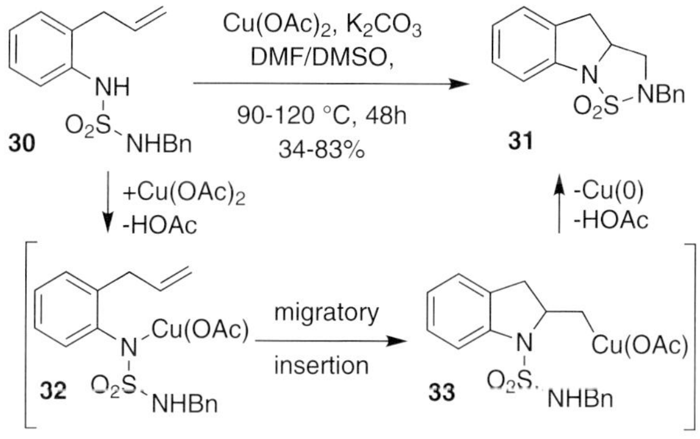

Scheme 10.13 Copper-catalyzed diamination of alkenes.

Widenhoefer and Han described a one-pot procedure for the synthesis of 2,3,5-trisubstituted furans from α-alkenyl β-diketones (Scheme 10.14) [19]. Treatment of 4-allyl-2,6-dimethyl-3,5-heptanedione **34** with a catalytic amount of PdCl$_2$(CH$_3$CN)$_2$ (5 mol%) and a stoichiometric amount of CuCl$_2$ (2.2 equiv.) in dioxane at 60 °C for 12 h gave 3-isobutyryl-2-isopropyl-5-methylfuran **35** in 77% isolated yield. A number of α-alkenyl β-diketones underwent oxidative alkoxylation under these conditions to form 2,3,5-trisubsituted furans in moderate to good yields. Most probably, the nucleophilic attack of enolic oxygen atom on the palladium-complexed olefin (cf. **36**) forms the palladium dihydrofurylmethyl intermediate **37** which undergoes β-hydride elimination and proton migration to afford furan **35**.

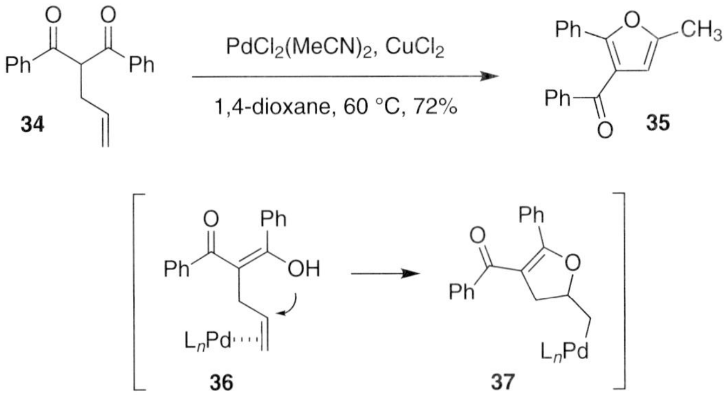

Scheme 10.14 Palladium-catalyzed synthesis of furans from 4-allyl-2,6-dimethyl-3,5-heptanedione.

The highly activated olefins are known to undergo bis-functionalization reactions with appropriate reaction partners in the presence of palladium [20]. The Yamamoto group reported the palladium-catalyzed reaction of the activated olefin **38** with the allylic carbonates of type **39** for the synthesis of cyclic ethers **40** (Scheme 10.15) [21]. The method is suitable for synthesizing five and six-membered cyclic ethers, however, lesser yields were observed for the formation of seven-membered cyclic ethers. Generally, the diastereoselectivities of the products were in the range of 60–70/40–30. The oxidative insertion of Pd(0) into the allylic carbonates **39** produces the π-allylpalladium complexes **41**, which on removal of ^{i}PrOH produce another π-allylpalladium complex **42**. The Michael addition of the oxygen nucleophile of **42** to **38** gives the intermediates **43**, which undergo intramolecular attack of the nucleophilic carbon on the π-allylpalladium complexes resulting in the formation of the cyclic ethers **40**.

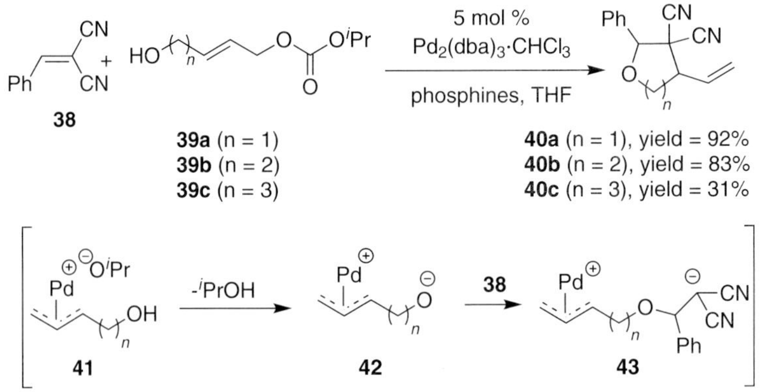

Scheme 10.15 Palladium-catalyzed reactions of activated olefins with allyl carbonate bearing hydroxyl group.

The palladium-catalyzed reaction of vinylic oxirane **44a** [22] and vinylic aziridine **44b** [23] with the activated olefin **38** for the formation of the five-membered cyclic ether **45a** and the pyrrolidine derivative **45b** was reported by Yamamoto's group (Scheme 10.16). Mechanistically, Pd (0) catalyst adds oxidatively to **44** to produce the π-allylpalladium complex **46**. The Michael addition of a hetero nucleophile in **46** to the activated olefin **38** gives **47**, which undergoes intramolecular nucleophhilic attack on the inner π-allylic carbon atom to give the cyclization products **45** with regeneration of the Pd (0) species.

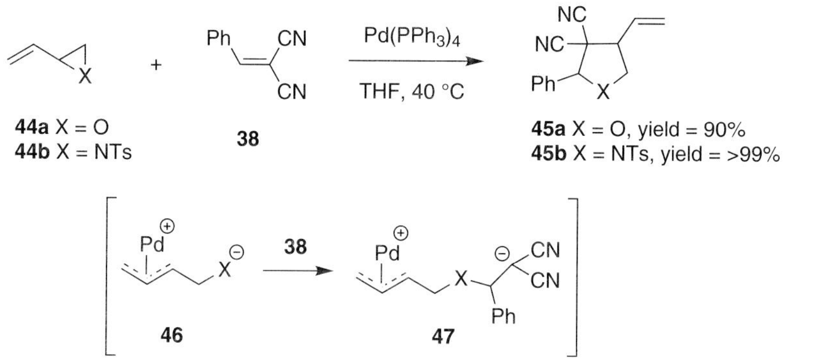

Scheme 10.16 Palladium-catalyzed reactions of activated olefins with vinyl epoxides/aziridines.

The reaction of olefin **38** with allene **48** bearing an active methine under palladium catalysis gave cyclopentanes **49** in good to excellent yields (Scheme 10.17) [24]. Mechanistically, the Pd(0) produced *in situ* adds oxidatively to an acidic C–H bond of **48** to give the hydridopalladium(II) intermediate **50**. This undergoes the carbopalladation reaction with **38** to lead to another hydridopalladium species **51**. Intramolecular hydropalladation of **51** gives the π-allylpalladium complex **52** which, on reductive removal of Pd(0), gives the carbocycles **49**.

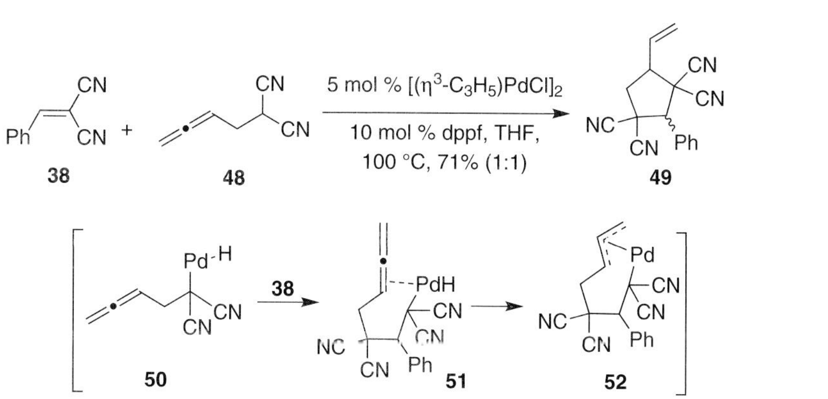

Scheme 10.17 Palladium-catalyzed reactions of activated olefin with allene bearing active methine.

10.3
Cyclization Reactions of Alkynes

10.3.1
Cyclization with Tethered Nucleophiles

The attack of various nucleophiles on an alkyne in the presence of a transition metal represents a powerful tool for accessing a number of cyclic products. Mechanistically, the coordination of metal catalyst to the π-system renders the alkyne moiety susceptible to nucleophilic attack, thus enabling intramolecular cyclization. In this section we will give a few examples of this category [25].

The silver-catalyzed cyclization of acetylenic alcohols **53** leading to 2-methyleneoxolanes **54** was reported by Pale and Chuche (Scheme 10.18) [26]. The Thorpe–Ingold effect is very influential in this case. The cyclization of the acetylenic alcohols **53a,b,e,f** required only catalytic amounts of Ag_2CO_3, while **53c,d** needed a stoichiometric quantity.

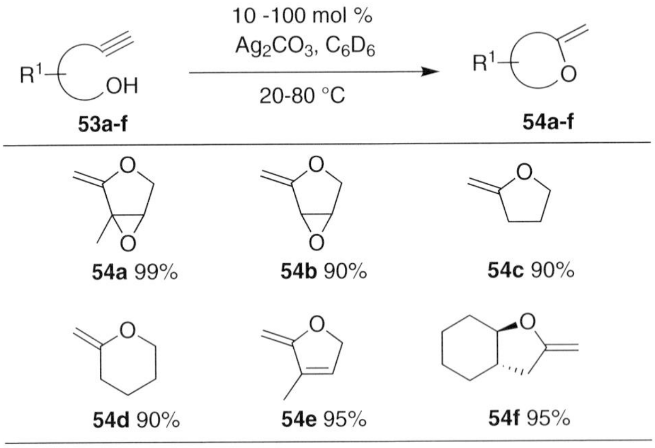

Scheme 10.18 Silver-catalyzed hydroalkoxylation of alkynes.

The gold-catalyzed cyclization of (Z)-enynols **55** afforded (Z)-5-ylidene-2,5-dihydrofurans **56** and fully substituted furans **57**, depending on the nature of the R^1 group (Scheme 10.19) [27]. For instance, when R^1 = alkyl or aryl, **56** was formed. On the other hand when R^1 = H, **57** was obtained. Two reaction conditions were employed; (i) $AuCl_3$ in CH_2Cl_2, (ii) $(PPh_3)AuCl/AgOTf$. Both generally gave products in high yields.

The synthesis of a new class of heterocycles that is, phosphaisocoumarins, was reported by Ding and Peng. They found that Cu(I) iodide efficiently catalyzed the cyclization of o-ethynylphenylphosphonic acid monoethyl esters **58** which led to the formation of phosphaisocoumarins **59** (Scheme 10.20) [28]. In all the cases only six-membered products were obtained via 6-endo-dig cyclization. The same research group developed a novel Ag_2CO_3-catalyzed cyclization of (Z)-2-alken-4-ynylphospho-

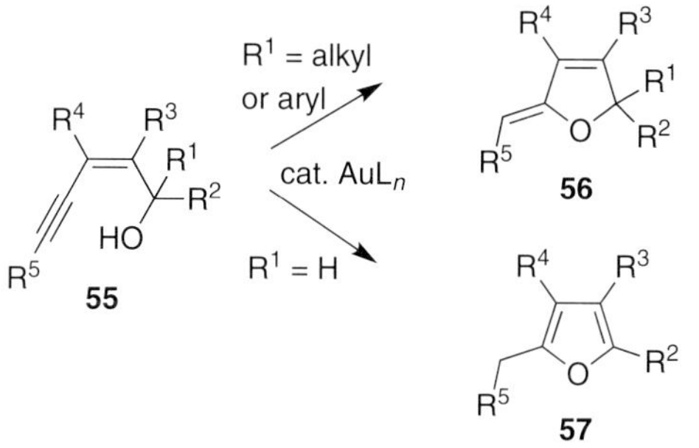

Scheme 10.19 Silver-catalyzed hydroalkoxylation/isomerization.

nic monoesters **60** to 2-ethoxy-2*H*-1,2-oxaphosphorin 2-oxides **61** in CH_2Cl_2 at room temperature (Scheme 10.21) [29]. Regioselective nucleophilic attack of the triple bond by the phosphonyl oxygen in the endo mode (cf. **62**) gives the vinyl silver species **63**, which subsequently undergoes proton transfer, with regeneration of the silver catalyst, to produce **61**.

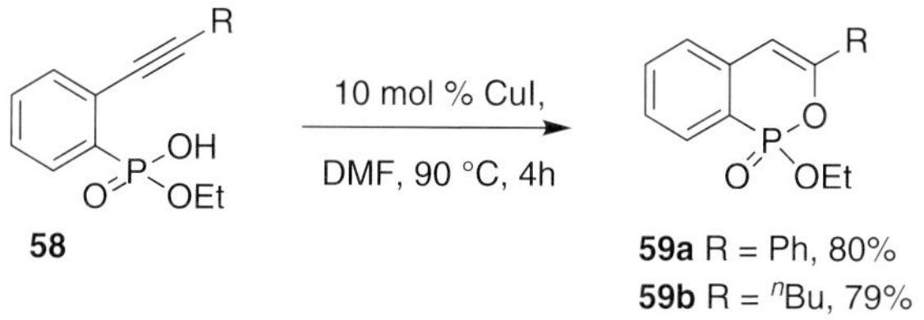

Scheme 10.20 Copper-catalyzed cyclization of *o*-ethynylphenylphosphonic acid monoethyl esters.

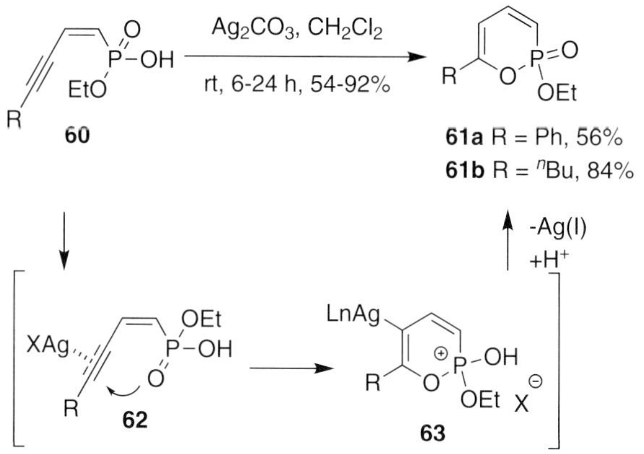

Scheme 10.21 Silver-catalyzed cyclization of (Z)-2-alken-4-ynylphosphonic monoesters.

Dalla and Pale reported the cyclization of alkynoic acids **64** which readily gave lactones **65** (Scheme 10.22) [30]. Negishi and Xu showed that Ag catalyzed the lactonization of alkynoic acid **66** for the synthesis of (Z)-γ-alkylidenebutenolide **67**, a precursor for the synthesis of lissoclinolide (Scheme 10.23) [31]. It is interesting to note that exo-dig/endo-dig cyclization depends on the nature of the catalyst [32]. Rossi and coworkers also cyclized bromoalkynylacetic acids **68** to the corresponding (Z)-3-bromo-5-ylidene-5H-furan-2-ones **69** (Scheme 10.24) [33].

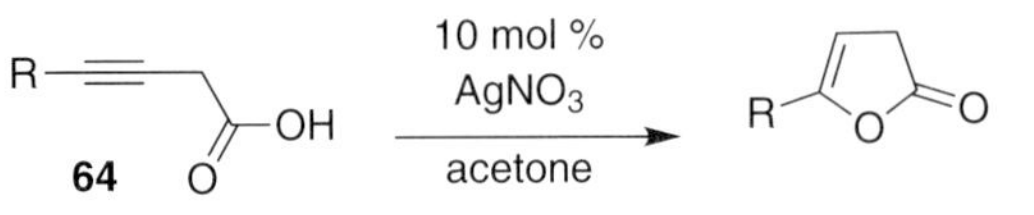

Scheme 10.22 Silver-catalyzed hydrocarboxylation of alkynes.

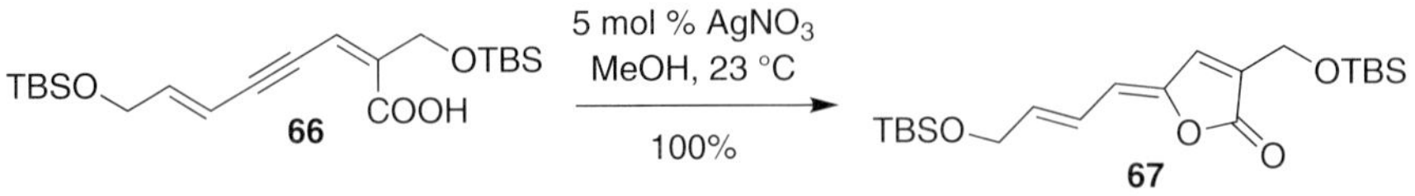

65a R = n-C_5H_{11} (95%)
65b R = CH_2=CH$(CH_2)_7$ (95%)

Scheme 10.23 Silver-catalyzed cyclization of alkynoic acid.

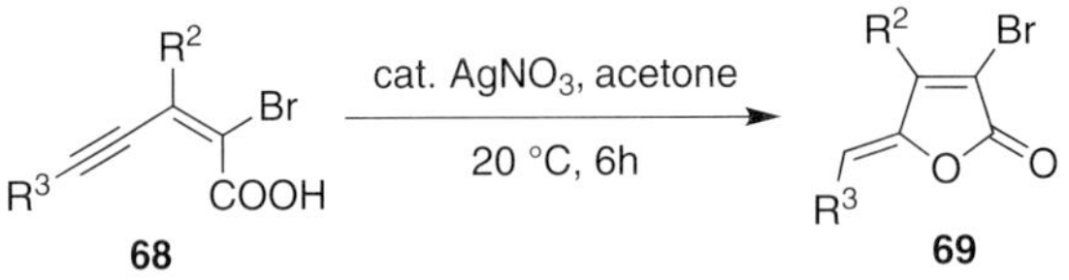

Scheme 10.24 Silver-catalyzed cyclization of bromoalkynylacetic acids.

Chaudhuri and Kundu reported the synthesis of benzodioxepinone **71a** and benzoxazepinone **71b** via cyclization of alkynoic acid **70a** and **70b**, respectively, in the presence of 5 mol% cuprous iodide and 1 equiv. of triethylamine in acetonitrile at 80 °C (Scheme 10.25) [34]. Though the substrates used were different, the efficiency of the gold catalysts can be judged by the recent report on the gold-catalyzed cyclization of alkynoic acids under extremely mild conditions (Scheme 10.26) [35].

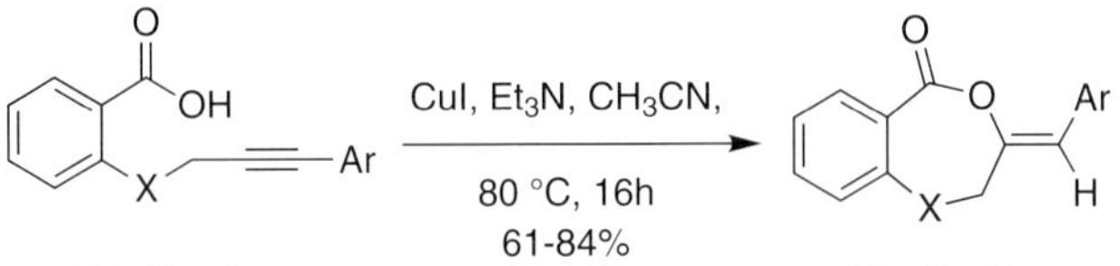

70a X = O
70b X = NBn

71a X = O
71b X = NBn

Scheme 10.25 Synthesis of benzodioxepinone and benzoxazepinone via copper-catalyzed hydrocarboxylation.

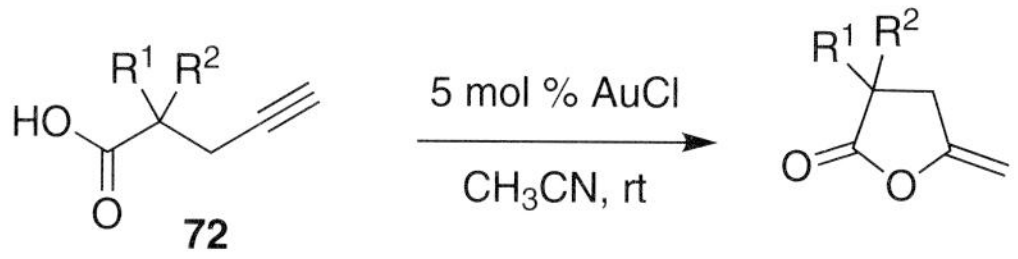

Scheme 10.26 Gold-catalyzed hydrocarboxylation of alkynes.

73a R¹ = COOMe, R² = allyl, 72%
73b R¹ = Ph, R² = propargyl, 98%
73c R¹ = COOEt, R² = Bn, 94%

Dyker and coworkers reported the AuCl$_3$-catalyzed intramolecular hydroamination of optically pure **74**, which gave chiral dihydroisoquinolines **75** and isoindoles **76** (Scheme 10.27) [36]. The former product was obtained via 6-endo-dig cyclization whereas the latter product was obtained from the 5-exo-dig cyclization/isomerization cascade.

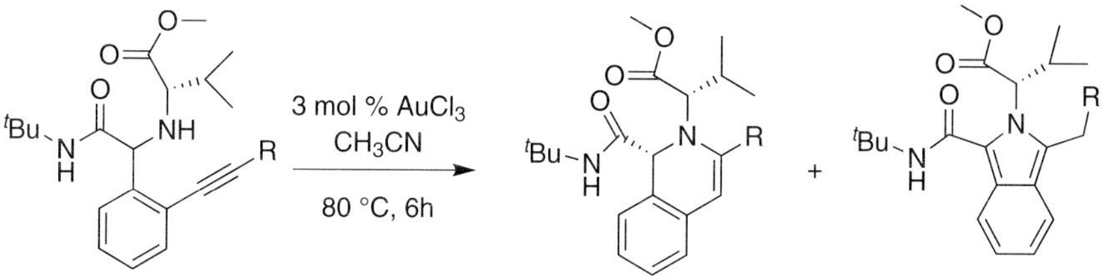

74a R = ^{n}Bu
74b R = Ph

75a R = ^{n}Bu, 23%
75b R = Ph, 35%

76a R = ^{n}Bu, 49%
76b R = Ph, 38%

Scheme 10.27 Gold-catalyzed hydroamination of alkynes for the synthesis of dihydroisoquinolines and isoindoles.

The cyclization of O-propargyl carbamates **77** afforded 4-methylene-2-oxazolidinones **78** via intramolecular nucleophilic addition of nitrogen atom to alkynes (Scheme 10.28). The reaction is highly dependent on the nature of the substituent (R¹) on the nitrogen. By the selection of either silver or copper catalyst, products were obtained in good yields. An illustrative example of the utility of a similar type of transformation can be found in the construction of a hexacyclic substructure of communesin B [37].

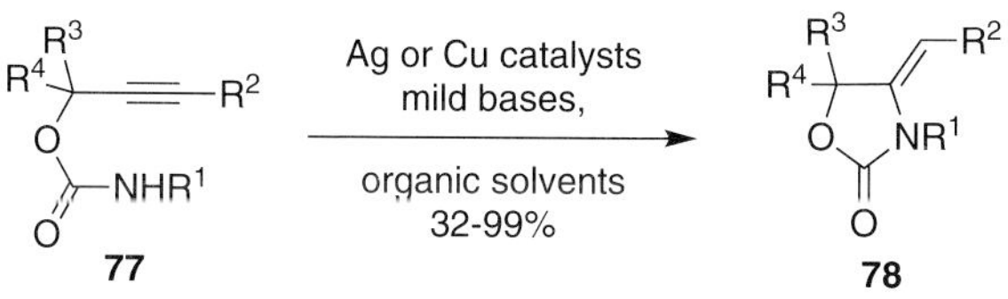

Scheme 10.28 Silver/copper-catalyzed hydroamidation of alkynes.

The tandem catalytic cyclization–hydroalkoxylation process of homopropargylic alcohols **79** was reported by Krause and Belting (Scheme 10.29) [38]. The catalytic system makes use of two catalysts that is, a metal catalyst and a Brønsted acid catalyst. The process is attractive and efficient for the synthesis of tetrahydrofuranyl ethers **80** in moderate to good yields. An elegant example of a similar strategy can be found in the synthesis of A-D rings of azaspiracid [39] and the spiroketal core of rubromycins [40]. The platinum-catalyzed tandem cyclization–hydroarylation of homopropargylic alcohols was also reported recently [41].

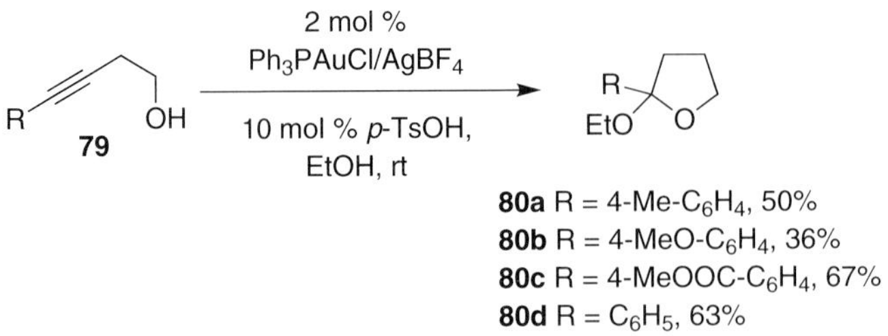

80a R = 4-Me-C_6H_4, 50%
80b R = 4-MeO-C_6H_4, 36%
80c R = 4-MeOOC-C_6H_4, 67%
80d R = C_6H_5, 63%

Scheme 10.29 Gold-catalyzed tandem cycloisomerization/alkoxylation of homopropargylic alcohols.

The bis-cycloisomerization of homopropargylic diols **81** was discovered by Genet and coworkers (Scheme 10.30) [42]. Such a cascade cyclization process provides an efficient and atom economical means of generating strained bicyclic ketals **82** from simple acyclic precursors. Lewis acid type activation of alkynes was reported in order to affect two intramolecular cyclizations, as shown in intermediate **83**.

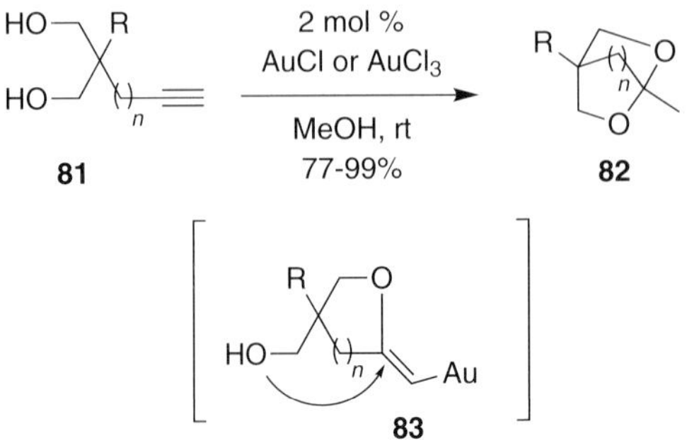

Scheme 10.30 Gold-catalyzed cycloisomerization of bishomopropargylic alcohols.

The tandem process for cyclization followed by oxidation of the resulting enol ethers has been reported recently. The process makes use of a cationic gold complex and converts (Z)-enynols **84** into butenolides **85** (Scheme 10.31) [43]. The cleavage of carbon–carbon triple bonds in (Z)-enynols under mild conditions is the important feature of these reactions. It was proposed that the intermediate **86** was produced, via gold-catalyzed intramolecular hydroalkoxylation, and this was further oxidized to butenolide by the reaction with dioxygen.

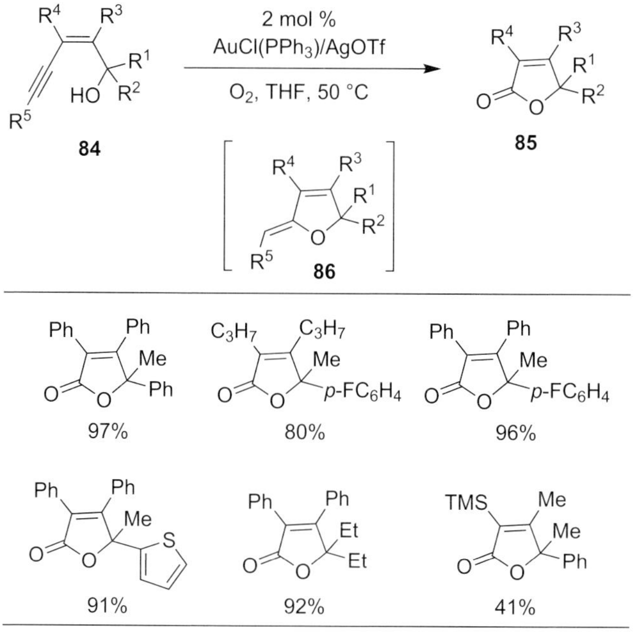

10.3 Cyclization Reactions of Alkynes

Scheme 10.31 Gold-catalyzed domino cyclilzation/oxidative cleavage reactions of (z)-enynols.

The copper-catalyzed CO_2 entrapment reaction of propargylic alcohols was reported by Deng and coworkers. For instance, the reactions of **87** with CO_2 in a [BMIm][PhSO$_3$]/CuCl catalyst system produced the corresponding α-methylene cyclic carbonates **88** in high yields (Scheme 10.32) [44]. It was reported that copper catalysts immobilized in ionic liquids can be reused three times without losing their activity.

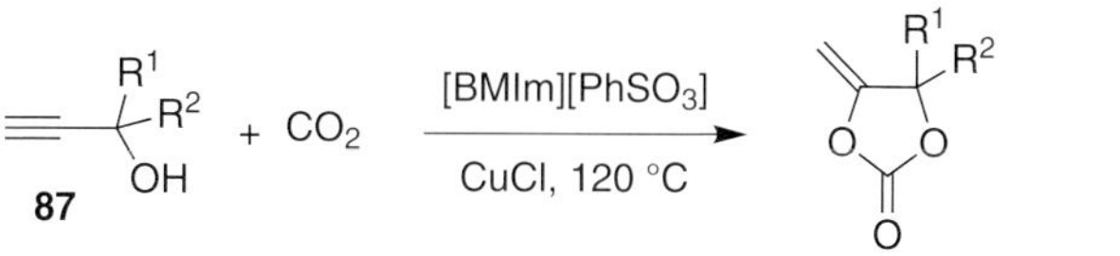

88a R^1, R^2 = Me, Et, 97%
88b R^1, R^2 = Me, ^{i}Bu, 94%
88c R^1, R^2 = Me, Me, 94%

Scheme 10.32 Copper-catalyzed reactions of propargylic alcohol with CO_2.

The alternative route for the synthesis of cyclic carbonate involves the use of alkynyl *tert*-butyl carbonates as starting material. Shin and Kang reported the gold-catalyzed cyclization of *tert*-butyl carbonates **89**, derived from homopropargylic alcohols, which led to the formation of cyclic enol carbonates **90** (Scheme 10.33) [45]. This reaction seems very general; however, internal alkynes were not viable substrates for this cyclization. Similarly, the cyclization of *N*-Boc protected alkynylamines in the presence of gold catalysts gave oxazolidinones with the liberation of isobutene. This cyclization was triggered by the addition of Au(PPh$_3$)Cl (5 mol%) and AgSbF$_6$ (5 mol %) to a solution of **91** in CH$_2$Cl$_2$. Various 2-oxazolidinones **92a** were obtained in high yields through the intermediacy of vinyl intermediate **93** (Scheme 10.34) [46].

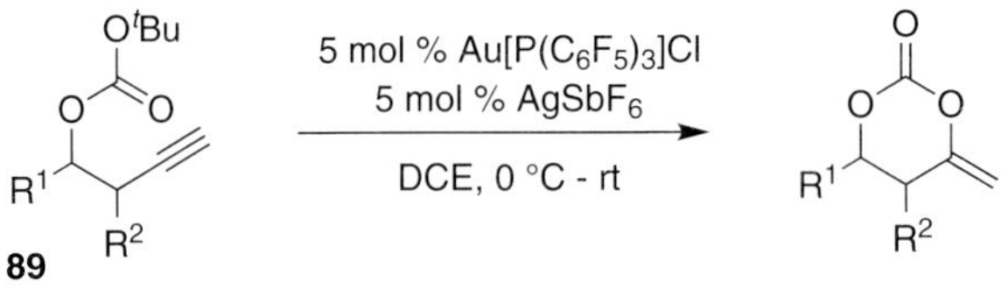

90a R^1 = p-Cl-C$_6$H$_4$, R$_2$ = H, 70%
90b R^1 = Cy, R$_2$ = H, 75%
90c R^1 = n-Pr, R$_2$ = H, 78%
90d R^1 = α-Cy, R$_2$ = α-Me, 71%

Scheme 10.33 Gold-catalyzed cyclization of alkynyl tert-butyl carbonates.

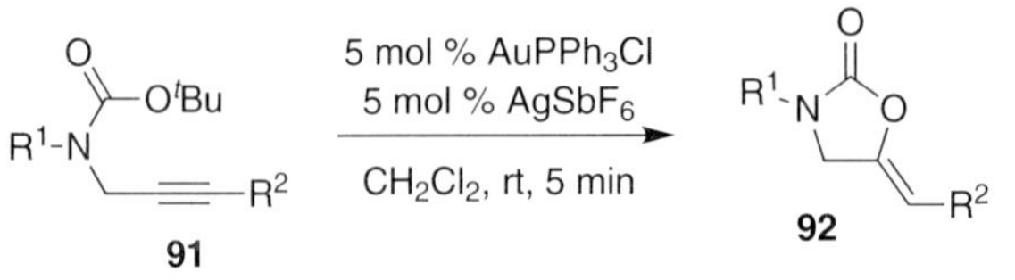

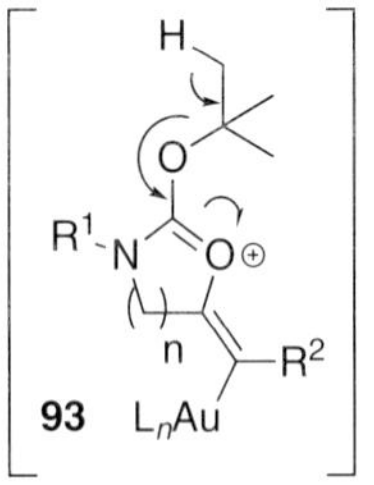

R^1 = 4-OMe-C$_6$H$_4$, R^2 = H, 80%
R^1 = 4-OMe-C$_6$H$_4$, R^2 = Me, 75%
R^1 = 4-OMe-C$_6$H$_4$, R^2 = Ph, 74%
R^1 = 4-OMe-C$_6$H$_4$, R^2 = COOEt, 95%
R^1 = H, R^2 = H, 92%
R^1 = Bn, R^2 = H, 95%
R^1 = allyl, R^2 = H, 80%
R^1 = Et, R^2 = H, 69%

Scheme 10.34 Gold-catalyzed cyclization of N-Boc-protected alkynylamines.

Gagosz and coworkers developed an efficient procedure for the synthesis of oxazolones **95** from tert-butyloxycarbamates **94** (Scheme 10.35) [47]. The starting material **94** could be easily obtained from bromoalkynes and carbamates via Cu(II)-catalyzed cross-coupling reactions.

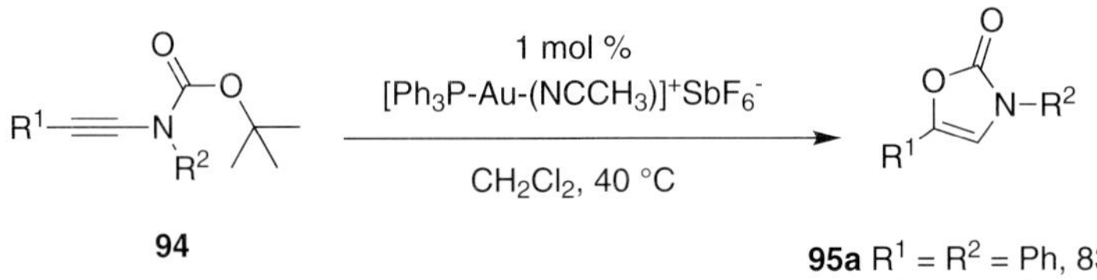

95a R^1 = R^2 = Ph, 83%
95b R^1 = Ph, R^2 = Bn, 78%
95c R^1 = ^{t}Bu, R^2 = Ph, 58%
95d R^1 = ^{n}C$_5$H$_{11}$, R^2 = Ph, 74%

Scheme 10.35 Gold-catalyzed cyclization of tert-butyloxycarbamates.

Not only hetero- but also carbon-nucleophiles are known to undergo addition across alkynes. The alkynylmalononitriles **96** underwent cyclization in the presence of $Pd(OAc)_2$-COD catalyst to afford the *Z*-isomers of the corresponding carbocycles **97** in all cases (except when R = TMS) (Scheme 10.36) [48]. Since the *Z*-isomers were obtained, the presence of the π-allylpalladium intermediate was ruled out completely. C–C activation by a metal catalyst as shown in **98** was proposed.

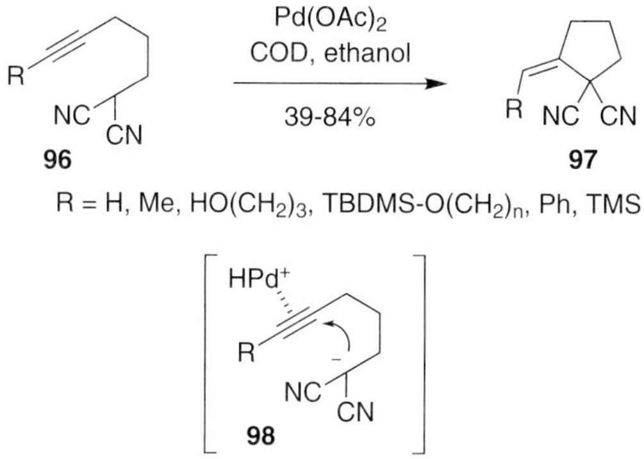
Pd(OAc)2
COD, ethanol
39-84%
R
NC
CN
96
97
R = H, Me, HO(CH2)3, TBDMS-O(CH2)n, Ph, TMS
HPd+
98

Scheme 10.36 Palladium-catalyzed hydrocarbonation of alkynes.

The Yamamoto group reported a novel approach for intramolecular hydroamination, hydroalkoxylation and hydrocarbonation reactions of alkynes. For instance, the reactions of the aminoalkynes **99a,b**, alkynols **99c**, and alkynylmethines **99d** in the presence of catalytic amounts of $Pd(PPh_3)_4$ and carboxylic acid (either acetic acid or benzoic acid) in 1,4-dioxane at 100 °C gave the corresponding nitrogen heterocycles **100a,b**, oxygen heterocycles **100c**, and carbocycles **100d** in good yields (Scheme 10.37) [49]. The mechanism of this reaction is depicted in Scheme 10.38. Hydropalladation of the alkynes **99a–d** with the H–Pd–OCOR, generated from Pd(0) and benzoic acid, most probably produces the vinylpalladium species **101** which, via β-elimination, gives substituted phenyl allene **102**. Subsequent hydropalladation of the allene **102** with H–Pd–OCOR gives the π-allyl palladium species **103**. Intramolecular nucleophilic attack of the pendant nucleophiles on the π-allylcarbon gives the products **100a–d** with the regeneration of active catalyst H–Pd–OCOR. The same research group could also make the reaction enantioselective, however, high catalyst loading was necessary in order to obtain good yields and ees (Scheme 10.39) [50]. The methodology has been extended for the intramolecular hydroamidation [51] hydroanilination [52] and hydrocarboxylation [53] to give tetrahydroquinolines, cyclic lactams and lactones, respectively.

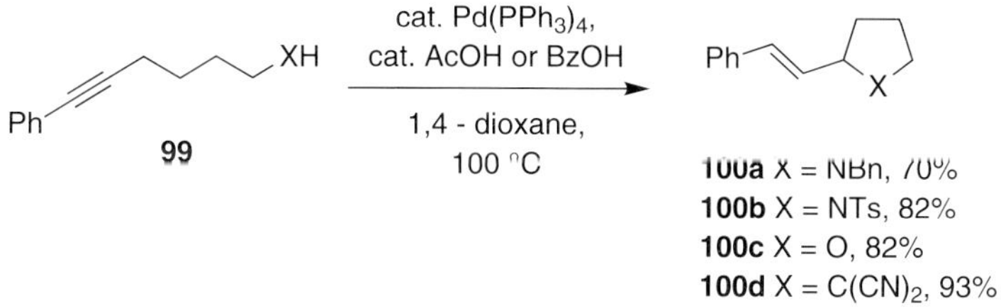
cat. Pd(PPh3)4,
cat. AcOH or BzOH
1,4 - dioxane,
100 °C
Ph
XH
99
X
100a X = NBn, 70%
100b X = NTs, 82%
100c X = O, 82%
100d X = C(CN)2, 93%

Scheme 10.37 Palladium-catalyzed addition of pronucleophiles to alkynes.

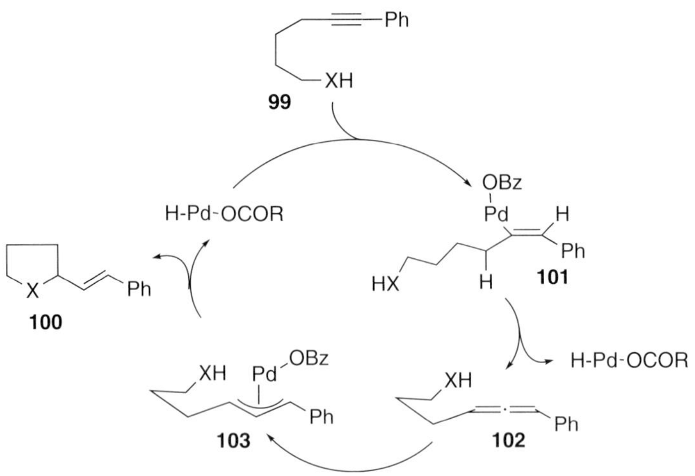

Scheme 10.38 Mechanism for palladium-catalyzed addition of pronucleophiles to alkynes.

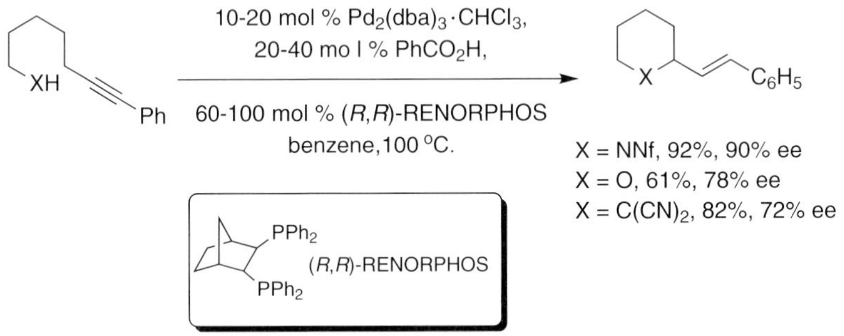

Scheme 10.39 Palladium-catalyzed asymmetric addition of pronucleophiles to alkynes.

X = NNf, 92%, 90% ee
X = O, 61%, 78% ee
X = C(CN)$_2$, 82%, 72% ee

Toste and coworkers reported the Au(I)-catalyzed addition of carbon nucleophiles across the C–C bond of alkynes (Scheme 10.40) [54]. The acetylenic dicarbonyl compounds **104** under experimentally simple conditions underwent cyclization in an exocyclic manner to give carbocycles **105**. However, for the cyclization to occur the alkynes must be unsubstituted. The mechanism involves nucleophilic attack on a Au(I)-alkyne complex by the enol form of the ketoester. The pioneering work from the same research group has shown that endocyclic cyclization is also possible if the carbon chain linking the nucleophile to the alkyne is short (Scheme 10.41) [55]. Interestingly, in this case both terminal as well as nonterminal alkynes **106** worked very well, giving cyclopentenes **107**.

The first example of gold(I)-catalyzed addition of silyl enol ethers to alkynes was reported by the same research group. The reaction allows for the diastereoselective synthesis of a variety of bicyclic frameworks **109** containing all-carbon quaternary centers from easily available TBS-protected enol ethers **108** (Scheme 10.42) [56]. An elegant example of this strategy can be found in the total synthesis of the natural products (+)-lycopladine A [56], (+)-fawcettimine [57] and plantecin [58].

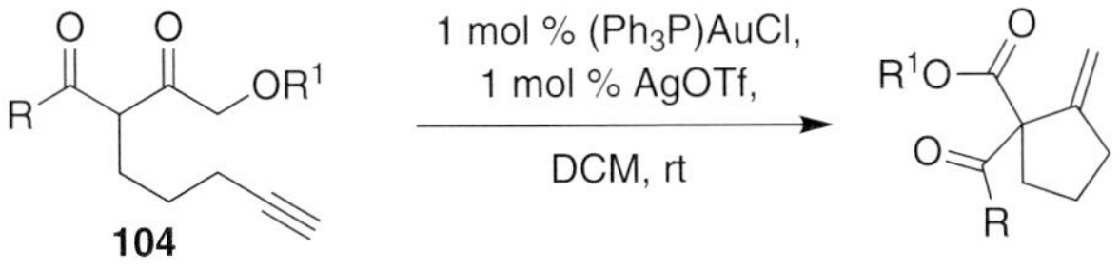

104

105a R = R[1] = Me, 94%
105b R = Ph, R[1] = Et, 93%
105c R = Me, R[1] = ^{t}Bu, 81%

Scheme 10.40 Gold-catalyzed exo-selective hydrocarbonation of alkynes.

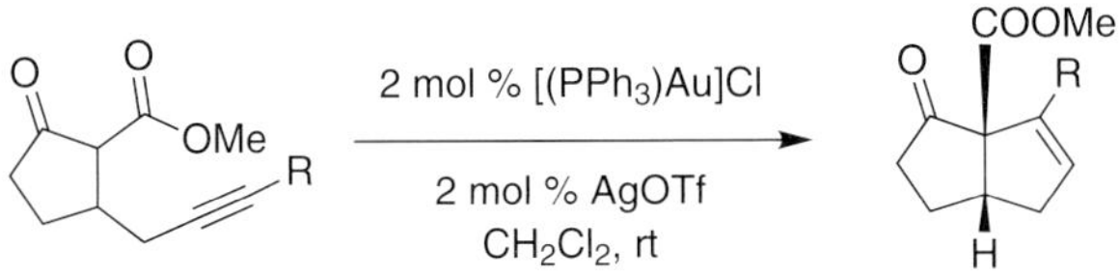

106

107a R = H, 83%
107b R = Me, 74%
107c R = Ph, 94%

Scheme 10.41 Gold-catalyzed endo-selective hydrocarbonation of alkynes.

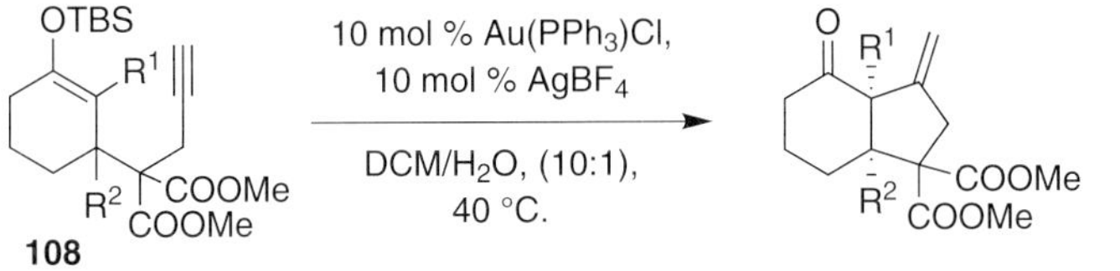

108

109a R[1] = Me, R[2] = Me, 90%
109b R[1] = Ph, R[2] = H, 83%

Scheme 10.42 Gold-catalyzed addition of silyl enol ethers to alkynes.

10.3.2
Domino sp–sp^2 Coupling and Cyclization

Sonogashira reaction between alkynes and vinyl/aryl halides followed by cyclization with pendant nucleophiles represents an excellent route to the synthesis of heterocycles. For example, the reaction of *o*-ethynylphenols **110** with a wide variety of unsaturated halides or triflates in the presence of $Pd(OAc)_2(PPh_3)_2$, CuI and Et_3N gives 2-vinyl and 2-arylbenzo[b]furans **111** (Scheme 10.43) [59]. This approach was extended to the synthesis of furanopyridines **113** from **112** as shown in Scheme 10.44 [60]. Similar examples were also reported by other researchers [61].

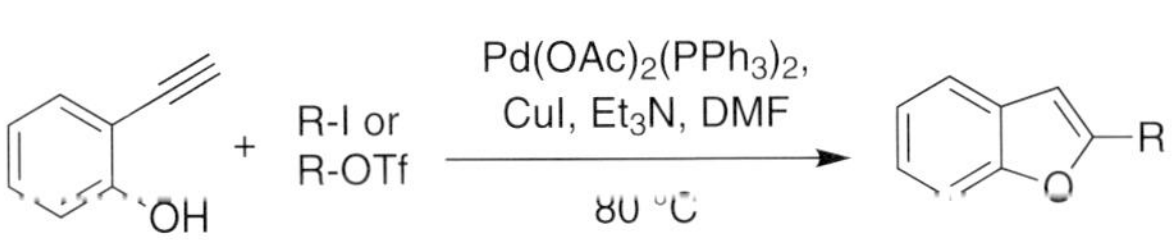

110 R = aryl or vinyl **111**

Scheme 10.43 Palladium-catalyzed cyclization of *o*-ethynylphenols with vinyl iodide/triflate.

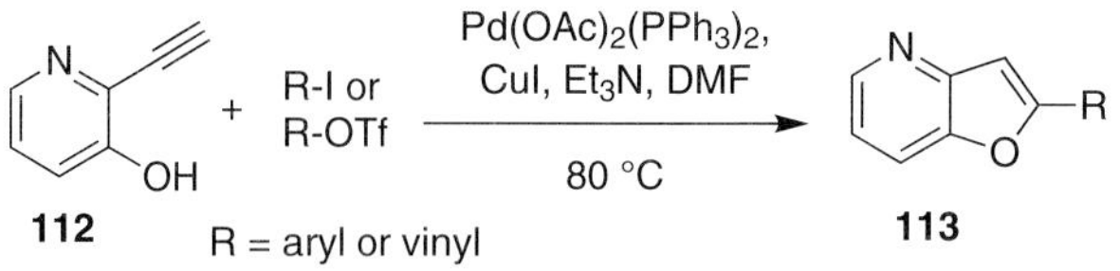

Scheme 10.44 Palladium-catalyzed synthesis of furanopyridines.

The one-pot reaction of 4-alkoxy-3-iodo-2-pyridone **114**, terminal alkynes and aryl halides gave furo[2,3-b]pyridones **115** (Scheme 10.45) [62]. Activation of the alkyne as shown in **116** was proposed. Kundu and coworkers reported the synthesis of (Z)-2,3-dihydro-2-(ylidene)-1,4-benzodioxins using palladium–copper catalysis (Scheme 10.46) [63]. For instance, aryl halides reacted with **117** in the presence of 3.5 mol% $(PPh_3)_2PdCl_2$ and 7 mol% CuI in triethylamine to afford products **118** in moderate yields. The reaction of 2-mercaptopropargyl benzimidazole **119** with various iodobenzenes catalyzed by Pd–Cu led to the formation of 3-benzylthiazolo[3,2-a]benzimidazoles **120** (Scheme 10.47) [64].

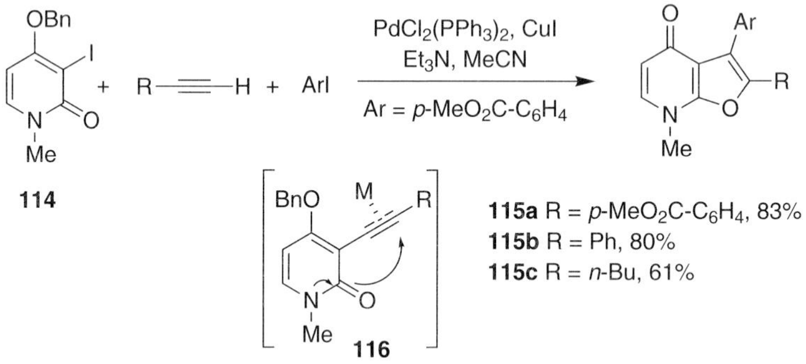

115a R = p-MeO$_2$C-C$_6$H$_4$, 83%
115b R = Ph, 80%
115c R = n-Bu, 61%

Scheme 10.45 Palladium-catalyzed synthesis of furo[2,3-b]pyridones.

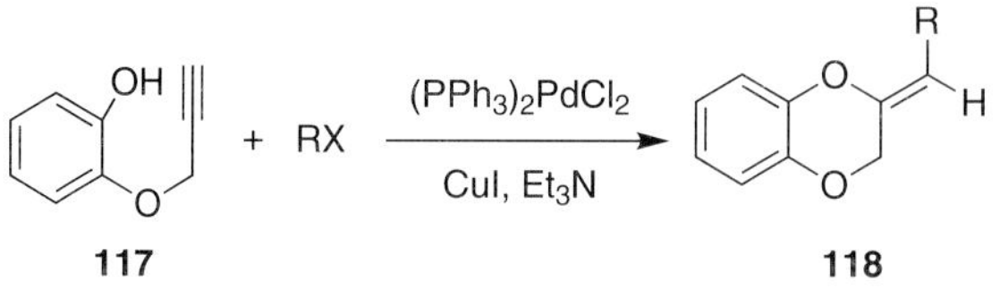

Scheme 10.46 Palladium-catalyzed reaction of o-propargyl catachol with aryl halide.

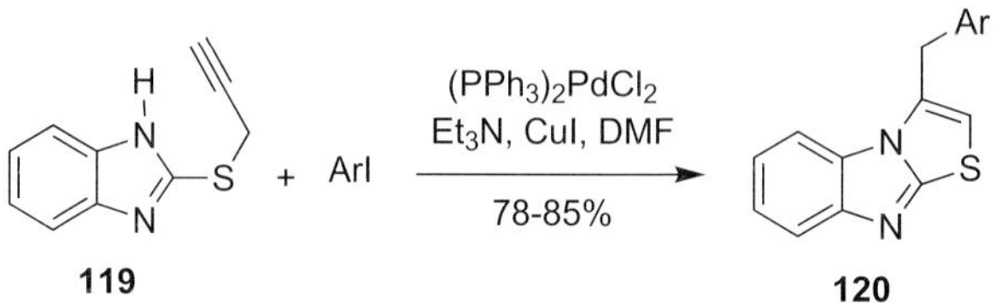

Scheme 10.47 Palladium-catalyzed synthesis of 3-benzylthiazolo[3,2-a]benzimidazoles.

Zhang and coworkers reported the Pd–Cu-catalyzed coupling/cyclization of the terminal alkynes with the resin bound *o*-iodoaniline **121**. Subsequent treatment of the reaction mixture with TBAF gave the indole derivatives **122** in high yields (cf. **123**) (Scheme 10.48) [65]. Similar examples were also examined by other researchers [66].

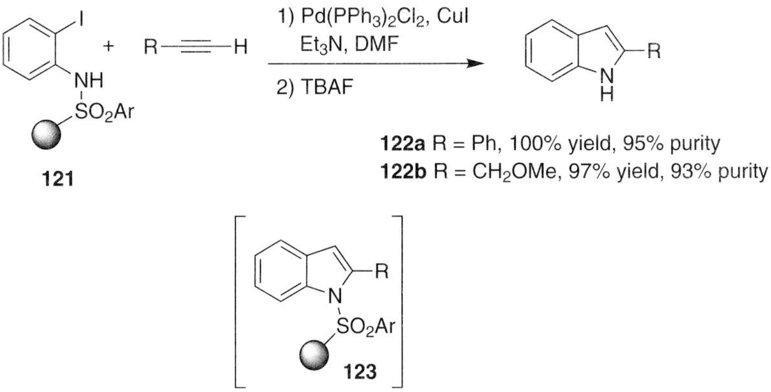
1) Pd(PPh3)2Cl2, CuI
Et3N, DMF
2) TBAF
121
122a R = Ph, 100% yield, 95% purity
122b R = CH2OMe, 97% yield, 93% purity
123

Scheme 10.48 Palladium-catalyzed synthesis of indoles.

Chowdhury *et al.* reported the synthesis of a novel class of compounds that is, isoindoline fused with triazole via tandem Sonogashira reaction/[3 + 2] cycloaddition reaction. For instance, treatment of *o*-iodobenzyl azide **124** with terminal acetylenes in DMF in the presence of Pd–Cu bimetallic catalyst and Et_3N gave isoindolotriazoles **125** in good yields (Scheme 10.49) [67].

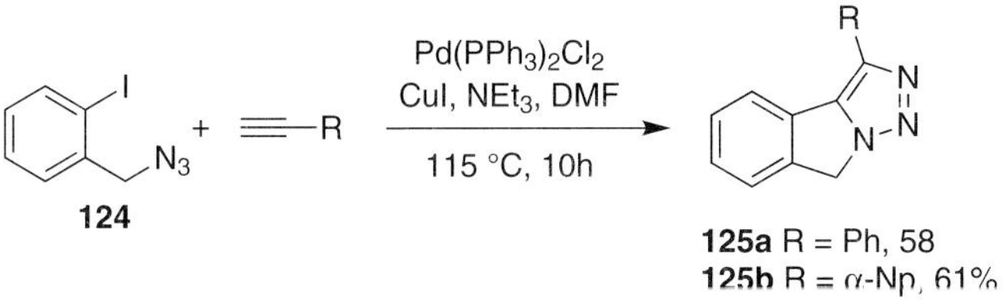
Pd(PPh3)2Cl2
CuI, NEt3, DMF
115 °C, 10h
124
125a R = Ph, 58
125b R = α-Np, 61%

Scheme 10.49 Palladium-catalyzed reaction of *o*-iodo benzyl azide with terminal alkynes.

Yamamoto and coworkers reported the synthesis of 6H-dibenzo[b,d]pyran-6-ones **127** via a tandem procedure using the Sonogashira coupling/benzannulation reaction of aryl 3-bromopropenoate **126** with terminal alkynes under Pd–Cu catalysis (Scheme 10.50) [68]. First the enyne derivatives **128** were formed and then these underwent benzannulation (see Scheme 10.133) under the present reaction conditions to afford the products.

The carboxylic acid **129** on reaction with phenyl acetylene under Pd–Cu bimetallic catalysis in the presence of 4 equiv. of Et_3N in acetonitrile gave the cyclized product *E*-**130** in 23% yield (Scheme 10.51) [69]. On the other hand it was reported that the use of 1-hexyne, instead of phenyl acetylene, afforded a mixture of *E* and *Z* isomers. A similar domino coupling/cyclization reaction was used for the total synthesis of melodorinol [70].

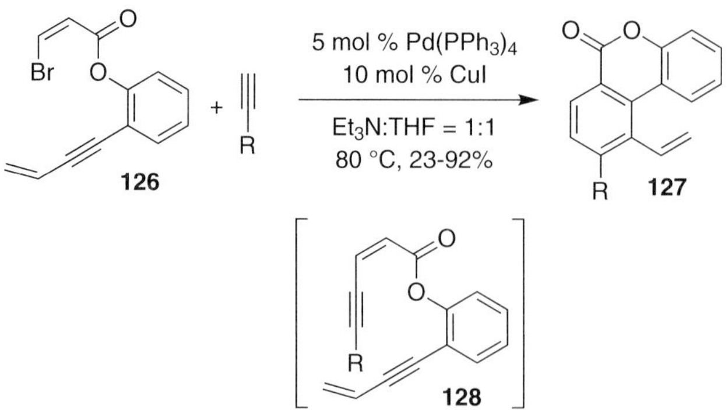

Scheme 10.50 Palladium-catalyzed tandem Sonogashira coupling/benzannulation reaction.

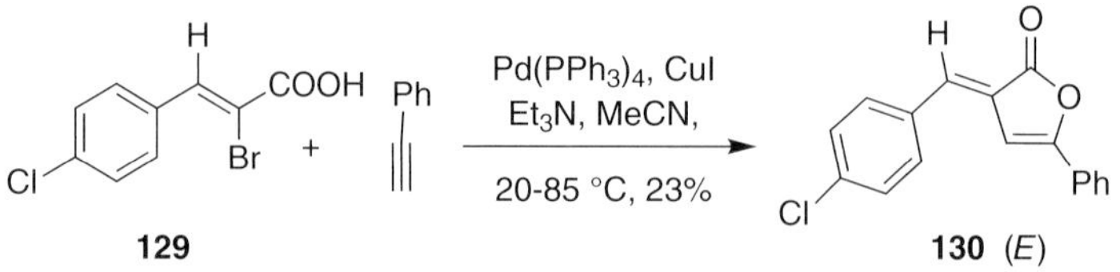

Scheme 10.51 Palladium-catalyzed tandem Sonogashira coupling/hydrocarboxylation reaction.

It is possible to perform similar reactions without using palladium. A copper(I)-catalyzed procedure for the synthesis of 2-arylbenzo[b]furans **132** from o-iodophenols **131** and terminal aryl alkynes was described by Venkataraman and coworkers (Scheme 10.52) [71]. However, a somewhat sophisticated copper source was required in this case.

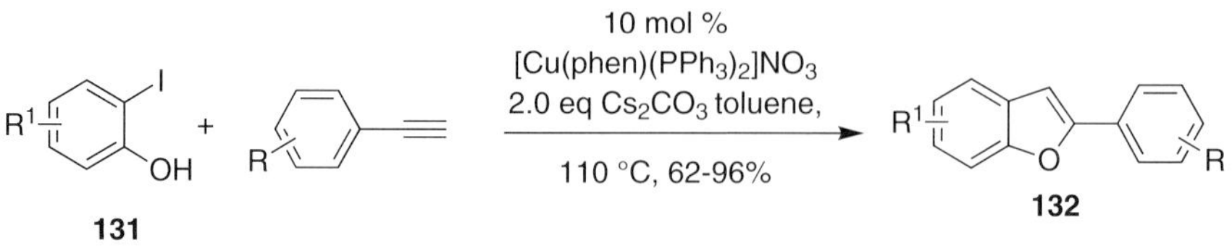

Scheme 10.52 Copper-catalyzed tandem Sonogashira coupling/hydroalkoxylation reaction.

The palladium-free method for the synthesis of 2-aryl and 2-heteroaryl indoles **134** from o-iodo anilines **133** and terminal aryl alkynes through a tandem copper-catalyzed process was reported by Cacchi and coworkers (Scheme 10.53) [72]. The best results were obtained with [Cu(phen)(PPh$_3$)$_2$]NO$_3$ in the presence of K$_3$PO$_4$ in toluene or 1,4-dioxane at 110 °C. They also extended this approach for the synthesis of quinoxalines **136** from 2-bromo-3-trifluoroacetamidoquinoxaline **135** adopting the same methodology (Scheme 10.54) [73].

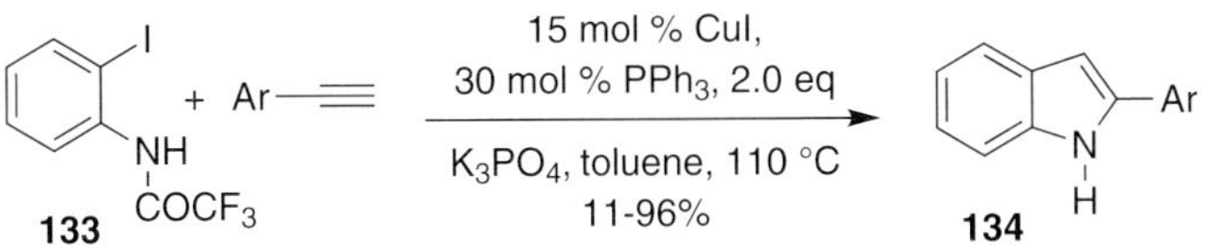

Scheme 10.53 Copper-catalyzed tandem Sonogashira coupling/hydroamination reaction.

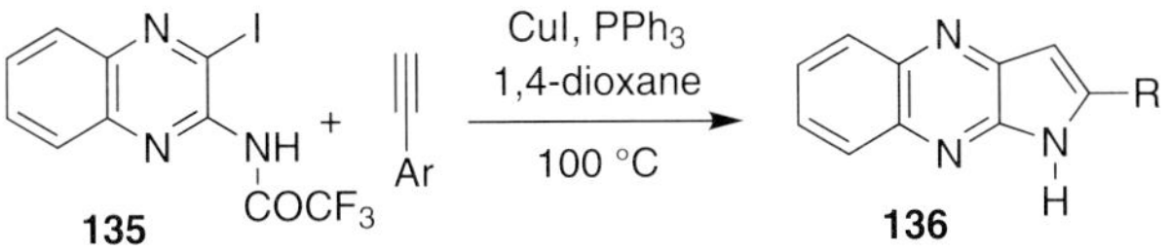

Scheme 10.54 Synthesis of quinoxalines via copper-catalyzed tandem Sonogashira coupling/hydroamination reaction.

The reaction of N-protected 2-iodoanilines **137** with 1-(tributylstannyl)-1-substituted allenes **138** in the presence of Pd–Cu bimetallic catalyst gave 2-methyl-3-substituted indoles **139** (Scheme 10.55) [74]. The reaction proceeded via a Stille coupling reaction between **137** and **138** to generate the intermediate **140** which underwent *in situ* cyclization.

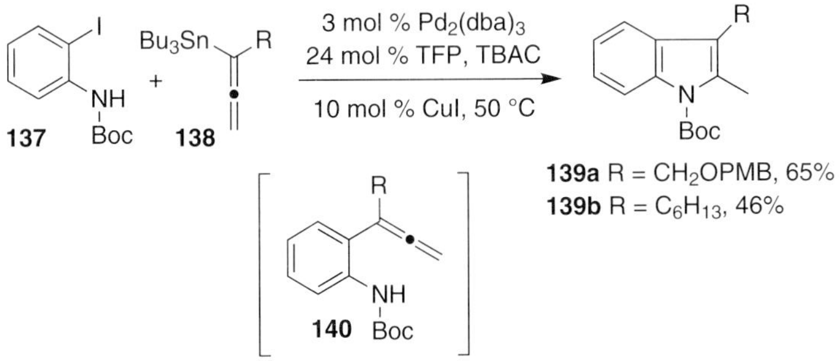

Scheme 10.55 Copper-catalyzed tandem Stille coupling/hydroamination reaction.

Buchwald and coworkers reported a domino process involving copper catalyzed C–N coupling and intramolecular hydroamidation for the synthesis of pyrroles **142** from haloenynes **141** (via intermediate **143**) (Scheme 10.56) [75]. The reaction of a substrate bearing a terminal alkyne does not give the product, however, this transformation can be accomplished by the use of a TMS group which masks the terminal acetylene and is deprotected *in situ*.

10.3.3
Cyclization/Functional Group Migration Reactions

Recently, the transition metal-catalyzed cyclization of aryl alkynes having an X-FG group at the ortho position has been gaining much interest. It was found that migration of FG from X to the C-3 position took place readily to produce the corresponding 2,3-disubstituted benzo fused heterocycles in an atom economical manner (Scheme 10.57). Representative examples are described herein.

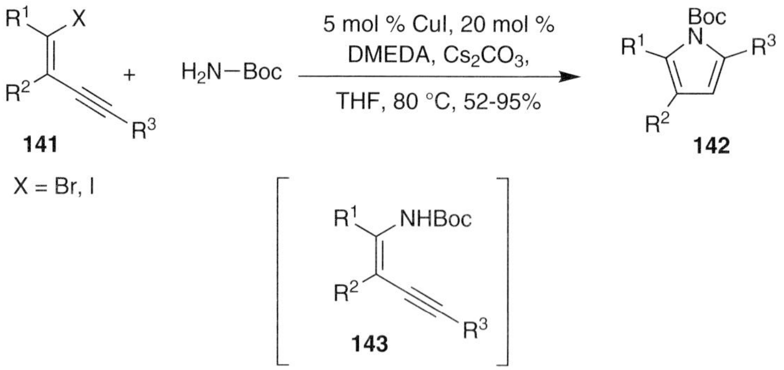

Scheme 10.56 Copper-catalyzed domino C–N coupling/hydroamination reaction.

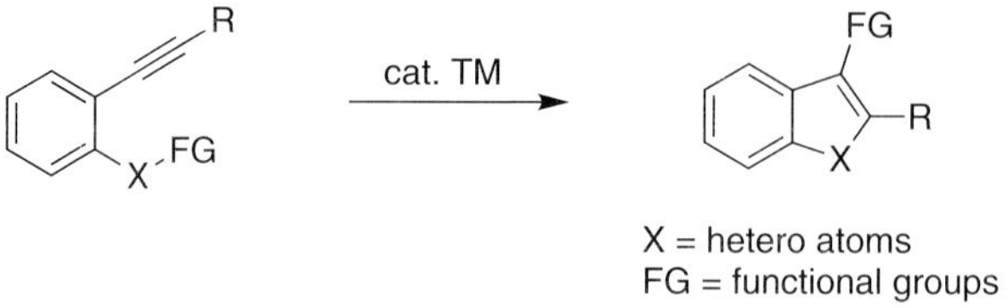

Scheme 10.57 Metal-catalyzed addition of X-FG across alkynes.

The gold-catalyzed intramolecular carbothiolation of alkynes for the synthesis of 2,3-disubstituted benzothiophene is known (Scheme 10.58) [76]. The reaction involves the migration of groups such as α-alkoxy alkyl, PMB and allyl from a sulfur atom to an alkyne. Various thiophene derivatives **145a–c** were obtained from readily available **144a–c**. A plausible mechanism is illustrated in Scheme 10.59. Nucleophilic attack of the sulfur atom of **144** on gold-coordinated alkyne (cf. **146**) most probably gives the vinylgold intermediate **147**. Migration of R groups in **147** to the carbon atom bonded to the gold atom produces the intermediate **148**. Elimination of gold chloride from **148** gives the products. Recently, the migration of silyl groups has also been reported [77].

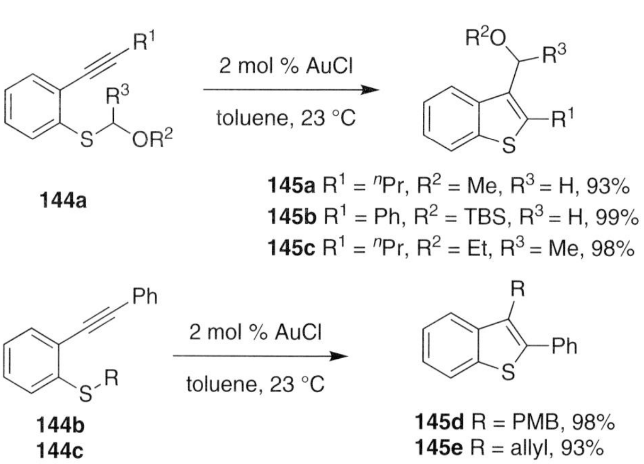

145a $R^1 = {}^nPr$, $R^2 = Me$, $R^3 = H$, 93%
145b $R^1 = Ph$, $R^2 = TBS$, $R^3 = H$, 99%
145c $R^1 = {}^nPr$, $R^2 = Et$, $R^3 = Me$, 98%

145d R = PMB, 98%
145e R = allyl, 93%

Scheme 10.58 Gold-catalyzed carbothiolation of alkynes.

Scheme 10.59 Mechanism for gold-catalyzed carbothiolation of alkynes.

We have also reported the gold-catalyzed intramolecular amino-sulfonylation (formal addition of N–S bond to a triple bond) for the synthesis of 3-sulfonylindoles **150** (Scheme 10.60) [78]. The procedure involved the treatment of *ortho*-alkynyl-*N*-sulfonylanilines **149** with catalyst AuBr$_3$ in toluene at 80 °C for 1 h. It is interesting to

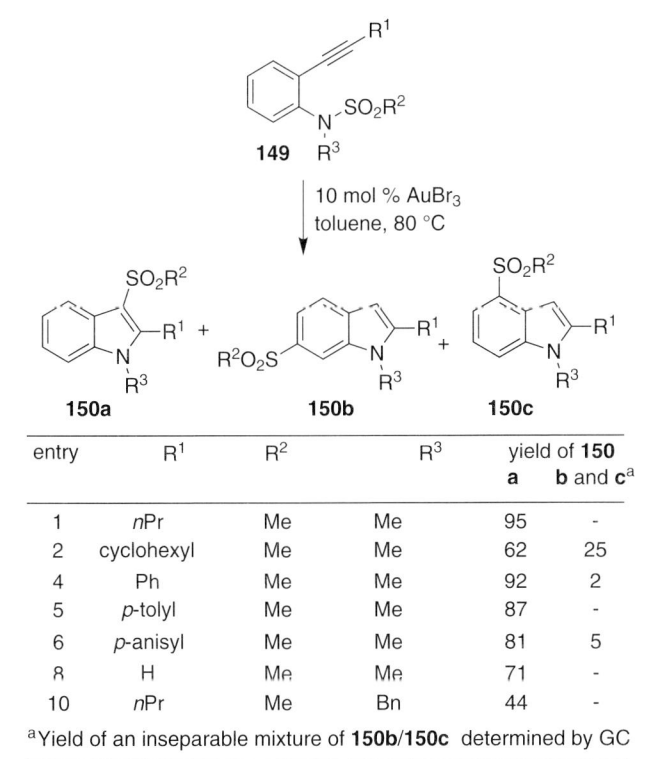

entry	R^1	R^2	R^3	yield of 150 a	b and c[a]
1	nPr	Me	Me	95	-
2	cyclohexyl	Me	Me	62	25
4	Ph	Me	Me	92	2
5	p-tolyl	Me	Me	87	-
6	p-anisyl	Me	Me	81	5
8	H	Me	Me	71	-
10	nPr	Me	Bn	44	-

[a] Yield of an inseparable mixture of **150b**/**150c** determined by GC

Scheme 10.60 Gold-catalyzed aminosulfonylation of alkynes.

note that the use of InBr$_3$ as a catalyst instead of AuBr$_3$ gave 6-sulfonylindoles as the major products.

The platinum-catalyzed cyclization of *ortho*-alkynyl aryl amides **151** for the synthesis of 3-acyl indoles **152a** has been reported (Scheme 10.61) [79]. In most cases, small amounts of deacylated indoles **152b** were obtained. This reaction is strongly influenced by the solvents and aromatic solvents containing an electron-donating group enhanced the rate of reaction. Mechanistically, the coordination of PtCl$_2$ to the alkyne (cf. **153**) increases the electrophilicity of the alkynes which in turn results in the attack of the tethered nitrogen atom to form the zwitterionic intermediate **154**. An intramolecular [1,3]-migration of the acyl moiety then yields the intermediate **155** which produces indole **152a** and PtCl$_2$ is regenerated.

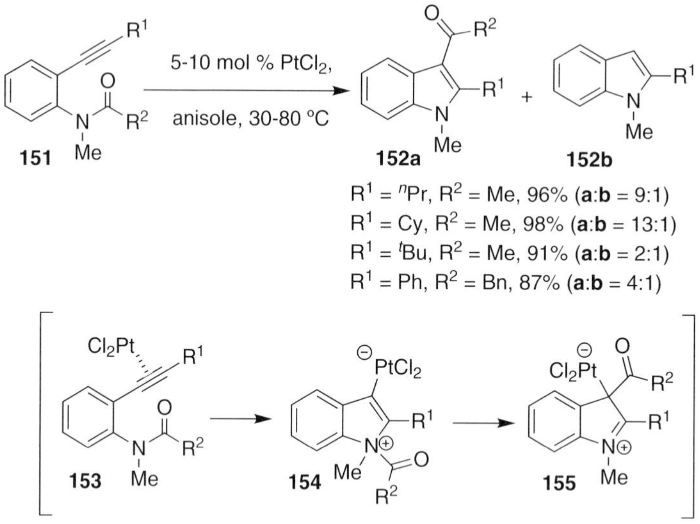

Scheme 10.61 Platinum-catalyzed cyclization of *ortho*-alkynyl aryl amides.

The Yamamoto group reported the PtCl$_2$-catalyzed cyclization reaction of *o*-alkynylphenyl acetals **156** in the presence of COD to give 3-(α-alkoxyalkyl)benzofurans **157** in good yields (Scheme 10.62) [80]. The mechanism is very similar to those reported for previous cases. Almost at the same time the Fürstner group reported similar results [81].

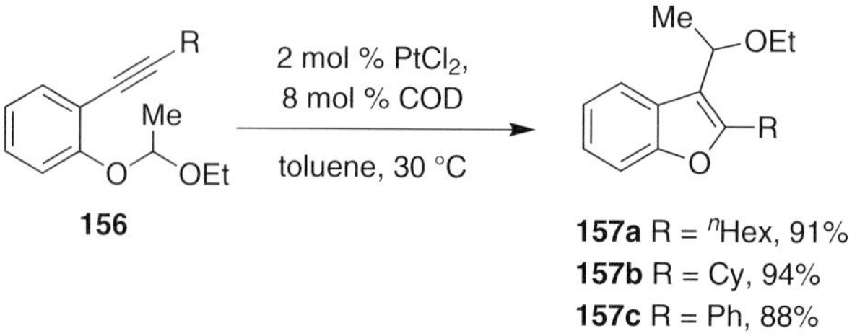

Scheme 10.62 Platinum-catalyzed carboalkoxylation of alkynes.

10.3.4 Cyclization with Carbonyls/Imines/Epoxides/Acetals/Thioacetals, and so on

10.3.4.1 Aliphatic Tether

The cycloisomerization of alkynyl ketone to furan was reported by Gevorgyan and coworkers. They developed an efficient method for the synthesis of 2-monosubstituted and 2,5-disubstituted furans **159** via the CuI-catalyzed cycloisomerization of alkynyl ketones **158** (Scheme 10.63) [82]. Mechanistic studies revealed that this reaction proceeded via allenyl ketone **160**. The logical extension of this method was elaborated for the synthesis of pyrroles **162** via copper(I)-catalyzed cyclization of alkynyl amines **161** (Scheme 10.64) [83].

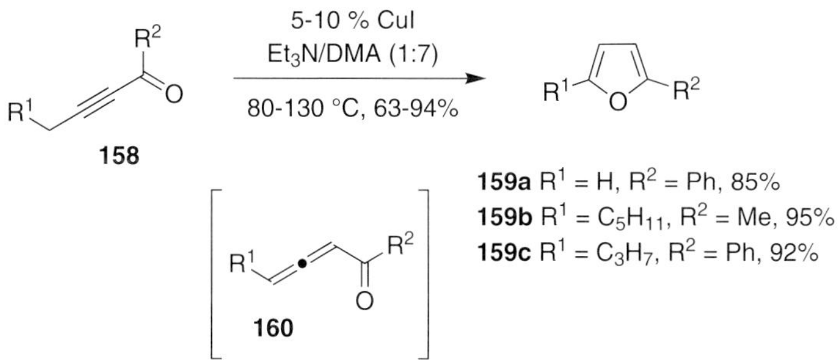
R2
R1
O
5-10 % CuI
Et3N/DMA (1:7)
80-130 °C, 63-94%
R1
O
R2
158
159a R1 = H, R2 = Ph, 85%
159b R1 = C5H11, R2 = Me, 95%
159c R1 = C3H7, R2 = Ph, 92%
R1
R2
O
160

Scheme 10.63 Copper-catalyzed cycloisomerization of alkynyl ketones.

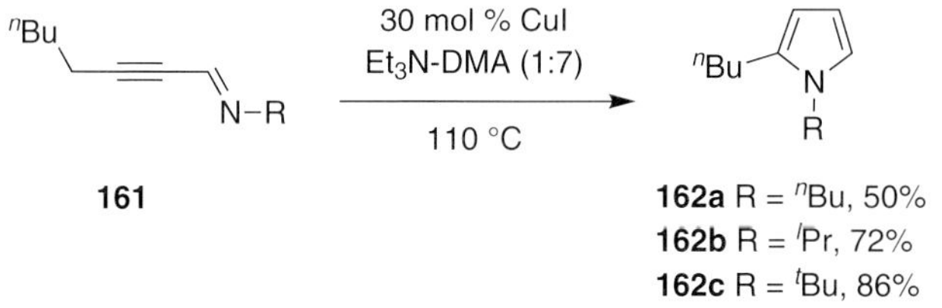
nBu
N–R
30 mol % CuI
Et3N-DMA (1:7)
110 °C
nBu
N
R
161
162a R = nBu, 50%
162b R = iPr, 72%
162c R = tBu, 86%

Scheme 10.64 Copper-catalyzed cycloisomerization of alkynyl imines.

Isomerization of alkynyl epoxides **163** into furans **164** in the presence of gold(III) chloride took place at room temperature (Scheme 10.65) [84]. The coordination of the triple bond of **163** to $AuCl_3$ enhances the electrophilicity of the alkyne (cf. **165**), and the subsequent nucleophilic attack of the epoxide oxygen at the distal position of the alkyne forms the species **166** which, on protodeauration, gives furans **164**. The gold-catalyzed synthesis of 2,5-disubstituted oxazoles **168** from the corresponding propargylcarboxamides **167** (Scheme 10.66) was also reported[85]. The addition of amide oxygen to alkyne is highly stereospecific to form the vinyl gold species **169** which on protonation, aromatization and regeneration of catalyst gives products **168**.

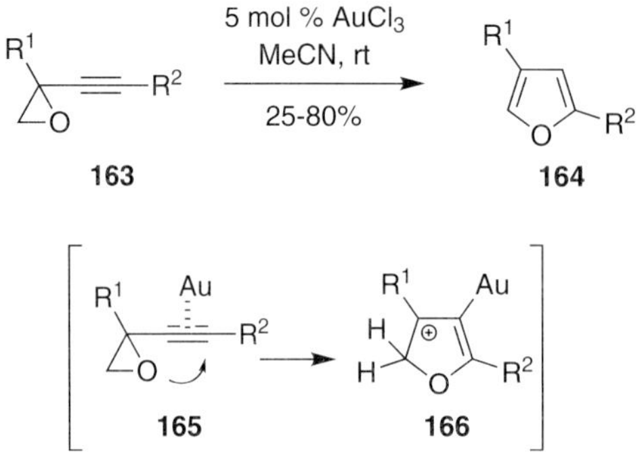

Scheme 10.65 Gold-catalyzed cycloisomerization of alkynyl epoxides.

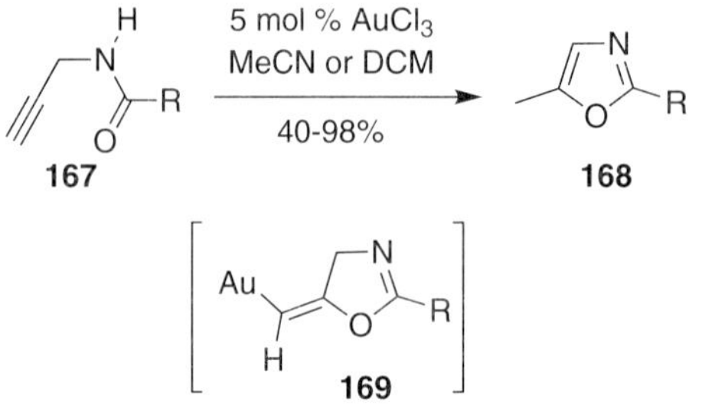

Scheme 10.66 Gold-catalyzed cyclization of propargylcarboxamides.

The Larock group reported a new approach for the novel cyclization of 2-(1-alkynyl)-2-alken-1-ones **170** with nucleophiles in the presence of catalytic amounts of AuCl$_3$ which led to the formation of highly substituted furans **171** (Scheme 10.67) [86]. Examples of nucleophiles include alcohols, activated methylenes and electron-rich arenes such as N,N-dimethylaniline and N-methyl indole. The coordination of the triple bond of **170** to AuCl$_3$ triggered the cascade, as shown in **172**. An inexpensive and air stable catalytic system that is, Cu(I) catalyst in DMF can also be used for this reaction [87].

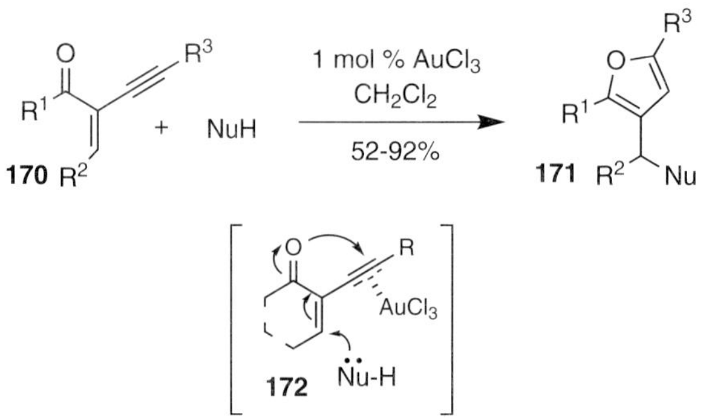

Scheme 10.67 Gold-catalyzed cyclization of 2-(1-alkynyl)-2-alken-1-ones.

The gold-catalyzed cyclization reactions of 4-phenyl-2-oxo-3-butynoic ester **173** with a variety of nucleophiles are reported by Liu and coworkers (Scheme 10.68) [88]. The method provided an efficient and general route to highly substituted 3(2H)-furanones **174a–i**. As shown in **175**, the reaction is triggered by the coordination of catalyst to the alkyne moiety.

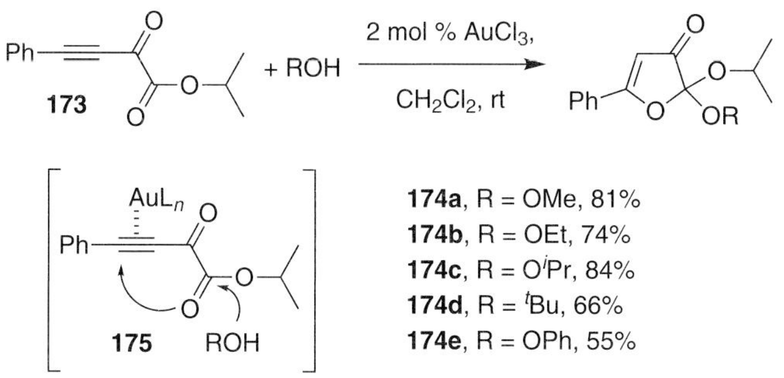
Ph
173
+ ROH
2 mol % AuCl3,
CH2Cl2, rt
Ph
OR
AuLn
Ph
175
ROH
174a, R = OMe, 81%
174b, R = OEt, 74%
174c, R = OiPr, 84%
174d, R = tBu, 66%
174e, R = OPh, 55%

Scheme 10.68 Gold-catalyzed cyclization of 2-oxo-3-butynoates.

The Yamamoto group developed an efficient method for constructing highly substituted cyclic enones **177** from alkynyl ketones **176** (Scheme 10.69) [89]. Coordination of the catalyst to the alkyne **176** generates the oxetenium intermediates **179** via the oxonium species **178**. Subsequent ring opening in **179** afforded the desired ketone **177**.

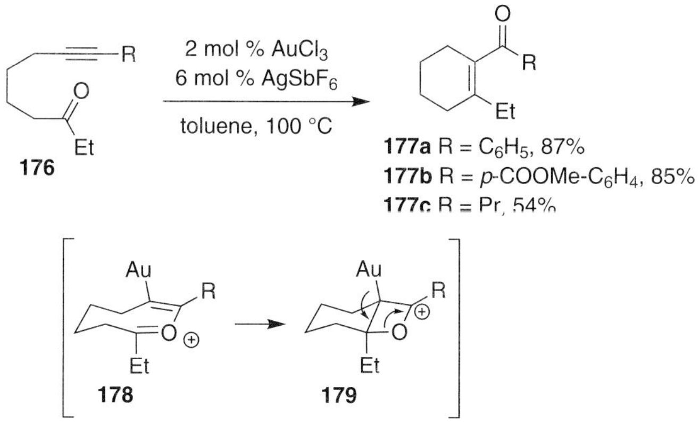
R
Et
176
2 mol % AuCl3
6 mol % AgSbF6
toluene, 100 °C
R
Et
177a R = C6H5, 87%
177b R = p-COOMe-C6H4, 85%
177c R = Pr, 54%
Au
R
Et
178
Au
R
Et
179

Scheme 10.69 Gold-catalyzed intramolecular cyclization of alkynyl ketones.

Dake and coworkers reported a domino process for the synthesis of pyrroles **181** starting from ketoalkyne **180** and amines. Either silver trifluoromethanesulfonate or a mixture of gold(I) chloride, silver trifluoromethanesulfonate, and triphenylphosphine catalyzes the formation of pyrroles from substituted β-alkynyl ketones and amines (Scheme 10.70) [90]. The reactions proceed by using 5 mol% of catalyst with yields of isolated pyrroles ranging from 13% to 92%.

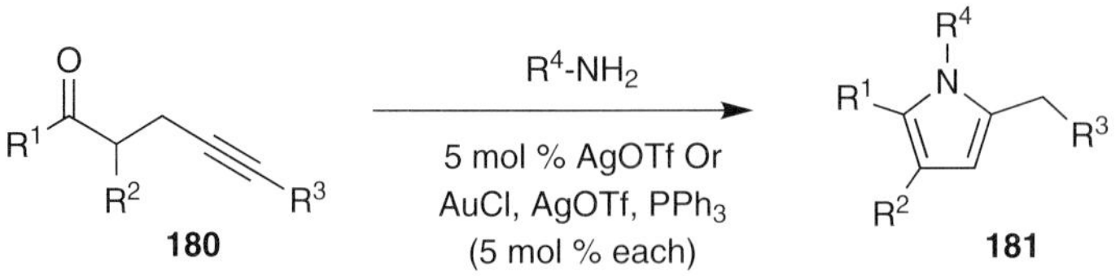

486 | *10 Transition Metal Catalyzed Cyclization Reactions of Functionalized Alkenes, Alkynes, and Allenes*

Scheme 10.70 Silver-catalyzed synthesis of pyrroles.

10.3.4.2 Aromatic Tether

Benzannulation Reactions The metal catalyzed [4 + 2] benzannulation between benzopyrilium species, formed from enynal or enynone, with alkynes and alkenes offers an attractive route for the synthesis of aromatic compounds [91]. Yamamoto et al. reported AuCl$_3$-catalyzed formal [4 + 2] benzannulation between o-alkynyl-benzaldehydes **182** and alkynes **183** (Scheme 10.71) [92]. Various naphthyl ketones **184a,b** were obtained in good to high yields. It was found that regioselectivity was strongly dependent on the substituent on the alkynes. For instance when $R^1 = C_3H_7$ and $R^2 = H$, the regioisomer **184a** predominates, on the other hand, when $R^1 = COCH_3$ and $R^2 = H$, the regioisomer **184b** predominates. As shown in Scheme 10.72 the intermediate auric -ate complex **186** is formed by the activation of alkyne by catalyst (cf. **185**). The Diels–Alder type [4 + 2] cycloaddition of **186** with an alkyne **183** occurs as shown in **187** to form the intermediate **188**. The subsequent bond rearrangement, as shown in **188** with arrows, affords the naphthalene derivatives **184** and regenerates AuCl$_3$. Recently, [4 + 2] benzannulation between benzopyrilium species and arynes has been reported [93].

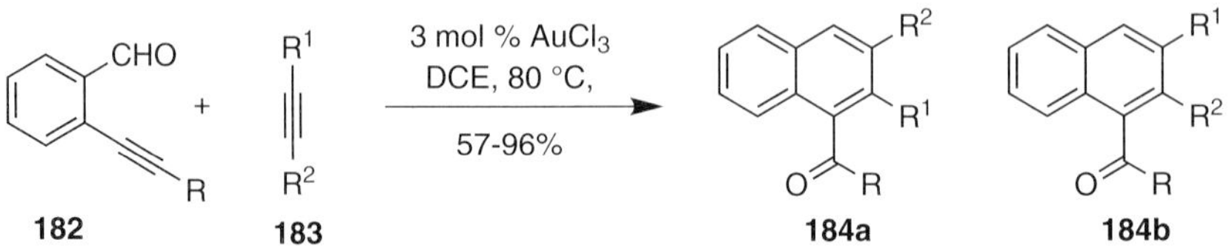

Scheme 10.71 Gold-catalyzed reactions of o-alkynylbenzaldehydes with alkynes.

The intramolecular version of this reaction was also reported (Scheme 10.73) [94]. Treatment of the tethered alkynyl enynones **189**, in which a carbon chain is attached to the carbonyl group, with a catalytic amount of AuBr$_3$ in DCE gave the naphthyl ketones **190** in good yields. Analogously, the AuBr$_3$-catalyzed benzannulations of **191**, in which a carbon tether was extended from the alkynyl terminus, also proceeded smoothly, and the cyclized naphthyl ketones **192** were obtained in high yields. Similarly, enynones **193** and **195** containing tethered alkenes underwent benzannulation in the presence of catalytic amounts of Cu(OTf)$_2$ to give **194** and **196**, respectively (Scheme 10.74).

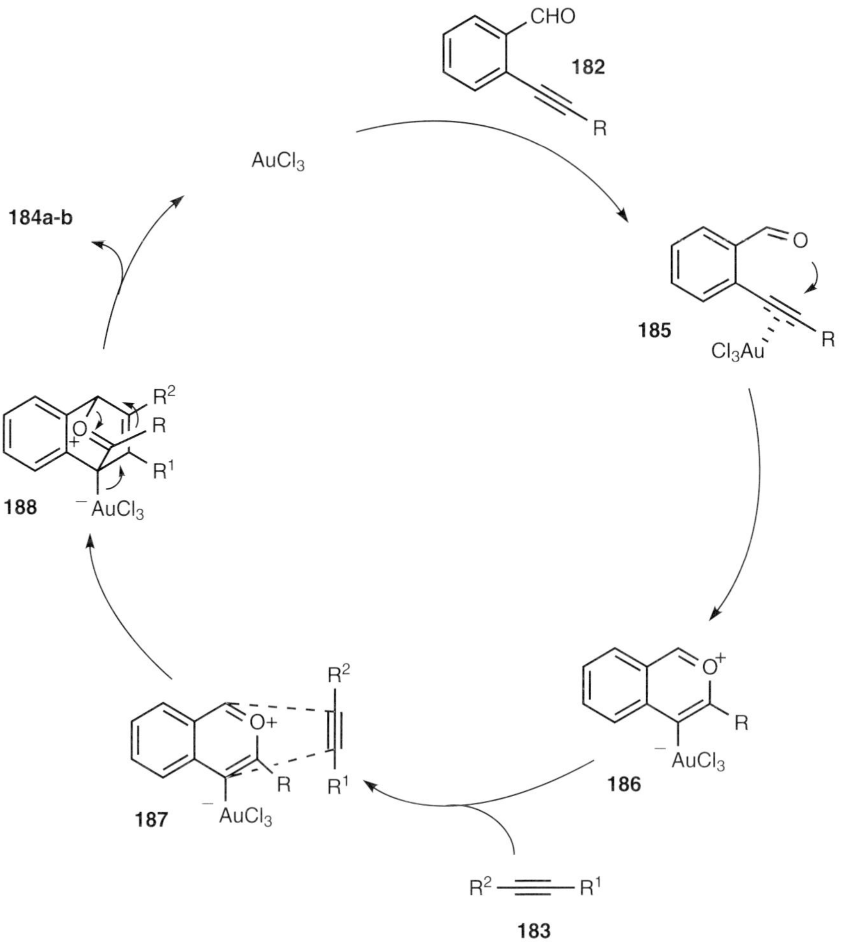

Scheme 10.72 Mechanism for gold-catalyzed benzannulation reaction.

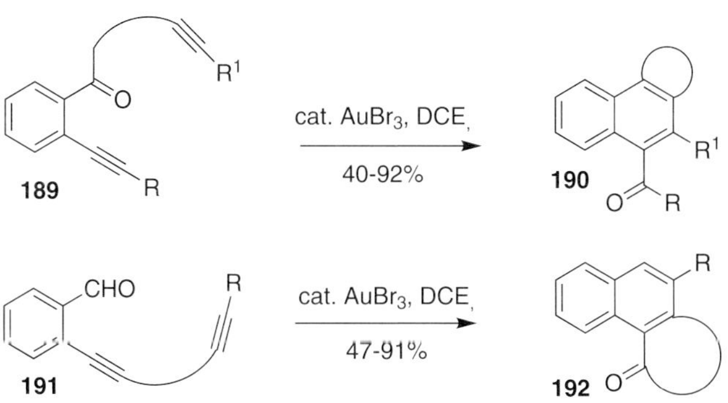

Scheme 10.73 Synthesis of naphthalenes via gold-catalyzed benzannulation reaction.

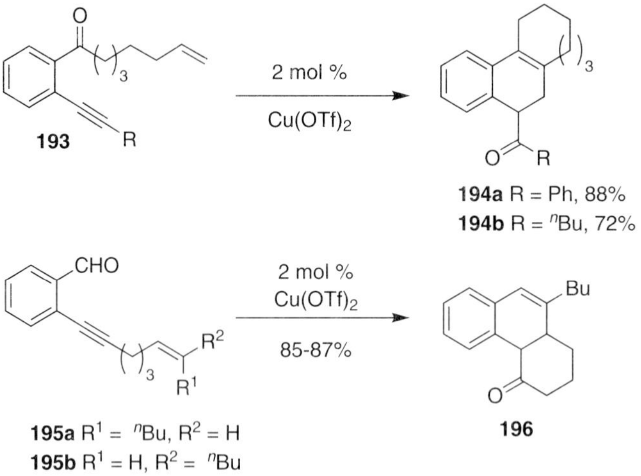

Scheme 10.74 Synthesis of 1,2-dihydro naphthalenes via gold-catalyzed benzannulation reaction.

A successful application of this methodology is the total synthesis of (+)-ochromycinone and (+)-rubiginone B_2 (Scheme 10.75) [95]. The chiral alkynal **197** was subjected to intramolecular [4 + 2] benzannulation reaction in the presence of 2 mol% $AuCl_3$ to afford (+)-Rubiginone **198**. (+)-Ochromycinone **199** was easily obtained via demethylation of the −OMe group by BCl_3. Another application [96] of this methodology has also been reported recently by Dyker et al. for the total synthesis of rac-heliophenanthrone **200** (Scheme 10.76). It was found that gold catalyst gave lower yield (34%) for the benzannulation of **201**, however, the use of $PtCl_2$ proved beneficial, giving the product **202** in 71% yield. The benzannulation product **202** thus obtained was converted into heliophenanthrone **200** through a series of steps.

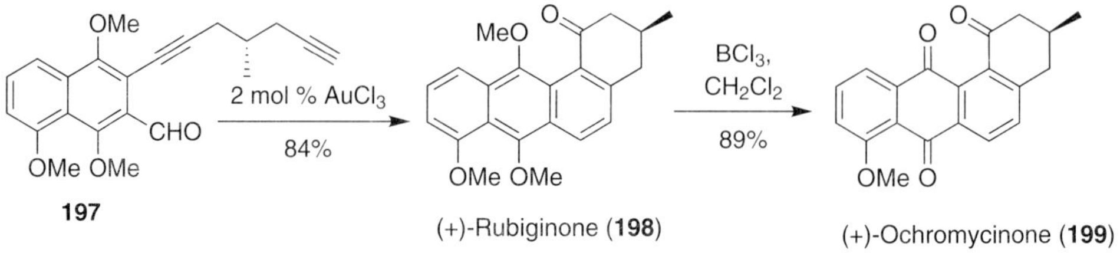

Scheme 10.75 Synthesis of rubiginone via benzannulation reaction.

An innovative extention of the above approach by Oh and coworkers involved the synthesis of [6,7,n] tricyclic compounds **204** via [3 + 2] cycloaddition of **203** (Scheme 10.77) [97]. It is interesting to note that only trace amounts of [4 + 2] benzannulation products were obtained in each case. Since the main structural feature distinguishing **203** from **191** is the gem-ester group, the author argued that the Thorpe–Ingold effect might be playing some role in favoring the formation of the five-membered ring products **204** via intermediates **205**.

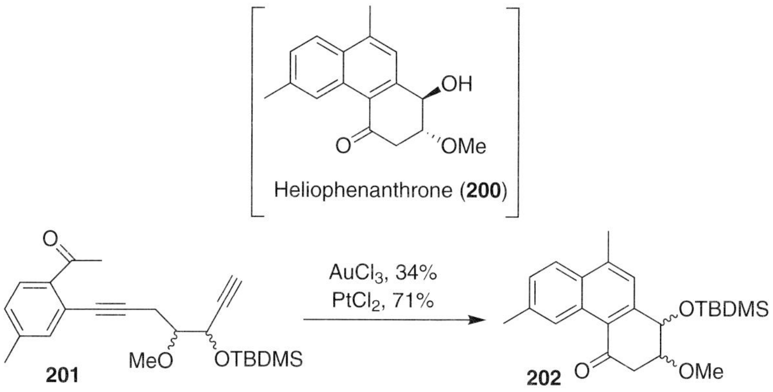

Scheme 10.76 Synthesis of *rac*-heliophenanthrone via benzannulation reaction.

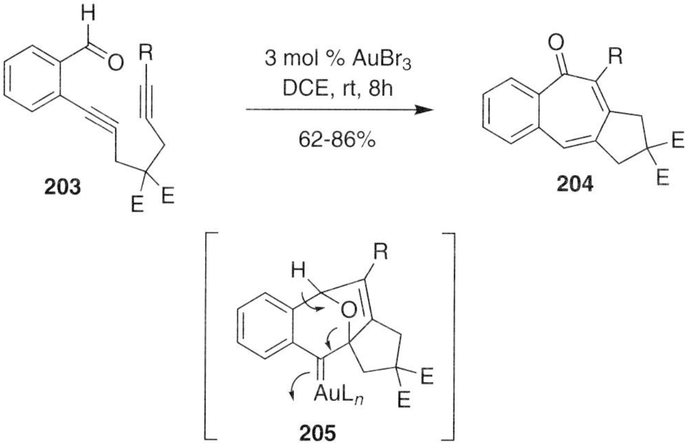

Scheme 10.77 Gold-catalyzed synthesis of [6,7,*n*] tricyclic compounds.

It is interesting to note that the reaction of enynals **206** with dimethyl acetylenedicarboxylate **207** in the presence of Pd(OAc)$_2$ and COD gives the bicyclic pyrans **208** in good yields (Scheme 10.78) [98]. The mechanism of the reaction is remarkable and intermediacy of the benzopyrylium intermediate **209** is proposed. The subsequent deprotonation afforded the bicyclic triene **210** and Pd(OAc)$_2$ was regenerated. The reaction of **210** with **207** gave product **211** which, on ring opening, gave pyrans **208**.

The [4 + 2] benzannulation between the pyrylium intermediates with carbonyl compounds is also reported. Functionalized naphthalenes **213a** or **213b** were obtained in good to high yields by heating the substrates **182** and ketones **212** at 100 °C in 1,4-dioxane in the presence of AuBr$_3$ (Scheme 10.79) [99]. A plausible mechanism is shown in Scheme 10.80. The gold -ate complex **186** was generated from **182** through intermediate **185**. The reverse electron demand-type Diels–Alder reaction of **186** with the enol **212a**, derived from **212**, followed by dehydration, generated the intermediate **215** through **214**. The subsequent bond rearrangement, as shown in **215** with arrows, afforded the naphthyl ketones **213a,b** with regeneration of AuBr$_3$.

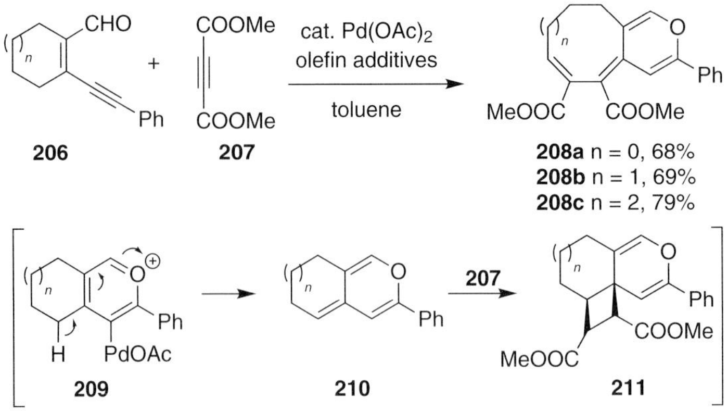

Scheme 10.78 Palladium-catalyzed reaction of enynals with dimethyl acetylenedicarboxylate.

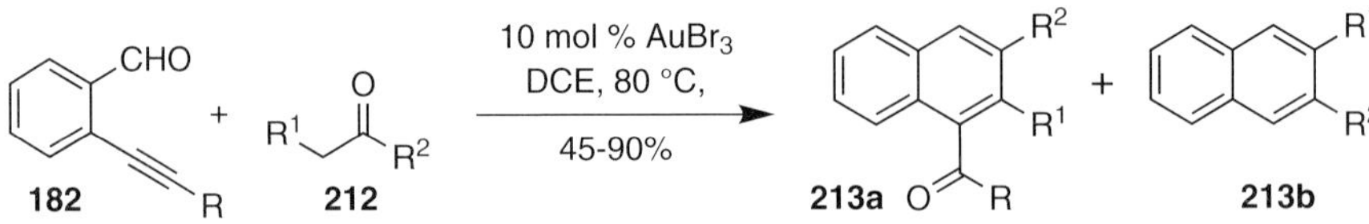

Scheme 10.79 Gold-catalyzed reactions of o-alkynylbenzaldehydes with ketones to form naphthalene derivatives.

Other Cascade Processes The synthesis of cyclic alkenyl ethers **216** via the Cu(I)-catalyzed intramolecular cyclization of O-alkynylbenzaldehydes **182** with alcohols has been reported (Scheme 10.81) [100]. A survey of metal catalysts and solvents revealed that the combination of copper(I) iodide and DMF was the best one, providing products in high yield. The superiority of this catalyst system is evident from the fact that even terminal alkynes give the desired products, in comparison with the previous catalyst system [Pd(OAc)$_2$] which gives the products in very low yields with the formation of unidentified by-products [101]. The benzopyrilium intermediate **217** was proposed as an intermediate which on trapping by the alcohol gave the products.

A tandem alkynylation–cyclization of terminal alkynes with *ortho*-alkynylaryl aldehydes **182** leading to 1-alkynyl-1H-isochromenes **218** by using a gold-phosphine complex as catalyst in water was developed by Li and Yao (Scheme 10.82) [102]. Mechanistically, the reaction of terminal alkynes with Me$_3$PAuCl in the presence of a weak base generates the gold acetylide, which then forms the chelating intermediate **219** followed by attack on the triple bond to give the vinylgold intermediate **220**. The intermediate **220** then afforded final products by protonolysis with regeneration of catalyst.

The silver(I)-catalyzed cyclization reaction of alkynones **221** with alcohols gave 1-allenyl chromenes **222** (Scheme 10.83) [103]. This is the best mehod for incorporation of the allene moiety into the isochromenes. The reaction was tolerated over a wide

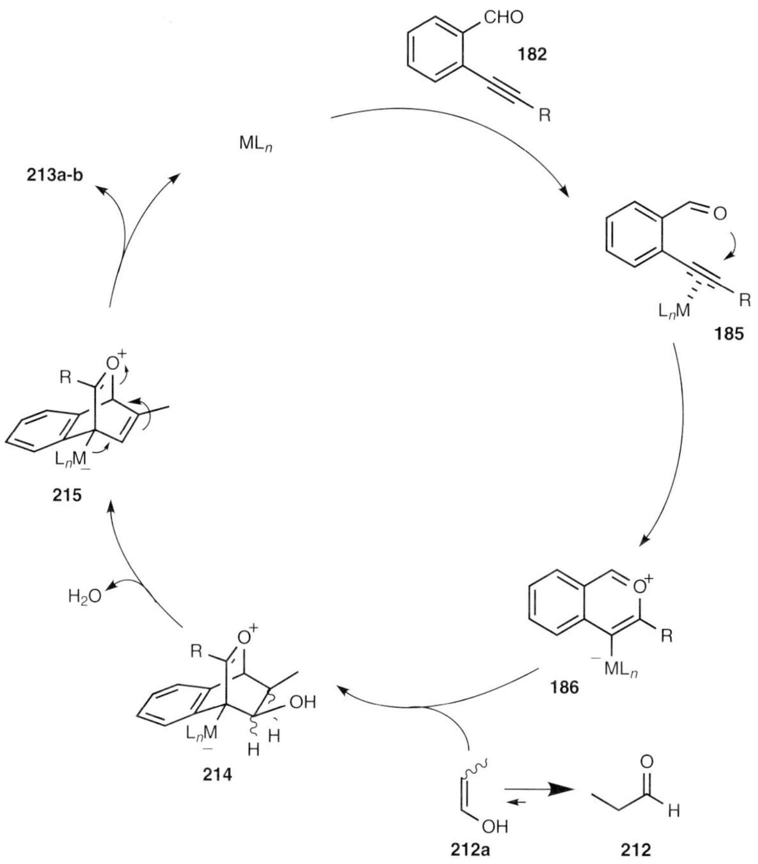

Scheme 10.80 Mechanism for the formation of naphthalenes from o-alkynylbenzaldehydes with ketones under gold catalysis.

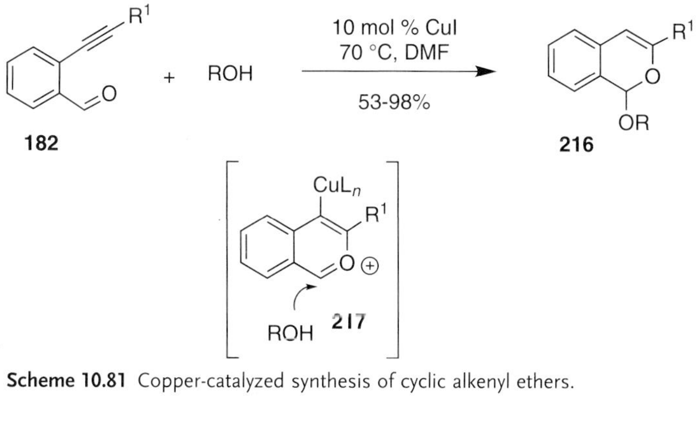

Scheme 10.81 Copper-catalyzed synthesis of cyclic alkenyl ethers.

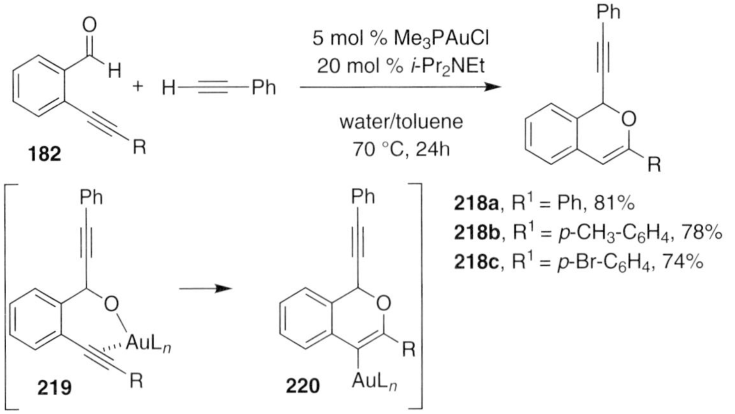
5 mol % Me3PAuCl
20 mol % i-Pr2NEt
water/toluene
70 °C, 24h
182
219
220
218a, R1 = Ph, 81%
218b, R1 = p-CH3-C6H4, 78%
218c, R1 = p-Br-C6H4, 74%

Scheme 10.82 Gold-catalyzed reaction of *o*-alkynylbenzaldehydes with terminal alkynes.

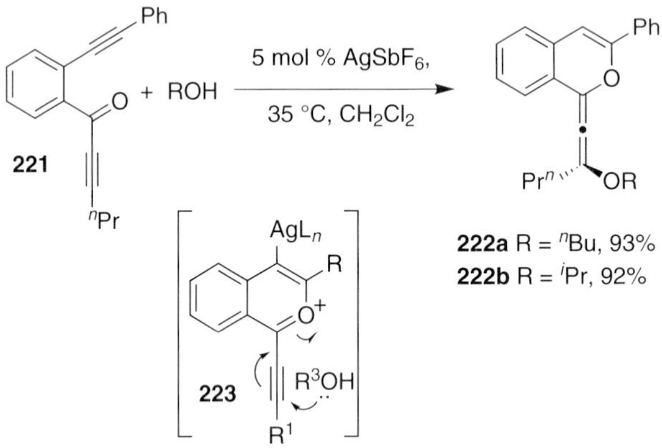
5 mol % AgSbF6,
35 °C, CH2Cl2
221
223
222a R = nBu, 93%
222b R = iPr, 92%

Scheme 10.83 Silver-catalyzed synthesis of 1-allenyl chromenes.

range of substrates except for the terminal alkynes and TMS-protected alkynes. The reaction most probably proceeded through the benzopyrylium cation **223**, which underwent subsequent trapping with alcohols to form the annulation products.

The silver-catalyzed tandem process for cyclization/Mannich reaction of *o*-alkynylaryl aldimines **224** with various nucleophiles gives 1,2-dihydroisoquinolines **225** in a highly atom economical manner (Scheme 10.84). [104] Examples of pronucleophiles include nitromethane, acetyl acetone, dimethyl malonate, malononitrile, acetone and acetonitrile. The reaction was carried out in the presence of 3 mol% AgOTf in dichloroethane as a solvent at 60–80 °C. Not only simple carbon nucleophiles but also terminal alkynes could be used as pronucleophiles for this reaction. The generation of isoquinolinium salts **226** was proposed as the key step for the reaction to occur.

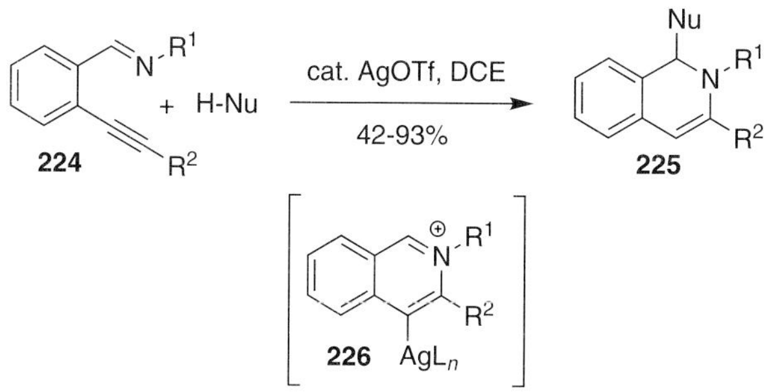

Scheme 10.84 Silver-catalyzed synthesis of 1,2-dihydroquinolines.

The domino allylation and cyclization processes of o-alkynylbenzaldehydes **182** catalyzed by Pd(II)-Cu(II) are reported. The reaction of **182** with allyltrimethylsilane under Pd–Cu catalysis gave the isochromene derivatives **227** in moderate to good yields (Scheme 10.85) [105]. In some cases, the formation of small amounts of chlorinated product, formed by the delivery of the −Cl group from $CuCl_2$, was observed. The activation of aldehyde, as shown in the σ–π chelating complex **228**, allows nucleophilic attack of allyltrimethylsilane leading to the formation of **229**. The coordination of the triple bond of **229** by $CuCl_2$ promotes the cyclization reaction. It was found that the present reaction can also be extended to domino cyanation and cyclization. For instance, when **182a** was treated with trimethylsilylcyanide in the presence of Pd(II)–Cu(II) catalyst, the cyclic alkenyl ether **230** was obtained in 94% yield (Scheme 10.86).

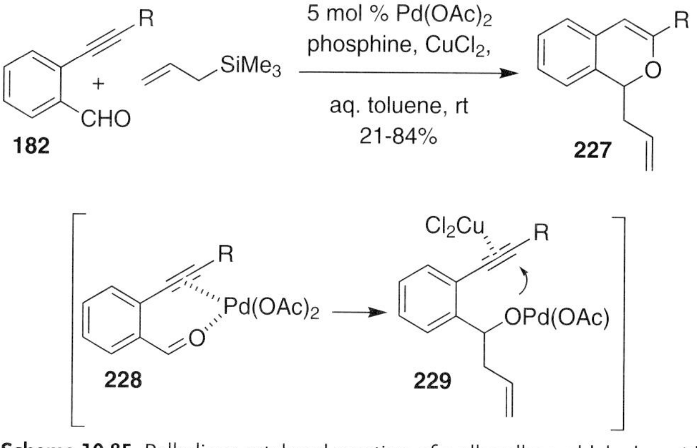

Scheme 10.85 Palladium-catalyzed reaction of o-alkynylbenzaldehydes with allyltrimethylsilane.

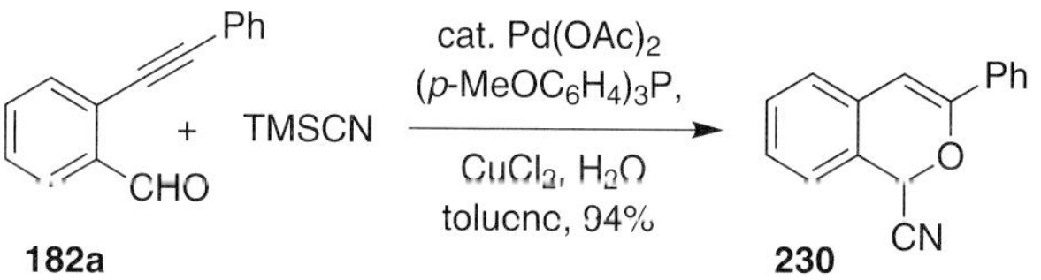

Scheme 10.86 Palladium-catalyzed reaction of o-alkynylbenzaldehydes with trimethylsilylcyanide.

A novel procedure for the tandem nucleophilic allylation–alkoxyallylation of alkynylaldehydes is known. The reaction of **182** with allyltributylstannane and allyl chloride proceeded in the presence of catalytic amounts of the allylpalladium chloride dimer at room temperature in THF to give the corresponding bisallylated 5-*exo-dig* cyclic ethers **231a** along with 6-*endo-dig* cyclic ethers **231b** (Scheme 10.87) [106]. The *in situ* generated bis-π-allylpalladium complex reacted with **182** to give the π-allylpalladium intermediate **232**. The attack of the alkoxy anion on the alkyne through path **a** or path **b** as shown in **233** afforded the 5-*exo-dig* products **231a** or 6-*endo-dig* products **231b**, respectively. The selectivity of 5-exo and 6-endo cyclization was dependent on the functional groups present on the alkyne. It was found that the alkynylbenzaldehydes having an electron-withdrawing group in R gave the 5-exo products predominantly, while those having an electron-donating group in R afforded the 6-endo products predominantly. Later, we extended this approach to the synthesis of 1,2-dihydroisoquinolines **234** from *o*-alkynylarylimines **224** (Scheme 10.88) [107].

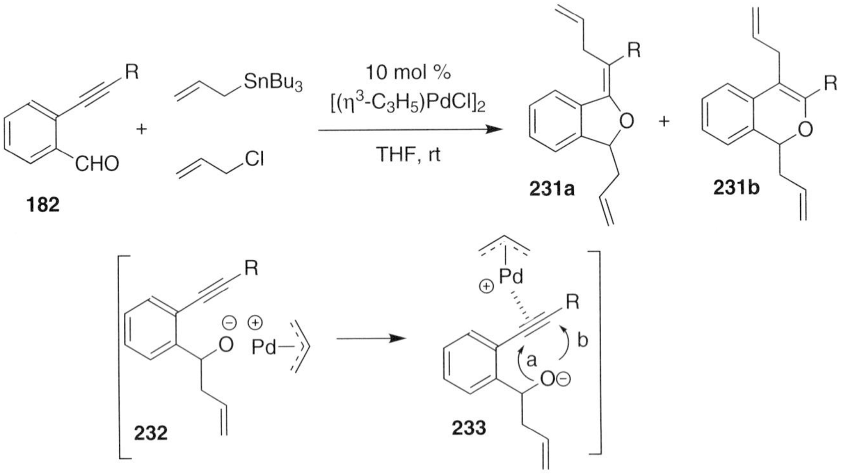

Scheme 10.87 Palladium-catalyzed reaction of *o*-alkynylbenzaldehydes with allyltributylstannane and allyl chloride.

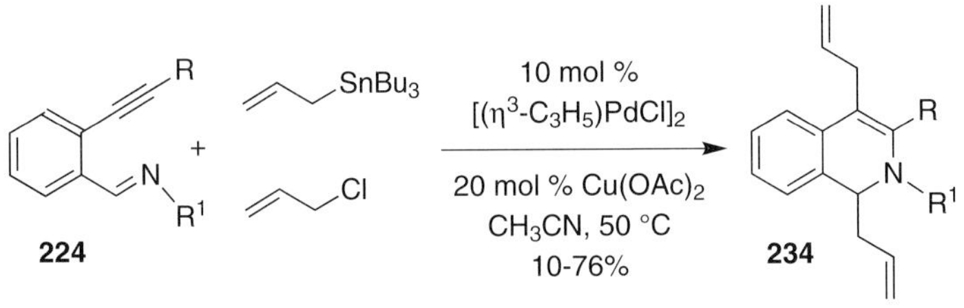

Scheme 10.88 Palladium-catalyzed reaction of *o*-alkynylbenzaldimines with allyltributylstannane and allyl chloride.

A general protocol for the synthesis of functionalized indenes **236** from *o*-alkynylbenzaldehyde acetals **235** has been developed by the Yamamoto group (Scheme 10.89) [108]. One of the proposed mechanisms for the cyclization of **235** involves the formation of the oxonium intermediate **238** through activation of the alkyne, as shown in **237**. The ring-closing reaction in **238** produced palladium carbene species **239**. Migration of R^1 in **239** followed by elimination of $PdCl_2$ produced the carboalkoxylation product **236**. Interestingly, the cyclization of the corresponding thioacetals **240** proceeds with thio group migration (not R^1 group migration) to afford 1,3-dithiosubstituted indenes **241** (Scheme 10.90).

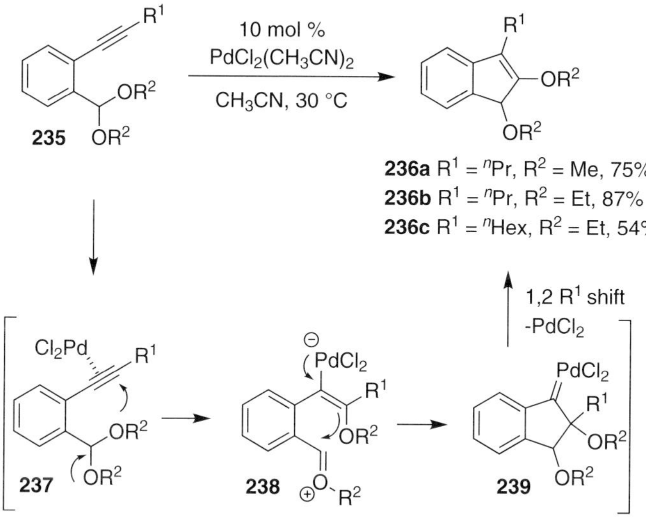

10 mol %
PdCl2(CH3CN)2
CH3CN, 30 °C
235
236a R1 = nPr, R2 = Me, 75%
236b R1 = nPr, R2 = Et, 87%
236c R1 = nHex, R2 = Et, 54%
1,2 R1 shift
-PdCl2
237
238
239

Scheme 10.89 Palladium-catalyzed reaction of *o*-alkynylbenzaldehyde acetals.

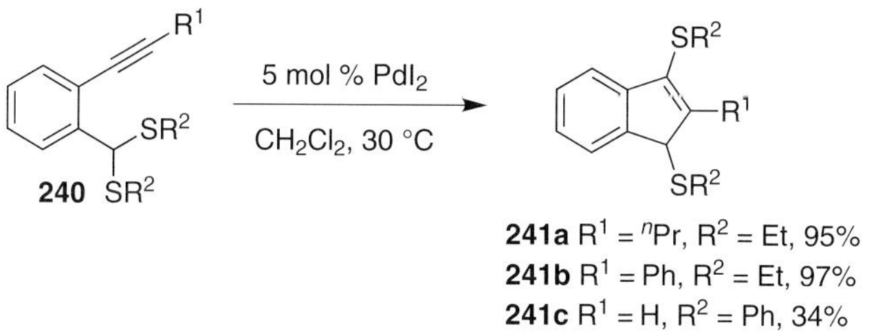

5 mol % PdI2
CH2Cl2, 30 °C
240
241a R1 = nPr, R2 = Et, 95%
241b R1 = Ph, R2 = Et, 97%
241c R1 = H, R2 = Ph, 34%

Scheme 10.90 Palladium-catalyzed reaction of *o*-alkynylbenzaldehyde thioacetals.

A new observation in the catalytic cyclization of *o*-alkynylbenzaldehyde acetals **235** to the functionalized indenes **242** has been reported (Scheme 10.91) [109]. It was found that the cyclization was strictly controlled by the number of triphenylphosphine ligands on the Pd catalyst. Only complexes with three available coordination sites on Pd catalyze this reaction. A mechanistic study suggested that π-coordination of Pd to the benzene ring was a key step, controlled by the number of vacant coordination sites.

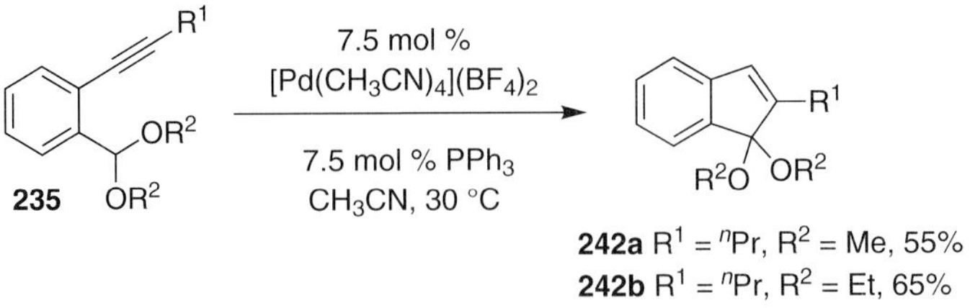

Scheme 10.91 Synthesis of indenes via palladium-catalyzed hydrocarbonation of alkynes.

242a R^1 = ^{n}Pr, R^2 = Me, 55%
242b R^1 = ^{n}Pr, R^2 = Et, 65%

The synthesis of *N*-(alkoxycarbonyl)-indoles **244** via AuCl$_3$-catalyzed cyclization of 2-(alkynyl)phenylisocyanates **243** in the presence of alcohols is known (Scheme 10.92) [110]. The product **244** was obtained in 57% yield along with side product **245** in 32% yield. Another gold catalyst, such as NaAuCl$_4$·2H$_2$O, gave **244** and **245** in 41 and 40% yields, respectively.

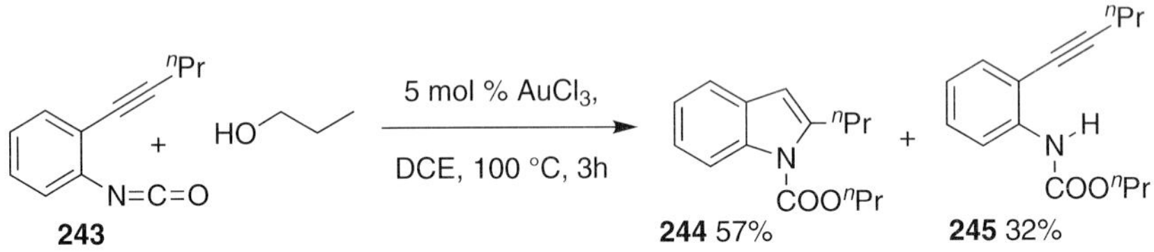

Scheme 10.92 Gold-catalyzed cyclization of 2-(alkynyl)phenylisocyanates.

The gold-catalyzed cascade cyclization of *o*-alkynyl nitrobenzenes **246** has been reported by Yamamoto and coworkers (Scheme 10.93) [111]. The reaction afforded isatogens **247** or anthranils **248** depending on the nature of R^1. For instance, when R^1 = Ar or cyclohexenyl, the corresponding isatogens **247** were formed in a major amount together with small amounts of anthranils **248**. However, when R^1 = Pr or ^{t}Bu, the corresponding anthranils **248** were obtained as sole products. The mechanism of the reaction is interesting and is depicted in Scheme 10.94. The common intermediates, **249** and **250**, are most probably involved for the formation of both types of products.

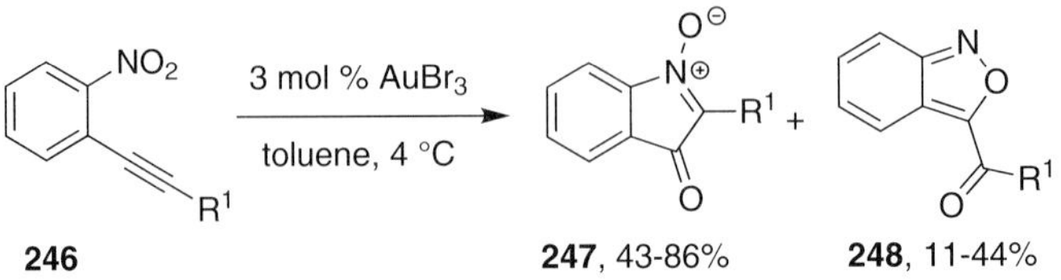

Scheme 10.93 Gold-catalyzed reactions of *o*-alkynyl nitrobenzenes for the synthesis of isatogens or anthranils.

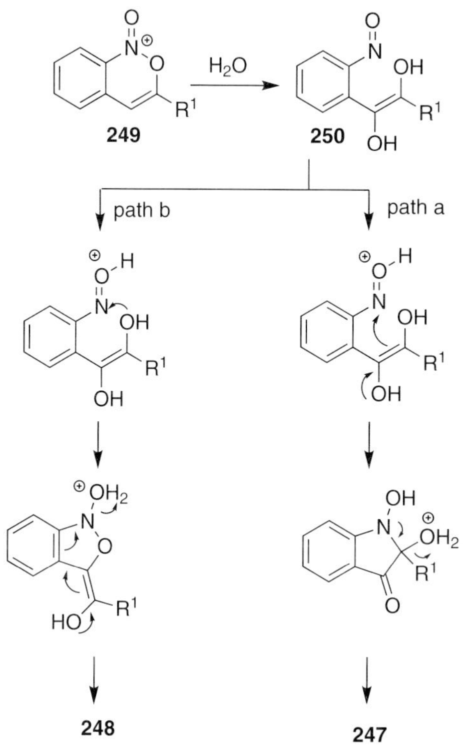

Scheme 10.94 Mechanism for the formation of isatogens or anthranils.

Palladium-catalyzed cascade three-component coupling reactions for the synthesis of the N-cyanoindoles 253 were reported by our research group (Scheme 10.95) [112]. The reaction takes place between 2-alkynylisocyanobenzenes 251, allyl methyl carbonate 252a and trimethylsilylazide in the presence of catalytic [Pd$_2$(dba)$_3$], with tris(2-furyl)phosphine as ligand, at 100 °C in THF. As described in the mechanism, first, Pd(0) reacts with allyl methyl carbonate and TMSN$_3$ to give π-allylpalladium azide 254 with the elimination of CO$_2$ and TMSOMe. The insertion of the isocyanide 251 into the Pd–N$_3$ bond in π-allylpalladium azide 254 then gives the π-allylpalladium intermediate 255. Elimination of N$_2$ followed by the 1,2-migration of the π-allylpalladium moiety from the carbon atom to the α-nitrogen atom in 255 gives the palladium-carbodiimide complex 256 which could be in equilibrium with the palladium-cyanamide complexes 257. Finally, the N-cyanoindoles 253 were formed via the insertion of the alkyne moiety into the Pd–N bond of this intermediate 257 followed by the reductive elimination of Pd(0).

The tandem procedure for the synthesis of 3-allyl-N-(alkoxycarbonyl)indoles 259 via the reaction of 2-(alkynyl)phenylisocyanates 258 and allyl carbonates in the presence of Pd(PPh$_3$)$_4$ (1 mol%) and CuCl (4 mol%) bimetallic catalyst (Scheme 10.96) [113]. Initially, the insertion of the isocyanates 258 into the π-

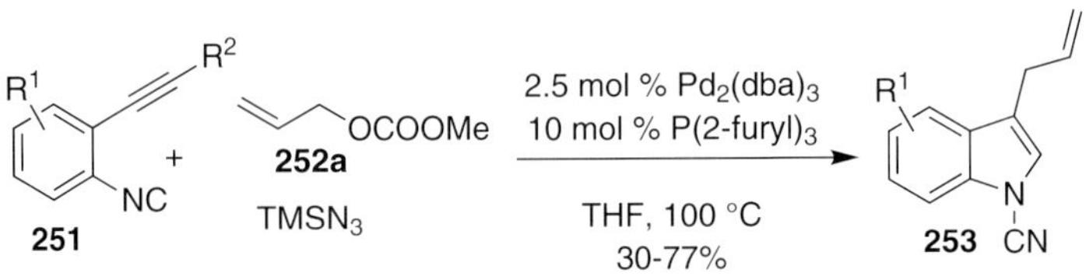

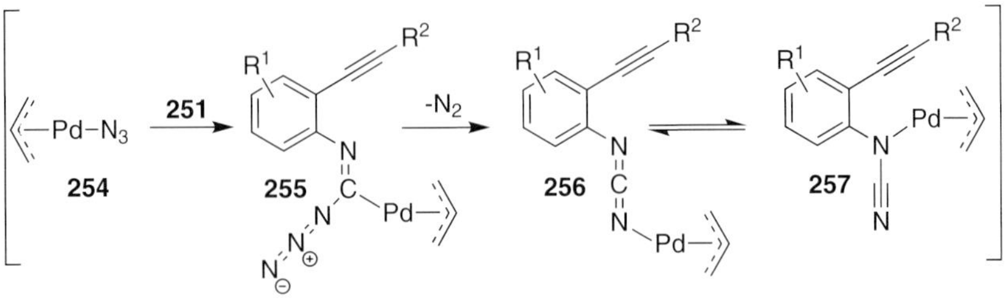

Scheme 10.95 Palladium-catalyzed reaction of o-alkynylisocyanobenzenes with allyl methyl carbonate and trimethylsilylazide.

allylpalladium alkoxide complex **261**, formed by the reaction of allylmethyl carbonate with Pd(0), produces the π-allylpalladium intermediates **262**. This intermediate, with Pd–N bonding, could be in equilibrium with the Pd–O bonded intermediates **118**, which should more probably be represented as the bis-π-allylpalladium analogue **119**. Insertion of the alkyne then occurs to form the indoles **259** and the Pd (0) species is regenerated. It should be noted that no carboamination takes place at all in the absence of CuCl, **260** being the sole product.

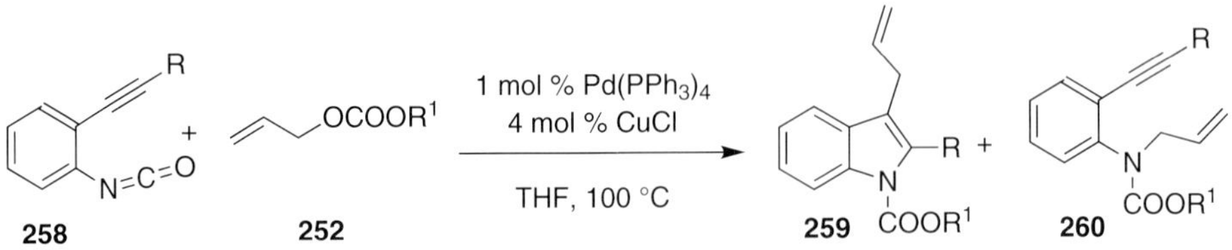

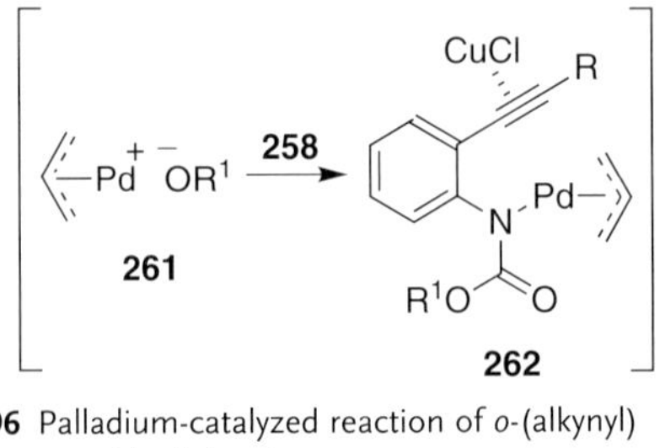

Scheme 10.96 Palladium-catalyzed reaction of o-(alkynyl) phenylisocyanates with allyl carbonate.

10.3 Cyclization Reactions of Alkynes

Yamamoto and coworkers reported the synthesis of oxindoles **264** from *ortho*-alkynylaryl isocyanates **258** and 1-pentyne **263** (Scheme 10.97) [114]. It was proposed that the formation of alkynyl palladium hydride takes place first and then inserts across the C–C triple bond of alkyne (cf. **265**) to form the vinylpalladium intermediate **266** (Scheme 10.98). Intramolecular nucleophilic attack of the vinylpalladium species on the isocyanato group produces the oxy-palladium intermediate **267** which is then transformed into Z-**264** after elimination of Pd(0) followed by isomerization. It is interesting to note that the Z isomer is formed first, and then isomerizes to the E-isomer under dppe catalysis.

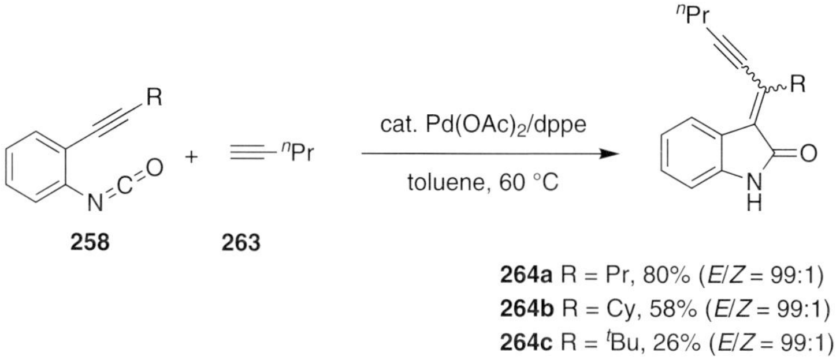

264a R = Pr, 80% (E/Z = 99:1)
264b R = Cy, 58% (E/Z = 99:1)
264c R = ^{t}Bu, 26% (E/Z = 99:1)

Scheme 10.97 Palladium-catalyzed synthesis of oxindoles from *o*-alkynyl arylisocyanates and 1-pentyne.

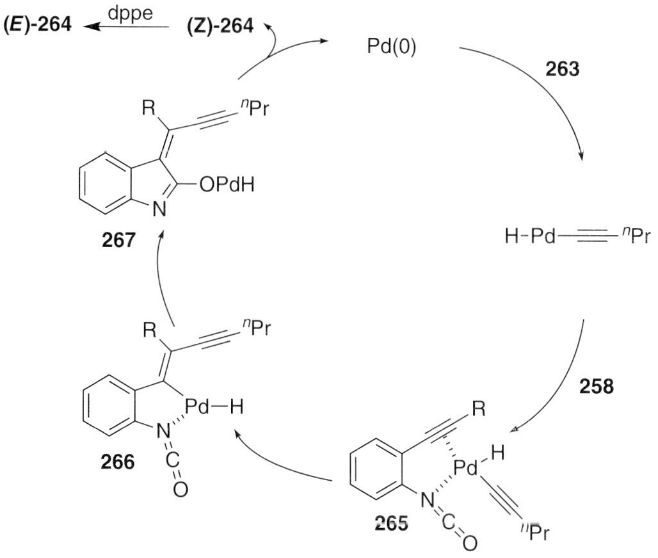

Scheme 10.98 Mechanism for the synthesis of oxindoles from *o*-alkynyl arylisocyanates and 1-pentyne.

The cyclization of iminoalkynes tethered with the aromatic ring under palladium catalysis gave 3-alkenylindoles. For instance, when iminoalkynes **268** were treated with 5 mol% palladium acetate and 20 mol% ^{n}Bu$_3$P in 1,4-dioxane at 100 °C, indoles **269** were obtained in good yields (Scheme 10.99) [115]. The mechanism of the reaction is quite interesting and is depicted in Scheme 10.100. First the formation of HPdOAc takes places in the presence of trace amounts of water which is supposed to be present in the reaction medium. The insertion of the palladium species into the C−C triple bond gives vinylpalladium **270** and then indolenine **273** via intermediates **272** or **271**. The indolenine **273** isomerizes to indoles **269** under the present reaction conditions.

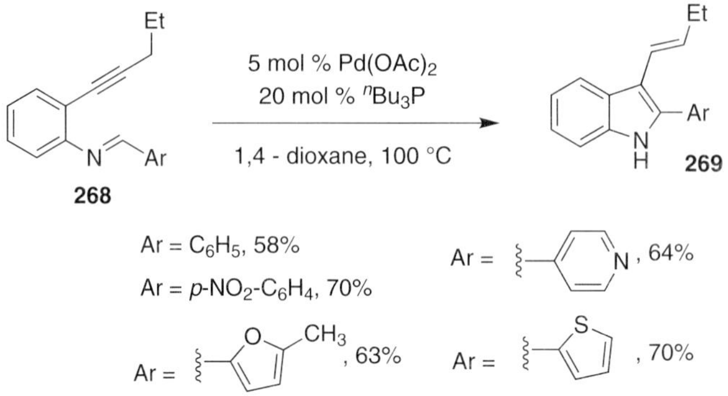

Scheme 10.99 Palladium-catalyzed synthesis of indole from iminoalkynes.

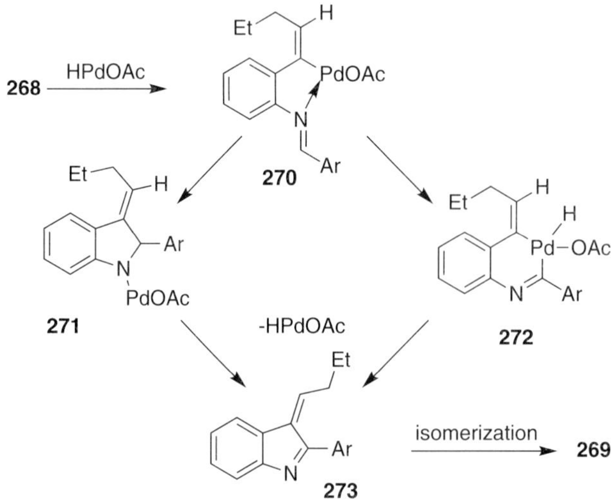

Scheme 10.100 Mechanism for the formation of indole from iminoalkynes under palladium catalysis.

10.3.5
Hydroarylation Reactions

Shi and He described the preparation of coumarins from aryl alkynoates (Scheme 10.101) [116]. The substrates **274** on treatment with gold catalyst in dichloroethane at an appropriate temperature gave the corresponding isocoumarins **275** in excellent to fair yields. The silver salts were added in order to increase the reactivity of the gold catalysts. The authors proposed that the key step was metallation of the arene which subsequently attacked the electron-deficient alkynes giving the products. The alternative mechanism which involved activation of the C—C bond by metal catalyst was ruled out. The 5-*exo-dig* hydroarylation of *o*-alkynyl biaryls via a C—H activation pathway has been reported recently [117].

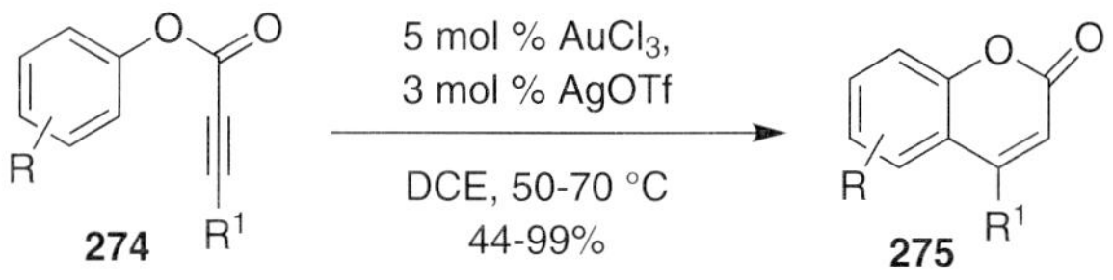

Scheme 10.101 Gold-catalyzed hydroarylation of alkynes for the synthesis of coumarin.

The cyclization of *N*-propargyl *N*-tosylanilines **276** was also catalyzed by the cationic complex produced from [Au(PPh$_3$)Me] and HBF$_4$ (Scheme 10.102) [118]. This intramolecular reaction gave 1,2-dihydroquinolines **277** and proceeded under milder conditions and with better yields than the cyclization catalyzed by Pt(II).

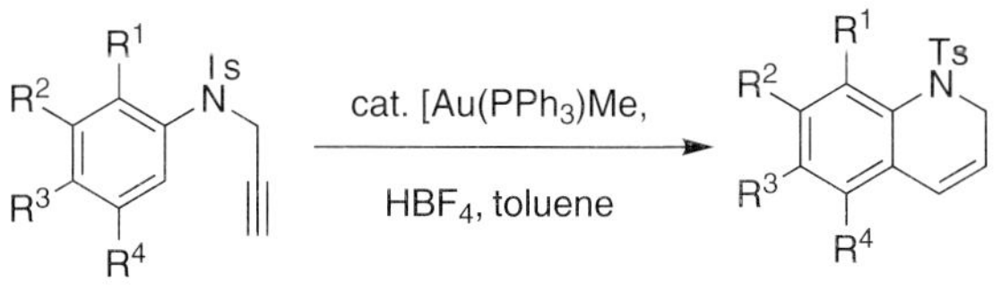

276a R^1 = R^4 = OMe, R^2 = R^3 = H   **277a** (23 °C, 4h, 71%)
276b R^1 = R^4 = H, R^2 = R^3 = OMe   **277b** (50 °C, 4h, 89%)
276c R^1 = H, R^2 = R^3 = R^4 = OMe   **277c** (50 °C, 17h, 92%)

Scheme 10.102 Gold-catalyzed hydroarylation of alkynes for the synthesis of 1,2-dihydroquinolines.

Echavarren recently reported a novel gold-catalyzed cyclization of the substrates of type **278** (Scheme 10.103) [119]. When gold(I) was used as a catalyst, azepino[4,5-*b*] indole derivatives **279** were obtained, whereas the use of Au(III) catalysts led to indoloazocines **280** by an 8-endo-dig process. This type of regiochemical control by the oxidation state of the metal catalyst is noteworthy.

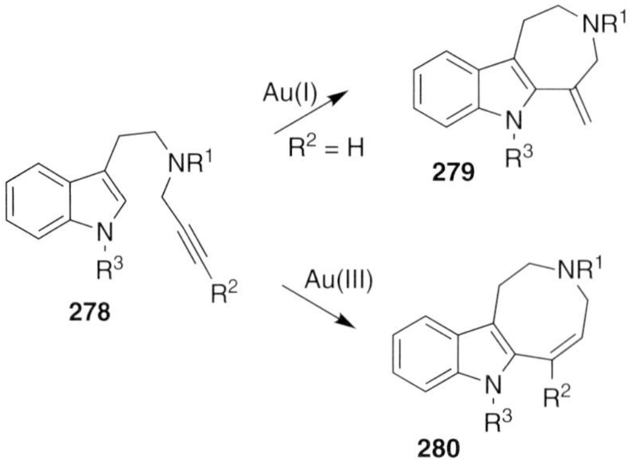

Scheme 10.103 Gold-catalyzed hydroarylation of alkynes for the synthesis of azepino[4,5-b]indole and indoloazocines.

10.3.6
Miscellanious Cyclization Reactions

The Yamamoto group reported that a palladium complex in the presence of a copper salt effectively catalyzed the three-component coupling of non-activated alkynes, allyl methyl carbonate and trimethylsilyl azide. Regioselective formation of 2-allyl-1,2,3-triazole **281** was achieved in the presence of catalytic amounts of $Pd_2(dba)_3 \cdot CHCl_3/P(OPh)_3$ and $CuCl(PPh_3)_3$ (Scheme 10.104) [120]. On the

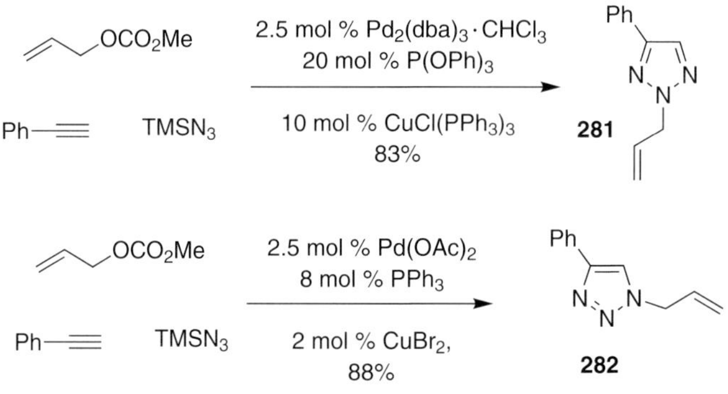

Scheme 10.104 Palladium-catalyzed reaction of phenyl acetylene with allyl methyl carbonate and $TMSN_3$.

other hand, under catalysis of Pd(OAc)$_2$/PPh$_3$ and CuBr$_2$, 1-allyl-1,2,3-triazole **282** was produced [121]. When 1-phenyl trimethylsilylacetylene was used as a starting material and the reaction was performed in the presence of Pd$_2$(dba)$_3$·CHCl$_3$/P(OEt)$_3$ and CuCl, 1,5-diallyltriazole **283** was formed selectively (Scheme 10.105) [122]. The formation of copper acetylide was proposed in all these cases. The [3 + 2] cycloaddition between the copper acetylide and active azide species (cf. **284**) afforded the triazole framework.

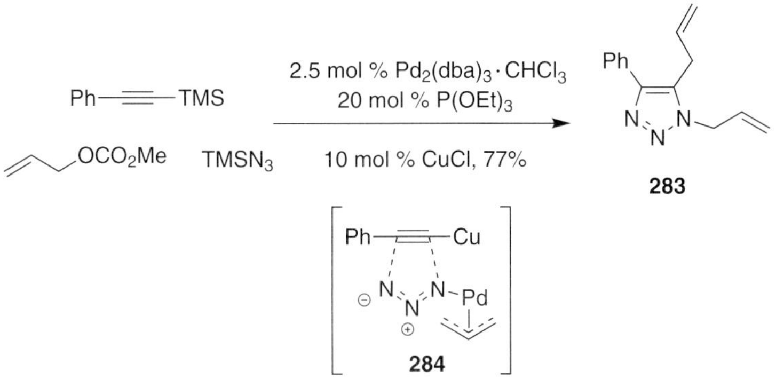

Scheme 10.105 Palladium-catalyzed reaction of 1-phenyl trimethylsilylacetylene with allyl methyl carbonate and TMSN$_3$.

The copper-catalyzed reaction of isocyanides **285** with electron-deficient alkynes **286** gave substituted pyrroles **287** (Scheme 10.106) [123]. A catalytic system made use of Cu$_2$O and the 1,10-phenanthroline ligand. A proposed mechanism is illustrated in Scheme 10.107. First, the 1,4-addition of the nucleophilic intermediate **288a** or **288b**, generated from **285** with the extrusion of H$_2$O, on the alkynes **286** takes place. The newly generated copper enolate intramolecularly attacks the isonitrile carbon to generate the cyclized intermediate **289**. The C–Cu bond in the intermediate **289** is protonated by isocyanide **285**, and the intermediate **290** is produced with regeneration of the copper-intermediate **288**. 1,5-Hydrogen shift in **290** produces the pyrroles **287**.

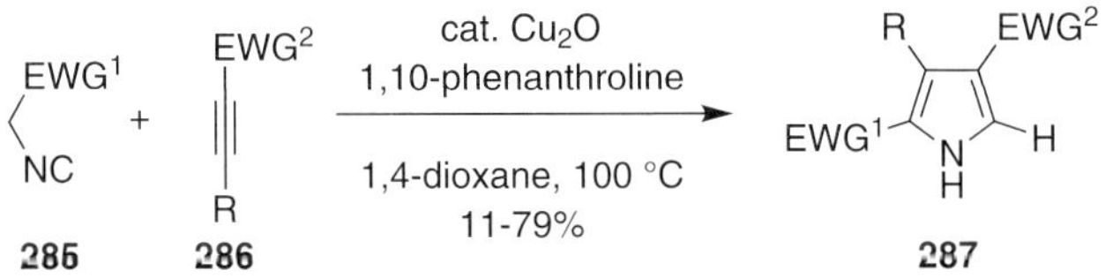

Scheme 10.106 Copper catalyzed synthesis of pyrroles from isocyanides and alkynes.

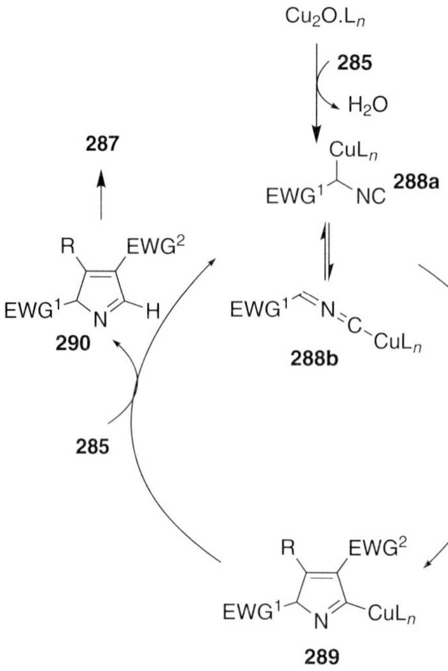

Scheme 10.107 Mechanism for the formation of pyrroles from isocyanides and alkynes under copper catalysis.

10.4
Cyclization Reactions of Allenes

10.4.1
Cyclization with Tethered Nucleophiles

Synthesis of heterocycles and carbocycles via the metal-catalyzed addition of nucleophiles across the C−C bond of allenes is one of the most important processes in organic synthesis [124]. Conventionally, the use of silver and mercury salts is well known to catalyze such transformations [125]. However, the Lewis acidity of the former catalyst and the toxicity of the latter restrict their use in organic synthesis. Recently, it has been reported that palladium and gold catalysts can also be used to afford these transformations. In this section the most significant examples including silver-catalyzed reactions are discussed.

In 1996, the Yamamoto group reported the palladium-catalyzed cyclization of allenes having carbon nucleophiles in the tether. The substrates **291** upon treatment with $[(\eta^3\text{-}C_3H_5)PdCl]_2$ – dppf catalyst, afforded carbocycles **292** in very good yields (Scheme 10.108) [126]. Although reactions were run at high temperature, relatively neutral reagents were employed. Ma and Zhao reported the palladium-catalyzed cyclization of 2-(2′,3′-dienyl)malonates **293** with organic halides to form vinyl cyclopropanes **294** (Scheme 10.109) [127]. The intermediacy of the π-allylpalladium complex **295** was proposed.

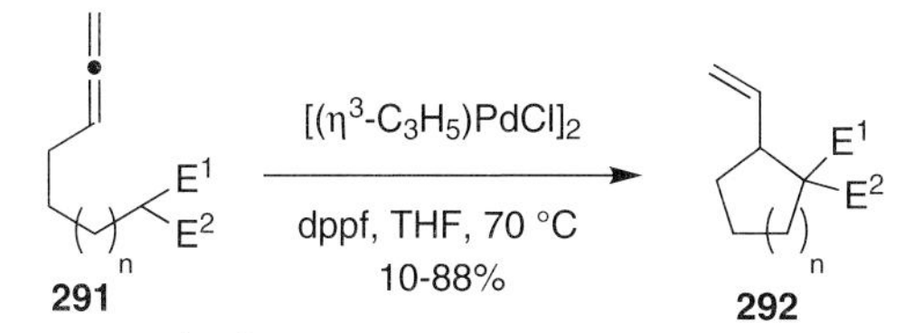

Scheme 10.108 Palladium-catalyzed hydrocarbonation of allenes.

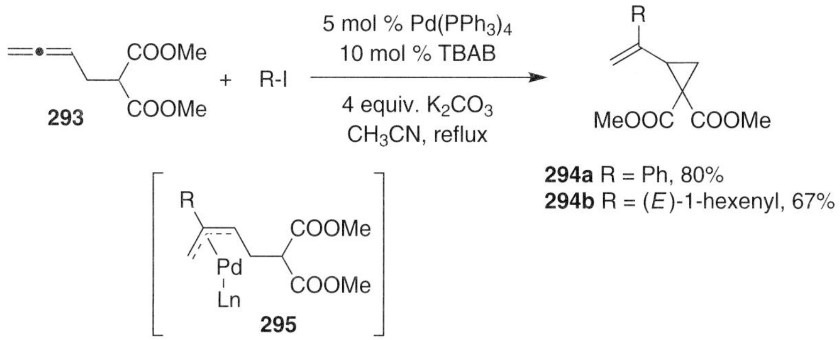

n = 1, 2; E^1, E^2 = CN, COOMe, SO_2Ph, Meldrum's acid

294a R = Ph, 80%
294b R = (E)-1-hexenyl, 67%

Scheme 10.109 Palladium-catalyzed reactions of allene bearing active methine with vinyl/aryl iodide.

The synthesis of nitrogen-containing heterocycles **297** from aminoallenes **296** could also be achieved (Scheme 10.110) [128]. It was observed that the addition of a mild proton source such as acetic acid increases the yield and rate of reaction. Recent research revealed that gold salts are very effective for activating allenes. The Yamamoto group reported the gold-catalyzed intramolecular hydroamination of allene **298** which gave access to five- and six-membered heterocycles **299** (Scheme 10.111) [129]. It is found that the chirality is transferred from the starting aminoallenes to the products under these reaction conditions. A few months later Widenhoefer et al. reported the synthesis of heterocycles via the gold-catalyzed activation of allenes with a much broader scope [130].

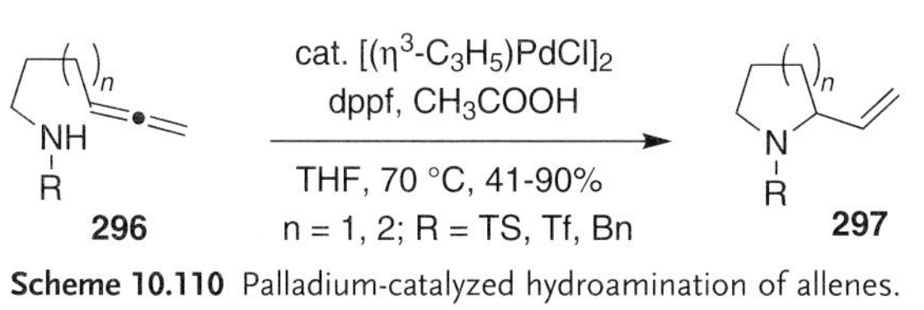

n = 1, 2; R = Ts, Tf, Bn

Scheme 10.110 Palladium-catalyzed hydroamination of allenes.

As shown in Scheme 10.112, the use of optically pure allenylcarbinol **300** gave cis-2,5-dihydrofuran **301** in high yield [131]. Furstner et al. also reported that allenyl alcohol **302** cyclized in the presence of $AgNO_3$ to afford dihydrofuran **303** with complete chirality transfer, however, a stoichiometric amount of the silver salt was needed in this case (Scheme 10.112) [132]. Marshall et al. later reported the cyclization of allenic acids **304** in the presence of catalytic amounts of $AgNO_3$ to give butenolides **305** (Scheme 10.113) [133]. High yields were generally obtained in all the cases.

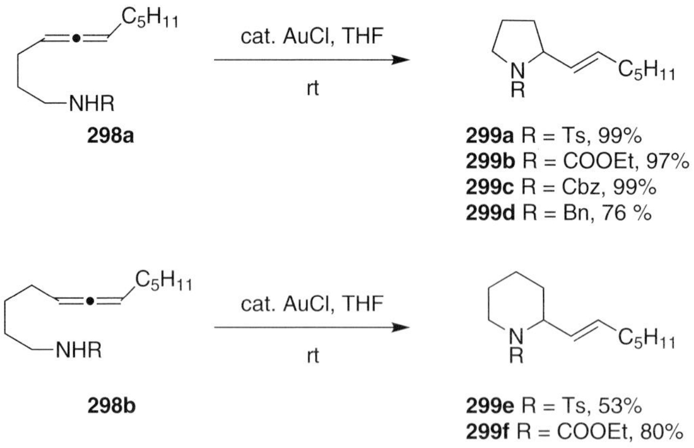

Scheme 10.111 Gold-catalyzed hydroamination of allenes.

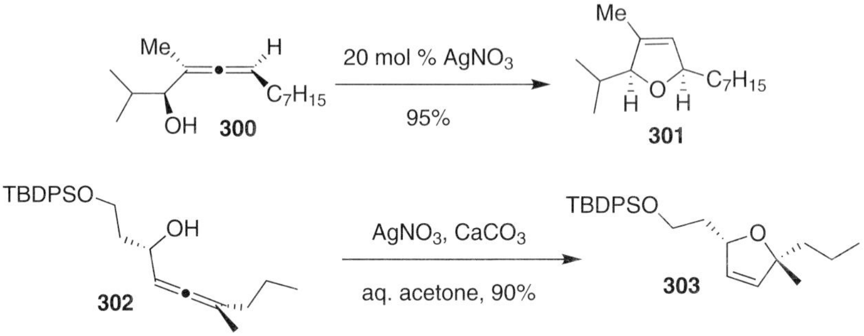

- **299a** R = Ts, 99%
- **299b** R = COOEt, 97%
- **299c** R = Cbz, 99%
- **299d** R = Bn, 76 %

- **299e** R = Ts, 53%
- **299f** R = COOEt, 80%

Scheme 10.112 Silver-catalyzed hydroalkoxylation of allenes.

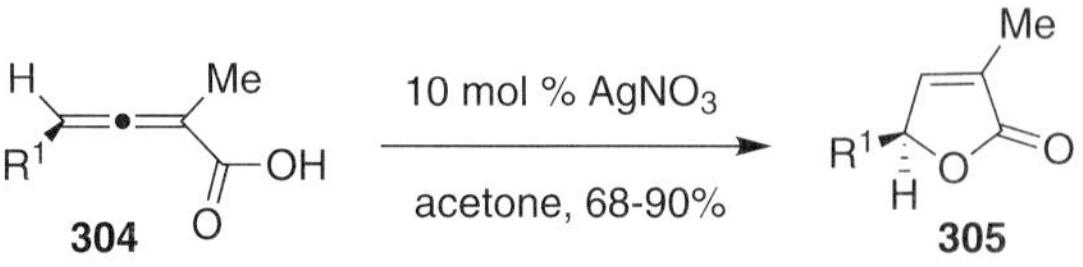

Scheme 10.113 Silver-catalyzed hydrocarboxylation of allenes.

R^1 = n-C_5H_{11}, CH_2=CH(CH_2)$_7$, c-C_6H_{11}

Functionalized α-hydroxyallenes **306** were smoothly converted into the corresponding 2,5-dihydrofurans **307** by using 5–10 mol% of gold(III) chloride as catalyst. This mild and efficient cyclization method was applied to alkyl- and alkenyl-substituted allenes which furnished tri- and tetra-substituted dihydrofurans in good to excellent chemical yields and with complete access to center chirality transfer (Scheme 10.114) [134]. Evidence for the *in situ* reduction of gold(III) during the cyclization of allenyl carbinols was reported recently [135].

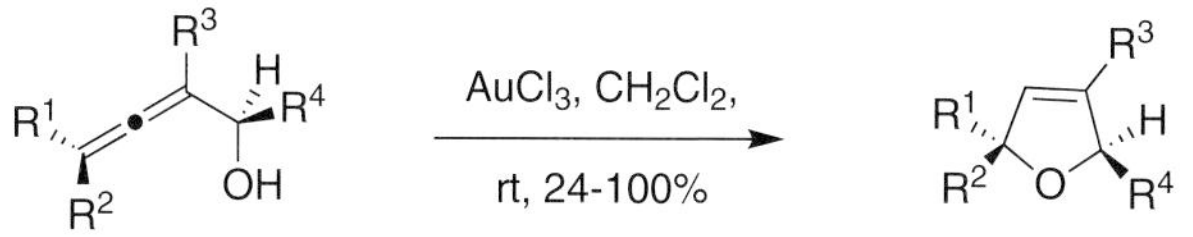

10.4 Cyclization Reactions of Allenes

Scheme 10.114 Gold-catalyzed hydroalkoxylation of allenes.

R^1 = ^{t}Bu, H, $H_2C=CH(CH_2)_2$; R_2 = Me, ^{n}Bu, ^{n}Hex, H;
R^3 = H, Me; R^4 = COOEt, COOMe, CH_2OH, CH_2OTBS, CH_2OMe

The gold(III) chloride-catalyzed cycloisomerization of various α-aminoallenes **308** gave the corresponding 3-pyrrolines **309** in good yields (Scheme 10.115) [136]. It was reported that the reactivity depended on the nature of the protecting groups. For example, when X = H, 5 days (room temperature) were needed to get 74% yield of the product. On the other hand, when X = Ts, the reaction afforded the corresponding product in 95% yield within 1 h at 0 °C. They also reported the cyclization of α-thioallenes **310** to 2,5-dihydrothiophenes **311** (Scheme 10.116) [137]. This is the first example of C—S bond formation in the history of gold catalysis.

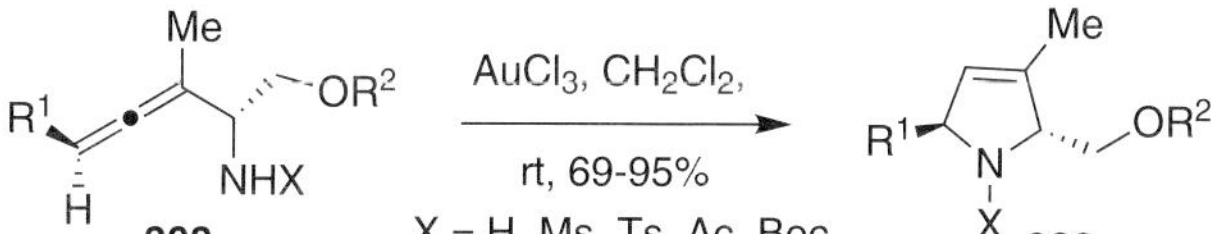

Scheme 10.115 Gold-catalyzed hydroamination of allenes.

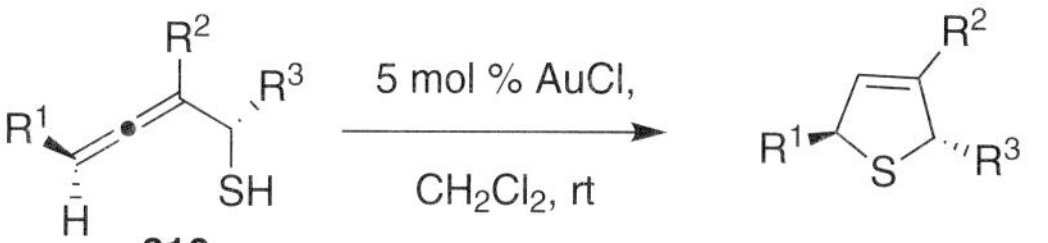

311a R^1 = ^{i}Pr, R^2 = Me, R^3 = p-CF_3-$C_6H_4OCH_2$, 67%
311b R^1 = ^{n}Hex, R^2 = Me, R^3 = CH_2OBn, 82%
311c R^1 = $H_2C=CH(CH_2)_7$, R^2 = H, R^3 = H, 43%

Scheme 10.116 Gold-catalyzed hydrothiolation of allenes.

Silver salts effectively catalyze the cyclization of O-(2,3-butadienyl) N-tosyl-carbamates **312** to provide the *trans*-oxazolidinones **313** predominantly or exclusively (Scheme 10.117) [138]. It was observed that the ease of cyclization depended on the kind of substituents on the nitrogen atom. Generally, electron-withdrawing groups facilitate the reaction.

$AuCl_3$ efficiently catalyzes the cyclization of *tert*-butyl allenoates **314** into γ-butenolides **315** (Scheme 10.118) [139]. Mechanistically, the authors believed that the formation of allenoic acid took place first, then subsequent cyclization afforded the products.

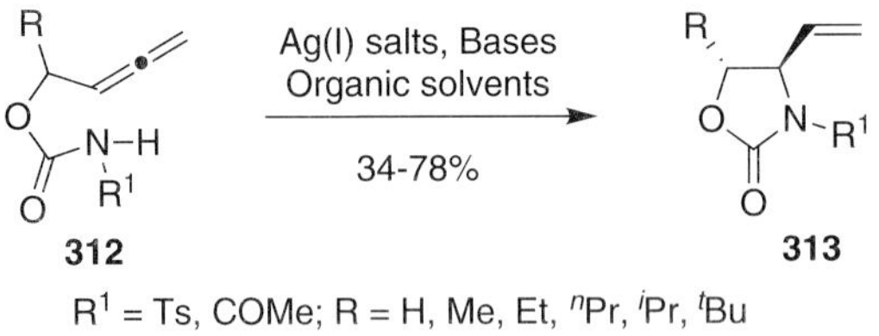

Scheme 10.117 Silver-catalyzed hydroamidation of allenes.

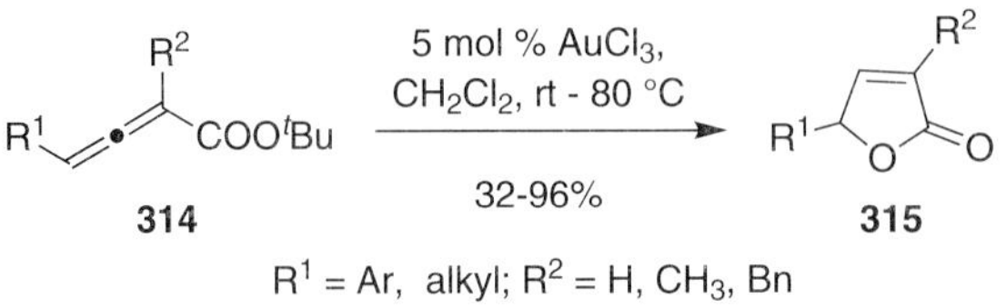

R^1 = Ts, COMe; R = H, Me, Et, ^{n}Pr, ^{i}Pr, ^{t}Bu

312 → 313, Ag(I) salts, Bases, Organic solvents, 34–78%

Scheme 10.118 Gold-catalyzed cyclization of t-butyl allenoates.

314 → 315, 5 mol % AuCl$_3$, CH$_2$Cl$_2$, rt - 80 °C, 32–96%

R^1 = Ar, alkyl; R^2 = H, CH$_3$, Bn

Although phosphine gold(I) complexes are the best catalysts for activating the C–C unsaturated bond of allenes, asymmetric versions of these processes are challenging because of the linear geometry of gold(I) complexes in which chiral phosphine ligand stays far away from the reactive center. Recent research showed that such a process involving gold-catalyzed asymmetric hydroamination and hydroalkoxylation was possible by using axially chiral phosphine ligands. Widenhoefer and Zhang reported gold(I)-catalyzed intramolecular enantioselective hydroalkoxylation of allenes **316** for the synthesis of optically active five- and six-membered cyclic ethers **317** (Scheme 10.119) [140]. The axially chiral dinuclear gold-phosphine complex **318** was found to exhibit best catalytic activity. This is the first example of catalytic enantioselective hydroalkoxylation of allenes catalyzed by transition metals.

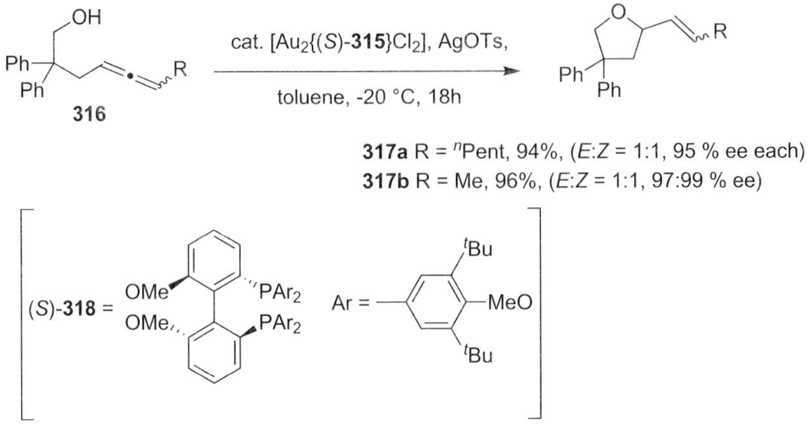

316 → **317**, cat. [Au$_2${(S)-**315**}Cl$_2$], AgOTs, toluene, −20 °C, 18h

317a R = ^{n}Pent, 94%, (E:Z = 1:1, 95 % ee each)
317b R = Me, 96%, (E:Z = 1:1, 97:99 % ee)

(S)-**318** = biphenyl with OMe, OMe, PAr$_2$, PAr$_2$; Ar = 3,5-di-^{t}Bu-4-MeO-phenyl

Scheme 10.119 Gold-catalyzed asymmetric hydroalkoxylation of allenes.

The first example of the dynamic kinetic enantioselective hydroamination of axially chiral allenes was reported by the same research group (Scheme 10.120) [141]. This remarkably rare transformation worked well only with trisubstituted allenes **319**, obtaining high ees of the products **320/321**. The disubstituted allenes **322** gave very low yield of products **323** under almost identical reaction conditions. The presence of two substituents on the terminal carbon of the allenes might attenuate substrate stereocontrol as well as retard the rate of C–N bond formation relative to racemization, leading to dynamic kinetic enantioselective hydroamination.

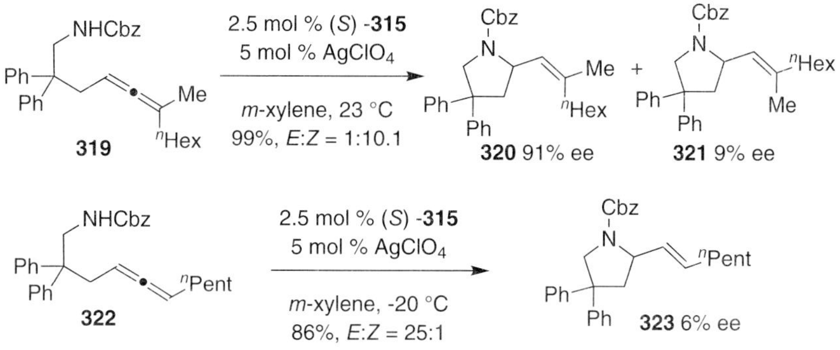

Scheme 10.120 Gold-catalyzed dynamic kinetic enantioselective hydroamination of axially chiral allenes.

The seminal publication from the Toste group reported the gold-catalyzed enantioselective intramolecular hydroamination of allenes **324**, for the synthesis of pyrrolidines **325a–d** and piperidines **325e–i**, (Scheme 10.121) [142]. This is the first

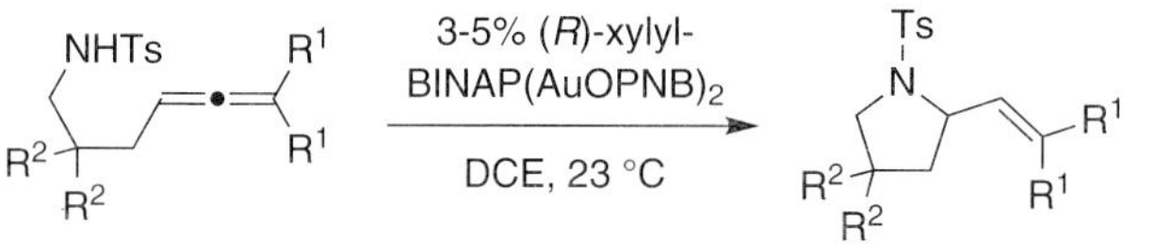

325a R^1 = Me, R^2 = H, 15h, 98%, 99%ee
325b R^1 = Et, R^2 = H, 17h, 90%, 99%ee
325c R^1 = -CH$_2$(CH$_2$)$_2$CH$_2$-, R^2 = H, 15h, 75%, 83%ee
325d R^1 = -CH$_2$(CH$_2$)$_3$CH$_2$-, R^2 = H, 17h, 88%, 98%ee

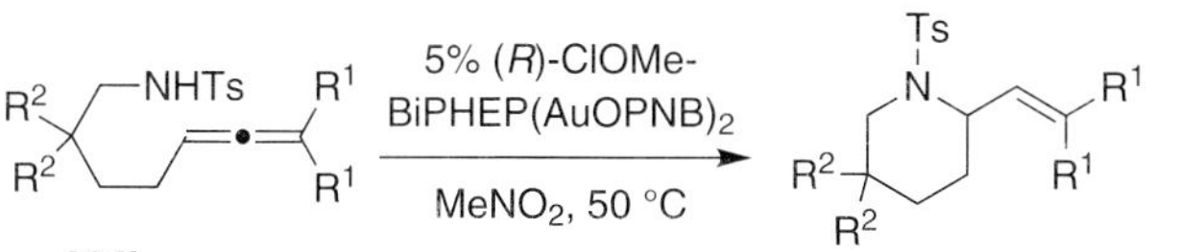

325e R^1 = Me, R^2 = H, 15h, 88%, 81%ee
325f R^1 = Et, R^2 = H, 24h, 41%, 74%ee
325g R^1 = Me, R^2 = Me, 24h, 70%, 98%ee
325h R^1 = Me, R^2 = Ph, 24h, 70%, 88%ee
325i R^1 = Me, R^2 = -CH$_2$(CH$_2$)$_3$CH$_2$-, 17h, 66%, 97%ee

Scheme 10.121 Gold-catalyzed asymmetric hydroamination of allenes.

10 Transition Metal Catalyzed Cyclization Reactions of Functionalized Alkenes, Alkynes, and Allenes

example of catalytic enantioselective hydroamination of allenes catalyzed by transition metals. Both research groups used chiral dinuclear gold-phosphine complexes to make this reaction enantioselective. Recently, Toste and coworkers reported a chiral counter anion strategy for asymmetric hydroamination and hydroalkoxylation reactions [143].

Ma et al. described the synthesis of butenolides **327** from aryl/alkenyl halides and 1,2-allenoic carboxylic acids **326** in the presence of a Pd/Ag bimetallic system (Scheme 10.122) [144]. In the absence of Ag_2CO_3, the reaction afforded the products in lower yields. On the basis of this observation the authors proposed the intermediacy of the vinyl silver intermediate **328**, which after transmettalation with R^2PdX (cf. **329**), followed by reductive removal, afforded butenolides. They also reported a halolactonization process for the synthesis of β-chlorobutenolides **331** from allenic acids **330** in the presence of excess amounts of $CuCl_2$ (Scheme 10.123) [145]. β-Bromobutenolides could also be obtained just by replacing $CuCl_2$ by $CuBr_2$. The reaction proceeds through the stereoselective halocupration of **330** to form the Cu-containing intermediate E-**332**. Intramolecular attack of the carboxylic group on the copper forms the six-membered intermediate **333** which provides the products after reductive removal of copper. Ma's research group also reported a halolactamization process that is, the cyclization of allenamides into cyclic haloamides [146]. Pioneering work from Ma's research group disclosed palladium-catalyzed coupling-cyclization of 2,3-allenamides **334** and aryl/vinyl iodides for the synthesis of iminolactones **335** (Scheme 10.124) [147].

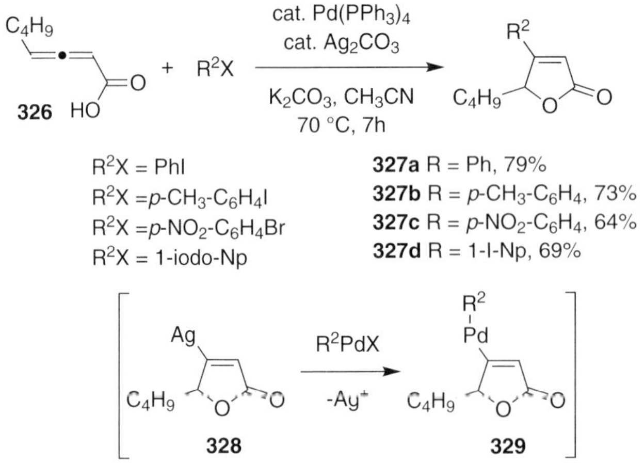

$R^2X = PhI$
$R^2X = p\text{-}CH_3\text{-}C_6H_4I$
$R^2X = p\text{-}NO_2\text{-}C_6H_4Br$
$R^2X = 1\text{-iodo-Np}$

327a R = Ph, 79%
327b R = p-CH_3-C_6H_4, 73%
327c R = p-NO_2-C_6H_4, 64%
327d R = 1-l-Np, 69%

Scheme 10.122 Cyclization of allenoic acids with aromatic halides in the presence of palladium and silver catalyst.

Pd–Cu bimetallic catalyst has also been used successfully for the intramolecular 1,2-addition (Scheme 10.125) [148]. Carboxylic acids **336**, alcohols **338a**, N-substituted amides **338b,c** and carbamates **340** were used as internal nucleophiles which afforded lactones **337**, tetrahydrofurans **339a**, pyrrolidines **339b,c** and oxazolidinones **341** in good yields.

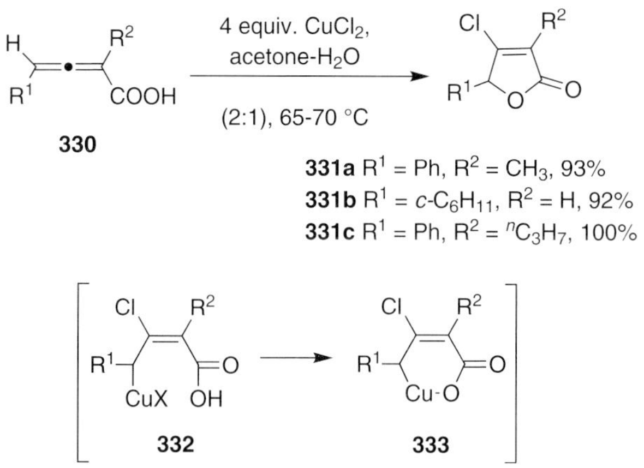
4 equiv. CuCl2,
acetone-H2O
(2:1), 65-70 °C
330
331a R1 = Ph, R2 = CH3, 93%
331b R1 = c-C6H11, R2 = H, 92%
331c R1 = Ph, R2 = nC3H7, 100%
332
333

Scheme 10.123 Copper-catalyzed halo-lactonization of allenoic acids.

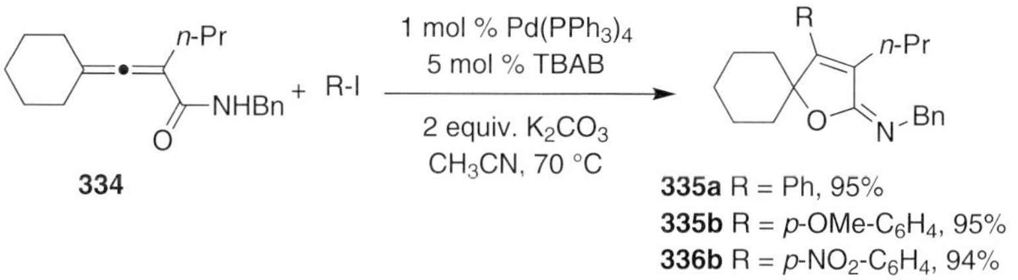
1 mol % Pd(PPh3)4
5 mol % TBAB
2 equiv. K2CO3
CH3CN, 70 °C
334
335a R = Ph, 95%
335b R = p-OMe-C6H4, 95%
336b R = p-NO2-C6H4, 94%

Scheme 10.124 Copper-catalyzed coupling/cyclization of 2,3-allenamides with aryl/vinyl iodides.

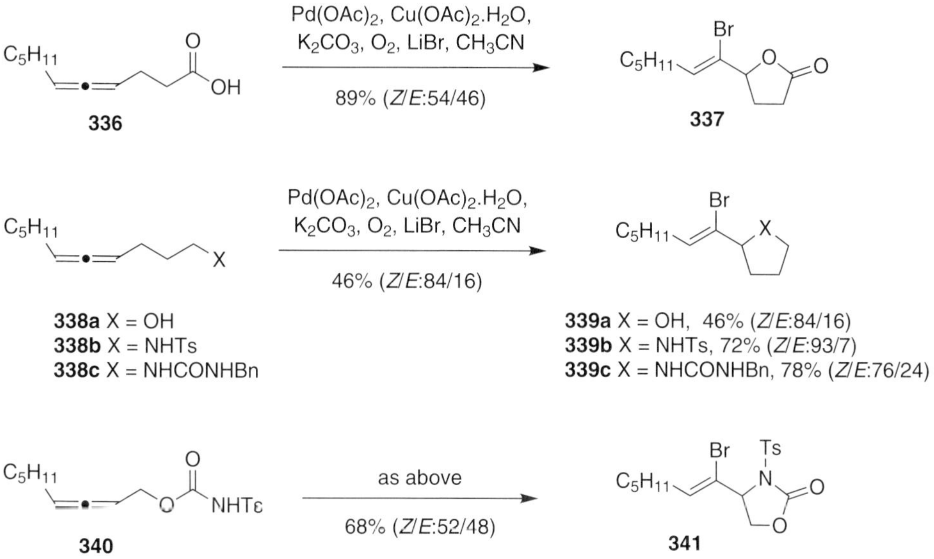
Pd(OAc)2, Cu(OAc)2.H2O,
K2CO3, O2, LiBr, CH3CN
89% (Z/E:54/46)
336
337
Pd(OAc)2, Cu(OAc)2.H2O,
K2CO3, O2, LiBr, CH3CN
46% (Z/E:84/16)
338a X = OH
338b X = NHTs
338c X = NHCONHBn
339a X = OH, 46% (Z/E:84/16)
339b X = NHTs, 72% (Z/E:93/7)
339c X = NHCONHBn, 78% (Z/E:76/24)
as above
68% (Z/E:52/48)
340
341

Scheme 10.125 Cyclization of allenes with tethered nucleophiles under palladium–copper catalysis.

10.4.2
Cyclization with Carbonyls

Cyclization of allenyl carbonyls represents a very convenient method for the preparation of multiply-substituted furans [149]. Marshall et al. reported a route for the syntheses of furans **343** involving Ag(I)-catalyzed cyclization of allenyl ketones/aldehydes **342** (Scheme 10.126) [150]. This group was the first to show that this type of cyclization is very useful for the synthesis of multiply-substituted furans. However, this procedure does not produce C-3 substituted furans. Ma et al. have reported a novel palladium-catalyzed process for the synthesis of C-3 substituted furans **345** from the reactions of allenyl carbonyls **344** with either aryl iodide [151] or allyl halides [152] (Scheme 10.127).

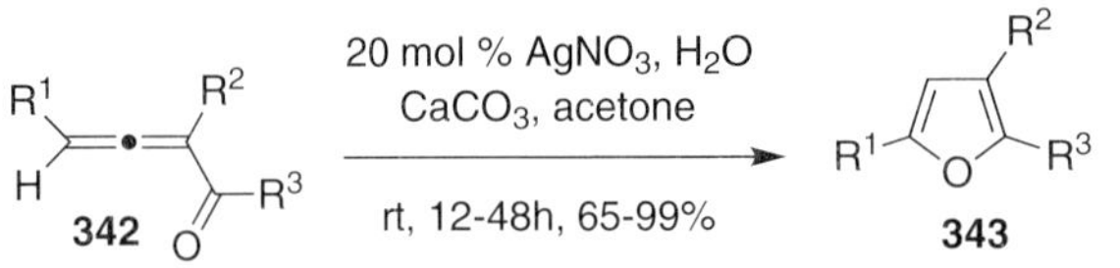

R^1 = H, CH_3, CH_2OBn, $(CH_3)_2CHCH_2$, $(CH_3)_3C$; R^2 = CH_3, C_7H_{15}; R^3 = H, $COOCH_3$, CH_2OAc, CH_2OTBS, CH_2OMOM

Scheme 10.126 Silver-catalyzed cycloisomerization of allenylketones.

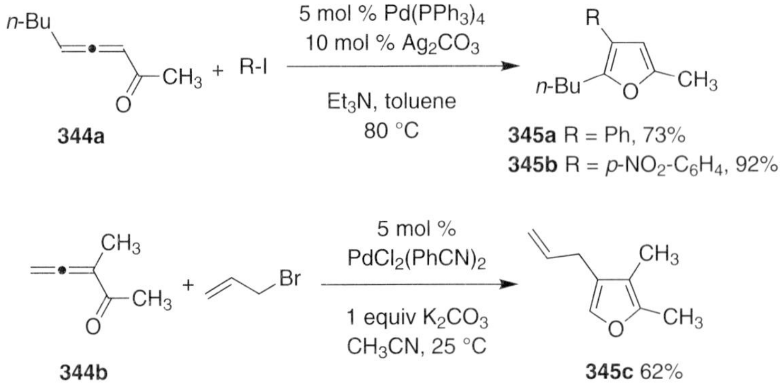

Scheme 10.127 Cyclization of allenylketones with aryl/allyl halide under metal catalysis.

Marshall and coworkers reported the activation of allenes by silver catalysts. It is reported that cyclization of allenyl ketone **346** provided furan **347** in the presence of catalytic amounts of $AuCl_3$. The reaction was extended also to one-pot cycloisomerization/hydroarylation of **348** with α,β-unsaturated ketones **349** to give C-2 substituted furans **350** (Scheme 10.128) [153]. Recently, gold(III) porphyrin complexes have been utilized for the cyclization of allenones [154].

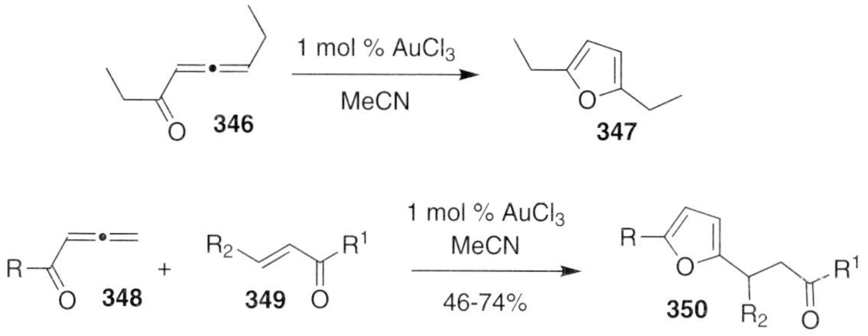

Scheme 10.128 Gold-catalyzed cyclization of allenylketones.

Gevorgyan et al. have shown that 1,2-iodine, -bromine, and -chlorine migration in haloallenyl ketones **351** takes place in the presence of AuCl$_3$ (Scheme 10.129) [155]. For this reaction iodo and bromo allenyl ketones gave the best results, compared to their chloro analogues. It was reported that the reaction proceeded through the halirenium intermediate **354**, formed by the intramolecular Michael addition of X to the enone moiety, as shown in **353**, which via subsequent addition–elimination furnishes 3-halofurans **352**.

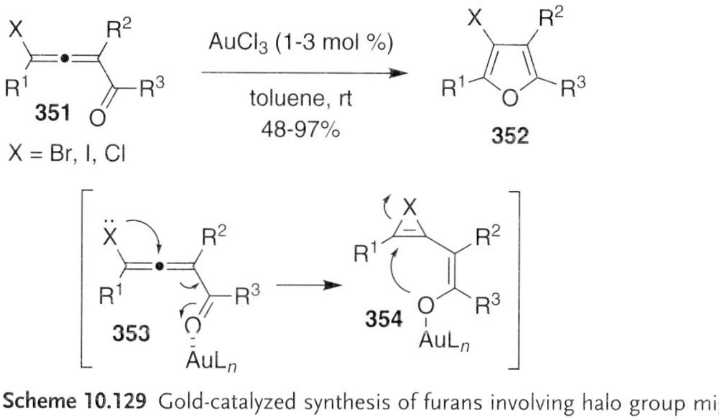

Scheme 10.129 Gold-catalyzed synthesis of furans involving halo group migration.

10.4.3
Hydroarylation Reactions

The first example of arylation of allenes was observed by Hashmi, where furan was used as the aromatic partner [153]. One can judge from the recent work published in the literature that the hydroarylation of allenes methodology became very general. The power and utility of the arylation of allenes in the C–C bond-forming process is dramatically manifested in the enantioselective total synthesis of (−)-rhazinilam by Nelson and coworkers (Scheme 10.130) [156]. An excellent diastereoselectivity was obtained for the cylization of **355** to **356**.

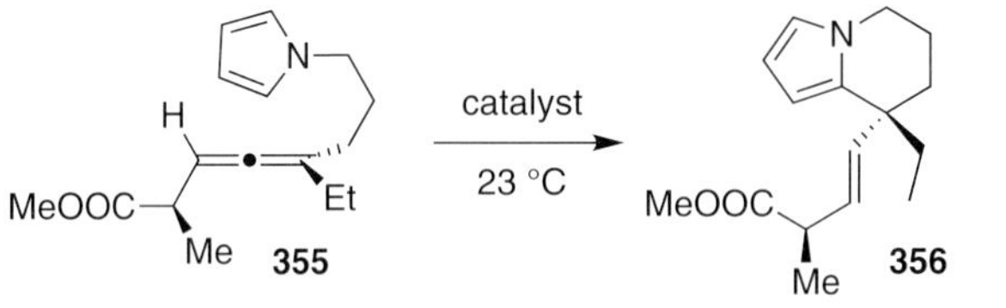

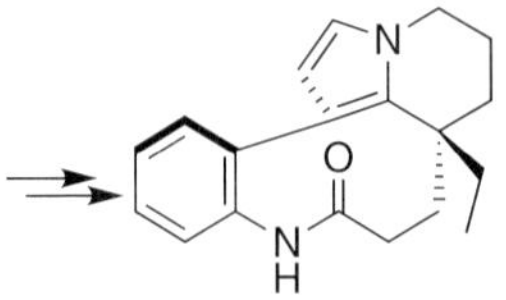

Scheme 10.130 Gold-catalyzed hydroarylation of allenes.

catalyst (mol%)	dr	yield (%)
AuCl$_3$ (10)	92:8	27
AuCl$_3$ (5); AgOTf (20)	92:8	82
Ph$_3$PAuOTf (5)	97:3	92

(−)-Rhazinilam

The Widenhoefer group found that Au[P(t-Bu)$_2$(o-biphenyl)]Cl activated by AgOTf is a highly active and highly selective catalyst for the intramolecular exo-hydroarylation of 2-allenyl indoles **357** for the synthesis of tetrahydrocarbazoles **358** (Scheme 10.131) [157]. The allenyl indoles **359** that possessed an axially chiral allenyl moiety after cyclization underwent complete transfer of chirality to give product **360** (Scheme 10.132). Such transfer of chirality from the allene to the newly formed stereogenic carbon atom with selective formation of the E-alkene is similar to that reported by Yamamoto for the hydroamination of allenes [129]. The gold(I)-catalyzed enantioselective variant of this method has also been documented [158].

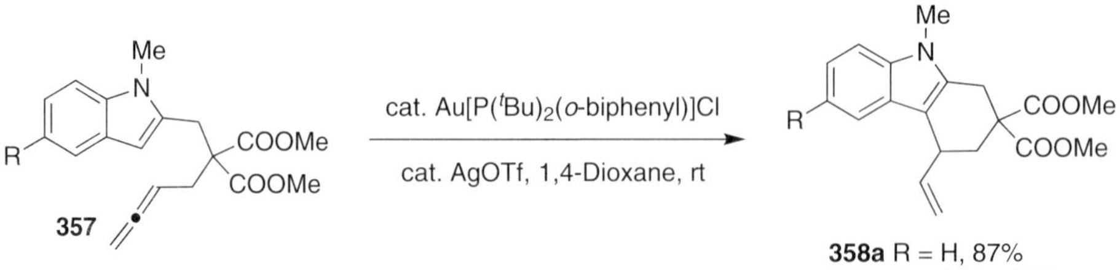

358a R = H, 87%
358b R = OMe, 89%
358c R = F, 91%

Scheme 10.131 Cyclization of allenylindoles under gold catalysis.

359 52% ee

360 52% ee

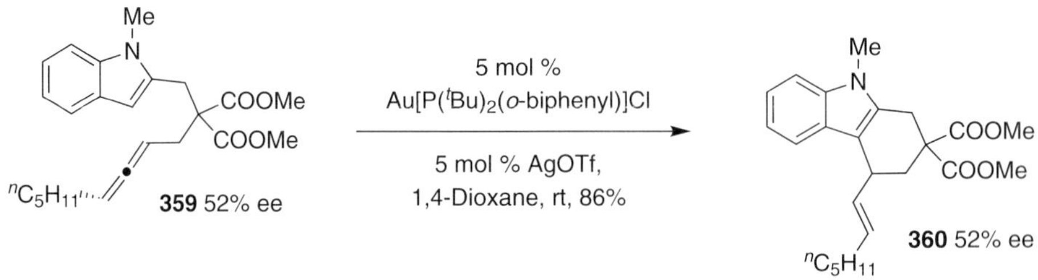

Scheme 10.132 Transfer of chirality during cyclization of allenylindoles under gold catalysis.

Not only heteroatom-containing heterocycles but also benzene rings bearing an electron-donating substituent can be used. Gagnè and Tarselli reported a highly electrophilic phosphite gold(I) catalyst for producing vinylbenzocycles **362** from allenes **361** in good to excellent yields (Scheme 10.133) [159]. The catalyst is tolerant of trace water and oxygen, it is bench-stable, and can be utilized in air with unpurified commercial solvent.

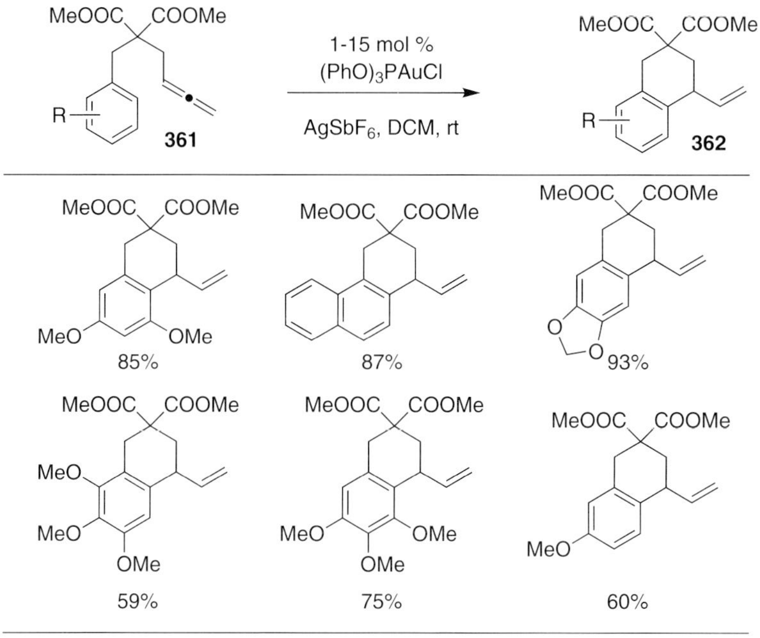

Scheme 10.133 Synthesis of tetrahydronaphthalenes via gold-catalyzed hydroarylation.

The cyclization of allenic anilines offers an efficient route to dihydroquinoline derivatives under the catalysis of gold. The hydroarylation takes place at the terminal carbon of allenes **363**, leading to a highly selective formation of six-membered rings **364** (Scheme 10.134) [160]. It is interesting to note that the product arising from the endo manner cyclization was not obtained at all.

10.5
Conclusion and Perspective

The transition metal-catalyzed cyclization of alkenes, allenes and alkynes bearing carbon-, oxygen-, nitrogen- and sulfur-containing functional groups represents a very valuable method for the synthesis of a wide variety of carbocycles and heterocycles. We have presented such very useful processes with emphasis on the work done in our laboratory. In the next few years we are likely to see many new catalytic systems developed for the cyclization of alkenes, alkynes and allenes. We hope that this review

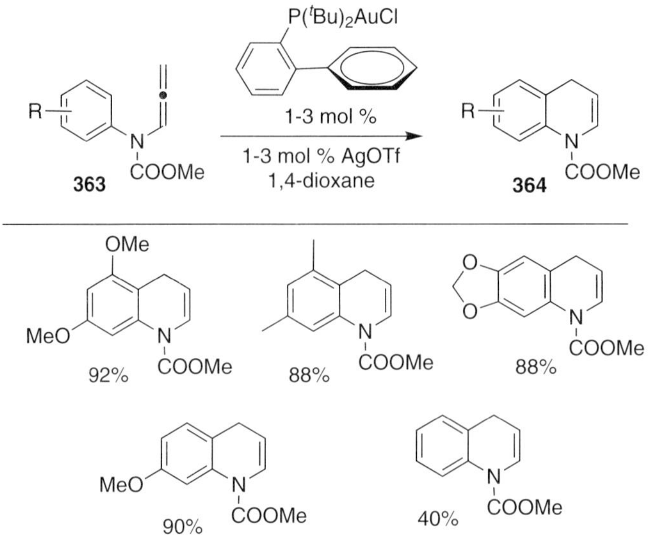

Scheme 10.134 Gold-catalyzed hydroarylation of allenic anilines.

has provided an appropriate background for the present topic and we can expect to see many more applications of these useful cyclization methodologies in the near future.

10.6
Experimental: Selected Procedures

10.6.1
Synthesis of Compound 13

To a mixture of chloro(triphenylphosphine)gold(I) (24.7 mg, 0.05 mmol) and silver triflate (12.8 mg, 0.05 mmol) in dry toluene (2.0 mL) were added the olefin (4.0 mmol) and TsNH2 (1.0 mmol) with stirring in a sealed tube. After the reaction was heated at 95 °C for 14–48 h, the mixture was directly loaded onto a silica gel chromatograph to afford analytically pure products.

10.6.2
Synthesis of Compound 20

A solution of alkenyl indole (0.50 mmol), PtCl$_2$ (0.010 mmol), and HCl (4.0 M solution in dioxane, 0.025 mmol) was stirred at 60 °C for 18 h, cooled to room temperature, and quenched with 1 N aq. NaOH. The layers were separated and the aqueous layer was extracted with ether. The combined organic extracts were washed with brine and concentrated under vacuum to obtain a residue which, on column chromatographic purification, gave the products.

10.6.3
Synthesis of Compound 40

To a solution of the activated olefins (0.5 mmol), cat. $Pd_2(dba)_3 \cdot CHCl_3$ and cat. phosphines in THF (5 mL) was added the allylic carbonates (0.75 mmol) at room temperature under Ar, and the mixture was stirred for an appropriate time at an appropriate temperature. The solvent was removed *in vacuo*, and the residue was purified by silica gel column chromatography.

10.6.4
Synthesis of Compound 82

A mixture of bis-homopropargylic diol, AuCl or $AuCl_3$ (2 mol%) in degassed methanol was stirred under an argon atmosphere at room temperature. After completion of the reaction, the mixture was filtered through a short pad of Celite (EtOAc), and the solvents were evaporated under reduced pressure to give the corresponding ketal.

10.6.5
Synthesis of Compound 100

To a mixture of substrates and cat. $Pd(PPh_3)_4$ in dry 1,4-dioxane was added cat. amounts of acetic/benzoic acid, and the mixture was stirred at 100 °C until reaction was completed. The reaction mixture was then filtered through a short silica gel column, using ether as an eluent, and the filtrate was concentrated. The residue was purified by a silica gel column chromatography to give the products.

10.6.6
Synthesis of Compound 134

To a solution of substrates (0.50 mmol) in 1,4-dioxane, cat. CuI (0.025 mmol), (±)-1,2-*trans*-cyclohexanediamine (0.050 mmol), and K_3PO_4 (1.00 mmol) were added. The mixture was stirred for 1.5 h at 110 °C under an argon atmosphere. After cooling, the reaction mixture was diluted with ethyl acetate, washed with water, dried over Na_2SO_4, and concentrated under reduced pressure. The residue was purified by chromatography to give the products.

10.6.7
Synthesis of Compound 145

To AuCl (0.005 mmol) in toluene (0.75 ml) was added the substrates (0.25 mmol) in toluene (0.5 ml) under argon atmosphere in a pressure vial. After stirring at 25 °C for 2 h, the reaction mixture was filtered through a short silica gel column using ethyl acetate as eluent. The residue was purified by silica gel column chromatography to give pure products.

10.6.8
Synthesis of Compound 184

To $AuCl_3$ (3 mol%) was added a mixture of substrate (103 mg, 0.5 mmol) and alkyne (0.6 mmol) in DCE (1.5 mL) at room temperature under Ar atmosphere. The resulting homogeneous solution was stirred at 80 °C for 1.5 h and then cooled to room temperature. The reaction mixture was transferred to a silica gel column, and the product was isolated using ether as eluent.

10.6.9
Synthesis of Compound 259

Allyl methyl carbonate (0.6 mmol) was added to a solution of isocyanate (0.5 mmol), $[Pd(PPh_3)_4]$ (0.005 mmol), and CuCl (0.02 mmol) in THF (0.5 mL) under an argon atmosphere, and the mixture was stirred at 100 °C for 1 h. The reaction mixture was cooled to room temperature, filtered through a short florisil pad, and concentrated. The residue was purified by column chromatography to afford pure products.

10.6.10
Synthesis of Compound 275

The substrate was mixed with $AuCl_3$ (5 mol%) and AgOTf (15 mol%) in 2.0 ml of DCE and the mixture was stirred at 50–70 °C. Stirring continued until the disappearance of the starting material was observed, as monitored by TLC. The final product was purified by flash chromatography.

10.6.11
Synthesis of Compound 287

To a 1,4-dioxane solution (1 mL) of Cu_2O (3.6 mg, 25 µmol) and 1,10-phenanthroline (50 µmol) were added ethyl butynoate (0.6 mmol) and ethyl isocyanoacetate (0.5 mmol) under an Ar atmosphere. The solution was stirred at 100 °C for 2 h. After consumption of the starting materials, the reaction mixture was cooled to room temperature and filtered through a short Florisil pad and concentrated. The residue was purified by column chromatography to afford the products.

10.6.12
Synthesis of Compound 325

To a solution of aminoallene in DCE or $MeNO_2$ (0.30 M) was added an appropriate gold(I) complex. The resulting homogeneous mixture was protected from ambient light and left to stir at the indicated temperature (23 or 50 °C). Upon completion, as judged by TLC analysis of the reaction mixture, the solution was loaded directly onto a silica gel column. Purification by flash column chromatography afforded the desired cyclized product.

Abbreviations

aq.	aqueous
Ar	Aryl
bipy	2,2′-bipyridine
Boc	*tert*-butoxycarbonyl
^{n}Bu	n-butyl
^{t}Bu	*tert*-butyl
cat.	catalytic
COD	1,5-cyclooctadiene
dba	dibenzylideneacetone
DCE	1,2-dichloroethane
DMA	*N,N*-dimethylacetamide
DMEDA	*N,N′*-dimethylethylenediamine
DMF	*N,N*-dimethylformamide
dppe	1,2-bis(diphenylphosphino)ethane
dppf	1,1′-bis(diphenylphosphino)ferrocene
equiv.	equivalent
EWG	electron-withdrawing group
Nf	nonafluorobutanesulfonyl
Phen	1,10-phenanthroline
PMB	*p*-methoxybenzyl
TBAB	tetra-n-butylammonium bromide
TBAC	tetra-n-butylammonium chloride
TBS	*tert*-butyldimethylsilyl
TDMPP	tri(2,6-dimethoxyphenyl)phosphine
TFP	tri-2-furylphosphine
THF	tetrahydrofuran

References

1 Reviews: (a) Yamamoto, Y. (2007) *J. Org. Chem.*, **72**, 7817–7831; (b) Nakamura, I. and Yamamoto, Y. (2004) *Chem. Rev.*, **104**, 2127–2198; (c) Kamijo, S. and Yamamoto, Y. (2004) *Multimetallic Catalysis in Organic Synthesis* (eds M. Shibasaki and Y. Yamamoto), Wiley-VCH, Weinheim, Chapter 1; (d) Nakamura, I. and Yamamoto, Y. (2002) *Adv. Synth. Catal.*, **344**, 111–129; (e) Patil, N.T. and Yamamoto, Y. (2006) *Top. Organomet. Chem.*, **19**, 91–113; (f) Patil, N.T. and Yamamoto, Y. (2007) *Arkivoc*, **10**, 121–141; (g) Patil, N.T. and Yamamoto, Y. (2007) *Arkivoc*, **5**, 6–19; (h) Yet, L. (2000) *Chem. Rev.*, **100**, 2963–3007; (i) Zeni, G. and Larock, R.C. (2004) *Chem. Rev.*, **104**, 2285–2309; (j) Alonso, F., Beletskaya, I.P. and Yus, M. (2004) *Chem. Rev.*, **104**, 3079–3160; (k) Zeni, G. and Larock, R.C. (2006) *Chem. Rev.*, **106**, 4644–4680; (l) Sato, F., Urabe, H. and Okamoto, S. (2000) *Chem. Rev.*, **100**, 2835–2886; (m) Widenhoefer, R.A. and Han, X. (2006) *Eur. J. Org. Chem.*, 4555–4563; (n) Fürstner, A. and Davies, P.W. (2007) *Angew. Chem., Int. Ed.*, **46**, 3410–3449; (o) Jimenez-Nunez, E. and Echavarren,

A.M. (2007) *Chem. Commun.*, 333–346;
(p) Marion, N. and Nolan, S.P. (2007) *Angew. Chem., Int. Ed.*, **46**, 2750–2752;
(q) Marco-Contelles, J. and Soriano, E. (2007) *Chem. Eur. J.*, **13**, 1350–1357;
(r) Hashmi, A.S.K. (2007) *Chem. Rev.*, **107**, 3180–3211; (s) Gorin, D.J. and Toste, F.D. (2007) *Nature*, **446**, 395–403; (t) Ley, S.V. and Thomas, A.W. (2003) *Angew. Chem., Int. Ed.*, **42**, 5400–5449; (u) Trost, B.M. and McClory, A. (2008) *Chem. Asian J.*, **3**, 164–194; (v) Kirsch, S.F. (2006) *Org. Biomol. Chem.*, **4**, 2076–2080; (w) Cacchi, S. (1999) *J. Organomet. Chem.*, **576**, 42–64; (x) Varela, J.A. and Saa, C. (2003) *Chem. Rev.*, **103**, 3787–3801; (y) Barluenga, J., Santamaria, J. and Tomás, M. (2004) *Chem. Rev.*, **104**, 2259–2283; (z) Shen, H.C. (2008) *Tetrahedron*, **64**, 3885–3903; (aa) Bongers, N. and Krause, N. (2008) *Angew. Chem., Int. Ed.*, **47**, 2178–2181; (bb) Crone, B. and Kirsch, S.F. (2008) *Chem. Eur. J.*, **14**, 3514–3522; (cc) Muzart, J. (2008) *Tetrahedron*, **64**, 5815–5849; (dd) Patil, N.T. and Yamamoto, Y. (2008) *Chem. Rev.*, **108**, 3395–3442.

2. (a) Wolfe, J.P. and Rossi, M.A. (2004) *J. Am. Chem. Soc.*, **126**, 1620–1621;
(b) Kang, S.H., Kim, M. and Kang, S.Y. (2004) *Angew Chem. Int. Ed.*, **43**, 6177–6180; (c) Kang, S.H. and Kim, M. (2003) *J. Am. Chem. Soc.*, **125**, 4684–4685; (d) Hori, K., Kitagawa, H., Miyoshi, A., Ohta, T. and Furukawa, I. (1998) *Chem. Lett*, 83–84; (e) Coulombel, L., Favier, I. and Dunach, E. (2005) *Chem. Commun.*, 2286–2288; (f) Zhao, P., Incarvito, C.P. and Hartwig, J.F. (2006) *J. Am. Chem. Soc.*, **128**, 9642–9643.

3. Yang, C.-G., Reich, N.W., Shi, Z. and He, C. (2005) *Org. Lett.*, **7**, 4553–4556.

4. Yang, C.-G. and He, C. (2005) *J. Am. Chem. Soc.*, **127**, 6966–6967.

5. Qian, H., Han, X. and Widenhoefer, R.A. (2004) *J. Am. Chem. Soc.*, **126**, 9536–9537.

6. Reich, N.W., Yang, C.-G., Shi, Z. and He, C. (2006) *Synlett*, 1278–1280.

7. Zhang, J., Yang, C.-G. and He, C. (2006) *J. Am. Chem. Soc.*, **128**, 1798–1799.

8. Han, X. and Widenhoefer, R.A. (2006) *Angew. Chem., Int. Ed.*, **45**, 1747–1749. Also see: Liu, Z. and Hartwig, J.F. (2008) *J. Am. Chem. Soc.*, **130**, 1570–1571.

9. Zhou, C.-Y. and Che, C.-M. (2007) *J. Am. Chem. Soc.*, **129**, 5828–5829.

10. Youn, S.W., Pastine, S.J. and Sames, D. (2004) *Org. Lett.*, **6**, 581–584.

11. Liu, C., Han, X., Wang, X. and Widenhoefer, R.A. (2004) *J. Am. Chem. Soc.*, **126**, 3700–3701.

12. Han, X. and Widenhoefer, R.A. (2006) *Org. Lett.*, **8**, 3801–3804.

13. Nguyen, R.-V., Yao, X. and Li, C.-J. (2006) *Org. Lett.*, **8**, 2397–2399.

14. Youn, S.W. and Eom, J.I. (2006) *J. Org. Chem.*, **71**, 6705–6707.

15. Adrio, L.A. and Hii, K.K. (2008) *Chem. Commun.*, **20**, 2325–2327.

16. Sherman, E.S., Chemler, S.R., Tan, T.B. and Gerlits, O. (2004) *Org. Lett.*, **6**, 1573–1575.

17. Zabawa, T.P., Kasi, D. and Chemler, S.R. (2005) *J. Am. Chem. Soc.*, **127**, 11250–11251.

18. Hövelmann, C.H., Streuff, J., Brelot, L. and Muñiz, K. (2008) *Chem. Commun.*, **20**, 2334–2336.

19. Han, X. and Widenhoefer, R.A. (2004) *J. Org. Chem.*, **69**, 1738–1740.

20. Patil, N.T. and Yamamoto, Y. (2007) *Synlett*, 1994–2005.

21. Sekido, M., Aoyagi, K., Nakamura, H., Kabuto, C. and Yamamoto, Y. (2001) *J. Org. Chem.*, **66**, 7142–7147.

22. Shim, J.-G. and Yamamoto, Y. (1998) *J. Org. Chem.*, **63**, 3067–3071.

23. Aoyagi, K., Nakamura, H. and Yamamoto, Y. (2002) *J. Org. Chem.*, **67**, 5977–5980.

24. Meguro, M. and Yamamoto, Y. (1999) *J. Org. Chem.*, **64**, 694–695.

25. For some examples, see: (a) Cacchi, S., Fabrizi, G. and Moro, L. (1998) *Synlett*, 741–745; (b) Gabriele, B., Salermo, G. and Lauria, E. (1999) *J. Org. Chem.*, **64**, 7687–7692; (c) Gabriele, B., Salermo, G., De Pascali, F., Costa, M. and Chiusoli, G.P. (1999) *J. Org. Chem.*, **64**, 7693–7699; (d) Kato, K., Tanaka, M., Yamamoto, Y. and

Akita, H. (2002) *Tetrahedron Lett.*, **43**, 1511–1513; (e) Li, C.C., Xie, Z.X., Zhang, Y.D., Chen, J.H. and Yang, Z. (2003) *J. Org. Chem.*, **68**, 8500–8504; (f) Ramana, C.V., Mallik, R., Gonnade, R.G. and Gurjar, M.K. (2006) *Tetrahedron Lett.*, **47**, 3649–3652; (g) Ramana, C.V., Mallik, R. and Gonnade, R.G. (2008) *Tetrahedron*, **64**, 219–233; (h) Liu, B. and De Brabander, J.K. (2006) *Org. Lett.*, **8**, 4907–4910; (i) Harkat, H., Weibel, J.-M. and Pale, P. (2007) *Tetrahedron Lett.*, **48**, 1439–1442; (j) Dieguez-Vazquez, A., Tzschucke, C.C., Lam, W.Y. and Ley, S.V. (2008) *Angew. Chem. Int. Ed.*, **47**, 209–212; (k) Trost, B.M. and Rhee, Y.H. (2002) *J. Am. Chem. Soc.*, **124**, 2528–2533; (l) Genin, E., Antoniotti, S., Michelet, V. and Genet, J.-P. (2005) *Angew. Chem. Int. Ed.*, **44**, 4949–4953; (m) Barluenga, J., Dieguez, A., Rodriguez, F. and Fananas, F.J. (2005) *Angew. Chem. Int. Ed.*, **44**, 126–128; (n) Barluenga, J., Dieguez, A., Rodriguez, F., Fananas, F.J., Sordo, T. and Campomanes, P. (2005) *Chem. Eur. J.*, **11**, 5735–5741.

26 Pale, P. and Chuche, J. (2000) *Eur. J. Org. Chem.*, 1019–1025.

27 Liu, Y., Song, F., Song, Z., Liu, M. and Yan, B. (2005) *Org. Lett.*, **7**, 5409–5412.

28 Peng, A.-Y. and Ding, Y.-X. (2003) *J. Am. Chem. Soc.*, **125**, 15006–15007.

29 Peng, A.-Y. and Ding, Y.-X. (2005) *Org. Lett.*, **7**, 3299–3301.

30 Dalla, V. and Pale, P. (1994) *Tetrahedron Lett.*, **35**, 3525–3528.

31 Xu, C. and Negishi, E-i. (1999) *Tetrahedron Lett.*, **40**, 431–434.

32 Anastasia, L., Xu, C. and Negishi, E.-i. (2002) *Tetrahedron Lett.*, **43**, 5673–5676.

33 Rossi, R., Bellina, F. and Mannina, L. (1998) *Tetrahedron Lett.*, **39**, 3017–3020.

34 Chaudhuri, G. and Kundu, N.G. (2000) *J. Chem. Soc., Perkin Trans. 1*, 775–779.

35 Genin, E., Toullec, P.Y., Antoniotti, S., Brancour, C., Genet, J.-P. and Michelet, V. (2006) *J. Am. Chem. Soc.*, **128**, 3112–3113.

36 Kadzimirsz, D., Hildebrandt, D., Merz, K. and Dyker, G. (2006) *Chem. Commun.*, 661–662.

37 Crawley, S.L. and Funk, R.L. (2006) *Org. Lett.*, **8**, 3995–3998.

38 Belting, V. and Krause, N. (2006) *Org. Lett.*, **8**, 4489–4492.

39 Li, Y., Zhou, F. and Forsyth, C.J. (2007) *Angew. Chem., Int. Ed.*, **46**, 279–282.

40 Zhang, Y., Xue, J., Xin, Z., Xie, Z. and Li, Y. (2008) *Synlett*, 940–944.

41 Bhuvaneswari, S., Jeganmohan, M. and Cheng, C.-H. (2007) *Chem. Eur. J.*, **13**, 8285–8293. Also see: Barluenga, J., Fernández, A., Satrústegui, A., Diéguez, A., Rodrguez, F. and Fañanás, F.J. (2008) *Chem. Eur. J.*, **14**, 4153–4156.

42 Antoniotti, S., Genin, E., Michelet, V. and Genet, J.-P. (2005) *J. Am. Chem. Soc.*, **127**, 9976–9977.

43 Liu, Y., Song, F. and Guo, S. (2006) *J. Am. Chem. Soc.*, **128**, 11332–11333.

44 Gu, Y., Shi, F. and Deng, Y. (2004) *J. Org. Chem.*, **69**, 391–394.

45 Kang, J.-E. and Shin, S. (2006) *Synlett*, 717–720.

46 Robles-Machin, R., Adrio, J. and Carretero, J.C. (2006) *J. Org. Chem.*, **71**, 5023–5026.

47 Istrate, F.M., Buzas, A.K., Jurberg, I.D., Odabachian, Y. and Gagosz, F. (2008) *Org. Lett.*, **10**, 925–928.

48 Tsukada, N. and Yamamoto, Y. (1997) *Angew. Chem., Int. Ed.*, **36**, 2477–2480.

49 For Hydroamination, see: (a) Kadota, I., Shibuya, A., Lutete, L.M. and Yamamoto, Y. (1999) *J. Org. Chem.*, **64**, 4570–4571; (b) Lutete, L.M., Kadota, I., Shibuya, A. and Yamamoto, Y. (2002) *Heterocycles*, **58**, 347–357; (c) Patil, N.T., Pahadi, N.K. and Yamamoto, Y. (2005) *Tetrahedron Lett.*, **46**, 2101–2103. For hydroalkoxylation, see: (d) Kadota, I., Lutete, L.M., Shibuya, A. and Yamamoto, Y. (2001) *Tetrahedron Lett.*, **42**, 6207–6210. For Hydrocarbonation, see: (e) Kadota, I., Shibuya, A., Gyoung, Y.S. and Yamamoto, Y. (1998) *J. Am. Chem. Soc.*, **120**, 10262–10263.

50 (a) Lutete, L.M., Kadota, I. and Yamamoto, Y. (2004) *J. Am. Chem. Soc.*, **126**, 1622–1623; (b) Patil, N.T., Lutete, L.M., Wu, H., Pahadi, N.K., Gridnev, I.D. and Yamamoto, Y. (2006) *J. Org. Chem.*, **71**, 4270–4279.

51 Patil, N.T., Huo, Z., Bajracharya, G.B. and Yamamoto, Y. (2006) *J. Org. Chem.*, **71**, 3612–3614.

52 Patil, N.T., Wu, H. and Yamamoto, Y. (2007) *J. Org. Chem.*, **72**, 6577–6579.

53 Huo, Z., Patil, N.T., Jin, T., Pahadi, N.K. and Yamamoto, Y. (2007) *Adv. Synth. Catal.*, **349**, 680–684.

54 Kennedy-Smith, J.J., Staben, S.T. and Toste, F.D. (2004) *J. Am. Chem. Soc.*, **126**, 4526–4527.

55 Staben, S.T., Kennedy-Smith, J.J. and Toste, F.D. (2004) *Angew. Chem., Int. Ed.*, **43**, 5350–5352.

56 Staben, S.T., Kennedy-Smith, J.J., Huang, D., Corkey, B.K., LaLonde, R.L. and Toste, F.D. (2006) *Angew. Chem., Int. Ed.*, **45**, 5991–5994. See also: Minnihan, E.C., Colletti, S.L., Toste, F.D. and Shen, H.C. (2007) *J. Org. Chem.*, **72**, 6287–6289. Allenes can also be used instead of alkynes, see: Miura, T., Kiyota, K., Kusama, H. and Iwasawa, N. (2007) *J. Organomet. Chem.*, **692**, 562–568.

57 Linghu, X., Kennedy-Smith, J.J. and Toste, F.D. (2007) *Angew. Chem., Int. Ed.*, **46**, 7671–7673.

58 Nicolaou, K.C., Tria, G.S. and Edmonds, D.J. (2008) *Angew. Chem., Int. Ed.*, **47**, 1780–1783.

59 Arcadi, A., Cacchi, S., Rosario, M.D., Fabrizi, G. and Marinelli, F. (1996) *J. Org. Chem.*, **61**, 9280–9288.

60 Arcadi, A., Cacchi, S., Giuseppe, S.D., Fabrizi, G. and Marinelli, F. (2002) *Synlett*, 453–457.

61 For the formation of benzofurans using a similar process, see: (a) Kabalka, G.W., Wang, L. and Pagni, R.M. (2001) *Tetrahedron*, **57**, 8017–8028; (b) Bergbreiter, D.E., Case, B.L., Liu, Y.-S. and Caraway, J.W. (1998) *Macromolecules*, **31**, 6053–6062.

62 Bossharth, E., Desbordes, P., Monteiro, N. and Balme, G. (2003) *Org. Lett.*, **5**, 2441–2444.

63 Chowdhury, C., Chaudhuri, G., Guha, S., Mukherjee, A.K. and Kundu, N.G. (1998) *J. Org. Chem.*, **63**, 1863–1871. See also: (a) Chowdhury, C. and Kundu, N.G. (1996) *J. Chem. Soc., Chem. Commun.*, 1067–1068; (b) Chaudhuri, G., Chowdhury, C. and Kundu, N.G. (1998) *Synlett*, 1273–1275.

64 Heravi, M.M., Keivanloo, A., Rahimizadeh, M., Bakavoli, M. and Ghassemzadeh, M. (2004) *Tetrahedron Lett.*, **45**, 5747–5749.

65 Zhang, H.-C., Ye, H., Moretto, A.F., Brumfield, K.K. and Maryanoff, B.E. (2000) *Org. Lett.*, **2**, 89–92.

66 For the formation of indoles using a similar process, see: (a) Torres, J.C., Pilli, R.A., Vargas, M.D., Violante, F.A., Garden, S.J. and Pinto, A.C. (2002) *Tetrahedron Lett.*, **58**, 4487–4492; (b) Flynn, B.L., Hamel, E. and Jung, M.K. (2002) *J. Med. Chem.*, **45**, 2670–2673.

67 Chowdhury, C., Mandal, S.B. and Achari, B. (2005) *Tetrahedron Lett.*, **46**, 8531–8534.

68 Kawasaki, T. and Yamamoto, Y. (2002) *J. Org. Chem.*, **67**, 5138–5141.

69 Rossi, R., Bellina, F., Bechini, C., Mannina, L. and Vergamini, P. (1998) *Tetrahedron*, **54**, 135–156.

70 Lu, X., Chen, G., Xia, L. and Guo, G. (1997) *Tetrahedron Asym.*, **8**, 3067–3072.

71 Bates, C.G., Saejueng, P., Murphy, J.M. and Venkataraman, D. (2002) *Org. Lett.*, **4**, 4727–4729.

72 Cacchi, S., Fabrizi, G. and Parisi, L.M. (2003) *Org. Lett.*, **5**, 3843–3846.

73 Cacchi, S., Fabrizi, G., Parisi, L.M. and Bernini, R. (2004) *Synlett*, 287–290.

74 Mukai, C. and Takahashi, Y. (2005) *Org. Lett.*, **7**, 5793–5796.

75 Martin, R., Rivero, M.R. and Buchwald, S.L. (2006) *Angew. Chem., Int. Ed.*, **45**, 7079–7082.

76 Nakamura, I., Sato, T. and Yamamoto, Y. (2006) *Angew. Chem., Int. Ed.*, **45**, 4473–4475.

77 Nakamura, I., Sato, T., Terada, M. and Yamamoto, Y. (2007) *Org. Lett.*, **9**, 4081–4083.

78 Nakamura, I., Yamagishi, U., Song, D., Konta, S. and Yamamoto, Y. (2007) *Angew. Chem., Int. Ed.*, **46**, 2284–2287. Also see: Nakamura, I., Yamagishi, U., Song, D., Konta, S. and Yamamoto, Y. (2008) *Chem. Asian J.*, **3**, 285–295.

79 Shimada, T., Nakamura, I. and Yamamoto, Y. (2004) *J. Am. Chem. Soc.*, **126**, 10546–10547.

80 Nakamura, I., Mizushima, Y. and Yamamoto, Y. (2005) *J. Am. Chem. Soc.*, **127**, 15022–15023. Also see: Nakamura, I., Chan, C.S., Araki, T., Terada, M. and Yamamoto, Y. (2008) *Org. Lett.*, **10**, 309–312.

81 Fürstner, A. and Davies, P.W. (2005) *J. Am. Chem. Soc.*, **127**, 15024–15025.

82 Kelin, A.V. and Gevorgyan, V. (2002) *J. Org. Chem.*, **67**, 95–98.

83 Kelin, A.V., Sromek, A.W. and Gevorgyan, V. (2001) *J. Am. Chem. Soc.*, **123**, 2074–2075.

84 Hashmi, A.S.K. and Sinha, P. (2004) *Adv. Synth. Catal.*, **346**, 432–438.

85 Hashmi, A.S.K., Weyrauch, J.P., Frey, W. and Bats, J.W. (2004) *Org. Lett.*, **6**, 4391–4394.

86 Yao, T., Zhang, X. and Larock, R.C. (2004) *J. Am. Chem. Soc.*, **126**, 11164–11165.

87 Patil, N.T., Wu, H. and Yamamoto, Y. (2005) *J. Org. Chem.*, **70**, 4531–4534.

88 Liu, Y., Liu, M., Guo, S., Tu, H., Zhou, Y. and Gao, H. (2006) *Org. Lett.*, **8**, 3445–3448.

89 Jin, T. and Yamamoto, Y. (2007) *Org. Lett.*, **9**, 5259–5262.

90 Harrison, T.J., Kozak, J.A., Corbella-Pane, M. and Dake, G.R. (2006) *J. Org. Chem.*, **71**, 4525–4529.

91 Asao, N. (2006) *Synlett*, 1645–1656.

92 (a) Asao, N., Nogami, T., Lee, S. and Yamamoto, Y. (2003) *J. Am. Chem. Soc.*, **125**, 10921–10925; (b) Asao, N., Takahashi, K., Lee, S., Kasahara, T. and Yamamoto, Y. (2002) *J. Am. Chem. Soc.*, **124**, 12650–12651.

93 (a) Asao, N. and Sato, K. (2006) *Org. Lett.*, **8**, 5361–5363; (b) Sato, K. Menggenbateer, Kubota, T. and Asao, N. (2008) *Tetrahedron*, **64**, 787–796.

94 Asao, N., Sato, K., Menggenbateer and Yamamoto, Y. (2005) *J. Org. Chem.*, **70**, 3682–3685.

95 Sato, K., Asao, N. and Yamamoto, Y. (2005) *J. Org. Chem.*, **70**, 8977–8981.

96 Dyker, G. and Hildebrandt, D. (2005) *J. Org. Chem.*, **70**, 6093–6096.

97 Kim, N., Kim, Y., Park, W., Sung, D., Gupta, A.K. and Oh, C.H. (2005) *Org. Lett.*, **7**, 5289–5291.

98 Sato, K., Yudha, S.S., Asao, N. and Yamamoto, Y. (2004) *Synthesis*, **2004**, 1409–1412.

99 Asao, N., Aikawa, H. and Yamamoto, Y. (2004) *J. Am. Chem. Soc.*, **126**, 7458–7459.

100 Patil, N.T. and Yamamoto, Y. (2004) *J. Org. Chem.*, **69**, 5139–5142.

101 Asao, N., Nogami, T., Takahashi, K. and Yamamoto, Y. (2002) *J. Am. Chem. Soc.*, **124**, 764–765.

102 Yao, X. and Li, C.-J. (2006) *Org. Lett.*, **8**, 1953–1955.

103 Patil, N.T., Pahadi, N.K. and Yamamoto, Y. (2005) *J. Org. Chem.*, **70**, 10096–10098.

104 Asao, N., Yudha, S.S., Nogami, T. and Yamamoto, Y. (2005) *Angew. Chem., Int. Ed.*, **44**, 5526–5528.

105 Asao, N., Chan, C.S., Takahashi, K. and Yamamoto, Y. (2005) *Tetrahedron*, **61**, 11322–11326.

106 Nakamura, H., Ohtaka, M. and Yamamoto, Y. (2002) *Tetrahedron Lett.*, **43**, 7631–7633.

107 Ohtaka, M., Nakamura, H. and Yamamoto, Y. (2004) *Tetrahedron Lett.*, **45**, 7339–7341.

108 (a) Nakamura, I., Bajracharya, G.B., Mizushima, Y. and Yamamoto, Y. (2002) *Angew. Chem., Int. Ed.*, **41**, 4328–4331; (b) Nakamura, I., Bajracharya, G.B., Wu, H., Oishi, K., Mizushima, Y., Gridnev, I.D. and Yamamoto, Y. (2004) *J. Am. Chem. Soc.*, **126**, 15423–15430. Also see: Nakamura, I., Bajracharya, G.B. and

Yamamoto, Y. (2005) *Chem. Lett.*, **34**, 174–175.

109 Nakamura, I., Mizushima, Y., Gridnev, I.D. and Yamamoto, Y. (2005) *J. Am. Chem. Soc.*, **127**, 9844–9847.

110 Kamijo, S. and Yamamoto, Y. (2003) *J. Org. Chem.*, **68**, 4764–4771.

111 Asao, N., Sato, K. and Yamamoto, Y. (2003) *Tetrahedron Lett.*, **44**, 5675–5677.

112 Kamijo, S. and Yamamoto, Y. (2002) *J. Am. Chem. Soc.*, **124**, 11940–11945. Also see: Kamijo, S., Jin, T. and Yamamoto, Y. (2001) *J. Am. Chem. Soc.*, **123**, 9453–9454.

113 Kamijo, S. and Yamamoto, Y. (2002) *Angew. Chem., Int. Ed.*, **41**, 3230–3233.

114 Kamijo, S., Sasaki, Y., Kanazawa, C., Schüβler, T. and Yamamoto, Y. (2005) *Angew. Chem., Int. Ed.*, **44**, 7718–7721.

115 Takeda, A., Kamijo, S. and Yamamoto, Y. (2000) *J. Am. Chem. Soc.*, **122**, 5662–5663.

116 Shi, Z. and He, C. (2004) *J. Org. Chem.*, **69**, 3669–3671.

117 Chernyak, N. and Gevorgyan, V. (2008) *J. Am. Chem. Soc.*, **130**, 5636–5637.

118 Nevado, C. and Echavarren, A.M. (2005) *Chem. Eur. J.*, **11**, 3155–3164.

119 Ferrer, C. and Echavarren, A.M. (2006) *Angew. Chem., Int. Ed.*, **45**, 1105–1109.

120 Kamijo, S., Jin, T., Huo, Z. and Yamamoto, Y. (2003) *J. Am. Chem. Soc.*, **125**, 7786–7787.

121 Kamijo, S., Jin, T., Huo, Z. and Yamamoto, Y. (2004) *J. Org. Chem.*, **69**, 2386–2393.

122 Kamijo, S., Jin, T. and Yamamoto, Y. (2004) *Tetrahedron Lett.*, **45**, 689–691.

123 Kamijo, S., Kanazawa, C. and Yamamoto, Y. (2005) *J. Am. Chem. Soc.*, **127**, 9260–9266.

124 (a) Zimmer, R., Dinesh, C., Nandanan, E. and Khan, F. (2000) *Chem. Rev.*, **100**, 3067; (b) Ma, S. (2005) *Chem. Rev.*, **105**, 2829–2871; (c) Krause, N. and Hashmi, A.S.K.(eds) (2004) *Modern Allene Chemistry*, Wiley-VCH, Weinheim.

125 (a) Arseniyadis, S. and Gore, J. (1983) *Tetrahedron Lett.*, **24**, 3997–4000; (b) Kinsman, R., Lathbury, D., Vernon, P. and Gallagher, T. (1987) *J. Chem. Soc., Chem. Commun.*, 243–244; (c) Kimura, M., Fugami, K., Tanaka, S. and Tamaru, Y. (1991) *Tetrahedron Lett.*, **32**, 6359–6362; (d) Ha, J.D., Lee, D. and Cha, J.K. (1997) *J. Org. Chem.*, **62**, 4550–4551; (e) Ha, J.D. and Cha, J.K. (1999) *J. Am. Chem. Soc.*, **121**, 10012–10020.

126 Meguro, M., Kamijo, S. and Yamamoto, Y. (1996) *Tetrahedron Lett.*, **37**, 7453–7456.

127 Ma, S. and Zhao, S. (2000) *Org. Lett.*, **2**, 2495–2497. Also see: (a) Ma, S., Jiao, N., Zhao, S. and Hou, H. (2002) *J. Org. Chem.*, **67**, 2837–2847; (b) Ma, S., Jiao, N., Yang, Q. and Zheng, Z. (2004) *J. Org. Chem.*, **69**, 6463–6466.

128 Meguro, M. and Yamamoto, Y. (1998) *Tetrahedron Lett.*, **39**, 5421–5424.

129 Patil, N.T., Lutete, L.M., Nishina, N. and Yamamoto, Y. (2006) *Tetrahedron Lett.*, **47**, 4749–4751.

130 Zhang, Z., Liu, C., Kinder, R.E., Han, X., Qian, H. and Widenhoefer, R.A. (2006) *J. Am. Chem. Soc.*, **128**, 9066–9073.

131 Marshall, J.A. and Bartley, G.S. (1995) *J. Org. Chem.*, **60**, 5966–5968. See also: Marshall, J.A. and Wang, X.J. (1991) *J. Org. Chem.*, **56**, 4913–4918.

132 Lepage, O., Kattnig, E. and Furstner, A. (2004) *J. Am. Chem. Soc.*, **126**, 15970–15971. For related examples, see: (a) Aurrecoechea, J.M. and Solay, M. (1998) *Tetrahedron*, **54**, 3851–3856; (b) Marshall, J.A., Yu, R.H. and Perkins, J.F. (1995) *J. Org. Chem.*, **60**, 5550–5555; (c) Marshall, J.A. and Pinney, K.G. (1993) *J. Org. Chem.*, **58**, 7180–7184.

133 Marshall, J.A., Wolf, M.A. and Wallace, E.M. (1997) *J. Org. Chem.*, **62**, 367–371.

134 Hoffman-Roder, A. and Krause, N. (2001) *Org. Lett.*, **3**, 2537–2538.

135 Hashmi, A.S.K., Blanco, M.C., Fischer, D. and Bats, J.W. (2006) *Eur. J. Org. Chem.*, 1387–1389.

136 Morita, N. and Krause, N. (2004) *Org. Lett.*, **6**, 4121–4123.

137 Morita, N. and Krause, N. (2006) *Angew. Chem., Int. Ed.*, **45**, 1897–1899.

138 Kimura, M., Fugami, K., Tanaka, S. and Tamaru, Y. (1991) *Tetrahedron Lett.*, **32**, 6359–6362.
139 Kang, J.-E., Lee, E.-S., Park, S.I. and Shin, S. (2005) *Tetrahedron Lett.*, **46**, 7431–7433.
140 Zhang, Z. and Widenhoefer, R.A. (2007) *Angew. Chem., Int. Ed.*, **46**, 283–285.
141 Zhang, Z., Bender, C.F. and Widenhoefer, R.A. (2007) *J. Am. Chem. Soc.*, **129**, 14148–14149.
142 LaLonde, R.L., Sherry, B.D., Kang, E.J. and Toste, F.D. (2007) *J. Am. Chem. Soc.*, **129**, 2452–2453.
143 Hamilton, G.L., Kang, E.J., Mba, M. and Toste, F.D. (2007) *Science*, **317**, 496–499.
144 Ma, S. and Shi, Z. (1998) *J. Org. Chem.*, **63**, 6387–6389. For a similar process involving the use of Pd(0)/Ag catalyst system, see: (a) Ma, S. and Li, L. (2000) *Org. Lett.*, **2**, 941–944.
145 Ma, S. and Wu, S. (1999) *J. Org. Chem.*, **64**, 9314–9317. For review on related reactions, see: Ma, S. (2004) *Pure Appl. Chem.*, **76**, 651–656.
146 (a) Ma, S. and Xie, H. (2000) *Org. Lett.*, **2**, 3801–3803; (b) Ma, S. and Xie, H. (2005) *Tetrahedron*, **61**, 251–258.
147 Ma, S. and Xie, H. (2002) *J. Org. Chem.*, **67**, 6575–6578.
148 (a) Jonasson, C., Karstens, W.F.J., Hiemstra, H. and Backvall, J.E. (2000) *Tetrahedron Lett.*, **41**, 1619–1622; (b) Jonasson, C., Horvath, A. and Backvall, J.E. (2000) *J. Am. Chem. Soc.*, **122**, 9600–9609.
149 For a general review on the activation of allenes by metal catalyst, see: (a) Zimmer, R., Dinesh, C., Nandanan, E. and Khan, F.A. (2000) *Chem. Rev.*, **100**, 3067–3126; (b) Ma, S. (2005) *Chem. Rev.*, **105**, 2829–2871.
150 (a) Marshall, J.A. and Wang, X-j. (1991) *J. Org. Chem.*, **56**, 960–969; (b) Marshall, J.A. and Bartley, G.S. (1994) *J. Org. Chem.*, **59**, 7169–7171, and references cited therein.
151 (a) Ma, S. and Zhang, J. (2000) *Chem. Commun.*, 117–118; (b) Ma, S., Zhang, J. and Lu, L. (2003) *Chem. Eur. J.*, **9**, 2447–2456.
152 Ma, S. and Li, L. (2000) *Org. Lett.*, **2**, 941–944.
153 Hashmi, A.S.K., Schwarz, L., Choi, J.-H. and Frost, T.M. (2000) *Angew. Chem., Int. Ed.*, **39**, 2285–2288.
154 Zhou, C.-Y., Chan, P.W.H. and Che, C.-M. (2006) *Org. Lett.*, **8**, 325–328.
155 Sromek, A.W., Rubina, M. and Gevorgyan, V. (2005) *J. Am. Chem. Soc.*, **127**, 10500–10501.
156 Liu, Z., Wasmuth, A.S. and Nelson, S.G. (2006) *J. Am. Chem. Soc.*, **128**, 10352–10353.
157 Zhang, Z., Liu, C., Kinder, R.E., Han, X., Qian, H. and Widenhoefer, R.A. (2006) *J. Am. Chem. Soc.*, **128**, 9066–9073.
158 Liu, C. and Widenhoefer, R.A. (2007) *Org. Lett.*, **9**, 1935–1938.
159 Tarselli, M.A. and Gagnè, M. R. (2008) *J. Org. Chem.*, **73**, 2439–2441.
160 Watanabe, T., Oishi, S., Fujii, N. and Ohno, H. (2007) *Org. Lett.*, **9**, 4821–4824.

11
Ring-Closing Metathesis of Dienes and Enynes
Miwako Mori

11.1
Introduction

Metathesis is one of the most important and useful reactions for the synthesis of cyclic compounds in modern synthetic organic chemistry [1]. The reaction is catalyzed by the transition metals. In the metathesis reaction of a diene, fission of a double bond occurs and a new double bond is formed simultaneously. Since the alkenes formed in this reaction react further with other alkenes, many alkenes should be formed by cross-metathesis (CM). Therefore, in the early days, only ring-closing metathesis (RCM) of dienes was investigated and has been a useful tool for the synthesis of cyclic compounds (Scheme 11.1).

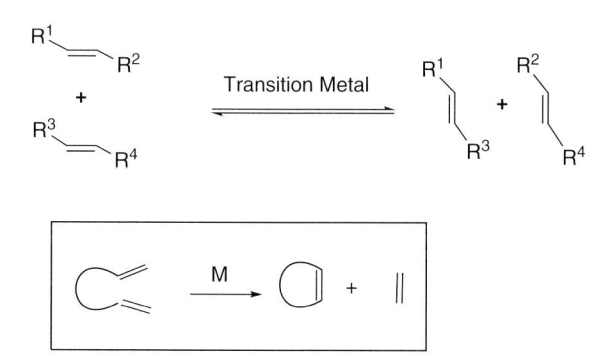

Scheme 11.1 Ring-closing olefin metathesis.

The pioneering work on olefin metathesis was undertaken by Villemina [2] and Tsuji [3], who synthesized lactones from the linear chain compounds using RCM of the diene with a tungsten complex (Eqs. (11.1) and (11.2))

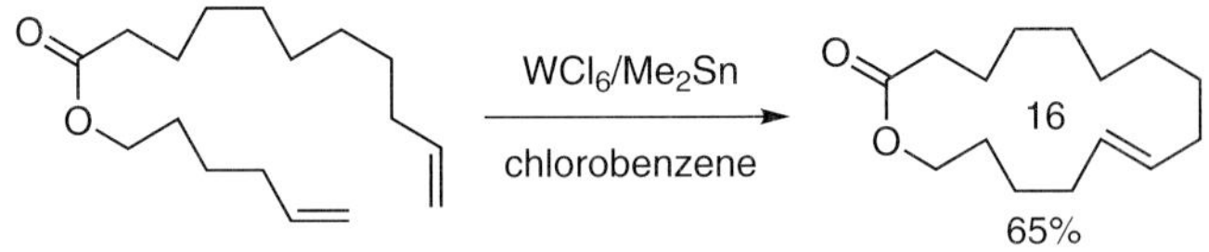

[Reaction 11.1: macrolactone diene → cyclic product 16, WCl₆/Me₂Sn, chlorobenzene, 65%]

$$\text{(11.1)}$$

[Reaction 11.2: macrocyclic diene (R = C₈H₁₇) → 19, WCl₆/Cp₂TiMe₂, benzene, 18%]

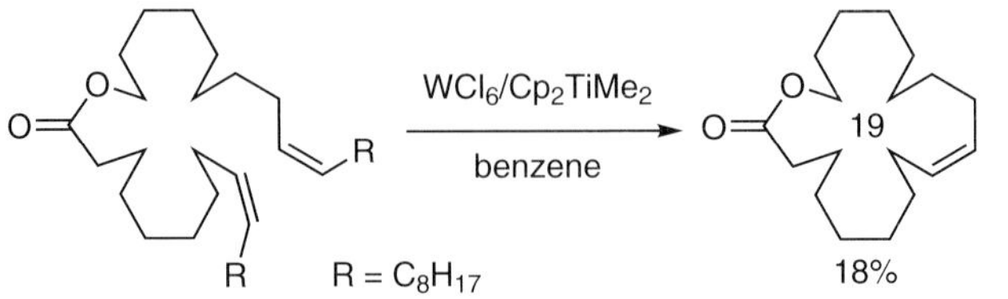

$$\text{(11.2)}$$

Schrock discovered the molybdenum catalyst **1a** for olefin metathesis in 1990 [4] and Grubbs used this complex for the synthesis of the cyclic compound **3a** from the diene **2a** (Scheme 11.2) [5]. This is the first example of transition metal-catalyzed olefin metathesis by a metal carbene complex. The reaction proceeds via [2 + 2] cycloaddition of an alkene part of the diene and the methylidene molybdenum complex. Then ring opening of molybdenacyclobutane **4** gives the molybdenum carbene alkylidene complex, which reacts with another alkene part of the diene intramolecularly to form molybdenacyclobutane **5**. Ring opening of **5** gives the furan derivative **3a** and the methylidene molybdenum complex is regenerated.

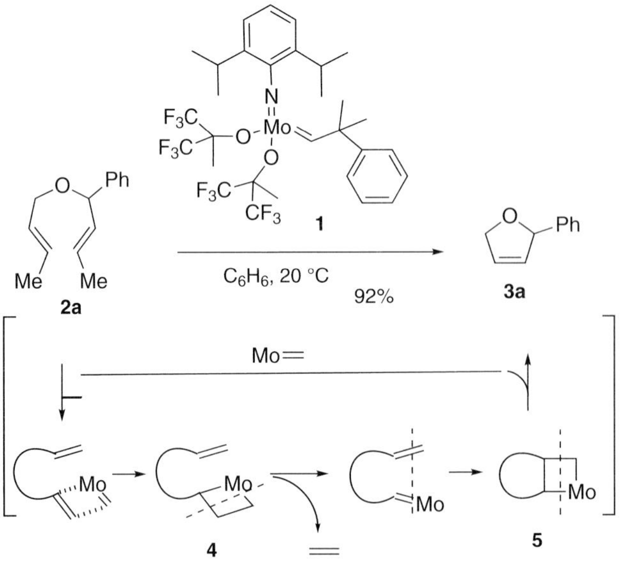

Scheme 11.2 Molybdenum-catalyzed alkene metathesis.

In the same year, Forbes reported the synthesis of seven- to medium-sized cyclic ketones **3b** and **3c** using this catalyst **1** (Scheme 11.3) [6]. In this case, they used no solvent and found that highly substituted dienes **2b** and **2c** showing the Thorp-Ingold

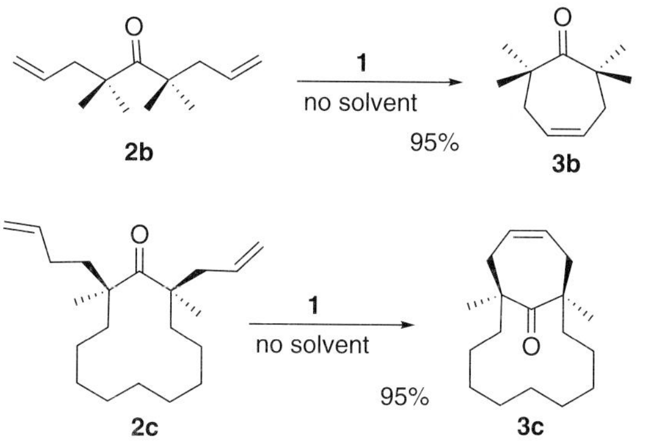

Scheme 11.3 Cyclization by Thorp–Ingold effect.

effect cyclized to give functionalized ring compounds. These results indicated that the molybdenum-catalyzed RCM of diene was very effective for the synthesis of carbo- and heterocycles.

Then Grubbs *et al.* discovered the ruthenium carbene complex **6a** [7] and synthesized many cyclic compounds using this complex. Furthermore, they found that the ruthenium benzylidene carbene complex **6b** had similar reactivity to **6a** (Figure 11.1) [8]. Complex **6b** is relatively stable and easy to handle and commercially available. Thus, many cyclic compounds may be synthesized from the diene moiety in the molecule using RCM.

In 1999, the second-generation ruthenium carbene complexes, having a heterocyclic carbene as a ligand were developed. These catalysts are very effective for olefin metathesis, especially, for the RCM of dienes with substituents, compared with the first-generation ruthenium catalysts **6a** and **6b**. Furthermore, it was clear that they could catalyze cross-metathesis (CM) of alkenes, asymmetric ring-closing olefin metathesis (ARCM), and ring-opening metathesis (ROM) of cycloalkene-ene. Metathesis reactions are now widely used in natural product synthesis. Novel retrosynthetic analyses were developed because a double bond constructed by the RCM of a diene could be converted into a carbon–carbon single bond after hydrogenation. Thus, the bond fission at the various positions in the molecule could be considered in the retro-synthetic analysis.

The metathesis reaction of enyne, which has an alkene and an alkyne part in a molecule, is a very interesting and useful reaction for obtaining cyclic compounds [9]. In this reaction, the double bond of enyne is cleaved and a carbon–carbon double bond is formed between the double and triple bonds, and the cleaved alkene part

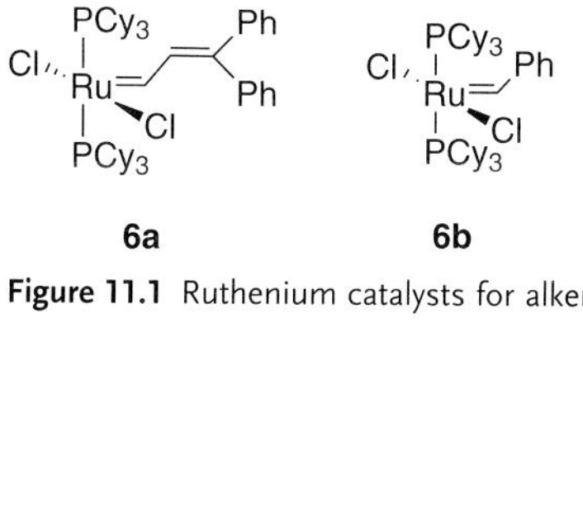

Figure 11.1 Ruthenium catalysts for alkene metathesis.

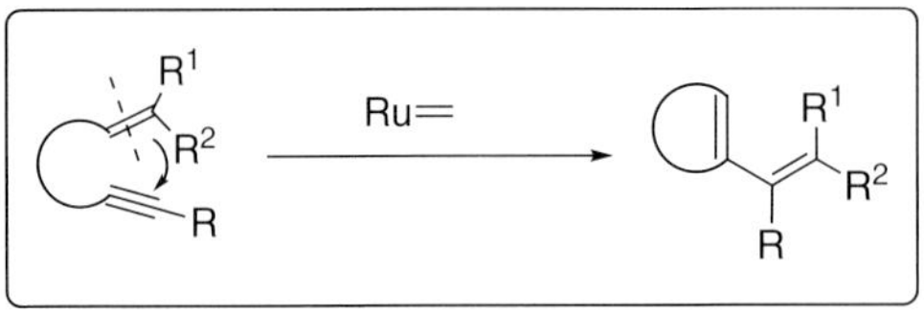

11 Ring-Closing Metathesis of Dienes and Enynes

Scheme 11.4 RCM of enyne.

migrates onto the alkyne carbon to produce a cyclic compound having a diene moiety (Scheme 11.4).

The first example of enyne metathesis was reported by Katz in 1985 [10]. He prepared the phenathrene derivative from an enyne conjugated with a biphenyl skeleton using the Fischer tungsten carbene complex (Eq. (11.3)). In this reaction, the real species is the tungsten carbene complex and the carbon–carbon double bond is formed between an alkyne and an alkene carbon. The cleaved methylene part of the alkene moves on the alkyne carbon to form the phenanthrene derivative.

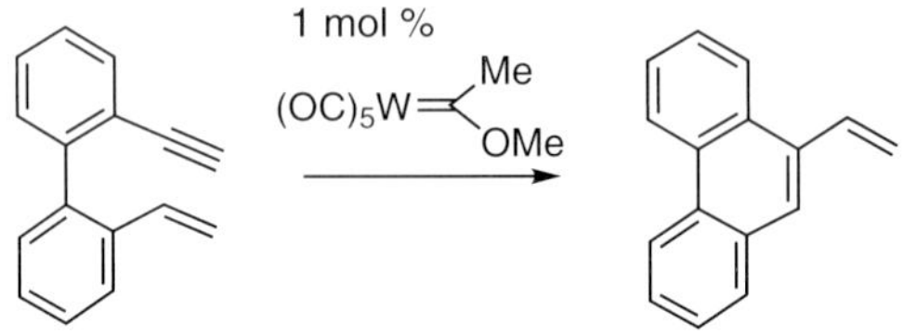

$$(11.3)$$

Mori reported the chromium-catalyzed enyne metathesis (Eq. (11.4)). In the case of an enyne having the same substituents on an alkene as those on the Fischer chromium carbene complex, the reaction proceeded with a catalytic amount of Fischer chromium carbene complex [11].

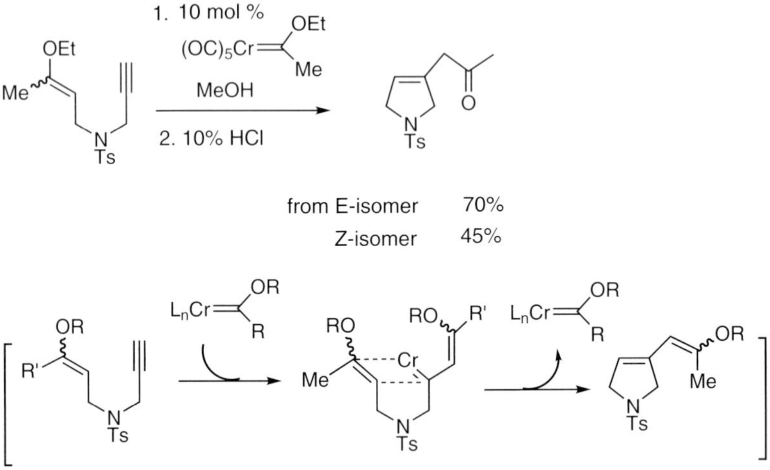

from E-isomer   70%
Z-isomer   45%

$$(11.4)$$

Since Mori observed that the ruthenium carbene complex **6a** or **6b** was very effective for enyne metathesis [12], enyne metathesis has been a useful tool for the synthesis of cyclic compounds having a vinyl group. Furthermore, it was found that in the enyne metathesis, the reaction proceeded smoothly under ethylene gas, and most alkene metatheses were carried out under ethylene gas [13].

Although many types of metathesis are now known, in this chapter, the RCM of dienes and enynes will be described.

11.2
Ring-Closing Metathesis of Dienes

11.2.1
Synthesis of Carbocycles and Heterocycles Using Ring-Closing Metathesis of Dienes

In 1992, Grubbs and Forbes reported the synthesis of carbo- and heterocycles using Schrock's molybdenum carbene complex **1** [5, 6]. Then Grubbs discovered the ruthenium carbene complex **6a** for the RCM of dienes [7] and synthesized many cyclic compounds using this catalyst. In 1995, he found that ruthenium carbene complex **6b** [8], which is now commercially available, is also effective for RCM and many cyclic compounds **8** could be synthesized from dienes **7** in high yields [14] (Scheme 11.5).

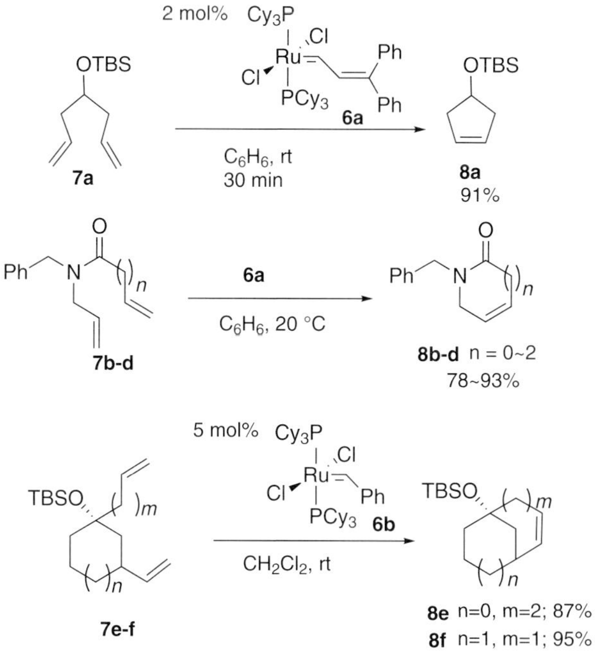

Scheme 11.5 Development of ruthenium-catalyzed alkene metathesis.

These results herald the dawn of RCM of dienes. Many researchers could obtain the ruthenium carbene complex **6a** or **6b** and various cyclic compounds were synthesized using RCM. The procedure of RCM is very simple and the reaction conditions are mild, that is, a solution of diene and **6a** in an appropriate solvent, usually CH_2Cl_2 or toluene, is stirred at room temperature, and the yields of the cyclic compounds are high.

Enantiometrically pure cyclopentene derivative **11** was synthesized from diene **9** using RCM followed by treatment with $LiBH_4$ by Crimmins *et al.* (Scheme 11.6) [15].

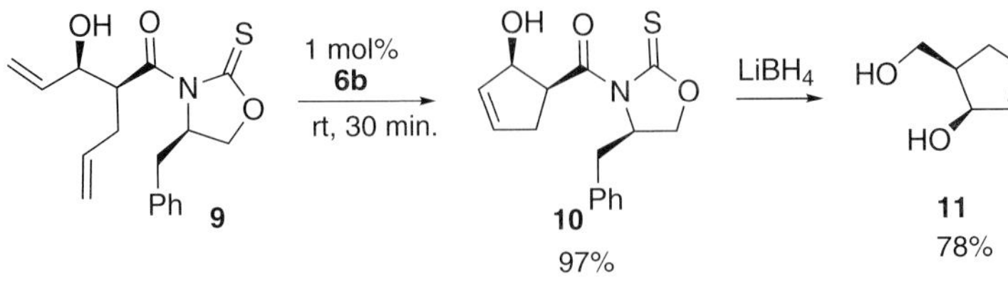

Scheme 11.6 Synthesis of enantioselective cyclopentenol derivative.

Although preparation of medium-sized ring compounds is difficult using the traditional synthetic method, the RCM of dienes using ruthenium catalyst was effective for the synthesis of these cyclic compounds (Scheme 11.7). RCM of diene **12a** afforded eight-membered compounds **13a** in good yields, however, diene **12b** bearing *cis* substituents on the cyclohexane ring gave the eight-membered compound **13b** in 33% yield [16].

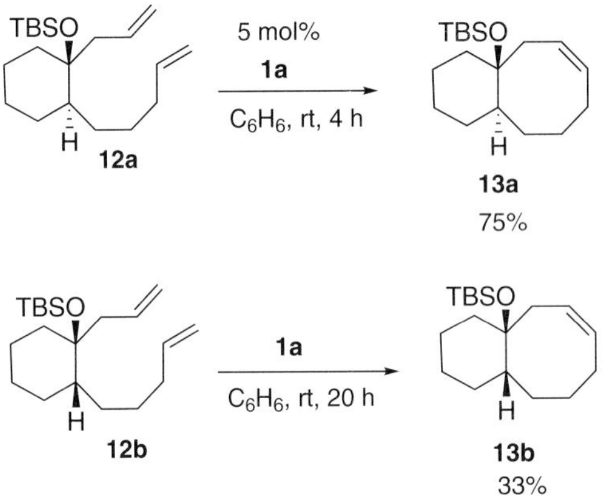

Scheme 11.7 Construction of medium-sized cyclic compound.

It is interesting that the macrocyclic lactone **15** could be synthesized from diene **14** using RCM. In each case, the substrate was introduced by slow addition of the CH_2Cl_2 solution of the ruthenium catalyst **6a** or **6b** and the yield of the macrocyclic compound was high (Scheme 11.8) [17]. Although the traditional macrocyclic lactonization is a condensation between an alcohol and carboxylic acid moieties in a molecule, RCM of the diene could be used as a novel tool for the synthesis of a macrocyclic lactone.

Hirama reported the synthesis of polyether **17** from diene **16** and succeeded in the synthesis of ciguatoxine using this method (Scheme 11.9) [18].

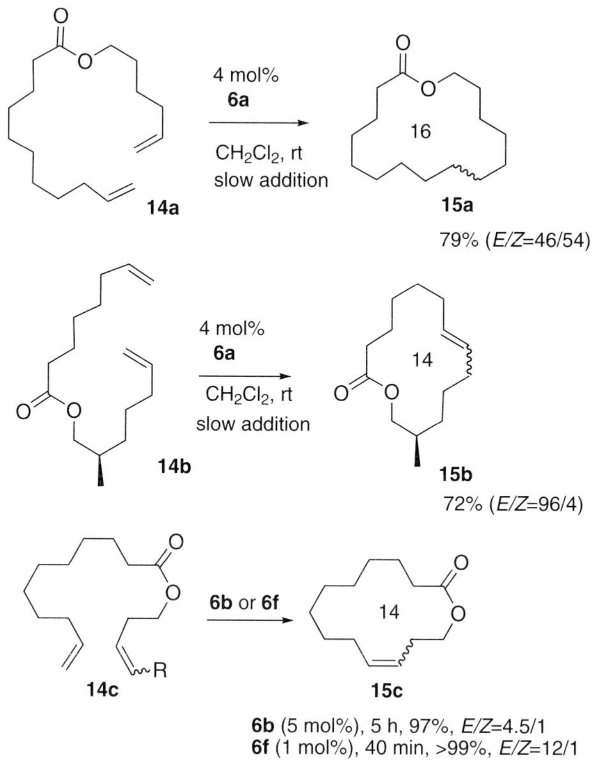

Scheme 11.8 Synthesis of macrocyclic lactone.

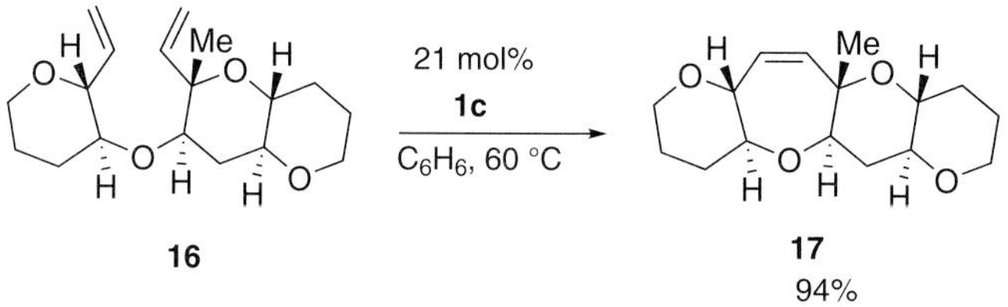

Scheme 11.9 Synthesis of polyether.

An interesting technique for the synthesis of crown ethers **19** has been developed by Grubbs using template-directed RCM or ROM [19]. Various crown ethers **19** have been prepared from the linear polyether **18** in the presence of an appropriate metal (Scheme 11.10). For the synthesis of the 14-membered crown ether **19a**, the Li cation gave good results. Surprisingly, polyether **18a** could be synthesized in high yield by ROM of the crown ether **19a** using **6b** at room temperature, and the crown ether **19a** was regenerated from polyether **18a** in the presence of the Li cation using RCM.

Although metathesis reactions are very useful for the synthesis of cyclic compounds from dienes, the reaction rate of metathesis of dienes having substituents on the alkenes are slow. In 1999, Hermann, Nolan and Grubbs independently found

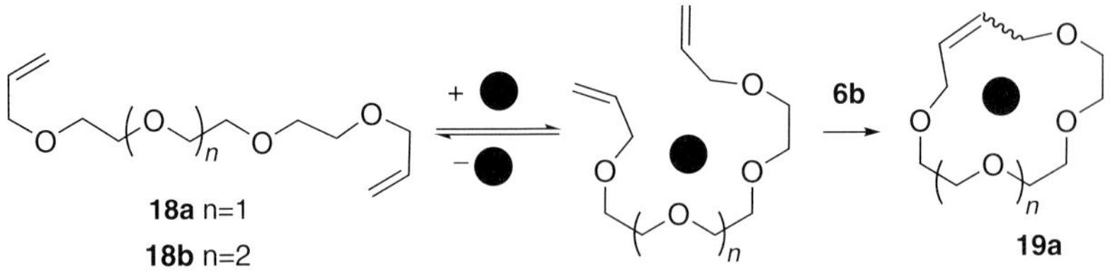

substrate	template	yield (%)	cis : trans
18a n=1	—	39	38 : 62
	LiClO$_4$	95	100 : 0
	NaClO$_4$	42	62 : 38
	KClO$_4$	36	36 : 64
18b n=2	—	57	26 : 74
	LiClO$_4$	89	61 : 39
	NaClO$_4$	90	68 : 32
	KClO$_4$	64	25 : 75

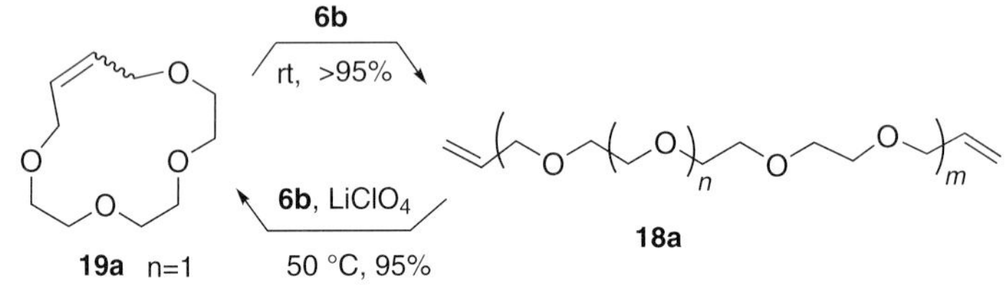

Scheme 11.10 Synthesis of crown ether.

the novel ruthenium carbene complexes **6c** [20] and **6d**, **6e** [21] and **6f** [22], which have a heterocyclic carbene ligand (Figure 11.2). It was observed that complexes **6d–f** were very effective for diene metathesis when there are substituents on the alkene (Table 11.1) [23]. Importantly, complex **6f** is now commercially available. Thus, many reports of the synthesis of carbocycles and heterocycles using alkene metathesis with **6f** as the catalyst have appeared. Furthermore, various ruthenium catalysts **6g** [24], **6h** [25], **6i** [26], **6j** [27], and **6k** [28] have been synthesized and are very effective for RCM. Since then cross metathesis (CM), ring-opening metathesis (ROM) and tandem metathesis were developed using these ruthenium carbene complexes.

Grubbs investigated the reason for the higher reactivity of second-generation ruthenium carbene complex **6f** in comparison to that of the first-generation ruthenium catalyst **6b** [29]. The initial step of this reaction is dissociation of a phosphine ligand from the ruthenium metal of **6b** or **6f**, and then an olefin coordinates to the ruthenium metal to give complex **21** (Scheme 11.11). Cycloaddition occurs to give ruthenacyclobutane **22**, whose ring opening gives an alternative olefin and ruthenium methylidene carbene complex **6l**. Surprisingly, the dissociation of the phosphine ligand from the ruthenium metal in **6b** having two phosphine ligands is faster than that in **6f** having a phosphine and N-heterocycliccarbene. However, coordination of the olefin to the resulting complex **20f** having an N-heterocyclic carbene ligand is overwhelmingly faster than that to **20b** possessing a phosphine ligand. Therefore, the overall reaction

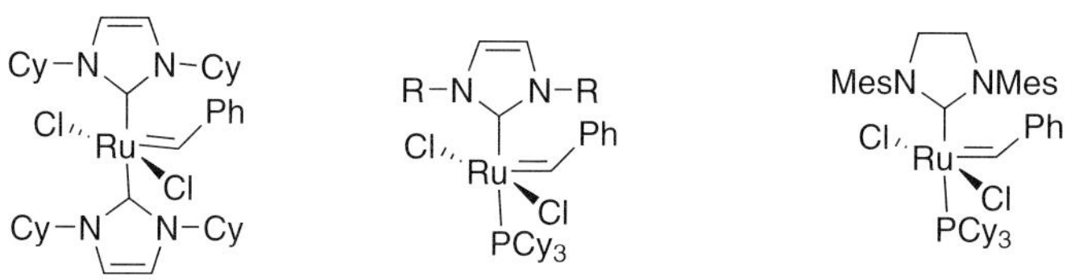

6c (Hermman) **6d** R = CHMePh (Hermann) **6e** R = Mes (Grubbs, Nolan) **6f** (Grubbs)

(Mes = 2,4,6-trimethylphenyl)

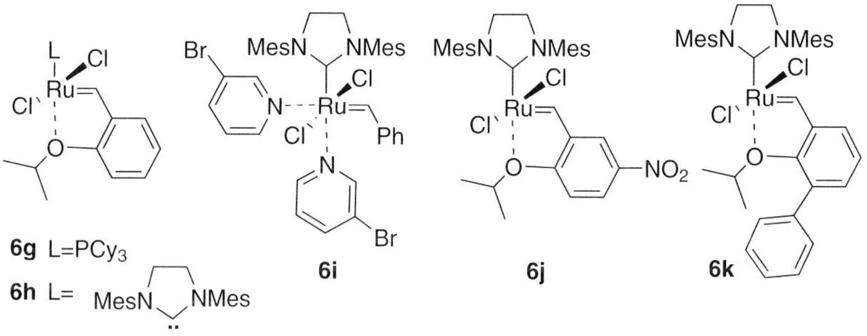

6g L=PCy_3 **6h** L= MesN NMes **6i** **6j** **6k**

Figure 11.2 Ruthenium catalysts for alkene metathesis.

rate using **6f** is faster than that using **6b**. A higher reaction temperature is required when the second-generation ruthenium carbene complex **6f** is used to promote dissociation of the phosphine ligand from **6f**.

Various poly-hydroxylated carbocycles could be synthesized from the corresponding dienes. A concise stereoselective synthesis of myoinositol derivative was achieved by RCM of diene **23a** prepared from readily available tartaric acid [30a] (Scheme 11.12). The poly hydroxylated cyclopentenol derivatives **24b** and **24b′** could

Table 11.1 RCM at 45 C utlizing 5 mol% catalysts **6b** and **6e**.

Entry	Substrate	Product	6b	6e
1	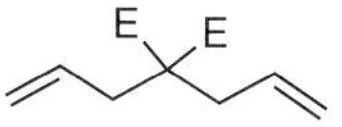	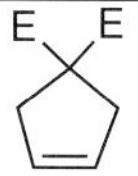	quant	quant
2	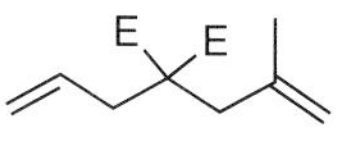	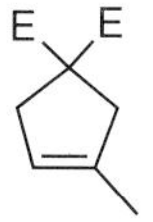	82	quant
3	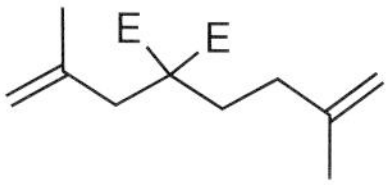	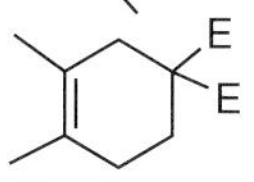	0	95
4	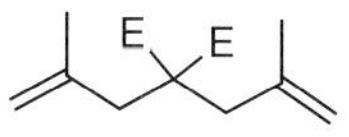 E=CO_2Et	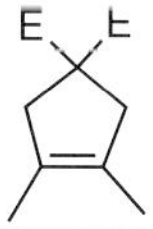	0	40

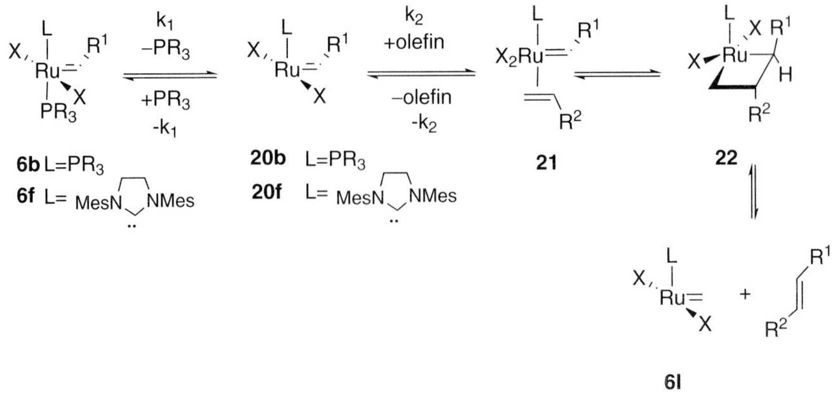

536 *11 Ring-Closing Metathesis of Dienes and Enynes*

Scheme 11.11 Reaction mechanism for alkene metathesis.

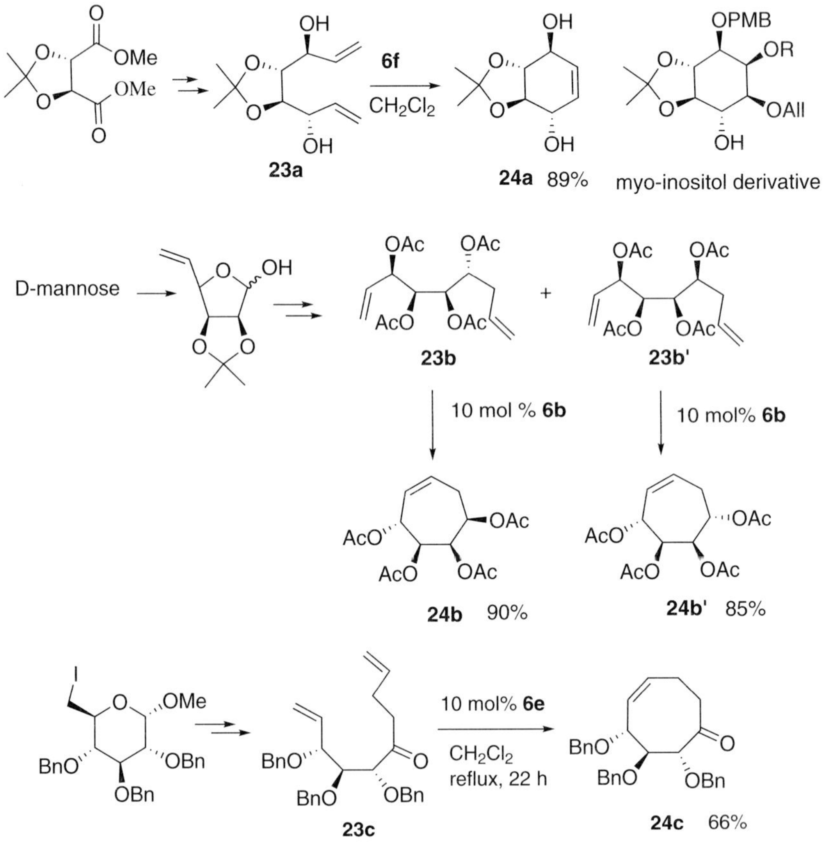

Scheme 11.12 Synthesis of cycloalkene derivative using RCM.

be synthesized in enantiomerically pure form with high chemical yields by RCM of acyclic, chiral polyhydroxylated 1,8-nonadiene precursors **23b** and **23b′** derived from D-mannose [30b,c]. From 6-deoxy-6-iodoglycoside, Nolan synthesized the functionalized eight-membered compound **24c** [30].

RCM of diene **25a** containing the silyl ether gave cyclic silyl ether **26a** in high yield. In a similar manner, cyclic silyl ethers **26b** and **26c** were prepared from dienes **25b** and **25c** and then these compounds **26b** and **26c** were treated by Tamao oxidation to give hydroxylated compounds **27b** and **27c** [31]. D-Altritol could be synthesized using this procedure. The double bond of the cyclic silyl ether **26d** prepared from the linear diene **25d** by RCM was treated with OsO_4 followed by treatment with TBAF and then NaOMe to afford D-altritol (Scheme 11.13) [32].

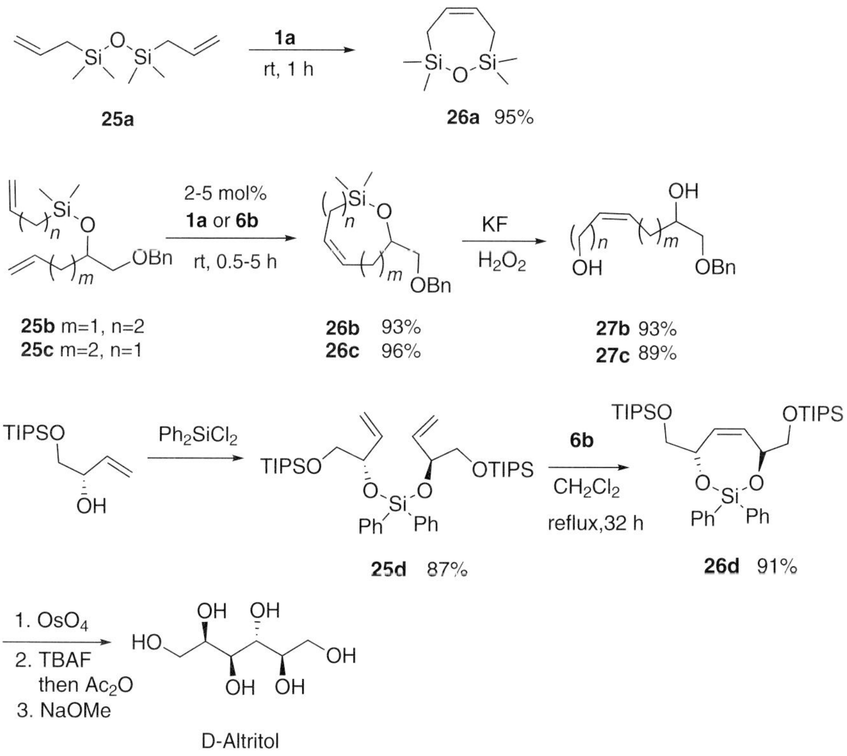

Scheme 11.13 Synthesis of cyclic compound containing silylether.

Although the first generation ruthenium catalyst **6a** or **6b** did not work in the RCM of a diene containing an enol ether, RCM between an enol ether and an alkene of **28a** using the second generation ruthenium catalyst **6f** proceeded smoothly to give cyclic enol ether **29a** (Scheme 11.14) [33]. In the case of the RCM of an alkene and the silyl enol ether, cyclic compound **29b** was formed in high yield. These results are useful because cyclic enol ethers **29a** and **29b** should give the cyclic ketones [34].

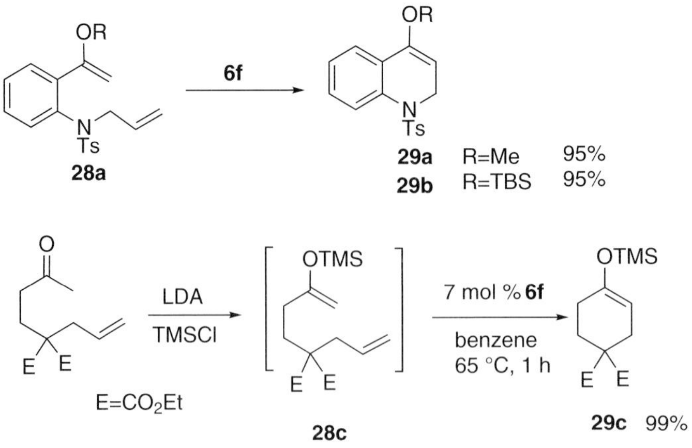

Scheme 11.14 Metathesis of diene containing enol ether.

The RCM of 2,2-divinylbiphenyl **30a** using a second-generation ruthenium catalyst **6f** led to the substituted phenanthrene derivative **31a** in quantitative yield under very mild reaction conditions [35]. Helicene **31b** was synthesized from **30b** in CH_2Cl_2 in a sealed tube using **6j** as catalyst (Scheme 11.15) [36].

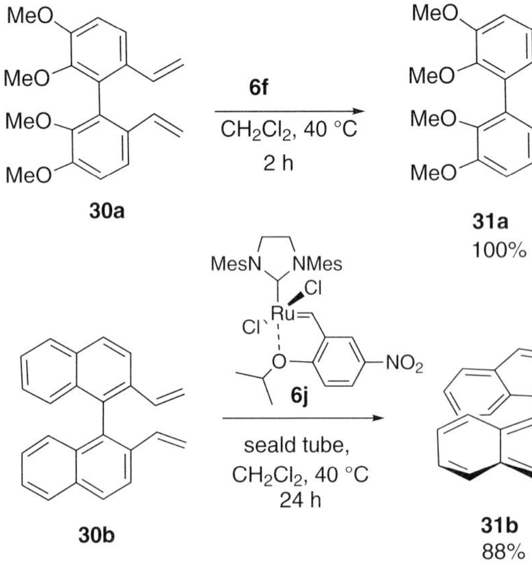

Scheme 11.15 Synthesis of aromatic compounds.

Imamoto reported a new synthetic approach to phenol derivatives **33**, which are most important classes of aromatic compounds, from linear compounds **32** using RCM [37]. Triene **32a** was treated with 7.5 mol% of **6f** in CH_2Cl_2 at room temperature for 2 h to give the phenol derivative **33a** in 93% yield. Polysubstituted phenol **33b** was also synthesized from triene **32b** in 84% yield (Scheme 11.16).

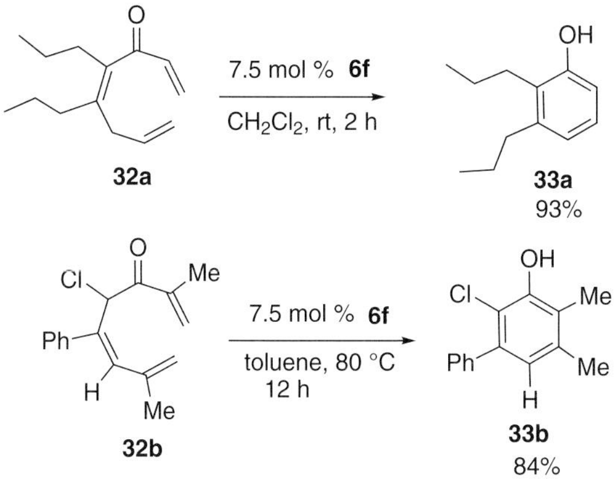

Scheme 11.16 Synthesis of phenol derivatives.

Double RCM of diene **34a** having a protected hydroxy group with **6b** gave *cis*-decaline **35a** as a major product along with a small amount of *trans*-decaline **35a'** in 80% yield in a ratio of 8 to 1. However, in the double RCM of diene **34b** having the hydroxy group, *trans*-decaline **35b'** was a major product [38]. Synthesis of (+)-epilupinine was achieved using double RCM as a key step by Ma. When **6f** was used as the catalyst, the double RCM reaction of (3S,4R)-**34c** afforded a mixture of (4S,5R)-**35c** and (4S,5R)-**35c'** in a ratio of 1.96 to 1 in 89% yield along with a dumbbell-type cyclization product **36** in 4% yield. Treatment of **35c** and **35c'** with Pd/C/H$_2$ and LiAlH$_4$ gave (+)-epilupinine in high yield [39a]. α-Isosparteine was synthesized from tetraene **34d** by double RCM followed by exhaustive reduction [39b]. Similarly, the triadjacent cyclic ether **35e** has been synthesized from hexaene **34e** using triple RCM [39c]. A quadruple RCM reaction of **34f** was reported for the synthesis of bis-spirocyclic compounds **35f**, **35f'**, and **35f''** [39d] (Scheme 11.17).

RCM of **37a** with **6b** gave lactone **38a**, but the yield was only 40%. It is reasoned that the carbonyl oxygen should coordinate to the ruthenium metal to give **39a** to decrease the catalytic activity. Thus, when the reaction was carried out using **6b** in the presence of Ti(O^{i}Pr)$_4$, the yield was improved to 72% [40]. Later, a new strategy to access the fumagillol skeleton was proposed. An Evans aldolization and a RCM involving enone **37b** in the presence of Ti(O^{i}Pr)$_4$ were used for the preparation of the key cyclohexenone intermediate **38b**, which was readily converted to fumagillol (Scheme 11.18) [41].

RCM of chiral diallylamine **40** in the presence of Lewis acid using the second generation RCM catalyst **6b** led to the enantiopure pyrrolidine derivative **41** in 93% yield under very mild conditions (Scheme 11.19). In the absence of Ti(O^{i}Pr)$_4$, the desired pyrrolidine derivative **41** was not formed [42].

There have been some reports of olefin isomerization of the product in RCM when the catalysts are stressed by high temperatures, high dilution, and forced high turnovers. While the exact mechanism responsible for this isomerization is unknown, recent results indicate that a ruthenium hydride species, such as **45**, formed from the decomposition of the ruthenium metathesis catalysts catalyzed the migra-

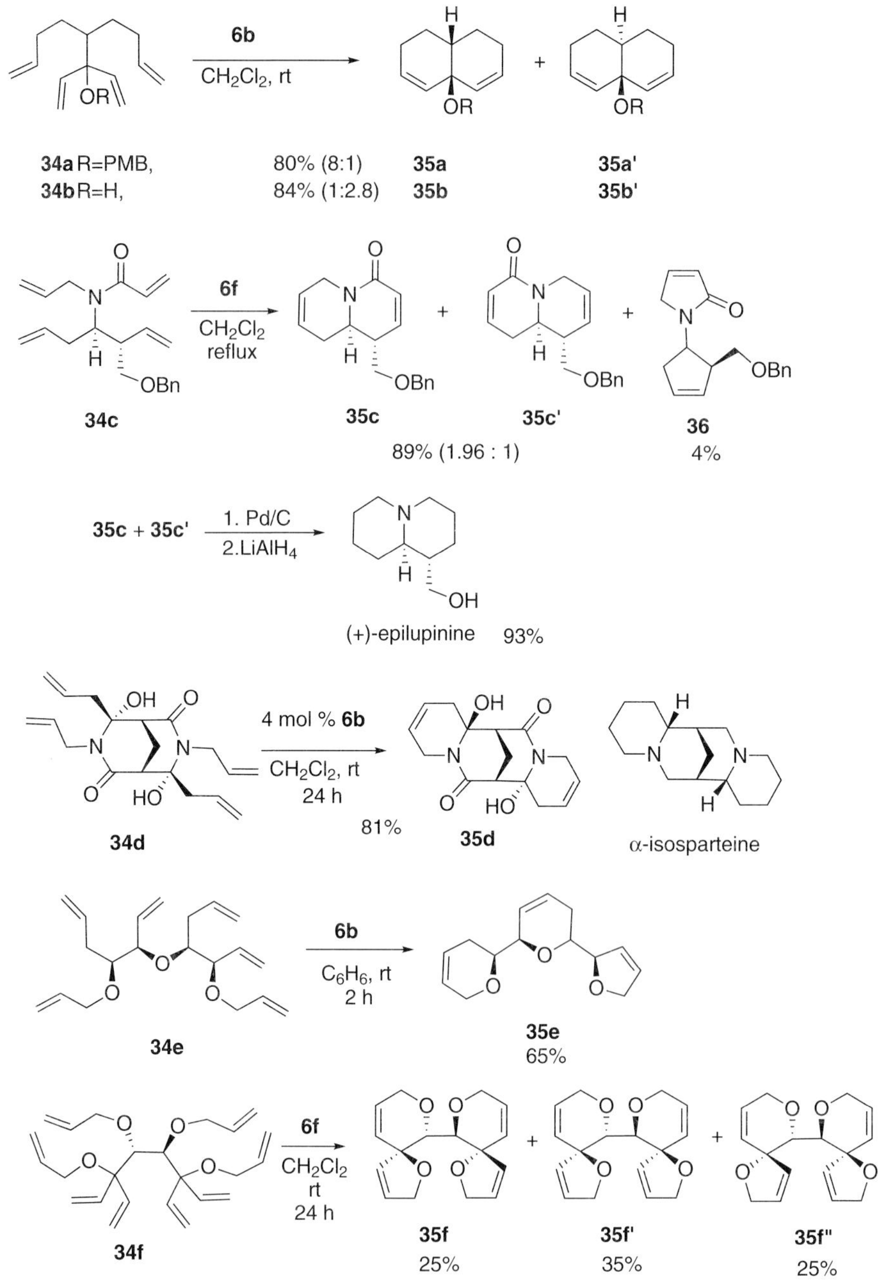

11 Ring-Closing Metathesis of Dienes and Enynes

34a R=PMB,	80% (8:1)	35a 35a'
34b R=H,	84% (1:2.8)	35b 35b'

Scheme 11.17 Double RCM of tetraenes.

(+)-epilupinine 93%

α-isosparteine

35e 65%

35f 25%   35f' 35%   35f" 25%

Scheme 11.18 Diene metathesis in the presence of Ti(O^{i}Pr)$_4$.

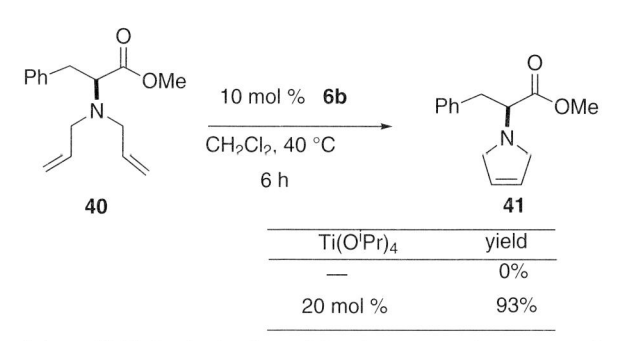

Scheme 11.19 Synthesis of pyrrolidine derivative in the presence of Ti(O^{i}Pr)$_4$.

tion of olefins under metathesis conditions [43]. To prevent isomerization, Grubbs examined the RCM of **42a** in the presence of various acids. Addition of acetic acid or 1,4-benzoquinone afforded the desired **43a** in higher yield (Table 11.2). RCM of **42b** in the presence of 1,4-benzoquinone gave the desired compound **43b** in 90% yield while, in the absence of additive, the isomerization product **44b** was obtained in 84% yield (Scheme 11.20) [44].

Table 11.2 Prevention of isomerization during RCM.

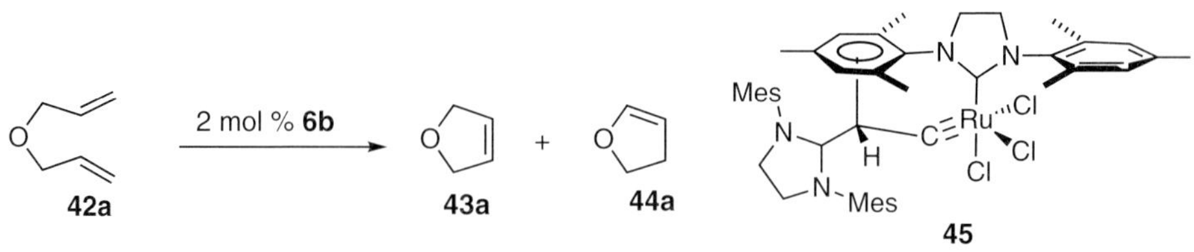

	Equiv.	43a (%)	44a (%)
none	none	<5	>95
CH_3COOH	0.1	>95	none
1,4-benzoquinone	0.1	>95	none
galvinoxy	0.2	80	20
TEMPO	0.5	7	93
4-methoxyphenol	0.5	17	83

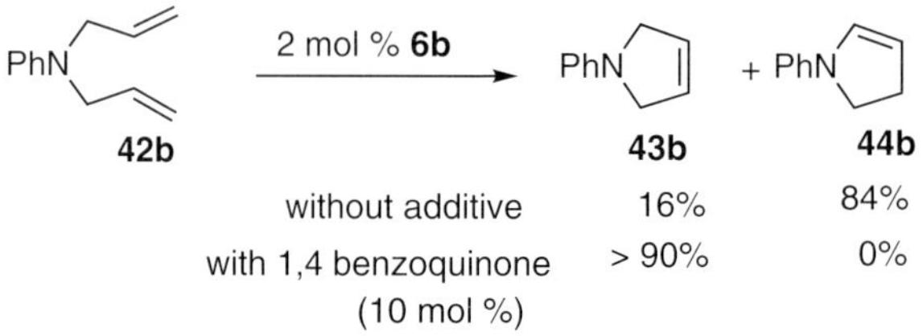

	43b	44b
without additive	16%	84%
with 1,4 benzoquinone (10 mol %)	>90%	0%

Scheme 11.20 Prevention of isomerization.

11.2.2
Synthesis of Carbo- and Heterocycles Using Ring-Opening Metathesis–Ring-Closing Metathesis of Dienes

Ring-opening metathesis of cyclobutene in the presence of an alkene was reported by Snapper in 1995 [45a]. ROM of cyclobutene **46a** with 1-octene gave the ring-opening product **47a** in 63% yield. As shown in Scheme 11.21, two possible reaction courses should be considered. One is the reaction of cyclobutene and the ruthenium heptylidene complex **48a**, generated from 1-octene and the carbene complex **6l**, the ruthenium carbene complex **49a** being formed as an intermediate. It reacts with 1-octene to afford **47a** and the ruthenium carbene complex **48a** is regenerated. The alternative reaction course is that ROM of cyclobutene with the ruthenium carbene complex **6l** gives the ruthenium carbene complex **49a'**, which reacts with 1-octene to afford **47a** and the methylidene ruthenium carbene complex **6l** is regenerated. ROM–RCM of cyclobutene **46b** and **46c** having a substituent at the C1 position gave the furan or pyran derivative **47b** or **47c** in high yield. It is interesting that ring cleavage of cyclobutene occurs by treatment with the ruthenium carbene complex **6f** in the presence of an intramolecular alkene [45b,c].

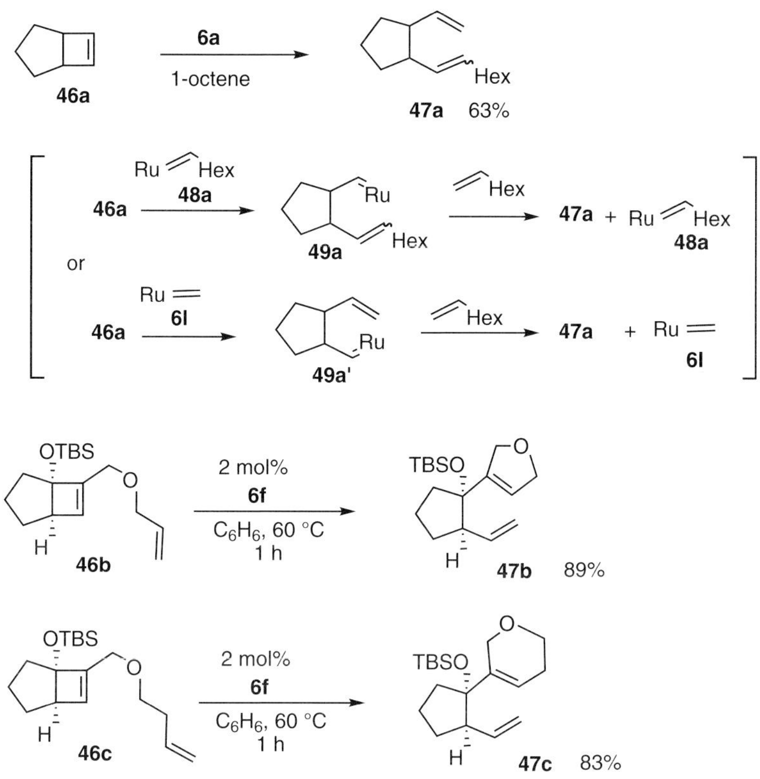

Scheme 11.21 ROM–RCM of cyclobutene-ene.

Metathesis of a substrate having more than three alkenes in a molecule proceeded via tandem reactions to give various cyclic compounds. It is denoted as ROM–RCM or ROM–CM. Grubbs reported the ROM–RCM of a disubstituted cycloalkene **50** having an alkene moiety in each tether (Scheme 11.22) [46]. Reaction of an alkene part in a tether of **50** with **6l** gave the ruthenium carbene complex **51**, which reacts with a cycloalkene to give the ruthenium carbene complex **52**, Then the ruthenium carbene intermediate **52** reacts with the cyclic C=C bond intramolecularly to afford the bicyclic compound **53**. Cleavage of the double bond of the four-, five- and six-membered cycloalkenes by ROM afforded the dumbbell-type bicyclic products **53a–c**. The ring size generated by ROM–RCM depends on the chain length of the tether.

Hoveyda *et al.* reported a novel method for the synthesis of chromene **55** by ROM–RCM of cycloalkene **54** bearing the phenyl ether at the 3-position (Scheme 11.23) [47]. The yield was improved when the reaction was carried out under ethylene gas. In the case of cyclopentene **54a** or cyclohexene **54b**, the yield was poor because the starting cycloalkene is in a state of equilibrium with the product and a thermodynamically controlled compound should be formed under these reaction conditions. They obtained the enantiomerically pure cycloheptene

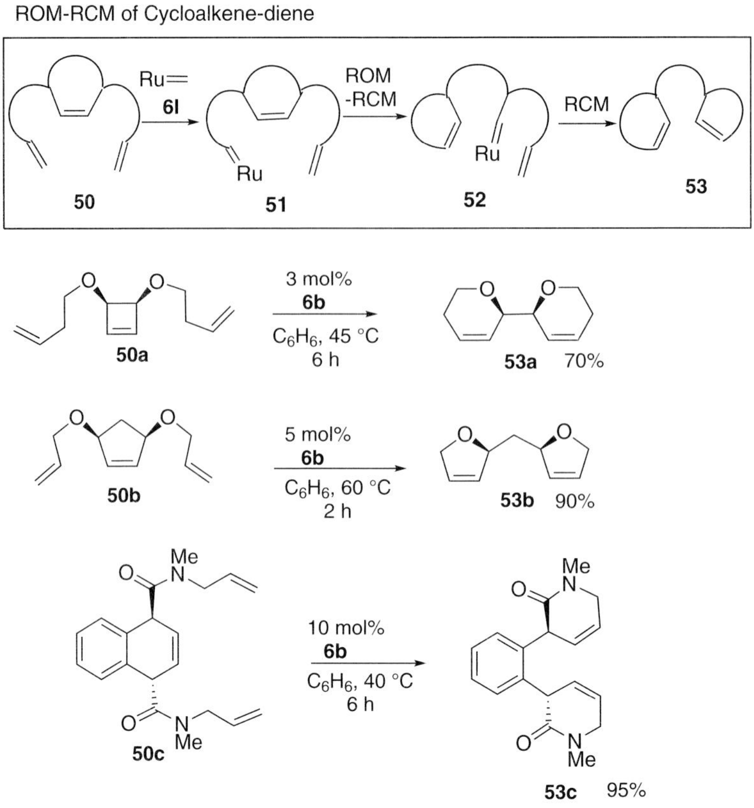

11 Ring-Closing Metathesis of Dienes and Enynes

ROM-RCM of Cycloalkene-diene

Scheme 11.22 ROM–RCM of cycloalkene-diene.

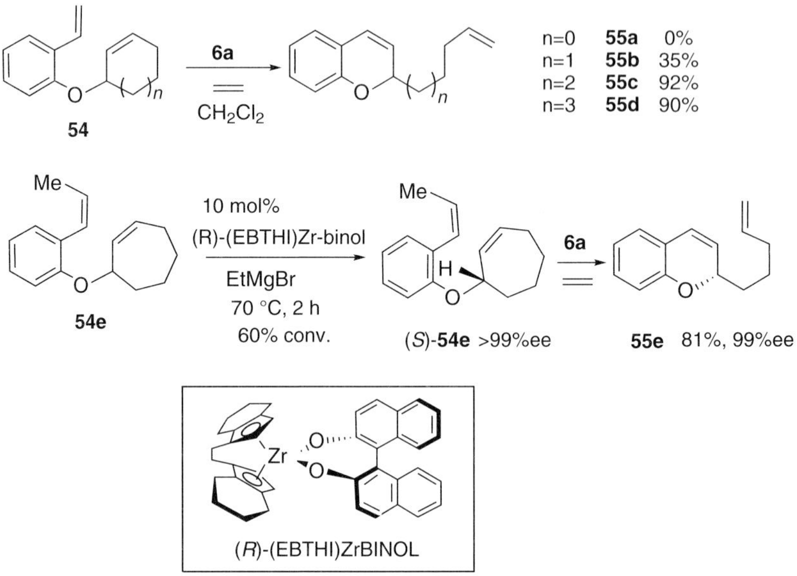

Scheme 11.23 Synthesis of chromen using ROM–RCM of cycloalkene-ene.

11.2 Ring-Closing Metathesis of Dienes

derivative (S)-**54e** using zirconium-catalyzed kinetic resolution of **54e** developed by themselves. Optically pure chromene **55e** was synthesized via ROM–RCM of (S)-**54e** using **6a**.

Since it was known that the cyclopentene ring of norbornene could be easily opened by ruthenium carbene complex **6l**, bicyclic compound **57** was synthesized from the norbornene derivative **56** having a terminal C=C bond in a side chain [48]. When compound **56a** was treated with **6b** under ethylene gas, the *cis*-pentalene derivative **57a** was obtained. In a similar manner, norbornenes **56b–d** were reacted with **1a** or **6b** in the presence of ethylene or terminal alkenes to give *cis*-indene derivatives **57b–d** in high yields. The ring size constructed in this reaction corresponds to the carbon chain length of the side chain on norbornene **56** (Scheme 11.24).

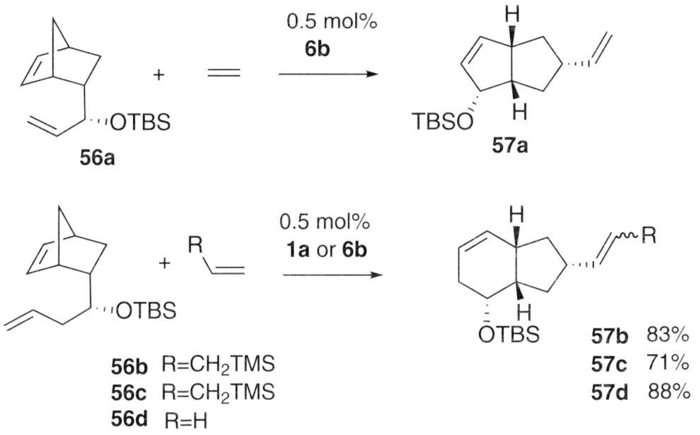

Scheme 11.24 Synthesis of bicyclic compounds from norbornene derivative.

Wright succeeded in the ROM–RCM of substrate **58a** having a bicyclo[3.2.1]ring system to form the spiro-bicyclic product **59a** in 82% yield [49]. Reaction of **58a** with the ruthenium carbene complex **6l** gives the ruthenium carbene complex **58a-Ru**, which reacts with the C=C bond in the furan ring to give the highly strained ruthenacyclobutane **60**. Ring opening of this complex **60** gives ruthenium carbene complex **61**, which reacts with a terminal C=C bond of **58a** to give **59a** and ruthenium carbene complex **58a-Ru** is regenerated. In a similar manner, **58b** having a bicyclo[2.2.1] ring system gave the cyclopentene derivative **59b** in high yield (Scheme 11.25).

Blechert reported that azacycle **63** could be synthesized by the ROM–RCM of the cycloalkenyl amine derivative **62** bearing an extra alkene moiety on the nitrogen substituent [50]. Various heterocycles could be synthesized from the disubstituted cycloalkene **62** having an alkene moiety on the hetero atoms. Double ROM of cycloalkene **62b** gave compound **63b** having two heterocycles. Treatment of **63b** with TBAF gave the pyrrolidine derivative having the hydroxy group and the terminal alkene in the tether. This compound **63b** was converted into the natural product, (−)-halosaline within a few steps. In a similar treatment of compound **62c**, the pyrrolidine derivative **64c** was obtained after treatment of the crude product **63c** with TBAF and was converted into indolizidine 167 B (Scheme 11.26).

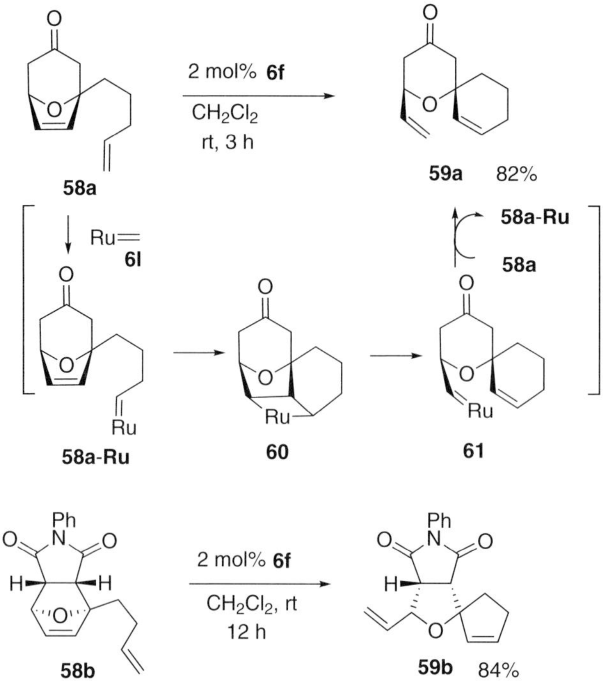

Scheme 11.25 Synthesis of spiro-compounds using ROM–RCM.

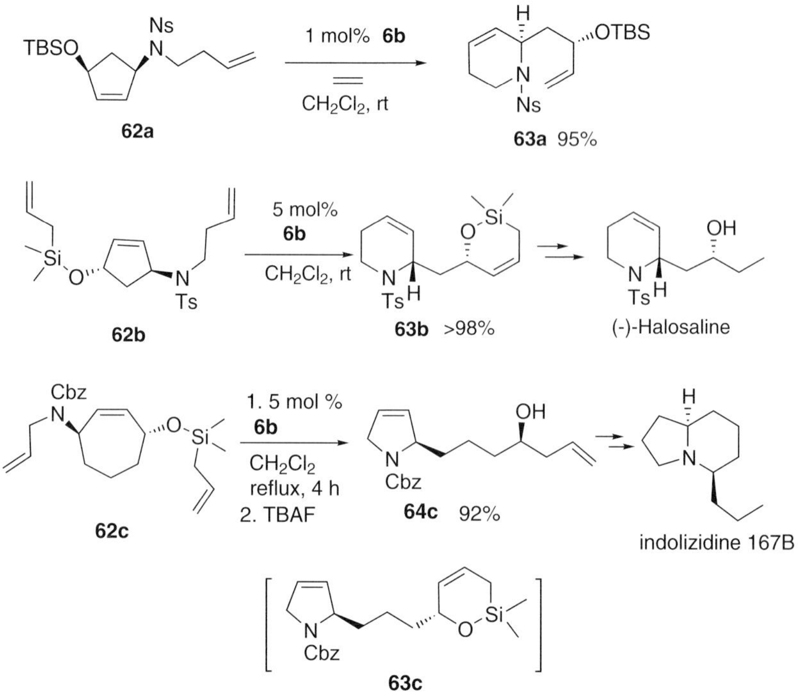

Scheme 11.26 One-step synthesis of nitrogen hetrocycles from cycloalkene by ROM–RCM.

11.2 Ring-Closing Metathesis of Dienes

Winkler reported the synthesis of stereochemically defined poly-heterocyclic ring systems (Scheme 11.27). Treatment of substituted oxanorbornanes **65a** with **6f** under ethylene gas gave the fused tri-heterocyclic ring compound **66a** in high yield. In a similar manner, compound **66b** having a penta-heterocyclic ring system could be synthesized from **65b**, albeit in low yield [51].

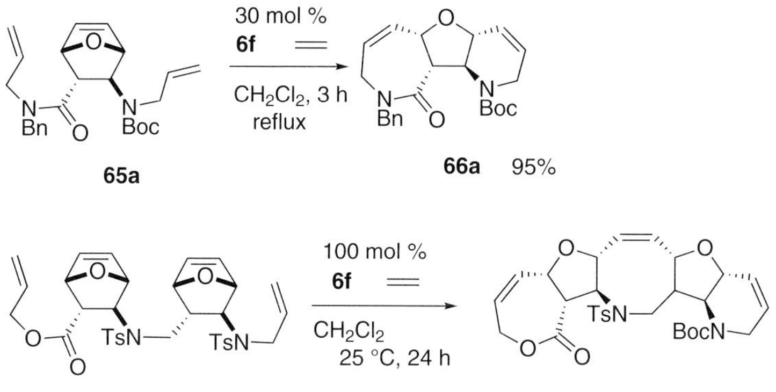

Scheme 11.27 Synthesis of polyheterocyclic compounds using ROM–RCM.

ROM–RCM of cycloalkene-ene proceeds in one operation via many steps to afford different cyclic compounds. It is not clear which alkene – the terminal alkene or the cycloalkene – reacts with the ruthenium carbene complex **6l** in the initial step, and it is difficult to determine the real reaction course. However, the result is interesting in that ring opening of the cycloalkene occurs and a complex molecule can be synthesized within one synthetic operation in high yield.

11.2.3
Catalytic Asymmetric Ring Closing Metathesis of Dienes

11.2.3.1 Asymmetric Synthesis Using a Chiral Molybdenum Catalyst

In olefin metathesis, a double bond is cleaved and a double bond is formed. Thus, no chiral carbon center is constructed in the reaction. To realize asymmetric induction by RCM, there are two protocols: kinetic resolution and desymmetrization of the symmetric prochiral triene (Scheme 11.28). Various molybdenum complexes have been synthesized for that purpose (Figure 11.3).

Grubbs synthesized the molybdenum catalyst **67** and reported the first example of asymmetric induction by kinetic resolution of the diene using RCM with complex **67** [52]. At 90% conversion of the diene **72** using the molybdenum complex **67**, the starting material **72** with 84% ee in *S*- configuration was recovered (Scheme 11.29).

Hoveyda and Schrock synthesized many chiral molybdenum catalysts and examined the effectiveness of these catalysts in asymmetric induction. In the kinetic resolution of **74** using a BIPOL-type molybdenum catalyst **68a**, cyclopentene

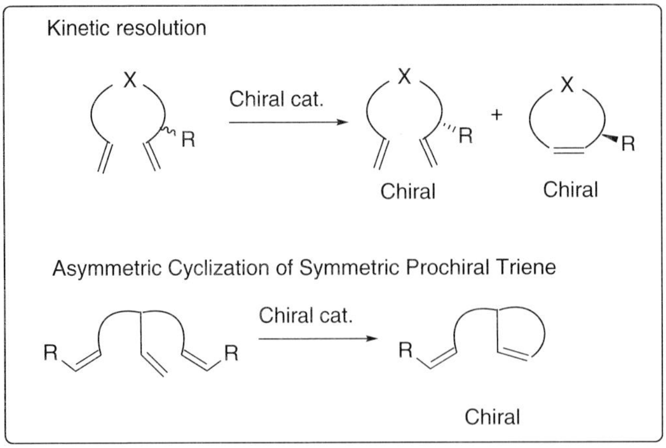

11 Ring-Closing Metathesis of Dienes and Enynes

Scheme 11.28 Procedure for asymmetric induction using olefin metathesis.

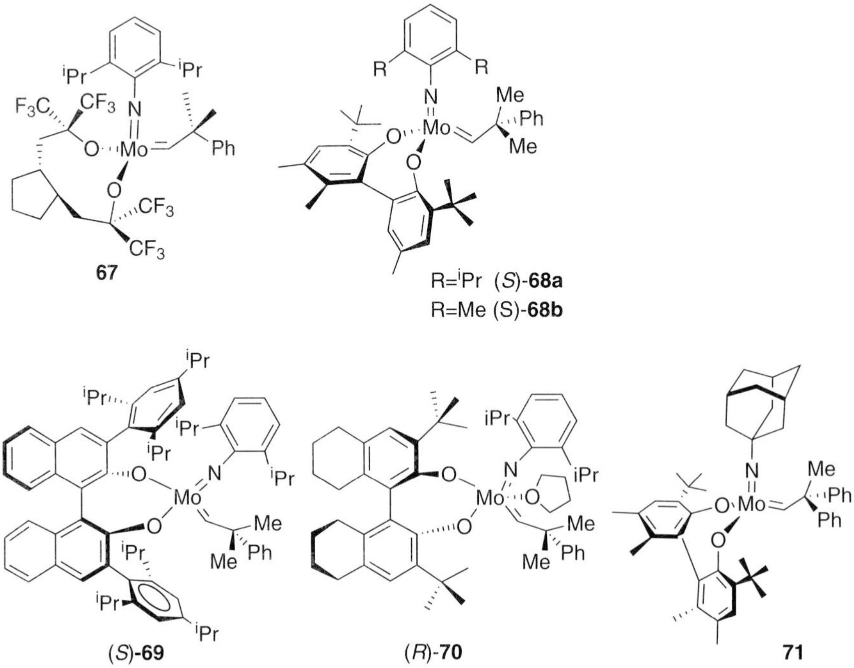

R=^{i}Pr (S)-**68a**
R=Me (S)-**68b**

(S)-**69**   (R)-**70**   **71**

Figure 11.3 Molybdenum catalyst for asymmetric synthesis.

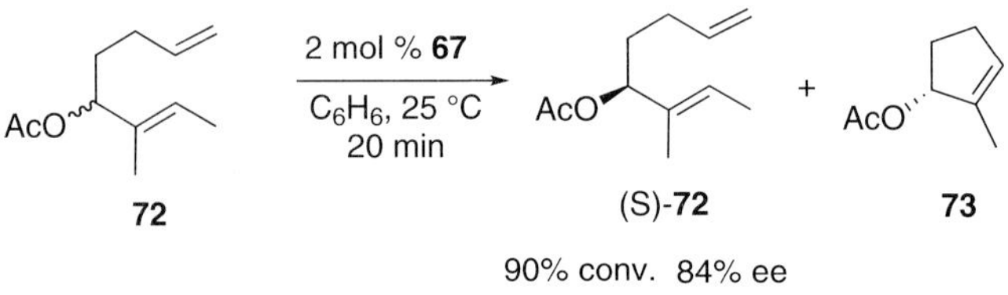

72 → (S)-**72** + **73**

2 mol % **67**, C$_6$H$_6$, 25 °C, 20 min

90% conv. 84% ee

Scheme 11.29 Kinetic resolution of diene using RCM.

derivative **75** with 93% ee was obtained at a conversion of 80% and the recovered diene (*R*)-**74** shows 99% ee [53a,b]. Desymmetrization of the symmetric triene **76** afforded the furan derivative **77** with 99% ee in 86% yield (Scheme 11.30) [53b].

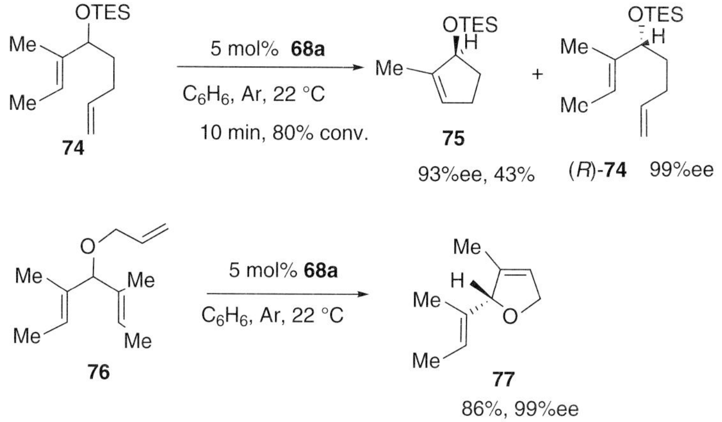

Scheme 11.30 Asymmetric synthesis of furan derivatives using ARCM.

Enantioselective desymmetrization of dienes by tandem Mo-catalyzed ROM–RCM of **78a** proceeded in the presence of diallyl ether to give cis-cyclopentapyrane **79a** with 92% ee in 54% yield. Addition of diallyl ether accelerates the formation of a chiral molybdenum carbene complex. On the other hand, the reaction of **78b** with **68a** gave oligomeric products. Thus, first, **78b** was treated with the ruthenium carbene complex **6b** under ethylene gas to give **80**, which was followed by the asymmetric ring-closing metathesis using **68a** to give *trans*-cyclopentapyrane **79b** with high ee (Scheme 11.31) [54].

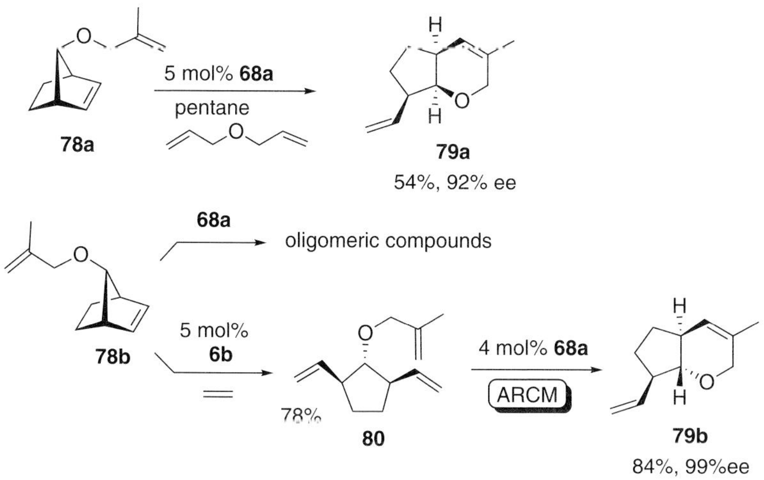

Scheme 11.31 Asymmetric ring-opening metathesis of norbornene derivatives.

Optically active cyclic ether may be synthesized by molybdenum-catalyzed ROM–RCM (Scheme 11.32). The cyclopentene derivative **81** was reacted with 5 mol% of chiral BINOL-type molybdenum catalyst **69** to give the pyran derivative **82** in high yield and high ee; this was further converted into the lactone **83** for the synthesis of anti-HIV agent tipranavir [55].

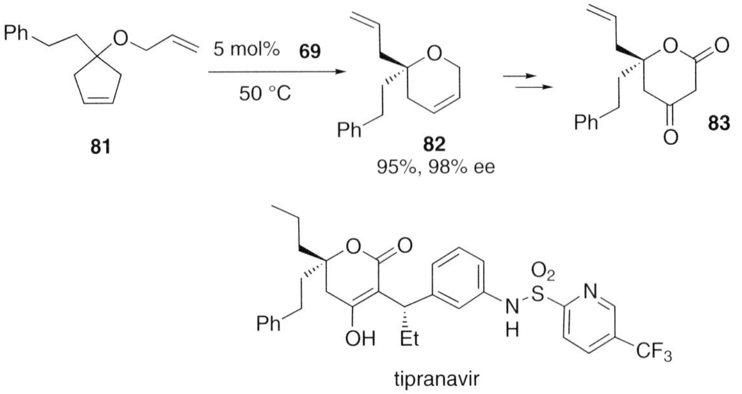

Scheme 11.32 Synthesis of pyrane derivatives using AROM.

Enantiomerically enriched nitrogen heterocycles have been synthesized by the same protocol (Scheme 11.33) [56]. Chiral piperidine derivative **85a** was obtained in high ee using modified BIPOL-type complex **68c**. Eight-membered nitrogen heterocycle **85b** with high ee could also be synthesized from prochiral triene **84b** in high yield.

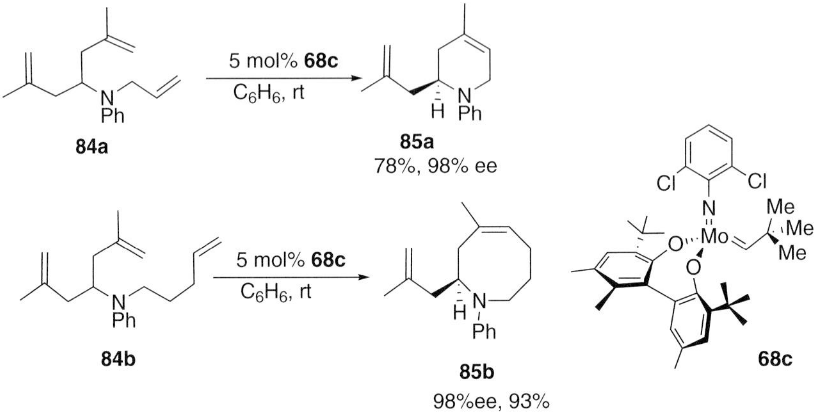

Scheme 11.33 Synthesis of nitrogen heterocycles using AROM.

Hetero bicyclic compounds **87** with high ees could be synthesized in high yields from trienes **86** using molybdenum catalysts (Table 11.3) [57].

11.2 Ring-Closing Metathesis of Dienes | 551

Table 11.3 Synthesis of azabicyclic compounds.

Entry	Substrate	Product	Cat (mol%)	Time (h)	Yield (%)	ee (%)
1	86a	87a	71 (5)	48	84	85
2	87b	86b	(S)-70b (15)	48	94	90
3	86c	87c	(S)-69 (10)	48	91	>98
4	86d	87d	(S)-69 (5)	12	92	88

The silicon-containing heterocycle **89a** with high ee could be synthesized from the linear compound **88a** containing silyl ether [58a]. The catalytic enantioselective RCM of polyene **88b** using the BIPOL-type complex **68b** afforded unsaturated siloxane **89b** efficiently. This was treated with K$_2$CO$_3$ in MeOH to give **90** (Scheme 11.34) [58b].

The 6,8-dioxabicyclo[3.2.1]octane skeleton is a common structural subunit in natural products. A conceptually new strategy affording these structures was described by Burke *et al.* (Scheme 11.35). For the syntheses of (+)-exo-brevicomin, they used desymmetrization of triene **92a**, derived from diol **91a** with C2 symmetry, via RCM. (+)-endo-brevicomin was synthesized by employing desymmetrization of trienes **92** derived from meso-diol **91b** using **68a** [59].

Preparations of molybdenum complexes are not easy because they are moisture sensitive. Thus, *in situ* preparation of the molybdenum catalyst for the catalytic

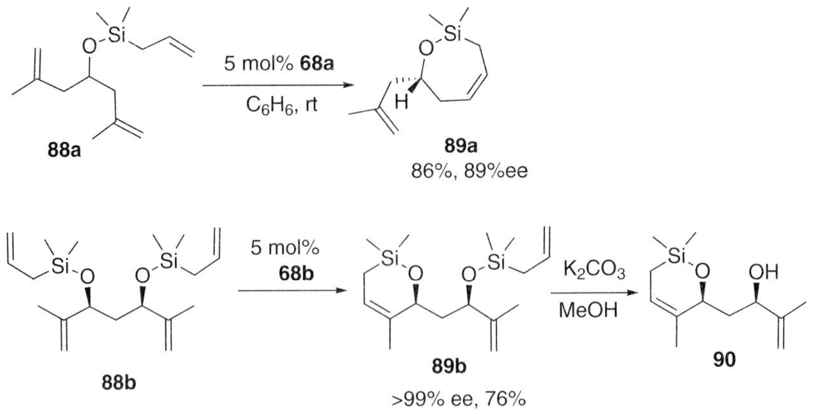

Scheme 11.34 Asymmetric synthesis of cyclic silyl ethers.

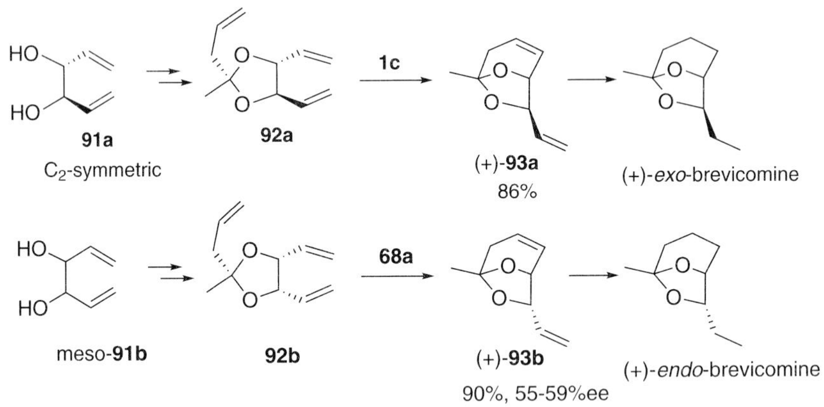

Scheme 11.35 Synthesis of brevicomine.

reactions has been reported. An active catalyst solution could be accessed by premixing of readily available (R)-**94** with a commercially available Mo-triflate **95**, which was ready for catalysis without further purification. This catalyst solution **70** promoted ARCM reactions with equal or higher levels of efficiency and selectivity as compared to **68a**: Desymmetrization of **76b** with the THF solution of **70** gave the furan derivative **77b** with 88% ee. Use of **68a** gave **77b** with 93% ee (Scheme 11.36) [60].

11.2.3.2 Asymmetric Olefin Metathesis Using Ruthenium Catalyst

Ruthenium catalysts for asymmetric synthesis have been developed (Figure 11.4).

Grubbs synthesized the ruthenium carbene complex **96a**, which catalyzed the ARCM of **76a** to afford (S)-**77a** with a high enantiomeric excess (up to 90%) (Scheme 11.37) [61].

They also synthesized various ruthenium catalysts **96b–f**. The addition of NaI to catalysts **96a**, **96c** and **96e** (to form **96b**, **96d** and **96f**) caused a dramatic increase

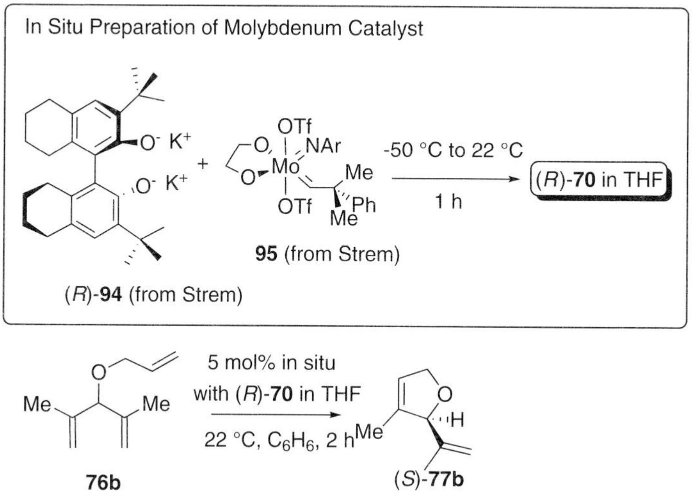

11.2 Ring-Closing Metathesis of Dienes

In Situ Preparation of Molybdenum Catalyst

(R)-94 (from Strem) + 95 (from Strem) → −50 °C to 22 °C, 1 h → (R)-70 in THF

Scheme 11.36 *In situ* preparation of molybdenum catalyst.

76b → 5 mol% in situ with (R)-70 in THF, 22 °C, C$_6$H$_6$, 2 h → (S)-77b

88% ee, 80% yield
with 5 mol% (S)-68a 93% ee, 86% yield

96a X=Cl
96b X=I

96c X=Cl
96d X=I

96e X=Cl
96f X=I

97

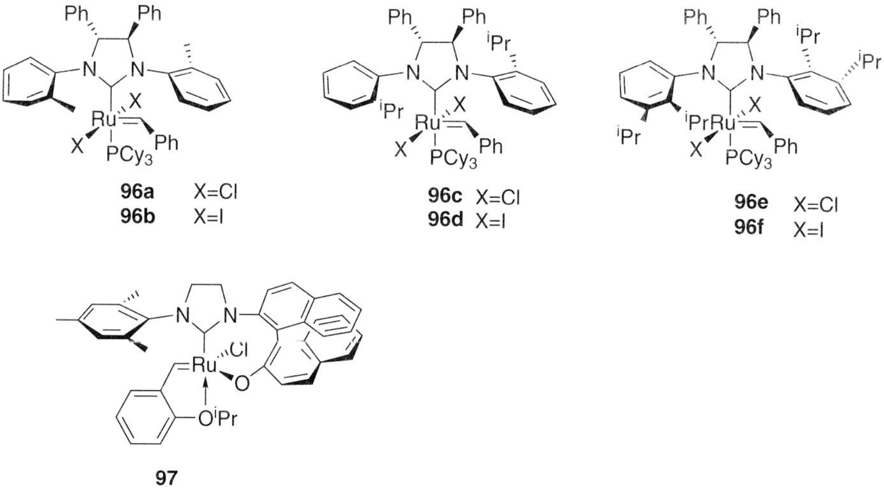

Figure 11.4 Ruthenium catalyst for asymmetric synthesis.

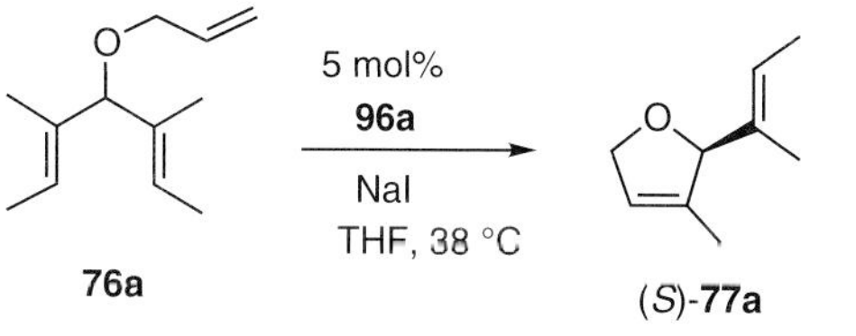

76a → 5 mol% 96a, NaI, THF, 38 °C → (S)-77a

82% conv. 90% ee

Scheme 11.37 ARCM using ruthenium catalyst.

11 Ring-Closing Metathesis of Dienes and Enynes

Table 11.4 Ruthenium catalyst for ARCM.

Entry	Substrate	Product	Catalyst	ee (%)	Yield (%)
1	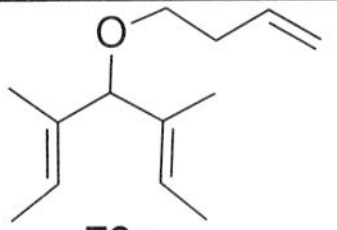 76c	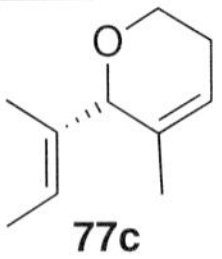 77c	96d	90	77
2	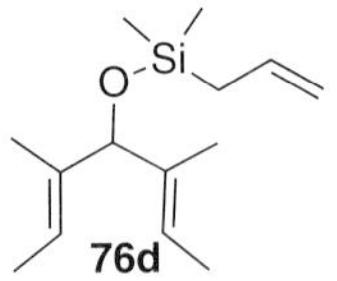 76d	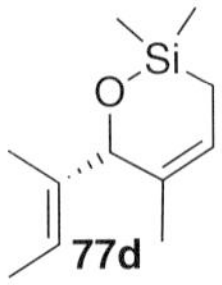 77d	96e	76	92
3	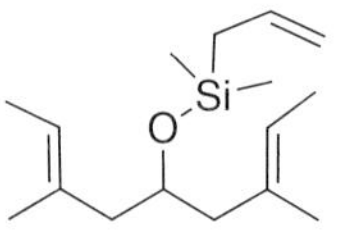 76e	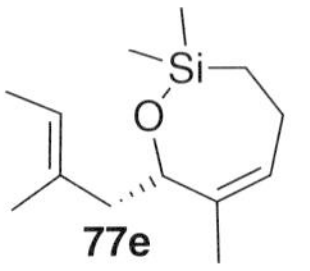 77e	96d	78	98

in enantioselectivity for the ARCM of trienes **76** to afford various sized heterocycles **77** with high ees (Table 11.4) [62].

Hoveyda synthesized recyclable ruthenium catalyst **97** for enantioselective olefin metathesis. This catalyst was very effective for AROM–CM and could be recovered after chromatography. The recovered catalyst could be reused without significant loss of reactivity and enantioselectivity. For example, the pyran derivative **99** with high ee was prepared by the ARCM–CM of **98** in the presence of styrene using ruthenium catalyst **97** (Scheme 11.38) [63].

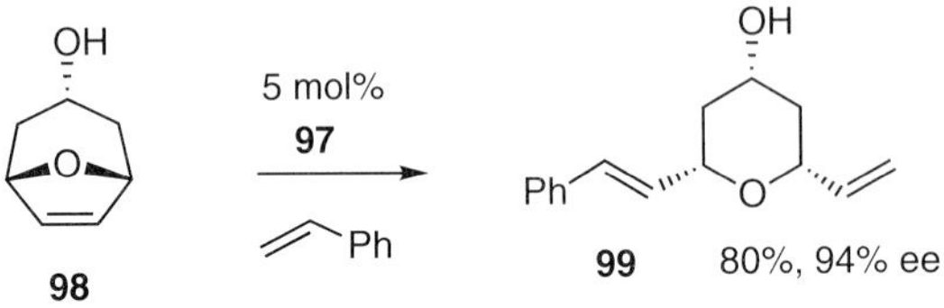

Scheme 11.38 AROM using ruthenium catalyst.

11.2.4
Polymer-Supported Catalysts

The development of cost-effective and environmentally benign catalytic systems has become one of the main themes of contemporary synthetic organic chemistry. In this context, efficient recycling and subsequent reuse of homogeneous catalysts are of prime importance. Development of the polymer-supported catalyst **100** (Figure 11.5) for olefin metathesis was performed by Grubbs first and it was used for polymerization of alkene [64].

11.2 Ring-Closing Metathesis of Dienes

Figure 11.5 Polymer-supported catalyst.

Barrett reported the recyclable, "boomerang" polymer-supported ruthenium catalyst **102** for olefin metathesis. The rate of reaction and the catalytic activity were found to be comparable to that of the homogeneous catalyst **6a**. The catalyst **102** could be recycled for RCM of **103** up to three times by simple filtration (Scheme 11.39) [65].

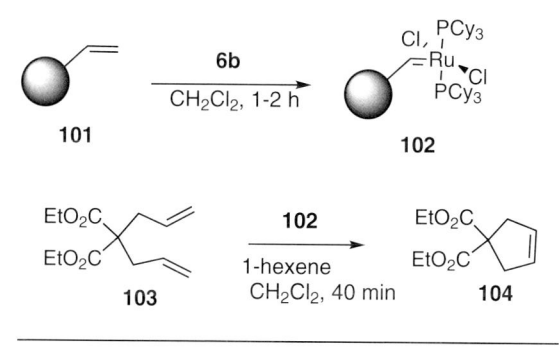

additive (mol %)	conversion (%)			
	run 1	run 2	run 3	run4
none	100	40	0	0
1-hexene (9)	100	100	50	10

Scheme 11.39 RCM using polymer supported ruthenium catalyst.

Hoveyda et al. synthesized catalyst **6h** for RCM and found the efficient formation of various trisubstituted olefins at ambient temperature in high yield within 2 h and it was re-used after purification. They also synthesized two dendritic and recyclable Ru-based complexes **105a** and **105b** (Scheme 11.40). Unlike monomer **6h**, dendritic **105b** could be readily recovered. Treatment of diene **42c** with 1.25 mol % **105a** (5 mol% Ru) led to efficient and catalytic RCM. The desired product **43c** was first isolated in 99% yield by silica gel chromatography through elution with CH_2Cl_2. Subsequent washing of the silica gel with Et_2O led to the isolation of the dendritic catalyst **105a**. Dendrimer **105b** effectively catalyzed tandem ROM–RCM of **50d** to afford **51d** in 94% yield. In contrast to the corresponding monomer **6h**, dendrimer **105b** could be easily separated from **51d** and recovered in 90% yield [25].

556 | *11 Ring-Closing Metathesis of Dienes and Enynes*

Scheme 11.40 ARCM using recyclable monomeric and dendritic Ru-based metathesis catalyst.

Polymer-supported second generation ruthenium catalyst **106** was developed by Blechert for substituted alkene **107a**. Although the initial catalytic activity of **106** was high, 2 days were required for completion of RCM of **107b** in the fourth cycle (Scheme 11.41) [66].

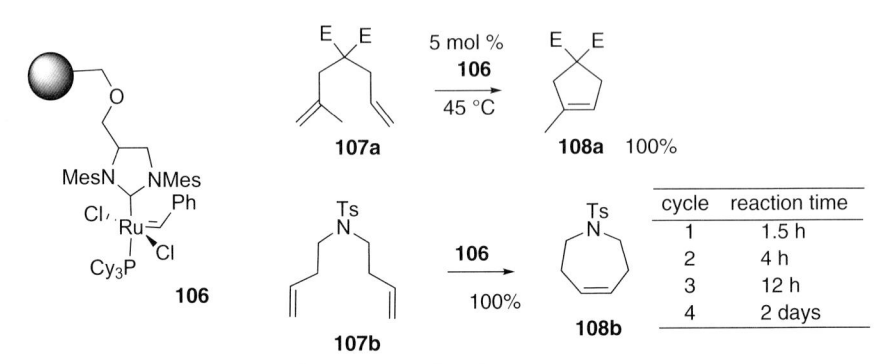

Scheme 11.41 RCM using polymer-supported catalyst.

Yao reported that polymer-supported ruthenium catalyst **110a**, prepared from **109** and **6b**, was recyclable and may be reused in 8 cycles of RCM of **107c** and the yield of **108c** is high [67]. He also synthesized the fluorous and recyclable catalyst **110b**, which was also very effective for RCM [68a]. RCM in the ionic liquid was examined using a recyclable ruthenium catalyst **110c** developed by himself [68b]. A mixture of 1-butyl-3-methylimidazolium hexafluorophosphate and CH_2Cl_2 (1:9) was used as the solvent. The catalyst could be re-used for 10 cycles of RCM of **107c** and the yield was high (Scheme 11.42).

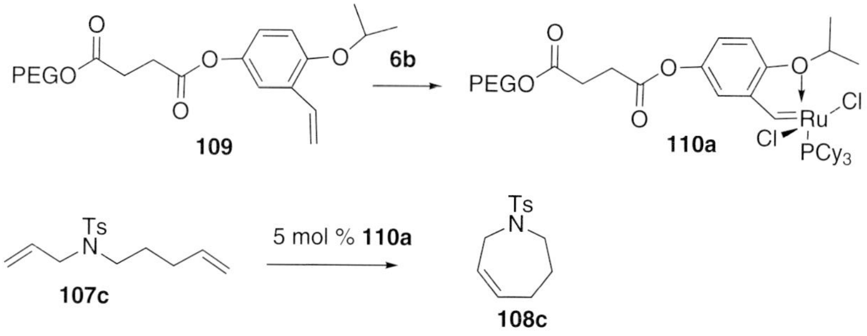

cycle	1	2	3	4	5	8
yield (%)	98	97.5	96.5	95	95	92

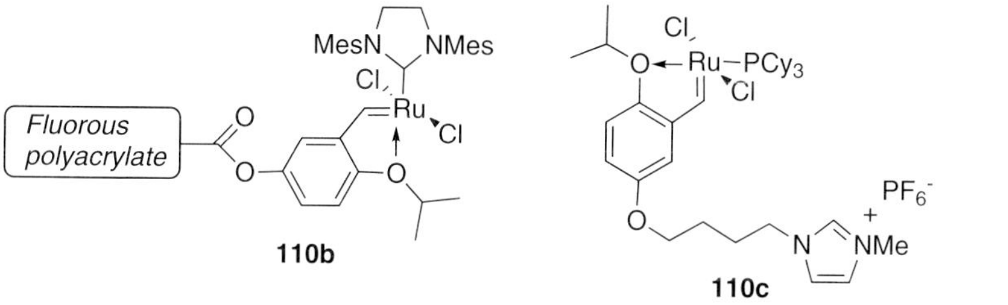

solvent: [Bmim]PF_6/CH_2Cl_2 (1:9 v/v, 10 mL)

[Bmim] 1-Butyl-3-methylimidazolium Hexafluorophosphate

Scheme 11.42 Polymer-supported ruthenium catalyst.

A chiral polymer-bound metathesis catalyst has been developed by Hoveyda et al. The supported chiral complex **111** showed appreciable levels of reactivity and excellent enantioselectivity (Scheme 11.43). This complex **111** could be recycled and easily removed from unpurified mixtures. In the first and second cycles of the recycle experiment, almost the same reactivity was shown. In the third cycle, high enantioselection and conversion were still obtained, but catalyst activity was notably diminished [69].

11.2.5
Synthesis of Natural Products Using RCM

RCM has had great influence on natural product synthesis since the cyclization occurs between dienes to form carbo- and hetero-cyclic alkenes easily. Because the double bond can be easily converted into the carbon–carbon single bond, retrosynthetic analysis is dramatically changed compared with the previous strategy.

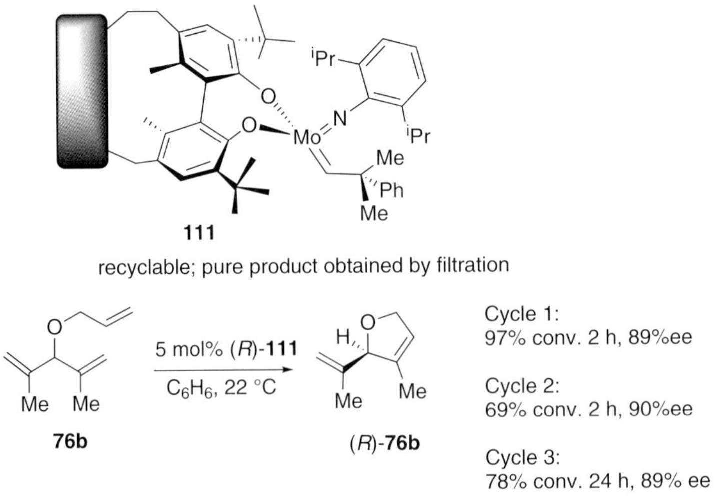

Scheme 11.43 Recyclable catalyst for ARCM.

Cycle 1: 97% conv. 2 h, 89%ee
Cycle 2: 69% conv. 2 h, 90%ee
Cycle 3: 78% conv. 24 h, 89% ee

Although there are many examples for the syntheses of natural products using RCM in the recent literature, in this chapter only representative examples are described.

Approaches to the total synthesis of complex natural products using RCM were first considered in 1994. Martin et al. succeeded in the construction of the core 13-membered ring from **112** and the 8-membered ring from diene **114** using RCM in the manzamine synthesis, and they then achieved the total synthesis of manzamine using RCM as the key steps (Scheme 11.44) [70].

Fürstner reported the concise, flexible, and high-yielding synthesis of the macrocyclic compound **117a**, which served as a key intermediate in the total

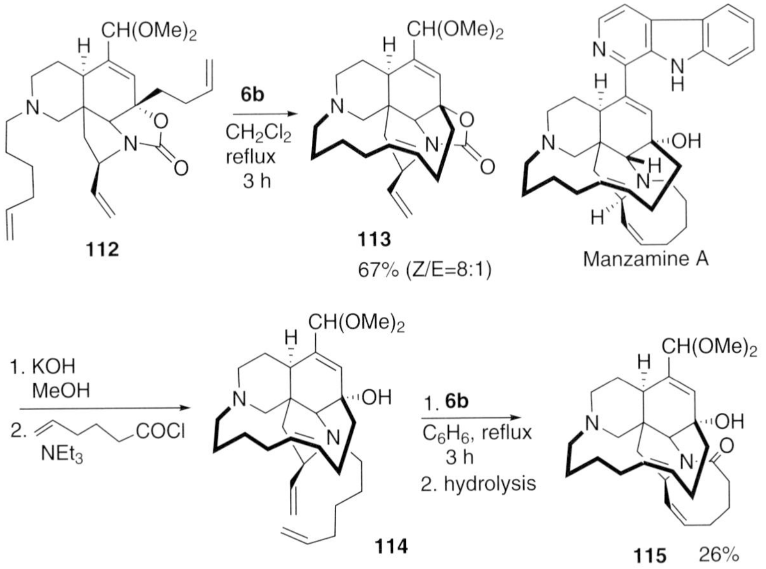

Scheme 11.44 Total synthesis of manzamine.

synthesis of alkaloid roseophilin with antitumor activity (Scheme 11.45). RCM for the formation of the 13-membered ring was catalyzed by the first generation ruthenium catalyst **6b**. Cyclization of diene **116b** required substantial amounts of catalyst **6b** (30 mol%) [71].

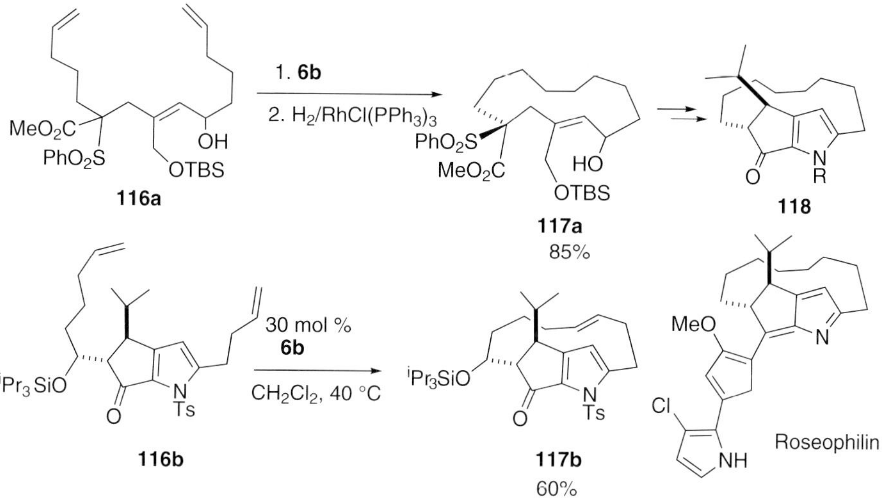

Scheme 11.45 Synthesis of roseophilin.

In the total synthesis of pinnatoxin A, a macro-RCM reaction of **119** was enabled by the action of Grubbs catalyst **6f**, resulting in the formation of the 27-membered carbocycle **120** with the E-olefin (75% yield) (Scheme 11.46) [72].

In the synthesis of nakadomarin A, for formation of the 8-membered ring, the alkyne in **121** was protected by a cobalt complex and RCM of **122** proceeded smoothly to give **123** in 83% yield. In the last stage of this total synthesis, the formation of the 16-membered ring F was again built by use of RCM of **124** to afford compound **125** in 72% yield ($E/Z = 1$ to 1.8) (Scheme 11.47) [73].

Total synthesis of kendomycin has been achieved by Smith III. RCM of **126** having a substituted alkene proceeded smoothly to afford **127** having a Z-olefin in good yield. Conversion of the Z–C=C bond in **127** to E-olefin followed by several steps led to the total synthesis of (−)-kendomycin (Scheme 11.48) [74].

The convergent total synthesis of brevetoxin B has been achieved by Yamamoto. RCM of **128** with the Grubbs catalyst **6f** formed the E-ring of the hexacyclic ether **129**, which was converted into the A-G ring segment through several steps. Furthermore, RCM of **130** provided the H-ring of the polycyclic ether framework **131**. A series of reactions of **131**, including oxidation of the A ring, deprotection of the silyl ethers, and selective oxidation of the resulting allylic alcohol, furnished brevetoxin B (Scheme 11.49) [75].

In Nicolaou's synthesis of abyssomicin C, the key step, RCM of compound **134**, proceeded smoothly using Grubbs catalyst **6f** in refluxing CH_2Cl_2 to afford a mixture of diastereomeric products **135** (about 3:2), in which the newly formed C=C bond is *trans*-configured (Scheme 11.50) [76].

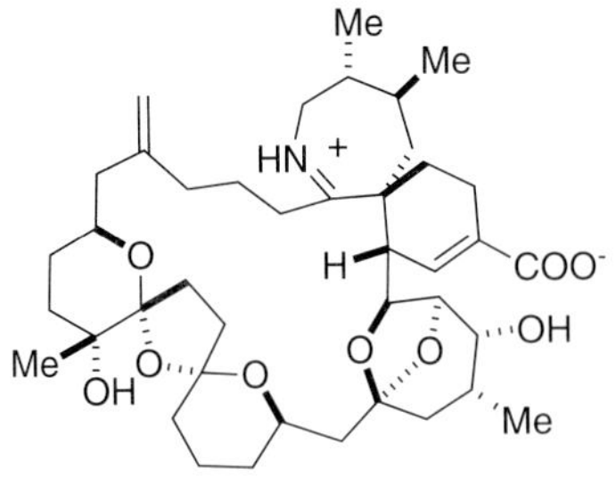

pinnatoxin A

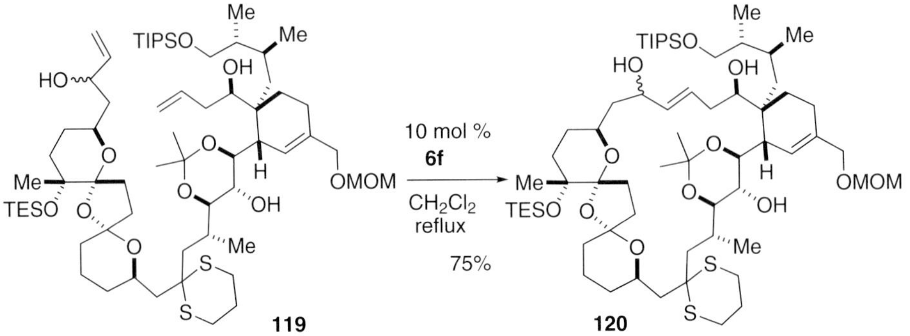

Scheme 11.46 Formal total synthesis of pinnatoxin.

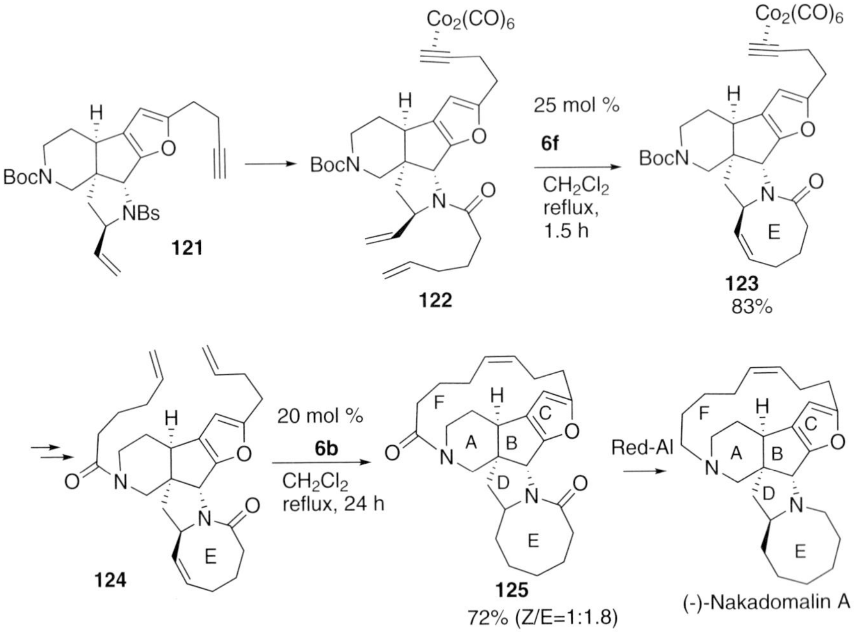

Scheme 11.47 Total synthesis of nakadomalin A.

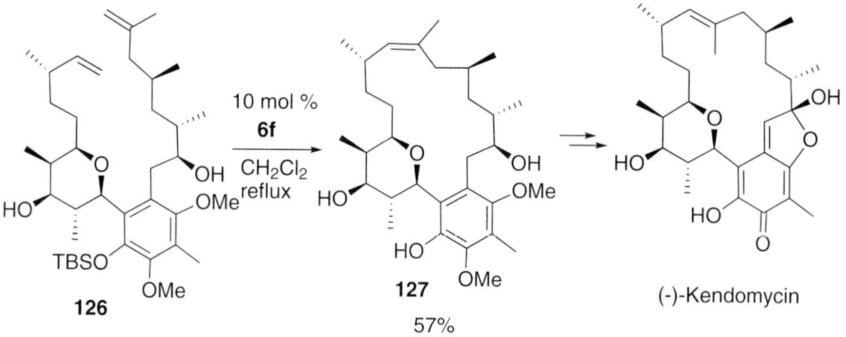

Scheme 11.48 Total synthesis of (−)-kendomycin.

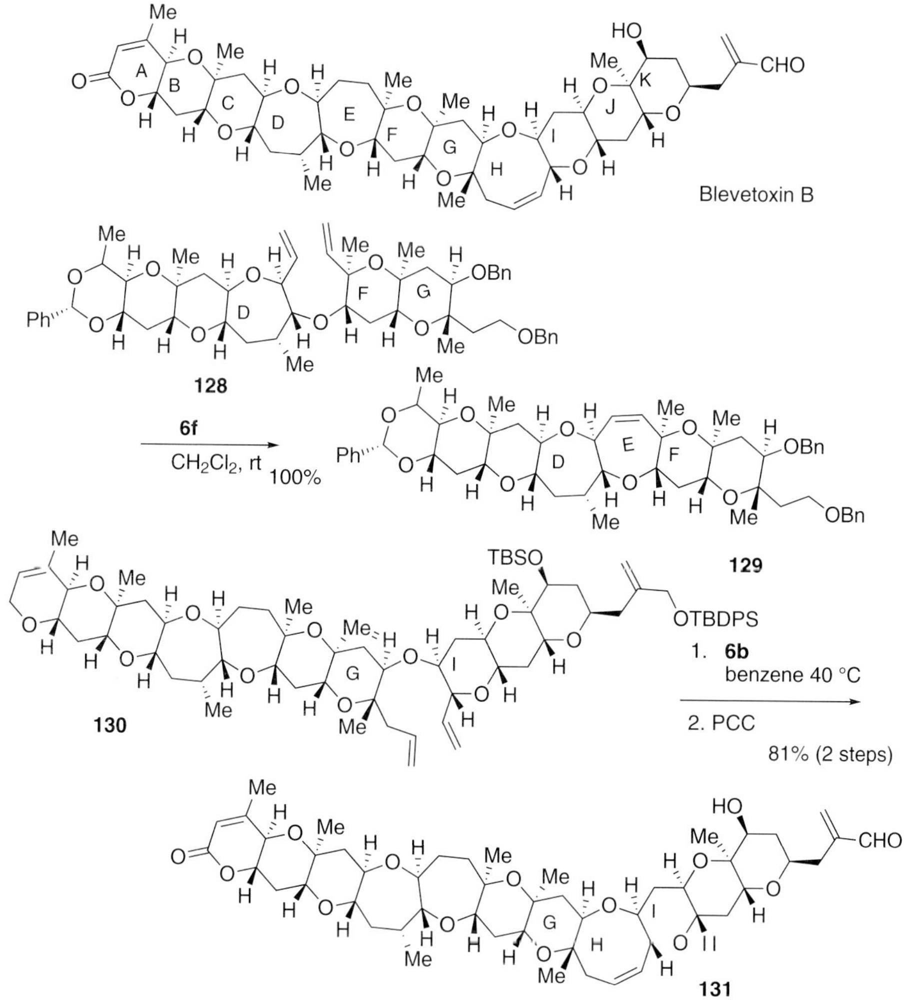

Scheme 11.49 Total synthesis of blevetoxin B.

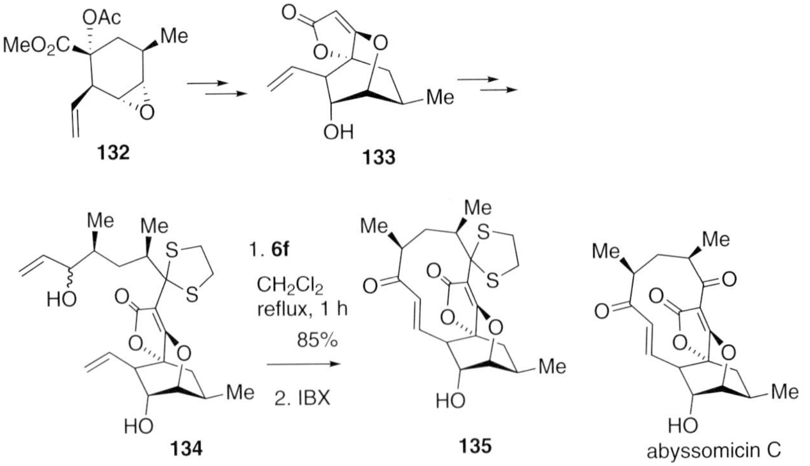

Scheme 11.50 Total synthesis of abyssomicin C.

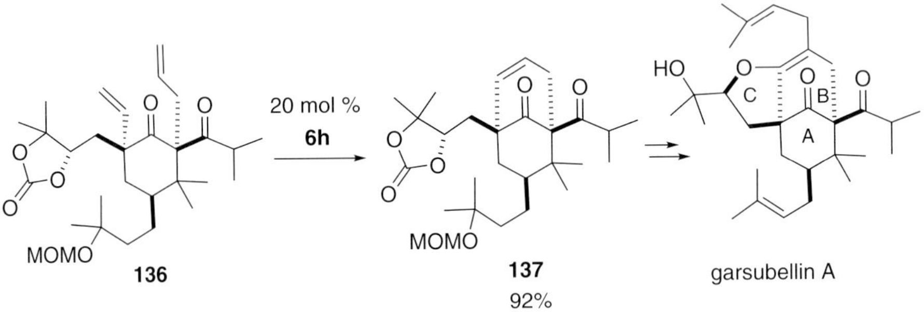

Scheme 11.51 Total synthesis of garsubellin A.

Garsubellin A was synthesized by Shibasaki by applying RCM of **136** with the Grubbs–Hoveyda catalyst **6h** to give **137** in 92% yield (Scheme 11.51) [77].

Fürstner succeeded in the total synthesis of ipomoeassin B. Exposure of diene **138** to catalytic amounts of the ruthenium carbene **6f** in refluxing CH_2Cl_2 afforded macrocycle **139** in excellent yield as a mixture of both geometrical isomers, which could be selectively hydrogenated with the aid of $RhCl(PPh_3)_3$ without affecting the lateral unsaturated esters (Scheme 11.52) [78].

11.3
Ring-Closing Metathesis of Enynes

11.3.1
Synthesis of Carbo- and Hetero-cycles Using Ring-Closing Metathesis of Enynes

Enyne metathesis was developed by Kats in 1985 using the Fischer tungsten carbene complex [10]. Then Mori reported the chromium-catalyzed enyne metath-

Scheme 11.52 Total synthesis of ipomoeassin B.

esis [11]. It was found that the ruthenium carbene complex **6a** or **6b** developed for olefin metathesis by Grubbs could catalyze RCM of enynes. Using this catalyst **6a**, Mori synthesized five- to nine-membered ring compounds **141** with a diene moiety from enynes **140** (Scheme 11.53) [12]. In each case, the yields are high. A possible reaction course is shown in Figure 11.6. The real catalytic species **6l** is generated by the reaction of an alkyne part in **140** and the carbene complex **6b**, as shown in

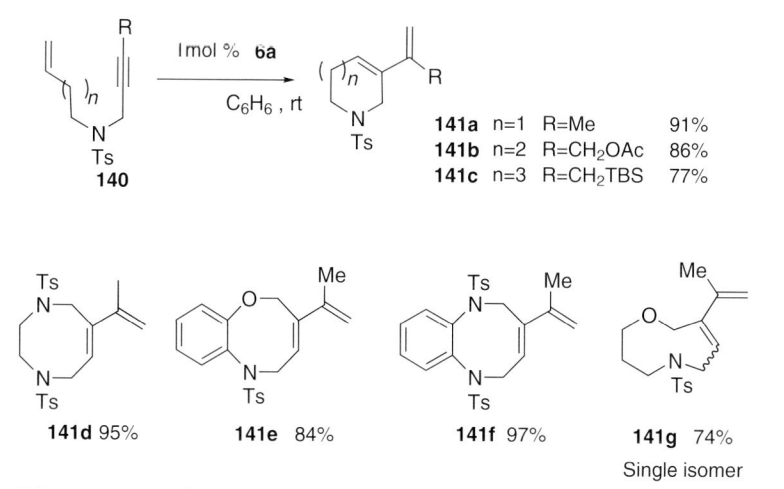

141a n=1 R=Me   91%
141b n=2 R=CH$_2$OAc 86%
141c n=3 R=CH$_2$TBS 77%

141d 95%

141e 84%

141f 97%

141g 74%
Single isomer

Scheme 11.53 Synthesis of heterocycles using ruthenium catalyst.

11 Ring-Closing Metathesis of Dienes and Enynes

Path a: Reaction of Alkyne with **6b**

Path b: Reaction of Alkene with **6b**

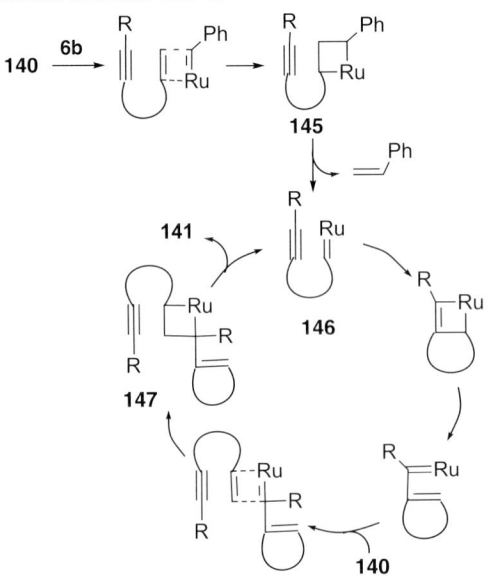

Figure 11.6 Possible reaction course.

Figure 11.6. The [2 + 2] cycloaddition of **6l** and **140** occurs to give ruthenacyclobutene **142** and ring opening of **142** gives the ruthenium carbene complex **143**, which reacts with an alkene part intramolecularly to give ruthenacyclobutane **144**. Subsequent ring opening affords cyclic compound **141** and the ruthenium carbene complex **6l** is regenerated (Figure 11.6, path a). The alternative mechanism (path b) is that ruthenium carbene **6b** reacts first with an alkene part of enyne **140** to produce ruthenacyclobutane **145**. Ring opening of **145** affords the ruthenium carbene complex **146**, which reacts with the alkyne part intramolecularly to give ruthenacyclobutene. Ring opening of this followed by the reaction of the generated ruthenium carbene and the alkene part of **140** affords ruthenacyclobutane **147**, whose ring opening gives **141** and ruthenium carbene complex **146** is regenerated. Later, a detailed study on the reaction mechanism was performed by Staub, who described that the reaction would proceed via path b and ruthenacyclobutene **142** generated from an alkyne part of enyne **140** and **6l** does not exist as a local minimum in the catalytic cycle [79].

In this reaction, enynes having a terminal alkyne do not give a satisfactory result. For example, RCM of enyne **148** afforded the cyclic compound **149** in only 21% yield. It is reasoned that an alkene part in product **149** further reacts with **6l** to afford ruthenium carbene **151**, which would be coordinated by the alkene part to produce **152**. Thus, the catalytic activity would decrease (Scheme 11.54). In fact, when the reaction of **148** was carried out under ethylene gas, the catalytic activity was much higher, affording **149** in 90% yield, even with the use of 1 mol% of the ruthenium catalyst **6b** [13].

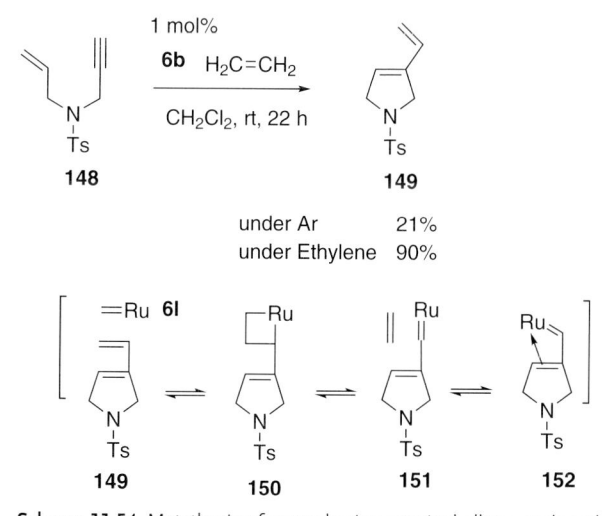

Scheme 11.54 Metathesis of enyne having terminal alkyne under ethylene gas.

(±)-Differolide could be easily synthesized by RCM of enyne **153** under the catalysis of **6b** to give **154** and the spontaneous dimerization of the product (Scheme 11.55) [80].

Isoquinoline derivative **157** was synthesized by enyne metathesis of amino acid derivative **155** followed by Diels–Alder reaction of the resultant diene **156** with dimethylacetylene dicarboxylate (DMAD) and then DDQ oxidation (Scheme 11.56) [81].

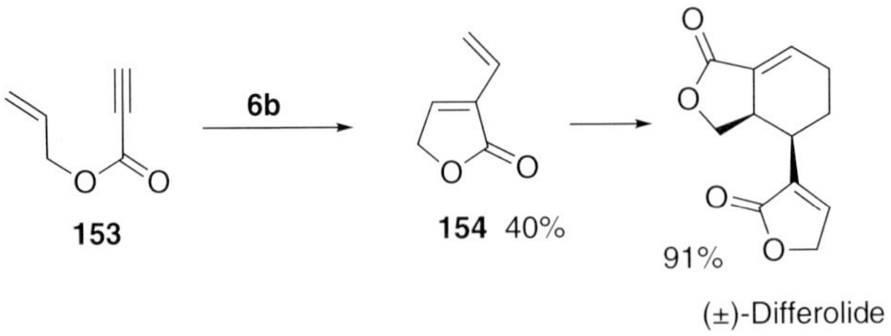

Scheme 11.55 Synthesis of (±)-differolide.

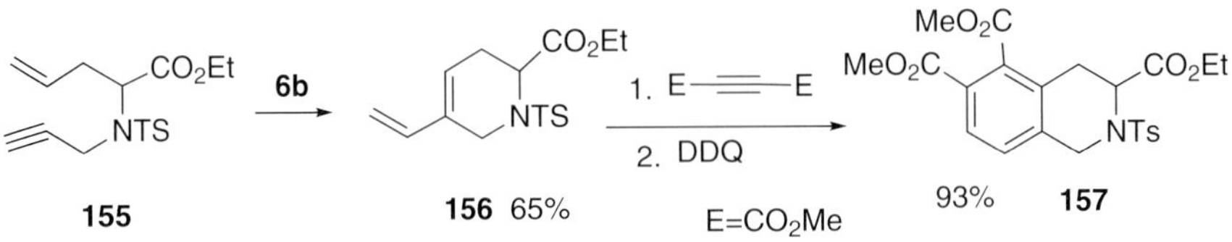

Scheme 11.56 Synthesis of isoquinoline derivative.

When enyne **158a** or **158b** was treated with 5 mol% of **6e**, the expected metathesis product **159a** or **158b** and the six-membered compound **160a** or **160b** were formed. Use of another second-generation ruthenium carbene complex **6f** gave a similar result. Presumably, when the ruthenium carbene complex **6l** reacts with the alkyne moiety in **158**, two regiochemically different pathways are possible (Scheme 11.57). Each gives two different products, **159** or **160**. However, it is not clear why the two products are formed when the second-generation ruthenium carbene complexes are used.

Carbacephems **162** and **164b** were synthesized from enynes **161** and **163b**, which were prepared from 4-acetoxyazetidinone, in high yields. It is interesting that carbapenem **164a** has been synthesized also in high yield by applying the same strategy when the second-generation ruthenium catalyst **6f** was used, though it is a highly strained structure. Diels–Alder reaction of **164a** with DMAD gave tricyclic compounds **165a** and **165a′** as a mixture of two isomers in a ratio of 1 to 2.8 (Scheme 11.58) [83].

Double RCM of the diene and enyne of **166** was reported by Ma, in which the quinolizidine derivative **167** was obtained in 75% yield. Diels–Alder reaction of **167** with maleic imide gave the tetracyclic compound **168** (Scheme 11.59) [84].

Macrocyclic compounds **170** or **171** were readily synthesized by RCM of the enyne of **169** (Scheme 11.60). In the formation of the macrocyclic compound, 10-membered and smaller ring compounds **170** formed from enyne **169** by RCM have invariably the exo-form, whereas the macrocyclic compounds **171** having 12-membered or larger rings have endo-form [85].

A new annulation reaction of unsaturated boronic esters with propargylic alcohol has been reported (Scheme 11.61). The reaction of propargyl alcohol and **172a** in the presence of **6b** gave the cyclic boronic ester **173a**. Transesterification of **172a** and propargyl alcohol is considered to afford a mixed organoboronic ester

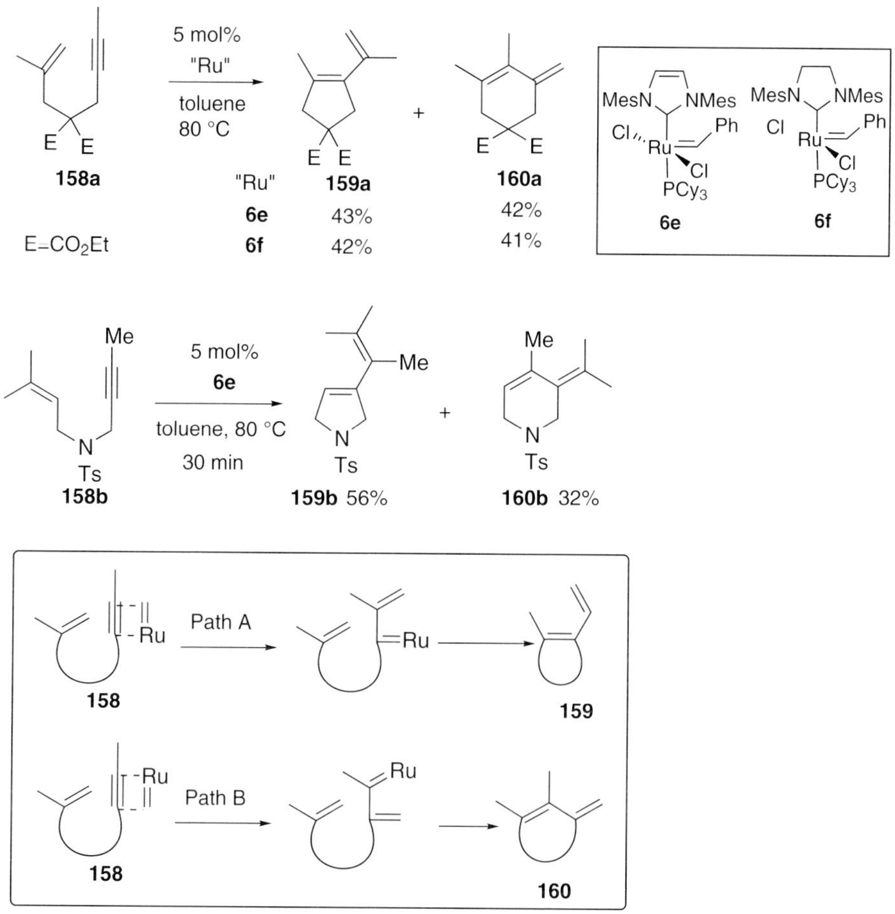

Scheme 11.57 Enyne metathesis using second generation ruthenium catalysts.

175a, which could undergo ring-closing enyne metathesis. Treatment of the cyclic boronic ester **172a** with H_2O_2 in aq. NaOH gave diol **174a** [86a]. Alkynylboronic ester **172b** reacted with homoallylic alcohol to finally provide functionalized 1,3-dienyl boronic acid **173b**, which was converted into ketone **174b**. A mixed organoboronic ester **175b** should be formed as an intermediate [86b].

A highly functionalized conjugated diene **178a** was synthesized through sequential silicon-tethered ring-closing enyne metathesis of silicon-tethers **176a** by the ruthenium carbene catalyst **6b** followed by Tamao oxidation [87a]. The metal-catalyzed dehydrogenative condensation between alcohol and silane using ruthenium catalyst, generating molecular hydrogen as the only by-product, allowed the subsequent enyne metathesis without isolating the intermediate silyl ethers **176b–d** to produce five-, six- and seven-membered cyclic silyl ethers **177b**, **177c**, and **177d** (Scheme 11.62) [87b].

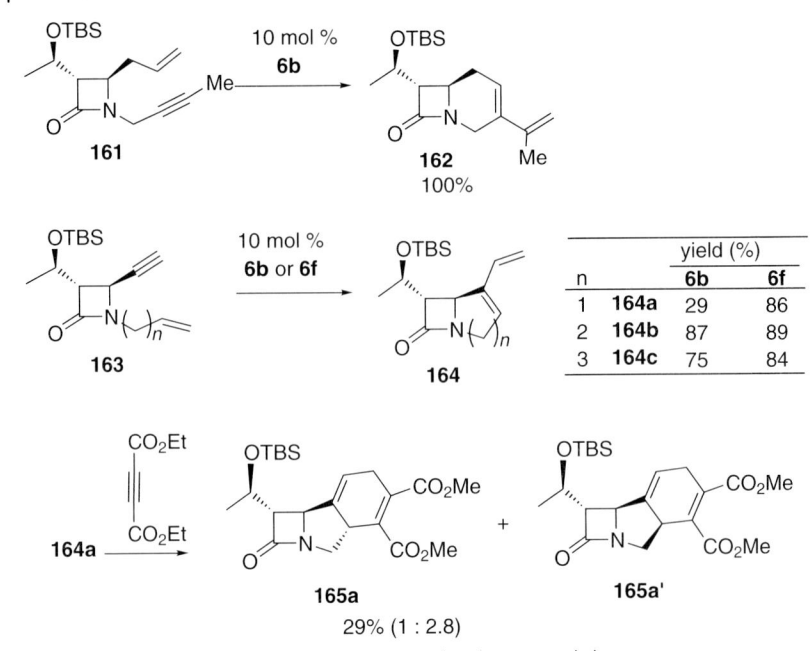

Scheme 11.58 Construction of carbacephem and carbapenem skeleton.

n		yield (%)	
		6b	6f
1	164a	29	86
2	164b	87	89
3	164c	75	84

165a + 165a'
29% (1 : 2.8)

Scheme 11.59 Double cyclization of diene and enyne.

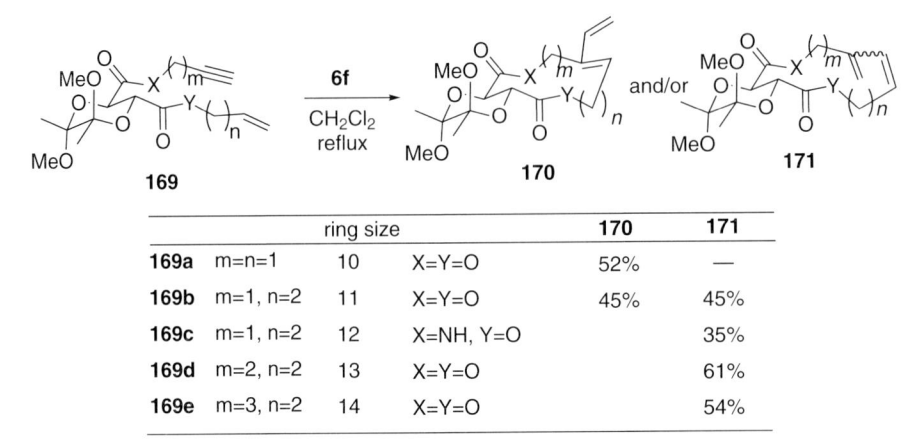

		ring size		170	171
169a	m=n=1	10	X=Y=O	52%	—
169b	m=1, n=2	11	X=Y=O	45%	45%
169c	m=1, n=2	12	X=NH, Y=O		35%
169d	m=2, n=2	13	X=Y=O		61%
169e	m=3, n=2	14	X=Y=O		54%

Scheme 11.60 Synthesis of macrocyclic compounds.

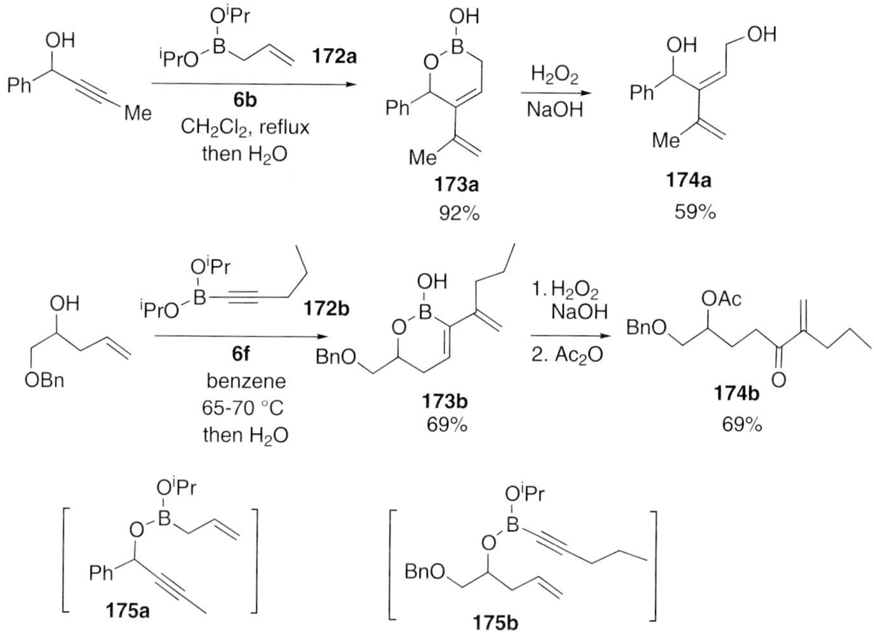

11.3 Ring-Closing Metathesis of Enynes | 569

Scheme 11.61 Boronic ester and alkynyl boric ester annulations.

Enyne **179a**, having a silyloxy group on the alkyne, gave cyclic compound **180a** having a vinyl silyloxy moiety, which was converted into methyl ketone **181a** by desilylation. In a similar manner, enyne **179b** afforded bicyclic methyl ketone **181b** in 68% yield after deprotection of the silyl group [88]. However,

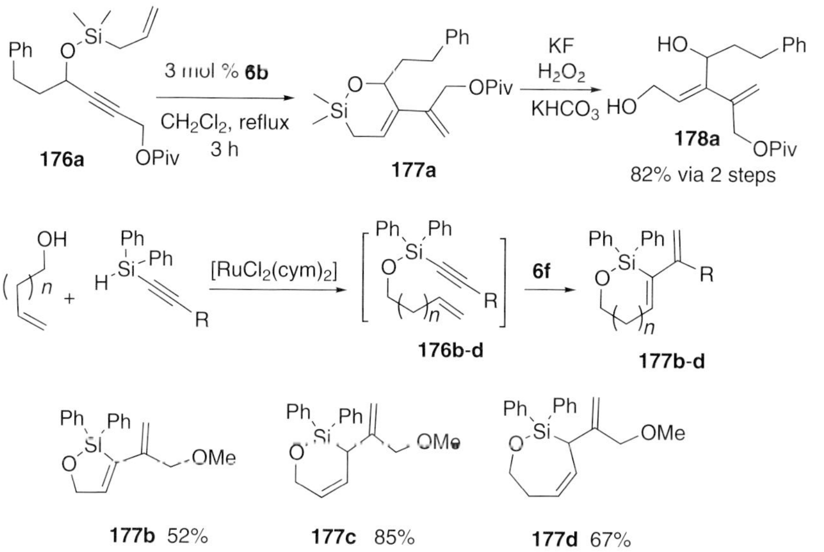

Scheme 11.62 Synthesis of cyclic silyl ether.

ynoate **179c** and yne-phosphonate **179d** did not give cyclized compounds. Ene-alkynyl ether **182a** or **182b** afforded cyclic enol ethers **183a** or **183b** in good to moderate yields (Scheme 11.63) [89].

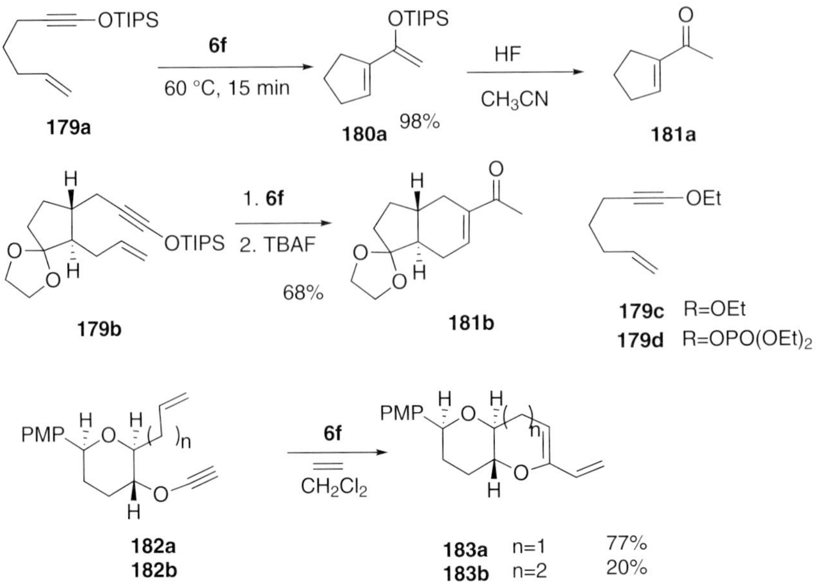

Scheme 11.63 Metathesis of ene-alkynyl ether.

RCM of ene-alkynyl amide **184a** using **6f** gave the pyrrolidine derivative **185a**, which afforded the indole derivative **186a** by Diels–Alder reaction. In a similar manner, one-pot RCM of ene-alkynyl amide **184b**, a one carbon-elongated homolog, followed by Diels–Alder reaction, gave the quinoline derivative **186b** in high yield (Scheme 11.64) [90].

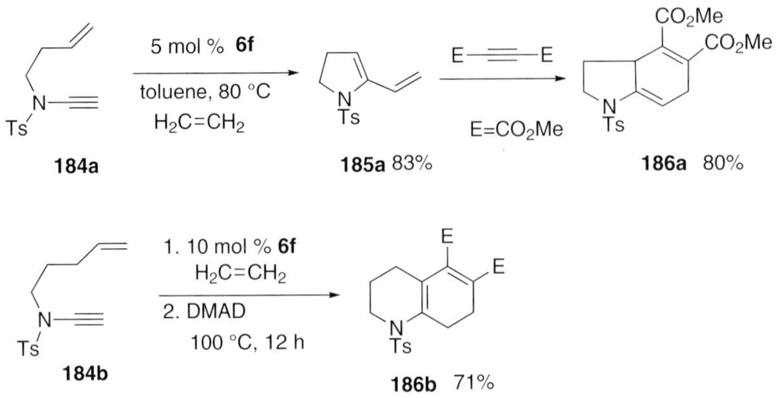

Scheme 11.64 Ring-closing metathesis of ene-alkynyl amide.

11.3 Ring-Closing Metathesis of Enynes

A one-pot RCM–CM reaction was carried out by Royer *et al.* (Scheme 11.65). The RCM of **187** in the presence of methyl acrylate gave cyclic compound **188** using the ruthenium carbene complex **6h**. The reaction proceeds via the intermediacy of **189** and subsequent CM with methyl acrylate to produce **188** (Scheme 11.65) [91].

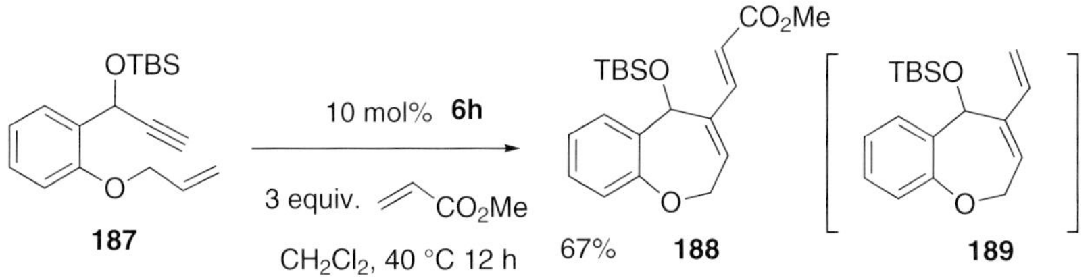

Scheme 11.65 Synthesis of benzoxepine by RCM-CM.

Metathesis of ene-enyne **190** using the ruthenium catalyst **6i** gave the cycloheptene derivative **191** having a triene unit. This reaction did not proceed with the first or second-generation ruthenium catalysts **6b** or **6f** (Scheme 11.66) [92].

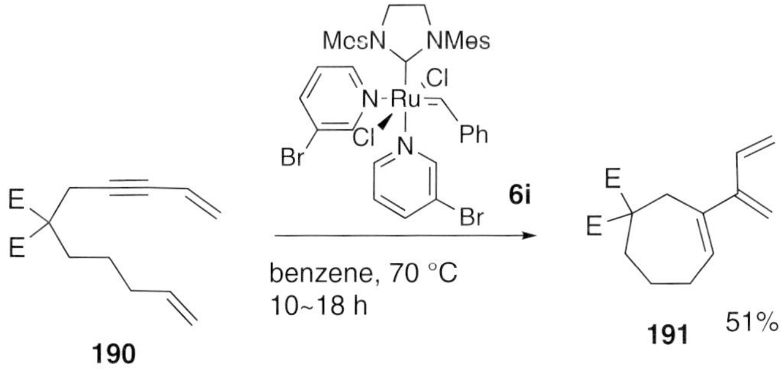

Scheme 11.66 Reaction of enyne containing conjugate ene-yne.

Metathesis of allene-yne **192** using molybdenum catalyst **1** gave the cyclic compound **193** having an allene moiety. RCM of **192-D** gave **193-D**, which indicates that the vinylallene skeleton was constructed by a metathesis-type reaction between the alkyne moiety and the proximal carbon–carbon double bond of the allene moiety (Scheme 11.67) [93].

The bond reorganization processes, defined as metallotropic shift, of various alkynyl carbene complexes with Rh [94a], Mn [94b], Re [94c], Cr [94d], Mo [94d], and W [94d] metals have been reported. The rearrangement involving Rh, Cr, Mo, and W is a [1,3]-shift, while that with Mn and Re is formally defined as a [1,1,5]-shift. However, the metallotropic shift of ruthenium alkynyl carbene complexes has not been observed until recently. The metallotropic [1,3]-shift of a transient ruthenium carbene complex is involved in the enyne RCM of diyne-containing substrates. On the basis of this concept, one-step construction of enediynes and oligoenynes was realized by the uniquely controlled repetitive metallotropic [1,3]-shift of ruthenium

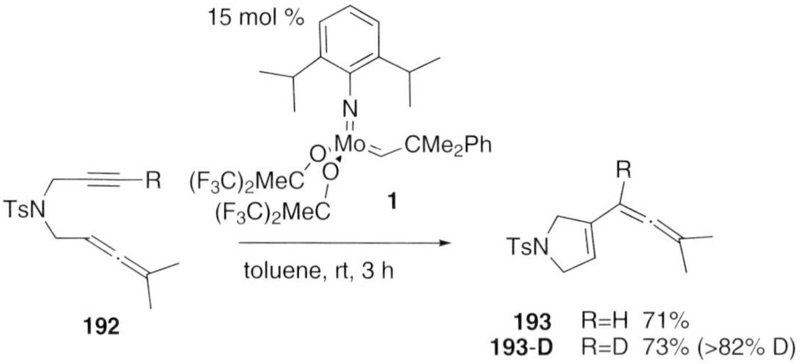

Scheme 11.67 RCM of allene-yne.

carbene species. Reaction of **194a** with **6f** gave the ruthenium carbene complex **195a**, which was converted into the ruthenium carbene complex **196a** via [1,3]-shift. RCM followed by a [1,3]-shift and then RCM gave the ene-diyne **197a** in high yield, as shown in Scheme 11.68. In a similar manner, **194b** and **194c** gave oligoenynes **197b** and **197c**, respectively, in one operation [94e].

A tandem ring-closing enyne metathesis/cyclopropanation reaction has been developed. 1,3-Dienes formed in a ring-closing enyne metathesis of **198a** or **198b**

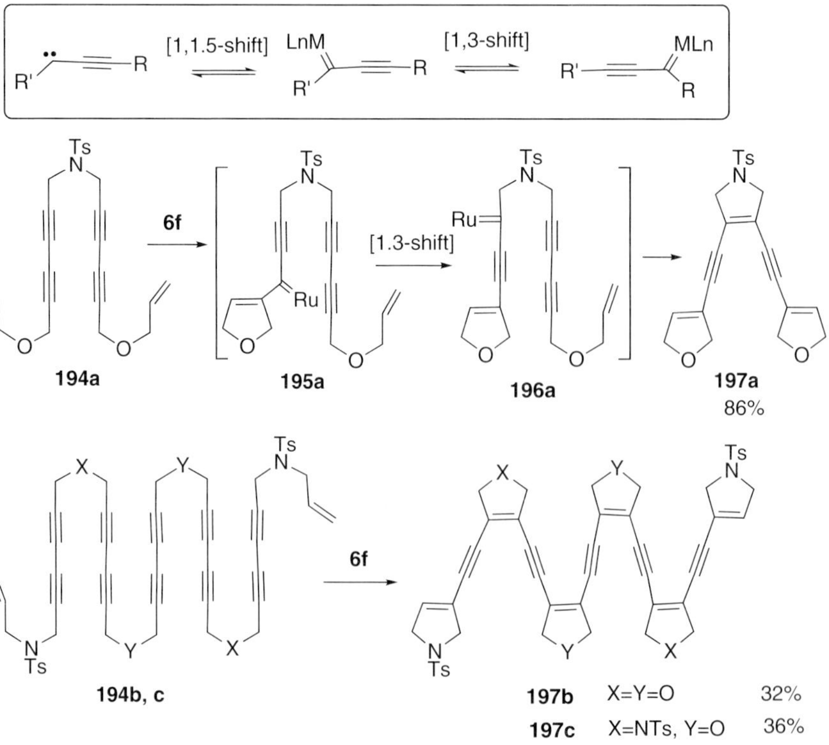

Scheme 11.68 Synthesis of oligoenynes by metathesis and metallotropy.

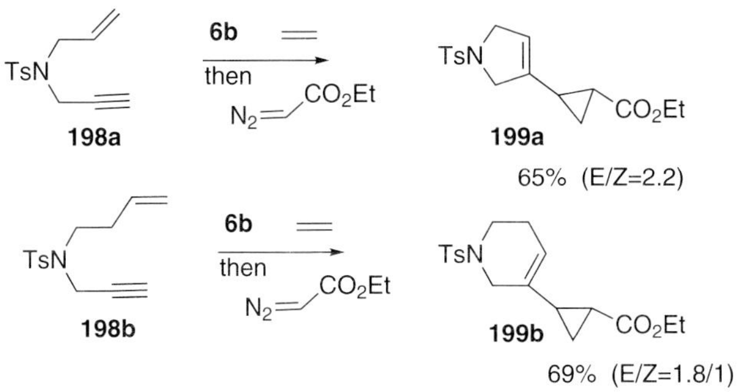

11.3 Ring-Closing Metathesis of Enynes

Scheme 11.69 Preparation of alkenyl cyclopropanes.

with Grubbs' ruthenium catalyst **6b** could be cyclopropanated to give **199a** or **199b** when diazoacetate was added to the reaction mixture at elevated temperatures (Scheme 11.69) [95].

It is generally considered that three- and four-membered cycles cannot be formed by RCM because such strained cycles may be reopened by the carbene catalyst with the driving force being the relief of ring strain. Various attempts for RCM of enyne **200** were made and the use of microwave irradiation proved beneficial, and the desired cyclobutene **201** was obtained in 58% yield (Scheme 11.70) [96].

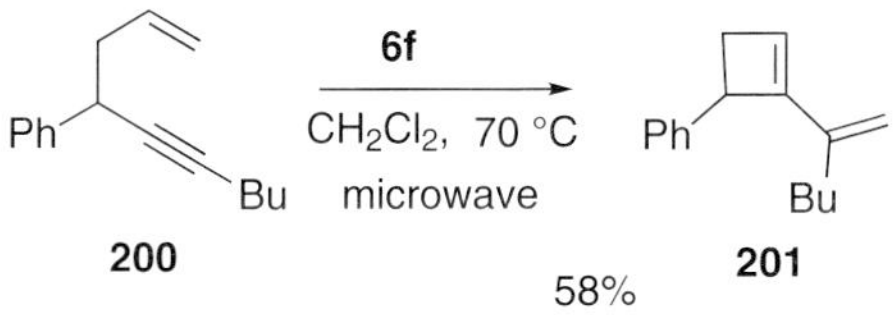

Scheme 11.70 1,5-Enyne metathesis.

11.3.2
Ring-Closing Metathesis of Dienynes

Grubbs reported an ingenious method for synthesizing bicyclic compounds from dienynes taking advantage of the metathesis reaction (Scheme 11.71). When a benzene solution of dienyne **202a** was stirred in the presence of 3 mol% of **6a**, the bicyclic compound **203a** was obtained in 95% yield in one operation. In the case of **202b**, two bicyclic compounds, **203b** and **204b**, were formed. Furthermore, dienyne **202c** gave the tricyclic compound **203c** in quantitative yield. Probably, the ruthenium carbene complex **6l** reacts with an alkene moiety in dienyne **202** to give the ruthenium carbene complex **205**, which reacts with an alkyne part to give ruthenacyclobutene **206**. Ring opening of this complex **206** gives the ruthenium carbene complex **207**, which reacts with an alkene part intramolecularly to give ruthenacyclobutane. Ring opening of this complex gives bicyclic compounds, **203** and/or **204** [97].

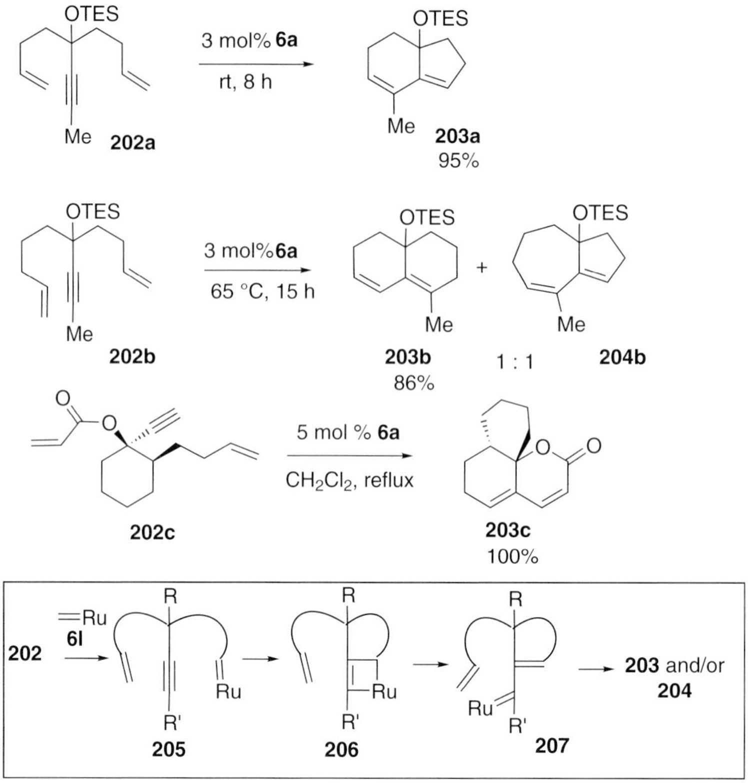

Scheme 11.71 Dienyne metathesis.

Grubbs demonstrated the synthesis of various fused polycyclic compounds using dienyne metathesis. Compound **209**, having a steroidal skeleton, could be synthesized from polyenyne **208**, whose double and triple bonds were placed at the appropriate positions in a carbon chain, in high yield in one operation, although the reaction includes many steps, as shown in Scheme 11.72 [98].

A new approach to the construction of a linearly fused 6-8-6 tricarbocyclic ring system was realized using dienyne metathesis. This ring system is a carbon framework analogous to the proposed transition state of isomerization of previtamin D3 to vitamin D3. The starting dienyne **211** prepared from indenone **210** reacted with **6b** to afford the target molecule **212** as a diastereomeric mixture at the C-10 position in 48% yield (Scheme 11.73) [99].

Synthesis of a polyoxygenated bicyclic compound **214** containing a medium-sized ring was achieved via tandem metathesis of dienyne **213** derived from D-ribose (Scheme 11.74) [100].

The transformation of dienyne **215** into the bicyclic vinylboronate **216** using **6b** was shown. The crude alkynylboronate **215** cleanly underwent ruthenium-promoted metathesis in 70% yield. The resultant bicyclic dialkenylboronic ester **216** was efficiently oxidized to the corresponding enoate **217** by treatment with Me$_3$NO in

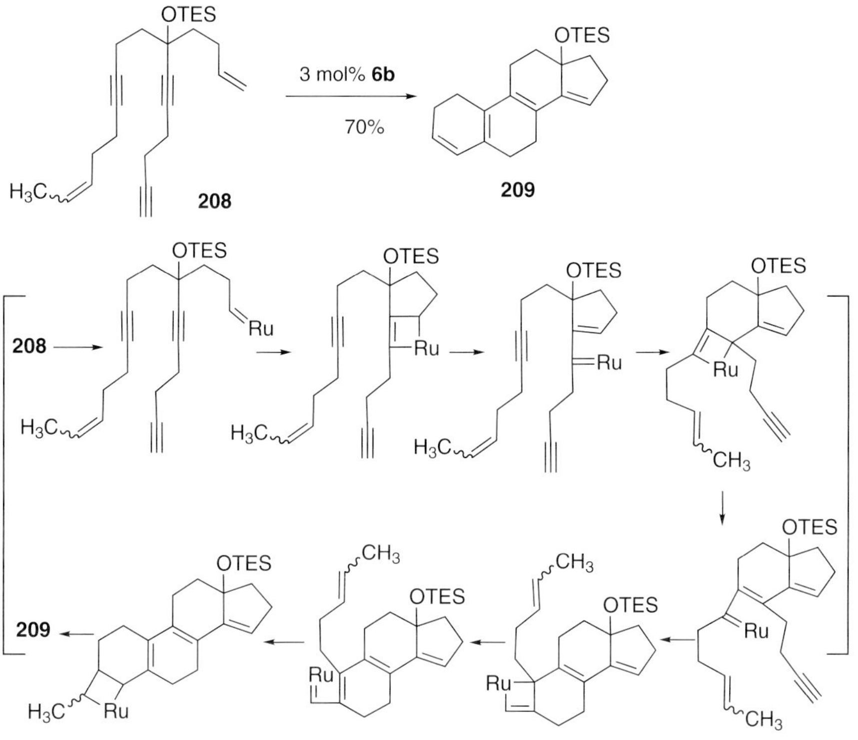

Scheme 11.72 Construction of a steroidal skeleton.

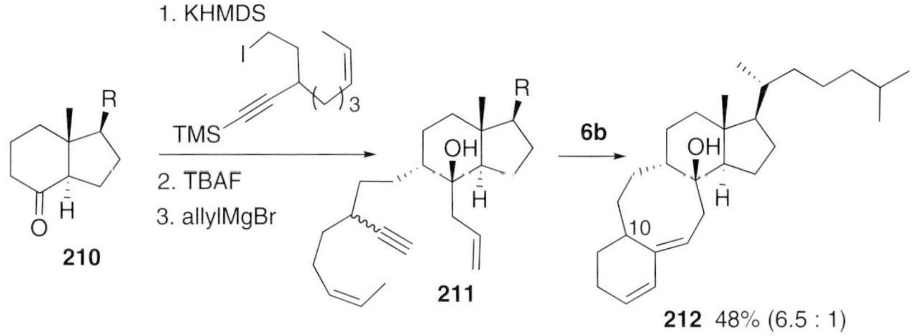

Scheme 11.73 Construction of 6-8-6 tricyclic ring system using dienyne metathesis.

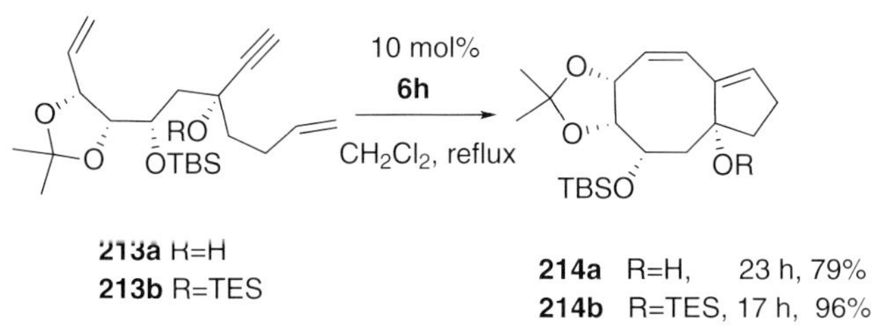

213a R=H
213b R=TES

214a R=H, 23 h, 79%
214b R=TES, 17 h, 96%

Scheme 11.74 Synthesis of bicyclic compounds.

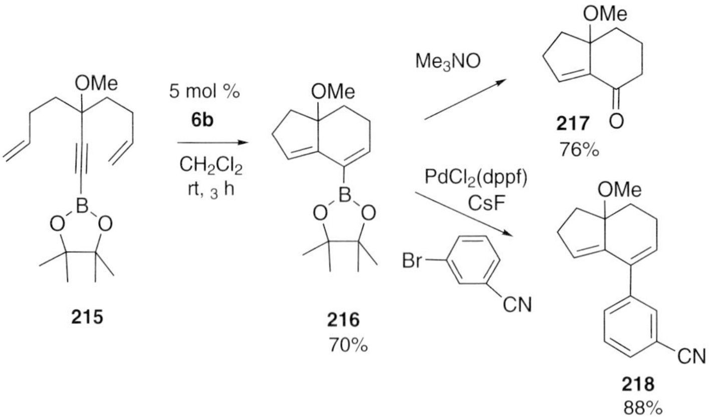

11 Ring-Closing Metathesis of Dienes and Enynes

Scheme 11.75 Reaction of enyne having alkenyl boronate.

refluxing THF. In the presence of CsF, reaction of **216** with 3-bromobenzonitrile under the catalysis of PdCl$_2$(dppf) furnished the cross-coupling product **218** (Scheme 11.75) [101].

A versatile route for the synthesis of a phosphorus oxide template was presented (Scheme 11.76). Ring-closing enyne metathesis using **6f** on substrates **219a** and **219b** led to the formation of mono- and bicyclic phosphorus heterocycles **220a** and **220b** [102].

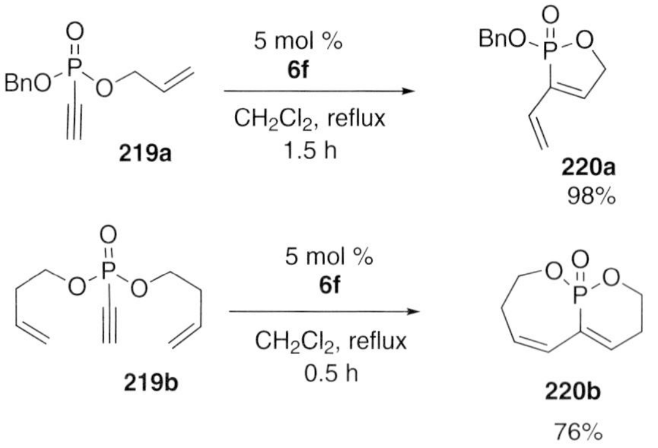

Scheme 11.76 Synthesis of phosphorus mono- and bi-cycles by RCM.

Base-promoted isomerization of propargyl amide **221a** gave the chiral alkynyl amide **222a**, which was subjected to RCM to afford the cyclic enamide **223a**. In a similar manner, propargyl amide **221b** was converted into the alkynyl amide **222b**, RCM of which gave bicyclic compounds **223b** and **223b′** in a ratio of 1 to 1 (Scheme 11.77) [103].

Dienyne metathesis of β-carboline derivative **225** afforded the pentacyclic compound **226** related to alkaloids containing a β-carboline unit. The starting material **224** was readily synthesized from tryptamine (Scheme 11.78) [104].

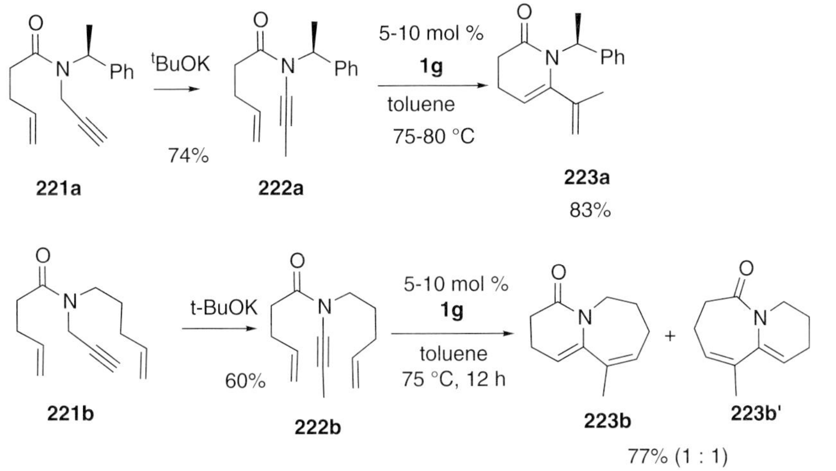

Scheme 11.77 Base-promoted isomerization of propargyl amide followed by RCM.

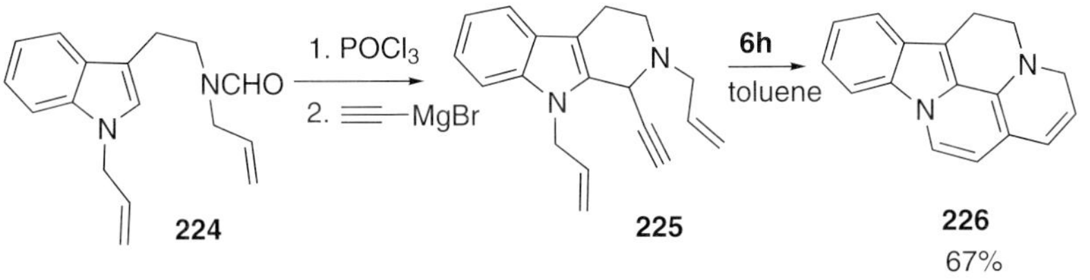

Scheme 11.78 Dienyne metathesis of β-carboline derivative.

Total synthesis of (±)-erythrocarine was achieved by Mori using dienyne metathesis. Regio- and stereo-selective insertion of carbon dioxide and an alkynyl group into the alkyne of **227** and hetero-Michael reaction gave the isoquinoline derivative, which was converted into dienyne **228**. Since the tertiary amine of **228** coordinates with the ruthenium catalyst and the catalytic activity is decreased, **228** was converted into **228·HCl** and dienyne metathesis was carried out using the ruthenium catalyst **6b** to afford the tetracyclic compounds **229a** and **229b** as a diastereomeric mixture in quantitative yield in a ratio of 1 to 1. From the α-isomer **229a**, erythrocarine was synthesized (Scheme 11.79) [105]. Hatakeyama succeeded in the total synthesis of erythravine using a similar procedure [106].

Honda *et al.* succeeded in a diastereoselective total synthesis of (−)-securinine in an optically pure form by employing RCM of the corresponding dienyne **230** as a key step (Scheme 11.80). To construct a fused-furan ring, dienyne **230**, which was synthesized from (+)-pipecolinic acid and has one terminal alkene and one internal alkene, was used, because the ruthenium-carbene complex **6j** would at first react with the terminal alkene part to form a furan ring. From **231**, (−)-securinine was synthesized. Honda *et al.* also synthesized viroallosecurinine in a similar manner [107].

Hanna developed a concise route to a key intermediate in the total synthesis of guanacastepene A using dienyne metathesis (Scheme 11.81). The main feature

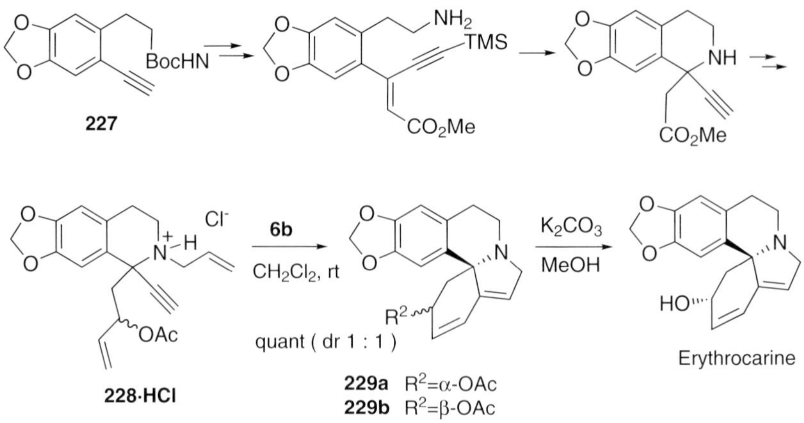

Scheme 11.79 Total synthesis of erythrocarine.

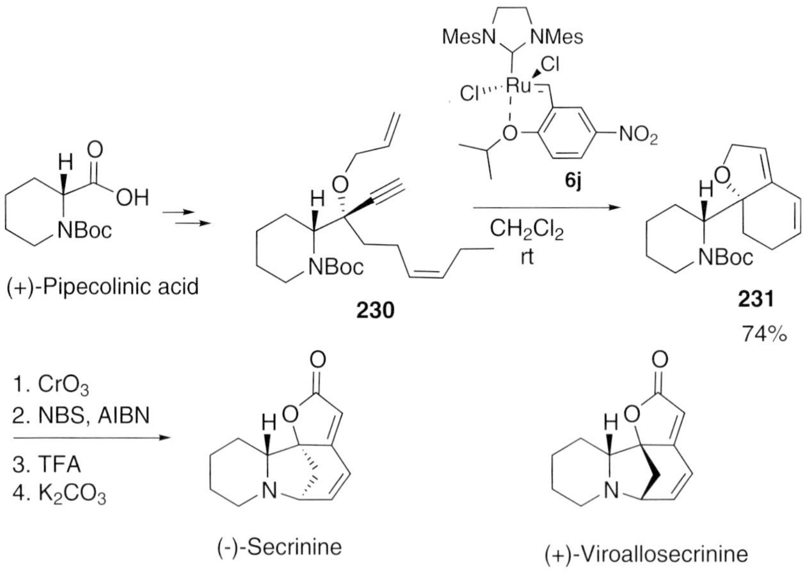

Scheme 11.80 Total synthesis of (−)-securinine.

includes the construction of fused seven- and six-membered rings. From the cyclopentanone derivative, the dienyne **232** was prepared. Dienyne metathesis of **232** was carried out using **6f** to give a mixture of tricyclic compound **233** in a ratio of 1 to 1. After a couple of steps, compound **234** was prepared to finish the formal synthesis of (±)-guanacastepene A [108].

11.3.3
Ring-Opening Metathesis–Ring-Closing Metathesis of Cycloalkene-ynes

Tandem cyclization of enyne, which proceeds via ROM followed by RCM and/or CM, is a very interesting and useful transformation, because different rings are formed

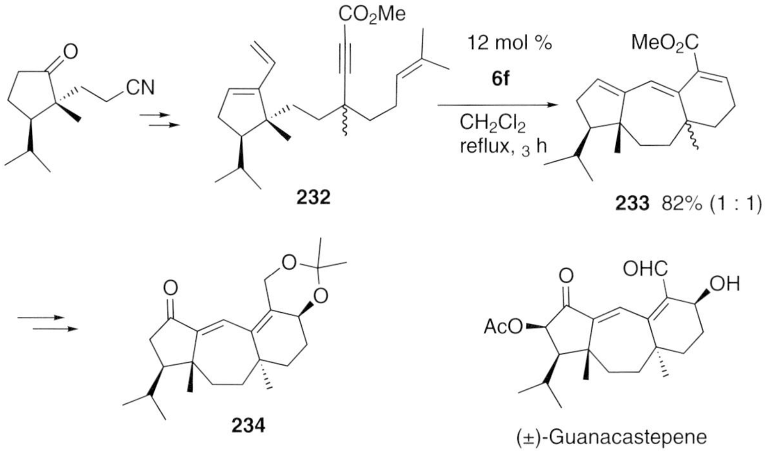

Scheme 11.81 Synthesis of (±)-guanacastepene.

from the starting cycloalkene via many steps by a single operation. Mori reported such tandem cyclization of cycloalkene-ynes. When cycloheptene-yne **235a**, whose alkynylic substituent was placed at the 3-position of cycloheptene, was reacted with the first-generation ruthenium-carbene complex **6b** in CH_2Cl_2 under ethylene gas at room temperature for 24 h, the pyrrolidine derivative **236a** was produced in 56% yield (Table 11.5, entry 1). Similar treatments of various cycloalkene-ynes **235** gave the pyrrolidine derivatives **236** in high yields. In each case, a pyrrolidine ring is formed and the carbon chain length of the substituent of the pyrrolidine ring depends on the ring size of the original cycloalkene (Table 11.5). The reaction is thought to proceed via cycloaddition of an alkyne part of **235** and **6l**. Ring opening of the generated

Table 11.5 ROM-RCM of cycloalkene-yne.

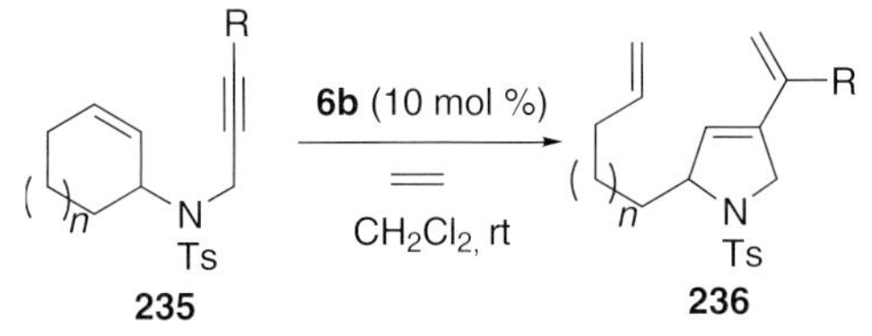

Entry		R	Ring size	n	Time (h)		Yield (%)
1	235a	Me	7	2	24	236a	56
2	235b	H	6	1	4	236b	78
3	235c	H	7	2	1	236c	70
4	235d	H	8	3	1	236d	75

ruthenacyclobutene gives the ruthenium carbene complex **237**. Then cycloaddition of **237** and an alkene part of the cycloalkene ring follows to afford highly strained cyclobutane **238**. Subsequent ring opening of this gives the ruthenium carbene complex **239**, which reacts with ethylene to give the cyclic compound **236** (Scheme 11.82). Formally, the double bonds of ethylene and cycloalkene are cleaved, and the carbon–carbon double bond is formed between the alkyne and cycloalkene carbons, and the two methylene groups derived from ethylene are introduced at the alkene and the alkyne carbons, respectively, to give the pyrrolidine derivative **236** (Figure 11.7). [109].

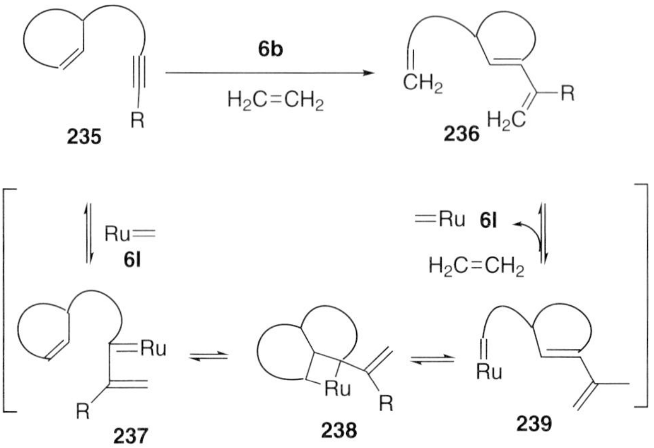

Scheme 11.82 Possible reaction course of ROM–RCM of cycloalkene-yne.

ROM–RCM of N-*trans*-1,4-disubstituted hexenyl-N-propargyl tosyl amide **240a**, having the alkene moieties in a tether, gave the pyrrolidine derivative **241a** in high yield. It is interesting that deprotection of the silyl group followed by the Dess–Martin oxidation gave the tricyclic compound **242** via the Diels–Alder reaction at room temperature. On the other hand, *cis*-disubstituted cycloalkene–yne **240b** afforded selectively triene **243**, which was a CM product of the alkyne part of **240b** and ethylene. Presumably, the ruthenium carbene site cannot approach the double bond in the cyclohexene ring due to steric hindrance and CM with ethylene occurs (Figure 11.8). ROM-RCM of *cis*- or *trans*-disubstituted cyclopentene–yne **240c**

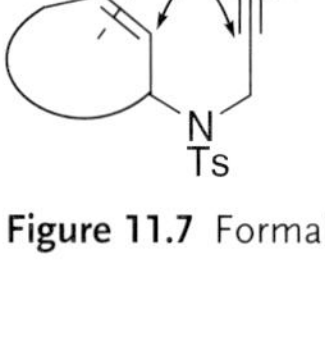

Figure 11.7 Formal reaction course for formation of **236** from **235** and ethylene.

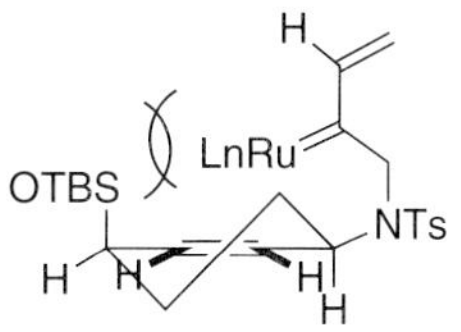

Figure 11.8 Steric repulsion in intermediary ruthenium carbene complex.

or **240d** with **6b** proceeded smoothly under ethylene gas to give the pyrrolidine derivative **241c** or **241d** in high yield, respectively. These reactions proceed in a highly stereoselective manner (Scheme 11.83) [110].

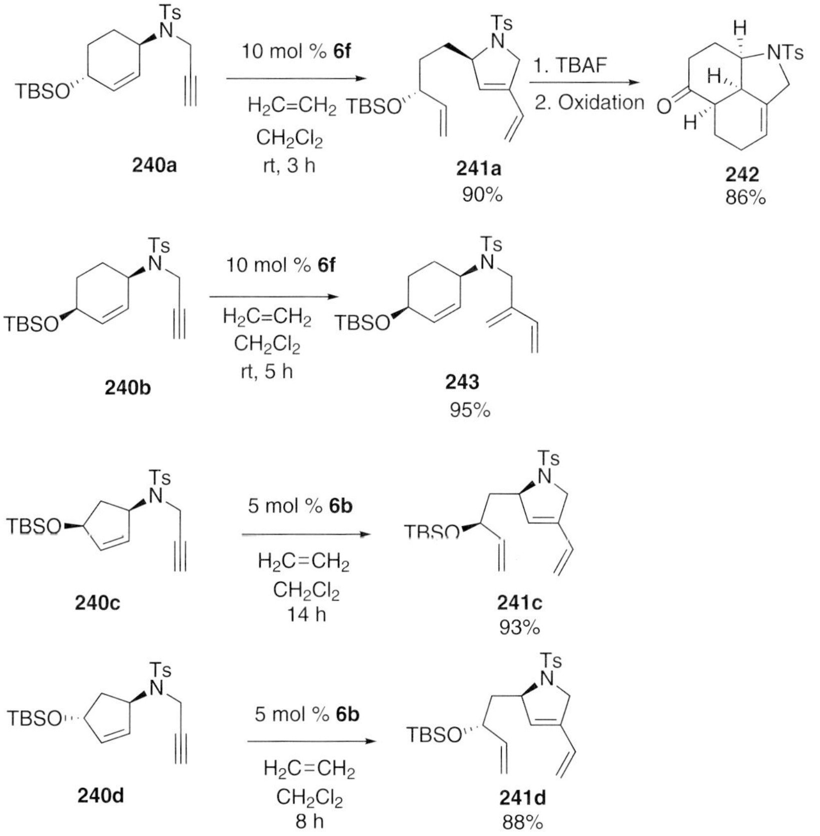

Scheme 11.83 ROM–RCM of cycloalkene-ynes.

A tandem reaction of enynes in the presence of olefins instead of ethylene was carried out by Blechert *et al.* Treatment of cyclopentenyl propargyl ether **244a** with **6f** in the presence of 2-allyl malonate afforded **245a** by ROM–RCM and CM. It is interesting that the D-ring of the estradiol derivative **244b** could be cleaved by ROM–RCM–CM of the propargylic side chain in the presence of an alkene (Scheme 11.84) [111].

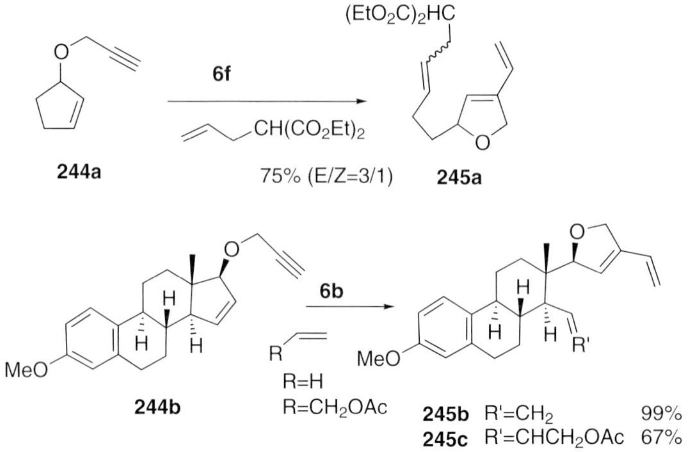

Scheme 11.84 ROM–RCM of cyclopentenol-yne.

If 1-alkynyl cycloalkenes **246** are treated with a ruthenium catalyst, the monosubstituted ruthenium carbene complex **248** will be formed from a highly strained ruthenacyclobutane intermediate **247**, as shown in Scheme 11.85. Subsequent reaction with ethylene would afford triene **250**, however, intramolecular RCM of the carbene moiety with the terminal C=C bond would afford the bicyclic compound **249**.

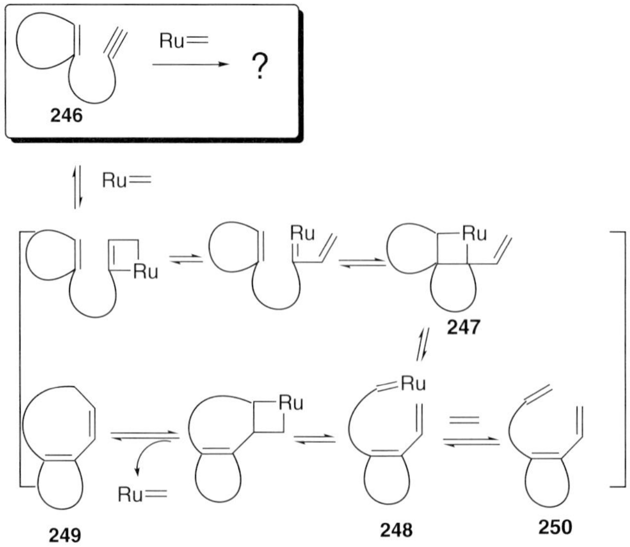

Scheme 11.85 Plan for ROM–RCM cycloalkene-yne.

When a cyclohexene-yne **246a** was treated with **6f** under an ethylene gas atmosphere, three products were produced. The expected bicyclic compound **249a** was obtained in only 14% yield, and the major product (57%) is an unexpected bicyclic compound which turned out to be **249b**. In addition, the dimeric cyclization compound **250** was formed in 26% yield. Presumably, the bicyclic compound **249b** would be formed by isomerization of the double bond of triene **250a** followed by the RCM of the diene moiety. Cyclopentene analog **246b** was found to give only the bicyclic compound **249b** in quantitative yield. Furthermore, N-cyclopetenylmethyl-N-homopropargyl tosylamide **246c** reacted with **6f** to give the bicyclic compound **249c** in quantitative yield (Scheme 11.86). These results indicate that, formally, the cyclopentene rings in **246b** and **246c** are cleaved and two carbon–carbon bonds are formed between the double and triple bonds, respectively, to produce **249b** and **249c** (Figure 11.9) [112].

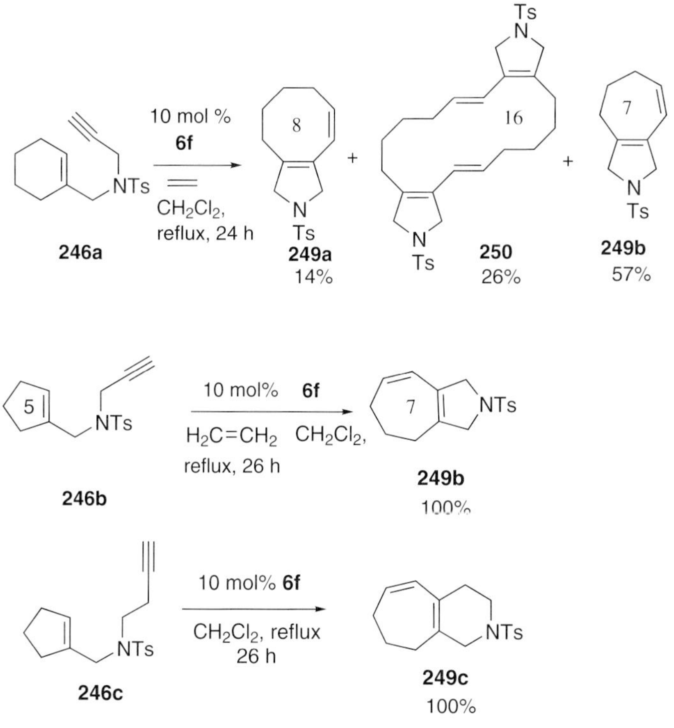

Scheme 11.86 ROM–RCM of cycloalkene-ynes.

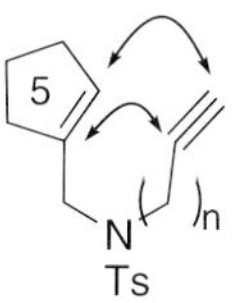

Figure 11.9 Formal reaction course in ROM-RCM of **246**.

In the above reaction, the ring size of the cycloalkadiene formed from **246** is the ring size of the initial cycloalkene plus two. Thus, for construction of the isoquinoline skeleton using this procedure, an initial ring size is four, and the chain length between the alkyne and the alkene carbons is four atoms. When the cyclobutene derivative **246d** was treated with **6f** under ethylene gas, the expected isoquinoline derivative **249d** was obtained in 60% yield. Furthermore, the cyclic amino acid derivative **249e** was obtained in 75% yield from the glycine derivative **246e** having a cyclobutene ring. The binaphthyl skeleton **249f** could be constructed in 76% yield when the naphthyl group was placed on the terminal position of the alkyne moiety (Scheme 11.87) [113].

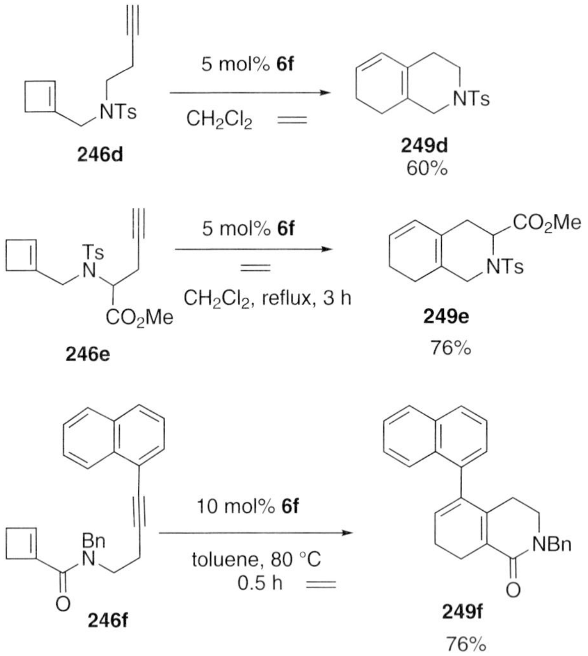

Scheme 11.87 Synthesis of isoquinolines by ROM of cyclobutene-Yne.

Plumet et al. described the domino metathesis of propargyl (2-endo-7-oxanorborn-5-enyl) ethers **251a–c** with allyl acetate in the presence of Grubbs' ruthenium catalyst **6b** (Scheme 11.88) [114a,b]. The reaction proceeded stereoselectively to produce substituted *cis*-fused bicyclic ethers **252a–c**. In a similar manner, the indolizidinone derivative **252d** was formed from azabicyclo[2.2.1]heptenone **251d** in high yield. Later, this reaction was further investigated and the pyrrolizidine derivative **253** was obtained as a minor product in 18% yield along with the indolizidine derivative **252d** in 50% yield when the toluene solution of **251d** and **6f** was warmed at 80 °C for 30 min under ethylene gas [114c].

North and Banti observed double ring-opening metathesis of dialkynyl cycloalkenes **254** affording the tricyclic compound **255** in high yield (Scheme 11.89). [115].

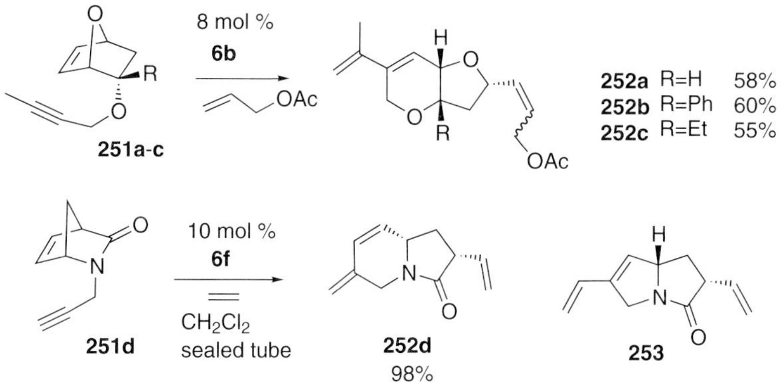

Scheme 11.88 ROM–RCM followed by CM of cycloalkene-yne.

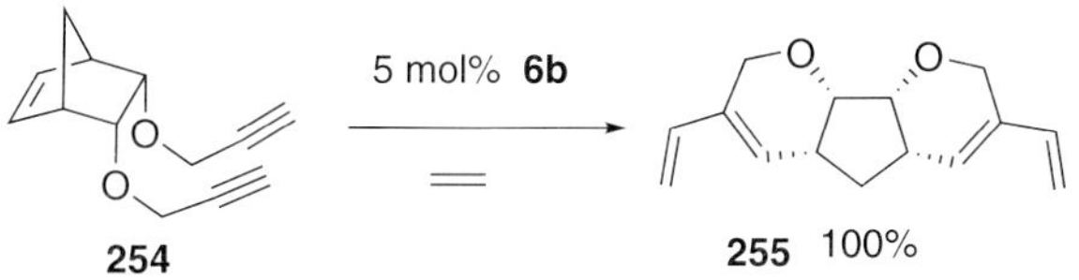

Scheme 11.89 ROM–RCM of norbornene derivative.

Diver found an interesting novel ring construction using CM of cycloalkene and the propargyl alcohol derivative (Scheme 11.90). CM of propynyl benzoate and cyclopentene gave the cycloheptadiene **256a** [116a]. Using 1,5-cyclooctadiene as the cycloalkene, the cyclohexadiene **256b** was formed [116b].

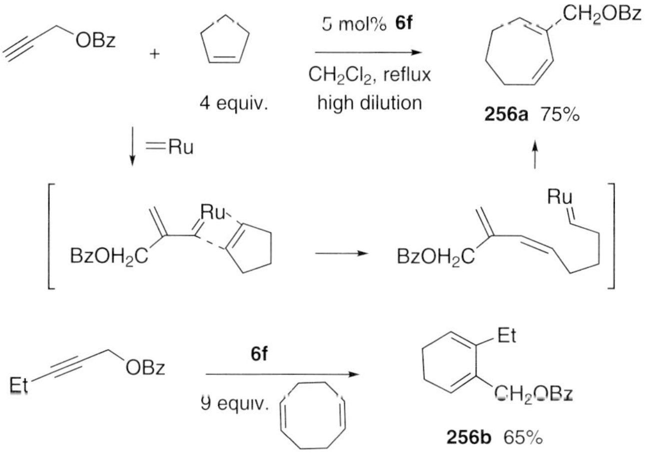

Scheme 11.90 Cyclization using CM followed by ROM.

11.3.4
Synthesis of Natural Products Using Ring-Closing Metathesis of Enyne

As shown in Section 11.2.5, many complicated natural products have been synthesized using RCM of dienes and many researchers know that the RCM of dienes is a useful tool for the synthesis of natural products. However, there are not so many reports of the total synthesis of complicated natural products using enyne metathesis [117]. Presumably, the use of enyne metathesis would be more complicated compared with that of diene metathesis because of the consideration of retrosynthetic analysis. The important point for the synthesis of natural products using enyne metathesis is how to use the diene moiety generated by enyne metathesis. The first such report is the synthesis of (−)-stemoamide reported by Mori, in which (−)-pyroglutamic acid was converted into the enyne **257** having an ester group on the alkyne, and the subsequent RCM was carried out using the ruthenium carbene complex **6b** to afford the bicyclic compound **258**, which was converted into carboxylic acid **259**. Halo-lactonization of **259** afforded **260**, which was converted into (−)-stemoamide by treatment with $NaBH_4$ under the catalysis of $NiCl_2 \cdot 6H_2O$ (Scheme 11.91) [118].

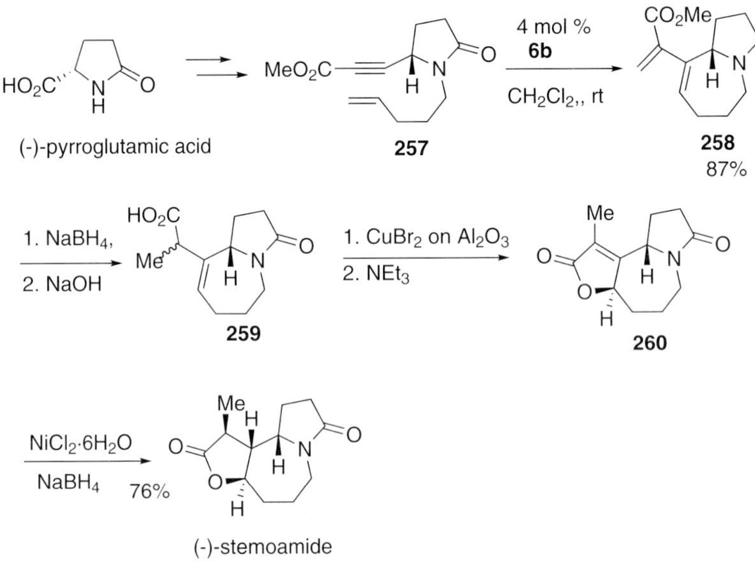

Scheme 11.91 Total synthesis of (−)-stemoamide.

An enantioselective biomimetic synthesis of longithorone A was accomplished on the basis of the proposed biosynthesis. The syntheses of two [12]-paracyclophanes **265** and **267** were realized by applying enyne metathesis macrocyclization of **264** and **266**, which were synthesized from the common substrate **263**. Synthesis of Longtholone A was achieved from **268**, which was formed by the Diels–Alder reaction of **261** and **262** followed by oxidation (Scheme 11.92) [119].

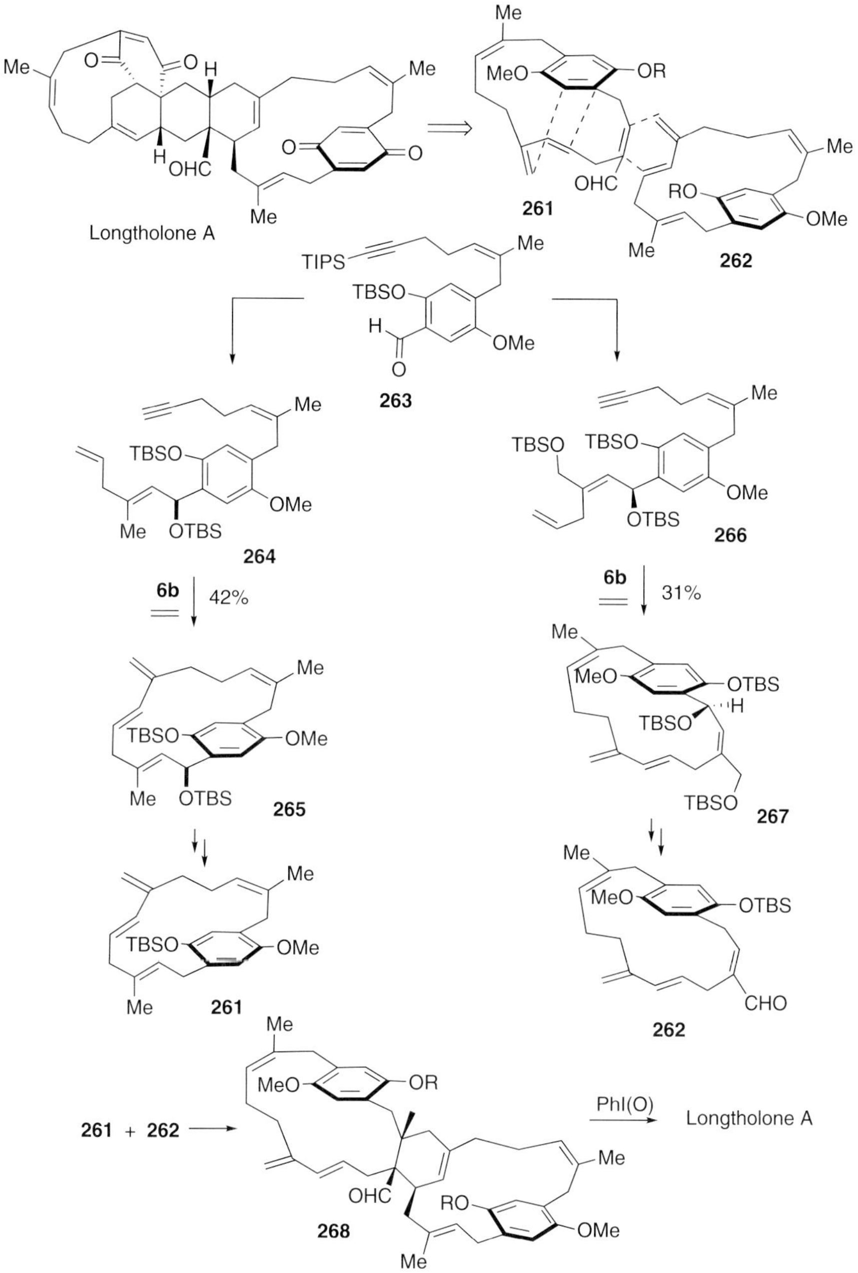

Scheme 11.92 Total synthesis of (−)-longthorone A.

Synthesis of the anthramycine derivative **273** was achieved by Mori using RCM and CM (Scheme 11.93). L-Methionine was converted into enyne **269**, and RCM of **269** using **6b** gave the pyrrolidine derivative **268**. Deprotection followed by condensation with acyl chloride gave **271**. Reductive cyclization of **271** followed by treatment with dil.

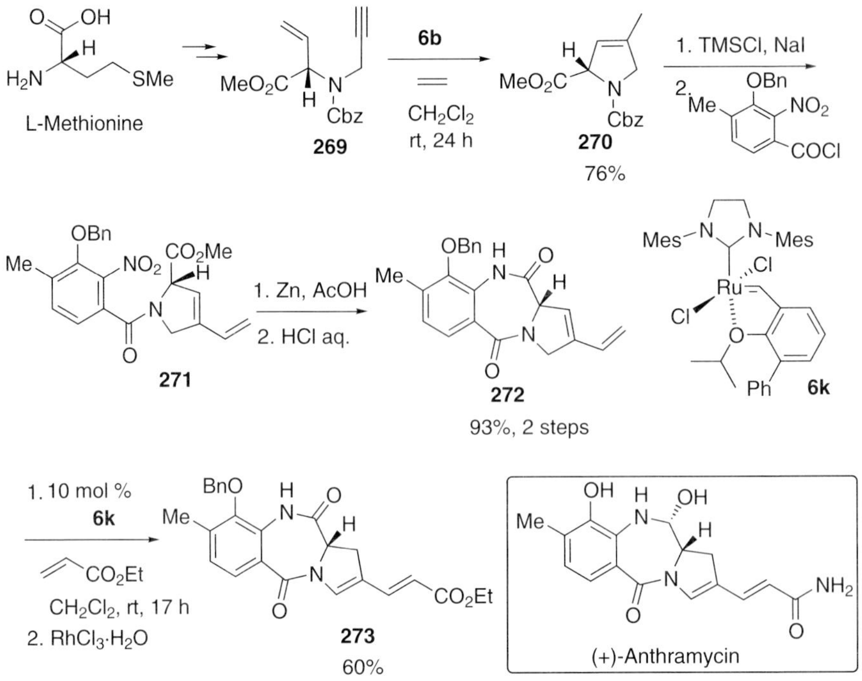

Scheme 11.93 Synthesis of anthramycin derivative.

HCl afforded pyrrolo-1,4-benzodiazepinone **272**. CM of **272** with ethyl acrylate using catalyst **6k** followed by isomerization of the double bond with RhCl$_3$·H$_2$O afforded the desired compound **273** [120].

The total synthesis of (+)-anatoxin-a was achieved by Martin [121] and Mori [122] by the same strategy. The key step is the construction of an azabicyclo[4.2.1]nonene ring system. For that purpose, pyrrolidine derivative **274** having *cis*-2,5-substituents was synthesized from (+)-pyroglutamic acid. Enyne metathesis of **274** was carried out using **6b** to form the cyclic unit **275** and anatoxin-a could be synthesized (Scheme 11.94).

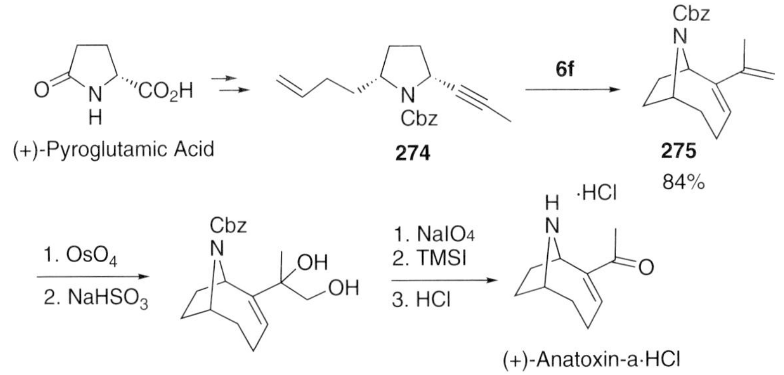

Scheme 11.94 Total synthesis of anatoxin-a.

11.3 Ring-Closing Metathesis of Enynes

By a similar procedure, (+)-ferrunginine was synthesized from (−)-pyroglutamic acid [123]. Construction of an azabicyclo[3.2.1] ring system was carried out using enyne metathesis of **276** using **6b**. Wacker oxidation to the resultant diene **277** afforded the methyl ketone. Subsequent deprotection followed by methylation was carried out to give (−)-ferruginine (Scheme 11.95).

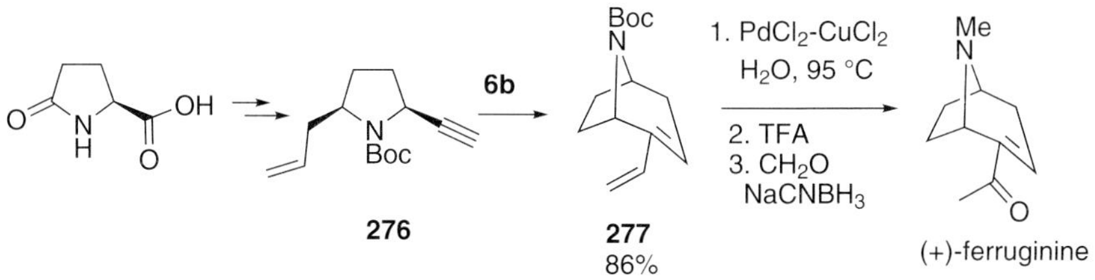

Scheme 11.95 Synthesis of (+)-ferruginine.

Kozmin developed a highly efficient synthesis of a cyclic compound bearing the methyl ketone from enyne using RCM [88]. As an application of this method, they succeeded in the synthesis of elemophilane [124]. RCM of enyne **278** having the silyloxy group on the alkyne, followed by treatment with HF, gave cycloalkene **279** having the methyl ketone functionality. From this compound **279**, α- and β-elemophilane were synthesized (Scheme 11.96).

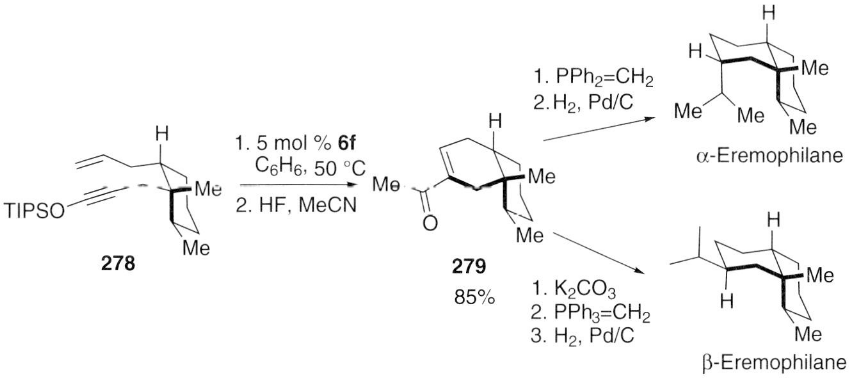

Scheme 11.96 Total synthesis of eremophilane.

Morken synthesized (−)-11a,13-dihydroxanthatin by RCM of enyne **280a** using **6f**, methylation, and CM with methyl vinyl ketone using **6f** [125]. On the other hand, Martin et al. succeeded in the first total synthesis of 8-epi-xanthatin using one-step RCM-CM of **280b** in the presence of an excess amount of methyl vinyl ketone (Scheme 11.97) [126].

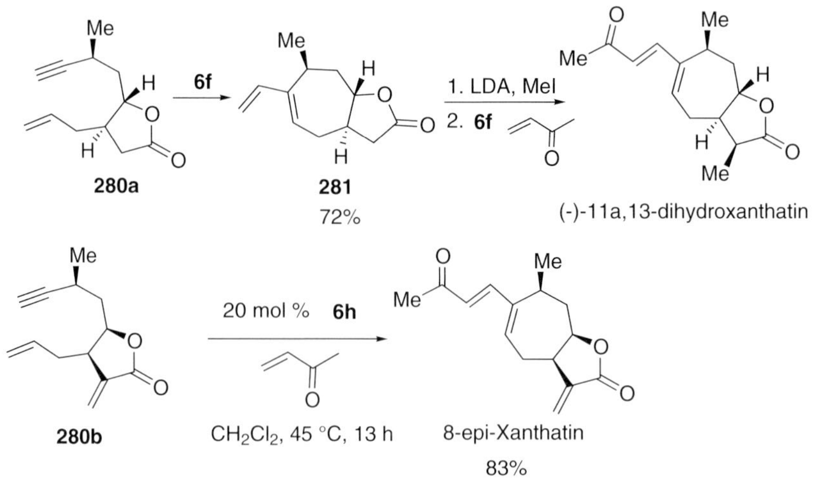

Scheme 11.97 Synthesis of dihydoxanthatin and 8-epi-xanthatin.

New allocolchinoids **285** and **285'** having functionality at position C10 or C11 in the C ring were synthesized using the RCM of enyne **282** for construction of the seven-membered ring. A Diels–Alder–aromatization sequence was carried out to form the aromatic C ring (Scheme 11.98) [127].

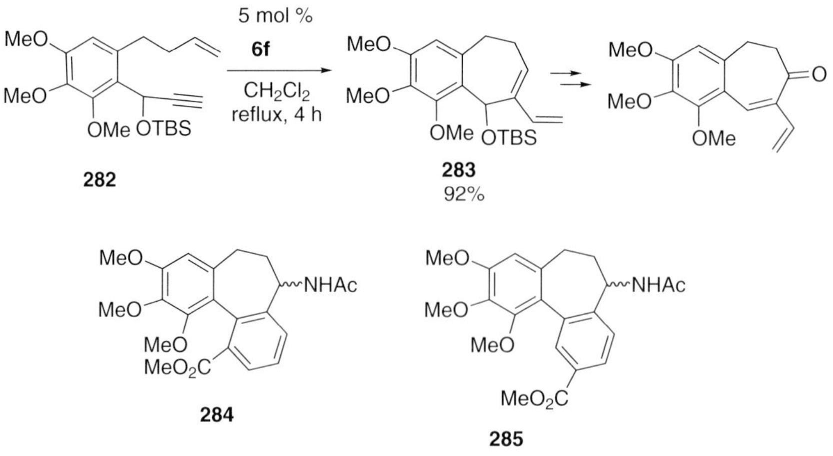

Scheme 11.98 Synthesis of allocolchicine.

The *agalacto*-spirolactone B subunit of quartromicins has been synthesized by Bedel *et al.* using the Claisen–Ireland/RCM of enyne approach [128]. Compound **286** was converted into enyne **287**, which was treated with **6f** to afford **288** in 73% yield. This was converted into subunit B of quartromicin (Scheme 11.99).

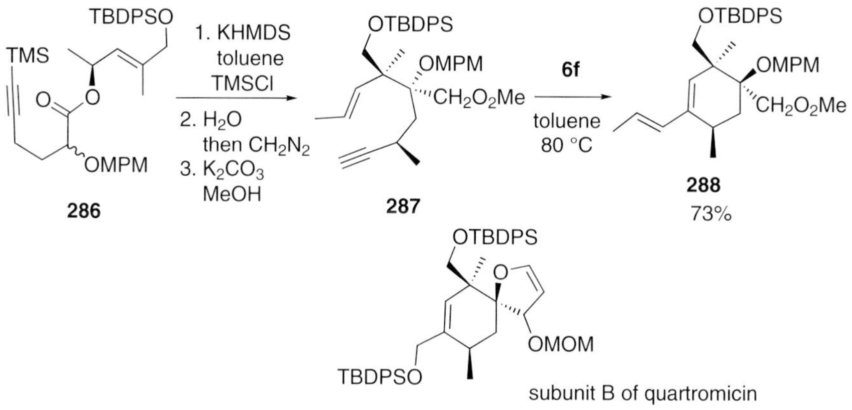

Scheme 11.99 Synthesis of subunit of quartromicin.

An enantioselective synthesis of (−)-galanthamine was realized in 11 steps starting from isovanillin (Scheme 11.100). The enyne **289** (92% ee) underwent an efficient RCM reaction in the presence of 3 mol% of **6b** to give diene **290** in 85% yield. Hydroboration and oxidation of an alkene part of **290** gave homoallylic alcohol in excellent yield, which was treated with sequentially the palladium catalyst and SeO$_2$ oxidation to give **291**. From this compound (−)-garanthamine could be synthesized via mesylation of the hydroxy group, deprotection of N-Boc, and intramolecular amination [129].

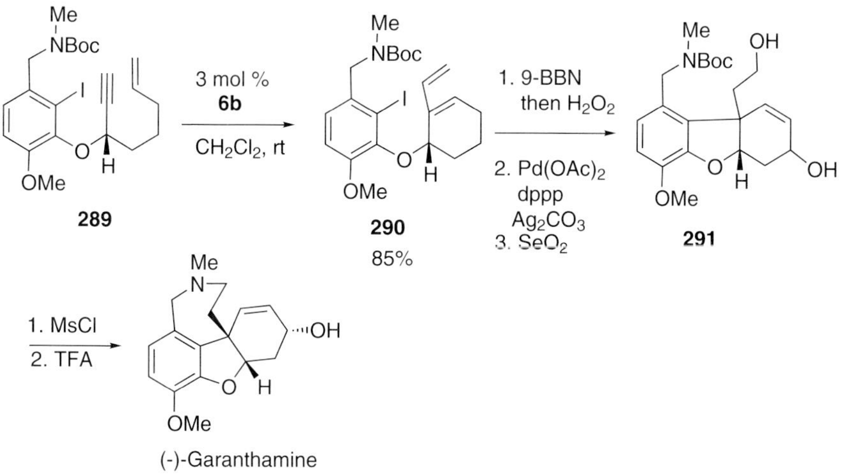

Scheme 11.100 Total synthesis of garanthamine.

A concise and highly enantioselective route was developed for the synthesis of angucyclinone-type natural products (Scheme 11.101). The chiral vinylcyclohexene derivative **293** was synthesized using RCM of enyne **292**. Diels–Alder reaction followed by aromatization gave the compound having a benz[a]anthraquinone skeleton. Utilizing this strategy, the total synthesis of natural product YM-181741 was easily achieved [130].

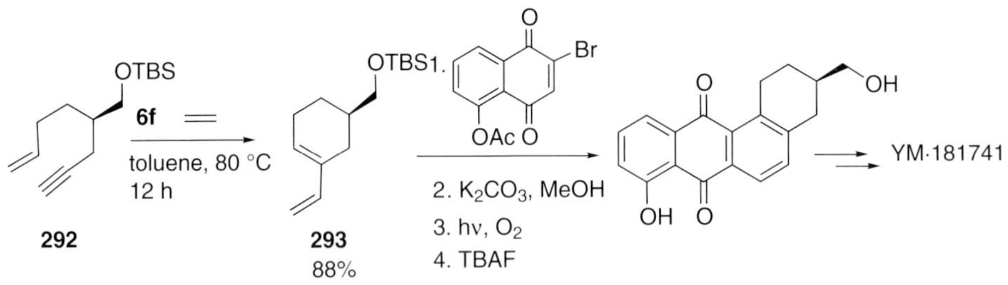

Scheme 11.101 Synthesis of YM-181741.

11.4
Perspective

Since the discovery of a stable and isolable catalyst for metathesis by Schrock and Grubbs in 1990 and 1993, a wide range of olefin metatheses, and diene, enyne and alkyne metatheses have been developed. The Nobel Prize 2005 in Chemistry was awarded to the pioneers, Professor Grubbs, Professor Schrock and Professor Chauvin, in this field, which acknowledged that the metathesis reaction has occupied an important position in recent synthetic organic chemistry. The remarkable features of these reactions are that double and triple bonds are cleaved and double and triple bonds are formed simultaneously. These reactions are unique and useful for the synthesis of cyclic compounds. Many cyclic compounds have been prepared efficiently by the metathesis reaction. Medium-sized, macrocyclic, and even five- and six-membered rings may be constructed by RCM instead of traditional methods. It should be noted that the catalysts of the metathesis reaction tolerate many functional groups in the molecule. The second-generation ruthenium catalysts can be used for metatheses of diene or enyne having an electron-donating group or an electron-withdrawing group on the multiple bonds. It is expected that metathesis will be further used for the syntheses of various complicated molecules and macrocyclic compounds because the reaction procedure is very simple, the yields are high, and transition metal carbene complexes are now commercially available or easily prepared from **6a** or **6b**.

11.5
Experimental: Selected Procedure

11.5.1
Typical Procedure for the Synthesis of a Cyclized Compound from Enyne Using RCM

A solution of **148** (74.8 mg, 0.30 mmol) in CH_2Cl_2 (10 mL) and benzylidene ruthenium catalyst **6b** (7.5 mg, 3 mol%) was stirred at room temperature under ethylene gas (1 atm, balloon) for 2.5 h. The solvent was evaporated and the residue was purified by column chromatography on silica gel (hexane–ethyl acetate, 5 : 1) to give a colorless oil of **149** (73.9 mg, 99% yield) Eq. (11.5)

$$\underset{\underset{\text{Ts}}{\text{148}}}{\text{allyl-propargyl-N-Ts}} \xrightarrow[\text{CH}_2\text{Cl}_2]{\underset{\text{H}_2\text{C}=\text{CH}_2}{3 \text{ mol } \% \text{ } \mathbf{6b}}} \underset{\underset{\text{Ts}}{\text{149}}}{\text{3-vinyl-pyrroline}} \quad (11.5)$$

90%

References

1. (a) Grubbs, R.H. (ed.) (2003) *Handbook of Metathesis*, Wiley-VCH, Weinheim; (b) Fürstner, A. (ed.) (1998) *Topics in Organometallic Chemistry*, vol. 1, Springer-Verlag, Berlin, Heidelberg; (c) Trunk, T.M. and Grubbs, R.H. (2001) *Acc. Chem. Res.*, **34**, 18–29; (d) Fürstner, A. (2000) *Angew. Chem. Int. ed.*, **39**, 3012–3043; (e) Mori, M. (2007) Recent progress in metathesis reaction, in *New Frontiers in Asymmetric Catalysis* (eds K. Mikami and M. Lautens), John Wiley & Sons, pp. 153–206.
2. Villemin, D. (1980) *Tetrahedron Lett.*, **21**, 1715–1718.
3. Tsuji, J. and Hashiguchi, S. (1980) *Tetrahedron Lett.*, **21**, 2955–2959.
4. Schrock, R.R., Murdzek, J.S., Dimare, M. and O'Regan, M. (1990) *J. Am. Chem. Soc.*, **112**, 3875–3886.
5. (a) Fu, G.C. and Grubbs, R.H. (1992) *J. Am. Chem. Soc.*, **114**, 5426–5427; (b) Fu, G.C. and Grubbs, R.H. (1992) *J. Am. Chem. Soc.*, **114**, 7324–7325; (c) Fu, G.C. and Grubbs, R.H. (1993) *J. Am. Chem. Soc.*, **115**, 3800–3801; (d) Fu, G.C., Nguyen, S.-B.T. and Grubbs, R.H. (1993) *J. Am. Chem. Soc.*, **115**, 9856–9857.
6. Forbes, M.D.E., Patton, J.P., Myers, T.L., Maynard, H.D., Smith, D.W. Jr, Schulz, G.R. and Wagener, K.B. (1992) *J. Am. Chem. Soc.*, **114**, 10978–10980.
7. Nguyen, S.T., Johnson, L.K. and Grubbs, R.H. (1993) *J. Am. Chem. Soc.*, **115**, 9858–9859.
8. Schwab, P., France, M.B., Ziller, J.W. and Grubbs, R.H. (1995) *Angew. Chem., Int. Ed. Engl.*, **34**, 2039–2041.
9. Reviews for enyne metathesis: (a) Mori, M., (1998) *Top. Organomet. Chem.*, **1**, 133–154; (b) Poulsen, C.S. and Madsen, R. (2003) *Synthesis*, 1–18; (c) Mori, M. (2003) in *Handbook of Metathesis*, vol. 2 (ed. R.H. Grubbs), Wiley-VCH, Weinheim, pp. 176–204; (d) Giessert, A.J. and Diver, S.T. (2004) *Chem. Rev.*, **104**, 1317–1382; (e) Mori, M. and Kitamura, T. (2005) Ene-yne and alkyne metathesis, in *Comprehensive Organometallic Chemistry III. Transition Metal Organometallics in Organic Synthesis* (ed. T. Hiyama), Elsevier, pp. 271–310.
10. (a) Katz, T.J. and Sivavec, T.M. (1985) *J. Am. Chem. Soc.*, **107**, 737–738; (b) Sivavec, T.M., Katz, T.J., Chiang, M.Y. and Yang, G.X.-Q. (1989) *Organometallics*, **8**, 1620–1625; (c) Katz, T.J. and Yang, G.X.-Q. (1991) *Tetrahedron Lett.*, **32**, 5895–5898; (d) Sivavec, T.M. and Katz, T.J. (1985) *Tetrahedron Lett.*, **26**, 2159–2162.
11. (a) Watanuki, S., Ochifuji, N. and Mori, M. (1994) *Organometallics*, **13**, 1129–1130; (b) Mori, M. and Watanuki, S. (1992) *J. Chem. Soc., Chem. Commun.*,

1082–1084; (c) Watanuki, S. and Mori, M. (1993) *Heterocycles*, **35**, 679–682.

12 (a) Kinoshita, A. and Mori, M. (1994) *Synlett*, 1020–1022; (b) Kinoshita, A., Sakakibara, N. and Mori, M. (1999) *Tetrahedron*, **55**, 8155–8167.

13 Mori, M., Sakakibara, N. and Kinoshita, A. (1998) *J. Org. Chem.*, **63**, 6082–6083.

14 Morehead, A. and Grubbs, R.H. (1998) *Chem. Commun.*, 275–276.

15 (a) Crimmins, M.T. and King, B.W. (1996) *J. Org. Chem.*, **61**, 4192–4193; (b) Crimmins, M.T. and Zuercher, W. (2000) *Org. Lett.*, **2**, 1065–1068.

16 Miller, S.J., Kim, S.-H., Chen, Z.R. and Grubbs, R.H. (1995) *J. Am. Chem. Soc.*, **117**, 2108–2109.

17 (a) Fürstner, A. and Langemann, K. (1996) *J. Org. Chem.*, **61**, 3942–3943; (b) Lee, C.W. and Grubbs, R.H. (2000) *Org. Lett.*, **2**, 2145–2148.

18 (a) Maeda, K., Oishi, T., Oguri, H. and Hirama, M. (1999) *Chem. Commun.*, 1063–1064; (b) Hirama, M., Oishi, T., Uehara, H., Inoue, M., Maruyama, M., Guri, H. and Satake, M. (2001) *Science*, **294**, 1904–1907.

19 Marsella, M.J., Maynard, H.D. and Grubbs, R.H. (1997) *Angew. Chem. Int. Ed.*, **36**, 1101–1103.

20 (a) Weskamp, T., Schattenmann, W.C., Spiegler, M. and Herrmann, W.A. (1998) *Angew. Chem. Int. Ed.*, **37**, 2490–2493; (b) Weskamp, T., Kohl, F.J., Hieringer, W., Gleich, D. and Herrmann, W.A. (1999) *Angew. Chem. Int. Ed.*, **38**, 2416–2419.

21 Huang, J., Stevens, E.D., Nolan, S.P. and Peterson, J.L. (1999) *J. Am. Chem. Soc.*, **121**, 2674–2678.

22 Scholl, M., Trnka, T.M., Morgan, J.P. and Grubbs, R.H. (1999) *Tetrahedron Lett.*, **40**, 2247–2250.

23 Scholl, M., Ding, S., Lee, C.W. and Grubbs, R.H. (1999) *Org. Lett.*, **1**, 953–956.

24 (a) Harrity, J.P.A., Visser, M.S., Gleason, J.D. and Hoveyd, A.H. (1997) *J. Am. Chem. Soc.*, **119**, 1488–1489;
(b) Kingsbury, J.S., Harrity, J.P., Bonitatebus, P.J. and Hoveyda, A.H. (1999) *J. Am. Chem. Soc.*, **121**, 791–799; (c) Harrity, J.P.A., La, D.S., Cefalo, D.R., Visser, M.S. and Hoveyda, A.H. (1998) *J. Am. Chem. Soc.*, **120**, 2343–2351.

25 Garber, S.B., Kingsbury, J.S., Gray, B.L. and Hoveyda, A.H. (2000) *J. Am. Chem. Soc.*, **122**, 8168–8179.

26 Love, J.A., Morgan, J.P., Truka, T.M. and Grubbs, R.H. (2002) *Angew. Chem. Int. Ed.*, **41**, 4035–4037.

27 Michrowska, A., Bojok, R., Harutyunyan, S., Sashuk, V., Dolgonos, G. and Grela, K. (2004) *J. Am. Chem. Soc.*, **126**, 9318–9325; Michrowska, A., Mennecke, K., Kunz, U., Kirschning, A. and Grela, K. (2006) *J. Am. Chem. Soc.*, **128**, 13261–13267.

28 Wakamatsu, H. and Blechert, S. (2002) *Angew. Chem., Int. Ed.*, **41**, 2403; Zaja, M., Stephen, J.S., Dunnen, A.M., Rivard, M., Buschmann, N., Jiricek, J. and Blechert, S. (2003) *Tetrahedron*, **59**, 6545–6558.

29 (a) Sanford, M.S., Ulman, M. and Grubbs, R.H. (2001) *J. Am. Chem. Soc.*, **123**, 749–750; (b) Sanford, M.S., Love, A.L. and Grubbs, R.H. (2001) *J. Am. Chem. Soc.*, **123**, 6543–6554; (c) Love, J.A., Sanford, M.S., Day, M.W. and Grubbs, R.H. (2003) *J. Am. Chem. Soc.*, **125**, 10103–10109.

30 (a) Conrad, R.M., Grogan, M.J. and Bertozzi, C.R. (2002) *Org. Lett.*, **4**, 1359–1362; (b) M-Contelles, J. and de Opazo, E.J. (2000) *J. Org. Chem.*, **65**, 5416–5419; (c) M-Contelles, J. and de Opazo, E.J. (2002) *J. Org. Chem.*, **67**, 3705–3717; (d) Boyer, F.-D., Hanna, I. and Nolan, S.P. (2001) *J. Org. Chem.*, **66**, 4094–4096.

31 Evans, P.A. and Murthy, V.S. (1998) *J. Org. Chem.*, **63**, 6768–6769.

32 Chang, S.B. and Grubbs, R.H. (1997) *Tetrahedron Lett.*, **38**, 4757–4760.

33 Arisawa, M., Theeraladanon, C., Nishida, A. and Nakagawa, M. (2001) *Tetrahedron Lett.*, **42**, 8027–8030.

34 Okada, A., Oshima, T. and Shibasaki, M. (2001) *Tetrahedron, Lett.*, **42**, 8023–8026.

35 Iuliano, A., Piccioli, P. and Fabbri, D. (2004) *Org. Lett.*, **6**, 3711–3714.
36 Collins, S.K., Grandbois, A., Vachon, M.P. and Cote, J. (2006) *Angew. Chem. Int. Ed.*, **45**, 2923–2926.
37 Yoshida, K. and Imasmoto, T. (2005) *J. Am. Chem. Soc.*, **127**, 10470–10471.
38 Lautens, M. and Hughes, G. (1999) *Angew. Chem. Int. Ed.*, **38**, 129–131.
39 (a) Ma, S. and Ni, B. (2004) *Chem. Eur. J.*, **10**, 3286–3300; (b) Blakemore, P.R., Kilner, C., Norcross, N.R. and Astles, P.C. (2005) *Org. Lett.*, **7**, 4721–4724; (c) Heck, M.-P., Baylon, C., Nolan, S.P. and Mioskowski, C. (2001) *Org. Lett.*, **3**, 1989–1991; (d) Wallace, D.J. (2003) *Tetrahedron Lett.*, **44**, 2145–2148.
40 (a) Ghosh, A.K. and Liu, C. (1999) *Chem. Commun.*, 1743–1744; (b) Ghosh, A.K. and Wang, Y. (2000) *J. Am. Chem. Soc.*, **122**, 11027–11028.
41 Boiteau, J.-G., Van de Weghe, P. and Eustache, J. (2001) *Org. Lett.*, **3**, 2737–2740.
42 Yang, Q., Xiao, W.-J. and Yu, Z. (2005) *Org. Lett.*, **7**, 871–874.
43 Hong, S.H., Day, M.W. and Grubbs, R.H. (2004) *J. Am. Chem. Soc.*, **126**, 7414–7415.
44 Hong, S.H., Sanders, D.P., Lee, C.W. and Grubbs, R.H. (2005) *J. Am. Chem. Soc.*, **127**, 17160–17161.
45 (a) Randall, M.L., Tallarico, J.A. and Snapper, M.L. (1995) *J. Am. Chem. Soc.*, **117**, 9610–9611; (b) White, B.H. and Snapper, M.L. (2003) *J. Am. Chem. Soc.*, **125**, 14901–14904; (c) Limanto, J. and Snapper, M.L. (2000) *J. Am. Chem. Soc.*, **122**, 8071–8072.
46 Zuercher, W.J., Hashimoto, M. and Grubbs, R.H. (1996) *J. Am. Chem. Soc.*, **118**, 6634–6640.
47 (a) Harrity, J.P.A., Visser, M.S., Gleson, J.D. and Hoveyda, A.H. (1997) *J. Am. Chem. Soc.*, **119**, 1488–1489; (b) Harrity, J.P.A., La, D.S., Cefalo, D.R., Visser, M.S. and Hoveyda, A.H. (1998) *J. Am. Chem. Soc.*, **120**, 2343–2351.
48 Stragies, R. and Blechert, S. (1998) *Synlett*, 169–170.
49 Usher, L.C., E-Jimenez, M., Ghibiriga, I. and Wright, D.L. (2002) *Angew. Chem. Int. Ed.*, **41**, 4560–4562.
50 (a) Voigtmann, U. and Blechert, S. (2000) *Org. Lett.*, **2**, 3971–3974; (b) Stragies, R. and Blechert, S. (1999) *Tetrahedron*, **55**, 8179–8188.
51 Winkler, J.D., Asselin, S.M., Shepard, S. and Yuan, J. (2004) *Org. Lett.*, **6**, 3821–3824.
52 (a) Fujimura, O. and Grubbs, R.H. (1996) *J. Am. Chem. Soc.*, **118**, 2499–2500; (b) Fujimura, O. and Grubbs, R.H. (1998) *J. Org. Chem.*, **63**, 824–832.
53 (a) Alexander, J.B., La, D.S., Cefalo, D.R., Hoveyda, A.H. and Schrock, R.R. (1998) *J. Am. Chem. Soc.*, **120**, 4041–4042; (b) Zhu, S.S., Cefalo, D.R., La, D.S., Jamieson, J.Y., Davis, W.M., Hoveyda, A.H. and Schrock, R.R. (1999) *J. Am. Chem. Soc.*, **121**, 8251–8259; (c) La, D.S., Alexander, J.B., Cefalo, D.R., Graf, D.D., Hoveyda, A.H. and Schrock, R.R. (1998) *J. Am. Chem. Soc.*, **120**, 9720–9721.
54 Weatherhead, G.S., Ford, J.G., Alexanian, E.J., Schrock, R.R. and Hoveyda, A.H. (2000) *J. Am. Chem. Soc.*, **122**, 1828–1829.
55 Cefalo, D.R., Kiely, A.F., Wuchrer, M., Jamieson, J.Y., Schrock, R.R. and Hoveyda, A.H. (2001) *J. Am. Chem. Soc.*, **123**, 3139–3140.
56 Dolman, S.J., Sattery, E.S., Hoveyda, A.H. and Schrock, R.R. (2002) *J. Am. Chem. Soc.*, **124**, 6991–6997.
57 Funk, T.W., Berlin, J.M. and Grubbbs, R.H. (2006) *J. Am. Chem. Soc.*, **128**, 1840–1846.
58 (a) Kiery, A.F., Jemelius, J.A., Schrock, R.R. and Hoveyda, A.H. (2002) *J. Am. Chem. Soc.*, **124**, 2868–2869; (b) Weatherhead, G.S., Houser, J.H., Ford, G.J., Jamieson, J.Y., Schrock, R.R. and Hoveyda, A.H. (2000) *Tetrahedron Lett.*, **41**, 9553–9556.
59 Burke, S.D., Mu1ller, N. and Beaudry, C.M. (1999) *Org. Lett.*, **1**, 1827–1830.
60 (a) Teng, X., Cefalo, D., Schrock, R.R. and Hoveyda, A.H. (2002) *J. Am. Chem. Soc.*,

124, 10779–10784; (b) Aelits, S.L., Cefalo, D.R., Bonitatebus, P.J. Jr, and Houser, J.H. (2001) *Angew. Chem. Int. Ed.*, **40**, 1452–1456.

61 Seiders, T.J., Ward, D.W. and Grubbs, R.H. (2001) *Org. Lett.*, **3**, 3225–3228.

62 Funk, T.W., Berlin, J.M. and Grubbbs, R.H. (2006) *J. Am. Chem. Soc.*, **128**, 1840–1846.

63 Van Veldhuizen, J.J., Garber, S.B., Kingsbury, J.S. and Hoveyda, A.H. (2002) *J. Am. Chem. Soc.*, **124**, 4954–4955.

64 Nguyen, S. and Grubbs, R.H. (1995) *J. Organomet. Chem.*, **497**, 195–200.

65 (a) Ahmed, M., Barrett, A.G.M., Braddock, D.C., Cramp, S.M. and Procopiou, P.A. (1999) *Tetrahedron, Lett.*, **40**, 8657–8662; (b) Barrett, A.G.M., Cramp, S.M. and Roberts, R.S. (1999) *Org. Lett.*, **1**, 1083–1086.

66 Schurer, S.C., Gesseler, S., Buschmann, N. and Blechert, S. (2000) *Angew. Chem. Int. Ed.*, **39**, 3898–3901.

67 Yao, Q. (2000) *Angew. Chem. Int. Ed.*, **39**, 3896–3898.

68 (a) Yao, Q. and Zhang, Y. (2004) *J. Am. Chem. Soc.*, **126**, 74–75; (b) Yao, Q. and Zhang, Y. (2003) *Angew. Chem. Int. Ed.*, **42**, 3395–3398.

69 Hultzsch, K.C., Jernelius, J.A., Hoveyda, A.H. and Schrock, R.R. (2002) *Angew. Chem. Int. Ed.*, **41**, 589–593.

70 (a) Martin, S.F., Liao, Y., Wong, Y. and Rein, T. (1994) *Tetrahedron Lett.*, **35**, 691–694; (b) Humphrey, J.M., Liao, Y.S., Ali, A., Rein, T., Wong, Y.L., Chen, H.-J., Courtney, A.K. and Martin, S.F. (2002) *J. Am. Chem. Soc.*, **124**, 8584–8592.

71 Fürstner, A., Gastner, T. and Weintritt, H. (1999) *J. Org. Chem.*, **64**, 2361–2366.

72 Sakamoto, S., Sakazaki, H., Hagiwara, K., Kamada, K., Ishii, K., Noda, T., Inoue, M. and Hirama, M. (2004) *Angew. Chem. Int. Ed.*, **43**, 6505–6501.

73 Ono, K., Nakagawa, M. and Nishida, A. (2004) *Angew. Chem. Int. Ed.*, **43**, 2020–2023.

74 Smith, A.S. III, Mesaros, E.F. and Meyer, E.A. (2005) *J. Am. Chem. Soc.*, **127**, 6948–6949.

75 Kadota, I., Takamura, H., Nishii, H. and Yamamoto, Y. (2005) *J. Am. Chem. Soc.*, **127**, 9246–9250.

76 Nicolaou, K.C. and Harrison, S.T. (2006) *Angew. Chem. Int. Ed.*, **45**, 3256–3260.

77 Kuramochi, A., Usuda, H., Yamatsugu, K., Kanai, M. and Shibasaki, M. (2005) *J. Am. Chem. Soc.*, **127**, 14200–14201.

78 Fürstner, A. and Nagano, T. (2007) *J. Am. Chem. Soc.*, **129**, 1906–1907.

79 Lippstreu, J.J. and Straub, B.F. (2005) *J. Am. Chem. Soc.*, **127**, 7444–7457.

80 Hoye, T.R., Donaldson, S.M. and Vos, T.J. (1999) *Org. Lett.*, **1**, 277–280.

81 Kohta, S. and Sreenivasachary, N. (2001) *Eur. J. Org. Chem.*, 3375–3383.

82 (a) Kitamura, T., Sato, Y. and Mori, M. (2001) *Chem. Commun.*, 1258–1259; (b) Kitamura, T., Sato, Y. and Mori, M. (2002) *Adv. Synth. Catal.*, **344**, 678–693.

83 (a) Barrett, A.G.M., Baugh, S.P.D., Braddock, D.C., Flack, K., Gibson, V.C., Procopiou, P.A., White, A.J.P. and Williams, D.J. (1998) *J. Org. Chem.*, **63**, 7893–7907; (b) Duboc, R., Henaut, C., Savignac, M., Genet, J.-P. and Bhatnagar, N. (2001) *Tetrahedron Lett.*, **42**, 2461–2464; (c) Savigna, M. and Genet, J.-P. (2004) *Eur. J. Org. Chem.*, 4840.

84 Ma, S., Ni, B. and Liang, Z. (2004) *J. Org. Chem.*, **69**, 6305–6309.

85 Hansen, E.C. and Lee, D. (2003) *J. Am. Chem. Soc.*, **125**, 9582–9583.

86 (a) Micalizio, G.C. and Schreiber, S.L. (2002) *Angew. Chem. Int. Ed.*, **41**, 152–154; (b) Micalizio, G.C. and Schreiber, S.L. (2002) *Angew. Chem. Int. Ed.*, **41**, 3272–3276.

87 (a) Yao, Q. (2001) *Org. Lett.*, **3**, 2069–2072; (b) Miller, R.L., Maifeld, S.V. and Lee, D. (2004) *Org. Lett.*, **6**, 2773–2776.

88 Schramm, M.P., Reddy, D.S. and Kozmin, S.A. (2001) *Angew. Chem. Int. Ed.*, **40**, 4274–4277.

89 Clark, J.S., Trevitt, G.P., Voyall, D. and Stammen, B. (1998) *Chem. Commun.*, 2629–2630.

90 (a) Saito, N., Sato, Y. and Mori, M. (2002) *Org. Lett.*, **4**, 803–806; (b) Mori, M., Wakamatsu, H., Saito, N., Sato, Y., Narita, R., Sato, Y. and Fujita, R. (2006) *Tetrahedron*, **62**, 3872–3881.

91 Royer, F., Vilain, C., Elkaim, L. and Grimaud, L. (2003) *Org. Lett.*, **5**, 2007–2010.

92 Kang, B., Kim, D., Do, Y. and Chang, S. (2003) *Org. Lett.*, **5**, 3041–3044.

93 Murakami, M., Kadowaki, S. and Matsuda, T. (2005) *Org. Lett.*, **7**, 3953–3956.

94 (a) Padwa, A., Austin, D.J., Gareau, Y., Kassir, J.M. and Xu, S.L. (1993) *J. Am. Chem. Soc.*, **115**, 2637–2647; (b) Casey, C.P. and Dzwiniel, T.L. (2003) *Organometallics*, **22**, 5285–5290; (c) Casey, C.P., Kraft, S. and Powell, D.R. (2002) *J. Am. Chem. Soc.*, **124**, 2584–2594; (d) Barluenga, H., de la Rua, R.B., de Sae, D., Ballesteros, A. and Tomas, M. (2005) *Angew. Chem. Int. ed.*, **44**, 4981–4983; (e) Kim, M. and Lee, D. (2005) *J. Am. Chem. Soc.*, **127**, 18024–18025.

95 Kim, B.G. and Snapper, M.L. (2006) *J. Am. Chem. Soc.*, **128**, 52–53.

96 Debleds, O. and Campagne, J.-M. (2008) *J. Am. Chem. Soc.*, **130**, 1562–1563.

97 (a) Kim, S.-H., Bowden, N. and Grubbs, R.H. (1994) *J. Am. Chem. Soc.*, **116**, 10801–10802; (b) Kim, S.-H., Zuercher, W.J., Bowden, N. and Grubbs, R.H. (1996) *J. Org. Chem.*, **61**, 1073–1081.

98 Zuercher, William J., Scholl, Matthias. and Grubbs, Robert H. (1998) *J. Org. Chem.*, **63**, 4291–4298.

99 Codesido, E.M., Castedo, L. and Granja, J.R. (2001) *Org. Lett.*, **3**, 1483–1486.

100 Boyer, F.-D., Hanna, I. and Ricard, L. (2001) *Org. Lett.*, **3**, 3095–3098.

101 Renaud, J., Graf, C.-D. and Oberer, L. (2000) *Angew. Chem. Int. Ed.*, **39**, 3101–3104.

102 Timmer, M.S.M., Ovaa, H., Filippov, D.V., van der Marel, G.A. and van Boom, J.H. (2001) *Tetrahedron Lett.*, **42**, 8231–8234.

103 Huang, J., Xiong, H., Hsung, R.P., Ramenshkumar, C., Mulder, J.A. and Grebe, T.P. (2002) *Org. Lett.*, **4**, 2417–2420.

104 Gonzalez, A., Dominguez, G. and Castells, J.P. (2005) *Tetrahedron Lett.*, **46**, 7267–7270.

105 Shimizu, K., Takimoto, M. and Mori, M. (2003) *Org. Lett.*, **5**, 2323–2326.

106 Fukumoto, H., Esumi, T., Ishihara, J. and Hatakeyama, S. (2003) *Tetrahedron Lett.*, **44**, 8047–8050.

107 (a) Honda, T., Namiki, H., Kaneda, K. and Mizutani, H. (2004) *Org. Lett.*, **6**, 87–90; (b) Honda, T., Namiki, H., Watanabe, M. and Mizutani, H. (2004) *Tetrahedron Lett.*, **45**, 5211–5214.

108 (a) Boyer, F.-D., Hanna, I. and Ricard, L. (2004) *Org. Lett.*, **6**, 1817–1820; (b) Boyer, F.-D. and Hanna, I. (2002) *Tetrahedron Lett.*, **43**, 7469–7472; (c) Boyer, F.-D. and Hanna, I. (2006) *Eur. J. Org. Chem.*, 471–482.

109 Kitamura, T. and Mori, M. (2001) *Org. Lett.*, **3**, 1161–1164.

110 Kitamura, T., Kuzuba, Y., Sato, Y., Wakamatsu, H., Fujita, R. and Mori, M. (2004) *Tetrahedon*, **60**, 7375–7389.

111 (a) Randl, S., Lucas, N., Connon, S.J. and Blechert, S. (2002) *Adv. Synth. Catal.*, **344**, 631–633; (b) Rückert, A., Eisele, D. and Blechert, S. (2001) *Tetrahedron Lett.*, **42**, 5245–5248.

112 Mori, M., Kuzuba, Y., Kitamura, T. and Sato, Y. (2002) *Org. Lett.*, **4**, 3855–3858.

113 Mori, M., Wakamatsu, H., Tonogaki, K., Fujita, R., Kitamura, T. and Sato, Y. (2005) *J. Org. Chem.*, **70**, 1066–1069.

114 (a) Arjona, O., Csaky, A.G., Murcia, M.C. and Plumet, J. (2000) *Tetrahedron Lett.*, **41**, 9777–9780; (b) Arjona, O., Csaky, A.G., Leon, V., Medel, R. and Plumet, J. (2004) *Tetrahedron Lett.*, **45**, 565–568; (c) Mori, M., Wakamatsu, H., Sato, Y. and Fujita, R. (2006) *J. Mol. Cat. A: Chem.*, **254**, 64–67.

115 (a) Banti, D. and North, M. (2002) *Tetrahedron Lett.*, **43**, 1561–1564;

(b) Banti, D. and North, M. (2002) *Adv. Synth. Catal.*, **344**, 694–704.

116 (a) Kulkarni, A.K. and Diver, S.T. (2003) *Org. Lett.*, **5**, 3463–3466; (b) Peppers, B.P., Kulkarni, A.A. and Diver, S.T. (2006) *Org. Lett.*, **8**, 2539–2542.

117 For a review of natural products synthesis using enyne metathesis: Mori, M. (2007) *Adv. Synth. Cat.*, **349**, 121–135.

118 (a) Kinoshita, A. and Mori, M. (1996) *J. Org. Chem.*, **61**, 8356–8357; (b) Kinoshita, A. and Mori, M. (1997) *Heterocycles*, **46**, 287–299.

119 Layton, M.E., Morales, C.A. and Shair, M.D. (2002) *J. Am. Chem. Soc.*, **124**, 773–775.

120 Kitamura, T., Sato, Y. and Mori, M. (2004) *Tetrahedron*, **60**, 9649–9657.

121 (a) Brenneman, J.B. and Martin, S.F. (2004) *Org. Lett.*, **6**, 1329–1332; (b) Brenneman, J.B., Machauer, R.M. and Martin, S.F. (2004) *Tetrahedron*, **60**, 7301–7314.

122 (a) Mori, M., Tomita, T., Kita, Y. and Kitamura, T. (2004) *Tetrahedron Lett.*, **45**, 4397–4400; (b) Tomita, T., Kita, Y., Kitamura, T., Sato, Y. and Mori, M. (2006) *Tetrahedron*, **62**, 10518–10527.

123 Aggarwal, V.K., Astle, J. and Rogers-Evans, M. (2004) *Org. Lett.*, **6**, 1469–1472.

124 Reddy, S. and Kozmin, S.A. (2004) *J. Org. Chem.*, **69**, 4860–4862.

125 Evans, M.A. and Morken, J.P. (2005) *Org. Lett.*, **7**, 3371–3373.

126 Kummer, D.A., Brenneman, J.B. and Martin, S.F. (2005) *Org. Lett.*, **7**, 4621–4623.

127 Boyer, F.-D. and Hanna, I. (2007) *Org. Lett.*, **9**, 715–718.

128 Bedel, O., Francais, A. and Haudrechy, A. (2005) *Synlett*, 2313–2316.

129 Satcharoen, V., McLean, N.J., Kemp, S.C., Camp, N.P. and Brown, R.C.D. (2007) *Org. Lett.*, **9**, 1867–1869.

130 Kaliappan, K.P. and Ravikumar, V. (2007) *J. Org. Chem.*, **72**, 6116–6126.

12
Ring-Closing Metathesis of Alkynes
Paul W. Davies

12.1
Introduction

Metathesis reactions for ring-closing processes have received tremendous attention over the last decade or so. Indeed, alkene metathesis is now considered a primary tool for cyclizations [see Chapter 11]. This chapter focuses on ring-closing metathesis of alkynes (RCAM) which has also been proven to be a valid synthetic strategy. Although it does not share the same extent of generality as ring-closing metathesis of alkenes (RCM), numerous applications are found in materials and organometallic chemistry as well as in the synthesis of complex organic molecules. This chapter describes the preparation of cyclic products by RCAM starting from variably substituted diyne and enyne-yne precursors. A brief introduction to the development of alkyne metathesis precedes discussion of the RCAM approach, the most common catalyst systems employed in RCAM, and applications in the synthesis of complex organic molecules, organometallics and materials chemistry. The synthetic advantages and some current limitations of the RCAM strategy will be highlighted and representative experimental procedures are given at the end of the chapter.

12.2
Alkyne Metathesis

Catalytic alkyne metathesis was first reported four decades ago over a heterogeneous mixture of tungsten oxides and silica at very high temperature (about 200–450 °C) [1]. This was followed by Mortreux's report of alkyne metathesis by a homogeneous catalysis mixture derived from $Mo(CO)_6$ (or related molybdenum sources) and simple phenol additives such as resorcinol, on heating in high boiling solvents [2]. Early mechanistic insight was provided by Katz and McGinnis who proposed a mechanism based on metal carbyne-promoted metathesis of alkynes [3]. The sequence of formal [2 + 2] cycloaddition and cycloreversion steps proceeding through a metallacyclobutadiene intermediate (Scheme 12.1) is analogous to the metallocyclobutane

Scheme 12.1 The acknowledged mechanism for alkyne metathesis.

invoked in alkene metathesis. This was subsequently verified experimentally by Schrock using high valent metal alkylidynes, with isolated metallocyclobutadiene intermediates shown to be catalytically competent species for alkyne metathesis [4].

The subsequent development of alkyne metathesis proceeded with refinement of the catalyst systems, the introduction of well-defined and isolable metal alkylidynes, and extension of the process toward more complex substrates. A number of reviews which cover these developments have been published recently [5]. A key contribution for the synthetic application of alkyne metathesis came in 1998, 30 years after the initial report of catalytic alkyne metathesis, when Fürstner and Seidel reported the first catalytic RCAM for the synthesis of functionalized macrocycles [6]. Both the Mortreux system and a well-defined metal-alkylidyne first prepared by Schrock in 1981 [4, 7] were employed in this chemistry.

12.3
Ring-Closing Alkyne Metathesis as a Synthetic Strategy

RCAM is the reaction of an acyclic diyne to afford a cyclic alkyne and an acyclic alkyne side-product (Scheme 12.2). The reaction is attractive in synthetic terms as it does not require the preparation and use of highly reactive functional groups. Furthermore, no reagents other than the substrate and catalyst system are required. A similar retro-synthetic logic to that used in RCM can be applied.

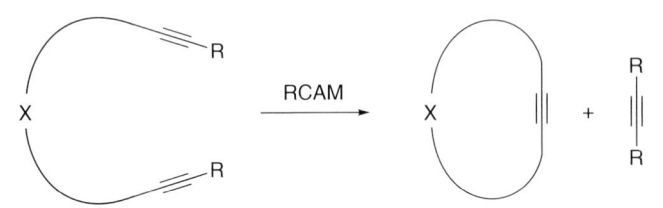

Scheme 12.2 The RCAM schematic.

For efficient RCAM, the substrates must bear end-capped alkynes (R = Me, Et) as terminal alkynes are incompatible with the catalyst systems currently used [6–9]. This has two synthetic implications; first, that any strategy incorporating RCAM must allow for substitution of the alkyne. Second, the alkyne side-product (but-2-yne or hex-3-yne) may not be readily purged from the reaction system, particularly under the milder reaction conditions. This can retard the rate of substrate conversion as the removal of the undesired metathesis side-product generally helps drive a reaction toward ring closure (cf. ethylene in RCM). Additionally, the polymerization of

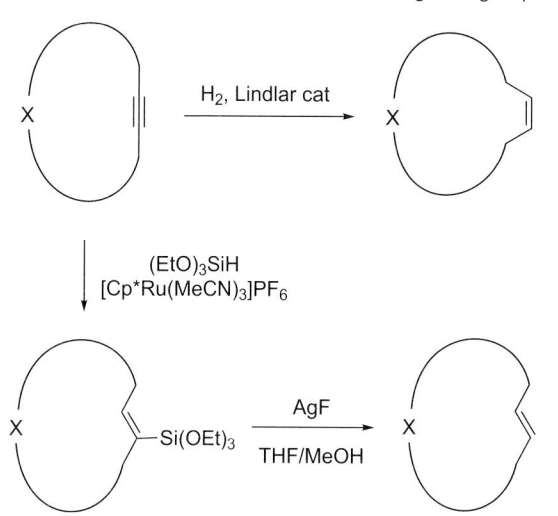

Scheme 12.3 Access to stereodefined macrocyclic alkenes post-RCAM.

but-2-yne by ring-expansion in the presence of Mo(VI) and W(VI) complexes is a possible catalyst deactivation pathway [5a, 10]. To minimize the impact of these issues, RCAM reactions are often run under reduced pressure to remove the side-product.

Although the scope and generality of many synthetic methods pale in contrast with alkene metathesis, the relationship between RCAM and RCM inevitably invites comparison. RCAM is limited due to the intrinsic requirement for the cyclic product to be of sufficient size to accommodate an alkyne. Therefore, a fairer comparison would be to consider macrocylic RCM and RCAM. While RCM is undoubtedly a superb macrocyclization strategy if the geometry of the resulting alkene is ultimately unimportant, it can be a risky strategy to employ when a stereodefined alkene is required. This predicament has been demonstrated in a number of cases where the undesired isomer has been formed preferentially [11].

In contrast, it was recognized at the outset of RCAM development that an alkyne offers an excellent opportunity to access alkenes with rigorous stereocontrol (Scheme 12.3) [5d, 6]. Indeed, the combination of RCAM and Lindlar reduction provides a stereoselective entry to (*Z*)-configured alkenes. Although inherently less attractive than a one-step strategy, similar retrosynthetic logic can be applied as with RCM whilst introducing a predictable component for sterocontrol.

When RCAM was introduced, the semi-reduction of acetylenes to (*E*)-alkenes was less well-established in the literature. However, more recent reports by Trost [12] and Fürstner [13] demonstrated that the hydrosilylation of alkynes, acyclic and cyclic respectively, proceeds with highly stereoselective trans-addition [14] using $(EtO)_3SiH$ and the cationic ruthenium complex $[Cp^*Ru(MeCN)_3]PF_6$. The reaction proceeds with good functional group tolerance. A subsequent protodesilylation step using AgF affords the (*E*)-alkene [13].

Despite there being relatively few final targets bearing alkyne moieties, at least in natural products synthesis, an alkyne provides numerous opportunities for synthetic manipulation. Although semi-reduction is to date the most widely used application, some alternative strategies are highlighted in Section 12.5. Furthermore, the rapidly growing interest in alkyne activation for a plethora of reactions suggests that processes to introduce alkyne moieties will become more important [15].

12.4
Catalyst Systems for Ring-Closing Alkyne Metathesis

RCAM applications have predominantly employed one of three catalyst systems; the "instant" Mortreux-type formed from the reaction of $Mo(CO)_6$ **1** and phenols; a well-defined tungsten alkylidyne **2** developed by Schrock; and the mixture of trisamido molybdenum complex **3** and CH_2Cl_2 developed by Fürstner (Figure 12.1).

12.4.1
The Mortreux "Instant" System

This "instant" system represents a very practical and easily applicable method which employs cheap, commercially available and stable reagents without the need to use rigorously purified solvents or an inert atmosphere. However, the reaction conditions are fairly harsh and the catalyst displays relatively low activity, limiting the general applicability of this system. Since its original discovery several modifications have been made to the reagents and reaction conditions employed in order to address these issues. These have recently been reviewed [5b,d].

A typical RCAM with this system will employ catalytic quantities of **1** alongside stoichiometric (or greater) quantities of an electron-deficient phenol, such as chlorophenol or trifluoromethylphenol, in a high boiling solvent, chlorobenzene for instance, at ~140 °C.

12.4.2
Tungsten Alkylidyne Catalysts

Part of a family of well-defined high valent metal alkylidyne complexes showing high metathesis activity [5c,h], the commercially available complex **2** is metathesis active under mild conditions, typically ambient temperature up to 90 °C. RCAM reactions

(Ar = 3,5-dimethylphenyl)

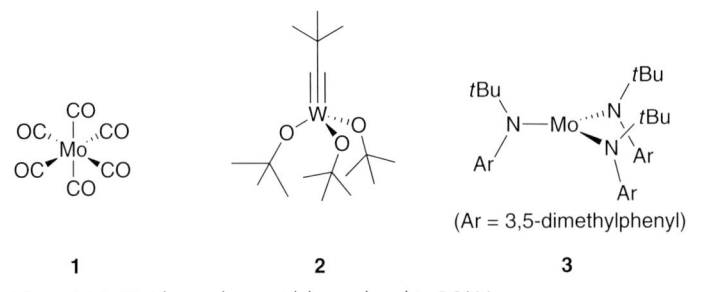

1   **2**   **3**

Figure 12.1 Metal complexes widely employed in RCAM.

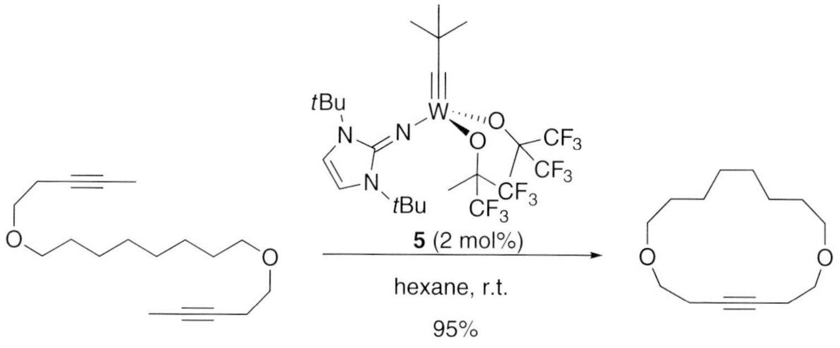

Scheme 12.4 Use of imidazolin-2-iminato-substituted tungsten alkylidyne **5** in RCAM.

with **2** are often run at 80 °C in toluene or a related aromatic solvent. In comparison with the "instant" system, rigorously inert (anhydrous and oxygen-free) conditions are required. Several modifications have been made to this structure to modify reactivity, and a recent report demonstrated that the effective RCAM of diyne **4** could be performed with imidazolin-2-iminato tungsten alkylidyne catalyst **5** at room temperature to afford cyclic alkyne **6** in excellent yield (Scheme 12.4) [16]. The same catalyst has also been employed in the preparation of cyclophanes [17].

12.4.3
Molybdenum-Based Catalysts

The apparent greater functional group tolerance of molybdenum species than tungsten ones provides an impetus to explore the metathetic reactivity of molybdenum alkylidynes [5c]. In 1999, Fürstner demonstrated that trisamido molybdenum species of the general type Mo[N(tBu)(Ar)]$_3$, developed by Cummins for dinitrogen activation [18], enabled metathesis in the presence of halogenated hydrocarbons such as CH_2Cl_2 [19, 20].

Analysis of the reaction between **3** and CH_2Cl_2 proved molybdenum chloride **7** and terminal alkylidyne **8** to be the major products (Scheme 12.5). Whereas **8** affected

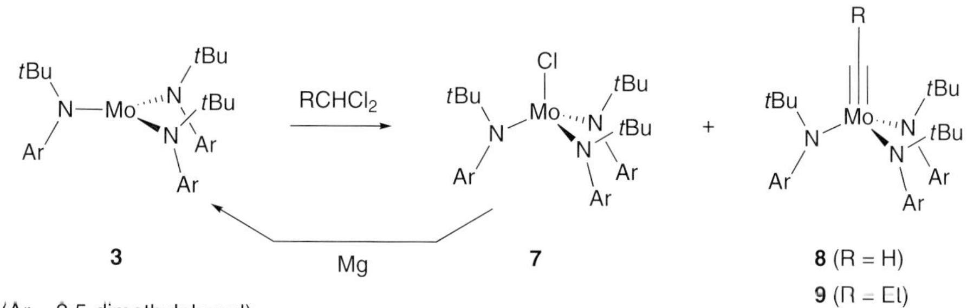

(Ar = 3,5-dimethylphenyl)

Scheme 12.5 Activation of the molybdenum trisamido complex **3** to form metathesis-active complexes and a "reductive recycle" strategy.

only one turnover in alkyne metathesis, **7** was catalytically active at 80 °C. Further refinement of this process was achieved by Moore using a "reductive-recycle" strategy whereby chloride **7** is reduced by magnesium back to **3** (Scheme 12.5) [10, 21]. The use of a higher order alkyl chloride gives rise to catalytically competent non-terminal alkylidyne **9**, which can be further modified by ligand exchange with alcohols into trisalkoxy molybdenum alkylidynes, known to be catalytically active [10, 21, 22].

While the metathesis active complex **7** can be isolated, applications have used *in situ* generation of the catalyst, typically **3** is reacted with the diyne in a toluene/CH_2Cl_2 mixture at ~80 °C under an inert atmosphere.

12.4.4
Comparison of the Reaction Systems

In the first report of RCAM a range of diynes, end-capped with methyl or ethyl units, were subjected to reaction with either the Mortreux "instant" system or tungsten alkylidyne **2** under high dilution conditions [6]. Whilst the "instant" system was effective for robust substrates, the reaction conditions were too harsh for more labile units. Complex **2** displayed significantly better generality, tolerating, for example, acetal, alkene, amide, carbonyl, enoate, ester, ether, furan, ketone, silyl ether, sulfonamide, and sulfone units [5d]. However, the high Lewis acidity of the tungsten center rendered substrates containing thioether, basic nitrogen groups and polyethers incompatible. In contrast, the combination of **3** and CH_2Cl_2 displayed wider applicability and tolerated functional groups which completely shut down the catalytic activity of **2** (see Table 12.1). Further comparative examples are given in

Table 12.1 A comparison of the effectiveness of the most common RCAM catalyst systems.

Product		Yield	
	"Instant"	Complex 2	Complex 3/CH_2Cl_2
(structure)	64%	73%	91%
(structure)	Decomposition[a]	55%[a]	74%[b]
(structure)	—	0	84%

[a] $n = 1$.
[b] $n = 2$.

12.5
Complex Molecule Synthesis using Ring-Closing Alkyne Metathesis

the target driven applications studied in Section 12.5. The enhanced reactivity profile of $3/CH_2Cl_2$ has been explained by a proposal that the close packing amido ligands efficiently shield the central molybdenum in the catalytically active template. By preventing coordination of potential donor substrates the effective Lewis acidity of the metal is thus attenuated [20]. In contrast, "acidic" protons such as those of secondary amides, which are tolerated by **2**, are incompatible with **3**, though tertiary amides are fully compatible. As with **2**, **3** is compatible with esters, isolated double bonds, silyl ethers, sulfones, aldehydes, nitro groups, ketones, alkyl chlorides, acetals and nitriles.

12.5
Complex Molecule Synthesis using Ring-Closing Alkyne Metathesis

12.5.1
Alkyne–Alkyne Metathesis

The RCAM strategy was migrated into target synthesis applications almost immediately after its initial demonstration as a viable methodology. The earliest examples of natural products prepared in this manner include the macrocyclic musks and fragrances, ambrettolide **10** [23], yuzu lactone **11** [23] and civetone **12** [24] (Figure 12.2), the olfractory properties of which depend on the double bond configuration.

RCAM of methyl-capped acyclic diynes was performed using the Mortreux "instant" system followed by near quantitative Lindlar hydrogenation to afford exclusively the required (Z)-alkene products [23, 24]. The preparation of **12** demonstrates the compatibility of both the user-friendly low-tech *in situ* approach of a **1**/phenol mixture and the more refined tungsten complex **2** with a carbonyl functionality (Scheme 12.6). These early examples provided a clear demonstration of the complementary nature of the RCAM/semi-reduction manifold to an RCM strategy which mainly led to the undesired (E)-isomer [25].

In an analogous manner, the insect-repellent alkaloids epilachnene and homologues, and the cytotoxic sponge extract motuporamine C (Scheme 12.7) [26] were prepared by an RCAM/Lindlar hydrogenation approach using either **2** or the "instant" system. The high yielding formation of the macrocyclic (Z)-alkenes in a

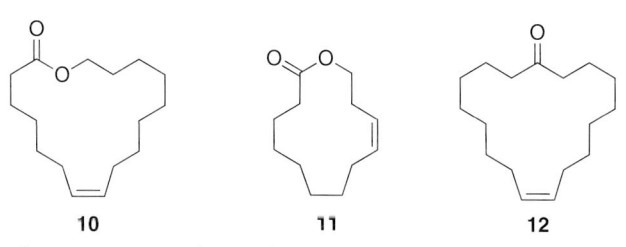

Figure 12.2 Some early natural product targets for the RCAM/Lindlar hydrogenation synthetic strategy.

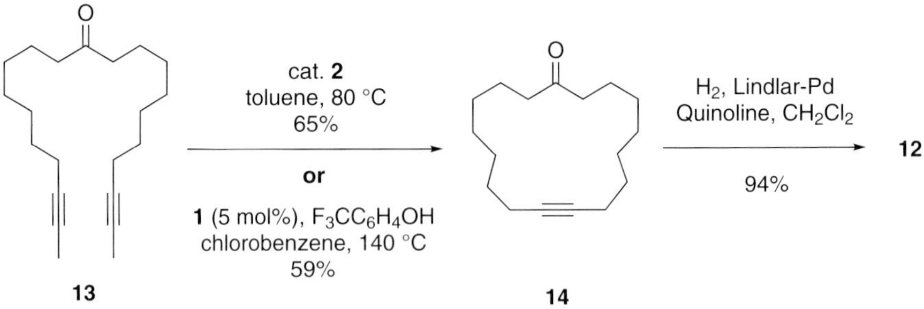

Scheme 12.6 The RCAM/Lindlar hydrogenation manifold in the synthesis of civetone.

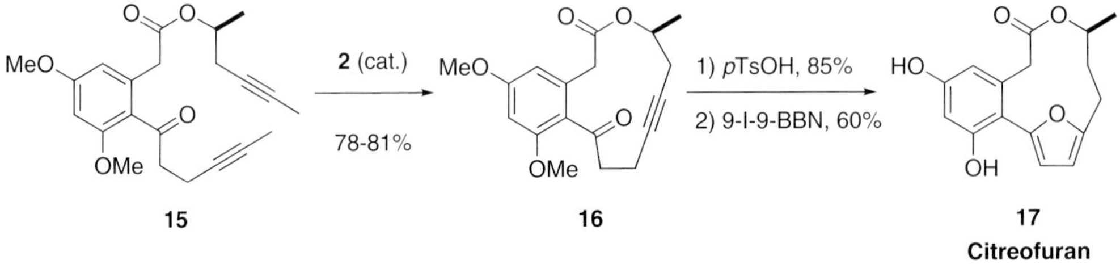

Scheme 12.7 A RCAM cycloisomerization approach to citreofuran.

fully stereoselective manner again compared well to related RCM approaches toward each of these molecules [25a, 27].

While macrocylic structures containing (Z)-olefins now suggest an obvious site for a RCAM-based disconnection strategy, the alkyne can also be employed for other means. In the synthesis of citreofuran, RCAM was incorporated into the retrosynthetic analysis as an alkyne-one precursor for the furan ring, bypassing a more obvious disconnection at the biaryl axis. RCAM using tungsten complex **2** worked well in the presence of the unprotected ketone although the importance of ensuring there were no catalyst poisoning impurities was noted (Scheme 12.7). Subsequently, under acidic conditions the yne-one cycloisomerizes to afford the furan [28].

Tungsten complex **2** has also been employed in the preparation of diaminosuberic acid derivatives (Scheme 12.8). Alkyne-bearing amino acid derivatives were linked via a dibenzyl unit in **18**. After RCAM, hydrogenation cleaves the benzyl unit and effects complete reduction of the alkyne [29].

A range of diyne-bearing oligopeptides have been tested under standard RCAM conditions with the alkylidyne **2** to access cyclic β-turn mimics [30] and mimics of the lantibiotic nisin Z [31]. Cyclization of **21** with **2** demonstrated how subtle changes can have severe impact on the outcome of the metathesis event (Scheme 12.9). While the secondary amide linkages in **21** presented no problem, the presence of an additional methyl group completely shut down the desired reaction [21].

Oligopeptide **23** bearing both alkyne and alkene side chains underwent selective metathesis to afford the cyclic alkyne **24**, albeit in low yield (Scheme 12.10) [22].

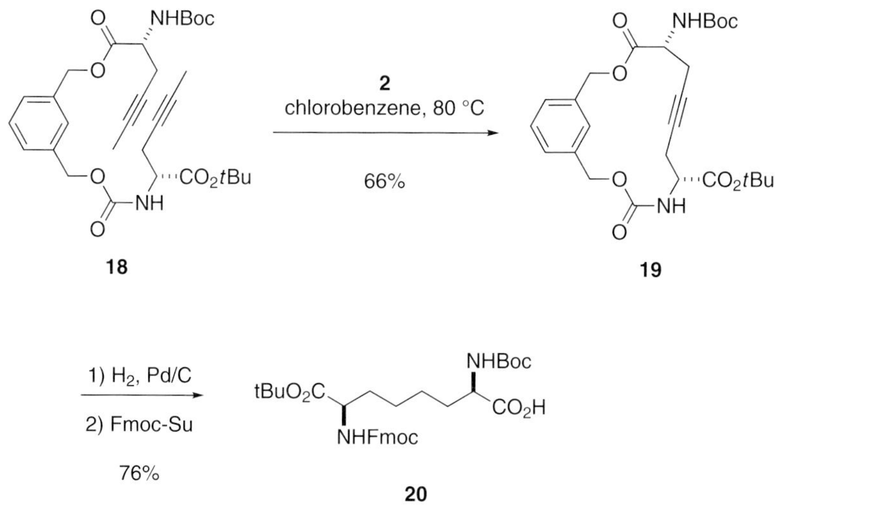

Scheme 12.8 Preparation of diaminosuberic acid derivatives by a RCAM hydrogenation approach.

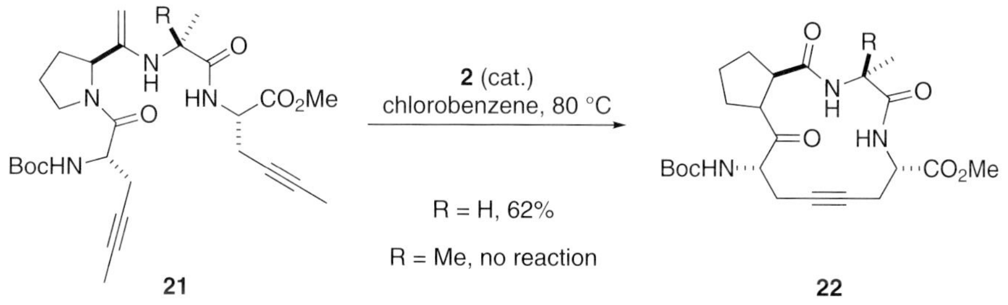

Scheme 12.9 RCAM on diynes with oligopeptides.

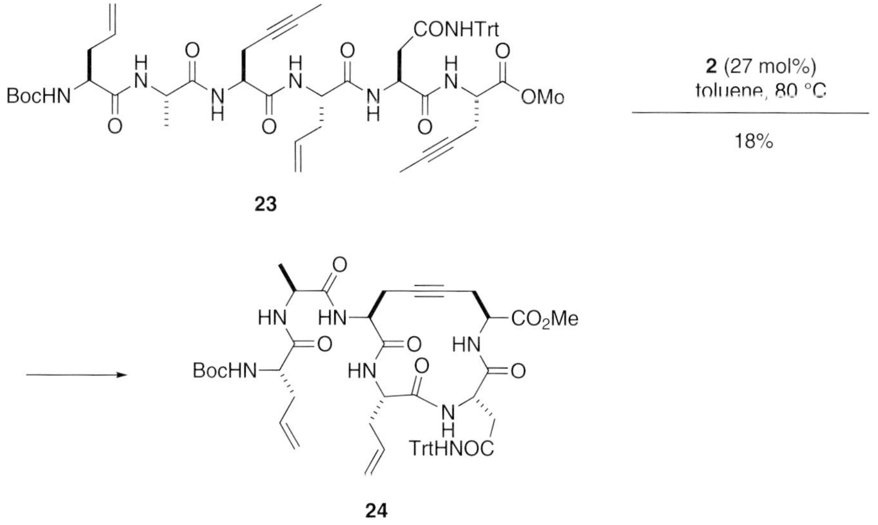

Scheme 12.10 An RCAM approach to lantibiotic nisin Z mimics.

Scheme 12.11 Microwave heating in RCAM.

In cases where the substrate is suitably robust, the "instant" Mortreux type system of **1** and phenols can be ideal. In the synthesis of turrianes, naturally occurring cyclophane derivatives, a modified Mortreux-type reaction system was employed. Heating diyne **25** under microwave conditions also allowed the overall reaction time to be reduced from 6 h to 5 min (Scheme 12.11). The cyclic product **26** was then subjected to Lindlar reduction and deprotection of the phenol units to access the target compound [32].

The "instant" system, however, was not suitable for the preparation of elaborate and sensitive materials such as those used in the synthesis of a prostaglandin lactone. In this case the molybdenum complex $3/CH_2Cl_2$ system proved to be most effective (Scheme 12.12), with tungsten alkylidyne **2** requiring longer reaction times and resulting in lower conversion. The "instant" system gave no product [33].

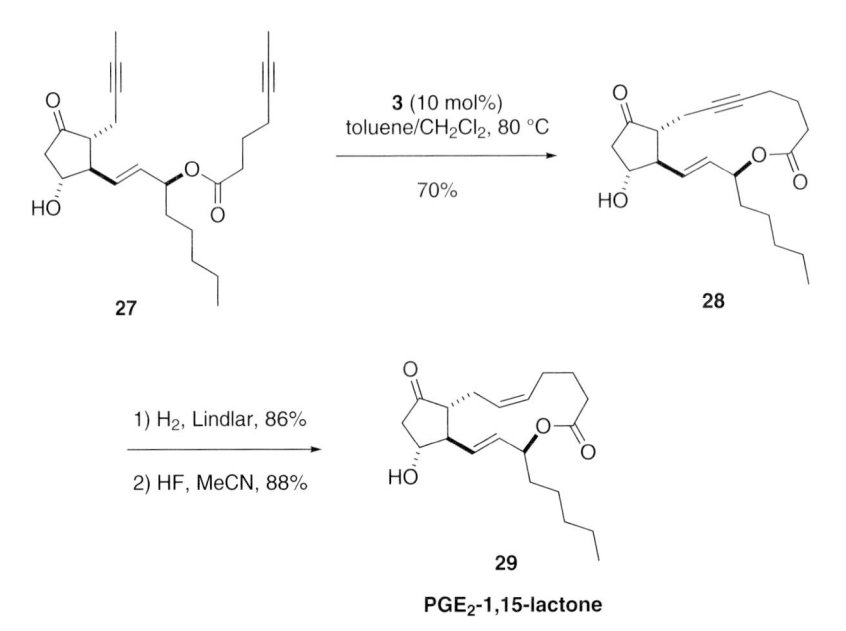

Scheme 12.12 Prostaglandin lactone synthesis.

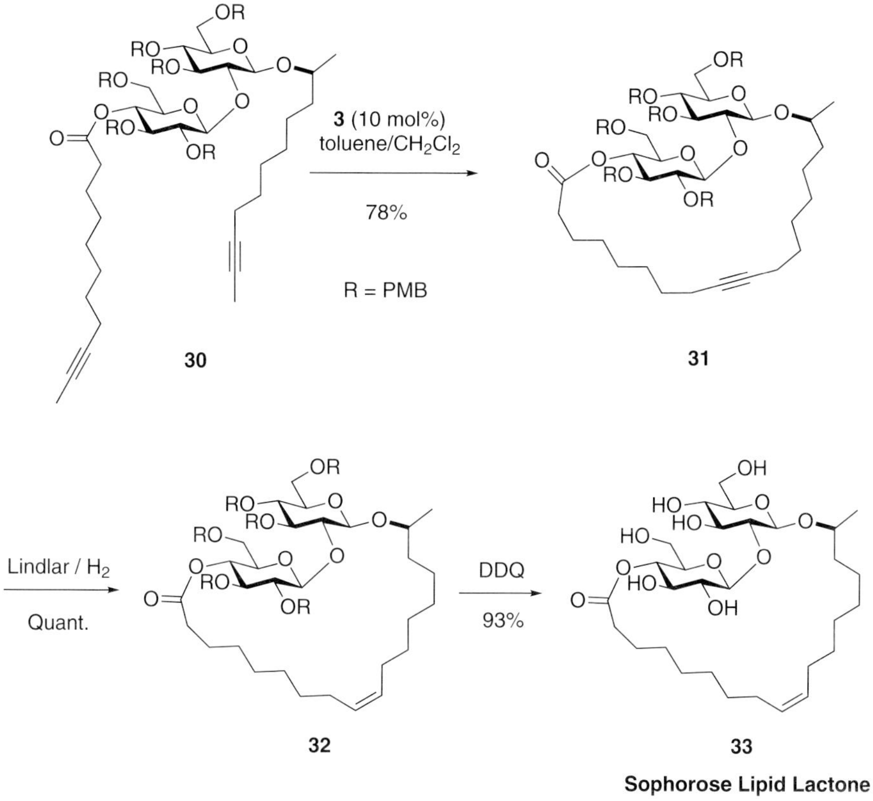

Scheme 12.13 Sophorolipid lactone synthesis.

The enhanced functional group tolerance of the 3/CH$_2$Cl$_2$ catalyst system was also applied to the synthesis of the secondary metabolite sophorolipid lactone **33**. Neither the acid-labile PMB ethers nor the glycosidic linkages were damaged by the Lewis acidic metal center of the catalyst during RCAM of diyne **30** (Scheme 12.13) [34]. RCAM has also been used to prepare alkyne-linked glycoamino acids by RCAM [35].

The syntheses of macrocyclic salicylate lactone cruentaren A **36**, reported independently by the groups of Maier [36] and Fürstner [37], highlighted the ability of both **2** and **3** to mediate complex RCAM processes, whilst demonstrating the improved reactivity profile of 3/CH$_2$Cl$_2$ toward acid labile units. The acyclic diynes **34** used in the two syntheses differ only by choice of protecting group and the extent of functionalization on the side chain (Scheme 12.14). Both RCAM processes proceeded smoothly although it was noted that complex **2** was incompatible with the OTHP functionality, necessitating the use of 3/CH$_2$Cl$_2$.

In the subsequent transformation to **36**, Lindlar reduction was carried out immediately after RCAM by Fürstner and coworkers, but as the final step of the synthesis by Vintonyak and Maier.

The synthesis of epothilones containing a (Z)-alkene was a particularly enlightening example of the potential of RCAM. The first three syntheses of epothilone A

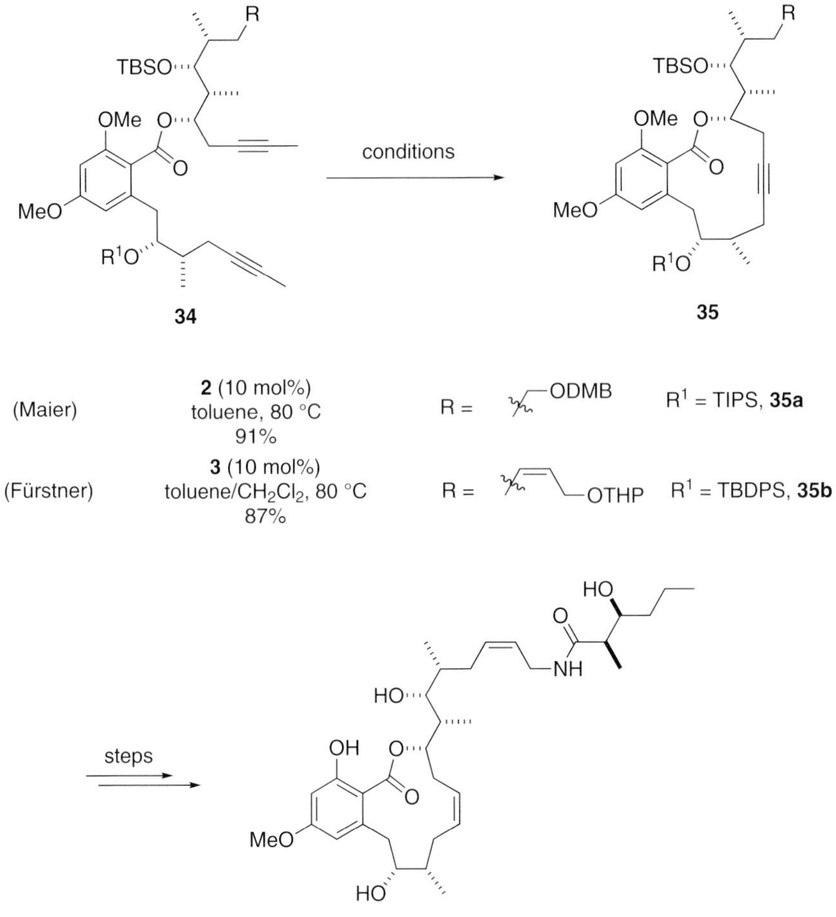

Scheme 12.14 RCAM/Lindlar reduction strategy toward cruentaren A.

used RCM to form the macrocycle with little, if any, selectivity toward the required alkene geometry, which represented a significant issue at the late stage of highly involved syntheses [11a-c]. In comparison the RCAM/Lindlar hydrogenation manifold provided selective access into the (Z)-alkene **39** in high yield (Scheme 12.15). Aside from the issue of alkene geometry, this synthesis highlighted the preparative relevance of the RCAM method by leaving labile functionality such as the aldol substructure intact, not effecting epimerization of the chiral center α to the carbonyl and proceeding even in the presence of the basic nitrogen and sulfur atoms of the thiazole [38].

The use of the RCAM/Lindlar manifold in the synthesis of epothilones represented a watershed moment in demonstrating that the strategy could be used for molecules of the same order of complexity that RCM was being applied to. The benign nature of the RCAM approach toward labile functionality has been further exemplified in the

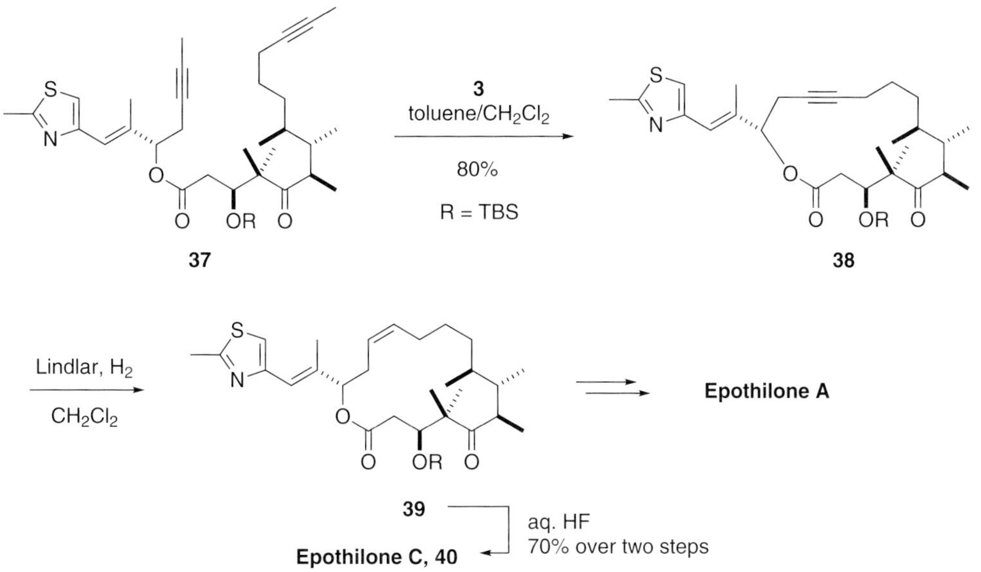

Scheme 12.15 RCAM-based synthesis of epothilones.

synthesis of what is believed to be amphidinolide V, where the strategy was coupled with an alternative use of the cyclic alkyne product (Scheme 12.16) [39]. RCAM of diyne **41** using **3**/CH$_2$Cl$_2$ proceeded smoothly despite the labile trans-configured vinyl epoxide and the allylic alcohol, to afford a particularly strained system **42**. Subsequent enyne metathesis under ethylene atmosphere with Grubbs' second generation ruthenium carbene installed the two exomethylene units in **43**.

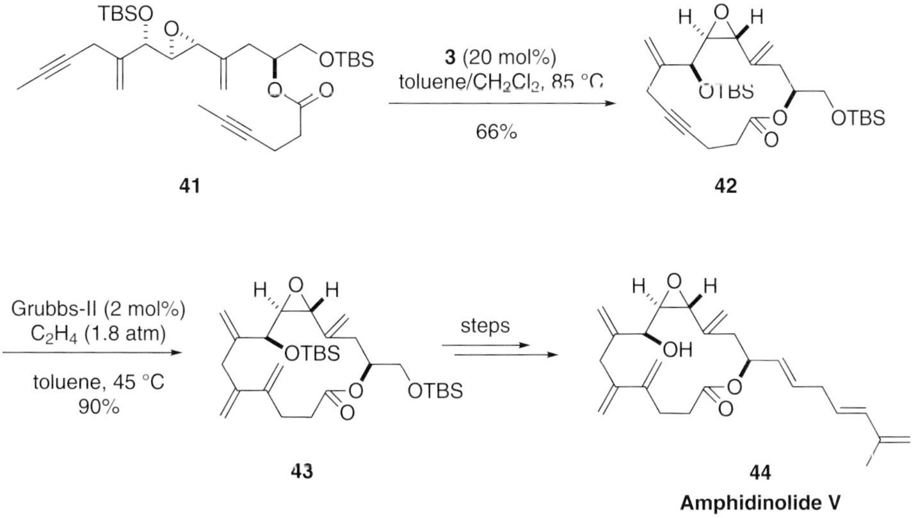

Scheme 12.16 The synthesis of amphidinolide V via an RCAM/enyne metathesis approach.

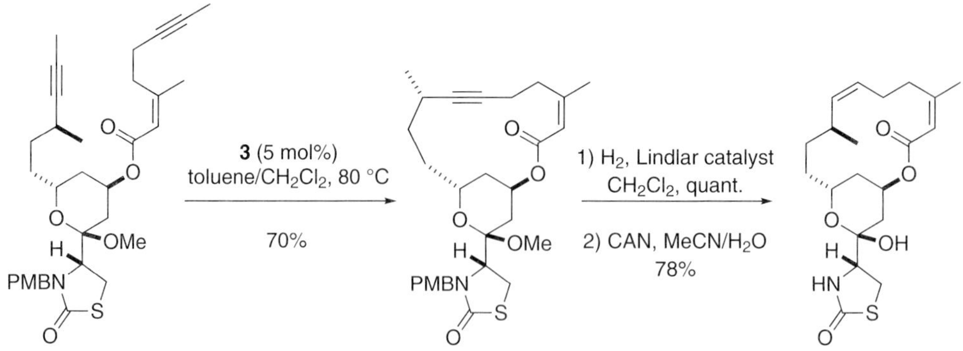

Scheme 12.17 Synthesis of latrunculin B via RCAM.

Completion of the synthesis gave a compound which corresponded to the NMR data of the isolated natural product with the exception of a singular resonance discrepancy. The RCAM/enyne metathesis synthetic approach was flexible enough to accommodate the preparation of isomers of the proposed structure, all of which were discovered to match the reported data less closely [39].

The syntheses of a range of actin-binding macrolides (latrunculins A, B, C, M and S) futher demonstrated the utility of both the RCAM/Lindlar hydrogenation strategy and the catalyst system derived from **3**. First employed to access latrunculin B (Scheme 12.17) [40], and similarly applied to latrunculin C, M and analogues [42], RCAM on diyne **45** worked well despite the dense array of functional groups, such as the mixed ketal and the thioazolidinone unit, and the branching α-substituent to one of the alkyne units. The tungsten complex **2** also worked but gave a slightly lower (63%) yield of **46**.

12.5.2
Enyne-Yne Metathesis

The most potent actin-binding macrolide of the latrunculin series, latrunculin A, represented a particularly interesting challenge for RCAM. The retrosynthetic strategy required a conjugated (Z)-olefin to be introduced and as such would represent the first use of an enyne-yne RCAM system in total synthesis. Prior to the report of this synthesis, a study employing the tungsten complex **2** had shown that the metathesis of model enyne-yne systems was viable (Scheme 12.18) [13b]. When the RCAM is coupled with reduction, the orthogonality of RCAM toward alkenes bypasses some of the difficulties associated with RCM of conjugated dienes, where reaction can occur at either the terminal or internal double bond, reducing the synthetic appeal of such an approach.

The synthesis of latrunculin A **52** not only required chemoselective enyne-yne metathesis but also led to a smaller and more constrained cyclic structure than previously attempted (a 16-membered ring). However, under the same reaction

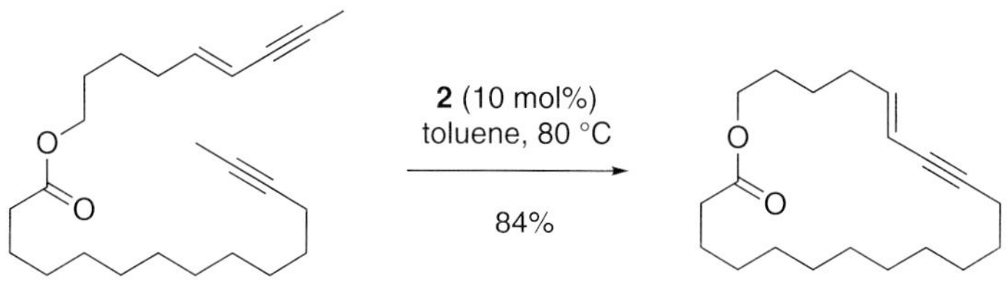

Scheme 12.18 A model study of enyne-yne metathesis to give conjugated cyclic enynes.

conditions employed for the other members of the latrunculin series the enyne-yne RCAM of **50** proceeded smoothly (Scheme 12.19). While a late stage protecting group exchange on the thiazolidinone from PMB to Teoc was required, this was a consequence of issues in the subsequent deprotection step, rather than with the RCAM. Lindlar hydrogenation of **51** thus installed the required (Z)-olefin [41, 42].

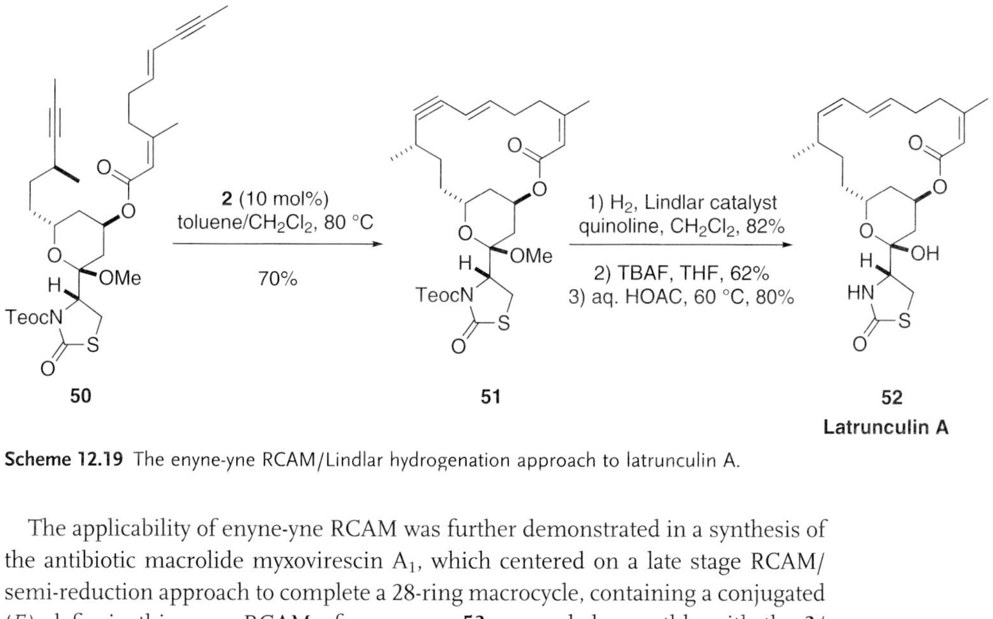

Scheme 12.19 The enyne-yne RCAM/Lindlar hydrogenation approach to latrunculin A.

The applicability of enyne-yne RCAM was further demonstrated in a synthesis of the antibiotic macrolide myxovirescin A_1, which centered on a late stage RCAM/semi-reduction approach to complete a 28-ring macrocycle, containing a conjugated (E)-olefin in this case. RCAM of enyne-yne **53** proceeded smoothly with the **3**/CH_2Cl_2 system despite the presence of the methyl ether close to the reaction site (Scheme 12.20). However, a high catalyst loading was required to achieve a high yield of the 28-membered cyclic alkyne **54**, probably due to the presence of a secondary amide linkage. The **3**/CH_2Cl_2 catalyst system was significantly more effective than the tungsten alkylidyne **2** in this case [43].

While the technology for semi-reduction to (Z)-olefins has proven to be robustly stereospecific, the semi-reduction to (E)-olefins has been less widely studied. Although model studies on cyclic enyne systems gave the desired outcome with very high selectivity [13b], application of the hydrosilylation technique to **54** gave

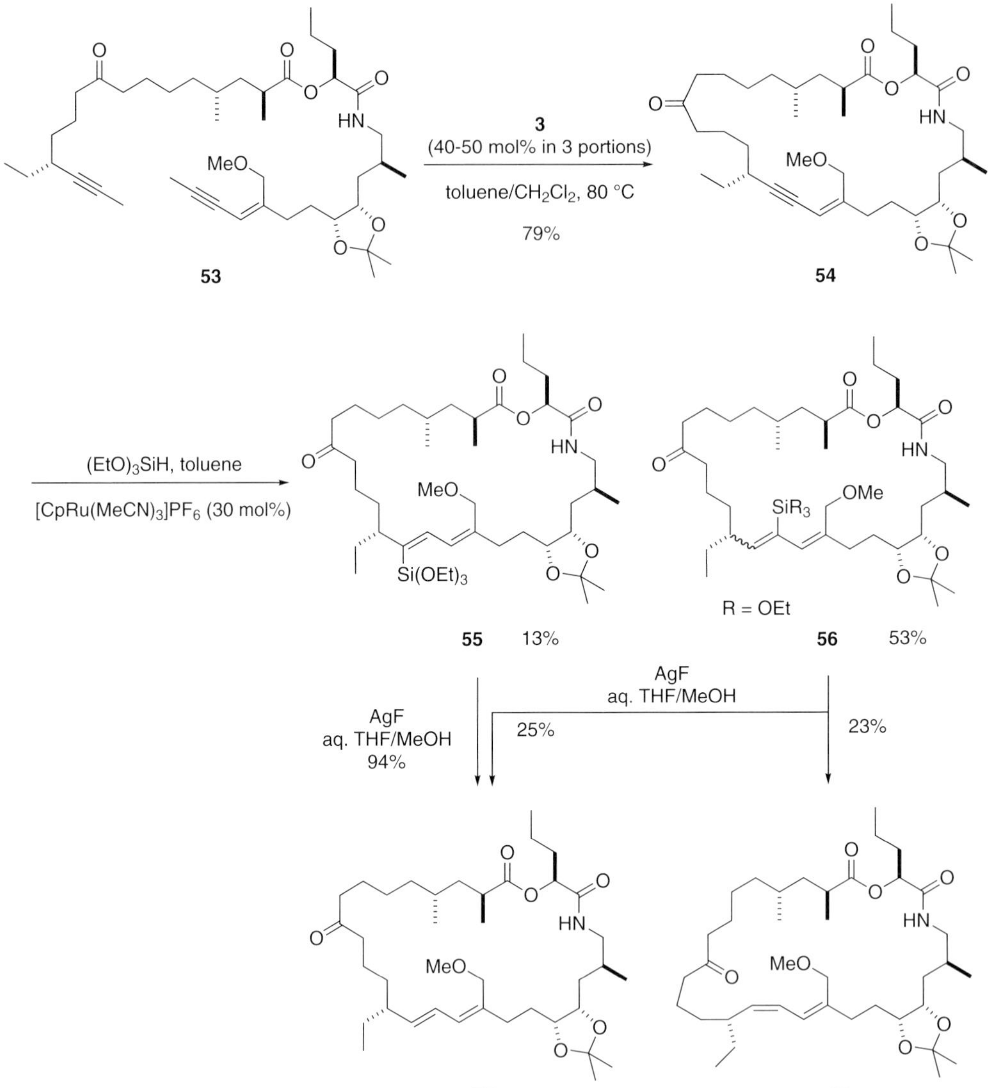

Scheme 12.20 The enyne-yne RCAM/hydrosilylation/
protodesiylation approach to myxovirescin A$_1$.

mixed results [12, 13]. The sterically less demanding [CpRu(MeCN)$_3$]PF$_6$ catalyst was ultimately used, affording 3 isomeric vinyl silanes **55** and **56** showing that the reaction had not been rigorously stereoselective on this substrate. Subsequent protodesilylation therefore gave a mixture of the desired (*E*,*Z*)-configured product **57** alongside the (*Z*,*Z*)-configured isomer **58**, demonstrating that this technology requires further improvement.

12.6
Ring-Closing Alkyne Metathesis within Transition Metal Coordination Spheres

An alternative application for RCAM has been reported by Gladysz and coworkers in the synthesis of metallamacrocycles by RCAM. Ligated alkyne-bearing species are metathesized to prepare chelated rhenium, ruthenium and platinum complexes. In the metathesis of diyne **59** the main product of the reaction is determined by the choice of catalyst system: tungsten complex **2** gives the *cis*-chelate **60**, whereas the "instant" system supplies the isomeric *trans*-chelate product **61** (Scheme 12.21) [44].

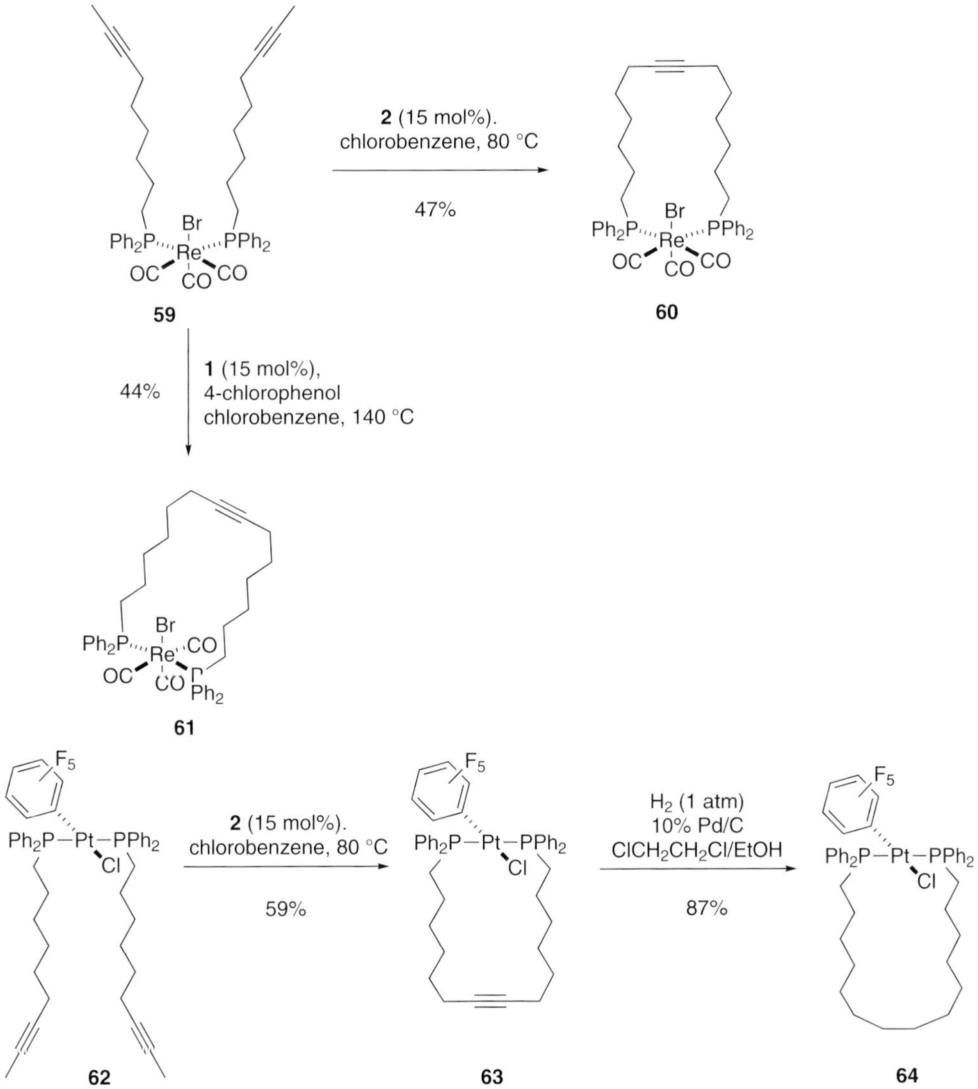

Scheme 12.21 RCAM within transition-metal coordination spheres.

12.7
Cyclo Dimerizations, Trimerizations, and Oligomerizations

Alkyne metathesis has been used widely in the area of functional materials preparation. The discussion in this section will be limited to a brief introduction to some recent examples in which an RCAM event eventually takes place [5, 45]. In comparison to the synthesis of complex natural products which predominantly employ the more well-defined metal complexes **2** and **3** in RCAM, the "instant" Mortreux-type systems enjoy wider application in this field as the relatively simple substrates employed are better able to tolerate the robust reaction conditions.

In their study toward substructures that would allow investigation of theoretical network properties of graphyne, the theoretical carbon allotrope consisting of sp- and

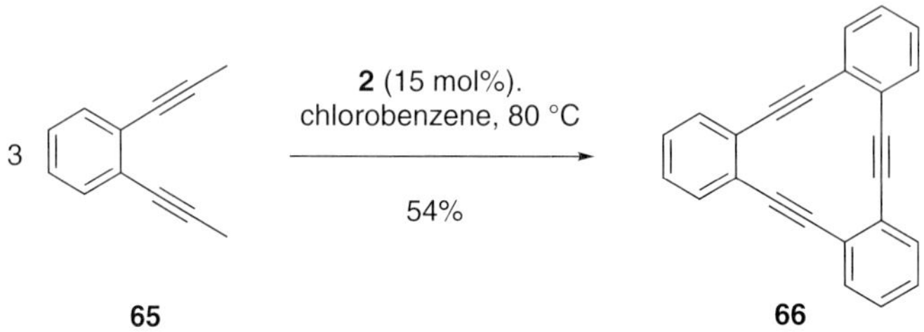

Scheme 12.22 Cyclotrimerization by alkyne metathesis.

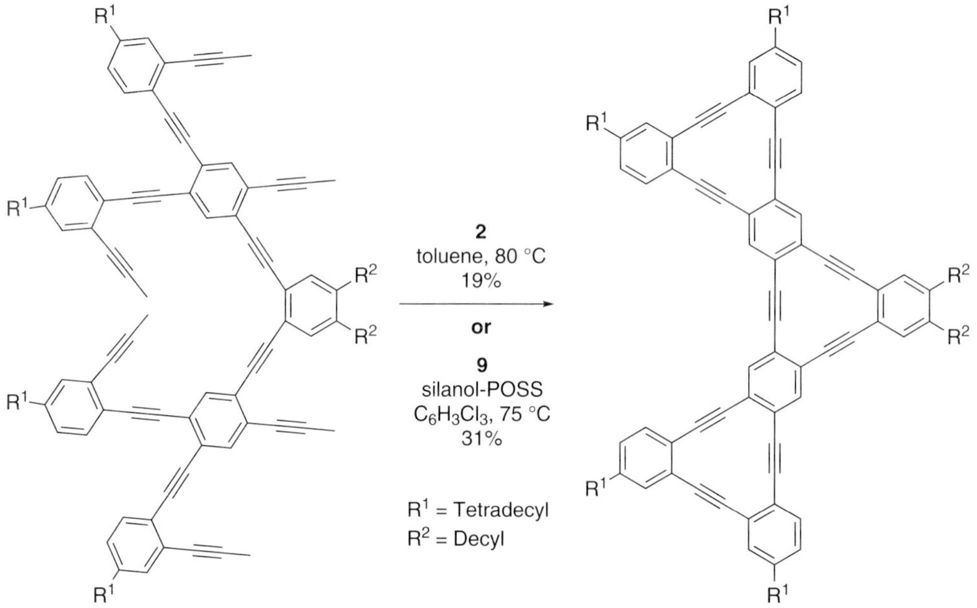

Scheme 12.23 Synthesis of a tri[12]cyclyne by RCAM.

sp²-hybridized carbon atoms [46], Haley and coworkers demonstrated an extension of the cyclodimerization and cyclotrimerization of acyclic diynes, which proceed smoothly, for example using the tungsten alkylidyne **2** (Scheme 12.22) [47]. Larger graphyne substructures were prepared by multiple RCAM of poly-yne **67** to afford the triscyclyne **68** in 19% yield with tungsten catalyst **2** or in 31% yield using molybdenum alkylidyne **9** (Scheme 12.23) [48]. The polyhedral oligomeric silsequioxane additives (silanol-POSS) employed in the latter system were used in place of phenols as they have been shown to limit polymerization of but-2-yne side-products [49]. The preparation of these synthetically challenging targets in such a rapid manner by RCAM further highlights the potential utility of alkyne metathesis.

The unique structures and novel properties of oligomeric macrocycles such as **70** have resulted in several synthetic approaches during which alkyne metathesis has been proven highly effective [50–55]. However, while the conventional vacuum-driven conditions have been employed successfully to access such compounds on a small scale, these processes do not scale-up well, possibly due to a "pseudo-poisoning effect" of the by-product but-2-yne on the metathesis catalyst. The preferential reaction of but-2-yne over aryl alkynes with the catalyst effectively reduces the efficacy of the catalyst toward the desired pathway. A technique to

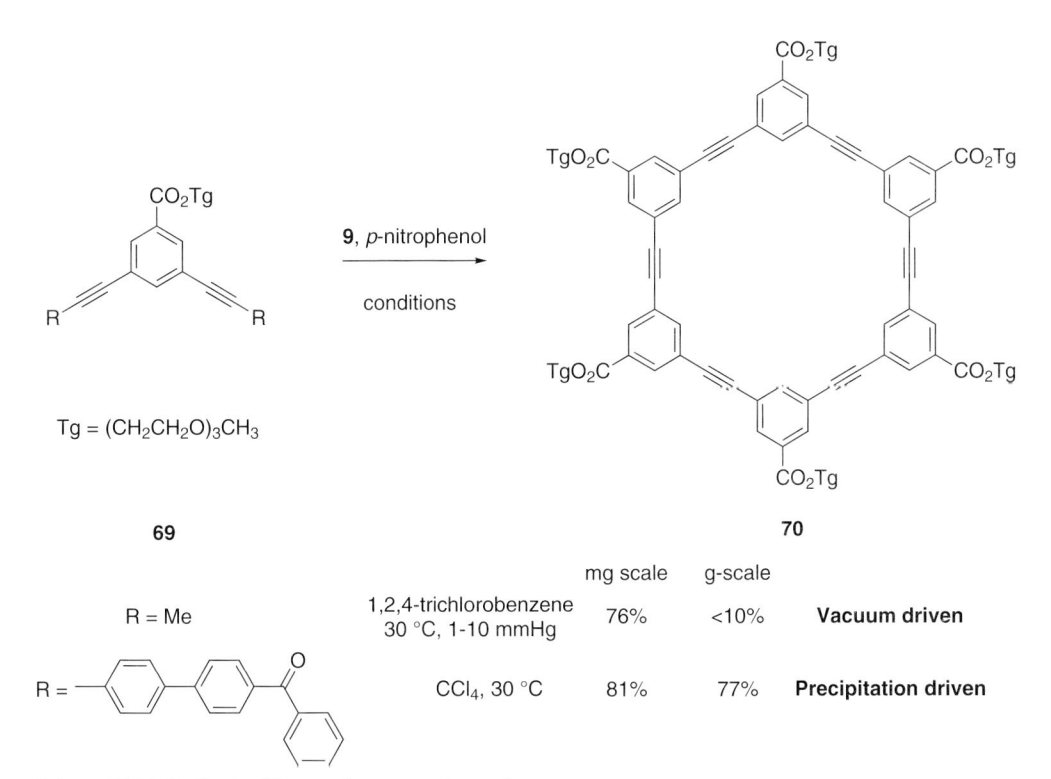

Scheme 12.24 Synthesis of hexameric macrocycles under vacuum-driven (non-scalable) and precipitation-driven (scalable) conditions.

bypass this issue has been developed by Moore and coworkers, employing substrates from which the acetylene by-product is poorly soluble and less reactive (Scheme 12.24) [56]. Precipitation of the by-product drives forward the metathesis equilibrium, thus removing any requirement for a vacuum. Although the atom economy with respect to substrate is diminished due to the sacrifice of a large unit, this approach has successfully allowed multi-gram quantities of hexakis(phenyleneethynylene) macrocycles to be prepared for the first time.

12.8
Conclusion and Perspectives

The examples shown throughout this chapter demonstrate that RCAM is a powerful methodology for the preparation of large rings in materials, organometallic, and synthetic organic chemistry. The technique is both robust and able to mediate transformations of complex structures bearing labile functionality. Many of the applications combine RCAM with a stereoselective semi-reduction to access defined alkene geometries. This strategy therefore presents a viable and predictable alternative to RCM macrocyclizations in the preparation of stereodefined alkenes and dienes.

Despite the clear synthetic utility, there is great opportunity for more widespread use of RCAM. However, as was the case in alkene metathesis, general uptake of the method appears dependent on the availability of a robust yet highly reactive catalyst system that shows good chemoselectivity. The current catalyst systems either display high reactivity and chemoselectivity but require rigorously inert conditions and careful handling, or the more robust systems display lower reactivity and/or functional group tolerance. The second key consideration relates to the current requirement for end-capped alkynes. A catalyst system that efficiently enabled RCAM of terminal alkynes would have significant synthetic benefit.

The rapid progress in the fields of alkyne and alkene metathesis and in the general area of catalyst design suggests that these are realistic goals which would have great effect on the synthetic impact of this already highly employable reaction.

12.9
Experimental: Selected Procedures

12.9.1
RCAM Synthesis of Compound 26 with Complex 1 [32]

A solution of diyne **25** (45 mg, 0.054 mmol), $Mo(CO)_6$ (1.4 mg, 0.0054 mmol, 10 mol %) and 4-trifluoromethylphenol (8.7 mg, 0.054 mmol) in chlorobenzene (30 mL) was refluxed for 6 h while a gentle stream of Ar was bubbled through the solution. After evaporation of the solvent, the residue was purified by flash chromatography on neutral alumina (hexanes/ethyl acetate 10 : 1 → 6 : 1) to give alkyne **26** (32 mg, 76%) as a colorless solid.

12.9.2
RCAM Synthesis of Compound 26 with Complex 1 under Microwave Conditions [32]

A SmithProcess vial (10 mL) containing a magnetic stir bar was charged with diyne **25** (27 mg, 0.032 mmol), 4-trifluoromethylphenol (5.2 mg, 0.032 mmol), Mo(CO)$_6$ (0.9 mg, 0.003 mol%) and anhydrous chlorobenzene (3 mL). The vial was sealed and evacuated through a cannula, and the resulting mixture was heated to 150 °C in a microwave oven (SmithCreator reactor) for 5 min. Work-up as described in Section 12.9.1 provides cycloalkyne **26** as a colorless solid (18 mg, 71%).

12.9.3
RCAM Synthesis of Compound 28 with Complex 3 [33]

An orange-brown solution of complex **3** (7.5 mol%) in anhydrous toluene (10 mL) was stirred at room temperature for 5 min under an argon atmosphere. A solution of diyne **27** (52.8 mg, 0.105 mmol) in anhydrous toluene (4 mL) was then added and the reaction mixture quickly warmed to 80 °C and stirred for 16 h at that temperature. The resulting mixture was cooled to room temperature, the solvent was evaporated and the residue was purified by flash chromatography (hexanes/ethyl acetate, 9:1) to yield **28** as a pale yellow syrup (32.8 mg, 70%).

12.9.4
RCAM Synthesis of Compound 35a with Complex 2 [36b]

To a solution of ester **34** (269 mg, 0.29 mmol) in anhydrous toluene (33 mL) was added a solution of tungsten complex **2** (13.8 mg, 29 µmol) in toluene (1.0 mL) and the mixture was stirred at 85 °C for 2 h under inert atmosphere. For work-up, the solvent was evaporated and the residue purified by flash chromatography (petroleum ether/EtOAc, 10:1) to give macrolactone **35a** as an amorphous solid (229 mg, 91%).

12.9.5
Enyne-yne RCAM Synthesis of Compound 54 with Complex 3 [43]

A solution of complex **1** (12.7 mg, 20.3 µmol) in anhydrous toluene (6 mL) and anhydrous CH$_2$Cl$_2$ (78 µL, 1.2 mmol) was added to a solution of enyne-yne **53** (31 mg, 41 µmol) in anhydrous toluene (35 mL) and the resulting mixture stirred at 80 °C for 14 h under an atmosphere of argon. A standard extractive work-up followed by flash chromatography (hexanes/EtOAc 4:1 → 3:1) gave cyclic enyne **54** as a pale yellow liquid (22.6 mg, 79%)

12.9.6
Polycyclization of 67 with Complex 2 [49]

A suspension of polyyne **67** (140 mg, 70 µmol) and (*t*-BuO)$_3$W≡C-*t*-Bu (100 mol%) in toluene was heated for 4 h at 80 °C under an Ar atmosphere. Concentration *in vacuo* and purification via Chromatotron (5:1 hexanes:CH$_2$Cl$_2$) afforded **68** (25 mg, 19%) as a bright orange solid.

12.9.7
Polycyclization of 67 with Complex 9 [49]

Reaction of **67** (40 mg, 20 µmol) with **9** (6 mg, 8 µmol) and silanol-POSS (49 mg, 50 µmol) in anhydrous 1,2,4-trichlorobenzene (5 mL) for 3 h at 75 °C and 1 torr afforded **68** (10 mg, 31%).

Abbreviations

Ar	aryl
BBN	9-borabicyclononane
Boc	*tert*-butyloxycarbonyl
CAN	ceric ammonium nitrate
Cat.	catalytic
Cp	cyclopentadienyl
Cp*	1,2,3,4,5-pentamethylcyclopentadienyl
DDQ	2,3-dichloro-5,6-dicyanobenzoquinone
DMB	3′,5′-dimethoxybenzoin
Fmoc	9-fluororenylmethoxycarbonyl
PMB	*p*-methoxybenzyl
RCAM	ring-closing alkyne metathesis
RCM	ring-closing (alkene) metathesis
TBAF	tetrabutylammonium fluoride
TBDPS	*tert*-butyldiphenylsilyl
TBS	*tert*-butyldimethylsilyl
THP	tetrahydropyran
TIPS	triisopropylsilyl
Trt	trityl

References

1 Pennella, F., Banks, R.L. and Bailey, G.C. (1968) *Chem. Commun.*, 1548–1549.
2 (a) Mortreux, A. and Blanchard, M. (1974) *J. Chem. Soc., Chem. Commun.*, 786–787; (b) Mortreux, A., Dy, N. and Blanchard, M. (1975/76) *J. Mol. Catal.*, **1**, 101–109; (c) Mortreux, A., Petit, F. and Blanchard, M. (1978) *Tetrahedron Lett.*, 4967–4968; (d) Bencheick, A., Petit, M., Mortreux, A. and Petit, F. (1982) *J. Mol. Catal.*, **15**, 93–101; (e) Mortreux, A., Delgrange, J.C., Blanchard, M. and Lubochinsky, B. (1977) *J. Mol. Catal.*, **2**, 73–82; (f) Mortreux, A., Petit, F. and Blanchard, M. (1980) *J. Mol. Catal.*, **8**, 97–106.
3 Katz, T.J. and McGinnis, J. (1975) *J. Am. Chem. Soc.*, **97**, 1592–1594.
4 Wengrovius, J.H., Sancho, J. and Schrock, R.R. (1981) *J. Am. Chem. Soc.*, **103**, 3932–3934; Pedersen, S.F., Schrock, R.R.,

Churchill, M.R. and Wasserman, H.J. (1982) *J. Am. Chem. Soc.*, **104**, 6808–6809.

5 (a) Zhang, W. and Moore, J.S. (2007) *Adv. Synth. Catal.*, **349**, 93–120; (b) Mortreux, A. and Coutelier, O. (2006) *J. Mol. Catal. A. Chem.*, **254**, 96; (c) Schrock, R.R. and Czekelius, C. (2007) *Adv. Synth. Catal.*, **349**, 55–77; (d) Fürstner, A. and Davies, P.W. (2005) *Chem. Commun.*, 2307–2320; (e) Fürstner, A. (2003) *Handbook of Metathesis*, vol. 2 (ed. R.H. Grubbs), Wiley-VCH, Weinheim, pp. 432–462; (f) Schrock, R.R. (2003) *Handbook of Metathesis*, vol. 1 (ed. R.H. Grubbs), Wiley-VCH, Weinheim, pp. 173–189; (g) Gradillas, A. and Pérez-Castells, J. (2006) *Angew. Chem. Int. Ed.*, **37**, 6086–6101; (h) Schrock, R.R. (2002) *Chem. Rev.*, **102**, 145–180.

6 Fürstner, A. and Seidel, G. (1998) *Angew. Chem. Int. Ed.*, **37**, 1734–1736.

7 (a) Schrock, R.R., Clark, D.N., Sancho, J., Wengrovius, J.H., Rocklage, S.M. and Pedersen, S.F. (1982) *Organometallics*, **1**, 1645–1651; (b) Freudenberger, J.H., Schrock, R.R., Churchill, M.R., Rheingold, A.L. and Ziller, J.W. (1984) *Organometallics*, **3**, 1563–1573; (c) Listemann, M.L. and Schrock, R.R. (1985) *Organometallics*, **4**, 74–83.

8 (a) Fischer, H., Hofmann, P., Kreissl, F.R., Schrock, R.R., Schubert, U. and Weiss, K. (eds) (1988) *Carbyne Complexes*, VCH, Weinheim; (b) Schrock, R.R. (1986) *Acc. Chem. Res.*, **19**, 342–348; (c) Schrock, R.R. (1995) *Polyhedron*, **14**, 3177–3195.

9 For a recent report on the use of terminal alkynes in alkyne metathesis see: Coutelier, O., Nowogrocki, G., Paul, J.-F. and Mortreux, A. (2007) *Adv. Synth. Catal.*, **349**, 2259–2263.

10 Zhang, W., Kraft, S. and Moore, J.S. (2004) *J. Am. Chem. Soc.*, **126**, 329–335.

11 (a) Bertinato, P., Sorensen, E.J., Meng, D. and Danishefsky, S. (1996) *J. Org. Chem.*, **61**, 8000–8001; (b) Nicolaou, K.C., He, Y., Vourloumis, D., Vallberg, H., Roschangar, F., Sarabia, F., Ninkovic, S., Yang, Z. and Trujillo, J.I. (1997) *J. Am. Chem. Soc.*, **119**, 7960–7973; (c) Schinzer, D., Limberg, A., Bauer, A., Böhm, O.M. and Cordes, M. (1997) *Angew. Chem. Int. Ed. Engl.*, **36**, 523–524; (d) Ono, K., Nakagawa, M. and Nishida, A. (2004) *Angew. Chem. Int. Ed.*, **43**, 2020–2023.

12 Trost, B.M., Ball, Z.T. and Jöge, T. (2002) *J. Am. Chem. Soc.*, **124**, 7922–7923.

13 (a) Fürstner, A. and Radkowski, K. (2002) *Chem. Commun.*, 2182–2183; (b) Lacombe, F., Radkowski, K., Seidel, G. and Fürstner, A. (2004) *Tetrahedron*, **60**, 7315–7324.

14 Trost, B.M. and Ball, Z.T. (2001) *J. Am. Chem. Soc.*, **123**, 12726–12727.

15 For an example, see the recent advances in platinum and gold activation of alkynes: Fürstner, A. and Davies, P.W. (2007) *Angew. Chem. Int. Ed.*, **46**, 3410–3449.

16 Beer, S., Hrib, C.G., Jones, P.G., Brandhorst, K., Grunenberg, J. and Tamm, M. (2007) *Angew. Chem. Int. Ed.*, **46**, 8890–8894.

17 Beer, S., Brandhorst, K., Grunenberg, J., Hrib, C.G., Jones, P.G. and Tamm, M. (2008) *Org. Lett.*, **10**, 981–984.

18 (a) Laplaza, C.E., Odom, A.L., Davis, W.M., Cummins, C.C. and Protasiewicz, J.D. (1995) *J. Am. Chem. Soc.*, **117**, 4999–5000; (b) Laplaza, C.E. and Cummins, C.C. (1995) *Science*, **268**, 861; (c) Cummins, C.C. (1998) *Chem. Commun.*, 1777–1786.

19 Fürstner, A., Mathes, C. and Lehmann, C.W. (1999) *J. Am. Chem. Soc.*, **121**, 9453–9454.

20 Fürstner, A., Mathes, C. and Lehmann, C.W. (2001) *Chem. Eur. J.*, **7**, 5299–5317.

21 Zhang, W., Kraft, S. and Moore, J.S. (2003) *Chem. Commun.*, 832–833.

22 (a) McCullough, L.G. and Schrock, R.R. (1984) *J. Am. Chem. Soc.*, **106**, 4067–4068; (b) McCullough, L.G., Schrock, R.R., Dewan, J.C. and Murdzek, J.C. (1985) *J. Am. Chem. Soc.*, **107**, 5987–5998; (c) Tsai, Y.-C., Diaconescu, P.L. and Cummins, C.C. (2000) *Organometallics*, **19**, 5260–5262.

23 Fürstner, A., Guth, O., Rumbo, A. and Seidel, G. (1999) *J. Am. Chem. Soc.*, **121**, 11108–11113.

24 Fürstner, A. and Seidel, G. (2000) *J. Organomet. Chem.*, **606**, 75–78.

25 (a) Fürstner, A. and Langemann, K. (1997) *Synthesis*, 792–803; (b) Fürstner, A. and Langemann, K. (1996) *J. Org. Chem.*, **61**, 3942–3943; (c) Fürstner, A., Koch, D., Langemann, K., Leitner, W. and Six, C. (1997) *Angew. Chem. Int. Ed. Engl.*, **36**, 2466–2469.

26 Fürstner, A. and Rumbo, A. (2000) *J. Org. Chem.*, **65**, 2608–2611.

27 Goldring, W.P.D. and Weiler, L. (1999) *Org. Lett.*, **1**, 1471–1473.

28 Fürstner, A., Castanet, A.S., Radkowski, K. and Lehmann, C.W. (2003) *J. Org. Chem.*, **68**, 1521–1528.

29 Aguilera, B., Wolf, L.B., Nieczypor, P., Rutjes, F., Overkleeft, H.S., van Hest, J.C.M., Schoemaker, H.E., Wang, B., Mol, J.C., Fürstner, A., Overhand, M., van der Marel, G.A. and van Boom, J.H. (2001) *J. Org. Chem.*, **66**, 3584–3589.

30 IJsselstijn, M., Aguilera, B., van der Marel, G.A., van Boom, J.H., van Delft, F.L., Schoemaker, H.E., Overkleeft, H.S., Rutjes, F.P.J.T. and Overhand, M. (2004) *Tetrahedron Lett.*, **45**, 4379–4382.

31 Ghalit, N., Poot, A.J., Fürstner, A., Rijkers, D.T.S. and Liskamp, R.M.J. (2005) *Org. Lett.*, **7**, 2961–2964; Ghalit, N., Rijkers, D.T.S. and Liskamp, R.M.J. (2006) *J. Mol. Catal. A.*, **254**, 68–77.

32 Fürstner, A., Stelzer, F., Rumbo, A. and Krause, H. (2002) *Chem. Eur. J.*, **8**, 1856–1871.

33 (a) Fürstner, A. and Grela, K. (2000) *Angew. Chem. Int. Ed.*, **39**, 1234–1236; (b) Fürstner, A., Grela, K., Mathes, C. and Lehmann, C.W. (2000) *J. Am. Chem. Soc.*, **122**, 11799–11805.

34 (a) Fürstner, A., Radkowski, K., Grabowski, J., Wirtz, C. and Mynott, R. (2000) *J. Org. Chem.*, **65**, 8758–8762; (b) Fürstner, A. (2004) *Eur. J. Org. Chem.*, 943–958.

35 Groothuys, S., van den Broek, S.A.M.W., Juijpers, B.H.M., IJsselstijn, M., van Delft, F.L. and Rutjes, F.P.J.T. (2008) *Synlett*, 111–115.

36 Vintonyak, V.V. and Maier, M.E. (2007) *Angew. Chem. Int. Ed.*, **46**, 5209–5211; Vintonyak, V.V. and Maier, M.E. (2007) *Org. Lett.*, **9**, 655–658.

37 Fürstner, A., Bindl, M. and Jean, L. (2007) *Angew. Chem. Int. Ed.*, **46**, 9275–9278.

38 Fürstner, A., Mathes, C. and Grela, K. (2001) *Chem. Commun.*, 1057–1059.

39 Fürstner, A., Larionov, O. and Flügge, S. (2007) *Angew. Chem. Int. Ed.*, **46**, 5545–5548.

40 Fürstner, A., De Souza, D., Parra-Rapado, L. and Jensen, J.T. (2003) *Angew. Chem. Int. Ed.*, **42**, 5358–5360.

41 Fürstner, A. and Turet, L. (2005) *Angew. Chem. Int. Ed.*, **44**, 3462–3466.

42 (a) Fürstner, A., Kirk, D., Fenster, M.D.B., Aissa, C. and De Souza, D. (2005) *Proc. Natl. Acad. Sci. USA*, **102**, 8103–8108; (b) Fürstner, A., De Souza, D., Turet, L., Fenster, M.D.B., Parra-Rapado, L., Wirtz, C., Mynott, R. and Lehmann, C.W. (2007) *Chemistry, Eur. J.*, **13**, 115–134; (c) Fürstner, A., Kirk, D., Fenster, M.D.B., Aissa, C., De Souza, D., Nevado, C., Tuttle, T., Thiel, W. and Müller, O. (2007) *Chemistry, Eur. J.*, **13**, 135–149.

43 Fürstner, A., Bonnekessel, M., Blank, J.T., Radkowski, K., Seidel, G., Lacombe, F., Gabor, B. and Mynott, R. (2007) *Chemistry, Eur. J.*, **13**, 8762–8783.

44 (a) Bauer, E.B., Hampel, F. and Gladysz, J.A. (2004) *Adv. Synth. Cat.*, **346**, 812–822; (b) Bauer, E.B., Szafert, S., Hampel, F. and Gladysz, J.A. (2003) *Organometallics*, **22**, 2184–2186.

45 Zhang, W. and Moore, J.S. (2006) *Angew. Chem. Int. Ed.*, **45**, 4416–4439.

46 Baughman, R.H., Eckhardt, H. and Kertesz, M. (1987) *J. Chem. Phys.*, **87**, 6687–6687.

47 (a) Hellbach, B., Gleiter, R. and Rominger, F. (2003) *Synthesis*, 2535–2541; (b) Miljanić, O.Š., Vollhardt, K.P.C. and Whitener, G.D. (2003) *Synlett*, 29–34; (c) Pschirer, N.G., Fu, W., Adams, R.D. and Bunz, U.H.F. (2000) *Chem. Commun.*, 87–88.

48 Johnson, C.A. II, Liu, Y. and Haley, M.M. (2007) *Org. Lett.*, **9**, 3725–3728.

49 Cho, H.M., Weissman, H., Wilson, S.R. and Moore, J.S. (2006) *J. Am. Chem. Soc.*, **128**, 14742–14743.

50 Ge, P.-H., Fu, W., Herrmann, W.A., Herdtweck, E., Campana, C., Adams, R.D. and Bunz, U.H.F. (2000) *Angew. Chem. Int. Ed.*, **39**, 3607–3610.

51 Bunz, U.H.F. (2001) *Acc. Chem. Res.*, **34**, 998–1010.

52 Zhang, W. and Moore, J.S. (2004) *Macromolecules*, **37**, 3973–3975.

53 Wilson, J.N., Bangcuyo, C.G., Erdogan, B., Myrick, M.L. and Bunz, U.H.F. (2003) *Macromolecules*, **36**, 1426–1428.

54 Brizius, G., Kroth, S. and Bunz, U.H.F. (2002) *Macromolecules*, **35**, 5317–5319.

55 Bunz, U.H.F. (2002) *Modern Arene Chemistry* (ed. D. Astruc), Wiley-VCH, Weinheim, pp. 217–249.

56 Zhang, W. and Moore, J.S. (2004) *J. Am. Chem. Soc.*, **37**, 12796–12796.

48 Johnson, C.A. II, Liu, Y. and Haley, M.M. (2007) *Org. Lett.*, **9**, 3725–3728.
49 Cho, H.M., Weissman, H., Wilson, S.R. and Moore, J.S. (2006) *J. Am. Chem. Soc.*, **128**, 14742–14743.
50 Ge, P.-H., Fu, W., Herrmann, W.A., Herdtweck, E., Campana, C., Adams, R.D. and Bunz, U.H.F. (2000) *Angew. Chem. Int. Ed.*, **39**, 3607–3610.
51 Bunz, U.H.F. (2001) *Acc. Chem. Res.*, **34**, 998–1010.
52 Zhang, W. and Moore, J.S. (2004) *Macromolecules*, **37**, 3973–3975.
53 Wilson, J.N., Bangcuyo, C.G., Erdogan, B., Myrick, M.L. and Bunz, U.H.F. (2003) *Macromolecules*, **36**, 1426–1428.
54 Brizius, G., Kroth, S. and Bunz, U.H.F. (2002) *Macromolecules*, **35**, 5317–5319.
55 Bunz, U.H.F. (2002) *Modern Arene Chemistry* (ed. D. Astruc), Wiley-VCH, Weinheim, pp. 217–249.
56 Zhang, W. and Moore, J.S. (2004) *J. Am. Chem. Soc.*, **37**, 12796–12796.